CHEMICAL INSTRUMENTATION: A SYSTEMATIC APPROACH

HOWARD A. STROBEL
Duke University

WILLIAM R. HEINEMAN
University of Cincinnati

CHEMICAL INSTRUMENTATION: A SYSTEMATIC APPROACH
THIRD EDITION

WILEY

A WILEY-INTERSCIENCE PUBLICATION

JOHN WILEY & SONS

New York • Chichester • Brisbane • Toronto • Singapore

Library of Congress Cataloging in Publication Data:

Strobel, Howard A.
Chemical instrumentation: a systematic approach / Howard A.
Strobel, William R. Heineman.—3rd ed.
p. cm.
"A Wiley-Interscience publication "
Bibliography: p.
Includes index.
ISBN 0-471-61223-5
1. Instrumental analysis. I. Heineman, William R. II. Title.

QD79.I5S76 1988
543′.07—dc19

88-11687
CIP

Printed in the United States of America

10 9 8

PREFACE

In the 15 years since the second edition of *Chemical Instrumentation*, the field of chemical methods of analysis has seen tremendous change. Measurements themselves have altered: sampling procedures have become more versatile, reliable, and automated and statistical and chemometric methods are being increasingly applied in order to obtain more information. In instruments, computers have become integral to the majority of models, advanced instruments increasingly are modular, and new sources such as lasers and new detectors such as photodiode arrays have come into more general use. Consequently, most established techniques have changed in methodology, instrumentation, and types of application. A parallel development has been that the central role of analytical chemistry in chemical measurements has been acknowledged.*

For the third edition of this book the goals have been to reflect these developments, include additional techniques, and present electroanalytical and column chromatographic methods more fully and authoritatively. In working toward these aims virtually all chapters in the second edition have been extensively revised. I have also developed new chapters and asked colleagues to write in other areas. The new chapters cover the following topics:

Microprocessors and microcomputers (Chapter 6)
Statistical control of measurement quality (Chapter 10)
Quantifying measurements and extracting information (Chapter 11)
X-ray fluorescence spectrometry (Chapter 21)
Surface spectrometric techniques (Chapter 22)
High performance liquid chromatography (Chapter 26)

Colleagues have contributed in several areas. Chapters 24–26 on chromatographic methods (Chapter 31, 2nd ed.) were written under the supervision of Dr. Ray P. W. Scott, who has recently retired as Director of Research, Perkin-Elmer Co. He wrote the chapter on chromatographic theory, and two of his colleagues, Dr. Kenneth Ogan and Mr. John D. Walters, wrote those on high performance liquid chromatography and gas chromatography, respectively. It has been especially satisfy-

*For example, the physical methods manual available to research personnel in the Dow Chemical Company is now entitled *Modern Methods of Analytical Chemistry*.

ing to have the strong contributions of my co-author, Professor William R. Heineman of the University of Cincinnati, in the electroanalytical chemistry chapters, Chapters 27–30 (Chapters 23–26, 2nd ed.). Their fine contributions as well as the new chapters greatly improve the quality of coverage of principal analytical methods.

Some chapters of the second edition have also been omitted and others condensed or consolidated. One major concession to length was reluctantly made, the omission of the chapter on nuclear magnetic resonance spectrometry. With the advent of a bevy of high-powered nmr methods mainly applied through software, this technique has seemed best left to monographs primarily for organic chemists.

In first developing this text, I was guided by the conviction that the modern methods of analytical chemistry that are used in analysis and research can best be mastered if chemical instrumentation is studied in its own right. Accordingly, after beginning with a critical examination of the process of measurement, this book turns to a systematic treatment of instrument *design* and instrumental *methods*. The adequate consideration of the first theme has dictated the strong undercurrent of physics, engineering, and physical chemistry that is evident on inspection of this book. These disciplines provide the fundamentals that are needed to understand design and function. The pursuit of the second theme, modern methods of analytical chemistry, has required a running discussion of physical properties and behavior throughout. Throughout the book the emphasis is on fundamentals. Above all, as both author and editor, I have aimed for a good balance between physical theory and design on the one hand, and chemical theory and methods on the other.

Because one focus is on instrumentation, a substantial amount of physical theory has been introduced, much of it through the sections Basic Electronics, Basic Optics, and Basic Quantification, that comprise the early part of this volume. The first section, Chapters 2–6, deals with basic electricity and electronics and microprocessors/microcomputers. These chapters furnish background for discussion of the processing and electronic aspects of instrumentation and electroanalytical instruments. In the section on basic optics, Chapter 7 deals principally with physical optics, Chapter 8 with optical sources and detectors, and Chapter 9 with monochromators and polychromators. This group of chapters sets the stage for dealing with optical spectrometric methods. The last basic division, Chapters 10–12, is new and introduces theory on statistical assessment of error, ways of quantifying measurements to extract information, and signal/noise considerations. This block of chapters was written in response to a consensus reached at a June 1982 curriculum workshop that instrumentation should always be introduced along with the general statistical and mathematical aspects of measurement.*

Although the later chapters on specific techniques are largely complete in themselves, it is not intended that they be studied without the collateral or prior reading of the appropriate parts of chapters in the basic sections, or without some accom-

*Workshop on Instrumentation in the Undergraduate Chemistry Curriculum sponsored by the NSF Project, Scientific Instrumentation Information Network and Curricula, in June 1982 in Lexington, VA, directed by Prof. Frank A. Settle, Jr.

panying laboratory experiments or at least study of manufacturers' brochures on instruments. It is hoped that the cross referencing ensures that the reader who begins a study with any one of the methods is promptly directed to pertinent background theory. The index may also be used to facilitate reference to background.

While the graduate–advanced-senior-level approach of the second edition has been retained, a serious effort has been made to improve the tutorial quality of the text by increasing the quality and range of difficulty of worked-out examples and providing a good selection of exercises at the ends of chapters. The modular approach to instrumentation initiated in the first edition and strengthened in the second has now been sufficiently developed that it is the integrating structure for the book. This approach helps the instrument user to understand:

How an instrument operates and how particular module(s) determine overall performance and specifications.

Limitations to measurement that grow out of instrument and measurement design.

Criteria by which one can select the model of instrument that will yield the desired measurements.

Criteria by which one can select appropriate modules to build an instrument for research uses.

In addition, the value of this modular approach to instrumentation is that it takes one basic way of looking at instruments in many different fields and, by repeatedly applying this perspective, it teaches the reader a skill, how to look at a new instrument with comprehension. This consistent perspective is often absent elsewhere. For example, manufacturers seldom describe how their instruments carry out measurements and tend to stress data acquisition and computer-related aspects.

STYLE AND NOMENCLATURE

Information that is more esoteric (either more advanced or more technical) is presented in smaller type to indicate that it is for a special audience. Worked out examples continue to be set in smaller type.

Most *references* appear in footnotes, or when cited in examples, in the body of the example. Those references that are valuable as a general bibliography have been numbered and collected at the end of each chapter; some of them are cited in the text by number.

Choices in *nomenclature* were made on the basis of good usage and elegance. Thus, the term *analyte* is employed throughout rather than the somewhat wordy "substance of interest." Similarly, the terms *quantify* and *quantification* are used to refer to processes in which amounts or concentrations of substances are measured rather than "quantitate" and "quantitation" since the latter terms appear to ignore accepted routes of derivation of new words. In addition, *interferent* is used to identify any nonanalyte that affects a measurement.

ACKNOWLEDGMENTS

I especially appreciated having Bill Heineman join me as co-author. At the outset we talked at length about the content and emphasis of the third edition and ways to make the text more valuable to users. His ideas have been valuable at many points. As he gradually became busier, however, he had to limit his contribution to writing lucidly on electroanalytical methods.

Of the several others who have contributed advice, I especially thank Charles H. Lochmüller for his significant help in organizing the chromatography section, for his reading of first drafts of those chapters and commenting on them critically, and for stimulating conversational advice at many stages as the chapters came to their final form. Peter C. Jurs helped by reading a late draft of Chapter 6 and recommending useful changes. Richard MacPhail read the section on lasers and provided good advice. George R. Dubay read through the mass spectrometry chapter and made suggestions about nomenclature and additional points that could be raised.

Darrel R. Wilder was kind enough to develop exercises for many chapters as well as answers, and John F. O'Keefe helped with some new exercises for the chapter on Raman spectrometry. And finally, I want to acknowledge the significant feedback from the many graduate students who used early versions of the electronics chapters and to undergraduates who used early versions of parts of Chapters 10 and 11. Student questions and suggestions helped materially in improving the approach and readability of the material. I wish also to acknowledge the real contribution of my wife, Shirley, for her forbearance during the long period of preparation of the third edition. Last, but not least, I am grateful to the many secretaries who were kind enough to enter original dictation in the wordprocessor and also to enter changes as editing occurred.

Bill Heineman wishes to acknowledge his indebtedness to Jon Kirchhoff and Hendrik Emons for reading Chapters 27–30 and commenting thereon, to Barbara Stallmeyer for typing these chapters, and to his wife, Linda.

HOWARD A. STROBEL

Durham, North Carolina
March 1988

PREFACE TO THE SECOND EDITION

Twelve years have passed since the appearance of the first edition of *Chemical Instrumentation*. Since the second edition represents a substantial rewriting, its main aspects deserve comment: its scope and aims, the ways in which the book has changed in response to developments in techniques and instrumentation, and some of the possibilities foreseen for use of this edition in courses.

Though a major revision has been carried out, the scope and aims of the book remain essentially the same. The volume treats a large number of the spectrometric, electrometric, and other physical methods that are important to chemists. What is lost by the omission of certain techniques (the book is long already) has—it is hoped—been compensated by the adoption of a consistent point of view.

The basic goal of the book is to offer a broad coverage of physical methods. The text reflects the conviction that such methods can best be mastered if the instrumentation necessary to measurement is studied in its own right. Thus equal emphasis is given to measurement principles and instrument design on the one hand and to techniques and their underlying theory on the other. The basic aim has also seemed best served by a semiquantitative approach, that is, by introducing mathematical expressions where necessary for good understanding but omitting most derivations of equations.

The other major goal is the development of a working mastery of the measurement process itself. It is easiest to describe the goal by suggesting criteria by which a working grasp can be recognized. Some of them are (a) facility in using instruments similar to or related to those one has studied; (b) an ability to devise and apply appropriate criteria for choosing an instrument (or a technique) for measurements on a given system; (c) facility in appreciating and mastering kinds of measurements that are unfamiliar. This aim has seemed best achieved by use of a modular approach* to instrumentation, a point that will be discussed below.

The second aim is actually also related to the time dimension of chemical measurement. One's grasp of physical methods must be sufficient to allow him to cope with the stream of new or modified instruments and techniques that appears. The questions that one must answer satisfactorily are: "What advantage(s) does the new instrument or technique offer over the present one?" and "How can the new

*A module is a subassembly or part that performs a particular function. Some examples are dc power supplies, monochromators, detectors, and amplifiers. When a modular point of view is taken, an instrument is represented by a block diagram.

device best be utilized in determinations of interest to me?'' Answering the queries should be a straightforward process, though not always an easy one, given a good grounding in measurement and techniques.

How does the second edition differ from the first? In the years that have elapsed since the publication of the first edition, there have been major changes in the field of measurement and analysis. Three of the most visible are (a) the well-nigh complete shift of electronics from vacuum tubes to semiconductor devices, (b) the increasing development of automatic instruments and tie-ins between instruments and computers, and (c) a growing acceptance of a modular or a ''systems'' view of instruments. All of these changes are reflected in this edition. Several chapters are given over to solid state electronics, and the older tube electronics is virtually excluded. Both automation and computer control are treated briefly and generally as they relate to instrument design. This discussion will be found in the parts of Chapters 1, 3, 9, and 17 that deal with the systems aspects of instruments. For reasons of brevity and because of rapid development in the field, little attempt is made to deal with automation and computer control in the discussion of individual instruments.

As noted above, a modular view of instrument design and function has been adopted. Its advantages deserve fuller comment. One is that it effects an economy both for the author and the reader. Modules are treated separately in early chapters where their characteristics can be explored with appropriate rigor and the most important different forms can be described and compared. Later, in chapters on techniques block diagrams are used for a particular instrument to show the modules that make it up and the pattern in which they are linked. For example, monochromators, optical sources, and detectors are discussed in Chapters 11 and 12, and the building up of spectrometers of various types is treated in Chapters 13–18, which deal with various spectroscopic methods. When an instrument is discussed, suitable choices of modules for it are suggested through examples and sometimes also through exercises at the ends of chapters. A very few modules that are peculiar to a single technique, such as burner-nebulizers in flame spectrometry, are discussed in chapters on techniques.

Some other advantages of a modular approach accrue to the instrument user. The most important is that he can understand an instrument on a ''macroscopic'' level. The types of components that make it up and their function can be perceived without the necessity of mastering a mass of detail. Further, manufacturer's specifications tend to become more intelligible since many aspects of the performance of an instrument are determined by the quality of particular modules. For example, the resolution of a spectrophotometer is mainly determined by the quality of its wavelength isolation device. A third advantage is that a ''systems'' approach is feasible. As a result, it becomes a fairly straightforward process to propose ways to optimize one or another aspect of instrument behavior, e.g., sensitivity, or to adapt it to new measurement situations.

The reader familiar with the first edition will observe other changes as well. He will find that the treatment of several topics has been expanded to chapter length (fluorometry, flame spectrometry, Raman spectrometry, and chromatography) and several new topics are included (NMR spectrometry, single sweep and pulse polarography, mass spectrometry, operational amplifiers, signal-to-noise optimization, digital electronics, and monochromators). Further, the theoretical background of most measurement techniques is presented in greater depth. Finally, SI notation and units have been generally adopted (see Appendix C).

As a result of the way the book has been revised, the second edition can be used in at least three kinds of courses:

1. A junior–senior analytical chemistry course emphasizing physical methods. An appropriate selection of material would be Chapter 1, followed by chapters selected from the blocks 13 through 16 and 18 through 31. The index should prove helpful in locating pertinent analytical information. In such a course it is recommended that a problem book appropriate to the aims of the particular course be used also.

2. A senior or first-year graduate course in instrumentation that stresses electronics and electrical methods. Appropriate material would be Chapters 1 through 9, 23, and chapters selected from the block 24 through 28. Chapter 19 and perhaps also Chapter 17 could be included.

3. A senior or first-year graduate course in instrumentation that treats optical and spectral methods. Appropriate material would be Chapters 1 and 8, 10 through 12, 17, and chapters selected from the blocks 13 through 16 and 20 through 22. Selected sections from Chapter 2 through 7 could be included.

When used in courses the book will need to be supplemented by a laboratory manual. Since courses in modern analysis and instrumentation differ appreciably in aim and scope from school to school, it seemed wise to omit laboratory experiments from this edition. The teacher is accordingly referred to the many fine laboratory manuals that have appeared in the past several years. Some are the following:

1. C. N. Reilley and D. T. Sawyer, *Experiments for Instrumental Methods*. New York: McGraw-Hill, 1961.
2. L. P. Morgenthaler, *Basic Operational Amplifier Circuits for Analytical Chemical Instrumentation*, 2nd ed. Danville, Calif.: McKee-Pedersen Instruments, 1968.
3. G. G. Guilbault and L. G. Hargis, *Instrumental Analysis Manual*. New York: Marcel Dekker, 1970.
4. A. James Diefenderfer, *Basic Techniques of Electronic Instrumentation*. Philadelphia: Saunders, 1972.
5. C. E. Meloan and R. W. Kiser, *Problems and Experiments in Instrumental Analysis*. Columbus, Ohio: C. E. Merrill, 1963.
6. R. W. Hannah and J. S. Swinehart, *Experiments in Techniques of Infrared Spectroscopy*, rev. ed. Norwalk, Conn.: Perkin-Elmer, 1968.
7. Issues of *Journal of Chemical Education* (this journal publishes suitable experiments from time to time).

The teacher will also wish to know that names of manufacturers of a particular type of instrument have been omitted in this edition. Now there is at least one instrument listing that appears annually, an August issue of *Analytical Chemistry*, entitled "Laboratory Guide." Usually a late fall issue of *Science*, "Guide to Scientific Instruments," is also published. These compilations are up-to-date and inclusive, and instruments appear to be classified with care.

An annotated bibliography of important references appears at the end of each chapter. An additional collection of references too recent to be listed there has been organized in the same way and appears in Appendix D. Both listings should be consulted.

ACKNOWLEDGMENTS

I especially thank two colleagues—Dr. Maurice Bursey and Dr. Charles H. Lochmüller—for writing Chapters 30 and 31 on mass spectrometry and chromatography, respectively. Their share in this enterprise is considerable.

In addition, I am grateful to the host of authors from whose works I have learned. The group numbers too many to name, but they are acknowledged in the many references at the ends of chapters.

I thank sincerely Drs. Leon N. Klatt and John T. Bowman for reading and commenting on the entire manuscript. Warm appreciation is also due to Dr. Marvin M. Crutchfield, who performed the same function for the chapter on nuclear magnetic resonance spectrometry, to Gray F. Crouse, who assisted greatly in the preparation of illustrations, to many students, who after using some of these chapters in preliminary draft, offered criticism and thoughtful suggestions, to Cathy Penny, Gloria Powers, and Sandy Parker for typing the numerous drafts and the final manuscript, and to Scott A. Miller for his generous assistance in reading proof.

Finally, I solicit the assistance of readers of this book in identifying errors and in offering suggestions.

H.A.S.

Leicester, England
March 1972

CONTENTS

CHROMATOGRAPHIC METHODS

CHEMICAL INSTRUMENTATION: A SYSTEMATIC APPROACH

Chapter 1

MEASUREMENT AND INSTRUMENTATION

1.1 INTRODUCTION

A chemical measurement begins when information is sought about a system. The simplest kind of measurement quantifies a property such as the absorption of light by a substance of interest using the technique of spectrophotometry. Nearly as easy is the extraction from that measurement of information such as the concentration of the substance. More complicated, since the process may require additional techniques, is establishing the substance's molecular structure and identity.

How does one select the physical property to be measured? Whether a single property of one *analyte*, our substance of interest, or a full description of all the analytes in a mixture is sought, the process usually calls for a comparison of the merits of several properties. Which properties are best suited to the kind of sample(s) available and the type of information sought? For example, some properties are highly *specific* and characteristic of particular substances, such as infrared absorption and X-ray fluorescence spectra. The information they furnish will contribute substantially to identification and quantitative determination on molecular and elemental levels, respectively. Other properties, such as refractive index, are characteristic but nonspecific. Nevertheless, they are well suited for quantitative determination in many simple systems. Finally, even properties like mass must be considered. Classical gravimetry and the thermal decomposition of compounds are commonly followed by changes in sample weight when known reactions occur.

Before actually making measurements of the selected property one must determine whether or not the results will actually address the questions needing answers. Measurements should always arise from a thoughtful analytical design that specifies the goals in mind, the appropriate set of measurements, and a host of ways to do the sampling and sample preparation that appear advisable. The design of a plan is crucial to economy of effort and even to clarifying the nature of the analytical problem. A good analytical design must be conscientiously pursued if any progress is to be made with more complex questions.

1.2 DESIGNING MEASUREMENTS

As research and industrial problems have become more complex, there is a new acceptance of developing *analytical designs*. A good plan calls for examining the basic questions that have prompted measurements and experimentation, identifying types of information that should lead to answers or solutions, and selecting methods that will yield this information. Even when one is planning a simple study involving chemical measurements, one engages in the process of design.

An elaborate but incomplete flow chart that suggests the progression of steps that may be taken in developing an analytical design is given in Fig. 1.1. How to raise questions that will aid the analytical design process is illustrated by examples below. Undoubtedly the full use of this process will be likely only for a new problem or piece of research. Even for routine measurements, however, any change in samples, instrumentation, or requirements on measurements, such as identifying trace constituents, may provoke systematic use of the chart beginning at the level of a selection of methods.

Example 1.1. An organic chemist intrigued by the excited states of a family of aromatic molecules decides to develop a research program to answer some questions. What molecules are most likely to have lifetimes that will permit useful measurements? What data should be collected besides lifetimes? Will fluorescence or phosphorescence spectrometry be a better technique? What experiments will yield the most information? What level of precision will be desirable?

Example 1.2. It is found that yttria-stabilized zirconia coatings on gas turbine blades in jet engines fail frequently after moderate operating times. What physical changes occur? Do blades lose their coatings and then fail? Is there any connection between failure times and the impurities in the kerosene burned in the engines? Are the stresses on blades (such as thermal stress as engines cycle from ambient temperature to 1050°C) contributing factors?

Example 1.3. In Western Europe and North America there has been great interest in lessening the damage caused by acid rain. Is it a single phenomenon? Won't the studies of acid rain have to deal with the mix of factors (atmospheric, biological, soil-related, climatic, geographic, industrial, etc.) that may be responsible? With what precision can the factors be learned and their impact assessed? What studies should be planned? Presuming that the factors can be identified, what measurements at what sites will serve to monitor them, and how long should they be monitored? What will be the most effective and least costly ways to diminish the impact of acid rain?

The process of analytical design calls for us to identify information that is likely to help answer the basic questions asked, select the methods that should provide the information, and develop a laboratory plan. The design should spell out necessary sampling of systems and modes of sample preparation that do not undermine the precision of the instrumental measurements. Often the steps with greatest error

Fig. 1.1 An outline of major processes involved in designing measurements for experiments or the pursuit of problems. Sequences of steps are not ordinarily as clear cut as suggested. Each box can in general also be expanded into a flow chart. Hatching indicates that statistical and mathematical procedures are necessary. [After D. L. Massart, *Z. Anal. Chem.* **305,** 113 (1981).]

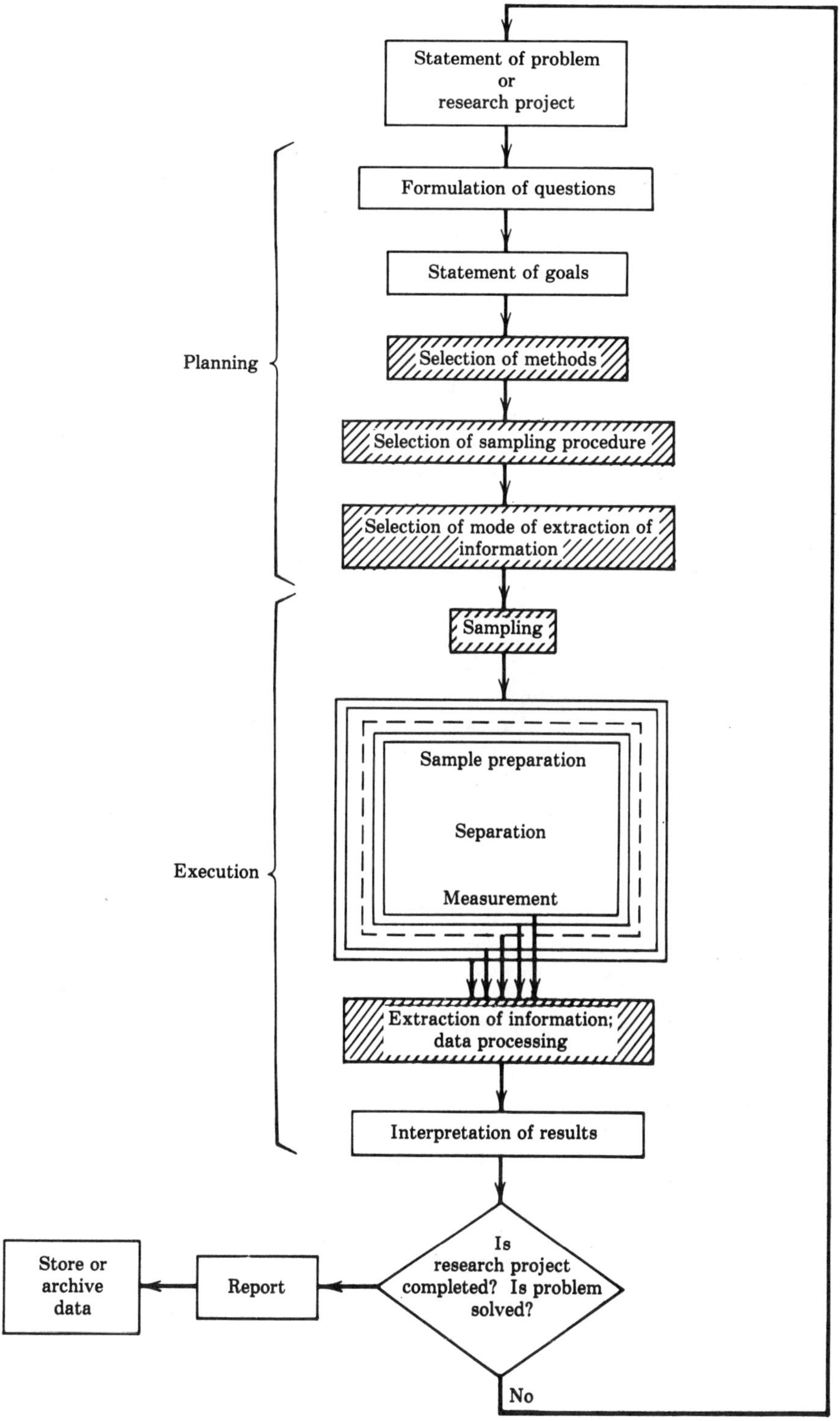
Statement of problem
or
research project
Formulation of questions
Statement of goals
Selection of methods
Selection of sampling procedure
Selection of mode of extraction of information
Planning
Sampling
Sample preparation
Separation
Measurement
Execution
Extraction of information; data processing
Interpretation of results
Is research project completed? Is problem solved?
No
Report
Store or archive data

are obtaining the minute samples needed for a sensitive technique such as mass spectrometry. The culmination of the design will be a clear statement of the actual measurements (determinations) to be made and the way in which information will be mathematically extracted from them and interpreted. *Analysis* is involved on many levels—conceptual, theoretical, and practical. Chemical analysis today is more than simple determinations.

In developing a design, asking good questions is probably the most crucial analytical technique involved. One function of the questions is to identify the techniques and types of measurements that will be useful in obtaining the desired results. Another is to assign priorities to the steps based on considerations of cost in time and money and of the probable value of the findings. The flow diagram in Fig. 1.1 suggests a systematic procedure for producing a measurement design.

Perhaps sequentially, perhaps simultaneously with mulling the kinds of chemical-structural-geometrical information that help resolve a problem, the research must select one or several techniques most likely to furnish information that will establish what chemical species or processes are responsible for the phenomena observed.

Example 1.4. The process of design of measurements is well illustrated by the evolution of procedures for predicting airplane engine failure. These procedures are based on the observation that a regular regimen of analyses of the metal content of spent lubricating oil from aircraft engines should provide information about their pattern of wear. Calibration data are obtained from samples of lubricating oil taken at times of engine overhaul when the extent of wear can be mechanically assayed. Those data about metal content can be correlated with observed wear and eventual engine failure to develop a schedule of warnings predicting the imminence of engine failure.

Considerable evolution of design has occurred over time. Both atomic absorption and atomic emission spectrometry with inductively coupled plasma excitation proved suitable analytical techniques. Metals in the spent oil appeared in suspended form, in chemical compounds, and in large particles that settled quickly. Samples that stood before analysis had to be mixed thoroughly by sonication (with ultrasound) before representative aliquots could be taken. Indeed, in order to include the large particles in sampling, small quantities of HCl and HF had to be added to prepare solutions for introduction into the spectrometer. [K. J. Eisentraut et al., *Anal. Chem.*, **56,** 1086A (1984).]

Developing and executing a good analytical design will also depend on having a clear picture of the connection between instrument design and the quality of measurements: this relationship will now be examined in terms of the modular approach.

1.3 THE MODULAR APPROACH TO INSTRUMENTATION

Basically, the modular approach is a conceptual view that calls for understanding an instrument as a system of modules, each of which performs a needed function, like amplifying a signal. Incidentally, the common use of block diagrams reminds us that instruments have long been conceived and explained in terms of modules.

Example 1.5. It may be anticipated by those unfamiliar with the modular way of viewing instruments that such an approach might be impossibly mathematical and abstract. A homely example will dispel that notion. We grow up familiar with a modular approach to our gasoline-operated automobiles: we understand the automotive system to be carburetor module through which a gasoline–air mixture is introduced to an internal combustion engine module whose crankshaft connects through a gear reduction module to a drive shaft–differential gear module that turns the wheels that propel the vehicle.

To develop the modular view it is useful to consider how a measurement is made on a sample. In the classical sense a measurement involves completing the following sequence of steps:

1. Generating the energy flow ("signal").
2. Impinging the energy flow on the sample.
3. Detecting the signal arising as the energy interacts with the sample.
4. Amplifying the signal as necessary.
5. Processing the signal.
6. Computing analytical information.
7. Displaying or reading-out of the results.

A modular approach works out best to mesh this description of measurement with actual instruments.

In this context a *module is considered to be the group of components that performs a particular function*, for example, detection or amplification. If we open an instrument case, we will always see a myriad of circuit boards, integrated circuits, wires, resistors, slits, mirrors, filters, photomultiplier tubes, and other components. But, by keeping the function of a module in mind, it should be possible to find its principal components even when they are not physically set off from the rest of an instrument. For a monochromator, we should look at least for its entrance and exit slits and wavelength dispersing device.

Only if we must calibrate or test a module and do not have access to information about test points or circuits from the manufacturer are we likely to have to identify many parts. In more modern instruments self-diagnostic programs are increasingly available to simplify this task. In the case of electronic modules our task is also eased since many appear on circuit boards and if failure of one is indicated, the whole board may be replaced and sent for repair.

Is it correct that most instruments are made from a standard collection of modules? Is it also true that only a fairly small set of modules is important enough to study extensively? If the answer to both queries is affirmative, an important question remains. Can one successfully trace instrument specifications to like specifications for individual modules? All these questions can be answered yes. It is even true that one module often sets an instrument specification.

Furthermore, nearly all modules are types that will appear in new instruments into the indefinite future. Improvements in design will enhance their operation but the modules will continue to perform the same function. In many cases both the original and improved version continue in use. A good example is the upgrading of the phototube to a photomultiplier tube; each type of detector has found its own area of application.

1.4 MODELING THE MODULE: GENERAL PROPERTIES

What sort of a subsystem is a module? In Fig. 1.2 a common *single-channel* spectrophotometer is pictured as a train of modules. The term channel refers to the signal path; it links all modules to make such an instrument a system. To interpret the figure note first that modules are named for the functions they perform. Their sequence should also conform with the measurement steps described earlier. Sometimes two modules are necessary for one step. For example, in Fig. 1.2 both a continuous wavelength source and a monochromator are necessary to generate the required monochromatic beam.

The striking difference in complexity of modules is worth noting. For example, if a spectrophotometer is to operate in the visible and near IR regions the continuous source of Fig. 1.2 may be a small tungsten-filament bulb like a light bulb. By contrast, the monochromator will consist, at least, of entrance and exit slits, a dispersing device such as a diffraction grating, several baffles, and a precision mechanism to vary the orientation of the dispersing device to the light beam. In general, it will also have collimating and focusing optics such as mirrors.

The term *signal* has been used up to this point in the general sense of a defined energy flow. In the early part of the channel the flow, for example, a light beam or a current, is only energy and it may be appropriate to call it ''signal''. After that flow has interacted with a sample, chemical information about analytes has been added to or *encoded* in it, by changing its amplitude, frequency, or phase. It is not difficult to picture changing these properties for 60 Hz ac in a wire, for example. The energy flow has truly become a signal.

Some modules such as a modulator or chopper may also add a carrier frequency to a signal, many will add background (e.g., a monochromator adds stray light), and all will add noise. It is the function of good design, which nearly always includes signal processing, to minimize or subtract such additions before the analytical signal is displayed.

The main question is still waiting to be asked. What sort of a system is a module? Note again by reference to Fig. 1.2 that for each module a signal appears as its input, is modified in some characteristic fashion, and then appears as its output.

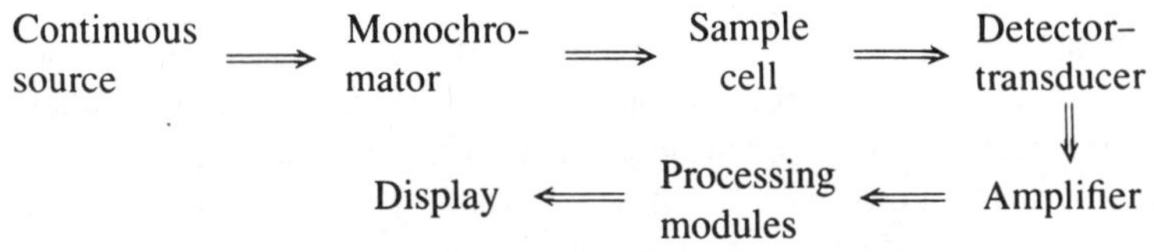

Fig. 1.2 Modular diagram of a single-channel spectrophotometer.

How can we think about this process in a consistent way so that we may easily characterize all modules, familiar and unfamiliar?

First, the type and intensity range of *input* a module can accept must be known. Examining the behavior of a photomultiplier tube as a detector will make the point clearer. Ultraviolet and visible radiation reaching the photocathode of the multiplier phototube is its input. The lower limit or *threshold* limit of light for which a response can be noted is usually set by internal noise, that is, random unavoidable fluctuations in the dark current. The upper intensity to which the detector can respond is reached when a further increase in input produces no additional growth in output. The module is said to *limit* or reach *saturation.*

If the light input is within a given range of intensities, it is "transduced" linearly to current by the phenomenon of photoemission. Figure 1.3 shows a representative plot of output as a function of input for a quite different type of detector, a photographic emulsion. The broken lines in the plot correspond to threshold and saturation values. The range of useful input, which is ordinarily called the *dynamic range*, lies between these limits. Commonly, it is taken to be the region of linear or near-linear response.

From the discussion it is evident that everything centers on the characteristic manner in which a module modifies the signal at its input, its *transfer function.* We are interested in modules precisely because of this ability. The *transfer function f* is defined by the expression

$$I_o = f(I_i), \tag{1.1}$$

where I_i is the amplitude of input and I_o that of output. If the function is linear, as in the central portion of the response curve in Fig. 1.3, a module will transform an input to an output without distortion.

In Table 1.1 several types of modules are listed to illustrate the variety of ways in which they interact with or transform signals. Often they are called *input transducers* if used as detectors and *output transducers* if used as source or readout

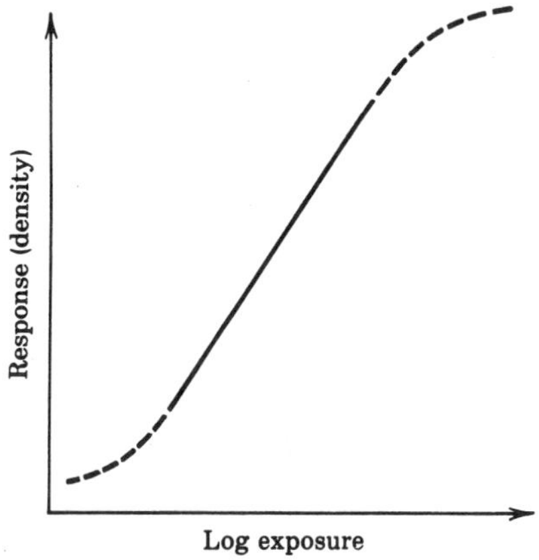

Fig. 1.3 Representative response curve: image density versus the logarithm of exposure for a photographic emulsion.

Table 1.1 A Few Properties of Some Major Modules

Module	*Function*	*Input*	*Output*	*Sensitivity of Module (usual name)*
Tungsten-filament lamp	Light emission (source)	dc current	Light	Efficiency
X-ray tube	X-ray emission (source)	dc current	X rays	Efficiency
Monochromator	Isolation of narrow λ band	Polychromatic radiation	Monochromatic radiation	Throughput
Photomultiplier tube	Detection	Light	Current	Efficiency
Thermocouple	Detection	Temperature	Voltage	Response
Ion-selective electrode	Detection	Ion activity (concentration)	Voltage	Response
Amplifier	Amplification	Voltage or current	Voltage or current	Gain
Counter	Counting (processing)	Voltage pulses	Binary or decimal sum	
Light-emitting diode (LED)	Emission (readout)	dc current	Light	Efficiency

modules. Note in the table that all the detectors have electrical outputs; this type of detector is greatly preferred in most instruments since the modules that follow a detector require an electric signal.

The rate of change of output signal with input signal is another important characteristic. This property is termed the sensitivity S and is defined in its precise sense* for a module or instrument as

$$S = dI_o/dI_i. \tag{1.2}$$

The magnitude of a module's response to a given increase in input (given a time longer than its response time) is thus measured by its sensitivity. It is the slope of a response curve such as that in Fig. 1.3.

*Sensitivity as applied to a whole instrument system does not by itself determine the smallest concentration of an analyte that gives an instrument response that may be differentiated from one caused by noise. This minimum concentration is properly termed the *limit of detection.* Clearly, the irreducible noise levels of each module ''combine'' to fix that minimum. The greater the slope of the response curve for very low input signals, however, the lower the concentration level that yields a discernable response.

There are many synonyms for sensitivity S (see Table 1.1). For an amplifier it is called gain, for a filter, attenuation factor, for an optical detector, response. Further, S is sometimes defined logarithmically (decibel notation). At other times it, or its reciprocal $1/S$, is used and termed a scale factor. The latter notation is common for a module such as an operational amplifier where sensitivity may be changed by altering the resistance in the feedback circuit incrementally, giving scale factors of $10\times$, $100\times$, and so on.

Does one need to examine all modules in such a detailed fashion? There is a simplification that merits exploration. Consider Fig. 1.2 again. In that instrument and most others two sequences of modules can be identified:

1. A set of *characteristic modules* that extends from the source through the detector. These modules relate to the physical technique on which the measurement is based and the nature of sampling it calls for.
2. A set of *processing modules* that completes the instrument. It includes an amplifier, processing modules (often only an AD converter and microcomputer), and display or readout.

While characteristic modules will vary considerably from method to method and with types of samples, modules in the processing cluster will be remarkably similar from instrument to instrument since data acquisition, data processing, and calculation of analytical quantities will most likely be centered around a microcomputer under program control regardless of physical technique. It is data capacity, types of calculation, and rates of processing that differ from instrument to instrument.

From this analysis there are two useful conclusions. First, in considering different instruments, it will in general suffice to deal with their characteristic modules. Second, we can treat processing modules once and expect insights to be generally applicable to all instruments.

1.5 BASIC INSTRUMENT DESIGN

Let us now investigate some of the good layouts for an instrument. How generally applicable is the single-channel design introduced in the last section? Given that the design is grounded in the sequential nature of steps in a measurement outlined earlier, it is of course the most widely used design.

Single-Channel Design. The single train of modules shown in Fig. 1.2 requires that for a given quality of output:

1. Every module should in general perform at least at the precision level desired, for example, at $\pm 3\%$ or $\pm 0.5\%$.
2. Drift in the performance of any module with time (aging) or with change in temperature or other environmental variable must be minimized since such changes are simply passed on to succeeding modules in the channel.

3. Coupling or interfacing of modules should be of good quality to enhance flow of signal.

Example 1.6. Given these requirements, what aspects of design and construction will be most important for a *general-purpose instrument* of moderate precision?

A single-channel instrument intended to operate at ordinary levels of concentration and precision can be designed with considerable simplicity and ruggedness. Ordinary concentrations are those at which changes in a measured property are easily detectable, and ordinary precision is assumed to be 1–3%. Unless there are special requirements on its range, such an instrument needs only a simple detector and a sturdy but sensitive output meter. It can be designed and built with attention to simplicity, ruggedness, and low cost. Common examples are low-cost pH meters and colorimeters. They are also to be preferred over more complex devices because they require minimal maintenance. The filter photometer or colorimeter shown in Fig. 1.4 is a good example of a single-channel instrument.

Example 1.7 What aspects of design and construction will be important for a *precision single-channel instrument*? Must all components be of high quality?

A precision instrument (precison >0.5% or better) will have to be more complex than a general-purpose instrument. Usually ruggedness will be sacrificed first. At the very least, a stable source, a sensitive detector of wide dynamic range, a stable, high-gain amplifier, a good precision processing system, and a digital readout will be needed. It is equally important that the performance of all components be good. Most will be operating close to their limits. If such an instrument is to be used at trace concentration levels, the detector and preamplifier must also have to have low noise characteristics.

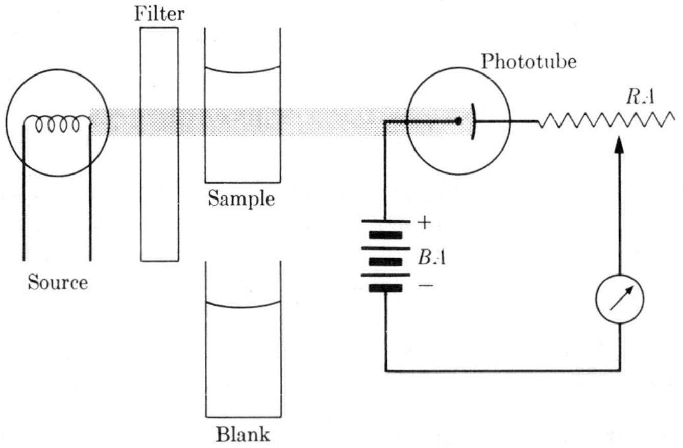

Fig. 1.4 Single-beam, direct-reading filter photometer. Readings are taken of the light over a small wavelength range that is transmitted by the sample and the blank. Battery BA supplies power to the phototube. The tap on resistor RA is varied to adjust the meter deflection to 100% when the blank is in the beam. After that step has been taken, when the sample is placed in the beam, the meter registers the percentage transmission of the sample.

Even precision single-channel instruments have become more common in recent years. Not only are modules that perform stably over considerable periods of time now more common but also program or computer control has added greatly to the stability of instrument operation. The latter is essential to assure that instrument registration holds constant in terms of wavelength, voltage, or another independent variable. More rapid instrument operation is also possible because of faster detectors and electronics. Together these aspects have considerably enhanced the attractiveness of single-channel design. Now signal baselines measured on blanks can be recorded in memory and simply subtracted from sample measurement responses in data processing.

Double-Channel Design. A common instrument modification is to divide the signal channel to effect a *double-channel* or dual-beam design. Ordinarily the division is made just before the sample module and the two channels are brought back together just afterward. This design is attractive for a technique such as spectrophotometry and is illustrated in Fig. 1.5.

Example 1.8. Is a second channel present in at least a figurative sense whenever a procedure calls for a blank?

Yes, as discussed in the legend of Fig. 1.4, even a single-beam spectrophotometer or filter photometer is used *twice* to obtain a single piece of information.

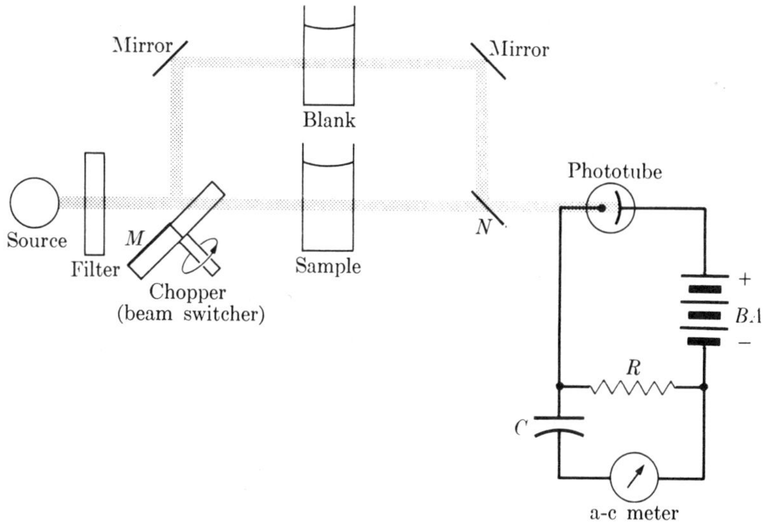

Fig. 1.5 One type of double-channel, direct-ratio filter photometer. Half of the rotating chopper wheel has been cut away entirely and the bottom half is covered by a front-surface mirror M. When the wheel is in the position shown, the beam travels through the sample cell and *through* mirror N, which is half-silvered, to the detector. If the chopper is driven by an ac synchronous motor, there will be 60 transmissions through each channel per second. The ac signal from the phototube passes through the *RC* filter to be read out on the ac meter as the ratio of the transmission of the sample to that of the blank.

A means to take the ratio of intensities transmitted in the two channels must also be added. The solution is to alter the dynamics of measurement. For example, a much more elegant solution to the problem of dealing with the signal in a double-channel instrument is outlined in the modular diagram of Fig. 1.6. Since the channels are involved alternately in the measurement, that is, the beam reaching the detector is first that from the sample channel and next that from the reference channel, the phasing of detector output furnishes a basis for making the comparison correctly. Expressed in other terms, the detector output can be sampled electronically in phase with the rotation of the chopper. In Fig. 1.6 the dynamics are also represented in the modular diagram of the instrument.

Under what circumstances are the added complexity and cost of double-channel design meritied? In general, a double-channel design represents a definite advance over a single-channel arrangement when

1. the many variables that affect a measurement are under only partial control and fluctuate in magnitude at a rate somewhat faster than can be accommodated by a procedure using blanks or standards or
2. a differential measurement appears desirable.

Further discussion of double-channel operation will be reserved for Section 17.5, when the topic will be explored as part of the discussion about spectrophotometers.

Sample Channel. Another major design modification of the basic channel pattern is attractive, the addition of a *sample channel.* In essence a separate train of modules is added to effect either

1. systematic sample preparation, for example, by automated dilution, addition of complexing agent, or extraction, to prepare a sample for the measurement, or
2. essentially complete resolution of a mixture, for example, by chromatographic separation.

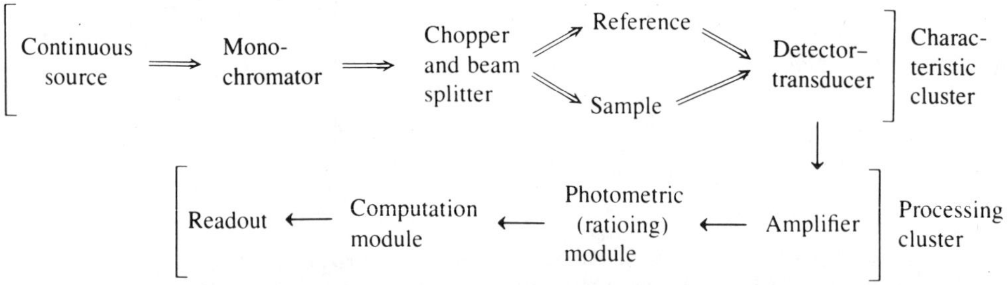

Fig. 1.6 Modular diagram of a double-channel spectrophotometer. Double and single arrows indicate movement of the "signal" from one module to the next between characteristic and processing modules, respectively.

While the idea is old, its implementation has required much development and thus has been commercially available only relatively recently. A sample channel is coupled into the signal channel of an instrument exactly where the sample must be introduced. One type of sample channel will be illustrated by Example 1.9 and still another described in Section 1.8. In all cases such a channel should be recognizable as a series of modules dedicated to carrying the sample through a prescribed series of steps.

Today sample preparation is becoming progressively more sophisticated through exciting developments in robotics and "hyphenated" instruments. The latter instrumentation will be further described in Section 1.8.

The Instrument as a System. The orderly flow of energy or signal in an instrument is characteristic of a good instrument. That being true, the instrument as a *system* must have designed into it arrangements, either in hardware or software, that keep background and fluctuations minimal. Only when this goal is realized can the encoded chemical information in the signal be reliably extracted at a later point.

One goal in design is to provide good *coupling of modules* to ensure minimum distortion of signal, maximum throughput of energy, and minimum noise. Example 1.10 below will clarify this idea.

Example 1.9. Probably the earliest example of a sample channel was in the Technicon Autoanalyzer, which performed automatic sequential analyses on physiological samples. Prior to the introduction of the device the necessary preparation steps were carried out by technicians with uncertain reproducibility. Its basic signal channel was that of a simple filter spectrophotometer. One sample channel was dedicated to each analysis, for example, for glucose or albumin. The channels were lengths of tubing; samples in a given channel were kept separate by air bubbles.

Thus necessary preparative steps, such as aliquoting, addition of reagents, dilution, mixing, and dialyzing (to remote protein), were carried out with good reproducibility in a maze of tubing that resembled the track pattern of a railroad freight yard. Each main sample was subdivided into as many aliquots as there were components to be determined. Amounts of chromogenic reagents or diluents added were regulated by proportional pumping. Finally, each channel incorporated a debubbler tee where air bubbles were removed. A sample then passed through a section of tubing of such length that it just filled the sample cell in the spectrometer. [L. T. Skeggs, Jr. and H. Hochstrasser, *Clin. Chem.*, **10**, 918 (1964).]

Example 1.10. Figure 1.7 shows two ways of coupling a sample cell and phototube (detector) in a spectrophotometer. All the optical signal from the cell should fall on the active area (photocathode) of the detector. Figure 1.7a thus illustrates poor coupling. In Fig. 1.7b coupling is good since a lens has been inserted to accommodate the finite optical aperture of the phototube. Alternatively, the detector might also be moved nearer the cell so that the beam would just illuminate its photocathode.

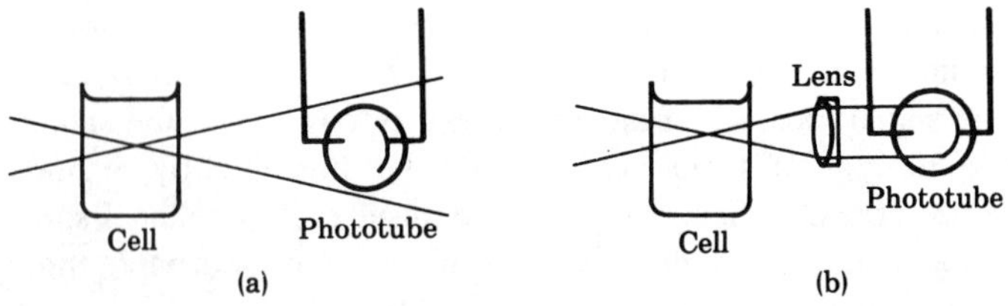

Fig. 1.7 Coupling of sample cell and detector in an absorption spectrometer. (*a*) Poor coupling: allowing divergence of beam from sample cell causes much of signal to bypass phototube. (*b*) Better coupling: insertion of lens before phototube directs the full beam to the phototube.

1.6 INSTRUMENT SPECIFICATIONS: WHICH MODULES DETERMINE?

How well can we relate instrument specifications, which represent a major part of the data furnished to a prospective instrument user or purchaser, to the behavior of modules? Or to put the question more sharply, how often is a specification fixed by a single module? Recalling the earlier discussion on this topic, the answers should be "often." Since we shall now look critically at these matters, it will be instructive to raise these questions with the basic instruments, a spectrophotometer and a pH meter. Their modules are listed in Table 1.2 and grouped according to whether they fall in the characteristic cluster or processing cluster. A detailed comparison of the specifications of the characteristic modules and of the instruments will be given for each.

First, we need to ensure that complexities of design which are beyond the scope of this introductory chapter can be ignored. For instruments of single-channel design we shall assume that manufacturers are employing modules of equal quality

Table 1.2 Modular Clusters in Representative Instruments

	Modules in Generalized Instrument	*Modules in UV–VIS Spectrophotometer*	*Modules in pH Meter*
Cluster of characteristic modules	Signal source	Continuous source and monochromator	pH Cell with reference and glass electrodes (self-source, sample cell, *and* detector)
	Sample module	Sample cell	
	Detector	Photomultiplier tube	
Cluster of processing modules	Amplifier	Amplifier	Amplifier
	Signal processing module	Processing module	Processing module
	Readout	Readout	Readout

for their instruments so that no module will limit performance as the weakest link in the chain.

Even where the design departs from the single-channel pattern, as in double-channel instruments, the modification is actually minimal. As just mentioned, to construct most double-channel layouts, a single-channel instrument is opened up at the sample module. Figures 1.5 and 1.6 show this arrangement. As evident in Fig. 1.5, the characteristic cluster of modules was left essentially unchanged and a dynamic aspect of chopping or modulation was introduced to aid signal processing.

Most other changes in design have to do with impositions of automatic control. For example, a feedback or control loop may be added to stabilize the output of an optical source. The general characteristics of the module are unaffected; the feedback loop ensures that fluctuations in output are minimized. A similar result is secured with computer control as will be discussed later.

To examine the degree to which instrument specifications derive from specifications of individual modules we can consider in some detail the part played by modules in the two instruments detailed in Table 1.2.

Example 1.11. For the UV–VIS spectrophotometer, a long list of specifications will be of interest. The most important of these are likely to be wavelength range, precision of wavelength setting, spectral slit width (resolution), precision in $\% T$ or absorbance, stray light, and scanning speed. What connections can we make between these instrument specifications and those for its characteristic modules? Other performance criteria seem to fall in the processing category: they are factors such as scale expansion, readout (in absorbance, $\% T$, or analyte concentration), and availability of first and second derivatives.

The first characteristic module is the continuous source. Clearly, the spectral range of the source will establish the maximum wavelength range of the instrument. Where higher intensity in part of the range or a still greater range is desired, additional sources will be required. Further, the more intense and the more constant its output, the greater the signal/noise (S/N) ratio of the spectrophotometer: precision in $\% T$ or absorbance will be enhanced.

The monochromator is next. The wider its entrance and exit slits (widths are usually equal), the greater the instrument S/N ratio and its precision in $\% T$. As is well known, in a monochromator a better S/N ratio is gained at the expense of resolution. The wider its slits, the greater also is its spectral slit width and thus the poorer its resolution. The quality and size of the monochromator dispersing element (usually a grating) and optics also directly enhance resolution and efficiency and reduce the level of stray light (at the exit slit). In essence, the monochromator will fix the instrument resolution, S/N ratio, and level of stray light. It may also reduce the spectral range provided by the source.

In the sample cell the essential interaction of energy flow and sample occur. If cell windows and walls reduce S/N and limit spectral range, other cells may be substituted. This module should affect no specifications.

The final characteristic module, the detector, plays a central role since it must not only respond to the intensity of the optical signal transmitted by the sample but also transduce it to an electrical output for passage to the train of processing modules. The response of the photomultiplier tube to light as a function of wavelength will be multiplied by (convoluted with) the intensity versus wavelength function available from the source–monochromator–

sample cell combination. It appears that it may reduce the wavelength range or cause the S/N to deteriorate. Clearly, choosing a photomultiplier tube with the broadest spectral response available and good sensitivity of detection will help preserve specifications the earlier modules have established.

Where is detector performance critical? It has the strongest influence on the limit of detection of measurement and on dynamic aspects of measurements. The higher its gain and the lower its noise, the better the possibility of working at trace concentration levels. If the spectrophotometer must scan rapidly or respond to rapidly chopped radiation, the detector must have a fast response.

A clear picture emerges that the main specifications of the UV–VIS spectrophotometer in fact result from the performance built into it by its characteristic modules. Only if processing modules cause deterioration in performance will the instrument as a whole fail to live up to this advance billing.

Example 1.12. The modular diagram of the pH meter in Table 1.2 is surprisingly simple. The cell is simultaneously sample module, a self-source (i.e., it generates its own signal), and the detector! Such a pile-up of functions is common with electrochemical instruments.

In essence the Nernstian response of the glass-membrane pH electrode relative to the reference electrode of the cell sets the sensitivity of measurement. Precision and accuracy are limited by the inherent difficulty in measuring cell emfs more precisely than 0.1 mV. At low (below 1) and high pH (above 12) interferences cause errors. For this instrument a consideration of characteristic modules is brief indeed. In any event the performance of characteristic modules determine the specifications of the overall instrument.

1.7 INSTRUMENT CONTROL

To operate an instrument requires control of its modular *parameters*, such as the current for a UV–VIS optical source and the operating potential for a photomultiplier tube detector. As different modules are taken up in the set of basic chapters that follow and later chapters on particular techniques, values of operating parameters will be defined. By what means can they be set for general use, adjusted for different samples, and controlled for constancy?

Three kinds of instrument control can be identified:

1. Manual Control. Either the user, or especially for simple instruments, the manufacturer, sets the working values for modules. For example, to operate a spectrophotometer values that must be set include source intensity, monochromator slit width, scanning speed and range, amplifier gain, and system time constant. In addition, the user is furnished sufficient information to permit using the manual controls provided for starting, inserting samples, stopping, and otherwise operating the instrument. If instrument performance varies during measurements, the user must attempt to reset values appropriately.

2. Automatic Control. In automatic operation the user selects the mode of operation, which includes provision for the needed operating values, and also starts and stops the device. To a considerable extent control is provided by feedback loops built in for selected modules by the manufacturer (see examples below). They establish and maintain stable operating conditions through the loops that incorporate servomotors or electronic circuits to ensure that modules are self-standardizing and/or operate automatically at the expected values. By such means automatic completion of a measurement is ordinarily assured during unattended operation.

3. Program Control. For nearly complete control and automation one or more microcomputers (microprocessors) is wired into an instrument (or connections are made to an outside computer). Nearly always programs for control are entered by the manufacturer into the read-only memory. Each set of computer programs includes provision for the user to select a measurement program appropriate to the samples and mode of measurement. For the program chosen, the user is prompted to supply information about the analytes, type of samples available, and other basic information. Together with the data supplied by the operator, the program calls on memory for operating values for modules and circuitry for appropriate data acquisition. Subsequently, the user, or more commonly, a second built-in microcomputer or a laboratory computer, uses other programs to process the analytical data and extract desired chemical information.

Some redesign of an instrument will, of course, be required to effect computer control. Sometimes sensing devices such as position encoders are inserted to furnish data about the state of each module to the computer. More commonly, devices that the computer can actuate are added, such as stepping motors and counters that keep track of the number of steps taken. Through these means program control of scanning and other operations is provided. A good account of the development of a sophisticated instrument with program control was given by Barnard.*

1.8 STATE-OF-THE-ART SAMPLE CHANNELS: TANDEM INSTRUMENTS

To understand the application of the *sample channel* concept introduced in Section 1.5, it is helpful to look closely at sample preparation. For a UV–VIS spectrophotometer, a sample is commonly dissolved. In this case preparation has at least involved a choice of solvent, dissolving process, and analyte concentration. Today dissolving or diluting is increasingly mechanized, especially when many samples are to be examined. With the use of robots and automatic sampling devices programmed for size of aliquot, kind and amount of solvent, and possible complexation with a reagent, one in effect establishes a sequence of operations that constitutes a *preparative* sample channel.

Yet another focus is warranted with the analysis of mixtures. In this case, what would be an ideal sample? In general, it would seem to be one in which its components have been separated in advance so that they can be presented one at a time for measurement. With other components eliminated, these individual analytes will suffer minimal matrix effects during measurement. Further, this arrangement

*T. W. Barnard, *Anal. Chem.*, **52,** 1172 (1979).

ensures that full information about the analyte will be obtained without having other responses show up as background that must be subtracted.

State-of-the-Art Sample Channels. The increasing number of hyphenated or tandem instruments developed in the past several decades offer sophisticated sample channels. The first such device may have been a gas chromatograph–mass spectrometer (GC-MS), but MS-MS, LC-MS, and many other hyphenated instruments have now appeared. The acronyms are believed self-explanatory. These instruments have almost twice the usual number of modules and at present are relatively expensive. In every case the first set of modules, for example, the characteristic modules of a gas chromatograph, *minus a detector*, serves to resolve the sample into separate components. This *separative* sample channel furnishes analytes one at a time to the sample module of the basic instrument.

Requirements for coupling a sample channel to a signal channel are sometimes demanding. The reason is that a flow of matter, for example, analytes in a carrier such as helium in a gas chromatograph, take the place of the flow of energy in a signal channel. If the carrier gas or liquid cannot be tolerated in the sample cell of the basic instrument, it must be separated. The answer is to insert a coupling module or *interface* which will

1. divert or vent the carrier to waste and
2. pass on with minimum loss the steady succession of analytes to the sample module of the basic instrument.

Example 1.13. The gas chromatograph–mass spectrometer design is shown in modular form in Fig. 1.8. Its interface is a jet device that takes advantage of the high-velocity carrier molecules (helium or hydrogen) to let them diffuse into a surrounding vacuum. The lower velocity and much larger momentum of the more massive analytes will cause them to enter the input port of the mass spectrometer, which is directly opposite the jet orifice. After species entering the mass spectrometer are ionized, the resulting ions are accelerated and separated to give a mass spectrum.

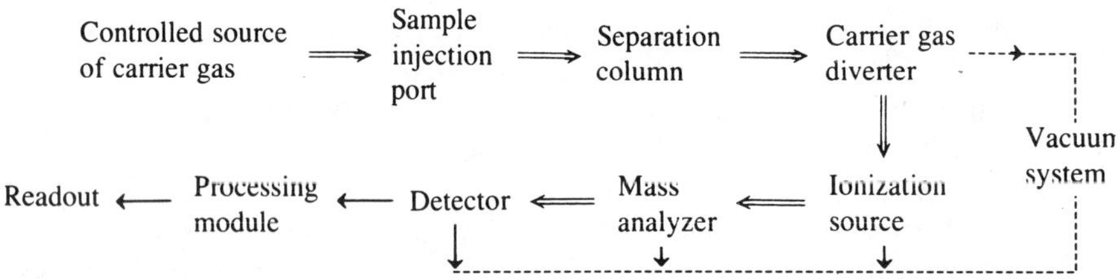

Fig. 1.8 Modular diagram of a GC–MS instrument in which a gas chromatograph sample channel prepares the sample (by resolving it into separate components) for the mass spectrometer. Note the interface in which the carrier gas of the gas chromatograph is separated and pumped out, allowing the sample components to enter the mass spectrometer at appropriate pressures.

When will we be well advised to select a hyphenated instrument with its state-of-the-art sample channel instead of a less complex instrument? Whenever interferences reduce the ability of an instrument to resolve signals from analytes or whenever the number of species in samples increases, some type of instrumental sample channel is needed. Recall that what may be termed extra-instrument methods, such as extraction of analytes into a favorable solvent, have long been employed and are still available for sample preparation. Where an instrumental solution is desired, use of robots to take aliquots, for example, is attractive. Many automatic sample preparation devices are also available. State-of-the-art sample channels enjoy application in research projects where the difficulty is not a high volume of samples but resolving sample mixtures into components. It is clear that the versatility and "sampling" efficiency of state-of-the-art channels is attractive, for tandem instruments are appearing in increasing numbers in laboratory settings.

By application of insights from a modular analysis we see clearly how great is the contribution of the new class of tandem instruments in enhancing analytical power. For a more complete discussion of the potential of this approach the reader is referred to Hirschfeld's paper. [T. Hirschfeld, *Anal. Chem.*, **52**, 297A (1980)].

1.9 PROCESSING MODULES

It is important now to look at the processing cluster. As a signal enters this part of an instrument, it will nearly always need to be further boosted in amplitude by a stable amplifier. In designing the cluster, a decision must be made whether after amplification processing will be carried out with the signal still in analog form (a signal of varying amplitude). Except in simple instruments the decision today is nearly always to convert it to digital form (a series of pulses in binary form, whose number indicates the signal amplitude). The conversion, termed a change from analog to digital domain, allows replacement of virtually all of the earlier analog processing modules by one or more microcomputers. As noted, these digital computers (microprocessors) bring the flexibility and versatility of program control to processing. A further reason for digitizing is the greater ease offered in providing control of modules by software executed by a microcomputer.

Even in an entirely analog instrument, it is the processing modules that will provide necessary filtering, taking off the ratio of the signal from the sample channel to that from the reference channel in a double-channel spectrophotometer, and providing other operations on a signal necessary to a particular type of technique.

In either digital or analog instruments, the readout may be a chart recorder, oscilloscope, or printed record. Some different types of display or readout module will be described in Chapter 3.

An insightful query is: How does the addition of microcomputers affect instrument performance? For an answer let us look both at the characteristic and processing clusters of modules. In the characteristic modules, currents, slit widths, etc. will now be set by digital devices actuated by computer program instead of the user or manufacturer. Yet the values will have been entered in the program by one of these persons. Further, every module will still interact with signals in the signal channel through its transfer function and input and output characteristics.

In the processing cluster the microcomputer will take the place of several modules. Important gains in the ease and quality of signal "massaging," data collection, and computation will result. Even in Fourier transform spectrometers a computer provides processing by shifting interferometer signals encoded with analytical information from the time domain to the frequency domain to allow them to be interpreted by the user.

With software control and processing is a modular interpretation of performance in the signal channel still valid? The signal containing the desired information about the chemical sample will still originate in the characteristic modules and it is they that will impose the performance criteria critical to the particular measurement technique. Thus, we can conclude that instrument specifications always depend ultimately upon performance of individual modules even in sophisticated instruments.

BASIC ELECTRONICS

Electronics and electrical circuitry enter the field of chemical instrumentation at three main points. First, many electrical and magnetic properties, for example, electrochemical cell potentials and NMR spectral transitions, require appropriate circuits in their measurement. Second, most properties of substances are sensed by detector–transducers like photomultiplier tubes that produce a current or voltage output signal. Not only do electrical circuits enable these detectors to operate stably, but use is made of the striking advantages of electronics to amplify detector outputs while introducing minimum noise. The third point of entry of electronics is through digital microcomputers which provide sophisticated processing and display of signals and extraction of chemical information. Through appropriate software microcomputers can also tailor the operation of instruments to measurements being made. Accordingly, Chapters 2–5, which comprise the bulk of this division of the book, attempt the development of a working knowledge of the basic electrical and electronic circuitry (both analog and digital) that finds use later in instrument modules. Chapter 6 concludes the division with a discussion of microprocessors and microcomputers.

Chapter 2

ASPECTS OF ELECTRICITY AND ELECTRONICS

2.1 INTRODUCTION

Chemical information obtained by an instrument is nearly always encoded as an electrical signal. Indeed, as has already been made clear, this type of encoding allows the power of electronics to be called on for useful processing. For example, electronic circuits can amplify weak signals, ensemble-average instrument outputs to extract information from a forest of noise, or divide one signal by another as in a measurement of sample light transmission.

It is the purpose of this chapter to review and reinforce the basic aspects of electricity needed for a working approach to instrumentation and to introduce the first solid-state device we shall deal with, the diode. A good place to begin is Table 2.1 with its compilation of essential electrical variables. At the top of the table are the variables most likely to represent chemical information in a signal: voltage, current, and frequency. We shall want to keep track of these aspects of signals. The next set of variables, resistance, capacitance, and inductance, are the most important properties for determining the behavior of a circuit. Yet they are called passive properties because they reduce the amplitude of signals. For completeness, electrical charge and power are also included in the table.

2.2 VOLTAGE AND CURRENT; OHM'S LAW

In dc *and* ac circuits Ohm's law interrelates current, voltage, and resistance quantitatively. Recall that current exists in a closed circuit that also incorporates a source of voltage such as a power supply. In dc circuits the variable relating voltage V and current I is resistance R. Ohm's law quantifies the relationship

$$V = IR \qquad \text{or} \qquad I = V/R. \tag{2.1}$$

Stated differently, resistance is a measure of the ability of a circuit to limit current. From the equation it is clear that resistance is also a linear property. For example, doubling the voltage in a circuit will double the current.

Ohm's law will stand as perhaps the single most valuable relationship for interpreting electronic circuits. Examples of its application are discussed below.

Table 2.1 Electrical Quantities[a]

Quantity	*Symbol*	*Unit (abbr)*	*Component/Symbol*[b]
Voltage, potential difference	V, E	volt (V)	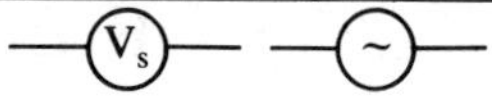
Current	I	ampere (A)	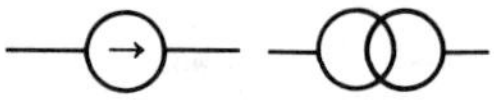
Frequency	f	hertz (Hz)	
Resistance	R	ohm (Ω)	Resistor 
Capacitance	C	farad (F)	Capacitor 
Inductance	L	henry (H)	
Impedance	Z	ohm (Ω)	Inductor 
Charge	Q	coulomb (C)	
Power	P	watt (W)	

[a]Some notes are in order.

SI prefixes are universally used to scale these electrical quantities up and down. For example, MW refers to a megawatt (10^6 W) and nA to a nanoampere (10^{-9} A). In the older literature prefixes such as $\mu\mu$ for p (pico), as in $\mu\mu$F, are also found.

Though names of units derive from proper names, they are not capitalized. Abbreviations of units, however, are capitalized. For example, hertz and megahertz are abbreviated Hz and MHz.

Even in current literature the omega is often omitted in stating the size of a resistor, for example, one sees 10k (or 10K) for 10 kΩ and 5 M for 5 MΩ.

When lowercase letters are used for current and voltage, they often stand for instantaneous values or ac values.

[b]Symbols given for voltage and current designate voltage and current sources.

Some Definitions and Conventions. Before pursuing applications of Ohm's law, recall that a *voltage* is always a difference of potential *between two points*. For example, a car battery produces a standard 12 V between its positive and negative terminals and the power line in households and apartments provides an alternating 115 V (usually 230 V outside the United States) between the terminals of power outlets. In the same sense, reference to the voltage at a point in a circuit will signify the value of the potential difference between that point and ground. Usually a *ground*, for example, one used for a circuit, is a connection to a metal water pipe or to the common terminal of a power line. If a circuit is not grounded, it can assume any potential relative to ground.

Further, the *current* in a circuit is taken by convention to be a *positive* current moving from a point of higher (more positive) potential to a point of lower (more negative) voltage. A convention is actually a necessity since the flow of charge that constitutes a current is borne by charge carriers that sometimes are negative in sign (electrons and negative ions) and at other times are positive (holes and positive ions). All that has been said applies also to alternating current, as will be made explicit below.

Another convention is that the loop character implied by the name circuit is often not shown in a drawing. The reason is straightforward: many circuits are completed through a metal chassis on which they are mounted and others are completed through ground.

If a dc source supplying power to a circuit is replaced by an ac generator, the current will constantly change in amplitude and reverse its direction periodically. To permit consistency in calculating power expenditure in ac and dc circuits an *effective or average ac current and voltage* can be defined. These root-mean-square (rms) values are by definition the square root of the average square of the instantaneous ac current or voltage over a cycle. For example, the effective ac current I is expressed by $I = I_p/\sqrt{2} = 0.707I_p$, where I_p is the peak value of the ac current. The effective voltage V for an ac circuit has a parallel definition: $V = 0.707V_p$. It is interesting to note that the peak voltage in an ac line rated at 115 V (its effective voltage) is just $V/0.707 = 162.6$ V.

The fact that most circuits are networks and contain at least one branch is a complication. In such a case Ohm's law is not immediately useful and interpretation must wait on more complex calculation or systematic reduction of the network in some fashion.

Kirchhoff's laws. Consider the example of a branching circuit given in Fig. 2.1. Even in this simple case, the circuit is properly called a *network*. It does not matter what the components in the branches are. Two simple rules for dealing with branching were developed by Kirchoff.

1. Currents in the leads going into a junction or *node* (f or g i Fig. 2.1) must sum algebraically to zero. In other words, charge must be conserved: all current "going into" a node must leave the node. The sign of a current will indicate its direction based on the convention for positive currents. Mathematically, $\Sigma I_j = 0$, where I_j is the current in each branch j at the node.

2. Voltages around any *loop* in a circuit must sum to zero. Mathematically, $\Sigma V_j = 0$, where V_j is the voltage drop across each component j in a loop.

It is beyond the scope of this book to deal more completely with networks than in the two examples that follow.

Example 2.1. What can we learn by applying Kirchhoff's rules qualitatively to the circuit of Fig. 2.1? The first law tells us that at point f the current must divide between components A and B and on emerging on the far side at g will again combine and be equal to its original value.

It is self-evident that the potential drops in the branches between points f and g in Fig. 2.1 must be identical. What does Kirchhoff's law add in this simple case? Basically, it calls attention to the loop that branches A and B form and emphasizes that voltages around the loop must add to zero, that is, $V_A + V_B = 0$.

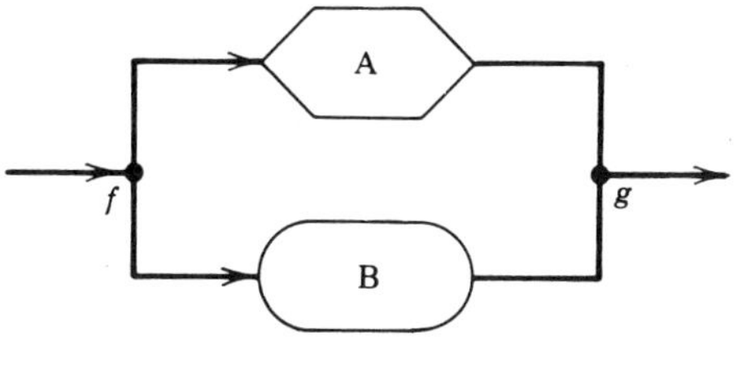

Fig. 2.1 A simple electrical network.

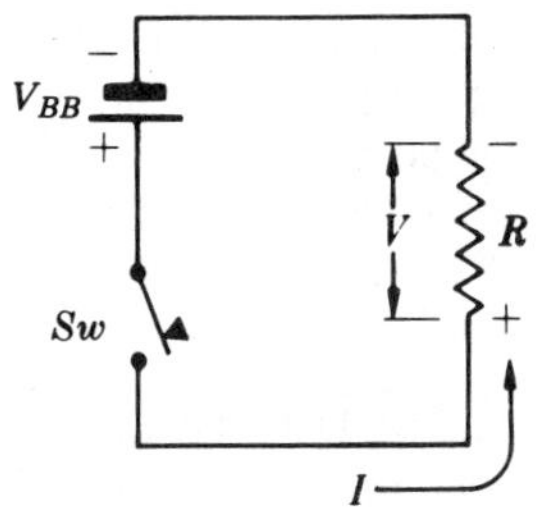

Fig. 2.2 Schematic diagram of a simple dc resistive circuit. When switch Sw is closed, current I exists in the circuit. The voltage drop across R is equal, and opposite in sign, to V_{BB}, the battery voltage.

Example 2.2. What can one learn by a detailed application of Kirchhoff's and Ohm's laws to the circuit of Fig. 2.2?

One develops a feeling for the magnitude of the basic variables:

1. Ohm's law defines the current in the circuit: $I = V_{BB}/R$.
2. Stating Ohm's law as $V = IR$ reminds us that there is a *voltage drop* of magnitude IR across the resistor.
3. According to Kirchhoff's first law, the same current must exist in all the components in a loop.
4. Kirchhoff's second law implies that the direction of the voltage drop across R must be opposite to that of the battery: the two voltages must add to zero algebraically.

Power Consumption. How much of the power in signals (power $P = VI$) is lost in the resistance of a circuit? Recall that power is given in watts (W) and represents energy expended per unit time. The definition of power $P = VI$ is valid for calculation of instantaneous power regardless of the type of signal. For example, by use of Ohm's law the power dissipated in a resistance R is found to be just

$$P = VI = V^2/R = I^2R. \tag{2.2}$$

A radical difference between the expenditure of power in resistance and in the two other passive qualities, capacitance and inductance, must be noted: all resistance degrades electrical energy into heat. In contrast, capacitances and inductances consume power to build up related electrical and magnetic fields. In all three cases electrical power has been instantaneously consumed; but insofar as capacitances and inductances are concerned, there is a reasonable chance of recovering it usefully. Only a resistor dissipates power.

Voltage and Current Sources. What is ideally sought in a dc *voltage source*? Surely, it should be able to deliver a constant voltage of reasonable amplitude regardless of loading, temperature, humidity, and so forth. Consider the voltage requirement first. Loading, that is, supplying power to a load, may well reduce the output voltage of a source, as will be discussed in the next section. The nature of the problem becomes clear by noting that if the resistance of a load changes, since $V = IR$, the current supplied to it by the source must vary in inverse proportion to keep V constant. Thus an ideal voltage source must be able to furnish whatever current is needed to keep its output voltage constant. The ability to do so is called *current compliance*. While sources can furnish small currents easily, few can supply the large currents necessary for loads with very small resistance, since they will approximate short circuits. It appears that real sources are most likely to have good current compliance for *loads of moderate to high resistance*.

How similar are the demands on a *current source*? Ohm's law will govern its current–voltage behavior also. It must supply any voltage needed to keep its output current constant ($I = V/R$) as the load changes. In this source, *voltage compliance* is needed. While a short-circuit load would offer no voltage compliance problem to a practical current source, only an ideal current source could still supply a constant current to a high-resistance load. By and large, good current sources are more difficult to design and more expensive than are good voltage sources.

Some *practical sources* should be mentioned here. Batteries, while mainly used in portable instruments, are a common voltage source. The difference in discharge characteristics of the standard types of dry cell used in smoke alarms and flashlights and the mercury cell found in automatic cameras and hearing aids is worth noting. A dry cell initially furnishes 1.5 ± 0.08 V and has an internal resistance of about $\frac{1}{4}\Omega$. Its voltage falls steadily toward 1V as it is used and resistance rises to several ohms. By contrast a mercury cell provides a constant 1.35 ± 0.01 V through 95% of its useful life.

Sources that give either a precisely reproducible voltage or current under fixed conditions are vital for calibration of instruments and measurement procedures. Zener diodes (see Section 2.11) are often used as practical constant voltage sources. Operational amplifier circuits provide still better constancy as analog current and voltage sources (see Section 4.4). Power supplies can be designed to have quite good characteristics, as will be seen in the next chapter. Finally, programmable digital sources are still more versatile, precise, and expensive.

While resistance will be discussed at some length in the next section, it is worth looking ahead and noting that in ac circuits the term impedance (Z) includes both the dissipative behavior characteristic of resistance and also storage aspects peculiar to capacitance and inductance.

2.3 RESISTANCE, RESISTORS, AND SOME APPLICATIONS

Resistors are among the most important components of electronic circuits. For example, they are the heart of most voltage dividers and potentiometers, provide start-up loads for power supplies, and are used with capacitors to develop a wide range of frequency-sensitive devices such as filters.

Resistors. A resistor is distinguished by its nominal resistance, design tolerance (common tolerances are ±5, ±10, and ±20%), power rating ($\frac{1}{4}$ W to many watts, the size increasing in proportion to the ability to dissipate power), and temperature coefficient of resistance (see discussion following). Since any resistor heats during use, the last parameter must be taken into account in precision applications.

Common resistors are fabricated by molding a weighed amount of graphite powder and resin into a cylinder, imbedding a long lead wire in each end, adding a color code on the jacket to identify the value of resistance and usual variability (design tolerance), and baking to form the finished product. What values of resistance are available? Common resistors are supplied in magnitudes from 0.01Ω to 100 MΩ in certain *preferred values*. As would be expected, many values of resistance are available in the series with ±5% tolerance and relatively few in the series with ±20% tolerance. Wire-wound resistors are available to take care of the need for high-wattage, low-resistance types. Many values are available and design tolerances are usually ±1% or smaller.

In the manufacture of *precision resistors* not only are closer tolerances maintained, but the amount of capacitance and inductance introduced during fabrication is kept to a minimum. Common design tolerances for precision resistors are ±1%, and in the best cases tolerances are ±0.005%. Wire wound resistors of ±1 and ±0.5% precision are sometimes employed. More commonly the same precision is obtained with less cost by depositing appropriate metal films on a nonconducting base. For resistance bridges like the Wheatstone bridge, high-precision resistors that have a low temperature coefficient of resistance are preferred. Wire of manganin, an alloy of Cu/Mn/Ni which has a coefficient of roughly $1 \times 10^{-7}\ \Omega\ \Omega^{-1}\ K^{-1}$ in the room-temperature range, sees many applications of this kind.

A range of variable resistors whose resistance can be altered mechanically is also available. The type called a *potentiometer* or *pot* is made by winding uncoated high-resistance wire on a circular nonconducting form. As its symbol  suggests, it has a wiper which is a flat contactor that can be moved along the wire by a turn of a knob. A variable resistance is obtained when one connection is made to the wiper and the other to either end of the winding. Pots see general use as smoothly variable but somewhat imprecise resistors. In video monitors for instance they are used as amplitude and position controls. Sometimes these variable resistors are used as resistance *trimmers*. For example, instead of placing a large, expensive precision resistor in series in a circuit, commonly one couples a regular resistor of approximately the right size to a small pot. Then the pot is adjusted during a calibration step to give the required total resistance. Finally, high-precision potentiometers can be made by winding a long precision resistance wire onto a helical insulator of 3–10 turns.

Since resistance is a temperature-dependent property, materials whose resistance is highly reproducible during repeated temperature cycling make good *resistance thermometers*. Platinum is such a material. Indeed, the platinum resistance thermometer is a secondary temperature standard over the range from −183 to +630°C. The average *metal temperature coefficient of resistance* is about $+0.003\ \Omega\ \Omega^{-1}\ K^{-1}$ at room temperature.

Mixtures of semiconductors that have large negative temperature coefficients of resistance are formed into beads and also used as resistance thermometers. They are commonly used under the generic name *thermistors*. Though not as reproducible in resistance, they are nonetheless attractive temperature sensors. The difference in thermal behavior of metals and semiconductors is shown in Fig. 2.3. Because of the large temperature coefficients of the latter a particular thermistor can usefully cover only a range of about 100–200°C near room temperature.

How troublesome is the heating caused by *power dissipation* in resistors? The accompanying rise in temperature will alter the value of resistances, increase currents in semiconductors, alter dimensions of components because of thermal expansion, and change alignment in layouts. Fortunately, with integrated circuit chips and other semiconductor devices, currents are small and dissipation is minimal. The packing density of integrated circuits, discrete transistors, diodes, and resistors on most circuit boards is kept deliberately low so that natural radiation or convection will provide sufficient cooling. Where packing density is high, however, forced cooling is required.

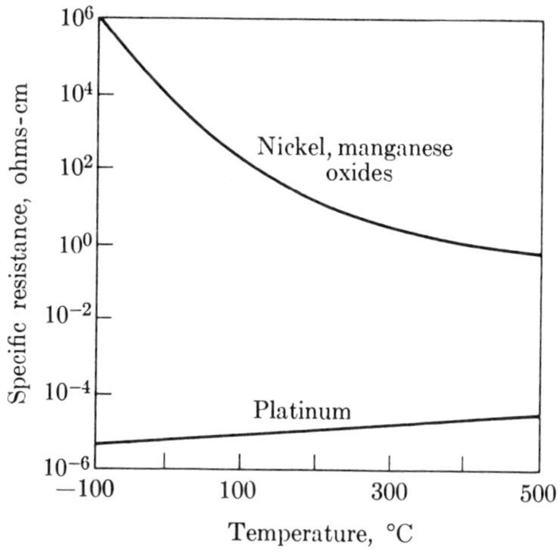

Fig. 2.3 Comparison of temperature coefficients of resistance for a representative metal (platinum) and semiconductor (oxide mixture).

In instruments the effect of temperature changes on operation is minimized by allowing time for preliminary warm-up to safe steady-state temperatures. Where levels of use are sufficient and power dissipation is small, it is common for instruments to be left on for long periods to avoid warm-up delays and wear. A third option for sensitive modules such as crystal oscillators that supply precise frequencies is to thermostat them.

Example 2.3. The schematic diagram of an amplifier circuit calls for a 5-kΩ resistor that will carry effective currents of up to 10 mA. How should one go about selecting an appropriate preferred value for the resistor? Assume that a resistor in the ±10% tolerance range will be adequate.

First, check a standard list of preferred resistors in the ±10% series. The value closest to 5 kΩ is 4.7 kΩ. Second, a safe wattage rating must be estimated. The maximum power dissipation expected is $I^2R = (0.01\ \text{A})^2 \times 4.7 \times 10^3\Omega = 0.5\ \text{W}$. From the resistor list we see that preferred values are available in $\frac{1}{4}$-, $\frac{1}{2}$-, 1-, and 2-W sizes. A 4.7-kΩ resistor with a 2-W rating would seem to be a good choice since this wattage would ensure a margin of safety.

Example 2.4. How are resistance wires used as thermal conductivity detectors in gas chromatography?

In the simplest case two matched sections of nichrome wire (a Ni/Cr alloy) are arms of a Wheatstone bridge. The wires are centered in separate compartments of the detector to permit the eluant from the column to flow past one wire and a stream of carrier gas (under the same conditions) to flow past the other.

A steady current is passed through the bridge, causing the wires to reach a steady-state temperature well above the temperature of the gas. The flowing gas cools each wire in proportion to its thermal conductivity. A carrier gas like helium, which has a very high thermal conductivity, will ensure a substantial drop in the temperature of both wires and a consequent drop in their resistances. Carrier gas diluted by an eluting component will almost certainly have a lower thermal conductivity. As a result, the reference detector wire bathed by pure carrier will be cooled more and its resistance will be lower, unbalancing the bridge. The extent of unbalancing will be the output signal.

The process of estimating currents and voltages in electrical circuits may be greatly simplified by using Ohm's law to develop a single *equivalent resistor* that will take the place of a set of series or parallel resistors.

Series Resistances. If two or more resistors are in *series*, as in Fig. 2.4, an equivalent resistor is determined as follows. There is a single current I in the circuit giving voltage drops IR_1 across R_1 and IR_2 across R_2. The total voltage drop, which of course equals the voltage of the source is $V = IR_1 + IR_2 = I(R_1 + R_2) = IR_{eq}$, where

$$R_{eq} = \Sigma R_i. \tag{2.3}$$

Thus, resistances in series add.

Parallel Resistances. How may one model resistances in parallel as in Fig. 2.4? Kirchhoff's laws tells us that there are equal voltage drops across the resistors and that the total current I in the main part of the circuit is the sum of the currents in each separate resistance, $I = I_1 + I_2 = V/R_1 + V/R_2 = V/R_{eq}$. The resistance equivalent R_{eq} is given by

$$1/R_{eq} = 1/R_1 + 1/R_2 \tag{2.4a}$$

On rearrangement, Eq. (2.4a) yields

$$R_{eq} = R_1R_2/(R_1 + R_2). \tag{2.4b}$$

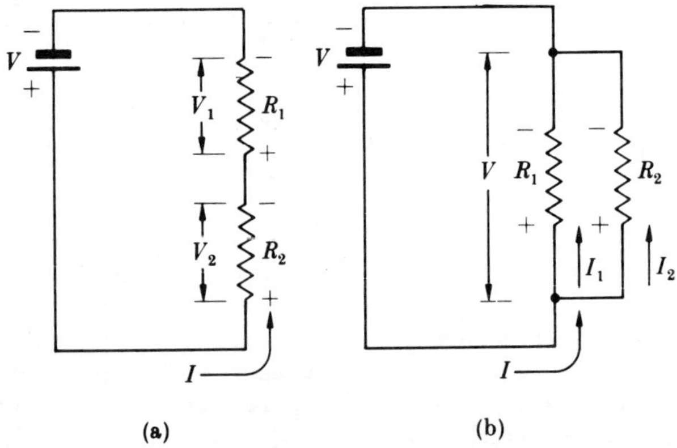

Fig. 2.4 (a) Resistors in series. (b) Resistors in parallel.

For the general case of resistances in parallel,

$$1/R_{eq} = \Sigma(1/R_i). \tag{2.5}$$

Example 2.5. If a 10-kΩ resistance is placed in parallel with a 100-kΩ resistor, can the equivalent resistance be approximated by the value of the smaller resistance?

Yes, the suggested approximation is valid. Applying Eq. (2.4b) gives an exact value of 9.09 kΩ.

Example 2.6. Attention should be called to a special case. When several parallel resistors have the same value, Eq. (2.4b) yields an equivalent resistance of $R_{eq} = R/n$, where n is their number. This result is so simple that it suggests a useful short cut to answering questions like the following: What is the value of R_{eq} if a 20-kΩ resistor is in parallel with one of 40 kΩ?

Consider the 20-kΩ resistor as a parallel pair of 40-kΩ resistors and the whole as three 40-kΩ resistors in parallel. Then $R_{eq} = 40/3 = 13.3$ kΩ.

Voltage Divider. Series resistors are often employed as a voltage divider. The pair in Fig. 2.5 serves as a good example. Voltage V is in effect divided into two parts. If output leads are attached to the top and bottom of resistor R_1, output voltage V_0 will be available:

$$V_0 = R_1V/(R_1 + R_2). \tag{2.6}$$

Note that the fraction of the supply voltage V appearing as V_0 equals the ratio of R_1 to the total resistance of the series. If the voltage across R_2 is of interest instead, the leads can be moved to that resistor. The output of any source of fixed or varying voltage can be so divided. Precision resistors will of course be necessary for a precision voltage divider.

Unfortunately, drawing power from one resistor of a voltage divider, that is drawing current as well as voltage, will affect the output voltage V_0. Only if minimal current is drawn can we expect the output predicted by Eq. (2.6). A way to calculate the precise voltage furnished at different currents will be developed in the next section using the Thevenin model.

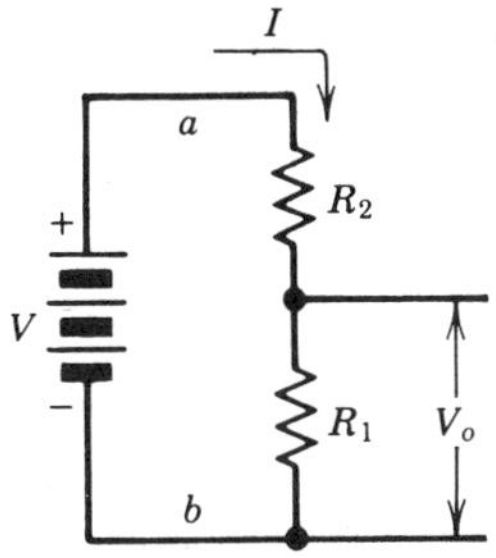

Fig. 2.5 A dc voltage divider circuit. The output V_o is the fraction of the total voltage V that exists across resistor R_1.

2.4 DEVELOPING MODELS OF CIRCUITS: THEVENIN'S THEOREM

Analyzing the behavior of a circuit can be simplified by developing a model of all or part of the circuit. Recall that this approach was used in treating groups of resistors as a single equivalent resistor R_{eq}. In many ways electrical modeling is analogous to using a molecular orbital model of a substance to simplify description of its energy levels. The circuit model, termed an *equivalent circuit*, behaves exactly like the original circuit in terms of the aspect of interest.

A *Thevenin equivalent circuit* is one of the most useful circuit models. The theorem undergirding the model states that any two-terminal network of resistors and voltage sources can be replaced by an equivalent circuit containing a voltage source and a series resistor. It proves easiest to explain this type of modeling by applying it to a common circuit.

Consider how the voltage divider pictured in Fig. 2.6a can be reduced to its Thevenin equivalent to develop a simple expression for V_0, the dc output of the voltage divider when a load is put across R_1. The equivalent circuit that results (shown in Fig. 2.6b) will deliver a *voltage and current identical to those furnished at the terminals of the original circuit.*

1. From the algebra of the Thevenin theorem it can be shown that the equivalent voltage V_{th} is the *open circuit potential* V_0. This value is just the portion of

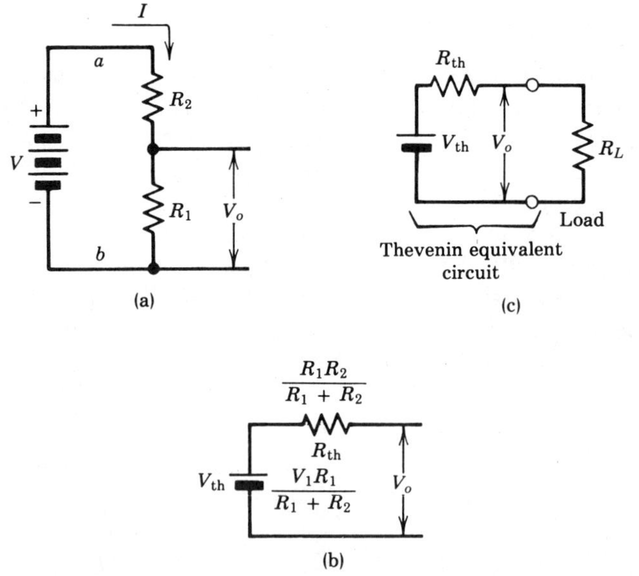

Fig. 2.6 A resistive voltage divider and its Thevenin equivalent circuit. (a) A dc voltage divider circuit. The output V_o is the fraction of the total voltage V that is dropped across R_1. (b) The Thevenin circuit equivalent to the voltage divider. (c) The Thevenin equivalent circuit connected to a load of resistance R_L. Note that R_{th} and R_L form a new voltage divider.

the total voltage V that appears as voltage drop IR_1. To calculate it, use Ohm's law to obtain current I in the voltage divider and then calculate IR_1. We have

$$I = V/(R_1 + R_2)$$

and

$$IR_1 = VR_1/(R_1 + R_2) = V_0 = V_{th} \tag{2.7}$$

2. Similarly, the Thevenin resistance R_{th} can be shown to be the *equivalent resistance* of the circuit. To calculate it, replace the voltage source by its internal resistance (often it can be assumed to be zero). In terms of the circuit of Fig. 2.6a consider that the battery is shorted by a wire. Since points a and b are now connected, R_{th} is the combination R_i and R_2 in parallel:

$$1/R_{th} = 1/R_1 + 1/R_2 \quad \text{or} \quad R_{th} = R_2R_1/(R_2 + R_1). \tag{2.8}$$

Example 2.7. In the circuit of Fig. 2.6a let $V = 45$ V, $R_1 = 800\Omega$, and $R_2 = 2200\Omega$. Calculate the values of V_{th} and R_{th}.

From Eqs. (2.7) and (2.8)

$$V_{th} = 45 \times 800/(800 + 2200) = 12 \text{ V}.$$

$$R_{th} = 800 \times 2200/(800 + 2200) = 587\Omega.$$

Example 2.8. Given the values of V_{th} and R_{th} just calculated, find the value of V_0 in Fig. 2.6c for different load resistances R_L. What is the value of V_0 when (a) $R_L = 1\text{k}\Omega$; (b) $R_L = 10\text{ k}\Omega$; (c) $R_L = 100\text{ k}\Omega$?

Note that R_{th} and R_L are in series and form a *new* voltage divider. For each value of R_L the current will be different. Nevertheless, in each case the *same* current will be in R_{th} and R_L and the resistance fraction $R_L/(R_L + R_{th})$ will determine the fraction of V_{th} that appears across the load. The values of V_0 are as follows:

(a) When $R_L = 1\text{ k}\Omega$, $V_0 = 12 \times 1000/(1000 + 587) = 7.6$ V.
(b) When $R_L = 10\text{ k}\Omega$, $V_0 = 12 \times 10{,}000/(10{,}000 + 587) = 11.3$ V.
(c) When $R_L = 100\text{ k}\Omega$, $V_0 = 12 \times 100{,}000/(100{,}000 + 587) \simeq 12$ V.

From the examples just worked out, it is apparent that a resistive voltage divider is a far from constant voltage source. Only when $R_L >> R_{th}$ does a source deliver to a load a voltage close to the Thevenin value. In Sections 2.12 and 3.8 more nearly constant voltage sources will be explored.

Output Resistance and Circuit Loading. From this discussion it is possible to develop a general conclusion about a voltage source. Look again at Fig. 2.6c. Load R_L forms a new voltage divider with R_{th} and part of the available voltage will appear across R_{th}. The problem lies in the finite *internal* or *source resistance* of the device. For any voltage source to deliver almost all of its output voltage to a load, the necessary condition is $R_{load} >> R_{internal}$. Fortunately, it proves possible to design voltage sources with equivalent internal resistance of only a few milliohms by use of active modules such as operational amplifiers (see Section 4.4).

2.5 ac AND SPECIAL SIGNALS

What kinds of periodic and special signals must be accommodated by instruments? The foundation being developed for dc and slowly varying dc signals, such as those provided by a phototube during a spectral scan or by an ion-selective electrode in an unknown solution, is now ready to be extended to the broader range of signals encountered in practice.

Sinusoidal Signals. The lower frequency part of the electromagnetic spectrum consists, of course, of sinusoidal or alternating current (ac) waves ranging from power frequencies through audio, radio, and microwave frequencies.*

Further, the modulation of instrument signals gives audiofrequency (af) signals ranging from 20 to 10,000 Hz. In NMR and ESR spectrometry radiofrequency (rf) signals range from 0.05 to 500 MHz and 1 to 20 GHz, respectively.

In an ac circuit the voltage may be represented as a sine wave (note the analogous treatment of electromagnetic optical radiation in Chapter 7) whose instantaneous value V_i is given by

$$V_i = V_p \sin 2\pi ft = V_p \sin \omega t. \tag{2.9}$$

Here V_p is the peak value (amplitude) of the voltage and ω is the angular frequency ($\omega = 2\pi f$). The period T (in seconds) of each cycle is, by definition, the reciprocal of the frequency f. If a circuit contains only resistance, the instantaneous current I_i is in phase with the voltage and may be expressed by a similar relation:

$$I_i = I_p \sin \omega t. \tag{2.10}$$

If current and voltage are not in synchronism, a phase angle ϕ will appear in the expressions; for example, Eq. (2.10) will become $I_i = I_p \sin(\omega t + \phi)$.

If many frequencies are present in a signal, they will add algebraically. A full treatment of such aspects is given in the analogous discussion of light waves in Chapter 7.

The amplitude of a sinusoidal or ac signal is commonly defined as the peak value V_p or I_p, as in Eqs. (2.9) or (2.10), but alternative definitions are valuable in particular applications. Recall that root-mean-square values of ac voltages and currents are ordinarily given where power concerns dominate. For example, ac power line voltages are always rms values. Since $V_{rms} = 0.707\ V_p$, conversions are easily made. Very occasionally peak-to-peak values such as $2\ V_p$ are cited.

*Electrical frequencies give way to optical frequencies at about 10^{11} Hz (100 GHz), the boundary between microwaves and long wavelength infrared waves.

A quite different approach is sometimes taken where a comparison is to be made between amplitudes of current or voltage in different parts of a circuit. Where attention is actually being directed to power levels, such a ratio will be defined in *decibels*. The ratio of power available in one part of a circuit, P_2, compared with that in another, P_1, is given by

$$\text{Power ratio (in decibels, dB)} = 10 \log_{10} (P_2/P_1). \qquad (2.11a)$$

If input or output voltages are to be compared, however, the equation becomes

$$\text{Power ratio (in decibels, dB)} = 20 \log_{10} (V_2/V_1). \qquad (2.11b)$$

Equation (2.11b) is derived below in Example 2.10.

Example 2.9. What is the power gain in decibels when the power level in a system is doubled? From Eq. (2.11a) the decibel gain is seen to be 10 log (2/1) = 3 dB.

Example 2.10. The voltage gain in an amplifier at 40 kHz is 200 but at 100 kHz the gain is only 100. By what factor is the power gain down at 100 kHz? Use Eq. (2.11a) for the calculation.

An additional assumption is needed before Eq. (2.11a) can be used, that the amplifier output voltage appears in both cases across essentially the same resistance R. Then $P = V^2/R$ can be substituted for P in Eq. (2.11a). The ratio P_2/P_1 reduces to V_2^2/V_1^2 and Eq. (2.11b) is obtained. Substituting the given output voltages in the latter equation, one finds that at 100 kHz the power gain is down by $20 \log V_2/V_1 = 20 \log 200/100 = 6$ dB.

Other Common Signals. A variety of other signals is shown in Fig. 2.7. A few interpretative comments about each type is in order.

A *square wave* has a frequency and period, as does a sine wave. A real square wave does not have infinitely steep sides, however, and its *rise time* t_r is an important specification. This interval is the time required for the signal amplitude to go from 10 to 90% of final value. Representative values of t_r for square waves are hundreds of picoseconds to a few microseconds.

A *sawtooth wave* is also periodic in form. As is evident in Fig. 2.7, a *ramp* is a linearly rising or falling voltage. It is characterized by a slope and an initial and final value. At some amplitude, usually just below the dc voltage that powers its generator, it limits, that is, levels off to a constant value. One can also have a current ramp.

Connecting a rising and falling ramp gives a symmetrical signal identified as a *triangular wave*. Slope and period will characterize such a wave.

A related kind of signal is the *pulse*. As shown in Fig. 2.7*e*, a pulse is distinguished by an amplitude and a width. Pulses need not be square as shown; many will be rounded. A periodic series of pulses can easily be generated. In this case pulses are described by their repetition rate (frequency), repetition interval (period), and a *duty cycle*. The last term gives the percentage of each period occupied by a pulse. For example, a train of pulses that are widely spaced might have a 5% duty cycle. In digital circuits, pulses whose amplitudes correspond to *logic states* are the basic signals.

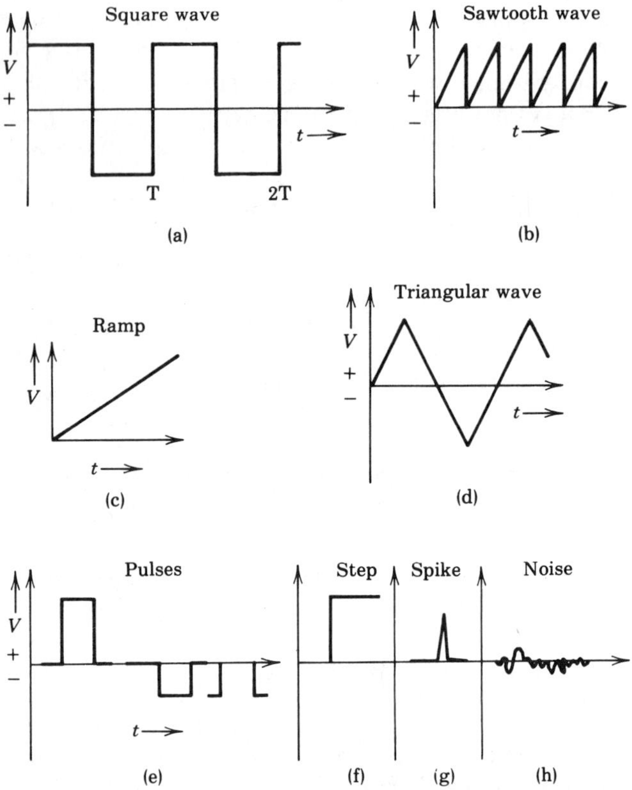

Fig. 2.7 Important types of signals other than sinusoidal waves encountered in electronics. (a) Square waves are generated by crystal clocks, whch are crystal-based oscillators, and by many relaxation oscillators. They are found in all digital circuits. (b) Sawtooth waves are commonly generated by partial charging of a capacitor followed by sudden discharging to return the voltage to the baseline. Sweep circuits in oscilloscopes and video monitors are driven by sawtooth waves. (c) Ramps are widely used in digital voltmeters and integrators. They are generated by constant-current charging of a capacitor in an operational amplifier feedback circuit. (d) Triangular waves result when a positive-going ramp is followed by a negative-going one. A common use is as a voltage sweep in cyclic voltammetry. (e) Pulses are generated by flip-flops in counting circuits and by one-shots and related devices to operate components for definite intervals. They also find use in measurement techniques such as pulse polarography. (f) A step signal is basically the leading edge of a pulse and sees use in testing circuits for rise time and so forth. (g) While spikes appear randomly as a part of larger-amplitude noise, they are commonly generated by capacitor discharge to trigger operation of a circuit. (h) Noise is ubiquitous. White noise (of all frequencies) may be generated by a random-frequency circuit.

In Fig. 2.7*f*, *g* and *h* there are also three less common signals. *Spikes* are generated as signals to trigger a circuit into activity. A *step signal*, which can be taken to be the initial rise of a pulse or square wave, makes a useful signal for testing the response time of a circuit or module. Finally, signals of random frequency and amplitude provided by *noise* generators are occasionally used for electronic purposes. For example, broad band noise, electronically generated, is used to decouple proton NMR transitions from ^{13}C NMR transitions.

2.6 CAPACITORS AND *RC* CIRCUITS

Capacitors are the second most widely used type of passive linear circuit component. In behavior they are somewhat like frequency-dependent resistors. This distinctive property causes capacitors to be valuable in filtering, blocking, and bypassing signals according to frequency. Less obviously, they are indispensable in the generation of waveforms and in differentiating and integrating signals.

To begin to account for their behavior recall that capacitors store charge Q on conducting plates and build up a potential difference V between the plates in the process. Capacitance is defined simply as the ratio of these two quantities:

$$C = Q/V. \tag{2.12}$$

Actually, some capacitance exists, often quite minute, between any two points of a circuit not directly connected by a conductor and affects circuit operation at very high frequencies. The unit of capacitance, the farad, is one coulomb per volt, as might be surmised from Eq. (2.12). Most circuit capacitances, however, are on the order of *micro*farads or *pico*farads.

What does one look for in a capacitor? Its capacitance and quality will depend on *geometry*, that is, shape, area, and spacing of its conducting plates (or sheets), and especially on the properties of the insulator layer between them. In electronic applications, insulators are usually called *dielectrics*. Their important properties are dielectric constant, resistance, and loss factor. In Table 2.2 information is given about most of the common types of capacitors.

How can the behavior of capacitors be described? Let us begin with current–voltage behavior as a function of time: first, only *changing* currents and voltages affect capacitors. This aspect is easily deduced by rearranging Eq. (2.12) as $Q = CV$ and taking the time derivative:

$$dQ/dt = I = C\,dV/dt. \tag{2.13}$$

Clearly, current in a capacitor is dependent on the *rate of change* of voltage, not just on voltage itself as in the case of a resistor.

Example 2.11. How much voltage will be developed on a 0.1-μF capacitor in 1 ms by a steady 0.1-mA current? Since variables are separable in Eq. (2.13), it can be rearranged to $I\,dt = C\,dV$. Assuming both time and voltage have begun from zero, on integrating between 0 and t we obtain $It = CV$ or $V = It/C$. On substitution of the values given, we have

$$V = 0.1\text{ mA} \cdot 1\text{ ms}/0.1\ \mu\text{F} = 1 \times 10^{-4}\text{ A} \times 10^{-3}\text{ s}/1 \times 10^{-7}\text{ F} = 1\text{ V}.$$

Charging and Discharging a Capacitor: V and I versus Time. The classic dc circuit shown in Fig. 2.8 lends itself to exploration of capacitor behavior in what

Table 2.2 Some Properties of Capacitors[a]

Type[b]	*Dielectric Constant*[c]	*Capacitance (range)*	*Maximum Voltage*	*Leakage*[d]	*Power Factor*	*Comment*
Mica	4–7	1 pF–0.01 μF	100–600	Low	Low	Good
Ceramic	1000	10 pF–1 μF	50–100	Low	High	Wide use; good; inexpensive
Polystyrene	2.5	10 pF–0.01 μF	100–600	Very low	Very low	High quality
Mylar	3.2	1 nF–10 μF	50–600	Low	Low	Wide use; good; inexpensive
Polycarbonate		100 pF–10 μF	50–400	Very low		High quality
Glass	7	10 pF–1 nF	100–600	Low	Low	Very stable
Electrolytic		0.1 μF–0.2 F	3–600	High		Polarized; short life
Oil		0.1 μF–20 F	$200–10^4$	Low	Low	High voltage use; long life

[a]To select a capacitor for a particular application the most important factors to take into account are the desired capacitance, maximum operating or working voltage, and design tolerance. The latter can be treated by the considerations discussed for resistors. Capacitor properties reflect the properties of the insulators or dielectrics used in their manufacture. The voltage rating is determined by the breakdown strength and thickness of the dielectric layer. At high frequencies, 50 MHz or above, the *power factor* of a capacitor also becomes significant. This quantity is defined as the ratio of power dissipated to that stored and is determined largely by the dielectric.
[b]Capacitors are usually classified by their dielectrics, the insulating material separating conductive layers.
[c]The dielectric constant ϵ or permittivity of a dielectric substance is defined by the ratio of capacitance C (when the dielectric is present between plates) to C_{vac}, its capacitance when there is a vacuum between them: $\epsilon = C/C_{\text{vac}}$.
[d]Leakage refers to the minute, though finite, conductance of a dielectric.

is termed the *time domain*. Consider that C has no charge initially and that at $t = 0$ switch Sw is moved to position 1. Almost instantaneously, current I exists in the circuit, and the capacitor begins to accumulate charge. Applying Kirchhoff's voltage law shows us that the battery voltage V_{BB} equals the sum of the potential drops across R and C. The former is just IR; from Eq. (2.12) the latter proves to be $V = Q/C$. We have

$$V_{\text{BB}} = IR + Q/C. \tag{2.14}$$

To deduce the behavior of the circuit as time changes, we differentiate Eq. (2.14) with respect to time and obtain $dV_{\text{BB}}/dt = 0 = R(dI/dt) + I/C$. The solution to

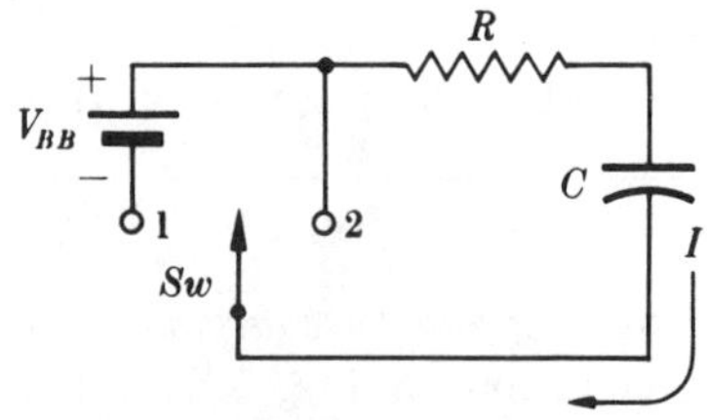

Fig. 2.8 Schematic of a circuit for charging and discharging a capacitor.

this differential equation proves to be

$$I = (V_{BB}/R)\, e^{-t/RC} \tag{2.15}$$

Two limiting conditions can be described at the outset. When charging begins at $t = 0$, the exponential term has the value one. At this instant, only resistor R limits the current: $I = V_{BB}/R$. After a relatively long interval, however, the current has fallen to zero and $V_C = V_{BB}$. At intermediate times while the capacitor acquires a charge, the current falls exponentially with time as predicted by Eq. (2.15). Examine the broken line plot in Fig. 2.9 to visualize this behavior. Further, note that at any intermediate time, $V_C = V_{BB} - IR$ for the RC circuit, where I is the effective current at that instant. The value of I that is defined in Eq. (2.15) may be substituted in Eq. (2.14) to give the capacitor voltage as a function of time, yielding finally

$$V_C = V_{BB}(1 - e^{-t/RC}). \tag{2.16}$$

The voltage–time curve is graphed as the solid line in Fig. 2.9.

Whenever desired, the capacitor may be discharged by moving switch Sw in Fig. 2.8 to position 2. Current and voltage will now fall exponentially along the broken curve of Fig. 2.9 as the discharge takes place.

Time Constant. A time constant $\tau_c = RC$ provides a convenient measure of the rate of capacitor charging or discharging. For example, during charging, at a time equal to τ_c, current will have fallen to 37% (1/e) of its initial value and voltage risen to 63% of its final value. When R is in ohms and C is in farads, the time constant product RC has the units of seconds. For example, the time constant for a circuit in which $R = 300$ kΩ and $C = 10$ pF is $3 \times 10^5\ \Omega \times 10 \times 10^{-12}$ F = 3 μs. A good rule of thumb is that after an interval of 5 RC seconds, capacitor values will be within 1% of their final magnitudes.

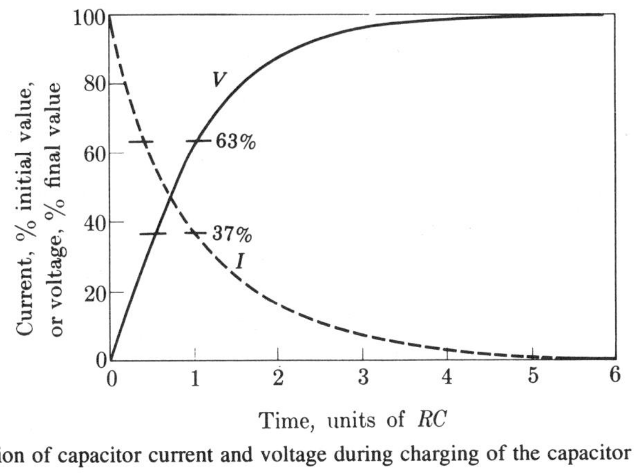

Fig. 2.9 Variation of capacitor current and voltage during charging of the capacitor in the circuit of Fig. 2.8.

ac Behavior. When energized by ac signals, a capacitor stores and releases charge in cycles. The process is complex. First, energy is drained from the circuit and becomes the energy of stored charge and then the process is exactly reversed. Second, a phase difference ϕ is introduced between the current and voltage. While there are instantaneous energy losses from the circuit, over the length of a whole cycle—one period—no energy is actually lost in a pure capacitance: what has been withdrawn is restored as the current reverses direction.

To deduce the magnitude of phase difference produced in a pure capacitance for a sine wave, we shift from the time to the *frequency domain*. Take Eq. (2.12) as $CV = Q$ and substitute for V the expression for amplitude of a sine wave $V_p \sin \omega t$ to obtain $Q = CV_p \sin \omega t$. By differentiating the expression for charge Q with respect to time, we obtain

$$dQ/dt = I = C\, d(V_p \sin \omega t)/dt = \omega C V_p \cos \omega t. \tag{2.17}$$

Note that the amplitude of voltage depends on a sine function and that of current on a cosine function. As a result, as the calculus shows, a 90° phase difference exists between current and voltage. Current is said to "lead" the voltage since the voltage rises as charge is stored.

To describe the impeding effect on current caused by the process of storing of charge, a new term, the *capacitive reactance* X_C, must be introduced. A useful definition of the magnitude is that X_C is the ratio of peak capacitor voltage V_p to peak capacitor current I_p. According to Eq. (2.17) the ratio is

$$X_C = (V_p/I_p)_{cap} = V_p/\omega C V_p = 1/\omega C = 1/2\pi f C. \tag{2.18}$$

Reactance is measured in units of ohms. As Eq. (2.18) suggests, X_C varies inversely as frequency and capacitance. For the reactance to be small, a condition often desired in a circuit, the product fC must thus be large. Because low reactance is often important it can be anticipated that at low frequencies circuits with large capacitances will be common while at high frequencies small capacitances will be needed.* Finally, Ohm's law becomes $I = V/X_C$ for a capacitor. Instantaneous voltage drops across capacitors can be calculated from it.

Shunt Paths in Circuits. In Fig. 2.10 a capacitor is wired in parallel with a resistor. Capacitors are often introduced in parallel to provide low reactance or *shunt* paths that cause ac components of a signal to bypass a main circuit. For example, undesirable ac might be shunted to ground. A shunt is seen to be nearly a short circuit. In Fig. 2.10 if the voltage between points A and B is V volts, the current in R will be V/R and that in the capacitor V/X_C. At low frequencies virtually all current will be in R unless C is large and its reactance X_C is thus small.

*For this reason also the small *distributed* or stray capacitance between wires are usually of consequence only at high frequencies.

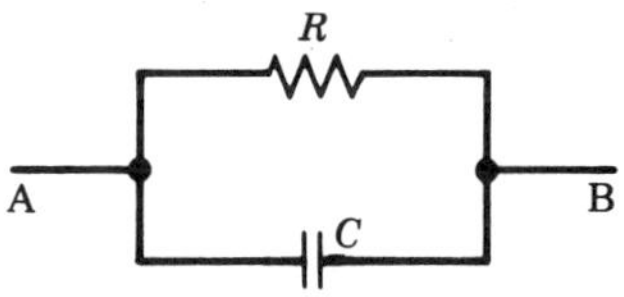

Fig. 2.10 A parallel *RC* circuit. The capacitor can be said to shunt the resistor.

Example 2.12. What is the reactance of a 20-pF capacitor at a frequency of 10 kHz? at 10 MHz? If 15 V rms is applied across the capacitor, what effective or rms current will exist in the capacitor at each frequency?

At 10 kHz, $X_C = 1/(6.28 \times 10 \times 10^3 \text{ Hz} \times 20 \times 10^{-12} \text{ F}) = 800 \text{ k}\Omega$. At 10 MHz, the reactance will be 10^3 times smaller, 800 Ω.

An Ohm's law calculation gives for the current $I = V/X_C$, where both I and V must be rms values. At 10 kHz, $I = 15/800 \times 10^3 = 19\ \mu\text{A}$, and at 100 MHz, $I = 15/800 = 19$ mA.

Example 2.13. If a leakage current must be kept below 1% in a circuit input with about 10 kΩ resistance, should one be concerned about a stray capacitance between an unshielded input and ground of about 10 pF for frequencies of 10 kHz? of 100 MHz?

At 10 kHz, the reactance attributable to stray capacitance is $X_C = 1/6.28 \times 10^4 \times 10^{-11} = 1.6\ \text{M}\Omega$. So large a shunting reactance can certainly be neglected in comparison with 10 kΩ. At 100 MHz, however, the reactance is 100 Ω, 10^4 times less, and the loss would exceed the named limit.

Series and Parallel Capacitors. Circuits containing two or more capacitors are usually simplified by calculation of an equivalent capacitor.

For *series capacitors*, as in Fig. 2.11a, the voltage V across the set must be the sum of the voltages across each capacitor. In terms of capacitances and charge accumulation on the capacitors we have

$$V = Q_1/C_1 + Q_2/C_2 + Q_3/C_3.$$

Since the charge for the right-hand plate of C_1 must have come from the left-hand plate of C_2, for example, the charge on each capacitor must be the same, $Q_1 = Q_2 = Q_3$. Factoring out the common charge gives $V = Q(1/C_1 + 1/C_2 + 1/C_3) = Q/C_{eq}$. Clearly, for capacitors in series it is the sum of the reciprocals of the individual capacitances that

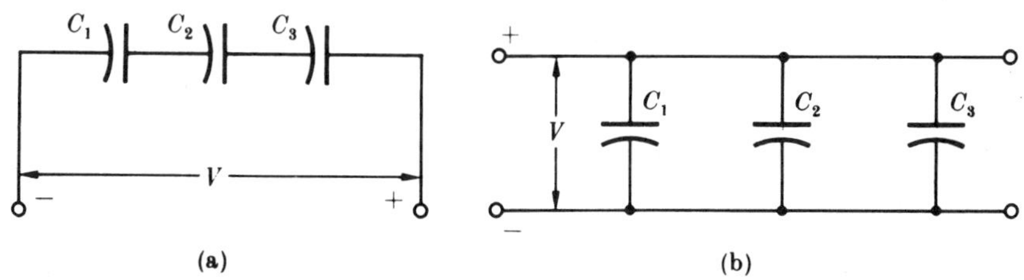

Fig. 2.11 (a) Capacitors in series. (b) Capacitors in parallel.

add to give the reciprocal of the equivalent capacitance

$$1/C_{eq} = 1/C_1 + 1/C_2 + 1/C_3 \tag{2.19a}$$

On the other hand, for *parallel capacitors*, as in Fig. 2.11b, the voltage across each is the same. Now the individual charges on the capacitors add. In this case, the total capacitance is $Q_1/V + Q_2/V + Q_3/V = Q_i/V$ and

$$C_1 + C_2 + C_3 = C_{eq}. \tag{2.19}$$

Thus, the capacitance of several capacitors in parallel is the sum of their different capacitances.

2.7 INDUCTORS; TRANSFORMERS

Inductors owe their usefulness as a type of passive linear component in alternating current circuits to still another type of current–voltage dependence. In inductors it is the rate of *current change* that is proportional to applied voltage:

$$V = L(dI/dt),$$

where L is the inductance (in henrys). For instance, if 1 V is imposed across a 1-henry inductor, current in it will rise at the rate of 1 A s^{-1}. Only brief attention will be given inductors since their use in electronic circuitry in instruments is relatively small.

An inductor is made by coiling a low-resistance wire, a mode of construction suggested by its symbol . For low-frequency use the coil is usually wound on a core of soft iron to increase the inductance. High-frequency inductors, which are often called chokes, by contrast have "air cores." Recall that the *solenoids* so often used in relays are basically inductors with moving iron cores.

When a circuit containing an inductor is energized there is an instantaneous withdrawal of energy to establish a magnetic field whose lines of force are concentrated along the axis of the coil. Current can build up only as fast as the magnetic field is established. This retarding effect on current is termed *inductive reactance*, X_L, and has units of ohms. Its magnitude is defined as

$$X_L = \omega L = 2\pi f L. \tag{2.20}$$

The property inductive reactance can be used in Ohm's law to deduce the connection between an rms alternating current and the voltage drop across an inductor: $V = IX_L$. In addition, an inductor causes current to be retarded 90° in phase relative to voltage. This aspect will be examined in more detail in the next section in connection with impedance.

Transformers. In a transformer a magnetic field set up by a primary coil is caused to couple efficiently with a secondary coil, usually by winding both on a common iron core that links the coils. The magnetic flux in the transformer is proportional to the number of turns n_1 in the primary coil; because in an ideal case all the flux loops through the secondary coil, the voltage induced in that coil is proportional to the turns ratio n_2/n_1 where n_2 is the number of secondary turns. The voltage V_2 that appears across the terminals of the secondary winding is the output of the transformer and is given by

$$V_2 = V_1(n_2/n_1), \tag{2.21}$$

where V_1 is the voltage imposed across the primary coil. Clearly it is possible to step up or step down voltages (and inversely, currents) in ac circuits by use of transformers of different turns ratios. A further discussion of transformers will be found in the next section when impedance matching is considered.

2.8 IMPEDANCE

Any practical circuit contains resistance, capacitance, and inductance. Can we simultaneously deduce the magnitude of voltage drops, phase shifts, power dissipation, and reactance in each of these types of component? Fortunately, Ohm's law is also applicable to ac circuits when a new quantity, the impedance Z, is defined. It is a comprehensive measure of the retarding effect of resistance, capacitance, and inductance on current. Ohm's law now reads*

$$I = V/Z \quad \text{or} \quad V = IZ. \tag{2.22}$$

A more useful definition of impedance that gives its magnitude will be given below. In the meantime, to develop a sense of how impedance relates to resistance and capacitive and inductive reactances, let us look again at definitions and especially at phase information for passive circuit properties. The results are collected in Table 2.3. A straightforward way to define impedance may be deduced from the vectorial representation of reactance and resistance in Fig. 2.12b. Note that angles are taken as positive in the counterclockwise direction. This manner of representation of the contributions to impedance and the general formula for impedance are generally valid. Actually, two defining equations result. The magnitude of Z as obtained vectorially is

$$Z = \sqrt{\mathbf{R}^2 + (\mathbf{X}_L - \mathbf{X}_C)^2} \tag{2.23}$$

*The magnitude and phase angle of reactances may be handled mathematically by representing reactance and impedance as complex numbers. The real part of each is associated with resistance, the imaginary part with reactance. In the complex plane, real parts of quantities are represented as distances along the x axis. Imaginary quantities, however, are plotted along the y axis. For reactances, graphing on the y axis will introduce the 90° phase differences that are observed when voltages in pure capacitances and inductances are compared with those in pure resistances. For example, a series combination of a capacitance and a resistance gives a complex impedance $Z = R + jX_C$, where $j = \sqrt{-1}$. Here the symbol j is substituted for i since the latter is used to identify current. The imaginary part, X_C, is graphed at right angles to the real part, which is R. See Fig. 2.12b.

Table 2.3 Phase Shifting and Power Properties of Passive Quantities

Passive Quantity	*Phase Angle*[a]	*Resistance or Reactance*	*Power Consumption*[b]
R	0°	R	I^2R
C	−90°	$X_C = -1/(2\pi fC)$	0
L	+90°	$X_L = 2\pi fL$	0

[a]The phase angle is that made by current vector **I** with voltage vector **V**.
[b]Here power consumption is calculated over a period equal to **VI** cos ϕ. The instantaneous power is always the product **VI**. For pure capacitances and inductances power consumption over a cycle is always zero, but instantaneous consumption will not be zero.

and the phase angle ϕ, as defined by its tangent in the figure, is

$$\tan \phi = (\mathbf{X}_L - \mathbf{X}_C)/\mathbf{R}$$

Resonance. It is apparent from Eq. (2.23) that under certain conditions the quantity $X_L - X_C$ must be zero and only resistance will oppose the current. By setting this difference equal to zero and substituting values for capacitive and inductive reactance we find this can happen at a particular frequency called the *resonant frequency*, equal to $1/2\pi\sqrt{LC}$. At this frequency the circuit is said to be in resonance since the apparent power I^2Z that must be supplied is a minimum. From Fig. 2.12*b*, the sum of the voltages across the reactive components must cancel at resonance. As a result of such impedance characteristics, *LC* circuits are often used as frequency-selective circuits, for example, in oscillators. References should be consulted.

Example 2.14. Consider the circuit of Fig. 2.12a. Take I = 10 mA. (a) What is the reactance and the voltage drop in each component if the effective current is 10 mA? (b) What must the applied voltage V_s be to develop the assumed current? Compare V_s and the

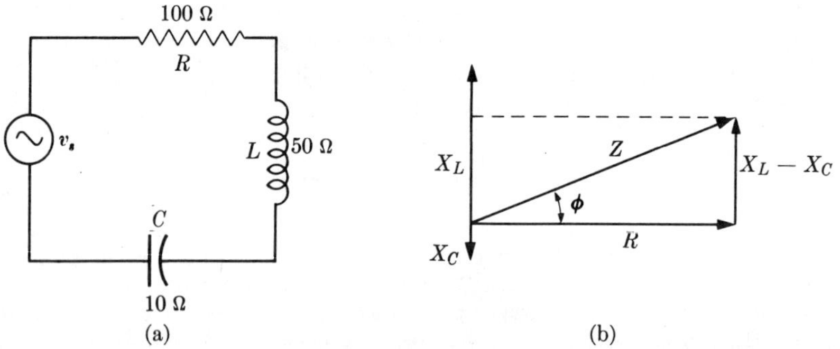

Fig. 2.12 A representation of the reactances and resistance for a capacitor, inductor, and resistance in series with an ac source *V*. (a) The circuit. (b) Vector diagram indicating the phase relationships of pure capacitive and inductive reactance relative to resistance. By the vectorial addition shown these components can be added to obtain the circuit impedance Z.

sum of voltages in the components. Does reference to Fig. 2.12b suggest a way to account for the difference?

(a) First, the impedance is

$$Z = \sqrt{R^2 + (X_L - X_C)^2} = \sqrt{(100)^2 + (50 - 10)^2} = 107.8\Omega.$$

Second, the voltage drop V_R across the resistance is $V_R = IR = 10 \times 10^{-3}$ A $\times$ 100 Ω = 1 V. The voltage V_L across the inductance is $V_L = IX_L = 10 \times 10^{-3}$ A $\times$ 50 Ω = 0.50 V. The voltage V_C across the capacitance is $V_C = IX_C = 10 \times 10^{-3}$ A $\times$ 10 Ω = 0.10 V.

(b) Finally, the applied voltage V_s must be

$$V_s = IZ = 10 \times 10^{-3} \times 107.8 = 1.08 \text{ V}.$$

Why does the sum of voltage drops (1 + 0.50 + 0.10 = 1.60 V) exceed the applied voltage? These voltages are out of phase. When they are added vectorially, they should sum to V_s.

Example 2.15. If in a series circuit containing capacitance *C*, resistance *R* and inductance *L*, values of the components are $L = 1.0 \times 10^{-3}$ henry, $C = 0.44$ μF, and $R = 150$ Ω, what is its resonant frequency? If a 20-V signal of this frequency is applied to the circuit, what is the current?

First, the resonant frequency

$$f = 1/(2 \times 3.14 \sqrt{1.0 \times 10^{-3} \times 0.44 \times 10^{-6}}$$
$$= 7.59 \times 10^3 \text{ Hz}.$$

Second, $Z = R$ at resonance and application of 20 V rms will give $I = V/R = 0.133$ A.

2.9 THE *p–n* JUNCTION AND THE DIODE

With this section the first solid-state circuit element, the diode, is introduced. Like the resistor and other *passive* circuit elements, the diode cannot enhance signal amplitudes. But while resistors, capacitors, and inductors all give doubled outputs when their inputs are doubled, diodes behave *nonlinearly*. The very different current–voltage response of a representative diode is shown in Fig. 2.13. Not only is the curve nonlinear in the upper right quadrant of the graph, but the current falls *nearly to zero* when the direction of external voltage is reversed. In essence a diode conducts in only one direction: it cannot conduct ac current.

Semiconductor Devices. Solid-state electronic devices are made from very pure single crystals (usually purified by zone refining) of semiconducting materials such as silicon, germanium, and gallium arsenide. A thin slab is first sawed from a single crystal, and its face is polished and then coated with a photoresist. Successive patterns for integrated cir-

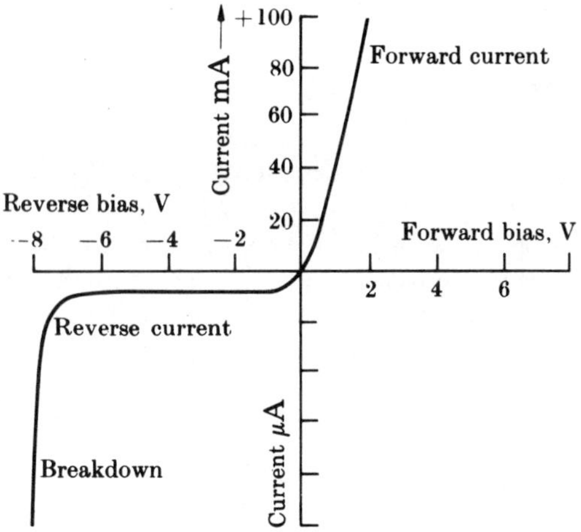

Fig. 2.13 Current–voltage curve for a representative semiconductor diode. Note that under conditions of reverse biasing the current is 1000 times smaller and the applicable scale is in microamperes. The breakdown voltage may be greater or smaller than shown.

cuits are photographed onto the surface with great precision, exposed areas are etched, electron donor or acceptor impurities are diffused into the etched regions, metal leads are deposited as required, the slab is cut apart to isolate individual circuits, and each is finally sealed into an air-tight package. The reader is referred to the literature for a current treatment of the fascinating art of the design and manufacture of solid-state electronic devices.

Semiconductors are crystalline materials that have an electrical resistivity intermediate between conductors and insulators. The class includes elements such as silicon and germanium that are in the region of transition between metals and nonmetals and compounds such as indium antimonide and silicon carbide. Their conductivity is a result of the continuous promotion by thermal or light energy of a tiny fraction of bonding electrons to the conduction band of nonlocalized orbitals.

Positive centers (holes) are also generated. A process of recombination of electrons and positive centers goes on continuously also, and thermal equilibrium is quickly reached. Pure substances of this type are classed as intrinsic semiconductors.

Control of Majority Charge Carriers. From an electronic point of view, the most important attribute of a semiconductor crystal is that the type of majority charge carrier in a given region can be controlled by adding a trace of another substance, a process termed *doping*. To fabricate a diode, for example, impurities that are stronger donors of electrons that the host crystal are introduced into one region and impurities that are better acceptors of electrons are diffused into an adjacent region. In the process it is arranged that a precisely defined junction will be formed between regions with different majority charge carriers. It then becomes possible to impose potential differences across such junctions and thus to control movement of majority charge carriers.

Consider the effect of introducing atoms of a donor element, for example, antimony (Sb) into a silicon lattice. Wherever an Sb atom with its five valence electrons substitutes in the lattice for an Si atom, a local region with an "excess" electron results. One antimony electron per atom needs only a small amount of energy to break away to the Si conduction band. This situation is illustrated schematically in Fig. 2.14. The doped region is termed *n type* to signify that its *majority carrier* of charge has negative polarity. In the doped substance there are still intrinsically produced electrons and holes (see discussion below), but the donated electrons outnumber them. Holes serve as *minority* charge carriers.

Conversely, introducing atoms of an acceptor impurity, such as gallium (Ga), results in the appearance of centers in the host lattice that are "deficient" by one electron. Transfer of an electron from a lattice atom to an impurity orbital costs little in energy. The holes that are left in the valence band are now the majority charge carrier. The doped region is labeled *p type* to denote its relative positive polarity. Conduction by holes can be roughly pictured as occurring by acceptance of valence electrons from neighboring semiconductor atoms. In this way holes can migrate, though with much lower mobility than electrons. Note that in hole conduction no electrons have been excited to the conduction band. Now electrons function as minority carriers.

The p–n Junction. How can we characterize such a junction? For example, the simplest bipolar solid-state device, the diode, is characterized by having one *p–n* junction. First, under quiescent conditions, as a result of normal thermal agitation, charge carriers diffuse across the boundary. Since concentrations of charge carriers differ on the two sides, a net number of holes diffuse out of the *p* region and a net

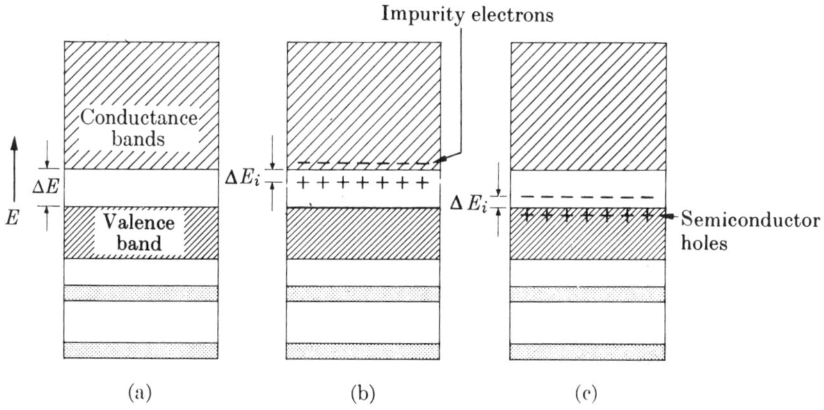

Fig. 2.14 Energy levels for electrons in semiconductors. ΔE is the band gap in the host semiconductor crystal ($\Delta E \simeq 1$ eV); ΔE_i is the excitation energy required to promote an electron or hole to a conduction level in the doped semiconductor ($\Delta E_i \simeq 0.01$ eV). (a) Energy bands in pure (intrinsic) semiconductors. (b) Highest electron energy level (+ + +) of added Group V impurity in relation to energy bands in host. In this n-type doped semiconductor conduction occurs by excitation of (impurity) electrons to the conduction band. (c) Highest electron level (– – – –) of Group III impurity in relation to energy bands in host. In this p-type doped semiconductor, conduction occurs by holes in the top of the valence band after electrons are excited to vacancies in the highest impurity electron levels.

number of electrons move out of the *n* region. In addition, holes crossing into the *n* region tend to recombine with electrons there and vice versa. As a result, on either side of the junction the number of charge carriers falls drastically and the area becomes a high-resistance *depletion region*.

The Diode. What occurs when an external dc potential is applied across the junction? Under the *forward biasing* shown in Fig. 2.15a, the lead to the *p* region is made positive and that to the *n* region negative. In the diode all majority charge carriers move toward the junction and current is substantial. (Refer again to Fig. 2.13.) With *reverse biasing*, however, polarities of the regions are reversed. Now majority charge carriers move away from the *p–n* junction, widening the depletion region such that only a minute *reverse* current exists.

Clearly, a diode behaves as a unidirectional valve. Any device that conducts in a single direction may also be termed a *rectifier* but in practice this term is reserved for the type of device which handles large currents and others are simply called diodes.

Since the forward resistance of a representative diode is only a few hundred ohms, substantial current flows when it is biased a few tenths of a volt (about 0.2 V for Ge, 0.7 V for Si) in the forward direction. When reverse-biased, such a diode has about a megohm resistance and the current across the junction is no more than a few microamperes.

If the reverse bias is increased sufficiently (5–200 V), the junction will "break down," and conduct excessively. Unless the current is limited externally, say by a series resistor, strong heating and junction damage will result. The relationship of the breakdown region to the normal operating region was illustrated clearly in Fig. 2.13.

In most instances an *avalanche mechanism* is responsible for the breakdown phenomenon. When a relatively high field exists across the broad depletion region

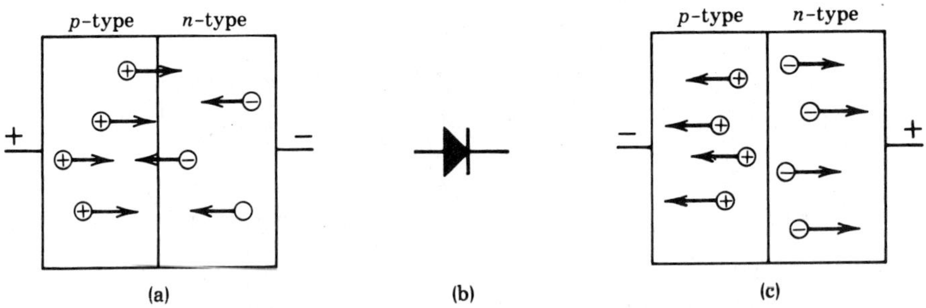

Fig. 2.15 Schematic illustration of movement of majority charge carriers in diode on application of an external potential difference. (a) Forward biasing: the *p* layer is made positive relative to the *n* layer, which is then negative. There is appreciable current since majority carriers cross the *p–n* junction. (b) Diode symbol. The arrowhead represents the anode and points in the direction of positive current. (c) Reverse biasing: the *p* layer is made negative relative to the *n* layer. The diode is nearly nonconducting. Now charge carriers are attracted away from the junction and only a minute current exists because of the movement of minority charge carriers.

created under reverse biasing, any electron crossing it gains sufficient energy to create electron–hole pairs by collision. Holes are also accelerated and produce further ionization. The process spreads in chain-reaction fashion.

Junction capacity. A reverse-biased *p–n* junction also acts as a capacitor. Recall that under such biasing most majority charge carriers move away from the junction, causing a substantial charge separation. The extended depletion region serves as a dielectric or insulator layer for the capacitor that results. Capacitance actually decreases with increasing reverse bias, since the layer widens rapidly as greater voltages are applied. Usual *p–n* junction capacitances are of the order of 1–15 pF. Commercial voltage-variable capacitors, called varactors, based on *p–n* junctions (often between aluminum and silicon) are also available and are often used for voltage control of frequency in resonant circuits. Their capacitance varies essentially as $1/\sqrt{V + 0.65}$, where V is the applied voltage.

2.10 THE DIODE IN CIRCUITS

It is the fact that a diode is a nearly ideal one-way conductor that mainly determines how it is used. Examine the specifications for a few types of diodes in Table 2.4. While fabrication has been varied to change voltage and current behavior, all the diode specifications in the table can be related easily to the representative I–V curve in Fig. 2.13. Some circuit applications will now be described. In examining how the diode in each responds to signals remember that a small forward voltage drop of 0.5–0.8 V occurs in each regardless of the current (Ohm's law isn't followed: the diode really has no conventional resistance) and that inverse voltages imposed must be kept below maximum rated values to prevent damage.

Table 2.4 Some Diode Specifications

Designation	V_{Rmax} [a] (V)	I_{Rmax} [b] (μA)	V_F [c] (V)	I_F [c] (mA)	Class
IN 914	75	5	0.75	10	General purpose signal diode
IN 4002	100	50	0.9	1000	Rectifier[d]
IN 1183A	50	1000	1.1	40,000	High-current rectifier
IN 6263	60	10	0.4	1	Schottky (low V_F)

[a]Maximum repetitive peak *inverse voltage* at 25°C, 10 μA leakage.
[b]The reverse or leakage current at V_R and 100°C ambient temperature.
[c]Paired values are effective forward voltage and current for continuous operation, for example, 0.75 V at 10 mA, and peak values are higher. As shown, the potential across the junction does increase slightly with increasing forward current.
[d]This rectifier is one of seven with identical specifications except for maximum inverse voltage ratings, which go up to 1000 V.

Rectification. This term is an old one meaning to convert ac to dc. Before rectification it is customary to use a transformer to step up or step down the available ac voltage. Then rectification can produce the desired dc voltages used to bias or operate most electronic devices and circuits. Figure 2.16a shows a *half-wave rectifier* and the waveform of the output voltage. While the circuit is simple, a disadvantage is that half the ac wave is lost.

A better arrangement is the bridge rectifier shown in Fig. 2.16c. It is a full-wave device that captures both positive-going and negative-going peaks. Note that the current through output resistor R in both circuits is unidirectional and the top of the resistor is at a positive potential. Bridge rectifiers usually are purchased as

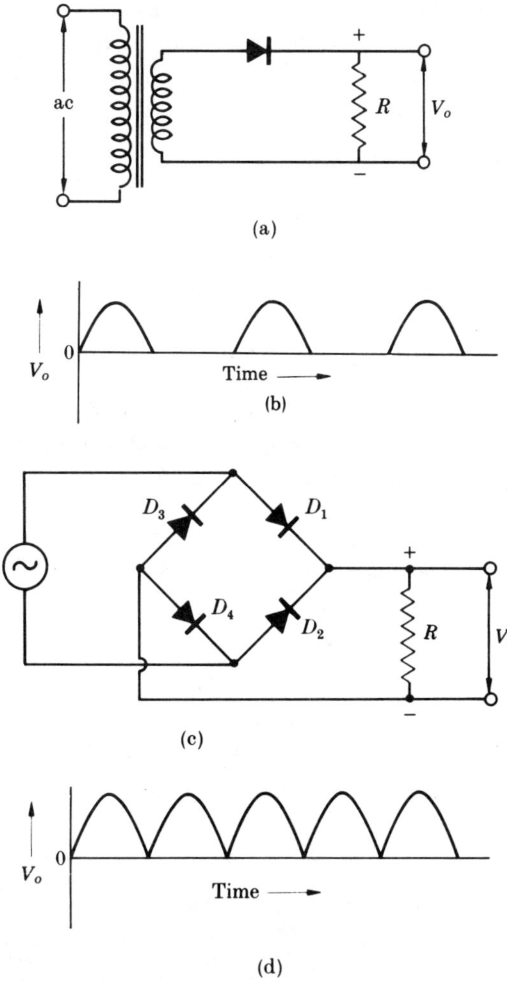

Fig. 2.16 Schematic rectifier circuits. (a) Half-wave type. (b) Output waveform of half-wave rectifier. Note that current exists in R_L only during half of each ac cycle. (c) Bridge type, which is a full-wave type. (d) Output waveform of bridge rectifier.

a unit. The smallest ones are rated at 1 A with breakdown voltages of 100–600 V. Larger ones are built for currents of 10–30 A and still greater inverse voltages. Rectifiers are the heart of dc power supplies, as will be seen in the following chapter.

Example 2.16. How does the half-wave rectifier circuit of Fig. 2.16a produce the fluctuating but unidirectional output V_0 shown in Fig. 2.16b?

To interpret this behavior requires combining an understanding of the variation of ac in the secondary winding of the transformer with diode behavior under different biasing. For 1/120 s the polarity of the winding from top to bottom will be + to − and positive current will move from the top of winding through the diode and R_L, causing the voltage drop across the resistor shown in Fig. 2.16a. During the next 1/120 s the secondary winding reverses polarity. Now the diode is reverse-biased and its current is miniscule. Indeed, during this part of the ac cycle V_0 is essentially zero according to Fig. 2.16b. During each succeeding cycle the pattern repeats.

Example 2.17. How does the bridge rectifier of Fig. 2.16c produce a full-wave rectified output? Why does it employ so many diodes?

In the bridge rectifier D_1 clearly plays the role that the single diode did in the last example. When the top of the secondary winding is positive, current conducted by D_1 will establish a + to − voltage drop across R. Why does D_3 not conduct? It cannot because its cathode is positive relative to its anode: it is reverse-biased. When current leaves the bottom of R, it reenters the bridge at the left and is conducted by D_4. Recall that the bottom of the secondary winding is negative and the current through D_4 will move in that direction. Why does not D_2 conduct? Again, it is reverse-biased since its cathode is also relatively positive.

When the polarity of the secondary winding changes direction, D_2 will be forward-biased and conducting, not D_4. Note that D_1 is now reverse-biased and cannot take the current from D_2. Current will again enter the top of R and move down it: during this half of the ac cycle the dc voltage across the resistor has the *same* polarity. Further, only D_3 is correctly biased for conduction, not D_4.

With the bridge rectifier no diode has to bear the whole reverse voltage; it is divided between a pair of diodes on each half cycle.

Example 2.18. The voltage–time curve is in Fig. 2.16d is idealized. How should it be modified to take into account the small forward voltage drops of the diodes? Assume the maximum voltage across the bridge is 18 V.

Since two diodes conduct simultaneously in the bridge rectifier the voltage drop across them must be about $2 \times 0.7 = 1.4$ V. Thus, the drop across R, which is V_0, will at maximum be at about $18 - 1.4 = 16.6$ V relative to ground. During parts of each cycle $V_0 = 0$. The curve will look like 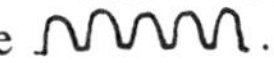.

Clipping Circuit. When it is desired to keep the amplitude of a signal below a predetermined maximum, diodes may be used. An example of a simple circuit called a clipping circuit that will limit amplitudes of both peaks and valleys of a signal is shown in Fig. 2.17. (If only ''tops'' were to be clipped, the right-hand diode and its bias supply would be removed.)

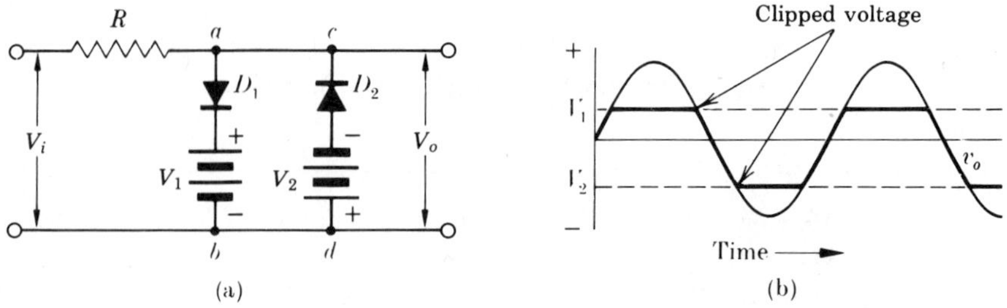

Fig. 2.17 Circuit for clipping or truncating peaks and valleys of ac signals admitted at the left. (a) The circuit. The dc voltages V_1 and V_2 applied in series with the diodes are shown as supplied by batteries but would usually be from a dc power supply. Diode D_1 clips peaks and D_2 clips valleys. (b) Waveforms at input (light line) and output (heavy line). Diode D_1 is biased by V_1 so that it will become conducting only if the input voltage exceeds $V_1 + V_d$, where V_d is the voltage drop across the diodes. When it conducts, the output voltage V_o remains constant since it cannot exceed the voltage across the arm *ab*. Any excess voltage in the signal appears as a voltage drop across *R*. In arm *cd*, diode D_2 is connected in the opposite fashion. When the input voltage is negative, D_2 will begin to conduct whenever is less than $V_2 + V_d$. Again, excess input voltage will be dropped across *R*.

How does clipping of peaks occur? The answer lies in the fact that D_1 can conduct only when the voltage to its anode (the arrowhead) is 0.6 V more positive than the bias applied to its cathode. Once D_1 begins conducting, the potential across it will remain at nearly 0.6 V even though the current through it increases greatly. Thus, the voltage across its branch circuit can be no greater than $V_1 + 0.6$ V. When V_i exceeds this value, the current in the diode and in *R* will increase. As a result, the voltage drop across *R* increases sufficiently to take up the excess. Similarly, when the voltage falls sufficiently that D_2 conducts, the bottoms of the signal peaks are clipped. The effect of clipping is evident in Fig. 2.17b and is further explained in the legend.

A clipping circuit is often used to limit the height of pulses that are to be counted to ensure a uniform response to all pulses. Similarly, a clipping circuit can be inserted before a gate or triggering circuit to limit voltages that might otherwise momentarily saturate the gate and slow its switching.

2.11 THE ZENER DIODE AS A SOURCE OF CONSTANT VOLTAGE

A diode specially designed to operate under breakdown conditions is termed a *zener diode.* As was evident from Fig. 2.13, under such conditions the voltage across a diode is nearly independent of current. For example, the voltage drop for a 653C4 zener diode increases only from 7.00 to about 7.15 V at 25°C as the (reverse) current rises from 5 to 20 mA.

Zener diodes are widely used as reference voltage sources. A representative circuit is illustrated in Fig. 2.18. Note that the diode is used with a current-limiting resistor *R* and that the input voltage must always be higher than the diode breakdown value for the circuit to operate. If the input voltage rises, there will be more current in the diode shunt without appreciably affecting the voltage across it or the load. Virtually all the voltage increase is reflected in a greater voltage across *R*.

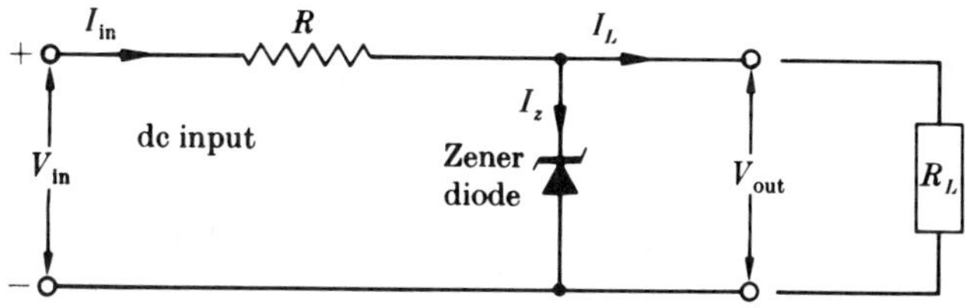

Fig. 2.18 Circuit incorporating a zener diode to achieve a regulated or constant output voltage.

The maximum diode current must, of course, not exceed the maximum rated value, $I_{z\,max}$.

The degree to which a zener diode approximates a constant voltage source is dependent both on its *incremental resistance* r_z, where $r_z = \Delta V_z/\Delta I_z$, and on its temperature coefficient of voltage. Representative values of r_z are from 10 to 100 Ω. To ensure precise reproducibility of breakdown voltage, zener diodes are often placed in a temperature-controlled oven and/or incorporated in an appropriate operational amplifier circuit (see Section 4.4).

Example 2.19. How can a suitable value of "dropping resistor" R for the zener diode circuit of Fig. 2.18 be determined?

First, note that $I_i = I_z + I_L$ and that $V_i = V_R + V_z$. Assume V_z is constant. It is easiest to proceed by use of limiting cases. Taking the smallest value of V_i and the largest value of I_L, R must be small enough to permit at least the minimum zener current $I_{z\,min}$ (one that ensures operation in the linear breakdown region) to exist. Thus, the resistance should be no larger than $R = (V_{i\,min} - V_z)/(I_{z\,min} + I_{L\,max})$. The value of $I_{L\,max}$ obtained from Ohm's law; $I_{L\,max} = V_z/R_{L\,min}$ can be substituted and $I_{L\,min}$ will be found in the zener specifications.

In addition, the resistor must ensure that at maximum input voltage and no load the maximum diode current is not exceeded. Thus, the equation

$$R = (V_{i\,max} - V_z)/I_{z\,max}$$

must also be satisfied.

Example 2.20. What increase will be noted in the voltage drop V_z in the regulator circuit of Fig. 2.18 if the diode current increases by 10 mA? Assume the incremental resistance of the zener diode is 20 Ω. Since $V_z = V_{z\,initial} + I_z r_z$, the increase will be $\Delta V_z = I_z r_z = 0.010 \times 20 = 0.20$ V.

REFERENCES

A lucid presentation of basic electrical measurements on the intermediate to advanced level is to be found in:

1. E. Frank, *Electrical Measurement Analysis*. New York: McGraw-Hill, 1959.
2. F. K. Harris, *Electrical Measurements*. New York: Wiley, 1952.

3. H. V. Malmstadt, C. G. Enke, and S. R. Crouch, *Electronic Measurements for Scientists.* Menlo Park, CA: Benjamin, 1974.

Other publications of interest are:

4. G. Klein and J. J. Zaalberg Van Zelst, *Precision Electronics.* New York: Springer-Verlag, 1967.
5. E. A. Boucher, Theory and application of thermistors, *J. Chem. Educ.*, **44**, A935 (1967).
6. R. Harruff and C. Kimball, Temperature compensation using thermistor networks, *Anal. Chem.*, **42(7),** 73A (1970).
7. J. E. Nelson, Oscilloscopes in chemistry, *J. Chem. Educ.*, **45,** A635, A787 (1968).

See also the references of Chapter 3.

EXERCISES

2.1 Derive the expression for average power dissipation $P_{av} = \frac{1}{2} RI_p^2$ for ac current in a resistance.

2.2 After standing for a year, a 1.5-V dry cell proves able to furnish only 5 mA of current. What is its internal resistance?

2.3 How can several 10-kΩ, 2-W resistors be connected to obtain an equivalent resistance of 1 kΩ with 20 W capacity? How many are required?

2.4 In the circuits of Fig. 2.4 assume $V = 15$ V, $R_1 = 10$ kΩ, and $R_2 = 5$ kΩ. (a) For the series circuit, what is the value of I, and of V_1 and V_2? (b) Show that in the series circuit Kirchhoff's voltage law $\Sigma V_j = 0$ holds. (c) For the parallel circuit what is the value of I_1 and of I_2? (d) Show that Kirchhoff's current law holds at the node where branching begins. What signs must be applied to I_1 and I_2? (e) Finally, what is the equivalent resistance of $R_1 + R_2$ in series? in parallel?

2.5 Use the Thévenin equivalent circuit developed for the voltage divider of Fig. 2.6*a* to calculate values for equivalent voltage V and resistance R. (a) What current will exist if a 2.5-kΩ load resistor R_L is across the output? (b) Apply Kirchhoff's laws to the circuit of Fig. 2.6*a*, with R_L attached, and develop an expression for the current in R_L. Show that it is identical to the expression we obtain when we use the Thévenin equivalent circuit.

2.6 The exploding-wire technique for obtaining metal vapor relies on discharging a capacitor of substantial charge through a metal wire of low resistance. For the case in which a 5-F capacitor charged to 100 V is connected across a wire of 0.01 Ω resistance, calculate the initial current and the energy dissipation.

2.7 Two capacitors of 1.0 and 0.01 μF capacity are connected in series across a 45-V battery. What is the voltage drop across each?

2.8 The charging of a capacitor is to be used to obtain an approximately linear rate of voltage increase. For this purpose a 0.05-μF capacitor is to be charged to only 20% of the applied voltage. What value of resistance must be used in series with it if the charging time desired is 10 μs?

2.9 An electric arc has a negative resistance characteristic, that is, its current increases as the voltage drop across it *decreases*. (See Fig. 8.5.) It may be brought into a condition of stable operation by placing a resistor R of proper value in series. Discuss the self-controlling features of this combination, which provides regulation when the current momentarily decreases or increases.

2.10 What time is theoretically required to discharge a 0.001-μF capacitor to 10% of its original value in a circuit with a resistance of 10 kΩ?

2.11 What will be the current in a circuit with a total resistance of 2.76 kΩ and a capacitance of 20 μF in series with a 50-V dc source (a) after 1 ms? (b) after 0.20 s?

2.12 There is an effective current of 40 mA at 50 kHz in an ac circuit. The total series resistance is 25 kΩ, and the impedance 30 kΩ. What amount of power is dissipated?

2.13 A resistor R, capacitor C, and inductor L are successively subjected to a 5-V square wave pulse. Sketch the voltage pulse (V versus time) and then draw the variation of current with time in each passive element during the period of the pulse.

2.14 A 10-μA ac signal appears in a 5-Ω resistor. It is desired to achieve as efficient a voltage transfer to the input of an amplifier as possible. Discuss whether a 100-Ω input impedance or a 1-M Ω input impedance would be preferable. (See Section 3.8 for additional information.)

2.15 Consider the diode circuit in Fig. 2.19 which ensures that the dc motor operates even when line power fails. Back-up power is available from a 12-V battery. Explain the role of the diodes, specifying which one or both are operating (a) when rectified dc from the line drives the motor and (b) when the line is dead and the battery is supplying power to the motor.

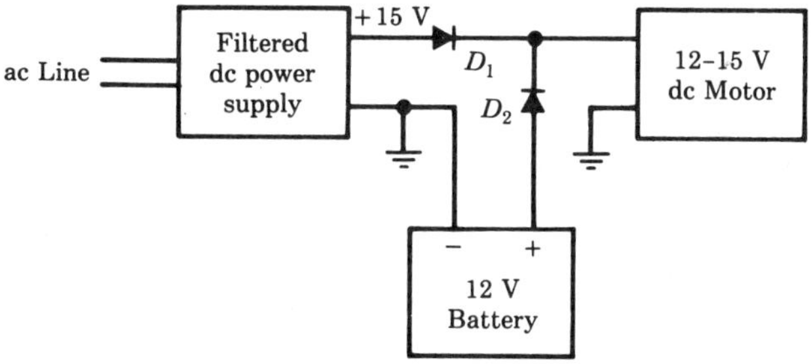

2.16 A germanium diode begins to conduct strongly when forward biased about 0.2 V. Often such a diode is wired across the terminals of a sensitive meter to protect it against an overload. (a) Draw a circuit that will protect a 50-μA meter of 2 kΩ resistance against both forward and reverse overloads. (b) Calculate the current at which diode protection begins. Is this protection sufficient?

2.17 In the bridge rectifier circuit of Fig. 2.15c a pair of diodes is always in series. In selecting suitable diodes for a particular rectifier, what diode ratings can be relaxed as a result, if any?

2.18 In Fig. 2.20 two diode clipping circuits are shown. For a sine wave input, sketch one cycle of the output waveform of each.

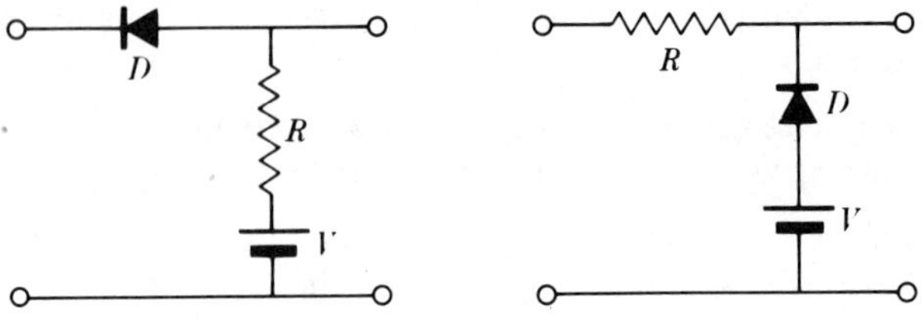

(a) (b)

2.19 In the circuit of Fig. 2.18 assume the zener diode provides 50 V at diode currents from 5 to 40 mA. The supply voltage V_i is 200 V. (a) What value of R will provide voltage regulation from $I_L = 0$ to I_{max}? What is I_{max}? (b) Let R have the value just found. If $I_L = 20$ mA, over what range of V_i will the circuit provide voltage regulation?

2.20 In a Zener-regulated supply like that of Fig. 2.17, the incremental resistance is 20 Ω. If the voltage is to be regulated to ±0.05 V, what is the tolerable range of diode current?

ANSWERS

2.2 300 Ω. *2.3* Connect in parallel; 10 required; *2.5* (a) 3.9 mA; (b) $I = V_1R_1/[R_1R_2 + R_L(R_1 + R_2)]$. *2.7* 0.446 and 44.6 V. *2.8* 895 Ω. *2.10* 23 μs. *2.11* 56 μs. *2.12* 40 W. *2.16* (b) 100 μA. *2.20* ±2.5 mA.

Chapter 3

ANALOG ELECTRICAL AND ELECTRONIC MODULES

3.1 INTRODUCTION

What electrical and electronic modules need discussion in a modular approach to instrumentation? Recall again the interrelationship of modules in an instrument summarized in Fig. 3.1.

This chapter introduces a few basic modules and their functions. Included are the dc power supply, Wheatstone bridge, and cathode ray tube or video monitor. Some electronic modules will be involved in signal processing and display, others will support the operation of characteristic modules as power supplies provide power to sources or detectors. To understand these devices as modules (see Section 1.4) means to see how they receive an input signal, transform it, and produce an output signal. Each module will be discussed in terms of basic concepts, modes of operation, and strengths and weaknesses.

Beginning with Section 3.7 the chapter moves to examining the central component of most electronic modules, the transistor. This revolutionary device is the heart of integrated circuits, the form in which nearly all electronic circuitry appears today. It is important for users of chemical instrumentation to understand how a signal existing as a current and/or voltage can be amplified by transistor circuits. With some simple rules of thumb about transistor operation, the application of electronic modules such as operational amplifiers can be made more effective.

In addition to the examination of modules, several operations that are central to electronics such as filtering, amplification, and feedback are described. Sometimes discrete modules will perform these functions, but more often, a module such as an amplifier circuit will perform several such functions. The optimum coupling of electronic modules is examined as well.

In these early electronics chapters the groundwork is developed for analog electronics in which modules receive and transform signals whose voltages and currents vary continuously in amplitude. Discussion of digital electronic modules in which pulses corresponding to binary bits (0 and 1) are handled will be reserved for Chapters 5 and 6. Modern instruments of course have both analog and digital modules; each type proves indispensable in certain applications.

Source ⟶ Sampling module ⟶ Detector ⟶ Amplifier ⟶ Processing modules ⟶ Readout

Fig. 3.1 Modular diagram of a generalized instrument.

3.2 *RC* FILTERS

It is essential in electronics to be able to reject unwanted frequencies or to favor desired ones. Sometimes electrical filters, which are frequency-dependent voltage dividers, are used for this purpose. For example, in any instrument noise can be minimized by filtering out broad bands of frequency that have no information. The simplest filters consist of a resistor in series with a capacitor, with the capacitor providing the necessary dependence on frequency through its reactance ($X_c = 1/2\pi fC$). Two simple filters of this type will be considered here.

Example 3.1. A 10-kHz signal containing chemical information about a sample is accompanied by 60 Hz power line hum and still lower frequency $1/f$ noise. How can the noise be rejected?

A high-pass filter whose cutoff frequency f_o is 4 kHz would reject strongly frequencies below 1 kHz while passing most frequencies above 8 kHz at their original amplitude.

High-Pass Filter. The capacitor–resistor voltage divider shown in Fig. 3.2*a* functions as a high-pass filter. Note that the output is taken across resistor *R*. Capacitor *C* will, of course, block dc and have its largest reactance at low frequencies. To find its relative output V_o/V_i, that is, its response, as a function of frequency it is legitimate to focus on peak values and use Ohm's law.

Assume that the current in capacitor *C* continues through *R*. This condition will be valid as long as very little current is taken from the output. Then, using Ohm's law, the peak current in the circuit is $I_p = (V_p)_i/Z$, or

$$I_p = (V_p)_i/\sqrt{R^2 + (1/\omega C)^2},$$

where $(V_p)_i$ is the peak input voltage, and the impedance of the circuit is $Z = \sqrt{R^2 + (1/\omega C)^2}$. Similarly, the peak value of the output voltage $(V_p)_o = I_pR$.

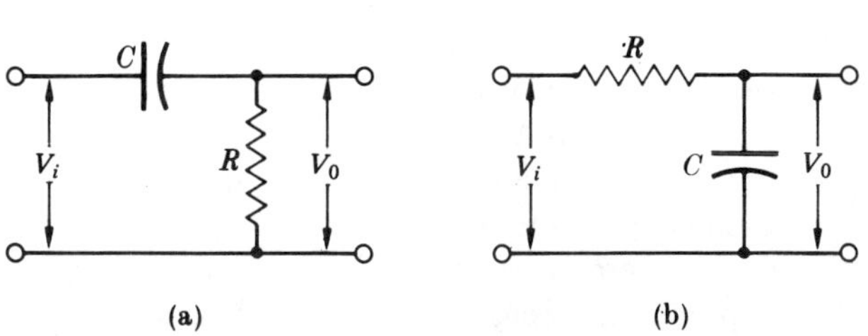

Fig. 3.2 Cutoff filters. (a) High-pass *RC* filter. (b) Low-pass *RC* filter. In each circuit the capacitor impedes low frequencies and blocks dc. Thus, low frequencies are attenuated in circuit (a) and passed in circuit (b).

Finally, the rms voltage ratio V_o/V_i can be expressed:

$$(V_p)_o/(V_p)_i = V_o/V_i = [R(V_p)_i/\sqrt{R^2 + (1/\omega C)^2}]/(V_p)_i$$
$$= 1/\sqrt{1 + (1/\omega RC)^2}. \qquad (3.1)$$

Recall that the peak values used in Eq. (3.1) are directly proportional to rms values of current and voltage. A graph of the ratio of rms values V_o/V_i for a high-pass filter is shown in Fig. 3.3.

It is useful to define a *half-power frequency* f_o as the frequency at which $P_o/P_i = \frac{1}{2}$ and $V_o/V_i = 1/\sqrt{2}$ (see Example 2.10).
From Eq. (3.1) it follows that $\omega RC = 1$ at this frequency. Since $\omega = 2\pi f$, the frequency f_o itself is defined as:

$$f_o = (2\pi RC)^{-1}. \qquad (3.2)$$

How sharply does this filter cut off? It can be shown that below f_o attentuation occurs at the rate of *20 dB per decade* of frequency.

The output ratio V_o/V_i for a high-pass filter is graphed in Fig. 3.3. Clearly, its characteristic *half-power frequency* f_o might also be termed a ''cut-on'' frequency. Since the filter is a voltage divider, it can be expected to function ideally only if the impedance of the load placed on its output is large compared with the size of resistor R. (Active filters greatly lessen this restriction; see Section 4.10.)

Low-Pass Filter. Conversely, taking the output of an *RC* circuit across the capacitor, as in Fig. 3.2b, yields a low-pass filter. Frequencies from dc up to about its half-power frequency f_o pass through with minimal attenuation. Its reactance is very low at frequencies above its cutoff f_o,* however, since as frequency rises the capacitor provides a ''shunt path'' to ground. For a low-pass filter also the half-power frequency f_o is given by Eq. (3.2). Naturally, the frequency pattern of the

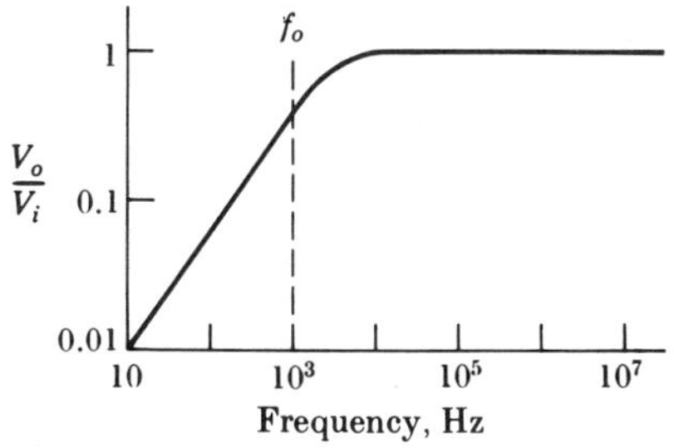

Fig. 3.3 Frequency response of a high-pass filter. The frequency f_o is called the characteristic frequency or cut-off point. It is defined by the relationship $f_o = 1/2\pi RC$. At f_o, $V_o^2/V_i^2 = 1/2$ and the power, which is proportional to V^2, has been halved. Its transmission falls off toward lower frequencies with a slope of 6 dB per octave or 20 dB per decade of frequency.

*In the time domain the perspective is different. Low frequencies fully charge the capacitor in Fig. 3.2b, yielding a voltage V_o equal to V_i. By contrast, high frequencies create their full voltage drop across the resistor, and fail to develop potential across the low reactance of the capacitor. The right-hand end of R is now at ground potential.

filter is the *reverse* of that in Fig. 3.3. Not far above f_o attenuation falls off linearly at 20 dB per decade of frequency.

Phase Angle. What happens to the phase angle ϕ of current relative to voltage in a signal as a result of passage through an *RC* filter?

In a high-pass filter it can be shown that ϕ is given by

$$\phi = \text{arc tan}(1/\omega RC). \tag{3.3}$$

Application of the equation shows that at the characteristic frequency f_o, current leads the voltage by 45°. When $f >> f_o$, however, $(1/\omega RC) \rightarrow 0$, and $\phi \rightarrow 0$. In other words, frequencies passed with little attenuation suffer little phase distortion, as would be hoped.

Equation (3.3) is also obtained for a low-pass filter. Is its phase behavior therefore similar? Only at the half-power frequency f_o is their phase behavior identical; here the phase angle is 45°. In this filter for frequencies $f << f_o$ which pass the filter freely however, $(1/\omega RC) \rightarrow \infty$, and $\phi \rightarrow 90°$.

3.3 *RC* DIFFERENTIATOR AND INTEGRATOR

The *RC* circuits just discussed respond very differently to signals they attenuate strongly: (a) the high-pass filter differentiates such signals with respect to time and (b) the low-pass filter integrates them with respect to time. Thus, these circuits are the heart of analog differentiators and integrators.

RC Differentiator. It is illustrative to work out differentiation by the RC high-pass filter in parallel ways. How the derivative dV_i/dt is obtained by a regular or *time domain* treatment will be shown first. Later, the *frequency domain* aspect of differentiation will be explored.

In Fig. 3.4 the *RC* circuit we are examining is shown again. Note that the voltage across *C* is just $V_i - V_o$. According to Eq. (2.13), we see that the output current *I* is just

$$I = Cd(V_i - V_o)/dt = V_o/R. \tag{3.4}$$

At frequencies for which signal attenuation is severe, it follows that $V_o << V_i$ and $dV_o/dt << dV_i/dt$. Substituting this information in Eq. (3.4) gives $C(dV_i/dt) \simeq$

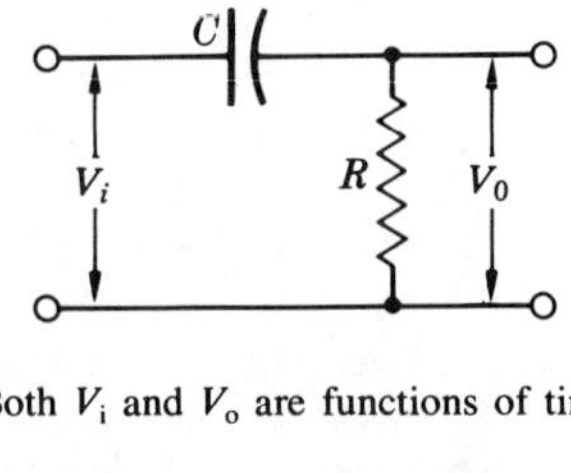

Fig. 3.4 An *RC* differentiator. Both V_i and V_o are functions of time. To secure differentiation it is necessary that $V_o << V_i$.

V_o/R and solving the equation for V_o gives

$$V_o = RC(dV_i/dt). \tag{3.5}$$

Equation (3.5) establishes that differentiation occurs. A "rule of thumb" covering the frequency range states that a high-pass *RC* filter will differentiate all frequencies f that meet the condition $f \ll f_o$, where f_o is the characteristic frequency of the circuit.

A Frequency Domain Treatment. How does electronic differentiation appear when we consider the frequencies that make up a signal? According to the Fourier theorem any input signal can be transformed into its sinusoidal frequency components. Each component frequency can be represented by an equation of the form $V_i = (V_p)_i \sin \omega t$. Its output V_o from the differentiator circuit of Fig. 3.4 will be the voltage drop across R:

$$V_o = RI = RI_p \sin(\omega t + \phi), \tag{3.6}$$

where ϕ is the phase angle between current and voltage in the output. Further, the time derivative of the input voltage is just

$$dV_i/dt = \omega(V_p)_i \cos \omega t. \tag{3.7}$$

We wish to compare Eqs. (3.6) and (3.7) under the conditions appropriate to differentiation, and substitutions for ϕ and I_p are required. The rule of thumb developed identifies the right conditions. A value for ϕ may be deduced from Eq. (3.3). In the frequency range for differentiation, $\omega RC \to 0$ and $\phi \to 90°$. Second, I_p can be found from Eq. (3.1). When $\omega RC \to 0$, $I_p = \omega C(V_p)_i$.* When the values for ϕ and I_p are inserted in Eq. (3.6), the equation becomes

$$V_o = \omega RC(V_p)_i \cos \omega t. \tag{3.8}$$

A comparison of Eqs. (3.6) and (3.8) shows that the output voltage indeed is a time derivative of the input voltage:

$$V_o = RC(dV_i/dt). \tag{3.9}$$

Example 3.2. What values of R and C are appropriate in the circuit of Fig. 3.4 if it is to differentiate signals of 10–100 Hz?

First, a characteristic frequency must be chosen. The rule of thumb developed to identify frequencies that will be differentiated with little error can best be applied to the highest frequency of interest, 100 Hz, since all lower frequencies will automatically be accommodated. Let $f_{highest} \ll 0.1 f_o$. Accordingly, we choose $f_o = 1$ kHz. Substituting this value in Eq. (3.3) will give the time constant of the circuit for the product *RC*, approximately 10^{-4} s.

*Multiply through the numerator and denominator of Eq. (3.1) by ωC to put it in a form where ωRC can be allowed to approach zero without creating difficulties.

Now appropriate values of R and C can be established. The resistance R can be taken as 5 kΩ since this value would probably not seriously load the circuit supplying the signal. When this value is substituted in $RC \simeq 10^{-4}$.

$$C = 10^{-4}\ \mathrm{s}/5 \times 10^3\Omega = 2 \times 10^{-8}\mathrm{F} = 0.02\ \mu\mathrm{F}.$$

Example 3.3. How is the input voltage to an RC differentiating circuit distributed between the capacitor and resistor? Use the circuit treated in Example 3.2.

An answer may be deduced by comparing the time constant τ calculated in Example 3.2 with the period T of the highest frequency of interest, which was 100 Hz. The period of a 100-Hz signal is $T = 1/f = 0.01$ s. Since that interval is 100 times longer than the time constant, the capacitor will be nearly fully charged at all times. As a consequence, there will be little current in R and very little voltage across it. Clearly, the assumption made to derive Eq. (3.5) holds. It also appears that the time constant τ of an RC circuit should be quite small relative to the period of any frequency that will be differentiated.

***RC* Integrators.** Consider now the other simple RC circuit, Fig. 3.5, which was discussed earlier as a low-pass filter. The output V_o is now taken across the capacitor. Again it is straightforward to proceed in the time domain and impose the condition of severe signal attenuation.

The voltage across resistor R in Fig. 3.5 is just $V_i - V_o$. On the basis of Eq. (2.13),

$$I = C(dV_o/dt) = (V_i - V_o)/R. \tag{3.10}$$

Whenever $V_o << V_i$, $C(dV_o/dt) \simeq V_i/R$. By separating variables, integrating, and simplifying, the value of V_o is found to be

$$V_o = (1/RC) \int V_i\, dt + \text{constant}. \tag{3.11}$$

The necessary condition to ensure that $V_o << V_i$ is that any frequency to be integrated be larger than cutoff frequency f_o. This limitation has the effect of requiring that the time constant RC of the circuit be large compared with the period of the signal to be integrated. If it is desired to treat the circuit in the frequency

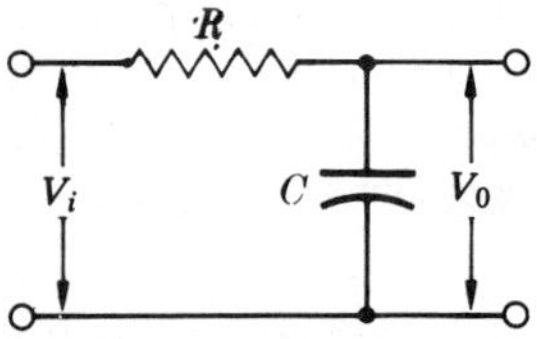

Fig. 3.5 An *RC* integrator.

domain, one should proceed in a manner parallel to that used with the differentiating circuit.

3.4 RATIOMETRIC DEVICES: THE WHEATSTONE BRIDGE

Signal measurements of the highest accuracy are available by comparing an unknown value with a standard. This strategy is usually called a *null method* or *ratiometric procedure* and involves varying the standard value systematically until it is identical with the unknown one. The comparison is best made by monitoring the difference in values and determining the point at which it becomes essentially zero; the process is a *null measurement*.

Example 3.4. A simple device for comparing voltages, the *potentiometer* is shown in Fig. 3.6. How does it apply the null principle of measurement? What advantages does the circuit have? How is it calibrated?

First, the resistor R, $R_A + R_B$ in circuit A, is usually a helically wound wire resistor with a take-off wiper designated by the arrow (see Section 2.3).

The voltage drop V_{R_B} in the upper circuit should be stable if voltage supply V_{BB} is stable, for the circuit resistance should be constant. Once this circuit has been calibrated (see below), voltage V_{R_B} is known precisely by the position of the wiper. During a measurement this known voltage is systematically varied while the difference between it and V_x is monitored in comparator circuit B by the difference detector M. The more sensitive this detector, the more precisely can the wiper position that corresponds to the null point $V_{R_B} - V_x = 0$ be ascertained. When the minimum or zero response is reached V_{R_B} can be read out.

An attractive feature of a potentiometer is that the unknown contact resistance at the wiper is not a part of the voltage divider and does not affect the accuracy of measurement. Only if $R_{contact}$ is so large as to reduce the sensitivity of the comparator circuit is precision affected.

Periodically, the voltage across R must be calibrated against a precise standard. This

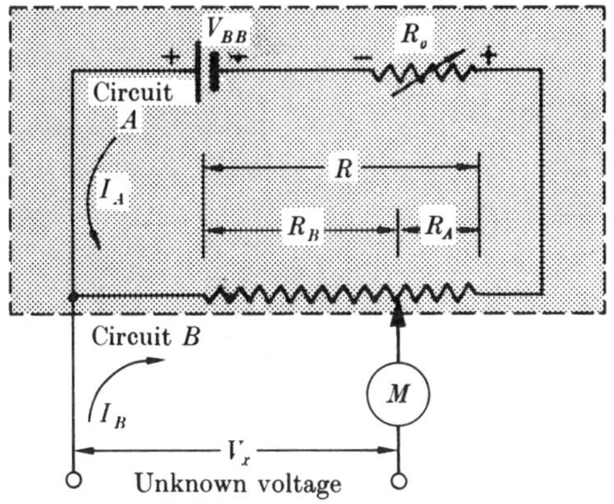

Fig. 3.6 Schematic of a simple potentiometer. V_x is an unknown voltage smaller than V_{BB}. R is a resistor (slidewire) whose resistance varies uniformly along its length. R_v is an adjustable resistor that permits the current in circuit A to be varied so that the voltage drop across R can be calibrated.

voltage V_{std} is substituted for V_x and the wiper is moved to a scale position R_{std} such that $I_A R_{std} = V_{std}$. Then the value of the resistor in series with V_{BB} is varied until I_A meets the null condition.

Wheatstone Bridge. When a precision measurement of a resistance or conductance is required, it is nearly always determined by a resistance bridge. The most widely used type is the Wheatstone bridge or a derivative of it. Its precision derives from the fact that it both relies on a voltage comparison and makes the measurement in a ratiometric fashion. Here the dc and low-frequency ($f < 2$ Hz) operation of the bridge is taken up; a discussion of its ac operation will be deferred to Section 32.5.

Its basic circuit is shown in Fig. 3.7. Two resistance branches, $R_1 + R_3$ and $R_2 + R_4$, are connected by a bridge or shunt BC in which a *null or difference detector* N (of resistance R_5) is located. By convention, arms R_1 and R_2 are called ratio arms, R_3 is usually the unknown resistance, and R_4 is a variable resistance. Measurements are usually made by balancing, that is, varying the resistance of at least one arm, for example, R_4, until there is no current in the shunt or bridging link. In other words, at balance the potential at points B and C must be equal.

What is the relationship of resistances at the condition of balance? First, consider that supply voltage V_{BB} appears across branch $R_1 + R_3$ and branch $R_2 + R_4$. Second, note that points B and C can be at equal potential only when the voltage drop across R_1 equals that across R_2. Consider that current I_1 exists in the first branch ($R_1 + R_3$) and I_2 in the second. At balance the following voltage relationships must hold: $I_1 R_1 = I_2 R_2$ and $I_1 R_3 = I_2 R_4$. Substituting for I_2 in the first equa-

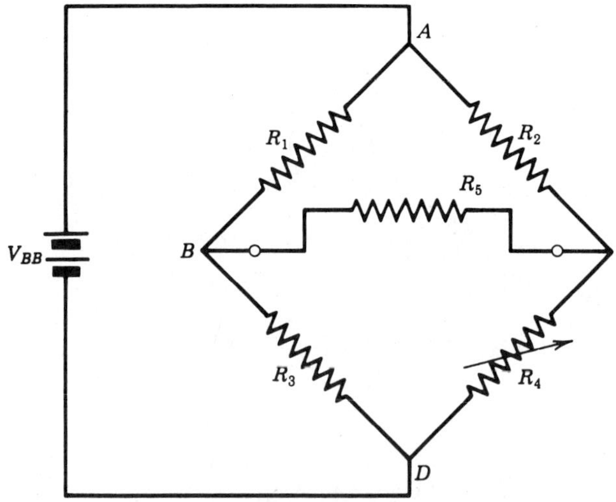

Fig. 3.7 A Wheatstone bridge. Supply voltage V_{BB} energizes the bridge. The null detector N has resistance R_5 and is connected to the output terminals of the bridge. Ratio arms are R_1 and R_2. R_3 is an unknown resistance and R_4 is a resistance that can be varied to balance the bridge.

tion gives $I_1R_1 = R_2(I_1R_3/R_4)$. Cancelling I_1 and rearranging gives

$$R_1/R_2 = R_3/R_4. \tag{3.12}$$

This equation defines the condition of balance.

The null procedure followed minimizes the need for precise, absolute detection, which is difficult, and calls for accuracy only in finding the minimum in the voltage difference, which is straightforward. As for the potentiometer, a major source of systematic error, contact resistance, is avoided. While the null detector must have sufficient sensitivity for the precision of balance required, if contact resistance greatly diminishes the difference signal, amplification will usually solve this problem.

An additional advantage of this bridge, and indeed all bridges, is that they are *ratiometric devices*. Their measurements are obtained in terms of ratios of quantities such as resistances, as in Eq. (3.12). It is these resistances that must be calibrated. Further, the balance point is independent of the energizing or source voltage V_{BB}. Quality Wheatstone bridges are constructed from resistors of high precision ($\pm 0.01\%$ tolerance) and with minimal temperature coefficients. For best precision R_1 and R_2 should be identical in value and construction so that they will drift in like fashion with time and temperature and maintain their resistance ratio of unity.*

If the unknown resistance is varying continuously, can a Wheatstone bridge still be used to measure it? For *continuous bridge measurements* the customary answer is to measure either unbalance currents $I_{\text{null det}}$ or *unbalance voltages* $V_B - V_C$. The following illustration and Appendix B pursue this possibility.

Example 3.5. A Wheatstone bridge configuration is employed when thermal conductivity detection is used with a gas chromatograph. As introduced in Example 2.4, let R_3 and R_4 be a matched pair of nichrome wire lengths and choose arms R_1 and R_2 to have equal resistance. By design R_3 is exposed to the chromatographc column eluant while R_4 is bathed by a flow of pure carrier gas. The bridge is placed in a thermostated oven along with the column. While current in the bridge raises the temperature of all the resistance arms, all should start at the same temperature. To be sure of this condition the bridge is balanced before a sample is introduced.

In the absence of eluting analyte, R_3 and R_4 are equally bathed by carrier gas and cooled by its thermal conduction. As components elute, the temperature of R_3 will change and thus also its resistance (according to its temperature coefficient of resistance) while R_4 will be stable. To detect the elution of components and respond to their relative concentration it is only necessary to measure the bridge unbalance current as a function of time (see Appendix B).

---

*In most semiprecision designs there is a much greater range of unknown resistance values since the ratio R_1/R_2 can also be varied by choosing fixed ratios of R_1/R_2 from perhaps 0.001 to 1000.0 Such bridges are versatile but seldom accurate to better than $\pm 1\%$.

3.5 VIDEO MONITORS AND OSCILLOSCOPES: CATHODE RAY TUBE DISPLAYS

The cathode ray tube (CRT) is one of the most versatile and widely used devices for displaying signals in analog form. It is the heart of video display units and oscilloscopes, two invaluable instruments for plotting the amplitude of a varying voltage or current y against time t or another variable x. These y–t or y–x plots appear on the face or screen of the tube. If yet a third variable requires display, the intensity at each distinguishable point (picture element or *pixel*), or the color hue if the option is available, can be varied.

In instruments and data stations a CRT also commonly serves for display of the analytical signal. Some examples of such displays are: observing NMR spectral traces during measurement to permit adjustment of instrument parameters for optimum performance; monitoring a biochemical reaction by display of kinetic measurements by a stopped-flow kinetic device to ensure quality while also storing them in memory; inspecting the evenness of distribution of platinum catalyst by inspecting the platinum microimage of the surface by a scanning Auger electron spectrometer.

In the CRT a focused electron beam is electrostatically deflected horizontally and vertically across a luminescent screen to trace signals of interest. Since electrons are charged and have negligible mass, an electron beam in a CRT can be moved or stopped almost instantly by electric fields proportional to signal intensity. Wherever the beam strikes the screen, luminescence is produced for 20 μs to perhaps 5 s, depending on the phosphor with which the screen is coated and its preparation. For a screen of standard persistence of about 20 ms, CRT displays have refreshing circuits that continually renew the display on the screen.

The general features of the CRT are shown schematically in Fig. 3.8. For details about the electron gun that generates the beam see Section 22.5. Suffice it here to say that control of beam intensity and focus determines the number of electrons in the beam and beam diameter.

How is a signal displayed? It must first be converted to a voltage, if not in that form already, and then applied across one of the pairs of deflection plates, usually the vertical pair, shown in the figure. As one plate is made positive with respect

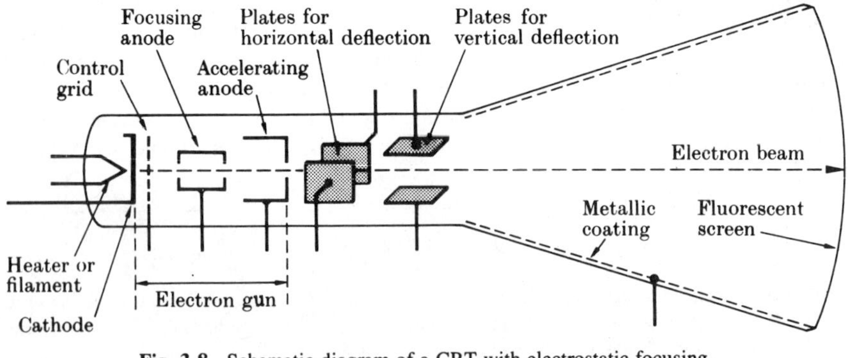

Fig. 3.8 Schematic diagram of a CRT with electrostatic focusing.

to the other, the electron beam is deflected toward it by an amount proportional to the applied signal. To accommodate a wide range of input amplitudes, signals are first passed through an attenuator and then through one or more precision amplifier stages. Thus the height of the trace on the screen quantitatively reports signal amplitude. Usually the screen is ruled so that amplitude can be read off directly in units denoted by the amplifier setting.

Actually, both periodic and transient signals are commonly displayed versus time. A voltage ramp whose slope is proportional to the desired scan rate (horizontal divisions per second) is applied to the horizontal plates. It sweeps the electron beam across the face of the tube while the field on the vertical plates produces a deflection proportional to signal amplitude. At the end of a sweep the field on the horizontal plates must suddenly be reduced to allow the beam to return to its starting point. The appropriate signal is a sawtooth wave (see Fig. 2.7b). Actually, most chemical data may usefully be displayed as *y–t* plots.

Cathode ray tubes also permit development of *indirect images of surfaces* and of *cross sections through complex structures*. In this case, actual image construction by computer (for the latter) will likely precede the display. If signals containing elemental or molecular information are obtained instrumentally by a point-to-point or multiple-angle scan, a CRT display can be coordinated with the scan. While this process will later be dealt with in detail, it is important here to examine the rastering process by which the image is formed.

In *rastering*, both horizontal and vertical deflection voltages that control the position of the electron beam are controlled by a raster program. The output of an instrument now modulates the beam intensity. One common raster program sweeps the electron beam across the full width of the screen again and again, beginning at the bottom and working by even increments to the top. Thus, a sawtooth voltage waveform is applied to the horizontal plates and a constant increment is added to the voltage across the vertical plates just after each horizontal sweep. (Recall that a similar scanning pattern is used in televisions!) Note that this program must also control the scan across a sample surface or through a sample if a cross-sectional image is sought. Again the term pixel identifies the smallest area on the tube that can give intensity information.

Oscilloscope. The dc-coupled, dual-trace, triggered oscilloscope represents one of the most dynamic display devices in the laboratory. The signal amplifiers in general-purpose scopes cover the range from dc to 10 MHz. Wideband scopes extend the range to 500 MHz and beyond. Generators (oscillators) provide fast-sweep frequencies for observation of waveforms.

If a periodic signal is to be displayed, an appropriate *time base* is one that sweeps the signal across once every period, every two periods, or every *n* periods. A stationary picture of one, two, or *n* periods of the signal is obtained. In a good oscilloscope the time base may be set precisely according to horizontal divisions per second.

Each sweep must also be synchronized to start at the same point in each cycle of a periodic signal to secure a stable display. To accomplish this result, the sweep is *triggered*. A fraction of the signal to be displayed is mixed with the time base in a circuit that generates a voltage spike; this spike initiates the start of the sweep.

x–y Plotting. A variable other than time can easily be displayed along the x axis. A classic example is the development of a Lissajous figure by comparison of a sine wave of unknown frequency with one of known frequency from a waveform generator. From the pattern the unknown frequency can be calculated.

Example 3.6. One cycle of a 5-kHz sine wave is to be displayed on an oscilloscope. Its period is $1/5000 = 200\ \mu s$. A 5-kHz sawtooth wave is generated as a time base. Assume that its ramp (rising) portion is 190 μs long and the return portion 10 μs long. Then only the last 5% of the cycle of the signal, which occurs during the beam return, will not be seen. To view two cycles of the wave on a CRT, the sweep frequency should be cut in half.

Example 3.7. In what uses will the limiting high-frequency performance of CRT amplifiers be most evident? The answer will perhaps depend on the application, but the nature of the limitation can be examined by displaying a square wave. The steep rise of the wave is, according to Fourier analysis, formed by superimposing very high frequencies. It should ideally be vertical. In practice an oscilloscope can display it only with a finite slope. The maximum rate of beam deflection is measured by its rise time t_r. If Δf is the stated frequency range, the rise time can be shown to be approximately the reciprocal of Δf, specifically, $t_r = 0.35/\Delta f$. For a dc to 10 MHz oscilloscope, for example, its rise time will be $t_r = 0.35/(10 \times 10^6) = 35$ ns even for a step function with an actual rise time of 1 ns.

Types of Oscilloscopes. In a *dual-trace* scope two signals can be simultaneously displayed as a function of time to allow the time relationship between them to be established. With only one electron gun (dual-beam oscilloscopes are also available but are more costly) time-sharing must be invoked. There are two input channels, each with its own amplifier and controls. Display is accomplished by multiplexing by means of a solid-state switch.

A dual-channel scope features different options. One is connection of channels A and B alternately to the deflection plates at a rate from 100 to several hundred kilohertz. Traces appear continuous unless a sweep rate faster than 100 cm μs^{-1} is used. An alternate mode allows switching only after a complete sweep for each channel.

A satisfactory display of very high-frequency signals is primarily limited by sweep rates. The *sampling oscilloscope* offers one ingenious way around this difficulty for repetitive signals. A signal that appears at its input is sampled at a slightly later point in the cycle each time it repeats. This point-by-point presentation continues until in a matter of perhaps 100 μs one complete cycle has been displayed. Then sampling begins again. This type of oscilloscope permits display of a periodic signal even of microwave (1–30 GHz) frequency.

Another solution is simply to increase "writing speed" and deal with the greatly diminished beam intensity by placing a microchannel plate intensifier (see Section 22.5) in front of the face of the tube so that the beam strikes it. The minute diameter of the channels ensures preservation of lateral coordinates. Then the actual stream of secondary electrons reaching the face is intensified by a factor of 10^5. Good brightness is ensured even for frequencies of 1 GHz.

Some of special types of CRT are the *storage tube*, in which an image is retained until erased, and the *digital readout* CRT, a modification that reads out the amplitude difference between two cursors that can be moved at will along a curve being displayed. In storage scopes the image is retained on a fine mesh wire grid next to the regular screen. Then a separate, unfocused electron gun floods the grid with electrons; they pass to the phosphor screen only where the trace has eliminated the repelling grid potential. Digital storage scopes are also available; these store data in semiconductor memory and convert it to deflection signals for display. A refreshing process continuously retraces the display. Clearly, storage in memory can also be made long term to provide for subsequent display whenever desired.

3.6 POWER SUPPLIES

Power supplies furnish constant voltage dc to electronic devices from amplifiers to microcomputers in most instruments. Usually the supplies derive their energy from ac lines (115 or 230 V, 60 Hz in the United States and 220 V, 50 or 60 Hz in most other countries). The series of operations performed by power supplies is summarized in Fig. 3.9. The important second step, rectification, was described in Section 2.10 and the widely used bridge rectifier circuit (see Fig. 2.16c) was taken up. To present the next step, filtering, it is worthwhile to apply it to a "worst case," the output of the half-wave rectifier, which was shown in Fig. 2.16a. Its output is a series of separated sine-wave peaks, an outcome far from the steady dc desired.

Filtering. If we model the rectifier output as dc on which ac has been superimposed, we can predict that smoothing will occur if a low-pass *RC* filter is added. Such a circuit is shown in Fig. 3.10. It should have a half-power frequency close to dc, which calls for a large value for *C* (0.1–20 μF). From a time-domain perspective, smoothing results because the capacitor stores charge when the diode is conducting. When the diode is cut off, the capacitor gradually discharges through the load represented by R_L. The time constant *RC* is always made as large as

ac Line ⟶ Transformer ⟶ Rectifier ⟶ Filter ⟶ Regulator ⟶ dc Output

Fig. 3.9 Modular diagram of a regulated dc power supply. First, a transformer steps up or down the voltage of an ac line. Then a rectifier converts the altered ac to dc. Ripple is subsequently diminished by filtering. Finally, a regulator is included to ensure an output constant within specified limits.

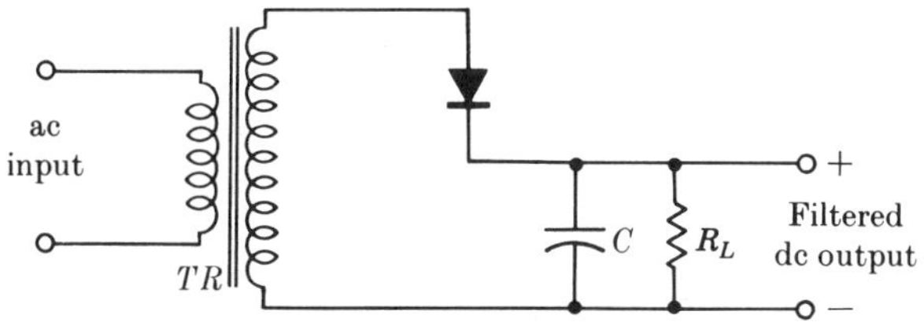

Fig. 3.10 Half-wave rectifier with simple *RC* filter.

practicable to minimize discharge. The output waveform is shown as curve V_c in Fig. 3.11.

What would happen if no load were present? Capacitor C would charge to the peak secondary voltage V_p and remain there. When current is drawn, however, $V_o < V_p$. It is convenient to return to modeling the output voltage as a dc component V_{dc} plus what is now a much smaller ac component called *ripple*. From Fig. 3.11 the dc output is seen to average $V_p - V_r/2$, where V_r is the largest amplitude of the ripple.

Clearly, in a smoothing circuit of the type shown, the greater C and R, the smaller is the ripple. If R is made large, the mean voltage obtainable at the output will depend greatly on the current drawn by the load. An additional factor that must be allowed for is suggested in Fig. 3.11a. If the capacitor has discharged appreciably, it may draw quite substantial currents I_c as it recharges. Since excessive currents that would damage diode or transformer must be avoided, C cannot be too large either.

Two other useful power supplies are the voltage doubler circuit and the split supply pictured schematically in Figs. 3.12 and 3.13, respectively. In cases where voltage but little power are needed, the voltage doubler offers a way to save on cost of components. By contrast, the split supply offers a pair of smaller, equal voltages of opposite sign. Many circuits require such voltages.

Regulators. The more current drawn by a load from a power supply of the type just described, the lower the voltage it furnishes. Where a constant output voltage is needed, as is true for most instruments, a regulator must be added following the filter. The addition of a regulator also insulates the output against line voltage fluctuations and further reduces ripple. Regulator specifications usually include either *regulation*, which is defined as the percent fluctuation in V_o, or the load stability factor $\Delta V_o/\Delta I$, where ΔV_o is the change in output voltage with a change ΔI in load current; the temperature stability factor $\Delta V/\Delta T$; and the amplitude of the remaining ac ripple and noise.

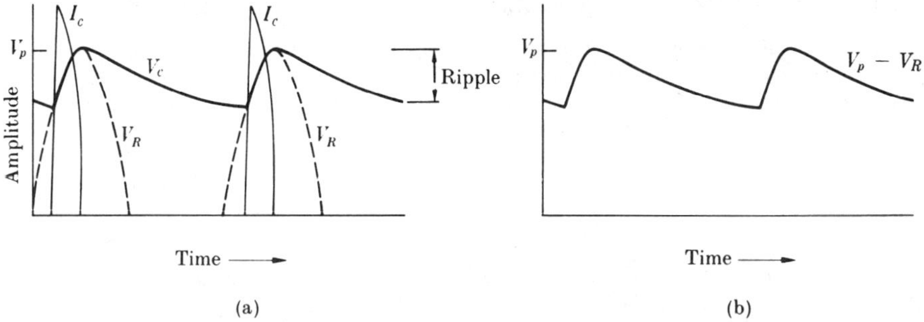

Fig. 3.11 Waveforms in the filtering circuit of a half-wave rectifier. The output voltage is V_C. Also shown are capacitor charging current I_C, and the output V_r that would appear across R if C were not present. (a) Actual waveforms. (b) Assumed triangular output waveform $V_p - V_r$.

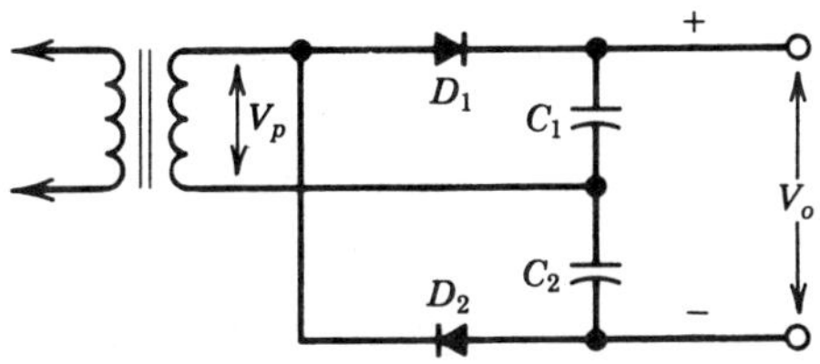

Fig. 3.12 Schematic diagram of a *voltage doubler power supply*. Each filtering capacitor will be charged to V_p if the load connected to the output terminals does not draw much current; thus, the total voltage available will be $2V_p$ at maximum. With diodes D_1 and D_2 oppositely connected to the top of the secondary winding, both halves of the power cycle will be rectified. For instance, on the ac half-cycles when the top of the transformer secondary winding is positive relative to the bottom, diode D_1 will conduct and charge capacitor C_1 as rapidly as its time constant will permit. If no charge is drawn from that capacitor on the opposite half-cycle, its voltage will gradually reach the peak ac secondary voltage V_p. On half-cycles when the bottom of the transformer winding is relatively positive, D_1 is cut off and D_2 will conduct to charge C_2. Note that current enters both C_1 and C_2 from the top and their voltages add.

A simple method of regulation is the series resistor–shunt zener diode arrangement described in Section 2.11. About 2–4% regulation is secured. A generalized regulated power supply is shown in Fig. 3.14 as a block diagram. Today a variety of packaged regulator chips that employ this approach are available for general use. For example, simple regulators of the three-terminal type (input, ground, and output terminals) are available for several voltages in the 5–24-V range. Each requires an input voltage 2–3 V greater than its output. Another series provides

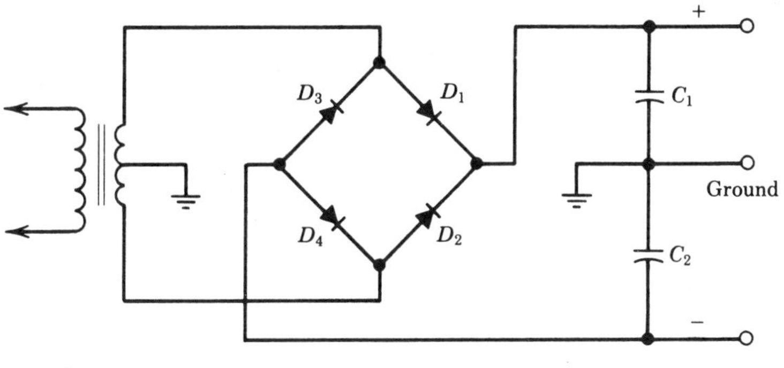

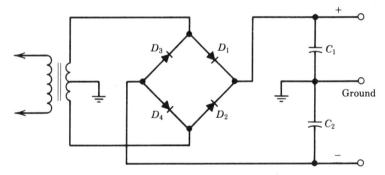

Fig. 3.13 Schematic diagram of a *split power supply* based on a bridge rectifier. This supply furnishes equal plus and minus voltages. If the load(s) attached to the output draw minimum current, each of these voltages will be $V_p/2$. The key to having equal positive and negative voltages available is that the centers of both the secondary transformer winding and the output are grounded. Both capacitors C_1 and C_2 are charged during each half cycle. Charging of C_1 occurs through diode D_1 during the half of the ac cycle when the top of the transformer is positive relative to the center tap and through C_2 and D_4 because ground is positive relative to the bottom of the winding. Because of the center tap, however, half of the ac voltage produced in the secondary winding appears across C_1 and the other half across C_2. Similarly, on the other half of the ac cycle, diodes D_2 and D_3 conduct, continuing the charging of the pair of capacitors to the same polarity. When outputs are taken with regard to sign, equal plus and minus voltages are available.

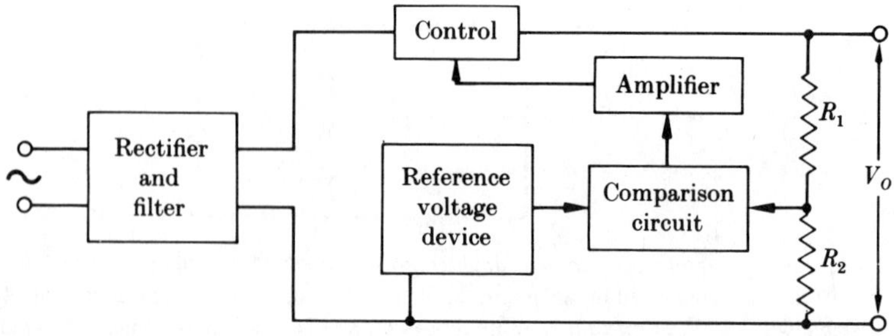

Fig. 3.14 Block diagram of a feedback-stabilized regulated power supply; V_o is the regulated output. Such a device is called a series regulator and acts as if it were a variable resistor. Its resistance is adjusted by the amplified output of the comparator in the direction needed to provide a constant current through $R_1 + R_2$ and thus ensure a constant output voltage. The comparator determines any difference between the voltage of a stable reference voltage and the IR drop across resistor R_2. For example, if IR_2 is smaller than V_{ref}, the comparator output, after amplification, causes the control device to pass more current.

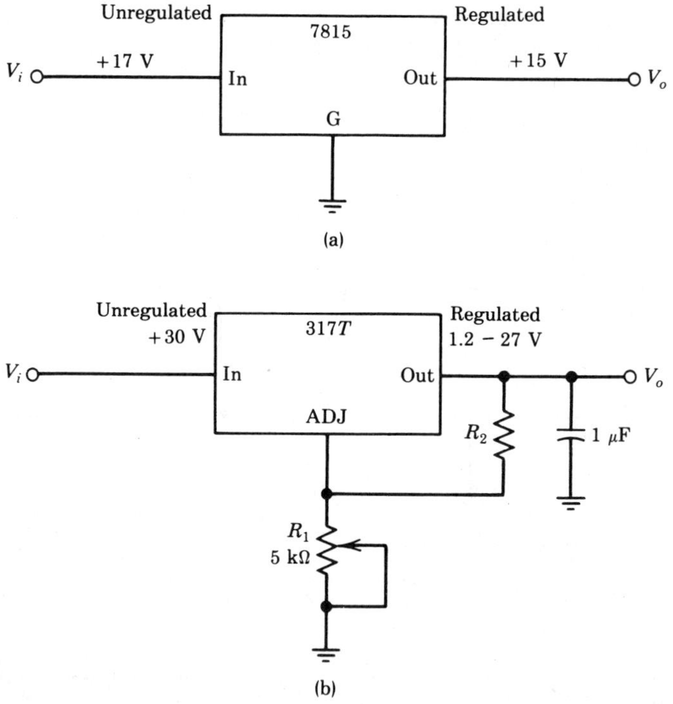

Fig. 3.15 Partial schematics for three-terminal regulators as the last stages of regulated power supplies. (a) Regulator providing a fixed voltage output. (b) Regulator furnishing voltages over prescribed range. This device maintains the voltage drop across R_2 (between output and ADJ terminals) at exactly 1.25 V. Thus, $V_o = 1.25\ (1 + R_1/R_2)$. If $R_2 = 250\ \Omega$, since R_1 can be varied from 0 to 5 kΩ by moving the wiper on the potentiometer, the regulated output can be anything in the range 1.2–26 V. The capacitor across the output provides a shunt path to ground for transient currents and might well be added to the output of the other regulator also.

regulation for negative voltages, and still others provide for adjustment of the output voltage over a small range; many of the latter are four-terminal devices. In Fig. 3.15 circuits for two representative regulators are shown. Some of their general characteristics are: maximum input voltage, 35 V; ripple rejection, 0.05–0.1%; voltage spike (sudden voltage swing in ac line) rejection, 0.1–0.3%; load regulation, 0.1–0.5% over maximum load change; temperature stability, 2% over full allowable temperature range; output impedance, about 0.02–0.05Ω.

3.7 ELECTRONIC MODULES AND THE TRANSISTOR

As suggested in Fig. 3.16, it is the presence of transistors or other *active components* that makes electronic devices substantially different from electrical circuits. This contrast is worth elaborating. In general, electrical circuits are able to modify and transform signals over a rather limited range and also attenuate signals. These disadvantages can be substantially lessened by adding transistors and other active devices so that the circuits can operate on signals while admitting energy from power sources. For example, a transistor behaves in an electronic circuit much as a sensor does in the steering wheel of a car with power steering. In such a car a driver need exert little energy to turn the front wheels: the sensor responds to a slight torque on the steering wheel by releasing a proportional flow of power to a motor to actually turn the wheels.

As instrument users what should we know about transistors? It will be important to gain a *working grasp of the varied ways in which they function*. Some practical reasons can be advanced for doing so even in a world where most electronic devices today have taken the form of integrated circuits. First, the input and output properties of integrated circuits, which are mainly comprised of transistors, are to a considerable extent defined by transistor behavior. Second, transistors usually are indispensible in interfacing integrated circuits with other circuits, as in coupling modules in instruments. Finally, users of instrumentation benefit from being able to apply their molecular insights to the operation of transistors and thus to the domain of electronics.

Accordingly, the properties and behavior of the bipolar transistor will be explored briefly here. In the next section the discussion will be extended to field-

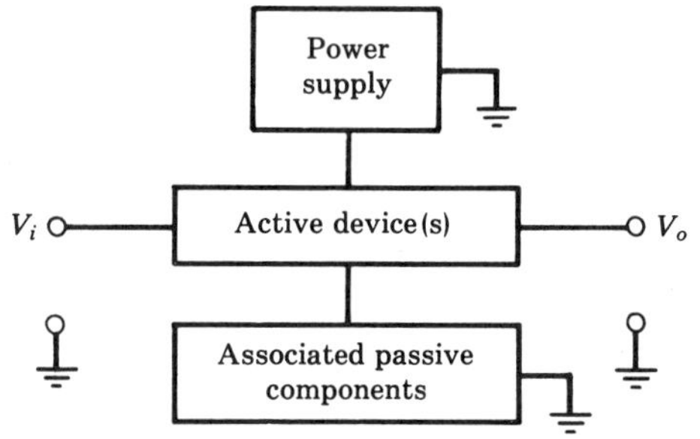

Fig. 3.16 Block diagram of a generalized electronic module.

effect transistors. Understanding these types of transistors will make the material of Chapters 4–6 more useful. Keep in mind the goal of developing a feeling for behavior and an understanding of results.

Transistors. *Bipolar* transistors were the first type developed commercially and are still widely used. They are three-electrode or three-region devices that have a pair of *p–n* junctions back to back, and are either *npn* or *pnp* types, as pictured schematically in Fig. 3.17. Their initial and final layers have the same polarity, that is, the *same majority charge carrier*. As shown, their electrodes from bottom to top are called the *emitter*, *base*, *and collector*, in that order.

In normal operation, majority charge carriers originating in the emitter cross the base (a region of opposite polarity) to the collector region, in which they are once again majority charge carriers. In the symbol for a transistor, the arrowhead identifies the emitter electrode and also indicates the direction of positive current. We shall deal only with *npn* transistors but everything said will apply also to *pnp* types if polarities and current directions are simply reversed.

Architecture, which is critical to transistor behavior, is shown in Fig. 3.18. The thinness of the central electrode, the base, is essential in ensuring a normal flow of charge carriers across its junctions. Whether such movement is possible depends almost exclusively on the base–emitter potential difference and the thinness of the base layer. Herein arises the possibility of control of the flow of power by a transistor. Clearly, the potential of the collector relative to the base must provide for the collection of these charge carriers.

Here are main facts about the behavior of *npn* transistors.

1. There are essentially two diodes present back to back, a base–emitter diode that is *forward-biased* and conducting, and a collector–base diode that is *reverse-biased*. (Recall that *p-n* junctions were discussed in Section 2.9.)

2. The collector must be biased positive relative to the emitter.

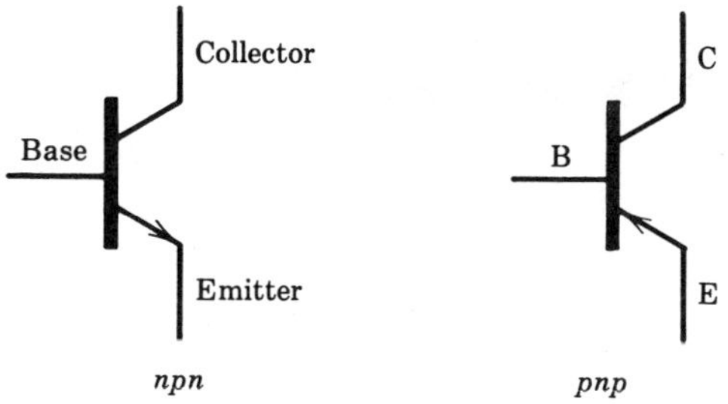

Fig. 3.17 Representation of bipolar transistors. The designations *npn* and *pnp* indicate nature of emitter, base, and collector areas, respectively. The emitter arrowhead indicates the direction of positive current; for example, in an *npn* transistor positive current moves from collector to emitter since the collector is always positively biased with respect to emitter. An *npn* transistor is said to be current-sinking since current goes toward ground. For an *pnp* transistor current direction and electrode polarities are opposite; this type of transistor is sometimes said to be current-sourcing. Older representations usually include a circle around the electrodes to identify the envelope.

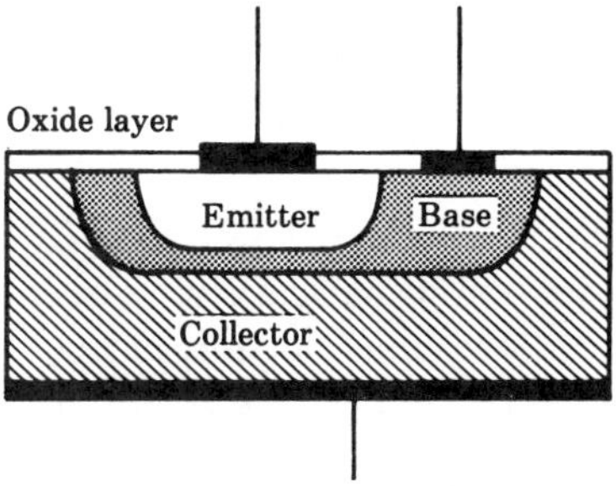

Fig. 3.18 Structure of an *npn* planar transistor.

3. Each type of transistor has specified maximum values of I_C, V_{CE}, I_B, power dissipation, temperature, and so forth that cannot be exceeded without damaging the device.

4. If these conditions are maintained, collector current I_C (current measured at that electrode) will be greater than the base current I_B by an approximately constant factor β termed the *current gain*: $I_C = \beta I_B = h_{FE} I_B$. Common values for beta are about 100.

How can there be a current gain? What actually happens is that the amplitude of the normal current through the transistor (from collector to emitter for an *npn* device) is controlled by the signal current into the base. There has been a transfer of signal from the base to the collector equivalent to amplification or gain since the base current is 100 times smaller. It is the collector current that powers any load. The following examples show applications of these rules.

Example 3.8. Bias voltages must be imposed on transistors for them to operate. If 20 V dc is imposed between collector and emitter of an *npn* transistor, can it be assumed to be dropped about equally at the two *p–n* junctions?

The first statement of fact about transistors leads us to the answer. If the base–emitter portion of a transistor is like a forward-biased diode, the base of an *npn* transistor will be about 0.7 V more positive than the emitter for all normal currents and voltages during analog or linear operation. Thus, most of the 20 V applied must be dropped not in the collector but across the collector–base junction, strongly reverse-biasing it. The diode model suggests for this junction that the reverse biasing causes extensive depletion of majority charge carriers and develops a high resistance layer that can sustain the large voltage drop.

Example 3.9. Why should a collector of an *npn* transistor be made substantially more positive than its emitter?

A relatively large positive potential on the collector ensures first that majority carriers in the collector will not enter the base. Second, and more importantly, it provides that majority charge carriers surviving the traverse of the base after entering it from the emitter will be collected. A third reason has to do with output signal and will be discussed below.

Example 3.10. Can the simple facts given about transistors provide a basis for knowing when the amplitude of collector current will be nearly independent of the collector voltage?

According to the last rule, I_C should be constant whenever I_B is constant. Such behavior will surely be observed only if biases of the three electrodes of a transistor as well as its temperature are held constant. Thus, for a constant collector current, a circuit that will hold biases constant must be sought.

In view of the care taken in transistor design and manufacture it may come as something of a shock to learn that there is great variability in transistor specifications. For instance, beta may range from 50 to 200 for a given type of transistor. Two main factors are responsible. Not only is much of the variability introduced in manufacture, but beta and other transistor properties also change with temperature and transistor parameters such as I_C and V_{CE}, the bias voltage between collector and emitter. Problems that might arise from such variability are avoided in circuits by designing for the worst case: most circuits using discrete transistors are built to operate satisfactorily using transistors off the shelf. As will be implied in Section 3.10, employing negative feedback skillfully is the best stratagem for making circuit behavior reasonably independent of the properties of individual active devices.

Consider now the two simple but representative linear transistor circuits shown in Figs. 3.19 and 3.20. Begin by identifying the place of entrance of *input signals* (the base) and of *output signals* (usually the collector). Remember that Ohm's and Kirchhoff's laws prove central in interpreting any circuit.. Electronic circuits can also be best used if voltages and currents are deduced at appropriate points by application of these laws.

A Current Source. In Section 2.2 it was stressed that good current sources are scarce. A practical current source design based on an *npn* transistor is shown in Fig. 3.19. A key aspect is providing a constant bias on the base regardless of local currents. Observe that a zener diode holds the transistor base at a fixed potential of 6.2 V above ground. Also note that the zener diode is part of a voltage divider that provides operating current for the diode. The right-hand voltage divider is, of course, the constant current path. The figure legend explains the rest. We shall see

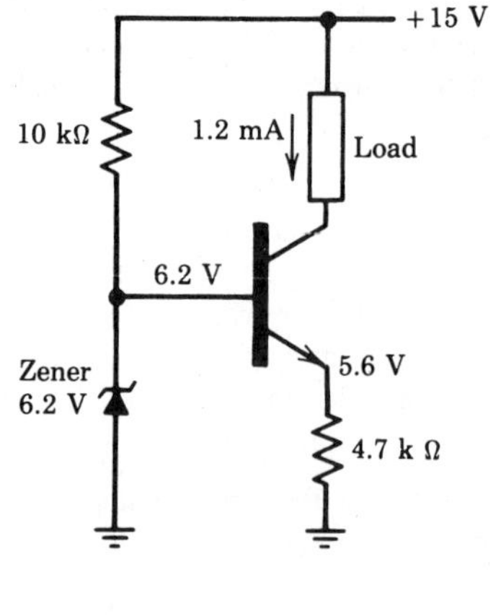

Fig. 3.19 Schematic diagram of a current source. A zener diode fixes the base potential (relative to ground) at 6.2 V. The forward-biased emitter junction of the *npn* transistor holds the emitter 0.6 V more negative, that is, at 5.6 V. Accordingly, current furnished to the load by this source is just that needed to produce a 5.6-V drop in the 4.7-kΩ emitter resistor.

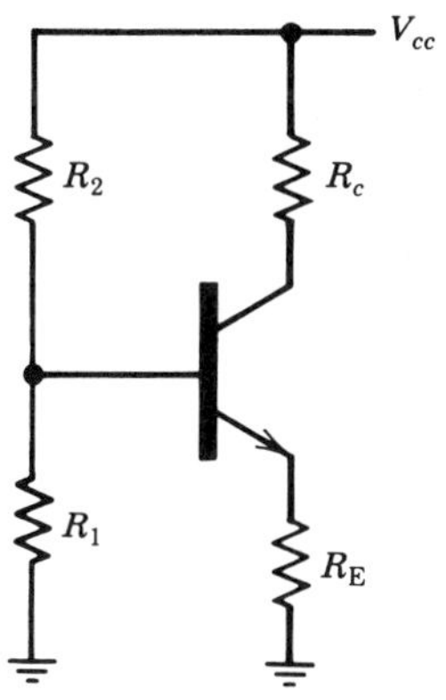

Fig. 3.20 A simplified diagram of an ac common-emitter amplifier circuit. The operation of the amplifier is greatly stabilized by using a voltage divider $R_1 + R_2$ to set the base voltage and by use of an emitter resistor (R_E).

the first applications of such sources in developing integrators and generators of sawtooth waves and ramps in the next chapter.

Example 3.11. How much voltage compliance does the current source of Fig. 3.19 show? Recall that compliance is defined as the ability of a current source to vary its output voltage as necessary to cope with different resistance (impedance) loads. According to Ohm's law, if the current I in loads is to be constant, the voltage must increase as load resistance rises.

The answer comes from noting the rule given above that the collector of the transistor must be positive relative to the emitter (and toward the base as well). To satisfy the rule, the absolute collector potential in the figure can range from +15 V down to a minimum potential of perhaps 7 V (0.8 V more positive than the base). Thus, at constant current the load could have a voltage across it that can vary from 0 V (nearly zero load resistance) to 8 V. There will be 8 V of compliance; if more is needed, a higher supply voltage could be used.

To convert the circuit into a precision current source, some method of temperature control for the transistor or other compensation would also need to be provided. Temperature coefficients are large for semiconductors and any current change will produce a small alteration in emitter voltage.

A Prototype Amplifier. An electronic amplifier circuit should produce an output that contains an enlarged version of the input signal. How does a device like a transistor make such a circuit possible?

The best answer is found by examining the operation of a representative circuit, the common-emitter amplifier shown schematically in Fig. 3.20. It will be explored as a voltage amplifier. Input in this design is always to the transistor base and output is taken from its collector. While the performance of the circuit will be analyzed briefly below in a series of examples, the results of analysis are summarized in Fig. 3.21. Note that the circuit as shown in Fig. 3.21 has been slightly modified (a) to exclude dc inputs and (b) to restrict its frequency range to audio

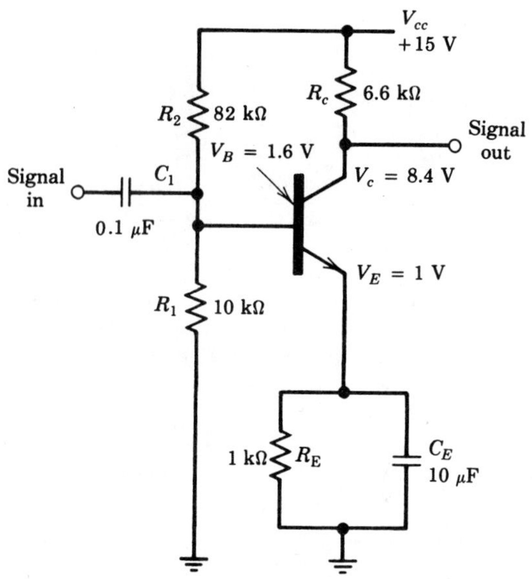

Fig. 3.21 An augmented version of Fig. 3.20. Calculated values of resistance have been added as well as an input lead with series capacitor C_1 to restrict input to ac; capacitor C_E to bypass emitter resistor R_E and improve signal gain; and an output lead (at the collector). In this circuit, the output signal is given by $V_o = V_{CC} - I_C R_C$. As the input signal to the base becomes more positive, increasing V_{BE} slightly, collector current I_C increases also. Since $I_C R_C$ is subtracted from the collector supply voltage V_{CC} to give output signal V_o, the output is 180° out of phase with the input.

frequency ac. Let us return to the figure after working through the following examples.

Example 3.12. In a basic circuit such as that in Fig. 3.20, too little information is provided to permit calculation of operating values without making some assumptions. What conditions and values may legitimately be assigned?

Some reasonable assumptions are

(1) that the supply voltage V_{CC} will be +15 V, a standard value;

(2) that the collector current of the transistor will assume a fairly standard quiescent value (with no input signal) of 1 mA;

(3) that $R_C >> R_E$ and $R_C + R_E < 10\ \text{k}\Omega$, which will ensure a good voltage gain in the amplifier. Further, it will be important that $R_1 + R_2$ be sufficiently large that minimum power be dissipated in this branch. Resistance values that fit these assumptions have been entered in Fig. 3.21.

Example 3.13. What dc voltages (biases) will be furnished the transistor in the circuit of Fig. 3.21?

To find bias values we use Ohm's and Kirchhoff's laws. The base bias V_B can be seen from the figure to be the fraction of the supply voltage V_{CC} furnished to the base by the voltage divider $R_1 + R_2$. Note that the bottom of R_1 is at ground potential, so the full

+15 V assumed for V_{CC} will be across the divider. Thus, $V_B = V_{CC}R_1/(R_1 + R_2)$, and $V_B = [10/(10 + 82)] \times 15 = 1.6$ V. No other bias will be as critical as this one.

To deduce the bias voltages for collector and emitter, we shall need to assume that the transistor can be considered mainly a part of the right-hand branch of the circuit in the figure. Accordingly, it is possible to calculate the voltage drops in R_C and R_E and subtract or add them to the fixed supply voltages brought into the circuit. The 1 mA of collector current that essentially exists in R_C, R_E, and the transistor will give rise to the voltage drops. (The amount of this current lost through the base and contributions to the current from minority charge carriers is insignificant and will be neglected.) By calculation the collector will be biased at $V_C = 15 - (1 \times 10^{-3}\text{A} \times 6.6 \times 10^3\Omega) = 15 - 6.6 = 8.4$ V. By similar calculation the emitter will be at $V_E = 0.0 + (1 \times 10^{-3}\text{A} \times 1 \times 10^3\Omega) = 1$ V. Note that all these values have been entered in Fig. 3.21, which mainly duplicates Fig. 3.20.

Example 3.14. What collector current *and* voltage will be obtained if a +5-μA input signal is directed to the base? Assume beta for the transistor to be 100.

Note that there must have been an earlier value for the base current that corresponded to the quiescent value of I_C. It can be calculated assuming that $I_C = \beta\ I_B$ holds down to very small base currents. For a 1-mA collector current, $I_B = 1 \times 10^{-3}/100 = 10\ \mu$A.

Now, if a signal increasing I_B by 5 μA is applied to the base, a new base current of 15 μA must exist. Similarly, the new value of collector current must be 15×10^{-6} A $\times$ 100 = 1.5 mA.

If I_C has increased, the collector voltage must have decreased. The new voltage drop in R_C will be $I_CR_C = 1.5 \times 10^{-3} \times 6.6 \times 10^3 = 9.9$ V. Thus, the new collector voltage will be $15 - 9.9 = 5.1$ V.

Clearly, there will be an upper limit to how much the collector current can increase before the collector bias is so low the transistor is cut off. Where a transistor is used in switching between logic states as in a computer, the cutoff state is regularly reached.

Example 3.15. Amplification of both ac and dc is possible with the circuit shown in Fig. 3.20. In this case, as in most integrated circuits, amplifiers are *directly coupled* to allow handling of both types of signals. In Fig. 3.21, however, the circuit has been modified to restrict its frequency range to ac signals by inclusion of the input capacitor C_1 and another capacitor in the transistor emitter circuit. Explain the role of the added capacitors. Assume the input impedance of the amplifier is 2 kΩ for ac.

First, the C_1R_1 combination will cause the input circuit to be a high-pass filter with a half-power frequency $f_0 = 1/(2 \times 3.14 \times 2 \times 10^3 \times 0.1 \times 10^{-6}) = 800$ Hz. Direct current will be blocked by the capacitor, which will allow the voltage divider $R_2 + R_1$ to determine the bias for the base without interference from external dc sources.

Second, the emitter capacitor will have a reactance $X_C = 1/(2 \times 3.14 \times 10^3 \times 10 \times 10^{-6}) = 15.9\ \Omega$ at 1000 Hz. This value is much less than 2 kΩ, indicating clearly that almost all ac will be shunted through the capacitor to ground and will bypass R_E. By this ingenious shunt, the dc bias on the emitter can be maintained with minimal ripple or ac contribution. Further, an amplified ac output signal is available from the circuit at the collector since ac contributes fully to the voltage developed across R_C.

In Fig. 3.21 values for transistor voltages have been entered and points of entrance and output of signals from the amplifier clearly identified. As can be deduced from Example 3.14 the *amplifier output signal* V_0 is just the difference

$$V_0 = V_{CC} - I_C R_C. \tag{3.13}$$

Since an increase in input signal leads to a rise in I_C and a *decrease* in V_0, a 180° phase change commonly occurs on amplification by a single transistor common-emitter amplifier. Finally, since a comparatively large change in I_C occurs for a small signal, a large replica of the input signal is created superimposed on the basic dc collector voltage. Amplification has occurred.

Many kinds of amplifier circuits and many additional stages like this one have been designed. The best designs are now provided in integrated circuits. For use in instruments high-performance, very high-quality amplifiers termed operational amplifiers now dominate the field. In the next chapter we shall look in detail at some of their applications.

Ebers–Moll Equation. This equation represents a valuable refinement of transistor behavior that is quite useful in circuit design. Collector current is not exactly proportional to base current as suggested earlier, but is more faithfully represented by the Ebers–Moll model, which give rise to the equation

$$I_C = I_s \left[\exp(V_{BE}/V_T) - 1\right] \tag{3.14}$$

Here I_s is the transistor saturation current, V_{BE} the base–emitter potential, and $V_T = kT/e$, where k is the Boltzmann constant, 1.38×10^{-23} J K^{-1}, and e is the basic unit of charge, 1.60×10^{-19} C. Thus, I_C is determined by V_{BE} rather than strictly by I_B. From the equation, which is valid for a diode as well as a transistor, base–emitter voltage rises 60 mV per decade of collector current at a fixed temperature. A large temperature coefficient for I_s makes difficult direct application of voltage signals to a base, however. Equation (3.14) actually describes I_C for a transistor over many orders of magnitude. It has also provided a theoretical basis for interpretation of analog logarithmic amplifiers based on transistors.

3.8 OPTIMUM COUPLING OF MODULES

Look again at Fig. 3.1 and the train of modules that make up a representative instrument. Electronic design also deals with the coupling between modules, represented only by arrows in the figure. Some criteria for good coupling are that

1. the signal should be transferred with little attenuation, and
2. coupling must introduce minimum distortion and noise.

Attention will be given to the first criterion here and to the second in Chapter 12.

The desired insight into coupling electrical and electronic modules can be gained by applying the Thevenin theorem (see Section 2.4). Two applications are shown

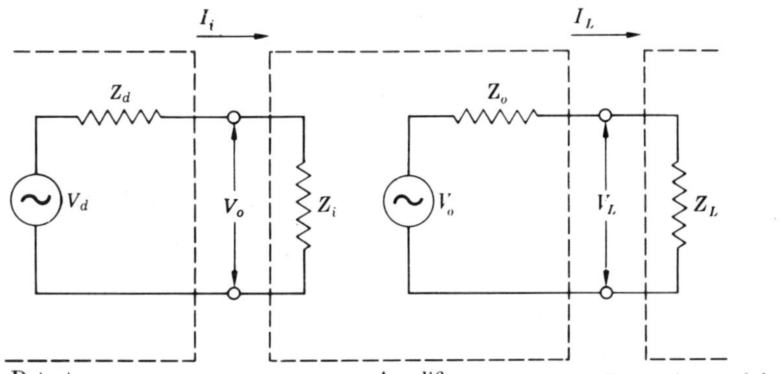

Fig. 3.22 Modular schematic of a train including detector, amplifier, and processing modules as part of a representative instrument. Thevenin equivalent circuits are used to represent outputs of both the detector and the amplifier. The input and output impedances of the amplifier, Z_i and Z_o, may be determined experimentally. To do so, one first disconnects all other modules and then applies varying values of V_i and V_{rev} to input and output. While doing so, one measures currents and voltages required by the following expressions and substitutes values $Z_i = (\delta V_i/\delta I_i)\, V_{rev}$ and $Z_o = (\delta V_{rev}/\delta I_o)\, V_i$.

in Fig. 3.22. In each the *output* of an electronic module is represented by a Thevenin equivalent circuit consisting of an output voltage in series with an output resistance or impedance.* When a detector is coupled to an amplifier, the detector output voltage appears across a *voltage divider* consisting of the output impedance Z_d and amplifier input impedance Z_i. The consequences are taken up in an example given below. As shown in Fig. 3.22, the pattern is repeated at the amplifier–processing module interface.

Some general principles governing coupling may be deduced. Whether a module should be designed with high or low input and output impedances will depend mainly on the type of electrical transfer of greatest interest.

1. Optimum *voltage transfer* will occur when $Z_i >> Z_o$.
2. Optimum *current transfer* will require that $Z_i << Z_o$.
3. Maximum *power transfer* will call for $Z_i = Z_o$.

Example 3.16. What fraction of the voltage developed in the detector of Fig. 3.22 will actually appear as an input to the amplifier?

According to Fig. 3.22 the detector output voltage appears across a voltage divider $Z_d + Z_i$. Neglecting any frequency-dependent behavior, the divider is identical to the resistive divider in Fig. 4.6. The fraction of V_o appearing at the amplifier input equals $Z_i/(Z_d + Z_i)$.

*It proves legitimate operationally to separate input and output aspects of electronic modules since one can in fact measure distinctive input and output signals and impedances.

The input voltage will thus be

$$V_o Z_i/(Z_d + Z_i).$$

Since the fraction of V_o appearing as input increases as $Z_i >> Z_d$, the condition for optimum transfer of voltage given above is confirmed.

Example 3.17. How can one match the output impedance of one module to the input impedance of a following module to ensure optimum power transfer?

It is often possible to match the impedances of standard modules by coupling them with a transformer. A partial diagram illustrating this use of a transformer is shown in Fig. 3.23. To interpret what is occurring we must describe the manner in which in the transformer secondary circuit impedance is related to that in the primary circuit. Such transformers are designed with a fixed turns ratio t (secondary winding turns to primary winding turns), frequency range, and wattage. Only the first need concern us now since it fixes both the voltage and current ratios in the transformer windings. For instance, we find that the voltage ratio is given by $V_4/V_1 = t$ and the inverse current ratio by $I_1/I_4 = t$. Further, from Ohm's law the effective current I_2 in the secondary circuit is just $I_2 = V_2/Z_2$, where V_2 is the effective voltage across the secondary winding and Z_2 is the impedance in the secondary circuit. Combining these three equations we obtain the equation for current in the primary circuit: $I_1 = V_1 t^2/Z_2$.

Now the impact of Z_2 on the impedance of the primary circuit can be learned by arranging the last equation in the form of Ohm's law. The result is

$$I_1 = V_1/(Z_2/t^2).$$

Its denominator can be taken as the equivalent impedance ($Z_{eq} = Z_2/t^2$) developed in the primary circuit. This behavior is commonly described by saying that the impedance of the secondary circuit has been *reflected* into the primary circuit by the transformer. Finally, the turns ratio to that will ensure that $Z_1 = Z_2$ can be found by letting $Z_1 = Z_{eq}$. Now $Z_1 = Z_2/t^2$ and $t = \sqrt{Z_2/Z_1}$.

Example 3.18. What turns ratio must a transformer have to transfer maximum power from a stereo amplifier of output impedance 100Ω to a 30-Ω load?

To achieve a match, we use $t = \sqrt{Z_2/Z_1}$ and find $t = \sqrt{30/100} = 0.55$. A step-down transformer with 55 secondary turns to each 100 primary turns is needed.

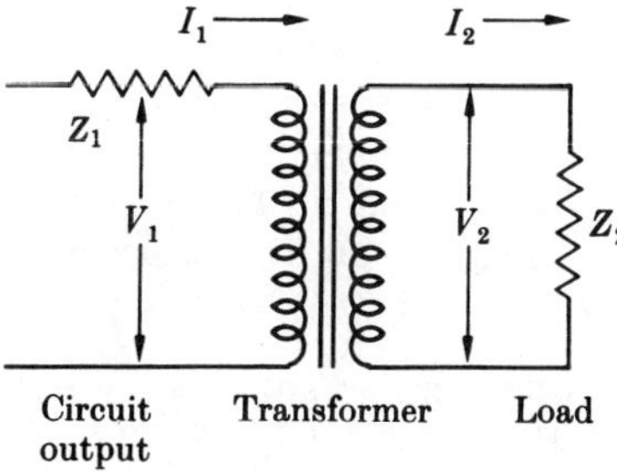

Fig. 3.23 Use of a transformer to match an input impedance Z_2 to an output impedance Z_1 of a preceding module of an ac circuit.

3.9 FIELD-EFFECT TRANSISTORS: JFET AND MOSFET

The development of field-effect transistors (FETs), the other main type of transistor, opened up new possibilities in electronics. In these devices there is a uniform channel of *n*- or *p*-type for majority charge carriers and its conductance is controlled by input signals. In one major type a *gate* is formed on one or both sides of the channel by diffusing in dopant that will give to a small area the opposite type of majority charge carrier. For example, in an *n*-channel device a gate can be set up by creating a *p*-type area. Gate regions of this type are pictured in Fig. 3.24*a*. The resulting device is called a junction FET or JFET. Its characteristic symbols are shown in Fig. 3.24*b* and *c*.

To bring an FET into operation, its gate is reverse-biased. The bias potential establishes a *field* that restricts channel conduction by widening the depletion layer at the *p–n* junction between gate and channel. The schematic for the JFET in Fig. 3.24 shows the channel narrowing. As noted above input signals are also directed to the gate. With FETs signal *voltage* will dominate in controlling current and the output signal. Recall that in the regular (bipolar) transistor conduction was mainly controlled by signal current. A second major kind of FET of basically similar behavior will be introduced later.

Majority charge carriers originating at the end of the channel called the *source* move toward the end called the *drain*. Actually, the two ends are often interchangeable.

What properties of FETs are distinctive? Since their operation as active devices depends upon an imposed field, the gate draws the most minimal of currents. Indeed, the input impedance can be as great as $10^{14}\Omega$, making it possible to develop a variety of circuits where input impedances are exceptionally high. Another at-

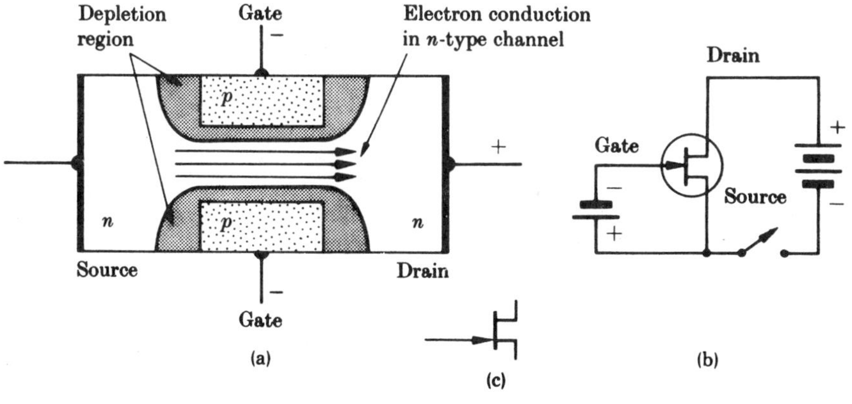

Fig. 3.24 Junction field-effect transistor (JFET). (a) Schematic of design. The channel is *n*- type and relies on electrons as majority charge carriers. Note that the gate is a *p*-type region. The reverse bias applied to the gate–channel junction causes the depletion layer to extend into the channel and channel conduction. (b) Symbol and simple biasing arrangements for an *n*-channel JFET. Biasing is also shown. Electrons, as majority charge carrier, must be attracted from source to drain. (c) Symbol with gate arrow placed to identify the source explicitly.

tractive feature is that FETs occupy an especially small area on a silicon chip. Thus, in large-scale integrated circuits FETs are often the transistor of choice.

In Fig. 3.25 a set of drain current (I_{DS}) curves are plotted versus the drain–source potential difference, V_{DS}. A family of curves is shown since each value of gate–source potential V_{GS} gives rise to a unique I_{DS} versus V_{DS} curve. Two aspects of the curves deserve attention. One is that the channel current is nearly constant as V_{DS} is varied at a given gate voltage. Recall that collector current behaved similarly with respect to V_{CE} for regular transistors. This is a fortunate result since it makes it possible for a large range of signal levels to be amplified linearly.

The other interesting aspect of the I_{DS} curves is that each *initial* slope depends upon the gate voltage V_{GS}. In this voltage range the channel resistance will vary linearly with the voltage applied to the gate. It should be possible in this range to use an FET as a *voltage-controlled resistor*.

There is a second major type of FET in which gates are insulated from the conduction channel. Since the usual insulating layer is SiO_2 and the gate is metallic, the prefix metal–oxide-semiconductor (MOS) is appended to FET to distinguish this type from junction devices. Earlier the alternate term ''insulated gate'' was occasionally used. The substrate (body) of a MOSFET is important electrically and is provided with a terminal also; thus, MOSFETs have four terminals. Again, channels may be either *n*- or *p*-type. Symbols for each type are given in Fig. 3.26. For the *p* type doping can also be varied to control within limits the gate–source voltage at which conduction begins in the channel. Thus, depending on the level of doping, there are depletion or enhancement *p*-type MOSFETs. Channel current curves as a function of gate–source voltage are also given for the two types in Fig. 3.26c. For several reasons it proves advantageous to use MOSFET transistors in large-scale integrated circuits.

What should be said to characterize the behavior of FETs in general? These transistors are brought into their active region by raising the gate voltage toward

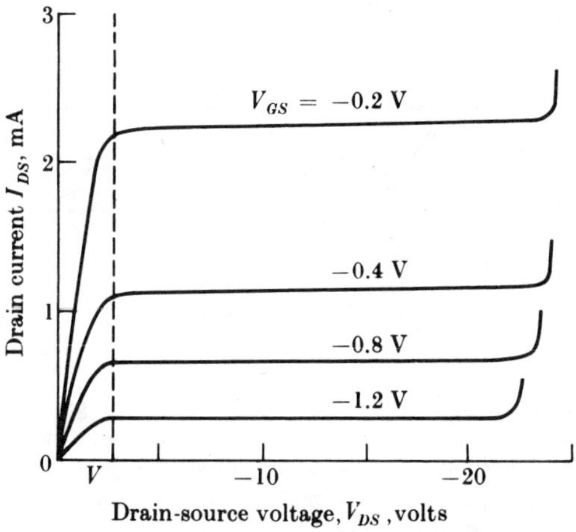

Fig. 3.25 Characteristic curves for a junction FET.

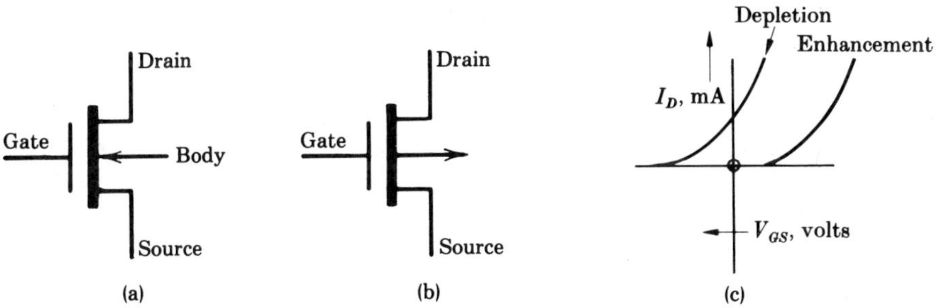

Fig. 3.26 Symbols for metal-oxide-semiconductor field effect transistors (MOSFETs). (a) An *n*-channel depletion type; (b) *p*-channel depletion type; (c) Drain current for both depletion and enhancement types as a function of gate bias.

that on the drain. For example, for an *n*-channel JFET, which requires a positive value for V_{DS} (Fig. 3.26c), the gate potential would gradually be increased from a negative voltage toward zero. Refer again to Fig. 3.25 for confirmation.

That figure certainly suggests that FETs would be good as central elements in all of the usual electronic circuits. A final comment is that there is even greater variability in device parameters for a given type of FET than was true for bipolar transistors. As a result, care must be taken in design to make allowance for the variability. Changes in manufacturing procedures have improved the situation somewhat.

An Analog Switch. A representative application in which FETs are especially valuable is electronic switching. What is most sought in an analog switch is a very high resistance (ideally, infinite) when it is OFF or open, and a low resistance when it is ON. A representative MOSFET analog switch with such properties is illustrated in Fig. 3.27. Its resistance is about 10^4 MΩ when the device is OFF and only about 50–100 Ω when it is ON.

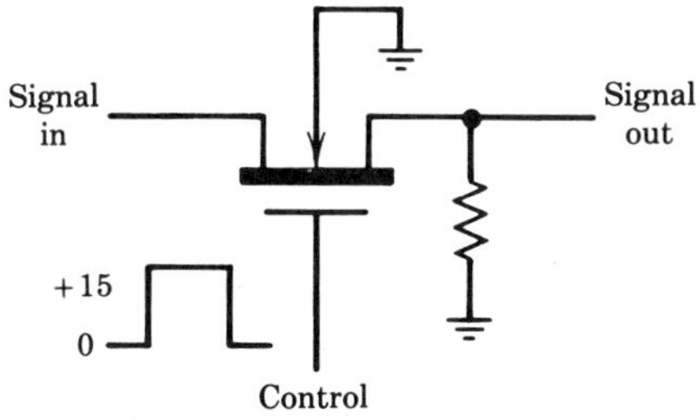

Fig. 3.27 Voltage-controlled FET analog switch. In the OFF state the resistance of the FET channel is extremely high (about 10^4 MΩ) and in the ON state only about 75Ω. To turn on the switch the gate (control lead of the enhancement type MOSFET) is raised in potential to +15 V. Then signals in the 0 to −10 V range pass through. To pass signals that are in the 0 to +10 V range also, the MOSFET body (substrate) should be held at −15 V (instead of ground) and the OFF pulse changed to −15 V (instead of 0).

To effect switching it is only necessary to change the bias on the gate. The switch is OFF with 0 V on the gate and ON when +15 V is applied. When this switch is ON, all frequencies in a signal, and different amplitudes of voltage will pass through.

In digital devices switching is the central operation. We can expect to see further development of FET switches as *logic* switches in the chapter on digital electronics.

3.10 FEEDBACK

Feedback is a process in which a portion of the output of an electronic circuit is added to the signal at its input. The addition may also occur within a single stage of a multistage device, for example, in a multistage amplifier between the collector of a particular transistor and its emitter, or across two or more stages. The characteristics of an amplifier are modified substantially by feedback, though the effect secured depends greatly on the phase of the returned signal relative to the incoming signal. Some feedback occurs naturally; most is arranged deliberately to achieve desired amplifier behavior.

An amplifier circuit that can be used with or without feedback is pictured in block diagram form in Fig. 3.28. The distinction between the circuit input V_s and the amplifier input V_i is important. The latter is the input to the amplifier itself. When the feedback loop is open (switch Sw open), the *open-loop* gain* A describes the output: $A = V_o/V_i = V_o/V_s$.

But when a fraction F of the output voltage V_o is fed back by use of the loop, the voltage appearing at the amplifier input becomes

$$V_i = V_s + FV_0. \tag{3.15}$$

It is also desirable to define a *closed-loop* or circuit gain G: $G = V_o/V_s$. Once the loop is closed, V_s and V_i are no longer equal, and $G \neq A$. Eliminating V_i from Eq. (3.15) gives an expression relating G to A:

$$G = A/(1 - FA). \tag{3.16}$$

The product FA is commonly called the *loop gain*.

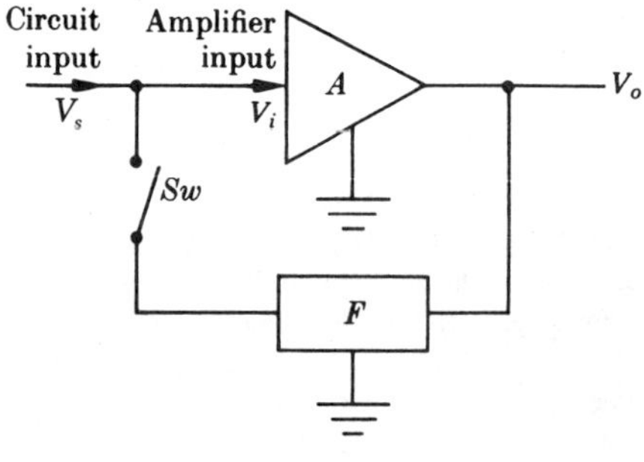

Fig. 3.28 Amplifier (shown as a triangle) with feedback network. A fraction F of output voltage can be fed back to its input by closing switch *Sw*. All voltages are taken relative to ground. The intrinsic gain of the amplifier is defined as $A = V_o/V_i$. When feedback occurs, the overall or circuit gain G experienced by the input signal is defined by the expression $G = V_o/V_s$.

*Only absolute values of gains will be considered.

Negative Feedback. This important process occurs when the feedback fraction F has a negative sign. What is fed back to the input subtracts from V_s. Whatever is necessary is done to ensure that the portion of the output fed back is 180° out of phase with the input. With such feedback $V_i \ll V_s$ and $G \ll A$. This drop in circuit gain is clearly evident in terms of Eq. (3.16). Its denominator is greater than unity when F is negative.

With negative feedback overall gain is sacrificed but impressive advantages are secured. In general, negative feedback brings about *greater stability, greater bandwidth*, and *reduction of any distortion and noise* that arise in an amplifier itself. *Bandwidth* is defined as the frequency range of an amplifier extending from its lower characteristic frequency f_0 (half-power frequency) to the upper half-power frequency f_0. A flat response curve and bandwidth refer to the same condition and indicate the frequency region where gain is relatively constant with frequency. For instance, stereo buffs seek audio systems with a flat response over a range from 20 to 20,000 Hz.

A case of great importance is the limiting situation in which the negative feedback fraction F is comparatively large. Then $|FA| \gg 1$ and $1 - FA \simeq |FA|$. Equation (3.16) reduces to

$$G \simeq -1/F \tag{3.17}$$

and the higher the amplifier gain, the better the aproximation. While the amplifier still provides power, the feedback circuit alone determines the gain! Stated differently, the character of the feedback circuit will determine both gain and *performance* for the overall circuit. This valuable result is the basis for the versatility of operational amplifier circuits, which will be discussed in the next chapter.

Example 3.19. If an amplifier has sufficient gain, some of it can be sacrificed to obtain a flatter response curve, that is, to reduce *frequency distortion*, as well as reduce other distortion in the output signal. Distortion other than the frequency type just referred to is measured by comparing output signals with input signals.

How feeding back a fraction of the output that is 180° out of phase with the input signal V_s accomplishes the improvements listed above may be sensed intuitively. The small fraction of output fed back adds a negative-going version of distortion to the input. When mixed with the distortion and noise normally generated in the amplifier, simple "cancellation" results. In any good amplifier, or indeed in any electronic circuit, it may be assumed that internal negative feedback has been provided.

Example 3.20. Calculate the value of circuit gain G for a case in which the amplifier gain A is 10^5 and the feedback fraction F of an external circuit equals -0.01. Does the result fit the case of Eq. (3.17)? If it does, what will be the relative roles of amplifier and feedback circuits in affecting the behavior of the circuit?

From Eq. (3.16) $G = 10^5/[1 - (-0.01)10^5]$. The circuit gain G is $10^5/1001 \simeq 100$. Clearly, a great deal of the output of the amplifier has gone into providing negative feedback! Equation (3.17) gives the same result: $G \simeq -1/F \simeq -1/(-0.01) \simeq 100$. We can conclude that a feedback of 1% is very large whenever the amplifier gain is high. With feedback of this magnitude, the performance of the overall circuit will be determined by the nature of the feedback circuit.

Positive Feedback. Whenever feedback is *in phase* with the input signal, it is said to be positive. Actually, some positive feedback occurs naturally in most amplifiers. Its effect is opposite to that of negative feedback. Now the factor F is positive, and the denominator of Eq. (3.16) approaches zero, causing the gain of the circuit to soar. In addition to this bootstrap effect a much narrower bandwidth is secured. This behavior is sought in introducing positive feedback in the input stage of a radio or TV. Positive feedback increases selectivity, making it possible to tune in the carrier frequency of each station or channel precisely. Frequencies on either side are virtually neglected.

With extensive positive feedback, the performance of an amplifier becomes unstable. Now any momentary fluctuation is likely to lead to oscillation, the amplification of the very narrow band of frequencies that experiences minimum loss to the exclusion of all others: the input signal is completely ignored. This behavior is totally unacceptable for an amplifier, but is put to good use in the design of ac signal generators and *oscillators*. In a high-quality oscillator a carefully cut, evenly thermostated, quartz crystal is installed in a positive feedback circuit and the output is a sine wave of precise frequency. References will provide information about oscillator design and behavior.

Control and Automation of Systems. Negative feedback also plays a central role in stabilizing the performance of many analog systems. For example, most analog recorders operate as servo devices employing negative feedback.

Analog Recorder. Many *analog recorders* employ negative feedback. In such a recorder a signal voltage or current is measured by a process that involves taking the difference between a variable standard voltage and the signal. The difference, sometimes called an error signal, is amplified and directed to a servomotor that systematically moves an output tap or wiper on the variable standard voltage supply. The negative feedback circuit continues to supply energy to the servomotor until the standard voltage tapped off matches the unknown voltage. The recorder display is ordinarily a pen tracing of the position of the wiper. Since the difference is driven to zero, the process is often called *null balancing*. Both potentiometric (resistive) and capacitive balancing are in use, with the latter used in analog recorders of better life and precision.

In negative feedback devices there may be several elements in the system *feedback loop* or *control loop* as just seen in the recorder circuit. Such systems are self-regulatory; they stand alone.* While self-regulating analog modules are common, where instrument versatility is important they have usually given way to program control. Microcomputers have increasingly been wired into instruments

*Self-regulating systems are studied in the discipline of cybernetics. Feedback loops are central to such systems.

to provide control. Flexibility is gained because the feedback circuit can be changed by the user by altering a program or a value. Even the style of control can be selected by the operator from a keyboard. See Chapter 6.

REFERENCES

Some general texts on an introductory to intermediate level are:

1. J. R. Cogdell, *Introduction to Circuits and Electronics*. Englewood Cliffs, NJ: Prentice-Hall, 1985.
2. W. D. Cooper and A. D. Helfrick, *Electronic Instrumentation and Measurement Techniques*. Englewood Cliffs, NJ: Prentice-Hall, 1985.
3. P. Horowitz and W. Hill, *The Art of Electronics*. Cambridge: University Press, 1980.
4. P. Kantrowitz, G. Kousourou, and L. Zuker, *Electronic Measurements*. Englewood Cliffs, NJ: Prentice-Hall, 1979.
5. H. V. Malmstadt, C. G. Enke, and S. R. Crouch, *Electronics and Instrumentation for Scientists*, Menlo Park, CA: Benjamin/Cummings, 1981.
6. B. H. Vassos and G. W. Ewing, *Analog and Digital Electronics for Scientists*, 2nd ed. New York: Wiley-Interscience, 1980.

Volumes dealing with a single topic are:

7. G. M. Ewing and H. A. Ashworth, *The Laboratory Recorder*. New York: Plenum, 1974.
8. S. M. Sze, *Semiconductor Devices—Physics and Technology*. New York: Wiley, 1985.

See also references listed for Chapters 3 and 4.

EXERCISES

3.1 It is desired to minimize a 60-Hz line voltage hum in an electronic circuit that will ordinarily carry 1–3 kHz signals. (a) Draw a simple *RC* filter that can be used. Choose a value for its cutoff frequency f_o and support your choice. Assume a roll-off of 20 dB per decade of frequency below f_o. (b) Select or calculate values for the filter components. Is there any advantage in picking $R = 1\ \text{k}\Omega$ instead of 10 kΩ? Explain. (c) For strong rejection of a 60-Hz hum what attenuation in decibels would be desirable?

3.2 In a simple *RC* low-pass filter what size capacitor will be needed with a 15-kΩ resistor?

3.3 What is the cut-on frequency f_o for the circuit of Fig. 3.2 if values of components are $C = 0.10\ \mu\text{F}$ and $R = 1\ \text{k}\Omega$?

3.4 Show that the response of a low-pass filter is given by the expression $V_o = V_i/[1 + \omega^2R^2C^2]^{1/2}$.

3.5 If Eq. (3.1) is reduced differently, one obtains $V_o/V_i = R/[R^2 + 1/\omega^2C^2]^{1/2}$. Draw an analogy between this expression and the one obtained in Chapter 2 for the output of a resistive voltage divider. Draw the circuits side by side and compare the equations for output voltage.

3.6 Assume a high-pass filter will be used to couple an ac circuit to an oscilloscope. (a) If it is desired to display signals of frequencies greater than 1 kHz, what half-power frequency would be reasonable for the filter? (b) For the filter if $R = 5\ \mathrm{k\Omega}$, and the oscilloscope input impedance is 100 kΩ, how much will the oscilloscope load the filter?

3.7 Is $V_o \ll V_i$ under the conditions worked out for an RC differentiating circuit in Example 3.2? (a) Calculate the ratio $(V_p)_o/(V_p)_i$ at 1 kHz. (b) Repeat at 100 Hz.

3.8 Construct a differentiating circuit that has an output of 2 V when the input voltage is changing at a rate of 5 V min^{-1}.

3.9 Assume values for the Wheatstone bridge of Fig. 3.6 as follows: $R_1 = R_2 = 2.00\ \mathrm{k\Omega}$; $R_3 = 1500\Omega$; $V = 10.00$ V. (a) Calculate the value of R_4 required for balance. (Do you need to choose a value for R_5?) (b) Assume $R_5 = 800\Omega$ and $R_4 = 1\ \mathrm{k\Omega}$. Calculate the unbalance current in R_5 using the Thevenin equivalent circuit. (See Appendix B.)

3.10 How can contact resistances be placed in series with the power circuit of a Wheatstone bridge? Why is the resistance error in a measurement minimized by this procedure?

3.11 Draw a schematic diagram of a recording type of Wheatstone bridge that can measure a resistance which is changing with time.

3.12 Show that power is transferred most effectively from one circuit (A) to another (B) when the output resistance of the first equals the input resistance of the second. (a) Draw circuit A as a Thevenin model (see Fig. 3.22) with R_A in series with dc voltage V. For circuit B show only the input resistance R_B. (b) Show that the power P_B expended in R_B, is given by

$$P_B = I^2 R_B = [V/(R_A + R_B)]^2 R_B$$

and that the total power P_T expended in A and B is just $P_T = I^2(R_A + R_B)$. (c) Assuming $V = 30$ V and $R_A = 50\Omega$, calculate the power expended in circuit B and the ratio of power expended in B to the total power when R_B has the following values: 10, 50, and 150 Ω. Show that the data support the opening statement.

ANSWERS

3.1 Choose $f_o = 600$ Hz. Power will be down 3 dB over midband. At 60 Hz, power will be down 23 dB. (b) For the filter, if $R = 10\ \mathrm{k\Omega}$, $C = 0.027\ \mu\mathrm{F}$. A large resistor will keep power losses low. (c) A more satisfactory attenuation figure would be 60 dB. *3.3* 1590 Hz. *3.12* (c) P_B has the values 2.5, 4.5, and 3.38 W, respectively, and P_B/P_T has the values 0.17, 0.5, and 0.75, respectively.

Chapter 4

OPERATIONAL AMPLIFIER CIRCUITS

4.1 INTRODUCTION

Negative feedback plays an important role in electronic circuits, as discussed in the last section. Operational amplifiers, which will be discussed in this chapter, are superior amplifiers that can provide the high gain required if there is to be strong feedback. Under these conditions, by appropriate design of a feedback circuit surprising results can be achieved. For example, with an operational amplifier (op-amp) it is possible to develop circuits that provide

1. nearly error-free amplification,
2. effective processing of signals, for example, precision differentiation and filtering, and
3. precise, stable values of voltages and currents.

While the circuits for such operations have long been known (e.g., some were described in Sections 3.3, 3.4, and 3.7), the introduction of operational amplifiers has transformed them. Their performance has been greatly stabilized, extended in frequency and power, and improved in precision. Indeed, op-amp circuits revolutionized chemical instrumentation in an earlier period much as the microcomputer has recently. Yet op-amps are analog devices while virtually all computers are digital in nature. Most signals containing chemical information, such as the potential of an ion-selective electrode, are in analog form. Today op-amp circuits continue to be used in analog modules such as sources and detectors as well as in the processing areas in which they offer special advantages.

4.2 THE OPERATIONAL AMPLIFIER

Fortunately, most high-performance amplifiers, regardless of design, qualify as op-amps. Properties of special importance are very high voltage gain (10^5 or higher), large input impedance, great stability, direct coupling internally (connections without capacitors), and a differential input stage. These properties make op-amps prime candidates for use in circuits with heavy feedback of the sort described at the end of the last chapter.

Since op-amp designs vary but properties are similar, it is convenient to represent them by a common symbol, which is universally the triangle shown in Fig.

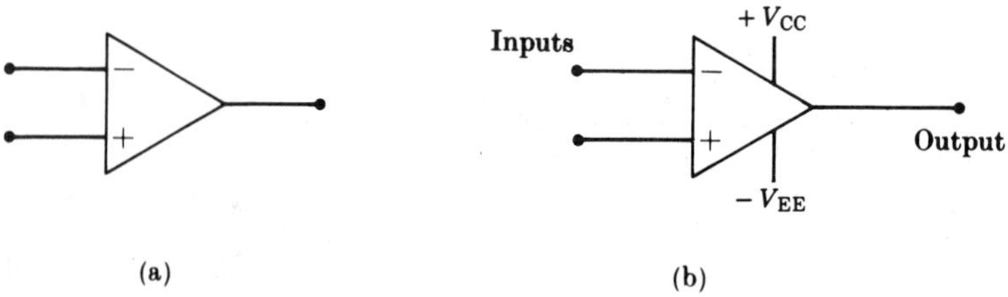

Fig. 4.1 Operational amplifier symbols and inputs. (a) The amplifier is symbolized by a triangle and its differential inputs identified as inverting (−) and noninverting (+). An incoming signal may be directed to either input or be connected differentially across both inputs. A third option is to apply two different signals to the inputs. The output (at the apex of the triangle) is single-ended. Inputs and outputs are taken relative to ground (ground leads are not shown). (b) Operational amplifier symbol with power supply connections $+V_{CC}$ and $-V_{EE}$ added.

4.1. Inputs are made to a side usually marked + and − and the output is taken from the opposite apex. In fact, the input signs serve only to identify the relative *phase of the output*. When a signal is directed to the − terminal, the output is an *inverted version* of the input signal (its phase is shifted by 180°). An input to the + terminal yields an output that is in phase (0° phase shift). It is customary to refer to the inputs as *inverting* and *noninverting* terminals, respectively.

Most op-amps come in integrated circuit (IC) form today. Some idea of what is involved may be gained by comparing the IC pin diagram in Fig. 4.2 of the model 741C, the long-time industry standard, with the diagram in Fig. 4.1*b*.

One of the main goals of this section is to develop a working approach to op-amp circuits with negative feedback. As discussed in the last section in connection with Eq. (3.17), negative feedback networks will themselves determine the be-

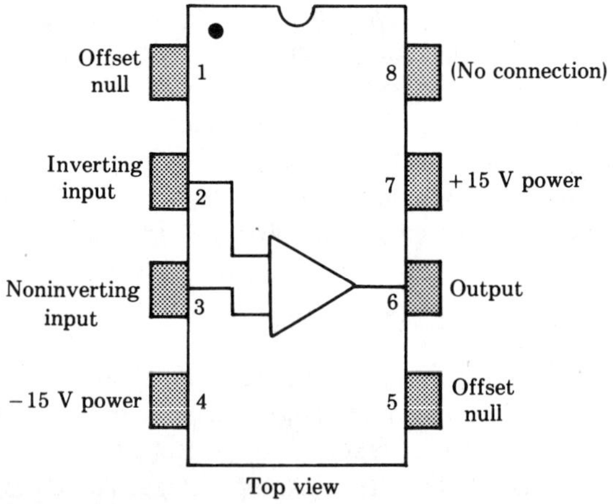

Fig. 4.2 Pin diagram of an integrated circuit (IC) Model 741C operational amplifier. The pins are numbered counterclockwise from the dot in the corner (or notch at the end) when the IC is viewed from the top. This industry standard op-amp contains 21 transistors and 11 resistors and is supplied as a mini-DIP (dual-in-line package) unit.

havior of an amplifier circuit if the feedback is sufficient. The amplifier simply supplies power! Note that the availability of an inverting input for an op-amp can make negative feedback straightforward. It is only necessary to insert a network with desired processing abilities between its output and *inverting input*. Such factors make a straightforward working approach possible.

What should be said about the op-amp itself? Why should there be a differential input? This aspect may be clarified by examining Fig. 4.3. It shows the first stage, which in this case is based on a matched pair of bipolar transistors, in some detail. The configuration is a *difference amplifier*, a circuit that faithfully amplifies the difference between the input signals to the bases of transistors Q_1 and Q_2. Drift in performance is greatly reduced because changes in conditions, for example, temperature, will affect both transistors equally. A second, crucial aspect is that the emitter leads are coupled and connected to a circuit that is not shown (between them and the power supply V_{EE}) that enforces a constant total current through the emitters and thus through Q_1 and Q_2. For instance, if the current in Q_1 falls by 1 mA, that in Q_2 must rise by 1 mA. Note also from the figure that at some point beyond the first stage of this amplifier, the differential character is dropped to provide the op-amp with a single-ended output. With a differential input, one input terminal leads to the inversion of the amplified signal while the other yields an output in phase with the input. This aspect will be dealt with below.

The formal expression for the output voltage V_o of an op-amp is just

$$V_o = A(V_- - V_+) = AV_s, \tag{4.1}$$

where V_- and V_+ are the voltages at inverting and noninverting inputs, respectively, V_s is the voltage difference or input signal, and A is the op-amp gain. In Fig. 4.4 the output of an op-amp with $A = 10^5$ is graphed as a function of input. Its gain is so large that only for differential inputs V_s within the range −0.1 to

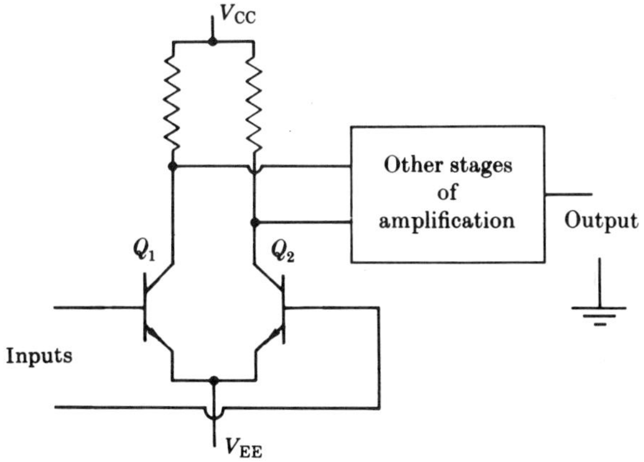

Fig. 4.3 Simplified diagram of a representative operational amplifier. Inputs are to the bases of transistor pair Q_1 and Q_2, which comprise a differential input stage.

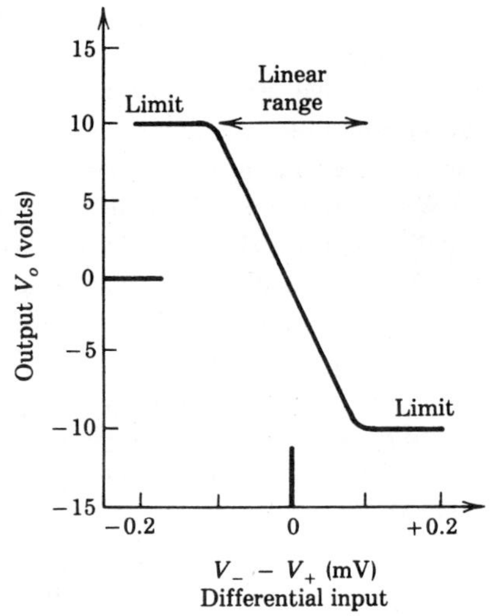

Fig. 4.4 Output voltage V_o as a function of differential input voltage, $V_- - V_+$, for an operational amplifier with a gain of 1×10^5. The amplifier operates in its linear range only for $V_i = -0.1$ to $+0.1$ mV. For smaller or larger differential input voltages, its output limits at about +10 or −10 V. While it might seem that V_o could limit at the supply voltages of ±15 V, some voltage drop always occurs internally during normal operation.

+0.1 mV will the op-amp give an output that is not a fixed value. In other words, the input range is less than a millivolt. Any differential input that is greater is said to cause the amplifier to *limit* and produce a value of minimal significance. Usually the limiting voltage of an op-amp is about 2–3 V smaller than the dc voltages of ±15 V or ±12 V that drive most operational amplifiers.

To understand and predict most of the behavior of op-amp circuits we need be concerned only with general aspects of op-amp behavior and Ohm's law. The important behavior can be summed up in two rules of thumb:

1. The input current I_A of an operational amplifier is minute, of the order of 5×10^{-8} to 1×10^{-14} A. Accordingly, take the input current of an op-amp as zero.
2. The potential at the inputs of a functioning operational amplifier can differ by no more than about 0.1 mV. Accordingly, for a functioning op-amp that has not limited take the voltage difference as zero, that is, $V_- - V_+ = 0$.

While rule 1 is a statement of a property, rule 2 expresses a dynamic relationship that is scarcely self-evident. How can an op-amp circuit enforce such a rule?

Rule 2 is enforced in part by the high gain, but mostly by the effect of negative feedback. Without such feedback, if the acceptable input range were exceeded, the amplifier output V_o would simply rise or fall to one of the limiting voltages. With feedback, the amplifier drives its output in the direction that brings the inputs to the same voltage. How closely it matches them will depend on amplifier gain A. Examples will help illustrate both aspects of behavior.

Example 4.1. Calculate the maximum potential difference that can exist between input terminals of an op-amp if it is to *operate* within its linear range. Assume its behavior is described by Fig. 4.4, that is, its limiting output voltages are −10 and +10 V and that its voltage gain is 10^5.

Equation (4.1) can be solved for V_s by substituting either limiting voltage for V_o. Rearranging the equation, one obtains $V_s = V_o/A = 10/(1 \times 10^5) = 0.1$ mV.

Example 4.2. How does feedback force the voltages at the input terminals of an op-amp to essentially the same value? Show by using the circuit of Fig. 4.5. Assume that initially $V_o = 10$ V and $V_i = 0$ V and that V_i is suddenly raised to 5 V.

Ohm's law can be used to estimate voltages at the op-amp inputs at different times. At the moment when V_i is changed to 5 V, it can be assumed that V_- is uncertain and that the network $R_1 + R_2$ will act as a voltage divider. With 5 V at the left and 10 V at the right, the voltage appearing at point A, V_-, will be 5 V + $(V_o - V_i)R_1/(R_1 + R_2)$ or 6.7 V. The consequence will be to drive V_o toward its negative limit. Assume as in Fig. 4.4 limiting voltages for the op-amp of +10 and −10 V. When V_o reaches 0 V, the potential delivered to point A will be 3.3 V. Finally, at $V_o = -10$ V, $V_i = 0$ V. In this case, the amplifier limited just as it enforced the second rule.

Inverting Amplifier. Connection of a circuit to the inverting input as in Fig. 4.5 yields an *inverting amplifier*. As in this representative circuit, the noninverting terminal is often grounded. An analysis of the behavior of this circuit by applying the two rules of thumb shows immediately that the difference in voltage at the input terminals of the op-amp must be essentially zero. Point A can be called a *virtual ground*. Often this behavior is important to instrumentation circuits.

The input impedance is another important property of the circuit that must also be taken into account. In Fig. 4.5, if point A is effectively at zero volts, an input signal "sees" only R_1 between it and ground; thus the *circuit* input impedance is that of R_1 or 10 kΩ. Note that the very high input impedance of the op-amp itself is unchanged; it still admits essentially zero current.

While an op-amp has a gain of 10^5 or more, what is the gain of the circuit in Fig. 4.5? It can be calculated in straightforward fashion. Since the second rule of thumb tells us that no current enters the amplifier, the current in R_1 must equal that in R_2. In this circuit, $I = V_i/R_1 = -V_o/R_2$. The voltage gain A_v can be expressed as

$$A_v = V_o/V_i = -R_2/R_1. \tag{4.2}$$

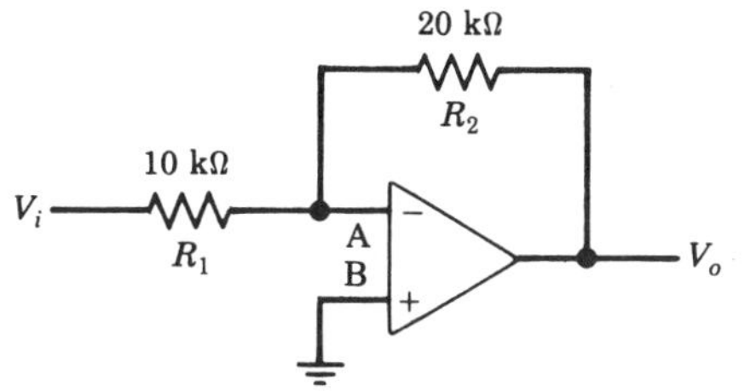

Fig. 4.5 An inverting amplifier. Input signal V_i is applied to the inverting input at A. Negative feedback through R_2 will drive point A as close to the potential at B (which is zero) as possible.

By rearranging Eq. (4.2) the output voltage is found to be

$$V_o = -V_i\,(R_2/R_1). \tag{4.3}$$

According to Eq. (4.2) voltage gain is in fact *independent of amplifier behavior*! With sufficient negative feedback and a powerful amplifier, this is always the outcome.

Noninverting Amplifier. The alternate circuit possibility is to direct an input signal to the noninverting terminal, as shown in Fig. 4.6. Its great attraction is that the input signal experiences a very high impedance; it must enter the amplifier itself. Note that negative feedback is still arranged through a feedback resistor R_2 between the output and the inverting input. High input impedance amplifiers like this one are essential in the measurement of signals from high impedance sources.

What is the gain of such a circuit? Consider first that inputs must be at the same potential, thus $V_i = V_B = V_A$. On the other hand, V_A is determined by the voltage divider that extends between V_o and ground. Thus,

$$V_A = V_B = V_i = V_o\,R_1/(R_1 + R_2). \tag{4.4}$$

Now the absolute value (see below) of the gain A_v is just V_o/V_i or

$$A_v = (R_1 + R_2)/R_1 = 1 + R_2/R_1. \tag{4.5}$$

Example 4.3. Consider the amplifier circuits of Figs. 4.5 and 4.6. Let V_i be 1 V. It is instructive to trace currents and dc voltages in each amplifier in turn and calculate voltages and amplification to gain insight into the behavior of the whole circuit.

(a) What currents exist in the different paths of the circuits and what is the value of V_o? First, for the circuit of Fig. 4.5, the rules of thumb tell us that the potential at input A will be essentially zero volts. According to Ohm's law, the input current in R_1 will be $I = (V_i - 0)/R_1 = 1/10\ \text{k}\Omega = 1 \times 10^{-4}$ A. Since the current entering the amplifier at terminal A is essentially zero, input current I must continue *through* feedback resistor R_2. This insight allows us to predict the output voltage V_o without using Eq. (4.1). First, note that V_o

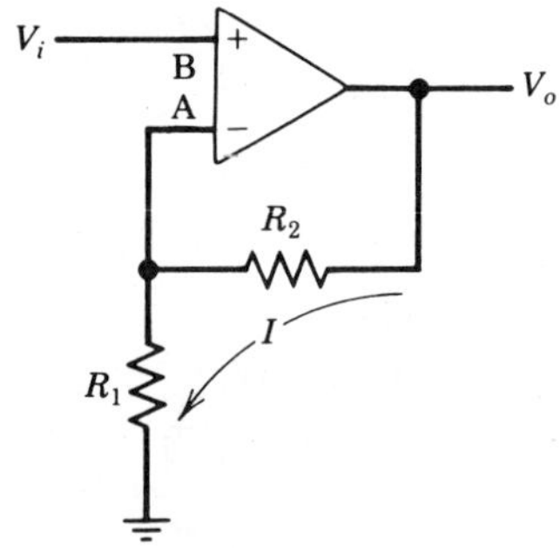

Fig. 4.6 A noninverting amplifier in which input signals V_i are directed to the noninverting input at B. Negative feedback through R_2 will drive point A as close to the potential at B as possible by supplying current I to make $IR_1 = V_i$.

must have a negative potential for current to go from A to the output through R_2, and a minus sign will be inserted. Ohm's law gives $V_o = -IR_2 = -1 \times 10^{-4} \times 20\ \text{k}\Omega = -2$ V.

Second, for the circuit of Fig. 4.6, the rules of thumb tell us that the input current I must be essentially zero since the input is directly coupled to an op-amp terminal that has no other connection. Since "no" current enters the op-amp, this input might be described as a voltage input. To calculate the output current, we use Eq. (4.5). By rearranging it, $V_o = -V_i(1 + R_2/R_1)$. On substitution, $V_o = -1(1 + 20/10) = -3$ V. Now I is just the current needed in the series combination $R_1 + R_2$ to give a 3-V drop or $I = 3/(10\ \text{k}\Omega + 20\ \text{k}\Omega) = 1 \times 10^{-4}$ A.

(b) What is the overall voltage gain? In the inverting amplifier of Fig. 4.5, the gain is $R_2/R_1 = 20\ \text{k}\Omega/10\ \text{k}\Omega = 2$. In the noninverting amplifier, the gain is $1 + R_2/R_1 = 1 + 20\ \text{k}\Omega/10\ \text{k}\Omega = 3$.

(c) How, if at all, is the output current related to the current in the feedback resistor? The output current to whatever load or additional circuit or components is attached at the output of either op-amp circuit is limited only by the current capabilities of the op-amp used. Indeed, as long as it can maintain the minimal difference of input voltages, all other available output current can be directed to a load. In practical cases, about 10–20 mA will be available to a load.

4.3 SOME BASIC OPERATIONAL AMPLIFIER CIRCUITS

The working approach just developed based on Ohm's law and some rules of thumb allows an examination of a large variety of op-amp circuits. A few basic ones will be taken up here.

Voltage Follower. An especially important application of an op-amp is as a *buffer* or isolation amplifier. For example, the noninverting amplifier in Fig. 4.6, which is commonly called a *voltage follower* since its output follows the phase of the input, plays this role well. It effectively isolates the preceding part of the circuit from the part that follows while still passing the signal between them. In Fig. 4.7 the simplest voltage follower is illustrated. Since the output of its op-amp is connected directly to its inverting input, feedback in this instance is total. As a result, the circuit voltage gain must be unity. Incidentally, by comparison the circuit of Fig. 4.6 can now be recognized as a voltage follower *with gain*.

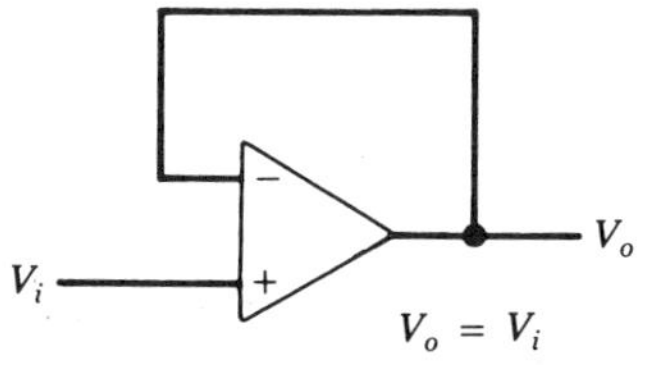

Fig. 4.7 Voltage follower circuit in which the output is fed back fully to the inverting terminal and the input is to the noninverting terminal. Since there is no other connection to the noninverting terminal, the circuit offers very high input impedance. This configuration is valuable with high-impedance sources and features a *voltage gain* of unity. The amplifier serves as a buffer.

Example 4.4. Show that the voltage gain is unity for the voltage follower in Fig. 4.7. Does this amplifier have power gain?

The approximate voltage gain can be found from the rules of thumb. Since $V_o = V_-$, V_+ ideally also equals V_o and the voltage gain V_o/V_i is unity. In fact, because of the small difference in input voltages (about 0.1 mV), the voltage gain will be a bit smaller than unity.

As for power gain, recall that an op-amp can supply currents of many milliamperes while maintaining the feedback. If the current drawn by a load is 10 mA, while the input current is 10 nA, the circuit power gain will equal $I_oV_o/I_iV_i = I_o/I_i = 10^6$.

Example 4.5. Show that the follower serves as a *buffer*.

From the first rule of thumb, a zero input current implies an infinite input impedance for any op-amp. The follower circuit input impedance will be very large, 10^9 Ω if a regular op-amp is used, as much as 10^{14} Ω if a MOSFET op-amp is used. With so large an input impedance, a follower will almost look like an open circuit to whatever preceeds it.

Inverter. A basic inverting amplifier circuit such as that in Fig. 4.5 in which the input and feedback resistors are equal will give an output in which $V_o = -V_i$. Since the magnitude of output is identical to the input but of opposite sign, the circuit serves to invert a signal, a transformation that is often of use.

The inverter also lends itself to change of scale by a constant factor. In Fig. 4.5 let the ratio of resistors R_2/R_1 be K, where K is a desired scale factor or multiplier. Then the output will be $V_o = -KV_i$.

Summing Amplifier. A more versatile use of an op-amp is shown in Fig. 4.8. This valuable circuit is an adder or summing amplifier. Since the inverting input is essentially at zero potential, currents in the two input leads sum without interfering with each other. Thus, $I_{sum} = V_1/R + V_2/R$. If the feedback resistor also

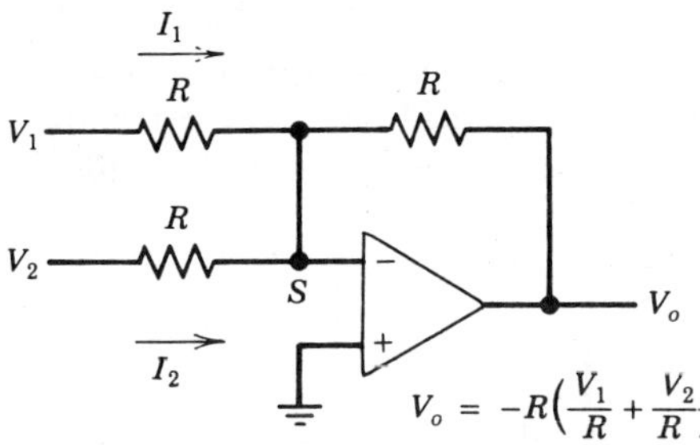

$$V_o = -R\left(\frac{V_1}{R} + \frac{V_2}{R}\right) = -R(I_1 + I_2)$$
$$= -(V_1 + V_2)$$

Fig. 4.8 Analog adder based on an operational amplifier circuit. All voltages, or currents, to be summed are taken in parallel leads to the inverting input. Often the point where such leads join is termed the summing point. For simple addition all resistances should be equal. If desired, a weighted sum may also be taken. In this case, one inserts appropriate values of input resistors R_i. Then one has an output equal to $V_o = -R_f[V_1/R_1 + V_2/R_2 + \cdots]$, where each signal V_i is weighted by $1/R_i$.

has the value R, the output of this circuit is simply $-I_{sum}R = V_o = -(V_1 + V_2)$. As suggested in the legend for the figure, weighted sums may also be taken. Subtraction can be arranged by a simple sign change, perhaps by insertion of an inverter in the appropriate input lead. Actually, this summing circuit is the basis for many digital-to-analog converters (see Example 4.7).

Example 4.6. Consider a practical application of the summing amplifier pictured in Fig. 4.8. Let $V_1 = 1$ V and $V_2 = -3$ V. Adding these voltages will of course result in a subtraction and should give the result -2 V. Since voltages are to be added, resistances in the input leads and R_f should have the same value, say 10 kΩ. Then output of the circuit V_o will be

$$V_o = -R(V_1/R + V_2/R) = -(V_1 + V_2) = -(1 - 3) = +2 \text{ V}.$$

Since the summing amplifier also inverts the output, if an accurate sign as well as the absolute magnitude is sought, this circuit should be followed by an inverter.

Example 4.7. Consider how we might design a simple *digital-to-analog converter* (DAC) based on a summing amplifier. Doing so will also give us a chance to look ahead to the next chapter on digital electronics. What is involved is constructing an op-amp summer that will convert a set of input voltages that represent a binary number (a digital signal) to an equivalent single voltage (an analog signal).

(a) Develop a design for a four-digit DAC based on the op-amp summer shown in Fig. 4.8. Assume the logic levels in the digital system furnishing binary numbers are as follows: 5 V for a binary 1 and 0 V for a binary 0.

(b) If the binary number 1010 is presented to the converter, what output will be produced? Does the output voltage calculated using the circuit agree with the decimal value of the voltage?

Some questions will guide the development of a design. How should the circuit of Fig. 4.8 be modified to accommodate four-bit numbers? It would seem there should be one input line for each bit; further, each line must be dedicated to a particular bit, for example, let the top line handle the least significant bit, the bottom line take the most significant bit, and so forth. By contrast no change need be made in the output since the DAC needs only a single analog output.

What values of input resistors will be needed? Recall that a four-bit binary number, reading from *right* to left, indicates *decimal* digits one, two, four, and eight. (See Appendix B for discussion.) The circuit can now be built.

(a) First, add two new input lines to the circuit, making a total of four. To estimate the value of resistance needed in each line, remember that the top input line represents the least significant bit. Then the four input resistors reading from top to bottom should have values which stand in reverse ratio to the magnitude of the digits: they indicate 1:2:4:8. Why? Assume a feedback resistor of 2 kΩ and an expectation of 1 V of output per decimal unit. Then a suitable resistor set will be 10, 5, 2.5, and 1.25 kΩ. Sketch the circuit.

(b) For the binary number 1010, 0 V will be received by the top (least significant digit) line and be across the 10 kΩ resistor, 5 V across 5 kΩ, 0 V across the 2.5 kΩ resistor and 5 V across the 1.25 kΩ resistor. From the equations given earlier, the output voltage will be $V_o = -2\,(0/10 + 5/5 + 0/2.5 + 5/1.25) = -10$ V. Since the decimal equivalent of the binary number input is 10, the output is as expected. Confirm that with a 2-kΩ feedback resistor and the logic levels selected that each decimal digit corresponds to 1 V.

Example 4.8. It should be evident that it is possible to develop an *analog computer* based on the use of op-amp circuits of appropriate types. Indeed, these offer fast and dependable ways to solve a variety of equations including complex differential equations. Nevertheless today's fast digital computers are so versatile and widely accessible that they are usually the computers of choice for such operations. It may be interesting to keep in mind this earlier use of op-amps as op-amp integrators and differentiators are discussed in Sections 4.7 and 4.8.

4.4 PRECISION VOLTAGE AND CURRENT SOURCES

Power supplies that furnish a precisely specified constant current or voltage can be prepared by using a reference voltage source with an op-amp circuit. It is only necessary that the design ensure that minimum current is drawn from the source.

Two constant voltage supplies based on this approach are illustrated in Fig. 4.9. The first is especially simple. Since the noninverting terminal has been

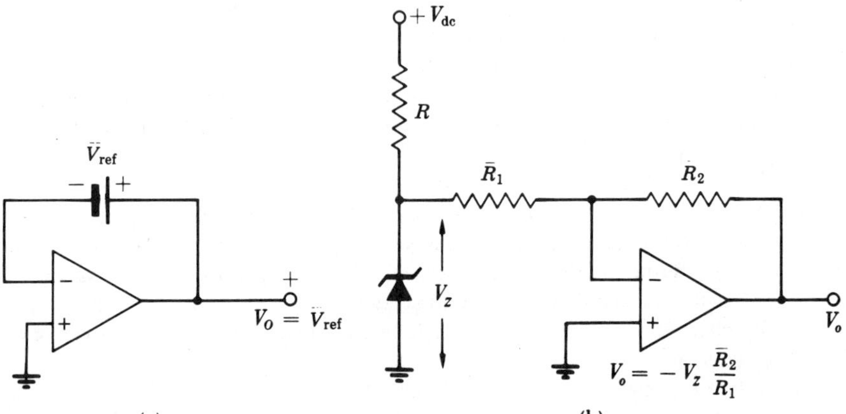

Fig. 4.9 Precision, constant voltage sources based on an operational amplifier and a reference voltage source. In both circuits the noninverting input of the op-amp is grounded and its output is taken relative to ground. Considerable current can be drawn from the output of either source without affecting the known voltage. (a) Standard cell or other low-current voltage reference in feedback circuit to the inverting input. (b) Voltage multiplying circuit supplying precise, constant voltage with a zener diode reference. The size of R_1 is chosen relative to that of R so that the current in the zener diode is controlled by the operational amplifier and therefore is constant. Clearly, precision resistors must be used and V_o cannot exceed the amplifier maximum.

grounded, V_- must also be essentially zero. Further, because $V_- = V_o - V_{ref}$, $V_o = V_{ref}$. Thus, the output voltage is fixed. The op-amp can provide up to nearly its maximum output current (ordinarily about 5–20 mA) while maintaining this constant value. Note that current cannot exist in the feedback loop, and thus be drawn from the reference voltage source unless it can enter or leave the op-amp (rule 1).

The schematic of a more versatile operational amplifier voltage source is pictured in Fig. 4.9b. When $R_1 < R_2$, the current through the zener diode is provided essentially by the amplifier. Thus the zener operates at constant current and furnishes a voltage to the op-amp constant to 0.01%. If more output current is needed (the zener diode uses part of the op-amp output), a booster amplifier can be used inside the feedback loop as in Fig. 4.10. For additional examples of constant voltage sources and their application in voltammetric measurements see Chapters 27–32.

Simple Potentiostat. The constant voltage source shown in Fig. 4.10 is ordinarily called a potentiostat. It is shown as it might be used with a three-electrode electrochemical cell, where the aim is to maintain the potential difference between working and reference electrodes constant. By use of Ohm's law, the value of $V_{ref} - V_{working}$ can be deduced. When the potentiostat is operational, both V_- and $V_{working}$ are zero, and V_{std} equals the potential difference between the working and reference electrodes.

The most evident difference between this circuit with a cell and one with only electronic components is that feedback resistance R_f is now the solution resistance between a pair of electrodes. Usually the reference electrode is placed very close to the working electrode to minimize this resistance.

Another difference is the presence of a booster amplifier, a standard noninverting amplifier that permits a much larger current than could be furnished by most

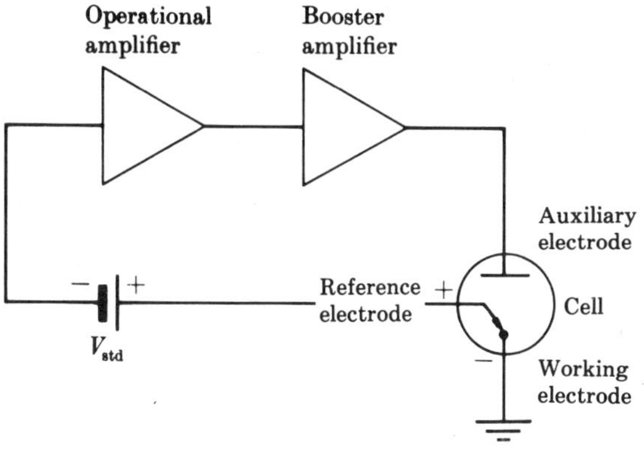

Fig. 4.10 A potentiostat, a constant potential device, for controlled-potential electrolysis. Since each noninverting terminal has been grounded, the sum of voltages in the feedback circuit must also be essentially zero. In other words, it must be true that $V_{std} = V_{ref} - V_{working}$. Usually V_{std} is supplied by a potentiometer circuit and can be set to a desired value. The booster amplifier ensures that much larger currents are available than the operational amplifier alone can supply.

op-amps, has been added to increase the current capacity (current compliance) of the cell. During an electrolysis, current through the cell may vary through several orders of magnitude. Regardless of the presence of the booster amplifier in the feedback loop, the potential drop between reference and working electrodes is maintained by the op-amp circuit.

Some Constant Current Sources. A generalized version of a *constant current source* employing an op-amp is shown in Fig. 4.11. In this case a constant current is developed *in the feedback loop*. To achieve this condition, a known current is developed at the inverting input and continued through the feedback loop. Some limitations of the design are that neither cell current I_{cell} nor voltage V_{cell} can exceed the rated output values of the amplifier and neither end of the load is grounded.

In Fig. 4.12 a more versatile constant current circuit is sketched. Though no current exists in the reference voltage source, currents as large as those a booster amplifier can supply will be held constant. Neither load terminal can be grounded.

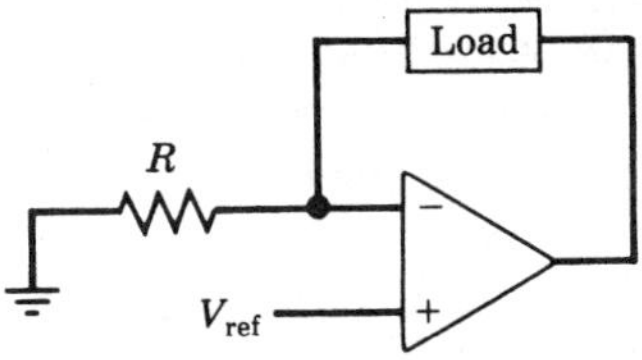

Fig. 4.11 A precision, constant current source based on a noninverting operational amplifier. In this case V_{ref} is furnished by a voltage divider associated with a reference voltage source. Feedback through the load develops voltage equal to V_{ref} at the inverting input. In response, I_i = constant = $(V_{\text{ref}} - 0)/R = I_{\text{load}}$. Variations in load resistance during measurements are provided for but the current compliance of the circuit cannot exceed the maximum current the amplifier can furnish. This circuit might be used for constant-current coulometry.

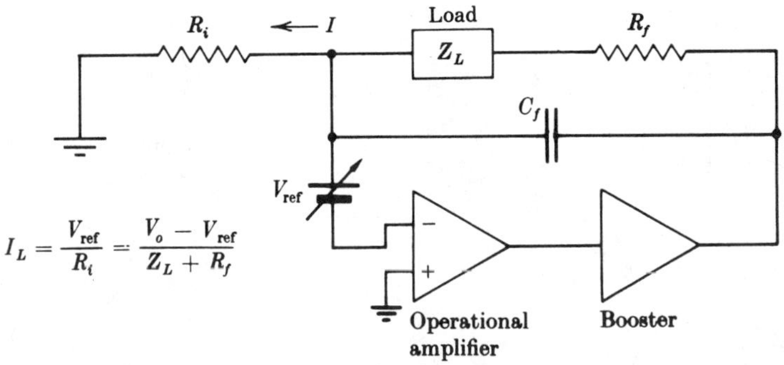

Fig. 4.12 A constant-current source supplying current to load without drawing current from the reference voltage source. The current may be adjusted to any level available from the booster amplifier as long as the voltage output of the operational amplifier is not exceeded; C_f and R_f should be inserted only if needed for stability.

4.5 MAXIMIZING AMPLIFIER PERFORMANCE

Today a wide range of operational amplifiers are available in inexpensive integrated circuit chips. Properties of a few of the most widely used op-amps are listed in Table 4.1. Note that op-amp voltage gains are indeed high: typical values are in excess of 10^5 or 100 dB. While input impedances (more about this later) are not listed, it can be expected that they are high. The two op-amps listed as having FET inputs have especially large input impedances. While limiting voltages are unlisted, they can be predicted to be 1–2 V less than the supply voltages. Limiting circuit currents, however, can only be predicted on the basis of circuit information. It will be useful now to take up aspects of input properties, output properties, and gain that deserve further attention.

Input Properties. For any design of op-amp there is a rated *maximum input voltage* that can be imposed either differentially, that is, between inputs, or simultaneously, on both inputs. Common-mode maximum input voltages, those imposed simultaneously, are usually at about the level of supply voltages. Maximum differential voltages are sometimes much smaller, for example, one is listed as 0.5 V in Table 4.1. Fortunately, most op-amps will tolerate 10–20 V. Actual maximum values must be observed to avoid damage. These values are very much larger than the useful differential voltage inputs, which can at most be about 0.1 mV. Recall that in practice feedback by an op-amp is instrumental in achieving the latter condition. Diode clamps are often used to protect inputs from inadvertently high voltages (see Section 2.11).

Ordinarily the *differential input impedance* Z_d, the impedance between one input and the other when it is grounded, is more important than any other impedance measure. From the circuits already described, such a parameter seems reasonable. This figure is seldom below 2 MΩ and with FET op-amps rises to as much as 10^6 MΩ. Actually, when negative feedback is provided by the usual circuit, it has the effect of boot-strapping or increasing the input impedance greatly (see Exercise 4.7 and also Section 4.6). Since inputs to op-amps are to the bases, or gates, of transistors, a minute *input bias current* I_B also exists in the circuit. With modern op-amps it is seldom greater than 50–80 nA and with FET op-amps it falls to the order of 20–30 pA. Interestingly, the effect of such a bias current is an additional voltage drop across the feedback network resistors or across any input or source impedance. (See Exercise 4.15 for a way to minimize the error.) As a result, the size of the input bias current basically establishes how large a feedback resistor can be; the smaller the current the larger the feedback resistor that can be tolerated and the larger the achievable circuit gain (see Section 4.6). Unfortunately, op-amps that are intended for very fast operations tend to have higher bias currents.

Perhaps unexpectedly, even with precision manufacture of integrated circuits, structural differences in the input differential pair of transistors leads to an output when there should be none. Consider first the *input offset current*. Usually its magnitude is about one-tenth the bias current. As a result of this offset current there will always be some output signal even when the two inputs are tied together.

Table 4.1 Characteristics of Several Operational Amplifiers[a]

Type	*Offset Voltage (mV)*	*Voltage Drift* ($V\,K^{-1}$)	*Bias Current (nA)*	*Slew Rate* ($V\,\mu s^{-1}$)	*CMRR (dB)*	*Gain* $A_v \times 10^5$	*Max. Output Current (mA)*	*Max. Diff. Inputs (V)*	*Comments*
308	2	6	1.5	0.15	100	3	5	0.5	
355	3	5	0.03	5	100	1	20	30	FET input
741C	2	—	80	0.5	90	2	20	30	Industry standard
TL081C	5	10	0.03	13	76	2	10	30	FET input
8017C	2	10	50	130	—	10	15	30	

[a]Representative values of important parameters are given. Slew rate is the rate of response of the output voltage of an op-amp to a sudden change in input. The common mode rejection ratio (CMRR) factor defines the ability of an op-amp to amplify a difference or differential signal at its inputs selectively as opposed to an identical signal applied to both inputs.

When this effect is a source of measurable error, it is customary to introduce the same resistance into the inverting and noninverting leads of an op-amp circuit.

Similarly, an *input offset voltage* exists for op-amps. By definition, the input offset voltage V_{os} is just the difference in input voltages that must be applied to bring the output voltage to zero when both inputs are grounded. The impact of an offset voltage is substantial. For example, a 3-mV offset voltage when the op-amp *circuit* gain is 200 will lead to an output voltage offset of $(200 \times 0.003) = 0.6$ V. On the basis of this result, we can predict that tying the inputs of an op-amp together will cause it to saturate!

Most op-amps include a circuit for trimming the offset voltage to zero (see Exercise 4.13). In practice trimming will need to be checked periodically because offset currents and voltages drift with temperature and time, the effect of temperature being much more important. For precision circuits, where offset voltages lead to measurable error, it is desirable to select op-amps with a lower offset value.

Output Parameters. In addition to the limitations on output voltages and currents already discussed, it is important to consider the useful *frequency range* of op-amps. In Fig. 4.13 the broken line gives the voltage gain for an op-amp as a function of the logarithm of signal frequency, a plot called a *Bode diagram*. Since most signals in instruments are dc or low-frequency ac (0–10 kHz), it appears that sufficient gain will be available. A slope of unity connotes a roll-off of gain of 6 dB/octave or 20 dB/decade. If a step change in voltage or a square-wave with its high frequencies had to be amplified, an op-amp could not be used.

Nevertheless, it proves useful to be able to measure the rate of response of an op-amp to a step signal. This value is termed its *slew rate*. As can be seen from Table 4.1, such rates extend from about 0.1 V μs^{-1} to more than 100 V μs^{-1}. Even faster slew rates are important in operational amplifiers used as comparators (see Section 4.9). In a real sense this is another aspect of the frequency-limited

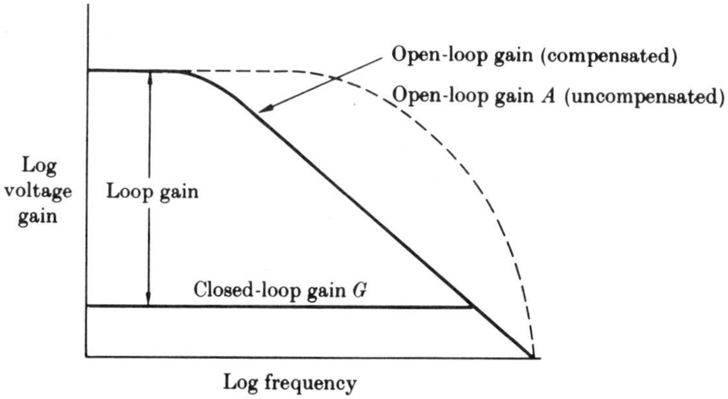

Fig. 4.13 Bode plot of amplifier response. The broken line represents the amplification of a normal operational amplifier. The full line represents gain in the same amplifier after a compensating *RC* circuit has been introduced to attain an ideal roll-off (attenuation) of 6 dB per octave or 20 dB per decade of frequency.

performance of the op-amps. Both the small base currents required to drive an op-amp and the existence of a compensation circuit limiting feedback at high frequencies (see below) tend to limit this rate.

Amplifier Instability. The reason for so restricted a frequency range must be sought in a *frequency-dependent phase shift* that can lead to amplifier instability. The phase of the output signal of an op-amp with respect to the input signal tends to increase somewhat with increasing frequency. Recall that the phase of the output will be zero or 180° at dc, depending on the input used. Actually, the slope of the plot in a Bode diagram gives the increment in this initial phase shift that is present at different frequencies. For example, a slope of -1 indicates an increment of 90°, or a total shift of 270°.

Suffice it here to indicate that the more the shift exceeds 270° and approaches 360°, the greater the possibility for instability. Note in Fig. 4.13 that the slope becomes almost infinite at the highest frequencies. The problem is that feedback at these frequencies changes from negative to positive. Sine waves 360° apart in phase will give constructive interference. In this situation an op-amp will favor production of the small set of high frequencies enjoying positive feedback and in fact becomes an *oscillator*.

It is customary to avoid such instability by introducing an *RC* network internally to provide frequency compensation. In the graph the full line indicates the response of an op-amp with frequency compensation. As would be expected, some gain is sacrificed at higher frequencies.

Summary. For convenience the limitations on op-amp performance are summarized here. They are:

1. Output limits on both voltage and current.
2. Input limits both on common-mode and differential signals.
3. A finite input impedance.
4. *Input bias currents* that vary with temperature, and occasionally time.
5. *Input offset voltages and currents.*
6. A voltage gain that is frequency-dependent.
7. Phase shifts that increase at "higher" frequency.
8. A finite slew rate (response time to a step signal).
9. Some internally generated noise.

The last two limitations will be further considered in Chapters 5 and 12, respectively.

4.6 PRECISION AMPLIFICATION OF CURRENT AND VOLTAGE SIGNALS

Operational amplifiers bring special assets to the process of precision amplification in instruments. They are differential or difference amplifiers, buffer later circuits from the modules where the signals originated, introduce little noise or distortion, and have very high gain. In most instruments we can expect that immediately following the detector there will be a precision amplifier to provide needed amplification of the analytical signal output. But what kind of amplifier circuits will be best adapted to detector outputs that are "mainly" currents and what kind to signals that are "mainly" voltages?

Current Amplification. For amplification of current signals an inverting amplifier circuit like that in Fig. 4.14 is effective. Such an amplifier may be identified in many ways. In a true sense it is a current-to-voltage converter, though when it handles low currents it may seem to be more of a charge amplifier or charge-to-voltage converter. It functions nearly ideally for current amplification since the input signal sees only the essential ground voltage maintained at the inverting input in this circuit. In other words, the signal is not opposed by a potential at the input. A further advantage is that the current does not enter the op-amp but must continue

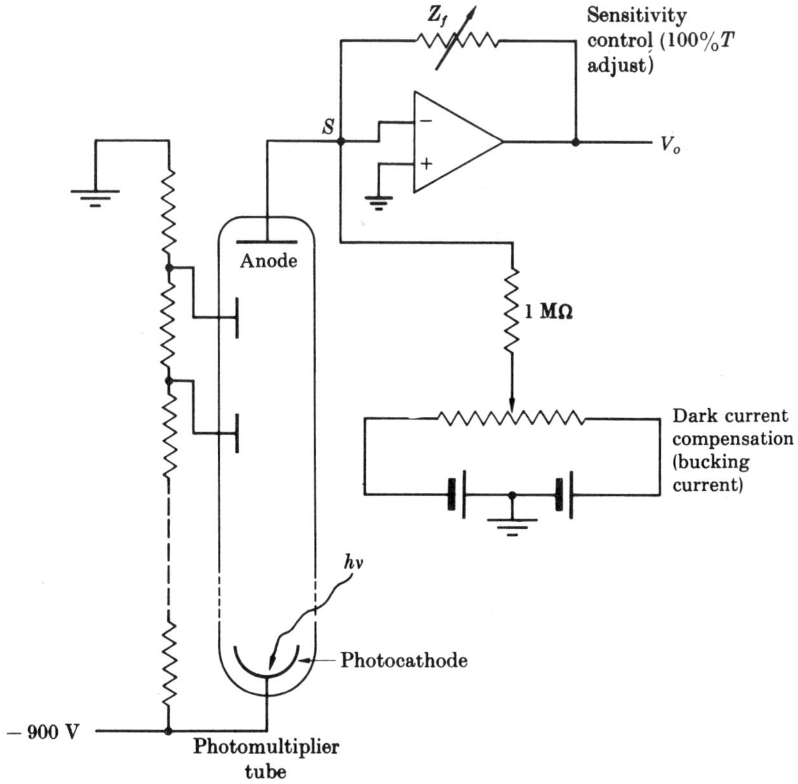

Fig. 4.14 An operational amplifier current-to-voltage converter applied to amplification of the output signal of a photomultiplier tube. Substituting a set of precision resistors for feedback impedance Z_f, allows one to select a multiplication factor appropriate to different levels of incident illumination. It is important that the op-amp offset voltage be trimmed to zero. The circuit also provides compensation of photomultiplier dark current. When the photomultiplier shutter is closed, the position of the wiper on the variable dc voltage divider is adjusted until the amplifier output V_o is zero. When a zero reading is obtained, a current equal to the dark current but of opposite sign is being added to it at the inverting input. This background current is subsequently subtracted from all amplified current signals.

into Z_f (rule 1). Thus, the output is $V_o = -I_i Z_f$. Note that amplifier gain may be varied by adjusting the value of feedback impedance Z_f. In this simple way a wide range of levels of light falling on the photocathode of the detector may be accommodated. Here the op-amp is clearly a buffer, as mentioned above. Any other detector–transducer with current output may of course be substituted for the photomultiplier tube shown in the figure.

Amplification Limits. How large a current gain may be provided in the circuit of Fig. 4.14 by varying feedback impedance Z_f? Assuming that noise is not limiting, the largest useful value may be calculated from amplifier characteristics taking particular account of the op-amp differential input impedance.

In first approximation, current I_a actually entering the amplifier at the inverting input is defined by the differential input potential across the two input terminals, $V_- - V_+$, and the *differential input impedance*. This impedance is the impedance *between* the input terminals with one terminal grounded. It should not be confused with the common-mode input impedance, which is always very much higher, and is also often quoted. Thus, we have

$$I_a = (V_- - V_+)/Z_d = V_-/V_d.$$

Since $V_o = I_i Z_f$, these two equations can be combined with an equation for the voltage gain $A_v = V_o/V_i$, to define Z_f in terms of other amplifier parameters: $Z_f = (I_a/I_i)A_v Z_d$.

A maximum value for Z_f can now be deduced as soon as the accuracy desired in current multiplication (gain) has been stated. Accuracy is mainly related to the ratio I_a/I_i since this factor expresses the fraction of input current lost to the op-amp. For 1% accuracy, $I_a/I_i \leq 0.01$; for 0.1% accuracy the ratio should be no more than 0.001.

Voltage Amplification. In circuits requiring precision amplification of voltage, a *differential amplifier* is always attractive to lessen dependence on semiconductor parameters. The simplest arrangement shown in Fig. 4.15 should be compared with that given earlier in Fig. 4.7. Not only is the inverting input connected in the usual fashion to an input voltage V_i, but a similar network is used for the noninverting terminal. Though the second resistor network ends at ground, the amplifier output V_o is defined by the surprisingly simple equation given with the figure. Though the voltage gain achieved is small, there is considerable power gain and

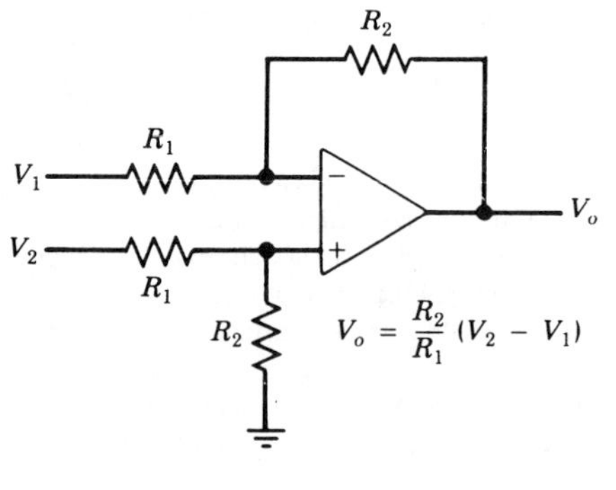

Fig. 4.15 A difference or differential amplifier. If high precision is required, 0.01% resistors must be used.

other stages of amplification may easily be added to achieve voltage gain as needed. Circuit specifications for precision amplification are discussed in Example 4.10. Another circuit with higher input impedance that is much better adapted to amplify signals from high impedance sources is illustrated in Fig. 4.16. For instance, the circuit has been used in several pH meters. Again, there are major differences between this arrangement and the usual op-amp circuit, as discussed in the legend. Either the output current, which is proportional to the cell voltage, or the voltage drop across R_f can be displayed as the output of the pH meter.

Example 4.9. Verify the equation for the gain of the differential amplifier in Fig. 4.15 by applying Ohm's law to all branch circuits.

First, calculate V_+ at the noninverting terminal; this voltage will be unaffected by operation of the amplifier. To do so, note that the first rule of thumb indicates all current caused by V_2 will continue to ground. Thus, V_+ can be calculated from the lower voltage divider $R_1 + R_2$. We find $V_+ = V_2R_2/(R_1 + R_2)$.

Second, let the current in the top half of the circuit be I_1. Applying rule 1, $I_1 = (V_1 - V_-)/R_1 = (V_- - V_o)/R_2$. But from rule 2, $V_- = V_+$ since the op-amp will drive the potential at V_- to the value at V_+, if it is within its capacity to do so. Combining the equations for the upper and lower voltage dividers and eliminating V_-, one obtains for the output voltage of the amplifier the equation $V_o = R_2(V_2 - V_1)/R_1$, as given in the figure.

Example 4.10. What circuit specifications will ensure that the differential amplifier in Fig. 4.15 will be a precision voltage amplifier?

To obtain the output voltage of the amplifier in the first example, a hidden assumption was made that the pair of resistors labeled R_1 were precisely equal, as were the pair labeled R_2. Clearly, precision amplification will depend on how well matched the pairs are. For $\pm 0.01\%$ accuracy, for example, resistors matched to 0.01% will be required. To minimize

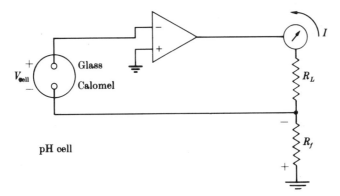

Fig. 4.16 Voltage-measuring circuit using series current feedback to the inverting input. A very high impedance is developed at the input of the operational amplifier by this design. In this circuit only the error signal ($V_{cell} - V_{Rf}$) reaches the amplifier input. Output meter I (calibrated in pH or mV) registers the current I needed to develop a voltage in feedback resistor R_f that is equal and opposite to V_{cell}. Alternatively, a voltmeter V can be employed in lieu of I as the readout device. Various modifications of this circuit are used in commercial pH meters.

power dissipation 100-kΩ resistors should be selected. Although they are expensive, they are well worth the cost. Equally important, the op-amp shown should have an FET input to guarantee that the differential amplifier will have high input impedance.

Rejection of Common-Mode Signals. Whenever a signal is applied between inverting and noninverting inputs or when a difference amplifier is used to compare signals, it is likely that some dc or ac will be applied in common to each input. How will the amplifier deal with this situation? If it amplifies both the difference voltage, which is of interest, and the common dc and ac voltage, which should be rejected, little will be gained. How can its ability to amplify only difference signals be evaluated?

It is desirable to proceed by defining these two types of voltages and their respective gains. The *difference voltage* V_d is of course just $V_+ - V_-$. It is more difficult to deal with the common input. By convention, this portion, V_{cm}, called the *common-mode voltage* is defined as the average of the signals at the inputs, $V_{cm} = (V_+ + V_-)/2$. Now gains for the difference and common-mode aspects of a signal can be compared. The *difference voltage gain* A_d is $A_d = \Delta V_o / \Delta V_d$, a dynamic version of the value stated in Eq. (4.1). Futher, the *common-mode voltage gain*, A_{cm}, will be dynamic gain $A_{cm} = \Delta V_o / \Delta V_{cm}$.

Ideally, the common-mode gain should be zero and the ratio of the difference and common-mode gains infinity. This ratio, A_d / A_{cm}, the *common-mode rejection ratio* (CMRR), is usually calculated for an op-amp as an index of its quality of performance. In a good op-amp the common mode rejection ratio should be at least 10^4 or 80 dB.

Follower Circuits for Voltage Amplification. An op-amp follower with gain (Fig. 4.6) is valuable whenever a signal from a high-impedance device must be amplified and a differential input is not needed. Recall that follower circuits have extremely high input impedances and that the output amplitude of the circuit of Fig. 4.6 is $V_o = -V_i(1 + R_2/R_1)$. It is possible to draw a current as little as 1 pA from a source with such an amplifier. Further, because of the feedback, its output impedance is very low (about 1Ω) and its output current nearly independent of voltage.

Instrumentation Amplifier. A variety of ideas were combined to develop the widely used three op-amp instrumentation amplifier shown in Fig. 4.17. This sophisticated circuit provides amplification even (a) for weak signals, (b) in noisy situations, and (c) in the presence of a large common-mode voltage. Examples are numerous. It is commonly used to amplify bridge–unbalance currents from a Wheatstone bridge thermal conductivity detector in gas–liquid chromatographs, the outputs of most biological "input transducers," and also the outputs of detectors in most instruments just before signal sampling and AD conversion. It combines three op-amps in an ingenious way to make a differential amplifier, as discussed in the figure legend. Its special assets are a high differential gain, high input impedance, and a large common-mode rejection ratio (100–200 dB or 10^5–10^6).

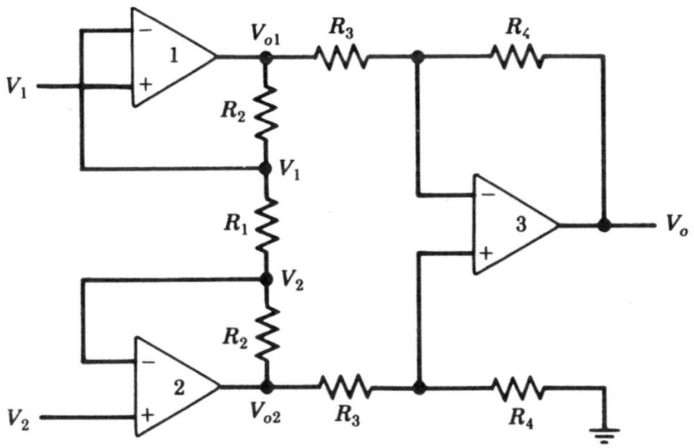

$$V_o = \frac{R_4}{R_3}\left(1 + \frac{2R_2}{R_1}\right)(V_2 - V_1)$$

Fig. 4.17 The standard instrumentation amplifier. Two op-amps are used in follower amplifier configurations at the input. They are cross-coupled to ensure high input impedance, a differential gain given by $1 + 2R_2/R_1$, and a common-mode gain of unity. Note that the input pair of op-amps has a differential output V_{01} and V_{02}. This output drives the straight difference amplifier, based on op-amp 3. Its differential gain is small, R_4/R_3, but its common-mode gain is nearly zero. For the entire circuit, the differential gain is given by the expression $A_d = (1 + 2R_2/R_1)(R_4/R_3)$. The overall common-mode rejection ratio (CMRR) is 10^5–10^6. A major advantage of this design is avoidance of the need for close matching of resistors with identical labels. The (differential) gain of the instrumentation amplifier can be adjusted by varying resistor R_1, which is attached externally when an integrated circuit form of the amplifier is used. Offset voltage can be compensated by trimming at the noninverting input of one of the input op-amps, for example op-amp 2.

For the entire circuit the differential voltage gain is $A_d = (1 + 2R_2/R_1)(R_4/R_3)$. Its advantage over the straight differential amplifier shown in Fig. 4.15 is that it does not require careful matching of resistors. Further, the device is available in hybrid form as an integrated circuit that provides external access to resistor R_1, which sets the gain.

4.7 ELECTRONIC INTEGRATORS

To integrate an electrical signal with respect to time, advantage may be taken of the ability of a capacitor to store charge. Recall that capacitors were used in this fashion as simple *RC* integrators (see Section 3.3). Unfortunately, the build-up of charge on the capacitor acted to limit performance and it was found that integration was accurately done only as long as the capacitor voltage V_o met the condition $V_o \ll V_i$. It is possible to add an op-amp to the *RC* integrator using the circuit shown in Fig. 4.18. Now an input signal sees only the virtual ground maintained by the op-amp at the inverting terminal and the limitation is removed. The electronic integrator ensures quality integration over a wide range of conditions.

Showing that integration occurs is straightforward. The current generated in R by the input is $I = (V_i - 0)/R$. Further, the current continues into the capacitor. At the end of an integration the capacitor charge Q will be $Q = \int I\,dt$. Its voltage,

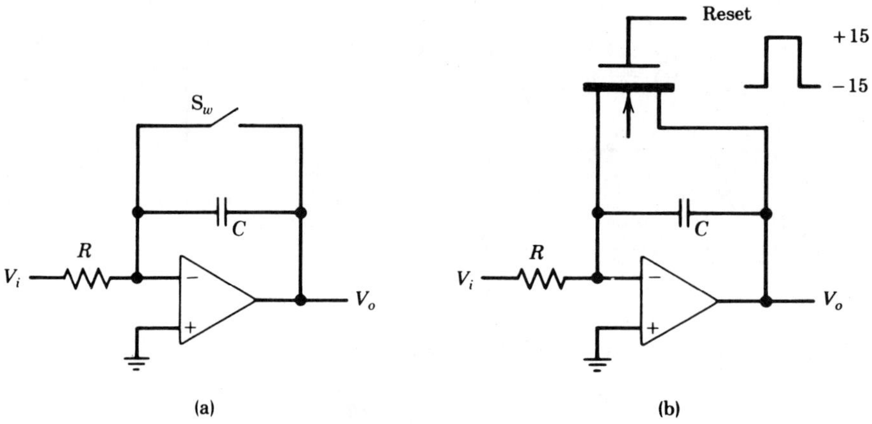

Fig. 4.18 Operational amplifier integrator. The output voltage V_o is proportional to the time integral of input signal V_i. Integration occurs during the interval of time t that the switch is open. (a) A mechanical switch is used. It is closed to reset the integrator to zero. (b) An n-channel MOSFET is substituted as switch. Resetting occurs when a +15-V pulse is applied to the gate. Only positive outputs can be accommodated with the circuit shown. A time diagram for a +15-V pulse that can be applied to the gate of the MOSFET to reset the integrator is shown.

which will also be the op-amp output voltage V_o, will be $V_o = Q/C$. Combining these three equations and inserting a negative sign on the right side to mark the inverting effect of the amplifier gives the expression

$$V_o = -(1/RC)\int V_i dt + \text{constant.} \tag{4.6}$$

Adding a "reset" switch across C allows the output to be reset to zero at any time by shorting the capacitor. Opening the switch will start integration. In Fig. 4.18 both mechanical and MOSFET switches are shown.

Example 4.11. How can an electronic integrator be used with a current input? For instance, the analytical signal of a liquid chromatograph with a UV detector is the current from its photomultiplier tube or photodiode.

In this case the signal should go directly to the inverting input without an intervening resistor. Equation (4.6) becomes

$$V_o = -(1/C)\int I_{\text{signal}}\, dt + \text{constant.}$$

Example 4.12. A voltage that increases linearly with time, called a *ramp*, can be generated by the circuit of Fig. 4.18. What change(s) in the circuit will permit the slope of the ramp output to be varied?

To generate a ramp, connect a voltage source as V_i. Since V_i will now be constant, Eq. (4.6) will become $V_o = V_{\text{source}}\, t/RC$. To change the slope of the ramp, either R or the value of V_{source} may be varied. Actually, if a current source is available, it may be substituted as the input to the circuit and resistor R removed.

Example 4.13. A portion of a chromatogram is given in Fig. 4.19 together with its time integral. How might the integral be obtained manually with the circuit of Fig. 4.18?

Integration would be started by opening the reset switch Sw at $t = 0$ when the onset of a peak is observed. Note that the integrator output first increases rapidly and then levels off gradually as a peak ends. To avoid drift from stray signals between peaks, the reset switch should be closed at the end of each peak and reopened only as another peak starts.

Example 4.14. In a program for automatic integration of peaks in a chromatogram what subroutines will be desirable?

Some important subroutines will allow for each peak detection of its

1. onset, which must be known to start integration;
2. position or time of its maximum, since this datum should be displayed to locate the peak;
3. termination (return to actual or projected baseline) so that integration can be stopped; and
4. approximate shape, in case of overlap by other peaks, to allow allocation of area to each one.

Subroutines relating to baselines will be especially important since they will establish both the onset and termination of integration. Often a menu will allow an operator to select whether a baseline is rising or falling. Clearly if the signal output fails to return to a recognized baseline, the integrator output V_o will continue to rise, introducing error.

Precision Analog Electronic Integration. How may electronic integration be made precise? The answer lies in minimizing at least three sources of drift in the output voltage.

1. *Capacitor leakage* may be minimized by use of a polystyrene or other high-quality capacitor (see Table 2.2) or by integrating for short intervals, resetting, and adding results.

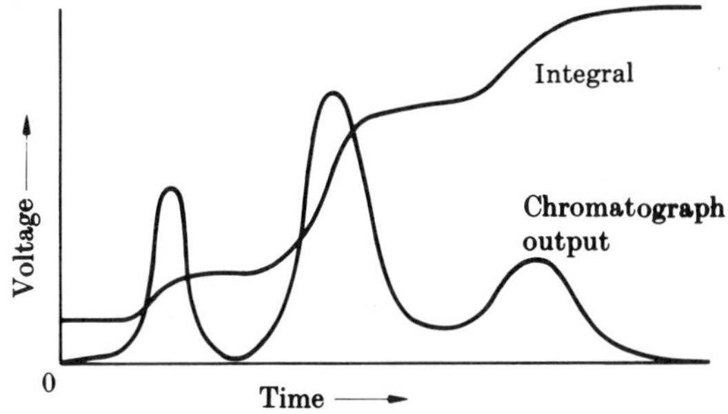

Fig. 4.19 A simple gas chromatogram and its time intergral.

2. *Input current, voltage offset,* and *bias current* of the op-amp may be reduced by employing an FET op-amp, careful initial trimming of offset voltage, and use of large R and C values. Integration for short intervals and adding outputs will also lessen error.
3. *Circuit leakage* to the op-amp can be minimized. The input lead should be well insulated as well as shielded against magnetic fields. Clearly, all signals, whether legitimate or stray, will be integrated.

If after such steps are taken the drift of an integrator remains too large for a given application, the insertion of a large resistor (several megohms) in parallel with C may be necessary to improve feedback. Clearly, it will also introduce error if there is some dc in the signal (by the voltage drop across it), or if too large, it may pick up induced current. Other strategies are available.

4.8 ELECTRONIC DIFFERENTIATORS

The performance of *RC* differentiators (see Section 3.3) can also be greatly enhanced by introduction of an op-amp. How this is done is shown in Fig. 4.20. The hallmarks of the circuit are the series capacitor in the input circuit and a resistor in the feedback circuit. The simple differentiator shown furnishes moderate current (2–20 mA) and has an extended frequency range. Later in the section some noise problems will be taken up.

To verify that the circuit differentiates a signal, consider first its input branch. A changing input signal, V_i, will cause a current I in capacitor C_i, as shown in Section 2.6, that will continue on through R_f. To find I, we differentiate the capacitor equation $C = Q/V$ with respect to time, obtaining $dQ/dt = I_i = C_i dV_i/dt$.

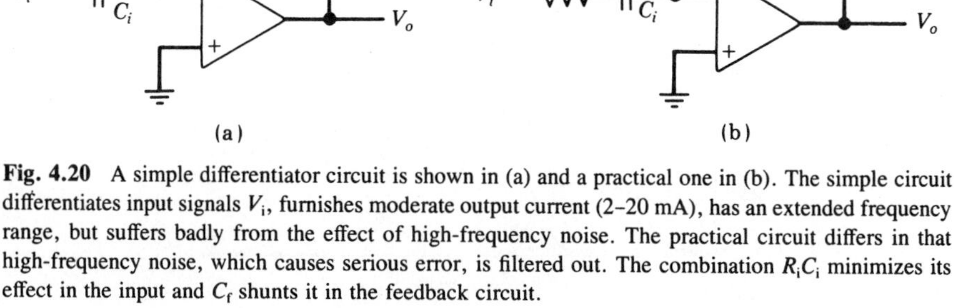

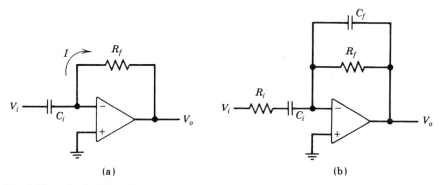

Fig. 4.20 A simple differentiator circuit is shown in (a) and a practical one in (b). The simple circuit differentiates input signals V_i, furnishes moderate output current (2–20 mA), has an extended frequency range, but suffers badly from the effect of high-frequency noise. The practical circuit differs in that high-frequency noise, which causes serious error, is filtered out. The combination R_iC_i minimizes its effect in the input and C_f shunts it in the feedback circuit.

As soon as the circuit of Fig. 4.20 is operating, $V_S = 0$ and all input current I_i will continue through the feedback resistor, that is,

$$V_o = -I_i R_f = -R_f C_i dV_i/dt.$$

Because of the ubiquitous presence of high-frequency noise in circuits, however, this simple differentiator proves inadequate. Though the amplitude of such noise may be minute, its rate of change of voltage with time is so large that it can contribute an appreciable fraction of the output of a simple differentiator.

A common strategy to minimize this error is to modify the circuit to block high frequencies as shown in the *practical differentiator circuit* in Fig. 4.20*b*. Note the *RC* filters formed, the series combination R_iC_i in the input circuit, and the parallel set R_fC_f in the feedback circuit. What characteristic frequencies f_o should they have? A good rule of thumb is for each to have an f_o ten times the highest *signal* frequency f_{max} expected. With this provision even a signal component with frequency f_{max} will be differentiated with an accuracy of $\pm 1\%$, which is sufficient for many applications.

The actual response of the practical differentiator as a function of frequency is shown in Fig. 4.21. Note that both scales are logarithmic. Differentiation results in a steadily higher output with increasing frequency. The peaking of output is traceable to the cutoff filters, which cause the differentiator response to drop rapidly at frequencies above f_o.

Example 4.15. It is desired to differentiate the output of a scanning UV–VIS spectrometer to obtain the extra sensitivity that derivative curves afford for recognition and quantification of components that give only shoulders on major absorption peaks. If a practical differentiator circuit is used, will its restriction to frequencies smaller than f_{max} pose any problems? Further, why is high-frequency noise seldom observed in recorded spectra?

To answer the first question, assume that the available spectrometer can be set to scan sufficiently slowly that there will seldom be frequencies greater than 100 Hz in its *absorb-*

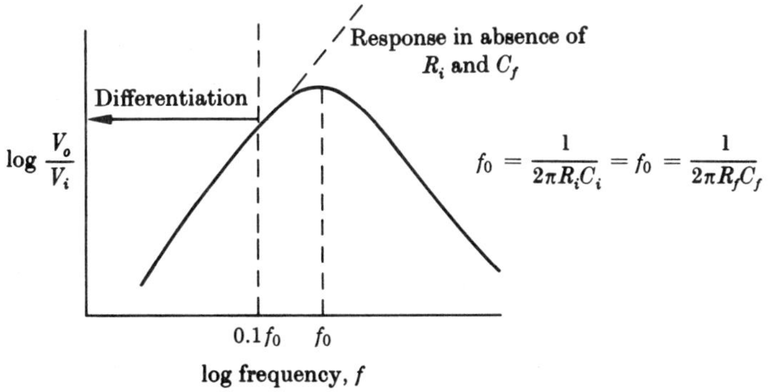

Fig. 4.21 Response curve of the practical differentiator of Fig. 4.20b.

ance signal. Thus, if $f_{max} \approx 1000$ Hz, essential frequencies will be satisfactorily differentiated. Usually high-frequency noise will not be observed since a fairly long time-constant is ordinarily selected for the amplifier, say 0.01 s, precisely to average out such noise before it is recorded.

Example 4.16. The titration curves produced by a potentiometric titrator are to be differentiated. What is the highest frequency likely to be encountered? What size resistors and capacitors should be used for the necessary filters?

For most titrations a full-scale response time of 1 s yields curves without distortion. Thus, one need handle signal frequencies no higher than about 1–2 Hz. From the rule of thumb given, filters with characteristic frequencies $f_o \approx 10f = 20$ Hz will be adequate.

For the basic differentiator a time constant $\tau = R_f C_i$ that is large relative to the period of the highest signal frequency is required, perhaps 10 s. Let $R_f = 1$ MΩ and $C_i = 10\ \mu$F.

The values of the other components are fixed by the rule of thumb for the characteristic filter frequencies of 20 Hz. Since $f_o = 1/2\pi RC$, we have $R_i C_i = R_f C_f = 1/2\pi f_o = 1/[2 \times 3.14 \times 20] = 8.0 \times 10^{-3}$. Then $R_i = 0.008/10 - 10^{-6} = 800\Omega$ and $C_f = 0.01/1 - 10^6 = 0.008\ \mu$F.

A second derivative can be taken if desired by placing a second practical differentiator in series with the first one. In this case, additional filtering is needed. Usually it takes the form of a low-pass filter inserted after both the first and second differentiators. The filter time constant may be as much as 100 τ.

Applications. It is valuable to add a differentiator to the processing modules of an instrument when a derivative

1. is a more sensitive measure of an analyte property than the variable being detected;
2. is required, as in measurement of a reaction rate; and
3. has a shape that lends itself better to actuating a control device or counting circuit than that of the variable itself.

Three important points about the use of differentiators must be noted. First, even if the independent variable in a measurement is not time, *as long as the actual independent variable is changing at a steady rate*, the time derivative of a variable will be proportional to the desired derivative. For example, in a potentiometric titration a *time* derivative of cell voltage is equivalent to one with respect to titrant volume as long as titrant is added at a steady rate. Second, a differentiating circuit can be expected to approximate closely the time derivative of a typical instrumental signal (a slowly changing dc voltage or current), except where it undergoes discontinuities in slope or very large slopes are involved. Here the trade-off arranged by excluding high frequencies and their attendant noise is experienced. By eliminating all such frequencies modest departures of the output from proportionality to the derivative (up to 20–40%) will be experienced whenever signals have very large slopes.

4.9 COMPARATORS AND SCHMITT TRIGGERS

There is often a need to determine whether an unknown voltage is larger or smaller than a reference voltage. A comparator is an op-amp circuit that furnishes such information by limiting at either its maximum positive or negative output.

Comparators. The most basic comparator is simply an operational amplifier, as shown in Fig. 4.22. One input is the unknown voltage V_i and the other a reference voltage V_{ref}. Note the absence of negative feedback in this circuit. Given gains over 10^5, whenever the difference V_s in voltages at its inputs exceeds a fraction of a millivolt, the comparator output will be at either its high (HI) or low (LO) limit. A sudden change in output will thus signal the matching of the input voltages or what is called a *zero crossing*. In other words, the time-based output wll resemble the op-amp transfer curve in Fig. 4.4. (See Fig. 4.30 for a zero-crossing level detector circuit based on a comparator and Exercise 4.12 for its explication.)

Instead of a regular op-amp, precision integrated circuit comparators with much faster response rates are ordinarily employed. They feature slewing at rates up to several thousand volts per microsecond. Integrated circuit comparators often have outputs of +5 and 0 V. How is this arranged? A simple way is to use an op-amp designed to operate with a +5 V power supply. Clearly comparators with this output range would supply voltages appropriate for a digital logic circuit of the TTL type, as will be seen later.

Comparators also have limitations. They have no external negative feedback that would raise their input impedance substantially. As a result, their input impedances are smaller, making efficient coupling difficult when a comparator is connected to a high-impedance signal source. Further, some comparators tolerate only a very modest range of input voltages, though many will accommodate a maximum range of ±30 V. A more serious disadvantage is that the comparison process is slow when the input signal changes only slowly. Finally, comparators have an extreme susceptibility to noise. Noise may either add to or subtract from an input signal and when the input signal essentially matches V_{ref}, the output of a comparator may swing several times from limit to limit instead of giving the single tran-

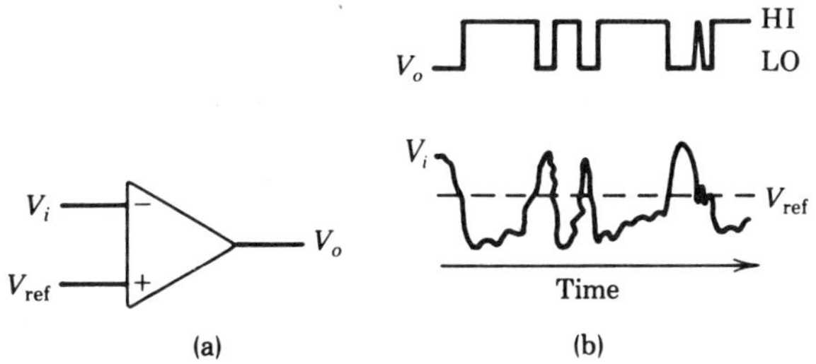

Fig. 4.22 Operational amplifier comparator. (a) Layout. (b) Input-output diagram. If V_i falls below V_{ref}, the output switches to HI. When the input signal $V_i > V_{ref}$, the output is LO. Because of the sensitivity of the comparator, noise in the signal is likely to cause false limiting whenever $V_i \simeq V_{ref}$ as shown.

sition expected. Clearly, it is inadvisable to use a faster or more sensitive comparator than is really needed as its speed would only exacerbate the problem.

Schmitt Trigger. The impact of noise on a comparator can be greatly reduced, however, by introducing *positive feedback* and converting the device to a Schmitt trigger. A common design is shown in Fig. 4.23a. Feedback is arranged through a resistor that couples the output to the *noninverting* input. The triggering pattern is illustrated in Fig. 4.23b. With positive feedback there is no longer one reference voltage or triggering *threshold* but two, one higher than the original V_{ref} and one lower. *Hysteresis* has been introduced, for the triggering from a LO to a HI output occurs only as the lower threshold is crossed from above, and a HI to LO output as the upper threshold is passed.* A second asset of the Schmitt trigger is that its speed depends only on the speed of the amplifier rather than on the rate at which the signal passes through the threshold region. With slowly changing signals, this property is a real advantage.

Example 4.17. How does the performance of a comparator contrast with that of an op-amp? Is a Schmitt trigger a comparator?

To compare an op-amp and a comparator, recall that an op-amp circuit is designed to operate within the linear region of its transfer curve (see Fig. 4.4). In most cases, feedback control is introduced to ensure that the value V_s is kept minute and Eq. (4.1) holds. By contrast, a comparator is designed (a) to report the sign of V_s and (b) to respond rapidly to changes in its sign. To accomplish the first the comparator assumes the appropriate limiting

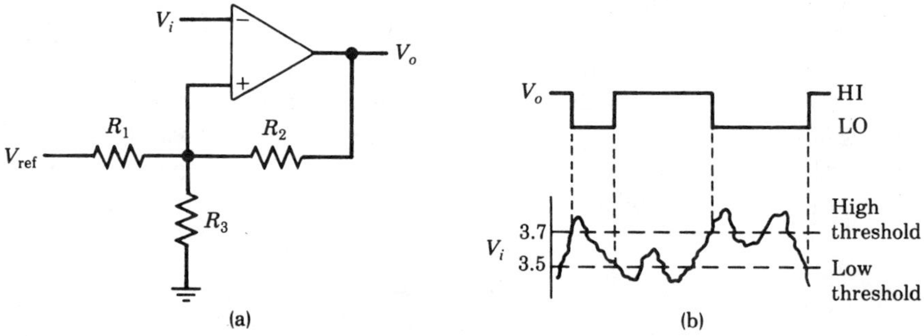

Fig. 4.23 A Schmitt trigger (comparator). (a) Circuit diagram. The distinctive feature is that it provides positive feedback (note the output connection to the noninverting input) through R_2. The effect is to speed transitions greatly and lessen the likelihood that noise will cause spurious switching. When the output is at its HI limit, the reference voltage at the noninverting input is higher because of the positive feedback and lower when the output is LO. As a result, there are two reference voltage thresholds separated usually by 0.2–0.3 V at which transitions are made. (b) Input–output diagram. Output switching for a particular input signal is shown. As is evident, the transition voltage now depends on both the amplitude of the input voltage and the current state of the output. The system is said to display *hysteresis*.

*The magnitude of separation of thresholds is a function of the feedback ratio, the comparator gain, and the difference between HI and LO output voltages.

state. To respond rapidly some internal circuit redesign may be required.

A Schmitt trigger may be described as a comparator in which the threshold for a HI–LO transition differs from that for the LO–HI transition.

Example 4.18. What is a common use of comparators?

Most digital voltmeters are based on comparators. The unknown voltage is directed to one input of a comparator and a linear ramp (see Example 4.12) to the other. During a measurement a clock provides square pulses. The pulses are counted as long as the ramp voltage is smaller than the unknown one and displayed as soon as equality is signaled by the comparator. For greater precision, a dual-slope integrator can be used with a comparator (see Section 5.3).

4.10 ACTIVE FILTERS

Active filters represent a happy combination of op-amps and *RC* filter circuits of the type discussed in Section 3.2. They are active because of addition of one or more op-amps. Actually, an op-amp furnishes not only circuit gain, but also provides the possibility of developing more ideal and more varied passband patterns. Today active filters are employed virtually to the exclusion of simple *RC* filters.

Two *high-pass* active filters are shown in Fig. 4.24. One provides an inverted output; the other is noninverting. As with simple *RC* filters, the resistor–capacitor combination determines the characteristic frequency ($f_o = 1/2\pi R_1C_1$) and the *series* capacitor identifies the filter as a high-pass unit. Further, the rate of attenuation or *roll-off* of output voltage V_o below the characteristic frequency f_o is the standard 20 dB per decade of frequency below f_o. Filters with one *RC* pair are also classified as first-order and are said to have "one pole" on the basis of a more sophisticated analysis.

A complementary *low-pass* active filter of the inverting type is shown in Fig. 4.25*a*. Its characteristic frequency f_o is also defined as $f_o = 1/2\pi R_1C_1$.

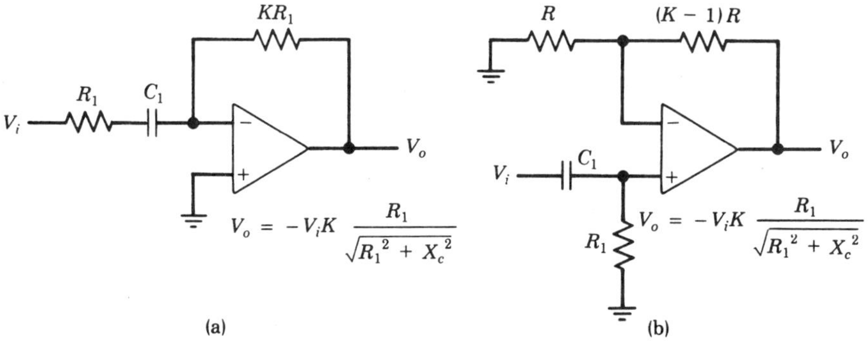

Fig. 4.24 Active first-order (one-pole) high-pass filters. Note the characteristic series input capacitor. These filters have typical 20 dB roll-off per decade of frequency at low frequencies. (a) Inverting type with gain. Since the input impedance is frequency-dependent, its gain is also. (b) Noninverting type coupling a regular *RC* high-pass filter to a voltage follower with gain.

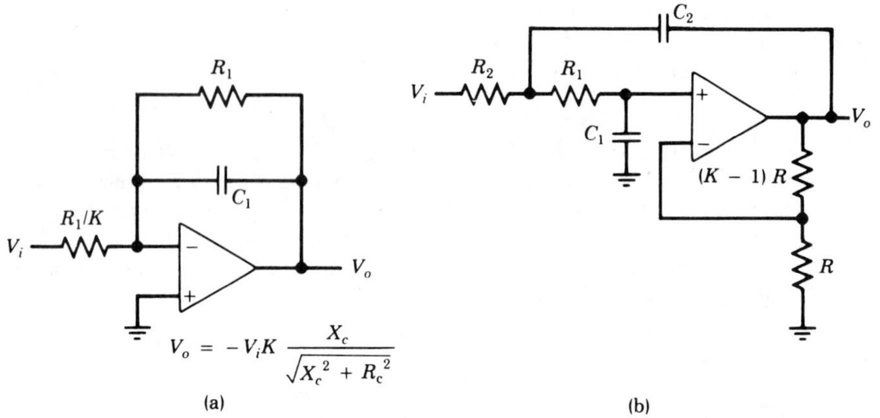

Fig. 4.25 Active low-pass filters of first and second order (one- and two-pole types). Note the characteristic resistive input. (a) Inverting first-order type. In this active filter the feedback impedance is frequency dependent. Roll-off at high frequencies is 20 dB per decade. (b) Noninverting second-order or two-pole low-pass filter. Note that if R_1 and C_1 were omitted, the filter would be the noninverting first-order counterpart of (a). When R_1 and C_1 are added, the roll-off with frequency doubles to 40 dB per decade. Further the active filter becomes a voltage-controlled voltage source (VCVS).

Example 4.19. Show that the R_1C_1 combinations in the inverting and noninverting versions of the active high-pass filters in Fig. 4.24a and b correspond to a regular passive *RC* filter of this type. Show also that these active filters have the output voltage given.

First, in the inverting active filter in Fig. 4.24a locate the *RC* filter elements themselves. They are the R_1C_1 pair in the input circuit. Voltage gain is provided by ensuring that the resistor in the feedback circuit is *K* times larger than R_1. In the noninverting version (Fig. 4.24b) of this active filter the R_1C_1 pair is directed to the noninverting input and the input voltage appears across resistor R_1. Compare this arrangement with the passive *RC* filter of Fig. 3.2a in which V_o is taken across *R*.

To determine V_o in these circuits, calculate input currents. For the circuit of Fig. 4.24a, $I_i = V_i/Z = V_i/(R_1^2 + X_C^2)^{1/2}$. Further, based on rule 1, $V_o = -I_iKR_1 = -KV_iR_1/(R_1^2 + X_C^2)^{1/2}$.

By contrast, for the circuit of Fig. 4.24b, consider that the voltage at the noninverting input will be unaffected by the gain of the op-amp. Thus $V_+ = V_iR_1/(R_1^2 + X_C^2)^{1/2}$. At the inverting terminal V_- will be driven to this value. The current I in resistor R must be $I = V_-/R$ and this current will also exist in resistor $(K - 1)R$. Thus $(V_o - V_-)/(K - 1)R = I = V_-/R$. Simplifying, we have $V_o = KV_-$. On substituting for V_- in this expression, the equation given in the figure is obtained.

Might additional RC circuits and op-amps increase the rate of signal attenuation? The idea is that successive units consisting of an *RC* circuit and op-amps could be added in such a way *that new units do not interfere with the others* but each adds 20 dB per decade to the previous roll-off. In other words, each unit increases the order of the filter by one and adds one pole. While a one-pole filter provides 20 dB roll-off per decade, a two-pole filter furnishes 40 dB roll-off, a three-pole 60, and so on.

A second-order low-pass active filter of the noninverting variety is shown in Fig. 4.25b. Note that it is easy to recognize the first pair R_1C_1 connected directly to the noninverting input. The legend of the figure provides some additional interpretation. Suffice it to say that it is possible to combine in series many circuits of the type shown in Fig. 4.25b and thereby increase roll-off drastically beyond the 40 dB provided by a second-order filter. Often sophisticated filtering must be used when it is necessary to filter a signal from nearby interfering signals. References should supply information about the wide variety of designs available.

REFERENCES

1. H. M. Berlin, *The Design of Operational Amplifier Circuits, with Experiments*. Derby, CT: EI Instruments, 1977.
2. J. G. Graeme, *Applications of Operational Amplifiers, Third-Generation Techniques*. New York: McGraw-Hill, 1973.
3. J. G. Graeme, *Designing with Operational Amplifiers, Applications Alternatives*. New York: McGraw-Hill, 1977.

See also references listed for Chapter 3.

EXERCISES

4.1 Develop an expression for the output voltage in each of the op-amp circuits of Fig. 4.26 in terms of the input voltage(s).

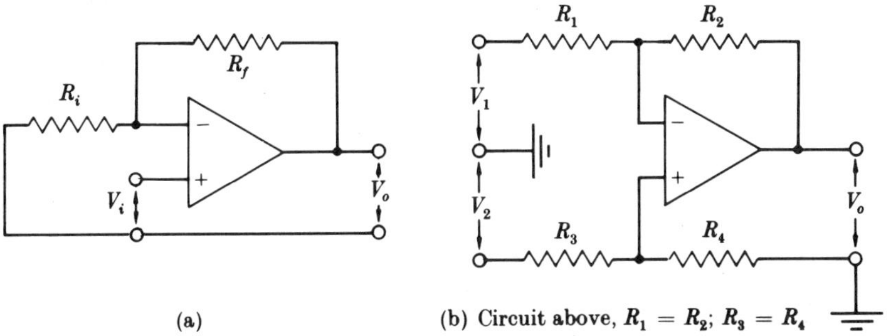

(a)

(b) Circuit above, $R_1 = R_2$; $R_3 = R_4$

(c) Circuit above, $R_1 \neq R_2$; $R_3 \neq R_4$

Fig. 4.26

4.2 Draw an op-amp circuit that will provide precision variable multiplication of an input voltage.

4.3 Show that the output voltage of the circuit in Fig. 4.27 is given by the expression $V_o = (1 + R_1/R_2)V_{ref}$.

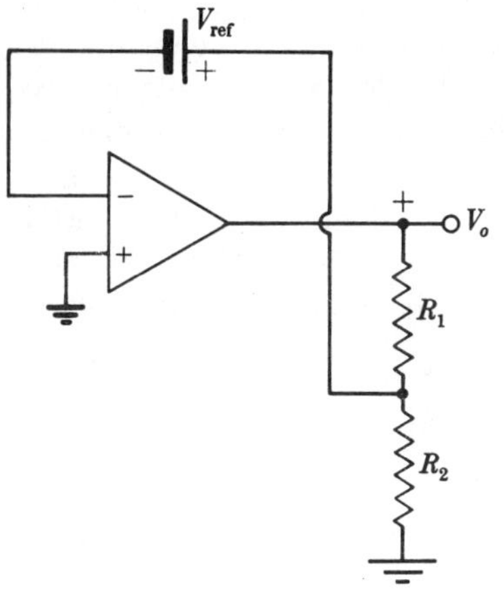

Fig. 4.27

4.4 Discuss the usefulness of the circuit of Fig. 4.5 as a constant voltage generator if a reference (standard) voltage source is substituted for V_i. How much current can the generator furnish if V_o is to be constant? To what extent is the choice of a value for Z_i limited?

4.5 Figure 4.9*b* depicts a constant voltage source. (a) What is the value of V_z? (b) Assume V_z is constant and that the current in the zener diode is I_z. Write equations for the current in R and in R_1. (c) What is the role of the op-amp circuit in keeping I_z and the current in R_1 and R_2 constant?

4.6 In the current-to-voltage op-amp circuit of Fig. 4.14 V_o is given by $V_o = -I_iZ_f$. When I_i is constant, does I_o have to be? Explain.

4.7 Calculate the effective input impedance of an op-amp circuit with negative feedback. Draw a basic op-amp circuit with a grounded noninverting terminal. Take the intrinsic op-amp gain as A. Assume the impedance of the op-amp without feedback is just the differential input impedance across the terminals, R_i. Since the voltage across the input terminals with feedback is reduced according to Eq. (3.15), calculate first I_i for the circuit and then $(R_i)_{effective}$.

4.8 Draw a schematic diagram of an op-amp circuit that will find the area under the peaks registered by a gas chromatograph with a flame ionization detector. The chromatograph output is a current.

4.9 Redraw Fig. 4.16 with a potentiometer in place of R_f. How can the potentiometer serve as the readout? What is the magnitude of amplifier current I in this case?

4.10 The op-amp circuit of Fig. 4.28 serves as a readout for the thermocouple which provides its input voltage. Assume the thermocouple has a resistance of 20 Ω. (a) Does the circuit successfully treat it as a voltage source? That is, does it transfer voltage without drawing current? Explain very briefly. (b) Develop an equation for V_o in terms of the thermocouple emf. (c) What advantage is gained by use of the potentiometer?

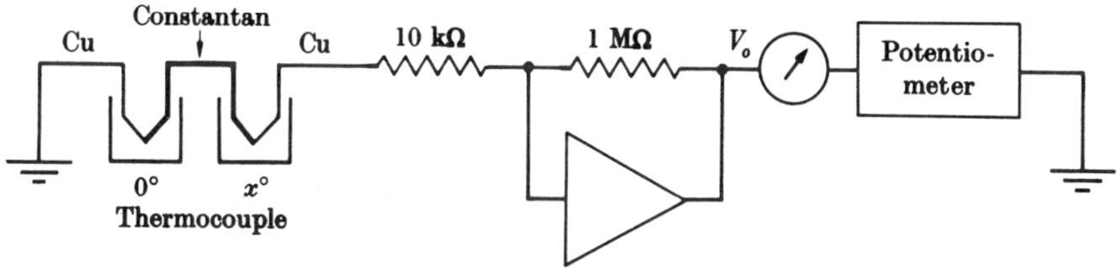

Fig. 4.28

4.11 (a) Develop an equation for the output voltage of the circuit of Fig. 4.29*a* in terms of the cell resistance *R*. What advantage does this circuit have over a Wheatstone bridge in monitoring a conductometric titration? (b) The oscillator can be represented by the Thévenin equivalent circuit in Fig. 4.29*b*. How will oscillator voltage V_i vary if R_{cell} varies from 1000 to 700 Ω during a titration? What relative error in V_o will result? (c) Sketch another circuit in which the error discussed in (b) is minimized by adding another operational amplifier. [See MPI Application Notes 2(6), 25 (1967).]

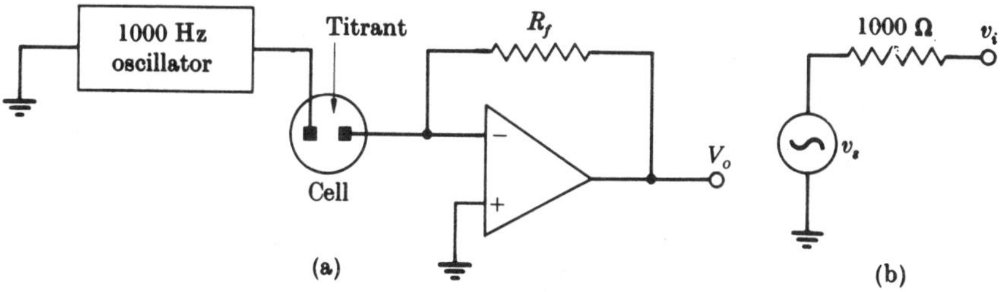

Fig. 4.29

4.12 In Fig. 4.30 a zero-crossing level detector using a Schmitt trigger is shown. (a) If diodes D_1 and D_2 are IN914 types, what approximate maximum input voltages can be tolerated by the circuit? Considering the behavior of the diodes under the biasing used, what voltage range is passed on to the comparator? Assume $R_1 \simeq 40$ kΩ and $R_2 = R_4 = 5$

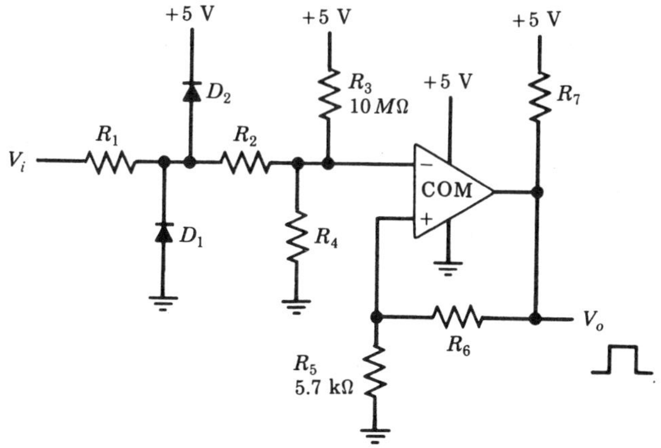

Fig. 4.30 A zero-crossing level detector.

kΩ. (b) It is desired that the feedback fraction be $\frac{1}{1000}$. What should the value of R_6 be? What maximum voltage can be imposed at the noninverting input of the comparator?

4.13 Sketch a simple potentiometer circuit to trim to zero the input offset voltage of the model 741C op-amp of Fig. 4.2. Wire a 100-kΩ potentiometer across pins 1 and 5 and connect its wiper to the −15-V pin. Describe the operation by which the offset voltage can be trimmed to zero.

4.14 A 741C op-amp is wired as the inverting amplifier of Fig. 4.5. Look up its input bias current in Table 4.1. Because the inverting input of the op-amp in this amplifier experiences an external impedance of R_1 in parallel with R_2, the bias current introduces a small input voltage. This error may be minimized by taking advantage of the existence of a nearly equal bias current at the noninverting input. If a resistor, R_3, is inserted between the noninverting input and ground, an equal input voltage can be developed at that terminal, thus cancelling this error. Draw the circuit just described. Derive an equation to calculate a value of R_3.

ANSWERS

4.1(a) $V_o = (1 + R_f/R_i)V_i$; (b) $V_o = V_2 - V_1$; (c) $V_o = [R_4/(R_3 + R_4)](1 + R_2/R_1)V_2 - R_2V_1/R_1$; *4.7:* $(R_i)_{\text{effective}} = (1 + AF)R_i$. *4.12* (a) About ±100 V, about −0.3 to 5 V; (b) $R_6 = 4.7$ MΩ. *4.14* $R_3 = R_2R_1/(R_1 + R_2)$.

Chapter 5

DIGITAL ELECTRONIC MODULES

5.1 INTRODUCTION

Since the mid-1970s digital circuits have become increasingly important in chemical instrumentation. Their use in microcomputers has propelled their acceptance in general. However, the characteristic modules of an instrument, such as source and detector, will be likely to remain analog in character. It is especially in the processing modules that digital electronics has had its greatest impact. In addition to the benefits gained from that development, digital circuits inherently offer the advantages of high speed, high precision, and virtual immunity from noise. When an instrument contains a microprocessor, if the characteristic modules related to the analytical method are analog, instrument signals are converted to the *digital domain* after detection and amplification. As a result, much modern instrumentation is half analog and half digital in character. Simpler instruments continue to be substantially analog in character.

A *digital signal* consists of binary digits or *bits*. Ordinarily, they are collected into a binary number. There is a review of the binary and other number systems in Appendix A. A number consists of digits *in particular positions*; the least significant bit (LSB) is ordinarily at the right end. An example is given below.

It can be expected that digital *circuits* will be functionally different since they will both *receive* and *produce* digital signals. This type of operation is possible only because the mode of operation of digital circuits is *switching* between HI and LO levels. As shown in Fig. 5.1, care is taken to identify each level by a voltage range and to separate the two ranges by far more than the amplitude of ordinary noise. Some examples of simple digital circuits are AND and NOR logic gates, which will be introduced in Section 5.4. In this chapter the most important types of digital circuits will be treated.

Example 5.1. What does a representative digital signal look like? Is it basically a discrete set of HI (current) pulses such as might be obtained by measuring the beta particle emission of a particular glucose sample labeled with ^{14}C and finding it to be 150 counts s^{-1}?

Actually, the train of current pulses is not a digital signal unless converted into binary code, perhaps by using a binary counter to obtain the count rate. The counter will yield the number 10 010 110. If these digits are sent with the least significant bit (on the right) first, this digital signal will appear in a circuit as . Note that the consecutive set of voltage amplitudes that indicate 1's and 0's are of constant width but are not separated.

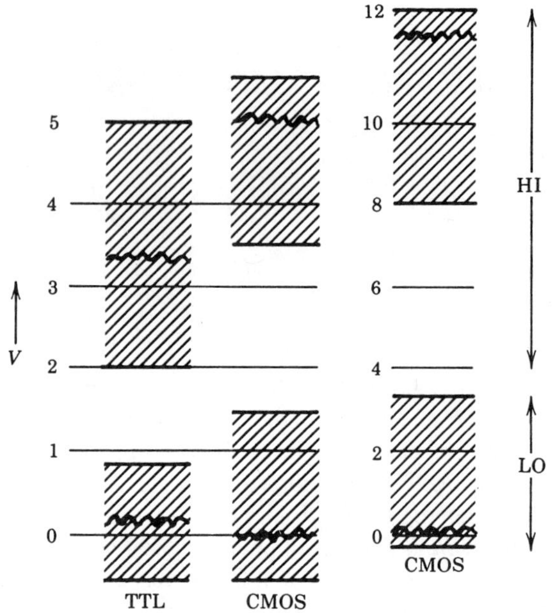

Fig. 5.1 Logic voltage levels for two digital logic families, TTL (transistor–transistor logic) and CMOS (complementary metal-oxide semiconductor) logic. Commonly, TTL circuits operate between ground and V_{CC} = 5 V while CMOS circuits most often operate between ground and V_{DD} = 5 or 12 V. *Input voltages* to circuits are acceptable as LO or HI if they fall in the hatched ranges. *Output voltages* from logic devices are guaranteed to fall in smaller ranges: in TTL, LO = 0–0.4 V and HI = 2.4–5 V; in CMOS, LO = 0–0.4 V and HI = 4.6–5 V or 11.5–12 V, with representative values shown by wavy lines.

Example 5.2 How are heights and widths of pulses formed in digital circuits based on transistor–transistor logic (TTL)? (Logic is discussed in Section 5.4.) Assume a 4-MHz clock (quartz crystal oscillator) is used to control switching. How much variation in pulse shape will be tolerated?

In transistor–transistor logic (TTL) circuits a binary 1 or HI level will nominally be 5 V and a 0 or LO level nominally 0 V. Each pulse height is allowed a voltage range as shown in Fig. 5.1. It appears that a binary 1 will average about 3.3 V and a 0 about 0.1 V.

Second, pulse widths will carry no information. Widths relate to the switching frequency in each digital device. For instance, if switching is routinely carried out by square wave pulses generated by a 4-MHz clock, each pulse will be 0.25 μs wide.

How do digital circuits transmit signals? As in Example 5.1, a several-bit signal may be sent *serially* along a single wire with the LSB arriving first. Alternatively, if it is arranged that all signals have the same number of bits, each digital position may be labeled by the bit it is to carry and provided an individual wire. Then all bits can be transmitted simultaneously in a mode termed *parallel* transmission. For instance, if all signals are 16 bits, 16 parallel lines will be needed. The latter set of conductors is called a *bus*.

5.2 SIGNAL SAMPLING

The process of ''data acquisition'' in an instrument seems a good starting point for dealing with digital electronics. What steps are necessary to convert analog data to the digital domain for processing by a microprocessor or microcomputer?

Initially amplification is nearly always required: many instrument detectors yield analytical signals of only a few millivolts, while AD converters are designed for inputs in the 1–10 V range. The detector signal should be amplified sufficiently to take advantage of most of the converter input range for efficient coupling and minimization of the digitization error. For this purpose an instrumentation amplifier (see Fig. 4.17) is a good choice because of its excellent stability and buffering.

In designing a sampling procedure for an analytical signal both its range of frequencies and noise level must be taken into account. Some general conclusions can be made about these factors. First, both line-frequency and high-frequency noise should be filtered out before sampling or spurious frequencies will be digitized as well as data frequencies. For signals that are nearly dc in character and have a good S/N one can simply make an AD conversion at regular intervals without other preparations. On the other hand, for a fast-changing signal samples must be taken rapidly and frequently and ''held'' while the signal amplitude is digitized. As discussed in Section 12.5, the number of samples taken per second, the sampling rate, must be twice the highest frequency in the analytical signal or chemical information will be lost.*

Sample-and-Hold-Systems. Usually a sample-and-hold circuit is employed for fast sampling. Both the sampling interval and rate can be selected. In this sampling process the analytical signal charges a circuit capacitor to a voltage proportional to its amplitude. After charging, an electronic switch isolates the capacitor at the input of a very stable, high impedance amplifier where its charge is held while the voltage is being digitized.

A precision sample-and-hold circuit is shown in Fig. 5.2. It has four elements: (a) op-amp 1, which provides at its output a low-impedance version of the input signal; (b) an FET switch (see Fig. 5.2b) which is opened and closed to establish a *sampling interval* on receipt of an external signal; (c) capacitor C, which is charged to a voltage representative of the analytical signal amplitude while the FET switch is ON; and (d) op-amp 2, which must isolate and maintain the capacitor voltage when the FET switch is OFF. Op-amp 2 provides a constant output signal for digitizing.

How can one guarantee that capacitor C will charge fully and rapidly? First, the circuit time constant RC should be kept small. In this circuit, resistance R is contributed mainly by the FET switch and is quite small (about 50 Ω). Selecting a value for capacitor C is more difficult. If its capacitance is small, its rate of charging will be faster (RC will be smaller). This matter must be taken up again

*Scanning with respect to wavelength or another variable in taking measurements basically modulates the intrinsic frequencies involved in the method as does direct modulation. It is the modulated frequencies in the analytical signal that must be captured in sampling.

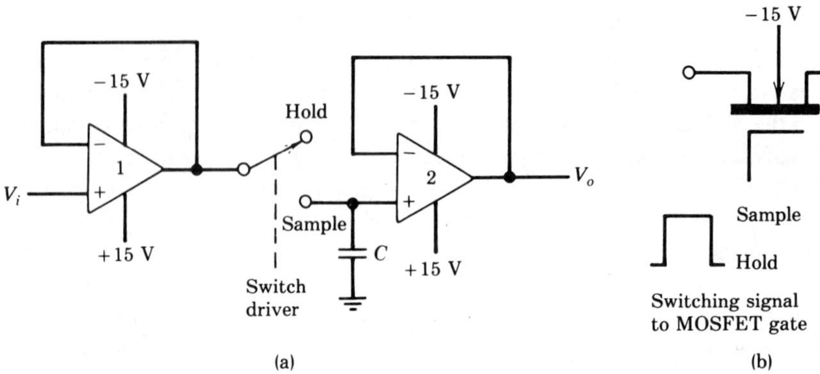

Fig. 5.2 Sample-and-hold circuit for high-speed analog sampling of a signal. (a) Schematic diagram. Op-amp 1 produces the current that charges capacitor C to the amplitude of the input signal. Then, on switching to the "hold" mode, op-amp 2, an FET amplifier or buffer, supplies a constant output voltage equal to the sample voltage being held. (b) A MOSFET gate that replaces the mechanical switch. Sampling occurs when the gate receives a +15-V pulse.

below. Op-amp 1 is also involved in this trade-off. Fortunately, if op-amp 1 is used in a voltage-follower configuration as shown, it will (a) isolate or buffer C substantially from the signal source, and (b) supply current for charging. Clearly, op-amp 1 should have a response time that is short relative to the anticipated rates of change of input signal as well as sufficient current capability to charge C quickly.

Op-amp 2 plays a crucial role during the holding interval. It should have very high input impedance which is secured both by use of the follower configuration shown and selection of a MOSFET input op-amp. Further, it should have minimal values of several types of error: voltage offset, gain error, input bias current, and deviation of amplifier gain from the expected value of unity for the follower configuration. In addition, its *settling time*, which is the interval required for its output voltage to become essentially constant after a step has occurred in its input voltage, should be minimal.

Given that these requirements are met, during the "hold" period, the rate of change in output voltage V_o, called *droop*, will depend mainly on the leakage current of the capacitor C both through the FET switch and to ground. The larger the value of C is, the less important the effect of leakage current. There is a trade-off, of course, for a larger capacitor will charge more slowly and require a higher current from op-amp 1 for charging. The capacitor should be a high-quality polystyrene, Teflon, or polypropylene to ensure low-leakage currents and minimum dielectric loss. The latter term defines the absorption of energy by the dielectric at high signal frequencies.

The *acquisition time* of the sample-and-hold module defines the minimum sampling time that will ensure an output of given accuracy. Note that uncertainty in the duration of sampling will arise mainly from what is called the *aperture time* of the FET switch, the delay between its receipt of the hold logic level, and the opening of the switch (perhaps 10–100 ns). In most instances the settling time of op-amp 2 is the largest factor in determining this value.

Integrated circuits as well as hybrid modules providing sample-and-hold capability are available. Most circuits are similar to that in Fig. 5.2.

High-Precision Circuits. In a high-precision sampling circuit, the sampling amplifier (op-amp 2) is usually included in the feedback loop of the input voltage follower and often an external hold capacitor is also required. Now op-amp 1 compares the device output voltage to the input signal. In response to the difference it charges capacitor C until the difference essentially vanishes. Both offset and common mode error are eliminated. The voltage-follower loop for op-amp 2 is retained, but its output is also connected to the noninverting input of op-amp 1. In such circuits accuracies of $\pm$ 0.002% are obtained and a converter of 12-bit or higher capability is employed.

Peak-Detector Device. It is possible also to modify the basic sample-and-hold circuit to operate as a *peak-detector device*. In this circuit C is charged continuously and the highest voltage reached by the input signal is presented to op-amp 2. Once op-amp 2 gives a steady output, which can be assumed to be a peak value, the switch can be reset in preparation for detection of a subsequent peak.

5.3 ANALOG-TO-DIGITAL CONVERTERS

Analog-to-digital converters (ADCs) may be classified by type roughly according to whether they *compare charges*, *compare voltages*, or fall into the category of the *flash converter*. One can infer that the module common to all converters must be a comparator. Recall that a comparator is based on an op-amp (see Section 4.9). Two modifications are called for: the op-amp is (a) diode-clamped to reduce its voltage swings to about the difference between logic levels and (b) provided with some positive feedback to allow it to swing between limits at high speed.

The *voltage-to-frequency converter* will serve as a good example of the charge-comparison or integrating type. A simplified block diagram is shown in Fig. 5.3. How does this converter function to measure V_x? During a conversion the op-amp integrator begins to develop an output proportional to V_x across its integrating capacitor. The integrator output voltage, call it V_C, will be compared continuously with a reference voltage V_{ref} by a comparator.

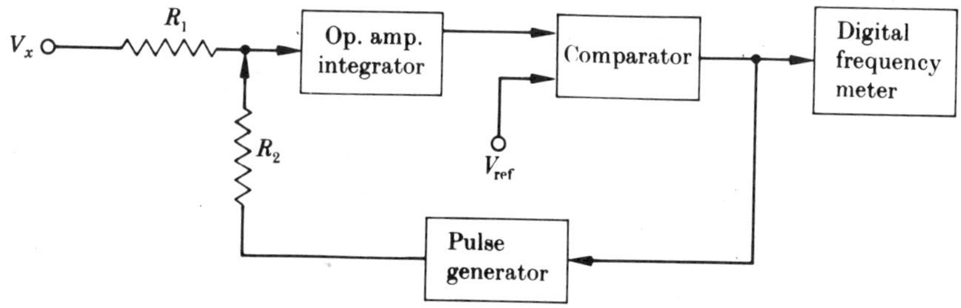

Fig. 5.3 Simplified block diagram of an AD converter of the frequency type.

Second, it is necessary to know that the pulse generator can produce pulses containing a small, fixed amount of charge repeatedly. The circuit directs them to the op-amp integrator. Each pulse is of a sign that acts to discharge the integrator capacitor partially. Pulse generation actually occurs, however, only when $V_C > V_{ref}$. It is thus the magnitude of V_C relative to the size of pulses that controls the rate of pulse generation. As soon as the integrator output falls below V_{ref}, the input signal begins to charge the integrator capacitor again. This type of converter is sometimes called a *switched-capacitor* converter.

Actually, the number of pulses produced per unit time, reported as a frequency by the final meter, is the output of this converter. Converters of this type with accuracies of $\pm 0.01\%$ are available. This type is attractive where an output must be sent some distance since frequency signals can be transmitted well over electrical or fiber optic lines. (This ADC can also serve as a *voltage-controlled oscillator*. For example, the integrated circuit voltage-to-frequency converter type LM331 will develop output frequencies from 0 to 10 kHz.)

Another charge-comparison or integrating type of ADC is commonly found in a *digital voltmeter*. A diagram is given in Fig. 5.4. Because of its integration patterns it is also called a *dual-slope* converter. First, a current I_x proportional to V_x (shown as supplied by an op-amp voltage-to-current device) charges capacitor C for a fixed time. If I_x is small, only a small voltage is built up. Then the capacitor is discharged by a constant, standard current until its voltage falls to zero.

The digital output of this ADC is the number of pulses provided by a precision

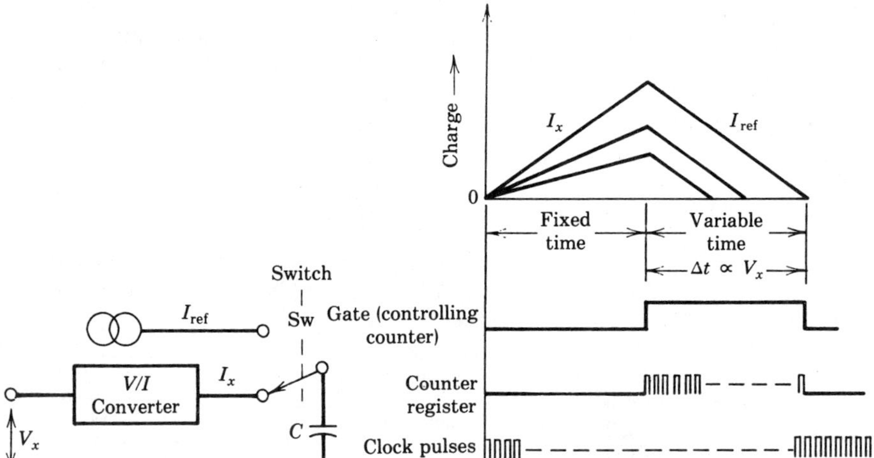

Fig. 5.4 A dual-slope AD converter or voltmeter. (a) An unknown voltage V_x is caused to furnish a proportional dc current I_x for a precisely timed interval while capacitor C charges. Then a constant current I_{ref} is used to discharge C completely. (b) During the period of discharge the number of pulses produced by a precision square-wave oscillator is counted. When the voltage of capacitor C falls to zero, switch Sw is opened and the counter read to obtain the total number of pulses, a value which is proportional to V_x.

square wave oscillator or clock (see Section 5.7) during the exact interval required to discharge the capacitor. This time Δt or number of pulses Δn will also be proportional to both I_x and V_x. Clearly, the voltmeter is also a precise ADC. The advantages of the dual-slope ADC are discussed further in Exercise 5.3. One virtue of the design is that by using a single capacitor for signal and reference it is not necessary that its capacitance be completely stable and accurately known. Further, these converters make few demands upon the comparator. While the integrating ADCs are inexpensive and accurate, they are relatively slow (conversion rate 1 kHz).

A quite different approach is represented by the second type of ADC, the voltage-comparison type known as a *successive approximation converter*. It is shown in block diagram form in Fig. 5.5. The principal of operation is to feed digital numbers that are successively better approximations to the amplitude of the input to a *digital-to-analog* converter (DAC) for comparison with the unknown signal. The program used calls first for delivery to the DAC of a logic 1 for the most significant bit with all other bits zero. If the DAC voltage is greater than V_x, the comparator changes output and causes the logic programmer to clear that bit, that is, set it to 0. The program then produces a 1 in the next most significant bit. But if the initial reference voltage generated is smaller than V_x, the comparator output will not change and the initial 1 will be saved in the register. In quick succession, each successive bit in the binary register can be tried as a 1.

The successive approximation converter is relatively fast, accurate, flexible, and widely used. Costs are higher than for integrating ADCs. Successive-approximation ADCs are fast because the time required for testing each bit is only that of the settling time of a 1-bit DAC. Representative conversion times are 1–50 μs for 8–12 bit converters. There is also a tracking ADC that is a variation of this device. A tracking device spends most time following input changes and a little "holding" a sample of the signal.

Example 5.3. An AD converter with $\pm$ 0.1% precision is needed. Should an 8, 12, or 16-bit device be selected?

Eight binary bits are sufficient to register the decimal number 256, 12 bits 4096, and 16

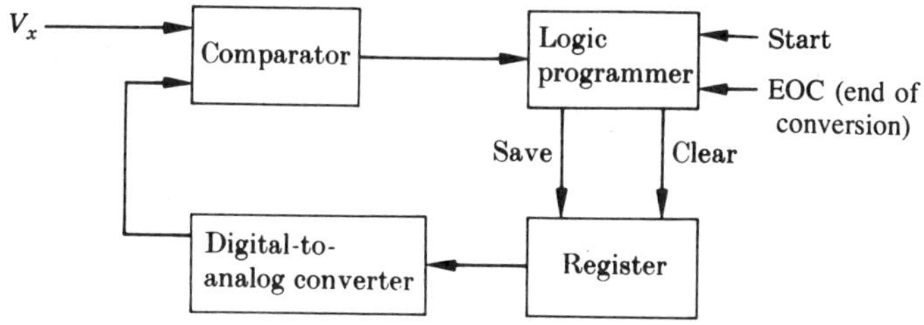

Fig. 5.5 Block diagram of an 8-bit AD converter based on a successive approximation approach to an unknown voltage V_x. The reference voltage is generated by the DA converter on instruction from a program. The digital equivalent of the analog voltage at the input is the reading of the register when the conversion is complete.

bits 65536. For $\pm 0.1\%$ precision an 8-bit ADC is inadequate; its precision is only $(1/256)$ times $100 = 0.4\%$. The best choice is a 12-bit device whose precision $(1/4096)$ times 100 $= 0.02\%$ surpasses the specifications.

Example 5.4. An 8-bit charge-comparison ADC is given an input range of -10 to $+10$ V by use of an external voltage offset. Describe its output.

Its output will be a series of pulses or bits up to 2^8. Thus, a -10 V input would correspond to 0 pulses, while a 0-V input would be equivalent to 128 pulses. One pulse or bit is about 0.08 V. The pulses may be tallied by a binary counter and its total then transmitted either serially or in parallel fashion over an 8-line output.

Finally, the *flash converter* is the sole representative in its category. It deserves brief mention because of its remarkable speed. In this device the input signal voltage is simultaneously directed to *n different comparators*. The other input of each comparator is connected to *n* equally spaced reference voltages (developed, for example, by using a voltage divider that uses a series of precision 1-kΩ resistors). The time of conversion is usually less than 20 ns. Commercial encoders of this kind are available with from 16 to 256 comparators, the lower number being equivalent to a 4-bit converter and the higher to an 8-bit one. The cost of flash converters is predictably greater than that of other converters of the same precision.

5.4 DIGITAL SIGNALS, SWITCHING, AND LOGIC GATES

The simplest way to develop electronic circuits that operate digitally is to design analogs of the electrical toggle switch. Recall that the switch provides two indefinitely stable states, ON (switch closed) and OFF (switch open), and it is easy to go from one to the other. If digital circuits are made from bipolar transistors, there are two fully reproducible states attainable by simple biasing: *cutoff* and *saturation*. The names describe the condition of the active device. In cutoff both *p–n* junctions are reverse-biased; in saturation, both are forward-biased. With other types of transistors comparable states may be generated.

Other requirements are also imposed on digital circuits. Of these the most important is that voltages supplied for the two states at the outputs be within the system range and be reproducible. Naturally, the devices should respond to input voltages at these levels as well. Recall that some standard HI and LO logic voltage levels were described in Fig. 5.1. Circuits with passive elements and two to four transistors can be designed so as to force each of the transistors into cutoff or saturation. Then, if power supply voltages have been chosen judiciously, when there are HI or LO voltages at the inputs, output voltages will be HI or LO also. Finally, since going from one level to another is accomplished by *switching*, it is possible to design circuits so that minimum power and time are required for switching. These points will be discussed again briefly at the end of the section.

Example 5.5. Some digital devices produce a HI or LO signal as an output. When does such a signal provide useful information for instrument control or signal processing?

A HI bit, as opposed to one that is LO, might indicate any of the following conditions:

That a given number is smaller than a reference value.

That data available from memory are ready to be received by a modular control device for control purposes.

That an action such as peak integration should be ended because the amplitude of an instrumental signal has returned to the baseline.

From these examples it appears that judicious combinations of logic-gate digital devices can also be used to verify that all necessary conditions are met before an operation such as adjustment of an instrument parameter or making a computation is begun.

Logic Gates. The simplest digital circuit is called a gate, or better, a *logic gate*. Conceptually, it has a direct tie to operations in logic in which there are only two states, TRUE and FALSE. Further, gate functions are related to Boolean algebra and the binary number code. Digital circuits are thus compatible with logic units and binary systems; each has two well-defined states. In this discussion, digital states will be identified as HI and LO except in tables, where 1 and 0 will be used.

Let us now consider some gates. Each will have two or more inputs, A, B,· · ·, and a single output O. As many inputs can be used as are needed, but only two, A and B, will be used here. The voltage at each input and the output can also be in two states and in no other. Some definitions are needed. Let A denote a HI state input that can (usually) be equated with the 1 state (binary), true state (logic), or closed state (switching). Similarly let $\overline{A}$ denote the LO state of an input.

AND, OR, and NOT Operations. Launching into a discussion of binary logic operations and logic gates may at first appear to be a diversion from the topic of digital electronics, but it proves instead to move us quickly to the heart of the matter. The three most fundamental logic operations that are related to gates are denoted as AND, OR, and INVERT (or NOT). They are easily defined by how they operate on variables A and B. Defining expressions are

$$A \cdot B = AB = O; \qquad A + B = O; \qquad \overline{A}$$

The first equation or statement is an AND statement and as a logic proposition should be read: when A is true AND B is true, the outcome or result O is true. This statement can also be translated electrically: if input A is HI and input B is HI there will be a HI output.

The second equation (A + B = O) is an OR statement and as a logic proposition should be read as follows: when A is true OR B is true OR both are true, output O is true. Usually this statement is described as an "inclusive or" concept.

The NOT or INVERT concept means to invert the condition suggested. For instance, $\overline{A}$ means that if A represents a 1 state, $\overline{A}$ represents the opposite. Consider the statement

Table 5.1 Some Logic Symbols and Truth Tables

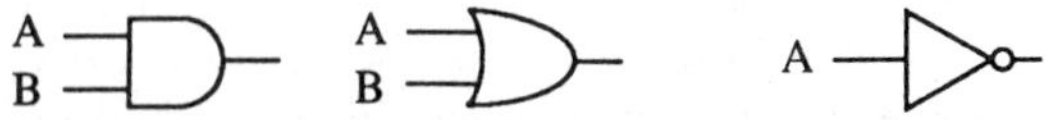

AND			*OR*			*INVERT*	
Inputs			*Inputs*			*Inputs*	
A	*B*	*Output*	*A*	*B*	*Output*	*A*	*Output*
0	0	0	0	0	0	0	1
0	1	0	0	1	1	1	0
1	0	0	1	0	1		
1	1	1	1	1	1		

$\overline{AB}$ = O. It should be read: when A is LO AND B is LO, the outcome is HI. Note the clear implication that other combinations such as A = HI and B = LO do not yield a result that is HI.

In Table 5.1 the operation of the basic logic gates is described. All possible combinations of inputs are given and the output for each, which is predetermined by the nature of a gate, is given. While such tables are often called *truth tables*, it is perhaps more informative to call them *input/output tables*. The first entries in Table 5.1 are for an AND gate, the second set are for an OR gate, and the third are for an INVERT or NOT gate. The key to using logic gates well is noting for each the one case for which a distinctive output is furnished.

In Table 5.2 the digital gate collection is extended. As the new acronyms NAND and NOR imply, these gates combine an INVERT function with either AND or OR functions. Again their possible input/output operations are tabulated. In Figs. 5.6 and 5.7 the actual layout of a NAND gate in two logic families is given and a fairly detailed explanation of their operation is provided.

The INVERT operation is symbolized by the *small circle* appearing at the out-

Table 5.2 Additional Logic Symbols and Truth Tables[a]

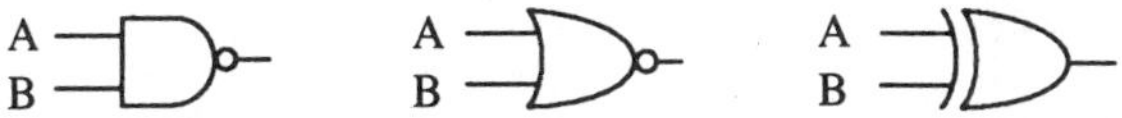

NAND			*NOR*			*XOR*		
Inputs			*Inputs*			*Inputs*		
A	*B*	*Output*	*A*	*B*	*Output*	*A*	*B*	*Output*
0	0	1	0	0	1	0	0	0
0	1	1	0	1	0	0	1	1
1	0	1	1	0	0	1	0	1
1	1	0	1	1	0	1	1	0

[a]The small circle appearing at the output in the symbols denotes inversion.

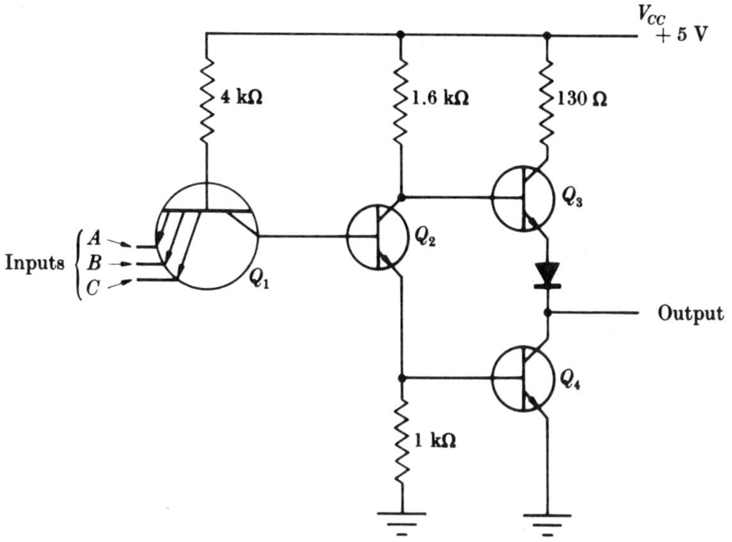

Fig. 5.6 A TTL-logic NAND gate. Some specifications are: speed, 6 ns; wattage per gate, 10 mW. Output voltages: logic 0, about +0.2 V; logic 1, about +3.3 V; noise immunity, 0.4 V. The operation of the NAND gate can be described by determining whether each transistor is in the state of being *cut off* or in *saturation*. During each change of logic level by a gate each of its transistors makes this type of transition. Consider first the case in which at least one input (A, B, or C) is LO, about 0.2 V. Then V_{BE}, the potential across the base–emitter junction, for transistor Q_1 will be sufficiently larger than 0.7 V to forward-bias this junction. Current will flow from V_{CC} out each input connection that is about 0.2 V. Since the base of Q_1 must be about +0.7 V, this potential is the high point *in the path to ground* that includes the following potential drops: V_{BC} of Q_1, V_{BE} of Q_2, and V_{BE} of Q_4. Each *p–n* junction can conduct only if the voltage across it is sufficient for forward-biasing, that is, if the sum of potentials exceeds $3 \times 0.7 = 2.1$ V. Here too small a voltage is available, and Q_2 and Q_4 must be cut off. Q_3 plays the roll of being an *active pull-up* device for the gate and ensures that its output will rise toward 5 V when the gate output is HI: when Q_2 and Q_4 are off, the base and collector of Q_3 rise in potential toward the 5 V supplied at V_{CC}. Q_3 conducts and saturates. The gate output will be HI (about 3.3 V). This is the expected result for a NAND gate with any input LO. Conversely, when all gate inputs are HI (or unconnected), there is no conduction from base to emitter of Q_1 and its base will be at a potential sufficiently high for the string of *p–n* junctions mentioned earlier to bring Q_2 and Q_4 into conduction and then into saturation. The gate output will be V_{CE} for Q_4 when it is in saturation, about +0.4 V. It is a LO, again the expected result for a NAND gate.

put. The new gates are useful because the outputs are opposite to those of AND and OR functions. The final new gate is an XOR gate, a design that gives the "exclusive or" function. (Refer again to Table 5.1 to compare this performance with the regular OR gate.) It is possible also to have an exclusive-NOR or *equality* gate. Its symbol would differ from that of the XOR gate by having a small circle at the output. An analysis of its behavior is called for in Exercise 5.10.

The several gates are compared in Table 5.3 to show that they perform sufficiently different functions so that all are useful: each gate gives a distinctive output pattern. Digital circuits of this kind that give a predictable output for a specified set of inputs are called *combinational devices.*

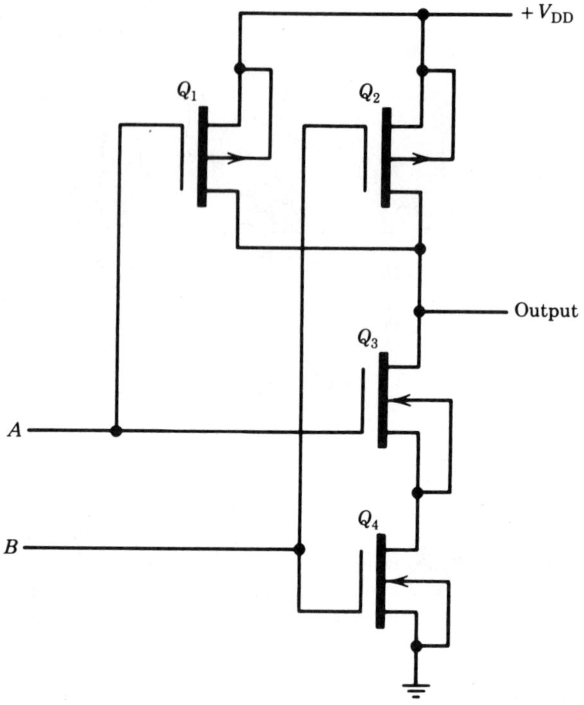

Fig. 5.7 A CMOS–NAND gate. Both polarities of enhancement-mode MOSFETS are part of any CMOS circuit. Circuit design takes advantage of this plurality. When both inputs of this gate are HI, Q_1 and Q_2 are nonconducting and Q_3 and Q_4 conduct. The expected LO is produced at the output.

Integrated Circuit Gates. Virtually all logic gates in use are designed as integrated circuits. Indeed, the scale of integration today can provide multitudes of gates on a single chip. In Table 5.4 chips commonly available for all the logic functions are listed for two logic "families." The table omits most chips with several inputs per gate and is intended to be illustrative rather than inclusive. It does provide symbols for both the regular HI-true system of gates and another not

Table 5.3 Comparison of Outputs of Several Kinds of Logic Gates

Inputs		*Outputs*					
A	*B*	*AND*	*NAND*	*OR*	*NOR*	*XOR*[a]	*XNOR*[a]
0	0	0	1	0	1	0	1
0	1	0	1	1	0	1	0
1	0	0	1	1	0	1	0
1	1	1	0	1	0	0	1

[a]The acronym XOR identifies the exclusive OR gate and XNOR the exclusive NOR or *equality* gate.

Table 5.4 Some Common Logic Gates in CMOS and TTL Logic Families[a]

Name	Statement	Symbol	LO-True Symbol	Part Number[b] CMOS	Part Number[b] TTL
AND	AB	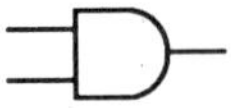	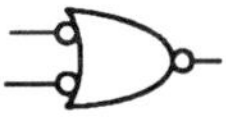	4081	7408
NAND	$\overline{AB}$			4011	7400
OR	A + B		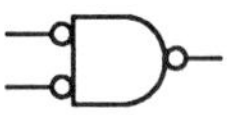	4071	7432
NOR	$\overline{A + B}$		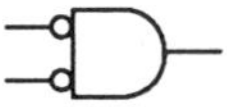	4001	7402
XOR	$A \oplus B$			4070	7486
XNOR	$\overline{A \oplus B}$			4077	74266
INVERT (NOT)	$\overline{A}$	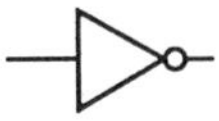	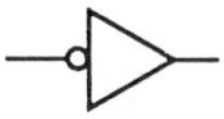	4069	7404

[a]Most gates are available with two, three, or four inputs and NAND and NOR gate chips with still more possibilities. Further, most are available in a low-power Schottky (LS) version that consumes much less (one-fourth) power and features somewhat more speed.
[b]Part numbers are given for ICs with 4 dual input gates per chip. All variations of basic gates in the 7400 TTL family are identified by insertion of letters into the number after its first two digits. For example, the low-power Schottky version of the NOR gate is listed as 74LS02. Other letters identify other subfamilies in the continuing upgrading of gate design. Integrated circuit packages of gates are commonly dual-in-line packages (DIP) which are rectangular in cross section (0.75 × 2 cm) and have 7–10 or more sturdy metal leads protruding downward along each side.

mentioned before, the LO-true family.* Note that LO-true logic gates are identifiable by *circles at the inputs.* Numbers for different types of gates have been standardized and are given in Table 5.4 as well. For example, the 74-series devices are in the widely used transistor–transistor-logic (TTL) family. New designs in the family are commonly identified by insertion of capital letters following the "74," as discussed in the table.

The translation of combinational logic concepts onto semiconductor chips is still an advancing technology. The emphasis in design is upon decreasing the size of circuit elements (present resolution is about a micrometer), increasing speed (present switching times are a few nanoseconds), and reducing power consumption

*LO-true devices give a LO output when all conditions for a gate or set of gates are met. Such devices are also known as "negative-true" gates and are described as obeying "assertion-level logic." In using LO-true devices it is helpful to label inputs and outputs with overbars and show the polarity of output signal.

per gate (presently as low as 1 mW though more representative values are 5–10 mW). Historically the main families of digital logic have been resistor–transistor logic, which is now obsolete, diode–transistor logic, which is seldom seen today, transistor–transistor logic (TTL), which is in widespread use, complementary and negative metal–oxide semiconductor logic (CMOS and NMOS), which are also in widespread use, and emitter-coupled logic (ECL), which finds applications in those specialized situations where ultrahigh speed is important.

May gates from different logic families be coupled? Yes, if differences in characteristic currents and voltages in the families are taken into account. For example, even though both TTL and CMOS gates operate with a supply voltage of +5 V, a TTL input in the LO state provides considerable current that must be accommodated by the device driving the gate. By contrast, a CMOS gate has nearly zero input current. Thus, coupling does require care. In general it is preferable to work with gates all in a given logic family.

Performance differs between families. TTL gates are faster but CMOS gates consume considerably less power. The power consumption of CMOS gates does rise as the frequency of switching (number of operations performed) goes up.

Open-Collector and Three-State Devices. In the majority of applications of logic gates, signal outputs are held HI or LO by a bipolar or MOSFET transistor. The term "active pull-up" describes this circuit arrangement. In some cases, such as connecting outputs to a data bus in a computer, it proves advantageous to redesign both TTL and CMOS gates: what they need is an open or LO output.

In TTL logic this can be supplied by simply having an open-collector output like the one in Fig. 5.8. It is basically the NAND gate of Fig. 5.6 in an open-

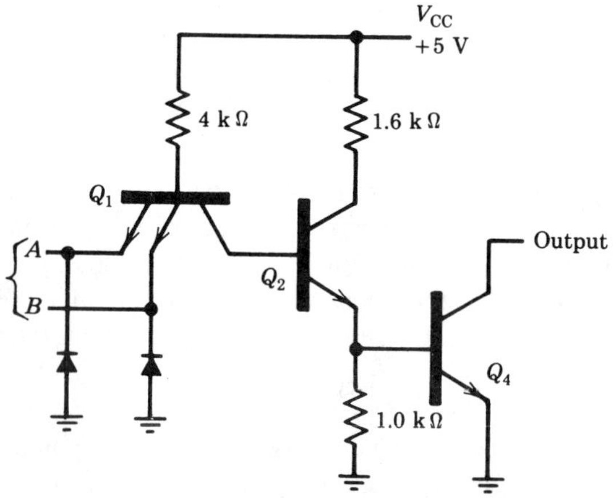

Fig. 5.8 A schematic diagram of an open-collector TTL-logic NAND gate. This circuit differs from that in Fig. 5.6 in two basic respects: the pull-up transistor Q_3 is missing and the inputs are no longer floating—each has a reverse-biased diode to keep it from going more negative than −0.5 V. The output of this NAND gate is simply *open*. It is employed in circuits that supply either a logic HI or LO as appropriate through a small external resistor (a pull-up resistor) to the collector of Q_4. This gate output is expressed or asserted only when it receives a HI at its output.

collector version. In this circuit the collector provides a LO response until an external connection to the output brings the potential up to a logic HI, as explained in the legend.

In the CMOS family this need was met by development of the *three-state gate*, a device that has a neutral state in addition to the two normal HI and LO logic states. The neutral state offers an open circuit and is made available when activated by an appropriate signal; the output still has the advantage offered by the "pull-up" output. As a result, three-state CMOS gates are often used in place of open-collector TTL devices in circuits related to microprocessors.

Example 5.6. A good illustration of use of three-state and open-collector gates is provided by the relationship maintained between logic gates and data buses in microprocessor systems. A data bus connects microprocessor, memory, and peripheral devices such as instrument modules and display units so that the microprocessor can be accessible to all devices. When all devices are attached to the bus, however, it is essential that not more than one impose its signals at any given time. That time is known because of receipt of a code signal and address identifying the particular device. Once addressed, a measurement device is able to draw the logic HI that it needs at its output in order to place its own information on the line. At all other times, however, it is advantageous to keep power drain down by having its gate output neutral, that is, with open collector or in the neutral state of three-state logic.

5.5 MULTIPLEXERS, DECODERS, AND OTHER COMBINATIONAL DEVICES

How easily can gates be combined in patterns that will make possible complex combinational tasks? Recall that a *combinational task* is one in which a given set of inputs will yield an entirely predictable output. It is the very predictability and precision with which the outcome occurs that makes such devices valuable. Recall the example of coupling a NOT and an AND gate to make a NAND gate. To make more complex arrays either discrete gates can be combined or a composite design can be laid as an integrated circuit. Some important combinational arrays will now be considered.

Multiplexers. A multiplexer is basically an integrated circuit switch which makes a connection between any of a large number (2–16) of inputs and a single, or perhaps two, outputs. A functional diagram of an eight-input TTL multiplexer chip is shown in Fig. 5.9. This multiplexer and others operate on receipt first of a binary code termed an *address* that indicates the particular input to be connected to the output followed by receipt of an enabling LO signal at the E input. Note that with a three-bit address any of eight inputs can be selected. In the figure the address inputs are labeled A_0–A_2 and the E input is also shown. The enabling pulse is sometimes termed a strobe pulse; if it must be a logic LO instead of a HI, there will be a circle at the input.

Not only are IC multiplexers inexpensive but their other advantages over mechanical switches are impressive. They switch nearly instantaneously and avoid

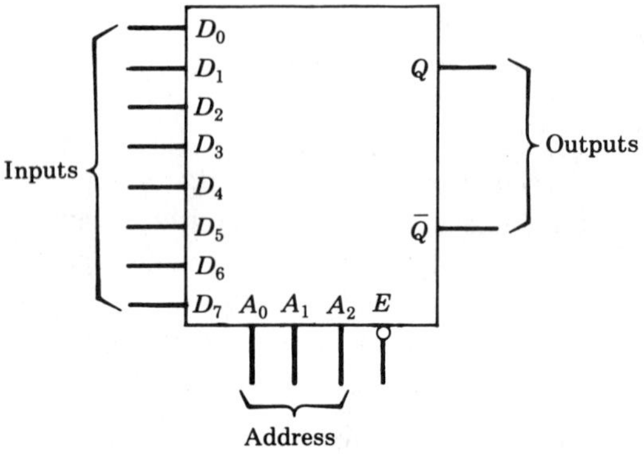

Fig. 5.9 A pin diagram of a 74LS151 Schottky TTL 8-input multiplexer. Note the data inputs D_0–D_7 and address inputs A_0–A_2. E identifies an enable input. The circle at E signifies that a LO pulse enables the device and *vice versa*. The outputs Q and $\overline{Q}$ are complemented, that is, when one is HI, the other is LO. When the multiplexer receives an address and a LO input to E, it will decode the address and connect the appropriate input from the set D_0–D_7 to the output, causing Q to go HI and transmission to take place.

the necessity of any external connections with all their susceptibility for pickup of noise and distortion. It is also straightforward to generate the address bits for a multiplexer in a digital circuit as we shall see in connection with microprocessors in the next chapter.

Both TTL and CMOS types of multiplexers are available. One type of CMOS multiplexer is strictly *digital*, as are all TTL multiplexers, and generates a digital output. Another type, however, offers bidirectional transmission gates and can *transmit analog signals* also. Clearly, since capabilities of such devices can vary considerably, it is essential to consider needs fully before setting up a circuit.

Is it possible to couple two 8-input multiplexers to obtain a 16-input device? The answer as shown in Fig. 5.10 is unequivocally yes. As the legend for the figure explains, their coupling is a straightforward process. Note the way in which simple logic gates help direct signals. An OR output gate is added at the output to ensure that the outputs of the two chips do not interfere. If three-state CMOS chips were used instead, they could be connected without the OR gate and a single output gate employed.

Decoder. An interesting but quite different combinational device termed a *decoder* is shown in Fig. 5.11. It is especially likely to appear in a system using a form of digital representation known as binary-coded decimal (BCD). This device, one stage of a decoder, translates or decodes a 4-bit binary as a decimal digit. In the BCD system, a decoder examines each 4-bit binary number received and determines which one of 10 different outputs, each standing for one decimal digit, should go LO. Its output might be used to drive a decimal printer, for example. A chip is also available for directly decoding binary numbers to obtain an LED (light-emitting diode) decimal display (see Fig. 5.27).

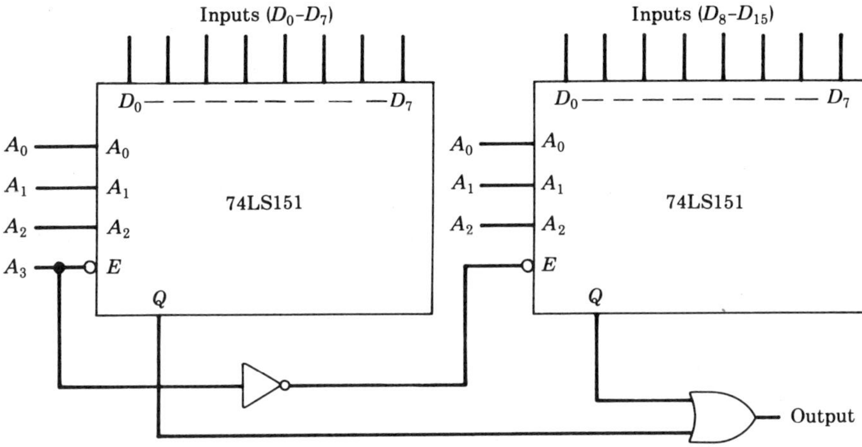

Fig. 5.10 A pin diagram of a 16-input multiplexer developed from two 8-input 74LS151 devices. In this case 4-bit addresses are needed since the 16 different inputs must have unique addresses. Note that the most significant address bit enables the appropriate chip and that address lines A_0–A_2 are connected to both chips. The circle at the E input indicates a LO bit is enabling. The left-hand multiplexer must handle inputs 0–7; it is enabled when A_3 is 0. At the same time, the inverter sends a HI input to chip 2, disabling it. The output of each chip goes to an input of an OR gate, which acts as a buffer. When the OR gate receives a HI from the Q of either chip, it transmits.

Other Combinational Devices. Chips that are the inverse of multiplexers, that is, which direct single inputs to one of many outputs see considerable use also. They are known by the somewhat infelicitous name *demultiplexers*. Some other combinational devices that are available are *adders*, other *arithmetic chips*, and *multiplier chips* (up to 16 bits times 16 bits). There is also the important combinational chip called a *magnitude comparator*. It provides an output that indicates whether two 4-bit input numbers, A and B, stand in the relationship $A > B$, $A = B$, or $A < B$.

A final example is a *parity-checker chip*. To ensure accurate transmission of binary words it is common to use a code bit with each word that causes the whole to have even or odd *parity*. A *word* is simply the conventional cluster of bits for a particular system, usually 4, 8, 16, or 32. Normally, a word is the number of bits that are processed together. By definition, for odd parity the number of 1 bits in each word is odd, and for even parity, even. The checker examines each incoming word for parity and asks retransmission of any that fails to meet the test.

Read Only Memory (ROM). With the advent of more elaborate techniques for making chips, it became possible to include a large number of combinational operations in read-only memories (ROMs). It is appropriate here to introduce the digital aspects of ROMs. In these super-combinational devices a single input provides an output that is a string of bits. For example, the sequence of bits may be an instruction for operation of an instrument module. Once an input/output table

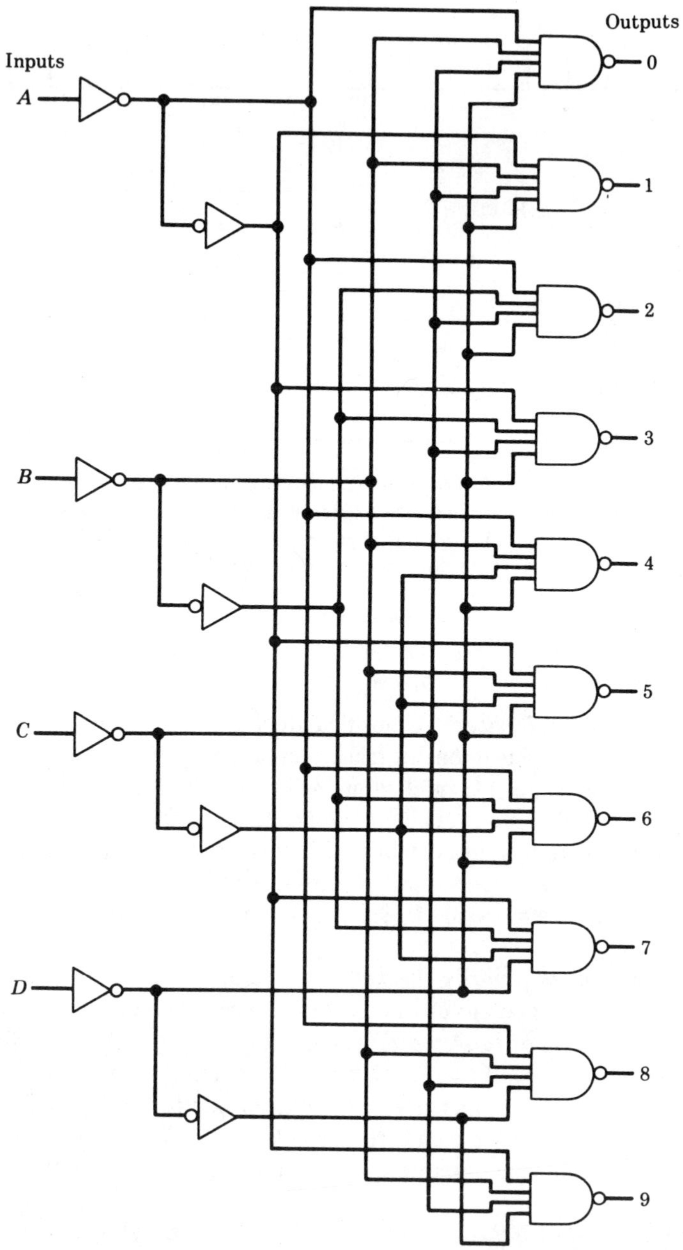

Fig. 5.11 A schematic of gates used in the TTL 7442 binary-coded decimal (BCD) decimal decoder. Inputs *A–D* receive least-to-most-significant bits, respectively, of binary-coded decimal input. This device furnishes a LO-true output. Note that the INVERT gates used with each input first produce a complement of the asserted bit and later reproduce the original bit. For every 4-bit binary input the appropriate NAND gate goes low to show the equivalent decimal digit. Circuitry ensures all outputs are HI (off) for any possible invalid input bit combinations, those that correspond to decimal numbers 10–15.

has been worked out for a module, groups of locations in ROM memory can be programmed with its essential control instructions. At the moment when each set of locations is accessed, it can be expected to provide the necessary digital signal.

Finally, programmable logic arrays also exist. In this case a chip provides many gates which the user can program by fusing internal links to connect them. These are like programmable read-only memories (PROMs). They provide more elaborate arrays of gates than would be available otherwise.

5.6 FLIP-FLOPS, MONOSTABLES, COUNTERS, AND OTHER SEQUENTIAL DEVICES

While the combinational operations provided by logic gates form one main pillar of digital operations, another is the predictable *sequential operations* provided by a different kind of digital circuit. In electronics terms, a sequential digital device must be based on a circuit whose voltage output reflects both the current input and the input that just preceded it. Such sequential devices can be expected to be indispensable along with combinational devices in the design of arithmetic–logic units in digital computers and for a variety of simpler devices such as counters, registers, and data latches. In arithmetic terms recall that operations such as adding and subtracting are procedures that call for a limited type of memory, that is, holding one datum and combining it in a predictable way with another.

The basic sequential device, a flip-flop, achieves this limited type of memory and the possibility of combination by providing extensive feedback between a pair of logic gates.* The name flip-flop is descriptive; the circuit has two stable output states and can be triggered from one to the other predictably. One of the simplest flip-flops, the *RS* type, is shown in Fig. 5.12. When a logic 1 is applied to *an*

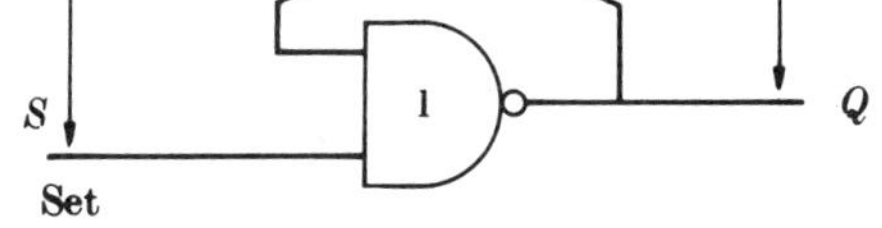

(a)

Truth table

Set	Clear	Q	$\bar{Q}$
0	1	1	0
1 (0)	1	1	0
1	0	0	1
1	1 (0)	0	1
0	0	1	1

(b)

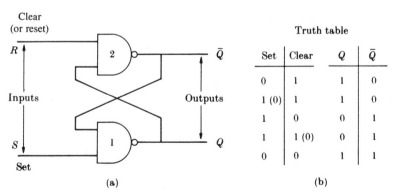

Fig. 5.12 Flip-flop of the *RS* (reset–set) type based on NAND gates. The unit is jam-loaded, that is, forced into one output state or the other. Only LO pulses at the input cause the output to flip from one stable state to the other. HI pulses cause no change. (a) Block diagram. (b) Truth table.

*Actually, there are many ways to construct a flip-flop. This type of device dates back to the early days of electronics. A less colorful name for it is bistable multivibrator.

appropriate input, each successive application of logic 1 and 0 to the other input produces a change in output. As a *bistable* device, it is ideally suited to the binary operations of digital electronics.

As the figure shows, the *RS* flip-flop consists of two cross-coupled NAND gates. Its operation can best be described by going line by line through the device truth table that shows all possible input and output states. First, assume the *set* (S) input is LO and the *clear* (R) or reset input is HI. Because one input of gate 1 is LO, output Q must be HI. Both inputs of gate 2 are HI and its output $\overline{Q}$ must be LO. This situation corresponds to the top line of the truth table. If the set input is now changed to HI, the output does not alter because the other input of gate 1 is still LO. (Its two inputs are HI and LO, still calling for a HI output.) The inputs to gate 2 are both HI, calling for a LO output.

When S is HI, however, the application of a LO pulse to the clear input R (third line of the table) causes the flip-flop to change state. (When the LO is applied, the inputs to gate 2 are HI and LO, causing its output to become HI. Since this change makes both inputs of gate 1 HI, its output goes LO.) If the clear input is next changed to HI (line four of truth table) the output does not alter. The reader should verify this situation and examine the change on returning to line one when the set input is next made LO.

The most important aspect of the flip-flop's behavior can now be deduced. Lines two and four of the truth table show its memory: its output remains fixed when either input is changed from LO to HI and shows which input was previously LO.

Nevertheless this simple circuit leaves much to be desired. One problem involves the timing of pulses. In the *RS* flip-flop an output transition tends to occur so quickly that it may alter the effect of an input signal. This property causes the RS type to be called transparent, since outputs follow inputs so quickly. Further, inputs of the *RS* type are called *jam-type*. Some way is needed in a flip-flop to hold an output state once it is reached. Further, a way to hold input states briefly is also required. They should be transferred to a flip-flop only after it has responded to a previous signal. If these changes could be made, new states would be transferred to a flip-flop only *after* it had responded to a previous input signal.

A second problem is that the output does not register the receipt of every pulse. From the truth table in Fig. 5.12 note that the *RS* flip-flop will respond to every zero pulse that arrives only when (a) both inputs R and S are HI and (b) the incoming pulse is directed to the proper input.

To avoid the first difficulty, most flip-flops are designed to be *edge-triggered*. Regular square clock pulses, which are generated by a crystal-controlled oscillator, are directed to an input gate in these flip-flops to trigger transitions. As one sees again and again, complex digital devices operate in the most orderly fashion when triggered by the arrival of regularly spaced square pulses at a clock input gate. This is certainly true for flip-flops, and digital computers operate in this manner also. In addition, a central timing device can synchronize changes of logic state throughout a complex of circuits. Such a *clock* is ordinarily a crystal-controlled oscillator that puts out a square wave of precise frequency (precision of one part in 10^6). Sometimes the clock is a separate (external) device. In any event, its standard signal is invaluable in providing control.

Since a pulse has two edges, logic gates can be designed to respond to either its positive (leading edge) or its negative edge (trailing edge). A wedge symbol is placed inside a chip diagram for each input for which a device is triggered on the leading edge (see Fig. 5.13b).

A positive-edge *D*-type flip-flop is shown in Fig. 5.13. In this design it is still possible to have the *RS* jam-type inputs. In the legend of the figure the operation of the device is discussed.

A *JK* flip-flop, whose operation is similar to that of the *D* type, has two data inputs. The device is a master–slave flip-flop in which inputs move from a master flip-flop to a slave flip-flop only on an edge of a clock pulse. Its pseudo pin diagram and truth table are given in Fig. 5.14. An interesting result obtains when both data inputs are HI. In this case the output "toggles" or alternates its level with each *positive edge* of a clock pulse. As a result, pulses at the output occur at half the clock frequency: the flip-flop divides the frequency by two. Incidentally, if the *D* flip-flop of Fig. 5.13 had its $\overline{Q}$ output coupled to its *D* input, it would also toggle.

Timing. With an eye to using a variety of combinational and sequential devices in a complex array, questions of switching times for transitions become important.

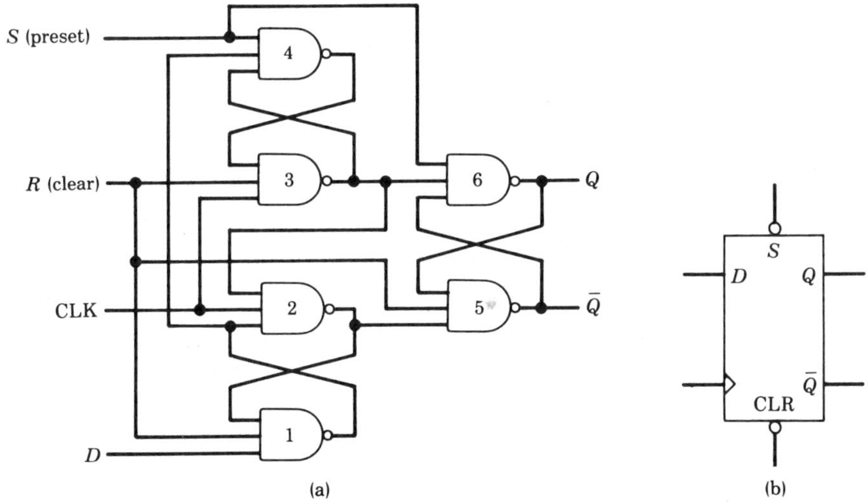

Fig. 5.13 A D-type, edge-triggered, flip-flop circuit based on NAND gates. (a) Schematic diagram. (b) Pin diagram of chip. Symbols used in the chip diagram have the following meanings: the wedge means edge-triggered, the small circles mean inversion. Thus this flip-flop is positive-edge triggered and its jam-inputs *R* (here labeled CLR) and *S* are LO-active. The circuit contains flip-flops based on gates 1/2, 3/4, and 5/6. When the CLK (clock) input is LO, outputs of gates 2 and 3 are HI, and *Q* and $\overline{Q}$ are stable. Assume *R* and *S* are HI for normal operation. When the CLK input is HI, outputs of gates 1 and 2 depend on input *D*, gate 2 taking the level of *D*. Further, since the output of 1 is an input to gate 4, its output takes the level of *D*. On the LO–HI transition of CLK, gate 2 transmits the *D* level to 5 and gate 3 transmits a $\overline{D}$-level to 6; at output *Q* the *D*-level also appears. Any later change in *D* has no effect until the next positive-edge of a clock pulse. LO levels at either *R* or *S*, since they are direct inputs to gates 5 and 6, take precedence over CLK and *D* inputs and establish the levels at *Q* (and $\overline{Q}$ immediately).

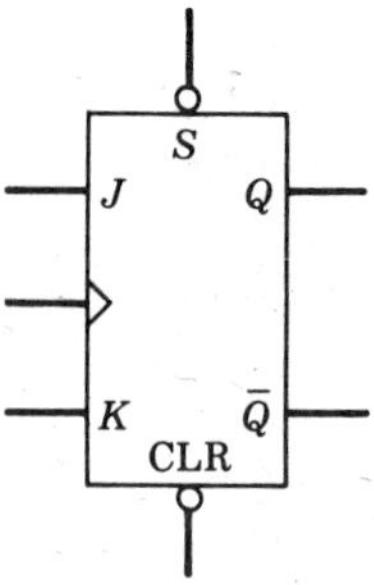

(a)

J	K	Q
0	0	no change
0	1	0
1	0	1
1	1	toggling

(b)

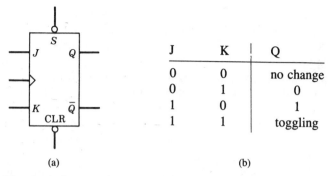

Fig. 5.14 (a) Pseudo pin diagram and (b) truth table for a *JK* flip-flop. The $\overline{Q}$ output is the complement of *Q*.

It is only necessary to return to the *D*-toggling arrangement just discussed to pursue this aspect. With the $\overline{Q}$-output coupled directly to the *D* input, the particular time at which a response is made to the *D* input becomes critical in terms of whether the flip-flop will behave erratically. With all flip-flops there is a specified set-up time (an interval during which data at inputs of gates must be stable for an accurate sampling to be made on the edge of the clock pulse). Earlier devices had a hold time of as much as 5 ns for the edge-triggering. Clearly, there is also a propagation time having to do with a signal moving from one device to another. It can be expected that timing will need close attention in any array.

Application: Relay Debouncing. When relays that can respond to high-frequency signals are used, it is observed that the metal reed closing the circuit is so flexible as to oscillate and make and break the connection a great many times during the first millisecond of closure. It is apparent that in a system of logic gates "havoc will be wreacked" by feeding in the voltage spikes from this type of behavior. The problem can be alleviated by introducing a flip-flop, either of the *RS* type or one that is more elaborate. A suitable circuit is shown in Fig. 5.15.

This circuit lets the NAND gate at the upper right transmit data pulses whenever "enabled" by a HI output at *Q*. Instead of relying on the bounce-prone relay to provide a HI, the relay now does nothing but change the output state of the flip-flop on the basis of its initial closure. Clearly, the flip-flop serves as buffer to ensure that the bouncing has no effect beyond the reed.

Enabling pulses will reach the NAND gate when the relay is in the up position. The S input is then grounded. Simultaneously the *R* input assumes a logic HI and *Q* goes HI. Note in the timing diagram (Fig. 5.15b) how "clean" the *Q* input is compared with the voltages at the *R* and *S* inputs.

Monostables. The monostable, also called a monostable multivibrator or *one-shot*, differs from flip-flops in having a single stable state together with a metastable, opposite-logic state. These circuits are valuable for timing since the duration of the metastable state is reasonably constant ($\pm 1\%$). When a monostable is forced

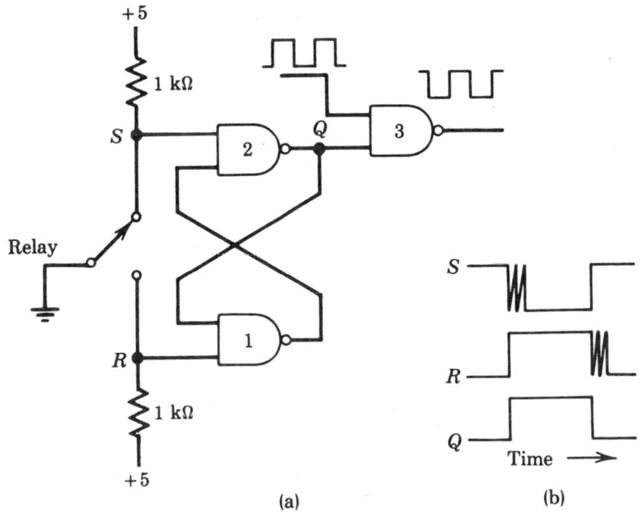

Fig. 5.15 A switch debouncing circuit. Debouncing is secured by employing an *RS* flip-flop. (a) Schematic diagram showing how the flip-flop is inserted between the relay and NAND gate 3 that the relay is to activate. NAND gate 3 transmits data pulses from its upper input when enabled by a HI at its lower input. The HI is furnished by the relay, which is debounced by the flip-flop consisting of NAND gates 1 and 2. (b) Timing diagrams showing voltages at inputs *S* and *R* and output *Q*.

into the metastable state by a trigger pulse, it will "relax" to its stable state after a reproducible time. Capacitor charging in a stable *RC* circuit often provides that interval. A variety of monostables are available in integrated circuit form.

In Fig. 5.16 a NOR gate monostable is shown schematically. Inspection of the diagram shows that the function of the trigger is to discharge capacitor *C*. This device and similar ones require that the trigger pulse exceed a minimum duration for the monostable state to be reached. Once the capacitor has discharged, there is a logic LO at the input of gate 2, causing its output, which is also the monostable's output, to switch to a HI. That output will endure until the external supply voltage recharges *C* to a voltage sufficient to return the input of gate 2 to HI status and cause the device output to fall LO again.

In Fig. 5.16b a partial diagram of an integrated circuit monostable is shown. In this case the trigger is applied to either input *A* or *B* of an AND gate that is connected to the clock input. Note also that the timing circuit is external to the chip. Otherwise the *IC* functions like a monostable made from discrete components.

Good *IC* monostables will provide intervals reliable to about $\pm 1\%$ if quality external components are used. Other widely used ways of generating timing pulses will be explored in the following section.

Pulse-Height Discriminators. Two Schmitt trigger circuits (see Section 4.9) and an exclusive-OR gate can be used to develop an *anticoincidence* circuit for pulse–amplitude discrimination. The circuit is shown in Fig. 5.17 and its behavior is explained in the legend.

Counters. It is time now to examine some sequential functions that may be performed by groups of flip-flops. Cascading several flip-flops to form a binary counter was perhaps their earliest application in digital electronics, and it is still a valuable one. Such a cascade acts as an accumulator that tallies at its outputs bit-places that give at each instant the total number of pulses or events that have been received at the input of the first flip-flop.

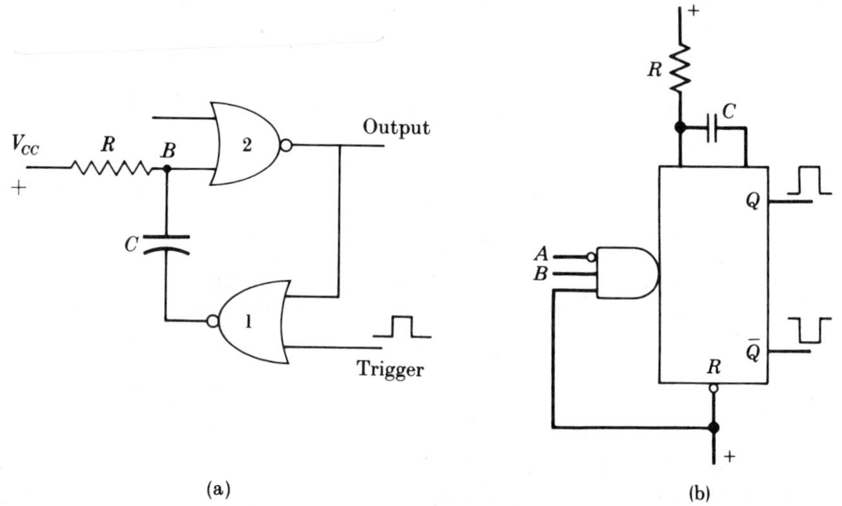

Fig. 5.16 Two versions of a monostable multivibrator or one-shot. (a) Schematic diagram of a NOR gate monostable. The output of gate 1 is capacitively coupled to the input of gate 2. The duration of a pulse at the output is about *RC* seconds. On receipt of a trigger signal (a 25–100 ns LO–HI pulse) outputs will change levels. In the process, capacitor *C* will be discharged. The levels at *Q* and $\overline{Q}$ exist for a period $\tau = RC$ during which *C* recharges sufficiently to recover the original logic level and return the monostable to its stable state. (b) An IC monostable with external resistor *R* and capacitor *C* forming a timing circuit. Its behavior parallels that of the NOR gate version.

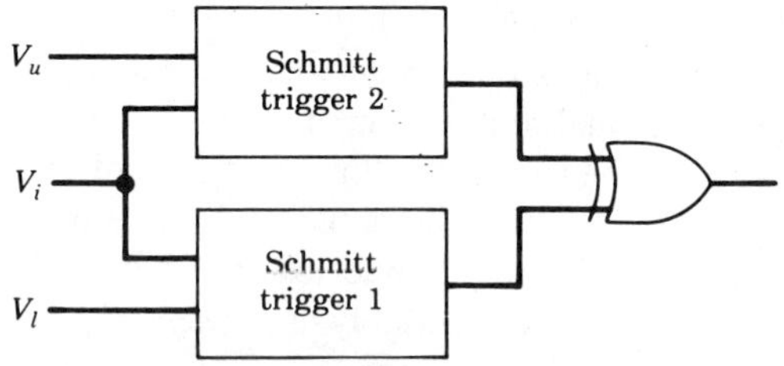

Fig. 5.17 A pulse-height discriminator of the window type. The firing voltages of the two Schmitt triggers are set to the upper V_u and lower boundaries V_l of the selected voltage window. Pulses whose height is below the lower edge of the window will trigger neither Schmitt and the output of the XOR gate will be LO. A pulse of amplitude greater than the top of the window will cause both devices to trigger, again giving a LO output. Only a pulse whose amplitude is within the window will trigger only one Schmitt and give a HI output. This device is said to operate as an *anticoincidence* device.

To make a counter of a flip-flop cascade it is only necessary to couple the Q output of each flip-flop to the clock input of the next. In Fig. 5.18a a cascade of four JK flip-flops is shown operating as a 4-digit binary counter. If all other inputs are at appropriate levels, each flip-flop will change its output on the arrival of the descending edge of a pulse at its input (note the wedge at each clock input symbolizing edge-triggering as well as the circle standing for descending edge).

The logic level at each Q output following the receipt of a pulse at the clock input of A is diagrammed in Fig. 5.18b. As a counter the output of flip-flop A registers the least significant bit, B the next more significant bit, and so forth. A study of the input/output diagrams in Fig. 5.18b shows that the output of B changes only half as often as that of A and that of C only one-fourth as often, and so on. It is because the factor 2 is involved that the output of each flip-flop registers one digit of a binary counter.

How can digital counters be made to register in other number codes? Two to n flip-flops can be grouped in clusters that correspond to one digit in the code of

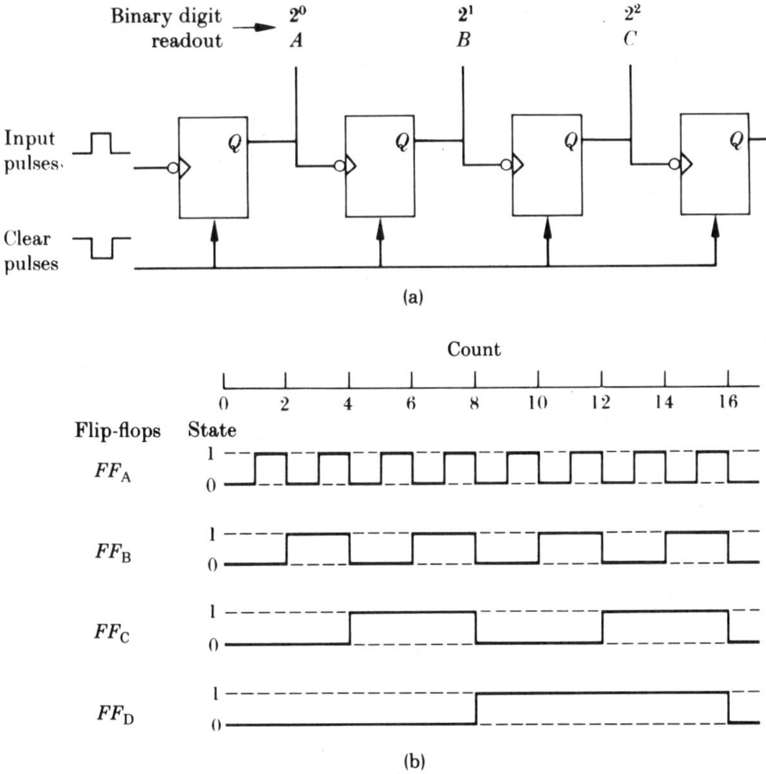

Fig. 5.18 (a) A 4-bit binary counter based on four JK flip-flops. The output state of each flip-flop is registered at A, B, C, and D. Pulses move through the counter by "ripple carry." J, K inputs (not shown) are kept HI or 1 for counting. (b) Output waveforms for the flip-flops that comprise the counter. Note that each flip-flop toggles only on the descending edge of a pulse.

interest and then attached to a decoder to translate the output into the new system. For example, to count in the octal code groups of three flip-flops are appropriate.

Decimal counting, however, requires groups of four. Perhaps the best known design is one called the *binary-coded decimal (BCD) counter*, which was introduced earlier. A pseudo pin diagram of a chip with a cascade of flip-flops that is interconnected to give the counter is shown in Fig. 5.19. Actually, this design is a dual-BCD counter with the top four flip-flops supplying one decimal digit and the bottom set a second. The interconnection must also cause all outputs to be cleared as soon as ten counts have been received. As discussed in the figure legend, this counter is unusually versatile since an external connection is made between Q_A and input $CLK_{B\text{-}D}$ in each of the four-bit counters and it offers other possibilities than BCD counting. It is also necessary to make an external connection between Q_D of the top set and CLK_A of the lower set.

How fast are digital counters? A striking aspect of flip-flop triggering is that a great many output transitions appear to be triggered simultaneously. Yet in fact each stage is clocked by the output of the previous stage. As a result, some time is required for the level change to ''ripple'' through a whole series. The *ripple counting* mode of propagation is unnecessarily slow. A further drawback at high rates of counting is that transient states sometimes appear and may cause spurious readings. For example, in going from a count of 3 to 4, transient states of 2 and 0 appear. In an alternative mode of coupling, *synchronous counting*, flip-flops are clocked simultaneously with a considerable gain in speed. The mode has the further advantage that no *intermediate states* appear before the final one.

Devices that count up to a number *n* and then return to zero are known as *modulo-n* devices. For example, a modulo-8 device would go from 0 through 7 and then return to zero. Decoding, that is, interpreting the output in the new code will be discussed shortly.

Many different types of counters are available as integrated circuit chips. In the TTL series there are four-bit ripple and synchronous counters available for BCD

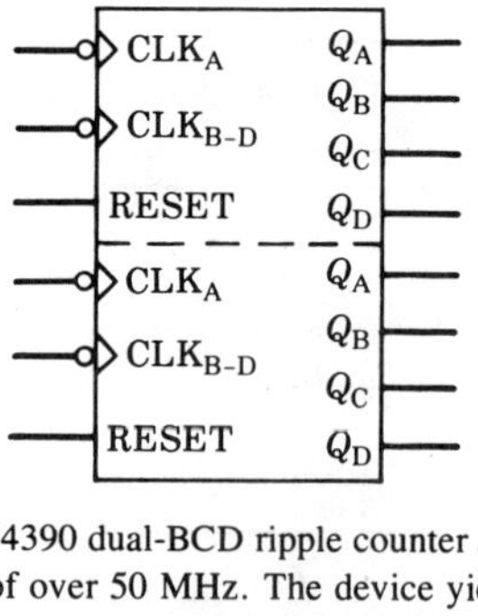

Fig. 5.19 Inputs and outputs of a 74390 dual-BCD ripple counter available in regular and low-power Schottky versions with count rates of over 50 MHz. The device yields two decimal digits and is also called an 8-bit counter. (It is possible, however to use flip-flop *A* independently of flip-flops *B*, *C*, and *D*. For example, the latter can serve as a scale-of-5 counter. In this case, separate use of the two clock inputs is required.) Each group of four flip-flops is termed a *decade counting unit* and is wired as a 1-2-4-8 counter: thus outputs for given binary inputs are: $1000 = 8_{10}$, $1001 = 9_{10}$, 0000 plus a carry of $1 = 10_{10}$. The comparable chip with an 8-bit binary counter is a 74393. Counters can be cascaded to expand a display to still more digits.

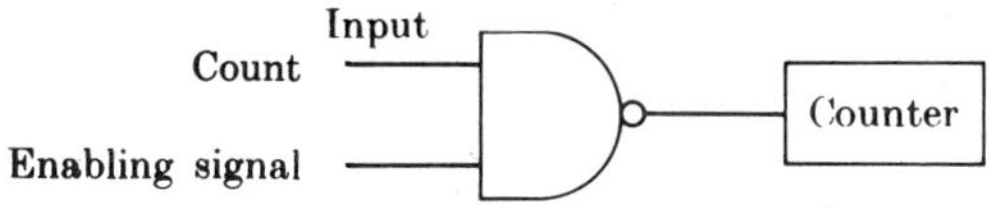

Fig. 5.20 Gating arrangement for a counter. The NAND gate transmits pulses only when the enabling input is HI.

and binary counting as well as 8-bit counters. In CMOS circuitry there are binary counters with 7, 12, and 14 stages as well as 24 stages. Further, counters, whether of the ripple or synchronous type, can always be cascaded to secure still more stages. Modulo-*n* devices are available for a range of *n* values.

Up/Down Counters. All counters that have been discussed fall into the category of up-counters: they count from 0 upward. A down-counter can be developed with ease. For example, to convert the binary counter of Fig. 5.18a to a down counter, the $\overline{Q}$ output of each flip-flop is simply connected to the clock input of the next one. Both types of counters are available on chips as well as up/down counters which have separate inputs for the two modes of counting.

Gating of Counters. In order that counting may be started at the time a measurement is to be made, it is necessary to gate the input. A simple gating arrangement based on the use of a NAND gate is shown in Fig. 5.20. The gate is "enabled" by a HI input. As long as the gate is enabled, the count passes through the gate to the counter.

It is also common that the count and control signals come from different origins. In this instance, the actual source of counts may be quite unsynchronized with enabling signals, which are likely to derive from an internal clock. In this case a circuit called an *asynchronous gate* is needed to ensure that only complete pulses are passed by the gate to the counter.

5.7 LATCHES, REGISTERS, AND DEVICES FOR BUS COMMUNICATION

In the ephemeral world of digital electronics in which device outputs so often change in accord with inputs, what kind of circuit can be developed to ensure that output data can be held for a time? This need is a common one; some examples are: (a) the output of a binary flip-flop counter must be available for a brief interval in order to be decoded in another number system by a second digital device; (b) the output of a decoder must be held sufficiently long for it to be displayed; (c) experimental data may need to be held until they can be appropriately processed. Probably the example most important for present concerns is that in an instrument with an embedded computer data that become available from an instrument module must be held by an output device before they can be placed on the bus that will take them at the appropriate time to a central processing unit. Data latches provide these very necessary functions.

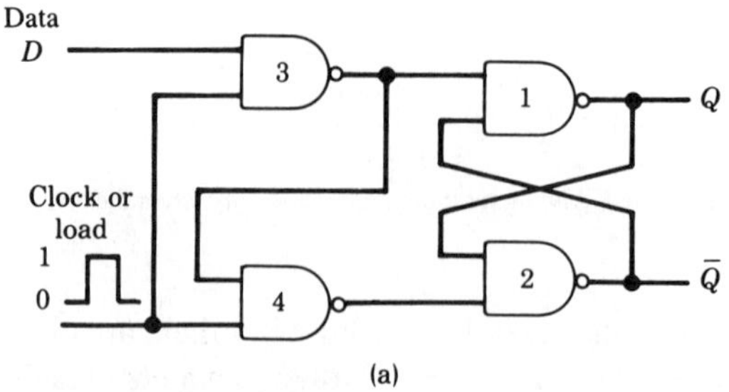

Inputs, t_n D	Output, t_{n+1} Q
0	0
1	1

(b)

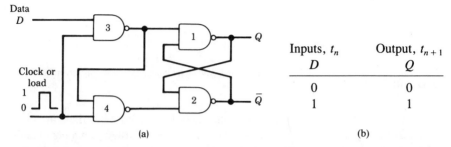

Fig. 5.21 The basic TTL data latch and its input/output table. This device is also called a transparent latch because when the clock or enable input is HI, the output changes directly (without a triggering edge) in response to the level at the *D* input. When the clock input is LO, the circuit latches or holds the last output even though the level at *D* changes. (a) The 1-bit gated data latch consisting of an *RS* flip-flop (NAND gates 1 and 2) and access gates 3 and 4. Clock and *D* inputs are marked. *Q* follows changes in *D* only when the clock input is HI. (b) Input/output table for the gated latch.

Data Latches. The simplest latch is just the gated *RS* flip-flop shown in Fig. 5.21a. Note that gates 1 and 2 form an *RS* flip-flop but that access is controlled through gates 3 and 4. As shown in the input/output table, the levels at *D* are transmitted to the output when the clock or load input is HI. When it is LO the output of the latch holds at the last reading obtained. Even though the *D* input will change subsequently the latch will hold that output reading. This type is called a transparent latch and is the kind ordinarily meant when the term data latch is employed. Its name derives from its direct-reading aspect, that is, jam-loading of the data onto the flip-flops. In order to latch a 4-bit or larger number a series of stages is needed. Chips provide such latches.

A simple example of the use of a data latch is shown in the Fig. 5.22. In this dual BCD counter arrangement, on receipt of a HI pulse the latch copies the output of the counter and holds it. The output is read by a BCD decoder that operates 7-

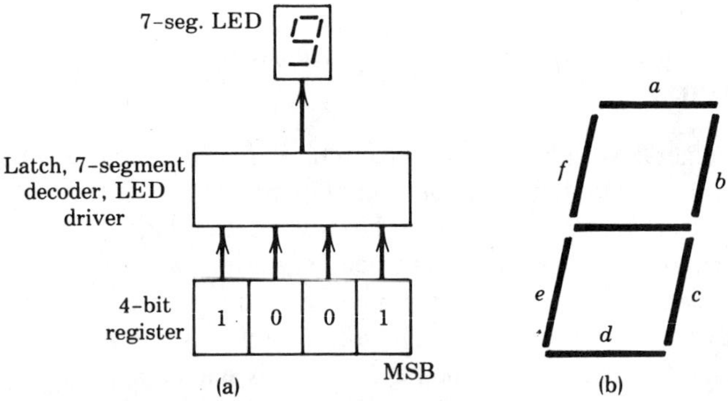

Fig. 5.22 Some components necessary to display a decimal digit. (a) The digit 9 is shown in a 4-bit BCD register. The bits are transferred in parallel fashion to a special chip that latches and then decodes them to drive a 7-segment LED display. (b) The seven segments of an LED digit are identified for an 8, which requires that all segments be lit.

segment LEDs so that the ultimate display is of the decimal equivalent of the total count. Transparent or regular latches have the virtue of allowing counters and other devices to operate without interference. Another example of the use of latches is given at the end of the section.

Registers. The type of register used in a microcomputer is very similar to a latch. For example, a set of D flip-flops can constitute a register. In the register chip shown in pin-diagram form in Fig. 5.23, four D flip-flops have been put on a chip. All "clock" inputs and all "set" inputs are connected so that there is a single input pin for each set. Since a register serves to store data, especially for computational or control purposes, it is often involved in additional functions.

Shift Registers. It is only necessary to take the chip discussed as a D register and arrange that each Q output drives the next D input and also clocks all of the flip-flops simultaneously as the chip would do, to use it as a shift register. What actually does it do? The data at the register outputs are shifted one stage to the right with each clock pulse. Any data present at the first D input is entered in the first flip-flop as well. In other words, if the initial levels at flip-flop outputs is 1001, and there is a 0 at the input of flip-flop A, after one clock pulse the outputs will show 0100, at the end of a second, 0010, and so forth.

Shift registers are employed to move data that is being held in a latch onto a bus in serial fashion. For example, a shift register can capture parallel data from a BCD counter and then output the data serially. Or a shift register may be wired to acquire serial pulses from a keyboard and later output them to a microprocessor in parallel fashion. Chips for each purpose are available.

Eight-bit shift registers are common and a few types are capable of operating with a shift to the right or to the left as determined by external control. Some shift

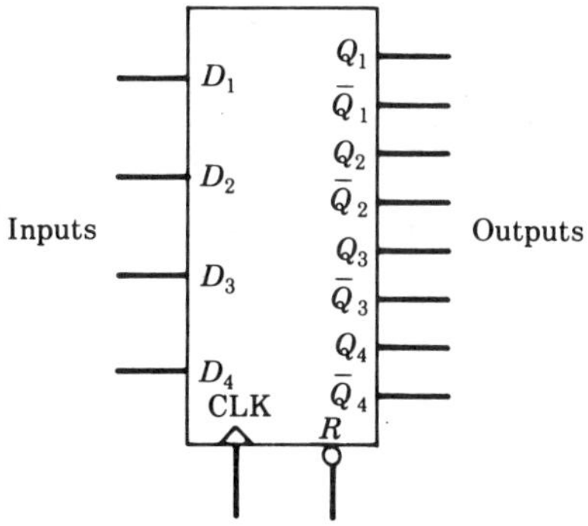

Fig. 5.23 Pseudo pin diagram of a 4-bit D register (74LS175) with regular or true Q outputs and complemented $\bar{Q}$ outputs. Though the chip contains four D-type flip-flops, a common "set," "clear," and clock input serves all four saving pins. Their D inputs *are* separate, of course. Often such a register is termed a latch because it holds the bits at its output, but the term is more correctly applied to what is often called a transparent latch.

registers are available with more than eight bits and usually provide access only to the input to the first flip-flop and the output from the last.

Coupling to a Computer Bus. In a computer all devices except the central processing unit, other processors, and memory are called input/output (I/O) devices. To ensure that they may be caused to operate and, when they have data, to transmit it to the CPU for processing, they must be connected to the data and control lines on the bus that starts with the CPU. This arrangement is suggested for an ADC in Fig. 5.24. Note that the ADC is coupled to a bus through an I/O port provided as a chip. Each port provides a buffer register with both a driver and a latch. The register (latch) holds data to be input to the CPU. A decoder on the chip ensures that the multibit address of the port on the control lines will be decoded when the CPU expresses on the bus the *device select signal* for that register or port.

Addressing of modules attached to the bus is done in a variety of ways depending upon the number of I/O devices and other requirements. An approach that is economical of bus lines as well as versatile is the *fully decoded system*. Only the most significant bits of the address are decoded by the I/O port and the three or four least significant bits are used within the I/O port to select particular registers or operations.

Real-Time Clock. To provide control intervals often needed by instrument modules, a real-time clock is valuable. A programmable version of such a clock is shown in Fig. 5.25. A precision 1-MHz oscillator is the basic clock. The other elements (and an attached computer that is not shown) make it possible to select a wide range of time intervals or clock periods. In this case divide-by-10 chips divide 1 MHz successively by 10 to supply the indicated frequencies to the multiplexer (MUX). For example, the MUX latch may be loaded with a binary address

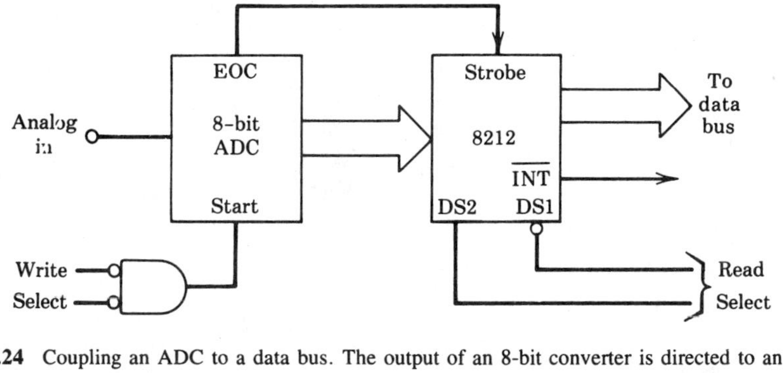

Fig. 5.24 Coupling an ADC to a data bus. The output of an 8-bit converter is directed to an 8212 buffer register, which serves as an input port on the data bus. Conversion is started by a program command to the start pin; when finished, data is strobed to the 8212 buffer by a signal over the end-of-conversion (EOC) line. Once the CPU has responded, transmission onto the bus occurs on receipt of sequential device select and READ signals from the CPU. (After H. V. Malmstadt, C. G. Enke, and S. R. Crouch, *Electronics and Instrumentation for Scientists.* Reading, MA: Benjamin/Cummings, 1981.)

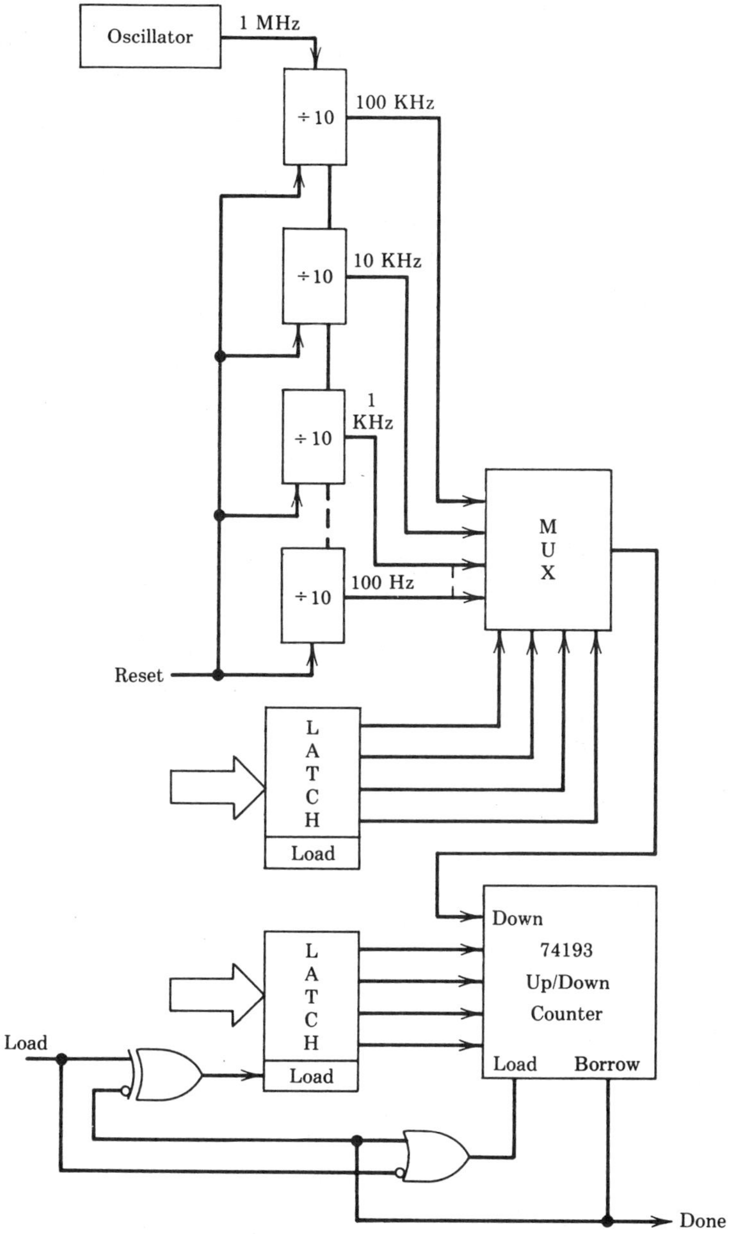

Fig. 5.25 Block diagram of a programmable real-time clock that can supply exact time intervals to modules in an instrument or be used to time an operation. A precision 1-MHz oscillator is the basic clock. The other circuits make it possible to generate a wide range of intervals. (After R. E. Dessy, *Laboratory Automation*, American Chemical Society, Washington, D.C., 1978.)

that will activate the MUX to pass the output from the 10-kHz chip. This output in turn will go to the down input of a 74193 up/down counter. One pulse from the 10-kHz chip will appear at the counter input every 0.0001 s. The counter latch may be set to 0000 and an output pulse will appear every 0.0001 s.

Example 5.7. Show how the programmable real-time clock in Fig. 5.25 can be set to provide 0.0012-s pulses, a more difficult task than just illustrated. Once set, how are the pulses generated? What changes would allow the real-time clock to cover a still greater range?

To obtain 0.0012-s pulses the computer program controlling the clock must set the MUX latch to pass the 10-kHz output. (Note that the computer may also be directed to preset the up/down counter to count up [connection not shown] and to start or finish counting at any number up to 15, binary 1111, by supplying the number to the lower latch that connects to it.) Since the MUX input here is to the down input of the counter, it will count down from the preset number. For 0.0012-s pulses, the latch must be loaded with binary code for 11, 1011. The counter will count down from this number, on the eleventh pulse reaching 0000. When the twelfth pulse is received, it will generate a borrow pulse at its output, which will be transmitted to the done line exactly 0.0012 s after its last one, as will become clear. Note that this pulse is also directed to the gates loading the counter and counter latch, so the counter will be preset again to 1011.

To extend the range of pulses providable, a counter with an 8-bit input and 8-bit latch could be used. This arrangement would make it possible to divide a decade by numbers from 2 through 256. A wide range of precise time intervals are thus potentially available from this programmable clock.

Couplers for Hostile Environments. On the output side it is sometimes possible that a signal will be transmitted from a bus to a difficult kind of environment. It may be one with high potentials or with minimal potentials. In either case transmission by an *optical coupler* will ensure sufficient isolation to protect both bus and device. A representative possibility is shown in Fig. 5.26 with further information provided in the legend.

5.8 LED DISPLAYS

Once a value has been obtained by a digital voltmeter or the results of a computation have been completed, some display of the digital result is needed. It has become conventional for numerical displays to use 7-segment light-emitting diodes (LEDs) or other graphic devices.

In Fig. 5.27, the process by which a 4-bit number is taken from a register and displayed by LEDs is suggested. On receipt of an enabling signal a 4-bit number is transferred in parallel fashion to a decoder (here a binary-coded decimal decoder) and thence to a display driver. What is necessary in decoding is translating the BCD output into signals for each of the seven segments of a digit. Those go through a single set of seven lines to the LED segments.

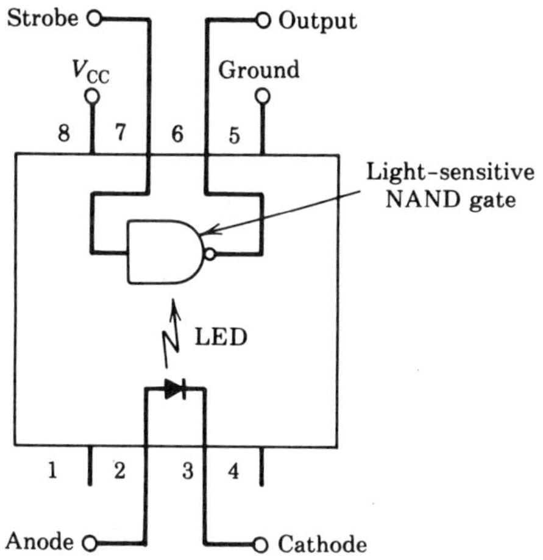

(a)

Input–Output Table

Input	Strobe	Output
0	1	1
1	1	0
x	0	1

(b)

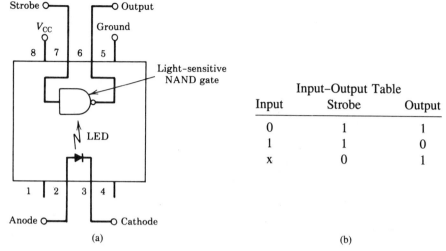

Fig. 5.26 A digital optical coupler. This chip is valuable for transmission between an instrument that must be operated in a difficult environment and a computer. The digital signal to be transmitted is directed to the light-emitting diode (LED) through pin 2. The LED sends a light pulse to a photoconductive NAND gate connected to pins 6, 7. From the input-output table, when a HI(1) pulse appears at the STROBE pin, the level of input (0 or 1) to the anode of the LED determines the output level: a HI input to the LED gives a LO(0) output from the chip and vice versa. When there is no strobing HI pulse, the output remains HI(1).

Most numbers have several digits requiring display. Having a decoder/driver for each digit to be displayed would be cumbersome, error-prone, and expensive. Instead it is common to use a single decoder/driver and install a multiplexing device to take successive digits from registers. This arrangement is used in Fig. 5.27 allowing each digit to be displayed successively rather than continuously. Nevertheless because of the high repetition rate of each digit (about 50 times per second) the output appears steady.

An important advantage of the multiplexing arrangement is that there is sufficient time to update the digits and registers during display. It would be possible, for example, to include in the microprocessor program the updating of the register at moments when no register is being enabled to put its digit bits onto the bus.

Interpretation of an LED Display. The multiplexer consists of a 1-kHz oscillator driving an octal counter/decoder whose output provides the successive HI pulses that will simultaneously activate an LED digit that is to be lighted and the register that is to transfer its data in parallel fashion onto a bus and to the decoder. Clearly, the internal bus connecting all of the registers and the decoder is an essential feature. The oscillator provides successive pulses to the octal counter/decoder, which in turn cycles a HI pulse to each of four output lines that are connected to the system.

Note that there is a parallel output from the BCD decoder that provides current to the several-digit display. Each line from the decoder could illuminate the segment with the pulse in any of the digits. Each line has a 320-Ω resistor, though for many lines the resistor

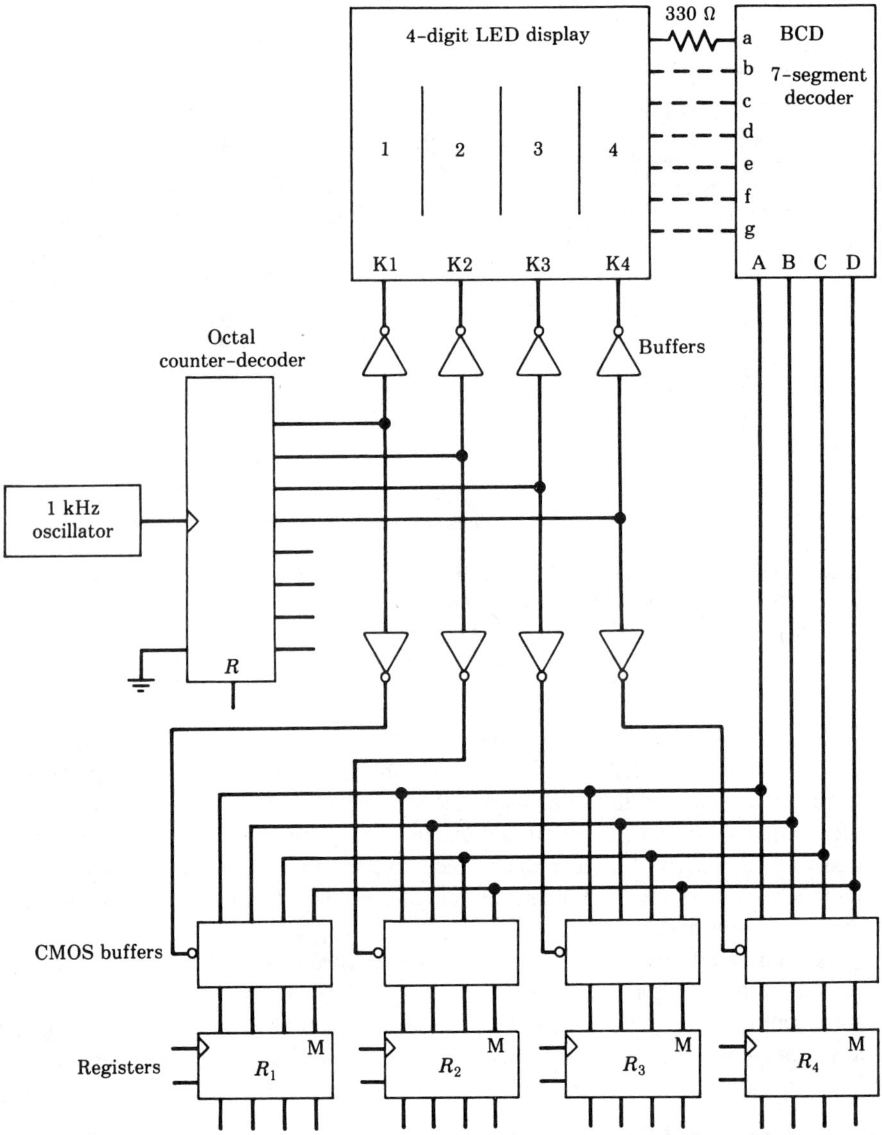

Fig. 5.27 Schematic diagram for a 4-digit multiplexed LED display. The digits are held in registers and successively "asserted" onto a short 4-bit bus that takes it to a decoder which selects the set of LED segments to be lit. The particular LED digit to be activated is determined simultaneously by that signal and a HI from one of the set of octal decoder/counters. Such a level enables (a) one digit to go on the bus through its CMOS buffer and (b) the particular LED units to be illuminated by generating a LO pulse at the output of the connected high-current buffer. The LO pulse takes the cathodes of its seven segments low, and the current supplied by the BCD 7-segment decoder/driver illuminates only that digit. Note that each line from the BCD 7-segment decoder to the LED set has a 330-Ω resistor and is connected at all times. Successive oscillator pulses turn on the other digits in turn. It is the repeated energizing of the LED digits (about 50 times a second) that gives seemingly steady illumination. (After P. Horowitz and W. Hill, *The Art of Electronics*, Cambridge, New York, 1980.)

is not shown. Note that the output pulse from the octal counter/decoder also determines which digit is to be illuminated. Simultaneously one of the cathode sets for a 7-segment digit is driven LO and data are transferred from a register onto the bus for decoding. The current signals from the BCD decoder reach the LED display while the cathodes of one set of segments are active and it is that digit that illuminates.

REFERENCES

Some volumes that present digital electronics on an introductory to intermediate level are:

1. J. J. Carr, *Digital Interfacing with an Analog World*. Blue Ridge Summit, PA: TAB Books, 1978.
2. L. S. Garrett, Integrated-circuit digital logic families, *IEEE Spectrum* **7**(10), 47; (11), 63; (12), 30 (1970).
3. E. R. Hnatek, *A User's Handbook of Integrated Circuits*. New York: Wiley-Interscience, 1973.
4. A. P. Malvino and D. P. Leach, *Digital Principles and Applications*, 3rd ed. New York: McGraw-Hill, 1981.
5. J. Millman, *Microelectronics: Digital and Analog Circuits and Systems*. New York: McGraw-Hill, 1979.
6. D. H. Sheingold, *Analog-Digital Conversion Handbook*, 3rd ed. Englewood Cliffs, NJ: Prentice-Hall, 1986.
7. H. Taub and D. Schilling, *Digital Integrated Electronics*. New York: McGraw-Hill, 1977.

Some volumes dealing with a single area are:

8. W.-K. Chen, *Passive and Active Filters*. New York: Wiley, 1986.
9. A. C. Dixon and J. Antonakos, *Digital Electronics with Microprocessor Applications*. New York: Wiley, 1987.
10. G. F. Weston and R. Bittleston, *Alphanumeric Displays—Devices, Drive Circuits and Applications*. Granada, 1982.
11. C. S. Williams, *Designing Digital Filters*. Englewood Cliffs, NJ: Prentice-Hall, 1985.

EXERCISES

5.1 Why must op-amp 2 in the circuit of Fig. 5.2*a* have a low input bias current?

5.2 Develop a design for an 8-bit DA converter based on an op-amp summing circuit (see Example 4.7). (a) Draw the circuit and label all components. Is this type of op-amp circuit appropriate to the conversion? (b) What digital precision can be expected? Explain. What precision input and feedback resistors should be used to obtain this precision? (c) If you build the circuit from a 741C op-amp chip and $\pm 0.1\%$ accuracy resistors, what values of resistors will ensure an output in the range of 1 to 10 V? Comment briefly on the choices for a TTL logic input. (c) Calculate the output of the converter when the binary number 11010011 is received.

5.3 The dual-slope digital voltmeter illustrated in Fig. 5.4 also serves as an AD converter. (a) How is it like the ratiometric devices discussed in Section 3.4? To what extent is it a null device? (b) Consider first the possibility that the clock or oscillator runs 10% faster than expected. How will this affect V_{max}? How will it affect the time required to discharge the capacitor from V_{max}? Since the pulses are 10% shorter than expected, how will the total discharge count be affected? (c) Will drift in the capacitor tend to cancel out? If so, explain the basis.

5.4 The AD converter shown in Fig. 5.3 is a voltage-to-frequency type. (a) Draw a simple op-amp integrator to replace the box so marked. Where might a gate be introduced to turn on the integrator each time a signal is to be digitized? (b) When $V_C = V_{ref}$, after time t, the comparator will change output and a charge pulse of opposite sign will be sent to the integrator input. Assume $V_C \leq 0.1\ V_{Cmax}$ in this converter. Is the fraction of capacitor charging small enough to give $t_{charging} \propto 1/V_i$ and $V_i \propto f_{pulses}$?

5.5 The pulse generator circuit in Fig. 5.3 is in a negative feedback loop. Show that this is true. How does feedback affect the accuracy of the device? Explain.

5.6 How many lines must there be for a truth table for gates with (a) three inputs and (b) four inputs?

5.7 Show how the BCD decoder in Fig. 5.11 responds when it receives a 1000 input at *ABCD*. Which input receives the least significant bit? A suggestion as to an approach: redraw the figure and indicate on it whether the logic level of every NAND input and output is 0 or 1. Thus, show that the decoder really decodes the binary input correctly and activates the correct NAND gate.

5.8 (a) Prepare a truth table or table of states for a two-input exclusive-OR (XOR) gate for HI-true signals. (b) What function does this gate perform on LO-true signals?

5.9 A digital parity checker can be made by connecting the outputs of two XOR gates to the inputs of a third XOR gate. (a) Draw the circuit. Label the inputs of one gate in the initial pair A, B and the inputs of the other C, D. (b) Show that the output of the circuit will vary according to whether the number of inputs that are 1 is odd. (c) How might the circuit be modified for a 6-bit input?

5.10 The exclusive-NOR (XNOR) gate has the symbol shown in Fig. 5.28. (a) How does its output indicate equality for an appropriate set of inputs? (b) In Fig. 5.29 a circuit that serves as a 4-bit equality detector is drawn. Show that its output is HI only when each bit of one number equals each bit of the second number. Direct the first number to inputs $A_3A_2A_1A_0$ and the second to inputs $B_3B_2B_1B_0$. Tabulate your results.

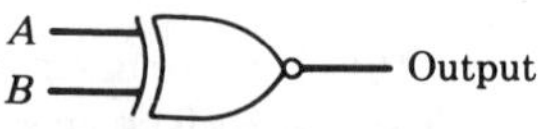

Fig. 5.28 An exclusive-NOR or equality gate. Its output is the inverse of the output of the XOR gate (see Table 5.3).

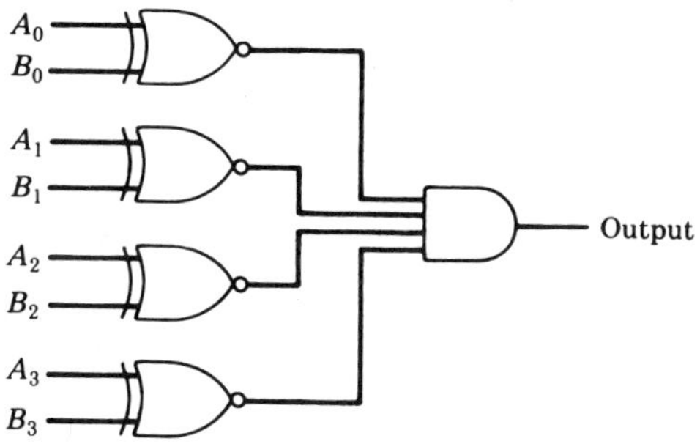

Fig. 5.29 A 4-bit digital comparator for comparing binary numbers $B_3B_2B_1B_0$ and $A_3A_2A_1A_0$. When the numbers or words are identical, the output of the comparator will be a 1.

5.11 Analyze the factors involved in selecting a capacitor for the sample-and-hold circuit of Fig. 5.2. What choices would give the best precision?

Chapter 6

MICROCOMPUTERS IN INSTRUMENTS

6.1 INTRODUCTION

After the survey of basic digital electronics in the last chapter the stage is set for discussion of digital microcomputers in instruments. We need to look at imbedded computers both as (a) modules playing a major processing role in the signal channel of such instruments and (b) controllers of instrument modules. When only minicomputers and mainframe computers were available, instrument users had the option of either recording data produced on tape for later processing by computer, an *off-line* arrangement, or direct connection to a computer through a suitable interface for real-time processing, an *on-line* arrangement. With the development of integrated circuit versions of central processing units or *microprocessors* in the early 1970s, the microcomputer became possible. Soon microprocessors began to be imbedded in instruments, that is, placed *in line* to give the high capabilities of modern instrumentation.

How does a microprocessor operate in an instrument? The best answer to the query can be obtained by seeing the digital microprocessor as a module functioning along side of instrument modules such as a source and detector. Further, since a microprocessor will usually be accompanied by the peripheral or modular parts of a computer such as keyboard, memory, input/output devices, and video display, hereafter this type of design will be identified as an in-line *microcomputer* in an instrument. (See Exercise 6.1 for further discussion.)

A microcomputer adds a programmable, high-powered calculating and logic device to an instrument and, because of software available to the user, in most cases provides both appropriate control of modules as measurements are made on different analytes and many types of data processing. For instance, one program will enable a user to carry out one-component measurements while another program will facilitate making multivariate analyses.

The nature of the digital domain in which computers operate must now be reviewed briefly. Persons familiar with the binary system will have a grasp of the number system employed in digital computers. Others may wish to consult Appendix A for a basic presentation of numbers and codes. In the digital domain all operations are described in terms of the two basic digital states, designated HI and LO or logic 1 and 0. The result is that all information and data appear as binary numbers. While software, at the machine level, is in binary code, virtually all

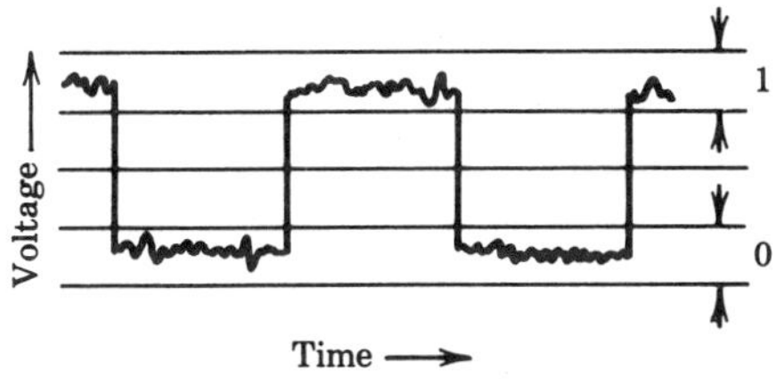

Fig. 6.1 A portion of a digital signal consisting of two bits. Digital logic levels are assumed to be 1 (HI) = +5 V and 0 (LO) = 0 V. Since the bit value can be only 1 or 0, a moderate amount of noise does not affect the value.

instrument programming involves higher-level languages such as PASCAL, BASIC, or FORTH.

Regardless of the language in which microcomputers are programmed, at the microprocessor level they accept and produce only binary signals; thus they are said to operate using machine code. An example of a signal in binary or machine code is provided in Fig. 6.1. Here two *bits* or binary digits of a digital signal are arranged serially in time for transmission. The exact interval of time allowed each bit is provided by an internal *clock* or oscillator. The clock furnishes square pulses whose width defines the lifetime of each bit in a signal. As one bit ends, the next one follows just as in the decimal number 738 one moves abruptly from digit 7 to digit 3. Though bits dominate within digital areas of an instrument, in virtually all instruments a final translation of binary code into decimal form will guarantee a familiar output for the user. Figure 6.1 also shows the main reason the error rate of digital devices is extremely low. As a result of the large separation of voltages defining logic 1 and 0, noise amplitudes can seldom obscure the identity of a bit.

6.2 COMPUTER ARCHITECTURE

In this discussion the term architecture connotes the general layout of a central processing unit (CPU) and the mode of coupling of computer and instrument modules. Figure 6.2 shows the architecture of an instrument with an in-line microcomputer; as is evident, not all modules are shown. Its microprocessor or CPU is evident in the upper left corner. A glossary of common computer acronyms is given in Table 6.1. Some modules serve both as parts of the computer and the traditional instrument. Additional memory chips and even extra CPUs may of course be incorporated to improve processing and a printer/plotter or other display device added for graphics capability.

Some ways in which an in-line microcomputer can interact with instrument modules are suggested in the examples that follow. Only for high-level instrumentation will designers be likely to decide to effect all the possibilities. In fact, as long as an instrument has an output such as a RS-232-C port (see Section 6.5), the user can easily gain additional processing capability by connection through the port to a stand-alone microcomputer or laboratory information system.

Example 6.1. What are some of the ways in which an embedded microcomputer is used in an instrument to enhance instrument capabilities?

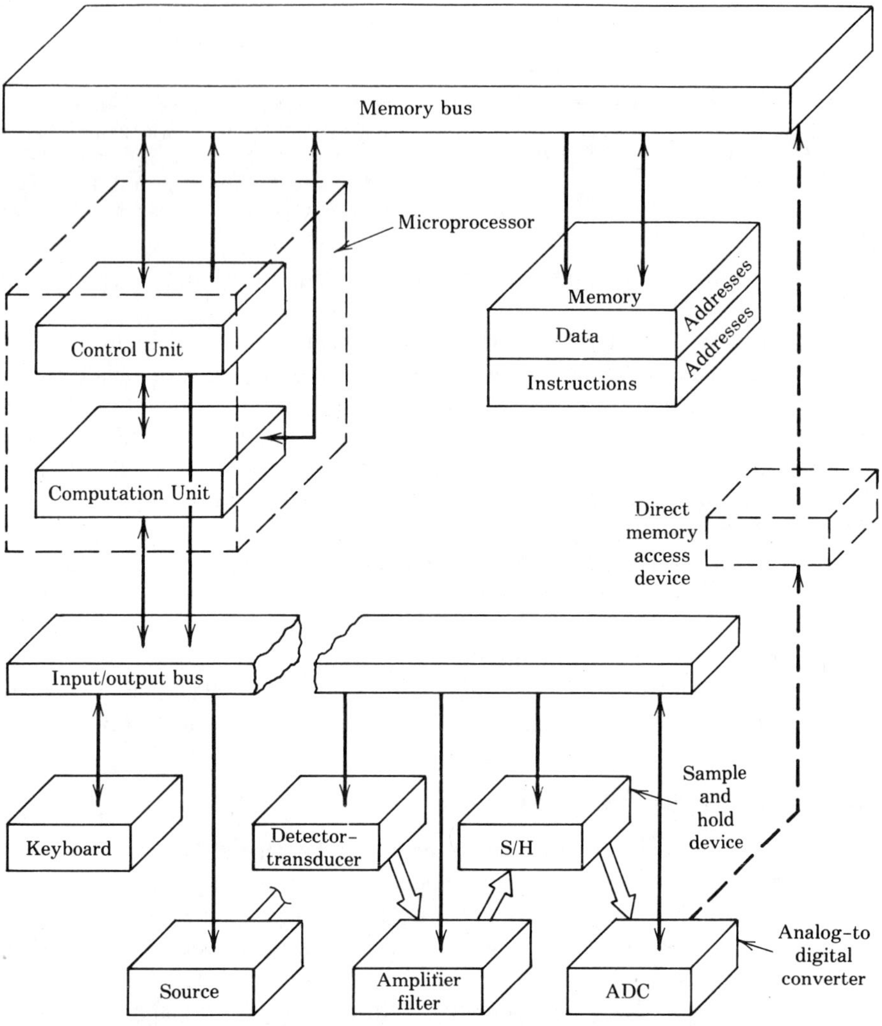

Fig. 6.2 Partial block diagram of an analytical instrument with a microcomputer built in. The microprocessor or central processing unit contains both a control unit and computation unit. Only one peripheral device and a few instrument modules are shown. All are interfaced to the data bus.

In some cases a computer serves only to process data: the computer takes the place of analog processing modules. Commonly, however, a computer also provides at least some instrument control (see Section 1.7) and directs proper currents, voltages, and operating signals to key modules. For instance, in a spectrophotometer a computer can effect a particular wavelength scan the user asks for by generating an appropriate number of pulses and directing them to a grating stepper motor. It is valuable to add the capacity for troubleshooting by computer by installation of appropriate transducers on instrument modules, adding software for an instrument operational check, and calling it up at startup and times

Table 6.1 Glossary of Common Computer Acronyms and Terms

ADC	Analog-to-digital converter
Address	A digital number used to establish contact with an individual memory location or an I/O device
ALU	Arithmetic-logic unit, a part of a CPU
Byte	A group of eight bits
CPU	Central processing unit or a microprocessor
DAC	Digital-to-analog converter
DMA	Direct memory access, a method of very fast transfer to or from memory without use of the CPU
IEEE-488	A standard for a general-purpose bus to link instruments and computer established by the Institute of Electrical and Electronic Engineers
I/O	Input/output, as in a module attached to a data bus
Peripheral	Any unit connected to the CPU such as keyboard, memory, or video display
RAM	Random access or read/write memory
Register	A digital device in the CPU that can hold an 8-, 16-, or 32-bit instruction datum or address
ROM	Read-only memory with permanently recorded data or program
RS-232C	A standard for a serial instrument output port developed by the Electronic Industries Association (EIA)
Word	The standard package of bits handled by a computer, usually 8, 16, or 32 bits

of component failure. Through the program it can be established which modules are functioning normally.

At the highest level, instruments are designed around the sophisticated control and information extraction capabilities of a computer. For instance, in a multidetector (diode array) dispersive UV–VIS spectrometer, a microcomputer can control several modules to achieve an optimum S/N. Then after computer-controlled observation of sample and standard transmission at several wavelengths, a matrix of Beer's law equations can be solved to determine concentrations of several analytes.

Example 6.2. What types of signal processing are common in instruments? In what sense can a computer become the main processing module of an instrument?

Recall that instrument processing begins with data-massaging operations that include analog and digital filtering of noise and the averaging of replicate runs. It also includes arithmetic operations such as the ratioing of signals (e.g., of reference and sample beam intensities in a spectrophotometer) and the subtraction of signals (e.g., the subtraction from the spectral curve of a sample of a background intensity curve previously obtained and stored in memory).

Other types of processing include differentiation, multiplication by a scaling factor, a search of an analytical curve for peaks, integration of area under peaks, taking Fourier transforms of data as is done in Fourier transform IR, NMR, and MS spectrometries, and the application of chemometric methods.

With access to software for all the desired processing operations and sufficient processing power and memory in an in-line microcomputer, a great many types of processing may be asked for by a user. As long as the original data are stored in memory, they may be called up again and again for different types of processing. In effect, each program chosen converts the computer to a new processing module to operate on the data.

These examples suggest the kinds of information an in-line microcomputer must provide or process. To carry out processing much storge and transmission is necessary in a microcomputer. Both instructions and data must be provided from its memory or external storage (diskettes or hard disks) by giving *addresses* that identify the locations where they will be found and addresses of the destinations to which the results of calculations are to be sent. Such calculations as are required will be made in an arithmetic/logic unit in the microprocessor following which results will usually be stored in memory.

Microprocessor. There is a partial diagram of the architecture of the microprocessor in an instrument in Fig. 6.2. Its *control unit* includes a register that holds the current instruction to be executed after it is obtained from a user-selected program that is in memory, a decoder that translates this instruction into specific switching operations, and another register called the program counter that holds the address of the next instruction. Elaborate wiring interconnects the myriad of gates, registers, and so forth that make up the control unit.* As the control unit executes an instruction, it issues necessary logic commands and directs them via wires such as those shown in Fig. 6.2. Wires will lead to all parts of the CPU, and via bus to units coupled to the buses, for example, memory units and instrument modules. The signals close some logic gates and open others, activate some registers, and so forth.

In addition the control unit holds a *stack*, a consecutive set of registers or, commonly, memory locations in RAM. They are called into use by instructions such as *jump to a subroutine* or *interrupt*. Whenever such a command calls for breaking the normal flow of a program, one or more stacks must store the next instruction in the normal program as well as any other information necessary to return to the program as soon as the diversion has been completed.

The heart of the *computation unit* is an *arithmetic–logic unit* (ALU). Most arithmetic and logic operations are carried out in all-purpose registers which are often called accumulators. These operations include: addition, subtraction, comparison of two data words bit-by-bit to see whether one digital number is equal to, larger than, or smaller than the other, multiplication, division, and many logic operations. Occasionally registers are also present that can only store numerical data. In architecture *memory* ranks second in importance to the microprocessor(s). It is essential for storing addresses, data, and instructions.

*The manner of connecting components is called a *microprogram*. If connections in a microprocessor cannot be altered, it is said to be "hard wired." In some designs connections can be altered by instructions, and the microprogram is said to be in *firmware*. Regular computer programs can thus be called *macroprograms*.

Buses. These packets of bidirectional (usually) shared conductors transmit data between parts of a microcomputer. Without buses there would be a plethora of private lines within a computer; with buses, architecture is much simpler. On the microprocessor chip itself a bus is a set of parallel conductors. At its edge they connect through pins to a set of flat parallel wires.

Most buses have a party-line structure: only one sender may put its message on a bus at a time. Control of access is appropriately exercised by the CPU. At any instant all devices are in effect disconnected from the bus except the one permitted to transmit data (talk) and the other to receive data (listen).

Input–Output (I/O) Ports. One important type of device invisible in Fig. 6.2 is an input–output port or interface connecting each I/O module to the bus. These ports are essential to the orderly transfer of data and instructions between the CPU and the modules. Often such exchanges are called input/output operations with the designation *IN* always referring to *transfer to the CPU*.

A port includes data latches (see Section 5.7), each of which holds a datum received at the precise moment that it is signaled. For an output latch the datum will be information from the CPU that is ready for transmission to a module; for an input latch, it will be data awaiting transmission to the CPU. Each I/O port also contains a decoder to identify the particular signal to which it alone must respond by getting ready to receive data in one latch or place data in the other latch on the bus for transfer to the CPU. Through I/O ports the precision of communication is ensured and only a single sender and single receiver are activated to transfer or receive data.

Communication with I/O Ports. For a CPU to contact a particular I/O port requires additional communication. Usually ports are furnished regular addresses by *memory-mapping* them, that is, treating each exactly like a storage site in memory. One entire block of addresses, usually at the upper end, is reserved for I/O ports alone. The CPU can address a port straightforwardly. Another approach, the *data-channel* method, calls for provision of a separate data channel or conductor to every port. In this system the CPU must issue for each datum a signal to indicate whether it is to be transferred to memory or an I/O port and must designate the destination. When the ability of CPUs to address memory was limited to 2^8 (256) locations, the data channel method was especially attractive since it saved memory. Today as many as 2^{16} memory locations can be addressed by more powerful microprocessors, and I/O ports are commonly memory-mapped.

A discussion of the timing of information transfer and details of CPU interrupts, setting flags, handshaking, and other aspects of data transmission is beyond the scope of the text. References should be consulted.

6.3 INSTRUCTIONS, DATA, AND ADDRESSES; PROCESSOR AND MEMORY

Given the architecture discussed in the last section, we can begin to examine the dynamics of microcomputers in instruments. The first step is to describe the nature of instructions made available through programs (software) for control of modules and data acquisition. Second, we must understand in a general way how CPUs decode and respond to the bits that comprise instructions.

Bits and Words. Binary language instructions are given in words. The size of a word depends on the computer and is the number of bits that can be handled in a single machine operation. In general, it is also the number held in each memory location and register. Most newer microprocessors use words of 16 or 32 bits. Historically, 8-bit words were common and were given the distinctive name *byte*, a term that is still widely used to identify different segments of 16- and 32-bit words, as shown in Fig. 6.3. Ideally, each instruction will be one word in length, though in a system where words are byte-length, two bytes are commonly required for an instruction. For each microprocessor, a word has a prescribed structure (see examples below). Word length has a direct effect on efficiency; for example, a 16-bit processor can execute a function on a 16-bit datum more than twice as fast as an 8-bit CPU. Nevertheless, a processor usually uses two or more different word lengths.

Programs. The strategy of making available to a computer the sequence of words that comprise the instructions of a program is ingeniously simple. Normally, except for jumps, instructions are entered via keyboard in a high-level programming language such as BASIC, FORTRAN, PASCAL, or FORTH. The reader is di-

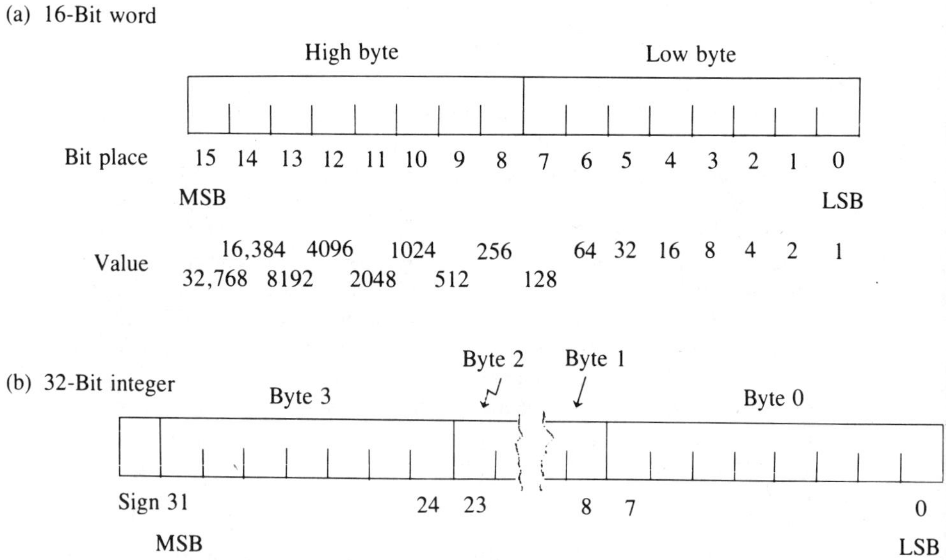

Fig. 6.3 Some digital signals in serial form. (a) A 16-bit word with bit places numbered. Its least significant bit is at the right end. The value of a 1 in each bit place is shown below the word. (b) A 32-bit *signed* integer or datum. If an integer is positive, the extreme left-hand bit is a 0, if negative, a 1.

rected to references for information about them. Where fastest execution is desired, assembly language (see below) is used. Instructions are stored in consecutive memory locations, usually beginning at a low address. For example, the first 16-bit instruction might be at address 1000, the next at 1002, and the third at 1004. When an instrument is turned on, its computer will *boot up* by reading the part of the ROM (permanent memory) that will ensure that the main part of the operating program is transmitted and stored in RAM (active memory). Then as the user selects a program for his measurement, that program will be called up by having the address of the initial instruction entered in the register called a program counter. After the instruction at the initial address has been fetched, the address in the program counter is automatically *incremented* by 2 (to 1002), to furnish the counter with the next address to be used. (In some 8-bit processors, and in instances when only bytes are being used, incrementing is only by 1.) This process will be repeated instruction by instruction.

What content can the words in a program have? Not every word will be an instruction for the microprocessor. They may also be addresses of modules of the instrument, data that are being taken from a sample, or any of a thousand other possibilities that will move the measurement process along. The collection of actual operations the microprocessor can perform is called an instruction set.

Instruction Sets. This set comprises particular instructions appropriate to each type of microprocessor and includes simple processes such as moving data, adding one number to another, and clearing a register, that are possible with a particular microprocessor. This set will be expressed both in machine language (bits) and in *assembly language* (comprised of mnemonics, e.g., ADD, for addition). A program called an assembler that is part of the operating system (Section 6.4) will translate each line in assembly language into one or more lines of machine language. The advantage of assembly language, rather than a high-level language like BASIC, is that decoding operations are performed much more quickly and microprocessors can work essentially in real time.* A list of essential categories of instructions for instruction sets is presented in Table 6.2. Attention to the design of a processor and its instruction set allows for greater power in a CPU since fewer instructions are needed to carry out a task.

Instruction Cycle. A CPU deals with an instruction by executing an elementary operation called an instruction cycle. In early microprocessors a linear sequence was followed in executing instructions: (a) the processor fetched from the next specified location in memory the contents entered there, (b) decoded the contents and executed the steps called for, and (c) incremented the address in the program counter so that the computer would be ready to fetch the next instruction as soon

*Real time is regular time. In instances where information from an instrument, perhaps experimental data, are available at a rate that exceeds the processing capabilities of a microprocessor it is common to store data and transfer them to the processor later when it will be less burdened. Delayed processing is said to be carried out in machine time.

Table 6.2 Essential Types of Instructions for Microprocessors[a]

Categories	*Nature of Operation(s)*
Transfer of information	Moving data between registers
Input/output	Moving data between an input or output device such as an ADC and the register termed an accumulator
Arithmetic	Adding, subtracting, and usually multiplying and dividing a data word by a word in a register
Logic	Comparison of each bit of a data word with the corresponding bit of a word by a logic operation such as AND, OR, or XOR
Increment/ decrement	Increasing or decreasing a data word by 2 or sometimes 1
Jump	Placing in the program counter a stated address (rather than the next address in the current sequence)
Call subroutine	Jumping to the address of the first instruction of a subroutine and storing the address of the next regular instruction on the stack (the last instruction of the subroutine returns to the program)
Processor	Resetting, halting, and other instructions that affect only operation of the CPU

[a]After H. V. Malmstadt, C. G. Enke, and S. R. Crouch, *Electronics and Instrumentation for Scientists*, Benjamin/Cummings, Reading, MA, 1981.

as the first had been executed. Today microprocessors usually fetch successive instructions from memory by a technique such as *pipelining*. The next instruction is stored in RAM in the CPU during the execution of the earlier instruction. Efficiency is significantly higher. For all processors the basic operation of the instruction cycle is repeated again and again during the execution of a program.

As described in Chapter 5 such switching between states is kept orderly and almost error free by being arranged to occur on the rising or falling edge of a digital clock pulse (see Section 5.4). During the design stage of a CPU a clock frequency is chosen on the basis of knowledge of the expected gate-switching speeds and of the largest number of gates likely to be involved in the simplest sequence. Sufficient time is allowed for complete transmission of instructions (different intervals are required for different commands) and execution of the operation. In cases where times needed are uncertain, for example, in obtaining complete information about an analyte from an instrument module such as a detector, the computer is programmed to wait for receipt of an *end of operation* signal on a designated line before taking up another instruction. In other cases it examines the status of digital *flags* before proceeding.

For timing, a clock or crystal oscillator of appropriate frequency is provided either on the basic microprocessor chip or as an auxiliary chip. Clock frequencies have risen as switching has become faster; today they are commonly 8 MHz or higher.

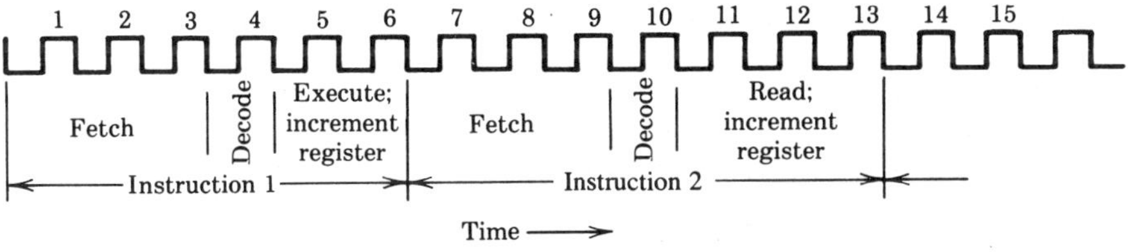

Fig. 6.4 Generalized timing diagram for a central processing unit. The numbers are the cycles of the square wave furnished by a clock (crystal oscillator) that times all operations precisely. The steps or operations making up machine (computer) cycles are listed below. A machine cycle includes fetching, decoding, and executing an instruction. It is simply labeled INSTRUCTION. The nature of an instruction determines the control signals sent during execution and the time required for execution. Two machine cycles are shown. Fetching an instruction from memory requires three clock cycles in this example. Decoding is the fastest operation since it involves only hardware.

A *timing diagram* for two instruction cycles is shown in Fig. 6.4. It is useful to think of a microprocessor as moving from instruction to instruction and allowing for each the number of clock cycles called for in microprogram for this process.

Execution of an Instruction Cycle. Once a microprocessor has been commanded by a program to start, instruction cycles begin. Its *microprogram* ensures that at each phase of a cycle its logic gates will receive necessary commands.

For example, at the beginning of an instruction-fetch cycle, signals set gates so that only the program counter can assert a signal, in this case the address of the instruction, onto the bus and only memory can latch (receive) the address. The decoding circuitry *in the memory system* that latches the address also provides a READ signal to the addressed memory location. The program counter and address decoder are disconnected.

On receipt of the READ, the contents of the memory location are latched (asserted) onto the bus lines. The instruction register and decoder in the control unit of the CPU are activated. When the instruction register has had time to latch the contents that are on the bus, it is disconnected. Decoding each instruction involves sending logic 1 or 0 bits on appropriate control lines. Usually decoding is hard-wired and each instruction activates a unique combination of gates. Finally, the program counter is incremented.

In addition to the versatility a good instruction set gives a microprocessor, the speed and efficiency of information transfers may be increased by having available a variety of *addressing modes*. When they are used a few bits of an instruction must be designated for identifying the mode of addressing. Programming is slightly more complex but this loss is compensated for by the considerable gain in versatility in execution. References may be consulted for further information.

Example 6.3. How are digits in a 16-bit word allocated among an instruction, the address where it will be found, and address identifying the place where it is to carry out an operation and where it is to leave the results? It will be useful to illustrate the mode of allocation for the DEC LSI-11 microprocessor in which 16-bit words are used.

First, the conventions with regard to numbering bits are that the first bit on the left is ordinarily the *most significant bit* (number 15), and the last on the right is the *least significant bit* (number 0). Second, in each word the most significant bits define the instruction or operation to be performed. Since the instruction is in code, it is referred to as the *op-code*. Third, the least significant bits give the location (address) where the data for the operation can be found. This information is called the *operand*.

The bits in a 16-bit word are allocated as follows in this system: (a) six bits beginning from the extreme right (bits 0–5) are saved for the operand that will give instructions for finding the first address [three bits for an address mode (3–5) and three for the designation of a register], (b) the next six bits (6–11) give a second operand when necessary (for an instruction such as MOVE or ADD that requires knowledge of both the location of data and the destination to which the result must be sent), and (c) the remaining bits (6–15 or 12–15) for the op-code define the process the computer is to carry out. Another example will show how coding is done.

Example 6.4. What is the computer word in assembly language for the instruction CLEAR? By definition, this operation calls for setting to zero all bits at the location addressed. It will of course be a single-operand instruction since it will prepare one site such as a register or a memory location to receive data.

Assembly language mnemonic		Word in machine language code
CLR	(binary)	0 000 101 000 001 011
	(octal)	0 0 5 0 1 3
		op-code / operand mode address

This op-code 0050 (octal notation) will be found in the instruction set for the CPU opposite the word "clear." To decode the address, first note that the *address mode* is 1. (The code 1 stands for a deferred mode in which the address will be found in a register.) In this case, the address is located in register 3.

Subroutines. Every program is likely to call more than once for common tasks such as scaling numbers (multiplying each by a constant) or averaging a set of data. Whenever feasible, a *subroutine* is written for such tasks. It can be called up when needed by placing a "jump to subroutine" instruction in the main program. Since subroutines are nearly always written in assembly language, they are executed hundreds of times faster than the operations in a high-level language in the main program. By use of subroutines one saves time in development of programs, saves memory space since a subroutine is entered only once, and considerably speeds the execution of programs.

Example 6.5. Show how a subroutine can be used in an *iterative operation*. Consider that replicate measurements of an analyte in a sample are to be continued until an average approximating the true mean is obtained. A microcomputer is to carry out this process, called taking a running average, by adding each new datum to the sum of all previous measurements and then taking a new average (see Section 12.8). Rather than take a great many data, a running average can be used to take just enough measurements to obtain an average of desired precision, for example, $\pm 1\%$.

The needed subroutine is a *loop program*. The nature of its operation is intuitively evident from the name. But how can the program be designed to exit the loop when the number of measurements is sufficient to secure the desired precision? To terminate the process one includes an instruction calling for a test of each new average against the previous one. The instruction will be to compare the difference between the average of $N + 1$ results and of N results with a reference value. As soon as this difference falls below the reference, the loop will be exited. If the difference is larger, averaging will be continued. With this process the final average will be a satisfactory approximation of the true mean.

Coprocessors and Array Processors. By closely coupling a second processor with the basic CPU, a microcomputer can execute an extended instruction set and the power of a computer can be considerably enhanced. The new coprocessor ideally may take on a special function such as mathematical operations or handling input/output.

Further, even in microcomputers computational tasks are increasingly being divided among special pipelined processors called *array processors*. These combinations are advantageous mainly when data arrays must be multiplied, inverted, or otherwise processed as, for example, in Fast Fourier Transformations.

Memory. A constant interplay between processing and the storage of data and instructions is characteristic of all computers. Memory devices in general use for storage can be subdivided into moving surface magnetic types and electronic types. Each type of memory sees use in instruments.

In *moving-surface* memory devices, storage occurs in thin films of a magnetic solid like iron oxide. It is coated onto a nonmagnetic plastic (usually Mylar) disk or tape. Recording or memorizing is done by moving the surface under a read/write head that incorporates an electromagnet. As data bits pass through its coils, the microscopic magnetization pattern it induces leaves permanent magnetic flux patterns detectable by a similar head. To permit fast access to all storage sites usually both the head and the surface move.

Bulk storage devices include regular magnetic tapes, *floppy disks* or diskettes, tape cassettes, and various hard disks. Floppy disks come inside a case in both 8-, $5\frac{1}{4}$-, and $3\frac{1}{2}$-in. diameter versions. The smaller ones have had wide use with microcomputers. Disks are formatted into files that are wedge-shaped sectors. About a third of the disk space available is given over to indices, gaps between sectors, and programs for identification of files and error detection. Storage capacity is substantially less on the $5\frac{1}{4}$-in disk but rates of information transfer are about 30 kbytes s^{-1}.

Hard disks are commonly provided in large-diameter Winchester type disks that have high recording densities. In a Winchester drive, the disk turns at high surface velocity under a fixed head and data transfer rates of up to 1 Mbyte s^{-1} are possible. Capacities of Winchester disks ranged from 4 to 14 Mbytes in 1987. As many as five disks can be placed on the spindle of the drive.

Rates of data transfer during writing and reading and of access to stored data are important. Where the density of data storage is high, as on a Winchester disk, and surface velocities are high, an average access time is 5 ms. Such block access devices are valuable for all kinds of storage where information is read often and changed infrequently. Data falling in this category include programs used occasionally, data stored pending processing, and data of archival value. The cost per bit of such storage is one to four orders of magnitude smaller than for purely electronic storage. Disk arrangements have virtually replaced tape storage for instrument use.

Optical disks are also being developed for storage. They promise to offer very high storage capacity (up to 10 Gbytes).

Electronic types of memory are mainly employed in semiconductor devices today. Both read-only memory (ROM) and read-and-write memory (RAM) are made up of individual elements or cells and are designed to be location-addressable. Usually each byte, or word, has a definite address and physical location assigned. Content-addressable memories would be useful but are quite uncommon.

ROM and RAM. What types of digital words should be stored in ROM and what types in RAM? If the words are the manufacturer's instructions for operation of a stepping motor used in wavelength scanning in a dispersive spectrometer or for acquistion of voltage readings from a potentiometric cell, in all probability they will be entered into ROM memory. They have been developed with great care for the particular instrument, will be employed again and again, and must never be lost.

Information or programs called up from a disk as one is using a computer should be put in the instrument's RAM memory.

Design of Memory. Most read/write (RAM) or volatile memory is designed in NMOS technology in rectangular arrays of rows and columns on a chip. Information is stored as an electric charge on a minute capacitor (50×10^{-15} F) that is switched on and off by a MOS transistor. A cell selected for reading or writing is one whose row and column are simultaneously activated. Charge is lost from these capacitors sufficiently rapidly by leakage, reading, and writing that it must be regenerated or refreshed as frequently as every 2 ms. Thresholding amplifiers are commonly used to refresh charge levels; they must, of course, first sample the cell content to determine its logic level (0 or 1). Chips with 1–256 K are presently available.

In most ROM memory the storage capacitor in series with the MOS transistor switch is replaced by either an open circuit (HI state) or a connection to ground (LO state). In PROM memories a fusible link connects the transistor to ground. Individual cells are programmed to a HI (open) state by a momentary strong signal that melts the link, leaving the requisite open circuit. Cells untouched are in a LO state. Information stored in ROMs and PROMs in nonvolatile. There are reprogrammable ROMs as well; references should be consulted for information.

Both RAM and ROM are random-access types of memory. Serial-access types, which are less expensive, are available too. These devices are either of charge-coupled or magnetic-bubble design. Bits are stored in them sequentially and are transferred cyclically through successive (64 or more) locations. Access time at its longest, if there are 64 locations, is 64 times the cycle time for RAM. Conventional video (cathode ray tube) display can be very satisfactorily refreshed by use of such memory.

6.4 OPERATING SYSTEMS

In earlier parts of this chapter microcomputer architecture, functioning, and the coupling or interfacing of the microprocessor to other modules was described succinctly. What ensures that the computer and instrument modules can operate compatibly and efficiently as a system? A set of programs called an *operating system* must also be available to the microcomputer to edit and save files and handle other elements in the system. In other words these programs manage CPU input and output as well as files. The rationale for such a system becomes clear on inspection of Table 6.3, which lists programs likely to be found in a representative operating system. It allocates the resources available in the computer among the programs that need as full access to them as possible.

Recall the procedure at startup of an instrument with a dedicated microprocessor: the microprocessor will read from its ROM an initial procedure, usually called a *bootstrap* program. It will automatically be entered into RAM in the CPU. This program allows the CPU to receive commands from the operator and to access additional portions of the operating system from ROM or from disk, tape, or another form of mass memory. Once the operator of an instrument selects an instrument program, the computer can draw on the assistance of the operating system to help with its execution. Then, as programs in the operating system are needed they can be called up by use of the program initially loaded from ROM.

Microcomputers will have the smallest operating systems and as one moves toward mainframe computers the size of systems will increase. Some common operating systems are designated by the registered trademarks CP/M, MS-DOS, and UNIX.

Table 6.3 Useful Programs for an Operating System

Program Name	*Content of Program*
Edit, load	Lists programs, provides for editing, and finds programs to permit them to be loaded
I/O device handlers	Reads and writes data files in instrument modules and other peripheral devices
File system	Creates data files in auxiliary storage devices such as disk and index files; usually also arranges clearing of files, loading of files, and controlling access to files
Language translator	Provides decoding or translating of languages in which programs are written into machine languages
Debugging	Provides analysis of program execution to facilitate finding errors
System fitting	Adjusts programs and operating system to conform to size, requirements of I/O devices, and operating system
Development program	Provides for changing the system as other computational facilities and peripherals are added
Diagnostic program	Provides for systematic use and checking of hardware items such as disk files and memory locations

Development System. In microprocessors dedicated to instruments a full operating system is seldom available. Instead a development system is commonly provided to allow a user to change the overall system, for example, by linking it to a laboratory information management system.

6.5 DATA COMMUNICATIONS; INTERPROCESSOR COMMUNICATION

The major flow of information involving the CPU, instrument and microcomputer modules, and external computer networks increasingly calls for standard methods of communication. The trend toward more generally accepted standards or protocols is growing steadily. Before briefly describing three widely used standards, we need to look at general aspects of data transfer.

Transfer Modes. In general, information is communicated 8 bits or 1 byte at a time. Buses within a computer or instrument and cables outside must include not only 8 conductors to carry the individual bits making up a byte but additional conductors to establish electrical ground and to announce the dispatch and receipt of information. This mode of transfer is called *parallel transfer* and offers advantages of less complex hardware, ease of accommodation to inputs of most CPUs, and shorter transfer times.

Example 6.6. Undoubtedly data acquisition is the most important data transfer in an instrument with an in-line computer. While the digital electronic aspects of these modules was described in Sections 5.1 and 5.2, we look now at the communications aspect. Figure 6.5 reminds us of the modules needed. As the figure makes clear, the microprocessor (CPU), under program control, must initiate each sampling of an analytical signal and its digitization. When the conversion of the signal to the digital domain is complete, the processor must send a signal to enable the ADC to transmit its datum to the CPU.

The bus (not shown) to couple the AD converter to the CPU should permit parallel transmission with its advantages of higher speed, greater clarity, and the efficiency of transfering an entire word at a time.

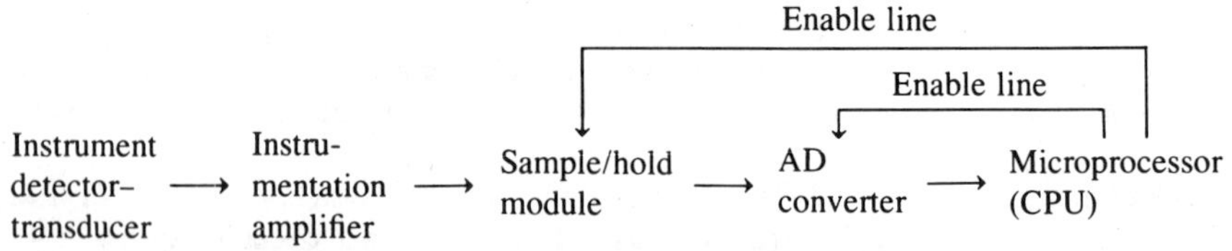

Fig. 6.5 Modular diagram of units active in acquiring data from an instrument with an in-line computer. The analog signal from the detector–transducer is first amplified and then filtered to reduce noise. Next the signal is sampled and each sample is digitized. The microprocessor, through its program, controls these operations.

The alternative to parallel communication is *serial transfer*. In this mode the bits making up a byte are converted to a sequence of bits for transfer; at the receiving end they are reconverted to a regular byte. Both conversions are made by digital integrated circuit devices. Serial transfer is advantageous for most instrument modules since measurements are made at particular times and for intervals determined by the requirements of an instrument. Another advantage of serial communication is that no more than a single pair of conductors is absolutely necessary, allowing the public telephone system to be used for transmission between widely separated computers. A disadvantage is that the timing of the communication is critical since it must not only be sent but received without confusion (see discussion on handshaking below).

Bytes are serially transmitted either *asynchronously* or *synchronously*. In asynchronous transfer a byte begins at any instant and the receiving device must detect and internally capture each byte independently. Asynchronous transmission is widely used in instruments and in any situation where each character of data must be sent as it is generated. Often this situation is described as an interactive environment. The reader is referred to references for discussion of synchronous transmission.

Standards. A standard or protocol to ensure standardized communication must define several types of variables. First, *output voltages and currents* in communicating devices must approximate actual driving voltages and currents on a CPU bus for electrical compatibility. Second, *digital levels and codes* for the communicating devices must fit the instruction set of the CPU. Third, a set of conventions relating to sending a byte are needed: there must be (a) a signal to indicate the beginning of a new byte, (b) a definition of the order of bits, for example LSB or MSB first, (c) a definition of the way to determine when the next bit arrives, and (d) a method to detect an error if noise in transmission is higher than expected. In general these functions are performed within an appropriate chip such as the universal asynchronous receiver–transmitter (UART).

Finally, it is essential that digital logic levels and mechanical requirements are defined precisely to ensure compatibility. In practice this means specification of voltages for logic 1 and 0 and defining the type of cable connector and its pin assignment.

ASCII. The protocol widely used for *alphanumeric* symbols, a category that includes letters, numbers, keyboard signals, and control signals, is the 8-bit *American Standard Code for Information Interchange* (ASCII, pronounced "as'key"). It is univerally used for serial transmission in computers and in most peripheral modules. In addition to the standard alphabet, it includes characters such as CR (carriage *return*), LF (line feed, a command calling to begin a new line of printing or a new instruction), and ESC (escape, a command delimiter that calls for moving from one screen to another).

A good example of ASCII code is shown in Fig. 6.6. The word begins at the extreme right with a "start" bit. The eleven bits that comprise a character word

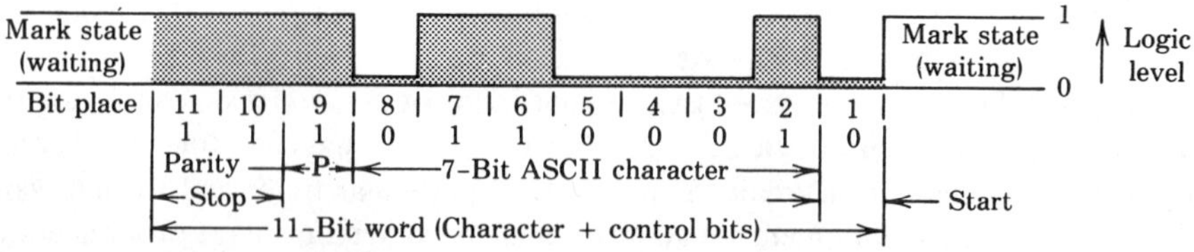

Fig. 6.6 An ASCII word for serial transmission. When a sending line is "quiet," it is in the mark (waiting) state. A "start" bit signals that the next seven bits will denote a character in ASCII code. In this example a "1" is being transmitted in code. The ninth bit is a parity bit. The encoding system controls parity also. If it is set for even parity, the number of 1's (in the ASCII bits plus parity bit) must be even. Odd parity implies an odd total number of 1's. The word ends with two stop bits.

are assigned as follows: one start bit, two stop bits, seven bits of code, and one error bit called a *parity* bit that follows the code bits. In general, parity bits allow a simple check for possible transmission error. If even parity is in use, the total number of 1's in the bit string including the parity bit must be even and vice versa for odd parity. Parity may be checked by the CPU, and if not met, a second transmission of a word requested.

RS-232 C Protocol. Instruments with built-in microcomputers nearly always provide one or more computer compatible digital outputs. A standard type of serial transmission output port designated RS-232 C and developed by the Electronic Industries Association (EIA) with some extensive suggestions by the International Organization for Standardization is included in most modern instruments. A LO-true set of logic levels applies. Voltages from −5 to −15 correspond to a 1 or ON condition; voltages from +5 to +15 to a 0 or OFF condition. Not only are bipolar logic level standards specified, but a standard 25-pin D-shaped connector is established, thereby ensuring physical compatibility. Commonly, eight pins are designated for operations such as transmitting and receiving. Other pins are for a hardware "handshake" to ensure that a message is ready to be transmitted or a bus is ready for receipt of a message.

A disadvantage is a susceptibility to noise that increases with an increase in baud rate and distance. Practically, good transmission is limited to a maximum of 15 m (50 ft). Baud rates for serial transmission (not part of the standard) on a line go from 110 to 19,200 (clock periods per second).

IEE-488 Protocol. More recently (1975) the IEEE protocol was adopted for parallel transmission by the Institute of Electrical and Electronic Engineers. It defined a standard general purpose interface bus (GPIB) that facilitates communication of instruments and computers. A standard 16-line bus is prescribed, eight lines for parallel transmission of data, three lines for handshaking, and five for bus management. In this standard the shape, size, and number of all pins in a connector, mode of handshaking, and method of data transfer are all specified. Up to 15 devices can be connected; only one device, such as a computer, can act as a controller that sends commands.

At least in the early years of having microcomputers in instruments, schematic diagrams and instruction codes or source codes have been virtually unobtainable. As a result, repair and adaptation of commercial instruments to new applications has been difficult. Even where information is lacking, it is still valuable to talk of the principles that apply and thus secure the kind of background that makes for effective use of instruments with microcomputers.

Multimicroprocessor Instruments. The low cost of microprocessors and their relatively limited capacity often makes it desirable to imbed two or more microprocessors in more complex instruments. For example, mass spectrometers and spectrofluorimeters are types that fall into this category. It is surprisingly easy to use two or more processors, given the modular nature of instruments.

For example, one microprocessor may control instrument modules and perform limited processing. Another may control display devices and provide interaction with an operator through a keyboard and video display. The latter processor will then almost certainly control the mode of instrument operation as the operator selects one from a menu. If some of its capacity is still unused, it may be desirable to devote the second processor also to some aspect of data processing. Here a microprocessor is dedicated to some cluster of modules.

Interaction between the processors must also be allowed. Given our discussion of the general complexity of communication, we can certainly look to ideas that have been developed in networks of computers and instruments for answers about how to proceed.

Somewhat surprisingly, a simple provision, sharing some part of common memory, is available. Processors will communicate with the shared memory locations, perhaps one adding acquired data and the other calling on it for processing. The sole problem is avoiding instances of simultaneous access. This can be done by developing hardware that assigns access on a first-time priority basis. The circuit will let processor 1 access the common memory the first time, and then immediately assign higher priority for the next operation to processor 2. With this type of rotating priority, there is a loss of speed in operation amounting to no more than one access cycle.

While the microprocessors may be different types, they need not even be programmed in the same language and need not operate synchronously, some minimal conformity will be necessary. They must have compatible word lengths and logic types. Clearly, their programs must use the same labels for accessing and the same conventions about location of flags.

Networks. In laboratories with many instruments it is common to tie several instruments into a computer network to pass information from one to others and to a minicomputer. The advantages secured from such electronic links are (a) both an in-line computer and the laboratory computer may deal simultaneously with aspects of the same experimental situation, (b) data keeping may be organized for the lab, and (c) resources may be shared. The reader is referred to references for general and specific information about kinds of networks in use and relative advantages.

6.6 GENERAL TYPES OF DIGITAL PROCESSING AND DISPLAY

After digitization and amplification of an analytical signal (see Section 12.5) it is of course the microprocessor and the associated memory that will store the data and subsequently interact with it. Programs will provide methods to massage, extract information, and calculate a variety of quantities from it.

Some possibilities of computer processing programs that will enhance the extraction of information are suggested in Table 6.4. Among the most important of the programs are those that allow the user to examine data from a variety of perspectives to decide finally how the results may best be displayed as well as how information may best be extracted from them.

Scale Expansion. Often a portion of a display with a high density of data will merit scale expansion in either the x or y dimension or both. With expansion the full features of the display will be evident. For example, only after expansion is it clear in many instances whether data will benefit from subtraction of a baseline or smoothing.

Baseline Subtraction. While subtracting a baseline or background has less relevance in a double-beam system, in a single-channel instrument substantial error may result unless the baseline is subtracted. For a successful subtraction it is necessary that the analytical curve be precisely referenced to the baseline and that scanning be linear and reproducible. Subtraction aids accurate data interpretation also. For instance, to check on resolution it might be useful to subtract the chromatogram of a pure substance from that of a mixture containing the compound at the same concentration in the same solvent system. If the chromatogram fails to fall to the baseline in the region of the analyte peak for the mixture, a second analyte must have failed to be resolved and thus eluted at the same time.

Peak Detection. In spectral and chromatographic records it is important to establish clearly the position and amplitude of peak maxima. Most peak identification is done on the

Table 6.4 Some Programs for Processing and Display of Data[a]

Display	*Data Processing*	*Information Extraction*
Raw point plot[b]	Baseline subtraction	Peak identification
Line plot (data points connected)	Least-squares regression plot	First and higher-order differentiation
Scale expansion (on x and/or y axis)	Multivariate regression	Peak area integration
Bar graph plot	Sliding average smoothing	Concentration calculation

[a]Programs for each processing option are called up from memory. They are generally applied to data points previously digitized and stored in memory and are tested by the user at a video monitor. The results are displayed in real time as obtained. Once a satisfactory approach is found, all data are usually treated in this fashion and results obtained are displayed with a printer with graphics capability. The ability to print out data or make a hard copy of a display is desirable. Manufacturer's literature and references should be consulted for information.

[b]Axes are labeled by use of a program that generates letters and numbers.

basis of a *peak track-and-hold* algorithm. A particular peak is tracked until a maximum is reached; then the position of each peak and its height are registered on a video monitor or other display device. To engage the algorithm a standard approach is to set a signal *threshold* after observation of the average noise level. Then four or five successive samplings of the signal must exceed the threshold before the peak program itself is engaged. This procedure ensures that momentary excursions caused by noise or baseline drift are ignored.

While most programs treat well-resolved peaks easily, a more sophisticated approach is required to cope with peaks that overlap badly. In a marginal case, the smaller of an overlapping pair may appear only as a shoulder on the other. Differentiation of such a signal can be a valuable way to see the onset of a peak and even provide a basis for extraction of quantitative information.

Bar Graph. A bar graph, the kind of display commonly used for a mass spectrum (relative peak intensity versus m/z), can also be used to present other types of data following peak identification. It is necessary to adjust the abscissa scale to the new units, say by entering the ratio of the sampling interval to the new scale units.

Quantification of Peaks. A great many types of chemical measurement yield curves in which peaks of finite width are observed as a function of an independent variable such as wavenumber. To determine the concentration of species responsible for the peaks either the location of the peak maximum, or more commonly, the integrated intensity of the peak must be calculated.

A peak identification program like the one described above can be used to establish the position of the peak maximum. A more precise procedure, however, is to determine the centroid or first moment of the system. For example, in Fig. 6.7 the necessary sampling of a mass spectral peak is shown. Its first moment or centroid $\bar{X}_m$ is given by $\bar{X}_m = \Sigma_{i=k}^{r} V_i X_i / \Sigma_{i=k}^{r} V_i$, where V_i represents the amplitude of the detector signal at the value of m/z denoted by X_i, and V_i is sampled at equal m/z intervals from $i = k$ to r. The effective integrated intensity or peak area is simply $\Sigma_{i=k}^{r} V_i$, the denominator of the above equation.

In order to translate the digital value of the centroid into the independent variable being

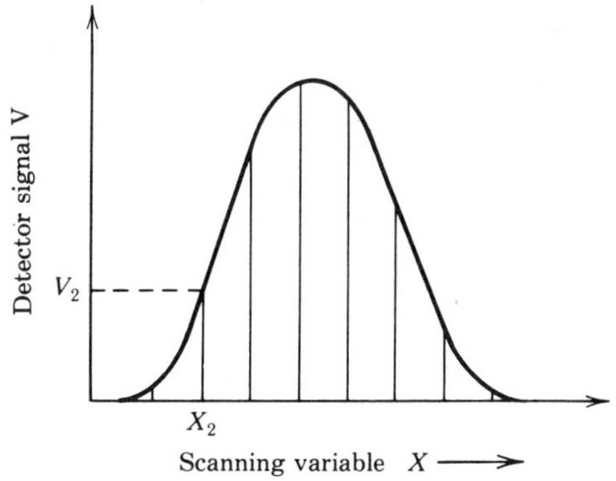

Fig. 6.7 Sampling of a mass spectral peak of an ion at equal m/z intervals to determine its centroid. Each sample of the detector voltage V is digitized to permit the actual calculations to be made by computer.

scanned, a calibration is required. In a spectrometer it is straightforward to use the wavelength scanning instrument to provide a value precise to within the specifications of the instrument. With a chromatograph a retention time for a standard will serve the same function. In mass spectrometry, a standard such as perfluorokerosene will produce fragment peaks across the range of values of m/z. As long as the first three peaks are entered, the program can be arranged to search for the additional peaks and include them as known values in the calibration that determines the m/z values for peaks in a mass spectrum of an analyte.

To obtain the integrated intensity or peak area is often more complex. In general, a valid baseline must first be established. Then noise must be minimized for both the baseline (the signal in the absence of analytes) and the sample. If $S/N < 100$, ways to diminish noise by signal averaging or other methods discussed in Chapter 12 will be important. If sampling to establish the peak is done at appropriate intervals, it should not be difficult to estimate the peak area as suggested above.

6.7 INSTRUMENT DISPLAY: GRAPHICS

With software providing graphics capability for an in-line computer the display of data in a variety of modes becomes feasible. A versatile instrument will provide a cathode ray tube (CRT) so that a user can interact with the modeling process and alter it within the limits of the software in use. Several benefits result:

1. Different ways to view experimental data can be tried visually in order to find the one that best presents the results.
2. The fitting of different mathematical models to data can be examined to test the appropriateness of different interpretations.
3. If graphics monitoring during the course of an experiment is possible, interactive adjustment of the experimental conditions may be feasible.

Recall that experimental results should in most cases be stored in memory as they are obtained and called up for these types of explorations.

Graphics software will furnish algorithms both for different display formats and for a variety of mathematical models useful in interpreting results. Formats for display may range from two- and three-dimensional plotting to bar graphs. Mathematical models will include linear least-squares plotting and may extend to multivariate analysis. A user will also need access to a printer/plotter for production of hard copies after deciding on the best type of display.

Example 6.7. Compare presentation of measurements by a chart recorder (potentiometric or capacitive type) with arrangements possible by a device with graphics capability such as a CRT or a digital printer/plotter.

Potentiometric or capacitive chart recorders display results in analog form. A pen traces the value of a dependent variable such as the absorbance of a sample while the chart moves at constant speed in synchronism with the scanning of wavenumber or other independent variable. The usefulness of IR and UV–VIS absorption curves for analytes scarcely needs comment. Similar displays are commonly obtained for structural studies by NMR and mass spectrometry.

With computer graphics capability many additional modes of presentation become available. Perhaps a bar graph display better shows the effect of a change in variable or an overlapped display of two curves illustrates their difference more clearly. Graphics capability also allows exploration of the information available in the data. Now baselines can be subtracted, derivatives of analytical curves may be taken up to any order provided for in the software, and so on. For example, one may learn by subtracting background whether or not a trend in a curve is an artifact of background and understand by differentiation whether a shoulder on an absorption curve represents the presence of a significant minor constituent in a mixture.

Example 6.8. How can computer graphics aid in interpreting supposedly linear data that actually seem to scatter considerably about a straight line?

Points will sometimes seem to deviate in a periodic fashion from a straight line. In this case, with graphics capability, fitting by a higher-order polynomial may be explored. The better fit may be desirable to facilitate interpolation of data. But a higher-order fit does imply a more complex dependence on the independent variable and the plausibility of that dependence must be questioned.

Alternatively, the user may view the scatter as indicative of error. In this case, it would be convenient to have measurements displayed as they are obtained to permit the user to adjust variables to secure better results as the experiment proceeds.

Once an appropriate representation has been developed one can obtain a hard copy from a plotter (see discussion below). Again, the ability of the computer to add alphanumeric characters in printing will ensure that curves are properly labeled. In the same manner it is often possible to develop block diagrams and charts for publications or oral presentations.

Graphics Software. Graphics software is available, both for particular designs of printer/plotters and for more general use. Hewlett-Packard has generated software called Plot-21 and Tektronics has similarly produced a package called Plot-10 for their own hardware. Graphics programs that are generally applicable include software for data plotting and spread sheets. Many software packages do not allow a user to add his own programs, however.

Standardization in this area has lead to the development of some graphics software independent of I/O devices. The Graphical Kernal System (GKS) intends to offer a "software interface" truly independent of hardware. For example, the interface must make it straightforward to draw a straight line in response to a simple command. Since plotting devices have varying numbers of resolution elements on the horizontal and vertical axes, such an interface must accommodate the different possibilities offered.

Three-Dimensional Images. While two-dimensional line graphs are in common use, to represent molecular structures spatially, to show the variation of chemical composition along the surface of a solid, or make 2D plots of NMR and other spectroscopies, it is valuable to be able to present a third dimension.

To represent such data, it is essential to decide on the perspective from which the surface is to be viewed or at minimum a viewing angle. Where a program allowing the choice of different angles is in use, one has an opportunity to select the angle that most nearly maximizes the ability to interpret the available chemical information. With a good choice *depth perception* becomes possible. The reader is familiar with contour plots in which lines of constant intensity of a dependent variable are plotted versus the other two variables to emphasize regions of large and small values.

The possibility of rotating a representation is especially appreciated. With molecular structures this proves important as one attempts to visualize properties that are dictated bv structure.

CRT Displays. A cathode ray tube (CRT) display unit is valuable not only for monitoring and possibly interacting with the measurement process but also lends itself to applications of computer graphics since by manipulating the electrical signals to the electron gun any portion of an image can be easily erased and replaced. As will be discussed below, color and shading may be introduced to include more information in such a display. Most displays are generated in raster fashion (see Section 3.5). Since in monocolor displays the electron beam intensity is modulated to represent either a 1 or 0 bit, the total image is often said to comprise a bit map. For such a display the image must be generated in memory and then displayed.

Resolution will depend on the bandwidth of the video monitor amplifier, with a practical limit for good monitors of 600–800 resolution elements per traverse of the screen. Vertically resolution is fixed by the number of scan lines; at present this number is about 240. Fortunately, it is possible to double the vertical resolution by using an *interlaced* CRT scan. In this type of display, horizontal lines are shifted slightly lower on alternate scans so that they appear between other scan lines. Separate data are displayed on the two adjacent scans so that half the points are displayed on each set of crossings. When interlaced scanning is used, longer persistence phosphors are ordinarily required since only in this way can screen flicker be minimized. Naturally, use of higher-performance video devices will secure greater resolution.

The positioning signals are usually received by a digital device called a graphics controller which directs them to a pair of DACs generating the horizontal and vertical deflection voltages that move the electron beam. When outputs are available from both DACs, a pulse turns on the beam of the appropriate electron gun and a spot or pixel is generated. Naturally, this type of screen image will be built from a series of spots.

A problem with a CRT display is that the persistence of phosphorescence on the screen is short. Commonly, refreshing of a display is performed by a circuit that operates directly from appropriate parts of memory rather than requiring any intervention by the CPU.

Color in displays ordinarily requires use of a different electron gun for each basic color—red, green, and blue. To obtain more than a basic set, individual colors must be mixed in varying intensity. With adequate memory, it is possible to assign to individual pixels on the screen as many as 6 bits or even a byte of memory. With this capability it is possible to call in a particular display for any 64 colors out of a total group of 256.

For a *vector* (or stroke) *graphics display* the arrangements are straightforward. The beam is turned on in one position and left on as it moves to the next, generating a stroke on the screen. If a straight line is needed, additional circuitry is required to ensure that the beam moves with uniform speed. In most instances a CRT will have a normal persistence phosphor. Resolution will be affected by the number of bits handled by the DAC, but usually the electron beam width is the limiting factor. Since the image must be constantly be refreshed, the refreshing electronics must have in internal memory the starting and stopping locations of each vector. To avoid flicker it is essential that the quality be sufficiently high for the number of pixels desired. If alphanumeric characters must also be displayed, their coding is usually provided by internal look-up tables in ROM. With 12-bit DACs, displays 4096 by 4096 refreshed at 30 times per second are possible.

Image Formation. As long as an instrument has generated the data necessary for an image, one can move with assurance to the production of a hard copy. In some instances data must also be entered from the laboratory. While it may be entered through a keyboard, efficient ways to avoid typing lists are available.

Choices for image formation involve scaling, location of position and so forth. It proves highly effective to be able to indicate choices by pointing to places on the screen rather than entering estimates of coordinates by keyboard. For example, a user can point to show the best position for the title of a figure, to suggest that a datum is to be deleted or modified, or to select an option from a menu of possibilities for types of display. The *light pen* is one device for pointing. It is a photodetector mounted in a probe that picks up the flash of light when a raster sequence moves the beam past the pen position. The timing of the electrical pulse can be used by the graphics controller to fix that beam location for acquisition by the computer. It is a disadvantage that the information obtained must be fed directly to the CRT controller, a facility not routinely available. Where a flat surface provides a spread from which data are to be entered into the computer, a device such as a *mouse* can be used to show choices. A mouse is a hand-held cursor that can be moved about on the table surface. When it reaches a point of interest, the position has already been noted and can be transmitted to the computer by pressing a button. In some versions additional buttons can be assigned functions according to the operation and process.

Printer/Plotters. Digital plotters offer considerable versatility. Often different colors may be used. With a *vector plotter* digitally encoded instructions move a pen in straight lines on a chart. Software will allow a choice of one or several colors of pens, if available, and instruct a pen tip whether to be up or down. Such a plotter moves a pen in straight lines from one stated position to a second. Since lines can go from any position on the plotter bed to any other, the plotters are essentially random-positioning devices. With microprocessor internal capability, one can call for an appropriate scale image and the computer will control drawing speed, and so forth. A plotter may use either a servo or stepping motor to position pens. The former is more expensive because of its versatility and speed while the latter tends to be more robust. With either device high resolution is obtainable (as high as 7650 × 10,300) because of mechanical precision.

Dot-matrix printers also are useful in graphics. Alphanumeric characters are easy to add since the software that fixed the necessary combination of dots can include ASCII characters. Where graphics are involved, each pin in the print head is independently controlled by the host microcomputer. An image is produced by a collection of dots which are placed on the chart as a slow raster. In the general case, since the paper will move only in a single direction, a map is formed in memory before a plot is begun. The raster video display program provides a computer with a bit map. Indeed, once a CRT display has been achieved, the formation of a hard-copy image is fairly simple: the data are simply dumped to the printer. A disadvantage of the arrangement is that resolution is limited to that achieved on the screen. With a laser dot printer, resolutions of 10 dots mm^{-1} can be realized. Nevertheless, such a printer requires nearly a megabyte of memory for a full page image.

REFERENCES

Some general references are:

1. J. T. Arnold, *Simplified Digital Automation with Microprocessors*. New York: Academic, 1979.
2. P. G. Barker, *Computers in Analytical Chemistry*. New York: Pergamon, 1983.
3. G. C. Barney, *Intelligent Instrumentation—Microprocessor Applications in Measurement and Control.* Englewood Cliffs, NJ: Prentice-Hall, 1985.
4. R. J. Bibbero, *Microprocessors in Instruments and Control.* New York: Wiley-Interscience, 1977.
5. K. L. Ratzlaff, *Introduction to Computer-Assisted Experimentation*. New York: Wiley-Interscience, 1987.
6. E. Sherman, *Data Communications—A User's Guide*. Englewood Cliffs, NJ: Prentice-Hall, 1985.

A volume on the introductory to intermediate level with emphasis on the digital electronics aspects of computers is:

7. D. J. Malcome-Lawes, *Microcomputers and Laboratory Instrumentation*. New York: Plenum, 1984.

Clear accounts of the incorporation of microprocessors/microcomputers in particular instruments are:

8. T. W. Barnard, Microprogramming, *Sci. Am.*, **248,** 50 (March 1983).
9. J. E. Wampler, Fluorescence spectroscopy with on-line computers: methods and instrumentation, in *Modern Fluorescence Spectroscopy*, E. L. Wehry (Ed.), Vol. 1. New York: Plenum, 1976.

Some volumes that emphasize specialized topics are:

10. R. Zaks, *Microprocessor Interfacing Techniques*. New York: Plenum, 1978.

11. P. He, J. P. Avery, and L. R. Faulkner, Cybernetic control of an electrochemical repertoire, *Anal. Chem.*, **54,** 1313A (1982).

12. W. A. Triebel and A. Singh, *The 68000 Microprocessor: Architecture, Software and Interfacing Techniques*. Englewood Cliffs, NJ: Prentice-Hall, 1986.

EXERCISES

6.1 Is it better to speak of imbedding a microprocessor or a microcomputer in an instrument? Discuss from a modular viewpoint considering (a) whether computer modules also serve as instrument modules and (b) any other ways to view the presence of the other computer modules in an instrument.

6.2 Compare the attractiveness of serial and parallel communication between instruments and an external (on-line) computer in terms of the connections (cable) required.

6.3 Describe as fully but as simply as possible what information must be communicated between particular computer modules when the next step in a program in assembler code is a CLEAR. (See Example 6.4.)

BASIC OPTICS

The design of instrumentation for optical spectrometry is informed by a general understanding of optical phenomena in all types of media. Accordingly, in Chapter 7 basic aspects of physical and geometrical optics are treated. These disciplines describe how light of varied frequencies (wavelengths) is refracted, reflected, polarized, diffracted, scattered, and undergoes interference in different media. This knowledge must be in hand to form and direct optical beams, separate beams into constitutent wavelengths, and perform other essential optical operations. In Chapter 8 the discussion is extended to the generation and detection of radiation as optical sources such as the hollow cathode lamp and detectors such as the photomultiplier tube are described. That discussion includes the electrical circuitry needed to establish dc biases for detectors and to couple them to preamplifiers or amplifiers. In the concluding chapter of this division, Chapter 9, monochromators and polychromators, the important wavelength dispersive modules, are examined. With the background of these three chapters the design of optical spectrometers and other optical instruments beçomes a matter of thoughtful selection of appropriate modules and their effective coupling.

Chapter 7

ASPECTS OF PHYSICAL AND GEOMETRICAL OPTICS

7.1 INTRODUCTION

The many types of interaction between matter and electromagnetic radiation in the optical range suggest the importance of this region of the spectrum. As may be seen in Fig. 7.1 the optical range covers essentially six decades of frequency from the far infrared (about 10^{11} Hz) through the vacuum ultraviolet (about 10^{17} Hz). The design of optical instrumentation arises out of a general understanding of the phenomena associated with the propagation of radiation in all types of media.

It is useful to classify optical phenomena according to the model employed to treat them. Thus, *physical optics* treats phenomena such as refraction that can be described by using a wave model for radiation. No picture of matter is necessary. *Geometrical optics* includes those phenomena where only the direction of energy transport needs to be accounted for, as in focusing a lens. Both these areas will be considered in this chapter. Interactions in which the quantized nature of radiation and atomic systems is essential to a description will, however, be deferred; they will be treated at modest length in the chapters on spectroscopy.

It should be emphasized that a completely rigorous treatment of any optical phenomenon requires quantum mechanics. Yet since wave theory provides a simple, rather elegant, model that yields widely valid equations, this more readily presented and interpreted approach will be employed. The "explanations" provided by this model are, of course, imprecise, as is known from a quantum-mechanical treatment.

7.2 ELECTROMAGNETIC WAVES

The classical model of light that will be used defines radiation as electromagnetic (EM) waves. Figure 7.2 pictures the simplest type of EM wave train, one which is linearly (or plane-) polarized and consists of a single frequency, that is, is *monochromatic*. Such a beam of radiation is best characterized as a moving force field with identifiable frequency, velocity, and intensity. Its frequency ν is truly constant, that is, independent of the medium. The radiation, of course, also has a wavelength λ, but it depends on the properties of a medium. It is defined for each

cycles per second

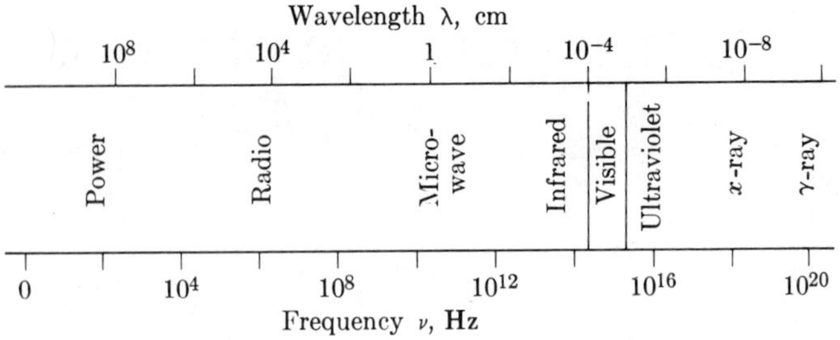

Fig. 7.1 Total spectrum of electromagnetic radiation. The optical region includes frequencies from the microwave to the X-ray range.

case by the fact that the product $\lambda\nu$ equals the velocity of radiation in the medium.* In a vacuum the product is just the speed of light c (2.998×10^8 m s^{-1}). This limiting case can be related to any other by use of the index of refraction n. By definition n is just the factor by which velocity is reduced if radiation traverses the medium instead of a vacuum. The general expression relating n and λ is

$$\nu\lambda_i = c/n_i \tag{7.1}$$

where i designates a particular medium. An example of the dependence of wavelength on velocity is illustrated in Fig. 7.3.

As shown in Fig. 7.2, the electric and magnetic aspects of EM waves may be identified with an electric field E and a magnetic field H. For a beam of radiation traveling in, for example, the z-direction, the variation of the two fields can be described by the expressions

$$E_y \sin 2\pi(\nu t - z/\lambda)$$
$$H_x \sin 2\pi(\nu t - z/\lambda). \tag{7.2}$$

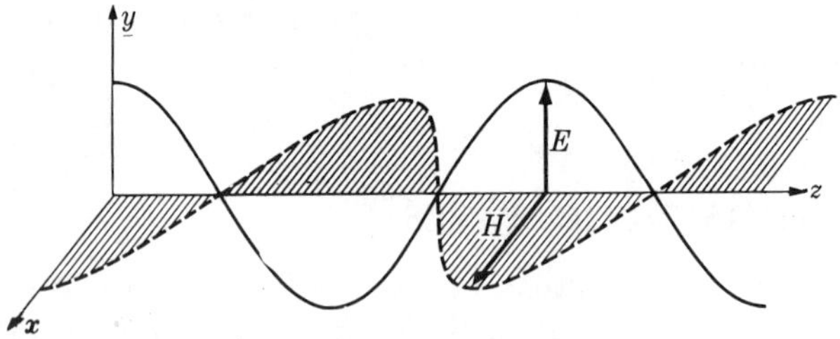

Fig. 7.2 Propagation of a linearly polarized electromagnetic wave train of a single frequency. The electric vector E and magnetic vector H represent the fields established at right angles to the direction of propagation.

*Strictly speaking, the velocity in this case is the phase velocity, not the actual rate of propagation, which is the group velocity. This matter will be examined in more detail in Section 7.6.

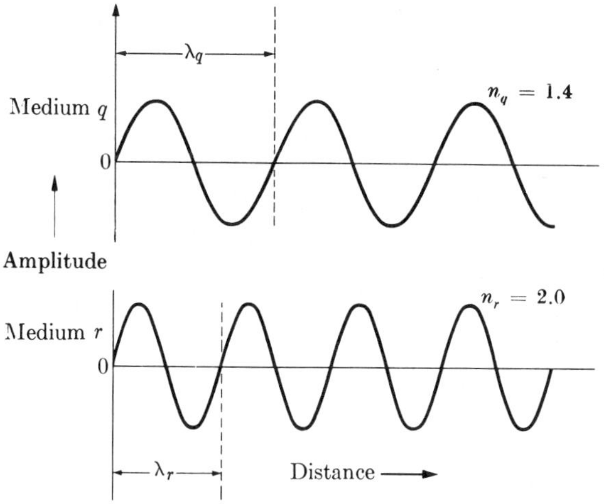

Fig. 7.3 Variation of wavelength of linearly polarized light with the medium through which it passes. Both its phase velocity and its wavelength are smaller in r, the medium of higher refractive index.

Note that the combinations νt and z/λ are dimensionless and define time and distance dependence, respectively. It should be apparent from the form of the equations that either the electric or the magnetic equation can be used in any situation in which *only* the wave character of radiation is important.

By convention, in the rest of the text the electric equation, that is to say, the electric vector, will be used exclusively unless otherwise noted. Since this aspect is the one active in quantum interactions with matter, the convention is especially useful here. It should be stressed that the usage does not in any sense imply that the magnetic aspect of radiation is less real.

The direction of vibration of the electric vector and phase relationships among the waves making up a wave train or beam are also important. In Fig. 7.2 a limiting case with regard to each was shown. The beam was *linearly polarized* since the electric vectors of all component waves were taken as vibrating in a single direction, parallel to the y axis. Other modes of polarization as well as unpolarized radiation will be treated in Sections 7.14–7.18. The beam in Fig. 7.2 also represented a limiting case in that it was *coherent*: the electric vectors of all the waves reached their maxima and minima at the same time and at the same point along the z axis. In short, all the waves were *in phase*. Commonly, however, as will be discussed below, waves are incoherent and originate either at different times or at different points in space. Thus, they show a phase difference.

Finally, it may be noted that the energy and intensity of an EM wave can be defined without reference to the particulate nature of light. The *intensity* I is a flux defined as the energy streaming through a unit area per second in the direction of wave propagation. Mathematically, it is just $I = E_y^2 c/8\pi$. Note that I is related to the square of wave amplitude and is not directly dependent on frequency or velocity.

7.3 WAVE INTERACTION

In optical systems, often by design, the paths of a great many EM waves intersect or are closely parallel. For example, refraction basically occurs because of the interaction of an incident wavefront with waves emitted by particles of matter that have been distributed by the front. How can nonabsorptive interactions like this be treated? In general, the answer is provided by the *principle of superposition*: force fields of waves add linearly in the region where waves overlap, whether they cross or actually coincide for a distance. At high intensities, however, linear superposition does not necessarily hold, but this is a topic beyond the scope of this book.

The application of the principle can best be pursued by a combined graphical and descriptive treatment. Again it will be convenient to use monochromatic, linearly polarized waves. For this purpose, Eq. (7.2) can be modified to focus attention on the phase or phase angle α:

$$E = E_y \sin(\omega t + \alpha), \tag{7.3a}$$

or in more general notation

$$y = A \sin(\omega t + \alpha), \tag{7.3b}$$

where y has now been substituted for E and A symbolizes the wave amplitude E_y. In both versions α includes the time aspect of *phase* and the aspect associated with distance and ω is the angular frequency ($\omega = 2\pi\nu$). Equations (7.3) are a statement of the magnitude of the electric field produced at any point by a passing EM wave train. The significance of the phase angle may be seen by setting t in Eqs. (7.3) equal to 0; then $E = E_y \sin \alpha$. Thus, α is identified as the constant that fixes the part of the cycle in which the wave disturbance is when timing starts.

In Fig. 7.4 two waves of different amplitude but of the same frequency are plotted as a function of time, first separately and then superposed. Their difference

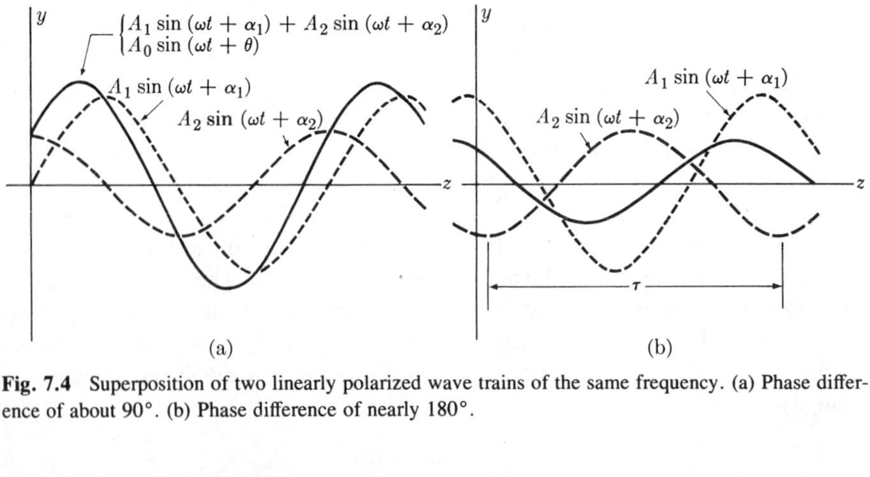

Fig. 7.4 Superposition of two linearly polarized wave trains of the same frequency. (a) Phase difference of about 90°. (b) Phase difference of nearly 180°.

is shown as about 90° in Fig. 7.4a and nearly 180° (or half a wavelength) in Fig. 7.4b. The result of superposition is the wave shown as a solid line. The expression for this result,

$$y = A_1 \sin(\omega t + \alpha_1) + A_2 \sin(\omega t + \alpha_2), \tag{7.4}$$

is obtained by vector addition of two equations like Eq. (7.3). Here A_1 and A_2 are the maximum amplitudes of the two waves. Trigonometric manipulation allows a reduction to

$$y = A_0 \sin(\omega t + \theta),$$

where A_0 is the maximum amplitude of the resultant and θ represents its phase angle.

Although practically it is impossible to produce a monochromatic wave like that described, theoretically it could be done by using an infinitely long train of waves. The short trains available experimentally can be shown to contain a finite spread of wavelengths. The matter is considered below in the discussion of Fourier analysis.

It is clear from Eq. (7.4) and Fig. 7.4 that the amplitude and thus of course the intensity of radiation resulting from interference depend strongly upon the difference in the phase angle of the waves. As $\alpha_1 - \alpha_2$ is increased, the amplitude of the resultant goes through a periodic variation. A maximum in *constructive* interference is secured when $\alpha_1 - \alpha_2$ equals 0°, 360°, 720°, ... Intensity minima, which mark the greatest degree of *destructive* interference, occur when the phase differences have the values of 180°, 540°, ...

An important question is the fate of the energy "lost" when waves interfere, since energy cannot be annihilated without some compensating effect. Clearly, no mass is being generated here, so the energy disappearing from one spot must be appearing elsewhere. This cannot be pictured by wave representation. Quantum mechanics is more successful with its probability function. According to it, the regions of destructive interference are those where there is a very low probability of finding photons. In other areas the probability of their being found is proportionately enhanced.

There is also the question of the *intensity* to be expected when a great many wave trains of the same frequency but of random phasing are superposed. All natural sources emit radiation of this kind in which the amount by which each train of waves is ahead or behind any other (either in time or in space) is purely a matter of chance. Analysis shows that n randomly phased wave trains of the same frequency give an array of waves that on the average has an amplitude increasing as the square root of n. The overall intensity must therefore increase *linearly* with the number of superposed waves, since intensity varies as the square of the amplitude. This result indicates, for example, that the intensity of radiation of 435.8 nm wavelength from a mercury lamp will on the average be the intensity from a single excited atom multiplied by the total number of atoms emitting. It is a somewhat striking result, since the emitting atoms are scattered over a large volume and emit quite randomly.

A more complicated instance is the interference of wave trains of different frequency as well as different amplitudes and phasing. If only two waves are involved, Eq. 7.4 may be modified as follows:

$$y = A_1 \sin(\omega_1 t + \alpha_1) + A_2 \sin(\omega_2 t + \alpha_2). \tag{7.5}$$

Regardless of how many waves are superposed, so long as the velocity of the component waves is the same, the resultant will move with this velocity. This behavior is illustrated in Fig. 7.5. For example, complex wave trains are found in the polychromatic radiation of hot tungsten filaments of incandescent lamps. If the radiation is moving through a vacuum, all the component frequencies will have the same velocity, but in air or any other material dispersion will complicate the superposition.

It is worth stressing that the basic equations of EM wave propagation are *linear* under all common conditions, whether the wave or the quantum character of light is predominant. In other words, two or more light waves can pass at the same time through the same section of a medium, each behaving independently of the others. The possibility of strict superposition of waves need only be abandoned when very intense radiation is encountered, as in the use of lasers. Similarly, the probability of absorption or emission involving a photon is independent of the number of other photons present under normal conditions, that is, there is no dependence on intensity.

The simplicity lent to EM processes by linearity is important, for nonlinear phenomena are well known in electronics (modulation, demodulation, parametric amplification) and in acoustics, where they give rise to beats between waves of

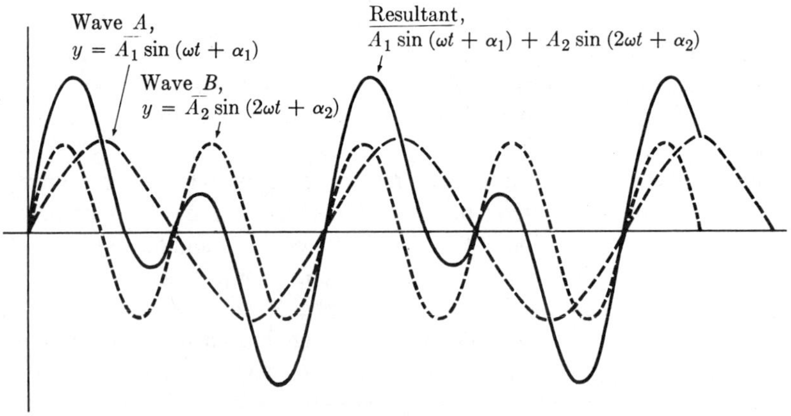

Fig. 7.5 Superposition of wave trains of different frequency and amplitude. Wave B has twice the frequency of wave A.

differing frequencies. Involved in these nonlinear phenomena are mixing of different waves of distinct phase and frequency by a medium or a device with a nonlinear behavior. Sum and difference frequencies are obtained and harmonics may also be generated.

A further contrast with electronics may be made in applications involving light from ordinary sources. Usually EM wave trains are the result of contributions from very many randomly occurring emission processes and are incoherent. Interactions of light of a particular frequency with a medium are in part hard to define because of such incoherence. It will be recalled that in electronics the phase of a signal is usually well defined.

7.4 FOURIER ANALYSIS

In some cases it is desirable to be able to reverse the process of superposition. Experimentally a prism or a diffraction grating is fairly successful in resolving polychromatic radiation, although neither device can produce wave trains of a single frequency. Only by mathematical analysis can a complex wave train be exhaustively resolved into a set of simple sine waves. The arbitrary nature of the process should be understood. The frequencies obtained in the analysis will depend largely on the assumption made about the constitution of the complex wave train.

In the most widely used method of resolution, Fourier analysis, the components are taken as monochromatic waves, each of an integral frequency. If ω is the basic frequency, then the waves will have the frequencies ω, 2ω, ... The complex wave is represented by an expression with a series of sine terms,

$$y = A_0 + A_1 \sin(\omega t + \alpha_1) + A_2 \sin(2\omega t + \alpha_2) + \cdots, \qquad (7.6)$$

where $A_0, A_1, A_2, \ldots$ are amplitude constants. Note that the terms are of the same form as those in Eq. (7.5).* In general, the coefficients $A_0, A_1, A_2, \ldots$ are evaluated by computer. For a discussion of Fourier series, Fourier integrals, and Fourier transforms, references should be consulted. (See, e.g., pp. 621–633 of ref. 9.)

The Fourier series representation is limited in that it represents in a satisfactory fashion only essentially infinite waves. When a Fourier analysis is to be performed on ordinary optical and electrical signals of limited duration, a Fourier integral is required. It differs from the series in that an infinite number of component waves that are only infinitesimally different in frequency are assumed. The application of the integral yields a continuous distribution of component waves rather than a series of separate ones, each with a discrete frequency and amplitude. Figure 7.6 shows the distribution of frequencies contributing to a spectral line, which always has a width. For example, the line might be the 435.8-nm mercury wavelength.

*The most general Fourier series also contains terms in the cosines of the frequencies.

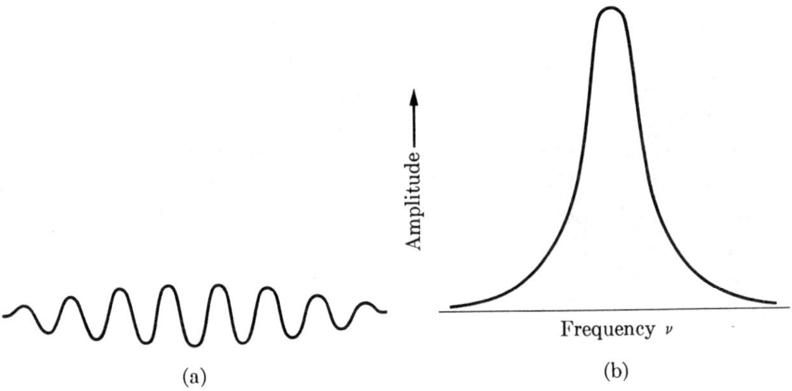

Fig. 7.6 The distribution of frequencies in a spectral line of finite width. (a) Wave contour of the "disturbance." (b) Amplitudes of the component frequencies.

The length of wave disturbance forming the image is shown at the left, and the amplitude of the component frequencies at the right.

7.5 SECONDARY EMISSION

When a matter particle—molecule, atom, or ion—is subjected to the periodically varying electric field of an EM wave, the particle is polarized, that is, a dipole is induced, usually one oscillating in phase with the incident wave. This process is suggested in Fig. 7.7. Assuming the impinging radiation is not of a frequency characteristically absorbed, the oscillating dipole instantaneously generates a secondary wave of the same frequency. In the wave model it is the interaction or interference of this secondary emission with the incident radiation that is the basis for the phenomena of refraction (transmission), reflection, and scattering. These will be treated in subsequent sections.

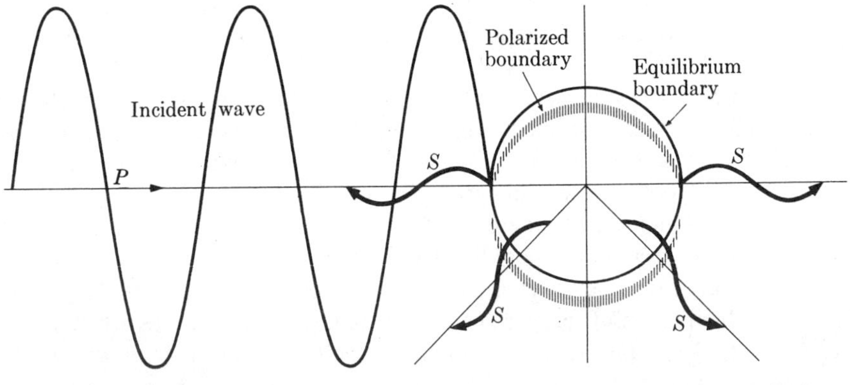

Fig. 7.7 Oscillative dipole induced in an atom by the passage of a monochromatic, polarized wave. The wavelength is assumed long with respect to atomic dimensions. Waves S are secondary waves produced by the oscillating dipole.

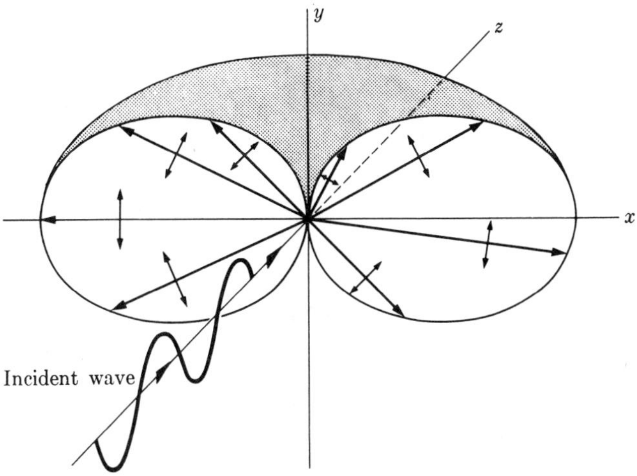

Fig. 7.8 Intensity pattern of the secondary emission of an oscillating dipole. The emitting particle is at the origin. The toroidal or doughnut-shaped envelope represents the intensity of the emission. For example, in the *xy* cross section shown the intensity of emission in a given direction is proportional to length of the radius vector. All secondary rays are linearly polarized.

In Fig. 7.8 the relation of secondary emission to the oscillating dipole is refined to permit graphing of intensities and angular distribution. Note (a) the relative amplitudes, (b) that the secondary waves are linearly polarized, and (c) that there is no emission perpendicular to the direction of travel of the primary wave. The greater the polarizability of a particle, that is, the larger the refractive index of the medium made up of the particles, the greater the amplitude of the secondary waves will be.

In quantum theory the secondary emission of radiation is regarded as a process of excitation to some unstable and unobservable state followed by an immediate reemission (in about 10^{-14} s). The incident light quantum is considered to be absorbed and reemitted.

Usually the excited molecule or other species drops back to the initial state. If the exciting frequency is in the visible or UV spectrum, however, molecules sometimes fall to an excited vibrational or rotational level, giving rise to secondary emission of different wavelength, a phenomenon termed the *Raman effect* (see Chapter 18).

The materials of greatest interest in physical optics belong to the class designated *dielectrics* or nonconductors. These have no, or at most very few, free electrons or ions and are usually optically transparent. Since electrically conducting solids either absorb radiation or reflect it strongly, they are optically of less interest and will be considered only very briefly in connection with reflection in Section 7.8. In addition, the present discussion will be concerned only with optically *isotropic* media, substances whose optical properties are the same in all directions. Anisotropic substances will be considered later in connection with polarized light in Section 7.15.

Whenever radiation passes from one dielectric material into another, it is partially reflected and partially transmitted. In the new medium it retains its characteristic frequencies but is propagated with a different velocity and thus, with a different wavelength. In general, the radiation also changes direction abruptly at an interface, that is, is *refracted.*

If the interface is small in extent, a condition that exists whenever a small particle is suspended in some other material, the radiation is scattered rather than reflected. As will be discussed in detail in Section 7.12, the division between reflection and scattering occurs when dimensions become of the order of a wavelength of the incident beam. Absorption may also occur at the interface and in the bulk of the second material, but since it is definitely a quantum effect it is considered separately in the chapters on spectrometry. In Fig. 7.9 the interfacial effects are shown schematically for unpolarized monochromatic radiation incident from air onto a piece of glass.

7.6 REFRACTION

The reader is probably familiar from a general physics course with the use of the Huygens wavefront construction to account for the angular relationships of refraction. (See pp. 25ff of ref. 1.) This approach is satisfactory for deducing the familiar statement involving velocities, Eq. (7.7), and Snell's law, Eq. (7.8).

$$\frac{\sin \phi}{\sin \phi'} = \frac{v_1}{v_2}; \tag{7.7}$$

$$\frac{\sin \phi}{\sin \phi'} = \frac{n_2}{n_1}. \tag{7.8}$$

In the expressions, the symbolism is taken from Fig. 7.9, that is, ϕ and ϕ' are angles of incidence and refraction measured relative to a perpendicular to the interface. The subscripts refer to the first and second medium respectively, and n is the refractive index. The ratio of sines becomes indeterminate for vertical incidence; in that case, experiment shows that the radiation is not deviated but makes the usual change in velocity.

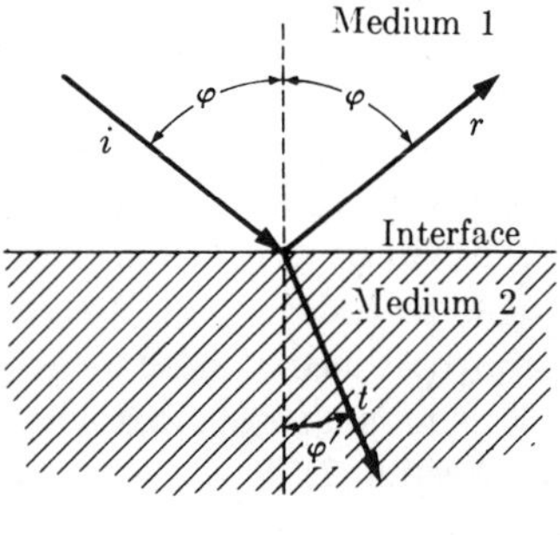

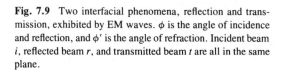

Fig. 7.9 Two interfacial phenomena, reflection and transmission, exhibited by EM waves. ϕ is the angle of incidence and reflection, and ϕ' is the angle of refraction. Incident beam i, reflected beam r, and transmitted beam t are all in the same plane.

For the common case of incidence from air, Eq. (7.8) can be simplified. Since n_1 is close to unity, we can use instead the nearly exact relation

$$n_2 = \sin\phi/\sin\phi'.$$

Were it possible to secure and work with truly monochromatic radiation, Eq. (7.7) would define a "true" velocity. But even the narrowest spectral emission line is actually a band of frequencies (Section 13.3), each of which will experience a different n and thus travel at a different velocity, as discussed in the next section. Accordingly, for a so-called monochromatic beam the velocity v of Eq. (7.7) represents a *phase velocity* at which a wavefront of a given frequency moves through a medium. The front is determined by combination of the secondary emission of atoms in the forward direction and the residue of the incident radiation (that propagates at velocity c). Thus v, the phase velocity of the combination wavefront, can exceed c if the phase of the secondary emission leads that of the incident radiation, or be less than c if it lags the incident light. (See ref. 4.)

In contrast to refractive index measurements, direct determinations of light velocities in matter involve observation of the rate of propagation of energy, thus a *group velocity* u. It is a "pulse-travel-time" velocity that makes allowance for the range of wavelengths present in a pulse and is defined relative to v by the expression $u = v - \lambda(dv/d\lambda)$. Differences in u and v may amount to several percent near an absorption band.

Example 7.1. Michelson measured the velocity of light in carbon disulfide by a rotating mirror device and found $c/u = 1.77$. He also showed by use of a refractometer that $n_2 = c/v = 1.64$.

A natural restriction on refraction that is of particular interest is the existence of a limit to the angle of refraction in the medium of higher refractive index termed the *critical angle* ϕ_c. If this is the second medium, radiation will be refracted in it at the critical angle when incident from medium 1 on the interface at 90°. Though this case of grazing incidence can be approached only as a limit, it leads to a mathematical definition of ϕ_c since, as $\sin\phi \to 1$, Eq. (7.8) becomes

$$\sin\phi' = \sin\phi_c = n_1/n_2. \tag{7.9}$$

Conversely, if radiation is incident on the interface from medium 2, total reflection (back into the medium) occurs when the angle of incidence exceeds ϕ_c (Section 7.8).

Two questions may now be posed: (1) if radiation is incident from a vacuum onto a medium, how is the wavelength decreased without any loss of energy, and (2) how are a refracted and a reflected beam formed, since secondary emission occurs in all directions in the plane formed by the incident beam and the interface?

The answer to the first may be expressed qualitatively in terms of the fact that the electric vector of a wave induces a dipole of the opposite sign. Consider a wave incident from a vacuum onto an isotropic medium. The positive end of the radiation vector effectively "moves" into an atomic region which has a net negative charge. The interaction will be much the same as if ones tries to push a positive charge through a medium of negative charges. The positive charge is slowed until some equilibrium velocity is achieved. The wave is slowed in the same way. Since the number of wavefronts arriving at the interface per second is not changed, it follows that the distance between successive fronts in the medium is smaller. The wavelength of the radiation has been shortened upon entering the medium.

To answer the second question, one must examine the interference or combination of the incident and secondary waves along the wavefront in the medium. In the refracting medium, the particles, whether atoms, molecules, or ions, are fairly regularly spaced in spite of small, random thermal movements. Therefore the lateral secondary radiation from the induced dipole destructively interferes. In the surface, the interference is complete for isotropic media except in the directions of (a) the refracted ray and (b) the reflected ray. If conditions are right for total reflection, only the latter appears.

Dielectric Constant and Refractive Index. It is of interest that the dielectric constant ϵ plays the same role in the electrical range as does the refractive index in the optical range, that is, serves as a measure of polarization. The quantities are very simply interrelated for substances whose magnetic susceptibility is unity, that is, for almost all transparent materials, by the equation $\epsilon_\nu = n_\nu^2$ where the subscript ν indicates that the expression holds when both are measured at the same frequency.

A representative picture of the dependence of n or $\sqrt{\epsilon}$ on frequency from dc through the optical range is shown in Fig. 7.10. Several types of polarization are operative in the radiofrequency range. As the frequency rises, each type in succession is "phased out." There is less time for it to occur and ϵ falls. For example, the decrease in n in the radiofrequency range is associated with increasing inability

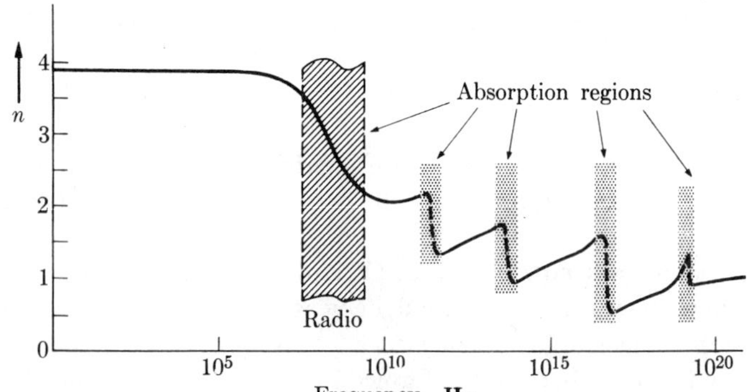

Fig. 7.10 Variation of n or $\epsilon^{1/2}$ with frequency for a polar dielectric.

of polar molecules to orient in phase with the incident radiation. Across the broad range of frequencies there is a progression toward lower and lower refractive indices until $n \simeq 1$ in the X-ray region. Here there is not even time for electronic polarization.

7.7 OPTICAL DISPERSION

It is a commonly observed phenomenon that the refractive index depends on frequency. For example, it is this dependence that allows a glass prism to spread a beam of white light into a spectrum. Thus the terms in Eqs. (7.7) and (7.8) should actually have a subscript designating the frequency.

The variation of refractive index with frequency is called *dispersion*. In order to describe the phenomenon more precisely, it is necessary to go beyond the wave model to one that will represent the interaction of bound electrons with radiation. For this purpose a simple classical model in which a dielectric is considered as a collection of bound electrons uniformly distributed in a lattice of atomic cores is adequate.

The incidence of light will cause a forced oscillation of each bound electron that will ordinarily be in phase with the periodic EM field. Since a nucleus is comparatively massive, it is reasonable to assume that in the short intervals before field polarity reverses it will not move. If radiation of ordinary intensity is incident, only small displacements of electrons from equilibrium positions need be considered, and a Hooke-law type of restoring force can be employed in developing an equation of motion.

To obtain an exact solution a convenient boundary condition to impose is that the system will have a natural or characteristic frequency, namely one at which it will absorb energy. For each "kind" of bound electron the value of natural frequency will differ. These different oscillations will not interfere, at least not at intensity levels below those of laser beams. The values of displacements obtained by solving the equation for one kind of electron lead to the following expression for polarizability in the y direction, α_y,

$$\alpha_y = \frac{e^2/m}{4\pi^2(\nu_y^2 - \nu^2)} \tag{7.10}$$

where e is the electronic charge, m is the electron mass, and ν_y is the natural frequency for the electron in the y direction. By introducing the expression for molar refraction, $M(n^2 - 1)/(n^2 + 2)\rho = 4\pi N_0\alpha/3$, the refractive index can be related to the polarizability. Here N_0 is Avogadro's number, M is the molecular weight, and ρ is the density. Since there are usually several electrons that can be moved with some ease, Eq. (7.10) is ordinarily a summation of several terms. Combining such a summation with the expression for the molar refraction yields finally the expression

$$\frac{M}{\rho}\frac{(n^2 - 1)}{(n^2 + 2)} = \frac{N_0 e^2/m}{3\pi}\sum_i \frac{f_i}{\nu_i^2 - \nu^2} \tag{7.11}$$

where f_i is the number of electrons of kind i in the atom (frequently termed the oscillator strength), and ν_i is the characteristic or natural frequency for the vibration of these electrons in the y direction. The type of interaction of radiation and matter described by Eq. (7.11) is termed *electronic polarization.*

The relationship between frequency and refractive index may be simplified for gases. It can be shown that for gases $n \simeq 1 + a$ where a is a constant much less than unity. Thus the term $n^2 + 2 \simeq 3$ and the term $n^2 - 1 \simeq 2(n - 1)$. With these approximations, Eq. (7.11) becomes

$$n - 1 = \frac{\rho N_0 e^2/m}{2\pi M} \sum_i \frac{f_i}{\nu_i^2 - \nu^2}. \tag{7.12}$$

A quantitative picture of the dependence of n on frequency may now be attempted. At low frequencies, where $\nu << \nu_i$, it is seen that n remains essentially constant and greater than unity. As ν approaches ν_i, the refractive index must become larger. Finally, when $\nu = \nu_i$, on the basis of this simple model, the refractive index becomes infinite. Because of damping effects that have not been considered the index actually remains finite. Note that the denominator $\nu_i^2 - \nu^2$ changes sign when $\nu > \nu_i$. Accordingly, as the frequency exceeds a characteristic absorption frequency, n must fall from a high value to less than unity. Here phase velocities will exceed c, the vacuum velocity of light. As ν increases further, normal dispersion is again observed. The anomalous behavior of n in the vicinity of absorption frequencies is shown in Fig. 7.11, as predicted by this simple model. It is significant that the behavior of n is intimately related to the existence of characteristic absorption frequencies.

Infrared Dispersion. At infrared frequencies the vibration of the electric vector of radiation is sufficiently slow that atomic nuclei can be displaced relative to one another before the direction of the imposed field reverses. In other words, in this

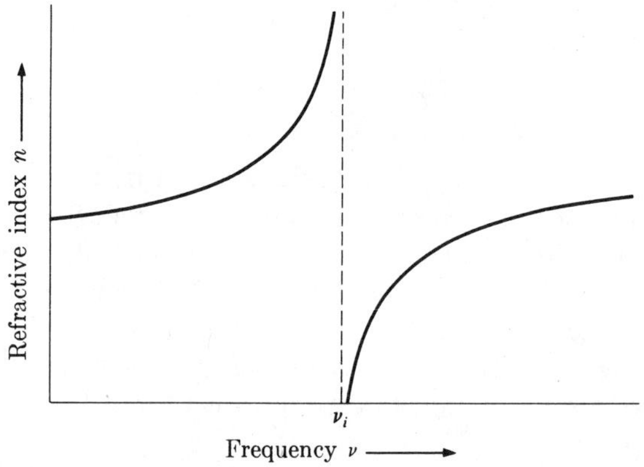

Fig. 7.11 Behavior of the refractive index n and the absorption index k for a dilute gas through the region of an electronic transition.

spectral range the field perturbs the normal vibrations and rotations of atoms and gives rise to *atomic polarization*. As a result of the greater magnitude of this form of polarization, the index of refraction is proportionately larger in the infrared. Again there is a sudden decrease in n at frequencies just larger than each characteristic vibrational absorption frequency.

7.8 REFLECTION

Reflection occurs whenever radiation is incident upon a boundary between dielectrics across which there is a change in refractive index. The quality of the surface has much to do with the nature of the phenomenon. The usual law of equal angles of incidence and reflection applies only in the case of regular reflection from a smooth surface, called *specular* reflection. Irregular surfaces give rise to diffuse reflection, of little interest in optical work. For most angles of incidence and types of interface, the reflected beam has a different total intensity, state of polarization, and phase from the incident one. In addition, if many frequencies are present there are usually chromatic differences in the reflected radiation as well.

The amount of reflection, that is the ratio of the intensity of the reflected radiation to that incident on the interface, is termed the *reflectance* or reflectivity. Since reflection can be considered as resulting from the secondary radiation of atomic dipoles induced in the surface, it is related to the polarizability of atoms. It seems reasonable to expect that for a given angle of incidence ϕ the reflectance will be greater the larger the refractive index of medium 2. Actually, allowance must be made for the interference of secondary and incident radiation, so that the refractive index of medium 1 is involved as well.

It can be shown for the simplest case, that of normal incidence onto the flat surface of a nonabsorbing medium, that the reflectance R is defined as

$$R = \frac{(n_2 - n_1)^2}{(n_2 + n_1)^2} = \frac{(n_{21} - 1)^2}{(n_{21} + 1)^2}, \tag{7.13}$$

where $n_{21} = n_2/n_1$. Clearly, the greater the difference in refractive index of the media, the larger the reflectance. Note also that the magnitude of reflectance at normal incidence is independent of which medium has the greater n.

Example 7.2. According to Eq. (7.13), visible light falling vertically from air onto a plate of ordinary glass of $n = 1.50$ would undergo about 4% reflection ($R = 0.04$). As light emerges from the far surface of the glass plate it will again be normally reflected with $R = 0.04$. The intensity of the transmitted radiation is thus approximately 92% of that incident. It may also be said that the plate has a transmittance of 0.92.

On the other hand, if $n_2 = n_1$ there can be no reflection. For example, X-rays cannot be reflected at all efficiently into air at "ordinary" angles, since $n_{\text{X-ray}} \sim 1$ for most substances. Note also the bearing Eq. (7.13) has on visibility: objects become invisible when surrounded by a medium of identical refractive index. A way to determine approximate refractive indices for glasses and other nonconducting solids such as minerals is suggested by this result (see Section 19.4).

When radiation falls obliquely on an interface, the reflectance varies with the angle of incidence. Polarization and phase changes also occur (see Section 7.14) but may be ignored if only the overall reflectance is of interest. It was shown by both Fresnel and Maxwell that the appropriate expression for the reflectance of a nonabsorbing medium is

$$R = \frac{1}{2}\left(\frac{\sin^2(\phi - \phi')}{\sin^2(\phi + \phi')} + \frac{\tan^2(\phi - \phi')}{\tan^2(\phi + \phi')}\right). \tag{7.14}$$

To solve the equation, the appropriate angles of refraction ϕ' are calculated from the known index n at the frequency of interest and for selected ϕ's from Eq. (7.8). The squares of the sine and tangent terms should be noted. These enter when amplitudes are converted to intensities. If only the ratio of wave amplitudes were of interest, unsquared trigonometric terms would be involved.

An illustration of the way reflectance changes with angle of incidence is given in Fig. 7.12. Although the graph plotted is for the reflection of unpolarized monochromatic light incident from air onto a glass of $n = 1.50$, the result is typical of optical cases. It is interesting that the reflectance is (1) approximately constant up to angles of incidence of 60°, and then (2) rises quickly to 100% at 90°, or grazing incidence. The reader may confirm the behavior by looking at this page at a low angle with respect to an overhead light.

Regardless of the angle of incidence, serious intensity losses occur when radiation must traverse many lenses, prisms, cells, and so on. For example, there need be only a few plates or lenses of $n = 1.50$ to reduce the transmitted power by 50%, since *each* surface has a reflectance of 0.04. Some ways of cutting reflection loss are suggested below.

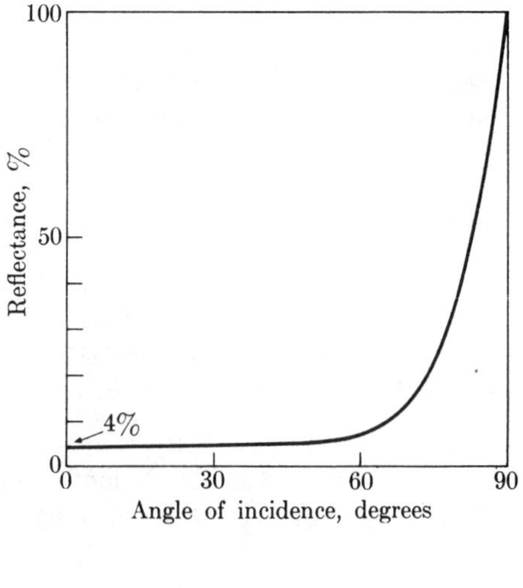

Fig. 7.12 Reflectance as a function of angle of incidence for unpolarized monochromatic light incident from air onto a glass plate of $n = 1.50$.

Multiple Reflection. An intimately associated phenomenon is that of multiple internal reflection. In a slab of dielectric internal reflections cannot be avoided although they may be minimized, allowed for, or utilized. The nature of these reflections may best be seen in the case depicted in Fig. 7.21 on page 215.

Some of the radiation reflected from the far face of the plate (ray 1′) will again be reflected from the front face (ray 2). In magnitude, ray 2 will be about 4% of ray 1′ (as before). The process will, of course, continue indefinitely. The radiation that actually emerges from the far side of the plate is the reduced "primary" beam plus a small supplement of radiation from the multiple reflections.

In the more common case of oblique incidence, the multiply reflected rays deviate, of course, from the primary beam. Together with the initially reflected energy, these rays constitute *stray radiation*. Since they will not undergo the same optical alterations, their reentry into the primary beam at any point will cause spurious results (see Section 9.7).

Reducing Reflected Losses. Two general means are used to reduce reflection when it is undesirable. Optical elements that are in contact, such as those in achromatic lenses, are cemented with a material of similar refractive index (rather than left with air gaps between them). This approach is also a basis for the oil-immersion technique used in microscopy, where a high-refractive-index oil fills the space between sample slide and objective lens.

The second method is the deposition of a low-reflection coating designed to secure destructive interference of light that would otherwise be reflected. The type of deposit and thickness of layer are decided on the basis of the following criteria:

1. The thickness of coating must be just one-fourth the wavelength in the center of the spectral band whose transmission is desired.
2. The intensity of radiation reflected from the outside of the coating must be equal to that reflected from the interface between the coating and the optical element. This condition is met if the refractive index of the coating, n_c, is given by

$$n_c = \sqrt{n_1 n_2},$$

   where the subscripts refer to the usual media 1 and 2, and light is vertically incident.

In Fig. 7.13 the manner in which the low-reflection coating reduces losses is shown diagrammatically. Any wave that emerges from the front of the coating after reflection at the coating–medium 2 interface will have traveled one half-wavelength more than a wave of the same frequency reflected directly from the coating.* On superposition, the two rays will destructively interfere: the energy that

*Actually, there is a phase change on reflection from a surface of higher refractive index. It need not be considered here, however, since the change is the same for each of the interfering rays.

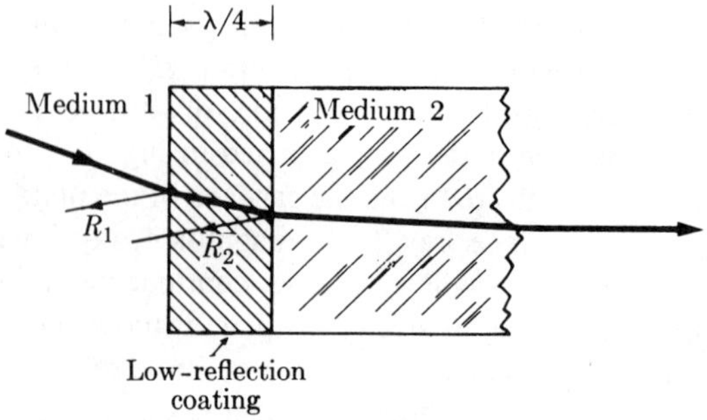

Fig. 7.13 The reduction of reflection by a low-reflectance coating. Although radiation is usually almost normally incident on the coated surface, it is drawn as obliquely incident to allow the reflected rays to be shown. Rays R_2 and R_1 superpose and destructively interfere, since they are nearly one half-wavelength out of phase.

would appear in reflected rays is shifted to the transmitted beam. If medium 1 is air ($n_1 = 1$), the value of n_c must be about $\sqrt{n_2}$. Thus, glass lenses of $n = 1.50$ to 1.65 require a coating of a transparent layer with n in the range of from 1.22 to 1.28.

Either magnesium or calcium fluoride is commonly used as a coating. Each is applied in layers of the desired thickness by vacuum evaporation techniques. The deposit has too high ($n = 1.35$) a refractive index, but is still quite effective. Most lenses intended for the visible region are coated so as to pass green light (550 nm) well. Even for red and blue light, their reflectance is no more than about 0.6%.

Reflective Coatings: Mirrors. Two other general means are used to increase reflection. For front-surface mirrors for use at normal or low-angle incidence a uniform layer of a metal of high reflectivity such as aluminum, or less commonly silver or gold, is vacuum deposited. An overlayer of SiO_2 is often added to lessen corrosion and increase durability.*

A second method permits development of reflectivities up to 95–99% over narrow wavelength ranges (10–30 nm). It involves deposition of a series of dielectric layers of alternating high and low index of refraction.

Total Reflection. Under certain conditions, radiation fails entirely to pass through an interface between transparent media and is totally reflected. (The trivial case of grazing incidence may be neglected.) Total reflection occurs when radiation strikes an interface with a medium of *lower* refractive index at an angle greater than the critical angle (Section 7.6). A graph of reflectance for this type of incidence is given in Fig. 7.14. It is noteworthy that the reflectance rises very steeply as the angle of incidence approaches the critical.

*Reflectivity will exceed 88–99% in general over the 180–3000 nm wavelength range. Below 180 nm the oxide layer on aluminum absorbs; if fresh aluminum is overcoated with a MgF_2 or CaF_2 layer, reflectivity remains at about 80–85% down to 120 nm. Usually a layer of SiO_2 is added also.

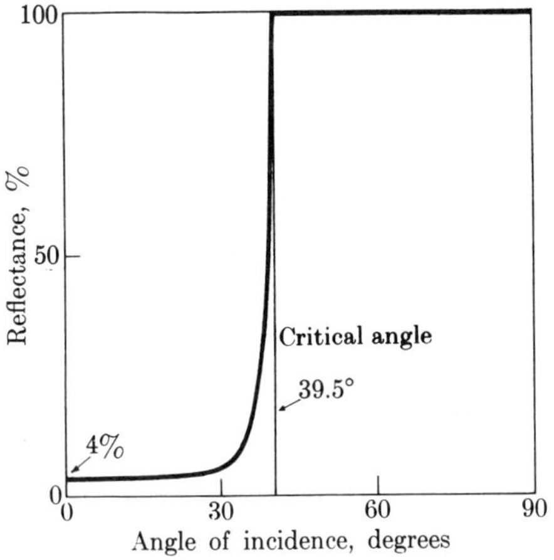

Fig. 7.14 Magnitude of internal reflection at various angles of incidence. Monochromatic radiation is incident from a glass of $n_D = 1.57$ into air.

In optical systems total internal reflection enjoys at least two applications that arise because all the power is reflected if the second medium is nonabsorbing.

1. *Totally reflecting prisms* such as are illustrated in Fig. 7.15 may be used instead of mirrors in situations where losses in intensity would be serious. Here the only losses occur at the entrance and exit faces.

2. *Fiber optics*, bundles of thousands of fine parallel fibers that are coated, can be used to transmit radiation into places that are difficult to access with very little loss of intensity. Multiple internal reflections will occur within each fiber as it transmits radiation.

If the second medium is absorbing, total internal reflection often represents a useful means of obtaining its absorption spectrum. This application will be considered further in the following section.

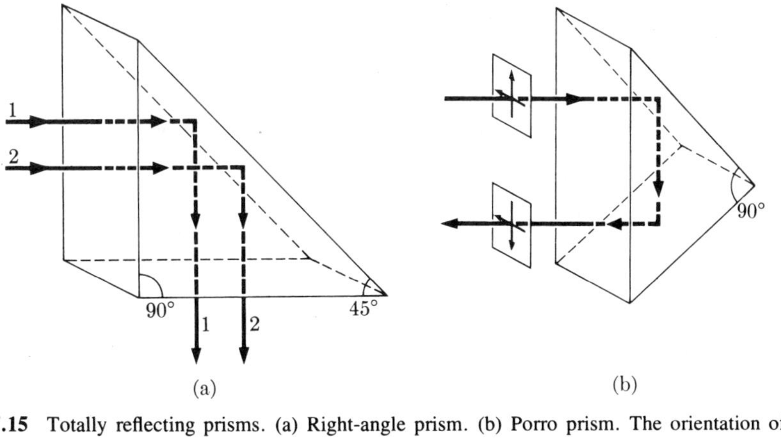

Fig. 7.15 Totally reflecting prisms. (a) Right-angle prism. (b) Porro prism. The orientation of an image on entering and emerging is shown by the arrows.

Aside from the obvious limitation imposed by a restricted range of transparency, there are other precautions that must be observed in using total internal reflection. Dispersion, polarization effects, and image rearrangements can occur under varying conditions and must be taken into consideration [4, 10]. For example, in the correct use of prisms shown in Fig. 7.15, radiation is always perpendicularly or nearly perpendicularly incident. Note also that the Porro totally reflecting prism brings about a reorientation of any image transmitted through it.

Reflection From an Absorbing Medium. If a medium is absorbing, for example, a red plastic or a metal, the reflectance from its surface at normal incidence can be calculated from Eq. (7.13) by use of the complex refractive index $\hat{n}$. By definition,

$$\hat{n} = n - ik, \tag{7.15}$$

where k is the attenuation index of the medium.* The reflectance equation for normal incidence becomes

$$R = \frac{(n_2 - n_1)^2 + k_2^2}{(n_2 + n_1)^2 + k_2^2}. \tag{7.16}$$

Except for strongly absorbing substances, the product k is small relative to the parenthetical term and can be neglected for normal (external) reflection from an absorbing medium. In the following section this discussion will be extended to internal reflection from an absorbing medium.

Example 7.3. What is the external reflectance in air of a gray glass plate of $\alpha = 10^4$ and $n_2 = 1.50$ at 500 nm? Since $\lambda = c/\nu$, $k = \alpha\lambda/4\pi = 10^4 \times 500 \times 10^{-9}$ m$/4 \times 3.14 = 3.98 \times 10^{-3}$. The magnitude of k^2 will be quite negligible. Thus, $R = 0.04$, as for a nonabsorbing plate.

7.9 OPTICAL MATERIALS

For substances to be suitable as transmitting walls of absorption cells, optical windows, lenses, and prisms, both high transparency and transmission are critical. The former depends on minimum spectral absorption and thus, on the absence of electronic and other quantum transitions in regions of interest. Absorption curves are graphed in Fig. 7.16 for several widely used optical substances. Note that all

*Alternative definitions are in use. The other definition is $\hat{n} = n(1 - ik)$; when used, the product nk appears in equations in place of k. The usual absorption coefficient α can also be used in place of k. For example, α appears in the Lambert exponential law for transmission $I/I_0 = e^{-\alpha d}$, where I and I_0 are the transmitted and incident intensities, respectively, and d the thickness of the medium. The substitution of α can be made by use of the expression $k = \alpha c/4\pi\nu$ where c is the vacuum velocity of light and ν its frequency.

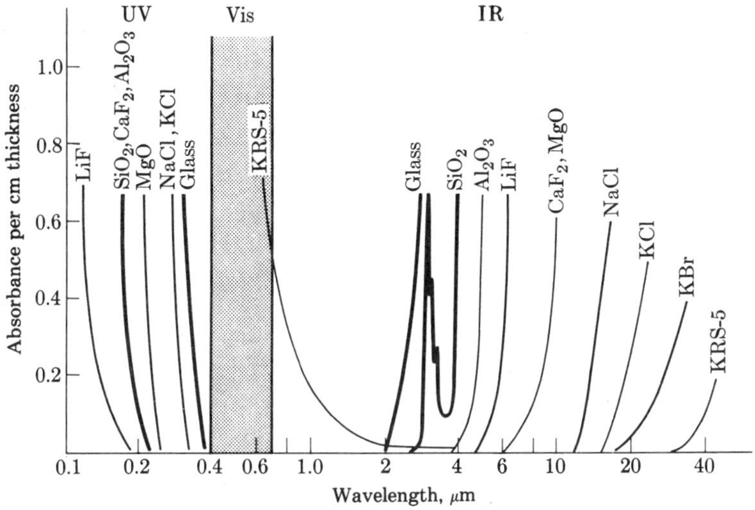

Fig. 7.16 Absorption curves for several optical materials. (Courtesy of Pergamon Press.)

are transparent in the visible except KRS-5. In general, effective optical use is precluded whenever the absorbance of a 1-cm slab is greater than 0.2.

Transmission, however, also decreases as reflection increases. Scattering will diminish transmission too. As just discussed, reflection increases both with increasing refractive index and angle of incidence. *Refractive index* values are plotted in Fig. 7.17 for many optical materials in their common regions of transparency. An average *reflection loss* of an uncoated window in the UV–VIS is 10–20% at normal incidence.

Useful Optical Ranges. In the UV simple inorganic substances, such as LiF, SiO_2, and diamond, offer sufficient transparency. Costs are generally high.

The materials in widespread use in the visible region, the silicate glasses, transmit well only from about 350 nm to 3 μm (3300 cm^{-1}). Many salts also offer good transmission in this region.

In the middle and far IR some relaxation of "mechanical" requirements is often made in the interest of adequate transmission. Many materials transparent in this region, like Groups IA and IIA metal salts, are relatively soft and susceptible to attack by water or lower alcohols. Careful handling, moderate humidity, and avoidance of samples that have more than a few tenths of a percent of water or alcohol will minimize surface erosion. Surfaces should be repolished at need.

The list of substances transparent in the IR also includes many insoluble, relatively hard materials. Both germanium and silicon, which are transparent from about 10,000 to 630 cm^{-1} (1–16 μm), make excellent single-element lenses because of their high refractive indices, 4.1 and 3.5, respectively. Since they reflect about 50% of incident radiation, they are often antireflection-coated for higher transmission in a particular range. Arsenic trisulfide is also often employed. Sev-

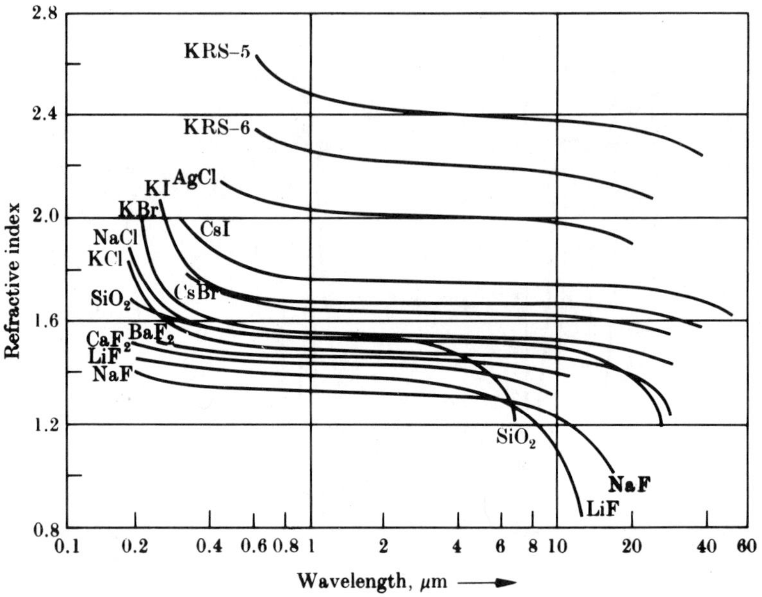

Fig. 7.17 Refractive index as a function of wavelength for many common optical substances. (Courtesy of Harshaw Chemical Company.)

eral water-insoluble, high-refractive-index, polycrystalline materials trademarked "Irtran" are also valuable in the near and middle IR. For example, Irtran-6 has a transmittance of about 60% from 10,000 to 350 cm^{-1} (1–28 μm). In the far IR polyethylene and crystalline quartz are useful as window materials.

The transparency of air must also be considered. Although it is nonabsorbing in the visible, two common components of air, CO_2 and H_2O, absorb in parts of the IR and all the gases absorb strongly at wavelengths shorter than 180 nm. In an IR instrument the reduction in beam intensity may be sufficient to warrant purging the interior of an instrument with dry nitrogen or evacuating it. For work in the far UV, evacuation is mandatory.

Lenses. Both good transparency and nearly constant refractive index are required over a reasonable spectral range for substances to be useful in lenses. Serious chromatic aberrations occur when the refractive index changes more than 1–2% with wavelength. For example, it is evident from Fig. 7.17 why lenses are seldom fabricated for UV use. Even in the visible region lenses are valuable mainly in applications in which imaging is sufficiently important to make the costs of overcoming chromatic aberration worthwhile (see Section 7.20). In the IR, however, the flat refractive index curves suggest that lenses will often be an attractive alternative to mirrors. For many applications, and especially those in the IR where refractive indices and reflectance are high, low-reflectance coatings will be desirable.

Fiber Optics. Bundles of flexible, transparent fibers of about 0.06 mm diameter make useful light "pipes." Transmission depends on total internal reflection at the walls. To maintain reflection, fibers are kept apart by cladding them with a coating of lower refractive index such as polyvinyl chloride. Numerical apertures of 0.50 are common for fiber optics.*

With glass fibers, transmissions in the visible and near IR range have steadily improved with purer materials. Use of arsenic trisulfide or other substances extends the capability of filter optics into the IR, and use of silica into the UV.

Not only are intensities transmitted, but images can also be carried through a fiber optics bundle. In this case bundles must be coherent, that is, the relative position of each fiber in a bundle must be identical at the two ends and the fibers must be fused at each end to eliminate dead space.

Representative applications of fiber optics are remote illumination, lighting without heat, telephonic transmission, remote sensing in optical methods of measurement, even, constant illumination of a large area such as a punched card (by splitting a bundle of fibers exposed to a small source into many subbundles), and remote viewing of images. Design considerations are often crucial [6].

7.10 OPTICAL FILTERS

An optical filter is a device that absorbs, reflects, or deviates all frequencies from an optical path except for one band or region. How well five representative types perform may be seen from transmission plots in Fig. 7.18. A band-pass filter like

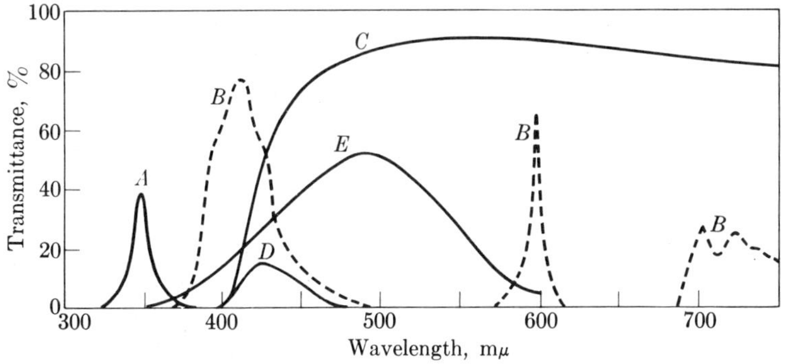

Fig. 7.18 Transmission of filters: *A*, Fabry–Perot or metal interference filter (The primary band at 700 nm has been blocked.); *B*, multilayer interference filter with primary band at 595 nm; *C*, Corning 3389 sharp-cutoff filter; *D*, Corning narrow band-pass filter made by combining filters 3389 and 5113; *E*, Corning 4060 glass filter.

*The numerical aperture (NA) is the sine of the maximum angle of incidence θ that permits transmission (thus also that permits total internal reflection). In this application it is defined as NA $= n_3 \sin \theta_{max} = (n_1^2 - n_2^2)^{1/2}$ where n_1, n_2, n_3 are the refractive indices for the fiber, coating, and external medium (usually air), respectively.

A, *B*, or *D* is valuable in separating an emission line or in isolating a narrow spectral band from the output of any continuous source, while a cutoff filter like *C* is useful in blocking a portion of a spectrum. To compare different band-pass filters, widths at half maximum transmittance, that is, spectral bandwidths (or band-pass), are conventionally stated. For interference filters, which will be discussed in the next section, bandwidths are about 10 nm (e.g., *A* or *B*) as compared with 40 nm for glass filter *D*.

Glass filters of the band-pass type transmit only a selected optical region. Other wavelengths are absorbed or scattered. Their absorbant is suspended in a glass plate or dissolved in a layer of gelatin between clear plates. Band-pass filters can also be made by cementing two filters that pass only a narrow range in common.

Interference Filters. Optical filters based on interference between multiple reflected rays can be made for the region that extends from about 2000 cm^{-1} (5 μm) through the near UV. Two types of filter, based on different methods of producing interference have been developed, the metal or Fabry–Perot filter and the multilayer dielectric type.

The *Fabry–Perot filter* consists of an extremely thin layer of dielectric sandwiched between semireflecting metallic films, usually of silver. The construction is shown schematically in Fig. 7.19 and is achieved by precise vacuum deposition. Suppose parallel light rays are perpendicularly incident on the left side. They suffer substantial reflection at each metal layer. Most of the radiation traversing the dielectric layer is reflected back. The interference that ensues as reflected rays superpose on incident rays, a phenomenon termed *Fabry–Perot interference*, causes the obliteration of all but several narrow bands of wavelengths. The transmission of such a filter is shown in Fig. 7.20.

In Fig. 7.21 an "exploded" picture is given as an aid in visualizing the interference. To separate the rays for the purpose of illustration, angular incidence is shown; actually normal incidence is required if rays are to interfere. The condition for transmission is the condition for constructive interference: path differences be-

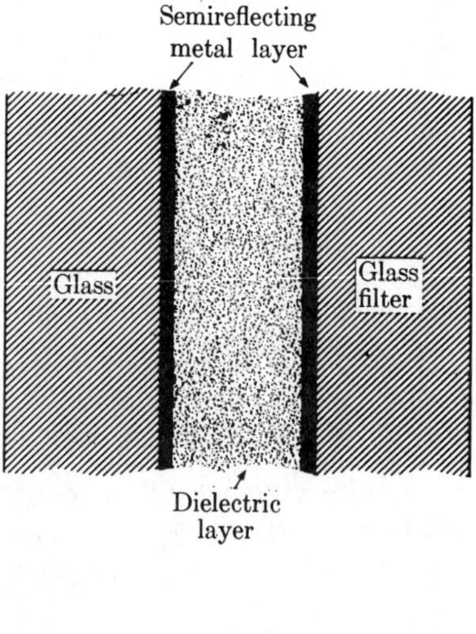

Fig. 7.19 Cross section of a Fabry–Perot type of interference filter. Thicknesses are exaggerated for clarity. The cover glass (outer layer) may be from 2 to 3 mm thick, the dielectric layer 0.4 μm, and the metallic films even thinner. Representative materials for the dielectric layer are calcium or magnesium fluoride and the metal is ordinarily silver. As shown, a filter may be used as a cover glass to pass an interference band of a single order.

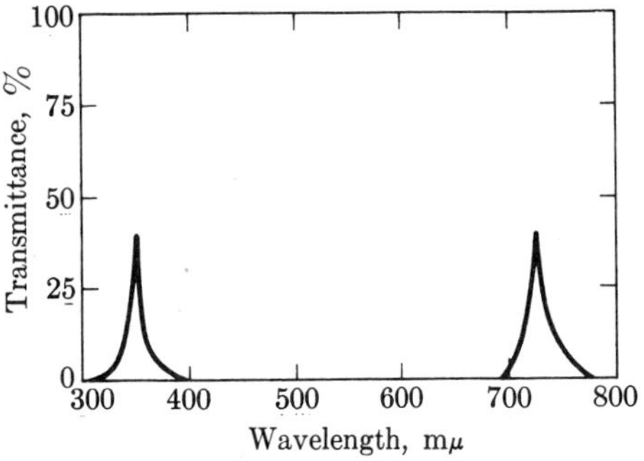

Fig. 7.20 Transmission curve of a Fabry–Perot interference filter.

tween superposed rays must be one wavelength or some multiple thereof. The expression for wavelengths at which full reinforcement will occur is

$$2d = m\lambda/n, \tag{7.17}$$

where d is the thickness of dielectric layer and n its refractive index at wavelength λ (vacuum), and m is the *order* ($m = 1, 2, 3, \ldots$), a factor that allows for path lengths that are a multiple of the wavelength. For example, rays 1 and 2 *completely* reinforce each other when $aa' + a'b$ is exactly one wavelength in the medium (λ/n), or a multiple. The same conclusion is reached for all unprimed ray pairs, for example, 2 and 3, 1 and 3.

Since partial reinforcement occurs for other path distances, the Fabry–Perot filter actually transmits a band. Equation (7.17) defines only the central wavelength. Intensity of transmission diminishes once path differences differ very much

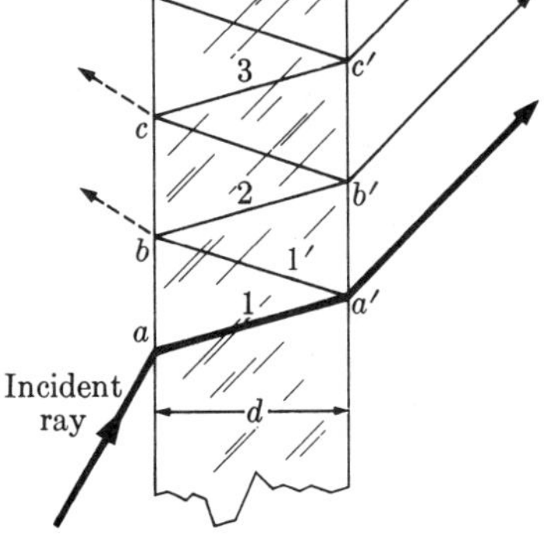

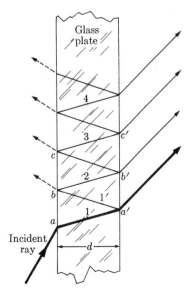

Fig. 7.21 An "exploded" sketch of the multiply reflected rays in an interference filter. Normal incidence is always used, although an obliquely incident ray is drawn here to allow the reflected rays to be shown.

from any multiple of the wavelength. The maximum transmission in the center is dependent on the extent to which the metal mirrors absorb. For about 5% mirror absorption, peak transmission is usually 40%, half-bandwidths 12.5 nm, 1/100 bandwidths 77 nm. By sacrificing intensity, still narrower bands can be achieved.

When is the *order* of interference of importance? It is convenient to rewrite Eq. (7.17) in the form

$$\lambda/n = 2d/m \tag{7.18}$$

to emphasize the influence of the order. Wavelengths that are one-half $2d$, one-third $2d$, and so on will also constructively interfere and be transmitted. The origin of the bands shown in Fig. 7.20 is now evident. The right band at about 720 nm must be first-order, and the left one at about 360 nm must be second-order.

Example 7.4. How thick a dielectric layer is required in a Fabry–Perot filter to secure a first-order band at 500 nm? Assume a calcium fluoride dielectric of $n = 1.35$ is used.

Solving Eq. (7.18) for d and substituting values gives

$$d = \frac{\lambda m}{2n} = \frac{500 \times 1}{2 \times 1.35} = 185 \text{ nm}.$$

Ignoring the small change of n with λ, we find that this filter also passes a second-order band centered at 250 nm and so on as the cover plates permit.

Second- and third-order bands are narrower, and most interference filters are arranged to transmit one of these. If a third-order 500-nm band is desired, it must be isolated from bordering transmission bands at about 750 and 375 nm. These "side" bands may be blocked by an additional filter. It may be substituted for one of the cover plates usually present. An ingenious choice of filter is a Fabry–Perot system whose first order matches the desired band.

What if the radiation deviates from normal incidence? In this case, the band transmitted shifts toward shorter wavelengths. While the shift is only about 2.5 nm for a 10° deviation from normal incidence, it increases rapidly. Clearly, the aperture angle of light to be filtered should be kept small if bandpass is to stay within specifications for a filter.

The variation in transmission wavelength when a filter is tipped does suggest a method of continuous tuning or scanning over a short wavelength range. By tilting, the band passed can be shifted smoothly as much as 10–30 nm toward shorter wavelengths.*

Finally, a continuously variable Fabry–Perot type of filter is also obtainable. A wedge-shaped slab of dielectric is deposited between semireflecting metallic layers. In this case a different wavelength is transmitted at each point along the length of the filter. Different frequencies may then be isolated by passing a slit assembly

*S. A. Pollack, *Appl. Opt.* **5**, 1749 (1966).

along the filter. Such a filter can be used in place of a prism or grating as the dispersing element in a monochromator.

Multilayer Filter. This kind of interference filter consists of alternating layers of high- and low-refractive-index dielectric materials of suitable thickness vacuum-deposited on a transparent slab. Layer thickness ranges between one-fourth and one wavelength, with 5–25 layers usually required for a filter. Although a detailed treatment is beyond the scope of this text, it should be noted that narrower half-peak widths (as small as 4–8 nm at present) and greater transmittances (40–70%) can be secured with multilayer dielectric filters. Very importantly, sharp cutoff filters are also possible. The disadvantages are greater cost and higher transmittance in unwanted parts of the spectrum. Generally, Fabry–Perot filters are more extensively used.

Christiansen Filter. This acts by deflecting all but a narrow band of wavelengths from the optical path. It consists of a suspension of tiny irregularly shaped particles of a transparent solid of a given dispersion $dn/d\lambda$ in a liquid of different dispersion. It is necessary that at one wavelength both have the same index of refraction. If a collimated beam is incident, only the wavelength at which particles and liquid have identical refractive indexes is transmitted freely; others are deflected. By choosing an appropriate binary mixture of liquids, composition can be varied to shift the wavelength of the band passed by the filter.

7.11 TOTAL INTERNAL REFLECTION

When radiation undergoes total internal reflection, it actually penetrates a fraction of a wavelength into the medium beyond the reflecting surface. In discussing the phenomenon it will be convenient to use the classical terms "denser," referring to a medium of higher refractive index, and "rarer," referring to one of lower index. Because of the superposition of incident and reflected waves at the boundary, a standing wave is set up normal to the reflecting surface in the two media [10]. In the rarer medium it is termed an *evanescent wave* since its amplitude falls off exponentially to zero within a fraction of a wavelength from the interface.

In Fig. 7.22 the nature of the standing wave is sketched for the case of incidence at an angle greater than the critical angle on a boundary between denser and rarer media where neither medium is absorbing. As the figure suggests, the electric field amplitude on either side of the interface may be substantial.* Its amplitude is strongly dependent on the angle of incidence, gradually falling to zero as the angle is increased toward 90°. Though not shown in the figure, the evanescent wave has field strength in all *three* directions because of the nature of the interaction.

*If reflection occurs from a metallic surface or a metallic mirror, a standing wave also is produced, but a node, that is, a region of minimum amplitude, occurs at the surface because of the high conductivity of the metal. In this case, the depth of penetration is also smaller (about 10 nm).

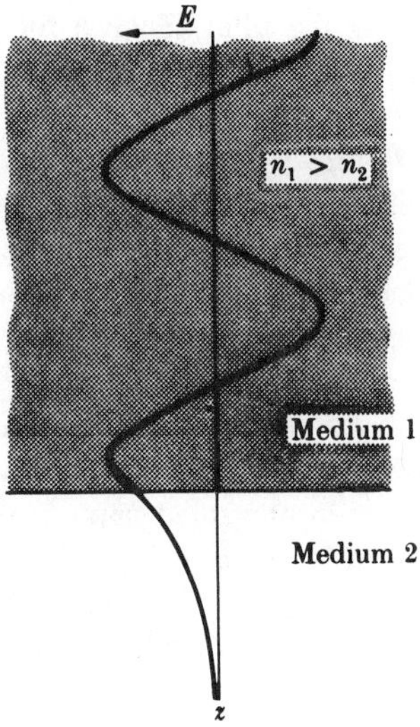

Fig. 7.22 Amplitude of standing electric wave formed on either side of an interface during total internal reflection. The superposition of incident and reflected waves in the upper (denser) medium causes the standing wave. The depth of penetration d_p into the rarer medium is the distance over which the amplitude falls to e^{-1} of its value at the reflecting interface.

Frustrated Total Reflection. Either a lossless mechanism or an absorption mechanism can be used to withdraw energy from the evanescent wave. Here lossless "coupling" implies withdrawal of energy in optical form. To do so it is only necessary that a second transparent element be positioned not more than a fraction of a wavelength from the interface. Then part of the optical energy is transmitted through the second element, and the rest is still internally reflected. The descriptive name, frustrated total reflection, is applied. A simple arrangement for securing the effect is shown in Fig. 7.23.

The intensity of the transmitted beam is strongly dependent on the distance of separation. Accordingly, the technique offers a method of *intensity modulation* of a beam of radiation. For example, with the arrangement in the figure a transmit-

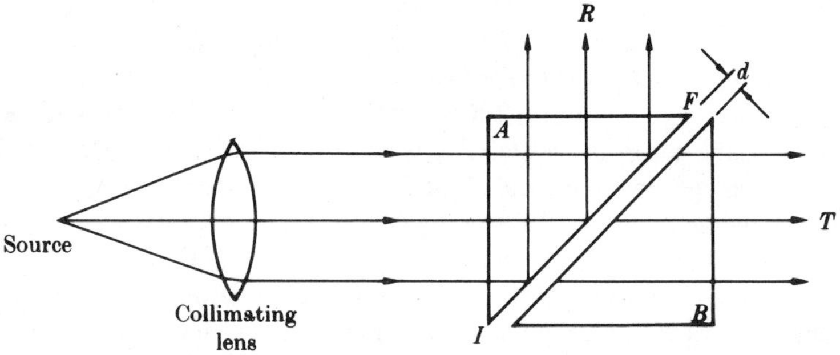

Fig. 7.23 Frustrated total internal reflection. Collimated radiation of wavelength λ (in denser medium) incident on the interface *IF* between prism *A* and air is either totally reflected (beam *R*) or transmitted through adjacent right-angle prism *B*. The transmission increases steadily as the distance *d* between prisms is decreased. It is zero for $d/\lambda = 1$, reaches 50% for $d/\lambda = 0.3$, and is 100% for $d/\lambda = 0$.

tance of 50% is attained with a separation of $0.3\lambda_1$, where λ_1 is the wavelength in the denser medium. At a separation of $0.8\lambda_1$, the transmission falls to zero. To modulate, the separation is varied at will. The transmitted intensity changes accordingly.

In addition, the intensity varies with the angle of incidence, decreasing steadily the more ϕ exceeds ϕ_c. Parallel polarized light is also favored in transmission relative to vertically polarized radiation (see Section 7.15). Finally, since the transmittance T and reflectance R prove to be *smooth functions* of the separation between prisms, it must be concluded that no interference phenomena occur in the gap between them. As will be noted below, their absence is characteristic of internal reflection.*

Attenuated Total Reflection. Where the rarer medium is absorbing, the intensity of radiation totally reflected also falls below 100%. In this case, energy is only absorbed at characteristic wavelengths and in general is then degraded to heat. The phenomenon is the basis for internal reflection spectroscopy (see Section 16.8), a widely used spectrometric technique.

Since the electric field amplitude in the rarer medium decreases exponentially, it is convenient to define a *penetration depth* d_p as the interval over which the field falls to $1/e$ of its value at the surface. This depth is given by the expression

$$d_p = \lambda_1/2\pi(\sin^2\phi - n_{21}^2)^{1/2}, \tag{7.19}$$

where λ_1 is the wavlength in the denser medium, ϕ the angle of incidence, and n_{21} the ratio of the refractive index of the rarer medium to that of the denser. It can be seen that penetration depth will decrease as the angle of incidence increases and as the two media become closer in refractive index. Since the extent of absorption depends on penetration, optimum conditions for attentuation of reflectance are an angle of incidence slightly greater than the critical angle and a small refractive index difference between rarer and denser media. The depth of penetration is not polarization-sensitive.

How great a penetration into the absorbing medium can be expected? Since both the angular and refractive index terms in the denominator of Eq. (7.19) will not differ markedly from unity, the denominator will in general be between 1 and 10. Thus the penetration is a fraction of the wavelength λ_1.

In most cases the rarer medium will be thicker than the depth of penetration. Thus it is also convenient to define an effective depth d_e of penetration in terms of the fraction by which the reflectance R is decreased by the presence of an absorbing medium. The expression used is $R = (1 - \alpha d_e)$ where α is the Lambert-law

*Another useful (lossless) method of frustrating total internal reflection is varying the refractive index of either medium. If incidence is at or just beyond the critical angle, decreasing, n_1 or increasing n_2 sufficiently will destroy total internal reflection. An electric field, pressure, or other appropriate variable can be used for this purpose. For example, if the second medium is a polar liquid, refractive index may be altered by application of an electric field.

absorptivity. The expression relates to a single reflection, of course. If the rarer medium is of low absorption, the effective thickness is given by the expression

$$d_e = \frac{E_0^2 n_{21} d_p}{2 \cos \phi} \tag{7.20}$$

where E_0 is the electric field amplitude at the interface.

Finally, a surprising aspect of absorption in the rarer medium under conditions of internal reflection is that to some degree it occurs in all directions. Because of the nature of the interaction, the evanescent wave has some electric field strength in all three directions. As a result, it is possible for an absorbing dipole in the rarer medium (connected with a quantum transition) to be oriented in virtually any direction and still absorb from the radiation incident on the interface. The intensity of absorption will, of course, vary with the orientation of the dipole.*

7.12 SCATTERING

Like reflection and refraction, scattering has its origin in the induced secondary emission of particles that lie in the path of radiation. Secondary radiation is scattered only if the particles

1. have dimensions about the order of magnitude or smaller than the incident wavelengths, and
2. are randomly distributed in a medium of refractive index different from their own.

In other cases the lateral rays originating from the secondary emission of atoms mutually cancel through destructive interference, leaving only a refracted and a reflected beam. The role of dimensions in causing reflection rather than scattering at a surface is illustrated in Fig. 7.24. In this figure a small particle is seen to give rise to a nearly spherical wavefront. If the radiation that the particle scatters is in the visible, an observer with a microscope or ultramicroscope could "see" the particle, but it would appear only as a point; no surface features would be resolved.

Approximate upper size limits of particles responsible for scattering in various regions of the optical spectrum are given in Table 7.1. Lower limits are more difficult to assign, since scattering efficiency of small particles varies about as the square of the volume at a given wavelength. For visible light, some familiar examples of scattering particles are those in smoke and freshly formed precipitates.

Small Particles. The simplest case of scattering is that involving (a) very small, (b) spherical, (c) optically isotropic particles. Their longest dimesion must be no

*This contrasts strongly with regular absorption processes in which a dipole can withdraw energy from the electric field only if perpendicular to the direction of propagation and parallel to the electric vector's vibration.

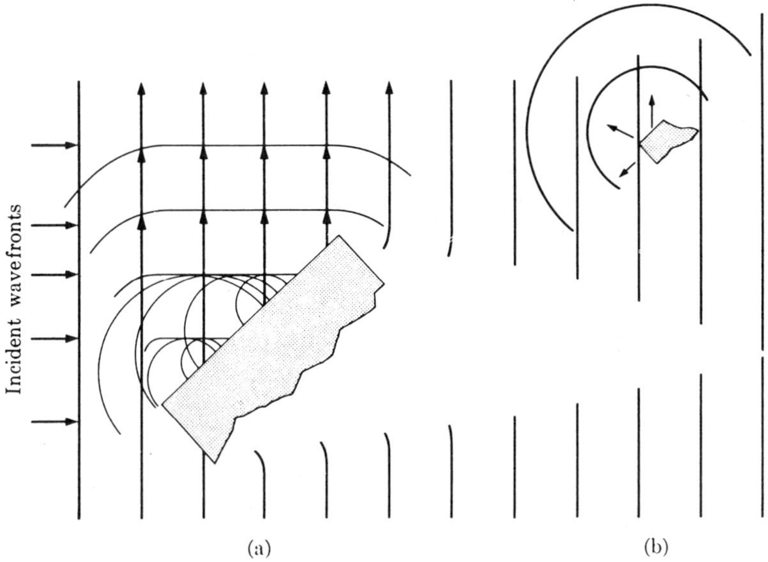

Fig. 7.24 Similarity of reflection and scattering of radiation. Monochromatic radiation is incident. (a) Reflection from particle whose dimensions are greater than 2λ. (b) Scattering by particle whose dimensions are smaller than $3\lambda/2$.

more than from 5 to 10% of the wavelength of incident radiation. Their scattering is not essentially different from that of the single atom shown in Fig. 7.8 and is known as *Rayleigh Scattering*. For a single particle, a symmetrical intensity pattern of the sort shown in Fig. 7.25 is obtained with unpolarized radiation. The intensity of scattering, I_s, is given in this case by the following expression:

$$I_s = \frac{8\pi^4\alpha^2}{\lambda^4 r^2}(1 + \cos^2\theta)I_0. \tag{7.21}$$

Here α is the polarizability of the particle, λ the vacuum wavelength, I_0 the incident intensity, θ the angle between the incident and the scattered ray, and r the

Table 7.1 Structures Producing Scattering

Incident Radiation		*Maximum Particle Size (approximate dimensions) (μm)*	*Type of Aggregate*
Wavelength (μm)	*Spectral Region*		
10	Infrared	15	Large colloidal particles
0.5	Visible	0.75	Colloidal particles, macromolecules
0.001	X ray	0.002	Small molecules, atoms

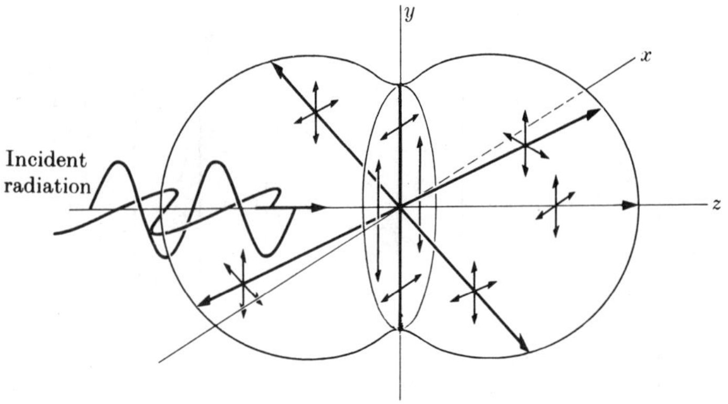

Fig. 7.25 Pattern of scattering of unpolarized monochromatic light by a small, spherical, isotropic particle. The particle is at the origin. The scattered intensity is indicated by the length of the vector. Only two cross sections of the scattering envelope are shown, that in the yz plane and that in the xy plane. The state of polarization of the scattered rays in those planes is indicated by the small vibrational vectors. Radiation scattered in either direction along the z axis (0° and 180° to the direction of propagation) is unpolarized; that scattered at 90° to the direction of propagation is linearly polarized.

distance from the center of scattering to the detector. It may also be shown that the polarizability varies roughly as the volume of a particle. Hence Eq. (7.21) predicts that the scattering will increase strongly as the particles become larger. In any collection of particles of different sizes the larger ones will contribute very heavily to the observed scattering.

Note also the inverse dependence of intensity on the fourth power of the wavelength. The intensity increases very strongly as the wavelength becomes shorter, that is as it approaches the size of the particle. For example, according to Eq. (7.21), violet light of wavelength 400 nm is scattered about 3.8 times better than green light of 550 nm. The blue color of the sky is a familiar result of this kind of discrimination. Dust particles of small dimensions, molecules of water, and so forth, cause the short wavelengths in sunlight to be scattered so efficiently as to make the predominant color blue. The red color of the sun at sunset and as seen through smoke or fog is confirmation that the long wavelengths (red) are much less effectively scattered and are more completely transmitted.

Liquids and Solutions. In pure liquids and in solutions of normal small particles, there is ordinarily a fairly regular array of "molecules." These systems are poor light scatterers, since here also lateral destructive interference ensures that virtually all the incident radiation is transmitted as the refracted beam (or reflected at the surface). There is a possibility, however, for very weak scattering. It originates in the inhomogeneities (small density and concentration fluctuations) constantly generated by thermal forces. Some volume elements will have more particles than the average, others less. Each nonaverage volume region acts briefly as a scattering center of polarizability $\Delta\alpha$, where

$$\Delta\alpha = \alpha_{\text{local}} - \alpha_{\text{av}}. \tag{7.22}$$

In this case $(\Delta\alpha)^2$, a very small quantity, replaces α^2 in Eq. (7.21), and the scattering is very faint.

Polymer Solutions and Sols. A markedly different result is obtained if the solute is composed of large molecules (15 Å < dimensions < 400 Å) of polarizability different from that of the solvent. Note that the scattering in the visible range will still be of the Rayleigh type. There is now a contribution to the scattered light by density fluctuations in the solvent, which is very small, and one due to the "concentration" fluctuations, which will increase with concentration.

Large Particle Scattering. Some other approach is necessary to handle particles whose dimensions are greater than 10% of λ. One difficulty is the fact that scattering centers in large particles are far enough apart for some interference from the secondary rays emitted from separate areas of the particle to be likely to occur. The situation is pictured for a large spherical particle in Fig. 7.26. The rays emitted in backward directions are highly susceptible to interference because of the large path differences possible. For example, compare distances for paths *SBA* and *SCA*. In the forward directions the path differences among rays are much less, and the interference is usually constructive. The result is a shift from the symmetrical intensity patterns of Figs. 7.8 and 7.25 to one showing much less scattering for angles greater than 90°. The new intensity pattern plotted in Fig. 7.27 shows the preponderance of forward scattering. Quite qualitatively, it may be seen that further increasing the size or bringing particles closer together so that there is lateral destruction would tend to give rise to a more and more clearly defined forward or transmitted beam. Eventually the ordinary situation of a single refracted beam would be attained.

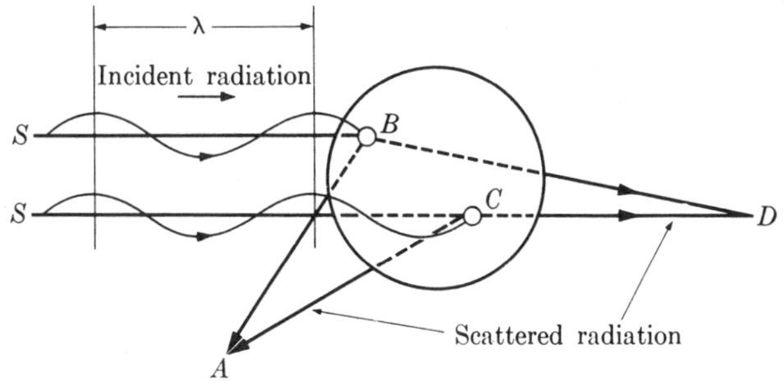

Fig. 7.26 Interference in scattering from a large molecule.

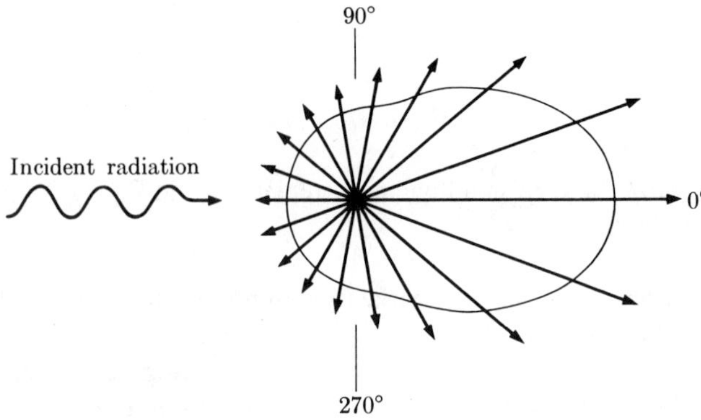

Fig. 7.27 Pattern of scattered intensities for a large molecule. Radiation is incident from 180°. Only a cross section of the scattered distribution is shown.

7.13 DIFFRACTION

Radiation is observed in most instances to travel in straight lines, a property termed *rectilinear propagation*. One exception has already been noted, that of scattering by small particles. Another, which is closely related, is what may be called an *edge-effect diffraction*. For example, the curved wavefront originating from the edges of the larger object in Fig. 7.24 is the result of diffraction. It is found that whenever a ray passes the edge of an obstacle or goes through an aperture it is somewhat "bent." As in Fig. 7.24, some radiation is spread into areas that would be in "shadow" if rectilinear propagation persisted.

To explain diffraction, the Huygens principle may be reintroduced. According to it, each point on a wavefront is potentially a new source of waves. The origin of a single wavelet at a slit is shown schematically in Fig. 7.28.

Diffraction gives rise to striking effects with narrow optical slits because of interference effects. One is the Fraunhofer pattern of alternating light and dark bands (fringes) produced by passing parallel illumination through a rectangular slit. To observe the pattern, the emergent radiation must be focused onto a distant plane surface. This pattern and its origin are discussed in Section 9.2. A very

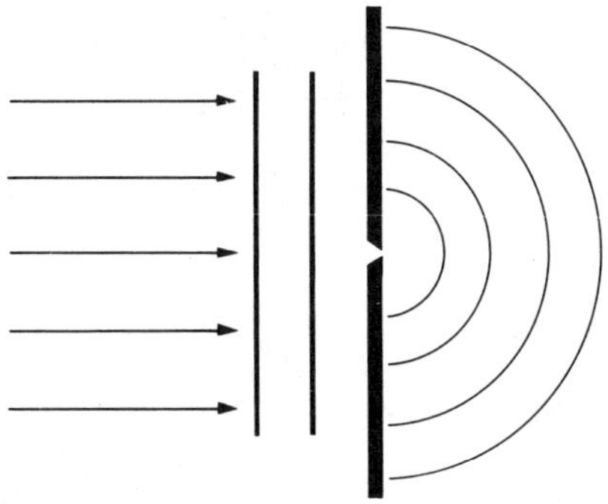

Fig. 7.28 A spherical wavelet spreading from a slit of infinitestimal width. If a very large number of these wavelets are produced side by side, as in a wide gap, the new wavelets interfere so as to give a plane wavefront and apparent rectilinear propagation. The analogy may be made with the production of a reflected image rather than scattered radiation (Fig. 7.24). For an ordinary narrow slit (Fig. 9.2), interference effects are appreciable.

different result is secured if many narrow, regularly spaced slits are used. The case of greatest interest, that of the diffraction grating, leads to the disposition of the incident radiation into a spectrum and will be considered in detail in Section 9.5. A second basic type of diffraction, the Fresnel class, covers cases in which the incident radiation is not parallel and the diffracted waves are not brought to a focus; since this is not so important in the optical systems to be encountered, it will not be discussed further.

7.14 POLARIZED RADIATION

Most sources produce EM disturbances in which the vibrations of electric and magnetic vectors occur with equal amplitude at all orientations perpendicular to the direction of propagation. Such radiation is said to be unpolarized. End on, a beam of this kind gives a vector picture like that of Fig. 7.29a, where only the electric vibration is shown.*

With few exceptions, contact with an interface between material media causes *preference* to be introduced into the vibrational pattern. The change may best be expressed if the representation of unpolarized radiation is resolved into two mutually perpendicular vibrations, as in Fig. 7.29b, by projection of every component vibration onto one or other of the rectilinear vibrations. Actually, the amplitude of the pair of vectors in (b) is much more enhanced by this process than is indicated in the sketch. The usefulness of this model will be seen shortly.

If a material medium affects the amplitude of these hypothetical wave trains differently, it is said to *polarize* the beam. When one vibration is completely eliminated, the radiation is said to be *linearly polarized*. It is also often described as plane-polarized.† In other cases the electrical vibration may continuously rotate in

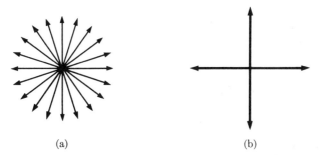

Fig. 7.29 End view of an unpolarized beam of radiation. The vibrations of the electric vector are shown (a) unresolved and (b) resolved into linear vibrations perpendicular to each other.

*This sketch falls short of a complete representation of the electrical vector. While there is no preferred direction of vibration, the constant changes in phase among the wave trains that make up an unpolarized beam give to it a predominantly elliptical character.

†In order to use the same "principle of projection" to represent all the kinds of preference, namely, that employed in Fig. 7.29a, the term linearly polarized will be used throughout [see ref. 13]. When the term "plane-polarized" is used, it is necessary to consider the plane formed by the vibration and the direction of propagation.

a clockwise sense, that is, the beam may be right-*circularly polarized*. The locus of the tip of the vector may for this kind of preference be portrayed in projection by a circle. The opposite result is also possible, that is, one yielding left-circularly polarized light. Finally, if the projection of the tip of the vector yields an ellipse, it is called *elliptically polarized*. The origins of these types of polarization will be considered in Section 7.17. Attention will be given in this section to linearly polarized light.

The relative intensities of the two orthogonal vibrational components in a beam may be determined by a second linearly polarizing device used as an *analyzer*. It also resolves light into orthogonal vibrations and selectively eliminates one. Accordingly, if linearly polarized light is incident, the intensity transmitted by the analyzer may be

1. 100%, if the direction of vibration of incident radiation is parallel to the characteristic (vibrational) direction of the analyzer;
2. between 100% and zero if resolved into a favored and a rejected component;
3. zero, if the direction of vibration of incident light is perpendicular to the favored direction of the analyzer. In this case the incident radiation generates *no* component in the vibrational direction passed.

Naturally, if unpolarized light falls on the analyzer, it should be transmitted at half intensity. Note that this is implied by Fig. 7.29b. These first and third orientations are pictured in Fig. 7.30. In general, a great deal about the state of polarization of radiation can be learned by observing the intensity transmitted by an analyzer as it is slowly rotated through 360°. It is convenient to define the degree of polarization P or *polarizance* as

$$P = \frac{I' - I''}{I' + I''} \tag{7.23}$$

where I' and I'' are the maximum and minimum intensities observed through an analyzer rotated through 360°. This equation is examined further in Exercise 7.21.

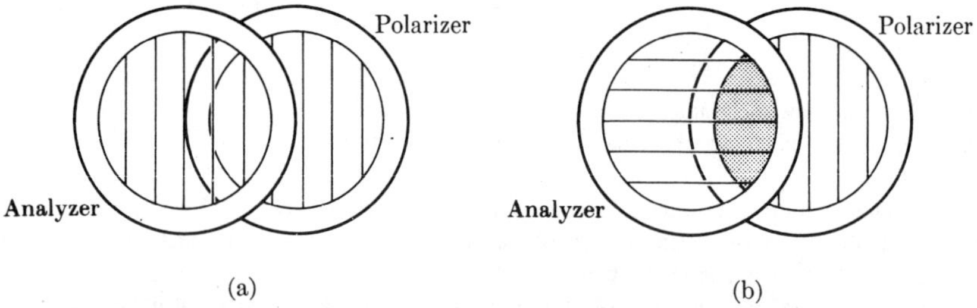

Fig. 7.30 Detection of linearly polarized radiation using Polaroid disks. (a) Analyzer oriented so that its vibrational direction is parallel to that of polarizer. (b) Analyzer aligned with vibrational direction perpendicular to that of polarizer (condition for extinction).

Example 7.5. To test the degree of polarization of a beam, it is allowed to fall perpendicularly on an analyzer. Regardless of the angular orientations of the analyzer, the beam is found to be transmitted with equal intensity. It can be concluded that the beam is unpolarized.

Selective Absorption. Some crystalline materials exhibit *dichroism*, a preferential absorption of radiation vibrating in one direction. Incident radiation is resolved accordingly, and the component vibrating in the preferred direction is reduced in intensity. The extent of reduction depends on distance of travel and absorptivity. The component vibrating orthogonally to the preferred direction is passed unaffected. Thus, such materials are polarizers. The phenomenon is illustrated in Fig. 7.31. The thickness of slab required for complete polarization and the frequency range over which strong polarization can be expected are, of course, functions of the nature of the material.

Example 7.6. The several varieties of Polaroid sheets undoubtedly provide the best-known examples of dichroism. The widely used type *H* and *K* sheets are prepared from polyvinyl alcohol films. *H* sheets are first stretched to align the molecules in the direction of extension and then impregnated with iodine. The I_2 molecules complex with the polymer and are thereby aligned. The component of incident light with vibrations parallel to the bond axis of I_2 molecules is absorbed; the component at right angles is transmitted. In type *K* film selective absorption is achieved in a film of polyvinyl alcohol by catalytic dehydration followed by parallel alignment of double bonds by stretching.

Reflection. Rays either reflected or refracted by a substance are found to be partially polarized except for normal incidence. A schematic illustration of this phe-

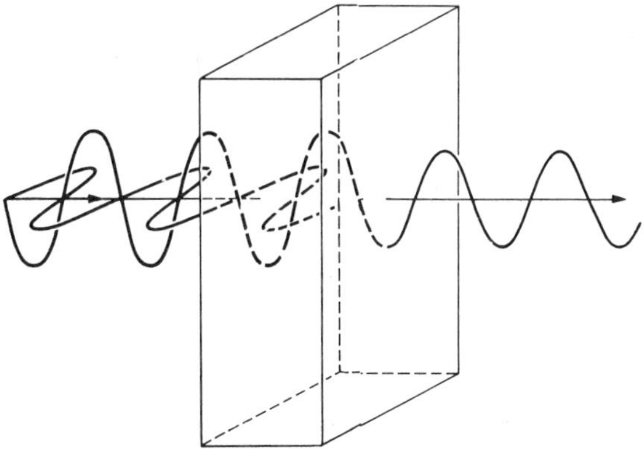

Fig. 7.31 Dichroic behavior by a crystal. As a result of parallel alignment of absorbing functional groups, the horizontal vibration is completely absorbed and the vertical vibration is absorbed only to a slight degree.

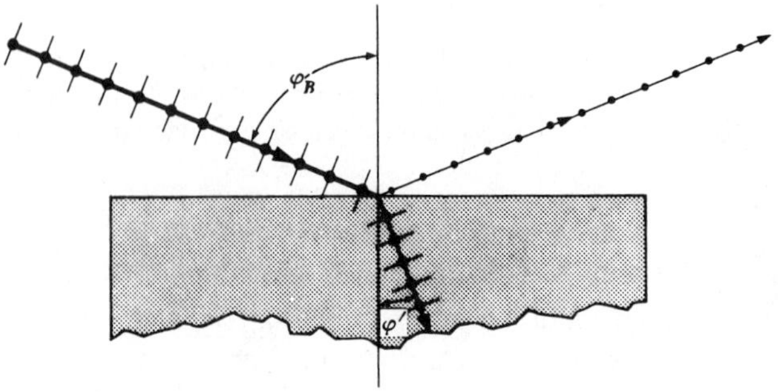

Fig. 7.32 Polarization by reflection from a dielectric. Radiation is incident at the polarizing angle ϕ_B, giving rise to a completely polarized reflected ray and a partially polarized refracted ray.

nomenon is given in Fig. 7.32. The plane of incidence is that of the page. The component vibrating perpendicular to this plane is the one that is more completely reflected; conversely, the component vibrating parallel to (or in) the plane of incidence is somewhat selectively refracted. As shown in the figure, for reflection from a dielectric there is a polarizing angle ϕ_B, for which the reflected beam is linearly polarized.

The reflectances of the two orthogonally polarized components are given by Fresnel's equations:

$$R_\perp = \frac{1}{2}\frac{\sin^2(\phi - \phi')}{\sin^2(\phi + \phi')} \tag{7.24}$$

$$R_\parallel = \frac{1}{2}\frac{\tan^2(\phi - \phi')}{\tan^2(\phi + \phi')} \tag{7.25}$$

where ϕ is the angle of incidence (and reflection) and ϕ' that of refraction. Unpolarized light is assumed incident. Compare these expressions with Eq. (7.14).

As may be seen in Fig. 7.33, the reflectance of the two polarized components varies considerably with the angle of incidence. This dependence is shown for both external reflection and internal reflection. The overall external reflectance is also shown as a dashed curve.

Do Eqs. (7.24) and (7.25) predict one or more polarizing angles? Yes, since Eq. (7.25) goes to ∞ when $\phi + \phi' = 90°$. Whenever this condition is met, $R_\parallel$ goes to zero. Two cases must be treated, $\phi' < \phi$ (incidence from the rarer medium) and $\phi' > \phi$ (incidence from the denser medium).

1. The polarizing angle for external reflection (into the medium of lower n) is termed *Brewster's angle* ϕ_B. It may be defined mathematically by substituting the condition $\phi + \phi' = 90°$ into Eq. (7.8) to give

$$\tan \phi_B = n_{21}. \tag{7.26}$$

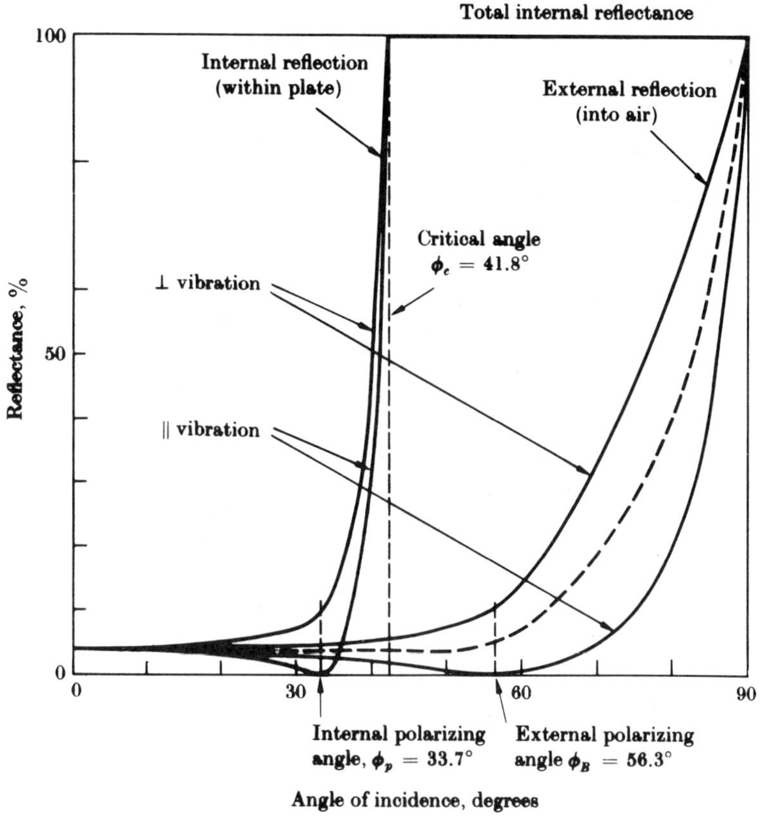

Fig. 7.33 Reflectance as a function of angle of incidence and direction of polarization [perpendicular (⊥) and parallel (||) to plane of incidence]. The interface is between dielectrics of refractive index $n_1 = 1.00$ and $n_2 = 1.50$. External reflection (from interface into medium 1) is characterized by Brewster's (polarizing) angle ϕ_B and internal reflection (from interface into medium 2) by principal (polarizing) angle ϕ_p and critical angle ϕ_c.

2. For internal reflection, the polarizing angle is called the principal angle ϕ_p. Since $\phi + \phi' = 90°$ still holds, ϕ_p must be just the complement of ϕ_B, that is,

$$\tan \phi_p = n_{12}. \tag{7.27}$$

Since the index of refraction varies with wavelength, the two polarizing angles and the reflectances at the angles defined by Eqs. (7.24) and (7.25) will be wavelength-sensitive. The higher the value of n_{21}, the greater is the intensity (reflectance) of the polarized radiation. For example, the intensity of the externally reflected polarized beam is low (5 – 15% of the incident "unpolarized" intensity) for most dielectrics in the visible region ($n \leq 2$), but is about 40% of that of an unpolarized incident beam for germanium in the infrared ($n = 4$).

Reflection by metallic surfaces also produces partial polarization. As with dielectrics, the perpendicular vibrations have the higher reflectance. An important difference is the absence of a polarizing angle, as might be surmised from the form of Eq. (7.16). While the intensity of the parallel vibrational component goes through a minimum, it is nevertheless strongly reflected at all angles.

7.15 ANISOTROPIC MEDIA

In the absence of orienting force fields, homogeneous liquids and gases are optically isotropic. Radiation coming into contact with them is affected in the same way regardless of direction of vibration or direction of propagation. Most crystalline substances, on the other hand, exhibit *optical anisotropy*. Sodium chloride and other compounds that crystallize in the cubic system are exceptions. In anisotropic substances, one direction is not optically equivalent to another because of differences in the numbers of each kind of atom encountered or in the spacing of units in the lattice pattern. It should be recalled that so-called solids like glasses and polymers (under some conditions) are noncrystalline. These are in fact supercooled "liquids" whose order is only of the short-range variety. A listing of substances according to their refraction behavior appears in Table 7.2.

Depending on its direction of vibration and propagation, a polarized monochro-

Table 7.2 Optical Classification of Substances

Type	*Physical State*	*Refractive Index*
Isotropic	Gases, liquids, solutions in absence of orienting fields	Single value of n
	Isometric crystalline solids, glasses, noncrystalline polymers	Single value of n
Anisotropic	Uniaxial crystalline solids (tetragonal, hexagonal, etc.), nematic and smectic liquid crystalline phases	Two values, n_o and n_E[a]: (1) $n_E > n_o$, uniaxial positive (e.g., quartz); (2) $n_o > n_E$, uniaxial negative (e.g., calcite)
	Biaxial crystalline solids (monoclinic, triclinic, etc.)	Three values of n
	Gases, liquids, solutions in strong dc electric field (Kerr effect)	Two or three values of n[b]
	Solutions of linear polymers flowing through narrow tube (flow birefringence)	Two or three values of n[b]

[a]Symbol n_o identifies the refractive index of the ordinary ray and n_E that of the extraordinary ray (see below).
[b]See Reference 6.

matic wave train travels through anisotropic crystals with varying speeds. The only exception occurs in a particular *direction* known as the *optic axis*. By definition, this direction (it is not really an axis) is the one in which monochromatic light travels with the same velocity regardless of the orientation of its vibrations with respect to the crystal. While certain crystalline materials show two optic axes, that is, are biaxial, this type of crystal will not be considered further. The optical character and constants of crystals reflect their degree of symmetry. Crystals belonging to the cubic system, such as NaCl, are optically isotropic. Those belonging to the tetragonal, hexagonal, or trigonal systems are uniaxial and the optic axis corresponds to the main or vertical axis of the crystals as they are usually represented. On the other hand, systems with lower symmetry such as monoclinic and triclinic are biaxial.

A schematic representation of the difference between wave velocities in a uniaxial crystal (one with a single optic axis) is drawn in Fig. 7.34. Suppose the crystal resolves the incident radiation into two rays with mutually perpendicular planes of vibration. One vibration occurs in a principal plane, that is, a plane formed by the optic axis and perpendicular to one of the main faces of the crystal. The other vibration is in the plane perpendicular to the principal plane. As stated above, the velocity along the optic axis establishes the ordinary index of refraction n_O. Since it is found that the ray with vibrations perpendicular to the principal plane travels at the same velocity, it is termed the *ordinary ray*. The other component (the ray with vibrations *in* the principal plane) is the *extraordinary ray* and has an index of refraction n_E that varies with the direction of propagation.* When traveling in a direction perpendicular to the optic axis n will exhibit either a maximum or a minimum, depending on the type of crystal [13].

Each of the refracted rays in an anisotropic substance may have a wavefront (Huygens construction) with a normal that makes an acute angle with respect to

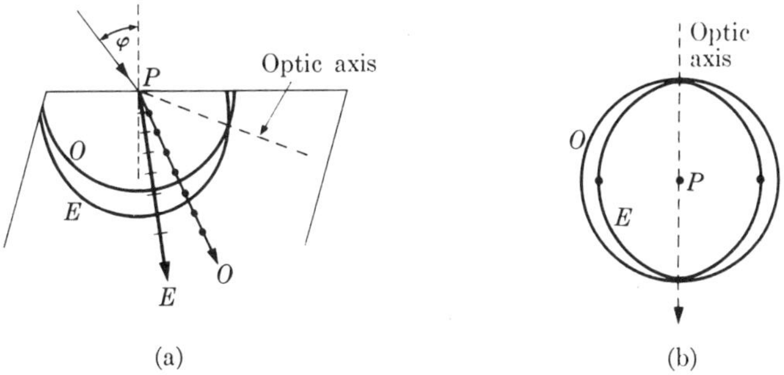

Fig. 7.34 Wavefronts in (a) calcite and (b) quartz. Note that the extraordinary wavefront travels more slowly than the ordinary one in quartz. In each drawing a monochromatic point source is considered to be placed at *P*.

*In crystallographic literature, ω and ϵ are commonly used to represent n_O and n_E, respectively, for uniaxial crystals.

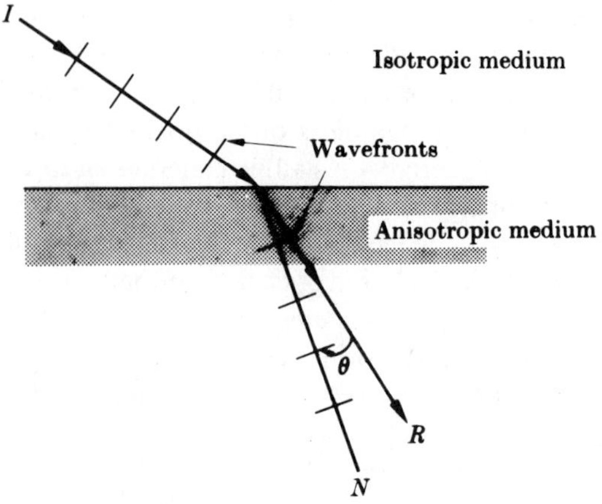

Fig. 7.35 Wavefronts of one of the refracted rays in an anisotropic substance at angle θ ($\theta < 90°$) relative to the direction of propagation. The wave normal direction N differs from the ray direction R.

the ray direction, as shown in Fig. 7.35. Two kinds of velocity may be defined: (a) the *velocity of the ray* V_r is the speed of propagation and (b) the *normal velocity*, the product of ray velocity V_r and cos θ. It is the latter velocity that defines the refractive index for the refracted ray, that is, $n_{ray} = c/V_r \cos\theta$. These *normal* parameters allow the most straightforward optical description of the crystal [see ref. 13].

Double Refraction. Now the actual results of anisotropy must be correlated. Unless radiation traverses an anisotropic substance along the optic axis, double refraction or *birefringence* results. What enters as a single ray breaks into two. The crystal resolves the incident radiation into beams that are transmitted at different angles because of their different indices of refraction, as shown in Fig. 7.34. Since dispersion also occurs, two complete spectral sets of rays appear. Each set is linearly polarized, as may be established by an appropriate examination of the spectra with an analyzing device.

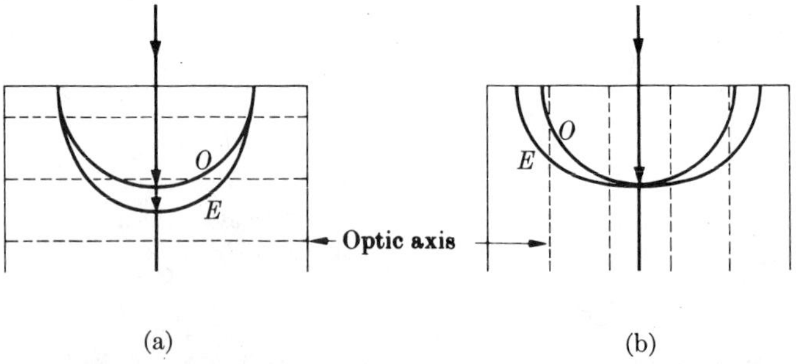

Fig. 7.36 Wavefronts in doubly refracting prisms of calcite ($CaCO_3$). (a) Optic axis (broken lines) perpendicular to direction of beams. (b) Optic axis along direction of travel.

Double refraction occurs even for normal incidence of parallel light except when a crystal face has been cut parallel or perpendicular to the optic axis, as depicted in Fig. 7.36. Interference takes place in the case shown in Fig. 7.36a, as will be discussed in Section 7.17.

7.16 POLARIZERS

Any one of the effects that produce linearly polarized radiation can be used as the basis of a polarizing device. The effectiveness of a polarizer can be judged by three criteria:

1. Efficiency
2. Spectral range of operation
3. Angular aperture

Reflection Polarizers. These make use of the partial polarization that results when radiation passes through an interface. Ordinarily, several properly oriented slabs of transparent material are used as shown in Fig. 7.37 to secure substantial polarization. Even though designed mainly as transmission devices, it is convenient to use the term "reflection" to identify this class since a reflection process produces the polarization.*

The degree of polarization can be increased by adding more plates, since each new one will eliminate more of the vertical vibration component. This type of polarizer is seldom used in the visible, but finds wide application in the IR. There, a set of four to six silver chloride, selenium, or germanium plates is employed.

Note that in the representative "stack-of-plates" polarizer in Fig. 7.37, the use of two oppositely sloping stacks avoids any displacement of the beam. Further, the design permits commercial, relatively thick plates to be employed. High efficiencies (polarizances of 99) can be achieved over the IR range.

Dichroic Polarizers. Such devices selectively absorb one of the orthogonal vibration components. Polaroid sheets (Section 7.14), frequently used in the visible

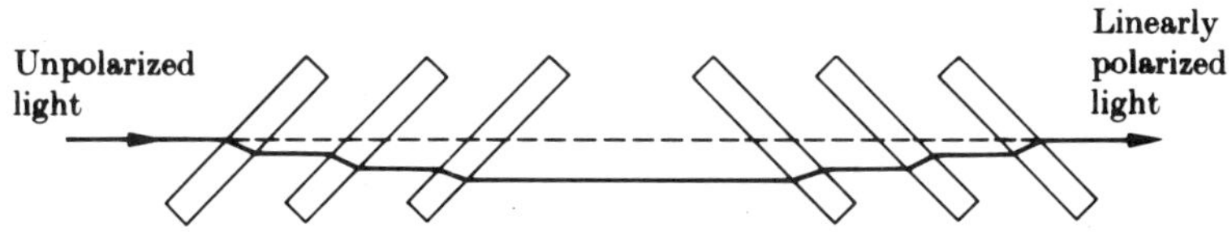

Fig. 7.37 A stack-of-plates reflection-type polarizer with plates mounted at the polarizing angle. The device makes use of selective transmission of light vibrating in the plane of incidence (the plane of the page).

*They are seldom used in the reflection mode, though a single reflection at Brewster's angle would yield linearly polarized light. Disadvantages of the method are that the angle varies with wavelength, reflectance at this angle is relatively small, and the incident beam is angularly displaced.

and near IR, illustrate the type. A second example is films of pyrolytic graphite (oriented microcrystalline graphite prepared by deposition at temperatures in excess of 2500°C), used for polarization in the IR from 2500 cm^{-1} to the microwave region.

Dichroic polarizers are still effective when incidence is off-vertical and polarize with nearly theoretical efficiency, that is, their polarized output has nearly 50% of the entering power. In addition they have the advantages of simplicity and, for those used in the visible, inexpensiveness.

Double-Refraction Polarizers. These separate the two orthogonal vibrational components in space. Each is refracted in a different direction as will be discussed in the following section. Polarization is essentially complete. As a result, these polarizers are without peer in the UV, and are commonly used in precision instruments in the visible. In the IR they see little use because of limited transmission. Their chief disadvantages are low angular aperture, small size, and high cost.

Most doubly refracting polarizers are constructed from sections of calcite ($CaCO_3$) crystals. The strong double refraction (birefringence) of this uniaxial crystal is suggested by the large difference between its refractive indices: n_O, 1.66; n_E, 1.49* (*D*-line). Its useful spectral range is defined by its transparency: from about 220 to 1800 nm (2200 Å to 1.8 μm). Polarizers using uniaxial crystals of the dihydrogen phosphate type have proved useful in the UV to about 185 nm.

Nicol Prisms. Nicol prisms or polarizers are prepared from calcite crystals with a minimum of alteration of natural crystal faces. The finished arrangement is pictured in simplified form in Fig. 7.38. An incident ray is resolved, as shown, into an ordinary ray vibrating perpendicular to the place of the paper, which is a principal plane, and an extraordinary ray. By splitting the crystal at a suitable angle and cementing the halves with Canada balsam, an isotropic substance of n_D = 1.55, the ordinary ray can be made to suffer total reflection at the interface. For this ray there is a decrease of refractive index at the interface. The other necessary

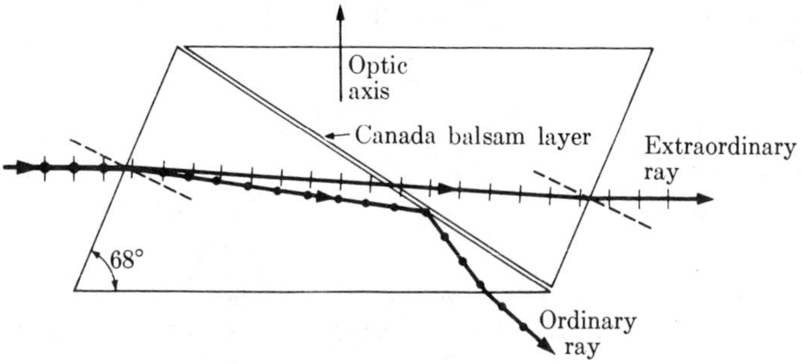

Fig. 7.38 Polarization by a nicol prism.

*The refractive index for the extraordinary ray ranges from this minimum value, applicable for a ray traveling perpendicular to the optic axis, to the value n_O.

condition, that the angle of incidence exceed the critical angle, can be ensured by restricting the initial incidence angle on the external face. Only the extraordinary ray emerges in the original direction. It appears as a beam parallel to the entering beam, but displaced laterally from it.

Some disadvantages of the nicol device must be noted. The two most severe ones are (a) that the emergent beam moves in a circle about its axis if the nicol is rotated and (b) that complete "extinction" cannot be obtained on "crossing" (see below) if an intense source is used. Use is also restricted to the visible, since Canada balsam is opaque to UV. In addition, an incident beam can be neither highly convergent nor divergent, since the range of angular approach for total reflection of the ordinary ray is only a few degrees. The range can be increased with a slight sacrifice in intensity by substituting an air gap for the Canada balsam film. A final disadvantage is that some dispersion occurs if polychromatic radiation is used.

Glan or Glan–Thompson Prisms. For these devices calcite is also used but crystals are trimmed so that the optic axis is perpendicular to front and rear faces. The Glan prism features an air gap; the parts of the Glan–Thompson prism are cemented either with Canada balsam or a resin of $n_O = 1.49$. In each of these polarizers the extraordinary ray emerges as the polarized beam and is achromatic, that is, suffers no wavelength dispersion. While the Glan polarizer is transparent in much of the UV, its air gap is responsible for loss of some intensity and it has a very small angular field ($\pm 4°$ deviation from parallel). The Glan–Thompson polarizer suffers no appreciable intensity loss. If a resin cement is used, virtually all internal reflection of the extraordinary ray is eliminated, ensuring a lower intensity of stray, unpolarized light. The Glan–Thompson prism also has a slightly greater angular field. Both of these types and many other polarizers are occasionally referred to as nicols.

Other polarizing prisms usable in the UV as well as the visible are the Wollaston and Rochon devices (Fig. 7.39). They are made from calcite or quartz pieces, which are cut and then cemented together with glycerine or castor oil (transparent down to 230 nm). As shown in Fig. 7.39*b*, parallel radiation enters the Wollaston prism perpendicular to the optic axis and leaves as two divergent, linearly polarized beams. The emergent beams are chromatic, that is, their component wavelengths are dispersed, if polychromatic radiation is used. Even when polarized radiation is not primarily sought, the Wollaston device is a useful means of obtaining two optical images side by side for comparison purposes. The Rochon prism is similar, but the transmitted beams do not emerge symmetrically; its advantage is that the undeviated beam is always achromatic.

Polarizer-Analyzer Pairs. Pairs of polarizers are employed in most experimental arrangements (see *Chapter 19*). One serves as a polarizer, the other as an analyzer. The paths of radiation in what are called the *uncrossed* and *crossed* positions are shown in Fig. 7.40. The amplitude of the light leaving the analyzer will be a function of its orientation relative to the plane of vibration of the incident radiation.

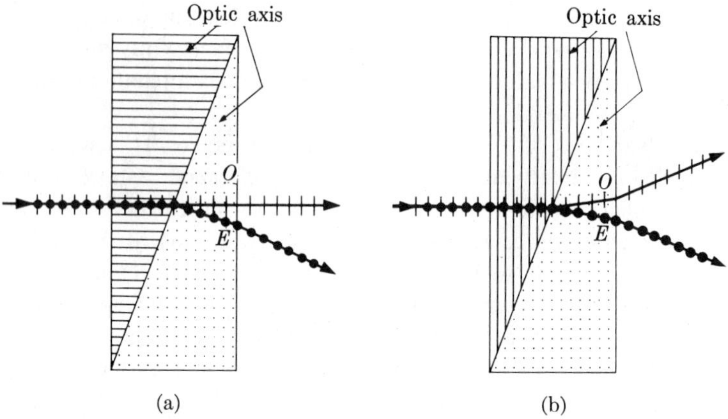

Fig. 7.39 Polarization by two types of compound quartz prism. Incident rays are monochromatic. (An optic axis perpendicular to the plane of the paper is indicated by dots.) (a) Rochon prism. (b) Wollaston prism.

From the vector diagram in Fig. 7.41 this amplitude EE is

$$EE = XX \cos \theta, \tag{7.28}$$

or, in terms of intensities,

$$\frac{I}{I_0} = \cos^2 \theta, \tag{7.29}$$

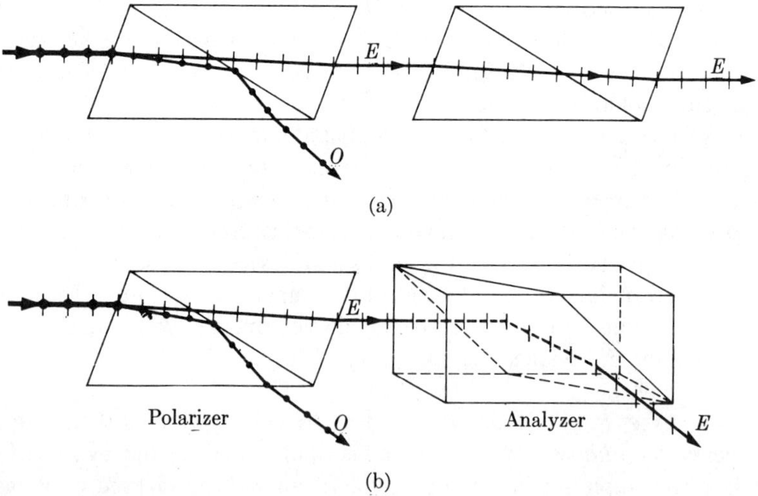

Fig. 7.40 Polarizer–analyzer pair. (a) Uncrossed pair. (b) Crossed pair; no light is transmitted by the analyzer.

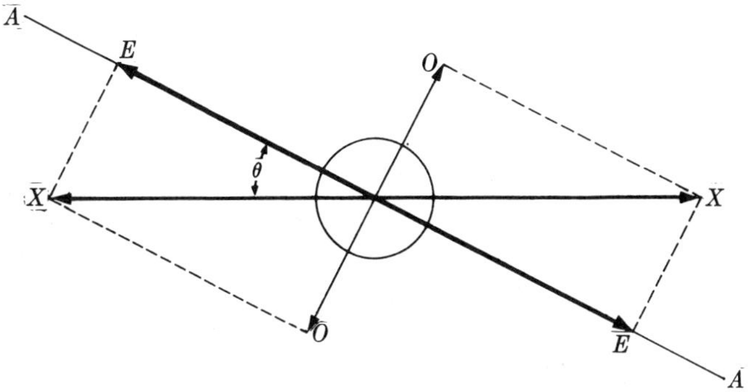

Fig. 7.41 Resolution of linearly polarized radiation by an analyzer in which the vibrational direction passed is *AA*. When radiation whose vibration is along *XX* is incident, it is resolved in the analyzer into components *EE* and *OO*, of which only *EE* is passed.

where I_0 is the incident and I the transmitted intensity. If $\theta = 90°$, corresponding to placing the polarizer and analyzer in the crossed position, no radiation will emerge from the analyzer. This angle is said to be one of extinction. The orientation of the vibrational plane of linearly polarized light can be most easily ascertained by setting an analyzer to extinction. The vibrational plane is then at right angles to its angular setting.

7.17 INTERFERENCE EFFECTS AND CIRCULAR POLARIZATION

In anisotropic media, the interference of EM rays can lead to the production of elliptically and circularly polarized radiation, as well as the linearly polarized type. These polarizations are closely related, as will be seen.

In Fig. 7.29b it was suggested that a beam of natural light can be resolved hypothetically into two vibrations at right angles to each other. If the incident light is incoherent, that is, the phase relationships among the waves comprising it are completely random, the emerging pair of orthogonal vibrational components will also be incoherent. As a result, no possibility exists for interference between them.

Consider now a simpler case in which the same representation applies. Let an unpolarized but coherent beam be perpendicularly incident on a uniaxial (anisotropic) crystal whose optic axis is parallel to its front face. This situation is pictured in Fig. 7.42. It is also assumed that $n_E > n_O$ as it is true for quartz. Accordingly, when the incident ray is resolved into two orthogonal components, the vibration along the direction of the optic axis (extraordinary ray) will travel more slowly because of the higher index for this vibration. Similarly, the vibration perpendicular to the axis (ordinary ray) will move faster. The greater the distance of travel, the more the phase of the extraordinary component will "lag" the ordinary component.

While these rays are out of phase at any point in their path through the slab, this fact cannot be detected conveniently until they emerge. At that point interference between the two components will be evident. It should be possible to calcu-

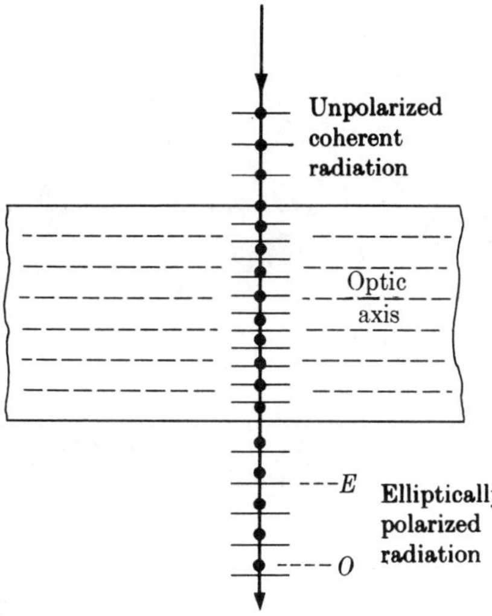

Fig. 7.42 Retardation of one set of polarized rays within a quartz crystal. The front and rear faces are cut parallel to the optic axis. After traversing the crystal, the ordinary wavefront is well ahead.

late the phase difference δ since it will reflect the difference in the number of wavelengths of each ray required to traverse the crystal. For the ordinary component $\lambda_O = \lambda_{vac}/n_O$, and the number of wavelengths across a slab is bn_O/λ_{vac}, where b is the thickness of slab. This expression is also termed the *optical path length* for the component. Taking the difference between this number and a similar value for the extraordinary ray and multiplying by 2π gives the phase difference in radians:

$$\delta = \frac{2\pi b}{\lambda}(n_E - n_O). \tag{7.30}$$

For example, the extraordinary ray may be $\pi/2$ radians (or 90°) behind. In this case, it is said to suffer a *quarter-wave retardation*.

Example 7.7. The thickness of quartz necessary to retard the extraordinary ray one quarter-wavelength for the D line can be calculated as follows from Eq. (7.30):

$$\delta = \frac{\pi}{2} = \frac{2\pi b}{589.3 \text{ nm}}(1.5553 - 1.5442).$$

Values of n_E and n_o for quartz have been substituted. Solving,

$$b = 13.3\ \mu\text{m}.$$

Such a plate would be very fragile as well as difficult to cut. If δ were 450°, 810°, and so on, the effect would still be that of one quarter-wave retardation and would allow more reasonable thicknesses to be used. For 810° ($9\pi/2$), b will be 119 μm or 0.119 mm. It is obvious that still greater thicknesses may be used if desired.

What is the effect of such interference on the character of the emergent light? If the vibrations of the two rays are in phase, they can be resolved into a single wave that is linearly polarized. Its vibrational plane will be at an angle of 45° to the vibrational planes of the rays, as indicated by the sketch in Fig. 7.43a. If the vibrations are out of phase by an angle between 0 and 90°, as in 7.43b, the resultant is a vector that does not lie in a plane but traces out an ellipse as the waves advance. A phase difference of 90° resolves into a circle as in Fig. 7.43c. It is common to describe the last two polarization phenomena by the figure traced by the resultant. Thus, they are *elliptically* and *circularly polarized radiation*. At greater phase differences the figures repeat. For example, a variation of 180° will again give linearly polarized radiation, but its vibrational plane will be perpendicular to that of Fig. 7.43a. Circular polarization will again appear at 270°, and so on. Note that at all but a few phase differences the output will be elliptically polarized light.

In practice, it is not necessary to use coherent light. Instead, any source of linearly polarized radiation can be employed. When its beam is resolved into two orthogonal components, each will be self-coherent and undergo interference when directed toward a birefringent crystal.

At least two kinds of applications of circularly polarized light deserve mention. One is the measurement of the circular dichroism of complex substances, a phenomenon which will be discussed in the next section and in Chapter 19. The other is the study of the optical properties of crystals [9]. For either, a combination of a (linear) polarizer and a quarter-wave retardation device is employed to generate circularly polarized light.

One form of quarter-wave retardation device is the so-called *quarter-wave plate*, which is usually a specially cut quartz or mica slab. Both the wavelength of quarter-wave retardation (recall that refractive index varies with wavelength) and the direction of vibration of fast and slow rays are engraved on the holder. Another more versatile device is the *quartz wedge*, which has a continuously varying thickness. Along its length it provides retardation of varying amounts (or at different wavelengths). Polaroid *circular polarizers* combine a linear polarizing dichroic layer with a quarter-wave retarding layer and are effective over different spectral ranges.

Finally, quarter-wave retardation can be secured by an electrooptical device. One type is the *Pockels cell*, which utilizes a uniaxial crystal of dihydrogen phosphate so oriented that light passes through it along its main or z axis. Some types

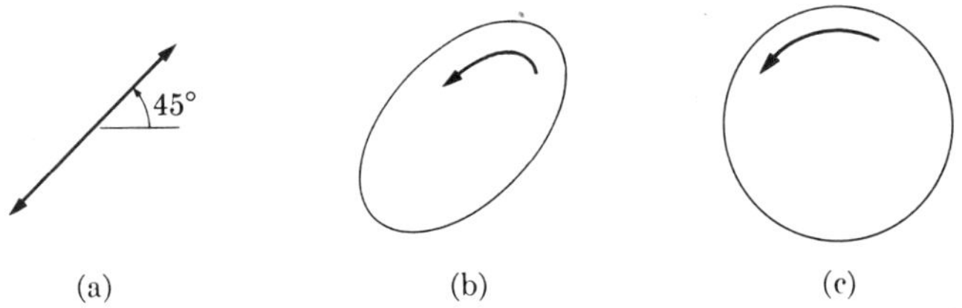

Fig. 7.43 Effect of phase difference δ on polarization. (a) Linearly polarized radiation, $\delta = 0°$. (b) Elliptically polarized radiation, $\delta = 40°$. (c) Circularly polarized radiation, $\delta = 90°$.

of piezoelectric crystals can also be used. Electrodes on either side of the crystal allow an electric field to be applied parallel to the z axis. Under the field, the crystal becomes biaxial. If the polarization of a preceding polarizer is parallel to the x or y axis of the crystal, the cell generates circularly polarized light, either left-handed or right-handed, depending on the field direction.

7.18 OPTICAL ACTIVITY

Substances whose molecules or crystals lack a plane or center of symmetry, that is, are asymmetric, are described by chemists as optically active since they have the ability to rotate the vibrational plane of linearly polarized radiation. Examples of optically active substances range all the way from sucrose to intricate complex salts of metals. If structural asymmetry occurs in molecules they are called chiral and the optical rotary power is evident in all physical states and in solution. There are some instances, however, in which it is a property of the crystal lattice and is shown only by the solid form.

Optical rotation can best be treated as a case of *circular double refraction*. Consider that incident linearly polarized rays are resolved into two circularly polarized rays as just discussed, one whose vector turns clockwise as the beam advances, the other, counterclockwise. In an optically inactive material their resultant is the original linearly polarized ray. In a material with optical rotatory power, however, these rays travel at different velocities. The material therefore has two unequal refractive indices, one for the left-handed circularly polarized rays n_L, the other for the right-handed rays n_R. As long as there is no absorption of either ray and they follow the same path, they have a resultant which is a linearly polarized ray, but one whose direction has been rotated with respect to that of the incident ray. This result is shown diagrammatically in Fig. 7.44. The plane of polarization of the resultant, of course, progresses in helical fashion through a steadily increasing angle as the beam traverses the medium.*

Optical Rotary Dispersion. For a sample the relationship between the difference in its refractive indices and the angle of rotation α' (in radians cm^{-1}) of the vi-

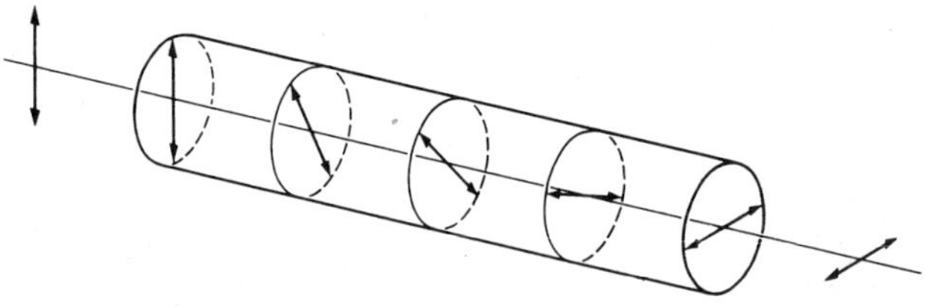

Fig. 7.44 Rotation of the (direction) plane of polarization in a medium with optical rotatory power.

*The phase difference between the two circular components is just twice the angle of rotation.

bration plane (direction) at a given vacuum wavelength λ is given by

$$\alpha' = (\pi/\lambda)(n_L - n_R), \tag{7.31}$$

where n_L and n_R are the refractive indices for the circularly polarized rays.

The form of the expression is identical with Eq. (7.30) for ordinary circular polarization. No more than a minute difference in refractive indices need exist for a rotatory effect to appear. For example, by solving Eq. (7.31) for the case where α is 10° per cm of travel and λ equals 500 nm (green light), one finds that Δn is just 3×10^{-6}. Since the indices are dependent on the wavelength, different wavelengths are rotated by different amounts, a phenomenon known as optical rotatory dispersion or ORD.

In regions of transparency the dependence of optical rotation of a substance on the wavelength is reliably expressed by the Drude equation:

$$\alpha = k_1/(\lambda^2 - \lambda_1^2) + k_2(\lambda^2 - \lambda_2^2) + \cdots = \sum_i k_i/(\lambda^2 - \lambda_i^2), \tag{7.32}$$

where k_i is a constant evaluated from experimental rotations and λ and λ_i are the vacuum wavelengths of the incident radiation and of a characteristic active electronic absorption peak, respectively. The k's may be either plus or minus. For many substances a single term of Eq. (7.32) is sufficient and almost all others require no more than two.

In an anisotropic medium like quartz, optical rotation is simply superposed on the dominant effect, ordinary double refraction. Such materials are usually crystalline and unambiguous information about their optical rotation may be obtained best by directing a beam along the optic axis of the crystal (Section 7.15). In other directions the differently circularly polarized rays would diverge, and there would be little opportunity to measure their relative difference.

Behavior in Electronic Transitions. For optically active substances behavior differs strongly in certain electronic transitions as shown in Fig. 7.45. Evidence indicates that one or at most three electronic transitions are optically active for most substances. The decrease in optical rotation with decreasing wavelength (increasing frequency) is sometimes described as anomalous dispersion of the optical rotatory power.

The true inport of optical activity is now evident. In an active electronic transition interaction with left- and right-circularly polarized light includes both

1. differential absorption, that is $\epsilon_L \neq \epsilon_R$, called *circular dichroism* (CD) and
2. strong *optical rotatory dispersion* (ORD), that is, n_L and n_R are substantially different and changing rapidly with wavelength.

As a result of differential absorption within an optically active absorption region, the two circularly polarized rays give a resultant that is elliptically polarized. The

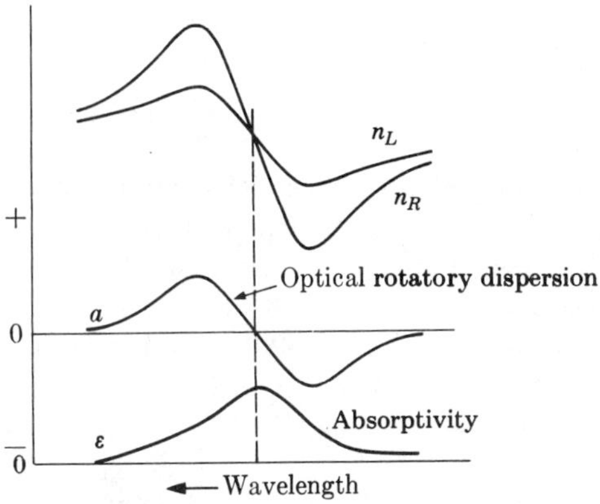

Fig. 7.45 Optical rotatory dispersion of a substance. The top set of curves refers to the refractive indices for left- and right-circularly polarized rays, the middle curve to the optical rotatory dispersion (ORD) of the substance, and the lowest curve to its regular absorption curve. If its differential absorption or circular dichroism (CD) when illuminated alternately by left- and right-circularly polarized light had been measured, a negative peak (below the axis) would have been recorded.

combination of optical rotatory dispersion and circular dichroism are often described as the *Cotton effect*. The two effects are connected by a Fourier transform relationship (the Kronig–Kramer transformation).

An example of these phenomena is illustrated in Fig. 7.45. If, as in this instance, right circularly polarized radiation interacts more strongly with the substance, it will be slowed or retarded more and also be more strongly absorbed ($\epsilon_R > \epsilon_L$). By convention this is a negative Cotton effect and the opposite situation a positive one.

In general, the wavelength of maximum circular dichroism corresponds to that of maximum absorption. This kind of differential absorption determination provides a measure of the asymmetry of an electronic transition and reflects a lack of symmetry in the molecular region in which the transition is localized. These points are expanded in Section 19.5.

7.19 IMAGE FORMATION

Lenses, curved mirrors, and aperture devices are indispensable in optical systems. Their role is the control of the flow of EM energy from one spot to another; more familiarly, they focus, collimate (render rays parallel), or diverge beams of radiation.

Focusing. Lenses and mirrors are characterized by two distinct sets of properties. One may be termed primary properties and includes refractive index, reflectance, absorbance, and curvature. The other is the set for which these optical elements are designed, definitely a drivative set. This group includes qualities such as focal length and focal point. Since these derive from the primary properties first cited, they will be as variable as are the primary properties.

As ordinarily used, the focal length and related qualities are considered constants. In this sense they are idealized concepts intended as standard. In general, they do define the behavior of optical elements precisely for rays that are *paraxial* in a narrow spectral region. The term paraxial designates rays (a) making very small angles with the transverse axis of the mirror or lens or (b) lying quite close to the axis. A good approximation to these conditions is possible in an optical system if highly convergent or divergent rays are blocked from access to lenses or mirrors. This can be done simply by the use of a variable aperture (a stop) or diaphragm on the axis immediately in front of a lens or mirror. An arrangement of this kind is pictured in Fig. 7.46. As expected, the smaller the opening, the more closely any lens approximates ideal focusing behavior.

The role of lenses and mirrors is the formation of optical images. A converging lens or mirror will form an image of an object placed beyond its focal point. Such optical elements are used to re-form an image of an object after some operation has been performed on the rays from the object. For example, a focusing lens is used in a spectrometer to produce images of the entrance slit after the radiation has been dispersed into the component wavelengths. As many images are produced as there are wavelengths present if these are not too close to each other.

When mirrors are used, it is desirable to eliminate the effects of having an object in the light path of the mirror. For example, Fig. 7.47 shows a conventional source–mirror arrangement used to obtain a collimated beam. Though effective, a portion of the aperture of the mirror is masked or vignetted by the source and is thus unused. Further, if an object with a sharp edge replaces the source, diffraction effects may result. These undesirable effects are eliminated if an *off-axis mirror* is substituted. As illustrated in Fig. 7.48, in this arrangement the source is no longer in the optical path of the mirror. The greater expense of manufacture of off-axis optics is often offset by their improved performance. For example, they are frequently employed in IR spectrophotometers where full use of beam energy is desirable.

In Fig. 7.49 the way in which rays from one point of an object define a point of an image is drawn for a simple convex lens. All the rays that travel from the object's point to the lens must be converged at spot T' to give the image point. As they converge they interfere. The result will be a spot of high or low intensity,

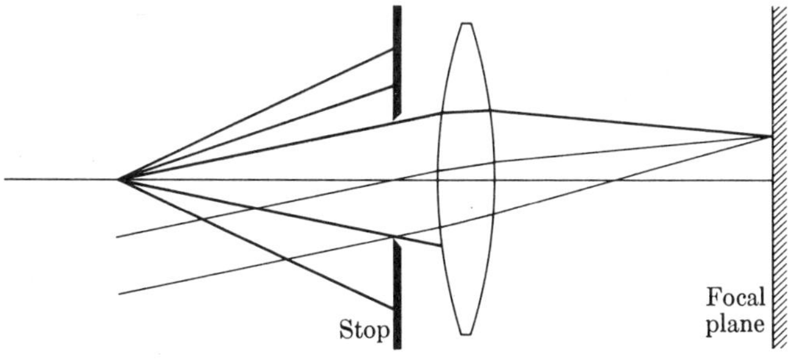

Fig. 7.46 A variable aperture or stop with a lens. Only paraxial rays are permitted to reach the lens.

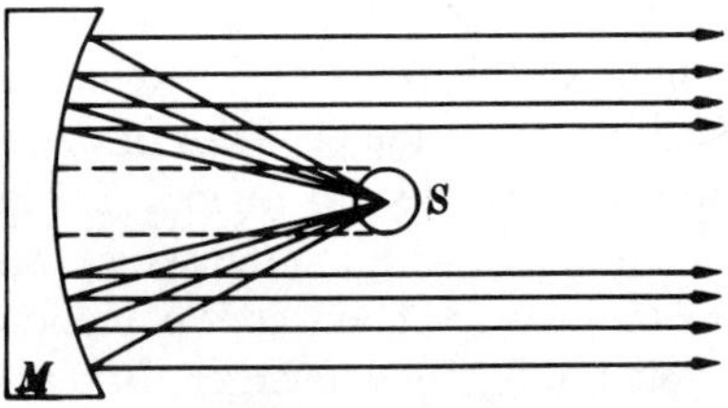

Fig. 7.47 Conventional placement of source S with respect to collimating mirror M. The reflecting surface directly behind the source is masked or vignetted by it.

depending on the brightness of the object spot. Since intereference forms an image point, the point must have a size related to the wavelength of the radiation. As a matter of fact, it can be shown that the point is a series of circular fringes with most of the intensity concentrated in the central spot. The greater the angle at which the rays from the lens cross and the shorter the wavelength, the more nearly does the image point duplicate the original one.

It is possible to set up several criteria that must be met for an image to be acceptable optically. Some of the more important of these are the following.

1. All wavelengths of a ray should come to a focus at the same point.
2. Each symmetrical pair of rays issuing from the image should focus in the same spot regardless of whether they are paraxial or whether they pass through a lens at an appreciable distance from its center.
3. Object points at a distance from the axis should be focused as points in the same plane as paraxial points.

If an appreciable portion of a lens or mirror is used, these criteria are not met in practice. Sources of the difficulty may be cataloged in terms of aberrations shown by the optical elements. In other words, the aberrations are types of departure from ideality. One way to compensate for them would be to sacrifice image intensity and restrict rays to the paraxial case. Fortunately, this is not necessary, since there are feasible means of minimizing one or more aberrations at the same time. Very seldom, however, is it possible to eliminate all aberrations in a given case.

Numerical Aperture. The solid angle over which light is "accepted" by a microscope, fiber optic, or other device is given in terms of its numerical aperture.

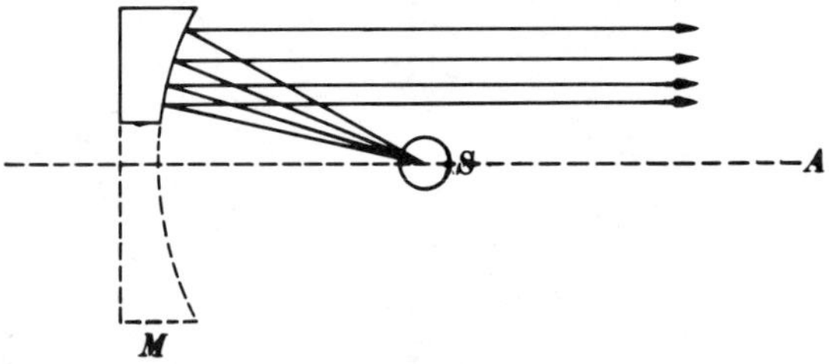

Fig. 7.48 Use of an off-axis mirror M, which is seen to be a portion of a full mirror of axis SA. By using it, masking of the reflecting surface is avoided.

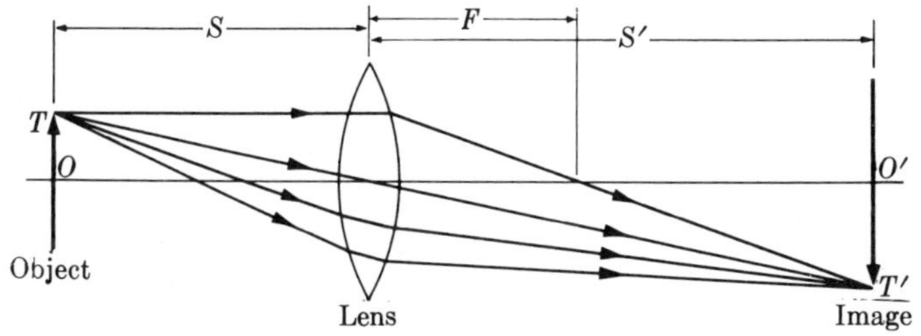

Fig. 7.49 Formation of an image point. The focal length F is defined as the distance from the center of the lens to the point at which incident rays parallel to the transverse axis of the lens are brought to a focus. The focal length, image distance S', and object distance S are related for a thin lens by the equation $1/S + 1/S' = 1/F$.

Only light falling within this cone is transmitted. The numerical aperture (NA) is simply the sine of the semiangle θ of the cone. Thus, a NA of unity implies acceptance over a solid angle of 180°.

Example 7.8. What is the NA of a magnifier that accepts light making an angle no greater than 30° with the normal? Since NA = sin θ, NA = sin 30° = 0.50.

---

f/Number. The geometrical light-gathering power of a lens is usually indicated by its f/number. By definition, f/(the / is part of the symbol) is the ratio of lens focal length f to lens clear aperture d. Light-gathering power increases as the inverse square of the f/number. For example, an f/2 lens passes four times more light than an f/4 lens when they are equally illuminated. If several lenses are involved in an optical system, the one of largest f/value will determine the light-gathering power. Thus, if both an f/2.8 and an f/16 lens are used in a beam, the latter effectively sets the light-gathering power.

An alternative index used in precise work is the T number. It is an f/number $\div$ (transmittance)$^{1/2}$. Note that the T number will always exceed the f/number for the same lens.

7.20 OPTICAL ABERRATIONS

At least five aberrations have been described for monochromatic light—spherical aberration, coma, astigmatism, curvature, and distortion. An additional one enters with the use of polychromatic radiation, chromatic aberration. Three of these will be described in detail, and a few common methods of correction will be suggested.

Spherical Aberration. This aberration is perhaps the least complicated departure from ideality. As illustrated in Fig. 7.50, it results in a blurred image at a point on the axis. Each symmetrical pair of rays from a given object point comes to a focus at a different distance from the center of the mirror. The reader may establish

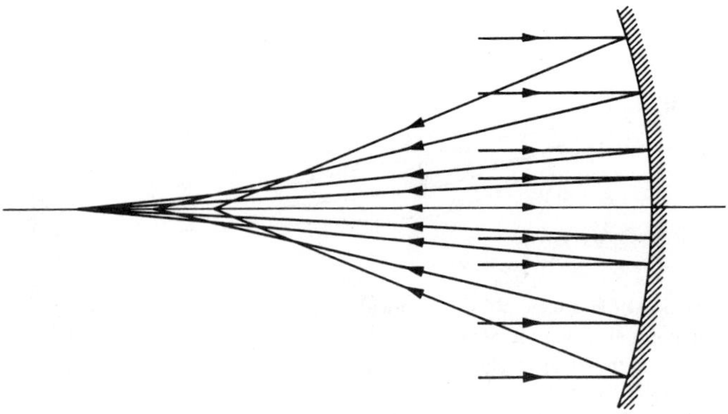

Fig. 7.50 Spherical aberration produced by a concave mirror.

the validity of this result for the mirror shown by applying the usual laws of reflection to various incident rays; those farthest from the axis will be found to focus nearest the mirror. A simple spherical lens produces the same kind of distortion. In this case the interference of the convergent rays produces not a bright point but a circle if some average focal distance is selected. The smallest circle is known as the *circle of least confusion*. Any detecting element or additional optical device is ordinarily placed at this distance from the mirror or lens. For example in a spectrophotometer, uncorrected spherical aberration causes a monochromator slit to appear as an enlarged spot on the detector. The resolution of different images is thereby limited.

Some common methods of reducing spherical aberration are

1. decreasing the aperture diameter with concomitant loss of intensity,
2. grinding the outer zone of the mirror or lens so that it is aspherical, or
3. using a correcting device such as an additional element of special shape and refractive index.

Since a concave parabolic mirror is free of spherical aberration, as shown in Fig. 7.51, it can be substituted to avoid elaborate grinding procedure if a mirror

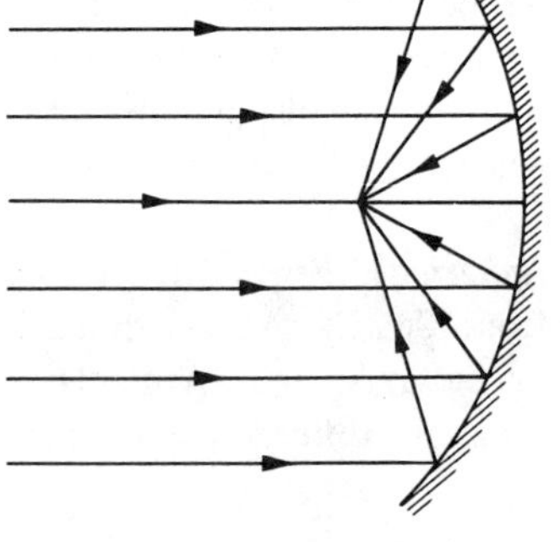

Fig. 7.51 Focusing for a parabolic concave mirror. It has no spherical aberration.

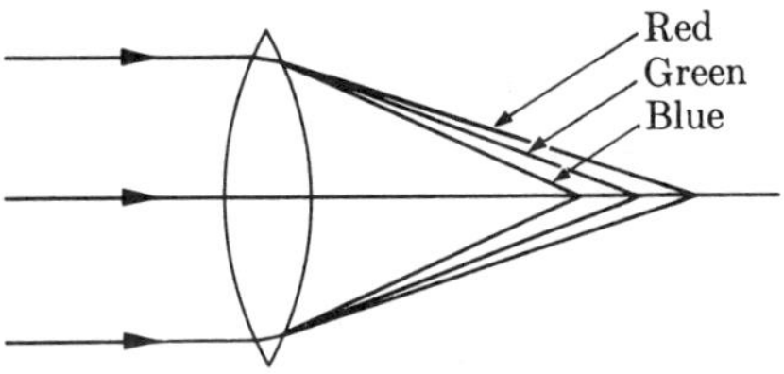

Fig. 7.52 Chromatic aberration produced by a simple lens.

is required. The third technique is frequently employed with lenses, so that compound lenses are the rule rather than the exception.

Off-axial rays give rise to *coma* and *astigmatism*. The first is closely related to spherical aberration and, like it, causes object points to become image areas. When object points are still farther from the axis, the distortion is termed astigmatism. In this case the blurring results from an object point being resolved only into two mutually perpendicular lines, though these exist in separate planes. It is best to intercept the focused beam at a distance intermediate between these planes. There the image is approximately disk-shaped and may again be said to be a circle of least confusion. A common measure taken to minimize astigmatism is the superposition of some cylindrical curvature on the basic spherical surface. Stopping down is obviously also effective.

Chromatic Aberration. Different wavelengths of light focus at different distances from a single lens because of the variation of the refractive index with wavelength. By the same reasoning, it is apparent that all wavelengths cannot be collimated by a simple lens. For example, an inexpensive spectroscope with a single-element lens ordinarily renders parallel only green light, which falls about in the middle of the visual spectrum. Instead of collimating it converges shorter wavelengths slightly and diverges longer wavelengths. In better instruments a two- or three-element lens is used to eliminate the chromatic error at *two or three chosen wavelengths*. Each part of the lens is a material of different refractive index and curvature. Compare the chromatic aberration of the simple lens of Fig. 7.52 with that of the two-element lens depicted in Fig. 7.53. At wavelengths intermediate to those for which the corrections have been made, there is some residual chromatic aberration.

The use of any correction scheme other than stopping down proves a compro-

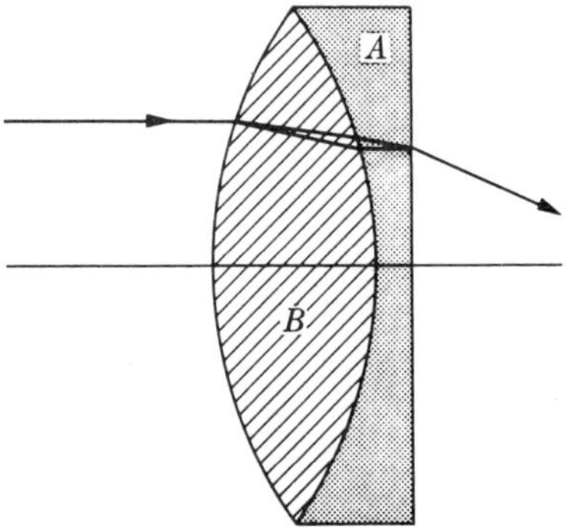

Fig. 7.53 A two-element collimating lens. *A* is quartz (n_D = 1.54). *B* is fluorite (n_D = 1.43).

mise. In other words, not all optical aberrations may be corrected simultaneously. When one understands the requirements of any particular optical system, one can decide among the many varieties of lenses and mirrors.

REFERENCES

A general introduction to many branches of optics is provided by:

1. J. K. Robertson, *Introduction to Optics, Geometrical and Physical*, 4th ed. New York: Van Nostrand, 1954.
2. J. P. C. Southall, *Mirrors, Prisms, and Lenses.* New York: Dover, 1964.

A clear detailed discussion of physical optics on an intermediate level is provided by:

3. W. H. A. Fincham and M. H. Freeman, *Optics*, 8th ed. London: Butterworths, 1974.
4. F. A. Jenkins and H. E. White, *Fundamentals of Optics*, 3rd ed. New York: McGraw-Hill, 1957.

An intermediate to advanced coverage of physical optics is given by:

5. R. W. Ditchburn, *Light*, New York: Wiley, 1964.
6. M. Garbuny, *Optical Physics*. New York: Academic, 1965.
7. J. R. Meyer-Arendt, *Introduction to Classical and Modern Optics*, 2nd ed. Englewood Cliffs, NJ: Prentice-Hall, 1984.
8. A. Nussbaum and R. A. Phillips, *Contemporary Optics for Scientists and Engineers*. Englewood Cliffs, NJ: Prentice-Hall, 1976.
9. J. Strong, *Concepts of Classical Optics*. San Francisco: Freeman, 1958.

Particular aspects of optics are treated comprehensively in:

10. N. J. Harrick, *Internal Reflection Spectroscopy.* New York: Interscience, 1967.
11. G. Kortüm, *Reflectance Spectroscopy: Principles, Methods, Applications*. New York: Springer-Verlag, 1969.
12. R. G. Newton, *Scattering Theory of Waves and Particles.* New York: McGraw-Hill, 1966.
13. W. A. Shurcliff, *Polarized Light, Production and Use.* Cambridge, MA: Harvard University Press, 1962.
14. W. H. Steel, *Interferometry.* New York: Cambridge University Press, 1967.
15. H. C. Van der Hulst, *Light Scattering by Small Particles*. New York: Wiley, 1967.

EXERCISES

A handbook listing physical constants will be needed in solving these exercises.

7.1 What are the velocity and wavelength (a) in water and (b) in benzene of the sodium emission line whose wavelength in air at 25°C is 589 nm?

7.2 Account for the fact that the dielectric constant of chloroform determined at 100 kHz is larger than the square of n_D for the substance.

7.3 In optical dispersion, does the rate of change of n with frequency or wavelength vary with wavelength? Where is it greatest for a substance?

7.4 Draw, without making any calculations, the dispersion curve in the visible and UV for a substance with characteristic electronic absorptions at 500 and 300 nm. Assume that $n_{600} = 1.50$ and that the 300-nm absorption is the stronger.

7.5 For a diamond ($n_D = 2.417$) illuminated with 589.3 nm light at perpendicular incidence, (a) what is the percentage reflected? and (b) if multiple reflections are ignored, what percentage will emerge from a thin slab of diamond? Assume air surrounds the diamond.

7.6 At 589 nm what is the reflectance in air of a polished plate of (a) silver and (b) aluminum? The pertinent constants are: for silver, $n = 0.1$, $k = 4$; for aluminum, $n = 1.0$, $k = 6$.

7.7 A layer of CaF_2 of 330 nm thickness is used as the spacer between semireflecting layers in an interference filter. At what wavelength is the second-order transmission region of the filter? Where do the first- and third-order bands lie?

7.8 What order interference will give a transmission band with a maximum at about 550 nm for an interference filter with a CaF_2 layer 0.61 μm thick? How close on the short wavelength side is the next transmitted band?

7.9 A Fabry–Perot interference filter that passes a second-order band centered at 600 nm is on hand. To eliminate the other bands an auxiliary Fabry–Perot filter is to be deposited on one side and the whole enclosed between glass plates. To secure optimum blocking of bands passed by the basic filter, what wavelength should be selected for the first-order band of the auxiliary filter?

7.10 Will light incident from air at 35° on a flat glass plate be totally reflected if the glass is extra-dense flint of index about 1.72?

7.11 The surfaces of a glass prism of refractive index 1.65 are coated with a CaF_2 layer of $n = 1.35$ to reduce reflection. (a) With light whose vacuum wavelength is 5000 Å (500 nm), how thick must the layer be for maximum effectiveness? (b) In this layer what is the phase difference for light of 4000 Å (400 nm) wavelength reflected from the upper and lower surfaces? Assume the refractive indices are constant over this range.

7.12 It is decided to construct a narrow-band IR reflection filter by making use of the change in refractive index across an absorption band. Mylar, which has an absorption band at 1740 cm^{-1} (5.75 μm), is selected and a layer of Mylar (average $n = 1.64$) is affixed to one of NaCl (average $n = 1.51$). Assume IR radiation is incident from the NaCl side. Discuss briefly how the device operates as a selective reflection filter in the region of the absorption band.

7.13 Determine an approximate reflectance versus wavelength curve for the following system over the range 550–450 nm; light is incident at a constant angle greater than the critical angle from a glass whose average n is 1.65 onto a liquid whose n is about 1.65 at 550 nm and 450 nm. The liquid has an absorption band at 500 nm whose half-width is 10 nm.

7.14 A bundle of 0.06 mm diameter glass fibers, packed coherently, is to be used for image transmission. What maximum resolution may be achieved? Consider that resolution involves seeing adjacent lines as separate objects.

7.15 In each fiber of the fiber-optic bundle total internal reflection must occur at the boundary with its coating layer. (a) On this basis suggest a minimum useful diameter for a single coated fiber optic that is to be used in the visible range. (Hint: consider the thickness of coating required.) (b) What is the maximum resolution that might be attained in a coherent bundle? (c) Would the transmission of a bundle be affected as the size of fibers that make it up is varied?

7.16 A beam of 589 nm light traveling in Plexiglas ($n = 1.50$) is found to be plane-polarized when reflected from a rare-earth glass at an angle of 50°. What is the refractive index of the glass?

7.17 What is the polarizing angle for silver chloride in the near IR? Assume $n = 2.10$.

7.18 Unpolarized light is to be perpendicularly incident on a slab of strontium chloride hexahydrate whose sides are parallel to the optic axis. The piece is ground to a final thickness of 0.545 μm. The crystal is uniaxial and has D-line indices of $\omega = 1.536$, $\epsilon = 1.487$. Will sodium light emerge elliptically polarized?

7.19 Calculate the maximum angle at which light can be incident on a nicol prism if the ordinary ray is to be totally reflected by the Canada balsam layer. (See p. 234 for diagram and refractive indices. Note that the balsam layer is perpendicular to the front and back faces.)

7.20 How does the intensity transmitted by a polarizer–analyzer pair, when their vibrations are at an angle of 45° relative to each other, compare with that when they are at 70°?

7.21 Show that Eq. (7.23) can also be expressed as $P = (I_{\parallel} - I_{\perp})/(I_{\parallel} + I_{\perp})$, when $I_{\parallel}$ corresponds to alignment of analyzer and polarized directions of preferred vibration and $I_{\perp}$ to the opposite condition.

7.22 Circularly polarized light is passed through a quarter-wave plate. What is the polarization of the emergent light?

7.23 The optical rotation of a 1-mm piece of quartz is 41.54° at 435.8 nm. What must be the difference in the refractive index for left- and right-handed circularly polarized light?

7.24 What effect will a half-wave plate have on the transmission of a beam of linearly polarized light? Will its action be dependent on wavelength?

7.25 In Raman spectroscopy it is necessary to detect weak, inelastically scattered radiation of many frequencies in the presence of intense monochromatic, elastically scattered light. To simplify instrumentation, a filter that *passes* all wavelengths but the elastically scattered one would be very useful. It has been proposed [M. Tobin, *J. Opt. Soc. Am.*, **49,** 850 (1959)] that two Fabry–Perot interference filters that *transmit* the elastically scattered light with 70% efficiency can be used for the purpose. Note that the light they do not transmit is reflected. Thus, their reflectance is about 97% except at the transmitted wavelength band, where it is only 30%. (a) Sketch a multiple reflection arrangement using a parallel pair of such filters. It should pass only the desired scattered frequencies from a Raman sample to the slit of a monochromator. (b) If four reflections from the filter occur, what intensity of Raman frequencies is passed? of elastically scattered light?

7.26 Sellmeir proposed the equation $n^2 = 1 + A\lambda^2/(\lambda^2 - \lambda_0^2)$ to represent optical dispersion in optically transparent regions. How closely is this related to Eq. (7.11)? [Hint: Start by substituting λ for ν in Eq. (7.11).]

7.27 Derive Eqs. (7.26) and (7.27), the expressions for Brewster's angle and the principle angle of a substance, from Snell's law, Eq. (7.8).

7.28 What are the advantages of using silver chloride plates over quartz plates in constructing a transmission polarizer for the IR region? Consult a handbook for properties.

7.29 It has been shown that the degree of polarization produced by passing a radiation beam through a stack of dielectric plates oriented at the polarizing angle is given by $m/\{m + [2n/(n^2 - 1)]^2\}$, where m is the number of plates and n the refractive index [p. 491 of Ref. 4]. (Multiple reflections have been considered.) A stack of six silver chloride plates is used as an IR polarizer. Assume $n = 2.10$. What fraction of the radiation emerging is polarized?

ANSWERS

7.1 (a) 2.25×10^8 m s^{-1} and 442 nm; (b) 2.00×10^8 m s^{-1} and 393 nm. *7.5* (a) 17.5%; (b) 68.1%. *7.6* (a) 0.98; (b) 0.90. *7.7* 445 nm, 890 nm, 297 nm. *7.8* Third, 413 nm. *7.9* 1800 nm. *7.11* (a) 92.5 nm; (b) 228°. *7.14* About 0.18 mm. *7.15* (a) The thickness of coating must be at least half a wavelength; thus in cross section a coating will contribute $2 \times 1/2 \times$ or 600-nm thickness for 600-nm light. If the glass fiber is the same thickness, a minimum diameter of 1.2 mm will be required. *7.16* 1.79. *7.17* 64° 33′. *7.19* 36 °. *7.20* at 45°, 50%; at 70°, 11.7%. *7.23* 5.03×10^{-5}. *7.25* (b) 0.8%; 89%. *7.29* 0.79.

Chapter 8

SPECTROMETRIC MODULES: SOURCES AND DETECTORS

8.1 INTRODUCTION

In this chapter and the next the groundwork for optical spectrometers is laid. The intent is to present the main aspects of the theory, operation, and properties of many of their characteristic modules. Optical sources and detectors are discussed here and two more modules, monochromators and polychromators, are described in the following chapter. Figure 8.1 suggests the range of optical spectrometers that use these. The supplying of dc operating voltages or currents for them and their coupling to preamplifiers or amplifiers will also be discussed; these aspects prove integral to understanding how the modules operate. Further, in discussing detectors, photon counting arrangements will be taken up as the preferred mode of operation at very low light levels.

With this background, discussion of spectrometers in later chapters on techniques can be focused on choosing modules to give spectrometers with desired properties. Each module makes a unique contribution to the performance of the system and the specifications of a spectrometer can usually be related directly to the optical properties of its characteristic modules. This modular approach will also allow spectrometers to be considered as instrument *systems* designed around a *signal channel*, as discussed in Chapter 1.

Example 8.1. Spectrophotometers require an optical source such as a tungsten lamp. How decisive are its properties in determining instrument specifications?

The spectral range of the source will of course establish the maximum wavelength range realizable for the spectrometer. Further, the more constant the output of the source, the greater the signal/noise ratio (S/N) of the instrument and the more accurately absorbance can be measured in highly absorbing solutions.

As illustrated in Chapter 1, these results will be secured (a) whether the spectrophotometer is single- or double-channel in design and (b) whether the manufacturer has set values of variables such as slit width or whether they are under the control of a computer program the user has chosen.

Example 8.2. In a spectrofluorometer for use at the trace-level, stray light often establishes the limit of detection for an analyte. What modular choices will reduce this problem?

Certainly the standard design (as in Fig. 8.1, the emission arm 90° to the exciting arm)

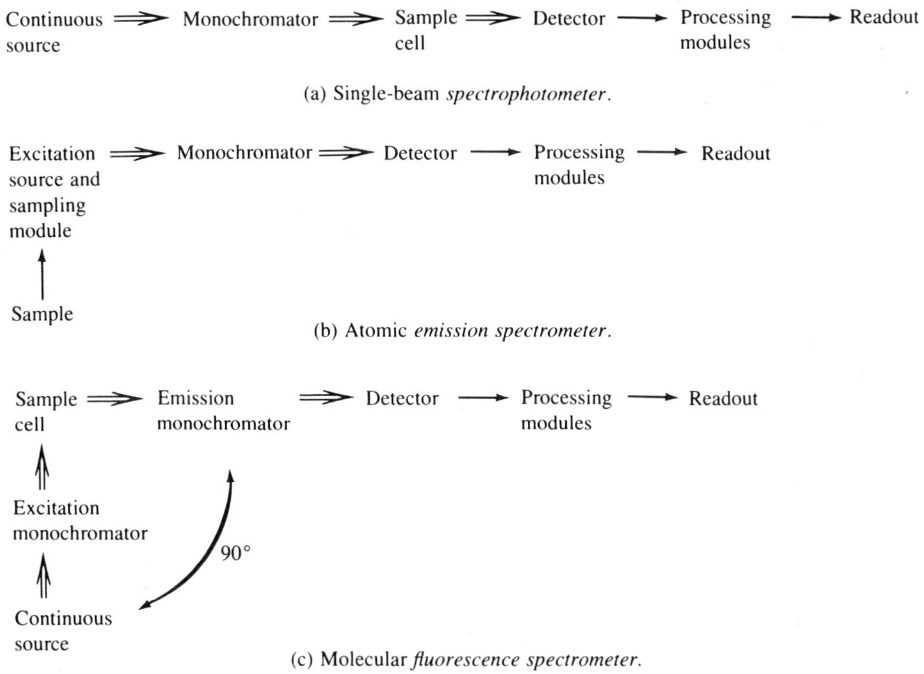

(a) Single-beam *spectrophotometer*.

(b) Atomic *emission spectrometer*.

(c) Molecular *fluorescence spectrometer*.

Fig. 8.1 Modular diagrams of some optical spectrometers.

will be essential in keeping unabsorbed exciting light, which may become stray light, away from the detector. Choosing a double monochromator to isolate *emission* wavelengths is also desirable to reduce stray light.

Other, more specialized optical modules will be described in chapters on individual methods. For example, the inductively coupled plasma torch will be discussed in connection with atomic emission spectrometry and the Michelson interferometer will be described in connection with Fourier transform spectrometers.

8.2 CONTINUOUS SOURCES

Continuous optical sources are those whose emission extends essentially without break over a broad spectral region. They are indispensable as monochromator sources. Ideally, such a source or emitter ought to provide an intense output that is (a) constant with time and (b) uniform with wavelength. How closely can ideal performance be approximated?

The first desired characteristic, constancy with time, can at least be closely approached. Some obstacles are that the output of all sources is subject to drift and fluctuations. Drift is a consequence of aging processes. Hot electrodes lose material by sublimation and envelopes become fogged, causing a reduction in intensity long before the filament fails. One method of slowing aging is suggested below. Heating and cooling cycles can also produce irreversible changes in output, effects best avoided by leaving a source on continuously during a working period.

Fluctuations in intensity occur during warmup as a source heats to its operating temperature, when the ambient temperature changes, or when the power energizing a source fluctuates. Continuous operation or allowance of an adequate time for warmup takes care of the first factor. The effect of ambient temperature changes can be minimized by insulation, for example, by use of a thicker envelope. Power fluctuations can be minimized by use of a regulated power supply (see Section 3.6).

Feedback control of output intensity is probably the best guarantee of constancy. It is only necessary to monitor the output by means of an appropriate detector, to compare its signal with that of a reference such as a Zener diode (see Section 2.11), and to use the difference signal to actuate a device that will provide reduced or extra power in an amount proportional to that signal.*

What of the second characteristic, equal output at all wavelengths? It cannot actually be attained by use of a source alone. The reason is to be found in the characteristics of continuous sources, which are best considered by type. The first of these will be discussed in terms of its ideal model, the blackbody.

Blackbody Radiation. The *blackbody* is a convenient model of a source like a heated solid, for example, a lamp filament or a SiC rod. The category includes most continuous wavelength IR and VIS sources. A nearly perfect blackbody can be built by developing a spherical furnace with a very narrow opening through which radiation is emitted and absorbed. Ordinary emitters can be related to a blackbody by their emissivity ϵ, the proportionality constant ($\epsilon < 1$) by which blackbody emission must be multiplied to give the emission curve of an actual source. The constant depends on both temperature and wavelength.

By definition, a blackbody absorbs all radiation falling on it, and is also a "perfect" emitter. Conversely, a good reflector is a poor radiator of energy. The emission spectrum of a blackbody is described by the following laws:

1. The *Wien* displacement law, which states that the wavelength of maximum emission λ_m of a blackbody varies inversely with absolute temperature, that is, the product $\lambda_m T$ remains constant. If λ_m is expressed in micrometers the law becomes

$$\lambda_m T = b = 2898 \tag{8.1}$$

or, approximately, $\lambda_m = 3000/T$. In terms of Ω_m, the wavenumber of maximum emission, Eq. (8.1) becomes $\sigma_m = 3.48T$. The frequency version giving ν_m is $h\nu_m = 5kT$.

2. The *Stefan law*, which states that the total energy J radiated by a blackbody

*For example, see H. L. Pardue and P. A. Rodriguez, *Anal. Chem.*, **39**, 901 (1967).

per unit time and area (power per unit area) varies as the fourth power of the absolute temperature:

$$J = aT^4, \tag{8.2}$$

where a is a constant whose value is 5.67×10^{-8} W m^{-2} K^{-4}.

3. The *Planck radiation law*, which shows that the monochromatic emissive power J_λ of a blackbody is proportional to the reciprocal of the fifth power of the wavelength:

$$J_\lambda = E_\lambda\, d\lambda = \frac{2\pi hc^2}{\lambda^5}\left[\frac{d\lambda}{e^{hc/kT\lambda} - 1}\right]. \tag{8.3}$$

Here J_λ is the power per unit area for radiation whose wavelength is between λ and $\lambda + d\lambda$; the other symbols have already been defined.

The source of radiation from hot bodies is the multitude of atomic molecular "oscillators" that are thermally excited in any solid. Since their thermal energies vary widely, the distribution of emitted energies is broad. Curves showing the distribution for a blackbody at several temperatures are given in Fig. 8.2. The shift in the wavelength of maximum emission λ_m should be noted. The curve for 2100 K is about what would be obtained for the tungsten filament in a light bulb. Actually all the solid curves are plots of Eq. (8.3).

To find the total energy emitted at a given temperature, we must determine the area under the curve for that temperature. We can easily do this by integrating the expression in Eq. (8.3) from $\lambda = 0$ to $\lambda = \infty$. The result is Eq. (8.2).

Infrared Sources. In the IR, heated inert solids ordinarily serve as continuous sources. Peak emission and the distribution of energy with wavelength can be

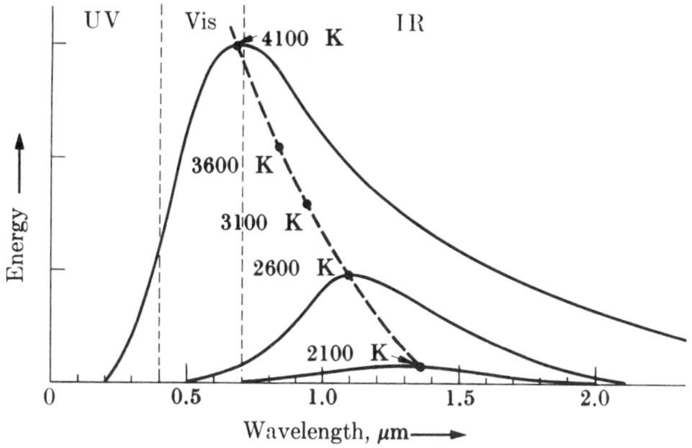

Fig. 8.2 Distribution of blackbody radiation at three temperatures. The broken line shows the dependence of the wavelength of maximum emission on temperature.

determined approximately by the use of Eqs. (8.1) and (8.2) for a blackbody. Emissivities are about 0.4–0.8.

How should a source operating temperature be decided? For example, is a good source for the middle IR one whose emission peaks at 1000 cm^{-1} (10 μm)? According to Eq. (8.1) such a source would be cooler than room temperature (290 K) and all parts of the instrument would radiate more energy than the source. A further problem is that the total energy it would produce—and thus also the energy it would emit at a given wavelength—would be quite modest. Contrast it with a source heated to 1000 K, which would radiate about 130 times as much energy. As hot a source as possible appears to be the best choice, since it furnishes better intensity over the whole range.

Practical IR sources operate at temperatures of about 1200–1800 K. Three factors preclude the use of still higher temperatures:

1. Source operating life would be shortened.
2. More heat would be dissipated.
3. Higher-intensity visible and UV light would be radiated, creating severe stray light problems in any IR photometric system (see Section 16.4).

Some commonly used IR sources are listed in Table 8.1 along with useful information about their nature, operating temperature, and useful wavelength range.

Table 8.1 Some Continuous Radiation Sources

Type	*Radiating Element*	*Envelope*	*Range (μm)*
Globar	6-mm diameter SiC rod at 1300–1500 K	None	1–40
Nernst glower	1–3-mm diameter ZrO_2, Y_2O_3 filament at 1200–2000 K	None	0.4–20
Tungsten lamp	W filament at 2300–3300 K	Glass	0.35–2.5
Tungsten–halogen lamp	W filament at 3500 K and halogen scavanger	Quartz	0.2–0.8
H_2 or D_2 lamp[a]	Arc discharge in gas	Fused silica	0.16–0.4
Xe lamp[b]	Arc discharge in Xe	Quartz, sapphire	0.23–2.1
Xe lamp[b]		LiF	0.15–0.20
Ar lamp[b]	Condensed discharge in low-pressure rare gas	LiF	0.10–0.15
He lamp[b]		None or Al foil	0.06–0.11
Hg lamp	Arc discharge in Hg vapor	Quartz	0.20–0.70

[a]For pulsed discharge a slightly modified lamp design is used.
[b]Y. Tanaka, *J. Opt. Soc. Am.*, **45,** 710 (1955); R. E. Huffman, J. C. Larrabee, and D. Chambers, *Appl. Opt.*, **4,** 1145 (1965).

These sources are maintained at operating temperature by passage of an electric current. Since the Nernst glower is a semiconductor, it must (a) be preheated to bring it to a conducting state and (b) be used with a series resistance or ballast to prevent burnout because of its negative resistance characteristics (see Section 8.4). The Globar has the advantage of greater ruggedness and better emissivity at shorter wavelengths.

In the far IR there is at present no fully suitable source. The best one available is a mercury lamp with a fused-quartz envelope. At frequencies below 150 cm^{-1} this source is considerably more efficient than a Globar. Below 35 cm^{-1} the mercury arc contributes most of the radiation, and above that wavenumber the hot envelope itself is the main emitter. Often this lamp is used for work up to 500 cm^{-1}.

Near IR, VIS, UV Sources. In the near IR and visible a tungsten lamp that consists of a tightly coiled tungsten filament in a glass envelope and operates at from 2600 to 3000 K is a widely used source. The transmission of the envelope limits the useful range to about 350–2000 nm.

The tungsten–halogen lamp is more efficient. It is designed with an internal chemical cycle that returns to the filament most tungsten that sublimes.* The lamp has a life more than twice that of a regular lamp and an extended spectral range.

In most filament lamps, useful life can be "traded off" for higher intensity by use of higher voltage. For example, an operating voltage 10% above the normal will elevate the luminous output of a filament lamp about 35% and reduce its average life to about 30% of the normal value.

The extent of the uniformly luminous area in a source greatly affects its optical applications. Lamps with vertical ribbon filaments are generally useful because they provide a large bright area. They are oriented so that the length of the flament is parallel to the slit to be illuminated. If very even illumination is required, the filaments should be considerably larger than the slit. In this connection arc lamps that provide an intensely illuminated circular spot are particularly valuable, especially the Pointolite lamps and the zirconium arc. By projection of their light a very uniform larger field can be secured.

Narrow-band semiconductor point optical sources are also available for the near IR and the longer-wavelength part of the visible region. For example, these diodes can be prepared from gallium arsenide or gallium phosphide–gallium sulfide combinations.

In the visible and UV, two high-intensity sources stand out: the high-pressure mercury lamp† and the high-intensity xenon lamp. In both the gas is excited by a

*A halogen (e.g., iodine), is added, and the envelope is fabricated of quartz to tolerate the higher operating temperatures involved. Near the hot (500–1000 K) quartz wall, gaseous iodine reacts with gaseous tungsten to form WI_2, a volatile iodide. When the diffusing gaseous iodide strikes the hot filament, it decomposes and tungsten is redeposited.

†A quartz-jacket, additive-type mercury arc light is also available. Its vapor is a mixture of mercury and iodine gas, the latter species originating from decomposition of an additive such as sodium iodide. (The gas is at about 6000 K.) It has an efficiency twice that of the regular mercury arc, but about half its life.

dc arc. Naturally discrete lines of the filling element also appear in the spectrum. Both the visible and UV regions are well represented in their discharge spectrum as is evident in Fig. 8.3.

While the output of the xenon lamp is smaller, it is more constant in intensity up to 450 nm. The emission continues up to 2.5 μm. Several very intense emission lines are superposed on its continuous spectrum between 830 and 1120 nm. The xenon lamp ordinarily has a ceramic–metal envelope, a built-in prefocused reflector that produces a very narrow uniform beam, and either a quartz or sapphire window, the latter giving superior UV transmission.

In discharge lamps run from a regulated power supply there are several sources of instability. One is arc wander, that is, random movement of the arc over the anode surface. Operation at a power above the specified average power level reduces wander.

Another cause of drift is change in the temperature of the envelope. Emission is affected because the internal pressure changes. Fortunately, such fluctuations are greatly reduced by the use of double-wall tubes or thicker windows. The output intensity of a lamp also diminishes gradually with time as a result of deposition of electrode metal on the lamp window.

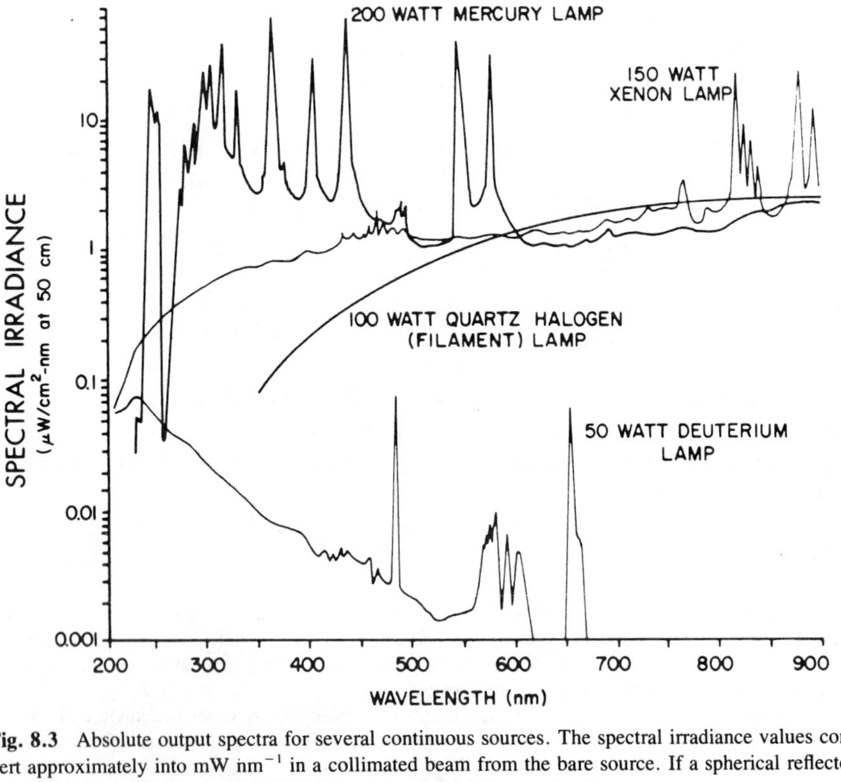

Fig. 8.3 Absolute output spectra for several continuous sources. The spectral irradiance values convert approximately into mW nm^{-1} in a collimated beam from the bare source. If a spherical reflector is added values will increase 25–30% (Courtesy of Oriel Corporation.)

For work only in the UV the continuous-emission spectrum of a low-pressure hydrogen or deuterium lamp is adequate down to 160 nm (1600 Å), the limit of transmission of a fused silica envelope. A higher-energy emission is attained with a deuterium filling. The lamp's upper limit of usefulness is about 370 nm, for at longer wavelengths neither spectrum is continuous. With a LiF window its usefulness can be extended downward to 125 nm, well into the vacuum UV.

Finally, lamps utilizing short-wavelength discharge continua of the rare gases allow work in the vacuum UV (200–50 nm). *Synchrotron radiation*, which is emitted by electrons, or sometimes positrons as a result of centripetal acceleration in the magnetic fields of a synchrotron or in a charged particle storage ring, is available at several national laboratories and their accelerator sites. It is especially attractive in the vacuum UV and X-ray regions (200–2 nm) for its high polarization, tunability, and intensity [11].

8.3 SPECTRAL STATES AND TRANSITIONS

In a generalized molecular system with two energy levels, E_1 and E_2, thermal equilibrium is established by collisions and redistribution of energy. If the levels represent a property that can be quantized, such as the energy of a molecular vibration or the electronic energy of an atom, particular frequencies of electromagnetic radiation can be emitted or absorbed by the system. Such processes are the basis of discrete wavelength sources.

A species in the upper state can emit a photon and enter the lower state. Because of quantization, Planck's law $\Delta E = h\nu$ must be satisfied for the transition and only frequency $\nu = \Delta E/h$ will be emitted. Ordinarly the Boltzman distribution can be used to calculate the relative populations of species in the two states:

$$N_2/N_1 = (g_2/g_1)\exp(-\Delta E/kT), \tag{8.4}$$

where N_2 and N_1 are the number densities, the number of species per cubic meter in states E_2 and E_1, respectively, g_2 and g_1 are the degeneracies of each, k is the Boltzmann constant (1.38×10^{-23} J K^{-1}), and T is the absolute temperature. The equation shows that the larger the energy difference between levels, the smaller the population of the upper level, and the higher the temperature, the *greater* the population on the upper level. Example 13.5 deals with relative populations of different states for two metallic elements.

The number of species making a transition during an interval of time dt is proportional to the population of species in the initial state and to the *Einstein probability coefficient* appropriate to the transition. Approximate values of such coefficients can in many instances be calculated. Three kinds of transitions and three probability coefficients can be considered for a two-level system where degeneracies are equal:

1. *Spontaneous emission* with probability coefficient A_{21}. The number of species dN spontaneously undergoing a transition from higher to lower level during an interval dt is just

$$dN = N_2 A_{21}\, dt, \tag{8.5}$$

where N_2 is the population of level 2. Representative values of the probability coefficient for allowed transitions are of the order of $10^8 s^{-1}$.

2. *Stimulated emission* with probability coefficient B_{21}. Such emission occurs where a population of excited species is irradiated by a uniform field of radiation of the frequency to be emitted. The number of species dN undergoing stimulated emission from level 2 during an interval dt is

$$dN = N_2 B_{21} \rho(\nu)\, dt, \tag{8.6}$$

where $\rho(\nu)$ is the volume density (m^{-3}) of the incident radiation.

3. (Stimulated) *absorption* with probability coefficient B_{12}. The number of species dN absorbing radiation from a uniform radiation field of the frequency required for the quantum transition in the interval dt is

$$dN = N_1 B_{12} \rho(\nu)\, dt. \tag{8.7}$$

During spontaneous emission each excited species emits independently and *incoherent* radiation is produced: amplitudes and phases of the emitted rays are unrelated in space or time. The discrete emission of gaseous discharge lamps and hollow cathode lamps is an example of spontaneous emission. By contrast, in stimulated emission a second photon is emitted exactly in phase at the very point where an arriving photon stimulates emission; further, the second photon moves in the same direction as the first. Thus, stimulated emission has both *temporal and spatial coherence*. Optical lasers, which are a prime example of stimulated emission, will be described in Section 8.5. It is important first, however, to introduce several simpler, widely used discrete sources.

8.4 GASEOUS DISCHARGE AND HOLLOW-CATHODE LAMPS

The transitions of species between excited states and a lower state just described are the origin of line spectra emitted by discrete sources (see Section 13.3 for a more detailed discussion). The thermal volatilization (if necessary) and excitation are secured by inducing energetic collisions, usually by means of an electrical discharge. Since many higher states are excited, emission occurs at a number of definite wavelengths. The chief value of such a source is that any of these wavelengths may be isolated by use of a filter or monochromator to furnish essentially monochromatic light.

A truly monochromatic line is not even ideally possible. Some causes of broadening are: natural broadening, which reflects the finite lifetime of species in excited states; Doppler broadening, which results from the motion of excited species relative to a detector; pressure broadening, which occurs because of collisions; and field broadening, which is attributable to the action of electric or magnetic fields. In low-pressure discrete sources lines have breadth mainly as a result of the Doppler effect. As a consequence lines have an essentially Gaussian shape and widths of the order 0.02 nm. Since pressure broadening can be very substantial, the emission of high-pressure gas sources is essentially continuous, as already noted.

The purity of spectral lines must also be taken into account in evaluating a discrete source. Factors such as nuclear spin can increase the number of lines relating to a ''particular'' electronic transition by removing degeneracy. The resulting pattern is described as *hyperfine structure*. The factors include nuclear spin, isotopic effects (variation in mass, spin, or other properties of a nucleus), the Stark effect (external electric field), and the Zeeman effect (external magnetic field). For example, the hyperfine structure of the mercury 253.7-nm line arising from isotope effects is shown in Fig. 8.4.

Vapor Discharge Lamps. Low-pressure (a few torr) gaseous discharge lamps are widely used as discrete sources. The charge carriers necessary to electrical conduction may be generated within the gas by ionization. Several modes are available. In practice, two or more are often effective simultaneously. Some are:

1. Exposure to cosmic ray showers and natural radioactivity: operative in all instances except where heavy shielding is employed (about 10 ions cm^{-3} s^{-1} are formed in a gas from these sources).
2. Illumination by energetic radiation: operative only if short-wavelength UV is used.
3. Exposure to electron emission from a hot electrode.
4. Heating to high local temperatures.

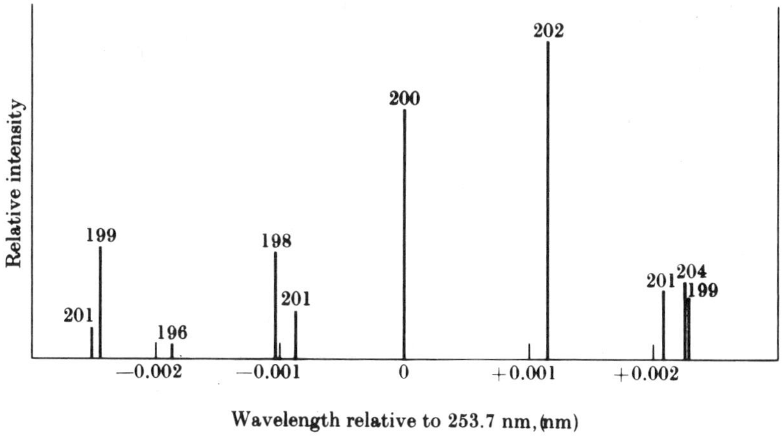

Fig. 8.4 Hyperfine structure of 253.7-nm emission line of a sample of mercury with normal isotopic abundance. The lines of each isotope are identified by mass number. In practice each line has a fine breadth. Note that there are several peaks for odd mass isotopes.

With any of these modes, conduction starts when electrodes are inserted into the gas and a potential difference is applied. As will be noted below, electrodes are unnecessary if a radiofrequency (rf) or microwave field is imposed.

In any method of ionization the necessary condition to produce charge carriers in a gas is that the energy U available be equal to or greater than the energy E_i of ionization of the gas molecules, that is, $U \geq E_i$. Values of E_i range from about 9 to 13 eV; those for the inert gases, nitrogen, and a few others are higher; those for olefins and less stable molecules are lower. The energy criterion is not a sufficient condition, however, since the method of energy transfer must also make ionization probable. Often energy is transferred mainly as translational energy, as is the case when molecules, or an ion and a molecule, of roughly equal mass collide, and little energy is available for ionization. By contrast, collisions between electrons and molecules are comparatively efficient in inducing ionization.

In general, molecular ions are *positively* charged since negative ions of this type are unstable relative to free electrons. Thus, the ion pairs formed on ionization commonly comprise an electron and a positive molecular ion.

Most discharge sources have cold electrodes* and rely on external agencies to ionize a few molecules and initiate conduction. To sustain it they use a voltage across the tube higher than the "breakdown value" of the system so that a self-sustaining discharge develops. It is made possible mainly by the bombardment of the cathode. Positve ions accelerated across the potential difference of the tube have sufficient energy as they strike the cathode to induce emission of secondary electrons. In turn, these electrons through collision with gas molecules maintain the level of ionization of the gas. Conventionally, a discharge of low current density is termed a *glow discharge* and one of large density an *arc*.

Actually, a tube with a self-sustaining discharge operates in an unstable current–voltage region. Unless an external resistance or inductance termed a *ballast* is placed in series, or other control is used, current will increase very rapidly (in milliseconds) until the tube "burns out," usually by cathode failure. The system is described as having a *negative resistance* characteristic, since a current increase causes the voltage to fall. The reason is that increased current leads to still greater ionization and easier conduction. In Fig. 8.5 this type of current–voltage behavior is illustrated. A tube with a ballast will maintain a nearly constant voltage drop across it in spite of current changes, as indicated in the figure by the combined characteristic curve.

Perhaps the two best-known vapor discharge lamps are those based on sodium and mercury. Initial heating is required to bring the lamp to operating temperature. At low internal pressure most of the output of the sodium lamp is concentrated in the D lines at 589.0 and 589.6 nm; a few weak sodium doublets and neon lines are also present. By contrast, the low-pressure mercury lamp is spectrally rich.

*An auxiliary heater may, however, be employed to volatilize an element, as in a sodium lamp. In this case, a low pressure of argon or neon is added to allow the discharge to be initiated easily. When the steady-state pressure of the element of interest has been built up (after 10–30 min) the emission of the starter gas is no more than 1% of the total.

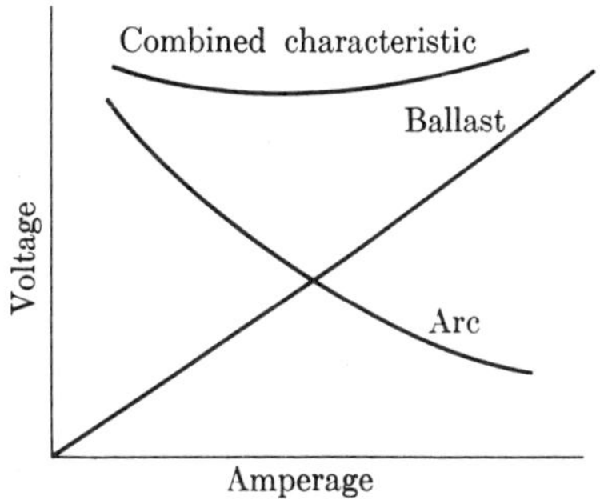

Fig. 8.5 Voltage–current curve for an electric arc. A series combination of arc and ballast has the characteristics given by the upper curve.

Though the mercury emission is concentrated in resonance lines at 253.7 and 184.9 nm, the latter line is blocked by the fused quartz envelope used.* The power supplies for these and other discharge lamps usually include the necessary ballast.

Raising the pressure of gas in a discharge lamp radically alters the emission. All lines are broadened and resonance line(s) in particular suffer self-reversal (see below). Note that each effect increases with pressure. When the pressure is high enough, an essentially continuous spectrum is secured as discussed in Section 8.2. High-pressure sodium, mercury, and xenon discharge lamps are presently available commercially.

The phenomenon of *self-reversal* of a line, that is, loss of its center, deserves comment. Resonance lines, which are mainly affected, involve an electronic transition to the ground state. Atoms in the ground state absorb such radiation in proportion to their population and an increase in gas pressure soon leads to virtually complete self-absorption of the characteristic (central) frequency. The ''wings'' of a line tend to remain.

rf and Microwave Discharge Lamps. An electrodeless gas discharge in which there is also no appreciable rise in pressure during operation (heating is localized) can be obtained by subjecting a gas to a radiofrequency or microwave electrical field. These ''cold plasma'' sources can achieve very high intensities and narrow line widths. A wide variety of commercial lamps is available, those for zinc, copper, thallium, mercury, and cadmium being of especially high intensity. The discharge occurs in a small capillary, perhaps 1 $\times$ 4 mm in diameter.

Hollow-Cathode Lamps. The commercial development of the hollow-cathode lamp made possible construction of discharge sources even for elements that were difficult to volatilize. Figure 8.6 shows representative designs. The key feature of this type of lamp is a cathode in the form of a hollow cup that is either made of, or more often lined with, the element or an alloy of the element to be excited to emission. About 10^3 Pa (7 torr) of a noble gas are added. In operation, voltage is applied accross the electrodes, and the positive noble-gas ions produced in the

*Substitution of a fused silica envelope makes this line available.

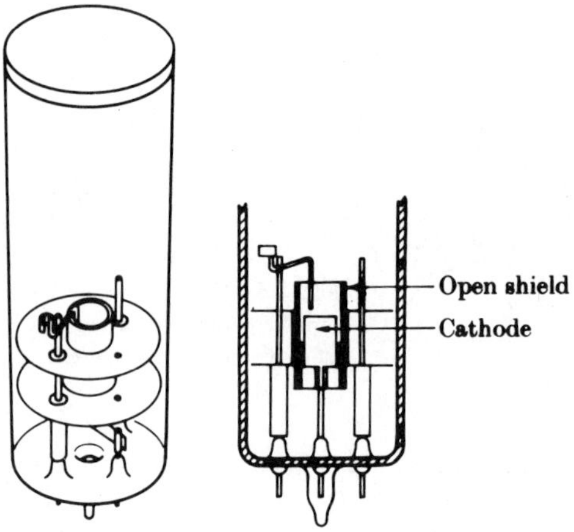

(a)

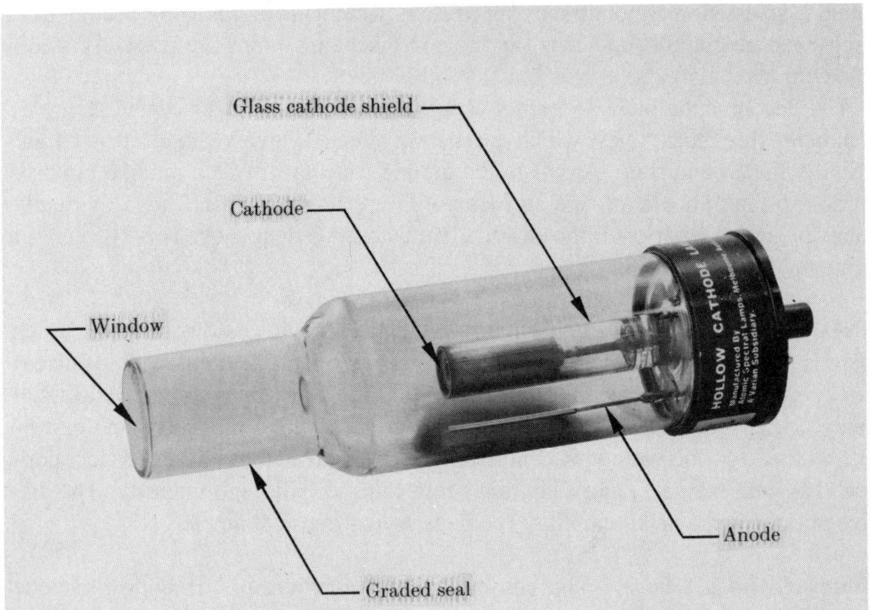

(b)

Fig. 8.6 Hollow-cathode lamps of shielded cathode design. [Courtesy of (a) Perkin–Elmer and (b) Varian Techtron.]

discharge strike the interior of the cathode and sputter out (i.e., volatilize) the desired element. Its atoms are then excited by collision with positive ions or secondary electrons (from the cathode) as in any vapor discharge lamp. After the lamp has warmed for 10–30 min, a stable output is attained. The lines emitted emerge through a window at the far end (see Fig. 8.6).

What operating current should be chosen for a hollow-cathode lamp? A high current excites more atoms per second and gives a steadier, more intense output. Unfortunately, a large current also leads to undesirable effects. First, it causes a large population of gaseous atoms in the ground state. Thus, there is appreciable self-absorption of the desired resonance line. Second, lamp lifetime is shortened. Third, there is more emission from excited *ions*. The optimum current is thus a value that ensures adequate intensity but minimizes undesirable effects. In general, recommended currents fall between 10 and 100 mA and potential drops between 100 and 200 V. Ordinarily a maximum safe value is also quoted; exceeding it subjects the cathode surface to destructive bombardment by cations.

Much thought has been given to ways to raise the fraction of atoms excited without increasing the emission of ions. The use of insulating sleeves (as in the design of Fig. 8.6) and of other shielding causes the ion current in the tube to be directed more efficiently *into* the cathode. Further improvements in design are to be expected. Commonly a bank of single-element lamps is mounted in a turret that provides easy interchange.

8.5 LASERS

In most types of sources producing discrete wavelengths, such as hollow cathode lamps, excited species in the active volume emit spontaneously and independently. As a result, even for a particular wavelength, the output is *incoherent*: emitted waves will be out of phase since they will be started by emitters at different times and places. By contrast, in lasers, each spontaneously emitted photon starts a ray that can *stimulate* like excited species in its path to emit identical photons in phase and in the same direction. Laser emission is thus both *temporally and spatially coherent*.

To gain an appreciation of the processes involved in establishing laser emission, we need to build on the discussion of spectral transitions in Section 8.3. Consider first the competition between the two ways in which emission takes place. Spontaneous emission leads to a decrease in the excited-state population N_2 according to the equation

$$N_2 = N_2 - 1 + h\nu,$$

where $N_2 - 1$ is the original population less one, and stimulated emission does so by the expression

$$N_2 + h\nu = N_2 - 1 + 2h\nu.$$

Clearly, stimulated emission produces amplification of the original photon.

Kinetic Aspects. Interaction of electromagnetic fields with a two-level molecular electronic system may be summed up by writing an equation based on Eqs. (8.5)–(8.7) for the rate of change of the excited state population with time:

$$dN_2/dt = -A_{21}N_2 - B_{21}\rho(\nu)N_2 - D_{21}N_2 + B_{12}\rho(\nu)N_1. \qquad (8.8)$$

The successive terms on the right represent the processes of spontaneous emission, stimulated emission, dark reaction losses, and absorption. The new term, $D_{21}N_2$, stands for all processes, such as collisions, by which excited species lose energy nonradiatively. These *dark processes* collectively cause the translational temperature of the system to rise. According to Eq. (8.8) a major obstacle to lasing is that photons of the lasing wavelength are as likely to be *absorbed* as to trigger stimulated emission, given that both ground and excited forms of the lasing species are ordinarily present. The importance of this competition can be seen by assuming for the moment that terms one and three, spontaneous emission and dark reactions, can be neglected, allowing Eq. (8.8) to be rewritten as

$$dN_2/dt = -B_{21}\rho(\nu)N_2 + B_{12}\rho(\nu)N_1$$

or

$$dN_2/dt = B_{21}\rho(\nu)[N_1 - N_2], \qquad (8.9)$$

since $B_{21} = B_{12}$ (see Section 8.3). According to Eq. (8.9), whenever $N_1 > N_2$ there will be only a net absorption of radiation and an increase in the excited-state population N_2. One can postulate that for stimulated emission to occur, a *population inversion* in which $N_2 > N_1$ must first be achieved. If that can be done, lasing, a net amplification, should be possible.*

3-Level and 4-Level Systems. Unfortunately, with a 2-level system there is little possibility of bringing about a population inversion. The reasons are many as will be elaborated below. With 3- or 4-level systems, however, practical lasers can often be developed. In Fig. 8.7 a simplified version of a 3-level system that will produce a population inversion is shown. Between levels 1 and 3 there is almost an equilibrium since both absorption (by forced excitation or *pumping*) and spontaneous emission are fast processes. The existence of an intermediate excited level E_2 to which access is gained from E_3 by a very fast nonradiative transition allows build-up of species in E_2. In this type of system a population inversion is possible if spontaneous emission from E_2 is relatively slow.

In Fig. 8.8 a ruby laser is used to illustrate a 3-level system that lases. It can be pumped to any of several excited levels E_3. Lasing is from level E_2, which has a lifetime of nearly 3 ms. A 4-level lasing system is also shown in simplified form in Fig. 8.9. Here inversion occurs between states E_3 and E_2.

*When $N_1 = N_2$, the system will neither produce nor lose characteristic photons and is said to be *bleached*.

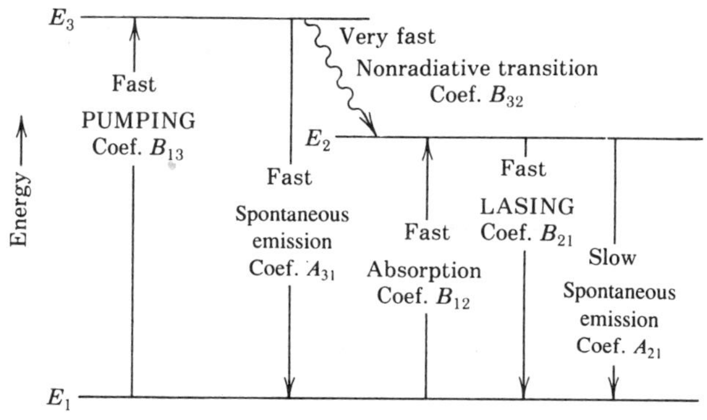

Fig. 8.7 A simplified three-level laser diagram. The system is pumped to excited state E_3 from ground state E_1 either by a flash lamp or an electric discharge. Relaxation from E_3 to E_2 occurs by a very rapid nonradiative reaction. Often the process is internal conversion between electronic states or vibrational relaxation within the excited electronic state. If $D_{32} > A_{31}$, most species will enter level E_2. Further, if $A_{21}N_2 < D_{32}N_3$, the population of E_2 may increase sufficiently to form a population inversion.

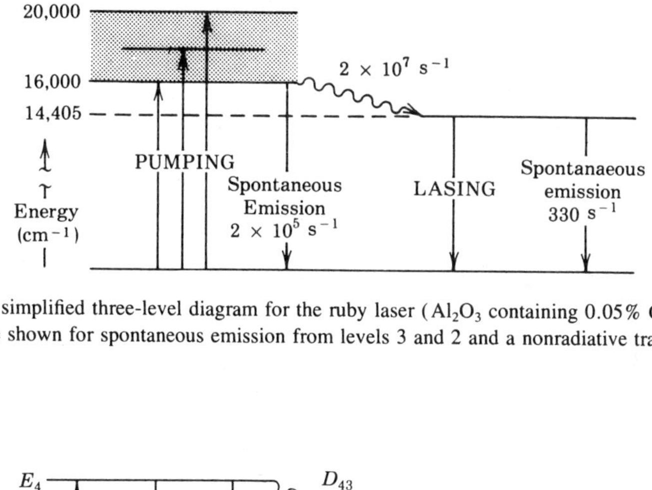

Fig. 8.8 A simplified three-level diagram for the ruby laser (Al_2O_3 containing 0.05% Cr_2O_3). Rate constants are shown for spontaneous emission from levels 3 and 2 and a nonradiative transition from 3 to 2.

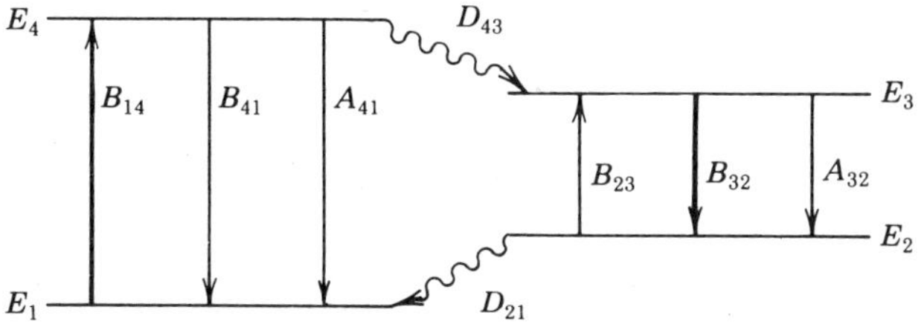

Fig. 8.9 A simplified four-level laser diagram. Pumping and lasing transitions are shown by heavy lines. Probability coefficient symbols are given for transitions. For kinetics to favor the buildup of population in level E_3, the lifetime E_3 should be relatively long and at least one fast mechanism should exist for relaxation from E_2. A nonradiative mechanism is suggested here.

Example 8.3. The neodymium YAG laser operates as a 4-level system. Its acronymn identifies the crystalline material, yttrium aluminum garnet, $Y_3Al_5O_{12}$, which has been doped with 0.5–2% Nd_2O_3. The lifetime of the state from which lasing occurs is about 500 μs. Pumping involves absorption at about 17,000 cm^{-1} and lasing occurs at 9434 cm^{-1} (1060 nm).

Example 8.4. The helium–neon laser is another well-known 4-level laser. Pumping is by an electrical discharge in a narrow tube that contains neon at about 0.1 torr and helium at 0.5 torr. For an average 30-cm laser tube, a pumping potential of about 2.2 kV is needed. Pumping accelerates electrons freed from helium atoms to high energies to excite other helium atoms to the $2s^1$ state.

A net transfer of energy from excited helium ions ($2s^1$ electronic level) to an excited level of Ne ($5s^1$ level) is facilitated by approximate energy equality of these levels. Subsequent relaxations and transfers in the Ne manifold lead to reasonable population of a number of longer-lived neon levels. The main lasing transitions and wavelengths are: 5s → 4p (3390 nm); 5s → 3p (632.8 nm); 4p → 4s (2040 nm); and 4s → 3p (1153 nm). Of this set the red (632.8 nm) line is most commonly used. Its emission can be favored by finding a way to remove species rapidly from its 3p lower terminus, a level which is also metastable. One method is to enhance its nonradiative loss of energy by use of a small-diameter laser tube that will increase collisions. With tubes longer than 1 m, the amount of lasing at 1153 nm becomes large and needs to be suppressed. Sometimes this is done by installing a small permanent magnet to induce Zeeman splitting of the line, which will minimize its amplification.

Optical Resonator. A standard way to increase the output power of a laser is to arrange for a wavetrain to travel through the lasing volume many times. Wavetrains can be folded back upon themselves by placing mirrors at either end and, as long as they move parallel to the axis, they will continue to increase in energy as they stimulate further emission. An example of an *optical resonator* is shown in Fig. 8.10. The resonant cavity extends from the total reflector on the left to the partial reflector on the right that allows some radiation to leave. One round trip of a wavetrain is shown. In this cavity the wavelengths that are enhanced are simply (a) resonant modes (b) that fall within the laser emission bandwidth, as will be discussed below.

The search for a good design for a resonator calls for examination of all its sources of gain and loss. In Fig. 8.10 an intrinsic laser amplification factor α is shown. In addition, there will be systematic losses δ since radiation of the lasing wavelength may be (a) absorbed, (b) imperfectly reflected at the mirrors, (c) scattered, (d) reflected by a mirror at an angle to the cavity axis and be lost in one or two reflections, or (e) lost at the edges as a result of diffraction. The last phenomenon is mainly responsible for the divergence of a laser beam as it travels (see Section 7.13). A solution to this problem, especially for narrow resonators, is substitution of focusing mirrors to define the cavity. Ideally, the focusing should just compensate for the diffraction. A *confocal* arrangement in which the mirrors

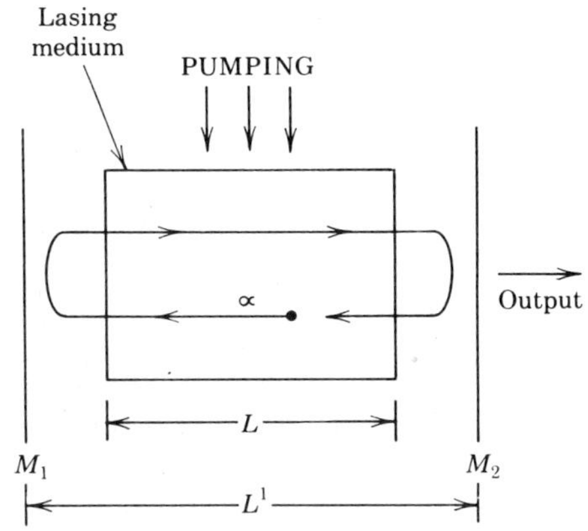

Fig. 8.10 For a lasing medium L meters long an optical resonator of length L' is defined by totally reflecting mirror M_1 and nearly totally reflecting mirror M_2. The mirror M_2 through which the laser beam emerges is also called a coupling mirror. Spontaneous emission by an atom at the dot starts the ray that is traced on a hypothetical round trip in the resonator. As the ray stimulates the emission of excited species along its path, gain α results. At the same time it suffers losses of magnitude δ. An axial standing wave is quickly created with distance between nodes equal to $cn_0/2L'$.

have a common radius of curvature that equals their separation, is particularly advantageous. This design constrains the beam to a small diameter "waist" in the resonator center. If the mirrors are moved farther apart, a limit will be reached when their common focus is at the center of the cavity. In general, their separation should permit the beam waist just to fill the volume of the laser medium.

Each wavetrain is known theoretically to have a finite bandwidth, as shown in Fig. 7.6. In the resonator there will probably be several wavelengths that correspond to standing waves; they are its longitudinal or axial modes.

To define the axial modes actually emitted by a laser requires an analysis of (a) the resonator modes and (b) the laser emission bandwidth, which is often called the laser gain curve. The axial modes are fixed by the resonator length L' and the speed of light. They are easiest to describe by their frequency separation:

$$\Delta\nu = n_0 c/2L', \tag{8.10}$$

where n_0 is the laser refractive index. Note that $2L'/n_0 c$ is the time required for light to make one round trip in the resonator.*

Second, the actual bandwidth of a lasing transition $\Delta\nu_R$ can be estimated by applying the Heisenberg uncertainty principle. The principle states that uncertainties in the complementary quantities time and energy are related by the expression $\Delta E \Delta t = h/2\pi$. For Δt, the average photon lifetime Δt_p, which is determined by

*A small correction in Eq. (8.10) will be required for the different refractive index in the nonlasing part of a resonator. The mode spacing will also be affected by the nonlinear nature of the amplification process and the fact that the refractive index of materials is strongly frequency-dependent near an optical transition.

the average number of rounds a photon makes in the resonator, can be substituted. Then by inserting $h\Delta\nu_R$ for ΔE in the equation, cancelling out h, and solving for $\Delta\nu_R$, we have $\Delta\nu_R \approx 1/\Delta t_p$. It appears that the bandwidth can be further narrowed by increasing the reflectivity of the partially transmitting output mirror since the lifetime of photons in the resonator will be increased.

In addition to falling into an axial mode, each wavetrain will also be characterized by a particular *transverse* electrical and magnetic (TEM) vibrational mode designated TEM_{mn}, where the m and n identify harmonics 0, 1, 2, ... In Fig. 8.11c,d two of these modes are shown in terms of the magnitude of the field strength of the electric vector across the diameter of the laser cavity. It is clear that the TEM_{00} mode represents a resonance with maximum amplitude along the axis of the laser cavity. It also has minimum divergence and for both reasons is ordinarily sought in laser design.* By contrast, the mode TEM_{10} represents a resonance whose amplitude falls to zero along the axis.

In Fig. 8.11a several resonator-stabilized axial modes are plotted as a function of frequency for a hypothetical laser. To establish the frequencies which actually are emitted by this laser, assume for it the emission bandwidth shown as a heavy curve in Fig. 8.11b. Also shown is the small set of axial mode frequencies that

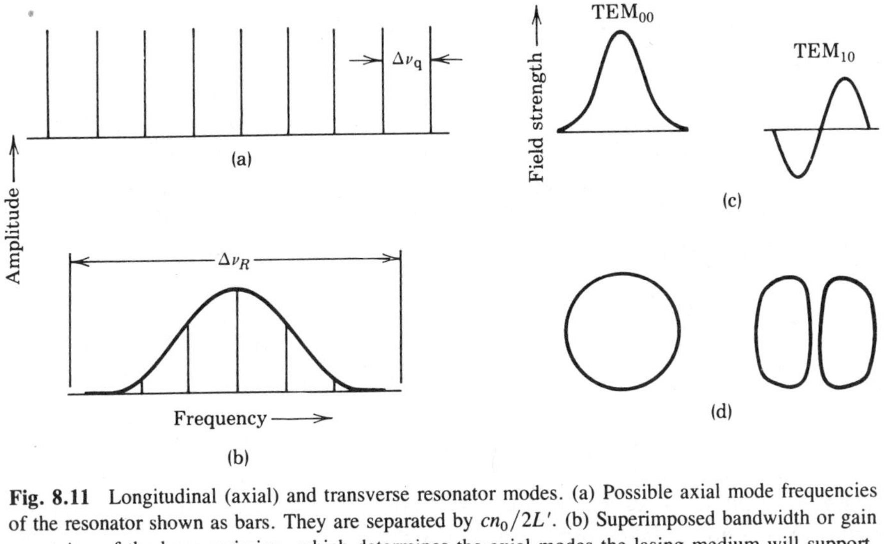

Fig. 8.11 Longitudinal (axial) and transverse resonator modes. (a) Possible axial mode frequencies of the resonator shown as bars. They are separated by $cn_0/2L'$. (b) Superimposed bandwidth or gain curve $\Delta\nu_R$ of the laser emission, which determines the axial modes the lasing medium will support. Laser output spectrum showing the relative amplitudes of the frequencies that fall within the gain envelope. (c) Transverse resonator modes TEM_{00} and TEM_{10}. The radiation field strength is plotted across the diameter of the resonator. (d) End view of the transverse modes. Clearly, the TEM_{00} mode gives the best beam shape for a laser source.

*Where it is difficult to maintain a laser output in the desired TEM_{00} mode, a variable iris diaphragm is often inserted in the resonator to force the laser into this Gaussian mode. The radius of curvature of the partially reflecting mirror can also be increased as a way of causing greater losses off-axis in the resonator and thus, reinforcement of the Gaussian mode.

fall in the allowed frequency range. Their transverse modes, which do not affect energy, are set by the laser design.

To secure a highly monochromatic output it becomes necessary to limit the laser to a single longitudinal mode. A simple way to do so is shown in the following example, and other methods are discussed under mode-locking at the end of the section.

Example 8.5. In principle, laser output can be limited to a single axial mode by simply adjusting the length of the cavity. Apply this idea to a particular He–Ne laser for which the characteristic 632.8-nm line has a bandwidth of 20 GHz (0.03 nm).

By making the cavity sufficiently short, the distance between modes will be greater than the gain bandwidth of 20 GHz and only a single axial mode will be allowed. In other words if mode q_0 is in the center of the lasing band, then modes on either side will be outside the band. To find an upper limit for L', solve Eq. (8.10). Assume n_0 is approximately 1.5. The maximum value of $L' = n_0 c/2\Delta\nu = 1.5 \times 3 \times 10^8 \text{ m s}^{-1}/2 \times 20 \times 10^9 \text{ s}^{-1} = 4.5/400 = 0.011$ m. Practically, $L' < 1.1$ cm. By this method we achieve a single mode and good monochromaticity, but lose output power since reducing the length also severely lowers the gain.

A noble gas ion laser is sketched in Fig. 8.12. Note that the optical resonator is formed by the combination of the planar Littrow mirror M_1 and the focusing mirror M_2, which serves as the "output coupler." The figure emphasizes that the laser beam tends to be linearly polarized in the plane of the paper. If the lasing area itself is capped with Brewster-angle windows, this polarization will be enhanced. The windows will ensure that it will be transmitted with minimum loss and any polarization at right angles will tend to be reflected out of the cavity. For many types of spectrometric measurement the availability of linearly polarized

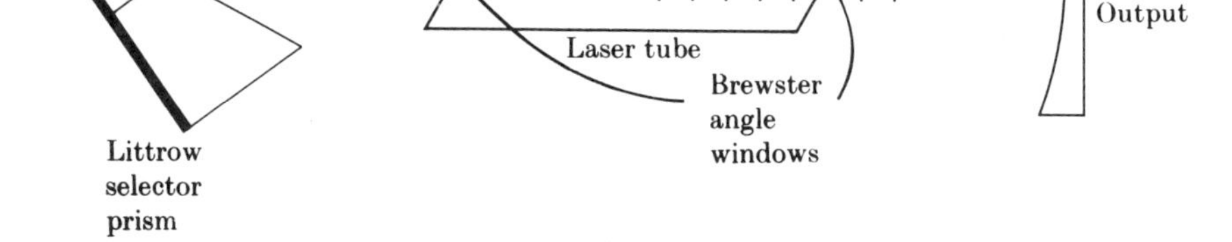

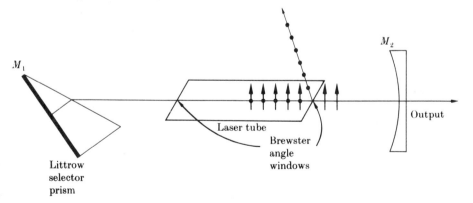

Fig. 8.12 Noble gas ion laser. Mirrors M_1 and M_2 define the resonator cavity. M_2 is not quite totally reflecting. The Littrow prism selects the lasing wavelength: it is the one returned parallel to the axis of the laser. Pumping is achieved by an electrical discharge across the cavity (electrodes not shown). The Brewster angle windows of the laser tube selectively pass radiation linearly polarized in the plane of this page, thus minimizing its reflection losses.

radiation is a considerable advantage. Several kinds of laser and some other types of discrete sources are listed in Table 8.2.

Tunable Lasers. Three types of tunable lasers have been developed to date. Dye lasers have proved eminently useful in the visible region and, with frequency doubling devices, in the UV as well. This type will be described below. For information about semiconductor diode lasers and F-center lasers, which provide limited but precise coverage of the IR, the reader is referred to general references.

Dye lasers are 4-level systems. Short wavelength radiation is used to pump molecules of a fluorescent dye such as rhodamine 6G in an organic solvent to an excited singlet electronic state (state 4). There follows a nonradiative vibrational relaxation to the ground vibrational state of the excited singlet (state 3). It is from this state that stimulated emission *molecular fluorescence* occurs. Emission occurs

Table 8.2 Intense Wavelengths of a Few Discrete Optical Sources[a]

Hydrogen Discharge Lamp (Quartz)	*Mercury Discharge Lamp (Quartz)*	*Sodium Discharge Lamp (Glass)*	*Lasers*	
			Wavelength (nm)	*Lasing Material*
			193	ArF (excimer)[b]
			249	KrF (excimer)
	253.6		308	XeCl (excimer)
	313.0		325.0	He–Cd
			337	N_2
365.7	366.3			
	404.7		400–800	Dye series
(G′) 434.0[a]	435.8		441.6	He–Cd
(F) 486.1			488.0	Ar^+ ion (ArII)
			514.5	Ar^+ ion (ArII)
	546.1		568.2	Ar^+ ion (ArII)
	577.0	(D) 589.0		
		(D) 589.6		
(C) 656.3	690.8		632.8	He–Ne
			694.3	Ruby: 0.05% Cr_2O_3 in Al_2O_3
	772.9			
		818.3	800–904	Diodes: GaAs, GaAlAs
		819.5		
	1014.0		1060	Nd/YAG: 0.5–2% Nd_2O_3 in
	1128.7	1138.2		yttrium aluminum garnet,
		1140.4		$Y_3Al_5O_{12}$
	3942.5		9200–10800	CO_2

[a]Letters in parentheses designate Fraunhofer lines identified orginally in spectrum of sun.
[b]The term here designates a heteronuclear excited complex though the name exciplex has been suggested for such a species.

during the fall to any of several excited vibrational levels in the ground electronic state (state 2). The final nonradiative relaxation to the true ground state (state 1) is a rapid process. Since solutions of dye molecules in organic solvents are involved, wide regions of emission characteristic of condensed-state fluorescence are obtained. How discrete frequencies are obtained will be addressed below.

Since the lifetime of the state from which molecular fluorescence occurs is of the order of 10^{-8} s, quite energetic pumping is required. About 35 kW cm^{-3} must be devoted to maintaining a population inversion. Suitable devices for pumping are the Ar^+ ion laser (488 and 514 nm), the N_2 laser (337 nm), the xenon flashlamp (giving a continuum from 340–840 nm), the Nd/YAG laser (second and third harmonics), and the excimer KrF* laser (249 nm).

In Fig. 8.13 a representative "jet-stream" dye laser is shown. The dye is actually a free-flowing laminar sheet. It is continuously circulated by a pump to prevent exposure leading to chemical degradation or "burning" of the dye. Pumping by the argon laser is approximately at right angles to the dye cell. The high pumping power density needed, among other factors, dictates use of a quite short dye pathlength.

Dye lasers can cover the wavelength range from about 325 to 1285 nm. Since most individual dyes provide coverage of a region of about 50–80 nm, a broadly tunable dye laser must be equipped for quick interchange of many dye solutions. Further extension of the overall range is hampered: at wavelengths shorter than 325 nm, pumping tends to lead to photodissociation of most dyes and at wavelengths greater than 1285 nm, energy transferred to a dye by pumping is mainly lost by fast nonradiative internal conversion processes. Such processes are much more probable in the IR than in the visible.

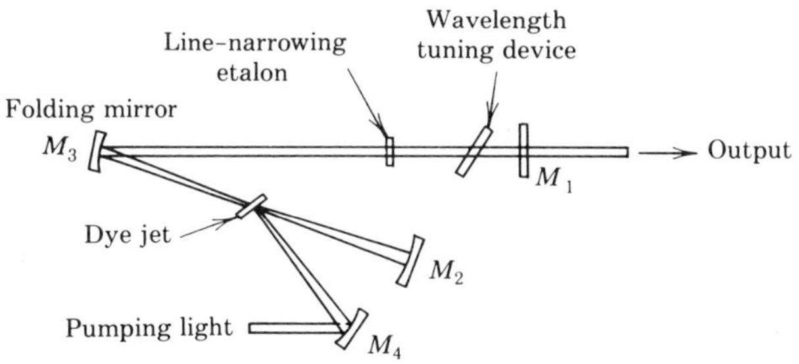

Fig. 8.13 Simplified diagram of a representative continuous-wave dye laser. The optical resonator extends from mirror M_2 (totally reflecting) to mirror M_1. The latter is sometimes known as the output coupler since it has a transmittance of about 0.03. This resonant cavity is said to be *folded*. The curvature of mirrors M_2 and M_3 is chosen to give a beam waist at the dye jet just equal to the thickness of the dye stream. The dye jet is at the Brewster angle with respect to the resonator axis. Further, the laser cavity is essentially stigmatic. The steady flow of dye reduces its decomposition under intense pumping. The pumping radiation from an Ar ion or N_2 laser (not shown) is focused on the dye jet by mirror M_4. An etalon whose thickness is just over half $\frac{1}{2}\Delta\nu_R$, the half-width of the frequency-gain band, is often inserted as a line-narrowing or mode-locking device. To tune the output frequency, a packet of perhaps three birefringent plates is rotated (or a tapered Fabry–Perot cell is partially inserted) in the cavity beam.

A dye laser can be tuned to a particular wavelength within the fluorescence band by pivoting a Littrow prism within the resonator (see Fig. 8.12). Other common laser tuning devices are: a stack of parallel birefringent plates, a diffraction grating (with achromatic beam expander), or a tapered Fabry–Perot interference cell. The first two devices must be rotated to effect tuning and the tapered cell must be moved in or out of a beam. Each device also produces some line narrowing. If a still narrower bandwidth is desired, a regular rectangular etalon can be inserted in the cavity also.*

Pulsed Lasers. Up to this point in the discussion, lasers that emit steadily have been described. They are said to have a *continuous wave* (cw) output. We turn now to a second valuable type, the *pulsed laser*. Since for many lasers of this type pumping is by exposure to brief bursts of high-intensity power, their output is also pulsed. An advantage of this mode of pumping is that it provides a straightforward way to achieve population inversions in less favorable cases. More importantly, pulsed lasers give output pulses of high power whose duration can be controlled. These lasers are uniquely valuable for

1. the study of nonlinear effects, which require high power, and
2. time-resolved spectroscopy.

They are also often used in lieu of cw lasers. A large percentage of lasers available commercially are of the pulsed type.

Pulsed laser systems vary widely in type; Table 8.3 describes many. Some emit in the UV. Some are pumped by an electrical discharge in which a bank of capacitors is discharged through a gaseous tube, others by a flashlamp or by another pulsed laser such as a pulsed nitrogen laser. Note that the pulses obtained are of modest energy, but since their lengths tend to be in the nanosecond range, pulse *powers* up to 200 MW can be secured. In many types of measurement and research still more powerful pulses and still shorter pulses are sought. How they may be achieved will be discussed below.

Q-Switching. For production of pulses of still higher energy a method called *Q-switching* that leads to greater population inversions is quite attractive. Recall that the quality factor Q for a resonator is the ratio of the energy stored in it to that being dissipated (or coupled to a following module). Q-switching calls for changing Q by a timed variation of loss in the resonator. Losses are first kept high (low Q) for an interval sufficient to achieve an unusually high population inversion. Then they are suddenly reduced nearly to zero (high Q). Just after losses are minimized, a single, giant pulse forms and is emitted. Q-switching makes gigawatt pulses possible. Care must be taken since such pulses stress laser optical materials severely.

*Etalons whose width can be precisely controlled by a piezoelectric device are useful for this purpose. With them emission bandwidths can be narrowed to a few megahertz (about 0.002 nm).

Table 8.3 Characteristics of Some Pulsed Laser Systems

Type	*Mode of Pumping*	*Wavelength (nm)*	*Pulse Energy*	*Duration*	*Repetition Rate*
CO_2	Electrical discharge	9.2×10^3–10.8×10^3	0.1–10 J	10^2–10^4 ns	
N_2	Electrical discharge	337	0.01 J	10 ns	
Ar ion (mode-locked cw)	Electrical discharge	514, 488	20 nJ	100 ps	80 MHz
ArF[a] KrF	Electrical discharge	193 249	0.2 J	20 ns	80 MHz
Nd:YAG (mode-locked cw)	Flashlamp	1064 532 (doubled)	130 nJ 13 nJ	100 ps 70 ps	80 MHz 80 MHz
Nd:YAG (mode-locked, *Q*-switched, cw)	Flashlamp	1064 532 (doubled) 355 (tripled)	80 μJ 50 μJ 10 μJ	100 ps 100 ps 100 ps	0–800 Hz 0–800 Hz 0–800 Hz
Dye	Flashlamp	400–1000	0.01–1 J	500 ns	
Dye	Pulsed N_2 laser	400–1000	10^{-3}–10^{-4} J		
Dye	Pulsed excimer laser	400–1000	0.01 J		

[a]An excimer or exciplex laser. For example, the KrF* exciplex can be formed by the reaction Kr* + F_2 = KrF* + F in a gas discharge-pumped laser.

To alter resonator losses we introduce some sort of optical switch. Many types have been used: saturable or bleachable dye cell, rotating mirror, and electrooptical modulator or shutter. Many of the lasers available commercially use an electrooptic Kerr cell and a design like that in Fig. 8.14. In this case a Glan–Thompson prism has also been introduced into the cavity as polarizer. The Kerr cell contains a polar substance such as nitrobenzene that has a high Kerr effect. In the absence of an electric field across this cell (electrodes not shown) the liquid is isotropic and polarized light passing through is unaffected. When a strong electric field E is imposed, the liquid becomes birefringent (see Section 7.15) and rotates the direction of polarization of light emerging from the Glan–Thompson prism. The field E is increased until the Kerr cell produces a quarter-wave retardation of the light that it transmits. Now losses in the optical resonator are nearly complete; any light reflected from M_2 will be directed outside the cavity on reaching the polarizer as shown. When a giant population inversion has been built up, the capacitors sustaining the cell electric field can suddenly be short-circuited to remove the applied voltage and allow the Kerr cell liquid to relax to its isotropic state. Since laser light will now remain in the cavity, stimulated emission will almost instantly build up a giant laser pulse that emerges through M_2.

If a shorter pulse than is permitted by *Q*-switching is desired, an alternative arrangement known as *cavity-dumping* is possible. This is described in the legend of Fig. 8.14.

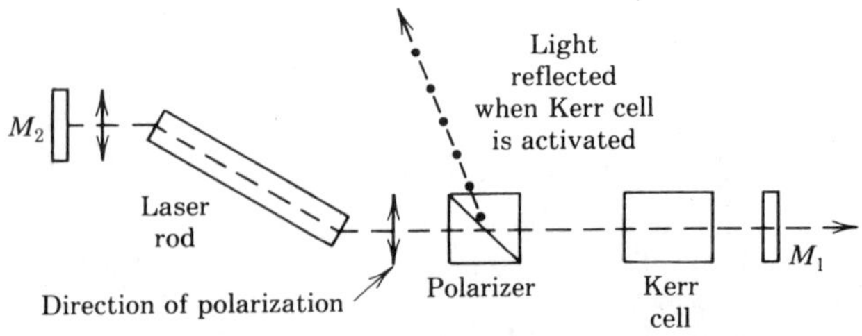

Fig. 8.14 Laser design for giant pulses. (a) For *Q-switching* a coupling mirror M_1 with $T = 0.05$–0.10 is selected. Either a Glan–Thompson or other type polarizer or the laser itself linearly polarizes radiation so that it vibrates in the plane of this page. The Kerr cell or other electrooptic modulator serves as a fast Q switch. A strong dc field applied across the cell rotates the direction of polarization 90° and causes light to be reflected out of the laser cavity. (b) If *cavity-dumping* is desired instead, a totally reflective mirror is selected for M_1. The application of a strong electrostatic field across the Kerr cell spoils the Q of the cavity as before and permits a large population inversion to form. Again the cell voltage is suddenly dropped leading to a buildup of an intense monochromatic standing wave. Now the high voltage is applied again, rotating the direction of polarization 90°, and the laser pulse is suddenly "dumped" from the cavity by reflection from the Glan–Thompson polarizer interface. If the second application of the field is sufficiently fast, a pulse as short as 3 ns can be achieved.

Mode-Locking. Most lasers have outputs that fluctuate continuously. The reason is to be found in the manner of origin of lasing modes. Each spontaneously emitted photon that differs in phase from the others will ordinarily give rise to a wavetrain of stimulated emission that is any one of the allowed frequencies or longitudinal modes. Since losses experienced by these wavetrains are also random, the amplitudes of modes will vary as randomly as their phases. The graph (shown in Fig. 8.15a, b) as compared with the idealized distribution of modes by frequency in Fig. 8.11b, illustrates this point.

Where the goal is production of ultrashort pulses one method is to "lock" the phase of all allowed frequencies (those separated by $\Delta\nu = n_0 c/2L'$ and under the laser gain curve). As these frequencies superpose they are in phase once within each period T, where $T = 2L'/n_0 c$, giving a single large-amplitude, short pulse. Its duration is $\delta t \simeq 2L'/Nn_0 c$, where N is the number of allowed axial modes. From this expression it is clear that for short pulses the laser cavity must be long and the laser transition broad so that many oscillating modes will superpose.

A simple, but impractical way to design a laser for mode locking was discussed in Example 8.5. A more common way is to add an etalon, which is a narrow Fabry–Perot resonator. Another is to insert an acoustooptic modulator that generates a high-frequency standing acoustic wave in a Bragg cell, leading to deflection of pulses that are not in phase with that wave out of the laser. Each type is illustrated in an example below.

Example 8.6. How does an *etalon* limit the axial modes of a laser? Where should it be inserted in the laser cavity for best effect? An etalon is simply a thin slab of glass with highly parallel, polished faces. Two optical properties of an etalon are important to its use in a laser. Like a Fabry–Perot interference filter (see Section 7.10) it will transmit certain frequency bands of light, but since its faces have relatively low reflectivity, the band passed will be quite broad. Second, because the thickness of an etalon is small compared with the length of a laser cavity, its bands will be much more widely separated. It will thus form a single axial mode of a laser by producing higher losses for all others. In practice about 75%

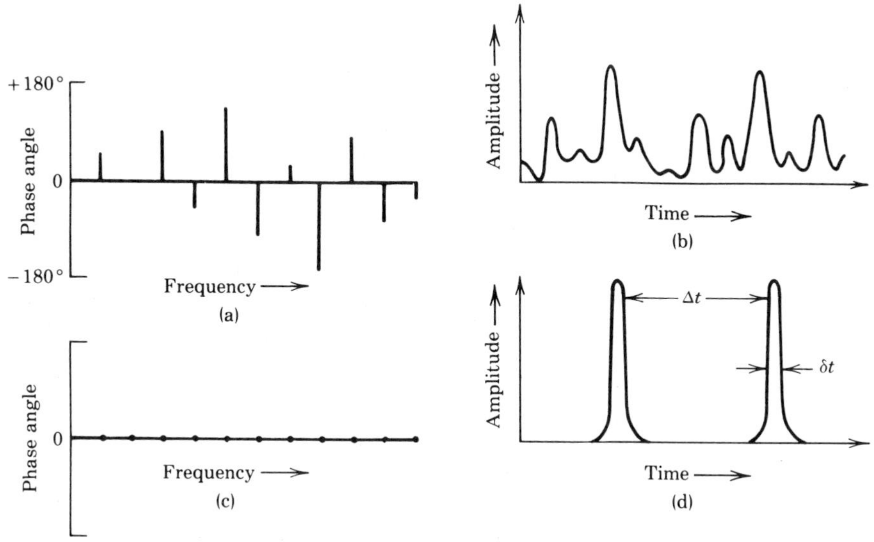

Fig. 8.15 Properties of axial modes in a laser. (a) Relative phase for normal output. In the absence of other factors, phases are usually random. (b) Time variation of amplitude of modes in normal output. (c) Locking modes to a single phase. (d) Time variation of output when mode-locked. Mode-locking gives a narrow signal-axial-mode pulse that is repeated every Δt seconds, where $\Delta t \simeq 1/\Delta\nu_R$.

of the output of a laser can be shifted to the axial mode an etalon favors. The etalon may be placed anywhere along the axis of the cavity. The frequency selected may be tuned over a small range by tipping the etalon a few degrees with respect to the laser axis (see Section 7.10).

Example 8.7. A properly designed acoustic cell or *acoustooptic modulator* can also be inserted to mode-lock a laser. How does this type of modulator produce mode locking? What frequency modulator should be used to mode-lock a particular laser that has axial modes separated by 88 MHz? What will be the temporal spacing of mode-locked pulses in its output?

An acoustooptic modulator consists of a transparent liquid-filled cell of rectangular cross section with a piezoelectric transducer firmly affixed to its lower end. An rf oscillator drives the transducer to set up a standing wave with planar wave fronts in the cell. Because this wave-front array is parallel to the axis of the laser, it affects passing optical frequencies or wavelengths exactly as planes of atoms do X-rays in a Bragg crystal. In this case, the modulator will deflect all but one longitudinal mode from the laser by diffraction and cause the laser to resonate mainly in the favored (nondeflected) axial mode. Most modes pass out through the sides of the laser.

In this simulated grating the grooves can be taken as the regions of higher refractive index at the compression maxima in the acoustic wave. Actually, the standing acoustic wave has two rarefaction and compression regions per cycle. If the laser mode spacing is 88 MHz, a 44-MHz transducer will provide a diffraction loss at twice that frequency and perform the needed mode-locking. Its output will be a train of pulses like those in Fig. 8.15d of duration 0.2 ns separated by 11.4 ns. For further information the reader is referred to references (8, 9).

Harmonic Generation. Some lasers generate intensities sufficiently high (about 1 MW cm^{-2}) to reach the region in which the refractive index of crystalline dielectrics varies with the optical electrical field strength. The equation obeyed is $n = n_0 + \pi\chi E$, where n_0 is the constant part of n, χ is the second-order nonlinear susceptibility of the crystal, and E is the electric field strength of the radiation. The significance of this effect is that monochromatic laser radiation will in many types of crystals generate appreciable intensities of the second and sometimes of higher harmonics as it passes through. This method of *frequency doubling* is commonly used to extend the frequency range of lasers.

The most useful crystals for doubling are phosphates, pentaborates, formates, or niobates with a single akali metal ion and usually a hydrogen (either ^1H or ^2H), such as KH_2PO_4 and $LiNbO_3$. They offer high efficiency because the refractive index for their second harmonic matches that of the fundamental, a condition that ensures conservation of momentum for the light.*

Applications of Pulsed Laser Radiation. Many new optical methods have been fostered by the availability of high-power coherent radiation available from pulsed lasers. The high flux of photons in a giant pulse makes possible 2-photon absorption and a technique known as saturation spectroscopy. In addition, a variety of types of Raman spectroscopy and other scattering processes that interact with vibrational transitions become accessible to study.

With pulses of the order of picoseconds and sometimes even shorter, it is possible to carry out time-resolved spectroscopy since the exciting pulse can be isolated from most subsequent molecular relaxation. For example, this approach permits fluorescence to be studied separately from the excitation that produces it. In particular, the nature of the excited state itself can be studied. Distance-measuring techniques such as argon ion laser "radar," which is known as LIDAR, become possible also.

8.6 OPTICAL DETECTORS

A wide range of detectors is required to sense radiation across the whole of the optical region, that is, from about 30 nm to 300 μm. The vacuum ultraviolet, UV, VIS, near IR (NIR), and middle and far IR are all included. Photon energies vary by a factor of almost 10^5 over this range, making possible many kinds of interactions with matter and thus several types of detector. *Quantum detectors* such as the photomultiplier tube and silicon photodiode respond to the absorption of photons by direct processes whose energy is characteristic of the detector material. On the other hand, *thermal detectors* simply respond to the temperature increase when the energy of photons is converted to heat. For example, a thermocouple or resistance thermometer detector can report the temperature increase in an attached, blackened strip that absorbs the radiation.

*T. E. Evans (Ed.), *Applications of Lasers to Chemical Problems*. New York: Wiley-Interscience, 1982, Chapter 1.

Most detectors are *single-channel*, *nonimaging* devices since they can neither discriminate between different wavelengths nor identify any pattern incident radiation may make on their sensitive surface. It is this type of detector, one that is sensitive over a broad band of wavelengths and whose response is essentially linear with intensity of light almost regardless of whether a rectangular or other shape of illumination falls on it, that will be discussed here. In the next section multichannel imaging detectors such as the linear diode array will be taken up.

Properties of many types of optical detectors are succinctly described in Table 8.4. While spectral range is a well-known property, others such as responsivity, rise time, and detectivity need definition. The spectral *responsivity R* is basically a measure of detector sensitivity. It gives the rms detector output current or voltage per unit of rms incident radiant power.* Since the signal for thermal detectors appears across a resistance, their responsivity is given in $V\ W^{-1}$. The *rise time* is the interval required for a detector's output to increase from 10 to 90% of its final response when a so-called step input signal reaches it and is a good measure of its response time.

The final detector property, the *detectivity D*, is a measure of what may be called a detector's limit of detection. Actually, this limit is basically established by a detector's noise level. It is important to know that the irreducible detector noise is usually stated in terms of the *noise equivalent power* (NEP). The NEP is calculated by measuring the watts of radiation which must be received by a detector to produce an output equal to its rms *noise voltage* under stated conditions (a given optical source, frequency bandwidth, and modulation frequency). The smaller the value of NEP is, the better the detector. As is logical, this term is the reciprocal of detectivity, that is, $D = \mathrm{NEP}^{-1}$.

More commonly, however, a specific detectivity D^* (pronounced dee star) is calculated from D. It is defined by the expression $D^* = DA_d^{1/2}(\Delta f)^{1/2}$, where A_d is the detector area and Δf is the bandwidth of detector electronics. The units of D^* are cm $\mathrm{Hz}^{1/2}\ \mathrm{W}^{-1}$.

Another useful specification for quantum detectors is the *quantum efficiency* η_q. Ideally, each incident photon will excite one charge carrier or photoelectron, but practically, losses of many types cause detector efficiencies to be less than one. The efficiency is calculated from the expression

$$\eta_q = (I/e)/(P_{opt}/h\nu),$$

where I/e is the number of charge carriers generated per second and $P_{opt}/h\nu$ is the number of photons incident per second.

Vacuum Phototube. The typical vacuum phototube is a photoemissive device whose cathode is coated by metals specially selected to yield a minimal work

*Here the watt, a familiar unit of power, is used rather than the lumen. The latter is the measure of luminance in the SI system. The lumen (lm) is the luminous flux per steradian from a source whose luminous intensity is one candela (the new candle).

Table 8.4 Characteristics of Some Optical Detectors[a]

Type of Detector (Output)	*Sensitive Element*	*Wavelength Range (μm) (Response Time)*	*Responsivity*[b] *(mA W^{-1} or V W^{-1})*	*Detectivity (D^*)*[c] (Type of Noise)[d]
		Quantum Detectors		
Photoelectric				
Photomultiplier or multiplier phototube (current)	Group I and V metals	0.16–1.2 (1–100 ns)	2–105	5×10^{14} at 1000 Hz (Shot noise)
Vacuum phototube (current)	Alkali metals	0.2–1 (<1 μs)		Shot noise
Solid state[e] (current)				
Silicon photodiode[f]	*p–n* junction in semiconductor crystal	0.16–1.1 (2–20 ns)	100–600	6×10^{13} (Shot noise)
Lead sulfide[g]	Polycrystalline layer PbS	0.7–4.5 (2–1000 μs)		1.15×10^{10} at 90 Hz (Johnson noise)
Doped crystal[h]	Germanium doped with Cu, Au, Zn	2–15 to 2–100 (0.1–1 μs)		10^{10} at 900 Hz (Johnson noise)

Thermal Detectors				
Thermocouple or thermopile, a multijunction thermocouple (emf)	Junction of dissimilar metals with blackened strip as absorber	0.2–40 (50–100 ms)	4–55	2×10^8 at 5 Hz (Johnson noise)
Bolometer (bridge detector) (resistance change)	Resistance wire or thermistor chip with blackened strip	0.25–40 (10 ms–5 μs^i)	5×10^{-5}–125	3×10^8 at 10 Hz (Johnson and $1/f$ noise)
Golay or pneumatic cell (capacitance change)	Gas chamber with blackened membrane	0.8–1000 (3–30 ms)		2×10^9 at 10 Hz (Johnson noise)
Pyroelectric[j] (voltage)	Ferroelectric crystal with absorbing film	0.1–1000 (2–100 ns)	15–2400	2×10^8 (Johnson noise)

[a]See also Table 8.5.
[b]Quantum detectors ordinarily are rated in mA W^{-1} since they develop currents while thermal detectors are rated in V W^{-1}.
[c]The specific detectivity D^* (dee star) has dimensions of cm $Hz^{1/2}$ W^{-1}. Values are valid at the cited chopping frequency for a 1-Hz bandwidth.
[d]See Section 12.3.
[e]These devices can be operated in either the *photovoltaic* mode (no external bias is used) or the *photoconductive* mode (reverse biasing is employed). For optimum performance, however, the construction will vary for the two modes. Data given are for use in the photoconductive mode, which gives shorter rise times.
[f]Used at 298 K. Other semiconductors such as InSb, InAs, and HgCdTe (MCT) are also employed, especially at 195 or 77 K.
[g]Chemically deposited in polycrystalline, granular films or layers. While it is basically a photoconductive detector, modeling of performance is difficult. Sometimes PbSe is substituted. Values given are for room-temperature operation. Detectivity improves with cooling. For example, greater sensitivity (up to 100-fold) and wider wavelength response (up to 7 μm) results on operation at 77 K.
[h]Generally used at 4–23 K.
[i]Shorter times are for a superconducting bolometer.
[j]Responds only to chopped or pulsed light.

function for photoejection of electrons. For any particular surface only a photon with energy greater than the work function will yield a photoelectron.

For safety of personnel, it is customary to ground the anode or collecting plate and to have the photocathode at a substantial negative voltage (-90 to -1200 V). The large gradient ensures that ejected electrons are quickly collected by the anode, which is a wire or sometimes a grid, to give a current. Vacuum phototubes see most use in measurement of higher light levels.*

Photomultiplier Tubes. In the multiplier phototube, or *photomultiplier tube* as it is commonly called, several stages of amplification are incorporated within the tube envelope. In addition to a photocathode, a train of several secondary emission electrodes or *dynodes* is built in before the anode is reached. Often the photocathode is only a semitransparent deposit on the inside of the tube window. Each dynode is maintained at about 90 V more positive than the last and is placed and shaped with care to provide between it and the next dynode a precise electrostatic field to guide the growing electron stream to the next dynode. One type of design is shown in Fig. 8.16. Note how different the dynode shapes and placements are. In other designs dynodes are arrayed in a venetian blind arrangement or a box-and-grid structure. In general, if δ secondary electrons result from every electron impact at a dynode, the tube current gain G is just $G = \delta^n$, where n is the number

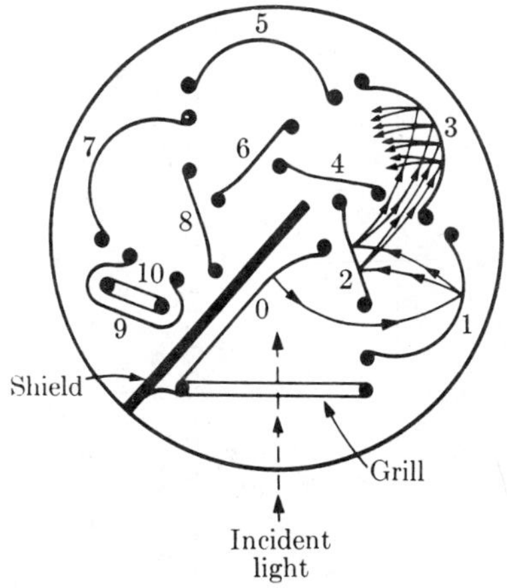

Fig. 8.16 A diagram of the RCA IP28 photomultiplier tube. Identification of electrodes; 0, opaque photocathode; 1–9, dynodes; 10, anode. Photoelectrons ejected from the cathode are accelerated toward the first dynode. A great many secondary electrons are ejected from dynode 1 as it is bombarded, and these are in turn beamed toward a second dynode, and so on. The strong electrostatic fields between dynodes result in effective focusing of the growing electron stream and short response times for overall response to a photoelectron. (Courtesy of Radio Corporation of America.)

*The addition of an inert gas at low pressure to a vacuum phototube permits internal current amplification through collision ionization, perhaps by as much as 10–20 times. Such gas-filled tubes are not used so widely as the vacuum devices.

of dynodes. If $\delta = 3$ and $n = 12$, then $G = 5 \times 10^5$. Actually, the number of dynodes can be as great as 14 and the gain as large as 10^8. A great virtue of secondary emission amplification is that it is almost noise free. In general, the maximum tolerable average anode current is 1 mA and the maximum anode power dissipation is limited to 0.5 W.

As many as two dozen different photocathode surfaces are presently available; most have been standardized by the Electronics Industries Association (EIA) and given "S" numbers, for example, S-1 and S-20. In Fig. 8.17 the responsivity and quantum efficiency of four cathode surfaces is shown. Not surprisingly, photocathode quantum efficiencies are high in the UV and fall off rapidly at longer wavelengths (and also at shorter ones, e.g., see p. 286). For the NIR, more efficient cathode surfaces consisting of Ga–As, Ga–In–As, and other combinations of Group III and V elements have also been developed. Yet quantum efficiencies and responsivities are low in this range and the photoelectric effect has not been observed above about 1.2 μm.

What arrangements must be made for power supplies? Commonly, potential differences between successive dynodes are precisely controlled by a resistive voltage divider circuit energized by a regulated dc power supply as illustrated in Fig.

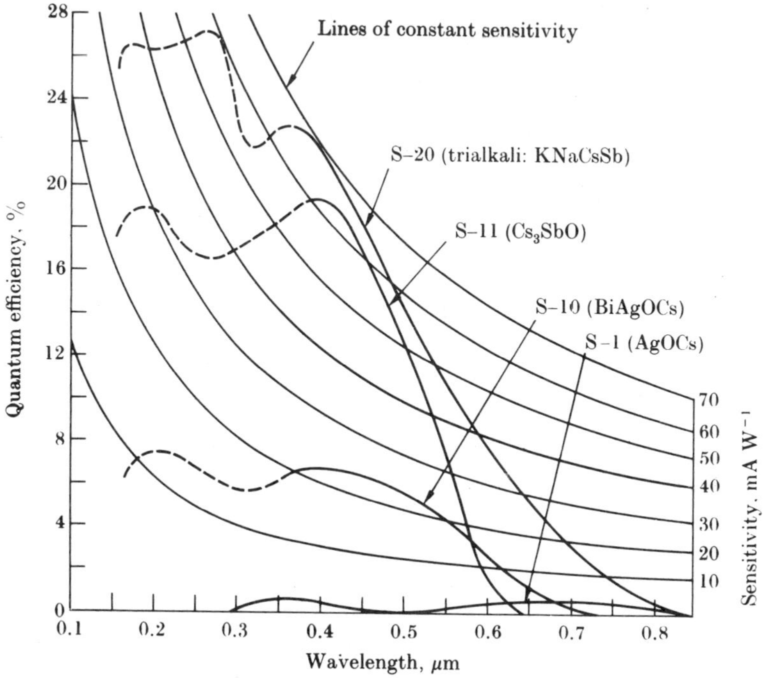

Fig. 8.17 Spectral response curves for several standard photocathode surfaces. Note the sensitivity scale at the right. Dashed portions of curves show response when UV-transmitting window is used in photomultiplier tube. In the IR some useful surfaces are Ga–As, Ga–In–As, and Na–K–Cs–Sb with 9791 glass windows.

4.14. Note that this op-amp circuit provides for measurement of the light intensity as a *dc signal*. Another method especially useful at very low light levels, a photon counting arrangement, will be discussed in the next section. Ordinarily manufacturers have installed useful divider circuits in the sockets into which tubes fit. When they must be modified for special applications, good information about choices to be made is available in handbooks and the literature [12, 13].

It is always important that the bias current in the divider be at least 10 times greater than the maximum average anode current expected to ensure linear operation. It is also necessary to limit current to keep heat dissipation low so that noise will be low.

Special Requirements. While equal resistors will ensure that the power supply voltage is divided equally among the dynodes, high and low light levels and especially fast signals present other requirements. High light intensities lead to large currents and a buildup of heavy space charge (the electron stream) between final dynodes and the last dynode and anode. As a result, the focusing electrostatic field is badly distorted and many electrons are not collected. This problem may be resolved by increasing voltage drops in this part of the tube. As shown in Fig. 8.18, higher potential differences are provided by use of larger resistors between later dynodes. With the greatest voltage between the last dynode and anode (up to perhaps 200 V), electrostatic focusing is preserved even as currents increase.

When a signal rise time shorter than 1 μs is desired, however, with pulsed or fast optical signals as in laser-based techniques or time-resolved spectrometry, still other changes may be necessary. In general, the anode current rise time should be no greater than 10% of the "step time" of the signal for accuracy. Specially designed photomultipler tubes offer rise

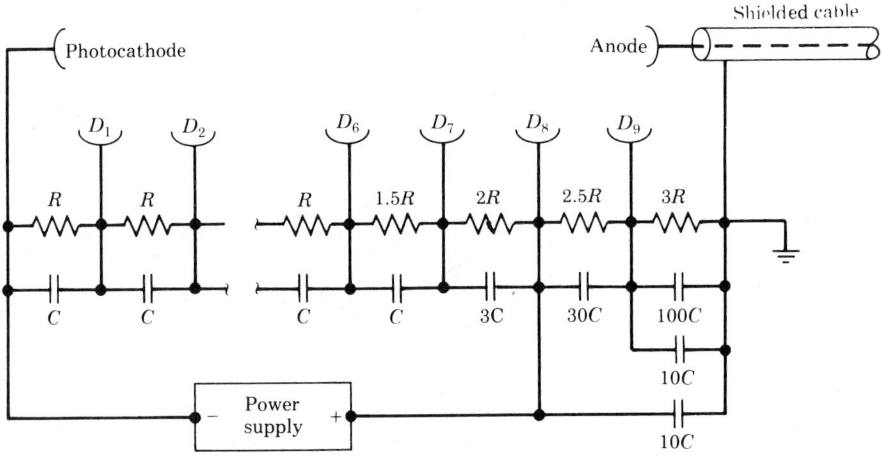

Fig. 8.18 Alternate circuit arrangement for a voltage divider for a photomultiplier tube that must handle larger and/or fast signals. It is wired to ensure that even at currents near the maximum tolerable value, an anode current is generated that is linear with light intensity and has a fast rise time. To cope with the field-distorting effect of higher space charge (electron density) between later dynodes the later resistors are gradually increased in size, thus increasing the potential drop between dynodes. Further, capacitors are inserted between dynodes to preserve fast recharging as current pulses pass.

times down to about 0.15 ns, though with reduced gain. Actually, as tube currents rise, rise times may be inadvertently lengthened if pulses become so large as to slow the recharging of dynodes. To diminish this effect capacitors are inserted between dynodes as shown in Fig. 8.18. Note that they increase manyfold in magnitude toward the end of the electrode chain. These capacitors, charged by the divider current, should hold ten times the charge needed to recharge the dynodes as a current pulse passes.

For still faster detection a *streak camera* based on an image converter permits resolution of events in the picosecond range. In the device electrons ejected from a photocathode pass through electrostatic deflection plates that rapidly sweep the beam across a sensitive detector surface, recording its amplitude as a function of time.

In photomultiplier tubes there is always background current to be reckoned with, even in darkness. A current called the *dark current* exists in the anode lead at all times. Its magnitude increases with signal current as shown in Fig. 8.19, and its value is predictable. It may be traced to factors such as ohmic leakage, thermionic and field emission, emission caused by radioactivity such as that traceable to the

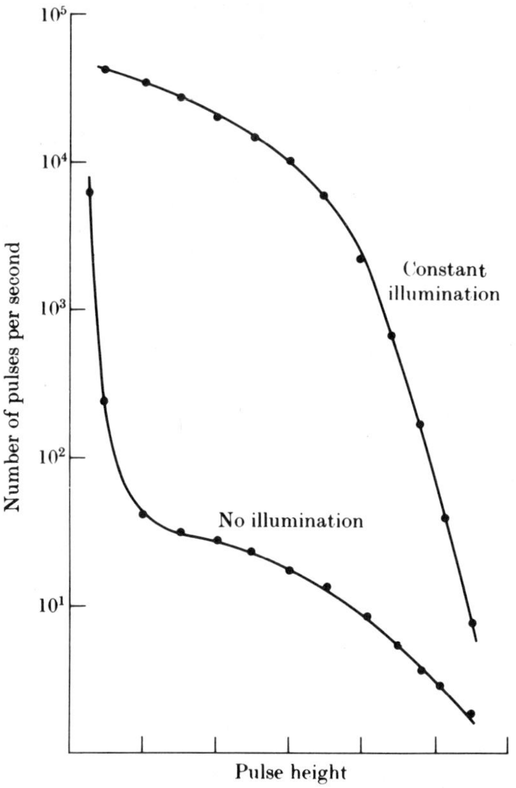

Fig. 8.19 Distribution of current pulses from a photomultiplier tube by amplitude. "Dark pulses" that originate when the tube is not illuminated are seen to be many fewer in number and much smaller in amplitude.

^{40}K in the glass envelope, and other factors. The ohmic leakage component is basically dc in character. By contrast, the other contributions are pulselike in character and consist of electron bursts that originate at any dynode or even the cathode and undergo partial amplification. Most "dark pulses" will be of lower amplitude than a photon pulse, as is evident in Fig. 8.19, and can be discriminated against on the basis of amplitude in a pulse-counting procedure. In standard dc current measurements an average dark current (background) can simply be subtracted. For example, a potentiometer zeroing arrangement for manual subtraction of dark current background was shown in Fig. 4.14.

Fortunately, some components of dark current can be minimized by thermoelectric or other cooling of a photomultiplier tube to 0 to −40°C (−185°C is required for a tube with an S-1 cathode). Other sources of dark current are less subject to control. In any event, dark current sets a lower limit to intensity measurements. Special attention will be given to measurements at low light levels in the following section.

Is noise related to dark current? Noise appears as a statistical fluctuation in the emission of electrons and gives rise to unsteadiness in both signal *and* dark current. If the time constant of the electronic circuit is long, noise will tend to average to zero. An excellent discussion of noise in photomultipliers can be found in ref. 13.

Usually the gain of a photomultiplier is best enhanced by use of improved dynode surfaces, better geometry, or additional dynodes. The lower the light level, the more important is a high first dynode gain. When accomplished, the transit time of photoelectrons is more uniform and stray magnetic fields have less effect. An increase in operating voltage at that stage will enhance the gain. Actually, such an overall increase is limited in practice by the possibility of tube damage by flashover or ionization of residual gas. Today the technique of photon counting provides the needed sensitivity to use photomultiplier tubes at very low light intensities as will be described in the following section.

As a result of large internal amplification, photomultiplier tubes can be used only at low power levels (about 10^{-14} to 10^{-4} lumens on a continuous basis). When power is on, exposure of a tube to even moderate levels of illumination can cause irreversible or at best slowly reversible changes in electrode surfaces. As a result these tubes are always mounted in light-tight housings with a shutter to control the entrance port.*

In the vacuum UV, from about 190 to 30 nm (300 Å) photoelectric efficiencies are generally low and radiation can be successfully detected by adding a *fluorescing plate* or scintillator as a converter in front of the photomultiplier window. A coating of sodium salicylate on a Pyrex of Vycor window serves the purpose. When radiation impinges, the compound fluoresces at about 400 nm, leading to a reasonably linear photomultiplier tube response across the spectral range cited.

Magnetic Electron Multipliers. While these detectors are similar in structure to photomultiplier tubes, by use of a different window and photocathode surface, they

*Accidental exposure to daylight when the power is off will not cause damage but will energize the photocathode and lead to excessive dark current for 24–48 hours.

permit detection from the far UV well into the X-ray region (150–0.2 nm). Photons enter through a thin metal film and on impact with a tungsten cathode release photoelectrons. A continuous dynode of coated glass with a potential of 2 kV across its length provides subsequent secondary emission. This geometry and crossed magnetic and electric fields accelerate electrons and constrain them to travel down the strip in a series of cycloidal jumps, producing an ever-growing avalanche of secondary electrons. The final electron burst is collected by an anode. The dark current of the multiplier is very small (0.1 electron s^{-1}) and gains of the order of 10^6–10^8 can be achieved.

Photodiodes. Silicon or germanium crystals are light-sensitive. When pure crystals are doped with elements from Groups 13 (III A) or 15 (VA) such as Ga and As, however, they show dramatically improved quantum detection well into the IR. Doping is always carried out so as to develop precise *p–n* junctions (see Section 3.7) and ensure fast rise times. Absorption of a photon leads to generation of a majority charge carrier in the conduction band by supplying the small amount of energy needed to create an excess electron (or hole). The new charge carriers are moved toward external electrodes by applying a dc bias (voltage). In the absence of light, dark current caused by minority charge carriers exists, but it may be reduced by cooling.

A variety of elegant, widely used photodiode detectors have been developed, a junction photodiode, a *p–i–n* diode, and an avalanche photodiode. While a photo transistor is also available, it has seen relatively little use.

Operation of Photodiodes. The *junction photodiode* is much like a regular solid-state diode. In the device the most efficient area for charge carrier generation is the depletion layer at the *p–n* junction (see Section 2.9). Photon absorption taking place away from the junction can be made small by making these regions thin. A photon that excites an electron to the conduction band leaves behind a hole; both charge carriers will migrate.

Two approaches to charge carrier collection are possible. If *no external bias is applied*, a photodiode will perform in the *photovoltaic* mode. As light is absorbed a small voltage is developed that causes a current. Two properties that must be taken into account are that the current developed depends logarithmically on light intensity and rise time is relatively long (order of milliseconds) in this mode. Alternatively, a photodiode can be operated in a *photoconductive* mode by simply reverse-biasing it as shown in Fig. 8.20. The op-amp circuit shown serves as an *I*-to-*V* converter for the detector. In this mode, the greater the applied reverse bias, the longer the linear range of operation and the shorter the response time. To some degree these advantages are offset at high reverse biasing by increased dark current and noise.

In the *p–i–n* diode, an *i* or intrinsic layer (i.e., native semiconductor) is left between the doped *p* and *n* layers. Because of this middle layer there is a better probability that photons will be absorbed, and because it also has higher resistance, it has a greater voltage drop and charge carriers will be swept more rapidly to an appropriate contact and response time will be good.

Finally, the *avalanche photodiode* is a semiconductor device providing internal amplification. It takes advantage of the possibility that charge carriers can be accelerated suffi-

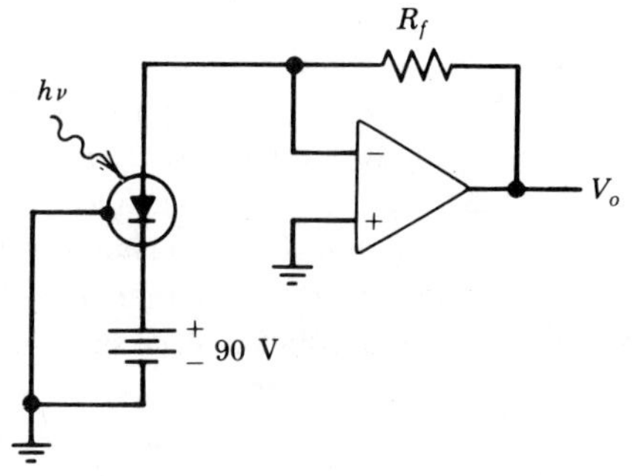

Fig. 8.20 A circuit diagram for a solid-state photodiode reverse-biased to operate in the photoconductive mode. This bias considerably extends the range of linear response. In the absence of light, reverse biasing limits current greatly; only minority charge carriers move. A photon transmitted by the window is absorbed in the diode to create an electron/hole pair. These charges are collected at the electrodes to give a signal current.

ciently to generate new electron hole pairs on collision. Multiplication of current by a factor ranging from 20 to 100 can be provided. Noise, however, rises proportionately.

Other Photodiodes and Photodetectors. In the NIR (12,500–4000 cm^{-1}; 0.8–2.5 μm) carefully deposited layers of PbS, and sometimes of PbSe and InI, make good quantum detectors (see Table 8.4). Their response drops rapidly at still longer wavelengths though it can be enhanced by cooling to 77 K. It has been difficult to develop this kind of device extensively because it is hard to model mathematically. Usually a PbS cell is made by depositing a thin film of semiconductor on glass either chemically or by a vacuum process. The larger the voltage across the cell, the stronger the output signal upon illumination. The voltage used is usually as high as possible without producing appreciable heating (less than 0.01 W mm^{-2} can be tolerated).

Low-temperature photoconductive diodes (see Table 8.4 again) are very sensitive, fast, broad-range detectors. These devices, usually doped germanium crystals, are mounted in the bottom of a liquid helium dewar to permit operation at 4 K. Rise times are less than 1 μs. Ranges in wavenumber for different dopings are; Au, 5000–1400; Hg, 5000–650; Cu, 5000–330; Zn, 5000–200. Intrinsic semiconductor diodes of InSb, HgCdTe, and other substances are also in use. Most of these detectors are background-limited devices whose D^* is enhanced by restriction of their entrance aperture angle.

Thermal Detectors. In the middle and far IR, radiation is ordinarily sensed by a thermal detector, a device that measures the temperature rise when radiation is absorbed. Most frequently used are the thermocouple or thermopile (a multiple junction thermocouple), bolometer (a resistance thermometer), and pneumatic or Golay detector (a gas thermometer). An important advantage of any thermal detector is that its response is independent of the wavelength of the radiation. Such a receiver responds to energy absorbed per unit area. The chief drawbacks of most thermal detectors are slow response, a moderate sensitivity, and a moderate D^* (see Table 8.4).

The typical absorbing device is a tiny (2 × 2 mm) blackened metal strip. A condensing mirror is used to reduce the much larger IR beam to this size. A re-

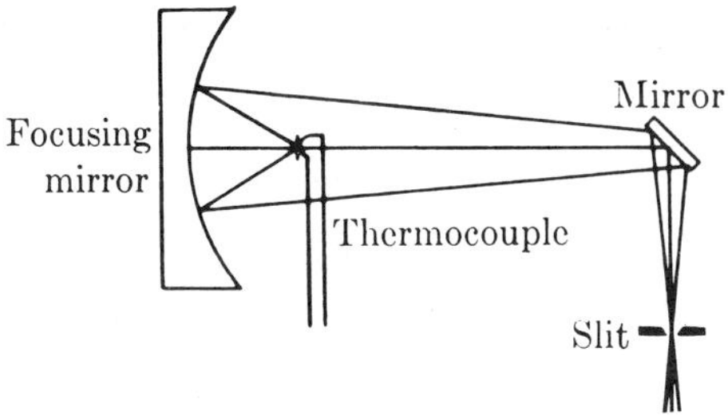

Fig. 8.21 A schematic diagram of a representative optical arrangement associated with a thermal detector. (Courtesy of Perkin–Elmer Corp.)

duction of 6 : 1 is usually feasible by an arrangement such as is shown in Fig. 8.21. In general, the detector is shielded by an evacuated holder to minimize noise (fluctuations in background temperature), for the temperature rise caused by the beam is minute.

A thermocouple, thermopile, or bolometer that will give a large output per degree is attached intimately to the absorbing strip. With a bolometer, a reference resistor or thermistor that is not irradiated by the beam must also be provided for its bridge circuit.

Thermocouple Response. What determines the magnitude of thermocouple response? Consider that incident radiation of power P is absorbed by the blackened strip affixed to the thermocouple junction. The rise in temperature is ΔT. Simultaneously, power is lost at the rate of L watts per degree rise, through radiation, conduction, convection, and the Peltier effect. The last may usually be neglected. The attainment of a final temperature is achieved exponentially, the response time t for 63% attainment being $t = C/L$, where C is the heat capacity of the blackened strip and thermocouple. For a short response time, C must be small and L large.

On the other hand, detector sensitivity is measured in terms of voltage E that is produced by the thermocouple per unit power P it receives. Let the thermoelectric power, that is, the voltage produced per degree rise in temperature, be Q. The output voltage is simply

$$E = Q\Delta T = QP/L. \tag{8.11}$$

From Eq. (8.11), it appears that high sensitivity calls for a small L.

The conflicting requirements for L can be resolved by making the absorber tiny: both the heat capacity and power loss will be small, yet the loss rate L may be large relative to C.

Pneumatic Detector. Unlike the detectors just described, whose response depends on the energy absorbed per unit area, the *pneumatic* or *Golay detector* is essentially a differential gas thermometer that responds to total energy falling on it. Either the filling gas itself or a blackened membrane making up one wall of the gas cell serves as the absorber. As radiation impinges, the gas is heated and expands against a second flexible membrane. In some designs membrane movement deflects a light beam that is incident, in others the membrane is part of a capacitor whose capacitance alters as a result of expansion. Simpler optics may be used with a pneumatic detector since the absorber is larger.

If an absorbing gas is used, the Golay detector has high sensitivity at wavelengths at which gas absorbs. A surprising degree of flexibility and specificity may be achieved by choice of a gas. As a result, this approach is widely used in nondispersing photometers (Section 17.10). If a blackened membrane is used, the device is termed a *Golay detector* and is equally sensitive at all wavelengths.

If a gas chamber is used, it may be large enough that the incident beam does not need focusing. Even the membrane type may be about 3 mm in diameter, allowing somewhat simpler optics.

The *pyroelectric detector* employs a thin slice of a ferroelectric crystal such as triglycine sulfate (usually deuterated) or $LiTaO_3$. Such materials have an internal electric polarization that varies with temperature. Even at room temperature they can be used as thermal optical detectors.* Any change in temperature causes a change in internal polarization and leads to current in the external circuit as long as is necessary to compensate for the shift in polarization. Somewhat anomalously therefore, to obtain a signal from the detector, incident light must be modulated. An ac output will of course be produced.

The resulting current I can be expressed as $I = A\Lambda dT/dt$, where A is the area of the crystal face, t the time, and $\Lambda = dP/dT$, the rate of change of polarization P with temperature. Clearly, the current in the external circuit exists to balance the polarization effects that occur as a temperature pulse is applied to the crystal.

A current-mode detector/preamplifier circuit is shown in Fig. 8.22. The detector window will establish the spectral range that will be received. In general an absorbing coating is added to the crystal if light of wavelength below 7 μm is of interest. For good response the minimum chopping rate is about 350 Hz. Most

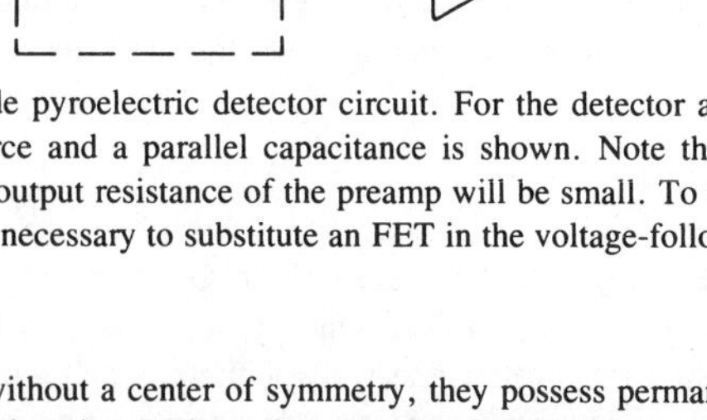

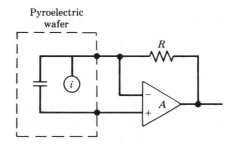

Fig. 8.22 A current-mode pyroelectric detector circuit. For the detector an equivalent circuit consisting of the current source and a parallel capacitance is shown. Note that this current-to-voltage converter ensures that the output resistance of the preamp will be small. To modify the circuit for the voltage mode, it would be necessary to substitute an FET in the voltage-follower configuration for the op-amp.

*Since these crystals are without a center of symmetry, they possess permanent dipoles, which form into microscopic polarized domains. By imposing a dc electric field, large numbers of domains become aligned in the direction of the field; the fraction so aligned is a function of temperature.

pyroelectric detectors have smaller responsivities than bolometers and thermocouples but their much shorter response times are a real asset.

8.7 PHOTON COUNTING: LOW-LIGHT-LEVEL DETECTION

Photon counting is perhaps the most sensitive of all techniques for measurement of low radiation levels. Indeed at these light levels, where the discreteness of photons dominates the scene, the output of a detector such as a photomultiplier tube consists of individual current pulses several nanoseconds long. Photon counting requires that pulses that are photon-induced be distinguished from dark-current pulses that originate within the tube from other causes. Fortunately, amplitudes of the former are substantially greater, as was shown in Fig. 8.19.

Figure 8.23 gives a block diagram of the configuration of a photon counter. Basically, a photomultiplier tube, amplifier–discriminator, and counter are required. The discriminator, which may be simply an operational amplifier with differential input, can be set to pass only pulses whose height exceeds an appropriate threshold, thus excluding most dark pulses. The readout is just the number of pulses counted. Instrumentation will be discussed further below.

Applications in which photon counting yields significantly improved measurements come readily to mind: low-level UV–VIS spectroscopy, spectrofluorometry, Raman spectrometry, and spectral measurements in astronomy. The method offers the advantages of improvement in S/N ratio, better long-term gain stability, a gain that is much less sensitive to dc supply voltage variations, better elimination of leakage and zero drift in photomultipliers, and a basically digital output which is less susceptible to noise.

Instrumentation. Many of the common photomultipliers are satisfactory in photon-counting circuits. For stable operation, the voltage is ordinarily set by finding a region in the count rate versus photomultiplier voltage plot in which the slope is smallest. Under these conditions every photon that yields a photoelectron at the photocathode produces a shower of 10^5–10^6 electrons at the anode lasting a few nanoseconds. Variations in the size of the burst can be traced to variations in the voltage supply, focusing, and in secondary emission thresholds of dynodes and to other random effects. Fortunately, much of the baseline "wandering" of photomultipliers can be eliminated by simply setting the discriminator to reject dark current bursts.

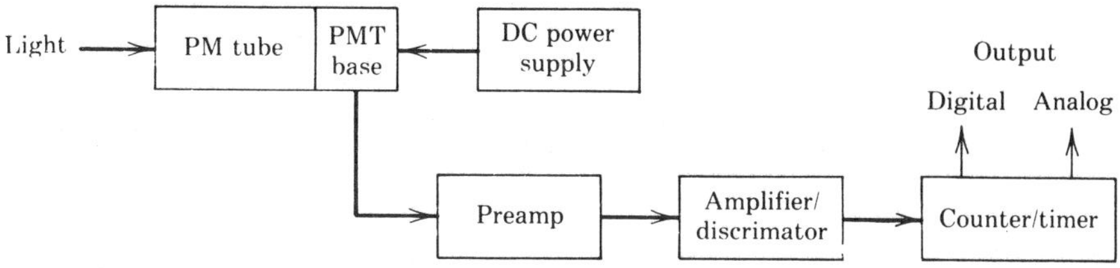

Fig. 8.23 Block diagram of a photon counting system.

It is also important that dark count rates be kept moderate by ensuring that the photomultiplier housing provides good shielding against external rf sources of radiation and other electrical interference. Good grounding practices must also be observed.

In order to ensure a fast rise time, and thus make possible counting rates up to 100 MHz, the anode of the photomultiplier is usually connected to a low-noise preamplifier of fast rise time through a 50-Ω coax cable that ends in a 50-Ω resistance. A representative photomultiplier pulse is about 20 μA and gives a 1-mV peak. Amplification is necessary to bring the pulse level to between 25 mV and 10 V for the discriminator.

Discriminators also contribute to the gain of the circuit. More importantly, they establish the maximum counting rate by the dead time they introduce. *Dead time* is the minimum time interval between two pulses that will pass the discriminator, for which it will still produce a pair of output pulses. Representative values are 10–40 ns, which correspond to 15–100 MHz counting rates. It is important that the output always be a pulse of a given amplitude to make subsequent counting or storage in memory straightforward.

Shielding will reduce noise. Amplifier noise in particular must be low so that it will not increase the amplitude of a dark pulse sufficiently to trigger the discriminator. Photomultiplier noise can be kept small by ensuring that the tube chosen has an active cathode area that is no larger than required for the optical beam and that the sensitivity of the cathode is as small in the red region of the spectrum as feasible.

Poisson Statistics. Randomness in the arrival rate of photons is subject to a statistical analysis similar to that used in radioactivity determinations and in counting random events generally. The signal is of course proportional to the total number of counts. Noise, the variation in rate of arrival of photons, falls in the category of *shot noise*, and is proportional to the square root of the number of counts. If measurements successfully exclude most dark counts, the S/N proves to be simply

$$S/N = (\text{total counts})/(\text{total counts})^{1/2} = (\text{total counts})^{1/2}. \tag{8.12}$$

Equation (8.12) can be expected to hold when the counting rate is high.

Example 8.5. What S/N ratio will be attained with a 20-s observation of a signal averaging 200 photons s^{-1}?

In a 20-s interval 4000 pulses are counted and $S/N = \sqrt{4000} = 63$.

Example 8.6. In spectrofluorometry light levels are frequently very low. Yet levels cover a range and it will be important that both photon counting and dc (analog) determination are provided. What differences must be made in the circuitry to switch from photon counting to dc determination?

For the former, the circuit is as just discussed. To switch to dc registration it is only necessary to select a current-to-voltage op-amp circuit and disconnect the preamplifier–discriminator–counter circuit. Naturally, the dc output will also generally be converted to digital form for processing.

Measurements at Very Low Light Levels. At exceptionally low levels of light the number of spurious nonphoton-induced pulses counted by the discriminator may be a significant fraction (greater than 1%) of the total. It can be shown in this case that

$$S/N = (r_s T)^{1/2}/(1 + 2r_b/r_s)^{1/2} \tag{8.13}$$

where r_s and r_b are total and background (dark) counting rates, respectively, at a given discriminator voltage and T is the counting interval [M. L. Franklin et al, *Anal. Chem.* **41**, 2 (1969)]. Note that this expression reduces to Eq. (8.13) when $1 \gg 2r_b/r_s$.

8.8 MULTICHANNEL DETECTORS

The development of fast, reliable multichannel (parallel) optical detectors in recent decades has added greatly to the versatility of spectrometers. Since such detectors are also sensitive to X rays, electrons, or ions over certain ranges of energy, they will be found in a variety of applications.

Probably the earliest type was the two-dimensional photographic emulsion, which was spread on a stiff backing such as film or a glass plate. While in less common use today, it still serves in many instruments and is described below. Somewhat later a new type appeared, arrays in which photomultiplier tubes were placed behind slits along a spectrometer focal curve. These devices will be taken up with atomic emission spectrometers in Chapter 13.

The advent of the solid-state detector arrays opened a new era. The first ones were two-dimensional and were basically adapted TV cameras. Good examples are the silicon vidicon and the silicon intensified target vidicon. Linear photodiode arrays designed especially for spectrometry and prepared by integrated circuit techniques are the most recent. These solid-state devices are discussed below. Other attractive multichannel detectors include the charge-coupled and charge injection solid-state devices and the image-dissector photomultiplier tube; references 17 and 19 should be consulted for information.

Multichannel detectors, because of their greater dimensionality, are inherently also *image sensors*. As a result, each must sense radiation, integrate the signal over time, and incorporate a readout system for systematic interrogation of its many light-sensing elements or pixels. Fortunately, with solid-state design the functions of sensing and integration are combined, and in linear photodiode arrays all three aspects are included in integrated circuits.

Photographic Emulsions. A photographic emulsion consists of minute silver halide crystals dispersed in a transparent, water-expandable medium such as gelatin which is spread on film or thin glass plates to provide dimensional stability. It is dried before use.

On exposure, the silver halide crystals that receive radiation build up a latent image. Subsequent development, a chemical process analogous to amplification, produces a black deposit of silver at the site of the latent image. Since the silver halide crystals are small, emulsions are capable of high spatial resolution. The useful spectral range is mainly the UV–VIS, and with special sensitizing, the NIR, X ray, and VUV.

The response of a representative emulsion is graphed in Fig. 8.24. Such a plot serves to calibrate the emulsion. As the figure shows, the density of deposit is related to the logarithm of *exposure*, the product of radiation intensity and time. The graph also makes clear

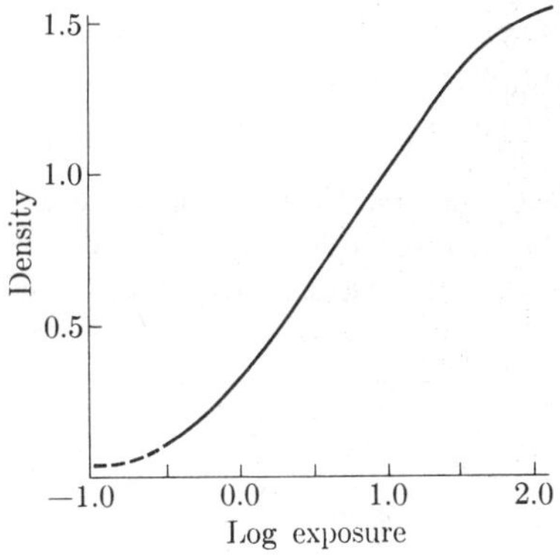

Fig. 8.24 Representative plot of image density versus the logarithm of exposure for a photographic emulsion.

the short dynamic range: a useful response is secured over about two decades of exposure. Given this short range, a user should make a preliminary survey of emulsions to select the one appropriate to the radiation level to be measured and to the longest feasible time of exposure. Stratagems allow extension of the range of response. For *very low intensities* such as originate from stars, prefogging of an emulsion by exposure to faint light serves to develop needed sensitivity, and for *very high intensities* inserting filters reduces sensitivity [22, 29].

Different *speeds* of response and *contrasts* (medium, high, very high) in emulsions are available. How are they related to the curve in Fig. 8.24? Contrast is proportional to the slope of an emulsion's calibration curve and speed to the range over which the curve is linear. For example, a fast, low-contrast emulsion will yield a curve of low slope whose linear range occurs at very low exposures. In general, speed and contrast cannot both be maximized.

Finally, the density of deposit in a developed photographic film or plate is ordinarily read out as a function of location by a photoelectric scanning procedure. The background density is subtracted to normalize the response.

Example 8.10. *For an elemental analysis* by atomic emission spectrometry the best lines of most elements, those with highest sensitivity of response, occur in the UV above 230 nm. Eastman Kodak SA No. 1 emulsion (useful range, 230–400 nm, speed low, contrast relatively high) is widely used for quantitative spectrographic analysis. It offers high contrast and freedom from graininess (caused by large AgX crystals) and thus gives sharp lines. Its cutoff at 440 nm ensures, even when no filter is used, that it can be used in a grating spectrometer at high orders in the UV with minimum overlapping of lines from the VIS and IR. Lines of certain elements are ordinarily used as internal standards to ensure still greater fidelity of emulsion response. Each analytical line is thus recorded as a ratio of its apparent response to that of a standard line in the same region of the spectrum.

Vidicons. The *vidicon* is a two-dimensional solid-state detector that is similar in shape to a small cathode ray tube (CRT). As stated above, it is basically a television camera with a face plate consisting of a slab of a single crystal of *n*-type silicon. On the inside of the plate an array of about 1000 × 1000 discrete photodiodes or pixels is developed by doping. During operation these diodes are reverse-biased by an electron beam (diameter about 25 μm) which continuously scans the back of the plate in a raster pattern, exactly as occurs in a CRT. During charging the depletion layer at the *p–n* junction in each diode widens and serves as a capacitor with a fixed charge.

Every photon that is incident on the front face and is absorbed generates electron-hole pairs. As generated electrons diffuse to the closest diode, they discharge it proportionately. Its exposure is read out as a charging current as the electron beam next scans across and recharges it.

The considerable advantages of the vidicon are a 2D display and ease of use because the device takes full advantage of developed CRT technology. Its limitations are a more modest dynamic range and resolution. For the vidicon dynamic range is diminished because of *lag*, a phenomenon resulting from the failure of diodes that have been substantially discharged to be recharged completely by one sweep of the electron beam. For these diodes a spurious charging will occur on the following sweep, giving rise to falsely high readout. Resolution is reduced by the phenomenon of *blooming*, a tendency of diodes adjacent to the one affected by incident light to be partially discharged as well. This effect is also described as cross-talk between pixels. In Table 8.5 the performance of the vidicon is compared with that of other array detectors.

Linear Photodiode Arrays. Thanks to the microartistry of large-scale integrated circuit design, linear arrays of photodiodes that are also self-scanning are now available. Diode arrays having numbers of elements ranging from 128 to 1024—and even up to 4096—are available. This multichannel detector makes an ideal sensor for an entire spectrum in a UV–VIS dispersive spectrometer. With that application in mind, newer arrays are made with adjacent diodes 2.5 mm long and spaced 25 μm on centers. This 100 : 1 aspect ratio ensures a rough correspondence between the dimensions of an individual diode and the image of a rectangular slit at the focal plane.

The *spectral response* of silicon photodiode arrays extends from about 170–1100 nm (the window admitting radiation must have suitable transmission) and reaches a peak quantum efficiency of about 0.7 in the 400–700 nm range. The overall responsivity varies by about a factor of 10. Diode response is rapid and linear. To minimize dark current, Peltier or thermoelectric cooling is commonly used to cool an array to about 0°C.

A diagram of the electronics of an array is shown in Fig. 8.25. Note that each diode–FET switch combination is poised between common and video lines and that a shift register is included to make self-scanning possible. Initially, each diode in the array is reversed-biased and its junction capacitance (see Section 2.9) is charged. Photons incident on these diodes create electron–hole pairs that partially discharge their capacitance, the actual loss of charge by each diode being proportional to the product of the number of incident photons and the diode quantum efficiency at that wavelength. This signal is stored in the diode until readout. In each diode readout occurs as it is recharged (see below.)

The duration and intensity of illumination determine both the final signal S/N and the exposure interval needed to acquire a spectrum. This interval is also the *integration time* for the signal. A longer integration time allows a higher S/N since the signal will be larger and noise averaged more completely toward zero. The upper exposure limit is set in practice by the time required for full discharge of

Table 8.5 Characteristics of Some Multichannel Detectors

	Photosensitive Elements	Wavelength Range (nm)[a]	Quantum Efficiency	Dynamic Range	Resolution Units or Pixels	Integration Time (s)[b]
Silicon vidicon	Silicon photodiodes on silicon face plate	250–1100	0.7 (400–700 nm)	500–1000	512 × 512	>1
Intensified or SIT vidicon	Multialkali on fiber optic array coupled to vidicon face plate	350–830	0.15 (480 nm)	500–1000	512 × 512	>1
Silicon photodiode array	Silicon photodiode array	<180–1100	0.7 (400–700 nm)	$10^3 - 2 \times 10^4$	256 × 1 up to 4096 × 1	1–500[d]
Intensified silicon photodiode array[c]	Microchannel plate intensifier coupled to photodiode array	165–900	0.12 (200–500 nm)	700 × 1		

[a] Maximum range given. In different models it may be reduced by windows (e.g., glass limits UV to 350–400 nm).
[b] Time during which element can accumulate electrical charge during exposures (limited by time required to saturate with dark counts).
[c] Can be modulated at periods as short as 5 ns.
[d] Shorter time at room temperature, longer at −40°C.

References: Y. Talmi, "TV-Type Multichannel Detection," *Anal. Chem.*, **47**, 697A (1975). Optical Multichannel Analyzer III, EG & G Princeton Applied Research, 1984.

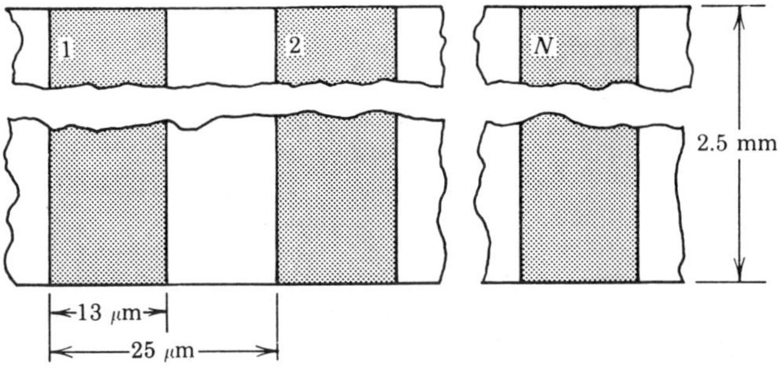

(a)

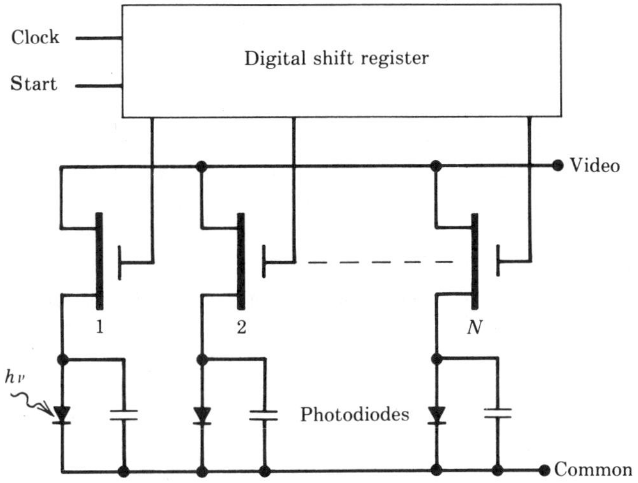

(b)

Fig. 8.25 Simplified schematic diagram of a self-scanned photodiode array. (a) Top view of array geometry as seen through the window through which light falls on the *top* of the array. Each diode is defined by a *p*-doped region 13 μm by 2500 μm (2.5 mm) that serves as the cathode. (b) Simplified electronic diagram. Diodes are reverse-biased by application of −5 V from the video line when their MOSFET switches are opened. A photon $h\nu$ is shown striking the sensitive *p*-doped region of a diode. The charge it generates in that diode will reduce its reverse bias. The digital shift register and MOSFET switches on the diodes allow the array to be self-scanning as a clock signal is received. (Courtesy of EG&G Reticon.)

the one or two diodes that receive the most light. For rapid spectroscopy a brighter source or wider slit or shutter will clearly be desirable as well as fast readout. Since a diode can be fully recharged in less than 1 μs, a 1028-element array should be scannable in no more than 1 ms. These times are short enough for fast spectrometric measurements.

The geometry of the sensitive surface of a diode array is shown in Fig. 8.26. Resolution will basically depend on the center-to-center spacing of its *p*-doped regions that locate the diodes. As the figure shows, the intermediate undoped re-

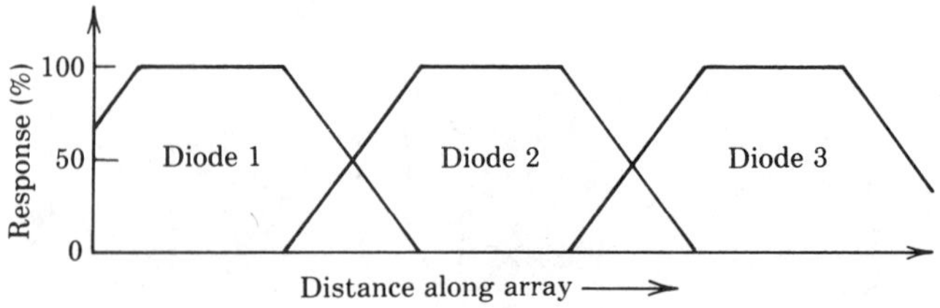

Fig. 8.26 Response function of diodes in a representative linear diode array as a function of distance along the array. Light incident directly on a *p*-type region produces a response normalized to 100%. Charge generated in the intermediate *n*-type regions, which are of reduced activity, will be split between adjacent diodes [18].

gions are also light-sensitive and they divide the charge generated in them by photons evenly between neighboring diodes.

Several different *modes of readout* have been developed for diode arrays. The regular or real-time mode in which all diodes are read out after a set exposure or integration time is described below; diode signals are stored in individual memory locations and can then be displayed. A variation of this mode would be to allow successive scans of a spectrum with readouts added to the same memory locations. A gain from this approach would be still greater dynamic range.

A variable integration time mode facilitates work with spectra that differ in intensity or the use of different exposures. The available range of an AD converter is best used by selecting integration times that are inversely proportional to the intensity of spectra.

For spectra that have broad features diode grouping is feasible. In this mode signals from adjacent diodes are summed after digitization and the sum stored in memory. The result is an improved S/N with some loss of resolution. In similar fashion diodes can be "skipped" by recharging them rapidly (0.5 μs or less) and ignoring their signals. This fast spectroscopy mode allows attention to spectral regions of interest for kinetic studies that involve rate determinations or the identification of intermediates.

Scanning a Diode Array. Light intensities received by each diode or pixel of an array during exposure are read out electronically by use of the circuit shown earlier in Fig. 8.25. When the MOSFET switch connected to a diode is closed by a clock pulse directed to it by the digital shift register, the diode is momentarily connected to the video line for recharging. The current pulse existing during recharging is integrated over the life of the pulse in a part of the circuit not shown, sampled by a sample-and-hold circuit, and digitized by an ADC. This signal goes to a microcomputer for processing and storage in an appropriate memory location. Subsequent display of the signal intensity across the array on a CRT will allow one to observe the spectrum.

In somewhat similar fashion, in charge-coupled devices, all capacitors are simultaneously discharged into corresponding stages of a charge-coupled analog shift register. Subsequently those charges are switched sequentially through the register by clock pulses and finally appear at its output for sampling and digitization.

Intensified Arrays. Both vidicons and diode arrays can be coupled to a microchannel plate intensifier (see Section 22.5) to improve their responses by a factor of 100. With a vidicon the combination comprises a *silicon-intensified target vidicon.* Similarly, with a diode array, one obtains an *intensified diode array*. Overall detector performance is now set by the more reduced performance of the intensifier: spectral range is reduced to about 165–900 nm and quantum efficiency to a maximum of about 0.1–0.2 from 200 to 500 nm.

Intensifiers that can be gated as useful in time-resolved experiments. The microchannel plate can be turned off by reducing its accelerating voltage and later restored for the period during which signal is to be developed. Current bursts as short as 40 ns (vidicon) or 10 ns (diode array) are possible.

IR Arrays. IR arrays have also been developed. Among the different types, the *pyroelectric array*, again of a self-scanning type, is particularly useful. The integrated circuit version is based on the use of single-crystal $LiTaO_3$ elements. This pyroelectric substance is highly sensitive and gives uniform response from element to element over the IR. Again, each detector element is connected through an FET switch to a digital shift register and video and common lines. Fewer light-sensitive elements are used than in UV–VIS arrays since features in IR spectra are usually broader-peaked. Modulation frequencies are about 50 Hz but the clock rate is of the order of 2 MHz.

In addition, arrays of germanium elements doped with the appropriate metals to give responsivities that are good in the middle and far IR are available. Another type is constructed from semiconductors such as InSb, PbSnTe, and HgCdTe. This set is usable in the near and middle IR and only the germanium-doped crystals are widely useful up to about 40 μm. Response times are fast and rapid reading is possible.

Applications. Multichannel detectors are widely used in instrumentation but the reader is especially referred to Section 9.9 where their main use in polychromators is described.

REFERENCES

Applications of physical and geometrical optics to design are treated in:

1. G. R. Elion and H. A. Elion, *Electro-Optics Handbook.* New York: Dekker, 1979.
2. R. D. Hudson, Jr., *Infrared System Engineering*. New York: Wiley-Interscience, 1969.
3. R. Kingslake (Ed.), *Applied Optics and Optical Engineering*. New York: Academic, Vols. I–V 1965–1969, (each devoted to a single topic).
4. L. Levi, *Applied Optics: A Guide to Optical Design*, Vol. 1, New York: Wiley, 1968.
5. W. J. Smith, *Modern Optical Engineering, the Design of Optical Systems*. New York: McGraw-Hill, 1966.
6. W. G. Driscoll and W. Vaughan (Eds.), *Handbook of Optics*. New York: McGraw-Hill, 1978.

Particular sources and detectors are treated on an intermediate or general level in:

7. W. Demtroeder, *Laser Spectroscopy: Basic Concepts and Instrumentation*. New York: Springer-Verlag, 1982.
8. G. M. Hieftje, J. C. Travis, and F. E. Lytle (Eds.), *Lasers in Chemical Analysis*. Clifton, NJ: Humana, 1981.
9. D. C. O'Shea, W. R. Callen, and W. T. Rhodes, *Introduction to Lasers and Their Applications*. Reading, MA: Addison-Wesley, 1977.
10. J. C. Wright and M. J. Wirth, Principles of lasers, *Anal. Chem.*, **52,** 1087A (1980).
11. C. Kunz (Ed.), *Synchrotron Radiation, Techniques and Applications*. New York: Springer-Verlag, 1979.
12. F. E. Lytle, Measuring fast optical signals: Detectors, *Anal. Chem.*, **46,** 545A (1974).
13. RCA, *Photomultiplier Manual.* Harrison, NJ: RCA Corporation, 1970.
14. RCA, *Electro-Optics Handbook*, 2nd ed. Harrison, NJ: RCA Corporation, 1974.
15. D. G. Jones, Photodiode array detectors in UV–VIS spectroscopy, Parts I and II, *Anal. Chem.*, 57, 1057A, 1207A (1985).
16. Y. Talmi, TV-Type multichannel detectors (and their) applicability to spectroscopy, *Anal. Chem.*, **47,** 658A, 697A (1975).
17. Y. Talmi (Ed.), *Multichannel Image Detectors*, Vols. 1 and 2. Washington: American Chemical Society, 1979 and 1983.
18. Y. Talmi and R. W. Simpson, Self-scanned photodiode array: A multichannel spectrometric detector, *Appl. Opt.*, **19,** 1401 (1980).
19. J. V. Sweedler et al., High-performance charge transfer device detectors, *Anal. Chem.*, **60,** 282A (1988).

and on an intermediate to advanced level in:

20. S. Caroli (Ed.), *Improved Hollow Cathode Lamps for Atomic Spectroscopy.* New York: Wiley, 1985.
21. G. Hernandez, *Fabry–Perot Interferometers*. Cambridge: University Press, 1986.
22. W. A. Hiltner (Ed.), *Astronomical Techniques.* Chicago: University of Chicago, 1962.
23. J. D. Ingle, Jr. and S. R. Crouch, Signal-to-noise ratio comparison of photomultipliers and phototubes, *Anal. Chem.*, **43,** 1331 (1971).
24. R. J. Keyes (Ed.), *Optical and Infrared Detectors*. New York: Springer-Verlag, 1977.
25. J. A. R. Samson, *Techniques of Vacuum Ultraviolet Spectroscopy*. New York: Wiley, 1967.
26. A. E. Siegman, *Lasers*. Oxford: University Press, 1986.
27. J. T. Verdeyen, *Laser Electronics*. Englewood Cliffs, NJ: Prentice-Hall, 1981.
28. A. Yariv, *Quantum Electronics*. New York: Wiley, 1975.
29. J. W. T. Walsh, *Photometry*. New York: Dover, 1969.

EXERCISES

8.1 To the degree that a carbon arc can be considered a blackbody emitter, what is the wavelength of maximum emission when its temperature is 5000 K?

8.2 Show by means of three sketches of intensity versus wavelength the appearance of the sodium D lines of sodium vapor lamps of (a) low pressure, (b) medium pressure, and (c) high pressure. Describe the significant effects.

8.3 Several longitudinal modes can coexist in a laser. (a) Show mathematically that these frequencies are basically harmonics. (b) What harmonic will exist in the optical resonator with a 0.1-m cavity for a lasing wavelength of 1000 nm? [Hint: Calculate the approximate multiple *m* of the basic wavelength that resonates in the cavity.] (c) What is the spacing between modes? What would be the spacing if the cavity were made 1 m long?

8.4 What is the wavelength of maximum intensity for a tungsten–halogen lamp in which the filament temperature is (a) 2600 K? (b) 3000 K?

8.5 Sketch a low-frequency version of laser mode-locking. For three waves of equal amplitude but frequencies of 30, 60, and 90 Hz, draw approximate amplitude–time plots for the individual waves and for their sum (a) when their phases are 0°, 60°, and 120° and (b) when they are in phase at 0°. Include four periods for each.

8.6 If a multiplier phototube with 10 dynodes has a grain of 10^7, what must the average ratio of numbers of secondary to incident electrons at each dynode be?

8.7 In scanning a photodiode array a current pulse is generated for each diode that has been exposed. (a) Show by sketch how the pulse might appear. Consult Fig. 3.11a. (b) Explain how a sample-and-hold circuit converts this into a square pulse whose height is proportional to the charge passed.

8.8 Both dynamic range and good S/N are sought with a linear photodiode array. (a) How does its dynamic range depend on integration or exposure time, dark current, and diode area? (b) In terms of S/N discuss the relative merit of increasing integration time and lowering diode temperature.

8.9 With an array of 512 diodes assume that a pair of diodes is the basic resolution unit. What resolution can be realized when wavelengths from 400–700 nm just cover its active length?

ANSWERS

8.1 580 nm. *8.3*(b) $m = 2 - 10^5/n_o$; (c) 1.5 GHz/n_o, 150 MHz/n_o. *8.6* Ratio = 5. *8.9* About 1 nm.

Chapter 9

MONOCHROMATORS AND POLYCHROMATORS

9.1 INTRODUCTION

Much versatile optical instrumentation is designed around a monochromator. It is a module that admits a wide spectral range of light and, by dispersing wavelengths in space, makes available at its exit slit only a narrow spectral band. Both its output wavelength, and in very many cases, spectral bandwidth, are tunable. In general, the optical system of a monochromator consists of

1. an *entrance slit* that provides a narrow optical image;
2. a *collimator* that renders the rays spreading from the slit parallel;
3. a component for *dispersing* this radiation;
4. a *focusing element* to reform images of the slit; and
5. an *exit slit* to isolate the desired spectral band.

For tuning to different wavelengths, a drive systematically pivots a ''table'' on which the dispersing element is held about an axis through its center. As a range of monochromator designs is explored in this chapter, basic quality requirements for the module such as resolution, dispersion, spectral range, stray light, and ruggedness will be related to design. Some brief definitions are provided in Table 9.1.

To illustrate the layout of a monochromator, a simple prism version together with a source are shown in Fig. 9.1. Note there is only a focal plane on which an exit slit might be located; indeed, Exercise 9.1 explores the problems of placing the slit. By using a tungsten lamp source and a monochromator with an exit slit one obtains a source of monochromator light for the visible and somewhat beyond.

Actually, the arrangement of Fig. 9.1 suggests a more elaborate but closely related double module, the *polychromator*. That module differs from the device shown only by having a multidetector array permanently mounted along the focal curve. Polychromators will be discussed in the last section.

9.2. THE SLITS AND RESOLUTION

The entrance slit of a monochromator is a narrow rectangular aperture* whose width, at least, is generally adjustable. When illuminated by radiation, the slit is

*Long slits may have curved sides. See p. 312.

Table 9.1 Some Monochromator Properties

Term and Symbol [dimensions]	*Definition*	*Determining Elements*
Resolution or resolving power $R = \lambda/\Delta\lambda$ or $R = \nu/\Delta\nu$ [dimensionless]	Potential to provide at the exit slit separate images of the entrance slit for nearly identical wavelengths	Slit widths, size of dispersing element, focal length of optics
Throughput [watts] [W mm^{-2} ster^{-1} nm^{-1}]	Ratio of radiant power output at exit slit (when its width yields a 1-nm bandpass) to the spectral radiance of the source	Slit widths, entrance aperture, dispersing element
Stray light	Radiation reaching exit slit that has not followed optical path	Baffles, nonreflecting walls
Dispersion	Spread of wavelengths in space	Dispersing element
Angular $d\theta/d\lambda$ [rad(nm)$^{-1}$)]	Angular range $d\theta$ over which waveband $d\lambda$ is spread by dispersing device	
Linear $dx/d\lambda$, D [mm(nm)$^{-1}$]	Distance dx over which a waveband $d\lambda$ is spread along the focal plane of a monochromator or polychromator	
Linear reciprocal $d\lambda/dx$, D^{-1} [nm(mm)$^{-1}$]	Range of wavelengths $d\lambda$ spread over distance dx along the focal plane of a monochromator or polychromator	
Spectral slit width bandwidth, or bandpass, S	Range of wavelengths included in monochromator output measured at half maximum intensity	Slit widths, dispersing element, focal length of optics

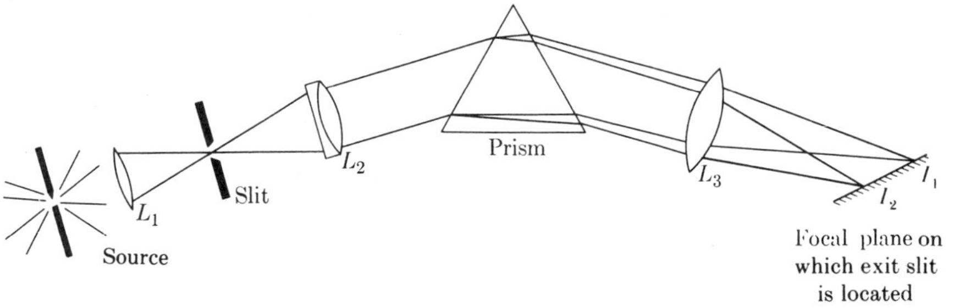

Fig. 9.1 Schematic diagram of a simple prism monochromator. Lens L_1 couples the source to the monochromator, directing light to its entrance slit. Lens L_2 collimates light and L_3 focuses the dispersed light. Each wavelength is brought to a separate focus along the focal plane of L_3 as an image of the entrance slit.

therefore a bright rectangle that can easily be reproduced by appropriate optics. But what is the role of the slit? Before tackling the question, the manner of operation of a monochromator must be described.

Inside the monochromator rays spread from the entrance slit as illustrated in Fig. 9.1 and illuminate the collimator, which renders them parallel. In effect, this parallel set of rays is a broadened version of the slit. This rectangle of light must become large enough to illuminate the full width of the dispersing device, a grating or prism. In turn, this component separates the incident polychromatic pattern into an array of monochromatic blocks, each of which leaves the dispersing element at a slightly different angle. The blocks overlap badly. Focusing by lens L_3 reduces each block to an image of the entrance slit. The collection of images falls in a plane termed the *focal plane* in which a stationary exit slit is located.

Ordinarily the exit slit is adjusted to the dimensions of the entrance slit so that it is just illuminated by an image of that slit. Different wavelength images can be brought to this slit by pivoting the dispersing element. No image is a single wavelength to be sure, a matter that will be considered further below.

In order to separate wavelengths with any efficiency, the slits of a monochromator must be closed as much as is feasible. There will actually be an optimum "narrowness" for a slit; closing it further will not produce an additional increase in resolution but will reduce intensity. The limit on resolution arises because of diffraction of the Fraunhofer type illustrated in Fig. 9.2. The drawing shows a horizontal cross section through a long, narrow slit. As pictured, a narrow slit gives not a single sharp image but several closely spaced ones.

Fraunhofer Diffraction. In Fraunhofer diffraction an aperture is illuminated by essentially parallel rays and the aperture is observed on the far side from a relatively distant point. The second condition is secured within a short distance in Fig. 9.2 by inserting a converging lens that brings sets of parallel rays from the slit to a focus at points along the line *RR*. When these conditions are not met, the diffraction is of the Fresnel type.

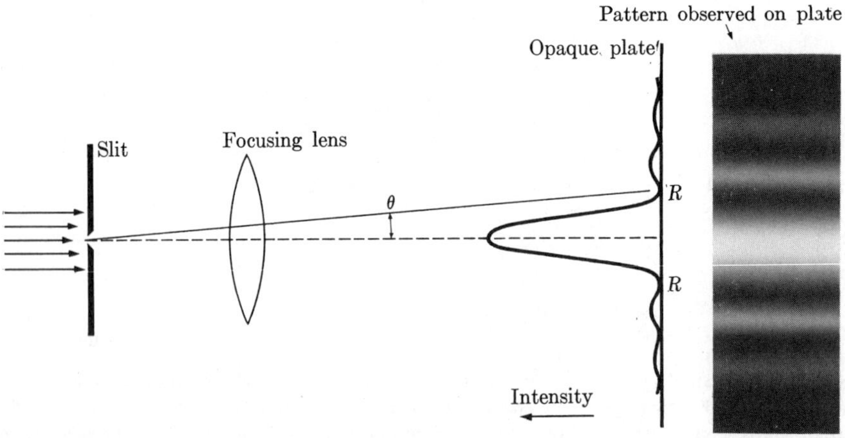

Fig. 9.2 Fraunhofer diffraction pattern of a single narrow slit. Illumination by monochromatic light.

The origin of Fraunhofer diffraction by a very narrow slit can be explained classically by use of the Huygens principle. The various points across the slit become sources of secondary wavelets and rays can be drawn in all directions from them. Consider first the rays arising along the slit and traveling parallel to the entering rays. The central member of this set is shown in Fig. 9.2 as a dashed line. All rays in this set will travel an identical distance as they are brought to a focus and will arrive at line *RR* in phase. There is reinforcement as shown.

Rays that leave the slit at angle θ, however, are shown to interfere destructively when focused. The one leaving the center of the slit is shown in oversimplified form as a solid line that terminates at *R*. Consider a ray parallel to it that originates from the "top" edge of the slit. For it to interfere destructively at *R* with the dashed ray, it must travel farther than the dashed ray by a distance equal to $\lambda/2$. The relation between slit width *W* and angle θ can now be deduced by constructing a right triangle whose apices lie at the top of the slit, the center of the slit, and the intersection of a perpendicular dropped from the top of the slit to the solid line. Applying trigonometry yields the equation $\sin\theta = \lambda/2 - W/2$, that is, $\sin\theta = \lambda/W$. According to this relation, once diffraction becomes a factor, further narrowing of slit width causes θ to increase and resolution cannot be improved further.

For experimental work it is fortunate that most of the energy passing through a narrow slit is concentrated in the central band. Ideally, the ratio of the intensity of this band to the one just to the right or left proves to be about 22. The optimum narrowing of the entrance slit (for maximum resolution) is attained in a monochromator when the central band fills the aperture of the collimating lens or mirror. Because it is impossible to attain this condition for all wavelengths simultaneously, adjustments should be made so that the condition will prevail in the center of the range of interest.

Resolution. The ability of a monochromator to separate nearly identical frequencies is described as its resolution or resolving power. In practice two operational definitions of resolution are used. The more widely used definition and the one that will be assumed unless a statement is made to the contrary is based on finding the closest-lying frequency lines or bands that are separated by a device. This definition leads to a quantitative expression for resolution *R*:

$$R = \nu/\Delta\nu = \lambda/\Delta\lambda = \sigma/\Delta\sigma \tag{9.1}$$

where ν and $\nu + \Delta\nu$ are the closest frequencies that can be regarded as having been separated. Note that comparable definitions exist in terms of wavelengths λ and wavenumbers σ. Differences $\Delta\nu$, $\Delta\lambda$, and $\Delta\sigma$ are measured between centers of lines or peaks.

How much drop in intensity must there be between adjacent spectral peaks for them to be considered resolved? In practice any of several answers may be given. The most liberal statement is that peaks are resolved when the intensity between them falls to at least 90% of the peak value. A much deeper valley is ordinarily insisted on, especially if intensities are to be read off.

The alternate definition of resolution is useful in a quite different context, that the resolution of a monochromator or spectrometer is the minimum peak width at half height achievable on measurement of an emission or laser peak of negligible intrinsic width. By this criterion a monochromator, for example, is said to have a resolution of 0.05 nm, if not only a 0.05-nm line but all narrower spectral lines emerge with a width of 0.05 nm.

Example 9.1. What is the resolution of a device that can separate two Gaussian peaks centered at 300.00 and 300.10 nm and have the signal drop to the baseline between them?

From Eq. (9.1) $R = 300/0.10 = 3000$. This value is clearly a minimum figure since still closer peaks might have been resolved.

Example 9.2. In making gas-phase UV measurements a spectroscopist is concerned that an available instrument may have insufficient resolution to yield true spectra. How may this spectroscopist use the alternate definition of resolution?

The spectroscopist should scan across any emission line of negligible width, such as the 632.8-nm line of helium neon laser (about 0.001 nm), at different slit openings. If the line is never displayed with a peak-width at half-height narrower than 0.15 nm, this figure can be taken as the instrument resolution.

Example 9.3. Will a monochromator of 0.15 nm resolution present a UV spectrum in which many absorption peaks have half-widths of about 0.45 nm with a precision of $\pm 1\%$? Assume that faithful rendition of peak heights is of interest.

In this case a useful measure of precision is the *peak-height test*. It can be shown that measured peak heights will deviate from true values by less than 0.5% if the ratio of the resolution of the spectrometer to the peak widths is 0.1 or less. For the UV peaks cited, the ratio is about $0.15/0.45 = 0.33$. The criterion is not met; it seems likely that precision will be poorer than $\pm 1\%$. Actually, it can be shown that the error will be about 3%.

The Slit Function. The width of slits plays a crucial role in determining resolution. Consider a representative monochromator that is illuminated with monochromatic light, whose entrance and exit slits are 1 mm wide, and whose optics permit the exit slit to be "filled" by the image of the entrance slit. Assume that monochromatic radiation of wavelength λ_0 is incident and the linear reciprocal dispersion D^{-1} (Table 9.1) of the instrument is 20 nm mm^{-1}. If the dispersing element is pivoted as in wavelength scanning, the image of the entrance slit will move along the focal plane and cross the exit slit opening at some time.

The dependence of intensity at the exit slit on wavelength setting may be seen from the successive images shown in Fig. 9.3a. The "front" of the image will appear at the edge of the exit slit when the dispersing element setting is $\lambda_0 - 20$ nm. In view of the dispersion given, this position should seem reasonable. At this setting, the intensity of light emerging must be essentially zero. If the setting is advanced to $\lambda_0 - 10$ nm, half the 1-mm-broad image should overlap the exit slit and 50% of its intensity be transmitted. At the setting λ_0, the exit slit aperture should be filled. The transmitted intensity is 100% of that available.

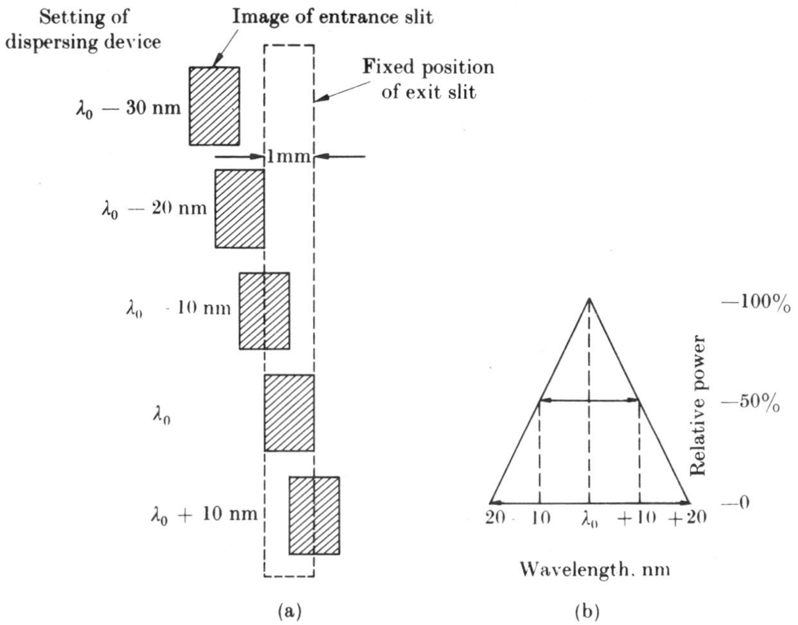

Fig. 9.3 Origin of triangular slit energy distribution. (a) Position of entrance slit image relative to exit slit during a wavelength scan. Monochromatic light of wavelength λ_0 is assumed incident. When the wavelength setting is λ_0, the exit slit is filled with light. (b) Intensity pattern at exit slit versus monochromator setting when monochromatic light is incident. This triangular pattern is also the intensity distribution at the exit slit as a function of wavelength when polychromatic light is present and the wavelength setting is λ_0.

When the dispersing element setting is increased to λ_0 + 10 nm, the image will have half departed and should again half fill the aperture, giving a relative intensity of 50%. The trailing edge will just disappear when the setting is charged to λ_0 + 20 nm. As graphed in Fig. 9.3b, a triangular plot of intensity versus wavelength setting is observed for monochromatic light on scanning.

Consider that the monochromator is next illuminated by a continuous source and that slits are still 1 mm wide. Now for *each frequency present* there should be a triangular distribution pattern like that of Fig. 9.3b. How wide a wavelength is passed at a given wavelength setting, for example λ_0? In accord with the earlier analysis, the exit slit should be fully illuminated only by the λ_0 wavelength image of the entrance slit. The images of wavelengths λ_0 − 10 nm and λ_0 + 10 nm should half illuminate it, and light from the λ_0 − 20 nm and λ_0 + 20 nm images should fall just short of emerging through the exit aperture. For continuous radiation a monochromator also gives a triangular intensity pattern like that in Fig. 9.3b, but the abscissa now represents the range of wavelengths passed at the setting λ_0. The pattern of Fig. 9.3b is usually called the *slit function*.

To define the width of wavelength band produced will require agreement on measurement at a particular intensity level. It is conventional to do so at half maximum intensity, sometimes called full width at half maximum (FWHM). This bandwidth is termed the *spectral slit width* or band pass. For example, a spectral slit width of 20 nm is found from the pattern of Fig. 9.3b. Spectral slit width S can be defined formally in terms of the physical width of the slits W and reciprocal linear dispersion D^{-1} of the device:

$$S = W \times D^{-1}. \tag{9.2}$$

The application of Eq. (9.2) to the monochromator just discussed gives a spectral slit width of 1 mm × 20 nm mm^{-1} = 20 nm, in agreement with the number read from Fig. 9.3. If only the angular dispersion $d\theta/d\lambda$ (rad nm^{-1}) is known, Eq. (9.2) can be rewritten as $S = W/[F_{\text{eff}}\, d\theta/(d\lambda)]$, where F_{eff} is the focal length of the focusing, and usually also the collimating, optic. Clearly, the longer the focal length of the optics is, the narrower S. For monochromators of high quality F_{eff} is usually at least 0.25 m. All else being favorable, increasing the focal length of the focusing optic is seen to be a good way to decrease the spectral slit width and increase resolution.

Limitations of the Triangular Slit Function. Equation (9.2) has two limitations worth noting. First, it is valid only under conditions such that slit width alone limits resolution. For example, diffraction effects, mismatch of slit curvature (see next section), and optical aberrations may also lead to widening of the spectral slit width S. For narrow slits, poor optics, or a monochromator of inherently low resolution, all effects may be the same order of magnitude, that is, no one will be limiting. Then the spectral slit width can be estimated by appropriate summing of the contributions or may be determined empirically. It is helpful to know that, for a grating monochromator, slit width is limiting if the instrument has a focal length of 0.5 m or greater.

Second, for narrow slits, the energy emerging from the slit of a monochromator is often best represented as a Gaussian curve. [A. Lempicki, H. Samuelson, and A. Brown, *J. Opt. Soc. Am.*, **51,** 35 (1961).] The triangular pattern of Fig. 9.3b is a good approximation but it truncates the tails of the distribution. Only occasionally does the triangular approximation lead to serious error as, for example, in measurement of an absorption that has a sharp edge, such as one in a semiconductor.

9.3 ENERGY THROUGHPUT

The amount of energy that enters and also leaves an optical module is called its *throughput*.* Few modules is an optical instrument will limit energy as much as a monochromator with its narrow entrance slit, and usually, multiple internal reflections from mirrors. Clearly, its throughput will need to be maximized if adequate energy, that is, a good S/N ratio, is to be maintained. Before looking at specific monochromators, the general aspects of transfer of optical energy need exploring.

*This property is proportional to the $f/$ number or speed of a monochromator; luminosity and *entendue* are synonyms.

Optical Coupling by Imaging. The transfer of optical energy between optical elements or modules is strongly facilitated by forming images. Recall that spatial integrity in an optical field can be preserved by this means, as in a camera or telescope. For an instrument, imaging is also important as a way of providing uniform intensity in a beam where needed. Thus, it is conventional to form an optical beam for an instrument by developing an image of a region of a source that is of uniform brightness. At points along the signal channel where uniform intensity is again needed, such as at a monochromator entrance slit, the image is formed again.

Thus, sources are needed that have a small, uniformly bright region. Two examples are the tungsten–halogen and gaseous-discharge lamps. In the first the region is provided by the incandescent filament coil and in the second, by the center of the gaseous discharge.*

Where will imaging be important? Certainly, internal imaging is essential in a monochromator, as discussed in the last section. Another instance follows.

Example 9.4. For quantitative work the *source must be coupled to a monochromator* in such a way that radiation illuminates the slit uniformly and just fills the acceptance *angle* of the monochromator. How may this be done effectively?

A good strategy is shown in Fig. 9.4. To secure uniform illumination, light emitted from a region of a source that is roughly rectangular and of high uniform intensity is collected over as large a solid angle as feasible and focused on the entrance slit. The beam from the source is reduced by a mask with a square aperture until the solid angle is sufficient to ensure that light entering the slit will just fill the acceptance angle of the lens or mirror

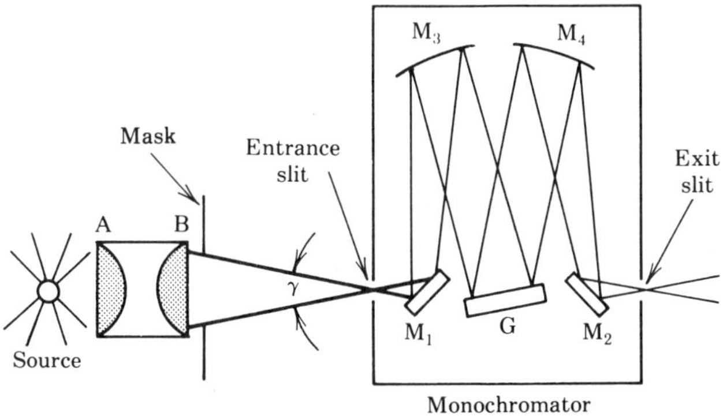

Fig. 9.4 A lens arrangement for coupling of source and monochromator. Lens A collects light and lens B forms at the entrance slit of the monochromator an image of the bright central region of the source. If desired, a transparent, heat-absorbing element may be added between A and B. The monochromator is a Czerny–Turner type employing off-axis spherical mirrors M_3 and M_4.

*A further reason for the approach arises from the optical principle that the energy flux of a beam may approach that of the source from which it originates, but cannot exceed it even if the beam is focused down. Since most slit apertures will be no larger than about 1 × 10 mm, the bright region of a source should also be small.

that illuminates the dispersing element. In Fig. 9.4 that element is a diffraction grating; it should be fully illuminated. Otherwise resolution and efficiency will be poor.

An especially effective arrangement for transfer of energy from a source is illustrated in Fig. 9.5. In this design good coupling is achieved with an off-axis elliptical mirror. Some other coupling systems based on lenses are shown in Fig. 13.11.

Within monochromators necessary collimating and focusing is usually performed by front-surface mirrors. By eliminating lenses, especially in the UV–VIS, where refractive index changes substantially with wavelength, chromatic aberration as well as other errors are minimized. Mirrors are also preferred because reflection is often more constant with wavelength than is transmission. Mirror and reflection grating surfaces are commonly of aluminum. Even in the vacuum UV (down to 100 nm) surfaces reflect satisfactorily if covered with a thin layer of dielectric like MgF_2 (see p. 320). Other metallic films (e.g., gold), are sometimes used in the IR to enhance reflectance.

It is helpful to estimate throughput at least semiquantitatively as different monochromators are examined. The following expression is useful:

$$P_0/N = D^{-1} l R_1 R_2 R_3 \cdots \zeta/4(f\text{-number})^2, \tag{9.3}$$

where P_0 is the radiant power leaving when the exit slit is set to give a 1-nm spectral slit width. Other new factors are: N, the spectral radiance (units: W mm^{-2} sr^{-1} nm^{-1}) of the source; l, the slit height; several R's, the reflectivities of the mirror surfaces; and ζ, the spectral efficiency of the grating. All these quantities are wavelength-dependent. An acceptable throughput is 0.10–0.20 or greater.

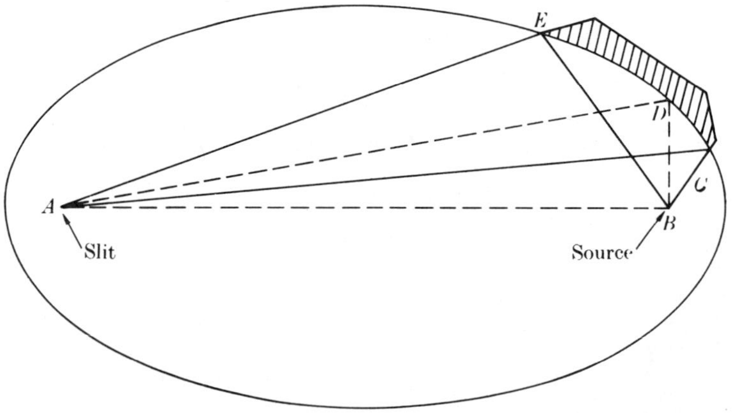

Fig. 9.5 Ellipsoidal mirror for gathering light from a source (not shown) at B, the near focus of the ellipse. Light is collected over a large solid angle (represented by plane angle *CBE*) and focused into a small solid angle (plane angle *BAE*) on a monochromator entrance slit (not shown) that is located at the distant focus of the ellipse at A.

The Slits and Throughput. In discussing the intimate connection between slits and resolution in the last section, the effect of slit width on beam energy was largely neglected. As a general rule, such considerations become of interest when a source of limited energy is involved, or when energy is unavoidably reduced in the process of securing high resolution.

When a monochromator is passing light from a monochromatic source, its energy throughput is measured by the product Wlp, where W and l are the width and height of the slit and p is the solid angle subtended at the slit by a preceding focusing element.*

When a continuous source is used, throughput still varies linearly with slit height and solid angle subtended, but now depends on the square of slit width. The reason is that both the energy flux and the spectral band passed by a monochromator increase as a slit is widened. As a result, the appropriate equation is now

$$\text{throughput} = W^2lp. \tag{9.4}$$

It is worth noting that the solid angle subtended will vary inversely with focal length. A short focal length will increase energy throughput. As will be noted, it will also diminish resolution and both cannot be maximized simultaneously.

Efficiency. The product *throughput* × *resolution* is termed the efficiency of an optical system. In many ways the efficiency is a useful criterion for comparison of different systems. The larger the product, the greater the power of the instrument in limiting cases when either resolution or energy throughput must be traded off to secure a desired level of the other. For the main types of sophisticated wavelength sorting devices, efficiency increases in the following order: prism monochromator < grating monochromator < interferometric sorter. It may be noted that cost also increases in this order while simplicity and ruggedness fall. The first two types of wavelength sorter are discussed in successive sections of this chapter; the last type is taken up in Section 17.8.

Slit Adjustments. To achieve fine performance by a monochromator, slits must be constructed with knife-edge jaws so that the aperture has smooth edges, can be closed to very narrow widths, and eliminates reflection from its edges. Variable slits are provided in a versatile monochromator. For example, widths may be continuously adjustable from 5 to 2000 μm. As noted earlier, a narrow slit will optimize resolution, while a wide one will optimize energy. In practice one must be traded off for the other.

For constant resolution, a spectrum should be scanned at constant spectral slit width regardless of the energy throughput. For example, such a constant-resolution

*A more general definition applicable to any optical system is that energy throughput is the smallest product of the following type encountered in the system: (area of a field stop or opening) × (solid angle subtended by the system at the field stop). Lenses, mirrors, prisms, or gratings used will, of course, affect throughput by their degree of transmission and reflection, but they are not ordinarily limiting.

scan with a prism monochromator would call for slit widths to be much narrower in the visible range than in the UV region to compensate for the variable dispersion of the prism.

On the other hand, when the available energy is limited, as in the IR region, securing an adequate S/N ratio at the detector calls for monochromator slits to be progressively opened in scanning toward longer wavelengths. The sources available simply do not yield enough intensity at long wavelengths. Some resolution is, of course, sacrificed in the process.

An alternative way to increase throughput is to increase the height of the slits. On first examination this choice seems preferable because it appears that resolution would not decrease; however, an unfortunate complication frustrates the possibility. The optical system of a monochromator produces an inherent curvature in slit image that becomes evident when slit heights exceed about 3 mm.*

As a result, a short slit height (3 mm or less) is preferred, except at very low energy. Where slit curvature must be allowed for, an additive correction to the spectral slit width can be estimated by running a calibration.

9.4 DISPERSION BY A PRISM; SOME PRISM MONOCHROMATORS

Dispersion of radiation in space involves angularly separating the different wavelengths present in a wavefront. While dispersion is commonly accomplished today by diffraction by a grating, there remain many instances in which refraction by a prism is a useful way to separate wavelengths.

A representative prism arrangement is depicted in Fig. 9.6. The two wavelengths present in the incident beam are refracted through different angles and are separated by the angular difference $\theta_2 - \theta_1$ on leaving the prism. As we would expect, the normal increase in refractive index n as wavelength *decreases* causes wavelengths in the blue range to be refracted most. The angular dispersion of the prism is defined as $d\theta/d\lambda$, where $d\theta$ is the angular separation of wavelengths differing by $d\lambda$ in magnitude. Actually, it is useful to resolve $d\theta/d\lambda$ into two factors as follows: $d\theta/d\lambda = d\theta/dn \cdot dn/d\lambda$, where $d\theta/dn$ is a geometric factor that includes the influence of prism shape and the angle of incidence and $dn/d\lambda$ is a specific factor characteristic of the prism material.

Geometric Factor. While dispersion will vary greatly with the angle of incidence, most symmetrical prisms are used at the *angle of minimum deviation D*. Under these conditions a ray passing through the prism traverses it parallel to its base and the angular deviation of

*In the Czerny–Turner or Ebert mounting (see Figs. 9.16 and 9.17), for example, the radius of curvature of the entrance slit image formed at the exit slit is half the straight-line distance between entrance and exit slits (usually no more than 10–15 cm). In precision instruments the loss of resolution on lengthening slits is sometimes diminished by giving the entrance slit a curvature that will oppose *the one produced* optically. Curvature is also somewhat dependent on wavelength and, even with compensating curvature, image superposition is completely possible at only one wavelength.

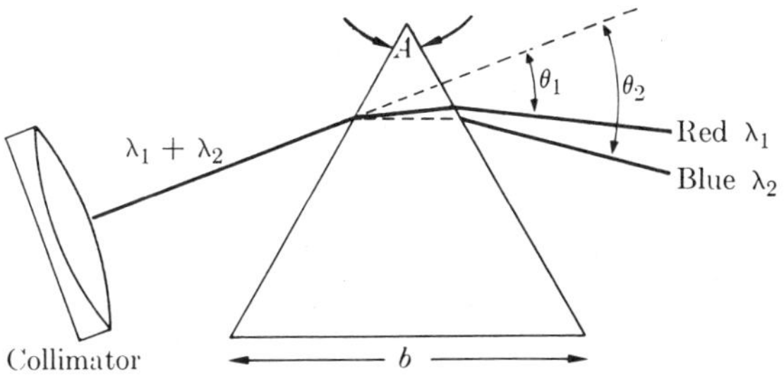

Fig. 9.6 Dispersion of light by a prism of apical angle A and base of length b. Blue wavelengths undergo a greater angular dispersion.

the emerging ray is a minimum. For example, this condition holds for the blue ray in Fig. 9.6. For a symmetrical prism, the angle D is defined by the expression (obtainable by use of Snell's law)

$$n = \frac{\sin (D + A)/2}{\sin A/2}, \tag{9.5}$$

where A is the apical prism angle and n is the refractive index of the prism material. For minimum deviation it can be shown that the geometrical factor in dispersion is given by

$$d\theta/dn = 2 \sin (A/2)/[1 - n^2 \sin^2 (A/2)]^{1/2}. \tag{9.6}$$

The angle of minimum deviation at a given wavelength can easily be found by (a) observing the emergence of the beam from a prism when a collimated monochromatic beam of light is incident, (b) rotating the prism slowly until a slight further rotation no longer produces a change in the angle of emergence, and (c) noting the angle of incidence, which is that of minimum deviation.

Using a prism at minimum deviation secures two important advantages. Optimum resolution is obtained because astigmatism is eliminated. In addition, internal reflection (at the far prism face) is minimized, leading to good transmitted intensity and minimum stray light generation by internal reflection.

What apical angle should be selected for a symmetrical prism? This angle influences dispersion and resolution. From Eq. (9.6) it can be shown that resolution increases as angle A increases. It may be appreciated intuitively that resolution will also improve linearly with the length of prism base b.

Specific Factor. How does dispersion depend on $dn/d\lambda$? In Fig. 7.17 refractive index was graphed against wavelength in the UV–VIS–IR range for several prism materials. On inspection of these curves it is apparent that the dispersion of any prism must vary dramatically with frequency since it depends on the slope $dn/d\lambda$. Near the center of each curve $dn/d\lambda \approx 0$, but at either end of a range the slope has a large value. This type of variation in dispersion with wavelength is contrasted in Fig. 9.7 with the nearly constant dispersion obtained from a diffraction grating.

In specifying dispersion, linear measures are also used as indicated in Table

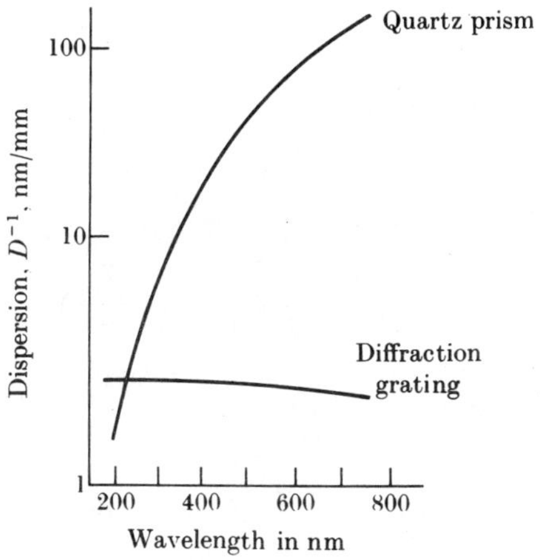

Fig. 9.7 A comparison of the linear reciprocal dispersion D^{-1} provided by a quartz prism and a representative grating over the UV–VIS range.

9.1. To relate linear dispersion D to the angular quantity $d\theta/d\lambda$, the following expression may generally be used:

$$D = F_{\text{eff}}\, d\theta/d\lambda \tag{9.7}$$

where F_{eff} is the focal length of the optical system. The inverse of Eq. (9.5), the reciprocal linear dispersion D^{-1}, was introduced in the last section and is also in common use. For example, a spectrograph with a linear dispersion of 0.0075 mm nm^{-1} at 400 nm has a linear reciprocal dispersion of 133 nm mm^{-1}.

Some Prism Monochromators. One of the most widely used prism monochromator designs is the *Littrow mounting* shown in Fig. 9.8. The characteristic feature is that the angles of incidence and emergence from the prism are nearly equal. It

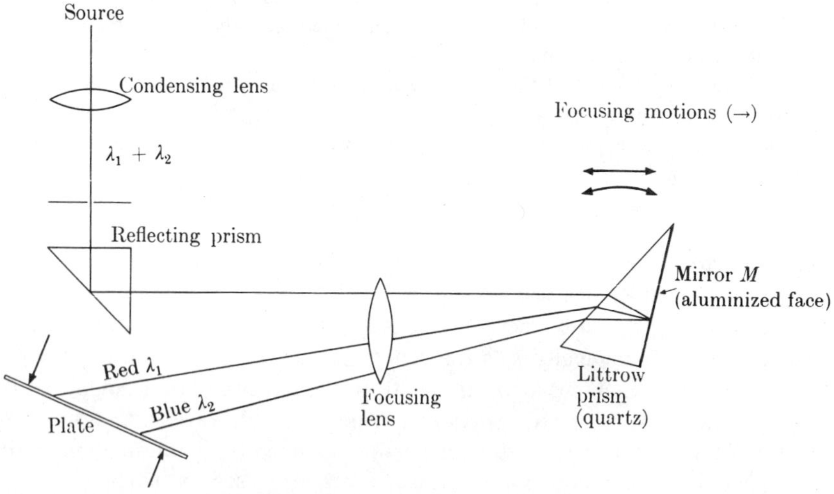

Fig. 9.8 Littrow prism and mounting.

should be noted that mirror M need not be integral to the prism, as shown, but may be separate. The particular advantages of the Littrow mount are: (a) a high degree of dispersion in a compact arrangement (each ray traverses the prism twice); (b) a single lens or mirror serves as both collimator and focusing device; and (c) avoidance of double refraction if an anisotropic material like quartz is used (light rays traverse nearly the same path in both directions). With this design, selection of wavelength is accomplished by turning the prism (or mirror M alone, if separate) through an appropriate angle.

In most other mounts there is only a single pass through the prism. As a result, to avoid birefringence either prisms must be made from an optically isotropic medium like fused silica or special pains must be taken in design.

A *constant-deviation* prism makes possible a monochromator with the advantage that the wavelength isolated by the exit slit can be easily identified from the setting of the prism table. For example, for the Pellin–Broca prism shown in Fig. 9.9, a 90° deviation occurs between entrance and exit beams. When it is used in a monochromator, the wavelength centered at the exit slit is the one that has been deviated exactly through 90°. An additional advantage is the preservation of beam alignment in an instrument for the isolated wavelength.

A *Cornu* prism is a symmetrical prism whose apical angle is 60°. It might be used in either a Littrow mounting with separate mirror or in the Wadsworth mounting to be considered below. If crystalline quartz is used, two 30° prisms, one of right-handed and the other of left-handed quartz, are cemented back to back along their long faces. The two pieces compensate each other and avoid the birefringence that would otherwise cause two images of the slit to appear for each frequency.*

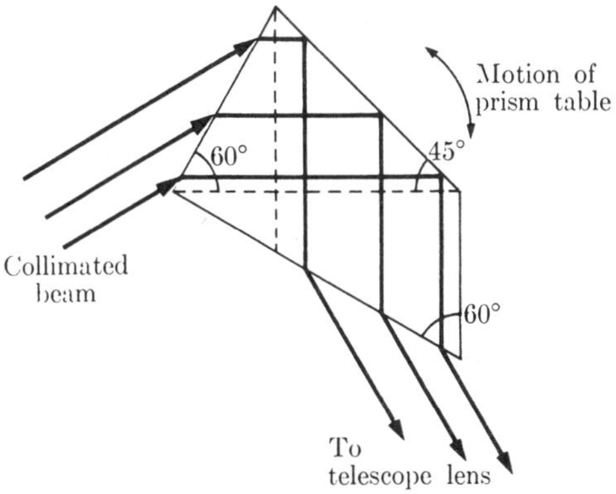

Fig. 9.9 Constant-deviation glass prism and mounting. This one-piece Pellin–Broca prism is equivalent to two 30° prisms cemented to a 45° totally reflecting prism. When a polychromatic beam enters at the angle of minimum deviation, only one of its wavelengths is incident on the back face at 45°. That wavelength emerges deflected through just 90°. By turning the prism table, successive wavelengths are brought into 90° deviation.

*Since rays traversing the prism at or near minimum deviation make at most small angles to the optic axis, there is no appreciable birefringence as a result of the anisotropy (Section 10.13) of quartz. The optical rotatory power (Section 7.18) would lead to birefringence in this case if there were no compensation.

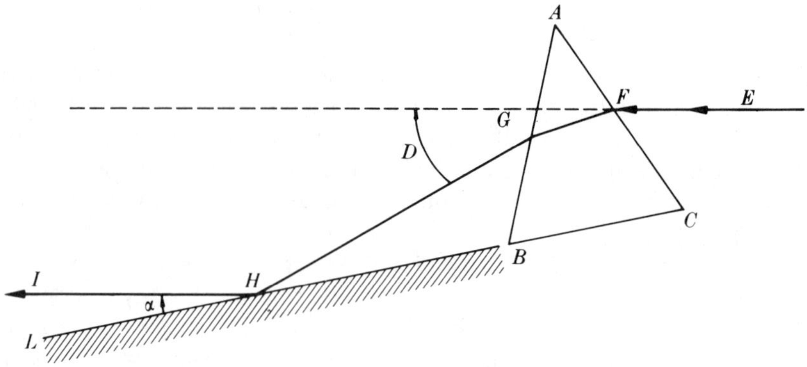

Fig. 9.10 Wadsworth prism mount. Entrance ray EF and exit ray HI are parallel for the wavelength that traverses the prism at minimum deviation. ABC is a symmetrical prism. Front-surface mirror LB is aligned with the prism so that LBC is a straight line.

Wadsworth Mounting. This design, illustrated in Fig. 9.10, operates at the angle of minimum deviation regardless of wavelength. A further advantage is that the ray leaves mirror *LB* (as ray HI) parallel to the entering ray *EF*. The deviation D is canceled by a reflection from the plane mirror at angle α, where $D = 2\alpha$. Further, by use of Eq. (9.6), the angle of reflection α for wavelength λ can be shown to be $\sin^{-1}[(n_\lambda/2) - 30]$ where a Cornu prism is used. Wavelengths are scanned by pivoting the entire mounting rigidly about *H*.

9.5 DISPERSION BY A DIFFRACTION GRATING

Where the ultimate in a dispersing device is needed, a diffraction grating is employed. Two outstanding advantages may be cited:

1. resolution and dispersion are very high for a long, fine grating, and
2. with proper arrangement of optics, a dispersion is obtained that is almost constant with wavelength.

The second factor makes it easier to establish the wavelength of radiation that appears at the output of a monochromator. In addition, it makes wavelength scanning simpler. For these reasons, today gratings are nearly always preferred as dispersing devices. Their main disadvantages relative to prisms are that they are less rugged, generate slightly more scattered light, particularly at shorter wavelengths, and are more expensive.

Gratings. A grating like that sketched in Fig. 9.11 may be designed either to reflect or to transmit light. Dispersion results from a complex process of diffraction and subsequent interference. So far as radiation is concerned, the rulings provide a series of slits (for transmission) or narrow mirrors (for reflection) uniformly placed along a smooth surface.

Fig. 9.11 Enlarged cross section of a representative echelette reflection grating. Since all its facets lie at the same angle, the grating is said to be blazed.

The Grating Equation. For an understanding of the fundamentals of grating behavior, consider first the simplest case, that of illumination of a *transmission grating* by parallel monochromatic radiation perpendicularly incident on the grating. This situation is shown in Fig. 9.12. As was done in Chapter 7, Huygens constructions will be used to show the effect of interference among diffracted rays. After transmission, interference among the wavelets from the slits gives rise to a series of very sharp wavefronts, each at a different angle relative to the grating.*

Some radiation passes straight through to give a parallel beam along the grating normal. If several wavelengths had been incident, this *zero-order* beam would contain all of them since in this direction all wavefronts constructively interfere. A second set of parallel beams, resulting in a plane wavefront, appears at angle of diffraction β. It is also formed as a result of constructive interference between wavelets originating at adjacent slits. For example, those from slit A travel just one wavelength further than those from slit B. This wavefront is the result of *first-order* diffraction. Similarly, another wavefront will exist (at an angle to the normal

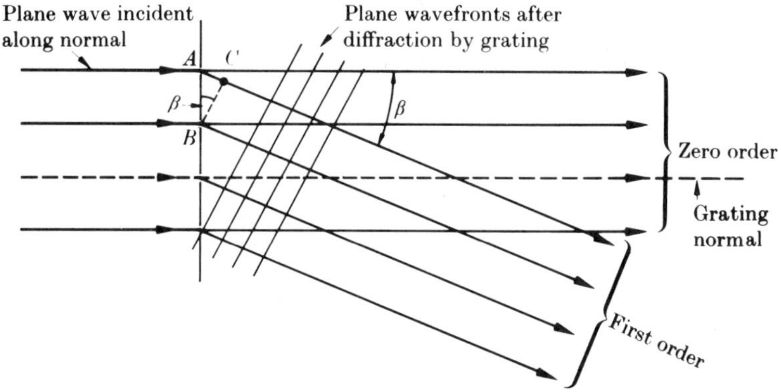

Fig. 9.12 Diffraction of monochromatic radiation by a transmission grating. A plane wavefront is incident from the left along the grating normal. Zero-order transmission results when transmitted light forms a new wavefront along the grating normal. A first-order plane wave appears at angle β to the normal. At this angle light from slit A has traveled one wavelength farther than that from slit B. A second-order plane wave appears (not shown) at a larger angle and still higher orders at still greater angles. Negative orders also exist on the opposite side of the normal.

*This pattern contrasts with the pattern from a single slit illustrated in Fig. 9.2. In that case, some transmitted radiation appeared in secondary or side images of the silt. With the many slits of a diffraction grating, the secondary maxima are weakened so greatly by multiple interference that they are seldom detected.

larger than β) in which light from slit A has traveled precisely two wavelengths farther than that from slit B. It will result from *second-order* diffraction. The process of finding other wavefronts can be continued though intensities usually fall rapidly with increasing order.

According to the geometrical representation of Fig. 9.12, the extra distance light from slit A travels for first-order diffraction is just AC, Since angle ABC, is also β, $AC/AB = \sin\beta$. But AB is the grating constant d so that $AC = \lambda = d\sin\beta$. This expression can be generalized. First, since the extra distance from slit A to any of the successive wavefronts that appear at even larger angles is just $m\lambda$ where $m = 1, 2, 3, \ldots$, a more general expression is

$$m\lambda = d\sin\beta. \tag{9.8}$$

The integer m may be positive or negative and is designated the *order*. Second, allowance must also be made for instances of oblique illumination of a grating. If light is incident at angle α, Eq. (9.8) becomes the general *grating law*

$$m\lambda = d(\sin\alpha \pm \sin\beta). \tag{9.9}$$

This expression holds for both transmission and reflection gratings. Note the possibility of a negative sign. It arises since light that is incident may be diffracted on either side of the normal. By convention, if angles α and β are on opposite sides of the normal, the negative sign applies in the equation; if both are on the same side, the positive sign is used.

Example 9.5. What wavelength of radiation is diffracted in the first order from a grating with 1000 grooves mm^{-1} when $\alpha = 30°$ (on one side of the grating normal) and $\beta = 5°$ (on the other side of the normal)?

The answer can be obtained by use of Eq. (9.9). The grating constant $d = 1$ mm/1000 $= 10^{-3}$ mm. Substituting in Eq. (9.9) gives $1\lambda = 10^{-3}$ mm$(\sin 30° - \sin 5°)$. Thus,

$$\lambda = 10^{-3}(0.500 - 0.087) = 413 \times 10^{-9}\text{ m} = 413\text{ nm}.$$

Example 9.6. If the sum $\alpha + \beta$ is fixed and both angles are on the same side of the normal, as in many plane grating monochromators, what first-order wavelength will appear for $\alpha = 40°$, $\beta = 5°$? Assume $d = 10^{-3}$ mm.

Equation (9.9) gives $\lambda = 10^{-3}(0.643 + 0.087)$ mm $= 730$ nm.

Example 9.7. What is the longest wavelength available with the monochromator just described?

From Eq. (9.9) and Fig. 9.13, it is clear that the longest wavelengths a grating will yield will be those in the first order. If $d = 10^{-3}$ mm, the long wavelength limit would be attained if $\sin\alpha = \sin\beta = 1$. Equation (9.9) gives a limiting wavelength $\lambda = 10^{-3}$ mm $\times 2 =$ 2000 nm.

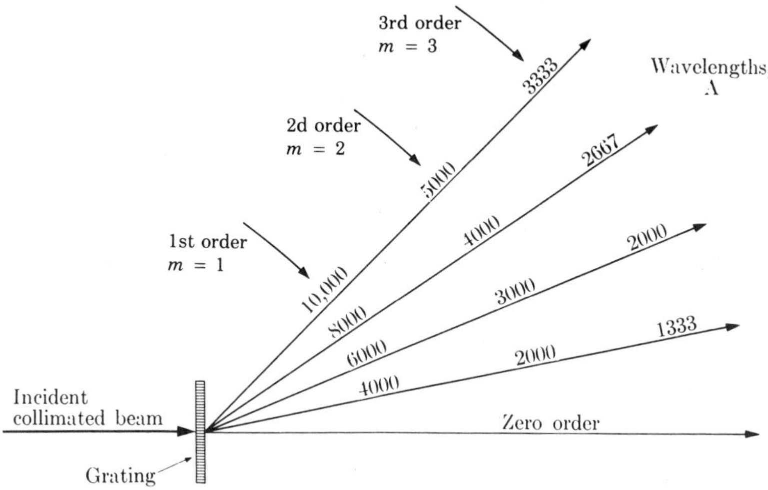

Fig. 9.13 Overlapping orders of spectra from transmission grating. A similar set of orders is formed at equal angles below the normal.

If parallel rays of polychromatic light are incident on a grating, each wavelength will, of course, appear at a different value of β. Actually, submultiples of this wavelength, each corresponding to an order higher than the first will also be found. The superimposed orders are illustrated in Fig. 9.13.

Although the overlap of orders is potentially troublesome, it can be avoided in straightforward fashion. Commonly, cutoff filters are used to exclude unwanted spectral ranges. Sometimes light is first dispersed by a prism so that no more than a narrow spectral region falls on a grating. The movement of the prism can be synchronized with the rotation of the grating so that both are responding in the same range. Some natural factors operate to eliminate spectral regions or to "sort orders" as well. For example, a detector insensitive to UV and still shorter wavelengths may be used if only the first-order visible region is of interest (see Fig. 9.13). In any case, the energy that has appeared in lower orders is unavailable for higher orders.

Reflection Gratings. Today virtually all gratings are of the reflection type. Groove densities range from 20 mm^{-1} for the far IR to about 3600 mm^{-1} for the VIS and UV ranges. Up to 6000 grooves mm^{-1} can be prepared! Replicas of planar ruled master gratings, interferometrically prepared (holographic) planar and concave gratings, and replicas of the latter types are in common use.

Example 9.8. How can a diffraction grating with an appropriate groove spacing be used efficiently to cover wavelengths from 200 to 800 nm? Assume cutoff filters may be used to block unwanted orders.

Recall that diffracted intensities are highest in the first and second orders. Further, since the longest wavelengths will be available in the first order, the grating can be used in the first order for the 400–800 nm range. For this range a filter with a cutoff below 400 nm can be inserted in the incoming beam to block UV and pass wavelengths above 400 nm. For the 200–400 range the grating can be used in the second order by use of the same range for β. In this case a new sharp cutoff filter should be inserted in the incident beam to block the visible region and pass only UV.

How serious a problem will the third-order 133.3–266.6 nm spectrum be? The third and higher order can be mostly neglected since their shorter wavelengths will be blocked by air and all their light will be comparatively low in intensity.

Ruled Gratings. Master gratings are cut by a properly shaped diamond tool directly into an aluminum film that has been deposited on polished glass (smoothed to $\lambda/10$). Interferometric controls and elaborate apparatus have been devised to ensure the necessary evenness of spacing of grooves over the usual ruling distances of from 10 to 25 cm.

Because masters take long to make, replica gratings are commonly used in their place. The process of replication involves (a) applying a film of "parting agent" to the master grating, (b) vacuum-depositing a layer of aluminum, and (c) attaching a glass or quartz base to the aluminum layer with epoxy cement. After an appropriate time interval the replica is separated.

Holographic Gratings. A polished glass block (smoothed to $\lambda/10$) is coated with photoresist. A regular array of interference fringes is formed on the light-sensitive coating by illuminating it with intersecting, coherent, collimated beams of monochromatic light (usually from an argon ion laser). The wavelength of light and the angle between the beams determines the fringe spacing. Development and processing produces (usually) grooves in the photoresist. Finally, a thin layer of aluminum or other reflecting coating is deposited.

Coatings. If a grating is to be used in the vacuum or far UV it is customary to add a layer of MgF_2 of appropriate thickness (about $\lambda/4$, where $\lambda = 120$ nm) to boost the reflectance in this region to around 70%. The refractive index of aluminum falls to unity and below in this region, giving rise to low reflectance. Magnesium fluoride has a higher index. Often a uniform protective layer of SiO_2 is vacuum-deposited on gratings whether they are to be used in the VIS, UV, or far UV.

Blazed Gratings. Gratings with flat, smooth facets instead of somewhat irregular grooves are said to be blazed and are highly efficient in diffraction at angles close to those for which specular reflection occurs. In Fig. 9.14 this case is illustrated for the incident and diffracted rays. The figure also shows the geometry of a blazed reflection grating of the *echelette* type. The *blaze angle*, that is, angle θ_0 of the broad facets relative to the grating normal is shown in relation to angles of incidence α and diffraction β. Note also the narrow steep faces that are seldom used. If ϕ is the angle of incidence and reflection, $\phi = \alpha - \theta_0 = \beta + \theta_0$. Rearranging gives $2\theta_0 = \alpha - \beta$, or since β may be on either side of the normal,

$$\theta_0 = (\alpha \pm \beta)/2.$$

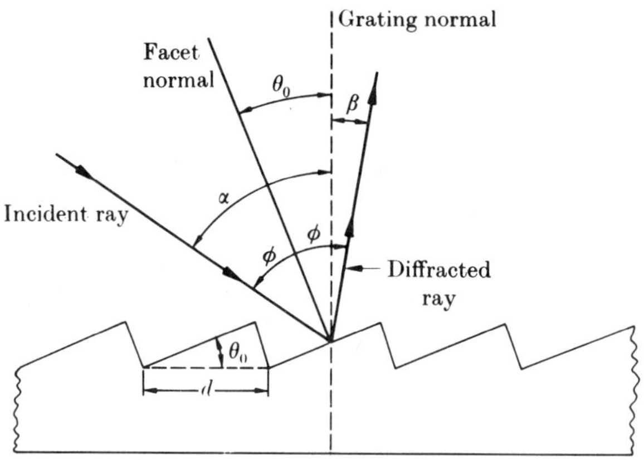

Fig. 9.14 Cross section of a blazed echelette reflection grating. Here α is the angle of incidence, β that of diffraction, and ϕ that of reflection. The blaze angle is θ_0.

In ruled gratings blazing is introduced by the shape of the diamond cutting tool. With holographic gratings an Ar^+ ion gun is used to cut away resist to shape grooves into flat blazed facets. Again the plus sign applies only when β is on the same side of the grating normal as α. The wavelength for which specular reflection and first-order diffraction coincide is called the *blaze wavelength* λ_B. With a blazed echelette grating nearly all the diffracted light appears on one side of the grating normal. As noted earlier light is concentrated in the first few orders. Further, diffracted light is centered about λ_B in the first order, $\lambda_B/2$ in the second, and λ_B/m in the mth order. Note that the wavelengths most favored in orders beyond the first follow from the grating law.

The *efficiency* of a reflection grating is one of its most important characteristics. It is defined as the percentage of monochromatic light diffracted in a given order relative to the percentage specularly reflected by a flat aluminized mirror. Clearly, a particular wavelength can be *efficiently* diffracted in only one order. For example, if 70% of the radiation of wavelength λ_B is diffracted in the first order, only 30% is left to be diffracted in other orders. The higher the efficiency, the more useful is the grating, since it can ensure a greater throughput and less scattered light in a monochromator. In the first order the efficiency of a ruled or holographic grating at the blaze wavelength may be 60–70%, but in the second order at $\lambda_B/2$ it is seldom more than 55%. In the first order a rule-of-thumb guide is that efficiency falls to half the value at λ_B when the wavelength becomes as short as $2/3\lambda_B$ and as long as $2\lambda_B$. For example, a grating blazed at 300 nm operates at greater than 30% efficiency in the first order from 200 to perhaps 600 nm. A similar rule relates to the blaze wavelength in other orders.

The stated or nominal blaze wavelength applies strictly to the Littrow configuration for which $\alpha = \beta$ and diffracted radiation returns along the path of incident radiation. The actual blaze wavelength is nearly the nominal value with most other mountings. The major exception arises when a grating is used at grazing incidence, as for spectral studies in the very far UV. In any case, the actual blaze value λ_A is related to the nominal value λ_B by the equation $\lambda_A = \lambda_B \cos \delta/2$, where δ is the difference between the absolute values of angle α and β.

Polarization. In general, it is only near the wavelength for which a ruled or holographic grating is blazed that it diffracts radiation without changing its polarization. At shorter λ's (low values of β) parallel light is most efficiently diffracted, at larger λ's (high angles), it is perpendicularly vibrating light. Polarization appears to be independent of groove spacing. At wavelengths greater than λ_B, however, the ratio of intensities of the two vibrational components, $I_\perp / I_\parallel$, is relatively flat. Corrections can be applied in this range with ease [p. 604 of 9, Chap. 7]. If polarized or partially polarized radiation is to be sent through a monochromator, it is often useful to pass it through a polarization scrambler to depolarize it prior to its entrance. Intensity effects are minimal for unpolarized light.

Stray Light. Radiation that a grating scatters for any reason may become part of the stray light that appears at the monochromator exit slit. Holographic gratings, with their greater uniformity of groove spacing usually have stray light levels about two orders of magnitude smaller than those of comparable ruled gratings.

Echelle Gratings. With greater precision in ruling, it became possible in the 1960s to make a new kind of grating called an *echelle*. While it had many fewer grooves ($8–150\ \text{mm}^{-1}$), even the steep faces were flat and and broad and could be used for diffraction.*

One is pictured in cross section in Fig. 9.15. As shown, diffraction is from the risers. Thus, echelles are effectively blazed at steep angles (about 63°). From Fig. 9.15 it is apparent that rays travel nearly parallel to the broader grating facets. In this *Littrow mode*, high efficiency is possible because angles of incidence and diffraction are those of specular reflection. Further, as will become evident from Eq. (9.12), if angles α and β approach 90° and are on the same side of the grating normal, resolution is maximized. Typically, echelle gratings are utilized in high

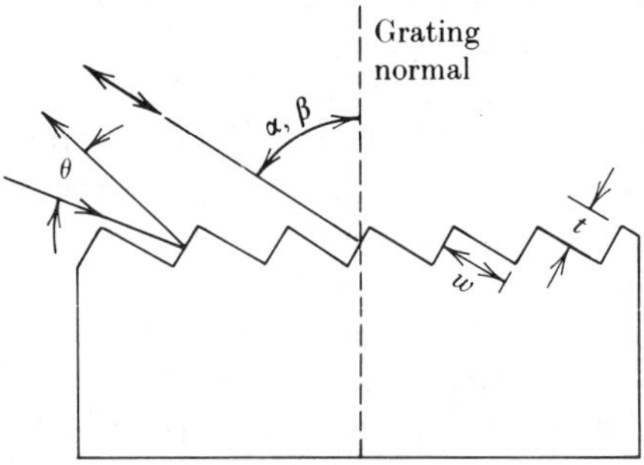

Fig. 9.15 An echelle, which is essentially coarsely ruled ($d_{\text{echelette}} \ll d_{\text{echelle}}$). The steep grating faces of width t are used in the Littrow mode. As a result, diffraction is based on highly efficient specular reflection. By using only the wavelength range in each order for which specular reflection obtains, high efficiency dispersion and resolution are secured.

*Risers in echelles are from 10 to 100 wavelengths wide in contrast to an average of 0.5–5 wavelengths in echellette gratings.

orders, 30 to more than 100, to take advantage of the high dispersion and high resolution provided.

For an echelle the distribution of energy into different spectral orders can be determined from what are called scalar theories of electromagnetic radiation. It can be shown that the wavelength at which efficiency peaks in order m is given by

$$m\lambda/d = 2 \sin \beta. \tag{9.10}$$

It is also true that the efficiency peaks in each order at the center of its free spectral range, a term which will be defined below. Since β is about 63°, the orders in which UV and VIS radiation appear are very large. For example, at about 200 nm, if $d = 0.0125$ mm, the order for peak efficiency is 112, while for 600 nm, the order is 37.

Since an echelle spectrum will contain a great many orders, special measures must be taken to ensure effective *order sorting*. Usually a small prism is mounted to provide cross dispersion, that is, dispersion by the prism at *right angles* to that of the echelle. Perhaps surprisingly, the spectrum appears in approximately parallel short segments. Each segment is a different order. For example, the highest order may be at the top and each successive lower row may represent one order lower. Since the output of this type of grating is a two-dimensional spectrum, echelle gratings are ordinarily incorporated in polychromator. They will be further discussed in Section 9.9.

Concave Gratings. It is also possible to rule the surface of a concave mirror. Indeed, this type of grating was one of the earliest developed, and until the 1950s was in common use. By virtue of the curvature of a concave grating, the ability to focus is combined with that of dispersion. The chief advantage of a concave grating is the high throughput it offers in a monochromator by eliminating two monochromator mirrors. It is this gain that has led to wide use of concave gratings in the far UV (below 170 nm) where reflectance is smaller.

Equation (9.7) also describes its diffraction behavior. A careful mathematical analysis is required, however, to determine optimum angles and distances at which to place slits in order to develop an image of the entrance slit at the exit slit with minimum aberration [14]. The standard configuration is that denoted as a Rowland circle and will be discussed in the next section.

The general use of ruled concave gratings in spectrometric work proved a mixed blessing, however. They introduce both *astigmatism* and *coma*, aberrations characteristic of spherical mirrors. Since the first produces images that are elongated and reduced in intensity, it is of greater concern. For example, in atomic spectrometric studies this aberration distorts the relation between element concentration and line intensity. A further disadvantage of concave gratings is that no successful process of making replicas exists, and each concave grating must be a costly master.

With the advent of holographic processes, it became possible to prepare concave gratings more easily. With computer-aided design an interference pattern was developed for exposure of the photoresist that reduces astigmatism at several angles.

Grating Characteristics. The theoretical *resolution* R of a grating may be shown to be defined by the relation

$$R = mN, \tag{9.11}$$

where m is the order and N is the *total number* of rulings illuminated. If we substitute for m from Eq. (9.9) this expression becomes

$$R = W(\sin \alpha \pm \sin \beta)/\lambda, \tag{9.12}$$

where W is the width of ruled area ($W = Nd$) and λ is the vacuum wavelength of incident light.

Equations (9.11) and (9.12) are based on the *Rayleigh criterion* for resolution by a multislit system like a diffraction grating. It defines resolution at the diffraction limit, which represents the ideal upper bound to resolution. By the Rayleigh criterion two spectral lines are said to be resolved when the maximum of the interference pattern of the first is incident on the first minimum (e.g., R in Fig. 9.2) of the interference pattern of the other line. There is about a 20% valley (above baseline) between peaks of lines so resolved.

These equations present key ideas. Equation (9.12) reveals surprisingly that, given the angles of incidence and diffraction, the resolution at a particular wavelength depends ideally on only the total ruled width of a grating. The order of spectrum does not appear in the equation. Clearly, wide gratings are much to be desired in precision work. The other main idea is the statement of the dependence of R on the angular configuration in which a grating is used and wavelength. Large angles of α and β are advantageous if they can be accommodated in a monochromator. For example, the ability to use angles as great as 63° is one attraction of the echelle grating discussed earlier.

Example 9.9. What is the maximum resolution of a grating for a particular wavelength? From Eq. (9.12), R will be a maximum when $\sin \alpha + \sin \beta = 2$. Thus, the maximum resolution possible is $R_{max} = 2W/\lambda$.

Example 9.10. Assume that a grating designed for work in the IR is blazed at 2500 cm^{-1} (4 μm), has a first-order resolution of 20,000 and has a ruled length of 10 cm. Calculate the grating constant d and its blaze angle.

The number of rulings N required can be obtained from Eq. (9.11), $R = mN$. Since $R = 20{,}000$, $N = 20{,}000/1 = 20{,}000$. The density of grooves is 20,000/100 mm = 200 mm^{-1}; its reciprocal is the grating constant and $d = 5$ μm. To calculate the blaze angle θ_0, assume operation in the Littrow mode where $\alpha = \beta = \theta_0$. Finally, substitution in the grating law gives $1 \times 4 = 2 \times 5 \sin \beta$ and $\sin \beta = 0.4$. Thus, $\beta = 23.5°$. This is also the blaze angle.

Free Spectral Range. In each order a grating has a wavelength band called the free spectral range F that appears only in that order. In this range there is no overlap of wavelengths from adjacent orders. Its breadth can easily be established mathematically. Let the free spectral range in the mth order be from λ_1 to λ_2, where $\lambda_2 > \lambda_1$. The shorther wavelength λ_1 must be just the wavelength that appears in the $(m+1)$th order at the same angle as λ_2 in the mth order. From Eq. (9.9), $(m+1)\lambda_1 = m(\lambda_1 + \Delta\lambda)$, where $\lambda_2 - \lambda_1 = \Delta\lambda$. Thus $F = \Delta\lambda = \lambda_1/m$.

Example 9.11. Light of wavelengths 800–200 nm is incident on a diffraction grating. What is its free spectral range in the first and second orders?

In a given order the shorter wavelength will also occur in the next higher order. Recall from Fig. 9.13 that when a grating is set at an angle at which 800 nm light is diffracted in order 1, it will diffract 400 nm light in order 2, 267 nm light in order 3, and so forth. The free spectral range must be 800–400 nm in order 1, 400–267 nm in order 2, and 267–200 nm in order 3. The equation given above confirms the length of the range as $400/1 = 400$ nm in order 1, $267/2 = 133$ nm in order 2, and $200/3 = 67$ nm in order 3. Such a range will exist in each order.

The *dispersion* of the grating, and thus of a monochromator or of a dispersive spectrometer, can be deduced by differentiating the grating law, Eq. (9.9). The angle of incidence being held constant, one asks what difference in angle of diffraction $d\beta$ should be observed for wavelengths that differ by $d\lambda$. Differentiating Eq. (9.9) gives for the angular dispersion,

$$d\beta/d\lambda = m/d\cos\beta, \tag{9.13a}$$

or for the linear reciprocal dispersion,

$$D^{-1} = d\lambda/dx = d\cos\beta/F_{\text{eff}}m \tag{9.13b}$$

where F_{eff} is the effective focal length of the system.

Example 9.12. Explain the nearly constant dispersion of planar diffraction gratings that was shown in Fig. 9.7.

Equation (9.13b) provides the basis for an explanation. A range of fairly small angles of diffraction is used in scanning in the low orders in which most monochromators operate. Since the cosine function is nearly constant at low angles, $\cos\beta$ and thus D^{-1} will change little during a wavelength scan with an echelette grating. For instance, an increase in β from 15 to 30° permits a doubling of the observed wavelength since $m\lambda \propto \sin\beta$, while dispersion decreases only about 10%, since it involves $\cos\beta$, which changes only from 0.95 to 0.85 over this range of β.

Example 9.13. An echelle with the following specifications is to be incorporated into a spectrometer: ruled area of grating 10 cm, 80 rulings mm^1, and steep facets 63° relative to the grating normal. What will be the central wavelength in its 40th-, 80th-, and 100th-order spectra? What maximum theoretical resolution can be attained when $m = 100$?

Equation (9.9) indicates that when $\alpha = \beta = \theta_0 = 63°$, $m\lambda = (2/80)\sin\beta = (2/80)0.89 = 0.0222$ mm $= 2.22 \times 10^{-5}$ m. Central wavelengths in the different orders as obtained from Eq. (9.9) are, 560 nm for $m = 40$, 280 nm for $m = 80$, 222 nm for $m = 100$.

According to Eq. (9.12) resolution when $m = 100$ will be $R = 10(2 \sin\beta)/222$ nm $=$ 20 cm $\times$ 0.89/222 nm $= 8 \times 10^5$.

9.6 SOME GRATING MONOCHROMATORS

Most early gratings were used in spectrographs and, as noted above, were of the concave variety. Unfortunately, appropriate mountings for concave gratings are bulky and require intricate movements of slits and grating to maintain imaging as the wavelength is changed (Section 13.6). They are especially valuable in the far UV region as noted. Today plane gratings are the most widely employed and monochromators based on their use will be described first.

The *Littrow mount*, like its prism counterpart, requires a single element for focusing and collimating. As a result, angles of incidence and diffraction are on the same side of the grating normal and are nearly equal. In terms of Fig. 9.14, $\alpha = \beta = \theta_0$. The grating equation becomes $m\lambda = 2d \sin\beta$. The Littrow design has been widely used in IR monochromators. It ensures high grating efficiency, good spectral purity, small aberrations, compactness, and economy.

Ebert Mounting. This mounting was historically the first compact, simple arrangement for a plane grating system that was substantially free of aberrations. The optical features of the design are shown schematically in Fig. 9.16. Entrance and exit slits are on either side of the grating and a single concave spherical mirror is used as a collimating and focusing element. Rays enter from slit 1, which is in the focal plane of the mirror, and are collimated as they are reflected to the grating. The diffracted radiation goes to the bottom half of the mirror and is focused on slit 2. The slit image suffers no aberration because two reflections occur off-axis. Since the entering and diffracted beams use different portions of the mirror, no scattering into the optical path results from the mirror. A further advantage is that a wavelength is selected by simple pivoting of the grating about the monochromator axis (shown as a dashed line).

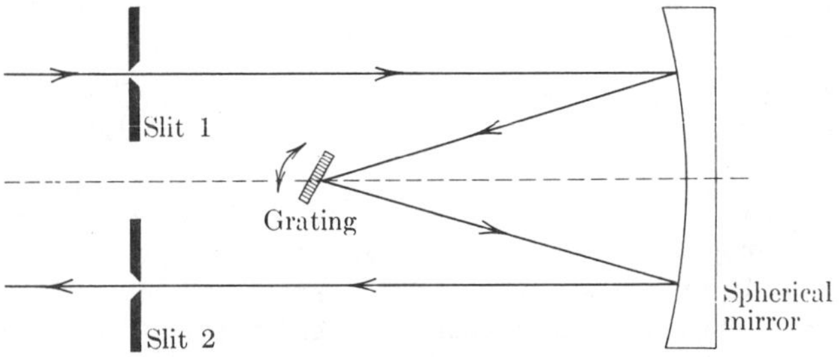

Fig. 9.16 The Ebert or Ebert–Fastie mounting.

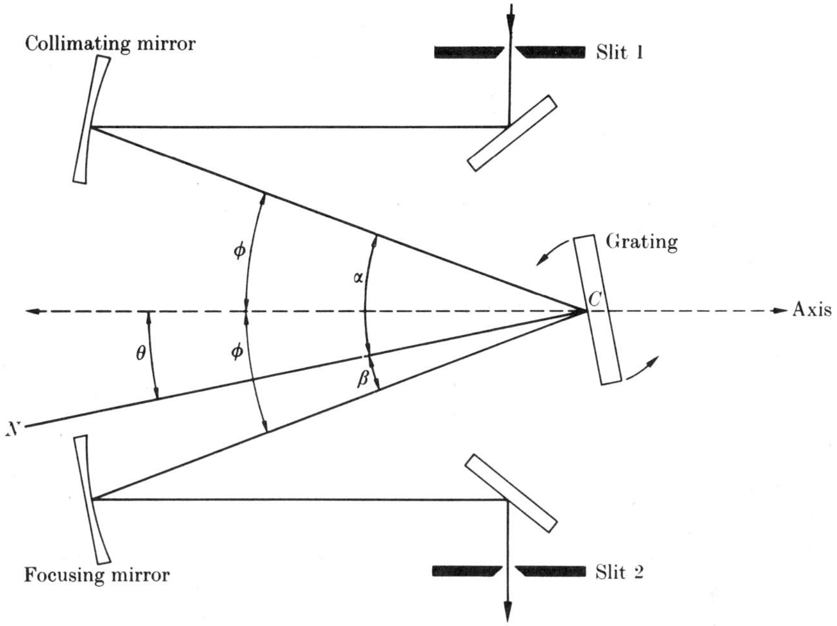

Fig. 9.17 Czerny–Turner grating monochromator. CN is the grating normal. Angle ϕ is determined by the manufacturer. To scan, the grating is pivoted about point C, increasing angle α from $\alpha = 0$ to $\alpha = 2\phi$.

Czerny–Turner Mounting. This widely used arrangement is shown in Fig. 9.17. It differs from the Ebert–Fastie mounting in that two smaller concave mirrors are used in place of a single large one. Parabolic mirrors may be used as well, though they introduce some astigmatism. All advantages of the Ebert mounting hold for this one also.

For the Czerny–Turner design Eq. (9.9) becomes

$$m\lambda = 2d \sin \theta \cos \phi \tag{9.14}$$

since the sum $\alpha + \beta = 2\phi$. Only grating angle θ is variable.

Rowland Circle. Several types of mountings have been developed for concave gratings. Nearly all are based on the Rowland circle, which is the circle drawn tangent to the center of the back of the grating and with a diameter equal to the radius of curvature of the grating. The entrance and exit slits of the monochromator or spectrometer are also located on this circle. The importance of the Rowland circle lies in the fact that the optical design worked out specifies precisely where spectrometer elements should be placed. While aberrations with this mounting cannot be completely avoided, the Paschen–Runge version of the mounting shown in Fig. 13.12 greatly minimizes them. Further, with the advent of holographic gratings it has become possible to produce gratings with minimum astigmatism. At least one design of compact UV–VIS monochromator using only a concave grating and two slits has become available.

Mountings for the Far UV. Monochromators for the vacuum and far UV, 180–50 nm, are nearly always based on diffraction gratings though prisms of lithium fluoride can be used down to about 120 nm. In this spectral region the chief design problems are adequate reflectance and compactness. Even with a MgF_2 overcoat, aluminum mirrors have only 70% reflectance at 120 nm and no more than about 15% reflectance at 50 nm. Thus, if three reflections are required (twice from mirrors and once from a grating) the beam intensity level falls to about 1–30% of the original figure. Compactness is important because of the need for evacuation of a monochromator to avoid absorption of radiation by air. While the Czerny–Turner mounting is often employed in the UV, its useful short wavelength limit is about 105 nm.

Both of the problems described above are surmounted by use of concave gratings. The *Seya-Namioka* and *Robin* mounts, usable to about 50 nm, and a *grazing incidence* mount,* for still shorter wavelengths, are available. Only the last is based on the Rowland circle and has a comparatively elaborate scanning–focusing mechanism. The other monochromators rely on particular angles and/or a special mechanism to maintain acceptable focusing. In the Seya–Namioka mounting [p. 460 of Ref. 14] a constant angle of about 70° is maintained between fixed entrance and exit slits and scanning is accomplished by rotating the grating about its vertical axis. This simple arrangement offers lower resolution than plane grating mounts, but features mechanical and economical advantages. By contrast, in the Robin monochromator [16], a constant, *small* angle (5–15°) is maintained between beams from the fixed entrance and exit slits. Each of these monochromators operates in the first order with a useful range of from about 50 to 300 nm at optimum resolution. With fixed slits, the volume to be evacuated can be kept small and source and detector can also be located in fixed positions. A further advantage is that monochromatic radiation always emerges from the monochromator at the same angle.

9.7 STRAY RADIATION

Radiation of wavelengths entirely different from those for which a monochromator is set will also be present in its output, though at low intensity. Assuming the compartment is lighttight, this unwanted or stray radiation originates from the fraction of the entering light that is lost from the optical beam by being

1. diffracted from a grating into unwanted orders and at unwanted angles;
2. reflected from surfaces of filters, lenses, or prisms present;

*For work below 50 nm (500 Å) only grazing incidence at present offers the possibility of sufficient reflectance, there being no suitable optical materials that do so. As will be recalled, nearly all materials show substantial reflectance at larges angles of incidence (80–85°).

3. scattered by imperfections, dirt, or scratches on optical components; or
4. diffracted from edges of slits.

Most but not all of this light will be absorbed or blocked from reaching the exit slit. The portion that emerges as stray light intensity S *adds* to the intensity T that has followed the optical path and the sum $S + T$ will be registered by the detector. Error will originate because the unwanted radiation will interact differently with a sample. The smaller T with respect to S, the more serious the resulting error. In a precision monochromator, stray light levels of 0.01–0.05%T are common in the UV; even in the IR the maximum level should be no more than 1%T. For a calculation of error attributable to stray light see Section 16.4. Here attention will be given to determining the intensity of stray light and ways to minimize it.

The level of stray radiation may be *estimated* quite simply. A straightforward method is to measure the apparent transmission of a substance at a wavelength where it is known to be opaque.

Example 9.14. A substance with a sharp transmission cutoff can be used to estimate stray light levels at frequencies at which it does not transmit. In the IR either quartz (opaque below 2300 cm^{-1}) or calcium fluoride (opaque below 900 cm^{-1}) can be employed. For a measurement, a slab of the substance is inserted in the optical path between monochromator and detector and the monochromator set to a wavenumber smaller than thc cutoff. Usually, a scale expansion is necessary to register stray radiation with moderate precision.

Alternatively, a polystyrene film can be used. If at least 0.07 mm thick, its absorbance is complete at characteristic IR wavelengths: 765 cm^{-1} or 705 cm^{-1}. A fast measurement is required to avoid having the film heat more than 1–10°C and become a blackbody source!

Example 9.15. A precise measurement of stray light can be made with aqueous solutions of KI, a standard reference material whose absorbance is certified at different concentrations at eight wavelengths from 240 to 275 nm. Since even concentrated solutions of KI have high transmission, how can additions of stray light be made reliably?

Two standard 10% transmittance filters can be inserted in the optical path as well, cutting the transmission to $0.1 \times 0.1 = 0.01$. Now stray light levels well below 1% can be measured with precision. The measured absorbance of the KI solution is compared with $0.01 \times T_{cert}$. The difference must result from stray light [ref. 28 at end of Ch. 16].

In spite of measures to reduce the intensity of stray radiation, its level may reach several percent of that of the optical beam, especially at monochromator settings far from those of *maximum emission of the optical source*. If wider slit settings are ordinarily employed at such wavelengths to compensate for inherently low intensities, more total radiation is admitted to the monochromator and more stray light generated.

Reducing Stray Light. Common measures to reduce stray light in a monochromator include design aspects such as the use of holographic gratings. These gratings have lower stray light levels because of the superior regularity of their grooves. Other steps routinely taken in monochromator design are to (a) paint all nonoptical surfaces a flat black; (b) insert baffles to obstruct radiation except of the optical beam; (c) use high-quality optical elements and slits that have knifesharp edges; (d) use windows in front of the slits to exclude dust and fumes.

Outside the monochromator other measures are valuable. Stray light is most severe in IR spectrometers. By simply placing the sample cell *before* the monochromator, light entering the monochromator is sufficiently limited to effect a great reduction in stray light (see Section 17.2). In addition, as discussed in the next example, a filter may be inserted in the signal channel.

Example 9.16. When would insertion of the filter in an optical beam help reduce stray light?

Dispersive spectrometers used in the *middle and far IR* (15–300 μm) are plagued by stray light of 1–5 μm from the region of peak emission of common IR sources. Usually two or three of the following types of filter are used: (a) a transmission filter which transmits only in the far IR, such as polyethylene with graphite dispersed in it or crystalline quartz, (b) a reflectance filter that will reflect the optical beam efficiently only over a narrow wavelength band that is of interest, or (c) scatter plates made by aluminizing a ground glass surface; they have a high reflectance only in middle and far IR. For instance, each alkali halide crystal can be used as a reflectance filter in its particular 10–15 μm-wide reststrahlen region ($NaCl$ crystals exhibit a high reflectance centered at 52 μm) in the middle or far IR. In this region their high reflectance (0.8 or higher) is the result of resonance between incident radiation and a natural lattice vibrational frequency.

Double monochromator. The most elegant and satisfactory approach to reducing stray radiation is to couple two monochromators in series, that is, use a *double monochromator*. With this arrangement stray radiation can be cut to less than 0.0001% of the intensity of the optical beam in the visible and UV and to about 0.1% or less in the IR at about 650 cm^{-1} (15 μm).

If desired, both dispersion and resolution can also be doubled, since dispersion of the two monochromators will add.

Alternatively, the greater resolution of the double monochromator can be "traded off" for a gain in energy throughput. A resolution comparable to that of a single monochromator can be secured by opening the slits to twice the width needed if only one monochromator were used. As a result, a quadrupling of energy at the exit slit is secured in the double unit, offset only slightly by reflectance losses in the second monochromator. An example of a double monochromator is shown in Fig. 9.18.

Somewhat the same effect as that of using two dispersing elements may be secured by simply arranging for a second pass through a single dispersing unit. Not all such arrangements reduce stray radiation or are compatible with double-beam design, however [19].

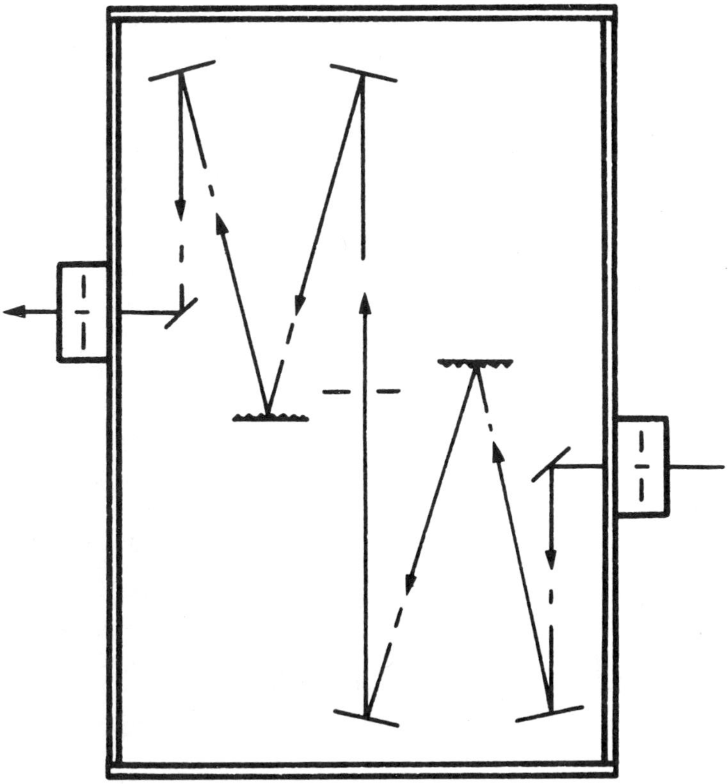

Fig. 9.18 A schematic drawing of a double monochromator. Note the internal slit separating the two halves. Baffling is not shown. Kinematic mounts allow a wide range of grating choices with respect to dispersion, blaze wavelength, and size. Slits may be fixed or continuously adjusted. (Courtesy of Spex Industries.)

9.8 SCANNING MONOCHROMATORS; CALIBRATION

While monochromators are sometimes used as finely tunable filters to isolate characteristic wavelengths, as in an atomic absorption spectrometer, they are more often employed for spectral scanning. Here the emphasis will be on scanning.

The choice of variable for the abscissa in scanning (e.g., wavelength or wavenumber) is seldom of special concern except in an entirely analog instrument. If a monochromator is part of an instrument with an imbedded computer, conversions between wavelength and wavenumber and the choice of scales for axes can be set by the operator. The user must also select a speed of scanning, but the selection is best made in terms of resolution, time constant of detector and associated electronics, and similar spectral considerations (Section 17.5) whether the user or an automatic program selects appropriate values. Here attention will be given to the instrumental arrangements for scanning.

Linear Wavelength Drive. The Czerny–Turner and Ebert mountings for gratings are ideally adapted to linear wavelength drives such as the one illustrated schematically in Fig. 9.19. A metal bar BC, termed a *sine bar* because of its use, is rigidly affixed (at its lower end) to the grating holder. Its upper end is held in contact with a drive plate P. This plate moves as the micrometer lead screw turns. The position of the sine bar when the grating angle θ is 0° is shown as a solid line. According to the sketch in Fig. 9.19, only zero wavelength could appear at this position. As the lead screw turns, it moves plate P to the left through a distance X that corresponds to rotating the sine bar and the grating through angle θ'. Note that θ' is just angle BCD and that $\sin\theta' = X/CD$. Since $CD = BC$ = constant, $X \propto \sin\theta$. Plate P advances a distance proportional to $\sin\theta$. With reference to Eq. (9.14), if the lead screw turns linearly with time, the sine bar drive will produce a linear wavelength scan.

Example 9.17 What form of mechanical linear wavelength drive can be used for a prism monochromator?

Such a prism drive tends to be analog in character. The challenge of a linear wavelength drive is that it must turn a prism through angular increments that vary in different parts of the spectrum because of the nonlinear fashion in which the wavelength of the refracted ray emerging from a prism depends on prism composition, shape, and size. A "program" incorporating that information can be cut into the circumference of a metal cam. During a scan, the cam is turned at constant speed and the lever that drives the prism table follows the circumference of the cam. The result is a systematic change in the angle at which light enters the prism.

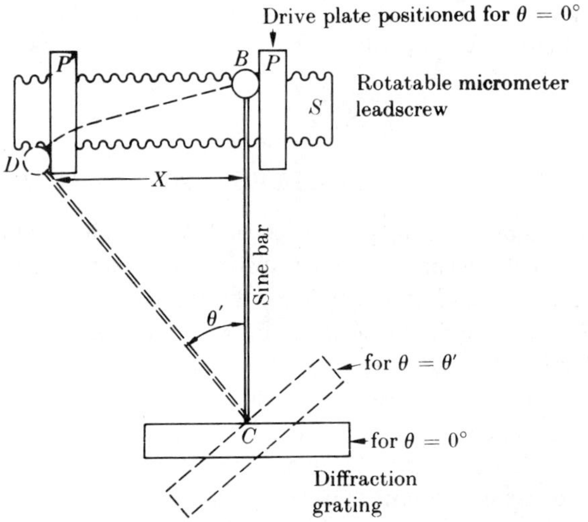

Fig. 9.19 Sine-bar grating drive for a linear-wavelength scanning unit. The sine bar is rigidly affixed to the grating holder. As the micrometer leadscrew turns, it drives threaded plate P. The plate pushes the sine bar, whose spherical end is held against it. When plate P has advanced distance X along the leadscrew, the sine bar and grating have shifted through angle θ'.

Computer-Controlled Scanning Drives. With microcomputer control, precision wavelength scanning becomes a matter of generating a digital voltage signal that will correspond to the desired final position of the monochromator grating or prism table. For instance, the final position will commonly be defined as an angle relative to an instrument axis. This position will provide a known angle of grating relative to the optical beam. The digital signal usually controls a linear wavelength drive; when the signal is converted to pulses, each pulse increases the wavelength at the monochromator exit slit by a precisely defined increment.

If the pulse is fed to a *stepper motor*, it turns the grating table a defined angular increment per pulse (digital signal converted to the total of bits it represents). A simple permanent magnet, multiphase stepper motor allows scanning in several modes that are defined by the way it is programmed. By appropriate patterns of pulses to the motor, the scan rate, acceleration and decceleration of scan, repetitive scanning, and ultraprecision wavelength indexing are possible. Motors with 200 steps per revolution are appropriate for such preprogrammed controllers. By contrast, a high precision motor (36,000 steps per revolution) with an appropriate grating can achieve a wavelength resolution of 7×10^{-5} nm in scanning. Corrections can be made for backlash in the drive shaft and other mechanical errors.

Other analog drives are also in use. In these drives a wavelength or *shaft encoder* must be added to provide wavelength indexing. One type of encoder consists of a lamp, a mirror (affixed to the drive shaft), and a detector/counter that will record light pulses reflected from the mirror as the shaft turns. A type of higher precision drive calls for a collar imprinted with a uniform coded pattern to be put around the drive shaft. A pencil of light is reflected from the collar and directed to an optical reader. As the grating table drive shaft turns, pulses are counted in the first case, and the scanner reads the pattern in the second; both identify the wavelength the dispersing element produces. If small wavelength increments must be registered, these arrangements have the disadvantage of complexity and cost.

Another type of drive features a magnetically driven *torsion bar* that turns the grating table directly. A servo system ensures that the final grating position is established with high precision. Through the controlling digital computer, the user enters the precise wavelength desired. It is converted by a program to a precise voltage designating the final grating table position and issued as a digital number to a DA converter. An 18-bit converter is necessary for high precision of setting. The voltage energizes a coil to turn the torsion bar. (The system is illustrated in Fig. 13.18.)

The final position of the torsion bar and table is established by servo control. As the table is turned, an analog transducer attached to the grating table shaft also generates a voltage that increases precisely with the shaft angle. Both this voltage and the original command voltage are fed to the differential inputs of an op-amp that drives the torsion bar. There is no further driving force for the torsion bar once it has reached its desired angle. If this op-amp is carefully adjusted for zero voltage offset, an exact final position of the grating table will be attained once the shaft transducer output just equals the command voltage. Fast slewing of the grating table is possible because the largest driving force is applied at the outset. Note the ingenious use of shaft position to provide a precise wavelength position.

Wavelength Calibration. Precisely known atomic emission lines and sharp molecular absorption peaks are nearly always used to calibrate the wavelength scale of a precision monochromator. In the visible and UV the best reference lines are in general those furnished by low-pressure elemental discharge lamps such as a mercury lamp, hollow-cathode lamps, or lasers (see Chapter 8). Lines used must be of good spectral purity and adequate intensity. The absorption peaks of rare-earth glasses are also useful. Thus, holmium oxide in a UV transmitting glass such as Corning CS3-138 provides eight sharp absorption peaks over the UV–VIS range.

In the IR, mercury emission lines are useful up to 4350 cm^{-1} (2.3 μm) but sharp absorption lines of common gases, such as water, carbon dioxide, and ammonia are used up to 250 cm^{-1} (40 μm). In addition, polystyrene films offer many sharp absorption bands in the near and middle IR. A good discussion of calibration techniques is available [15].

The wavelength calibration of monochromators with broader spectral slit widths, for example, 2–20 nm, must take into account the problem of which wavelength values best represents spectral features. Often calibrated absorption filters such as didymium (a mixture of Nd_2O_3 and Pr_2O_3) in glass are used. A wide variety of other methods has also been developed [11].

9.9 POLYCHROMATORS

The advent of multichannel solid state detectors like the vidicon and linear diode array made possible the rapid development of *polychromators* beginning in the 1970s. As suggested by the sketch in Fig. 9.20, a polychromator is an enhanced monochromator. To modify a monochromator the diffraction grating must be locked in place, the exit slit removed, and a multichannel detector installed permanently along the focal plane. In a monochromator, wavelength scanning calls for mechanically rotating the dispersing element about its vertical axis; in a polychromator it is accomplished by electronic scanning of the multichannel detector. Because a polychromator intimately links two functions, dispersion of wavelengths and their detection, it is really a *double module*. Not only is speed in readout gained but also, by permanent mounting of the grating and detector, a precise, stable wavelength calibration is assured.

As discussed in Section 8.8, the major types of multichannel detector that permit electronic scanning are solid state detectors such as the vidicon and photodiode array. These detectors are ordinarily flat and are best used with a dispersing arrangement which yields a flat focal plane. Under optimum conditions they can

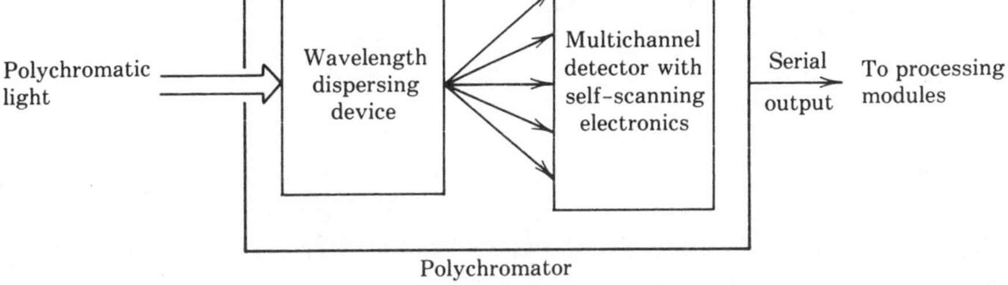

Fig. 9.20 A polychromator module.

detect as many wavelengths simultaneously as their number of resolution elements, pixels, or individual diodes. When coupled with a microchannel plate intensifier, they have a responsivity comparable to that of a photomultiplier tube. Stray light and background per element are of course not lessened by the fact that they are arrays, but charge transfer devices have very low dark currents. These detectors also require that a computer be added to provide needed electronic control, data storage, and data manipulation.

Multiwavelength Measurements. The correspondence between wavelengths and particular detector channels in a polychromator facilitates nearly simultaneous measurement of intensities of many wavelengths. In analogy to digital devices, it is usually said that signal information is available from a polychromator essentially in *parallel* fashion rather than in serial form as from a monochromator.

Rapid Scanning. Before multichannel detectors all rapid spectral scanning was done with a monochromator in which a scanning mirror could be rotated quickly. The mirror determined the angle at which radiation was incident on the dispersing device, and thus the wavelength appearing at its exit slit.

By employing a polychromator some distinct advantages in rapid scanning are gained. There are fewer alignment/registration problems since the grating (or prism) is locked in position. Further, a polychromator avoids the variations in optical performance with wavelength and time that are introduced in a scanning monochromator by moving the grating. Indeed, in a polychromator no mechanical movement is required except perhaps the opening of a shutter at the entrance slit.

In addition, fast reading (perhaps 10 μs per channel) of the output is possible. From 5 to 900 ms may be taken to sample the output of all the elements. Usually the time of observation at a given wavelength (particular detector element) can be adjusted to ensure that a signal can be integrated over a short period of time to smooth spurious fluctuations in signal intensity before it is read. When intensities are low, integration is especially valuable.

Advantages and Disadvantages. Since wavelengths over a considerable spectral range (the range is established by the dispersion of the grating and effective focal length of the optics) can be detected simultaneously, a polychromator enjoys a multiplex or Fellgett advantage. It is possible to observe an entire range in $1/N$ the time required to bring the N wavelength bands successively to a single detector in a monochromator and yet secure an identical S/N. Alternatively, observation for N units of time with a polychromator will ensure an improvement in S/N equal to $N^{1/2}$. A consequence of both capabilities is fine potential for acquiring dynamic information about systems.

In addition, wavelength accuracy is high since no mechanical or optical changes occur during electronic scanning of the detector channels. For the same reason calibration can be quickly accomplished and is stable with time in the absence of strong shocks or vibration. Changes in source intensity will also simultaneously affect all elements of the detector; as a result, sources with less constant output like lasers also become acceptable spectrometric sources. Changes in sample properties during spectrometric observation will also be sensed by all channels. Some illustrations of applications of polychromators will further show their advantages.

Example 9.18 A polychromator can assist considerably in finding the best wavelengths for a spectrophotometric titration when high precision is desired. In Fig. 9.21 an automatic titrator with a polychromator is shown.* The figure shows the modules involved and emphasizes the communication to and from the microcomputer. Note that the output of the diode array in the polychromator is immediately converted to the digital domain for handling by the microcomputer.

The polychromator offers the possibility of straightforward dual-wavelength titration. As a result, it facilitates titration in the presence of turbidity such as occurs in the titration of sulfate ion with Ba^{2+}. One wavelength is selected on the basis of approximately constant absorbance over the course of the titration and the other for high sensitivity to the component being monitored as a function of the progress of the titration.

A further application is to the titration of diphenylphosphate and *p*- and *m*-nitrophenol in an isopropyl alcohol–water mixture with tetrabutylammonium hydroxide solution. The

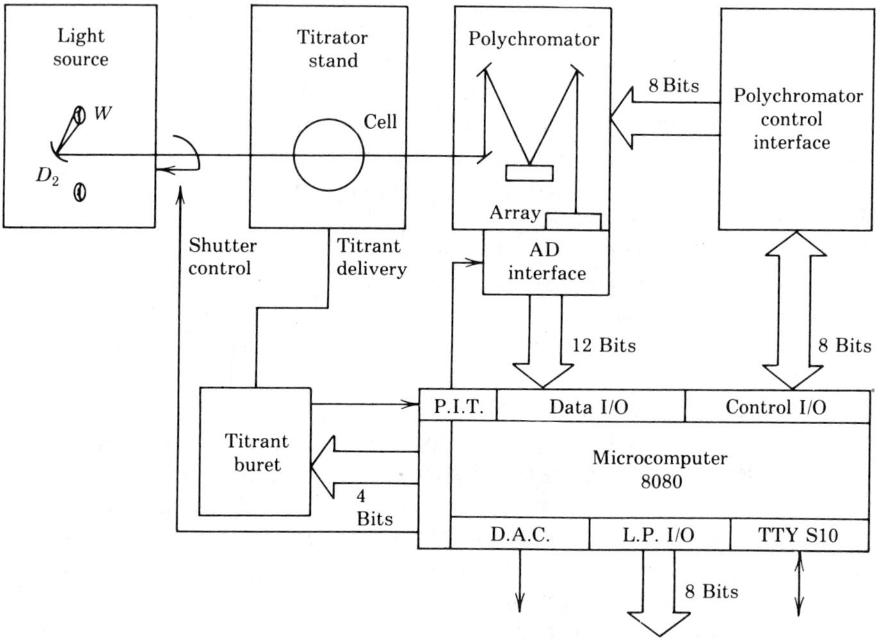

Fig. 9.21 Block diagram of a titrator that is based on spectrophotometric end points. As shown, the polychromator module includes a Czerny–Turner-mounted grating and a diode array detector along its focal curve. As rapidly as the self-scanning array produces an output it is digitized and stored in memory in the microcomputer. For display a recorder, line printer, or video monitor can be connected to the lower outputs. (Courtesy of *American Laboratory*.)

*A. H. B. Wu, T. Rotunno, and H. V. Malmstadt, *Am. Lab.*, **15**, 16 (September 1981).

pK_a's for the three analytes in water are 3, 7.1, and 8.3, respectively. It proves possible to obtain all of the information needed by simultaneously monitoring the absorption of the mixture at 475 and 550 nm; the end point for the diphenylphosphate is picked up at the first wavelength and those of the other two analytes, in the order listed, at the second.

Example 9.19. The Hewlett–Packard 1040A liquid chromatography ''detector'' features a UV polychromator built around an aberration-corrected holographic grating and a 211-element photodiode array detector. A deuterium lamp provides high output through the UV, and lower intensity up to about 600 nm. Since the sample cell *precedes* the polychromator, the spectrometer system is sometimes described as having reversed optics. The polychromator registers a first-order spectrum. The user can select instrument bandwidths from 4 to 400 nm, permitting measurements exclusively in the UV or from 190 to 600 nm. A microcomputer provides necessary control and data processing.

During the appearance of most chromatographic peaks successive spectra can be taken. If the spectrum observed on the chromatographic peak's forward edge differs from that during its middle portion and again from that on its trailing edge, it is likely that column resolution is inadequate to separate these analytes. Further, spectra can provide the confirmation necessary to establish the identity of an analyte believed to be present on the basis of retention times.

Fast scanning is critical to obtaining good spectra for rapidly eluting peaks. Securing a spectrum within 10 ms permits accuracy in peak height within 1% for even for a 1-s peak.

Example 9.20. Multiwavelength analyses are greatly facilitated with a polychromator. Systems specially designed for this purpose are called *optical multichannel analyzers* or *multichannel spectrum analyzers*. Software makes possible a variety of modes of extraction of data. Once a spectrum is known for a sample, a scanning and extraction program can be tailored. It may incorporate a fast scan mode to skip over individual elements of no interest at a rate of 0.5 μs per element. In a normal mode all elements will be read at 10 μs per element and the signal placed in memory. Or adjacent elements can be grouped to provide both an increase in sensitivity and speed.

For instance, if the only features in the spectrum of interest are a broad absorption peak at 480 nm and a sharp peak at 580 nm, one might program the analyzer to skip over detector elements registering wavelengths below 420, group the elements from 420 to 560 nm in pairs, scan through 600 nm in a normal mode, and skip all elements beyond 600 nm.

REFERENCES

Lucid, brief accounts of diffraction grating theory and design are given on an introductory to intermediate level in:

1. Bausch and Lomb Staff, *Diffraction Grating Handbook*. Rochester, NY: Bausch and Lomb, 1970.
2. S. P. Davis, *Diffraction Grating Spectrographs*. New York: Holt, Rinehart, and Winston, 1970.
3. M. C. Hutley, *Diffraction Gratings*. New York: Academic, 1982.
4. J and Y Diffraction Gratings Staff, *Handbook of Diffraction Gratings, Ruled and Holographic* (1980).

Monochromator design is presented on an intermediate to an advanced level in:

5. G. S. Hayat, J. Flamand, M. Lacroix, and A. Grillo, Designing a new generation of analytical instruments around the new types of holographic diffraction gratings, *Opt. Eng.*, **14,** 420 (1975).
6. J. F. James and R. S. Sternberg, *The Design of Optical Spectrometers*. New York: Barnes and Nobel, 1969.

Polychromator design is treated on an intermediate level in:

7. R. Kingslake, *Optical System Design*. New York: Academic, 1983.
8. P. M. Epperson et al., Applications of charge transfer devices of spectroscopy, *Anal. Chem.*, **60,** 327A (1988).
9. Y. Talmi (Ed.), *Multichannel Image Detectors*, Vol. 2. Washington: American Chemical Society, 1983.

Other references of interest are:

10. P. P. Acarnley, *Stepping Motors: A Guide to Modern Theory and Practice*, 2nd ed. London: Peregrinus, 1984.
11. D. H. Alman and F. W. Billmeyer, Jr., A review of wavelength calibration methods for visible-range photoelectric spectrophotometers, *J. Chem. Educ.*, **52,** A281, A315 (1975).
12. G. R. Harrison, The production of diffraction gratings. II. The design of echelle gratings and spectrographs, *J. Opt. Soc. Am.*, **39,** 522 (1949).
13. P. N. Keliher and C. C. Wohlers, Echelle grating spectrometers in analytical chemistry, *Anal. Chem.*, **48,** 333A (1976).
14. T. Namioka, Theory of the concave grating, *J. Opt. Soc. Am.*, **49,** 446, 460, 951 (1959).
15. K. N. Rao, C. J. Humphreys, and D. H. Rank, *Wavelength Standards in the Infrared.* New York: Academic, 1966.
16. S. Robin, Method of focusing of a far UV monochromator with fixed and distant slits, *J. Phys. Radium.*, **14,** 551 (1953) [French text].
17. M. R. Sharpe, Stray light in UV-VIS spectrophotometers, *Anal. Chem.*, **56,** 339A (1984).
18. Walter Slavin, Stray light in ultraviolet, visible, and near-infrared spectrophotometry, *Anal. Chem.*, **35,** 561 (1952).
19. A. Walsh, Multiple monochromators. I. Design of multiple monochromators, *J. Opt. Soc. Am.*, **42,** 94 (1952).

EXERCISES

9.1 The prism monochromator of Fig. 9.1 needs to be completed so that its output beam can be optically coupled to a solution cell or other module. Consider the following ways to add an exit slit in terms of advantages and disadvantages: (a) an exit slit is added that can be moved to any point along the focal curve to obtain different wavelengths; (b) the prism is placed on a small platform that can be rotated and an exit slit is placed in a fixed location along the focal curve.

9.2 A 453.8-nm ray is incident on a glass Cornu prism for which $A = 60°$ and $n_{453.8} = 1.63$. (a) What is the angle of incidence for minimum deviation? (b) For this prism what is the mathematical relation between the angle of incidence and the angle of minimum deviation?

9.3 On the basis of transparency and the variation of n with λ given in Fig. 7.17, (a) select two substances that should be good prism materials in the UV and two good in the visible; (b) indicate the optical range(s) where the following materials are most valuable for use as prisms in monochromators: quartz, lithium fluoride, and silver chloride.

9.4 Symmetrical prisms are usually employed at the angle of minimum deviation. Show that internal reflection (back into the prism) increases steadily when an angle of incidence smaller than the angle of minimum deviation is used.

9.5 Why are most mirrors used in monochromators front-surfaced?

9.6 It has been stated that a monochromator reimages its entrance slit at every wavelength along the focal curve. Yet when a spectrograph, which contains a polychromator, is used with a fixed width of entrance slit to record emission spectra on a photographic plate, some lines are wider than others. Explain this phenomenon.

9.7 For a monochromator (a) define throughput, (b) give at least two changes in the module that will increase throughout and explain why each is effective, and (c) define stray light.

9.8 Consider the Czerny–Turner monochromator design shown in Fig. 9.17. (a) Show that the grating law becomes $m\lambda = 2d \sin \theta \cos \phi$. (b) Derive an equation for its angular dispersion.

9.9 How does dispersion in a Czerny–Turner monochromator vary with angles of incidence and diffraction? Using the equations derived in the last exercise, when θ is varied from 15 to 35°, what change in dispersion can be expected?

9.10 A planar holographic grating is to be inserted in place of the Littrow prism in the cavity of an Ar^+ ion laser used in the configration of Fig. 8.12. A holographic grating with 1800 grooves mm^{-1} that is blazed at 550 nm is available. Efficiencies for the grating for different linear polarizations at several wavelengths are as follows: 550 nm, 80% $\perp$, 50% $\parallel$; 500 nm, $\perp = \parallel$ = 65%; 450 nm, 7% $\perp$, 65% $\parallel$; 400 nm, 7% $\perp$, 75% $\parallel$; 350 nm, 10% $\perp$, 60% $\parallel$; 300 nm, 10% $\perp$, 20% $\parallel$. For which of the Ar^+ ion laser wavelengths listed in Table 8.2 would the grating be satisfactory? Discuss.

9.11 For the light of wavelengths 1–5 μm (10,000–2000 cm^{-1}) normally incident on a reflection grating whose constant $d = 10$ μm: (a) calculate the range of angles of diffraction; (b) tabulate λ, $\sin \beta$, β, and $\Delta\beta$ for λ = 1–5 μm. (c) What variation is $\Delta\beta/\Delta\lambda$ can one expect over this wavelength range?

9.12 How can one tell that a grating is blazed in the visible range by visual inspection of the diffraction pattern?

9.13 (a) At what angle must a grating be blazed for it to have a peak efficiency in the first order at 1000 cm^{-1}? Assume $d = 5$ μm. (b) If a grating operates with 50% or more of its peak efficiency from 2/3 λ_B to $2\lambda_B$, over what frequency range will it achieve such efficiency? (c) Using the diagram of Fig. 9.17 and assuming that ϕ is constant, over what range of grating angles θ will a grating operate with 50% or greater efficiency?

9.14 Show from Eq. (9.9) that when a grating is used in the Littrow mode, its theoretical resolution is given by the equation $R = 2W \sin \beta / \lambda$, where β is the angle of incidence and diffraction and W is the ruled width of the grating blank ($W = Nd$).

9.15 Consider the echelle shown in Fig. 9.15. (a) Show that the grating law given in Eq. (9.9) can be rewritten as $m\lambda = 2w - t\theta$, where $\theta = \alpha - \beta$. (b) Also show that its angular dispersion can be expressed as $d\theta / d\lambda = m/t$. (c) What percentage change in dispersion will it undergo over the orders 80–105?

9.16 It can be shown that the actual blaze wavelength λ_B' for a diffraction grating is related to the nominal value λ_B by the equation $\lambda_B' = \lambda_B \cos(\alpha - \beta)/2$. What nominal value of blaze should be selected for a grating to be used at grazing incidence ($\alpha = 85°$, $\beta = -85°$) in a monochromator intended for work centered around 80 nm in the first order?

9.17 Comment on the following statement: "Historically, the first polychromator was in the spectrograph since a photographic plate (detector) appeared along the focal curve of its monochromator." Discuss whether it was actually a polychromator. Whether it was or not, did it secure the advantages listed for a polychromator?

9.18 Show in a sketch how the Czerney–Tuner monochromator of Fig. 9.17 is modified (a) when a linear diode array detector is added to make it a polychromator. (b) What advantages and disadvantages does such an array offer?

9.19 Explain the role of the monochromator in an atomic absorption spectrometer. Recall that needed monochromatic wavelengths are produced by a hollow-cathode lamp.

ANSWERS

9.2 (a) 54.6°; (b) $\sin \frac{1}{2}(D + A) = \sin i$. *9.3* (b) quartz: UV, LiF: UV, AgCl: IR. *9.7* (a) Absolute throughput will be the energy that emerges from the exit slit at a given wavelength setting. (b) Increase slit widths; increase the efficiency of the grating; increase the reflectance of mirrors. (c) Stray light is the radiation of all wavelengths that has not followed the optical path and still emerges from the exit slit and to which a detector is sensitive. *9.8* (b) $d\theta / d\lambda = m/2d \cos \theta \cos \phi$. *9.11* (b) $\Delta\beta / \Delta\lambda$ is nearly constant at 6° per μm. *9.15* (c) About 25%. *9.18* (b) Advantages are wavelength accuracy since geometry fixes the wavelength registered by each diode; S/N is higher since all wavelengths are registered during the entire time of observation. Disadvantages are that resolution is poorer, unless the wavelength range is limited; the response from diode to diode may vary by as much as 3% (a calibration will overcome this limitation).

BASIC QUANTIFICATION

Mathematical fields such as statistics also play major roles in chemical measurements; increasingly these are collected under the relatively new heading of *chemometrics*. In the first chapter of this division, Chapter 10, statistical modes of assessment of the precision and accuracy in measurements are discussed; their application to all steps in chemical measurements, from taking initial samples through chemical preparative procedures and instrumental determinations of a variable to the processing of instrument signals to reduce noise, makes it possible to ensure measurements of high quality. Such an approach allows the setting of limits of detection and determination for methods. In Chapter 11 we look at ways to quantify the analytical signals instruments yield. Throughout their existence in an instrument, they are generally voltages or currents. It is the application of mathematical models such as the use of calibration curves that quantifies these signals. After discussion of many one-component methods some chemometric strategies for multicomponent systems are presented briefly. Chapter 12 concludes the division by returning to the physical as well as the computational level to examine useful ways to enhance the instrument S/N ratio and extract chemical information.

Chapter 10

STATISTICAL CONTROL OF MEASUREMENT QUALITY

10.1 INTRODUCTION

A rich body of statistical procedures that focus on the reduction of both random and systematic error has been developed to improve the quality of measurements. Where should attention be given to quality, that is, at what points should the level of error be ascertained? To seek an answer consider the generalized diagram in Fig. 10.1. One's experience with measurements may suggest that error should be mainly assessed when one reaches the readout or display step. Yet if high quality is to be achieved this view is anomalous: each step in a measurement can benefit from an assessment of error as the overall process is set up. In this chapter we consider both the assessment of error in measurements and some ways of minimizing it to ensure that chemical information extracted from samples is (a) high in quality and (b) presented in a manner that makes its level of reliability clear.

As Fig. 10.1 implies, the application of statistics should ideally begin in the analytical design process introduced in Chapter 1. To keep error appropriately low, one should start by outlining the goals of measurement for a new problem or research project. Statistics and other mathematical procedures will aid in defining and optimizing some of the steps, such as the design of a sampling process for a given system, the conditions of measurement, and the number of replicate measurements desirable. Naturally, other aspects of measurement such as cost, time expended, and the amount of information that can be obtained from a particular method, also come into consideration in analytical design.

10.2 ERROR IN MEASUREMENTS; PRECISION AND ACCURACY

As the term quality is applied to measurements it means *reliability*. Reliable results must on inspection prove *valid*, *precise*, and *accurate*. A first step in assessment will be to define these criteria clearly. *Validity* is a statement of the degree to which a measurement by a particular instrumental method *and* analytical procedure yields the property (e.g., molecular weight) that it is supposed to measure. An example is given in Section 10.5. *Precision* is basically a measure of reproducibility and is specifically defined by observing the scatter in replicate measurements. Thus, poor precision results from random fluctuations. Much further attention will be given here to this topic. Finally, *accuracy* is defined as the degree of agreement between

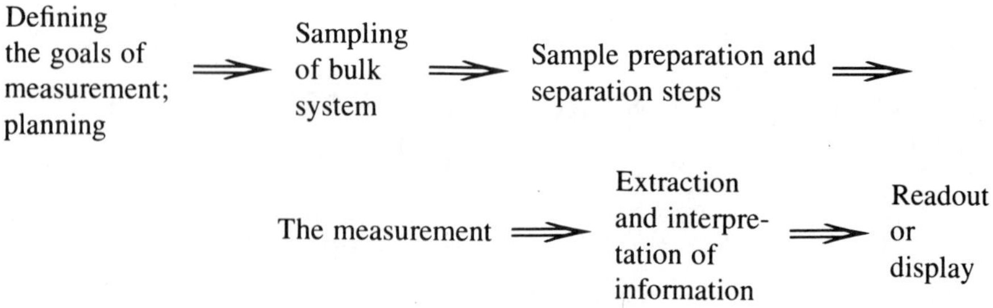

Fig. 10.1 Modular diagram of a measurement process.

a measured property such as concentration and the "true value." A dilemma is immediately evident. Unless the true value is available, accuracy cannot be calculated. For that reason we shall concentrate instead on ways to discover error and to minimize it in all the steps of a measurement system (see Fig. 10.1).

Metrologists, persons who make measurements with ultimate accuracy, seem to have the most philosophically grounded interpretation of reliability. In their determination of fundamental constants such as the speed of light, they seek to approach a true value as nearly as is physically possible. By contrast, persons using chemical instrumentation in a laboratory or in the field, tend to take an *operational approach*: they seek a *best value*. Best implies simply that uncertainty for a particular measurement has been reduced until it is smaller than a prescribed figure. As will be seen below, the uncertainty is usually reported as the range, or confidence limits, within which it is highly probable that future measurement results will fall.

There is another way to approach reliability in a set of replicate measurements: to focus on error. The total error e in a measurement can be defined by the expression

$$e = \Delta + \delta.$$

Here the total error is made up of Δ, *systematic error*, which causes *bias*, and δ, *random error*, which contributes to imprecision. The lack of validity is included implicitly through possible contribution to bias.

Random error is traceable to small changes in variables such as temperature and mass of reagent; it varies unpredictably between measurements. By contrast, *systematic error* or bias influences repeated measurements equally. Such errors are commonly *fixed* systematic errors that are constant in magnitude from measurement to measurement regardless of analyte concentration. They may also be *proportional* in magnitude to concentration or other variables. Unless stated otherwise, we shall deal with fixed errors.

An interesting conclusion about total error can be drawn. As is well known, the averaging of replicate measurements ensures that random error will tend toward zero. The equation above makes clear that as $\delta \to 0$, $e \to \Delta$: the total error may become almost entirely *systematic error*. It appears that in procedures of quality control and quality assurance much attention must be given to minimizing systematic error. A wealth of statistical tools will help us establish a framework for examining error generally. Fortunately, their power is such that they will allow us to determine whether a lack of reproducibility observed in the various stages of measurement can be interpreted (with high probability) as resulting mainly from (a) random error or instead from (b) a mixture of random *and* systematic error. For that reason we shall focus first on precision. Then in Sections 10.3 and 10.4 tests that help uncover bias will be reviewed.

Precision Measures. Some basic definitions from statistics need to be introduced to express the idea of precision clearly. First, a *population* is any complete set of data, for example, a set of n replicate measurements $x_1, x_2, \ldots, x_n$. Second, the *mean* of this population is the arithmetic average symbolized by $\bar{x}$ and defined by the equation $\bar{x} = \Sigma_i x_i/n$. An *arithmetic mean* is always more reliable than any individual result or some other kind of average in the cases we shall deal with.

The most widely useful measures of precisions are the *standard deviation* σ and its square σ^2, which is called the *variance* V. The latter is defined for a very large (essentially infinite) population of replicate measurements of x by the equation $V = \sigma^2 = \Sigma_i (x_i - \bar{x})^2/n$. The term in the denominator represents f, the *degrees of freedom* of the population. For small populations, however, $f = n - 1$, and $\sigma^2 = \Sigma_i (x_i - \bar{x})^2/(n - 1)$. The expression for the standard deviation is simply

$$\sigma = \sqrt{\Sigma_i (x_i - \bar{x})^2/(n - 1)}. \tag{10.1}$$

The calculation of either standard deviation or variance can be simplified as outlined in Example 10.1. Variance is the most common measure of precision, mainly because contributions to variance are additive. For that reason it is easier to use in comparing the precision of different methods or different steps in a procedure.

Equation (10.1) for quite small populations, for example, the usual three or four replicate measurements, yields only an *estimated standard deviation* s and not σ. As will be seen below, this estimate proves quite useful as long as the number of replicates n or degrees of freedom f is also stated. In practical situations a new set of terms is developed from the standard deviation. The *relative (estimated) standard deviation* s_r where $s_r = s/\bar{x}$ proves widely valuable. When multiplied by 100 to convert to percentage, it is called the *coefficient of variation* (cv) of a measurement.

Example 10.1. How can Eq. (10.1) be simplified? Actually, the quantity in the numerator, $\Sigma_i (x_i - \bar{x})^2$ can be shown to equal the difference $\Sigma_i (x_i)^2 - (\Sigma_i x_i)^2/n$. Less time is required for this calculation and once this difference has been found, division by the degrees of freedom yields the variance.

Example 10.2. A chemist analyzes for lead in three groups of samples (10 samples in all) taken from a production lot of brass utilizing a standard emission spectrometric procedure for lead and obtains the percentages listed below. What are the variance and estimated standard deviation in the results? Note that the modified equation from the first example is used for the calculations.

Results

	Group A	*Group B*	*Group C*
Number of determinations, $n = 4 + 3 + 4 = 11$	1.25	1.31	1.26
	1.33	1.35	1.38
	1.16	1.28	1.24
	1.27		1.31
Sum	5.01	3.94	5.19
Sum of squares	6.2899	5.1770	6.7457
$(\text{Sum})^2/n$	6.2750	5.1745	6.7340
Difference	0.0149	0.0025	0.0107

Sum of differences $(S) = 0.0149 + 0.00256 + 0.0107 = 0.0291$
Degrees of freedom $(f) = (4 - 1) + (3 - 1) + (4 - 1) = 8$
Variance $= (S)/f = 0.003637$; estimated standard deviation $= \sqrt{(S)/f} = 0.0603$

What statistical implications do groups A, B, and C have? In cases such as this one in which no information about differences in treatment of the samples in each group is given, it is simply necessary that when the data are combined, the total degrees of freedom reflect the existence of the groups. Since $f = n - 1$ for each group, these subtotals must be summed to give the overall degrees of freedom of the entire set.

If, however, the treatment of each group had differed in some way, for example, if each were analyzed by a different instrument, different chemist, different laboratory, or even on different days, it would be desirable to calculate the variance of each group separately and then add them to obtain the overall total. Often such differences in procedure are introduced deliberately to assess their possible significance. The separate variances are compared by tests that will be discussed in the next section.

The Normal Distribution. In most cases when a very large (essentially infinite) number of replicate measurements subject only to random fluctuations is made, the data they yield fall into the pattern shown in Fig. 10.2 and called a normal or Gaussian distribution. The ordinate denotes the frequency with which departures from the mean occur. Individual values scatter around the true mean, μ. If precision is high, the scatter of values is small and is described by a curve that peaks sharply like *A*; with poor precision, scatter is large and results distribute themselves into a flatter curve like *B*. Other important properties of a normal distribution are described in the legend for Fig. 10.2. As noted in Fig. 10.2, measurements do distribute themselves symmetrically around the arithmetic mean.

The normal distribution can be described mathematically by just two parameters, its mean and standard deviation. That we have defined these two properties of populations at the outset is clearly no accident.

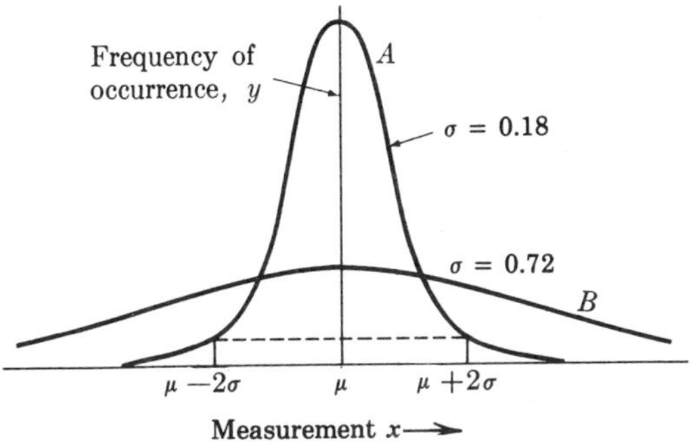

Fig. 10.2 A normal or Gaussian distribution of replicate measurements about their mean value μ. Curves A and B represent results of high precision and low precision, respectively; their standard deviations differ by a factor of 4. The curves are also plots of the normal error law defined by the equation $y = (1/\sigma\sqrt{2\pi})e^{-1/2[(x-\mu)/\sigma]^2}$. Several features of the curves are of interest: (1) their symmetry indicates that a negative error of particular size occurs just as frequently as a positive one; (2) their shape makes clear that very small deviations from the mean occur most frequently and conversely, that large errors are uncommon.

Since the ordinate in Fig. 10.2 is a frequency of occurrence, it is proper to ask "What is the probability P that replicate measurements will fall within a given range centered about the mean?" The answer can be obtained by finding the area under the curve between the chosen limits. A classic example in which the limits are just $\pm\sigma$ is shown in Fig. 10.3. An important virtue of working with a distribution is that calculated probabilities can also serve as measures of confidence and the limits involved as confidence limits with respect to random error. It can be said about the range in Fig. 10.3, for example, that the limits serve to define the range of random error in measurements at the 68% confidence level.

Extensive tables of calculated probabilities for normal distributions are available and may be consulted to obtain probabilities for almost any limits of interest. Fortunately, most other distributions of data reduce to the normal distribution un-

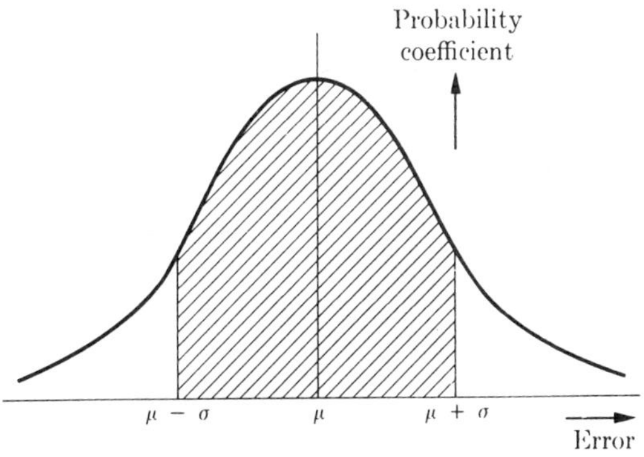

Fig. 10.3 The area under the curve between limits $\mu - \sigma$ and $\mu + \sigma$ defines the probability that measurements will fall within these limits 68% of the time.

Table 10.1 Probability P that Measurements will be Within Specified Limits[a]

Limits	P
$-\sigma$ to $+\sigma$	0.682
-2σ to $+2\sigma$	0.954
-3σ to $+3\sigma$	0.997

[a]For very large populations.

der selected conditions so that such a table does yeoman service. A very abbreviated set of values is listed in Table 10.1. For example, there is a probability of 0.954 (95.4%) that replicate measurements will have values falling between the limits $\mu - 2\sigma$ and $\mu + 2\sigma$. It can also be said that for these confidence limits only 5% of the time will a replicate measurement fall outside the range.

Precision of Means. The measures of precision defined above referred, of course, to individual determinations of a variable. How will the *precision of a mean* be affected by the number of measurements? To provide an answer requires at least two steps. First, consider that a large number of means will itself comprise a population, in this case a population of normally distributed means.*

Fortunately, the mean of individual measurements is equal to the mean of the population of means for the measurement, that is, $m_x = m_{\bar{x}}$. From this distribution, the standard deviation of a mean is found to be $\sigma_{\text{mean}} = \sigma_x/\sqrt{n}$. If σ_{mean} is unknown but a value of s for the measurement is available, an estimated deviation of the mean can be calculated:

$$s_{\text{mean}} = s_{\bar{x}} = s/\sqrt{n}. \qquad (10.2)$$

This value is often called the *standard error* of x. The inverse dependence of s_{mean} on $n^{1/2}$ defined by the equation of course provides the rationale for making replicate measurements to improve precision. Example 10.4 will pursue this idea.

Confidence Limits for Small Populations. It would be especially helpful to be able to define confidence limits for small populations consisting of the usual two to four replicate laboratory measurements. The statistics necessary to these conditions were developed by Student.† He calculated extensive values of a factor t by which the estimated standard deviation of the mean of a small population could be multiplied to obtain confidence limits for the related population of means. Quite explicitly, he sought to define confidence limits that would contain, to the desired probability, the elusive but desired true value, which statistically is called the true population mean μ. A brief set of values for Student's t is given in Table 10.2.

*Means are nearly always normally distributed even when individual measurements may not be.

†W. S. Gossett adopted the pen name "Student" in publishing his important work on small populations in 1908.

Table 10.2 Some Values of Student's t[a]

Degrees of freedom f $(n - 1)$	$\pm t$ $\alpha = 0.10$	$\alpha = 0.05$	$\alpha = 0.01$
1	6.31	12.71	63.66
2	2.92	4.30	9.92
3	2.35	3.18	5.84
4	2.13	2.78	4.60
9	1.83	2.26	3.25
∞	1.64	1.96	2.58

[a]Values of t having f degrees of freedom and the probability α of being exceeded.

Student showed that the *confidence limits* for a small set of replicate measurements can be expressed as

$$\bar{x} \pm t_{f,\alpha} s / \sqrt{n}. \tag{10.3}$$

Here subscript f is just the number of degrees of freedom $(n - 1)$ of the population n, and α is the probability that the population mean will fall *outside* the stated range. How is α related to the probability P of Table 10.1? They are related by the expression $\alpha = 1 - P$. Equation (10.3) provides the statement of the best value that we sought. The best way to report results when replicate measurements are made will be in the form of confidence limits given by Eq. (10.3), $\pm t_{f,\alpha} s / \sqrt{n}$.

Example 10.3. For the following set of replicate measurements of the concentration of an analyte in a sample what is the mean, estimated standard deviation, and percent relative standard deviation: 46.25, 46.40, 46.36, and 46.28%. How should these data be reported to convey their reliability at the 99% confidence level?

The mean and estimated standard deviation of an individual measurement are calculated to be $\bar{x} = 46.32\%$ and $s = 0.0695$, respectively. The percent relative standard deviation or coefficient of variation is just 0.15%.

In establishing the confidence limits, we translate 99% probability into a value of α of 0.01. From Table 10.2 the appropriate value of Student's t for three degrees of freedom is seen to be 5.84. Equation (10.3) gives the confidence limits for the best value or population mean: $46.32 \pm 5.84(0.0695)/\sqrt{4} = 46.32 \pm 0.20$.

Example 10.4. How many replicate measurements are worth making on a sample in the interest of better precision?

Table 10.2 can provide information about the trend in precision. As seen in the table, Student's t values at first fall rapidly with an increasing number of replicates and then decrease more slowly. This beneficial shift toward smaller values is opposed in practice by increasing analysis time and costs with each additional measurement. Inspection of the data suggests limiting replicates to three or four.

Example 10.5 How many times should a sample of gasoline that has been blended with methanol be analyzed to obtain a relative precision of ±2% in the determination? Assume the gasoline averages 4 wt.% methanol.

Since $\bar{x} = 4\%$, the precision called for will mean that a minimal number of results should fall outside the range $\bar{x} \pm 0.02\bar{x}$. Clearly, the range can be regarded as confidence limits allowing Eq. (10.3) to be used. We obtain

$$0.02\bar{x} = t_{f,\alpha}s/\sqrt{n} \quad \text{or} \quad \sqrt{n} = t_{f,\alpha}s/0.02\bar{x}. \tag{10.4}$$

We must select values for t and the estimated standard deviation on the basis of experience. Let $t = 3.0$ and $s = 0.05$. Then, $\sqrt{n} \approx 2$ and $n = 4$. Four replicates will be needed.

In spite of validated methods, calibrated instruments, and careful measurement procedures, results are sometimes obtained that are so far from the mean as to be quite improbable. These errors are neither random nor systematic but may more properly be called gross errors. Such data are commonly called *outliers*. If one drops outliers from a set of data, it is important that they be identified and the reasons for rejection given. Statistical criteria are useful in making such decisions [1, 5]. One method, the Q test, is examined in Exercise 10.4.

10.3 PROPAGATION OF ERROR

Each module and procedural step contributes some error in making measurements. Viewed as a dynamic system, a measurement can be no more reliable than its least accurate step. Similarly, an instrument can be no more reliable than its least precise module. Under some conditions it may actually be much less accurate. Knowing the most important sources of error and considering the manner in which they propagate thus is highly important in using and designing instruments and procedures.

The propagation of random error may be traced mathematically by using a precision measure such as the standard deviation and drawing on equations that describe the dynamics of the system. Systematic error may also be followed through a system by relying on the latter equations. Only the general case of propagation of random error will be discussed. Let R indicate a measurement. It will depend in some known manner on several variables, x, y, z, which are the outputs of instrument modules and of steps in a procedure such as extraction of an analyte into a solvent, that is $R = f(x, y, z)$. The error or uncertainty in R may then be found by taking the total differential:

$$dR = \frac{\partial R}{\partial x}dx + \frac{\partial R}{\partial y}dy + \frac{\partial R}{\partial z}dz. \tag{10.5}$$

In Eq. (10.5), the differentials dx, dy, and dz are to be regarded as fluctuations in the response of modules or steps that combine to produce dR, the error in the measurement. Note that the fluctuations in the response of components do not add directly but are first modified by their functional relationship to R. For example, dx is multiplied by $\partial R/\partial x$ before addition.

Equation (10.5) must be altered before it can be applied conveniently to experimental situations. This involves transforming the differentials dx, dy, and dz into standard deviations and eventually into variances. It can be shown that the variance V or σ_R^2 in the response R is given by the expression

$$\sigma_R^2 = V = \left(\frac{\partial R}{\partial x}\right)^2 \sigma_x^2 + \left(\frac{\partial R}{\partial y}\right)^2 \sigma_y^2 + \left(\frac{\partial R}{\partial z}\right)^2 \sigma_z^2. \qquad (10.6)$$

Equation (10.6) expresses the propagation of error. While σ^2, the variance, is often used in comparing experimental procedures, it is not as useful as σ in situations where bounds are to be reported for measurements, as is the case here. An adaptation to small sample statistics, and estimated standard deviations, if important, would be straightforward. An illustration of the application of Eq. (10.6) follows.

Example 10.6. The power W dissipated in a resistor must be estimated to within $\pm 1\%$ by a calculation based on $W = I^2R$. Will it be adequate to measure both resistance and the circuit current to $\pm 1\%$ (percent relative estimated deviations)?

A satisfactory answer may be obtained by evaluating the propagation of error for this case. Equation (10.6) becomes

$$V_W = s_W^2 = (\partial W/\partial I)^2 s_I^2 + (\partial W/\partial R)^2 s_R^2.$$

The values of the partial derivatives are $\partial W/\partial I = 2IR$ *and* $\partial W/\partial R = I^2$. Substituting these values and dividing through by W^2 will give for the relative estimated deviation:

$$\begin{aligned} s_W^2/W^2 &= 4I^2R^2 \cdot s_I^2/I^4R^2 + I^4 \cdot s_R^2/I^4R^2 \\ &= 4s_I^2/I^2 + s_R^2/R^2 \qquad (10.7) \end{aligned}$$

at the level of precision of $\pm 1\%$ in both I and R, the estimated relative deviation in the power will be

$$\sqrt{s_W/W^2} = \sqrt{4 + 1} = 2.2.$$

The answer makes clear that more precise measurements will be required. It also confirms one's intuitive expectation that a special effort should be given to improving the measurement of current.

In building or designing a measurement system, costly changes can frequently be saved by a preliminary analysis of the errors contributed by the steps in the overall procedure and the instrument modules. Their precision is usually known from previous experience or readily obtained from available data. Then, by applying Eq. (10.6), the reliability with which the measurement can be made is readily ascertained. If this precision is insufficient, the steps or modules that must be modified or replaced will be those whose individual terms contribute most to the total error as seen from the equation.

10.4 SYSTEMATIC ERROR: STATISTICAL ASSESSMENT

Systematic error or bias is by its nature difficult to identify and as a result ordinarily limits accuracy far more than random error, as noted earlier. Fortunately, straightforward statistical procedures are available to test for its presence even though they cannot identify its source.

The possibility of bias must always be sought in a new method, whether in research or analysis. The simplest way to check on a method is to use it to measure analyte in an appropriate *standard reference material.* By definition, such a material is homogeneous and contains one or more analytes at a concentration that is known with high accuracy and precision.* After replicate measurements are made on a reference material using the new method, the possibility of bias is assessed by applying a statistical test such as the Student t-test. While it is necessary to assume that data produced by the method are distributed normally, more often than not this assumption will be valid.

Student's t-Test. The t-test allows one to learn whether all the variability in a set of replicate measurements can be ascribed to random error. One begins the test by developing a clear *null hypothesis* for the measurement. Only if the statement is precise will the outcome be definitive. For example, if a new method is used on a standard material a good null hypothesis will be

> The difference between the mean $\bar{x}$ of replicate measurements of analyte in the standard material by the method being tested and the true mean μ for the material is attributable entirely to random error.

The t-test establishes whether the hypothesis holds. If it stands after a test, one can conclude

> *there is no significant systematic error in the new method.*†

A finding that it is false, however, is equivalent to demonstrating the presence of bias or systematic error in the method. In this event, one is entitled to make what is called the alternate hypothesis:

> *the difference between means must be at least in part attributable to bias.*

In applying the t-test, an "experimental" value of Student's t is first calculated for the set of replicate measurements by the equation:

$$t_{calc} = (\bar{x} - \mu_0)/(s/\sqrt{n}). \tag{10.8}$$

*The certified concentrations in standard reference materials may be assumed to represent a true mean μ with a zero standard deviation. Both the U.S. National Bureau of Standards (NBS) and the American Society for Testing Materials (ASTM) provide such materials in the United States.

†While this hypothesis cannot be *proved* true, as is true of most hypotheses, it may be shown to be false.

Here μ_0 designates the mean certified for the standard material. Next, t_{calc} is compared with values of t_{stat} from Table 10.2 or a more comprehensive table. Recall that the table will give statistically determined values of t for a population with the same number of degrees of freedom as the measurements and for several probabilities. The user must choose the probability. Selecting a probability $\alpha = 0.05$ will usually suffice, though for a critical case a value of $\alpha = 0.01$ will provide rigor. If $t_{calc} < t_{stat}$, the null hypothesis holds and the scatter in results can be attributed to random error. On the other hand, if t_{calc} is larger, it is probable that systematic error has occurred. To estimate the amount of bias and where it entered additional measurements would have to be devised (see Example 10.13).

When a standard reference material is not readily available, a common situation, a carefully standardized local reference material can be substituted. An example based on this approach will be given below.

Alternatively, the method to be tested can be compared with the performance of a "reference method" on a homogeneous but not necessarily standardized local material. This comparison is called a *two-group* experiment because it involves results from replicate measurements by two different methods on a common sample. The reference method may be well known to the users, or perhaps one that is familiar only through the literature. Naturally, the latter possibility will produce uncertain results at best.

The t-test needs to be applied to a different hypothesis for the two-group case. It will now read

> The difference between $\bar{x}_1$, the mean of measurements of a sample with method 1, and $\bar{x}_2$, the mean with method 2, is attributable entirely to random error.

An example of a two-group experiment will be included below.

What kind of judgment is involved in choosing a value of α for a t-test? The choice will greatly affect the value of t_{stat}. A good rule of thumb is that a conclusion made at the 0.01 probability level will be more convincing than one at the 0.05 level.

Example 10.7. How does the t-test relate to the use made of Student's t values in developing Eq. (10.3) to express confidence limits for the mean of small numbers of measurements?

Actually, the t-test exactly reverses that procedure. The numerator of Eq. (10.8) expresses the difference between the mean for the new method and the "true" value μ_0 while the denominator is simply the estimated standard deviation of a population consisting of means. Basically, Eq. (10.8) is a reworking of Eq. (10.3).

Example 10.8. A new analytical method is to be tested. A standard reference material with an analyte concentration of 4.41% is available. With the test method replicate determinations are carried out on four samples of the reference material, giving a mean analyte concentration of 4.83% with an estimated standard deviation of 0.38. Is the null hypothesis stated in the text for a similar case acceptable?

A value of t is soon calculated: $t_{calc} = (4.83 - 4.41)/(0.38/\sqrt{4}) = 2.21$. On examining Table 10.2, a t-value of 3.18 is quickly located for data with three degrees of freedom at the 95% confidence level ($f = 3$ and $\alpha = 0.05$). Since $t_{calc} < t_{stat}$, it can be concluded that random error will successfully account for all of the scatter.

Example 10.9. If only local standard materials are available for testing new methods, how should the t-test be modified?

The main difference will be that the estimated standard deviation of the local material must be included in Eq. (10.8). Let $\bar{x}_1$, s_1, and n_1 be the mean, estimated standard deviation, and number of replicate determinations made with the new method on the laboratory standard, respectively. Further, let μ', s_2, and n_2 be similar quantities for the laboratory standard as established over the years. Equation (10.8) must now be modified as follows:

$$t = (\bar{x} - \mu')/(s_1^2/n_1 + s_2^2/n_2)^{1/2}. \qquad (10.9)$$

Equation (10.9) may be applied as Eq. (10.8) was, by comparing the t_{calc} it yields with a value t_{stat} from the tables.

Two-Group Measurement. A method for determination of chromium that has just been published is to be tested. To establish its reliability, it is compared with a method long in use in the laboratory by a *two-group experiment*. Both methods are applied to a particular chromium sample. Let $x_1, x_2, \ldots$ be results from method 1 and $y_1, y_2, \ldots$ be results from method 2. The hypothesis will be:

> The difference between the means of results from the two methods can be entirely attributed to random error.

The calculation of t will involve the following expression:

$$t_{calc} = [(\bar{x} - \bar{y}) - (m_x - m_y)]/s_{\bar{x}-\bar{y}}, \qquad (10.10)$$

where $s_{\bar{x}-\bar{y}} = s_{comb}/\sqrt{n_x + n_y}$, and s_{comb} is the pooled standard deviation (see below). Since the hypothesis is equivalent to assuming that data from the two methods of determination belong to the same population of results, it must follow that $m_x = m_y$. Thus, Eq. (10.10) reduces to $t_{calc} = (\bar{x} - \bar{y})/s_{\bar{x}-\bar{y}}$.

The results obtained were as follows: percentages of chromium by method A were 27.54, 27.56, 27.51, 27.56, 27.52, and by method B, 27.28, 27.31, 27.30, and 27.32. The means are $\bar{x}_A = 27.54$; $\bar{x}_B = 27.30$; and $\bar{x}_A - \bar{x}_B = 0.24$. Further, the pooled variance, which will have $(n_x - 1) + (n_y - 1)$ degrees of freedom is

$$\begin{aligned} s^2_{comb} &= [\Sigma(x_i - \bar{x})^2 + \Sigma(y_i - \bar{y})^2]/[(n_x - 1) + (n_y - 1)] \\ &= [2.1 \times 10^{-4} + 9 \times 10^{-4}]/(4 + 3) = 4.3 \times 10^{-4}. \end{aligned}$$

The standard deviation of the combined *mean* will be

$$\sqrt{s^2_{comb}(1/n_x + 1/n_y)} = \sqrt{4.3 \times 10^{-4}(1/5 + 1/4)} = 0.014.$$

We find $t_{calc} = 0.24/0.014 = 17$. But t_{stat} with seven degrees of freedom and $\alpha = 0.05$ is only 2.4. Since $t_{calc} > t_{stat}$, the hypothesis is disproved; thus the two methods do not yield statistically similar results, and it is probable that the new method has systematic error.

Limitations of the t-Test. Basically, the *t*-test applies to data that are normally distributed. In the event data do not fit such a distribution, other nonparametric tests may be applied to assess whether systematic error is probable [2, 8].

Second, in such comparisons the test establishes not absolute accuracy but only accuracy in *terms of a particular reference material* or *reference method*. To avoid the latter limitation it is clearly preferable to analyze a range of samples, applying the *t*-test to each comparison. If this strategy is used, a further problem may emerge, that the *t*-test will be in error because a systematic error or bias will arise in only one or two samples because of an interfering substance in them but not in all samples.

Third, the *t*-test also includes an unrevealed assumption that the difference between a result obtained with the test method and the standard method used with the reference will be a true systematic error that is constant, and thus, independent of analyte *concentration*. The case of an error proportional to the concentration of analyte will be taken up below.

Regression Plots. When two methods are being compared, a good approach is to apply each method to a series of samples in which the analyte concentration is increased by increments and to plot the results by one method against those by the other. The data obtained from the reference method, which presumably has little systematic error, should be plotted along the abscissa and those from the test method along the ordinate. A regression equation should also be calculated (see Section 11.3).

What type of plot should result? The power of this type of visual analysis is illustrated in Fig. 10.4. If each method is both precise and accurate, a straight line with a slope of unity should be secured as in Fig. 10.4a. Furthermore, and perhaps less immediately obvious, the intercept of the curve should be the origin. As a contrast, Fig. 10.4b shows data for a case in which precision is poor but accuracy is good. Note that points scatter, but the line goes through the origin. A straight line of nonzero intercept, as in Fig. 10.4c, signals probable systematic error or bias in the test method. Indeed, the intercept gives an estimate of the bias in the method.

How would Fig. 10.4a be changed if the test method were subject to *proportional bias*? Up to this point a bias that is independent of the concentration of analyte has been assumed. In many situations systematic errors actually change with analyte concentration. For example, in X-ray fluorescence bias occurs because emitted X rays are absorbed to some degree by the analyte, and the absorption error is proportional to its concentration. In this case a curve like that in Fig. 10.4a but of lesser slope would be obtained. To overcome this type of error the procedure of standard addition (see Section 11.3) is often used.

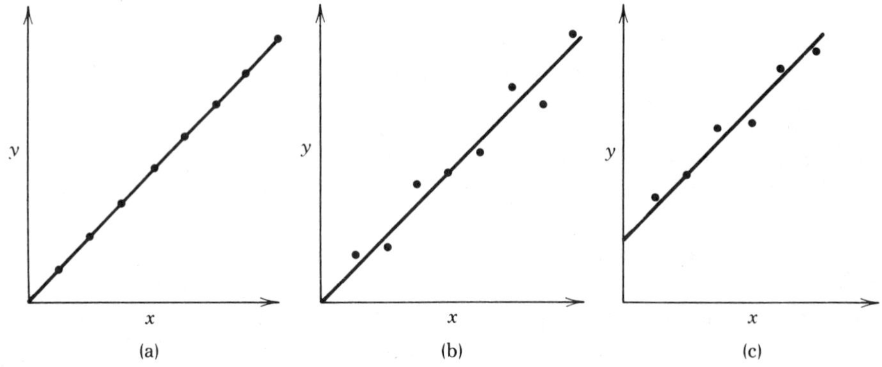

Fig. 10.4 A regression or least-squares comparison of a reference method with a new or test method. It is assumed that the reference method is well established and yields both precise and accurate results. Each point represents both methods: its x coordinate is the value obtained for the analyte by the reference method and its y coordinate that by the test method. (a) A representative plot for a test method which proves both precise *and* accurate. (b) A plot secured for a test method that is of low precision but good accuracy. (c) A plot obtained for a test method of moderate precision that introduces a systematic error (note the nonzero intercept).

The reader is referred to one of the references to pursue suitable mathematical calculations in connection with regression analysis and estimation of slope and intercept. Many programs available with instruments will provide necessary algorithms.

Naturally, regression analyses are also subject to error. Since the type of plot suggested is based on assumption of a straight line, it will force such a relationship even between methods that are nonlinearly related. Another likely source of error is that too small a range of concentrations may inadvertently have been available in samples, thus denying adequate estimates of precision and bias. A large range is preferable. It may also be the case that the method in use is plagued by larger errors at higher concentrations. If this is true, a weighting factor proportional to concentration needs to be multiplied by the values measured. Further, if a plot of response as a function of concentration is involved, it is important to force the plot to go through the origin by reducing the equation to $y = ax$. An example would be a Beer's-law blank for a spectrophotometric determination.

Finally, whenever data scatter badly (e.g., worse than in Fig. 10.4b) a value of the *correlation coefficient* r should be calculated to establish whether there is a significant linear correlation between the sets of x and y values. References should be consulted for details [1, 9]. If the value of $r_{calc} > r_{stat}$, as given in tables, the correlation is significant and a useful regression line can be calculated.

***F*-Test.** If two analytical methods are under comparison, it is reasonable to consider whether their relative variances or estimated standard deviations may not be compared for statistical significance. The Fisher F-test is such a method. Since the variances cannot really be compared directly, the test examines the ratio of their variances. A factor F is defined as

$$F = s_1^2/s_2^2 = V_1/V_2, \qquad (10.11)$$

where by convention the larger value is in the numerator. The test can be applied widely. Since a comparison of procedures is being made, the ratio F should ideally be unity for methods of comparable precision. As with all statistical tests the hypothesis to be tested must be stated unambiguously at the outset. A brief set of F values at the 95% confidence level is given in Table 10.3.

Extensive F-function tables have been developed. Since standard deviations compared may be based on different numbers of replicate measurements, the columns in F tables are by degrees of freedom for method 1 and rows for the degrees of freedom for method 2. Thus, separate tables are required for each probability level of interest, for example, there will be one for $\alpha = 0.05$ and another for $\alpha =$ 0.01.

Example 10.10. Replicate measurements carried out by using method 1 are to be compared with results obtained by using method 2. Assume all measurements are made on the same sample. Show how the F-test can be applied.

F_{calc} should be compared with the value in the table for the chosen value of α, say 0.05, and for the appropriate number of degrees of freedom. If $F_{calc} > F_{stat}$ it is probable at the selected level that the procedures compared are significantly different in bias. Conversely, if the observed value is smaller than that in the table, precision may be assumed to be comparable in both procedures.

Analysis of Variance (ANOVA). Actually Student's t-test and Fisher's F-test are both special cases of application of a powerful statistical technique called the analysis of variance (ANOVA). It allows an assessment of contributions from one or many factors believed to be influencing the outcome of a measurement. For example, this approach allows the comparison of methods to be extended from a pair or a group of two, as in the discussion of the Student t-test, to several. The F-test is often applied in this fashion.

Table 10.3 Values of F at the 0.05 Probability Level

Degrees of Freedom of the Lesser Variance	*Degrees of Freedom of the Greater Variance*					
	1	2	3	4	8	∞
1	161	199	216	225	237	254
2	18.5	19	19.2	19.2	19.4	19.5
3	10.1	9.55	9.28	9.12	8.85	8.53
4	7.71	6.94	6.59	6.39	6.04	5.63
8	5.32	4.46	4.07	3.84	3.44	2.93
∞	3.84	3.00	2.60	2.37	1.94	1.00

ANOVA is also routinely used in connection with "round-robin" interlaboratory testing of analytical procedures. In this case two variables are usually examined: the dependence of results obtained on the laboratories and the dependence on the samples used with the methods.

Like the other hypotheses tested in this section, the analysis of variance is applied to establish the factors that produce systematic error in results and the ones that simply contribute to the random error. When several kinds of effects are operating in a system a proper experimental design and use of ANOVA will help assess their relative importance and estimate their magnitude [6–8].

Example 10.11. A collaborative test of a new method is to be carried out in two or more different labs. What types of error should be considered?

Systematic error inherent in the method itself will need to be distinguished from that attributable to the differences in instruments, reagents, and ways of doing things in the two laboratories. Since it is only error in the method that is of interest, a complex experimental design will be necessary to provide a basis for separation of the two types of systematic error and of both from random error. It would be more satisfactory to assess the new method in the lab where it will be used. References should be consulted for more information on interlaboratory tests.

Example 10.12. In an attempt to reduce the estimated standard deviation found in an atomic absorption (AA) spectrometric determination of magnesium in soil, the broad sources of error were traced. Earlier work had shown that that the method itself had little error in the concentration range of interest. Accordingly, the possible sources of major error explored experimentally were the instrumental, sampling, and chemical (an acid extraction) aspects.

Data were collected systematically to evaluate each of the terms in the following equation for the total variance:

$$V_{\text{total}} = V_{\text{instr}} + V_{\text{samp}} + V_{\text{chem}}.$$

To simplify the problem of units, all data were calculated as coefficients of variation (see p. 345). How could data be collected to estimate these quantities? To obtain V_{total} after several samples were taken, preparative steps (mainly an extraction) were performed, and AA measurements made. A value of V_{instr} was obtained from several replicate AA measurements on a single sample after magnesium was extracted in the regular way. Rather than seek the variance of sampling directly, it was calculated by difference. For this purpose a series of replicate extractions and AA determinations were made *on one sample* and the resulting variance $V_{\text{instr}} + V_{\text{chem}}$ was subtracted from V_{total}. The results were

$$V_{\text{total}} = 275 = 12 + 235 + 28,$$

where the subvariances are listed in the original order. Clearly, sampling has contributed strongly to the variance and ways to improve this step require attention. Indeed, it appears that effort spent refining other aspects would yield relatively little improvement. [Suggested by the work of A. B. Calder, *Anal. Chem.* **36**(9), 26A (1964).]

10.5 CONTROL OF SYSTEMATIC AND RANDOM ERROR

Once the probability of bias or of substantial random error has been established, how can the source(s) of error be found and controlled?* A good way to identify sources of error is to list possible systematic and random errors in each step of a measurement. An example of such a listing is shown in Fig. 10.5 for mass spectrometric measurements. How many of the possible errors are occurring and their magnitude can be established by careful observation. Other possibilities will probably come to mind also. If plausible magnitudes of error can be established, their contribution to measurements can be estimated by a study of propagation of error (see Section 10.3). Another gain from making up a list is that concrete ways to deal with each type of error will often come to mind.

A second valuable approach to control of error is to set up categories into which errors commonly fall. Here are some common types and some approaches to reducing them.

1. *Methodological error,* especially in new methods, may result from a lack of validity of a technique for the measurement at hand or from its failure to give quantitative results. For example, a trace analysis may have too small a percent recovery of the analyte (see Section 11.6). Nevertheless if the amount of recovery is reproducible, the method may be corrected by incorporating that figure in the final calculations. Or, a redox titration may be based on an incomplete oxidation. In that case, if the problem is a slow reaction, the temperature can often be raised to permit a complete reaction in a reasonable time.

2. *Sampling errors* will often occur, especially in seeking representative samples of *heterogeneous masses* such as ore deposits or the atmosphere. Section 10.7 will deal with sampling theory.

3. *Procedural errors* often result from unclear protocols, departures from the procedures, or operator carelessness. To minimize error in routine measurements, detailed instructions must be drawn up to ensure that every sample is handled alike in all steps. An attractive way to reduce this source of error is to employ robots for sample preparation and other steps involving manipulations.

4. *Instrument error* must always be investigated.

Example 10.13. When should one be concerned with the validity of a measurement? Most new procedures or modifications of methods call for careful examination. For instance, applying a pH measurement that was developed for aqueous solutions, and is based on the

*Uncertainty in measurements in an earlier period was sometimes categorized as determinate or indeterminate error. *Determinate errors* were said to arise when the use of different modules, reagents, or processes gave differing values, while *indeterminate* or unpredictable errors were attributed to random excursions in temperature, humidity, line voltage, pressure, and a dozen other parameters. Commonly, closer study of indeterminate errors would uncover many determinate errors. In recent years the categories of systematic and random error have tended to replace the earlier ones since the latter terms are clearer and also are firmly grounded in statistical assessment.

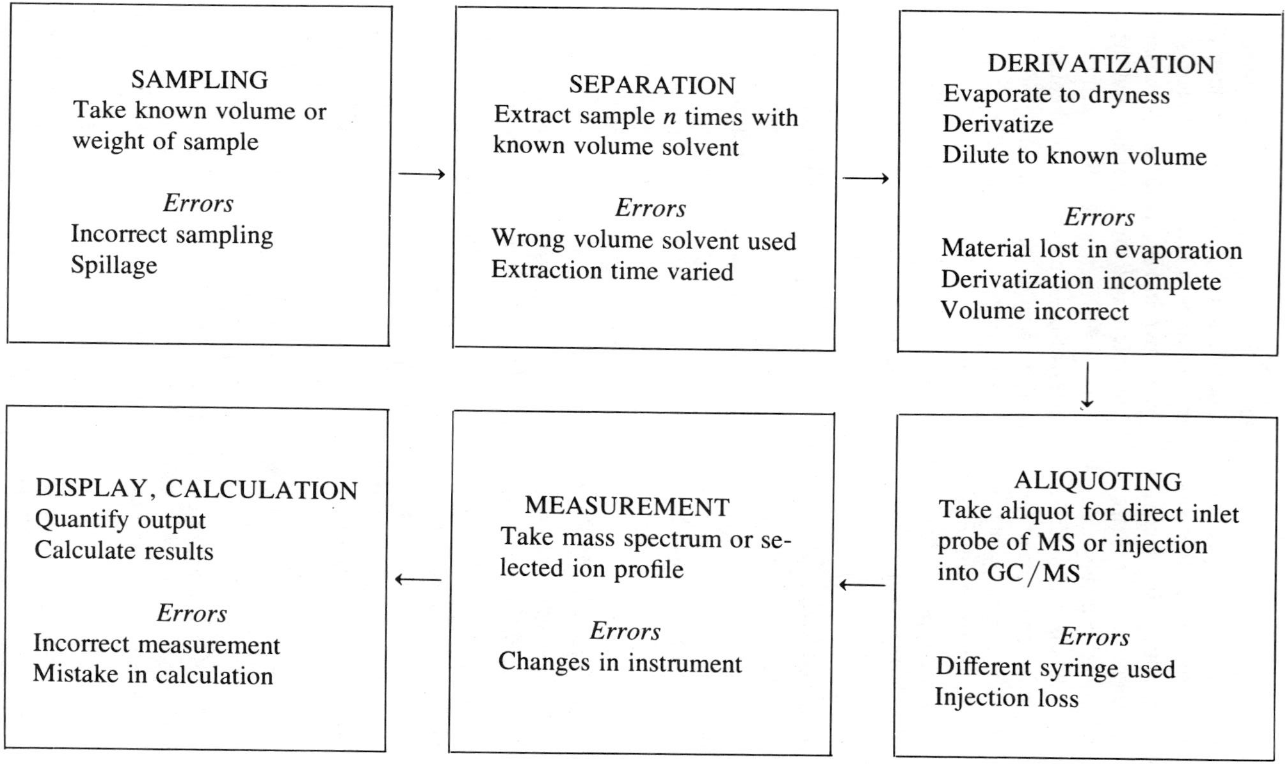

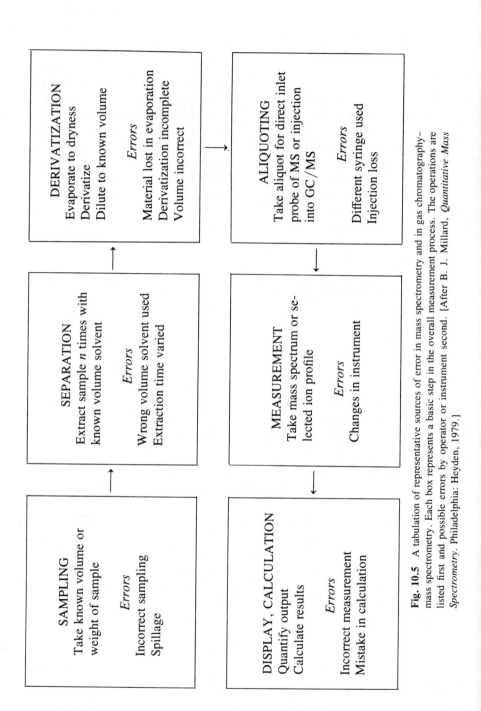

Fig. 10.5 A tabulation of representative sources of error in mass spectrometry and in gas chromatography–mass spectrometry. Each box represents a basic step in the overall measurement process. The operations are listed first and possible errors by operator or instrument second. [After B. J. Millard, *Quantitative Mass Spectrometry*. Philadelphia: Heyden, 1979.]

use of a glass electrode and pH meter, to acetone solutions is a such a modification. The system may yield stable readings in acetone but it cannot be assumed without careful examination that its readings are still a valid measure of hydrogen ion activity related to similar measurements in water (see Chapter 28).

As sources of error are identified, their contribution to overall error can be calculated by an analysis of propagation of error, as discussed earlier. Special attention can then be given to major sources of error.

To maintain quality control the statistical assessments of error discussed in the last section should be applied in laboratories to all new measurements and instruments and also to routine procedures that are used month in and month out whenever

new lots of reagents are bought, and *new persons* become involved in analysis;
instruments age or are modified;
part of a procedure is altered;
a *new* or *modified instrument* is put into use.

In a larger laboratory where routine samples are frequently analyzed, a sequential record of the deviations from the mean obtained whenever a control sample is analyzed, for example, a *control chart* such as that shown in Fig. 10.6, is valuable. The graph provides an immediate record of analyses for which error is larger than

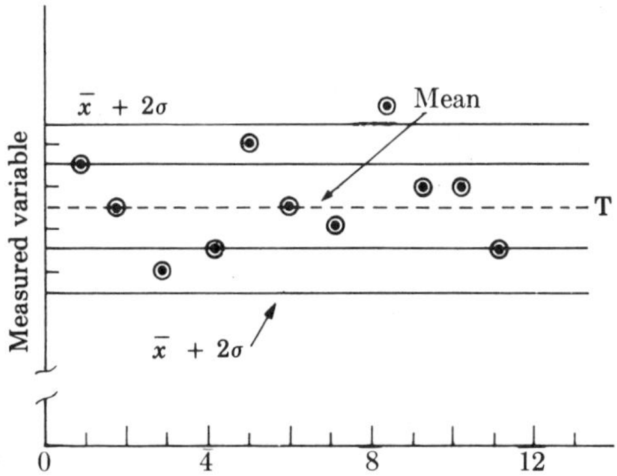

Fig. 10.6 Measurements on control samples plotted in the order obtained. When a routine analysis is to be carried out on similar samples over a long time it is valuable to chart the results with control samples in this way as a *control chart* for the measurements. When the method is first introduced, a sufficient number of samples are measured to obtain a mean and standard deviation. The mean becomes the target value T on the chart. Lines are also drawn to include specified portions of the total population of measurements with the method as standardized. Note the $\bar{x} \pm 2\sigma$ lines, which should include 95% of the measurements between them. They are called *warning limits* since few measurements will fall outside unless systematic error begins to occur. Though not shown above, ordinarily $\bar{x} \pm 3\sigma$ lines are added as action lines. Whenever a control sample gives a value this large or larger, a reassessment of the method is indicated. Some action must be taken.

expected random error, or of persistent departures from randomness such as "runs." Often $\bar{x} \pm 2\sigma$ limits are set on such charts as "warning limits" and $\bar{x} \pm 3\sigma$ as "action limits."

Minimizing Bias. What general procedures exist for minimizing systematic error? Three will be discussed, calibration, use of blanks, and use of controls. They are attractive because they can be used without a detailed study of a system. If procedures specific to a particular method seem necessary as well, they can be added.

Calibration or standardization is a valuable routine method for controlling bias. It is almost always applied when new pipets and other volumetric equipment, chemicals, and instruments are introduced. By direct testing or by comparing the performance of new units with that of a standard, corrections or calibration curves are developed. Ordinarily, calibrations are organized in tabular or graphic form or, even better, entered into memory and called up as part of control programs for instruments. Thus, during actual measurements these measures provide compensation for irregular scale graduations, chemical impurities, and untheoretical responses of instruments.

Blanks are frequently effective where many sources of bias must simultaneously be kept in check. A blank is a sample identical in makeup with the unknown except that the constituent of interest is omitted; in measurement it is treated in the same way as the unknown. If a blank and sample are affected alike by bias, an appropriate comparison of the results should produce data free of systematic error. In most cases values for blanks are simply subtracted from sample values.

A particularly good illustration of the usefulness of blanks is provided in spectrophotometric determinations. Recall that in this case the ratio of the value for the sample to that for the blank is taken. Because of ratioing, blanks in spectrophotometry can be used to eliminate the effect not only of solvent and other nonanalyte absorption and scattering, but also of reflection from and absorption in cell windows.

Finally, *controls* are more commonly used in noninstrumental systems that are subject to so many variables that not all can be regulated. When controls are used in analytical work, they are usually samples of known concentration in x as much like the unknown as possible and are used as a running reference. A general example is provided by the common method for testing the resistance of a new metal alloy to salt water. In this instance a few known samples are designated as controls and are treated in the same way as samples of the new alloy. The corrosion of the latter samples can then be compared with that of the controls to eliminate bias arising from variation in procedures.

The more empirical the procedure, the greater the need for controls. Thus, in flame emission and absorption analyses, controls or bracketing of samples by similar standard solutions are very valuable. Nebulization and evaporation rates of solution samples in flames depend on variables that can be precisely controlled only with great difficulty. By noting the results obtained with known samples just before and after analyzing an unknown, variation in operating conditions is compensated and any adjustments necessary for precision and accuracy may be made.

Control of Variables. Often error enters when parameters such as pH and temperature drift or vary randomly. The body of theory associated with a method usually identifies variables that will affect a measurement and must be held constant. For example, close thermostating must be provided when conductometric detection is used in ion chromatography because values are strongly temperature-dependent. In some cases good insulation may provide the degree of constancy needed.

Though constancy of a crucial variable may minimize random error, the question arises of whether an *optimum value* of the parameter may not profitably be sought. For example, in a kinetically controlled measurement involving an enzyme, what value of pH will give maximum sensitivity of measurement? This sophisticated aspect of control called optimization will be explored in the next section.

10.6 OPTIMIZATION PROCEDURES

When all variables affecting a measurement are set at optimum values, that measurement certainly will be made under conditions of highest S/N and sensitivity. In addition, the higher S/N secured will enhance accuracy and precision and thus the quality of the measurement.

What variables require optimization? Clearly the independent variables, for example, wavelength and concentration in spectrometric measurements, are excluded: the theory of each technique yields equations that define the central role of these variables. By contrast, few guidelines exist to show the contribution of less important variables such as temperature. Best values for these minor factors have often been found by changing the value of one factor at a time, but this process is not only slow but will give less than optimum results if the variables interact. The usual approach today, however, is to employ experimental *optimization methods* that allow one to plan and carry out the minimum number of experimental trials necessary to identify optimum values. Because these methods bring mathematics to bear on the measurement system, they can also be termed chemometric techniques.

Optimization methods can probably be best approached by examining the representative *response surface* illustrated in Fig. 10.7. In general this type of surface shows the dependence of a particular property of an analyte on two or three minor variables. For example, Fig. 10.7 might represent the fluorescence of a solute at a given concentration and at selected (excitation and emission) wavelengths as solution temperature and pH are varied over reasonable ranges.

Example 10.14. Interpret Fig. 10.7 by assuming that it represents measurements of the rate of an enzyme reaction at different pHs and concentrations of an impurity.

Though these kinetics measurements seem to require a three-dimensional response sur-

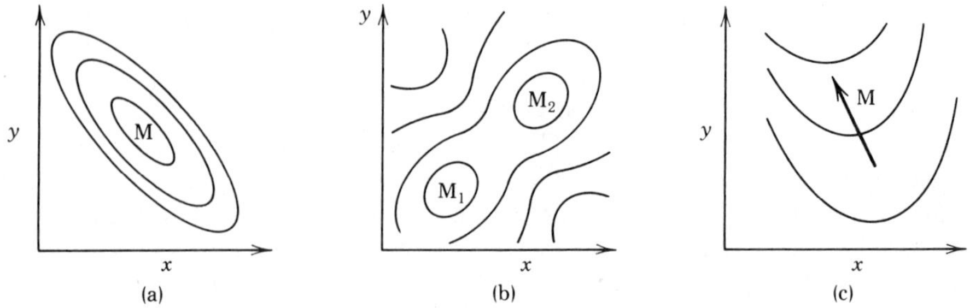

Fig. 10.7 Some two-dimensional measurement-response surfaces. The value of one important parameter such as pH is plotted on the y axis and that of another parameter on the x axis. Contours represent lines of equal response. (a) A surface with a single maximum (or minimum) response M, (b) one with two maxima (or minima), M_1 and M_2, and (c) a sloping surface. The goal of optimization is to find the values of x and y that yield a maximum response or will give highest sensitivity of measurement.

face, Fig. 10.7 involves only two. It is clear that Fig. 10.7 uses a common plotting strategy to project three dimensions onto two: it is a *contour diagram* made up of equiresponse lines that connect values of pH and impurity concentration that yield equal reaction rates. On this diagram optimum rates corrspond to peaks whose coordinates identify the optimum values of the minor variables. One does need to decide whether a central loop defines a valley or a peak. To find the overall or global maximum for the system requires a choice among competing local maxima if more than one appears.

Example 10.15. What minor variables are likely to be important in measurements?

Usually factors such as concentrations of interfering substances, temperature, and pH should be optimized. Interferents will usually compete with analytes when chemical preparative steps are involved in a measurement, pH will determine the position of any acid–base equilibrium that involves an analyte (as temperature may also), and temperature will affect solution properties such as density and refractive index.

It now becomes straightforward to relate optimization methods to response surfaces. Calculus or the mathematics of finite differences can be used to track changes in slope on a mathematical response surface and locate the global maximum. Complications are likely to arise mainly for cases in which there may be three or more important minor variables. It should be easy to identify them. Assuming different kinds of experimental situations can be ranked in importance, what different types of optimization processes can be identified?

The simplest case is that in which minor variables have a sufficiently small influence on the analytical signal that their effect can be *approximated as a random one*. Clearly, no attempt need be made to find their optimum values. For instance, temperature has little effect on UV–VIS spectrophotometric measurements.

In a sizable number of cases there are only two important minor variables. It is principally this type of case that will be described below. Where three factors are important, ways are indicated to extend the approaches that are described.

Simultaneous Optimization Techniques. In this approach a set of values is given to all variables of interest. Either values within the common or meaningful range are assigned randomly or by factorial design. Since the pattern of influence is usually of as much interest as the magnitude of optimal values, it is common to limit the initial exploration to two factors. Other factors are held constant and contribute only their random (background) fluctuations. This approach yields a two-dimensional response surface. It is also possible to deal with many variables by collecting them into just two *factors* [8].

To set up a *random design*, a large number of pairs of values that differ by increments can be entered in a table. Values must of course be realistic; for example, with biological macromolecules there are definite limits on temperatures and pHs that can be tolerated. By then consulting a random number table, a small number of sets can be selected for trial. Once the experiments have been made, the set giving the best response is to be taken as identifying optimal values of these variables. Usually, it is valuable to repeat the procedure near the optimal value to improve the selection of the final set of values. In general this type of approach functions most successfully where many variables are involved and experimental error is modest.

Alternatively, a *systematic* design can be used. One type is developed by incrementing each principal variable by the same amount to develop a set of values within an appropriate range. For example, for variable X, the values chosen might be $X_1, X_2, \ldots, X_n$. For parameter Y a similar set of incremented values is selected. All possible combinations $X_1Y_1, \ldots, X_n\,Y_n$ are judged worthy of investigation in the search for an optimal set. Clearly this method can lead quickly to an untenably large number of experiments.

A simultaneous, systematic design with more strategy is that of *factorial design*. This approach provides experimental designs that economically reveal the manner in which factors or variables influence a measurement. Two types of design are outlined in Table 10.4. In the single-factor design two factors, A and B, are tested

Table 10.4 Experimental Designs to Test Effect of Variables on Measurements and Their Interdependence

	Single-Factor Design			*Three-Factor Design*[a]			
Design Number	*Factor A*	*Factor B*	*Response*	*Factor A*	*Factor B*	*Factor C*	*Response*[b]
1	Average	Low	g	A	b	c	m
2	Average	High	h	a	B	c	n
3	Low	Average	j	a	b	C	o
4	High	Average	k	A	B	C	p

[a]Capital letters in the body of the table denote high values for the designated factors, lower case letters, small values.

[b]Average results are calculated in the three-factor design as follows: $\bar{a} = (n + o)/2$; $\bar{b} = (m + o)/2$; $\bar{c} = (m + n)/2$; $\bar{A} = (m + p)/2$; $\bar{B} = (n + p)/2$; $\bar{C} = (o + p)/2$.

one at a time. Each is varied individually; for example, both a low and a high value for B are used while A is kept at an average level. Four experiments suffice to ascertain the *independent* effect of the two variables on the measurement. Note that the importance of variable A will depend on the percentage change represented by results h–g and that of B by the change k–j. Additional experiments would be required to discover whether the factors are interdependent.

A still more economical, *multifactor design* is suggested in the right-hand portion of Table 10.4. The role of three factors, A, B, and C, at two levels is ascertained in only four experiments. At the same time their interdependence is checked. Each run differs from another in that two of the factors are changed in level. Thus, the experiments involving C (each capital letter indicates a factor at a high level) also involve the other factors first at high levels A and B and then at low levels a and b. Accordingly, if the average result labeled $\overline{C}$ (see footnote on Table 10.4) is significantly different from the result labeled $\bar{c}$, it must be factor c that has produced the difference. Factorial design can be extended to many more variables effectively [9].

As has been suggested [6], such studies are especially valuable to ascertain fairly small effects on measurements. A minimum number of experiments is required and averages are used to determine differences, giving more precision.

Applications of factorial design to measurement are numerous. One is ascertaining probable kinds of variations that may come up in interlaboratory round-robin measurements on a test series of samples. Another application is developing a measure of the ruggedness of a particular analytical technique. A rugged method provides a response that is largely independent of changes in the common variables. A third example of factorial design is to develop a design for sampling bulk materials, a subject that will be discussed in the next section.

Sequential Techniques. Sequential techniques are designed to permit one to start with a few initial experiments at most and vary conditions for later experiments as suggested by the response obtained. Three procedures will be described briefly and the simplex procedure, which is most widely used in chemistry, will be developed more fully.

1. *Single-Step Procedures.* In this design one factor is varied while the others are kept constant. Then another is varied in the same fashion. Experiments are done with each set of factors. Each new set is selected on the basis of judging whether response is increasing until an optimum is obtained.

2. *Steepest-Ascent Method.* Either by a single-step method or by a factorial scheme, experiments are set up that allow an estimation of the response gradient. New experiments are designed to move the system in the direction of steepest slope.

3. *Simplex Methods.* A *simplex* is a geometric figure defined by the number of main variables plus one. For example, a system with two main variables would have an equilateral triangle for a simplex and one with three variables, a regular

tetrahedron. An experiment is carried out at the conditions indicated at each vertex of a simplex. With the completion of the experiments a prescribed operation is invoked to move the figure toward a region in which an optimal response will be found.

Simplex Optimization. This method has gained wide acceptance since it requires a relatively small number of experiments and a minimum of calculations. As would be expected, two-dimensional simplexes that employ equilateral triangles (usually) are the most common. Two disadvantages should be noted. When the procedure yields a maximum, it cannot be known with any certainty that the result is a so-called ''global maximum'' rather than one of several local maxima. Actually, this disadvantage is also likely to obtain in other methods. It also proves difficult with three or higher dimensional simplexes to recognize patterns.

To illustrate the simplex technique, consider its application to a two-factor system in the examples that follow. A more detailed treatment is beyond the scope of the text.

Example 10.16. From information about the variables that may have the greatest effect on a system of interest, assume two have been selected for study with a view to optimizing system response. First, scale a two-dimensional coordinate space to include the entirety of the experimentally reasonable range of values of the two parameters. Let the variable X_1 be along the abscissa and variable Y_1 be along the ordinate. Select three sets of values of the parameters that seem favorable for good response in such a way that X_1, Y_1, X_2, Y_2, and X_3, Y_3 form an equilateral triangle. Draw in the points and the triangle they form. It should occupy only a fraction of the coordinate space. Carry out the measurement being studied for the three selected sets of parameters and tabulate the responses.

In this sequential technique examine the first simplex and choose the next point by the following procedure. Discard the point that gave the smallest response. Draw a second triangle the same size by using as its base the side that remains and enter a point as apex that is opposite the discarded point and an equal distance from the base. The second triangle or simplex is said to be formed by ''reflecting'' the first triangle through the side with the original points of higher response. In Fig. 10.8 the first triangle is ABC and the second BCD.

In Fig. 10.8 equiresponse lines are shown, but they were actually drawn in after the fact by using the actual response values which were tabulated but are not included here. After BCD note the development of two more simplexes. In terms of the contour lines it appears that good progress toward the optimum value was made in this sequence.

Example 10.17. In Fig. 10.9a the simplex process appears to stall. Simplex DFG was reached by reflection from DEF because point E was the low vertex. From the response surface it is clear that point G must also be a low vertex. Thus, continuation without employing a new strategy will lead to repeated reflection back and forth across line DF.

A standard strategy in this situation is to reject the second lowest point in the last simplex, DEF. As shown in Fig. 10.9b this step leads to triangle FGH, and now the process has moved on in the right direction. Observe also that as simplexes approach the response surface maximum, they begin to encircle it. The goal has been attained.

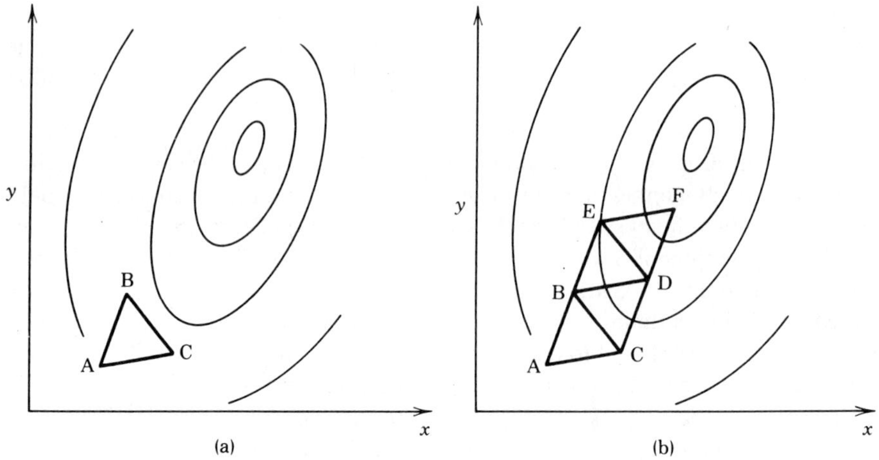

Fig. 10.8 Initial steps in a simplex optimization on a two-dimensional response surface. To show the logic of the process, contours (iso-response lines) have been drawn in after-the-fact. (a) The initial simplex (triangle ABC). (b) Successive replacements of the first simplex by the second one, BCD, third, BDE, and a fourth, DEF. [After S. N. Deming and S. L. Morgan, *Anal. Chem.*, **45,** 278A (1973).]

Example 10.18. Still another strategy in simplex optimization is to increase or reduce the size of the simplex itself. This approach must be used judiciously. Remember that the initial simplex size was chosen arbitrarily. A rule of thumb is to use a large simplex at the beginning because one may be far from an optimum set of values. The closer one appears to be getting to an optimum, the smaller the simplex can be made.

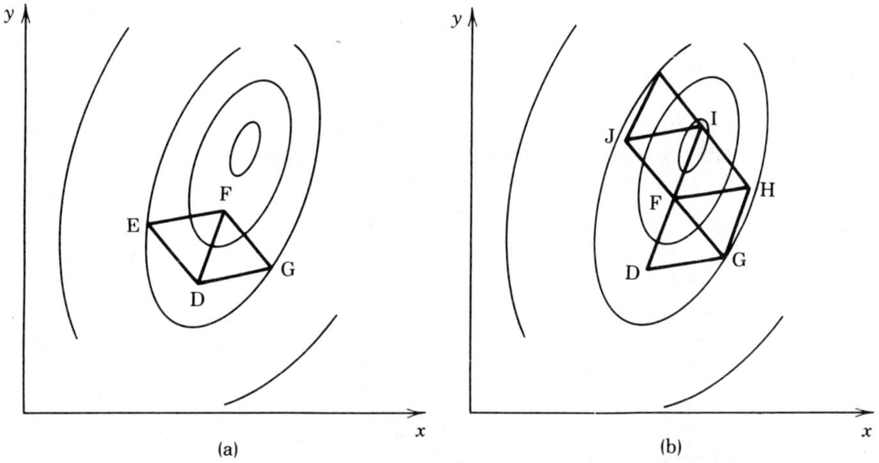

Fig. 10.9 Later steps in the simplex process begun in Fig. 10.8. (a) Development of a block to further progress. (b) Use of new strategy allows location of optimum response.

10.7 SAMPLING

Sampling must necessarily precede most analytical measurements. In research measurements, however, the homogeneity of experimental systems is usually carefully established so that accurate measurement of the behavior of interest can be made and sampling questions do not arise. *Sampling* is defined as the orderly selection of one or several representative portions of a larger system for analysis or measurement. The importance of good sampling must be stressed. An *invalid* sampling procedure will introduce a major systematic error in measurements on a system. Unfortunately, neither blanks nor controls can minimize such error; only a better sampling design will improve the situation.

Planning. What in general will constitute a reliable *sampling procedure*? Often the process is difficult because of the sheer size and complexity of the system. For example, how should one sample a forest for measurements that will provide information about acid rain? The essential first step is to develop a clear statement of the goals of measurement for the system of interest. Then each step in the measurement process outlined in Fig. 10.1 should be described as fully as possible and the desired precision and accuracy defined. Most of the information for this second step will be found in chapters that describe the measurement techniques to be used. Consideration must also be given to the nature of the bulk system. But it is the goals of measurement that will define both what is to be measured and how precisely it must be known.

Fortunately, analytical measurement goals tend to fall in categories. Some measurements intend to *describe an object* either (a) on the molecular or atomic level by identifying the species present or (b) on the microphysical level by locating each species on a surface or in the bulk of a specimen. For instance, the determination of the composition of a coal deposit would probably fall under the first, and determining the distribution of heterogenous catalyst particles on an inert support under the second. In other instances the goal may be *control*, as in monitoring a process stream to ensure a quality product. In a relatively small number of cases, measurements are intended for *threshold control*, for example in monitoring the concentration of a carcinogen such as ethylene oxide to ensure that it never exceeds a specified limit in a workplace.

The next step in planning is to identify the *parent population*, the overall aggregate of material or collection of items that make up the set about which information is sought. Once the parent population is described, a sampling plan can be developed that will secure samples for analysis that are *representative* of the aspect of a system that is under examination. If only selected members of a population are to be analyzed, the object will be to ensure that all parts of the system have an equal opportunity to be selected for measurement. Statistical techniques already developed will prove indispensable in relating desired precision to the questions raised above.

If the bulk aggregate of material is inhomogeneous there are two ways to proceed. One is to develop a *gross* or *composite* sample first by taking small *increments*, grabs, or samples from many portions of the body and mixing them thoroughly. From a homogenized gross sample small laboratory *subsamples* can be taken for analysis. It is essential that subsamples have the composition of the composite sample. The other technique is to analyze each increment separately. Relative advantages determine which method is followed.

Other requirements influencing the development of a plan will be the need for a sample of certain size, possible speed of sampling, equipment and personnel available for sampling, and cost. Because there are many criteria to be met as sampling is done and because different people will often be involved in sampling and measurement, it is essential to develop a careful written protocol listing all details. Peculiar care is required to ensure that steps in sampling are systematically followed.

Assessment of Sampling Variance. If at all possible, during the latter stages of planning a separate *assessment of variance* in the measurement process and in sampling should be carried out experimentally. It must be determined whether the sampling error that will result from a particular protocol is tolerable.

This assessment can often best be made indirectly. One set of samples can be taken through all the steps to obtain the total variance σ^2_{tot} of the method. Then a single sample can be subdivided and taken through only the process and measurement steps to obtain the variance contributed by those steps σ^2_{m}. With these data, the variance attributable to sampling, σ^2_{s}, can be obtained by subtraction: $\sigma^2_{\text{tot}} - \sigma^2_{\text{m}} = \sigma^2_{\text{s}}$. Example 10.12 illustrates this approach.

By comparing error in measurement and process steps with that in sampling, it is possible to decide whether to sample according to the initial procedure. The alternative is to try to refine the process. The question is important because sampling is often the step that contributes the *greatest* error in measurements.

Example 10.19. A coal seam in rock overburden is to be sampled in preparation for a series of analyses. It is found that an analytical precision of about 10% will be satisfactory. Since the rock will be ground coarsely, two questions become obvious. What increments should be taken to secure a representative gross sample? How can statistics be applied to ensure that the sampling makes sense? Before these questions can be answered some further information is needed.

In the planning stage, analysis of cores drilled from the seam show that (a) the coal and overburden (rock) differ substantially in hardness, and (b) during grinding of cores large chunks fracture mainly into pieces of rock or coal. Assume these preliminary studies show the seam to be 95% coal and 5% rock and that after grinding most pieces weigh about 50 g.

In this case experience suggests that a sample should be made up by taking many increments or grabs and combining them to make a gross sample. For instance, if only two particles (increments) are selected randomly either both will be coal, both rock, or one coal and one rock. Combining many increments will, however, give a gross sample approximating the composition of the seam. The gross sample will need to be ground finely and subsamples taken for analysis.

Since the analyst can assume only two types of solid in the sample, simple statistics provides a theoretical framework for sampling. The question asked is the fewest pieces (increments) one must take for a representative sample. From statistics the estimated standard deviation of mean composition for this system is given by

$$s = (1/n)\sqrt{P_{\text{coal}}P_{\text{rock}}n} \qquad \text{and} \qquad n > P_{\text{coal}}P_{\text{rock}}\,10^4/s^2, \tag{10.12}$$

where P_i is the fraction of component i in the coal seam and n is the minimum number of particles that can be in the gross sample if it is to meet the required precision (10%). After substitutions are made in Eq. (10.12), one has

$$n > 0.95 \times 0.05 \times 10^4/100 \geq 5.$$

The number of particles (increments), 5, is quite modest, though the weight of gross sample will be 0.25 kg. If a still more representative sample, one precise to $\pm 1\%$, is needed, more than 450 particles (about 22 kg) would need to be included in a sample. Finer grinding should of course greatly reduce the number of particles needed. Further, it will avoid samples of enormous size.

Example 10.20. Some of the ways in which sampling may fall short of the ideal are worth listing.

1. As discussed in the preceding example, too few increments of material may be taken to duplicate the diversity of an actual system. For instance, how many increments are needed at an abandoned chemical dump site?

2. The sampling procedure may inadvertently introduce preference with respect to some aspect of the system. Thus, fine particles in a solid sample may escape as dust unless the possibility is noted in advance. Further, the description of a system comprised of a mass of uniform crystals will unwittingly be skewed by regular sampling of pieces fractured at grain boundaries.

3. Changes in a sample may occur as it is taken or while it is stored awaiting measurement. Thus, oxygen may cause chemical changes in an organic sample, trace components in a sample of river water may be adsorbed onto the walls of a container, and in a slurry from an industrial process stream any liquids present may evaporate.

Random Sampling. For heterogeneous systems random sampling is the method more likely to ensure a low bias error, as seems clear intuitively. Nevertheless, a general answer cannot be given as to whether a random selection scheme or a uniform sampling procedure (e.g., selecting every *n*th increment) will be the better approach to sampling in general.

The *standard method for random sampling* is to divide the bulk of material into many segments, usually by imposing an imaginary three- or two-dimensional grid. Each unit or cell should then be assigned a number in sequence. After selecting the number of samples (see examples below), they are taken randomly. A random number table is used to identify cells that should be sampled. One starts in an arbitrary place in the table and moves down a column, or from left to right, to find the numbers that will identify "cells" on which measurements are to be made. The method of movement through the table must be selected in advance and continued until one has the requisite number of samples.

Example 10.21. What should be done to improve precision of a measurement if the sampling error is large and cannot be easily reduced? A rule of thumb is suggested by Youden: If the variance of the instrumental and procedural steps is no more than about one-third the variance of sampling, there is no reason to seek any greater precision in the instrumental and procedural steps.* Even if instrumental and chemical procedural steps could be made more precise, the overall measurement would show no *appreciable* gain in precision.

An interesting, perhaps unexpected, inference can be drawn. If sampling error seems certain to remain large, it makes eminent sense to seek a less precise analytical procedure that is faster and cheaper. For instance, sampling is at present the major source of error in determinations of aflatoxins (highly toxic compounds from molds) in peanuts. As a result, a fast method of low cost and only moderate precision is usually selected to determine the compounds. The tradeoff of precision, which could not be attained anyway, for time and cost is an easy one.

Example 10.22. Probably the most difficult question is how to sample when the properties of a material are not well known. When little is known about a bulk material, one simply takes several apparently representative samples. They are analyzed for the constituents of interest and confidence limits for the mean calculated from Eq. (10.3).

These results should permit a tentative decision about the *number of increments* needed to obtain a representative gross sample and how large each increment should be. The number must be large enough to give the desired level of confidence in the measurement. Since it is reasonable to assume that data will be distributed normally, the number of increments n can be found from the expression

$$n = t^2 s^2 / \sigma_r^2 \bar{x}^2, \tag{10.13}$$

where σ_r is the relative standard deviation desired in the measurement and s and $\bar{x}$ are the values determined above. One assumes a value for t, which depends on n, and uses an iterative procedure to obtain n. Good initial choices are the value of t for $\alpha = 0.05$ and $n = \infty$. The calculated value of n is now used to obtain a new value for t, and iteration is continued until n reaches a constant value. If data are distributed other than normally (Gaussian distribution), slightly different expressions for n may be necessary.

Defining the *size of increments* that is appropriate is also a crucial aspect of sampling. A useful approach has been developed by Ingamells.† In this approach a set of determinations is made on increments to develop a sampling constant K_s for each system, whether it is a biological system like a muscle, in which concentrations of certain elements are of interest, or a mountain. Basically, K_s is the minimum weight of sample (increment) necessary to insure that sampling uncertainty is no more than 1% at the 68% confidence level. Ingamells developed an equation for K_s and showed that it fits the relationship $WR^2 = K_s$ in many cases. Here W is the weight of sample (increment) taken for analysis and R is the relative standard deviation of sample composition.

*W. J. Youden, *J. Assoc. Off. Anal. Chem.*, **50,** 1007 (1967).
†C. O. Ingamells and P. Switzer, *Talanta*, **20,** 547 (1973); **23,** 263 (1976).

10.8 THE LIMIT OF DETECTION

At the very small concentrations that characterize *trace* and *ultratrace analysis*, and with the minute samples employed in *microanalysis*, the *limit of detection*, a new criterion of performance, becomes important. Because noise itself now becomes limiting, many questions arise. Should a minute instrument response be interpreted as arising solely from random fluctuations in one or more variables? And more importantly, how large must the reading from a sample be, compared with that from a blank, for a person to report that analyte *is* present?

At the detection limit statistics and chemistry become intimately mixed and statistical procedures must be used to interpret chemical results. The detection limit is simply the smallest concentration of analyte that can be certified as statistically different from a blank. To determine the limit two sets of measurements are necessary. In addition to a regular set, many measurements (10–20 are recommended) must be made on a blank under the same conditions. It must simulate regular samples but be without analyte.*

To calculate the limit of detection one begins by observing the *limiting response* x_L. For a given measurement procedure, it is the smallest response that can indicate with reasonable probability that a particular analyte is present. This response is expressed as

$$x_L = \bar{x}_{bl} + ks_{bl} = \bar{x}_{bl} + 3s_{bl} \tag{10.14}$$

where $\bar{x}_{bl}$ and s_{bl} are the mean and estimated standard deviation of the blank, respectively, and k is "a numerical factor chosen according to the confidence level desired."† The rationale for choosing $k = 3$ in Eq. (10.14) will be taken up below. It is important not to confuse the standard deviation s_{bl} with the standard deviation of the blank mean and other similar measures.

To obtain the limit of detection from x_L requires a calibration curve or a standard addition curve. A usable curve will be one based on quite pure standard solutions of extremely low concentration. The alternative, a long extrapolation of a calibration curve beyond the lowest points for standards, even in those favorable cases in which the curve is linear, will invite substantial error (see Section 11.3).

A good illustration of the operations involved in deciding the limiting response and limit of detection is provided in Fig. 10.10. Note that the calibration curve in the figure intersects the ordinate at a height corresponding to the mean blank response $\bar{x}_{bl}$. The limiting response x_L is represented by a point $3s_{bl}$ above that. If S is the slope of the curve, the limit of detection c_L is then simply $c_L = 3s_{bl}/S$. Details are worked out in the following example.

*Sometimes background or baseline levels are also of interest (e.g., in spectral measurements). For calculation of limits of detection, however, such results do not have the status of measurements on blanks. As might be expected, baseline values are often higher than those on a blank [R-P Stoessel and A. Prange, *Anal. Chem.*, **57**, 2880 (1985)].

†*Pure Appl. Chem.*, **45**, 99 (1976).

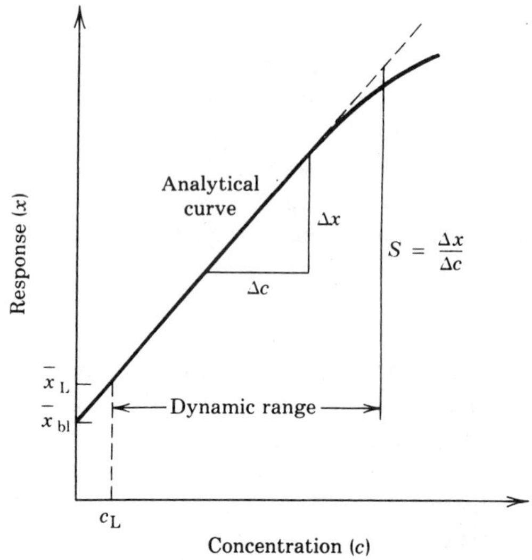

Fig. 10.10 An analytical curve showing the response x of a measurement system (full line) as a function of the concentration (or amount) of an analyte. The slope dx/dc is by definition the *sensitivity* of the measurement. By definition, the y intercept of the analytical curve is the mean response of a blank, $\bar{x}_{bl}$. The smallest significant response or limiting response x_L must be higher. It is defined by the expression $x_L = \bar{x}_{bl} + ks_{bl}$, where s_{bl} is the estimated standard deviation of the blank and k is a constant; usually $k = 3$.

Example 10.23. How does one calculate the limit of detection or limiting concentration c_L from Eq. (10.14)?

First the background or blank response must be subtracted from the limit of detection: $x_L - \bar{x}_{bl}$. The difference, according to Eq. (10.14), is just $3s_{bl}$. Note that this step is also evident in Fig. 10.10. Then to convert to c_L, divide by the slope, which is the limiting sensitivity S:

$$c_L = 3s_{bl}/S. \qquad (10.15)$$

Example 10.24. From a statistical perspective, how valid is a detection limit based on the confidence limits set by choosing $k = 3$? Assume the measurements follow a normal distribution.

One way to answer is to study the distribution of the measurement responses of a blank, as given in Fig. 10.11. Note the position of the limit x_L within the distribution. If as many as 20 replicate measurements have been made on a blank, s_{bl} can be assumed to approach the standard deviation. When these conditions hold, $k = 3$ corresponds to working at the 99% confidence level according to Table 10.1: a response this size or larger will result from background alone, less that 1% of the time. The hatched area in Fig. 10.11 represents this possibility. Stated differently, there is a nearly negligible chance that the signal as large as x_L would occur when analyte is absent.

As has been pointed out [8], with the IUPAC definition there is some probability that a

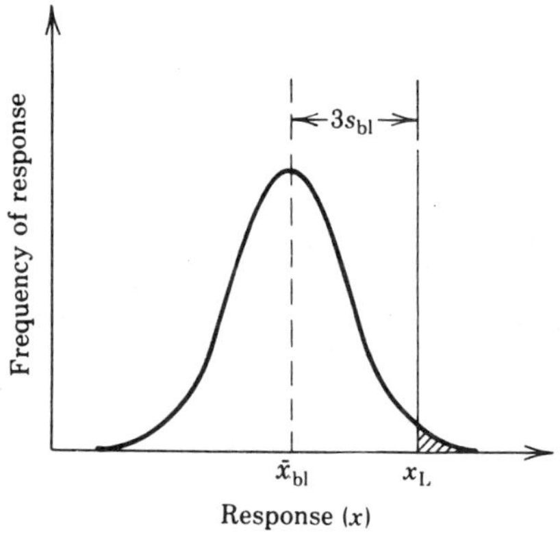

Fig. 10.11 Predicted distribution of measurements of the blank for a method whose responses are assumed to be distributed normally. The limiting response x_L is shown.

trace of analyte is present when measurements fall between $\bar{x}_{bl}$ and x_L. Setting $k = 2$ would virtually eliminate an error of this type. Most chemists agree, however, that reliability of detection is especially important. For example, certainty of the presence of an analyte is especially important in environmental studies. When $k > 3$, detection is virtually unquestionable.

Limit of Quantification. Actually, as quantitative measurements at the trace level have become more important than simple detection, a limit of quantification or *limit of determination* x_Q is recommended by IUPAC for comparison of the relative merits of different trace methods rather than the limit of detection.* The proposed definition is shown in Fig. 10.12 as the dashed vertical line at the far right.

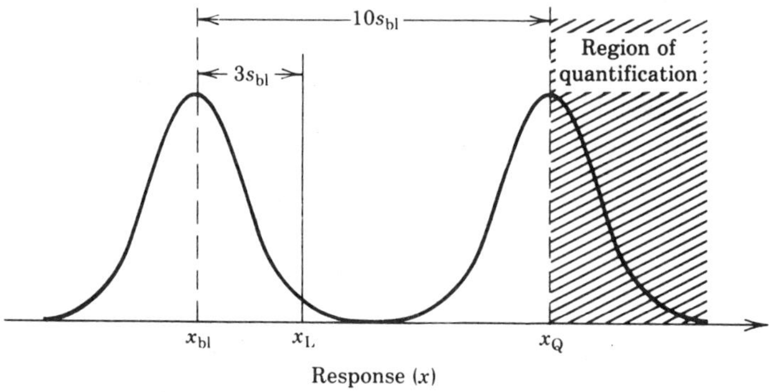

Fig. 10.12 The relative position of the limiting response x_L and the limit of quantification x_Q. Normal distributions are shown both for the measurement of blanks and of values of x_Q. Both distributions are assumed to have the same standard deviation. Note that x_Q is fully 10 estimated standard deviations above the mean value for blanks. The region of quantification begins at concentrations corresponding to x_Q ($10s_{bl}/S$, where S is the slope of the calibration curve) and extends upward indefinitely.

Pure Appl. Chem.* **54, 1566 (1982) and L.H. Keith *et al.*, *Anal. Chem.*, **55,** 2210 (1983).

By definition, x_Q is

$$x_Q = \bar{x}_{bl} + 10s_{bl} \tag{10.16}$$

A distribution of values of x_Q has been included in the figure. The limits of confidence $x_Q \pm 3s_{bl}$ may be added. Whenever a response is sufficiently great to fall an average of 10 standard deviations to the right of the mean value of the blank, it may be presented in standard form as a mean with confidence limits: a respectable quantitative determination can be made.

Example 10.25 What account, if any, should be taken of possible errors in the slope and intercept of a calibration curve at trace levels? Will such errors be likely to raise or lower detection limits?

To allow for the possibility that an extrapolated linear calibration curve may depart from linearity at small concentrations, the curve may be allowed to have its own uncertainty. Its slope can be expressed as a confidence interval $S \pm t_{f,\alpha}s_S$ where s_S is the standard deviation of the sensitivity and $t_{f,\alpha}$ is Student's t chosen for a particular number of degrees of freedom and level of confidence. This point can now be pursued formally.

Propagation of Error Approach to the Limit of Detection. A supplementary approach to calculations at trace levels is to consider the calibration curve for a measurement procedure as a straight line obeying equation $y = mc + i$ where m is the slope (sensitivity) and i the intercept. Solving for c gives $c = (y - i)/m$. The possibility exists that each of these terms has errors associated with it. A *propagation of error* approach (see Section 10.3) may be taken to derive an overall error to be expected in the concentration here expressed as an uncertainty or standard deviation s_c. An illustration by Long and Winefordner indicates the significance of such a consideration of error [G. L. Long and J. D. Winefordner, *Anal. Chem.*, **55**, 712A (1983). Three sets of their results for atomic fluorescence measurements of ICP-excited samples containing calcium, copper, and cobalt are presented in Table 10.5. The second column simply gives the limit of detection using the IUPAC value, the third lists the limit of detection in which the slope is also assumed subject to some uncertainty, and the fourth gives the error, taking into account propagation of uncertainty from both intercept and slope. Including the possibility of error in the calibration curve substantially increases the limit of detection as would be expected.

Table 10.5 Limits of Detection $c_L(k = 3)$ in ppm for ICP-Excited Fluorescence

Species Determined	$\frac{3s_{bl}}{m}$	$\frac{3s_{bl}}{m - t_\alpha s_m}$	$\frac{3\left[S_{bl}^2 + S_i^2 + \left(\frac{i}{m}\right)^2 S_m^2\right]^{1/2}}{m}$
Ca(II)	0.002_4	0.002_7	0.05
Ca(II)	0.002_1	0.002_4	5
Cu(I)	0.03	0.3	5
Co(II)	0.5	−2	6

In conclusion, it is important to note that the limits of detection (Eq. (10.14)) and quantification (Eq. (10.16)) are perhaps the most readily ascertainable values for a method at the trace level. Certainly, these limits have the virtue of a clear statistical basis. Trace measurements are best compared in terms of their limits of quantification or determination. Finally, where data are available about the nature of the sensitivity curve at low concentrations, it appears valuable to make the kinds of corrections indicated in the third column of Table 10.5. Needless to say, it will seldom be legitimate to extrapolate sensitivity values estimated at much higher concentrations to the trace level.

REFERENCES

Clear introductory volumes on statistical methods in chemistry are:

1. R. Caulcutt and R. Boddy, *Statistics for Analytical Chemists*. New York: Chapman and Hall, 1983.
2. J. C. Miller and J. N. Miller, *Statistics for Analytical Chemistry*, 2nd ed. New York: Wiley-Horwood, 1987.
3. L. A. Currie, Sources of error and the approach to accuracy in analytical chemistry, in *Treatise on Analytical Chemistry*, 2nd ed., I. M. Kolthoff and P. J. Elving (Eds.), Part 1, Vol. 1. New York: Wiley-Interscience, 1978.
4. J. K. Taylor, *Quality Assurance of Chemical Measurements*. Chelsea, MI: Lewis, 1987.
5. H. L. Youmans, *Statistics for Chemistry*. Columbus, OH: Charles Merrill, 1973.

Some general volumes on statistical methods are:

6. J. R. DeVoe (Ed.), *Validation of the Measurement Process*. Washington, DC: American Chemical Society, 1977.
7. G. Kateman and F. W. Pijpers, *Quality Control in Analytical Chemistry*. New York: Wiley, 1981.
8. D. L. Massart, A. Dijkstra, L. Kaufman, *Evaluation and Optimization of Laboratory Methods and Analytical Procedures. A Survey of Statistical and Mathematical Techniques*. New York: Elsevier, 1978.
9. G. E. P. Box, W. G. Hunter, and J. S. Hunter, *Statistics for Experimenters*. New York: Wiley, 1978.

A clearly presented specialized volume is:

10. American Society for Testing Materials, *Sampling, Standards, and Homogeneity*, Spec. Tech. Publ. 540. Philadelphia: ASTM, 1973.

Also of interest are:

11. G. L. Long and J. D. Winefordner, Limit of detection: A closer look at the IUPAC definition, *Anal. Chem.*, **55,** 712A, 1983.
12. B. Kratochvil and J. K. Taylor, Sampling for chemical analyses, *Anal. Chem.*, **53,** 924A, 1981.

See also references listed for Chapter 11.

EXERCISES

10.1 Group the following errors into systematic and random types: (1) instrument corrosion and wear, (2) solubility loss of barium sulfate in precipitation, (3) operator carelessness in measuring pH, and (4) measurement of solution absorbance with an unstabilized tungsten lamp. Justify your assignments.

10.2 A measurement of analyte concentration has a mean value of 10 units and an estimated standard deviation s of 1 unit. (a) Calculate the $s_{\bar{x}}$ or standard error for sets of 4, 8, 16, and 20 replicate measurements. (b) Compare the improvement in the $s_{\bar{x}}$ achieved by increasing the number of replicates from 12 to 16 with the improvement gained in going from 1 to 4 repetitions.

10.3 Interpret the following statement: The holographic monochromator is accurate to 1 nm and precise to ± 0.5 nm. Is the latter specification equivalent to reproducibility?

10.4 When measurements are replicated an occasional result will appear to differ substantially from the rest. A Q-test may be applied to ascertain whether it may be discarded before a mean and standard deviation are calculated.* To apply this statistical test, a rejection quotient Q, defined as Q = (outlying result − nearest result)/(range of results), can be calculated to see whether it exceeds the appropriate datum in the following statistical table of data obtained for such quotients at the 90% confidence level:

Number of Replicates	Q	Number of Replicates	Q
3	0.94	7	0.51
4	0.76	8	0.47
5	0.64	9	0.44
6	0.56	10	0.41

If $Q_{calc} > Q_{stat}$, the outlier may be discarded.

Apply the Q-test to the following data (in μg mL^{-1}) for the strontium content of a sample to see whether the outlying value may be discarded: 1.15, 1.02, 1.10, and 1.88. Interpret your decision statistically.

10.5 Comment on the statement, "Fig. 10.2 fails to include the possibility of bias in the measurements." If necessary, alter the figure to represent a case with bias.

10.6 In a given measurement if the bias in each step of a measurement is constant, will it be detected by a t-test or analysis of variance? Discuss.

10.7 If the bias in a measurement process varies in some way over time, is it likely to be discovered by a t-test or analysis of variance? How was the bias in radiocarbon dating found and a calibration curve developed? What factors were responsible for the systematic error? (See I. N. Olsson, (Ed.), *Radiocarbon Dating and Absolute Chronology*, New York: Wiley-Interscience, 1970.)

10.8 The estimated standard deviation for a method of measurement of CO in automobile exhaust gases is found to be 0.80 ppm. Using the t-table, calculate the 99% confidence limits for a quadruplicate analysis when the mean value is 1.02 ppm. Comment on the precision of the results.

*R. B. Dean and W. J. Dixon, *Anal. Chem.*, **23,** 636 (1951).

10.9 The reliability of a new method for the determination of nickel is evaluated by applying it to a standard reference material that contains 8.35% Ni. The results observed in three replicate analyses are 8.21, 8.15, and 8.13%. (a) Calculate the mean and estimated standard deviation of the mean. State the results of the analyses as you would report them. (b) It is decided to check for systematic error in the new method. Apply the *t*-test. State the hypothesis you are testing and report your findings at both the 95 and 99% confidence levels.

10.10 For a new method of measurement the limit of detection is to be established. (a) Write an equation for the limit of detection. Define terms. (b) For this method measurements on blanks yield an average response $\bar{x}_{\text{bl}} = 1.7 \times 10^{-8}$ with estimated standard deviation $s_{\text{bl}} = 0.6 \times 10^{-8}$. Calculate the limit of detection. (c) The slope of the calibration curve for the method at trace levels is found to be 2.50 μg^{-1} mL. Calculate the limiting concentration c_{L}. (d) Also calculate the lowest concentration that can be measured with reasonable quantitative reliability.

10.11 A stripping voltammetric study of cadmium levels in a river based on fast linear sweep is to be undertaken. Several replicate measurements on a blank as much like the composition of the river as possible, but without cadmium, yield currents that average 1.5×10^{-8} A with $s_{\text{bl}} = 0.8 \times 10^{-8}$ A at the voltage at which a cadmium peak is expected. When four samples of river water are tested, currents averaging 3.0×10^{-8} A are obtained at the same voltage. (a) Should a trace of cadmium be reported? (b) What level of response will be required to obtain a quantitative measurement?

10.12 A toxin *T* can be measured spectrophotometrically by complexation to form a species with an absorption maximum at 450 nm. Optimum conditions for the variables that influence the method need to be found to establish a protocol or set of procedures for regular measurements. Three variables are investigated for their importance using a factorial approach: reaction time, X_1; pH, X_2; and temperature, X_3. Nine measurements of absorbance of the complexed toxin are made with variables at appropriate low (LO) and high (HI) levels. For the measurements $s_{\text{absorb}} = 0.005$ AU.

Relative Level of Variables			Absorbance at 450 nm
X_1	X_2	X_3	
LO	LO	LO	0.115
HI	LO	LO	0.130
LO	HI	LO	0.120
HI	HI	LO	0.135
LO	LO	HI	0.130
HI	LO	HI	0.165
LO	HI	HI	0.125
HI	HI	HI	0.160

Which variables are significant factors in the determination of the toxin?

10.13 Assume a new enzyme assay is under development. Standard substrate and enzyme is prepared so that parallel trials can be run. One person does five assays and gets a mean value of 235 units with a mean standard deviation of 35 units. Another does five assays and gets 265 with a mean standard deviation of 25 units. Are the two people getting similar results?

ANSWERS

10.4 $Q_{calc} = 0.85$. It may be discarded on the basis that only once in 10 times will a result that different from the mean be likely in a set of four replicates. *10.8* Confidence limits will be $\pm ts/\sqrt{n} = \pm 5.84 \times 0.80/\sqrt{4}$ ± 2.34 ppm. The precision is poor. *10.9* (a) $\bar{x} = 8.16\%$; $s_{mean} = 0.024\%$. Report $8.16\% \pm 0.10\%$. (b) $t_{calc} = 7.9$ while $t_{stat} = 4.30$ for 95% confidence, $= 9.92$ for 99% confidence. It can be concluded at the 95% confidence level that systematic error has occurred, but not at the 99% level. A study of possible sources of systematic error should be made. *10.10* (b) $x_L = 3.5 \times 10^{-8}$, (c) $c_L = 0.14\ \mu g\ mL^{-1}$. *10.11* (a) By definition, the limiting response is $x_L = \bar{x}_{bl} + 3s_{bl} = 3.9 \times 10^{-8}$ A. Since the average response of river water samples fell below this limit, it should be reported that cadmium is absent. (b) For quantification of measurements, responses should be at least as large as $\bar{x}_{bl} + 10s_{bl} = 1.5 \times 10^{-8} + 10 \times 0.8 \times 10^{-8} = 9.5 \times 10^{-8}$ A. *10.12* Time and temperature of reaction are significant.

Chapter 11

QUANTIFYING MEASUREMENTS AND EXTRACTING INFORMATION

11.1 INTRODUCTION

In addition to the procedures for control of measurement quality introduced in the last chapter, we must deal with the processes by which chemical information is obtained from an instrument's analytical signal. What strategies exist for extracting information under the whole range of experimental conditions that may be encountered? Analyte concentrations may range from nearly 100% to ultratrace levels and they are often in complex mixtures.

In general, information is obtained from an analytical signal by applying at least three relationships, one physical and two analytical. For example, in a spectrophotometric measurement, the *physical* relationship is the functional connection by which the amplitude of the instrument signal is related to the amount of light absorbed by the analyte solution, which is expressed as either its transmittance T or absorbance A. The *analytical* steps, however, actually yield the desired chemical information, by application of

1. the theory of the analytical technique, which in this example gives Beer's law, and
2. a *mode of quantification* of that law, such as a calibration curve.

It is with the second analytical step of quantifying measurements that this chapter is mainly concerned. In later chapters signal processing and a wide range of techniques will be taken up.

The main modes of quantification include *use of a calibration curve*, *standard addition*, and *titration*. Each mode finds use under different experimental conditions with analytical techniques as different as direct potentiometry and mass spectrometry. Which mode is chosen substantially affects the reliability of measurements on a particular sample by a given technique. Therefore, the selection of the mode of quantification becomes a strategic choice. For this reason, even though programs (software) for quantification are included with instruments, we need to know the procedures well to choose the one most appropriate for the system of interest.

Some procedures that provide support for quantification also are examined in this chapter. Examples are physical methods such as *preconcentration*, which can enhance measurements of analytes at trace levels, and chemical methods such as *derivatization* and *complexation*, which will convert species to forms more tractable to measurement. Where chemical preparation is added, as in the last cases, one or more stages of sample preparation or separation become integral to a procedure.

Concluding the chapter will be a brief presentation of some chemometric procedures including simultaneous multicomponent analysis, factor analysis, and pattern recognition, which are multivariate procedures or methods for extraction of information on a level of higher dimensionality.

11.2 ONE-COMPONENT MEASUREMENTS; SAMPLE CHANNELS IN INSTRUMENTS

At present most analytical measurements are designed for determination of one species at a time. The attractiveness of developing one-component strategies is that quantifying a single-analyte method nearly always offers the simplicity of a linear calibration and that such calibrations are less sensitive to error.

It is simple enough to employ this approach by utilizing methods with sufficient selectivity for particular analytes. For example, a glass electrode responds selectively to the hydrogen ion in aqueous solutions under most conditions. To analyse more complex samples such as commercial herbicides or blood serum by one-component methods, each analyte can either be determined by a procedure selective for it or the several analytes can first be separated chromatographically or electrophoretically and then measured individually.

Selectivity describes the ability of a technique to respond specifically to one analyte. Standing against such a possibility in most samples are *interelement effects* in which other substances may respond to some extent as well or may interfere with the development of an analytical signal for the substance of interest. These effects are also called matrix effects or interferences (this last term is used mainly when influences can be traced to certain components as in many atomic absorption measurements). Some examples of interelement effects are (a) in pH measurements Na^+ and K^+ ions reduce the response of a glass electrode to H^+ ions in strongly basic solutions; (b) in IR spectrometry the overlap of absorption bands of many molecular species introduces uncertainty in the response of an analyte; and (c) in X-ray emission spectrometry of a solid sample, many analyte X-ray photons fail to reach the detector because they are absorbed and/or scattered out of the beam by other atoms in the sample.

In most cases some initial sample preparation will lessen interelement effects. The process is sometimes described as matrix substitution. For example, simply dissolving a solid sample substitutes the uniformity of a solvent for the diversity of a specimen. Nearly always sample preparation steps for particular techniques also reduce interelement effects. Preparation may include the use of selective chemical reactions, resolution of mixtures into individual components by separative procedures like gas or liquid chromatography, and use of spectroscopic methods like emission spectroscopy and nuclear magnetic resonance spectrometry that give separate analyte signals.

Both sample preparation and separation processes have become increasingly automated to improve their reliability and precision. As a result the series of steps they require have taken the form of a sample channel that is coupled to the measuring instrument. The main instrument, of course, provides the *signal channel* in which chemical information about the analyte is obtained (see Sections 1.5 and 1.8). As a consequence the overall instrument becomes both more complex, more versatile, and more reliable. How can we best understand what is gained by addition of a sample channel?

Quality criteria by which analyte sample channels can be evaluated are:

1. *Resolution*. Is an analyte presented for measurement under conditions that ensure at least that its response will have minimum overlap with the responses of other constituents?
2. *Recovery*. Is all the analyte that entered the sample channel presented for measurement?
3. *Reproducibility*. Are conditions and procedures repeated in precisely the same manner for each sample and analyte?

Preparative Sample Channels. In general, the goal of a preparative sample channel will be to isolate an analyte physically or to convert it to a chemical form that will afford selectivity. For instance, in spectrophotometry a metal ion of interest can often be selectively complexed by a reagent to intensify its absorption and also shift the absorption band to an ''open'' spectral region.*

Two other common approaches should be cited. One is a reverse approach in which the usual effect of interferents is *masked*. For this purpose a substance is added that will complex (sequester) the interferent and minimize its interaction. When masking can be used, it is valuable for analytes present at the trace level (see Section 11.6). Another common option is to extract an analyte quantitatively and selectively into another medium by dissolving a selective complexing agent in the extractant. Many examples of such reactions will be found in discussions of sample preparation in later chapters on particular techniques and in references.

Example 11.1. How do laboratories set up preparative sample channels for determination of a few analytes with good precision (ideally $\pm 1\%$) in a great many samples?

Clinical and environmental control laboratories provide automated sample preparation by installing *automatic analyzers*. These devices provide precise pipeting, pH adjustment, chemical complexing, and other necessary steps as needed whenever 5–10 analytes must be determined in a great many samples. A selective procedure is provided for each analyte consisting of an automated and programmed series of steps that begin with taking an aliquot of a sample.

Very often the sample channel consists of a manifold of tubes in which controlled pumps

*Visual spot tests are ultimate examples of selectivity; indeed, there are elegant collections of reactions that are specific for different analytes in the complex samples often encountered in field work. See F. Feigl, *Spot Tests in Organic Analysis*, 7th ed., New York: Elsevier, 1966; and F. D. Snell, C. Snell, and C. A. Snell, *Colorimetric Methods of Analysis*, Vol. 2A, Princeton: Van Nostrand, 1959.

and valves direct flows appropriately. Sample integrity is maintained during passage. Converging side streams add reagents, loops in the tubing allow time for reaction, and special portions of tubing permit a variety of physical operations such as dialysis. Finally, the sample moves into the sample module of the main instrument for measurement.

Three approaches are in use. Two use a manifold of tubes. In one, segmented samples move through the system with samples separated by air bubbles. Alternatively, a continuous, nonsegmented flow system known as *flow injection analysis* has also become important as a sample channel [J. Ruzicka and E. H. Hansen, *Flow Injection Analysis*. New York: Wiley-Interscience, 1981.] Pumping pressures in both types of sample channel are low.

A third option is to use a programmed robot to transfer an aliquot from a sample vial to a new vial, dilute this aliquot, add to it appropriate quantities of reagent, and see that it makes contact with a heater or other devices to carry out a host of desired steps. When the sequence of steps is completed, the vial is moved into position as the sample cell in the main instrument.

An automated sample channel repeats its reactions precisely in the same way for every sample and thus ensures a quality performance while relieving the user of considerable tedious work. A good preparative channel should ensure maximum and reproducible development of the reaction.

Separative Sample Channels. Gas, liquid, or supercritical fluid chromatographs and many other systems such as a capillary electrophoretic or mass spectrometric system will separate analytes and permit them to be measured individually. Thus these systems minus detector and display modules are commonly used as *separative sample channels*. The power of such a channel is clearly the special attraction of tandem or hyphenated methods such as gas chromatography/mass spectrometry that will be taken up in chapters dealing with measurement techniques. Needless to say, the appropriate choice of a separative column should ensure that even complex mixtures are resolved and the possibility of the precision and simplicity of a one-component analysis is available.

At the point of connecting a separative sample channel to the first module of a measuring instrument a specially designed *interface or coupling device* is necessary. This device provides for

1. an orderly flow of sample into the sampling module even though there may be change of pressure or concentration,
2. venting or diverting any carrier, such as helium in a gas chromatograph, and
3. adjusting conditions such as pressure and the rate of appearance of analytes to allow the sample to enter the instrument sample channel easily and effectively.

A great deal of ingenuity has gone into designing interfaces that ensure that the quality requirements listed above are met, and spray or jet devices have been especially successful.

Sequential One-Component Methods. In many instances the responses of individual analytes are resolved and information may easily be extracted about them from the available signal. For example, elements give characteristic emission and absorption wavelengths in atomic methods and some molecular methods such as raman and IR spectrometry afford separated wavelengths. In this situation *sequential* or even simultaneous one-component analyses of the components, whether atoms or molecules, can be made in the frequency domain of the measurements. Further, with sequential measurements, even where interelement effects are marked, effective ways to correct for these mathematically are commonly developed.

Example 11.2. How is it possible to obtain quantitative analyses of solid samples by X-ray emission spectrometry when intensities are strongly affected by interelement effects?

First, note that elemental peaks are ordinarily well resolved. That being true, intensities at expected peak wavelengths for an analyte can be corrected with high precision. The interelement effects, which are strong in solid samples, are mainly absorption and scattering by other atoms in the path of an emitted analyte X-ray. Because these phenomena are well understood, precise values of mass absorption coefficients and scattering coefficients are available in tables. Using such data and modern computer programs, corrections can be made with precision.

These procedures are carefully used in X-ray fluorescence spectrometry and in scanning of large objects where elemental determinations are obtained simultaneously with scanning transmission electron microscopy. Thus, even these methods can be said to devolve into essentially single-component methods.

Multicomponent Analyses. True multicomponent analyses are also common. For mixtures it is often possible to tackle the problem of unknown interelement effects by multivariate regression procedures. Other chemometric procedures such as factor analysis and pattern recognition are also important (see Section 11.9). Naturally, for effective use, adequate software for these procedures must be available.

How does one choose between a one-component and a multicomponent method? Given sufficient selectivity for the analytes, a one-component method offers easier quantification. With appropriate software multicomponent analyses become quite attractive, especially for complex samples. Such analyses seem sure to become steadily more widely used [5, 7, 8].

11.3 USE OF CALIBRATION CURVES

The most common way to quantify instrument signals for a single-component measurement is by using standards to develop a *calibration curve*. It is also known as an *analytical curve* or *working curve*. As shown in Fig. 11.1, the curve is basically a plot of signal amplitude x_i as a function of analyte concentration. Linear curves are sought since their plotting from calibration data is less sensitive to error. Further, the *dynamic range* of a measurement by a given method is taken to be the number of decades of concentration over which the calibration curve is essentially linear. The slope of the calibration curve is simply the *sensitivity*. The greater the sensitivity, the higher S/N, and the more precise and accurate the measurement in general.

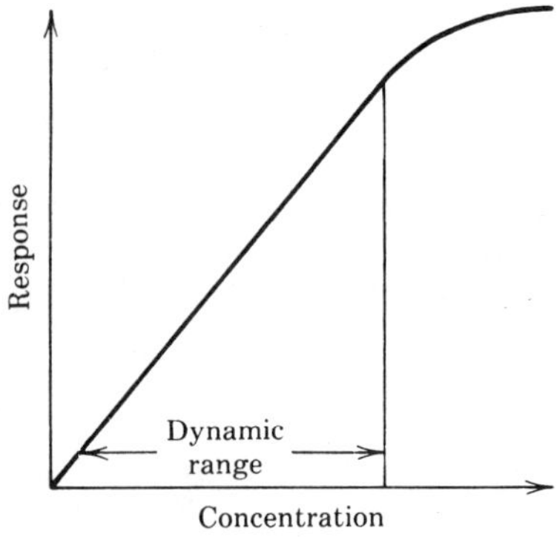

Fig. 11.1 A representative calibration curve for a one-component measurement.

For reliability in a calibration curve it is also necessary that *interelement* or *matrix effects* be minimal, as discussed earlier, or at least constant. Further, the reliability of a concentration curve depends on developing an equation that fits the analytical signal values (data points). In most cases a least-squares or regression process is employed. With an equation available, analyte concentrations can be readily calculated from measurements on samples.

Example 11.3. Spectrophotometric measurements on solid and liquid mixtures commonly require dissolution of the sample. Is the main reason that dissolving reduces matrix effects?

Actually, putting a sample in solution does allow substitution of the uniform solvent background of a solution for the possibly large interelement effects in a complex solid or liquid matrix. Further, standard solutions can be made up more simply if interelement effects are known to be small and duplication of components besides the analyte becomes unnecessary.

Example 11.4. How can interelement effects be reduced for solids that are difficultly soluble?

In this case the matrix can be simplified by grinding to a powder and diluting with an inert solid. For instance, in arc or spark atomic emission spectrometry, samples are often mixed with pure graphite powder to give a uniform material background or matrix before volatilization and excitation. In IR spectrophotometry, KBr, which is transparent through the middle IR, is often ground with solid samples to make a mull, and for near IR diffuse reflectance spectrometry, KCl often plays the same role.

Example 11.5. A good way to obtain a calibration curve for a given analyte by a given technique is to obtain measurements on three or more standards. Each standard will give a pair of X_i, C_i values. The slope and intercept (if not zero) are best determined by application of a least-squares procedure (to be described below).

Once the slope and intercept are known, the concentration of a sample C_i can be found by substituting the response x_i in the calibration equation and solving for C_i. Either the response x_i from a single sample, or a mean response, if the time and cost has been taken to measure a set of replicate samples, can be substituted.

Example 11.6. Once data for a calibration curve are available, how may they be fitted by a least-squares process to develop a regression equation?

In most modern instruments a mode of operation may be selected in which it is only necessary to let the instrument automatically acquire calibration data during measurements on standards. An operator using such an instrument need only identify the data from standards so that it is used for calculation of the slope and intercept values for the linear calibration curve. Once obtained, the equation of the calibration curve can be saved in memory.

Surely the simplest case is that in which theory requires a calibration curve to go through the origin, as in Fig. 11.1. Its equation will take the form

$$x_i = mC_i, \tag{11.1}$$

where x_i is the signal amplitude, C_i, is the analyte concentration, and m is the slope. Beer's law is a good example of such a relationship.

Least-Squares Analysis. The "least-squares" principle can be stated as follows: in replicate measurements of a variable the sum of squares of deviations from the arithmetic mean is a minimum. It derives from the normal law of error discussed in the last chapter. Since deviations from a mean represent random error, this principle is a statement about the manner in which error is distributed. We shall use the approach to obtain the equation of the curve which will connect the set of mean values deducible from measured points. This principle can best be presented by an illustration.

Example 11.7. Apply the least-squares procedure to find the best values of slope m and intercept b for data expected to be fit by the linear relationship:

$$y = mx + b \tag{11.2}$$

For example, such a curve would be sought in direct potentiometry with an ion-selective electrode. How can one proceed?

First, consider how to apportion experimental error between x and y in a consistent way. Though both will be uncertain to a degree, it is customary to assume that all error is in y since it is the dependent variable. This process is shown in Fig. 11.2. Thus, the deviations d_i of each measurement of y from the best value, that is, one taken from the yet unknown calibration curve, must be of the form

$$d_i = y_i - y_i' = y_i - mx - b, \tag{11.3}$$

where y_i' is the mean value of y for each particular value of x and the values of parameters m and b are mean values that must be calculated. Note that y_i' can only be algebraically found at this time by solving the equation for it.

Next the coordinates of each experimental point y_i', x_i' are substituted in Eq. (11.3). Recall that the sum of squares of the deviations is to be minimized according to the least-squares principle. Thus, each value of d_i must be squared and a sum of all values must be taken. The minimum of the sum is secured by taking its partial derivative with respect to

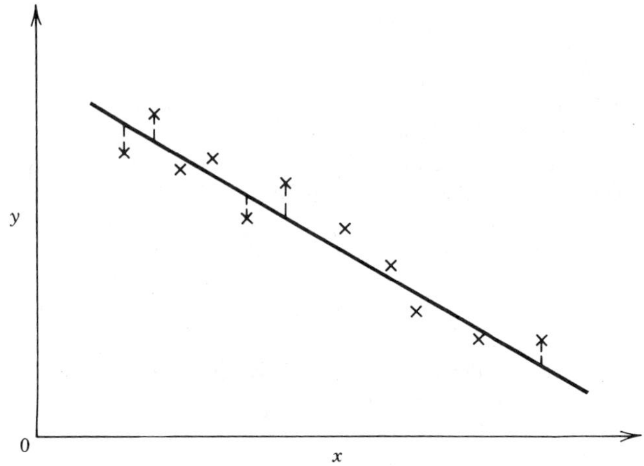

Fig. 11.2 A least-squares "best" fitting of data by a straight line. All random error is taken as occurring in the output response or signal, and thus in the y direction. Deviations in the y direction are minimized to secure the "best" value of slope and intercept.

m and setting it equal to zero. The mean slope found is given by Eq. (11.4). This value is substituted in Eq. (11.2) to give the mean value of intercept b in Eq. (11.5). A mathematical treatment of least-squares analysis will be found in the references [1, 9, Chapter 10].

Equations obtained for the slope and intercept in the case of a straight line, rewritten to make calculation easier, are

$$m = \frac{n \sum_{i=1}^{n} x_i y_i - \sum_{i=1}^{n} x_i \sum_{i=1}^{n} y_i}{n \sum_{i=1}^{n} (x_i)^2 - (\sum_{i=1}^{n} x_i)^2} \tag{11.4}$$

and

$$b = \bar{y} - m\bar{x} \tag{11.5}$$

Example 11.8. It is decided to fit a general straight line ($A_i = mC_i + b$, where A_i is the absorbance of component i at concentration C_i and a given wavelength) to spectrophotometric data and see whether an intercept is found. If one is found, should it be tolerated?

As suggested in Section 10.4, a nonzero intercept may indicate systematic error. Thus, if $A_i = mC_i$ was simply fit to spectrophotometric data, insight about systematic error might be lost. Once advantage has been taken of such information to reduce bias in the measurement, the final calibration equation developed must be the simple one that goes through the origin.

The Nonlinear Case. Fortunately, the least-squares principle can also be used to fit nonlinear equations to data. Often the appropriate type of curve, for example, logarithmic or hyperbolic, will be known or may be suggested by a rough plot of the data. References will give details.

When standard mathematical forms seem inadequate, an nth order orthogonal polynomial of the form $y_i = a_0 + a_1x + a_2x^2 + \cdots + a_nx^n$ can ordinarily be fit to data. For a good fit it is wise to make as many measurements as convenient; only in this way can the curvature be followed with some precision. It is customary to try to fit data with a low-order polynomial (e.g., $n = 3$) and move to larger values of n only if the fit seems unacceptably poor. Software that is readily available for computation of orthonormal polynomials by least-squares analysis will yield best values for the coefficients a_i.

One other case should be discussed here, that of the calibration curve that would be straight except that it is being applied over a concentration range longer than the acceptable dynamic range. Then curvature will be observed at higher concentrations, and precision will diminish in the region of curvature. This type of curve is often labeled a *working curve*; though it is still valid as a basis for quantifying the results, it is clear that compensation is being made for systematic error. There is every reason to apply a polynomial least-squares analysis to improve the precision.

Sources of Error in Least-Squares or Regression Equations. To minimize errors in setting up calibration curves by least-squares it is especially important to check whether in unfamiliar cases a nonlinear equation may best fit the data. The power of the least-squares approach is such that it will fit any type of curve to the data one has. More than once data that would have been best fit by a parabolic relationship have been reported as linear in form!

A second common error is taking measurements over too narrow a concentration range. The result is an equation that fits only over the narrow range. If an abbreviated range has been used because error increases at either high or low values of the independent variable, a better strategy is to employ a *weighting factor*. Commonly, the factor is a multiplier of the concentration that decreases in value with analyte concentration.*

Another error occurs when care is not taken to ensure that a calibration curve goes through the origin when it is so compelled by theory. This possibility was treated in Example 11.8.

Error in Calibration Curves. Let us look at two common calibration procedures with a view to estimating their reliability. The *single-point calibration* method is valid, in principle, for calibration curves that are required to go through the origin on theoretical grounds. Actually, the intercept may not be zero because of systematic error. When that is the case, the error can be shown to be

$$\text{error} = \left[(S_s - S_x)/S_s\right](I/m), \tag{11.6}$$

where S_s and S_x are responses for the standards and unknown, respectively, I is the intercept on the ordinate, and m is the slope of the curve.†

To minimize error in calibration curves, *multiple-standard curves* are usually prepared. To minimize error from this source, one senses that there should be one or more calibration points in each expected analyte concentration region. Confir-

*S. D. Christian and E. E. Tucker, *Am. Lab.*, **16**(2), 18 (1984).
†M. J. Cordone, P. J. Palermo, and L. B. Sybrandt, *Anal. Chem.*, **52**, 1187 (1980).

mation of this idea will be found in the discussion about predicting confidence levels for calibration curves given below. An example of its application is the well-known rule of thumb in precision atomic absorption spectrometry of taking readings on standards just below and above the concentration of a sample, a strategy known as *bracketing* the unknown concentration.

Predicting Confidence Levels. It can be shown that confidence levels available from a regression equation decrease rapidly the further analyte concentrations are from the concentrations of standards. The estimated standard deviation of the mean or standard error $s_{\bar{x}}$ in a linear calibration relationship based on a regression equation can be shown to be*

$$\text{est. std. error in } S_0 = s\left[(1/n) + (C_0 - \bar{C})^2/\Sigma(C_i - \bar{C})^2\right]^{1/2} \qquad (11.7)$$

where the estimated error for predicted response S_0 is *calculated* from the least-squares line for a solution whose concentration is C_0. In the equation s in the estimated standard error of the regression equation, n is the number of measurements on standards, and $\bar{C}$ is the mean concentration of standards whose concentrations are $C_1, C_2, \ldots, C_i$. It is possible to calculate a confidence band ΔS around a response S_0. Similarly, for that confidence band the regression curve allows one to establish a confidence band ΔC around the concentration C that is predicted when response S is observed. A confidence band calculated from Eq. (11.7) is shown in Fig. 11.3. From Eq. (11.7) we can deduce that the confidence band is narrowest at the mean of the set of standards. It widens slowly in either direction and becomes strikingly broader outside the region defined by standards.

Example 11.9. What confidence limits apply to a linear calibration curve obtained with four standard solutions when two measurements are made on each? The analyte concentration in the standard solutions is precisely 0, 2, 4, and 10×10^{-3} *M*.

It should be possible to calculate the confidence limits by applying Eq. (11.7). The limits will be just the estimated standard error in the instrument signal, which can be calculated as a coefficient of s. First, the mean value of standard solutions is 4×10^{-3} *M*. Then, at 4×10^{-3} *M* that error is $s[\frac{1}{8} + (4-4)^2\, 10^{-6}/56 \times 10^{-6}]^{1/2} = 0.35$ s and at 0 *M* and 1×10^{-2} *M*, 0.64 s and 0.88 s, respectively.

Example 11.10. Which factor will be most important in keeping precision high in measurement of a property: the precision of the method, the number of standards, or the concentration of the standards? Use Eq. (11.7) as necessary to evaluate these possibilities.

It is common sense that one must choose an inherently precise technique for good results. Note that in Eq. (11.7) the precision measure multiplies all terms. According to that equation, employing a larger number of standards will enhance precision by increasing the value of n. As discussed above, choosing standards that bracket anticipated sample concentrations is important. In Eq. (11.7) this factor enters in terms of the value of $C_0 - \bar{C}$. Ideally, the mean value of the standards should be close to the most common sample concentration. All listed factors will be important.

---

*D. G. Mitchell, W. N. Mills, J. S. Garden, and M. Zdeb, *Anal. Chem.*, **49**, 1655 (1977).

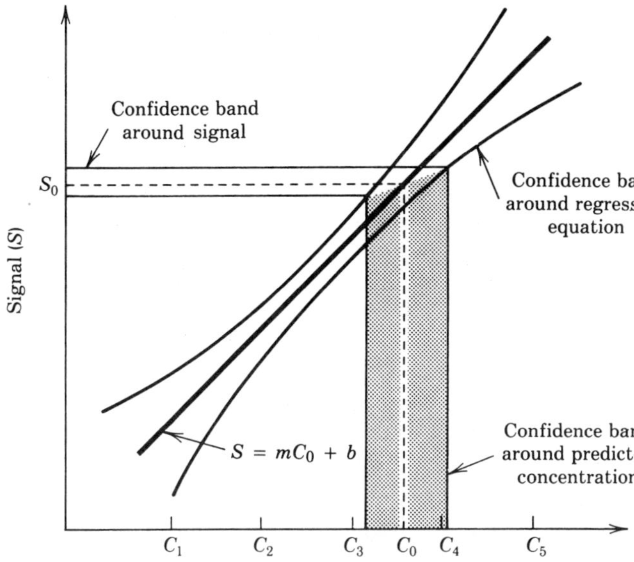

Fig. 11.3 A calculated least-squares or regression calibration curve to which has been added confidence limits. In most cases confidence limits for the signal S differ from those that will apply to the concentration of analyte. The narrowest confidence limits for both are at the mean concentration of each standard solution used to determine the curve; at concentrations below or above these values limits widen. Note that a wider confidence band exists for the analyte concentration. Its lower limit is determined by the intersection of the $S_0 - \frac{1}{2}\Delta S$ line with the confidence band for the regression equation and its upper limit in similar fashion by the intersection of the $S_0 + \frac{1}{2}\Delta S$ line with the band. These differences for the analyte concentration will be still greater if the slope of the regression line differs from 45° [After Mitchell et al, op. cit.].

11.4 STANDARD ADDITION METHOD

The method of standard addition is an especially useful mode of quantification for samples that have substantial matrix effects. Often the method is called *spiking*. In essence, it allows a calibration curve to be made up in each sample solution. First, a measurement is made on the sample. Next, a known amount of the analyte is added and a second measurement made. After as many as two or three more standard additions and measurements, the data are fitted by a linear regression equation.* What is important is that the added analyte is also subjected to the matrix effects characteristic of the sample.

Standard addition gives an analytical curve like that in Fig. 11.4. Its equation is

$$x_i = k_j(\Delta C + C_u), \tag{11.8}$$

*Standard addition procedures should be seen as clearly different from the use of internal standards. The latter are based on adding a known concentration of an easily measured substance not normally present to serve as a reference.

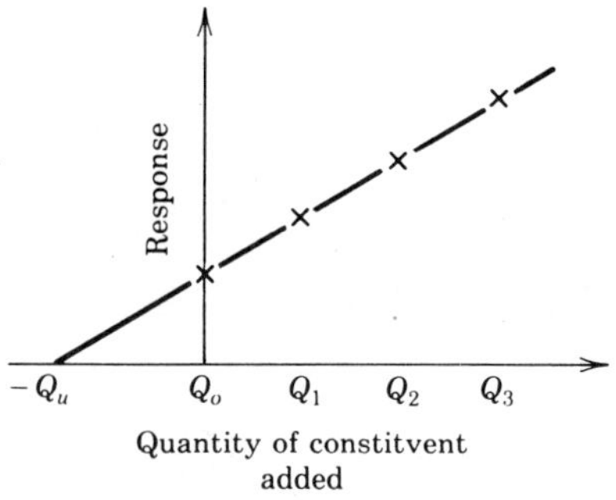

Fig. 11.4 A representative standard addition plot. A zero response is assumed when the concentration of analyte is zero. The initial response is at Q_0, corresponding to the response from the sample solution itself. Responses after three standard additions labeled Q_1, Q_2, and Q_3 together with the initial response should give a linear curve as shown. The intercept of this curve with the abscissa gives the concentration in the unknown sample.

where x_i is the response, ΔC the increment in analyte concentration with each standard addition, C_u the initial concentration of analyte, and k_j the sensitivity or proportionality constant relating the measured property j to concentration. A linear response to the analyte is assumed and as shown in Fig. 11.4 only the intercept on the x axis need be estimated with precision to find the concentration of analyte in the sample.

If only one addition can be made easily, precision is best served by making its amount greater than the estimated original amount or concentration. In this way two points quite different in concentration form the plot, ensuring that a reasonable value of the slope is secured. With a single addition, special care must, of course, be taken to avoid error. With regard to such an addition, prudence suggests that the standard be about 10 times the approximate concentration of the samples and that a check be made that the highest concentration expected will still fall within the linear part of the curve. The use of multiple standard additions is preferable.

While standard addition is used with essentially all techniques when concentrations are low and interelement significant effects are likely, in trace-level measurements it is usually *the* method of choice. At such low concentrations, most ways of minimizing matrix effects would be likely to lead to sample contamination or loss of analyte (see Section 11.6). In systems showing proportional systematic error (see Section 10.4) in which bias is proportional to the concentration of analyte, standard addition also sees regular use.

Some disadvantages of standard addition are the extra time required in sample preparation and the fact that it is easily applied only to solutions and powdered solids since all additions must be mixed completely with the sample system.

Example 11.11. Neutron activation analysis is a valuable trace technique. In this method an (n, γ) nuclear change is induced in most elements in a sample by neutron bombardent. Subsequent monitoring of the γ-ray emission leads to a quantitative determination (see Section 23.8). Yet with many samples, the γ-ray emission of a particular isotope is found to be consistently smaller than for a standard sample of the isotope. Can this type of systematic error be compensated for by standard addition?

Actually, standard addition is a good way around the problem. The effect is usually traceable to the presence of a neutron-absorbing isotope. Standard amounts of the analyte affected are added to powdered samples and the "spiked" samples are subjected to activation analysis. Extrapolation of a curve for the isotope like that in Fig. 11.4 allows the estimation of the analyte concentration in the original sample.

Example 11.12. Will standard addition minimize interelement effects in direct potentiometry with an ion-selective electrode?

Each such electrode also has a small selectivity for interfering ions. When present at sufficient concentration these interferents will alter both the slope and intercept of the calibration curve for the selected ion. The higher their concentrations (activities) are, the more serious their effect. Fortunately, the errors are automatically corrected by developing a standard addition plot. After each addition of standard solution to a potentiometric cell with such an electrode the potential is again measured.

Example 11.13. A zinc solution was analyzed by atomic absorption spectrometry using the method of standard additions. The data taken were tabulated as presented below. What is the equation of the standard addition curve and the concentration of zinc in the solution?

Step Number	Concentration of added Zn (ppm)	Absorbance
1	0.0	0.196
2	0.5	0.289
3	1.0	0.383
4	2.0	0.555

Several readings of absorbance were taken on each addition. Interestingly, it was found that optimum precision was secured by making a single large standard addition that just fit within the linear range of the curve.

A linear regression procedure used on the data yielded for the absorbance the equation $A_i = 0.199 + 0.179\Delta C_i$. The initial Zn^{2+} ion concentration was calculated to be 1.11 ppm. Applying statistics (Section 11.2) gave for the 95% confidence range for the original solution $C_0 = 1.110 \pm 0.0237$.

Method Development Based on Recovery Procedures. This section would be incomplete without a description of a procedure for testing new methods for trace analysis by standard addition. The percentage of analyte found or "recovered" is easily related to the standard amount added. For example, if a standard addition T is made to distilled water and Y is the concentration determined by the test method, its percentage recovery equals $100Y/T$.

The statistics of recovery should be carefully studied to ascertain the variability resulting from laboratory performance and/or from the test method. In particular, changes in the ratio of amount of standard added to initial (background) concentration will affect the variability greatly.

If a single standard addition gives incorrect recovery, interferences may be present. Nevertheless, if subsequent uniform additions give uniform increments in response or analytical signal, two conclusions may be drawn: (a) it is likely that interferences are operating and (b) extrapolation of the curve to its intercept will give a correct result.

11.5 TITRIMETRIC PROCEDURES

Titration is a venerable procedure that continues to be used extensively when a precision of ± 1 to $\pm 0.05\%$ is required. In the method a reactant solution of known concentration, the *titrant*, is systematically added to and reacted with a solution of the analyte. Both the *equivalence point* or *end point* of the reaction and the precise amount of the titrant added by that point are measured.

An important characteristic of this method of quantification is the subdivision of the measurement. One part of the instrumentation reports only whether or not an *end point* of the reaction has been reached, sometimes with the use of graphical methods. Because only a yes or no answer is required, this step need not be precise in the usual sense. Good quantitative precision is achieved instead by independently measuring with a device such as a buret the volume or weight of titrant reacting and can afford precision substantially higher than by other modes of quantifying.* The combined steps are like a ratiometric or null procedure.

The categories of chemical reaction to which titration is commonly applied are listed in Table 11.1. Within each category only those reactions are suitable that

1. proceed to completion or whose point of completion can be determined graphically;
2. display exact stoichiometry; and
3. are fast.

Where one criterion is not fully met, it may be possible to develop a strategy that will still allow use of the reaction. When these conditions are met and concentrations of solutions are sufficient (greater than 0.001 *M* usually), end points are sharply defined. Examples of titration curves are given in Fig. 11.5.

Example 11.14. What is the basis for the different requirements that reactions must meet to be suitable as a basis for a titration?

The rationale of the first requirement is nearly self-evident. If one is to titrate *to an end point*, accuracy will scarcely be possible without completeness of reaction. For reactions that are complete except in the region of the end point, that point can usually still be detected by graphical procedures. Data are taken well before and after the end point and plotted versus the volume of titrant added. The resulting lines are extrapolated to an intersection that identifies the end point. This approach is a valuable way to locate an end point even for reactions that are complete at all times, as is evident in the representative curves of Fig. 11.5.

When a nonstoichiometric precipitate (of variable formula) is formed, accuracy of determination is uncertain. It is possible to include under this heading also cases of precipitation reactions that proceed with coprecipitation or entrainment of foreign compounds in the precipitate and other forms of irregular behavior. In such cases other substances are trapped as a precipitate forms, and unreacted analyte may be trapped as well.

*In addition to the common type of titration with a *volumetric* buret, it is possible to use a *weight* buret. Its precision and accuracy are greater than that of the volumetric device by a factor of 10 because of the inherent precision of weighing.

Table 11.1 Principal Types of Titrimetric Methods

Type of Reaction	*Types of Titrant*	*Principal Product*[a]	*Examples of Titrimetric Reactions*[b]
Acid–base	Strong acid or strong base	Very weak acid or base or amphiphilic species	$\underline{H^+} + OH^- \rightarrow H_2O$ $\underline{H^+} + A^- \rightarrow HA$ $B^+ + \underline{OH^-} \rightarrow BOH$
Complexation	Salt of chelating or complexing agent (Na_2H_2Y, NaCN)	Complex ion	$\underline{H_2Y^{2-}} + Zn^{2+} \rightarrow ZnY^{2-} + 2H^+$ $\underline{2CN^-} + Ag^+ \rightarrow Ag(CN)_2^-$
Precipitation	Salt containing precipitating ion	Slightly soluble salt	$\underline{Ag^+} + Cl^- \rightarrow \underline{AgCl}$ $SO_4^{2-} + \underline{Ba^{2+}} \rightarrow \underline{BaSO_4}$
Redox Chemical	Acid or salt form of strong oxidizing agent or reducing agent	Reduced form of oxidant or oxidized form of reductant	$\underline{2MnO_4^-} + 5H_2O_2 + 6H^+ \rightarrow 2Mn^{2+} + 5O_2 + 8H_2O$ $\underline{2S_2O_3^{2-}} + I_2 \rightarrow S_4O_6^{2-} + 2I^-$
Coulometric	Electrons, which generate the actual titrant by electrolysis	Any of above	$(Br^- \rightarrow \frac{1}{2}\underline{Br_2} + e^-)$ $\underline{Br_2}$[c] $+ As_2O_3^{3-} \rightarrow AsO_4^{3-}$

[a]Product that is stable under conditions of titration such as a slightly ionized electrolyte.
[b]Conventions for writing ionic equations are followed except that solvation of ions is omitted. In each reaction the titrant species is the underlined.
[c]The chemical titrant Br_2 is generated by electrolysis at a platinum anode from a dissolved bromide salt.

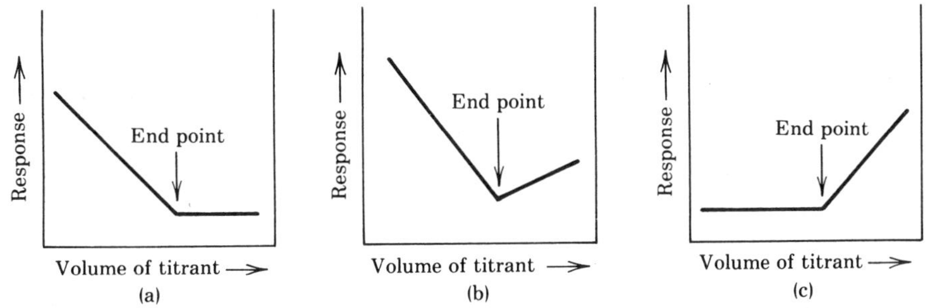

Fig. 11.5 Representative titration curves. For each titration reaction a mode of measurement is chosen that is selective for one or more reactants or products. Measurements are taken after increments of titrant solution are added both before and after the end point in order that a graphical determination of the end point can be made. (a) Only a reactant is detected. (b) Both reactant and a product are detected. (c) Only a product of the reaction is detected.

The final criterion of a sufficiently fast rate implies that a titration reaction must be complete in a few seconds. If a reaction is slow, not only will a titration take longer, but its precision may be lower. The process of locating an end point is most precise when a expected changes in concentration occur rather quickly.

Example 11.15. The requirement of a fast reaction in a titration is surprisingly limiting. (a) Can one predict the types of reactions that will be slow? (b) What strategies may be used to circumvent a slow rate?

(a) In general, the more complex its mechanism is, the slower a reaction. Thus, most redox reactions are slow because of the necessity of transferring many electrons. Precipitation reactions in which a substantial number of solvating molecules must be stripped from an ion before it can form a complex ion, precipitate, or species of changed oxidation state are also slow. Further, if a reaction involves fitting an asymmetrical species such as the thiocyanate ion SCN^- into lattice sites, it will tend to be slow. It can be anticipated that in all titrations rates will be slowest near the end point where the concentration or activity of both the analyte and titrant will be small.

(b) In redox titrations reaction flasks are often heated to *accelerate rates*. A common general strategy is to add an excess of titrant immediately and then *back titrate* by adding extra analyte of known concentration from another buret until the end point is reached. With a considerable excess of reagent rapid conclusion of the titration reaction is usually possible.

End-Point Detection. In most titrations, the end point is detected by monitoring the concentration of one or more components of the reaction by potentiometry, amperometry, or another instrumental procedure. Recall that in lieu of an instrumental measurement, an indicator may be added and its change of color at the end point detected visually. Yet instrumental monitoring gives desirable flexibility since for a typical reaction

$$\underset{\text{(titrant)}}{\mathrm{A}} \quad + \quad \underset{\text{(analyte)}}{\mathrm{B}} \quad = \quad \underset{\text{(product 1)}}{\mathrm{C}} \quad + \quad \underset{\text{(product 2)}}{\mathrm{D}}$$

it allows the concentration of any of the species to be followed. In general the species followed is the one whose optical absorption or other property allows it to be distinguished with minimum interference.

The major methods of end-point detection are listed in Table 11.2 and will be further discussed in chapters on each of the main techniques. The literature should also be consulted.*

Preparation of Standard Solutions. Part of the higher accuracy available from titrimetric procedures stems from the possibility of using primary standards and even standard reference material to calibrate or standardize titrant solutions.

*A good example of the rich sources of information available is I. M. Kolthoff and P. J. Elving (Eds.), *Treatise on Analytical Chemistry*, 2nd ed., Part I, Vol. 2, New York: Wiley-Interscience, 1979, Chapters 18–21; Vol. 3, 1983, Chapters 24 and 28.

Table 11.2 Some End-Point Detection Techniques

End-Point Technique (Signal Monitored)	*Nature of Titration Cell*	*End-Point Signal*
Potentiometry (Cell voltage)	Potentiometric cell with ion-selective or general-purpose indicator electrode	Voltage change
Voltammetry (Cell current)	Voltammetric cell with microelectrode	Current change
Coulometry (Cell current)	Coulometric cell with inert working electrode	Current change
Spectrophotometry (Cell absorbance)	Titration cell with light pipes to (a) optical source and (b) photomultiplier tube	Current change
Thermometry (Cell temperature)	Titration cell with thermocouple on wall	Voltage change
Conductimetry (Cell conductance)	Conductance cell with inert electrodes	Current change

---

Example 11.16. How useful are commercially available standard solutions for titration? How should titrants be precisely calibrated or standardized?

Many standard solutions whose concentrations are guaranteed precise to $\pm 0.1\%$ or better are commercially available. If the solutions are used with care, they retain their original concentration through long periods. Naturally, occasional restandardization will ensure that systematic error will not occur from this source.

In addition *primary standards* are often used in the laboratory. These are substances or solutions that can be prepared reproducibly with only moderate care. One example is constant boiling HCl, which can be prepared simply by distillation of reagent grade dilute HCl. Distillation is begun and about the first 75% of the distillate is discarded; then the distillate from 75–90% is saved as constant-boiling HCl. To obtain the exact concentration the barometric pressure is measured and the concentration at that pressure is obtained from the literature.

11.6 SEPARATION AND PRECONCENTRATION IN TRACE ANALYSIS

Trace analysis, the analysis for analytes present at concentrations below 0.01%, began to gain importance with the rise of the nuclear and semiconductor industries, because of their needs for analyses of low levels of impurities. More recently concern over environmental issues related to health protection and air and water quality and many research needs for lower limits of detection have spurred further development.

It is true that nuclear activation analysis, stripping voltammetry, and some other methods with low limits of detection allow elemental analysis of mixtures at the trace level. Yet if analytes present at the trace level in a sample can be separated or concentrated in a separate phase, a whole range of other valuable techniques can also be used. For example, procedures such as extraction, ion-exchange chromatography, and electrodeposition (in stripping voltammetry) are commonly used in this way.

Such preliminary steps are also attractive because they tend to minimize (a) matrix effects and (b) separability problems (measuring a trace amount of one constituent in the presence of large concentrations of similar substances). In this section the focus will be on separation and preconcentration techniques as important parts of the quantification strategy for trace analytes.

A number of common techniques for separation or preconcentration are described in Table 11.3. They may be compared by information in the table, but by reference to the literature their relative effectiveness will become much clearer. Criteria important in characterizing them are properties such as selectivity, relative speed, ease of use, coefficient of enrichment, and effect on further treatment. In addition, the degree to which a separation is quantitative is critically important: not all techniques provide 100% recovery. A method for estimating recovery is given at the end of the section.

In some separation procedures emphasis is placed on isolating the majority of trace constituents as a group. In other instances, however, the need for selectivity for a particular substance may dictate that it alone be isolated or that that substances which may interfere with its separation or determination be masked. In *masking* a complexing agent is added which will selectively tie up interferents.

Contamination; Loss of Trace Constituents. Every manipulation of a trace sample offers the potential of adding impurities. For example, carrying out standard operations in a regular laboratory can lead to serious contamination by airborne dust. Not only should a particular room be designated for trace-level work if many measurements are to be made, but a "clean-room" condition should be maintained. Alternatively, individual operations should be carried out in glove boxes or bags. Plastic equipment is greatly preferable to metal equipment. Among plastics, vessels of Teflon are found to have the smallest tendency to contaminate. Samples that must be ground should be handled in mortars of the highest hardness. For example, single-crystal Al_2O_3 mortars are available that are 1500 times harder than tungsten carbide. Clearly, great caution must be exercised in handling samples.

It is also important that all chemicals used in such analyses be ultrapure.* Where even these chemicals are insufficiently pure or chemicals not stocked are needed, a whole literature exists to assist the chemist in purifying his reagents.

At times the loss of trace constituents through sorption on walls of vessels becomes a severe problem. Teflon containers are suitable where those of Pyrex or other glass are not. Quartz vessels keep sorption to a minimum but clearly are unduly expensive.

*Some manufacturers that supply ultrapure chemicals and report the level of remaining impurities are J. T. Baker ("Ultrex" brand), E. Merck ("Suprapur" brand), and BDH Chemicals, Ltd. ("Aristar" brand).

Table 11.3 Some Separation and Preconcentration Techniques Commonly used in Trace Inorganic Analysis[a]

	Ion-Exchange Chromatography	*Extraction (Liquid–Liquid and Other Types)*	*Ashing or Oxidation of Organic Matrices*	*Electrodeposition*	*Precipitation and Coprecipitation*	*Volatilization*
Principle of method	Species may be chromatographed on appropriate cation-exchange or anion-exchange resin. A collector may be added for species at ultratrace level.	Extraction from aqueous into nonaqueous solvent on addition of chelating agent or a species that will associate with inorganic constituent.	Organic material is oxidized and driven off as gases (CO_2, H_2O, N_2, other) by careful burning in air (or oxygen-enriched air) or by "wet" oxidation by concentrated $HClO_4$ and/or H_2SO_4, and HNO_3.	Ionic species is deposited on working electrode by long reduction. Deposition usually followed by voltammetric stripping analysis.	An inorganic collector (carrier) is added. The trace element is precipitated or coprecipitated with the carrier.	Either an existing volatile compound is distilled or such compound is synthesized. Volatilization at 200–450°C.
Requirements and advantages	Selectivity of resin for all trace species must be high or differ substantially from one trace constituent to another. Traces can be isolated from hundreds of liters of solution on a small column. Only a small volume of eluent is required. Enrichment by 10^3 to 10^5. Very widely used.	Extracting solvent must offer favorable solubility (solvation) to species to be extracted. By use of range of complexing agents, method is widely applicable.	Organic matter usually interferes and must be removed. In dry burning after organic matter is charred, an oxidant like HNO_3 may also be added to avoid formation of thermally stable compounds. Additives may be necessary to avoid losses of constituents. With oxidation concentrated acid is a common alternative.	Potential on working electrode must be sufficient to deposit trace constituents in 10–30 min. Valuable for metal ions at 10^{-6} to 10^{-10} *M*.	Precipitate should be easily filtered and washed and not interfere in subsequent analyses. Amount of collector should be small. Substantial enrichment (10^3) possible. Method sees limited use.	Volatile compound of trace element must be easily formed. Commonly, chlorides and bromides in higher oxidation states are volatile. Method sees limited use.
Example(s)	The first traces of newly synthesized transuranium elements are usually mixed with a collector and isolated on an ion-exchange column.[b]	U(VI) extracted into tri-*n*-butyl phosphate from 2–6 *M* HNO_3 solution.[c]	Wet ashing of organic matrices may be started with HNO_3 and finished with $HClO_4$ as temperature is steadily increased.[d]	Traces of Co, Cr, Cu, Hg, Ni, and Zn were concentrated from water by electrodeposition on a graphite electrode.[e]	Al as collector can be coprecipitated with Co, Cu, Mo, and Zn traces from soil extracts by 8-hydroxy quinoline, thionalide, and tannic acid mixture.[f]	As (III), Os, Re, Sn(IV), chlorides and/or bromides volatilizable from $HClO_4$ and H_2SO_4 solutions at 200–220°C.[g]

[a]Processes should be quantitative, feasible, as simple as possible, and fast. They should introduce no extraneous impurities. Note that they may be used to separate either the macrophase (matrix) or trace elements. Once trace analytes have been separated, standard technique and measurement strategies will be used for quantification and/or identification.

[b]H. M. Leicester in M. E. Weeks, *Discovery of the Elements*, 6th ed., Chapter 31, Easton, PA, *J. Chem. Educ.*, 1956.

[c]C. J. Rodden, *Analysis of Essential Nuclear Reactor Materials*, Washington, DC: Division of Technical Information, U.S. Atomic Energy Commission, 1964.

[d]G. F. Smith, *Anal. Chim. Acta*, **8,** 397 (1953).

[e]B. H. Vassos, R. F. Hirsh, and H. Letterman, *Anal. Chem.*, **45,** 792 (1973).

[f]E. E. Pickett and B. E. Hankins, *Anal. Chem.*, **30,** 47 (1958).

[g]J. I. Hoffman and G. E. F. Lundell, *J. Res. Natl. Bur. Stand.*, **22,** 465 (1939).

Unmeasured Blank. As discussed in Section 10.5, blanks are a general means of avoiding types of systematic error not easily controlled otherwise. Stress has been placed on having pure chemicals for use in measurement steps in trace analyses, but far less control can be imposed on the steps of sample preparation. Contamination may occur in any step. In the not unusual case that a sample, after preparation, contains unidentified contaminants at the trace level, it is described as having an unmeasured blank: these substances will not have been added to blanks (see example below). Discovery of such contamination is often difficult. If it is believed to be occurring, the best way to check may be to subject a standard reference material to all steps.

Estimating Recovery. The percentage ''recovery'' of trace elements in preconcentration steps must be assessed to ensure precision. Ordinarily a kind of ''inverse standard addition'' procedure is valuable. Often radioactive tracers are used to assess percent recovery. A small quantity of a radioactively labeled substance that is otherwise identical to the anlyte is added to the initial sample as a tracer. Once it has been mixed with the sample, the next step is to measure the level of radioactivity in the sample. Then the concentration process, perhaps an extraction, is performed. As analyte is extracted, the tracer follows. The final measurement is of the radioactivity remaining in the sample. If all the analyte has been successfully extracted, no activity will remain. The ratio of the final activity to initial activity, after correction for decay and any volume effects, should give an estimate of the percent recovery. In particular, changes in the ratio of amount of standard added to initial (background) concentration will affect the variability greatly.

Example 11.17. If lead in the environment is being studied, even trace concentrations of lead in standards should be avoided. If lead is found, what steps can be taken to deal with it? What additional sources of contamination should be systematically checked for lead to avoid unmeasured blanks?

One approach to contaminated standards is to remove lead from them by an electrolytic or other procedure. Alternatively, if the concentration of lead impurity in standards can be accurately assayed, blanks may also be made up to contain that level.

In the study of trace levels of lead in whole blood an unmeasured blank may enter if an unanalyzed buffer or stabilizing agent is added in the field as samples of blood are taken. If the buffer contains a trace of lead unknown to the analyst, all blood lead analyses will by systematically high. Interlaboratory testing of whole blood from different sources in a community should show such bias.

11.7 OTHER STRATEGIES

Only principal strategies of quantification have been considered up to this point. A few of the other possibilities need brief mention to illustrate the range of possibilities.

Use of an *internal standard* is sometimes an attractive alternative to standard addition. An internal standard is a species that either has been added or is present naturally but is not to be measured. Each measurement on an analyte is made relative to a similar determination on the internal standard. The essence of the method is that matrix effects for analyte and reference be similar. For instance, in atomic emission spectrometry the intensity of an analyte line is often referenced to that of another metal that is naturally present in a set of samples at about a constant level. Indeed an analyte/reference pair is sometimes called an homologous pair.

Differential procedures are often used to improve precision of determination of species present at very low or very high concentrations. For example, in the determination of molecular weights of polymers by light-scattering measurements the small change in refractive index of a solvent on dissolving a solute (a few parts in 10^6) must be found with precision. It is straightforward to use a differential refractometer in which the index of the solution is compared with that of the solvent. Differential spectrophotometry is a similar procedure. In this case either the 0 or 100% transmission readings or both are made on standard solutions (see Section 16.6). Such "bracketing" by standard solutions is also common in atomic absorption spectrometry.

Derivative spectrometry, which involves taking the first or higher derivative of an analytical signal becomes useful in cases where analyte peaks overlap. For example, this approach is valuable in estimating the concentration of a component giving a shoulder on a main peak in an IR spectrum. The derivative spectrum provides sufficient additional detail to locate the minor peak more clearly (see Section 16.7).

Derivatives may be taken electronically or by computer program. The first or second derivative can be approximated by a ratio. For example, the first derivative of absorbance versus wavelength will be approximated by $\Delta A/\Delta\lambda$, where the change in absorbance over a small, fixed increment in wavelength is divided by the wavelength increment.

Finally, in instances where an analyte property cannot be conveniently determined directly, it often proves feasible to employ an *indirect measurement*. Some species of enzymes, for example, can be determined indirectly by determining their interference with the main reaction catalyzed by the enzyme. Similarly, fluorescence can be used to determine not only the fluorescing substance but sometimes that of an efficient quencher of the fluorescence of a common aromatic.

11.8 EXTRACTING INFORMATION: HOW MANY LEVELS?

As discussed in the opening section, extracting information from an instrument signal ordinarily begins with its translation into a physical response x_i such as an intensity of light. That operation is followed by a pair of analytical steps involving both an analytical relationship such as Faraday's law that relates the coulombs of charge used in coulometry to the concentration of an analyte and a mode of quantification that establishes the proportionality constant between that relationship and actual analyte concentration. But as made clear in Section 11.2, a one-dimensional operation is involved in each case. The following example will recall these operations.

Example 11.18 What are the relationships involved in determining the concentration of an aqueous solution of cobalt(II) ion by UV–VIS spectrophotometry?

To determine Co^{2+} in water, an excess of 8-quinolinol solution may be added to form a complex with cobalt(II) ion whose absorbance peaks at 365 nm. All measurements are made at 365 nm. The instrument signal is secured by appropriate ratioing of the transmittance of a blank for this measurement (solvent plus all reagents but with zero analyte) to that of the sample and taking of the logarithm to obtain the absorbance A. We have $A_i = \log(I_0/I_i)$.

Next, absorbance values at 365 nm are translated to Co^{+2} concentration by means of Beer's law ($A_i = e_i b c_i$), where b is the solution cell thickness and ϵ_i and C_i are the molar absorptivity and molarity, respectively, of the complexed ion. A calibration curve is obtained by plotting measured values of the absorbance of complexed standard cobalt(II) solutions versus cobalt concentrations.

A further step must be taken when the identity of a substance is sought. Now several procedures are ordinarily required, implying movement to a yet higher level. Another example will clarify the matter. Consider that an IR spectrum is examined to furnish information about the identity of a substance. The spectrum will feature two-dimensional encoding: it is a record of absorbance (dimension 1) versus wavenumber or wavelength (dimension 2). In comparison with extracting a concentration from a measurement, the entire absorption spectrum must be examined in a complex way to identify the substance. Perhaps the extraction process could be represented by an expression like

$$\text{identity of substance} = f(\text{IR absorption spectrum}). \tag{11.9}$$

Yet it may be argued convincingly that at present we simply cannot develop from quantum mechanics the kind of functional relationship required by Eq. (11.9). Indeed, more than one type of spectrum is often needed to establish an identity unambiguously. Nevertheless, a chemist, using ingenuity as well as mathematical procedures, can identify substances from spectra. Ordinarily he or she will use a library spectrum, if available, to validate the identification.

In the absence of a direct procedure, is any machine approach possible? Identities are nearly always established by comparing a given spectrum with spectra in libraries. If a library spectrum matches the one measured, identity is established. Excellent search or comparison methods have been developed. If a match is not secured, the program can be asked to furnish a list of possible candidates or a listing of highly probable functional groups.

11.9 SOME CHEMOMETRIC METHODS

When samples to be analyzed have several analytes, we begin to move away, at least in principle, from simpler measurement systems in which accuracy and sensitivity are high and optimum response is secured. The more elaborate systems of interest in research today call for monitoring a greater number of variables in complex samples. Examples are studies of catalytic systems, the earth's ozone layer, and neurotransmitters in the brain. In such systems one must take a greater number of measurements in fairly complex matrices. Elaborate procedures and complex evaluational techniques characteristic of chemometrics will usually be required.

Multicomponent Analysis. Simultaneous multicomponent analysis is a sophisticated alternative to the approaches discussed thus far that offers considerable power to deal with complex systems. Any suitable analytical procedure sensitive to all the components of interest can be used. Here the discussion will be mainly of absorption spectrometric or spectrophotometric methods. Recall the familiar spectrophotometric procedure of measuring light absorption at two wavelengths to analyze a two-component system. To extend the method to a great many components it is only necessary that Beer's law describe the absorption process (see Section 16.2) of each analyte. Each absorbing molecule must act independently and monochromatic light must be used. When these conditions are met, the method is applicable from the far IR through the far UV.

How is the complex series of absorptions described? If m components of a sample are to be measured, the absorbance of a dissolved sample must be observed at a minimum of m different wavelengths. Mathematically, this process yields a set of linear equations, one for the optical absorbance y_i at each wavelength i of each component j at concentration x_j:

$$\begin{aligned} y_1 &= S_{11}x_1 + S_{11}x_2 + \cdots + S_{1m}x_m \\ y_2 &= S_{21}x_1 + S_{22}x_2 + \cdots + S_{2m}x_m \\ &\vdots \\ y_m &= S_{m1}x_1 + S_{m2}x_2 + \cdots + S_{mm}x_m. \end{aligned}$$

Here each component j has a characteristic molar absorptivity S_{ij} at wavelength i. For example, the absorbance of component j at wavelength i is the product $S_{ij}x_j$. No cell path length appears since it is a constant factor multiplying every term.

It is important to recognize that this set of absorbance equations can be written in matrix form as

$$\begin{bmatrix} y_1 \\ y_2 \\ \vdots \\ y_m \end{bmatrix} \begin{bmatrix} S_{11} & S_{11} & \cdots & S_{1m} \\ S_{21} & S_{22} & \cdots & S_{2m} \\ \vdots & & \vdots & \vdots \\ S_{m1} & S_{m2} & \cdots & S_{mm} \end{bmatrix} \begin{bmatrix} x_1 \\ x_2 \\ \vdots \\ x_m \end{bmatrix}$$

Or each term can be represented by a symbol and the entire array shown as $\mathbf{Y}_m = S_{m\times m} \cdot \mathbf{X}_m$. Here $\mathbf{Y}_m$ and $\mathbf{X}_m$ are column vectors and $S_{m\times m}$ is a square matrix serving as the *calibration matrix*.

There is, of course, no reason that additional measurements cannot be made at wavelengths $n, o, p, q, \ldots$, which would add new equations to the above set. In this case $i > j$, and the system is said to be *overdetermined.* The advantage is a considerable gain in precision of measurement.

Before seeking a mode of solution of this large set of simultaneous linear equations (or its matrix representation), the procedure of measurement needs to be designed. The main choice will be the wavelengths. Presumably those chosen should give high values of molar absorptivities of components to ensure high sensitivity and thus, precision and accuracy. Ideally, each component should be represented by a wavelength at which it and no other components absorbs strongly. Since that situation will be difficult to realize, the practical choice will be more complex since the measurements should give optimum results for all of the components.

A reasonable strategy is to use an analogy to the calibration function (see Section 11.3) in which molar absorptivity plays the role of a sensitivity. Then the molar absorptivity of a component at each wavelength can be regarded as a *partial sensitivity.* Kaiser has defined the sensitivity of a multicomponent measurement as the absolute value of the determinant of the calibration matrix for the case illustrated in which the number of measurements is the minimum required for the number of components (square matrix).* In the matrix representation given above, the calibration matrix has been labeled S. To maximize S, the diagonal elements should be made large and the off-diagonal elements small. The challenge is to maximize the sensitivity and in that way ensure a simultaneous maximization of precision. Such a procedure is relatively simple but requires the calculation of a large number of determinants!†

As the number of wavelengths needed grows larger, whether one has found the "optimum value" becomes less important. Finally, it is worth noting that for X-ray fluorescence spectrometric measurements it is necessary to use a nonlinear type of model (see Chapter 21) and that programs giving good results have been developed.

Solving for Concentrations. Once wavelengths have been selected, and measurements made, the set of equations must be solved for component concentrations. If a single component were involved, the equation describing the absorbance would be basically the calibration function involving the sensitivity or molar absorptivity for the individual component at the wavelength presumably of its maximum absorbance: $A = \epsilon bc$. Reversing the relationship to calculate the concentration from the absorbance would be straightforward.

In this multicomponent system the set of equations must be solved. If represented as matrices, the equations reduce to the matrix form $\mathbf{Y}_m = S_{m \times m} \cdot \mathbf{X}_m$. But what is desired is

*H. Kaiser, Z. *Anal. Chem.*, **260,** 252 (1972) (German text).

†An alternative definition of sensitivity is $|S| = \sqrt{S' \cdot S}$. This expression regards sensitivity as the square root of the determinant of the product of the calibration matrix and its transpose. This procedure deals with what are called singular values of S and leads to a singular value decomposition. It provides a guide to varying the wavelength selection in the right direction. (A. Junker and G. Bergmann, Z. *Anal. Chem.*, **272,** 267 (1974) (German text).

the solution of the inverse relationship given by $\mathbf{X}_m = T_{m \times m} \cdot \mathbf{Y}_m$. Here $T_{m \times m}$ is the transpose of the sensitivity matrix (see below). This model implies that the *matrix of molar absorptivities* uniquely relates to concentrations and signals, but that each detailed composition of the mixture of components should also be related in a precise way to a particular set of measurements.

Two types of solutions have been suggested. One is the least-squares optimization procedures proposed by Kaiser and discussed above in which a set of equations involving ratios of determinants is developed for calculation of concentrations. The second involves using the matrix equation. In terms of the matrix approach the inverse relationship involving the matrix T must be worked out. It is possible to show that matrix S converts to matrix T by the process $T = (S' \cdot S)^{-1} \cdot S'$. In other words, T results when S is premultiplied by its transpose S'. This product is inverted and then postmultiplied by the transpose S'. Then the equation $\mathbf{X}_m = T_{m \times m} \cdot \mathbf{Y}_m$ can be solved.

Kalman Filter. Situations often arise in which the number of components in a sample is uncertain. If many possible components are known, the mathematical procedure called the Kalman filter can be used to elucidate simultaneously the number of components and their concentrations [18]. It uses an algorithm with a recursive structure and is ideally suited to measurements with considerable noise. As an on-line computational method, it can be stopped when results of a desired accuracy are attained. The Kalman filter offers at least as much power in determining the number of components in a system as does the mathematical technique of principal component analysis [3].

This section will conclude with a brief description of additional chemometric procedures that permit extraction of as much information as possible from large amounts of data furnished by complex systems. Factor analysis of data about a complex system identifies by matrix algebra variables that cluster into larger *factors* which can account for much of the behavior of the system. By contrast, pattern recognition begins with established categories such as molecular identities and establishes from the data the species that are present.

Factor Analysis. Methods such as factor analysis examine large amounts of data from many samples relating to many variables and identify a few fundamental *factors* [3, 9]. This process is also described as representing the information available in a smaller number of dimensions. Often many different instruments or sensors (detectors) provide data. Each factor normally includes at least two of the original variables. Other similar or equivalent methods are techniques known as *principal component analysis*, *eigenvector analysis*, and *Karhunen–Loewe expansion*.

Factor analysis is a widely used procedure that is also applied to problems arising in pattern recognition (see below), curve resolution, and calibration. An example of a complex curve that was resolved by factor analysis is given below.

In Fig. 11.6 the process of factor analysis is diagrammed. It calls for "decomposition" of a data matrix such as the absorbance matrix into two submatrices. Actually, there are two levels to the kind of factor analysis useful in chemical situations. An initial step yields an *abstract solution* that gives the correct number of abstract factors or *eigenvectors*.

Next, a chemically significant set of factors is sought. Insight and intuition about physical aspects that may be important are invaluable at this stage. In recent years, the devel-

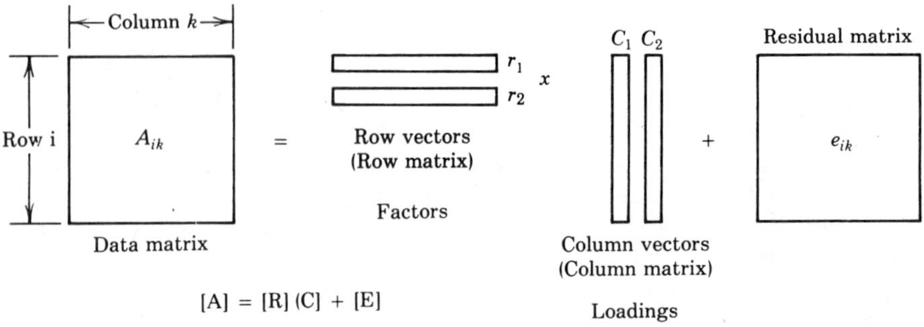

Fig. 11.6 A decomposition of a data matrix into submatrices. For an absorbance matrix Beer's law and a knowledge of the chemistry of the system are invaluable in identifying the factors. For a successful resolution it is necessary that the data matrix be complete, that is, that all solutions be made up the same way and that the residual matrix be as small as possible. Precision in measurement must be good.

opment of a technique called *target testing* has made it possible to test one by one those parameters that are believed physically significant [9]. The method makes it possible to extract a few significant factors from bewilderingly large amounts of data and numbers of potential variables.

Example 11.19. Kankare applied factor analysis to UV spectrophotometric measurements on acidified, aqueous bismuth choride solutions to deduce the probable number of bismuth–chloride ion complexes.* The other approaches reported in the literature had yielded a range of answers. He determined the rank of the UV absorbance matrix, and thus the number of factors or species of bismuth ion, by comparing residual standard deviations of absorbances for several solutions of different bismuth concentration with the residual absorbance error of the spectrophotometer used, which was 0.00368 absorbance units. As implied, he assumed successively larger number of components and made a comparison after each increase. Kankare found that the residual deviation reduced to 0.00379 assuming six components and 0.00231 assuming seven. Thus, since with seven absorbing species the estimated residual deviation was for the first time less than experimental error, he concluded seven species were present. To support this conclusion he calculated formation constants for the assumed species Bi^{3+}, $BiCl^{2+}$, $BiCl_2^+$, $BiCl_3$, $BiCl_4^-$, $BiCl_5^{2-}$, and $BiCl_6^{3-}$. Values obtained were in agreement with others found in the literature.

Pattern Recognition. This quantitative strategy deals with more diffuse kinds of systems by plotting multivariate (many-dimensional) data in two or three-dimensional fashion in order to emphasize patterns. Often the kind of question asked of a chemist is: From the data available can you tell whether (a) this sample of milk is from a cow or a goat? (b) the sweetner in this food is based on corn starch or starch from another plant? (c) this oil spill at sea represents oil dumped from this tanker or from that one? The point being made is that with multivariate analysis it is often possible to identify the three or four factors that contribute most to identifying all members of a class of substances. Pattern recognition provides techniques for classifying data into a set of categories [3, 8, 10, 12]. For example,

*J. J. Kankare, *Anal. Chem.*, **42,** 1322 (1970).

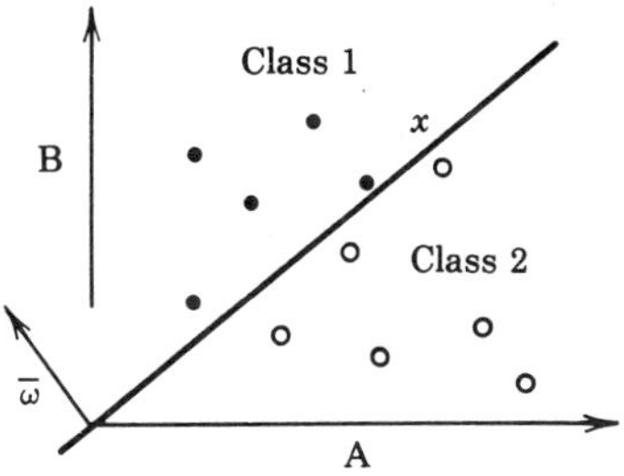

Fig. 11.7 The clustering of points indicates that there are two classes of similar molecules or systems in this two-dimensional space. Since the new molecule *x* falls into the cluster of dots, it can be identified as in their class. Where a great many properties are required to establish patterns it is possible to use factor analysis to reduce variables to principal sets of variables so that the graphical representation can be kept to two or three dimensions whenever possible.

the many peaks in the mass spectrum of a mixture of three substances might be grouped into categories consisting of peaks of the three molecules, thus identifying the species.

Methods are conceptually simple and can be treated essentially geometrically. Assume molecules are to be classified on the basis of one or more properties or *features*, A, B, . . . , N. If two properties A and B are sufficient to identify the molecules, to represent the set a two-dimensional coordinate system is used. Each molecule is represented by a point whose coordinates are the two properties, that is, by a *pattern vector* that extends from the origin to the point. For example, in very simple systems molecules might be characterized by m/e ratios of their *two most abundant mass spectrometrical* fragments.

For pattern recognition to be possible, however, it is essential that points that represent similar molecules tend to cluster as in Fig. 11.7. Here one set of points is represented by dots and a second by circles. The value of the approach is that by recognizing the pattern of properties of a new point *x*, it is seen to fall in the cluster of dots. For example, dots might represent methyl ketones and circles ethyl ketones.

Unfortunately, in many instances points representing pattern vectors do not cluster as closely as would be desirable and a variety of *classification procedures* or discriminant functions have been developed to separate classes. The predictive ability of a good classifier (usually 75–95%) is a measure of its usefulness. The class of molecules designated dots, for example, in Fig. 11.7 can be distinguished from the class of those represented by circles by the line that begins at the origin and divides the figure. Actually, as shown in the figure, classification is established in this case by setting up a *decision plane* defined by weighting vector to and the heavy diagonal perpendicular to the AB plane.

To develop a set of decision planes for more complex cases, a fraction of the molecular pattern vectors is often used as a *training set* to establish a classifier. Alternatively, molecular clusters can themselves be modeled and used to develop vectors without a training set. Reasonably compact clusters can be represented by their centers of gravity. A new molecule would then be classified by establishing to which center of gravity it was closest.

REFERENCES

Chemometric approaches and methods of quantification are presented in:

1. S. N. Deming and S. L. Morgan, *Experimental Design: A Chemometric Approach*. New York: Elsevier, 1987.

2. P. J. Elving and H. Kienitz, Methodology of analytical chemistry, in *Treatise on Analytical Chemistry*, 2nd ed., Part I, Vol. 1, I. M. Kolthoff and P. J. Elving (Eds.), New York: Wiley, 1978.
3. D. L. Massart, B. G. M. Vandeginste, S. N. Deming, Y. Michotte, and L. Kaufman, *Chemometrics: A Textbook*. New York: Elsevier, 1988.
4. J. Minczewski, J. Chwastowska, R. Dybezynski, *Separation and Preconcentration Methods in Inorganic Trace Analysis*. New York: Halsted (Wiley) 1982.
5. M. A. Sharof, D. L. Illman, and B. R. Kowalski, *Chemometrics*. New York: Wiley, 1986.
6. G. W. Small, Automated spectral interpretation, *Anal. Chem.*, **59,** 535A (1987).

More advanced chemometric topics are clearly discussed in:

7. K. Eckschlager and V. Stepanek, *Information Theory as Applied to Chemical Analysis*. New York: Wiley-Interscience, 1979.
8. B. R. Kowalski (Ed.), *Chemometrics: Theory and Application.* Washington, DC: American Chemical Society, 1977.
9. E. R. Malinowski and D. G. Howery, *Factor Analysis in Chemistry*. New York: Wiley-Interscience, 1980.
10. D. L. Massart and L. Kaufman, *The Interpretation of Analytical Chemical Data by the Use of Cluster Analysis*, New York: Wiley, 1983.
11. O. Strouf, *Chemical Pattern Recognition*. New York: Wiley-Research Studies Press, 1986.
12. K. Varmuza, *Pattern Recognition in Chemistry.* New York: Springer-Verlag, 1980.

Also of interest are:

13. J. J. Breen and P. E. Robinson (Eds.), *Environmental Applications of Chemometrics*. Washington, DC: American Chemical Society, 1985.
14. T. F. Brown and S. D. Brown, Resolution of overlapped electrochemical peaks with the use of the Kalman filter, *Anal. Chem.*, **53,** 1410 (1981).
15. J. R. Franke, R. A. de Zeeuw, and R. Hakkert, Evaluation and optimization of the standard addition method for absorption spectrometry and anodic stripping voltammetry, *Anal. Chem.*, **50,** 1374 (1978).
16. N. A. B. Gray, *Computer-Assisted Structure Elucidation.* New York: Wiley-Interscience, 1986.
17. D. L. Massart, A. Dijkstra, and L. Kaufman, *Evaluation and Optimization of Laboratory Methods and Analytical Procedures*. New York: Elsevier, 1978.
18. H. N. J. Poulisse, Multicomponent-analysis computations based on Kalman filtering, *Anal. Chim. Acta.*, **112,** 361 (1979).
19. B. E. H. Saxberg and B. R. Kowalski, Generalized standard addition method, *Anal. Chem.*, **51,** 1031 (1979).
20. C. Veillon, Trace element analysis of biological samples, *Anal. Chem.*, **86,** 851A (1986).

See also references listed for Chapter 10.

EXERCISES

11.1 Define the process of quantification of an analytical signal in a sentence or two.

11.2 Give a specific example of quantification for an electrochemical technique with which you are familiar. Very briefly describe your method in words, write any equation that is applicable, describe the kind of sample to which it will be applied, and illustrate any kind of graph that will be used.

11.3 A single-point calibration method is employed in a spectrophotometric determination of Ni^{2+} ion. For a 1.53×10^{-4} M Ni^{2+} standard an absorbance of 0.578 is obtained at the selected wavelength. (a) What is the molar absorptivity? Write the equation for a calibration curve based on this point. (b) The curve is used to determine the concentration of a Ni^{2+} unknown for which $A = 0.318$. Calculate its concentration.

11.4 From a consideration of Eqs. (11.6) and (11.7) what choice of concentration(s) would be most sensible for establishing a calibration curve?

11.5 Standard addition procedures are often used in atomic absorption spectrometry since they provide a calibration curve under the conditions of measurement. In atomic absorption, solutions are converted to an aerosol of tiny droplets that are carried into a flame. Would standard addition be able to cope with unpredictable variations in surface tension, viscosity, solvent burning (when an organic solvent is used), as well as interelement or matrix effects?*

11.6 In elemental analysis of powdered samples by atomic emission spectrometry with a dc arc and graphite electrodes, standard addition often ensures better results. (a) Describe briefly the sort of interelement effects that can occur. (b) Commercial mixtures of relatively pure inorganic compounds that can be added to powdered samples to allow for matrix effects are available. After addition of a weighed portion to the sample, the concentration of element x in the diluted sample is the unknown concentration C_x sought plus the known addition C_s. Concentrations are given as weight fractions, for example, C_x = wt. x in sample/wt. sample. Spectra are observed initially and after the known addition, and analyte line intensities are precisely measured. For the elements, we have for the initial sample, intensity I_x, and for the sample mixed with standard mixture, the intensity I_m. If w_x is the sample weight and w_s the weight of added standard, show that the concentration of species x in the sample is given by†

$$C_x = (I_x/I_m)\, C_s/\left\{1 + (w_x/w_s)\,(1 - I_x/I_m)\right\}.$$

11.7 Copper was determined in a stream below the point at which runoff water from mine tailings had mixed with it. Atomic emission measurements were used with quantification by the method of standard addition. In the procedure three 10-mL samples of river water were taken by pipet and amounts of a standard solution of 5.7 ppm Cu added by pipet.

*R. Klein, Jr. and C. Hach, *Am. Lab.*, **9**(7), 21 (1977).
†J. A. Larson et al., *Anal. Chem.*, **45,** 616 (1973).

Then deionized water was used to bring the total volume of each solution to exactly 50 mL. These solutions gave intensity readings (arbitrary units) as follows:

Flask Number	Added Standard	Intensity
1	0	12.52
2	10 mL	17.84
3	20 mL	23.13

(a) Calculate the Cu concentration in the water both in parts per million and in molarity. (b) Why is standard addition attractive as a method of quantification in this instance?

11.8 An unknown cadmium(II) solution is to be analyzed by a single-sweep voltammetry using the method of standard addition. In this method $I_p = kc$, where I_p is the peak current for an analyte whose concentration is c and k is a proportionality constant. A 25.00-mL aliquot of the unknown yields a peak current of 0.80 μA. After 10.00 mL of a 2.10×10^{-3} M standard cadmium solution is added by pipet and mixed in thoroughly, a measurement gives a value $I_p = 2.15$ μA. Calculate the cadmium concentration in the original solution.

ANSWERS

11.7 (a) 13.5 ppm, 2.12×10^{-4} M. *11.8* 3.12×10^{-4} M.

Chapter 12

SIGNAL-TO-NOISE ENHANCEMENT

12.1 INTRODUCTION

After the statistical and chemometric treatment of the whole sample measurement system just completed, we now focus on the physical aspects of instrument signal channels. What are the best ways to enhance the instrument signals? Since the chemical information in a signal will be present mainly as the amplitude of different frequency components, improving their S/N should greatly facilitate the task of extracting information.

In the physical domain it is noise, the random fluctuations always present, that tends to obscure the information for which a measurement is made. Noise ultimately establishes the lowest concentration of a species that can be quantitatively estimated, just as it fixes the precision of determinations at higher levels. Noise will appear in all modules of an instrument, signal source, sample module, detector, amplifier, processing modules, and display device. Nevertheless, attention will be focused on noise originating in the detector and amplifier, for, as will be seen, the behavior of these modules is usually crucial in extending the limit of detection or improving the precision of a measurement.

Developing ways to deal with noise has become increasingly important as the demand for lower detection limits and for greater precision in routine analysis has grown. Several strategies exist for improving the quality of measurements made under conditions of significant noise. One is to enhance the signal by all possible means, an approach that will be of interest in treating each technique in later chapters. A second is to reduce noise, a topic that will be explored here at some length. A third is to optimize S/N by techniques that discriminate between signal and noise, an approach that will be considered in the last sections of this chapter.

Since the concern in instrumentation is with the final result of a measurement, attention must be given not only to noise originating in a particular module but also to its transfer from one module to the next. For that reason we must not lose sight of the influence of a given module upon the preceding and succeeding ones. Questions of transfer of voltage, current, and power are thus relevant. Often, by judicious arrangement of transfer conditions, noise that has originated in a module may be selectively discriminated against in a transfer of signal to the next module.

12.2 THE ANALYTICAL SIGNAL AND ITS S/N RATIO

After the directed energy beam from the source has interacted with a sample and been detected, the output signal is properly called the *analytical signal*. It contains the desired *chemical information* (the true signal), *background* (almost inevitably), and *noise*. Since the goal of measurement is the chemical information, that information must be extracted. How difficult will the process be?

Two *limiting cases* will help illuminate what is involved. Consider first measurements that give strong analytical signals containing little noise or background. For these simple cases $I_{\text{anal sig}} \simeq I_{\text{information}}$ and the total signal may be used directly to calculate a desired property of the analyte or its concentration. Depending on the percentage of the total signal that is noise and uncorrected background, the relative standard deviation of the result may be 1%, 2%, ... Examples are environmental assays of copper(II) or cyanide ions in water. After appropriate complexation they are determined spectrophotometrically by filter spectrometer or colorimeter.

In the other limiting case $I_{\text{anal sig}} \simeq I_{\text{bgd}} + I_{\text{noise}}$ and the total signal may be almost entirely background and noise. The statistical aspects of this case were treated in Section 10.8 in the discussion of limits of detection and determination for measurements. The fluorometric determination of analytes at the trace level would be a good example.

Background. The term background can best be defined by examples. For a photomultiplier tube the background is its dark current (current when no light is incident). In a gas chromatograph with a flame ionization detector the background is the response observed when only carrier gas is eluting from a column. In measurements background is usually called the *baseline*. Random fluctuations or noise will be especially evident in the background and its *average value* must be obtained under each set of conditions. Once obtained, the average background must be subtracted from the total signal in either analog or digital fashion.

Example 12.1. In atomic emission spectrometry an excitation source such as an inductively coupled plasma torch produces luminous background that varies with wavelength. For an analyte emission line the observed analytical signal line equals the true value I_{line} plus I_{bgd}.

Two measurements are required to obtain I_{line}. Initially, the peak signal intensity at the characteristic analyte wavelength (it may need to be tuned in) is integrated over a short time interval. The spectrometer is then reset to a nearby wavelength at which only background radiation is observed. This reading, I_{bgd}, is also recorded over a short interval and subtracted to give I_{line}. Undoubtedly, I_{line} will also include noise!

Signal-to-Noise Ratio (S/N). For an analytical signal that is either close to the limit of detection or has a noisy baseline, the *signal-to-noise ratio* is a useful measure of quality. As just defined, signal is the analytical signal less background.

Noise can be defined as the *root-mean-square average* of fluctuations in the signal being measured.

In a particular system S/N is defined as the ratio of power in the signal relative to that in the noise:

$$S/N = 10 \log (V_s^2/V_n^2), \quad (12.1)$$

where V_s is the signal voltage and V_n the noise voltage.* Actually, there is an intimate connection between noise and uncertainty or random error. As shown in the following example, S/N essentially expresses the relative standard deviation or precision of the analytical signal.

Example 12.2. Whenever an analytical signal is essentially dc in character, its relative standard deviation describes both its precision of measurement and its noise. Now V_N, the rms noise voltage will be given by $V_N = [\Sigma (\bar{x} - x_i)^2/i]^{1/2}$, where $\bar{x}$ is the mean analytical signal, and x_i is its value after interval i. Whenever an instantaneous value of an analytical signal deviates from $\bar{x}$, noise is responsible. Note that this equation is the same as Eq. (10.1), that for the standard deviation of measurements in a set for the case where i is large. By substituting, S/N may be rewritten [6] $S/N = \bar{x}/V_N$ = mean/standard deviation = 1/(relative standard deviation) = $1/\sigma_r$.

12.3 NOISE

All extraneous and essentially random fluctuations added to a signal are termed noise. They originate ultimately in the discrete, noncontinuous nature of matter, charge, and energy. Noise has a statistical origin and may therefore be minimized but not eliminated. As a consequence, noise establishes both the limiting sensitivity of a module and the limit of detection of a technique or instrument. For convenience the present discussion will be restricted to electrical signals. The treatment is therefore directly applicable to most modules used in instruments. In other instances, as in treating lamp emission, an adaptation can be made [1].

Measurement of Noise. In analytical signals from instruments or their components noise is commonly evident, at sufficient amplification, as fluctuations in the baseline as a function of time. The peak-to-peak value of fluctuations may be translated to an rms measure of noise: the square root is taken of a sum of squares of several $V_{p\text{-}p}$ values.

Generally, noise is plotted as *noise power density* (W Hz^{-1}) versus frequency. Figure 12.1 shows such a graph of different kinds of instrument noise. Alterna-

*Voltages are squared since the power developed in a resistance or impedance is just V^2/R or V^2/Z. In Eq. (12.1) R or Z cancels since it can be assumed that both voltages are developed in the *same* resistance or impedance.

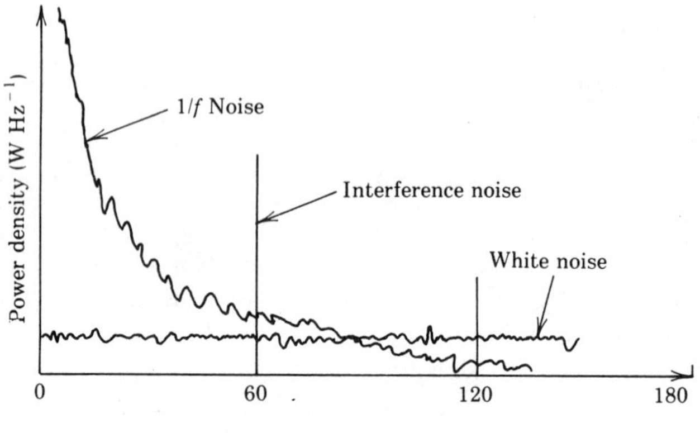

Fig. 12.1 Some representative power density spectra for noise. In most instruments $1/f$ noise predominates at low frequencies and white noise at high ones. Here types of noise are defined by frequency dependence.

tively, if a spectrum analyzer adapted for low frequencies could be used, it would provide a bar graph representation of power at discrete frequencies, for example, at 1-Hz intervals from dc to 10^3 Hz.

Environmental or Interference Noise. It is also possible to classify noise roughly according to its origin. The surroundings are always one source of noise. Light reflected by objects in a room, energy radiated by 60-Hz electrical lines in walls (fundamental frequency and harmonics), and mechanical vibrations transmitted through a floor are common examples. Much of this kind of interference noise appears as spikes, as shown in Fig. 12.1. The "pickup" of such random energy from surroundings may be minimized by techniques like shielding, thermostating, and filtering.

Satisfactory electrical shielding is perhaps hardest to devise, especially protection against the 60-Hz power distribution system in a building. Two steps should be taken. First, all shielding for an instrument should be grounded to the same *spot*, whether on a pipe or ground "terminal." It can be shown that there are differences of hundreds of millivolts between different places on most grounds. A spurious input signal through leads to ground can be avoided when a single connection is used. Second, for work at very low noise levels, differential-input amplifiers that have a high common-mode rejection ratio (Section 4.6) should be used. Any 60-Hz pickup will be common to both inputs of the circuit and will cancel when only the difference in signals at the two inputs is amplified.

Example 12.3. How can interference noise be minimized? In a sensitive amplifier or oscilloscope, leads should be kept short and the input to the first stage should be well shielded, especially if the source of the signal has a high impedance. These steps will minimize pickup of 60-Hz radiation. Adequate insulation or thermostating must be incorporated to protect against temperature and humidity changes.

Johnson Noise. It is possible to classify noise roughly according to its origin. One very basic type is noise that originates in a resistor (or conductor). Recall that all electrical signals are an ordered drift of electrons superposed on a basic random fluctuation. The noise associated with such fluctuations is called Johnson noise or resistor noise, since its magnitude is proportional to $R^{1/2}$. It has a root-mean-square voltage component given by the expression

$$v_{rms} = \sqrt{4kTR\Delta f}, \tag{12.2}$$

where k is the Boltzmann factor, T is the absolute temperature, R is the resistance, and Δf is the frequency bandwidth in hertz over which a measurement is made. Since the dependence is on Δf and not on frequency, Johnson noise may be called "white" noise: the contribution per $Hz^{1/2}$ is constant across the spectrum. Note that this noise exists whether there is a flow of current or not.

Example 12.4. A resistance of 300 kΩ at 300 K (about room temperature) in the input of an amplifier whose bandwidth is 10^5 Hz contributes an rms noise voltage of

$$\begin{aligned} 7.4 \times 10^{-12}\sqrt{RT\Delta f} &= 7.4 \times 10^{-12}\sqrt{3 \times 10^5 \times 300 \times 10^5} \\ &= 20 \times 10^{-6}\ \text{V} \quad \text{or} \quad 20\ \mu\text{V}. \end{aligned}$$

Ways to minimize Johnson noise are implicit in Eq. (12.2). It will be worthwhile in some instances to lower the temperature; in others, to narrow the amplifier bandwidth. For example, photomultipliers and other radiation detectors are often cooled to minimize Johnson noise.

Shot Noise. This is associated with variations in number of charged particles crossing a boundary or arriving at an electrode.* It arises from the essential randomness of such events. Shot noise originates in transistors as electrons or holes cross the depletion layer to the collector and in vacuum and photomultiplier tubes when electrons arrive at the anode. Such shot noise can be described by the following formula:

$$i_{rms} = \sqrt{2Ie\Delta f}, \tag{12.3}$$

where e is the charge on the electron and I is the collector current or other current due to some "single-event" type of process. Clearly, shot noise is also white noise

*The name is meant to suggest its resemblance to the discreteness of a stream of small shot.

since it depends on bandwidth Δf. Shot noise is a significant source of noise in spectrometric measurements at low light levels.

Flicker Noise. A third major type of noise is that termed flicker or excess noise. It is also called $1/f$ *noise* since it contributes an rms voltage proportional to $1/f^n$ where f is the frequency and n is a constant near unity. The mechanism by which it originates is not well established. In general it arises where granular material is present, as in certain types of carbon resistors, or where clusters of atoms may be involved, as in photocathode emission. What is clear is that flicker noise is likely to predominate at frequencies somewhere below 300 Hz. In this regard electron tubes are "quieter" than transistors, FETs are quieter than regular (bipolar) transistors. Flicker noise is a serious source of uncertainty in small signals of dc or very low frequency, such as a spectrophotometer signal.

Example 12.5. One form of $1/f$ noise is *drift*, the tendency of the output of a source or other module to vary slowly with time. Aging, temperature changes during warm-up of an instrument, and similar factors are responsible. Sometimes it is termed baseline drift. In many instruments it is a major factor in defining the detection limit of measurement. Drift can seriously affect the precision and accuracy of results. It may be noted that dc offset (see Section 4.5) is an error related to that of drift, in that it too is capable of amplification as a signal passes through a system. Both drift and dc offset may be minimized by modulation techniques (see Section 12.7).

12.4 MINIMIZING NOISE IN A SYSTEM

Fortunately, in those measurements in which the magnitude of the signal from an instrument is large compared with that of noise, noise may be ignored. Noise may, however, be significant in instances in which:

1. Source energy is limited as in IR spectrometry. Since the analytical signal derives from that energy, its amplitude and that of S/N will also be small.
2. There is a desire to increase the precision of a measurement. Often the easiest strategy will be to reduce the noise level.
3. Analyte concentrations are near the limit of detection. Because noise will actually be the limiting factor, all reasonable means of reducing it will ordinarily be employed.

How does noise add in a system like an instrument? We need now to examine the role of modules.

When noise signals from independent sources combine, the resultant noise V_N is the sum of the squares of rms noise voltages:

$$V_N^2 = V_a^2 + V_b^2 + \cdots + V_n^2,$$

where V_i designates the rms contribution by each source. The contributions to V_N^2 also depend on the frequencies sensed

$$V_N^2 = \sum V_{Ni}^2 \delta f_i = V_{N1}^2 \delta f_1 + V_{N2}^2 \delta f_2 + \cdots ,$$

where V_{Ni} is the noise associated with each small bandwidth δf_i. If the noise is white and all bandwidths δf_i are 1 Hz, $V_{Ni} = V_{N2} = \ldots$, and

$$V_N^2 = V_{Ni}^2 \sum \delta f_i = V_{Ni}^2 \Delta f,$$

where Δf is the system bandwidth and V_{Ni}^2 is the noise power per hertz or *noise density*.

What strategies exist for improving the quality of measurements made when the noise level is moderate to high? One is to enhance the signal by all possible means, a type of approach that will be valuable regardless of the type of measurement. For instance, Fourier transform techniques ensure stronger signals by allowing all frequencies of interest to be observed at all times.

A second strategy is to improve instrument design. In instrumentation attention must be given not only to noise originating in a particular module, but also to its transfer and accumulation from one module to the next. Note that only the detector generates the analytical signal, though all modules will introduce noise. The *noise figure* F of a module is a convenient measure of the effect of a module on the instrument S/N and is defined by the expression

$$F = (S_i/N_i)/(S_o/N_o), \qquad (12.4)$$

where the subscripts i and o refer to module input and output, respectively. When noise figures are known for modules that are coupled as shown in Fig. 12.2, an overall noise figure can be calculated. If the modules are of sensitivity or gain S_1, S_2, . . . , and noise figure F_1, F_2, . . . , the system noise figure will be

$$F = F_1 + (F_2 - 1)/S_1 + (F_3 - 1)/S_1 S_2 + \cdots . \qquad (12.5)$$

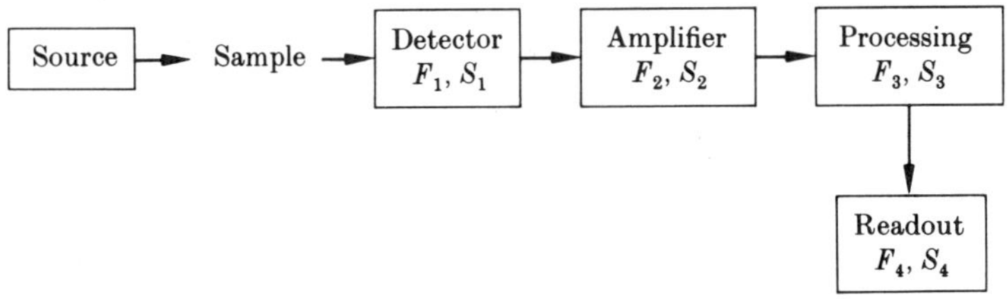

Fig. 12.2 Contribution of modules comprising a single-channel instrument to noise. Symbols are: sensitivity or gain, S_i and noise figure, F_i. The analysis is begun with the detector since chemical information from analytes is first added to the signal channel in the sampling module.

Note that the product of all preceding sensitivities or gains forms the denominator of each term; this factor causes noise contributed by modules to become progressively less critical toward the output of an instrument.

Example 12.6. What should the noise figure be for an ideal amplifier? Since this amplifier should contribute no additional noise and should amplify the signal and noise voltages at its input equally, it should have an F of unity.

Example 12.7. The S/N at the input of a voltage amplifier is 50 and at the output 25. Its noise figure in decibels is $20 \log 50/25 = 20 \log 2 = 6$ dB.

Example 12.8. Consider that in a representative instrument the detector is the first module to contribute appreciable noise and the amplifier is the second. How much attention should be given (a) to the quality of the detector? (b) to the quality of the amplifier?

(a) On the basis of Eq. (12.5), if F_1 refers to the detector, keeping its noise minute will be crucial to producing a favorable S/N at the readout. A quality detector of high gain and low noise should be used. If it has considerable Johnson noise, it should be cooled, probably at least to 0°C.

(b) The amplifier will contribute F_2 in Eq. (12.5) and also be important. In practice, two amplifiers are commonly used. The first, a low-noise, high-gain *preamplifier* is placed very near the detector. The shorter its input lead is, even if shielded against pickup, the smaller the noise. It is worth lavishing special care on the design of preamplifiers. The regular instrumentation amplifier that follows must be stable, but may be much noisier without diminishing precision since its denominator in Eq. (12.5) will probably be larger by two orders of magnitude.

Shifting the Channel Frequency. Attention to instrument characteristics such as the central frequency of measurement and system bandwidth will be decisive in limiting noise. The chemical information about an analyte in the energy beam falling on a detector is often at frequencies of 0–2 Hz. For example, this situation holds for most measurements with pH meters, gas chromatographs, and dispersive spectrometers.* For low-amplitude and low-frequency signals $1/f$ noise or flicker noise may dominate. In the IR $1/f$ noise is a problem because sources have low power, while this type of noise is not as important with higher source intensities in the UV–VIS.† Fortunately 1/f noise can be substantially reduced by (a) shifting the frequency of measurement from dc to a higher value by modulating or chopping the energy in the signal channel, and synchronously demodulating it. This process will be described in Section 12.7.

*In dispersive instruments one frequency band at a time is brought to the detector. Since the intensity of radiation to which the detector responds is constant, the signal frequency is 0 Hz and not that of the light.

†J. D. Ingle, Jr., *Anal. Chem.*, **49,** 399 (1977); Y. Talmi and R. Crosmun, *ibid.*, 340; J. D. Ingle, Jr., *Anal. Chim. Acta*, **88,** 131 (1977).

Narrowing the Bandwidth. By narrowing the system bandwidth Δf as much as possible, both Johnson and shot noise can be greatly reduced. Equations (12.2) and (12.3) make this clear. To carry information, the instrument channel must have a finite bandwidth Δf. While some modules, for example, a photomultiplier tube, may have a large bandwidth, the effective value for an instrument will be the smallest bandwidth for any of its modules. Indeed, the bandwidth is often set by adjusting *the time constant* of an active low-pass filter associated with an amplifier to a value from 0.01 to 10 s.

The *effective bandwidth* of a module or an instrument is basically defined in the time domain by its time of response t_r to a step signal, that is, an abrupt change in input. The height of the step is not critical as long as the signal does not overload the system. A convenient approximation to the response time is the *rise time*, the time for the output to increase from 10 to 90% of its final value. It can be shown that the effective bandwidth Δf is related to t_r by the expression

$$\Delta f = (1/2) \int_0^\infty \left[a(t)^2\right] dt \simeq 1/t_r, \tag{12.6}$$

where $a(t)$ is the slope of the relative time-response curve. Bandwidth is in hertz and time in seconds. The approximate relation stated in Eq. (12.6), $\Delta f \simeq 1/t_r$, is quite adequate for present purposes.

Example 12.9. What is the approximate bandwidth of a system whose rise time in response to a step change in input is 0.01 s?

Since bandwidth $\Delta f \simeq 1/t_r$, it should be $1/t_r = 1/0.01 = 100$ Hz.

Example 12.10. Will noise be a greater problem in a system with a 0.1-μs or a 0.1-s response time?

Again, the approximate bandwidths for the systems are 10 MHz and 10 Hz. With Johnson and shot noise both proportional to $\Delta f^{1/2}$, the faster system is more noisy by a factor of $(10^7/10)^{1/2} \simeq 10^3$.

Example 12.11. An IR spectrometer whose beam is chopped at 13 Hz by a rotating sector employs a thermocouple detector with a 35-ms response time, has an amplifier with a 1-Hz bandwidth, and displays spectra on a chart recorder with a 1-s full-scale response time. (a) Are the modules compatible in terms of bandwidth? (b) If the sharpest absorption peak is likely to be 5 cm^{-1} wide, how fast a spectral scan can be made without losing precision?

(a) If it is assumed that the chopper produces one square pulse of light per channel with every revolution, each pulse will have a length of $(\frac{1}{2}) \times (\frac{1}{13}) = 39$ ms. The light pulse lengths appear to be matched to the thermocouple response.

(b) The system time constant or bandwidth is fixed by the smallest bandwidth and will be 1 Hz. Noise will be strongly reduced by this narrow bandwidth. It appears that the scan rate cannot be faster than 5 $cm^{-1}\ s^{-1}$ (see Section 17.5).

The S/N ratio may be enhanced considerably by techniques such as digital filtering, lock-in amplification, and multichannel averaging that reduce the system bandwidth by extending the total observation time. Their importance warrants coverage in separate sections after an examination of signal sampling.

12.5 SIGNAL SAMPLING AND DATA ACQUISITION

To take advantage of the possibilities of signal processing that in-line microprocessors provide an instrument, the analog signals furnished by most modules such as optical detectors must be digitized, that is, converted to the digital domain. While the digital aspects of this process have already been described in Chapters 5 and 6, we return to the process to deal with the sampling step in more detail. In doing so the focus will be on passing the digital version on to the processing modules with precision and speed and with the chemical information intact. The *data acquisition* modules responsible for conversion are shown in Fig. 12.3. The amplifier ensures that the analytical signal is large enough to use the full input range of the AD converter and in that way ensure precision of digitization, as will be taken up below. Accuracy also depends on the proper choice of the frequency of sampling of the analog signal.

Sampling Frequency. The sampling theorem, which is a part of information theory, yields the result that the sampling frequency must be *at least* twice that of the highest signal frequency with chemical information, f_{max}. For example, sampling at 3,000 Hz would be appropriate for data containing signal frequencies up to 3,000/2 = 1,500 Hz. The highest signal frequency that will be correctly digitized is termed the *Nyquist frequency*. Once f_{max} is set, a sufficiently fast AD converter can be selected and the microcomputer program responsible for control of data acquisition can be set to provide control signals at appropriate rates to the sample-and-hold module and AD converter.

Because the sampling process involves mixing the frequency of sampling, f_s, and its harmonics with the frequencies f_i in the analytical signal, sampling will

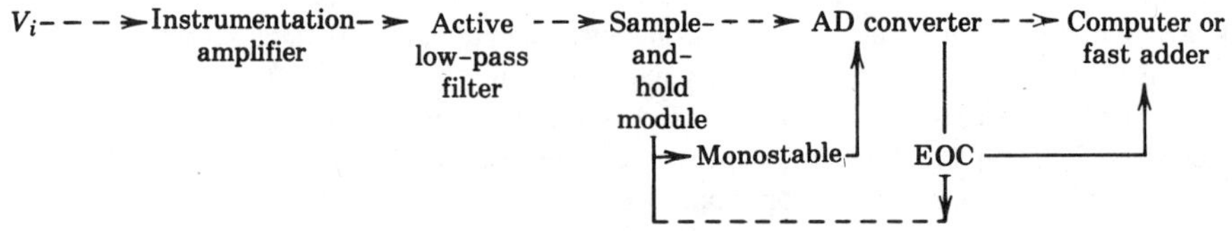

Fig. 12.3 A modular diagram of a fast-data acquisition system coupled to a digital integrator. The instrumentation amplifier raises the level of the signal from the detector until it matches the input range of the sample-and-hold device. The low-pass filter limits the bandwidth to that required for accurate sampling. The monostable (multivibrator) is selected to provide a delay time about twice that of the average settling time of the sample-and-hold module. An appropriate AD converter would have a still longer conversion time. Both times should be 25–100 times shorter than the smallest time interval of significance in the measurement system to permit at least 25–100 points to be added (integrated) for S/N enhancement. Note that the starting and stopping of the modules is controlled by signals in the lower channel: the monostable output pulse starts the converter. When it is finished, its end-of-conversion (EOC) line carrier a pulse to the sample-and-hold module to begin a new cycle.

produce sum and difference frequencies such as

$$f_i, \quad f_s \pm f_i, \quad 2f_s \pm f_i, \quad \cdots .$$

While amplitudes of harmonic sum and difference frequencies will be smaller, it is clear that filtering will be necessary to block unwanted frequencies. A good choice is to include a cutoff filter that will block frequencies above the Nyquist value, f_{max}.

Without the cutoff filter, when $f_s - f_i$ falls within the spectral range of interest, *aliasing* or folding is said to be taking place. Both f_i and $f_s - f_i$ will now appear in the spectrum though only f_i is valid.

Example 12.12. Since there are no sharp "cutoff" filters in electronics, what can be done to block frequencies above the Nyquist limit? When is it wise to increase the sampling rate to minimize the problem?

A complex filter may be a good choice for blocking them, for example, a 6-pole, low pass, active filter. If its half-power frequency f_o is set at the upper limit of the analysis range f_{max}, frequencies that are greater than f_{max} will not be strongly attenuated and will appear as differences $f_s - f_x$. A better decision would be to set the sampling rate at about three times, or in some instances 10 times, f_{max}. This rate takes into account the actual falloff rate of available filters. Sampling at still higher rates is unattractive because the memory requirement for storage of data becomes very substantial.

Example 12.13. In the far IR, if only the spectral range from 1 to 500 cm^{-1} is sought, how many f_{max} be best set?

In this case an optical filter that would drastically attenuate wavenumbers higher than 500 cm^{-1} should be introduced. If the filter has too gradual a cutoff above 500 cm^{-1}, the Nyquist frequency should be raised to 1000 cm^{-1} or higher.

Example 12.14. The proton NMR spectrum of 1,3,5-trimethylbenzene is observed at 100 MHz using a Fourier transform spectrometer. The data acquisition rate is set for 2000 Hz. When the transmitter frequency was set at frequency f' the trimethylsilane (TMS) (reference) resonance appeared at 176 Hz. For a second run, the transmitter frequency was increased by 200 Hz to f'', causing the TMS peak to register at 376 Hz. The sampling rate was not altered. Was the sampling (data acquisition) rate equally appropriate for both runs? The data obtained were as follows:

	Aromatic Protons	*Methyl Protons*	*TMS Protons*
Run 1			
Resonances observed at frequency f' (Hz)	852	401	176
Chemical shift (ppm)	6.76	2.25	0
Run 2			
Resonances observed at frequency f'' (Hz)	948	601	376
Apparent chemical shift (ppm)	5.72	2.25	0

In the first run with a sampling rate of 2000 Hz, all resonances up to the Nyquist frequency (1000 Hz) should have appeared correctly, including the aromatic proton peak at 852 Hz (676 Hz relative to TMS). In the second run, all peaks should be 200 Hz higher and the aromatic proton peak should appear at 1052 Hz. Because of aliasing it in fact shows up incorrectly at $f_s - f_1 = 2000 - 1052 = 948$. Increasing the sampling rate to 2400 Hz would have reduced this problem.

Quantization Error. Note that the precision with which analog signals are converted to digital form is basically set by the digital *voltage increment* that corresponds to the least significant bit. For example, an 8-bit AD converter with a 10-V output will have a least measure or minimum output increment of $10/256 = 0.04$ V; its *best* precision is 0.4% of full output.

More subtle errors enter during conversion, such as errors attributable to zero-offset, scale, nonlinearity, and nonmonotonicity, and must also be taken into account (7). The zero-offset error may be minimized by appropriate trimming of the converter. Similarly, calibration will allow for a correction of other errors.

Example 12.15. It is sought to obtain a resolution of 10,000 in observation of mass spectra. A mass spectrometer with a data digitization rate of 50,000 Hz is available. Can the desired resolution be secured if a spectral scan rate of $100 m/z$ min^{-1} is used? What precision per point will be obtained if a 14-bit AD converter is employed?

When a mass spectrum is scanned at rate of $100 m/z$ min^{-1}, a resolution of 10^4 will mean that peaks at say 100.00 and $100.01 m/z$ will have to be distinguishable. Assume a minimum of three points (samplings) are required to define each peak. Given the specified resolution, the total time for sampling per peak can be no greater than roughly 0.01 or 0.003 min per sampling. The AD converter must produce a count of 1.6×10^4 in 0.003 min or 1.6×10^4 counts/0.003×60 s or nearly 10^5 counts s^{-1}. The digitization rate is inadequate.

If the rate is increased to 10^5 Hz, however, by installation of a faster AD converter, spectra can be obtained with the desired precision. Alternatively, precision can be traded off for speed. If a 12-bit converter (precision better than 0.1%) is substituted for the 14-bit unit (precision better than 0.01%), fewer counts will be produced per second and the digitization rate will be adequate.

12.6 DIGITAL INTEGRATION AND SMOOTHING

It is possible to gain significant S/N enhancement by means of integration connected with the AD conversion. While an op-amp integrator with time switching* might be used to achieve averaging of noise toward zero, digital counterparts are more common. There are integrating AD converters, for example, that use either the dual-slope or the voltage-to-frequency approach. They are able to reject noise

*It is necessary that the integrator operate between sampling points, which presumably would be evenly spaced.

at the frequency corresponding to the integration time and to multiples of that basic frequency. The striking advantage of this mode is its rejection of much line-voltage interference when the integration time is set to one period of 60 Hz (13 ms). Nevertheless the disadvantage of having the measurement centered at 0 Hz where $1/f$ noise is greatest and limited to slowly changing signals is severe.

Digital Integration. For signals with no change occurring in times shorter than 1 ms, it is possible to take advantage of digital integration by using a fast data acquisition system. As was shown in Fig. 12.3 the crucial operation of integration is taken care of in the computer or in a fast adder in the hardware version. Note that provision is made for settling of the output of the sample-and-hold module by inserting a monostable unit that controls the start of data conversion.

Digital Filtering. In the time domain digital filtering becomes quite feasible because it basically can be regarded as a process of *weighting* or convolution. The approach is most successful if all of the frequencies in the signal are higher than most of the $1/f$ noise. By digital filtering procedures it is possible to remove a great deal of the random error in signals measured as a function of time without degrading the information appreciably. Recall that wavelength scans in a dispersive spectrometer are basically in the time domain since wavelength is being varied linearly with time.

Sliding or Moving Average. The sliding average is perhaps the most common type of smoothing. In this method each raw data point is replaced by an average of a prescribed number of preceding and succeeding points. It is important to note two criteria that must be satisfied: (a) sampling must be at fixed, uniform intervals and (b) the relationship between variables must be continuous and more or less smooth.

Example 12.16. How does one take a moving average? Must weighting always be uniform?

In a moving average "smooth" weighting is uniform. The averaging function joins seven points and is moved along the data array. Points 1–7 from the raw data are first averaged and the mean is substituted for point 4. Next the average of points 2–8 replaces point 5, and the process is repeated until the end of the array.

Many other types of weighting can be and are used. A polynomial, for example, a quadratic function, can be also applied to parts of a raw data array for a more sophisticated type of smoothing.

In all forms of smoothing the signal bandwidth is narrowed. The price paid will of course be some loss of information; usually peaks are broadened and lose height. For that reason smoothing functions and the number of points included in the smooth must be chosen with care. The user is always in a good position to evaluate visually how smoothing is affecting his data. Further, whenever measurements are being made relative to standards and both are treated in the same fashion, precision is not lost even if some signal distortion occurs. For absolute measurements, a program-controlled least-squares fitting of a curve to data is preferable (see Section 11.2).

The greatest advantage of digital smoothing is probably in visual interpretation of data. Whenever parameters are to be extracted manually, it is valuable to have "cleaner" curves to inspect.

Savitzky-Golay Convolution. A least-squares mode of weighting of points in the sliding average was proposed by Savitzky and Golay [8]. They pointed out that even the rectangular sliding average follows an equation like:

$$Y_j^* = \Sigma\, C_i Y_{ji}/N, \qquad (12.7)$$

where j is an index relating to the particular point around which the average is performed, C_i is what is termed a convoluting factor, and N is the number of points averaged. A convolution is thus a process in which one set of values is multiplied times a corresponding set of ordinate values. In the case of rectangular smoothing $C_i = 1$ for all values. For an analog RC filter, values of C will follow an exponential function. The Savitzky–Golay smoothing is performed by using a least squares procedure to develop appropriate values of C_i based on fitting a polynomial to the data and developing a set of integers which provide the appropriate weighting function. For either a cubic or a quadratic function the set of integers is the same.

The digital smoothing process and a recursive method of smoothing that is computationally more efficient have been described [5].

Example 12.17. Can signal averaging (see Section 12.8) and digital smoothing be usefully combined?

Following a digital averaging of a set of runs by a least-squares method, smoothing will be quite effective. Signal averaging of a set of data can be shown to result in a gain in S/N of $\sqrt{100} = 10$ if 100 runs are averaged. A comparable improvement could be achieved by taking only 16 sets of data for the average and then applying a 9-point least squares smoothing to the average.

Running Average. In finding the arithmetic mean of a noisy steady-state signal a type of calculation known as a running average, and the process as *exponential averaging*, may effect a significant saving of time. The *running average* A_i is the average up to and including point i and is usually calculated on a batchwise basis, as illustrated in Example 12.18. If the signal is expected to be a steady-state dc voltage on which random noise is superimposed, this average will gradually approach the dc signal expected. Readings can be taken very quickly if the noise is low. If noise is high, precision can be traded off for speed. As few as one-fourth of the total of results that might initially be assumed necessary have proven in general all that are needed to get an average. An advantage of the method is that neither the noise level nor an approximate time constant or bandwidth need be known in advance. A digital voltmeter is a convenient display device.

Example 12.18. How should an efficient batchwise running average be designed? One way is to sample a steady-state analytical signal at about a 1-kHz rate and to average all readings.

When 256 have been averaged, the mean is displayed. Successive batches of 256 readings are averaged and each new batch is mean-averaged with the earlier overall mean. Unless the noise level is high, the grand mean will quickly approach the true value.

12.7 CORRELATION TECHNIQUES, MODULATION, AND LOCK-IN AMPLIFIERS

Most important methods for extraction of signals from a great deal of noise derive from modern information theory. Here the method called *cross correlation* will be briefly described because of its importance and attention focused on lock-in amplifiers, one of its main applications.

Cross correlation is a mathematical operation that requires taking the product of two different but coherent signals A and B from a process such as a spectrophotometric measurement. The product is integrated over as long a time as possible. The integral equals what is called the cross correlation function $C_{AB}(\tau)$ of A and B, which is the limit approached as the period of integration $-T$ to $+T$ approaches infinity. This process is expressed by the equation

$$C_{AB}(\tau) = \lim (1/2T) \int_{-T}^{T} f_B(t)\, f_A(t - \tau)\, dt, \qquad (12.8)$$

where f_B and f_A are the coherent signals and τ is equivalent to a phase difference. Function C_{AB} contains information about all the frequencies in the two signals that are correlated and about the phase difference between them. This information may be extracted from C_{AB} by a Fourier transformation.

Like correlation functions in statistics, Eq. (12.8) describes the degree of coherence that exists between the functions in the time domain. For the enhancement of S/N it is important that (a) cross correlation of noise with a periodic signal yields a value that approaches zero and (b) cross correlation of a signal with a replica of itself yields a result that approaches a definite value. It is beyond the scope of this volume to develop the theory of correlation further. Further information may be found in references [7]. Before describing the lock-in amplifier, modulation of signals needs discussion.

Modulation. To minimize $1/f$ noise the central frequency of a signal that is basically dc can be moved upward into a region with less noise. In terms of Fig. 12.1, when the analytical signal is centered around 0 Hz, it can be moved to a substantially higher frequency at which there is little interference noise or $1/f$ noise. Shifting the central frequency can be accomplished simply by the technique of *amplitude modulation*. Before discussing modulation, however, we need to remember that the process must be reversible: the goal is extracting the signal from noise. Other steps must follow modulation: (a) selective amplification of the modulated signal (frequencies not immediately around the modulating frequency must receive little amplification), (b) synchronous demodulation, in which the modulating frequency must be removed along with much noise, and (c) filtering, in which residual carrier frequency and all other frequencies except a very narrow band that contains the chemical information desired are removed. These steps will be described when the lock-in amplifier is taken up below.

Modulation involves mixing or multiplying the signal with a carrier wave that is appreciably higher in frequency in a way that causes the carrier *amplitude* to vary with the amplitude and frequency of the signal.* In instruments sinusoidal modulation is sometimes used but modulation employing a train of square waves known as a *pulse train* is more common. For example, in a spectrophotometer, modulation is often achieved by interrupting the optical beam periodically by a rotating sector mirror.

How should the frequency of modulation f_0 be chosen? In general, f_0 should be a frequency that is

1. high enough to be above most $1/f$ noise,
2. free of line-voltage and other kinds of interference noise,
3. low enough compared with the response frequency $(1/t_r)$ of modules so that it is well transmitted through the instrument, and
4. larger than f_{max}, the highest frequency containing chemical information about the samples.

Observing a *noise density spectrum* for the instrument is helpful in meeting the first two criteria. It should be possible to estimate the response time of instrument modules. The last requirement arises because the modulation frequency in effect sets the highest frequency the carrier wave can carry. Clearly, the highest signal frequency will vary with the speed of measurement and the complexity of chemical information encoded. Trade-offs must be worked out in advance. Since noise at the ac line frequency and its harmonics and subharmonics may have high amplitudes, it is common to place notch filters after the amplifier to attenuate those components strongly. Otherwise the lock-in amplifier may be overloaded with noise.

A second important question is where along the signal channel should modulation take place? Locating the place requires analysis of the kind of noise or background to be minimized. Usually it proves advantageous to modulate as early in the signal channel as possible: only drift and $1/f$ noise that enters after the point of modulation can be minimized.

Example 12.19. In a double-beam dispersive UV–VIS spectrophotometer, Johnson and shot noise will usually be less than 1% of signal amplitude, but drift and other types of $1/f$

*Modulation may also be arranged to alter the *frequency* or the *phase* of the carrier wave. In addition to its use in improving S/N, modulation also serves other valuable purposes such as facilitating transmission of a signal through a medium (e.g., a telephone wire or interstellar space) that will attenuate the basic signal. In this use, the signal is borne on the carrier. The transmission of radio and television signals on carrier waves are good examples. A third application is to permit a fiber-optic link or coaxial cable to carry many channels of information (e.g., several conversations or digital data transmissions).

noise may be more substantial. The shifting of the central signal frequency to a higher frequency by modulation thus proves valuable in eliminating the major noise component in this type of instrument.

In this instrument and in others, modulation nearly always serves other purposes too. For example, it provides a way to arrange time-sharing of a single detector with both the sample and reference beams. Subsequently, with a modulated wave it is easy to preserve the phase relationship of the two parts of the signal during signal processing and then take the ratio of sample and reference energies electronically.

Example 12.20. From Fig. 12.1 it appears that selecting a carrier of 10–35 Hz would meet both criteria if modular response times are of the order of milliseconds and frequencies in the signal are less than 2.5 Hz. When response times are 10^{-6} s or smaller, a good choice for f_0 would be 250–10^5 Hz.

Example 12.21. Consider the consequences of two different choices about where to locate the modulator in a spectrophotometer.

If a mechanical chopper is placed between the source and sample, changes in the intensity of the source are part of the "signal." That being true, it is essential that a stable source be used. (Actually, in double-beam instruments, since changes in source intensity affect both sample and reference beams, they can be subsequently eliminated by ratioing the beams.) Noise originating in the sample module and detector, however, can be reduced greatly by this modulation.

On the other hand, if the modulator (perhaps an FET) is placed in the sample channel between the detector and the processing modules, the noise from the characteristic modules will be part of the signal and be difficult to remove by modulation.

Lock-in Amplifier Enhancement of S/N. To apply Eq. (12.8) to a lock-in amplifier, let $f_B(t)$ be a function such as the square wave that is modulating the energy flow from an instrument source. The other function $f_A(t - \tau)$ will then be the detector output signal, which will be coherent, though perhaps out of phase by τ, with the output of the modulator. The device performing the cross correlation will be called either a *phase-sensitive detector*, or more commonly, a lock-in amplifier. Fortunately, even when $S/N \ll 1$, information can still be extracted by this cross-correlation process.

A lock-in amplifier, in general, consists of a low-noise selective amplifier, a multiplier, and a low-pass filter. In Fig. 12.4a a modular diagram indicates the general arrangement. The output of the lock-in amplifier (Fig. 12.4b) is ideally only the narrow band of frequencies Δf that contains the chemical information provided by the measurement.

The role of the three parts of the lock-in amplifier can be simply described.* The selective amplifier passes a band of frequencies somewhat broader than the carrier plus the band of information frequencies Δf. As shown in Fig. 12.5 the

*Many lock-in amplifiers also have an internal oscillator (sine-wave generator) that can provide a modulation frequency, if needed.

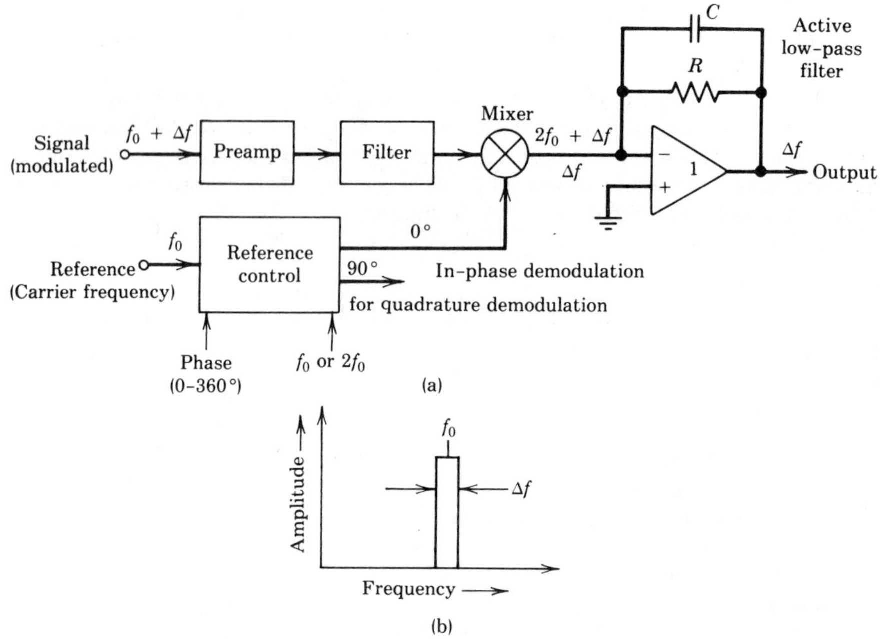

Fig. 12.4 A block diagram of a generalized lock-in amplifier arrangement for an instrument and an idealized version of its output. (a) A simplified block diagram. A stable carrier frequency f_0 modulates the input signal of frequencies Δf, giving frequencies that include $f_0 + \Delta f$. A high-quality, low-noise preamplifier boosts the amplitude of the modulated signal sufficiently that noise will not further degrade it. Some noise can be removed after the preamp by switching in a filter appropriate to the type of noise in the system being used. In the mixer the amplified signal is multiplied by a reference signal also of frequency f_0 that is coherent with the modulating frequency. The mixing and subsequent low-pass filtering causes synchronous or phase-sensitive demodulation of the input signal. The active low-pass filter passes only the narrow bandwidth Δf where $\Delta f = (4\pi RC)^{-1}$ and removes the rest of the carrier frequency. [A reference signal 90° out of phase is also available as shown. It may be introduced into a separate channel (not shown) to give a quadrature output.] (b) The ideal output from the lock-in amplifier circuit is the narrow band Δf that contains the desired chemical information. The lock-in amplifier has permitted Δf to be centered at the modulating frequency f_0 rather than at 0 Hz (dc) where $1/f$ noise was strongest.

multiplier or mixer module receives two inputs, the modulated signal and the carrier frequency f_0. The latter, as a reference, can be adjusted in phase and also doubled. Phase-sensitive demodulation occurs in the multiplier plus low-pass filter. The mixer output includes two terms, one represents the sum of frequencies in the channels $f_0 + f_0 + \Delta f$, which will need to be rejected. The second, Δf, which represents the difference of frequencies, will need to be saved and presented as the output. For synchronous demodulation it is not essential that the reference be in phase with the carriers but it must be coherent with it. If the phase difference is zero, the output signal is a maximum. As shown in Fig. 12.5, the phase difference at the mixer can ordinarily be adjusted to an optimum value either manually or automatically.

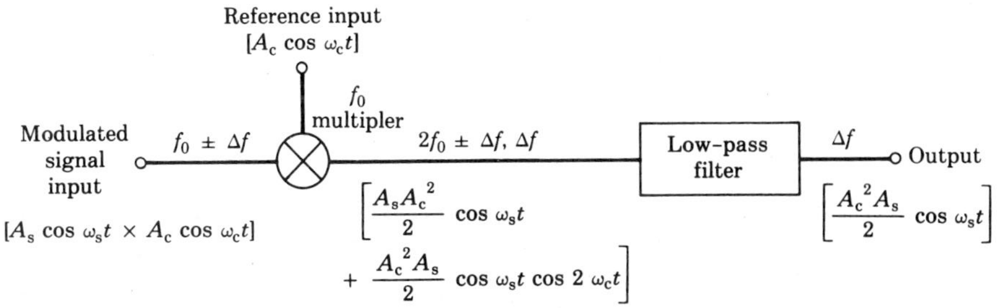

Fig. 12.5 A synchronous demodulation circuit annotated to show frequency components and their time dependence. The output of the multiplier consists of two terms, a component of information frequencies Δf (angular frequency ω_s) which represents the signal, and another dominated by the modulating frequency f_0 (angular frequency ω_c). Only the former component is passed by the low-pass filter.

In addition, it is important that the phase-sensitive detector be able to carry large amounts of noise, that is, asynchronous signals, without distortion. Its *dynamic reserve* is defined as the maximum tolerable ratio of nonsynchronous rms voltage to synchronous voltage at the input when the lock-in amplifier is set at maximum sensitivity or gain. Representative ranges for reserves are 40–80 dB. Output stability is usually proportional to dynamic reserve.

In the low-pass filter the remnants of the carrier are easily separated from the signal information. Its output is Δf, the dc plus low-frequency information, at amplitudes that are representative of their amplitudes in the carrier wave. In most cases the time constant of the filter is adjustable to allow a trade-off between the speed of the system and the amount of averaging over cycles of the carrier. In other words, the bandpass can be narrowed very substantially below 1 Hz as appropriate.

Lock-in amplifiers allow signal information to be recovered when it is buried in noise several orders of magnitude greater. A wide variety of versatile lock-ins is available; however, there are some situations in which the amplifier cannot be successfully used. The approach is unsuccessful with transient signals that cannot be effectively triggered for repetition, very fast signals, signals with extremely high reptition rates, and discrete signals consisting of a few counts per minute. It all these cases modulating a carrier wave with the signals is a difficult process.

Boxcar Integrator. A device known generally as a boxcar integrator or averager is attractive as a way of measuring repetitive signals such as those generated in systems using pulsed laser sources. After a precise delay the signal can be said to be cross-correlated with a rectangular pulse of unit amplitude. A modular diagram of the integrator is shown in Fig. 12.6. In this gated device the channel width or aperture of its single channel can be varied from about 100 ps to 5 ms. As repeated signals arrive, during this interval, their information components add, while noise averages toward a uniform background. Further, the observation interval or aperture can be begun after a precisely controlled delay.

The boxcar integrator also has an advantage for cases in which almost any portion of a

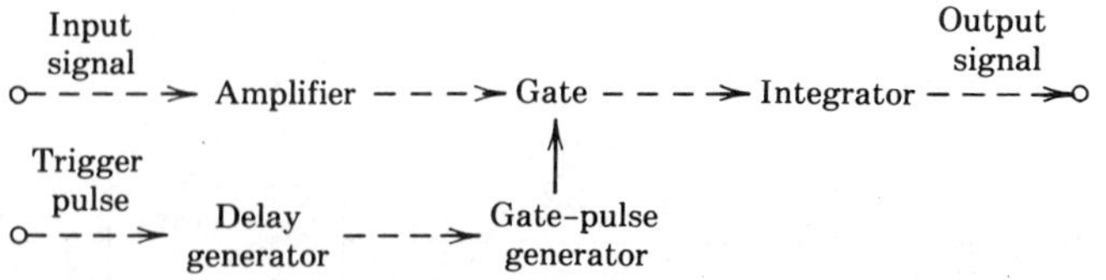

Fig. 12.6 A modular diagram of a boxcar integrator. Any portion of a repetitive signal can be observed by setting (a) the delay generator to operate at the desired interval after the signal begins and (b) the gate generator to produce a pulse of the desired width when the delay ends. A trigger pulse inputs the signal and starts the delay generator simultaneously. The signal pulse is integrated over the period when the gate is open. To improve S/N, the same portion of the input waveform can be observed repeatedly. To observe more of the waveform the delay generator can be scanned across the input signal.

pulse may be of interest. Flexibility is provided by having both a variable delay and a precisely variable aperture. A scan mode is often provided in which the delay is systematically increased with each repetition while keeping a constant aperture to permit a survey of an entire pulse before one decides which portion to record precisely.

Some examples of valuable applications of boxcar averagers are in coherent Raman spectrometry and in time-resolved spectroscopic studies using pulsed lasers. For example, consider how a boxcar integrator might be used with the fluorescence decay curve of Fig. 15.5.

With pulsed signals, the boxcar averager or integrator offers a great deal of flexibility relative to the multichannel analyzer. The user can operate the gate (or aperture) of a boxcar integrator in quite sophisticated fashion. The flexibility and speed of the boxcar are also strong advantages in characterizing signals with high repetition rates, as from a pulsed laser, and very fast signals. By contrast, the multichannel analyzer is much more effective for the routine observation of entire pulses.

12.8 MULTICHANNEL AVERAGING; MULTICHANNEL ANALYZERS

Perhaps the most widely used strategy to improve S/N in spectral measurements is the collection of techniques known as *signal multichannel averaging*. Recall that these measurements typically involve intensity versus wavelength plots. Both analog and digital types of averaging are used, though the latter dominate because of the relative inexpensiveness of computer memory chips. Averaging involves algebraic summing of repetitive signals. Where an instrument such as a Fourier transform NMR spectrometer behaves reproducibly, it is straightforward to call for precise repetition of an NMR spectrum as many times as desired. The multichannel averager samples the basic signal at equal intervals. The sampling in repeated measurements must be at corresponding times to permit coherent addition of information, usually by having a trigger activate the rf pulse and sampling procedure.*

*Usually the free-induction decay curve is averaged and a Fourier transform of the final curve taken to obtain the conventional frequency-domain spectrum.

Table 12.1 Beneficial Effect of Signal Averaging Using a Multichannel Averager. Counts in One Channel of a Multichannel Averager[a]

Number of Repetitions	*Background*[b] B	*Signal*[c] S	*Analytical Signal*[d] $M = S + B$	*Fluctuations in Background* $B^{1/2}$	S/N[e] $(M - B)/B^{1/2}$
1	10^3	10	1,010	31	0.32
100	10^5	1000	101,000	316	3.2
1,000	10^6	10,000	1,010,000	1,000	10

[a] Any system employing pulse-counting electronics will produce in a given wavelength or scanning time interval a number of counts proportional to the intensity. Examples are an X-ray fluorescence spectrometer or optical spectrometer operating at low light levels.

[b] *Average* number of counts received in absence of analyte.

[c] Counts attributable to analyte, that is, the true signal.

[d] Total counts received in the channel.

[e] Since the signal $(M - B)$ is a small number obtained by subtracting two large numbers, an AD converter of high resolution must be used or the quantization error will be significant.

In Table 12.1 the gain in S/N that results from averaging is shown for one channel of a multichannel averager. Noise and background B are large in this example, Since Poisson statistics apply to counting, the fluctuation in B is proportional to $B^{1/2}$. In each channel the true signal adds coherently with each sweep and is proportional to N, the number of repetitions. Background also accumulates as N, as the table indicates. Noise accumulates as $N^{1/2}$. Thus, the S/N ratio increases as $N^{1/2}$. In Table 12.1 noise is not shown explicitly, but background is. As is shown, the true signal is $M - B$, and the steadiness in B and in M increases with the number of repetitions. The greater the number of repetitions accumulated, the more clearly the signal will rise above the noise.

It is helpful to recall that the overall time constant of the system also keeps rising with each addition to the multichannel averager. Observation of a spectrum for $T \times 10^3$ s provides 1000× more time for noise to average to uniformity. By analogy to Eq. (12.6), the system bandwidth Δf has diminished in proportion since $\Delta f_{sys} \simeq 1/t$, where t now identifies the total time of observation.

Both hard-wired multichannel scalers and microcomputers functioning as *multichannel averagers or analyzers* (see below) are available. The scaler version will count pulses and requires for inputs both the particular pulses and (on a separate line) a channel-advance signal (a pulse) or a channel address. Each signal input pulse causes the scaler to increment the count and the channel signal to advance the memory channel currently being addressed.

Example 12.22. Normalized averaging is also available in some types of *signal* analyzers. Each datum I_i for a channel produced on the ith sweep is not entered in the channel but combined with the average value from the previous sweep according to the algorithm $A_i = A_{i-1} + (I_i - A_{i-1})/2^J$, where A_i is the average channel value after i sweeps, A_{i-1} the average after $i - 1$ sweeps, and J a positive integer keyed to the number of elapsed sweeps. The normalized average is saved at the address of the channel and may be displayed to give the shape of a curve nearly from the beginning of scanning, however noisy it may be. The advantage is that averaging may be discontinued whenever the S/N appears satisfactorily large.

Multichannel Analyzer. A modular diagram of a multichannel analyzer is given in Fig. 12.7. All modules may be in hardware or most may be in software. If a microcomputer is functioning as an MCA, memory locations comprise the required set of channels and are sequentially addressed during each scan. Channel capacities are generally 10^6 (20 bits binary) or more. Multichannel analyzers usually have sampling rates up to 200 kHz with fast sample-and-hold devices and AD converters. Accurate operation of the sampling device (see Section 5.2) is critical to performance. It is also important that the AD converter have sufficient resolution, especially to handle demanding signal processing such as subtraction of one large signal from another. Averaging over many scans will improve S/N as just discussed.* Program control of sampling will ensure that repeated signal traces are coherent. That will not be true if the signals of interest must be triggered, as with animal nerve impulses. In these cases an external trigger will be required to start both the nerve discharge and MCA cycle.

If analog signals are received from the outside and are simply directed to a multichannel averager, it is possible to make the necessary conversion to digital form by using a voltage-to-frequency AD converter. Its output is a pulse rate or number of counts that is proportional to signal amplitude. Quite similar is the direct reception of a pulsed signal. This case was illustrated in Table 12.1. Alternatively, it is possible to use a multidetector array, for example, a multichannel diode array or vidicon detector, and arrange to have the output of each unit directly assigned

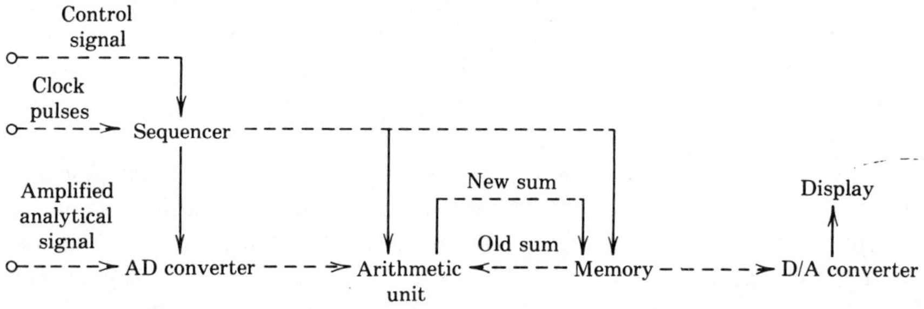

Fig. 12.7 A modular diagram of a multichannel averager. The sequencer is in hardware or software (computer) and controls sampling (not shown) and digitization as well as the addition of new values and their storage in memory. Note that the sequencer arranges on a continuing basis that memory furnishes an old sum to the arithmetic unit to be added to a new value received from the AD converter. The resulting new sum is then placed in memory. The possibility of display of analog values is provided for by a DA converter.

*Noise resulting from quantization steps can be reduced by signal averaging, however, only if the noise in the signal is larger than the quantizing interval.

to a particular memory channel. Only the rate of reading then needs to be under control to synchronize it with the period of interest.

Example 12.23. A portion of an optical spectrum is to be recorded by a multichannel analyzer. A rate of electronic scanning is set that calls for completing a scan every T seconds to allow a rate of digitizing without loss of precision. If the spectrometer detector is a linear diode array of 1024 channels, the analyzer should have the same number of channels. Every T seconds the signal will swept across the 1024 channels, transferring each digitized output to its memory channel. In each channel the signal will add its numbers to the number already in memory.

REFERENCES

1. W. R. Bennett, *Electrical Noise.* New York: McGraw-Hill, 1960.
2. A. Bezegh and J. Janata, Information from noise, *Anal. Chem.*, **59,** 494A (1987).
3. R. J. Bibbero, *Microprocessors in Instruments and Control.* New York: Wiley-Interscience, 1977.
4. R. N. Bracewell, *The Fourier Transform and Its Application*, 2nd ed. New York: McGraw-Hill, 1978.
5. M. U. A. Bromba and H. Ziegler, Digital filter for computationally efficient smoothing of noisy spectra, *Anal. Chem.*, **55,** 1299 (1983); *ibid.* **53,** 1583 (1981).
6. G. M. Hieftje, Signal-to-noise enhancement through instrumental techniques. Part I. *Anal. Chem.*, **44**(6), 81A (1972).
7. G. Horlick and G. M. Hieftje, Correlation methods in chemical data measurement, in *Contemporary Topics in Analytical and Clinical Chemistry*, Vol. 3, D. M. Hercules, G. M. Hieftje, L. R. Snyder, and M. A. Evanson (Eds.), New York: Plenum, 1978.
8. A. Savitzky, and M. J. E. Golay, Smoothing and differentiation of data by least-squares procedures, *Anal. Chem.*, **36,** 1627 (1964).
9. R. Swanson, D. J. Thoennes, R. C. Williams, and C. L. Wilkins, Determination of the Nyquist frequency, *J. Chem. Educ.*, **52,** 530 (1975).
10. M. R. Thompson and R. E. Dessy, Use and abuse of digital signal processors, *Anal. Chem.*, **56,** 583 (1984).
11. F. E. Woodard, W. S. Woodward, and C. N. Reilley, Microprocessor-based laboratory data aquisition systems, *Anal. Chem.*, **53,** 1251A (1981).

EXERCISES

12.1 Develop an equation for the rms *current* contributed by Johnson noise. (Hint: Use Ohm's law).

12.2 What connection is there between the number of stages, modules, and components in an instrument and the S/N of an instrument? In an instrument is it better to use a one-stage amplifier with a microammeter readout or a two-stage amplifier with a milliammeter readout? On what basis would you choose one or the other?

12.3 Discuss the connection between the S/N of an instrument and the precision and frequency of its calibration. What is the connection between the standard deviation of measurements on a species to which an instrument responds sensitively and its S/N? Consider measurements both at high levels of concentration and at trace levels.

12.4 A sample-and-hold and low-pass filter to minimize aliasing are to be set up to sample the spectral range 1–100 cm^{-1}. (a) What should the sampling frequency be? (b) What gains and costs would result from selecting a higher frequency? (c) If a low-pass active filter with an $f_0 = 1500\ cm^{-1}$ and a fall-off of 60 dB per octave is used, what attenuation will occur at 1000 cm^{-1}?

12.5 Should one attempt to use a simple low-pass filter to lower the noise level in a steady signal such as the output of an ion-selective electrode or of a photomultiplier tube receiving radiation transmitted by a highly absorbing solution at a particular wavelength.

12.6 Give examples of factors that contribute to noise in an optical system.

12.7 In Section 12.4 it was stated that the intensity of light and not its frequency is measured in a dispersive spectrophotometer. Does this accord with the response curves of common optical detectors such as photomultiplier tubes?

12.8 Discuss the best way to couple a thermocouple detector and preamplifier in an IR spectrophotometer. The thermocouple has a resistance of 30 Ω. The noise voltage and current of the preamplifier (taken at its input) is 1×10^{-7} V and 1×10^{-11} A.

12.9 (a) If an available multichannel averager offers either a 14-bit or a 10-bit converter, why would one choose one converter over another? (b) For the measurements reported in Table 12.1, what precision of AD converter will be advisable, 10-bit, 12-bit, or 14-bit?

12.10 Why can quantization error not be reduced by multichannel averaging?

12.11 Draw a block diagram of a lock-in amplifier arrangement for a Wheatstone bridge. Describe it briefly also. What advantage does it offer over the usual energizing and detection arrangements?

12.12 Draw a time diagram (see Fig. 15.5) for a boxcar integrator that is to record a 1-ns interval of a fluorescence decay curve. Show on separate lines (a) the decay curve, (b) the trigger pulse, (c) the delay pulse, (d) the gate interval, and (e) the output pulse.

12.13 Assume the gated portion of a signal yields 100 counts s^{-1} and the background 10^3 counts s^{-1}. (a) How many repetitions of a portion of a pulse waveform whose signal-to-background ratio is 0.01 will be needed to achieve a S/N of 10? (b) Describe the role an oscilloscope might play in choosing the length of the delay interval and the gate interval.

OPTICAL SPECTROSCOPIC METHODS

Chapter 13

ATOMIC EMISSION SPECTROMETRY

13.1 INTRODUCTION

The first type of optical spectroscopy to be treated is atomic emission, perhaps the oldest chemical technique now widely used. The main steps in an atomic emission measurement are as follows:

1. Conversion of a sample to a free-atom gas.
2. Excitation of the atoms and their ions to higher electronic states.
3. Measurement of emission wavelengths and intensities.
4. Extraction of qualitative and quantitative information from these signals.

Emission spectrometry is widely used because wavelengths and intensities characteristic of electronic transitions in atoms of all elements can be made with high precision under appropriate conditions.

Atomic emission spectroscopy offers extreme specificity and good sensitivity as an elemental technique. A nearly complete elemental analysis can usually be made on a single sample of a complex material in the UV–VIS. All ranges of concentrations can be handled and absolute limits of detection are of the order of nanograms generally and are substantially smaller for certain elements.

Example 13.1. What elements can be determined by emission spectrometry in the UV–VIS? How complicated is sample preparation?

All metals and some nonmetals such as boron and phosphorus can be determined in any kind of sample. For many other nonmetals, including the halogens, even the least energetic electronic transition falls in the vacuum UV region ($\lambda < 190$ nm), which requires the use of somewhat more complex instrumentation. The emission spectrometry of lanthanides and a few heavy elements is complex because of the sheer number of their electronic transitions. For the same reason, it is hard to determine other elements in their presence.

Samples must be volatilized and atomized to a free-atom gas; those atoms must in turn be excited electronically. It is possible to aid volatilization and excitation by first grinding solid samples, and especially by dissolving them whenever feasible. Solution samples not only have fewer interelement or interference problems, but can be atomized and excited more easily. Further, quantification of spectra by use of solution standards can be carried out more readily.

Probably most emission spectrometric measurements are quantitative determinations. By appropriate techniques, relative standard deviations of $\pm 1\%$ are obtainable for many elements at both major ($> 1\%$) and minor concentrations (about 1%). As sample size decreases or as concentrations fall to the trace level, precision is at best $\pm 5\%$.

Emission spectroscopy is a valuable technique for the study of excited states of all atomic species, many ions, and a number of small molecules and radicals like C_2N_2, C_2, NH, or OH. In *atomic* emission spectrometry, procedures are of course designed to minimize the diffuse spectra of such ''molecular'' species.

In addition, emission spectroscopy offers a means of determining temperature (and electric and magnetic field strengths) in ''hot'' systems. Line intensities are exponentially dependent on temperature and intensities of particular lines can be measured relative to those of other lines to determine the temperature of emitting species up to about 10^9 K. Further, by analyzing the radiation from a system, that is, measuring wavelengths, line widths, and intensities for many species, a spectroscopist can ascertain the electric and magnetic field strengths (from splitting of lines), pressure, and degree of ionization.

13.2 ATOMIC SPECTRA

The next two sections will develop background important to atomic techniques. Recall that free atoms absorb or emit radiation of characteristic frequency ν or wavelength λ as they undergo electronic transitions. Indeed, the close relationship between ΔE, the energy change in transitions, and electromagnetic waves is given by the expression $\Delta E = h\nu$, where the product $h\nu$ also defines the energy of a photon or quantum of radiation absorbed or emitted.

Many nonradiative mechanisms also promote electronic transitions. For example, atoms may be excited by collision when a gas is heated. Collisions also result in formation of ions and excited ions. Similarly, excited species may transfer energy by collision and return to the ground state.

When a large number of atoms of an element is excited, many different states will be reached. As each atom returns to the ground or to a lower state, its emission frequency is observed; the set of all such frequencies is the *emission spectrum* of the species. Wavelengths of excited *atoms* are often designated by a Roman numeral I. Excited ions are said to give second or higher-order spectra and are labeled II, III, ... For example, Fe III will indicate a third-order spectral line arising from states of Fe^{2+}. As might be expected from the close relationship between electronic structure and the periodic table, the spectra of an isoelectronic sequence of atomic species resemble one another. Thus, similar spectra are shown by Sc III, Ca II, and K I. The sharply defined wavelengths emitted by free atoms or monatomic ions are often designated *line spectra*.*

*The term originated because in a dispersive spectrometer discrete wavelengths are images of the spectrometer entrance slit and thus appear as bright rectangular lines.

In contrast, excited molecules give rise to *band spectra*. The term molecule includes transitory species such as OH and C_2 that exist at high temperatures or in very dilute gases. Their emission spectra are complex since their electronic changes generally occur simultaneously with probable (allowed) rotational and vibrational transitions. As a result, each electronic change is a collection of closely spaced lines which appears as a band unless dispersed by a spectrometer of high resolution. Nearly all atomic spectra excited in air or other gases contain bands as well as lines. Regular line spectra broadened by pressure in a concentrated gas (see Section 13.3) should not be confused with band spectra.

Finally, condensed systems like solids emit continuous spectra on excitation. Such blackbody radiation originates from thermal vibrations in the solid and thus depends on temperature rather than characteristic energy levels of a particular species (see Section 8.2).

Units. While frequency is the variable related to quantum energy changes and is invariant even through changes of medium, it is not always the quantity most precisely determinable. In the electrical part of the electromagnetic spectrum, it is very precisely measurable by comparison with the frequency of a standard oscillator. In the optical part, however, wavelengths can in general be more precisely observed.* Partly for this reason, in the optical range the radiation accompanying a transition is often given in terms of $1/\lambda$, a quantity called *wavenumber* (units, cm^{-1} and symbol, σ). The usefulness of the wavenumber is that it is proportional to the energy of transition. Unless otherwise specified, wavelength and wavenumber refer to the vacuum wavelength.†

Example 13.2. Calculate the factor for conversion of wavenumber to joules per mole.

Given that $E = h\nu$ and $c = \nu\lambda$, from the definition $\sigma = \lambda^{-1}$, we have $E = hc\sigma$ joules per quantum. The energy of 1 mole of such quanta will be $hc\sigma N_0$, where N_0 is Avogadro's number. Thus,

$$\begin{aligned} E &= 6.63 \times 10^{-34}\ \text{J s} \cdot 3.00 \times 10^{10}\ \text{cm s}^{-1} \cdot 1\ \text{cm}^{-1} \cdot 6.02 \times 10^{23}\ \text{mol}^{-1} \\ &= 11.96\ \text{J mol}^{-1}. \end{aligned}$$

Atomic Spectra. Emission and absorption spectra of hydrogen-like atoms can be successfully accounted for by the use of the Bohr model. Summing the potential and kinetic energy of the single electron possessed by such atoms gives the following expression for the nth electronic energy level:

$$E_n = -\frac{2\pi^2 m_e e^4 Z^2}{h^2} \cdot \frac{1}{n^2}, \qquad (13.1)$$

*Frequencies might then be precisely calculated by means of the expression $\nu = c/\lambda$, but the velocity of light c is not known with comparable accuracy.

†Recall that λ, the vacuum wavelength, is defined by $\lambda = c/\nu$ and that the wavelength in air, λ', is shorter. It can be shown that $\lambda' = \lambda/n$, where n is the refractive index of air. At room temperature the difference is of the order of three parts in 10^4.

where m_e is the mass of the electron, e is the electronic charge, Z is the atomic number, h is Planck's constant, and n is the principal quantum number for the level.* For spectroscopic purposes (and historically) Eq. (13.1) is often rewritten in terms of the wavenumber σ_n of the nth level. When this is done, the constant quantities are grouped together in the *Rydberg* constant for hydrogen, R_H. One has

$$\sigma_n = -\frac{E_n}{hc} = -\frac{2\pi^2 m_e e^4}{h^3 c} \cdot \frac{1}{n^2} = \frac{R_H}{n^2}, \tag{13.2}$$

where $R_H = 1.0967759 \times 10^5\ \text{cm}^{-1}$ and $R_\infty = 1.09737318 \times 10^5\ \text{cm}^{-1}$.* It will be seen below that R_H/n^2 is called the term value for the nth state.

According to the lower state to which transitions are made, a line spectrum for a given atom can be analyzed into different series. This procedure is clearly illustrated for hydrogen in Fig. 13.1. Note, for example, that the Balmer series for hydrogen represents transitions to the second electronic level ($n = 2$).

For any set, the *series limit* represents the greatest energy (shortest-wavelength line) possible for the series. Such a line would be associated with a downward

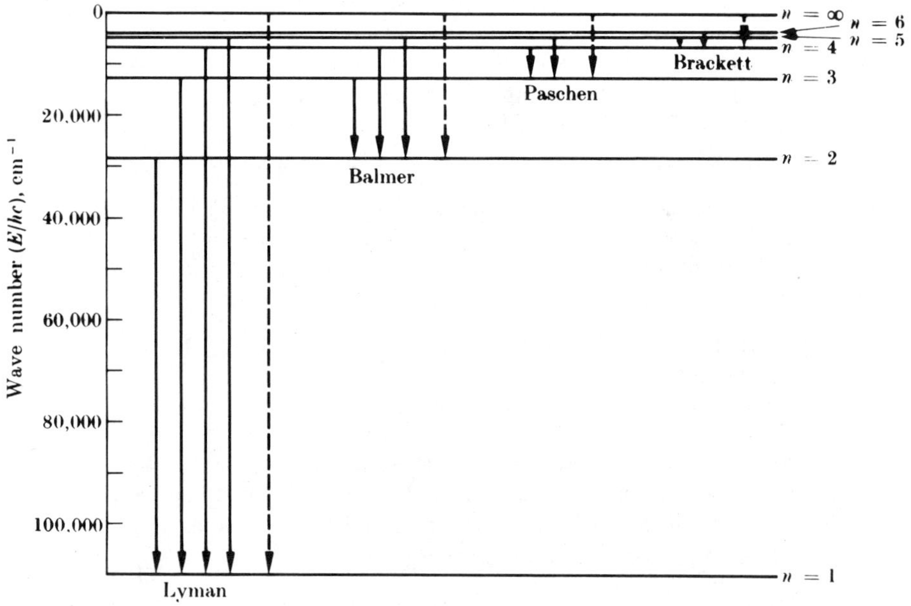

Fig. 13.1 Spectroscopic emission series for the hydrogen atom. The broken-line electronic transition in each case indicates the series limit. The wavenumber of each emission line is the difference between the values at each end of its arrow. Principal quantum numbers n are given at the right.

*Actually, the reduced mass μ of the system should be used in lieu of the electronic mass. This quantity is defined as $\mu = m_e M/(m_e + M)$ where M is the mass of the nucleus. For hydrogen μ is 0.07% less than m_e; for heavier atoms the difference is still smaller and becomes insignificant as M increases.

transition from the ionized state. For example, the limit fot the Balmer series of hydrogen would be observed when a separated electron ($n = \infty$) of zero kinetic energy was captured in an $n = 2$ hydrogen level. These limits are indicated in Fig. 13.1.

Three other properties of series may be noted. First, there is an increasingly close spacing of levels as the principal quantum number increases, thus a convergence of emission lines toward a *series limit*. Second, different series of lines overlap in an actual spectrum, which causes most emission spectra to be complex. Third, since most separated electrons have finite translational kinetic energies, their capture will give rise to a spectral continuum that extends from the series limit toward shorter wavelengths.

For multielectron atoms both electrostatic repulsion between electrons and the effect of electron spin enter into the determination of energy levels. As a result energies differ for different patterns of electron probability density. In describing the emission spectra of these elements it is conventional to assign a term or *term symbol T* to each different atomic state. Electronic transactions are then described by a difference such as $T_q - T_p$, where T_q is the term symbol of a particular upper state and T_p that of a particular lower state. Each term is also associated with a wavenumber value marking its position relative to a common reference energy. Thus, a downward transition gives rise to an emission line (in cm^{-1}) given by $\sigma = T_q - T_p$. The formal statement of this relationship between levels is called the *Ritz combination principle*.

A term diagram, often called a Grotrian diagram, after its initiator, represents graphically the simpler electronic transitions for a given atomic species. It should be understood that not all differences between terms correspond to actual spectral lines and that whether transitions are "allowed" must be deduced from quantum–mechanical principles.

Figure 13.2 shows the term diagram for sodium. Note that values of the (orbital) angular-momentum quantum number l are given across the top and values of n downward. Allowed transitions are shown as diagonals that include the wavelength of the emission line.

To unravel the complexity of this diagram, consider the formal scheme used to construct it. Both angular momentum and electron spin can be represented as vector quantities. The summing of such motions becomes vector addition. When a species has a single valence electron, as sodium does, coupling can be represented by $l + s$ when the vectors are parallel and by $l - s$ when they are oppositely directed. The coupling is described by the so-called inner quantum number j. When the valence electron of sodium is in an s-level ($l = 0$), $l + s$ and $l - s$ give a single value, 1/2; when it is in a p-level ($l = 1$) the coupling produces j values of 1/2 and 3/2. Now the term for an atom is symbolized by the capital letter describing the angular momentum of its highest-energy electron. Thus S is the symbol for the ground state of sodium in an s-state. the value of j is given as a right subscript of the term, e.g., $S_{1/2}$ or $P_{1/2}$ or $P_{3/2}$. Finally, a left superscript is used to indicate multiplicity, which is just $2s + 1$, where s is the resultant electron spin quantum number. Since sodium has a single valence electron, its multiplicity

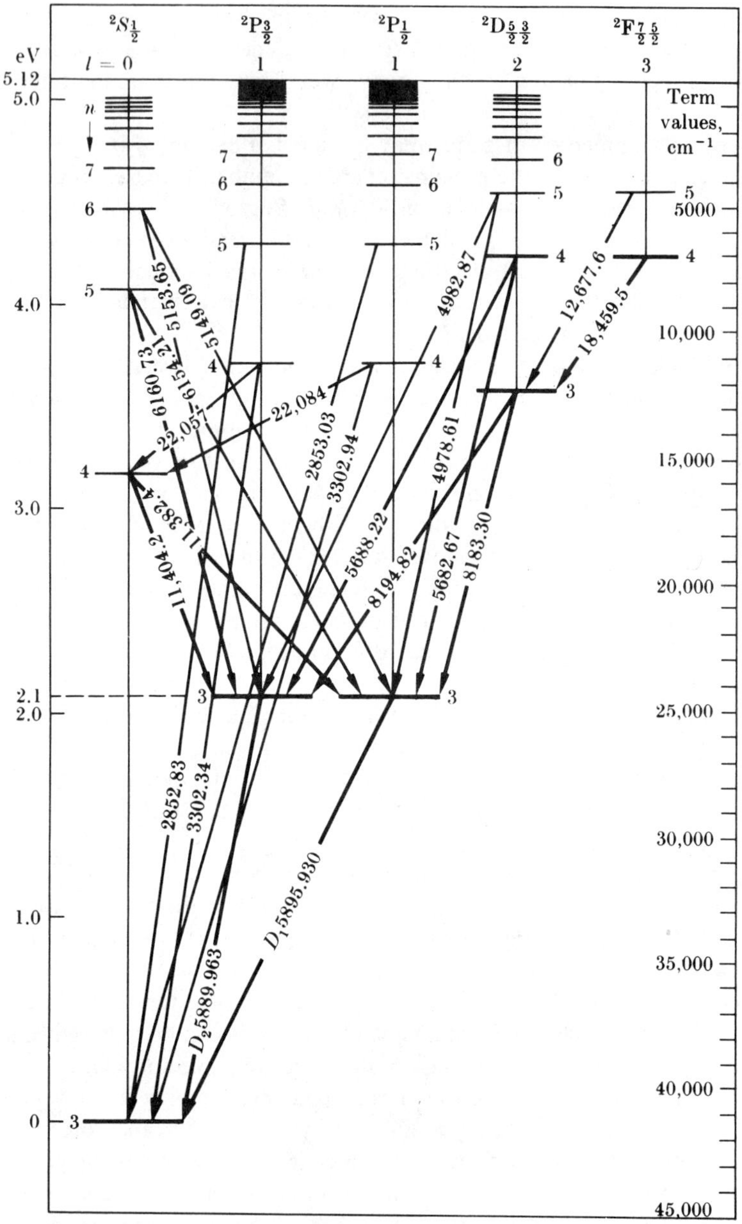

Fig. 13.2 Term diagram for the sodium atom. Emission transitions are shown by arrows with wavelengths in angstroms. These values may be converted to nanometers by dividing by 10. (After Grotrian. Courtesy of Springer-Verlag, Heidelberg.)

Table 13.1 Values of *j* Possible for Sodium

Term	*l*	*Value of j*
S	0	1/2
P	1	1/2, 3/2
D	2	3/2, 5/2
F	3	5/2, 7/2

is $2(\frac{1}{2}) + 1 = 2$ for all states. In Table 13.1, the term and the values of the inner quantum number are given for sodium for each value of l. The complete term symbols in Fig. 13.2 have now been identified. A more detailed treatment is beyond the scope of this text.

In Fig. 13.3 the so-called principal series spectrum for each of the alkali metals is shown. Each series marks transitions from a higher electronic level to the ground state for that atom. In all cases the ground state is the same, $^2S_{1/2}$. Thus, the lines are described by a term combination $T_q - {}^2S_{1/2}$. For example, the doublet comprising the sodium *D*-lines is the pair $^2P_{3/2} - {}^2S_{1/2}$ and $^2P_{1/2} - {}^2S_{1/2}$ at about 17,000 cm^{-1} (589 nm). It is instructive to identify the origin of many of the lines.

In the emission spectrum of an element it is found that its strongest lines are those for transitions terminating in the ground state. Such lines are called *resonance lines*. The most intense of these originates from the first excited state that can "combine" with the ground state to given an allowed transition. For example, in the case of sodium the brightest "line" is actually the *D*-doublet, and the 589.0 and 589.6 nm pair are resonance lines.

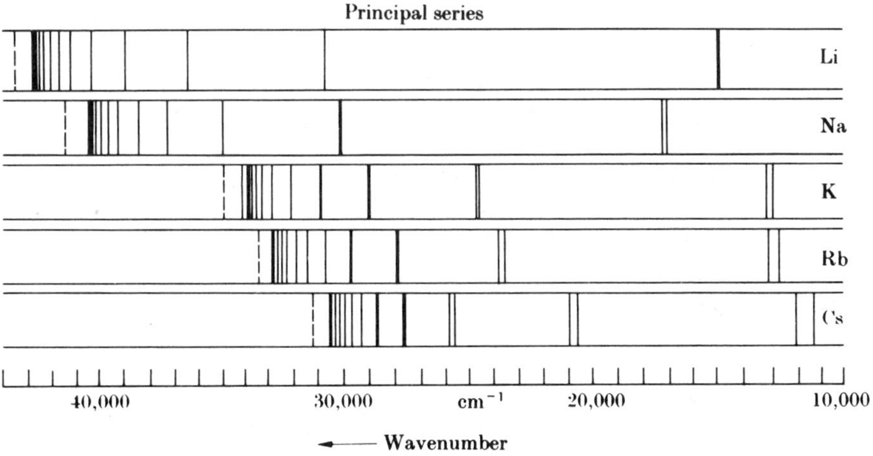

Fig. 13.3 The so-called principal emission series for the common lkali metals, that is, the set of transitions to the ground electronic state (2s for Li, 3s for Na, etc.). Since only transitions for which $\Delta l = \pm 1$ are allowed by selection rules, all these transitions are from upper p-states ($l = 1$) to the ground s-state ($l = 0$). Each transition gives rise to a doublet since the spin ($s = \frac{1}{2}$) can couple either $l + s = \frac{3}{2}$ or $l - s = \frac{1}{2}$ in the p-state. The separation of lines in the doublets of sodium and potassium has been exaggerated.

While the fine structure of atomic spectra can be accounted for quantitatively by term diagrams, so-called *hyperfine structure* cannot. These very slight (hundreds of wavenumbers) splittings of many lines have two nonexclusive origins. There is a definite isotope effect for any element that is traceable to isotopic variations in nuclear volumes. The energy of very penetrating electrons such as those in s-orbitals is affected. The other source of hyperfine structure is nuclear spin. It couples with orbital motion in vector fashion to give sums and differences for all isotopes except those of zero spin.

13.3 INTENSITIES AND SHAPES OF SPECTRAL LINES

Whether for atomic or molecular absorption or emission, transition line intensities are proportional to the population of the initial level and to the *Einstein probability coefficient* for the transition. Approximate values can in many instances be calculated. The possibility of absolute spectroscopic measurements clearly depends on knowledge of such values.

To describe transitions between two nondegenerate states 1 and 2 of a system that have energy E_1 and E_2, three probability coefficients were introduced in Section 8.3:

1. A_{21}, the probability coefficient for *spontaneous emission*.
2. B_{21}, the probability coefficient for *stimulated emission*.
3. B_{12}, the probability coefficient for (stimulated) *absorption*.

In defining these coefficients, Einstein showed that $B_{12} = B_{21} = 8\pi^2|R|^2/3h^2$ and $A_{21} = 8\pi h\sigma^3 B_{21}$, where R is a *transition dipole moment*.* Unfortunately, since it is difficult to formulate exact wave equations for the states involved, few transition probabilities have been calculated reliably. Ordinarily probabilities are measured.

A distinctive difference between spontaneous and stimulated transitions must be noted. The probability of an induced transition, B_{12} or B_{21}, is independent of frequency (or wavenumber), but the probability of a *spontaneous* transition varies as the third power of the frequency or wavenumber. For this reason, emission lines from electronic transitions, which are of high frequency, are very much more intense than absorption peaks in the IR that correspond to vibrational changes.

Most transitions that occur in the optical range depend upon an electric dipole moment. Such a moment exists if there is a nonsymmetrical charge distribution about the center of mass. Since this condition is met only for transitions between

*Thus, $|R|^2$ is the square of the absolute value of the moment between the states 1 and 2. In general R is resolved into components along a set of axes so that

$$|R|^2 = |R_x|^2 + |R_y|^2 + |R_z|^2.$$

certain states of an atomic system, probable or *allowed* transitions can easily be specified. The specifications are called *selection rules*.* Further, high-order effects may also give rise to such transitions. One is a magnetic dipole transition, which is roughly only 10^{-5} times as likely as an electric dipole transition. Another is an electric quadrupole transition, which is only 10^{-8} times as probable as an electric dipole transition.

When there is an unsymmetrical charge distribution in a single direction, as along a bond axis, the transition is well adapted to investigation by use of linearly polarized radiation. For example, a stretching vibration of a heteronuclear bond might be so studied in a crystal.

In still other instances charge distributions that are symmetrical around a bond axis may not be symmetrical in space. These cases give rise to an electric *quadrupole moment*, defined by the expression

$$Q = \Sigma_i q_i x_i^2.$$

Here q_i is a particular charge and x_i is its distance from the center of mass. There is a distinct difference between dipole and quadrupole transitions. As noted above, the latter are of very low intensity. Such transitions are among those observed in Raman spectroscopy.

Example 13.3. The basis for selection rules in atomic transitions may be appreciated qualitatively by considering the case of the hydrogen atom in its ground state. On the average its electronic charge is distributed symmetrically around the proton, that is, the electron is in a 1s state. A transition to a 2s configuration, which is also spherically symmetrical, would fail to yield a transition dipole and would be "forbidden." However, if the incident radiation were of proper energy, the atom could undergo a transition to a $2p_x$, $2p_y$, or $2p_z$ configuration. If linearly polarized light of this energy with its electric vector in the y direction impinges on a hydrogen atom, the only change that can occur is that from the 1s to the $2p_y$ configuration. Only this change can yield a transition dipole. The field intensities of the $2p_x$ and $2p_z$ configurations along the y axis are zero; while they would yield a transition dipole, they have no projection along the direction of the incident radiation.

Line Intensity. The intensity I_{21} of an emission line is defined as the energy radiated per second. It is the product of the number of atoms spontaneously undergoing a given transition in a second and the energy of the photons released. The number of atoms (or molecules) dN spontaneously undergoing a transition from higher state 2 to lower state 1 during an interval dt is

$$dN = N_2 A_{21} dt, \tag{13.3}$$

*Transitions that are improbable or forbidden may occur when a molecule is perturbed by a collision or interaction with near neighbors, though they will be much lower in intensity.

where N_2 is the population of state 2. Since the energy released per photon is just $h\nu_{21}$ for $dt = 1$ s, the resulting expression for I_{21} is

$$I_{21} = h\nu_{21}A_{21}N_2. \qquad (13.4)$$

How shall N_2 be found? If thermal equilibrium has been attained, it can be defined in terms of the Boltzmann distribution. In a representative system where few atoms are in an excited state, the equation is

$$N_2/N_0 = [g_2/g_0]\,e^{-(E_2-E_0)/kT}, \qquad (13.5)$$

where g_2 and g_0 are statistical weights in the excited and ground states, respectively, and E_2 and E_0 are energies in those states. Finally, an expression for line intensity in representative systems is obtained by substituting Eq. (13.5) in (13.4):

$$I_{21} = h\nu_{21}A_{21}N_0[g_2/g_0]\,e^{-(E_2-E_0)/kT}. \qquad (13.6)$$

Example 13.4. How may statistical weights be found so that relative populations in excited states can be calculated at different temperatures? Calculate the ratio g_2/g_0 for sodium atoms for the $^2P_{1/2}$ excited and $^2S_{1/2}$ ground states.

The statistical weights can be found from the expression $g = 2j + 1$. Note that j values are tabulated in Table 13.1 for several species. Using the formula, the ratio g_2/g_0 for sodium in the state is $[2(1/2) + 1]/[2(1/2) + 1] = 1$.

Example 13.5. What expression can be used to calculate the population of an excited state E_2 when an appreciable fraction of atoms are excited?

In this case the population N_2 can be related to the total population of the species in all states N by the more general form of the Boltzmann expression $N_2/N = [g_2/Z(T)]\,e^{-E_2/kT}$, where $Z(T)$ is the quantum mechanical partition function for the species and is defined by the equation $Z(T) = \Sigma_j g_j e^{-E_j/kT}$, where the sum is taken over all states. The equation for N_2/N reduces to Eq. (13.6) when few atoms are in an excited state.

Radiative Lifetime. Spontaneous emission is by nature a first-order kinetic process, that is, one whose rate is dependent only on the concentration of excited species at a given time. If there is an initial population $N_2(0)$ in an excited state, and additions to the population are not made, an exponential decay rate will be noted:

$$N_2(t) = N_2(0)\,e^{-t/\tau}, \qquad (13.7)$$

where $N_2(t)$ is the population at time t and τ is a constant termed the radiative lifetime of the excited state. Its value will be principally determined by the Einstein coefficient A_{21}; by comparison of Eqs. (13.3) and (13.7) it is seen that this coefficient and τ are reciprocally related. Since most values of A_{21} are of the order 10^8 s^{-1}, lifetimes tend to be in the range of 10^{-8} s.

It is clear that spontaneous emission may also be viewed as a relaxation process. The usual exponential type of decay of the excited state is observed.

Shape of Spectral Lines. All emission and absorption lines have a finite width. There are several causes. A so-called *natural width* is defined by the mean lifetimes of initial and final states involved in a transition, that is, the average length of time a species remains in each state *i* before undergoing a transition. According to the Heisenberg principle, the uncertainty in energy of each state ΔE_i is complementary to this lifetime τ_i and is given by $\tau_i \Delta E_i \simeq h/2\pi \simeq \tau_i h \Delta\nu_i$, where h is the Planck constant. If atoms could remain an essentially infinite time in excited and ground states, the uncertainty in the energy of the states would become vanishingly small. Then a transition from excited state E_2 to lower state E_1 would yield a single transition frequency ν_{21} of magnitude $(E_2 - E_1)/h$. But only ground-state lifetimes are "infinite," and actual transitions lead to the emission of a band of frequencies that has a breadth $\Delta\nu_N$ at half-maximum intensity. It can be shown that if τ_2 and τ_1 are lifetimes in the two states, the natural line width of the emission (or absorption) is given by

$$\Delta\nu_N = (1/2\pi)(1/\tau_2 + 1/\tau_1). \quad (13.8)$$

In general, natural line widths are minute. The narrowest lines are resonance lines, those for which state 1 is the ground state. For example, the mercury resonance line at 253.7 nm has a natural width of 3×10^{-5} nm.

Example 13.6. What fraction of atoms will be excited in a flame at 3000 K for a sample of gaseous sodium atoms? For a sample of gaseous zinc atoms? Assume that local equilibrium is attained.

Under these conditions Eq. (13.5) should predict the fraction of excited atoms. For sodium and zinc g_2/g_0 has values of 2 and 3, respectively, and values of N_2/N_0 of 6×10^{-4} and 6×10^{-10} are calculated. Since the figure for zinc is six orders or magnitude smaller, for reasonable emission intensity it is clear that a much hotter excitation source is needed.

Example 13.7. Once excitation is stopped in a system undergoing resonance emission, what fraction of the original number of excited atoms remains after a time in seconds equal to $1/A_{21}$?

The fraction can be found from Eq. (13.7). It is just $N_2(t)/N_2(0) = e^{-1} = 0.37$.

Example 13.8. What is the natural width of a representative resonance line of wavelength 250 nm?

Assume $\tau_2 \simeq 10^{-8}$ s. For a resonance line, the equation defining $\Delta\nu_N$ reduces to $\Delta\nu_N$

$= 1/2\pi\tau_2$ since $\tau_1 = \infty$. Substituting in the equation gives

$$\Delta\nu_N = 1/(6.28 \times 10^{-8}) \simeq 10^7 \text{ s}^{-1}.$$

This breadth may be converted into a wavelength by use of the wave relationship $c = \lambda\nu$. Since $\Delta\lambda$ is desired, this expression must be differentiated. One obtains

$$d(c/\lambda) = d\nu = (c/\lambda^2)\,d\lambda = d\nu \quad \text{or} \quad \Delta\lambda_N = \lambda^2(\Delta\nu_N/c),$$

when increments are substituted for differentials. At 250 nm,

$$\Delta\nu_N = (250)^2 \times 10^{-18} \text{ m}^2 \times 10^7 \text{ s}^{-1}/(3 \times 10^8 \text{ m s}^{-1}) = 2 \times 10^{-6} \text{ nm}.$$

Other influences also act to broaden lines. In Table 13.2 the effects are summarized. *Doppler broadening* results because wavelengths increase if an emitting (or absorbing) species is moving away from a detector, and vice versa. The effect is sufficiently large for Doppler broadening usually to define the lower limit of widths that can be observed. Collisions or near collisions of species undergoing transitions also increase line widths. The phenomenon is termed *pressure broadening*, but is also known as Lorentz and/or Holtsmark broadening.

At low pressures the Doppler effect is the main cause of broadening and lines have an essentially Gaussian shape. At flame temperatures, values of line half-breadths are of the order of 5–50 $\times$ 10^{-4} nm. Pressure broadening can be very substantial indeed [9].

Table 13.2 Factors Determining Width of Spectral Lines of Gaseous Species

Factor	*Mechanism by Which Influence is Exerted*	*Variables Affecting Influence*	*Designation of Contribution to Width and Its Value for 253.7-nm Hg line*
Finite lifetime of states	Heisenberg uncertainty principle	Nature of initial and final states of species	Natural width 3×10^{-5} nm
Motion of emitting or absorbing species relative to detector	Doppler effect	$(\text{Temperature})^{1/2}$ and $(\text{Mass})^{-1/2}$	Doppler width ~0.02 nm
Molecular collisions involving emitting or absorbing species	Intermolecular forces causing "shift to red" or nonradiative transfer of energy	Nature, density, and temperature of surrounding gas	Pressure-broadened width Very variable

This chapter has dealt with general aspects of absorption and emission. Yet a consideration of molecular transitions has been deferred to Section 16.3.

An accurate rendering of a narrow line will require a quality spectrometer. With a spectrometer of low resolution, narrow lines will have an apparent breadth that is too great and an apparent intensity that is too small. In most such cases the *integrated intensity*, which is given by

$$I = \int_{\nu_1}^{\nu_2} I(\nu)\, d\nu, \tag{13.9}$$

where ν_1 and ν_2 are the limits of the line, better expresses line intensity as a function of frequency. In general, Einstein transition probabilities are determined from such measurements.

Self-Absorption. The radiation from excited atomic gases is also subject to self-absorption, a phenomenon which diminishes the intensity of the central frequency, especially for intense lines with high transition probabilities. The origin of self-absorption is apparent on examination of the spatial distribution of species in an excited gas. While the center will usually be a hot, strongly emitting region, in the peripheral region temperatures are often low enough to permit ground-state atoms to exist. They can reabsorb the characteristic wavelengths produced in the core and afterward will be likely to lose the excitation energy by simple collision processes that are nonradiative. In addition, the smaller velocities characteristic of cooler atoms will reduce the Doppler broadening and further concentrate self-absorption at the center of the radiation band. Figure 13.4 shows both a line whose

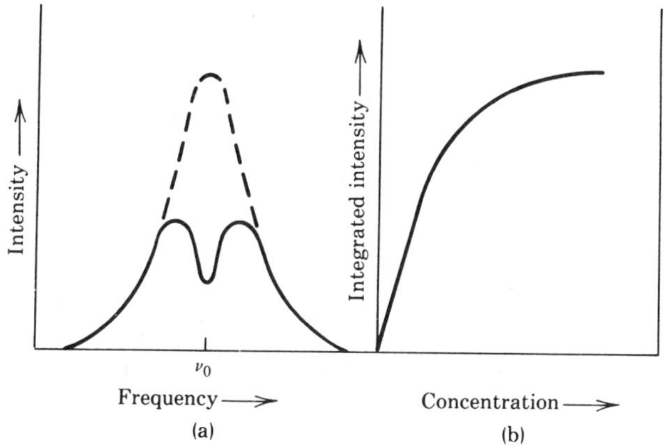

Fig. 13.4 Effect of broadening influences and self-absorption on an atomic emission line. (a) The broken-line curve shows the shape of the peak. It is widened by the Doppler effect and pressure broadening. Few ground-state atoms are present. The full line shows the effect of self-absorption on the peak. Ground-state atoms of the emitting species have reabsorbed some of the photons of the characteristic, central frequency and its amplitude is now considerably lower. (b) Calibration curve for the species based on integrated intensity. As concentration increases self-absorption leads to a loss of linearity.

peak intensity is considerably reduced by self-absorption and its effect on a characteristic calibration curve.

Background Radiation. Since line intensities are always measured against a background, that intensity must be known and subtracted. Probably the most important source of background is continuum radiation. It may result from blackbody radiation from incandescent particles, conversion of variable electron kinetic energies to radiation as electrons recombine with positive ions, and free-electron transitions. The last involves changes in kinetic energy of electrons when they decelerate in local electric fields associated with ions and other electrons. (Note the similarity to Bremsstrahlung radiation in X-ray emission.) Another contribution may be molecular emission, which ordinarily is a closely spaced series of lines unresolved by a spectrometer. Clearly, low background will be essential for precision observation of spectral lines. Low-intensity lines may often be indistinguishable from background fluctuations.

13.4 EXCITATION SOURCE AS SAMPLING MODULE

The module most characteristic of emission spectrometers, the *excitation source*, will be introduced here. It is also the most critical module because it must volatize and atomize samples—literally separate atom from atom—as uniformly as possible since concentrations of atoms in the gas should be representative of those in a sample. The source should also excite atoms of as many kinds as possible. Since the dissociation processes are inherently easier for solutions than for solids, the dissolution of samples is the first step in much atomic emission spectrometry.

The many kinds of excitation sources in use may be compared on the basis of the following criteria:

1. high energy flux;
2. reproducibility in sample introduction and energy transfer;
3. stability of excitation;
4. high sensitivity; and
5. ease of use.

It is clear that an excitation source must be designed to transfer a great deal of energy very quickly and smoothly to carry out the necessary processes with efficiency. Yet the source best suited to excitation of one type of sample may differ from that for another. Thus, solids can be more easily excited in electrical arcs and sparks and solutions in inductively coupled plasma sources. Where intensities of spectral lines must be measured precisely, as in quantitative measurements, stability and reproducibility in a source are clearly valuable. Trade-offs among criteria are common, for some performance criteria are incompatible with others.

Flames. Sufficiently high temperatures can be attained in flames (up to 3000–3400 K in the hottest, the N_2O–C_2H_2 flame) to determine some 70 elements quantitatively. The temperature of the gas in flames is, of course, limited by the heat of combustion of the fuel. If stable oxides of elements tend to be formed, the use of very hot and/or reducing flames that have excess fuel will usually lessen such formation. Flame sources will be considered in Chapter 14 in connection with the discussion of atomic absorption and fluorescence methods in which they find widest use. At the same time attention will be given in Section 14.8 to *flame* emission spectrometry. Both high precision and accuracy as well as lower instrumentation costs are obtainable in many instances in flame emission spectrometry for many elements.

Electrical Discharges. Energy may be efficiently transferred from an electrical discharge to a sample. Discharges are classified according to the manner in which they are generated and their specific properties. Conventionally, *arcs* are recognized as discharges that (a) must be initiated by an auxillary spark or momentary mechanical conduction across a gap and (b) produce minimal ionization. *Sparks*, on the other hand, will (a) jump their gaps unassisted and (b) give spectra richer in frequencies corresponding to highly excited atoms and ions. Some discharges may be classified as either arcs or sparks. *Plasma discharges* are electrical discharges in which ionization is sufficient to affect the properties of the gas in the discharge. They must be constrained by a magnetic field, a temperature inversion, or other means.

The stability of discharges generally increases in the sequence

$$\text{arc} < \text{spark} < \text{plasma discharge} < \text{glow discharge}.$$

In general, only the first three types have high current densities. The requisite sensitivity of excitation was first obtained easily from arcs and sparks.

Electric Arc. The dc arc is perhaps the simplest and least costly high-energy excitation device in use. Its best application is in the qualitative analyses of solids, and if its current is well stabilized, it is also satisfactory for some quantitative determinations. Figure 13.5 shows a standard dc arc circuit.

Two aspects of the discharge deserve emphasis. The plasma energy is highest near the cathode where the temperature is about 4700 K as a result of a concentration of high-velocity ions, electrons, and atoms. Thus, samples to be examined are placed on the cathode where they are quickly vaporized and excited by bombardment. Second, in operation the discharge is between the anode and a small spot on the cathode. It tends to wander over the cathode surface, causing the length of the arc and its temperature to change. Resulting current fluctuations are opposed by the RL circuit and feedback circuits can add further stabilization. In any event, arc wander leads to fluctuations in emission intensities. With an open arc, gases from the air interact with analyte atoms somewhat giving rise to small populations

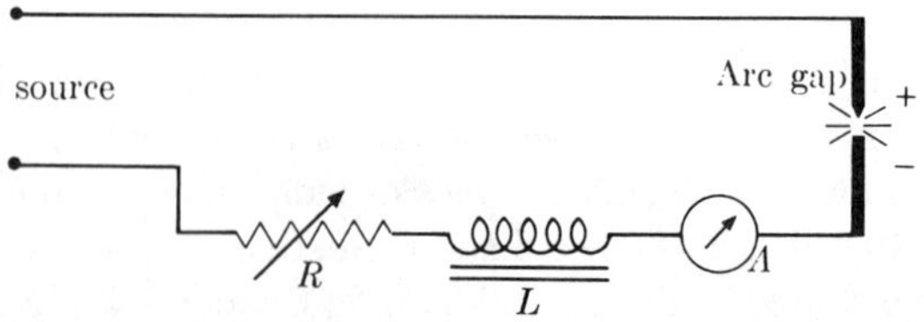

Fig. 13.5 Circuit for a dc arc. The dc source is usually a regulated power supply that furnishes 110–220 V at 3–30 A. Ammeter *A* and variable resistor *R* allow the current to be adjusted after the arc has been started by closing the electrode gap momentarily. Iron-core inductor *L* and resistor *R* stabilize the arc somewhat against fluctuations. The arc has a negative resistance characteristic and its resistance *falls* as current increases. Resistor *R* is present specifically to limit the current. (See also Fig. 8.5.)

of gaseous "molecular" species such as MO, diminishing the population of excited M atoms and the intensity of its lines.

Stallwood Jet. The excitation sensitivity of a dc arc using graphite electrodes can be enhanced by a Stallwood jet, a device that surrounds the arc with a concentric curtain of gas such as an Ar–O_2 mixture. It serves both to cool the electrodes and the outer layer of the arc plasma and to exclude atmospheric gases. By cooling the discharge it slows sample volatilization and keeps a sample in the arc a longer time. Thus, it permits more efficient excitation. Further, by reducing pressure it narrows line widths. By excluding the atmosphere, the jet eliminates the formation of cyanogen (C_2N_2), whose intense bands often obscure or provide a heavy background for emission lines of interest.

Electric Spark. Spark discharges have found their best application in quantitative work. They furnish high precision rather than great sensitivity. Basically, spark generation is simple and requires a circuit much like that for an ac arc, although elaborate units have been devised to provide constancy and intensity of excitation as well as reasonable flexibility. Sources differ principally in the arrangement used to initiate breakdown in the spark gap. Figure 13.6 illustrates a circuit for a condensed spark. Each spark is in fact an oscillating electrical discharge.

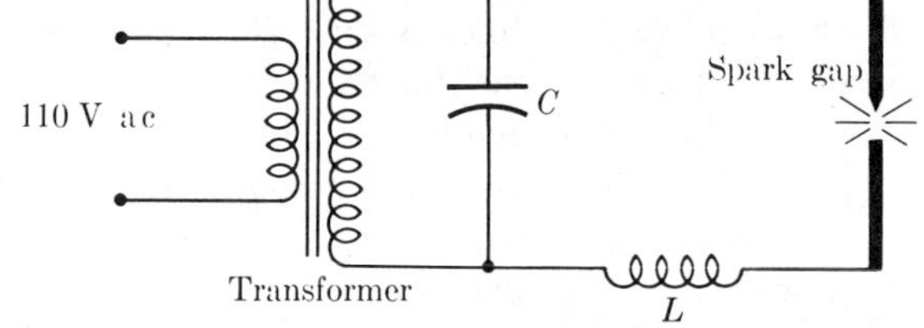

Fig. 13.6 Simplified circuit for a spark excitation unit. The transformer output may range from 10,000 to 100,000 V, depending on the application. Capacitor *C* and inductor *L* bring about a series of sparks at the gap, each of which is in fact an oscillating electrical discharge. The frequency, usually 1–2 kHz, duration of a spark, usually 10^{-4}–10^{-5} s, and amount of current will depend on the magnitudes of *L* and *C* and on electrode separation.

Multipurpose Sources. Several commercial excitation sources provide for dc and ac arcs and a mechanically interrupted, high-voltage damped spark. The variety of discharges is sufficient to accommodate most types of qualitative and quantitative analyses.

Plasmas. A significant advance in emission spectroscopy occurred in the 1960s when high-temperature plasma sources were developed. A plasma is simply an electrically neutral gas whose ionization has been carried to the level ($>1\%$) where gas properties are appreciably affected by the population of electrons and ions. For example, a plasma interacts strongly with electric and magnetic fields. Yet even modest concentrations of free charge carriers can exist only at quite high temperatures. Application of the Boltzmann equation to one of the most stable molecules, N_2, shows, for example, that at 6700 K the gas is sufficiently dissociated that equal amounts of N_2 and N are present, but only at 14,500 K are equal amounts of N and N^+ predicted.

To be effective in atomizing and exciting samples, it is important that plasmas attain at least local equilibrium (see discussion below). Without it a plasma is likely to be unstable and its production and excitation of free-atom gases will not be really predictable for most elements. Fluctuations in intensity will be likely and the rate of collisions energetic enough to ensure efficient atomization will no longer be certain.

Equilibrium in Plasmas. It is useful to describe two possible levels of equilibrium that may be attained in a plasma. In *thermodynamic equilibrium*, both the distribution of particle velocities and intensities of different wavelengths of radiation can be theoretically described. Boltzmann's law would describe the former and Stephan's law the latter. Almost never does this condition hold.

Fortunately, many plasmas at least have a distribution of particle velocities that represents rough equilibrium even though radiation does not. Collisions between species are mainly responsible for energy transfer between particles. In this instance a plasma is said to be in *local thermodynamic equilibrium* (LTE). As a result, the description of populations in Section 13.2 will be applicable for plasmas. The most likely departure from local thermodynamic equilibrium is an overpopulation of the ground state. A system showing this behavior is sometimes described as in "partial LTE."

Measuring Temperatures in Excitation Sources. How can one estimate the temperature of a plasma? of a flame?

The *line-reversal method* is a common null procedure in which the emission of a continuous source, for example, an incandescent strip tungsten filament is compared with a slit image of a wavelength in the plasma or flame. To make the comparison the bright filament is focused on a region of interest in a flame and both are focused on the entrance slit of a spectrometer of moderate dispersion. If necessary, a sample is introduced in the source that contains the comparison element, for example, sodium. The spectrometer is set to the expected wavelength. The filament is made steadily brighter by increasing *its* temperature. The intensity of the flame line from the spectrometer will fall to zero and then *reverse* (drop below background). Immediately before reversal the flame temperature equals that of the source: incoming radiation is being absorbed by some flame atoms (for a 1 → 2 transition) while an equal number are returning from the upper state (a 2 → 1 transition). From the calibration of the continuous source the flame temperature is known.*

For a plasma or flame in LTE, temperatures can be measured by observing the *ratio of integrated intensities* of two or more spectroscopic lines of an element in the sample. Transition probabilities must be known. In the absence of self-absorption the ratio is given by $I_1/I_2 = (A_1 \lambda_2 g_1 / A_2 \lambda_1 g_2) \exp[(E_2 - E_1)/kT]$, where the A's are Einstein spontaneous transition probabilities, λ_i is an emission wavelength, g_i is a statistical weight, and E_i is an energy of transition. The uncertainty in transition probabilities will limit accuracy.

Developing an Analytically Useful Plasma. Physical constraints as well as "pinching" effects and other special properties of plasmas are employed to constrain a plasma for analytical use. Its volume is confined by a quartz tube just inside of which a vortex of inert gas is swirling. Both are *relatively* cold. The lower temperature causes ionization and conductivity to fall at the edge of the plasma, an effect often termed a "thermal pinch." As the current is constrained to the hotter central region, there are additional consequences: (a) both the plasma temperature and current density increase, (b) there is a greater "magnetic pinch," and (c) the volume of the discharge collapses further. The more dense plasma that results is considerably more stable and has a higher temperature.

In addition, a striking and valuable result, the formation of an *annular plasma*, is secured in the ICP. When a plasma has a torroidal or doughnut shape, the problem of sample introduction is greatly eased. Current in conductors, and in the plasma, is localized near the surface *when high frequencies are used*, a phenomenon called the *skin effect*. The eddy currents caused by oscillations of electrons in the rf field form circular paths localized near the *outer* regions of the plasma. Thus, heating is correspondingly greater in these regions, and the plasma takes on an annular shape.

There are two additional requirements. First, it is essential that the plasma have a region in which temperatures will still be very high (6000–6800 K), but that is free of intense continuous emission. Fortunately, such a band exists about 10–20 mm above the luminous region. Sometimes this observation region of analyte emission is called a plasma jet or a plasma flame. Second, reaction of analyte atoms with the plasma gas should be minimal. This goal is secured by use of argon or another inert gas for the plasma.

Several modes of plasma generation are in use. One design of a *dc plasma jet* is shown in Fig. 13.7. Note that the analytical region is just below the luminous central region. Samples are introduced into the inert gas that chills the outer portion of the plasma as it passes through the ring. The mode of sample introduction will be described below as nebulizers are taken up. Advantages of the dc jet are a very large dynamic range for excitation intensities, low cost, precision, and basically high sensitivity. Limitations are that electrodes may contaminate samples and that some arc wander occurs (just as in a regular dc arc) with resulting instability. It also proves difficult to inject samples other than by the direct means.

Microwave-generated plasmas have also been explored as emission sources. They usually employ frequencies greater than 300 MHz. Most offer high electron

*A. G. Gaydon and H. G. Wolfhard, *Flames, Their Structure, Radiation, and Temperature*, 4th ed. New York: Wiley-Halsted, 1979.

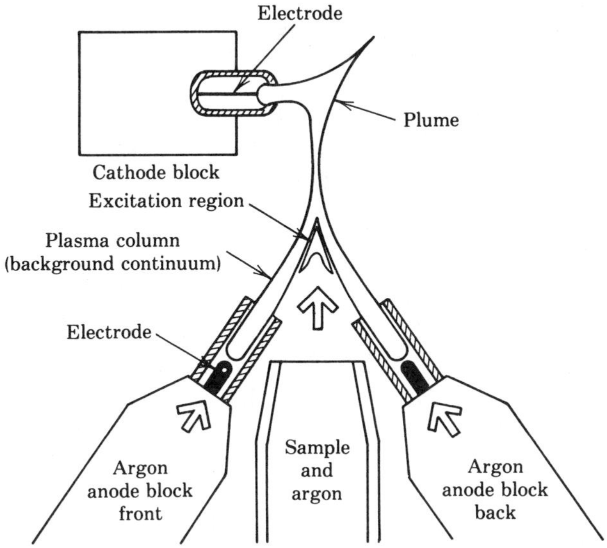

Fig. 13.7 A schematic of a dc plasma jet. Anode and cathode blocks provide sleeves that shelter carbon electrodes. Argon flows out around the dual anodes. The intricately shaped dc plasma provides a high temperature region just below its luminous central portion for introduction of the sample aerosol. (Courtesy of Applied Research Laboratories, Inc.)

temperatures but fail to give local thermodynamic equilibrium. It appears that the microwave plasma devices are most attractive as detectors for gas chromatographs where one component is detected at a time.

The *inductively coupled plasma torch* has proved to be the most valuable plasma source. Radio frequency (rf) power (in the United States a frequency of 27.12 MHz is commonly chosen since it falls in a band set aside for commercial use) is coupled electrically to the plasma either inductively or capacitively. In Fig. 13.8 a common version of the inductively coupled plasma torch (ICP) is illustrated.

An important question with any excitation source is whether a sample can be introduced with minimum disturbance. Early attempts to inject samples into plasmas as aerosols were minimally effective: the aerosol tended to be repelled by the steep thermal gradient at the plasma surface. With an annular or ring-shaped plasma an ingeniously effective injection into the hole in its center is possible. An aerosol is simply propelled down the axis of the plasma where it meets little resistance. While in the plasma, aerosol is heated to 5700–6300 K by *radiation and convection from the plasma*. In the roughly 2 ms at these high temperatures droplets lose solvent, chemical structures are destroyed, and excitation and ionization occur. Samples are more uniformly atomized and excited than they would be if introduced into the plasma itself. In other types of excitation it is often necessary to readjust the coupling to ensure continuing efficient transfer of energy. Further, since the plasma loses heat mainly indirectly, the nature of analytes in the sample affects the plasma relatively little. Axial sample introduction ensures a very stable plasma.

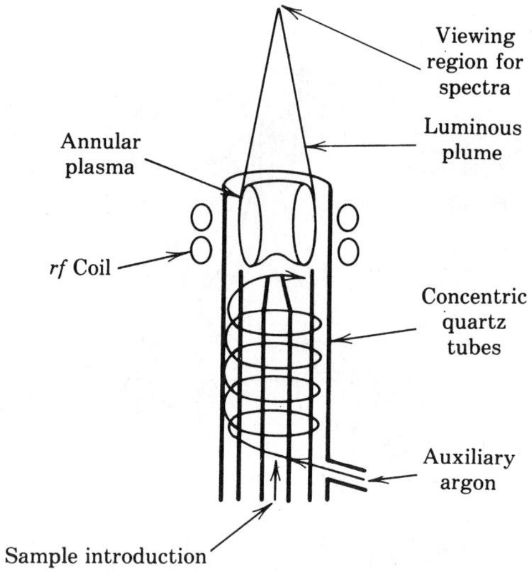

Fig. 13.8 Cross-section of an inductively coupled plasma torch. The external water-cooled coils through which energy is coupled to the torch are also shown. A coaxial cable (not shown) transmits power from an rf oscillator to the coils. Three concentric quartz tubes shape and constrain the plasma. The sample travels up the central tube from the bottom as an aerosol in a stream of argon (flow of 1 mL min^{-1}) and traverses the plasma axially. Note the annular nature of the plasma. To confine and ensure a stable plasma, auxiliary argon flowing at a rate of 10–15 mL min^{-1} swirls around the plasma as a helical vortex. The initial ionization needed to form a plasma is created by a brief application of a Tesla coil.

To introduce samples with precision, automatic devices are used whenever possible. For solid samples that are to be examined in that state after grinding and other laboratory steps, procedures outlined in Section 13.7 are employed. Most samples can be dissolved, however, and the solution can be introduced automatically into the plasma in the form of an aerosol.

Aerosols are formed mainly by *pneumatic nebulizers* and sometimes by *ultrasonic nebulizers*. One of these devices accompanies the ICP excitation source. In Fig. 13.9 a crossed-flow type of pneumatic nebulizer is shown. It has the advan-

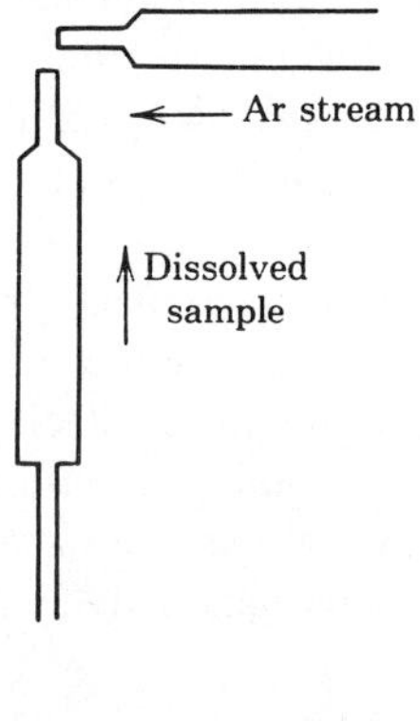

Fig. 13.9 A crossed-flow nebulizer. A pump produces a constant stream of solution flowing through the capillary tip at 1 mL min^{-1}. An argon stream directed toward the sample solution at right angles nebulizes it into an aerosol.

tage that a continuing series of samples can be introduced by a peristaltic pump, which produces a steady flow essentially regardless of solution viscosity. Fortunately, as each new solution is introduced, it also cleanly displaces the previous one in a short interval. The rate of flow of new solution can be increased briefly to speed rinsing or rinsing can be done with solvent. Note that argon is used only to form the aerosol. At present ultrasonic nebulizers see only specialized use since they almost always require thorough cleaning between use with different kinds of samples.

The significant advantages of both dc and ICP plasma sources result from their high temperature and inert argon atmosphere. Even "refractory" elements such as tungsten and zirconium appear in free-atom form. Further, some working curves cover as many as 5 decades of concentration, for example, from 1 ppb to 10^2 ppm. The emission region is also small, ensuring higher intensities. Self-absorption effects are also much reduced: most species are in an excited state rather than the ground state. Both standard and minitorches that consume less argon and require less power are in use. They offer nearly the same limits of detection.

The very success of the source in excitation is also the cause for spectral interferences by elements such as iron or transuranium elements that give rich emission spectra. Capital and operating costs are also higher for the ICP torch than for the dc plasma jet. As a result, if no more than one or two elements are to be determined, the dc plasma jet may be the more attractive type of excitation source. Once many elements are to be determined one must upgrade optics, computer, and software, and the lower cost of the dc plasma jet appears less a factor. Finally, while solids may be volatilized into an ICP torch, no satisfactory way to introduce powder directly into the center of an annular plasma has yet been developed.

Other Sources. Some other types of discharge source that have utility are the demountable hollow-cathode lamp, the glow discharge source, and the laser microprobe. The demountable hollow cathode lamp operates exactly as the regular one does (see Section 8.4). The sample is placed in the cup of a hollow metal cathode, the lamp evacuated, and an inert filler gas (usually Ne or Ar) added. Inert gas ions created by an electrical discharge are accelerated toward the cathode interior. When they impact, atoms of the sample are sputtered out where they may also be excited by collision with accelerated ions. With hollow-cathode excitation the limit of detection is usually quite low and when sputtering atomizes a sample matrix effects seem minimal.

Microsampling of the surface of any specimen is possible by use of a focused laser beam (a microscope is used) or a microspark. In this way an analysis can be performed on an area whose dimensions are of the order of 10 μm. A laser microprobe accomplishes the vaporization of the spot and sometimes also its excitation. If necessary, an arc between auxiliary electrodes placed just above the target spot can excite the vapor. A spark microprobe makes use of an electrode whose tip is polished to a 1-μm point. After it is positioned about 25 μm from the sample (which must either conduct electricity or have a thin conducting layer deposited, an ac spark is used to volatilize and excite to emission an area just below. Such microprobes allow effective sampling of small areas of any material such as individual cells in biological samples or inclusion areas in alloys.

13.5 CRITERIA FOR SPECTROMETER DESIGN

Now that excitation sources have been described, all modules necessary to construct atomic emission spectrometers have been presented. Recall that monochromators and polychromators were described in Chapter 9 and detectors in Sections 8.6 and 8.8. On the basis of the functions to be performed a generalized emission spectrometer can be blocked out in modular form as shown in Fig. 13.10. It will be a goal of this section to examine briefly the choices made in developing the spectrometers in general use.

How important is each module to spectrometer performance? We need an overview of the contribution of the different modules to the development of the spectrometer analytical signal. It can be shown* that the analytical signal obtained from a spectrometer at a single wavelength is related to the performance of characteristic modules by the expression

$$\text{signal} = \left[\frac{(1-\alpha)MC}{\psi}\right]\left(A_{21}h\nu_{21}\frac{g_2}{Z(T)}\,e^{-E_{21}/kT}\right)f(\theta,\lambda)\,g(\lambda). \qquad (13.10)$$

The terms are identified from left to right as follows. First, the bracketed quantity contains: α, the fraction of analytical species that is ionized; M, the weight of sample atomized (volatilized as atoms) per second; C, the concentration of analytical species in the sample; and ψ, the velocity of transport of sample through the excitation zone. Thus, the bracketed quantity represents the concentration of volatilized atoms in the region of excitation. Second, the parenthetical quantity is seen to be the product of the fraction of species excited [Eq. (13.6)] and all factors but concentration necessary to convert this value to an emission intensity. Third, $f(\theta, \lambda)$ represents the geometrical factor that defines the solid angle of the emission observed by the monochromator and its transmission efficiency with wavelength. Fourth, $g(\lambda)$ represents the efficiency of the detector as a function of wavelength. Since the signal is integrated over a period of time, $\int_0^t Q\,dt$ should be substituted for MC, where Q is the rate of volatilization and dt is the interval.

It becomes evident on inspecting Eq. (13.10) that the effectiveness of atomization or volatilization and of excitation is of paramount importance. Ideally the

Sample
↓
Excitation source ⟹ Wavelength isolator ⟹ Detector ⟶ Processing modules ⟶ Readout

Fig. 13.10 Modular diagram of a generalized atomic emission spectrometer.

*M. L. Parsons, W. J. McCarthy, and J. D. Winefordner, *J. Chem. Educ.* **44,** 214 (1967).

procedure used should maximize the rate of volatilization M, provide as slow a rate of transport ψ (through the excitation zone) as possible, and maximize the temperature T without unduly increasing the degree of ionization α. As some of the modules are individually examined in the next section, these criteria should be kept in mind. Since standard procedure discussed in the preceding chapters can ensure an efficient performance by the monochromator, detector–transducer, amplifier, integrator, and readout, these optical modules will receive little attention here. The reader should consult earlier chapters for information.

Are matrix effects that are so characteristic of solids and to a much lesser extent of solutions, evident in Eq. (13.10)? It will be useful to think of the expression as dealing with each single component. Rates of atomization, for example, will be quite dependent on the enthalpy of sublimation of particular constituents from a sample quite as much as on the rate of energy delivery. In the excitation source, the temperature experienced by a component will determine the effectiveness of its excitation. Yet there is little opportunity for equilibrium to be attained and it is likely that the temperature in the locality of particular component atoms will be dependent on the composition of the sample. It is clear that provision in the processing modules for integration of output with time will be important in reducing both types of matrix effects.

13.6 EMISSION SPECTROMETERS

A variety of spectrometers has been designed because of varied needs in different wavelength regions as well as to provide levels of dispersion, resolution, and other characteristics desirable in actual applications. It should be kept in mind that certain characteristics may be mutually exclusive. For example, resolution is nearly always achieved by sacrificing "speed." Dispersion and wavelength range are another such pair: a high dispersion instrument inherently has a short spectral range. For convenience, definitions of resolution and dispersion for grating systems are summarized in Table 13.3.

The optical system of a dispersive emission spectrometer* is designed around a monochromator or polychromator. The latter is a double module in which the exit slit of a monochromator has been replaced by a multichannel detector (or an array of slits coupled to single-channel detectors) mounted in the focal plane. The basic theory for these modules was given in Chapter 9. Here spectrometers based on these modules will be examined. It should be noted also that the emphasis will be on grating spectrometers; in recent years prism spectrometers have become relatively rare.†

*Historically, visual devices for observing emission spectra have been classified as *spectroscopes*, devices in which spectra are recorded on photographic film or plates as *spectrographs*, and scanning devices which can be used in several wavelength regions as *spectrometers*. Today these distinctions are less important and only the general term *spectrometer* will be used in this volume.

†Prism spectrometers suffer substantial limitations compared with grating instruments: a spectral dispersion that is inherently smaller and strongly nonlinear with wavelength, lower resolution and optical efficiency, and marked curvature in images for long entrance slits.

Table 13.3 Summary of Several Performance Parameters for Grating Spectrometers

Resolution or resolving power $R = \lambda/\Delta\lambda$ where nearly identical wavelengths λ and $\lambda + \Delta\lambda$ are just separated. Units: dimensionless	$R_{max} = mN$, where N is the total number of rulings and m the order of spectrum observed, or $R = W(\sin\alpha + \sin\beta)/\lambda$, where W is the width of ruled surface, α and β are angles of incidence and diffraction, respectively.
Dispersion. Angular dispersion $d\beta/d\lambda$ is the angular separation $d\beta$ achieved for a wavelength interval $d\lambda$. Units: rad nm^{-1}	For a fixed angle of incidence: $\frac{d\beta}{d\lambda} = \frac{m}{d\cos\beta}$, where d is the grating constant and β the angle of diffraction. For the Czerny–Turner mounting: $\frac{d\Theta}{d\lambda} = \frac{m}{2d\cos\phi\cos\Theta}$, where the angles are identified in Section 9.6
Linear reciprocal dispersion $d\lambda/dl$ is a measure of the distance dl along the focal curve over which the wavelength interval $d\lambda$ is spread. Units: nm mm^{-1}	$\frac{d\lambda}{dl} = \frac{d\cos\beta}{mr}$, where r is the focal length if a lens is used, or the distance from grating to focal curve, if a mirror is used.

The reader interested in prism instruments is directed to Section 9.4 for coverage of prism monochromators, which are essentially spectrometers, and to references at the end of this chapter.

Coupling of Excitation Source. To secure reasonable line intensities, light from the excitation source must be collected over as large a solid angle as feasible and focused efficiently on the entrance slit. One or more lenses or mirrors between the two will serve this purpose. Two useful arrangements are illustrated in Fig. 13.11. For good coupling it is basically required that the solid angle of the lens, L_1 in Fig. 13.11a, be equal to or greater than the solid angle of the spectrometer. As long as photoelectric detection is employed, the intensity of light will be integrated over the entire height of the spectral line and uniformity of intensity will be unimportant. If photographic detection is in use, lens L_1 should be placed much nearer the entrance slit to ensure uniform illumination. While each point in the source now furnishes light equally to each point in the slit image at the focal plane of the spectrometer, the entrance slit may not be completely filled. What must be

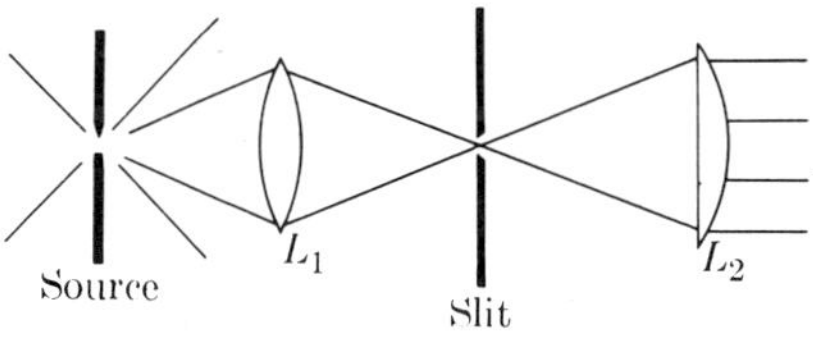

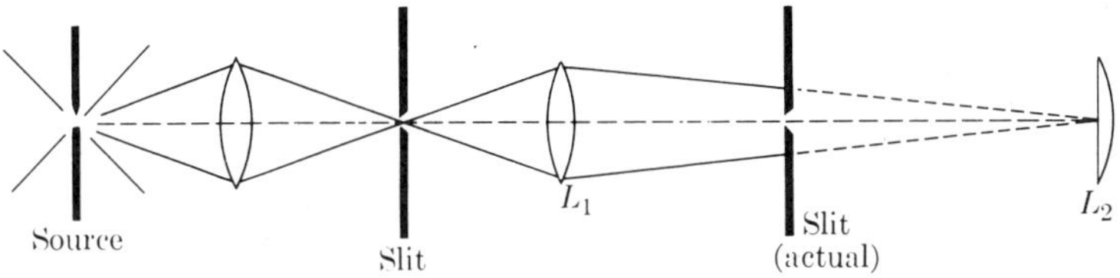

Fig. 13.11 Schematic diagram of coupling of an excitation source to a monochromator. (a) An arrangement providing maximum intensity at the slit. (b) An arrangement providing uniform illumination of a slit at moderate intensity.

ensured is that the collimating optic in the spectrometer is completely illuminated by the image of the source.

In Fig. 13.11b an alternate coupling arrangement is shown. Radiation is observed only from a selected portion of the source, an approach that would allow study of some part of the source, for example, to observe species as a function of height in a plasma or flame or to find a position at which the signal/background ratio was suitably low. Finally, as illustrated in Section 9.3, concave mirrors can also couple efficiently.

Concave Grating Instruments. The special attraction of a concave grating spectrometer lies in the curvature of its grating, which allows it to provide both dispersion and focusing. Thus, such spectrometers in principle need no collimating or focusing lenses or mirrors. With fewer optical elements they have enhanced throughput, and if a holographic grating is used, considerably reduced stray light, astigmatism, and coma.

Most of these spectrometers have their slit, grating, and focal plane along the circumference of a *Rowland circle*, which is a circle whose diameter equals the grating radius of curvature. These spectrometers are inherently bulky since long focal length gratings are ordinarily used to secure high dispersion and resolution and are of necessity also designed for very fine adjustment to realize the high-quality spectra of which they are capable.

The *Paschen–Runge mounting* (spectrometer) shown in Fig. 13.12 illustrates a common placement of spectrometer elements on the Rowland circle. The long wavelength range available in this spectrometer (190–800 nm, and with purging down to 165 nm) is a considerable asset and this design is widely used for direct-

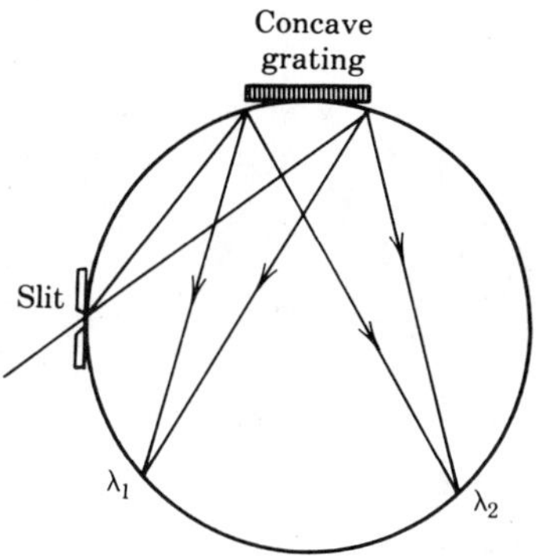

Fig. 13.12 A Paschen–Runge mounting for a concave grating. Three different designs that are in use, all based on focal lengths of 1 m or longer, will be described. (1) In the direct or *simultaneous* design, the Rowland circle holds an entrance slit mounted in a chosen position, a concave grating, which is fixed in position, and a series of about 60 exit slits precisely set at positions of particular elemental lines. Detectors are then located behind the slits. (2) In a second design, essentially a *sequential* type of spectrometer, there are two or more entrance slits positioned along the circle and a curved photographic plate in a holder placed to capture the wavelength range chosen. The slit giving the least aberration and the desired range is used. (The plate must be scanned after chemical development.) (3) A third, *sequential* Paschen–Runge design is implemented for the observation of a set of known wavelengths, which the operator must specify. Now an array of more than 200 exit slits is placed on 2-mm centers along the Rowland circle between λ_1 and λ_2. To obtain a spectrum, a slit in about the position of a particular line is selected and one of two photomultiplier tubes provided is moved behind it. In this design the entrance slit can be precisely positioned, and is moved until the center of the chosen line falls on the selected exit slit.

reading (simultaneous) spectrometers. At angles of incidence and diffraction greater than about 45° optical aberrations begin to increase and throughput to decrease.

A more compact Rowland circle instrument, the *Eagle design*, is shown in Fig. 13.13. In the Eagle layout all components are on one side of the Rowland circle, giving a layout reminiscent of a Littrow prism spectrometer. Since rays from the

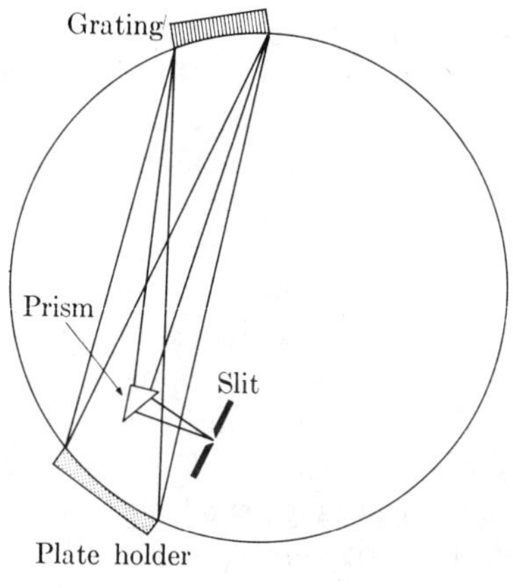

Fig. 13.13 The Eagle mounting for a grating spectrograph.

grating return over a path nearly the same as that taken in incidence, astigmatism is slight. Interestingly, by use of a totally reflecting prism, the slit can be located within the Rowland circle. A tradeoff of elaborateness in adjustment has been made for compactness: three adjustments are required for a change of wavelength range: a rotation of the grating, a rotation or tilting of the plateholder or other detector, and movement of the plateholder or grating to alter the distance between them. As is evident from the figure, the angle of diffraction β is fairly large.

A common concave grating arrangement that is not based on the Rowland circle is the *Wadsworth mounting* shown in Fig. 13.14. By having an extra mirror to collimate light from the entrance slit, the design almost completely eliminates astigmatism. The speed and light-gathering power of this design are greater than for many others since the grating is used at about half the image distance. But bulk is again a disadvantage, and the focal curve is not circular.

Plane Grating Instruments. Most modern UV–VIS spectrometers employ planar diffraction gratings. Two developments have made these gratings attractive, the formulation of processes for precision replication of gratings, and the development of holographic procedures for "ruling." Important advantages are that planar gratings are substantially free of optical aberrations and permit more compact spectrometric systems. A representative Ebert–Fastie design is shown in Fig. 13.15. By comparison with the Ebert monochromator (see Fig. 9.16), it is apparent that it is possible to use either a side-by-side positioning of slit and detector or an over-under arrangement as shown here. These arrangements basically provide for off-axis use of a spherical mirror to avoid aberrations.*

A plane-grating instrument offers several advantages over one with a concave grating. Greater spectral brightness, or grating efficiency, can be achieved since a plane grating can be blazed for peak efficiency in the spectral region of interest

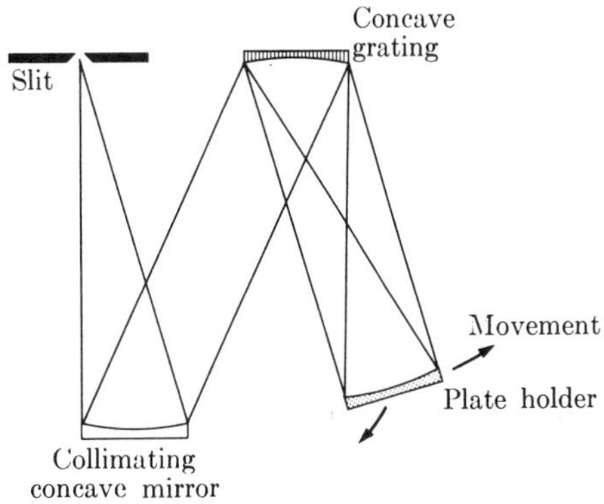

Fig. 13.14 The Wadsworth mounting for a concave grating.

*In the side-by-side Ebert monochromator mounting aberrations are minimized only at the exit slit, making it attractive as the main module of a scanning spectrometer. On either side of the slit, astigmatism and coma occur and soon become appreciable. The *over-under arrangement* of Fig. 13.15 achieves relative freedom from these aberrations even at the ends of a 75-cm plate centered as shown

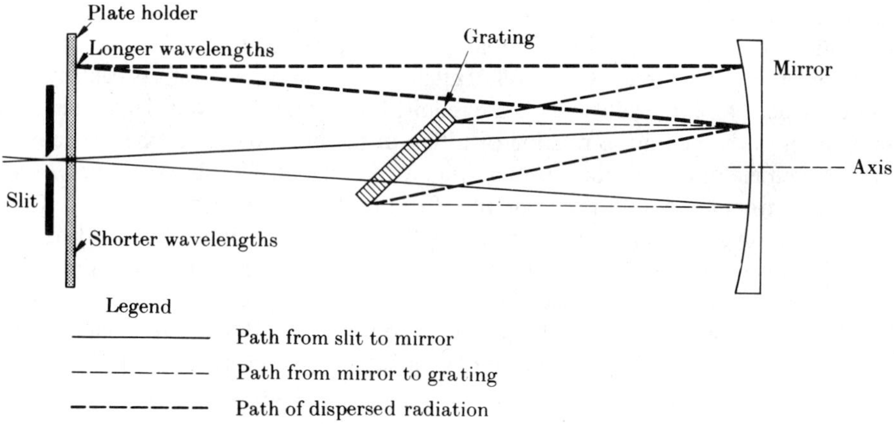

Fig. 13.15 Top view of an Ebert-Fastie spectrometer. This modification features an over-under arrangement in which the slit and plate holder are centered on the mirror axis below and above the grating, respectively. The entrance slit is directly below the plate (though it is here shown farther from the mirror for clarity). It is necessary to tilt the slit slightly to overcome a slight tilting of lines on the plate. The focusing action of the mirror is shown for a single, relatively long wavelength. No provision is shown for separating orders; a fore-prism or filter might be so used. (Courtesy of Jarrell-Ash Co.)

(see Section 9.5). Further, the grating permits a mechanically simple arrangement for wavelength adjustment, makes possible regular, aberration-free use of large angles of incidence, and features a virtually flat focal curve. As a result a planar instrument has the potential of:

1. Considerably greater wavelength range. Since large angles of incidence can be used without difficulty, access is gained to the long-wavelength region ($\lambda >$ 700 nm) in the first-order spectrum.

2. High dispersion and resolution in a moderately compact instrument. High-order visible and UV spectra can be efficiently observed with an appropriate planar grating and a large angle of incidence.

Example 13.9. An Ebert–Fastie spectrometer with a 3.4-m focal length mirror and a 900-line mm^{-1} grating blazed at 330.0 nm gives a useful range of 210–470 nm in the first order when set for incidence at 5.8°. The linear reciprocal dispersion is 0.5 nm mm^{-1}. The second-order UV spectrum (105–235 nm) will overlap unless blocked.

Example 13.10. A 2950 groove mm^{-1} grating blazed at 4 μm is used in an Ebert–Fastie spectrometer with a 37.6° angle of incidence. At 37.6° in the 18th-order the 219–241 nm band is centered at the detector. Recall that λ_B when $m = 18$ will be 4 μm/18 = 222 nm. A high value of D^{-1}, 0.044 nm mm^{-1}, and a theoretical resolution of 725,000 are secured. All the lower-order, longer-wavelength bands that also appear can be separated by an order-sorter arrangement (see below).

Echelle Spectrometers. Instruments with dispersion greater by a factor of 10 than that of regular spectrometers can be achieved by using an echelle grating (see Section 9.6). Recall that its wide rulings permit use of angles of incidence and diffraction as large as 63°. Since orders ranging from 35 to more than 100 are common, often a prism is employed as an *order sorter* to make it possible to extract spectral information successfully. The prism is oriented so that its dispersion is at right angles to that of the grating, an arrangement described as *cross dispersion*. In the resulting two-dimensional spectrum the abscissa represents wavelength and the ordinate, the diffraction order. A representative echelle spectrometer is shown in Fig. 13.16.

To scan in an echelle instrument it proves easier to move the detector than the grating. The resulting spectrum is often detected by moving a photomultiplier tube to appropriate positions. Computer control of the positioning of the photomultiplier tube is a straightforward process. It is also important that the entrance slit be only 1–2 mm high to allow the display of many orders. The vertical separation of orders is determined by the prism dispersion. If its dispersion is small, there will be little separation, but a wide range of wavelengths can be registered in the detection area. Conversely, a large prism dispersion will lead to a few well-separated orders, each of which registers a short wavelength range.

Because high orders are used, resolution is unusually high with an echelle device, as noted earlier. Spectral efficiency (throughput times resolution) is also high because observations are made close to the blaze angle, and stray light is low since

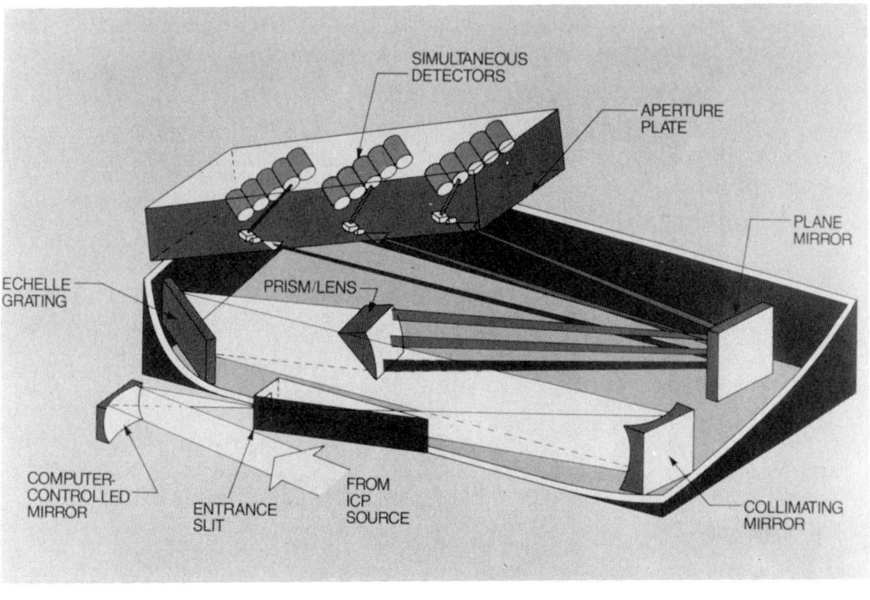

Fig. 13.16 A scanning echelle spectrometer. An echelle grating provides the dispersion. Since diffraction is into high orders (m = 40–110), it is essential to have a prism of low dispersion at right angles serve as an order sorter. (Courtesy of Leeman Labs, Inc.)

the small amount present is widely dispersed. Finally, the wavelength coverage in a single spectrum may also be from 190 to 800 nm by virtue of display of the spectrum in two dimensions.

Example 13.11. A fore-prism is used as an order-sorter for an echelle grating by being turned at an angle of 90° to the axis of the echelle. Describe the appearance of its spectral display. Assume that the fore-prism causes "red," or longer, wavelengths to appear at the top of the detection area and "blue," or shorter, wavelengths to appear at the bottom.

Recall from the grating law that a grating must display longer wavelengths in lower orders. With this echelle, lower-order spectra should appear at the top of its detector field. Successively higher orders will be arranged roughly parallel to the first at lower and lower positions. Any wavelength that is imaged simultaneously, say in the 50th and 51th orders, will appear at the same horizontal place in each spectrum, but in the higher order it will be a line directly below the same line of lower order.

Direct-Reading Spectrometers. In *multichannel spectrometers* many wavelengths are measured simultaneously by employing a polychromator (see Section 9.9). A common polychromator design based on a regular (echellette) grating calls for placement of a series of photomultiplier tubes, each behind a slit positioned along the focal curve of a Rowland circle spectrometer where a wavelength of interest will appear as shown in Fig. 13.17. When several wavelengths are close together, a small mirror or prism can be placed to "fold" the path of the beam to permit a detector to be located to one side where there will be more room. Some

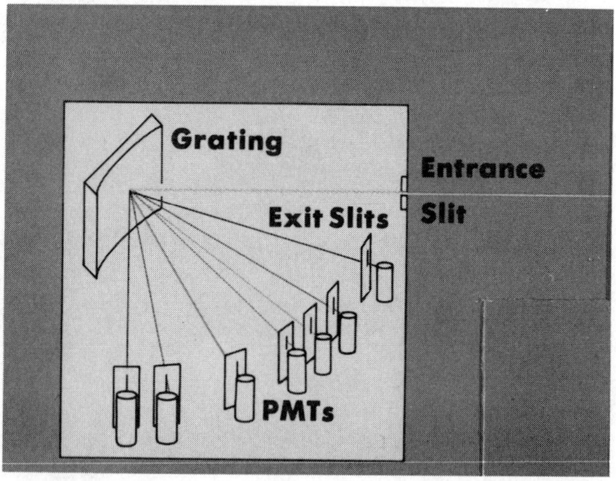

Fig. 13.17 The Paschen-Runge polychromator of a direct-reading emission spectrometer. The slits that isolate lines of elements of interest are precisely mounted along the instrument focal curve for the grating being used by attachment to a metal base at pre-located positions and each slit is accompanied by a photomultiplier. Photomultiplier outputs are directed to integrating circuits (not shown). Note that the accompanying spectrometer excitation source may be an inductively-coupled Ar plasma torch or an arc-spark source. The spectrometer permits measurement of many elements simultaneously. (Courtesy of Thermo Jarrell Ash Corp.)

slits will inevitably be used to monitor background, others, elemental lines, and still others, elemental lines being used for reference. Up to 55–60 slits can be placed and their detectors monitored simultaneously. About 35 elements can be determined in a favorable case.

Direct-reading echelle (two-dimensional) polychromators or spectrometers have a series of precisely positioned slits. Indeed, the slits may be etched in a focal plane plate. One design calls for positioning a detector behind each slit. Another calls for registering the coordinates of spectral lines in computer memory and moving the closest detector to the correct position for each wavelength of interest. Clearly, more wavelengths can be accommodated in the second design since the size of detectors will not be limiting. With these instruments it is desirable to calibrate with several lines of a mercury arc lamp rather than one, since two coordinates are required to specify each line.

While direct-reading spectrometers are too expensive and inflexible (slits must be securely anchored) for general use, they are ideally suited to the analysis of large numbers of similar samples for a specific set of elements. Thus the steel, aluminum, and other materials industries make extensive use of this type of instrument.

In operation, the intensities of lines are compared by determining the ratios of the time-integrated outputs of the properly positioned detectors during an interval of from perhaps 2 to 40 s. Each detector's output is integrated by a suitable capacitor–resistor circuit. The capacitor voltage is proportional to the time integral of the line intensity. Both noise and interelement effects are reduced by this integration.

Maintaining Alignment. It should be emphasized that a very stable mounting and constant room conditions will be necessary to maintain wavelength calibration in a spectrometer for any length of time. For example, changes in temperature will affect the grating constant d, the refractive index of air, and in more subtle ways the alignment of spectrometer parts.

Use of servo-control to maintain exact alignment is therefore well worthwhile. In a direct-reading echellette spectrometer one procedure is to place an auxiliary source such as a mercury lamp near the excitation source and employ one channel to monitor a strong line from the lamp. A quartz slab called a refractor plate is placed behind the entrance slit. To effect realignment, the slab is rotated to displace all wavelengths; rotation is continued until the mercury reference line is again received at peak intensity.

Some disadvantages of direct-reading spectrometers are the difficulty of adding slits and detectors for new elements, the low signal level occasioned by the high dispersion of the usual spectrometer of this type, and high cost. Use of high-energy excitation sources such as inductively coupled plasma torches is desirable. Securing reliable background values also offers problems; seldom will more than two or three exit slits be used for measuring background, and using these corrections far from the wavelength of measurement may introduce error.

Scanning Spectrometers. To scan spectra, a spectrometer with a scanning monochromator and detector is used. Controlling scanning precisely is desirable. Wavelengths are impinged successively on the exit slit by rotating the grating table. Wavelength drives for scanning are readily available, as described in Section 9.8. Often a stepper motor rotates the grating table. Linear wavelength scanning occurs if the lead screw of a sine-bar mechanism is driven by the motor. Though precise, this type of scanning is slow and wear will gradually impair accuracy. For greater accuracy and speed, a more direct drive can be used and the shaft encoded with an optical pattern that can be read and continuously interpreted in terms of wavelength. Cost rises with this arrangement, however, especially if small increment scanning is required. Another design will be discussed below.

How are diffraction orders separated in scanning? Usually, a set of bandpass filters, each of which passes a selected wavelength band, is mounted in a filter wheel. As the grating is turned, the wheel can be turned from position to position to bring each new filter into position. Clearly, filters will also block wavelengths outside the band passed and thus reduce both background and stray light.

There will usually be provision for rotation of the grating table at high speed, a process called "slewing." Slewing may be invoked between lines of interest or to reach a spectral region of interest. An elegantly designed, high-accuracy slewing drive is shown in Fig. 13.18. With program control one can slew to an exact wavelength. Then a standard slow scan can be called for to scan through a peak to register intensity and background on either side.

The advantage of a linear scan rate is the steadiness of the drive mechanism and perhaps the simplicity of the program and electronics. Because analysis time will be a product of the scan rate and wavelength range, scanning can be as rapid as the minimum integration time and detector response rate permits. Too slow a

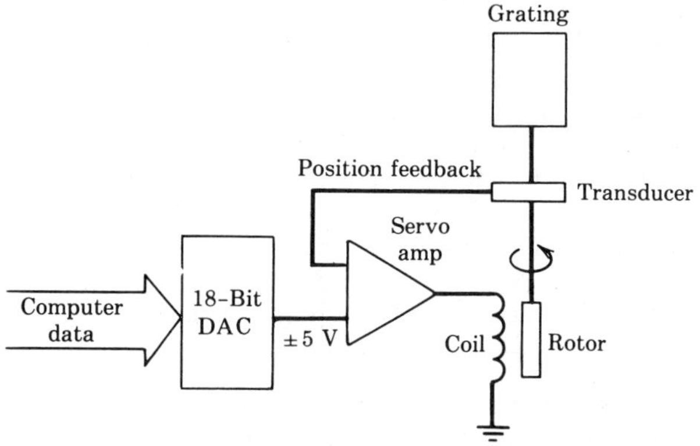

Fig. 13.18 A grating drive based on a magnetically driven torsion bar. The arrangement ensures that fast slewing can be achieved. To drive it to a new position, a digital command is converted to analog form by an 18-bit DAC and becomes one input of the differential amplifier. As the grating table turns, a transducer provides a second digital signal proportional to the table position. The position of the grating table can be prescribed to one part in 500,000. The servo-system controls the magnetic field in the coil, producing extremely fast rotation. When the transducer on the drive shaft feeds back a signal equal to that of the command input, the rotor stops and the final position is obtained. (Courtesy of Thermo Jarrell-Ash Corp.)

scan rate will result in excessive measurement time and consumption of a considerable amount of sample as well as possible drift in the intensity of the excitation source.

A marked advantage of a scanning monochromator is the ease of making corrections for background. It will be straightforward to enter into the program a measurement of background intensity on the side of a line with the lower baseline. Wavelength and intensity measurements may be made at several positions before, during, and after a peak in order to identify the best background reading as well as the true peak. Further, possible spectral interferences can be identified.

13.7 PREPARATION OF SAMPLES

It is nearly always advantageous to begin a study of a sample with an elemental analysis by emission spectrometry. Since the analysis should not only identify the elements in the sample but estimate their concentrations semiquantitatively, the information will guide a decision about other measurements. Thus, in emission spectrometry a wide range of sample types must be handled. The method of sample preparation should ensure:

1. homogeneity;
2. minimal matrix or interelement effect; and
3. sufficient preconcentration to ensure that elements at the trace level are present above their limits of detection.

Metals are often sampled by preparing them as self-electrodes, especially in metallurgical industries where samples are taken frequently for purposes of quality control. Samples of melts may be taken by casting rods. Rapid cooling is essential to minimize fractional crystallization. Solid samples may be machined into rods. All these samples would be excited by high voltage spark or perhaps dc arc.

For minerals and heterogeneous inorganic and organic solids sampling must minimize what would otherwise be severe matrix effects. Volatilization, atomic dissociation, and excitation will occur at different rates for different substances. Thus, some kind of "matrix exchange" is desirable. Substitution of a solution matrix would probably be best, but requires the availability of sufficient sample and a dissolving process that is not excessively difficult. For example, refractory fuel ash from fossil fuel power plants can be dissolved by a two-step process: fusion with lithium metaborate at 1050°C for 20 min followed by dissolution in HCl.

Dissolution. A solution matrix is probably the best choice for most samples. As discussed earlier, solutions may be nebulized and introduced into a plasma for efficient excitation. A very occasional alternative for solutions is introduction into graphite cups, followed by solvent evaporation and electrical excitation. An older method still in use is spark excitation of solution residues left on a copper rod.

Geological and metallurgical samples are ordinarily dissolved in a strong mineral acid. If the materials are siliceous, treatment with hydrofluoric acid is nearly always required.

Organic and biological samples ordinarily require prior destruction of the organic matrix. Sometimes this involves dry ashing by heating under conditions such that the vapor and smoke released can escape, but burning is avoided. If volatile elements such as arsenic and mercury are present, "wet-ashing" or oxidation is a better procedure. Nearly always there is an initial heating in concentrated nitric acid followed, if needed, by a second heating in a nitric–perchloric acid mixture.

Since even trace elements will ordinarily be of interest, it is essential that mineral acids and solvents have minimal impurities. Ultrapure acids (see Section 11.6) should be used. In addition, preconcentration can be ensured if volumes are controlled.

During preparation of a sample it may be desirable to remove (a) elements whose spectra are line-rich, such as iron or the lanthanides, perhaps by ion exchange, and (b) alkali metals present in large concentrations, since their easy ionization will lower the excitation temperature for other elements.

Grinding. Heterogeneous solid samples that are not to be dissolved will require grinding or crushing. Once the particle size is reduced sufficiently to ensure homogeneity, a common approach is to mix the powder with an excess of an excitation *buffer* such as ultrapure powdered graphite. The mixture is introduced into a hollow that has been machined in the cathode or lower electrode. Ultrapure graphite is ideal for preparation of cathodes of this type; it machines well, introduces a very little background in the emission spectrum, and is very high-melting. The buffer determines the rate of volatilization, and with its low ionization energy, the temperature and the composition of plasma of an arc or spark. As a result, matrix effects are "leveled" or held constant in a series of samples. It also proves easier by this method to introduce an internal standard (see Section 13.9) if needed.

13.8 QUALITATIVE ELEMENTAL ANALYSIS

Sensitive elemental analysis can be made by emission spectrometry. Once a spectrum of a sample is in hand, the first step is to establish with precision the wavelength of all lines that are believed to be of interest. Usually this is done by comparison with a simpler reference spectrum, as will be discussed below. Then wavelengths are assigned to particular elements. When three or more lines of an element are identified, the presence of an element can be taken as confirmed. Actually, since emission wavelengths are unique, if measurements have been made with high accuracy, determining a single line will be sufficient. In qualitative analysis it is also important to confirm expected intensities approximately.

Wavelength Identification. In the case of a *direct-reading spectrometer*, exit slit positions for known lines are carefully set by the manufacturer. Subsequently, if periodic recalibration is maintained, the appearance of intensity at a slit can be taken, in the absence of interference, to be the desired wavelength. For spectra taken with other spectrometers the amount of effort required for identification will vary with the type of wavelength isolation device.

In a *scanning spectrometer*, a continuous registration of wavelength is obtained from a sensor on a stepper motor that drives the grating (or prism) table; wavelength values of lines are obtained by interpolation and their intensities by the detector output. Results are stored in computer memory and plotted on command. The instrument wavelength scale must, of course, be calibrated, and periodically recalibrated against a standard. A common standard in the UV–VIS is the mercury spectrum, which has a relatively small number of bright lines.

Photographed Spectra. Generally, with a *spectrograph* spectra are recorded photographically. To facilitate wavelength identification of lines, known spectra are ordinarily photographed next to unknown spectra. The spectral range recorded is also noted. Wavelengths are established by interpolation between lines of the known spectra. In the case of a grating instrument, one must watch for overlapping orders. Sometimes a composite master spectrum containing three or four of the most persistent lines of each element may be furnished in such a way that it appears opposite the unknown spectrum. Though convenient, accurate alignment of spectra must be checked in this case.

To begin identification it is best to use relatively simple spectra. More complex spectra can be more easily compared to a standard like iron, which has a line-rich spectrum. For example, in Fig. 13.19, portions of a simple copper and complex iron spectrum have been photographed beside an unknown spectrum. From inspection of the figure, it is clear that magnification (scale expansion) must be employed in the comparison to realize the resolution inherent in a photographic plate or film. It is customary to project enlarged images of the several spectra onto a flat screen to permit use of a scale or ruler for distance measurements. A scanning microdensitometer may also be used to provide a wavelength scale, again by comparison with a standard for its own calibration.

Assignment of Lines to Elements. Once wavelengths have been identified in a recorded spectrum, the task remaining is assignment of lines to elements. Wavelength tables are invaluable in this regard [18, 19]. For clear identification of each element believed present, at least its principal lines should be found unless spectral interference is evident. It is unnecessary to assign every line present: surprisingly often, lines that will not be found in tables may appear, perhaps as a result of intrusion of other orders or grating ghosts.

Line intensities also are valuable in making wavelength assignments. Since intensities may vary with excitation conditions, how can they be employed reliably? One way may be to examine intensities of multiplets in the spectrum since roughly correct intensities will be known for such lines. Expected multiplets should be located as soon as an element is tentatively identified. If they can be verified, it will be strong evidence of the identity of the element.

Sensitivity and the Choice of Lines. Actually, lines of a given element range in sensitivity, that is, in their tendency to be excited. Resonance lines have the highest sensitivity and ion lines, the lowest. In trace analysis it is useful to seek very sensitive lines.

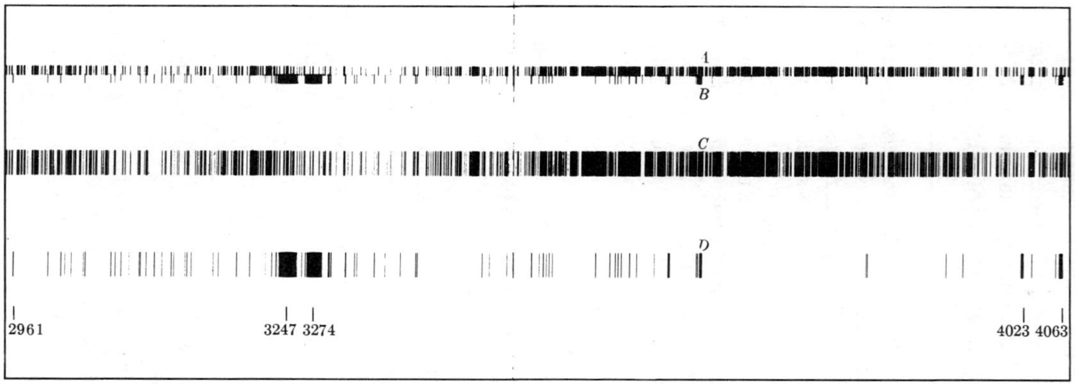

Fig. 13.19 Portions of a copper and an iron spectrum produced by a high resolution spectrometer. Spectra A and B are iron and copper, respectively. They are repeated at greater slit height in curves C and D. This first-order grating spectrum was obtained at a linear reciprocal dispersion of 5.4 A mm^{-1} on glass plates. Conditions: 10 μm slit width, 3 A dc arc, 10 s exposure for iron, 35 s for copper. (Courtesy of Jarrell-Ash Co.)

Background intensity is also critical. Whether a faint spectral line can be detected at all will depend on the natural background as well as on stray light in the spectrometer system.*

13.9 QUANTITATIVE ANALYSIS

Emission spectrometry is deservedly popular as a means of rapid quantitative elemental analysis. The dynamic range of emission methods is so great that concentration levels are identified by the terms *major* ($>10\%$), *minor* ($\approx 1\%$), and *trace* ($<0.01\%$).

Analytical Curves for Samples. In emission spectrometry the dominant strategy for quantification is the use of an analytical curve. A well-selected set of standards is employed to obtain plots of relative intensity of selected emission lines versus concentration. As would be expected, accuracy and precision are dependent on the close regulation of excitation conditions, including the selection of an appropriate integration time. Most sources are given to fluctuation and times of volatization and excitation of elements will vary.

Selection of an Analytical Line. How should emission lines be chosen for determination of an element? Once its presence is established, by procedures just discussed, a single line will usually suffice for its quantitative determination. Factors of importance in the selection of the line are lack of interferences with other species, the internal standard chosen, if any, and a native line intensity sufficient for accuracy at the concentration expected.

For a sample in which analytes are present in major or minor levels, lines ending in an excited state, or ion lines, are usually preferred since they will not suffer from self-absorption. If an analyte is present over a wide range of concentrations in a series of samples, it will be desirable to use two or more lines in most cases to be sure precision is maintained. For elements at trace levels, neutral atom resonance lines are the best choice.

A final criterion is that lines be chosen in the same wavelength region if at all possible. Most detectors are somewhat wavelength dependent.

Solid Samples. If solid samples are being examined, special care must be taken in selecting the analytical standard. If the sample compositions are known approximately, it will be easier to select standards that will match roughly in terms of analyte concentration and matrix. Without such a match, matrix effects may overwhelm the analysis. A variety of standard reference materials are available from the U.S. National Bureau of Standards and secondary standards are available from many companies.

*To the extent that stray light is a dominant factor, the greater the dispersion of the spectrometer, the lower the background intensity at a given wavelength: the stray light must be spread over a greater area in an instrument of high dispersion.

In many instances use of an *internal standard* is desirable to reduce the effect of fluctuation in source intensity resulting from changes in temperature as sample composition varies as well as some unsteadiness in all excitation sources. Adding a standard amount of an uncommon element such as scandium is often attractive. This approach will also greatly reduce matrix effects. Where samples are ground, crushed, or present as powders, samples can be mixed with appropriate standards as they are prepared for excitation. With an internal standard present, analytical curves are plots of the ratio of the intensity of the line for the analyte to the line for the internal reference versus concentration of the analyte. Such a curve is shown in Fig. 13.20.

When only moderate precision (5–20%) is needed, a semiquantitative analysis will suffice. This type of analysis can be carried out more quickly. One method is the dilution technique described below.

Example 13.12. As shown in Fig. 13.20 a suitable internal standard line is one that is close to the analyte line, about equal to it in excitation energy, and influenced similarly by variations in procedure or conditions. Sometimes the pair is termed a homologous pair. Suitable pairs may be selected by experiment, but extensive data are available in the literature and should be consulted. It is entirely possible to use an element as an internal standard that is present originally in high concentration.

Example 13.13. In the dilution method of semiquantitative analysis the following sets of mixtures is made up from graphite and the analytical sample: (1) 100 mg of sample; (2) 10 mg sample and 90 mg graphite; (3) 1 mg sample, 99 mg graphite; (4) 0.1 mg sample and 100 mg graphite. Each mixture is a 10-fold further dilution of the original sample. This set and a 100-mg standard that contains known concentrations of elements thought to be present (in a graphite base) is completely "burned."

Relative line intensities are then observed and approximate concentrations of elements are determined by simple comparison. For example, if the intensity of a sensitive copper line in mixture (3) approximately matches that in the standard, the copper concentration in the original sample must be about 100 times that in the standard. Similarly, an intensity

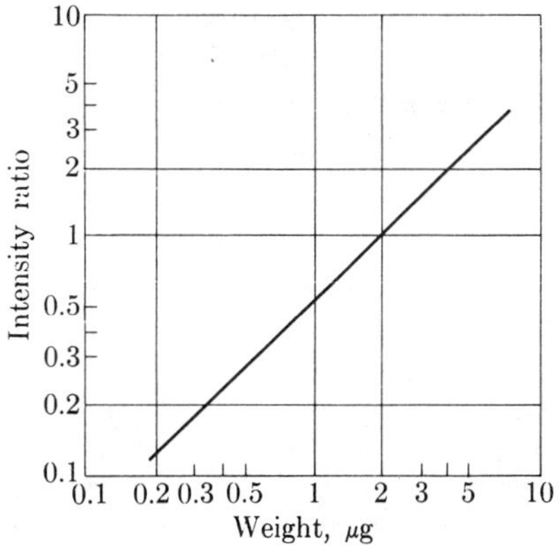

Fig. 13.20 Plot of spectrographic intensity ratio for Ni 2632.9/Mo 2775.4 as a function of the concentration of nickel. (By permission, from N. H. Nachtrieb, *Principles and Practice of Spectrochemical Analysis*, New York: McGraw-Hill, 1950.)

match between a lead line in the standard and the lead line from mixture (2) indicates the lead is present at 10 times its concentration in the standard.

Solution Samples. For an inductively coupled plasma source or other type of system that will admit solution samples in straightforward fashion, development of calibration curves is as easy as for other analytical methods. Clearly, the solvent, usually water, provides a standard matrix. The calibration curve will be a simple linear plot given by the equation $I_a = mC$, where m is the slope of the curve, I_a the intensity of the analytical line, and C the concentration. In every case, a background correction I_b should be made if background is greater than the order of magnitude of acceptable error. A more correct equation for the working curve would then be $I_a - I_b = mC$, where I_a and I_b are intensities for the analytical line and background, respectively.

Solution Residue Methods. In cases where traces are being determined from solution samples, it is also possible to use microsolution samples by initially evaporating the solvent. Such procedures are not in wide use, but do give possibilities for high sensitivity and low limits of detection. The substrate on which to place the solution sample prior to evaporation should, of course, make minimal contribution to the analysis. A rotating-platform method has been commonly employed where there is no real limitation in sample size, and a copper or silver disk is used with arc or spark excitation. Alternatively, a hollow-cathode lamp may be used. For microsolution samples, the sample may be vaporized in a small furnace from a graphite crucible, loop of refractory metal, or strip of graphite yarn. Often the furnace is placed under the excitation source in a way that will allow the plasma gas to carry the sample vapor into the plasma.

REFERENCES

Some general references on emission spectrometry are as follows:

1. P. W. J. M. Boumans (Ed.), *Inductively Coupled Plasma Emissions Spectroscopy. Part 1: Methodology, Instrumentation and Performance.* New York: Wiley, 1987.
2. E. L. Grove (Ed.), *Analytical Emission Spectroscopy*, Vol. 1, Parts I and II. New York: Dekker, 1971 and 1972.
3. R. D. Sacks, Emission spectroscopy, in *Treatise on Analytical Chemistry*, 2nd ed. Part 1, Vol. 7, P. J. Elving (Ed.), New York: Wiley-Interscience, 1981, Chapter 6.

Detailed practical information on emission spectrometric analysis is given by:

4. American Society for Testing Materials, (E-2 Compilation), *Methods for Emission Spectrochemical Analysis*, 6th ed. and suppl. Philadelphia: ASTM, 1971 and 1977.
5. P. W. J. M. Boumans, *Theory of Spectrochemical Excitation.* New York: Plenum, 1966.
6. M. Thompson and J. N. Walsh, *A Handbook on Inductively Coupled Plasma Spectrometry.* New York: Chapman and Hall, 1983.

Atomic spectra are discussed on an introductory to intermediate level in:

7. B. P. Straughan and S. Walker (Eds.), *Spectroscopy*, 2nd ed. New York: Wiley-Halsted, 1976, Vol. 1, Chapter 1.

An advanced treatment of atomic spectra may be found in:

8. G. Herzberg, *Atomic Spectra and Atomic Structure.* New York: Prentice-Hall, 1937; also Dover, 1944.
9. G. H. Kuhn, *Atomic Spectra*, 2nd ed. New York: Academic, 1969.

Instrumentation is emphasized in:

10. R. A. Sawyer, *Experimental Spectroscopy*, 3rd ed. New York: Prentice-Hall, 1951; also Dover, 1963.
11. K. I. Tarasov, *The Spectroscope.* New York: Wiley-Halsted, 1974.

Other publications of interest are:

12. R. M. Barnes (Ed.), *Developments in Atomic Plasma Spectrochemical Analysis.* Philadelphia: Heyden-Wiley, 1981.
13. S. Bashkin and J. O. Stoner, Jr., *Atomic Energy Levels and Grotrian Diagrams*, Vol. 1. New York: Elsevier-North Holland, 1976.
14. S. Greenfield, H. McD. McGeachin, and P. B. Smith, Plasma emission sources in analytical spectroscopy I, II, III, *Talanta*, **22,** 1, 553 (1975) and **23,** 1, (1976).
15. V. A. Fassel, Quantitative elemental analyses by plasma emission spectroscopy, *Science*, **202,** 183 (1978).
16. K. Fuwa (Ed.), *Recent Advances in Analytical Spectroscopy.* Oxford: Pergamon, 1982.
17. B. Magyar, *Guide-Lines to Planning Atomic Spectrometric Analysis.* New York: Elsevier Scientific, 1982.

Compilations of spectral lines of elements are available in standard handbooks as well as in Ref. [1] and:

18. G. R. Harrison, *Wave Length Tables*, 2nd ed. Cambridge, MA: MIT Press, 1969 (wavelengths given for 109,000 lines).
19. W. F. Meggers, C. H. Corliss, and B. F. Scribner, *Tables of Spectral Line Intensities*, Parts I and II, National Bureau of Standards, Monograph 32. Washington, DC: U.S. Govt. Printing Office, 1961 (wavelengths given for 39,000 lines).

EXERCISES

13.1 For sodium atoms calculate the ratio N_2/N_0 in an acetylene–air flame at 2500 K assuming the excited state of interest is $^2P_{3/2}$.

13.2 The effect of self-reversal on line intensities leads to departures from linearity in calibration curves. Consider that the phenomenon affects principally intense emission lines ending at the ground state. (a) In what analyte concentration ranges might calibration be usefully based on such lines? (b) At concentrations affected by self-absorption, what other types of lines could be used for determination of species?

13.3 Calculate an intensity for the emission line of hydrogen gas at 656.3 nm. Assume the Einstein probability coefficient for spontaneous emission has an approximate value of $10^8\ s^{-1}$ and that N_2 for the hydrogen excited state $= 10^{-18}$ mol m^{-3}.

13.4 In a Paschen–Runge spectrometer, for a given grating and slit width, what will be the effect of increasing the grating order on (a) dispersion, (b) resolution, (c) precision of wavelength setting (if an exit slit must be moved), and (d) throughput? Explain answers briefly. [See M. W. Routh and K. J. Paul, *Am. Laboratory* **17**(6), 84 (1985).]

13.5 Consider that an order-sorter prism unit produces a spectrum along the length of the entrance slit of a grating spectrograph that it accompanies, with 260 nm appearing at the bottom of the entrance slit and 500 nm at the top. Lower and higher λs fail to enter the slit. Assume the grating angle is such that spectra of fifth and higher orders appear on the photographic plate of the spectrograph. (a) How many orders will be seen on the photographic plate? Note that the largest pertinent value of the product $m\lambda$ is $5 \cdot 500 = 2500$. (b) Sketch in rough fashion the several sets of spectra as they would appear vertically. [See R. F. Jarrell, *J. Opt. Soc. Am.*, **45,** 259 (1955).]

13.6 A grating spectrograph has a linear reciprocal dispersion of 0.5 nm mm^{-1} in the first order. The iron triplet at 310.00, 310.03, and 310.07 nm is photographed in two orders. What is the separation of the extreme lines in the first order? In the second order?

13.7 A concave reflection grating with 300 grooves mm^{-1} and a focal length of 1.5 m is available. A collimated beam containing wavelengths 550.0 and 540.0 nm is normally incident. In the curved focal plane at a distance of 1.5 m, what is the separation (in millimeters) of these wavelengths in (a) the first-order spectrum, (b) the fourth-order spectrum?

13.8 Is a spectrometer of resolution 10,000 capable of resolving the 5538.57- and 5539.28-Å (553.857- and 553.928-nm) iron lines?

13.9 What is the theoretically attainable resolution of a grating with 600 grooves mm^{-1} whose ruled area extends 10 cm?

13.10 Self-electrodes of brass are arced by a spectroscopist interested in determining their lead content. He finds that a spectrum photographed when a dc arc is first struck shows good contrast throughout. One recorded after the arc has operated 2 min shows appreciable continuous background, especially in the red portion of the spectrum. Offer a possible explanation for the change.

13.11 To establish the approximate concentration level of aluminum, silicon, and copper in an unknown, the technique of successive dilutions of unknown S is to be used. Let S-1 and S-2 identify the unknown after 10-fold and 100-fold dilution with pure graphite. Let G, G-1, and G-2 designate standards containing 0.1, 0.01, and 0.001% concentrations of each of 40–50 elements. From the data in Table E13.11 determine approximate concentrations for each element sought.

Table E13.11 Relative Intensity[a]

Sample	*Aluminum (309.2 nm)*	*Silicon (288.1 nm)*	*Copper (324.7 nm)*
S	7	1	2
S-1	6	0	1
S-2	4	0	0
G	4	4	4
G-1	3	3	3
G-2	2	2	2

[a]Range: 7 indicates a maximum response and 0 means undetected.

13.12 What type of spectrograph should be used for precision quantitative analyses of steel samples? List reasons for the choice and cite data where possible.

13.13 Suggest suitable excitation methods for the following quantitative spectrographic determinations: (a) lead in chamber-process sulfuric acids, (b) calcium in blood, (c) zinc in a brass casting, (d) traces of silicon in germanium powder, (e) potassium in porcelain chips. Consult references if necessary.

ANSWERS

13.1 1×10^{-4} s. *13.6* 0.16 mm and 0.32 mm. *13.7* Approximately (a) 6.6; (b) 21.7. *13.9* 60,000. *13.11* Aluminum, 10%; silicon, $<10^{-3}$%; copper, 10^{-3}%.

Chapter 14

ATOMIC ABSORPTION, ATOMIC FLUORESCENCE, AND FLAME EMISSION SPECTROMETRY

14.1 INTRODUCTION

Two further elemental spectroscopic methods, atomic absorption and atomic fluorescence spectrometry, will be treated in this chapter as well as a related technique, flame emission spectrometry. Atomic absorption spectrometry is by far the most widely used of the three.

Development of Atomic Absorption Instrumentation. Up to this point only the emission of a cloud of *excited* atoms has been considered. Since ground-state atoms are also present in an atomic gas, might not their optical *absorption* also be usefully measured? In the simplified modular diagram of an atomic absorption spectrometer in Fig. 14.1, only the flame sampling module seems out of the ordinary. Yet the appearance of working absorption spectrometers was delayed more than a century after the development about 1830 of the first atomic emission instruments because of crucial instrumentation problems. A good bit of insight into how effective instrument design enables good measurements is afforded by studying how these problems were resolved.*

One problem was the narrowness of absorption lines of *gaseous atoms* (invariably < 0.005 nm) as compared with the spectral slit widths obtained with many monochromators. If the latter width is broad compared with the former, the sensitivity of measurement of absorption is poor. This situation is shown in Fig. 14.2. Actually, if the slits of a high resolution monochromator are narrowed sufficiently and the device is coupled to an intense source like a xenon lamp, the bandwidth and height of curve B can be approximated. Stability of output would still be a problem as well as precision of wavelength setting. The development of modern hollow-cathode lamps for individual elements solved this problem: their very narrow emission lines (0.001 nm width) provide precisely the wavelengths that will be absorbed by gaseous analyte atoms of the same element.

A second challenge was to find a convenient, reproducible way to convert samples into free-atom gases. While flames can perform this function, many flames are noisy and their moderate temperatures lead to incomplete atomization of samples. It took the development of slot burners that use premixed gases to secure flames of low intrinsic luminosity and minimal noise (fluctuation). Further, new fuel–oxidant combinations such as N_2O–C_2H_2

*J. W. Robinson, Atomic absorption spectroscopy, in I. M. Kolthoff and P. J. Elving (Eds.), *Treatise on Analytical Chemistry*, 2nd ed., Vol. 7, Part I, New York: Wiley Interscience, 1981.

Monochromatic source → Flame sampling module → Wavelength isolation device → Detector → Processing modules → Display

Modulator ↑ (to Monochromatic source); Sample ↑ (to Flame sampling module)

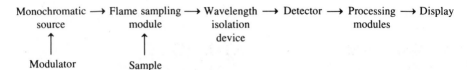

Fig. 14.1 Modular diagram of a simple atomic absorption spectrometer.

were found to give hotter flames that atomized most refractory samples efficiently. The second problem was resolved.

With the hollow-cathode lamps currently available about 70 metals and metalloid elements today can be sensitively determined by flame atomic absorption. The method is of necessity mainly a quantitative technique since ordinarily a different hollow-cathode lamp is required for each analyte. Spectral interferences in which a hollow cathode line is overlapped by a characteristic line of another element are rare because the lamp lines are so narrow. Some matrix effects do, however, occur.

Two other atomic methods, *flame emission spectrometry*, which is readily accessible with most flame atomic absorption spectrometers, and *atomic fluorescence spectrometry*, a trace method, will also be treated. As a fluorescence method, the latter also requires a bright monochromatic source to excite the fluorescence that is measured.

This set of atomic techniques has several attractions. These methods are straightforward, highly selective, and can be used for most elements. All excel as trace methods for metals and some nonmetals, and instrumentation costs are moderate.

14.2 NEBULIZER–BURNER AS SAMPLING MODULE

The dissociation of analytes into free (gaseous) atoms, a process discussed in Section 13.4 for atomic emission spectrometry, is also required in atomic absorption

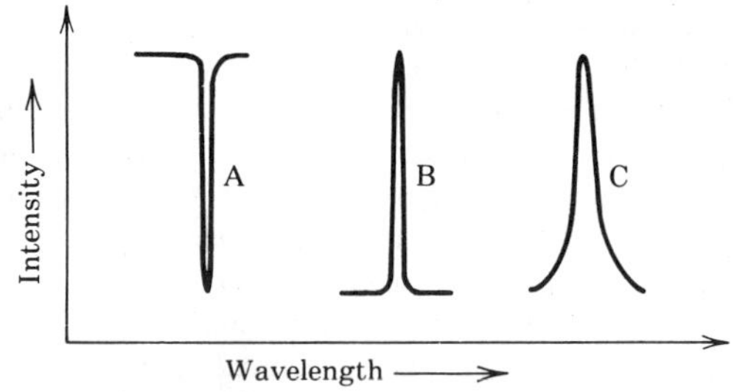

Fig. 14.2 Relative bandwidths of absorption and emission bands. Curve A, an atomic absorption band of element X; curve B, a hollow-cathode lamp emission band; curve C, bandwidth of output of a monochromator. By superimposing curves A and C it is clear that only a fraction of a monochromator band will be absorbed by gaseous atoms of element X, while essentially all of the hollow-cathode lamp emission will be so absorbed.

and fluorescence spectrometry. Since these new methods require atomization with minimal excitation or ionization, lower temperatures of atomization are sufficient and a nebulizer–burner sampling module is often used.

To reduce interelement effects, solutions are widely used as a way to present analyte samples. Solutions are then introduced into a *nebulizer* to produce a spray of fine droplets called an *aerosol*, which can be directed to a burner. The *pneumatic nebulizer* is the simplest and most common type. It generally uses the oxidizing gas for the burner to aspirate the solution and form a spray. Such an arrangement is evident in Fig. 14.3. Unfortunately, *at most* 10% of a sample appears in the flame! As will be seen, special steps can be taken to improve the efficiency of this process. Alternatives such as ultrasonic nebulization also find use.

The sample aerosol usually has an undesirably wide distribution of drop sizes. Larger droplets seldom have time to evaporate in a flame and can be removed by obstructing the path of the aerosol with baffles. Larger droplets will collide with them, coalesce, and can be drained away. A further problem is that all the solvent in drops must be evaporated. Anything hastening this process, such as heating the nebulizer chamber, automatically ensures that a larger fraction of a sample will interact with a flame.

An ideal nebulizer–burner combination should:

1. Convert in the nebulizer as large a fraction of a sample solution into small droplets as possible.
2. Eliminate large droplets of solution and preferably evaporate much of the solvent in other droplets before they reach the flame.
3. Be designed so that cleaning and adjustment are easy, solution viscosity effects are small, there is minimal crust or residue left by volatizing samples, and corrosion by even HF–HNO_3–HCl mixtures is small.
4. Respond rapidly to new samples.
5. Produce a stable flame in the burner, that has minimal luminosity, absorption, and flicker and a well-defined internal structure and shape.

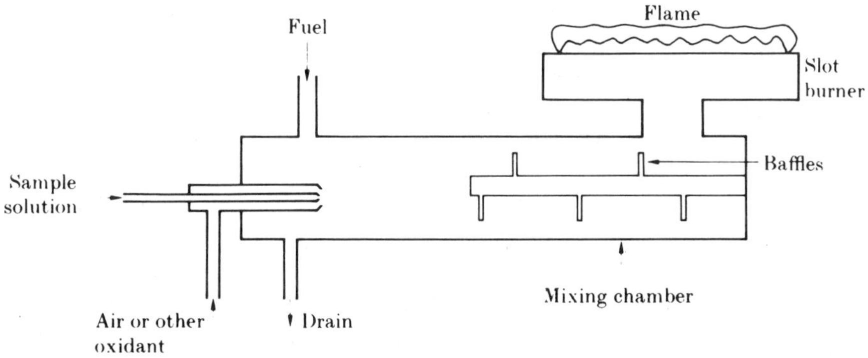

Fig. 14.3 A premix nebulizer-burner arrangement. Baffles in the mixing chamber collect larger drops that strike their surfaces. The burner face is usually flat with a slot perhaps 10 cm long down its center.

6. Prevent flashback.*
7. Produce a quiet flame that provides large populations of atoms from species introduced.

The separate aspects of nebulizer–burner design may now be considered.

Gas Regulation. Both fuel and oxidant gases, including air, are supplied from cylinders. A two-stage regulator for each, today often computer controlled, ensures a steady gas flow. Once a burner has been chosen, a process discussed below, two properties of flames affect the selection of appropriate gas pressures. The concentration of fuel gas in the mixture must fall between the upper and lower limits of flammability with the oxidant to sustain combustion. For example, the volume percent of fuel in an air–acetylene mixture must be between about 10 and 90%. Unless otherwise stated, it can be assumed that nearly stoichiometric mixtures of combining gases are used.

Second, gas pressures must be appropriate to the size of the burner channel and the *burning velocity* of the mixture. The term defines the speed of propagation of a flame front in a gas mixture, a value that depends somewhat on burner design and conditions such as pressure. Burning velocities are usually of the order of 1–10 $m\ s^{-1}$ at atmospheric pressure. (Detonation velocities, realizable when reaction is initiated by a shock wave, are usually about 100 times greater.) It is found that a flame can be maintained at a burner head only if the fuel gas or combustible mixture moves at a speed 3–10 times the burning velocity. Higher gas velocities cause a flame front to blow away, lower ones can lead to flashback which, unfortunately, can lead to explosion.

Flashback is most likely during changes of conditions, for example, during ignition, alteration of pressures, or extinguishing a flame. Precautions are straightforward—with C_2H_2–N_2O flame flashback can virtually always be avoided if on starting, the C_2H_2 is ignited first and then the N_2O turned on; on shutting down, the N_2O is first turned off and then the C_2H_2 flame is extinguished. Mixing is usually governed by a microcomputer today.

Burner. In the widely used *premix burner* shown in Fig. 14.3, gases are blended prior to emergence from the burner head. The burner yields a *laminar* (i.e., layered) flame that has low luminosity and little flicker (noise). Most theoretical studies have been made on such flames since thermodynamic equilibrium is attained in their central zone, as discussed in the next section.

Other advantages of the premix burner are that solution viscosity is no particular problem since the solution enters through a large tube, and big droplets can be eliminated before they enter the burner slot. Because of the mixing chamber there is only modest dependence of signal on sample flow rate. The device is also mechanically simple.

A premix burner head usually is about 10 cm long and has one flame slot, as shown in Fig. 14.3. The design gives absorption intensities proportional to flame length. Air is entrained in controlled fashion and carbon build-up when fuel-rich flames are used is avoided. In the Boling type of premix burner three closely ad-

*Flashback occurs when flame propagates within the burner.

jacent, parallel slots are used in a flat surface; this has the advantage that a large central portion of the flame is protected against the unsettling diffusion of air, the outer slots providing a type of peripheral flame curtain.

There are two possible disadvantages of a premix burner. When a solvent mixture such as alcohol–water (see Section 14.7) is used, selective evaporation may cause sample to remain in the unevaporated portion of the drops. This potential difficulty is avoided if almost all solvent is evaporated from the drops prior to their entering the flame. There is also a minor possibility of flashback explosion. To lessen this danger, oxidants such as air or nitrous oxide (with acetylene) are employed instead of oxygen. Good design can minimize the volume occupied by the mixed gases and further lessen the danger. Automatic optical monitoring of the flame and use of a mixing chamber with a blow-out diaphragm that ruptures at a low excess pressure are worthwhile additional safeguards.

Another early type of burner-nebulizer, the total consumption burner, in which solution samples are aspirated directly into the flame, is today in occasional use in flame emission work. It is simple mechanically and introduces all of a sample into the flame. The latter feature would be highly advantageous were it not for the burden of evaporation and atomization it puts on the flame. When massive sample introduction occurs, one obtains a cooler, turbulent flame that is both instrumentally and acoustically noisy, and somewhat luminous.

14.3 FLAMES AND SAMPLE ATOMIZATION

In Table 14.1 several analytically useful flames and their approximate maximum temperatures are listed.

Table 14.1 Some Analytically Useful Flames[a] and Maximum Temperatures[b]

Fuel and Oxidant	T_{max} (K)[c]
Hydrogen–air	2300
Hydrogen–oxygen	2950
Methane–oxygen	2950
Acetylene–air	2450
Acetylene–oxygen	3400
Acetylene–nitrous oxide	3200

[a]Oyxgen is seldom used as the oxidant to lessen possibilities of explosion. For minimum flame luminosity, however, the oxy-hydrogen flame is preferable to the air–acetylene or oxy-acetylene flames.

[b]Maximum experimental temperature determined in the interzonal region of premixed flame formed from stoichiometric mixtures at 25°C and burning in the atmosphere at 1 bar total pressure. Stoichiometric mixtures give the hottest flames.

[c]Temperature measurements are made on flames (see Section 13.4) in which gases have been well dispersed before burning, although there is no limitation in principle.

Atomization. For sensitive spectral measurements atomization should be as complete as possible. Collisions with energetic gas molecules are principally responsible for atomization. Thus, the hotter a flame, the more effective the process. Flames must also be stable and something must be known about the background levels of chemical species derived from the flame gases as well as those derived from the sample. Not all parts of a flame will be equally valuable for atomization or for observation of an atom cloud spectrometrically (see below). Small concentrations of radicals and molecules such as OH and C_2 will exist at various heights in flames along with analyte atoms, ions, excited atoms and ions, and even analyte monoxides.

For a given flame at what height and to what extent can the concentration of analyte atoms be optimized? What variables should be controlled? Usual thermodynamic criteria can be applied to atomization processes to the extent that equilibrium is obtained. For example, in the atomization of NaCl any enhancement in the concentrations of counter-atom species such as chlorine will tend to decrease dissociation. While a high temperature will ensure that atomization is as complete as possible, the greater the temperature, the stronger the tendency for excited species and ions to be produced. A compromise is indicated. Finally, formation of analyte monoxides can be reduced by increasing the proportion of fuel (a reductant) in the gas mixture, though a lower temperature will result.

Flame Structure. Against this background, it is important to ask how closely concentrations of chemical species are related to flame "structure." A premix flame is sketched in cross section in Fig. 14.4. Given its reasonably stable structure, it is possible to appraise different regions in terms of atomizing efficiency and concentrations of interfering species.

Extending upward from the burner slot there is a familiar A-shaped flame front in which most chemical combustion occurs. This region, which is about 0.1 mm thick is called the *primary combustion zone*. It appears as a blue–green or blue–white layer. Its color and luminosity result from both band (closely spaced lines) and continuous emission by molecular fragments such as C_2, CH, and CHO. The strong release of heat during reaction means that most expansion and heating of the flame gas also takes place in this thin zone. As an analytical sample (introduced into the gases earlier) reaches the primary zone, it is vigorously atomized as its molecules are bombarded by highly energetic gas molecules.*

As shown in the figure, above the blue primary zone and extending upward from it from one to several centimeters is an *interzonal region*. Its height depends mainly on burner design and rate of gas flow. Sometimes the terms interconal zone or central zone are also applied to this region. The middle part of the interzonal region along the length of the flame (usually about 10 cm) is commonly used for flame spectroscopic measurements since in addition to a long path length, the region offers a maximum concentration of sample atoms (or excited atoms), low

*Excited species also form in the primary zone. The region is little used in flame emission spectroscopy, however, because of the difficulty of detection of emission against the strong background.

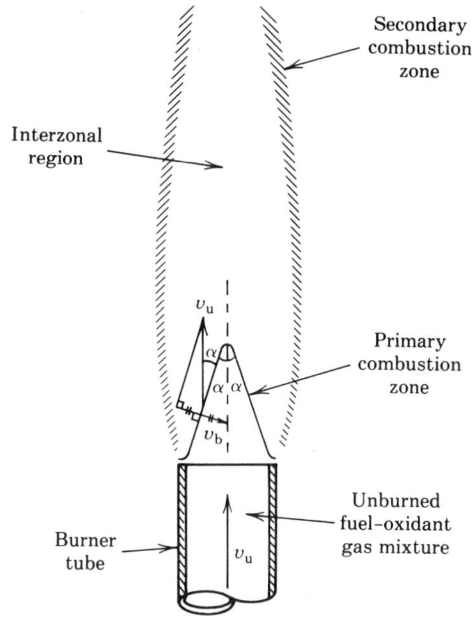

Fig. 14.4 A cross section of a premixed laminar flame taken perpendicular to the slot of the burner. The characteristic regions are discussed in the text. The velocity of the unburned fuel–oxidant gas mixture v_u must be greater than the burning velocity v_b to avoid flashback. For a given flame, the value of v_b may be found by taking the component of v_u that is perpendicular to the flame front or primary combustion zone. (After Alkemade et al. [9].)

luminosity, steady-state conditions, and reasonable homogeneity.* For lower-temperature flames a serious drawback is that O and OH radical concentrations in this region are sufficient to cause many metal atoms to appear as monoxides (MO). By contrast, the high-temperature nitrous oxide–acetylene flame greatly reduces this problem.

Finally, the external border of the flame is called the *secondary combustion zone* or outer zone. Further reactions ($CO \rightarrow CO_2$, etc.) occur and release heat that may offset the loss that occurs as cold surrounding air penetrates this region by diffusion and turbulent mixing.

Lengthening the Central Region. Two ways to lengthen the interzonal region and make it more useful analytically have been investigated. The first approach is to use a *fuel-rich premix flame*. In this case molecular fragments such as C_2 persist in the flame, lending it a reducing character. As a result, the concentration of oxides of elements present in the flame is lower and populations of both free atoms and excited atoms is higher. Another approach used mainly for cylindrical burners is the formation of a *separated flame* by fitting a tubular

*Near-equilibrium is attained among molecular motions in the roughly 100 μs required for species to move 1 mm through this region because there is time for many elastic and inelastic collisions.

silica sleeve of 2–4 cm length to the top of the burner, or by using a surrounding sheath of flowing gas (nitrogen or argon). The flame temperature drops somewhat in a separated flame, but background emission from burning gases is reduced one or two orders of magnitude and oxide and hydroxide formation is lowered by forestalling the diffusion of atmospheric oxygen into this region. Compare this approach with that of the Stallwood jet in Chapter 13, which yields improved results by the same technique. Both methods can lead to a significant improvement in specificity and sensitivity.

Interferences. The effectiveness of flame spectroscopic procedures for a particular element depends only in part on external variables such as the fuel gas pressure and width of the burner slot. It is also determined by processes called *interferences* or *interelement* (*matrix*) *effects* already described as occurring in the flame. As implied, these effects diminish the concentration of analyte atoms or lead to absorption or emission at analytical wavelengths. The main types of interferences and general ways of combating them are summarized in Table 14.2. Some effects (especially 3–5) are specific for a given element; others are nonspecific. Still others (1, 2) affect a blank as well as a sample.

The interferences affecting atomization are the most troublesome. The category includes chemical (so-called cation or anion) interferences and the tendency of some elements to form stable monoxides in a flame that affect all flame techniques. Interferences that alter the fraction of ionized atoms are important for easily ionized analytes such as alkali metals.

Example 14.1. In air–acetylene and other lower temperature flames aluminum sulfate and phosphates suppress the production of gaseous calcium atoms. Suggest the causes of this effect. How may the effect be reduced?

Both species appear to contribute to poor atomization of calcium through a chemical effect. Aluminum does so by forming $CaAlO_2$, which is difficult to dissociate, in the flame. Similarly, phosphate ions depress the atomization of calcium and other alkaline earth metals through phosphate formation. The effect may best be minimized by using a higher temperature N_2O–C_2H_2 flame.

Example 14.2. How troublesome are ionization interferences in the quantitative measurement of alkali metals by atomic absorption? How may this type of error be reduced? Will it be more serious for an air–C_2H_2 or and N_2O–C_2H_2 flame?

If α represents the fraction of analyte atoms ionized and β the fraction excited, $(1 - \alpha - \beta)$ will represent the fraction available for atomic measurement. Anything that changes this fraction between the standards used for calibration and samples will produce this sort of error. Almost always ionization is by far the larger factor and excitation may be neglected.

As Table 14.2 suggests, a large amount of an easily ionized element that does not interfere spectrally can be added to samples as a buffer or *ionization suppressant* to stabilize the percent of ionization. Its presence at about 100–1000 $\mu g\ L^{-1}$ concentration becomes especially important in hotter flames that produce more ionization, such as N_2O–C_2H_2.

Table 14.2 Sources of Interference and Error Associated With Flame Atomization[a]

Type of Error	*Cause of Error; Technique Affected*	*Method of Reducing Error*
1. Chemical (or cation and anion) interference	Formation of stable compounds of analyte that are atomized with difficulty in flame; AA, FE, AF	Use hotter flame or fuel-rich flame; avoid certain anions; use solvent extractions or ion exchange to effect prior separation of offending ion; chelate the analyte
2. Background (or nonspecific) interference	(a) Emission by flame gases; FE (much less important for AA and AF); emission by analyte; AA and AF	Change flames or use other analytical line for FE; modulate source emission in case of AA and AF and selectively detect analytical signal
	(b) Absorption and light scattering by molecular species in flame; AA, AF (important at low concentrations)	Correct for background by deuterium arc, Zeeman, and other methods; between 190 and 175 nm use separated flame with Ar shielding
3. Ionization interference	Ionization of analytes; AA, FE, AF	Add easily ionized nonanalyte; lower flame temperature
4. Excitation interference	Flame temperature change as solvent composition changes or high analyte concentration appears; possible encrustation of solids on burner; AA, FE, AF	Ensure constant solvent composition; wait for steady state; reduce concentrations
5. Spectral (or radiation) interference	Emission or absorption of another atomic species or its oxide within waveband passed; FE (much less important for AA, AF)	Use narrower slit or select another intense analytical line in a spectral region free of such interference

[a] It is assumed that instrumental variables such as spectral slit width and gas pressures will be set appropriately. Abbreviations used are: AA, atomic absorption spectrometry; FE, flame emission spectrometry; AF, atomic fluorescence spectrometry.

Example 14.3. Sodium and potassium are to be determined in a series of samples by atomic absorption and cesium is added at about $100–10^3$ μg L^{-1} as an ionization suppressant. How will it achieve this effect?

If sodium and potassium are present, a certain fraction of each determined by their relative ionization energies will appear in a flame as ions and be unavailable for flame spectroscopy. When a more easily ionized element like cesium is added, it adds to the population of electrons and thus depresses the ionization of the other two elements. Thus, the atomic absorption or flame emission of these elements is enhanced, though to different degrees. Since potassium also has this effect on sodium, and vice versa, the presence of cesium will ensure minimal ionization-induced error if the relative concentrations of potassium and sodium in samples vary considerably.

14.4 ELECTROTHERMAL ATOMIZER AS SAMPLING MODULE

For the lowest possible detection limits the electrothermal or resistive furnace atomizer was a second exciting development in sampling modules for atomic absorption spectrometry. The heart of this device is most often a hollow graphite rod or cup in which one or more drops of solution sample can be volatilized by resistive heating. When the temperature is carefully programmed, ultratrace samples may usually be analyzed without preconcentration. In addition, microsamples can be easily handled. Since electrothermal atomizers are complex to use, they find employment mainly when samples are too small for flame atomizers and when their extra-low detection limits are valuable. Other distinctive modes of atomization such as the decomposition of hydrides of nonmetals and some low-activity metals are treated in Section 14.7 as a part of the discussion of quantitative absorption methods.

Figure 14.5 shows a representative adaptation of the furnace atomizer devel-

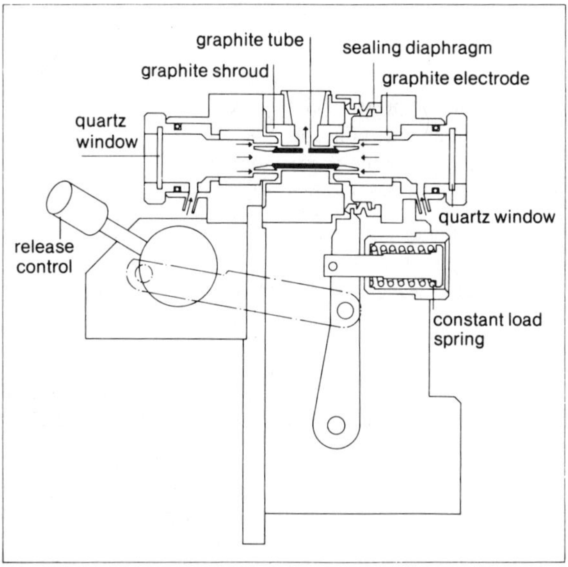

Fig. 14.5 A representative electrochemical atomizer. The carbon tube in which a sample is ashed and then atomized is a simple hollow graphite cylinder open at both ends. The small hole at the top is the injection port for samples and, during operations such as ashing, it vents gaseous products. Note that the tube is held by graphite end cones, which are supported in water-cooled blocks. Electrical current resistively heats the tube. An inert gas, usually argon, enters from the ends of the tube and flows through and around the graphite tube to protect it from burning and remove ashing products. A hollow cathode source directs a beam of monochromatic light down the axis of the assembly. (Varian Model GTA-95 graphite furnace, courtesy of Varian Instrument Group.)

oped by Massmann. The furnace itself is a graphite tube of high purity that is held between two spring-loaded clamps to ensure good electrical contact. End windows transmit the optical beam. Samples of about 1–5 μL are introduced into the center of the furnace, often into a graphite boat. During observation the tube is bathed in a flow of inert gas (usually argon) to exclude the atmosphere.

Other high-melting, electrically conducting substances can also be adapted as atomizers. A V-shaped tungsten filament serves well. Some means of guiding the micropipet by which a solution sample is deposited on the filament is needed.

In electrothermal atomization a temperature program is developed for each analyte and kind of sample. A high degree of automation is required to achieve reproducibility. To be certain that desired temperatures are reached, the furnace temperature is monitored, often by transmitting its radiation through a fiber optic channel to an outside optical pyrometer. Comparison of the furnace radiation with that of an internal heated filament in the pyrometer provides current adjustment through a feedback circuit.

Example 14.4. In the usual furnace *temperature program*, an initial period of slow warming to evaporate solvent is succeeded by a high-temperature step for sample ashing, followed by a second much higher temperature step for atomization (up to 2200 K). During atomization the flow of gas is reduced or stopped so that analyte atoms will remain a maximum time. A representative temperature program and an accompanying optical absorption curve for a sample are shown in Fig. 14.6. Note the brevity of the dissociation period; sometimes there is no more than a 10-ms window in which to measure analyte absorption or the absorbance will be spuriously low. The program concludes with a very high-temperature "heating out" step to clean out the furnace and then returns the temperature to ambient for long enough to return to the original baseline.

Example 14.5. Where a *preconcentration of sample* is sought, programming can be arranged so that analyte from several drops may be accumulated in the furnace. For this purpose, a modified temperature program is used that provides only steps through "ashing" while drops of sample are deposited. Once sufficient solid sample has been accumulated, a single dissociation step produces the desired stronger analytical signal. In Section 14.7 the adaption of this procedure for quantification by standard addition is described.

The electrothermal atomizer offers high analytical sensitivity and very low limits of detection. Measurement of analyte amounts of the order of nanograms is possible. Some disadvantages are the virtual necessity of operating such atomizers automatically, the need to have an instrument designed to accommodate such controls, and a greater measurement time.

14.5 CRITERIA FOR SPECTROMETER DESIGN

The modular diagram of Fig. 14.1 suggests a place to begin an analysis of the characteristics of modules that will most affect the performance of an atomic ab-

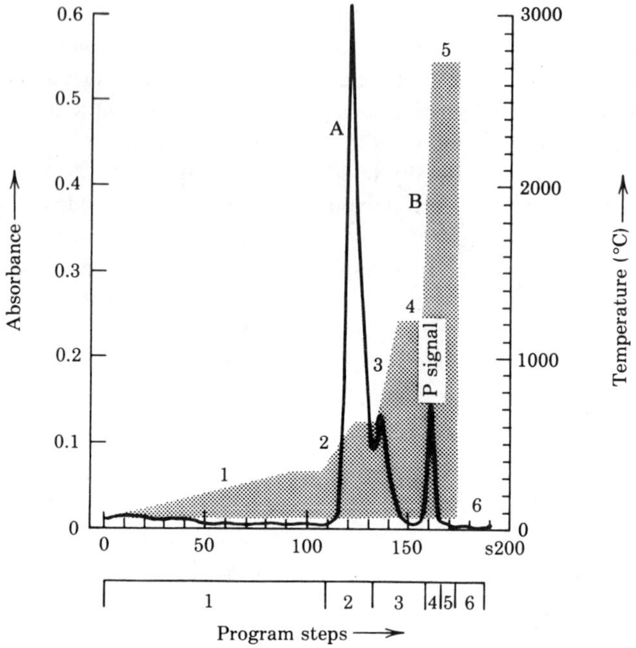

Fig. 14.6 Temperature program for an electrothermal atomizer used for determination of phosphorus in soya oil. Correlation of optical absorbance, curve A, and programmed furnace temperature, curve B, as a function of time. The six steps in the temperature program are as follows: (1) a long, slow-temperature ramp for drying a sample; (2, 3) faster ramps for sample decomposition to remove matrix without "explosions"; (4) brief steady sample atomization at 1200°C during which the internal furnace gas flow is cut to 50 mL min^{-1}; (5) a steady heating out at 2700°C with full gas flow of 300 mL min^{1}; (6) return to ambient temperature. Note that the sample peak appears *as the temperature is increased* to the final plateau. (Courtesy of Perkin-Elmer Corporation.)

sorption spectrometer. The goal will be stating criteria by which to select modules that are appropriate for measurements on different kinds of samples. Flame emission and atomic fluorescence spectrometers will be examined later in the section.

Atomic Absorption Spectrometer. Where absorption is being observed, Beer's law (see Section 16.2) will describe the process as long as the light irradiating the free atom gas is monochromatic and the analyte atoms absorb independently: $P = P_0 e^{-abc}$, where P_0 and P are the incident and transmitted radiant powers (or intensities), a is the analyte absorptivity, b is the pathlength, and c is analyte concentration. It will be assumed that (a) a hollow-cathode lamp (see Section 8.4) or other narrow-line monochromatic source provides the incident P_0 and (b) concentrations of analyte atoms in the burner flame are sufficiently small that they absorb independently. Further, it will be assumed that transmitted light power P will be corrected for atomic emission at the wavelength of the analyte absorption line. First, the constant a and c from Beer's law must be defined for a gaseous system and the effect of the monochromator on the power from the optical source deduced.

It can be shown [17] that the signal produced by the detector is

$$\text{signal} = P_0 \exp\left\{ -\left[0.037\lambda_0^2(g_2/g_1)\, A_{21}(\delta/\Delta\nu_D)\right] L(c\phi\epsilon\beta/e_f Q)\right\} \cdot f(\theta, \lambda)\, g(\lambda). \qquad (14.1)$$

The exponential term, which appears in braces, has been subdivided into factors that represent the original Beer's law variables. These terms will be discussed below. The first *post*exponential term $f(\theta, \lambda)$ defines the monochromator throughput (see Section 9.3) in terms of the solid angle θ of the incident beam appearing at its entrance slit and monochromator response as a function of wavelength; the second term, $g(\lambda)$, represents the efficiency of the detector as a *function* of wavelength.

Exponential Terms. In the exponential the long-bracketed factor represents a, the analyte "atomic absorptivity." Several special points must be made. First, a is defined at λ_0, the wavelength at the *center* of the atomic absorption line for the transition from electronic level 1 to 2. Clearly, a could also be given for any other atomic transition. Second, the width of the source line (band width of incident light) is assumed negligible. Third, the absorptivity has been defined in terms of spontaneous emission from state 2 to 1 as is suggested by the appearance of terms such as A_{21}. The lone numerical coefficient includes a variety of factors. Finally, two variables relating to the width of the absorption line are included. The first is $\Delta\nu_D$, the Doppler half-width (in s^{-1}) of the absorption line, that is, the broadening due to movement of atoms relative to a fixed point. This term appears in the denominator since, the wider this line, the smaller is the degree of absorption of the incident radiation. The second factor, δ, which is always a fraction, represents a broadening due to other causes such as increased partial pressure.

In the rest of the exponential expression we find L, the *path length* in the flame, and $(c\phi\epsilon\beta/e_f Q)$, a factor defining the concentration of analyte atoms in the flame. Each of its variables requires definition. First, c is the analyte concentration in the sample solution. Second, ϕ is the rate of introduction of that solution into the flame. Third, ϵ is the efficiency of that introduction, a combination of nebulization and evaporation efficiencies. Fourth, β is the efficiency of atomization. Fifth, the product $e_f Q$ in the denominator expresses the flow rate of unburned gases, and the expansion factor e_f takes care of the increase in volume on combustion and expansion in the flame.

Several assumptions have been made about the burner flame: (a) that thermal and chemical equilibrium obtain, a condition that holds reasonably in the central region of the outer core of a premix flame; (b) that the flame is optically thin, that is, does not self-absorb or fluoresce (clearly, the former assumption is valid only at low concentrations); (c) that is will be possible instrumentally to correct for analyte emission; (d) that flame luminosity and scattering can be neglected.

In absorption measurements it is customary to take the ratio P/P_0, the transmittance, or obtain log P_0/P, the absorbance. A full discussion will be given in Section 16.2. The conversion of Eq. (14.1) to either form is straightforward. For example, since $A = abc$, the absorbance equals the exponent in eq. (14.1).

While flame absorbances should vary linearly with solution concentration, linearity is usually obtained only over one or at most three orders of magnitude. At increasing partial pressure of atomic vapor, the absorption line broadens, resulting in a smaller δ and reduced absorption at the center of the line. While temperature does not appear explicitly in the equation, the probability coefficient A_{21} varies as $T^{1/2}$. A great source of variability in signal is the set of factors ϵ and β, which depend mainly on chemical interferences as considered in Section 14.3.

Flame Emission Spectrometer. In similar fashion inquiry can be made as to the influence of characteristic modules in a flame emission spectrometer on overall performance. The modular diagram in Fig. 13.10 can serve as a basis for this analysis. It can be shown [17] that the following expression describes their influence if assumptions cited in the next paragraph are valid:

$$\text{signal} = (c\phi\epsilon\beta/e_f Q)L\left[h\nu_0 A_{21}(g_1/g_2)\exp(-E_{21}/kT)\right] f(\theta, \lambda)\, g(\lambda). \tag{14.2}$$

Most of the terms have just been defined, for example, the concentration of ground-state atoms in the flame (in parentheses) is multiplied by flame path length L. The following bracketed term gives the intensity of emission by an excited atom in going from state 2 to 1. Note $h\nu_0$, the size of the quantum emitted; A_{21}, the Einstein coefficient for spontaneous emission; and the terms defining the fraction of atoms excited.

Basically, Eq. (14.2) predicts a linear dependence of signal on concentration and flame path length at a given flame temperature and a low solute concentration. What can be learned from Eq. (14.2) about measurement procedures or modular performance? It is found that for a given element ϵ and β vary with chemical environment, that is, specific "chemical effects" alter response, as might be expected. Such factors affect procedures and will be treated in Section 14.7. Second, the exponential dependence on temperature suggests that the choice of flame will be a crucial "instrumental" decision. Third, background intensity will have to be corrected for whenever it is appreciable. Finally, the central role of burner and nebulizer is emphasized. Flame emission spectrometers will be discussed in Section 14.8.

Atomic Fluorescence Spectrometer. Optical fluorescence spectrometers, whether intended for atomic or molecular fluorescence measurements, are sufficiently similar that a single treatment of design criteria will be found in Section 15.3.

14.6 ATOMIC ABSORPTION SPECTROMETERS

Figure 14.7 gives a modular diagram of a single-beam atomic absorption spectrometer.* It will now be valuable to examine choices for each module in terms

*The term spectrophotometer can also be used for this instrument but seems not to connote its versatility (it can easily be adapted to emission measurements) or the careful attention to wavelength it provides.

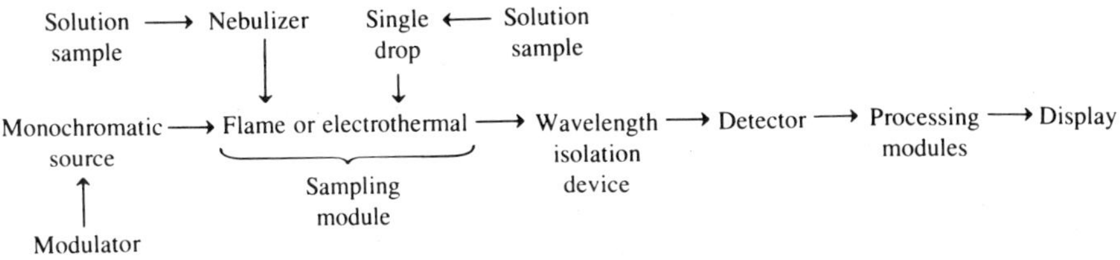

Fig. 14.7 Modular diagram of a single-beam atomic absorption spectrometer. For example, the modulated output of a hollow-cathode lamp can be directed through the interzonal region of a burner flame containing an atomized sample to a Czerny–Turner monochromator. Its first mirror will present a collimated beam to a planar grating and the second collect light diffracted from the grating and focus it at the exit slit, through which it will pass to the detector.

of the needs of different applications. Along the way means of diminishing systematic error will be fully treated. The section will conclude with a brief discussion of double-beam atomic absorption spectrometers and flame atomic emission spectrometers.

Sources. Hollow-cathode lamps for individual elements are commonly used as optical sources. For certain sample types multielement hollow-cathode lamps are attractive. In some instances, a brighter source such as a microwave-excited or electrodeless lamp, may be preferred. Whatever the source, its output is modulated for reasons that will be discussed below in connection with detection and processing. A square wave may be imposed on the dc power energizing the lamp or the beam may be chopped. It is important that the lamp current be large enough to give sufficient intensity for good sensitivity in measurement but not so high as to cause self-absorption and reduced lamp life.* Extraneous emission lines of the analyte element are subsequently blocked by the monochromator. The striking advantages of these sources are high brightness, establishing the wavelength precisely, high monochromaticity, and freedom from wavelength drift.

Monochromators. Since atomic spectra supplied by hollow-cathode lamps are not line-rich, a fairly low dispersion monochromator is often adequate. It must have sufficient resolution (0.2–0.5 nm in most systems) to pass only the selected hollow cathode line. Note that the actual optical bandwidth will really be *determined by this line*, which will be as narrow as 0.002 nm, guaranteeing minimum spectral interference. Neither is it essential that wavelength accuracy be superb in absorption work; the monochromator can be directed by a microcomputer program to scan through a range centered on the selected wavelength of a hollow cathode lamp and then select the peak wavelength for measurement.

To maximize instrument throughput monochromator entrance and exit slits can be kept moderately wide for the reasons just outlined. In some designs achromatic lenses can also be substituted for mirrors with some further gain.

*Appropriate working (and maximum) currents are specified on lamps for at least a pair of useful analytical wavelengths. Because lamps are modulated they do not run constantly at the specified current and life is extended.

When a particular instrument is to be used both for atomic absorption and flame emission measurements, better monochromator resolution (about 0.05 nm) will be desirable since emission spectral interferences are more numerous. Good wavelength precision in setting will also be needed.

Sampling Modules. Fortunately, a nebulizer–burner sampling module without background correction proves satisfactory for the large majority of samples. For measurement of trace-level constituents or for microscopic samples, an electrothermal atomizer and background correction will be needed.

An appreciable systematic error can, however, arise from losses caused by the presence of larger molecules or particles in the optical path. If the *unwanted absorption* and *scattering* they cause reduces beam intensity appreciably relative to analyte absorption in a flame, graphite furnace, or other atomizer, the losses must be corrected for. In general, correction is needed in measurements made

1. at the trace or ultratrace level;
2. on samples containing high concentrations of "inert" salts, which will vaporize as clusters of ions;
3. under conditions such that other background or absorption will occur at an analytical wavelength.

Though devices to correct for background differ in particulars, most employ time-sharing to permit two different beams to traverse the sampling module and yet be distinguished at the detector: the flame or furnace analyzer cell serves as a blank as well as a sample cell. One part of each modulation cycle is given to the (measuring) beam from the hollow cathode and another to a different beam that responds mainly to background.

Example 14.6 How important will background correction be in the UV compared to the visible? Will correction be more important for flames or electrothermal sampling modules?

Background corrections are much more important in the UV. One reason is that scattering increases as the fourth power of the light frequency. Further, molecular absorption is likely to be greater in the UV. Correction in the visible is seldom necessary.

Since total samples are vaporized in electrothermal or furnace atomizers, background correction is a virtual necessity. Smokes of solid particles as well as vapor with significant molecular density are common. These will scatter and absorb light.

Deuterium Arc Corrector. One of the simplest arrangements for subtracting background is to direct light from a deuterium arc (hydrogen continuum lamp) through the sampling module along the path taken by radiation from the hollow cathode lamp. A representative dual-beam instrument incorporating this type of background correction is shown in Fig. 14.8. When a background correction is to be made, the deuterium arc lamp is turned on and modulated at the same frequency as the hollow cathode lamp, but 180° out of phase with it. Clearly, corrections are possible only in the UV range where the arc provides intense radiation.

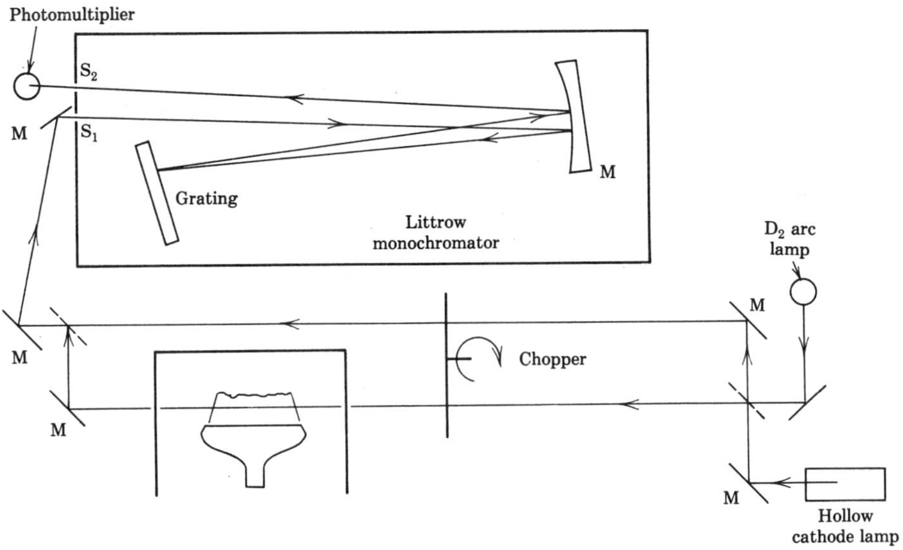

Fig. 14.8 Optical schematic of a double-beam atomic absorption spectrometer with a D_2 background corrector. The beam splitter (broken line) establishes the sample beam and reference beam. The chopper blocks the sample beam while light is going through the reference beam and vice versa. The beams are recombined just before entering the monochromator. Only a single ray is shown, though actual beams illuminate fully the mirrors M, grating, and other optical elements. Slits are designated S. When background correction is sought, the deuterium lamp is turned on, its intensity is adjusted to equal that of the hollow-cathode lamp, and it is pulsed out of phase with the hollow-cathode lamp. The intensity of the deuterium beam through the flame cell is first subtracted from that of the hollow-cathode beam for both sample and reference beams before the ratio is taken. (Model 2380 Atomic Absorption Spectrometer, courtesy of Perkin-Elmer Corporation.)

Radiation reaching the detector at different times comprises

1. the modulated (hollow cathode) analytical wavelength, whose amplitude is its initial value less that absorbed by analyte and scattered or absorbed by background;
2. the modulated broad-band deuterium signal, whose amplitude is the initial value less background loss;
3. and when no sample is present, the incident amplitude from the hollow-cathode lamp or deuterium arc less background loss.

The difference between the hollow cathode absorbance and that for deuterium is the absorbance due to the analyte. Clearly, the deuterium beam intensity must be matched to that of the hollow cathode beam to ensure a valid correction.

Advantages are low cost and simplicity. Disadvantages of the deuterium lamp corrector are that it averages a correction over the spectral slit width of the monochromator, not just the hollow cathode line width. If the width is as great as 1 nm for greater throughput, it may fail to correct accurately for a background varying strongly with wavelength, which is called a structured background. Further, throughput of the hollow cathode wavelength is reduced 50% by a necessary beam splitter.

Smith–Hieftje Corrector. This approach relies on development of the modulated correction beam by short pulses (duration, ~5 ms) of high current through the hollow cathode lamp itself. Such pulses do not appreciably reduce the life of the tube. Under high current the hollow cathode line broadens greatly, permitting correction for background immediately adjacent to the analytical wavelength. Since broadening is accompanied by self-absorption, the background beam will have a minimum amplitude *at the central wavelength*. In this procedure also, corrections are obtained as shown above. There is no difficulty in keeping track of the phase since the high-current pulses are precisely timed.

The advantages are that corrections are made simply and very near the analytical wavelength whether in the UV or VIS, and costs are smaller since fewer power supplies, beam-splitters, and so on are required. The disadvantages are that the measurements are made with less sensitivity than with a deuterium lamp system and hollow-cathode lamps of slightly different design (to avoid arcing) must be used.

Zeeman-Effect Corrector. The Zeeman splitting of spectral lines is the basis for another type of corrector [11]. The imposition of a dc magnetic field across the sampling module or the hollow-cathode source causes each absorption line to split into several components of different wavelength and optical polarization.* The central component (line) retains the original wavelength and is polarized parallel to the magnetic field ($\parallel$ component), while the new lines on either side are polarized perpendicular to the direction of the magnetic field ($\perp$ components). Different beams traveling the same path can be created on a time-sharing basis by inserting a rotating linear polarizer that will pass in turn the two components. Subtraction of the intensity of the $\perp$ component from the $\parallel$ one gives the desired corrected intensity. It is also possible to omit the polarizer entirely, place the electrothermal atomizer between the pole pieces of an electromagnet, and simply cycle its magnetic field. When it is OFF, the $\parallel$-component appears, and when ON, the $\perp$-component.

The advantages of the Zeeman arrangement are that the basic beam intensity remains constant between the main measurement and the background correction measurement and that correction is possible at any analytical wavelength. The system is also reproducible and precise. Its disadvantages are that space in the sample compartment is reduced to accommodate the magnet if a polarizer is used; throughput, and thus sensitivity, is reduced and costs also tend to be higher. A modest disadvantage is the bending back of the calibration curve toward the concentration axis at high analyte concentrations. With increasing concentration, the side $\perp$-bands broaden and begin to absorb some of the central (analyte) wavelength. Thus, the correction beam begins to include a fraction of analyte absorption. As a result, the working curve is overcorrected and soon turns downward. The effect is especially pronounced at the level of concentrations likely in a flame atomization cell and may limit the upper useful absorbance to about 0.5.

---

*Electronic transitions in atoms are degenerate magnetically and a steady magnetic field causes them to have different energies and polarizations.

Detectors. Recall that UV and VIS wavelengths are to be detected. Photomultiplier tubes with their high gain and sensitivity in these ranges are the logical choice in a majority of cases.

Processing. A dynamic aspect of detection must, however, be taken up. *Emission* at the analytical wavelength by excited analyte atoms and luminosity of the flame will spuriously increase the signal intensity unless prevented. Fortunately, modulation of the dc energizing the hollow cathode with either a sine or square wave of constant frequency, and subsequent lock-in amplification only of in-phase signals of that frequency will cause rejection of analyte emission which is out of phase.

Double-Beam Atomic Absorption Spectrometers. To compensate for drift as well as slower fluctuations (noise) in output of the hollow cathode lamp, the detector, and the amplifier, a double-beam spectrometer design is often used. Both drift and much noise are eliminated and stability of operation is considerably enhanced. Yet with the employment of modern modules of increasing stability, double-beam arrangements are less needed today.

A double-beam spectrometer is shown schematically in Fig. 14.8. Note that the second beam is not through a blank but through air; the design is valued for the dynamic aspects mentioned earlier. The use of a pair of different beams through the flame takes care of the usual compensation aspects of a second beam through a blank.

Atomic Emission Spectrometer. Atomic absorption spectrometers can easily serve as either emission or absorption devices. For example, to convert the single-beam spectrometer shown in Fig. 14.7 to an emission spectrometer, it is only necessary to inactivate the hollow-cathode lamp and insert a chopper between flame and detector to modulate the emission signal at the frequency to which the amplifier is tuned. An alternative is to modulate the detector output. All of the virtues of a single-beam device are available to this instrumentation: high beam intensity, ruggedness, and simplicity. Flame emission spectrometers and methods will be discussed in Section 14.8.

14.7 QUANTITATIVE ATOMIC ABSORPTION SPECTROMETRY

Atomic absorption spectrometry is a widely applicable and generally sensitive technique for the quantitative determination of elements. Most metals, metalloids, and some nonmetals (about 70) can be determined by atomic absorption. Hollow-cathode lamps are available for all and microwave electrodeless lamps for many. Spectrometric aspects of elemental determinations are simplified in atomic absorption spectrometry because the lamps provide exactly the wavelength needed for each analyte at high intensity. Thus, measurements are especially convenient and there are also well-defined analytical procedures for most elements.

For quantitative determination of elements well above the limits of detection, atomic absorption is capable of a precision of ±2%. When a double-beam procedure and background correction is employed, precision can be better than ±1%. For trace determinations approaching the limit of detection and work with minute samples, precision is expectedly poorer. Not only are sampling procedures taxed when determinations are made close to the limit of detection; instrumentation is taxed as well. The technique is excellent for elemental measurements at the trace level.

Perhaps the principal limitation of atomic absorption is the lack of direct procedures for nonmetals. Oxygen and nitrogen (*in the flame* and all beam paths) begin to absorb strongly at wavelengths shorter than 185 nm. Recall that the region is called the vacuum UV for this reason. It is in this range also that the most sensitive lines of other nometals occur. Use of argon plasmas as "flame cells" will permit some nonmetals to be determined in the 190–175 nm region even without providing vacuum paths for optical beams, though it may be helpful to purge them with argon or helium.

The major sources of error arise from the sampling provided in the flame cell or furnace atomizer. Random changes occur in the formation of aerosol, transfer of droplets, volatilization, and atomization. Systematic errors arising from interferences and background will also be involved unless standards faithfully model samples in content. Often pure standards are used, necessitating concerted attempts to reduce interferences as samples are run. Many suggestions to reduce systematic errors were made earlier in Table 14.2 and in the last section. Background effects include (a) absorption of radiation by flame radical species such as OH as well as a variety of gaseous species, (b) scattering of radiation by flame gases, and (c) emission by excited species other than the analyte. Finally, noise contributed by instability in the source and in the flame will be increasingly important as analyte concentrations decrease toward limits of detection. The contribution of hollow-cathode lamps, electrodeless discharge lamps, and slot burners to noise is today much reduced because of their greater stability, allowing measurements at lower concentrations.

Selection of Analytical Wavelength. The choice of wavelength for a determination is straightforward in most cases. Spectral interference with lines of other elements will be unlikely given the narrowness of hollow-cathode lamp lines. In general, the resonance line of an element is the best choice since it offers the highest intensity. When it is desired to avoid regions of excessive flame background or close-lying lines of other elements that can be excluded only partly by the monochromator, other intense lines are possible choices. Another choice is posed when sample concentrations of an analyte will vary over many orders of magnitude and dynamic ranges are short. In this situation, it may be necessary to select for an analyte as many as three lines with different sensitivities and prepare as many analytical curves.

Sample Preparation. In the vast majority of cases samples are presented in solution. Some (e.g., physiological fluids and drinking water samples) are furnished as liquids. For samples that require dissolution, acid digestion or solvent extraction may be the first step. A common sequence is to dry a sample, to free it of organic material by ashing (see Section 11.6), and then to dissolve it.

Liquids with high salt content, such as sea water and blood, offer problems as the solution is evaporated. Encrustation of salts around the slot of a burner when a sample is introduced quickly leads to unsteadiness of the flow of sample and of fuel and oxidant gases and introduces serious errors. In electrothermal atomization salts give rise to smokes and high background correction. A way around the difficulty may be to complex the analyte in the sample solution and extract the complexed species. Actually, extraction offers the dividends of concentrating the analyte, eliminating interferents that are not extracted, and, if an organic extractant is used, enhancing the sensitivity of the measurement. The complexation and extraction procedures need not be selective, only effective for the analyte(s). Hollow-cathode lamps provide the needed selectivity.

Many metalloids and low-activity metals in nonmetal families are most sensitively determined by chemical conversion to hydrides, which are then volatilized and decomposed by heat. The group of elements includes mercury, arsenic, selenium, tellurium, antimony, bismuth, germanium, tin, and lead. Because hydride formation is a selective reaction, they are also effectively concentrated by removal from the original matrix.

Complexing Agents and Solvents for Extraction. Some common complexing agents for metals and solvents are ammonium pyrrolidine dithiocarbamate, which is often used with methyl isobutyl ketone, 8-hydroxyquinoline (with ethyl acetate or butanol), and ethylenediaminetetraacetic acid. In addition, methyl pentyl ketone and pentyl acetate are also useful extractants.

Example 14.7. For the determination of zinc in a sample of brass the optimum zinc concentration level is 0.5–5.0 $\mu g\ mL^{-1}$ if the resonance line (213.9 nm) is used, and 2000–20,000 $\mu g\ mL^{-1}$ for another possible line (307.6 nm). An air–hydrogen flame (2000°C) provides the lowest detection limits and is preferred to the hotter air–acetylene flame. A bandpass of 5 nm in the spectrometer can be tolerated if a zinc lamp is used, but it must be reduced to 1 nm for a brass hollow-cathode lamp, which also produces copper lines (at 216.5 and 217.9 nm).

While interferences are slight even with a 200-fold excess of other elements, zinc is usually complexed with ammonium pyrrolidine dithiocarbamate when determined at the trace level and then extracted into a solvent such as methylisobutylketone. Constant flame conditions are important if zinc is determined at 213.8 nm because of appreciable absorption by flame gases. The extent of absorption should be ascertained and corrected for if significant.

Hydride Generator. This separate, small module, provides chemical conversion of the compounds of metalloids and many low-activity metals to hydrides. For hydride preparation sodium borohydride is added to an acidified sample. For instance, arsenic will combine in such a reaction to form arsine AsH_3.

In practice, the generator cell in which the reduction occurs must first be flushed with Ar or N_2 to remove oxygen. When hydride formation is complete, the hydride is carried slowly on a further quantity of flushing gas into a tube of silica or other high-temperature glass, decomposed by heat, and measured by atomic absorption. Usually the silica tube is placed on top of the regular slot burner.

For mercury, it is customary to arrange reduction of bound mercury by a reducing agent such as $SnCl_2$ in acid solution. The elemental mercury formed may then be volatilized easily by bubbling an inert gas through the solution and passing the gas through a heated tube as discussed above. Detection levels are below 1 ppb. For mercury best results are obtained by circulating the vapor until an equilibrium is attained.

Clearly, the determination of any of these elements in the presence of considerable amounts of organic material or background substances will require special attention. Often organic material must be first digested with a strong oxidizing agent. Interfering substances may be masked if necessary.

Solvent Enhancement of Sensitivity. Sensitivities of flame techniques can be surprisingly enhanced by the use of at least a partly nonaqueous solvent. Both the possible contribution to flame temperature of the enthalpy of combustion of a solvent and facilitation of nebulization and evaporation because of lowered suface tension and viscosity contribute. Care should be taken to ensure that there is not an accompanying increase in interferences.* To retain good solubility for samples, alcohols or ketones are ordinarily chosen.

Calibration Curves. Ordinarily, analytical signal readings are quantified by reference to a working curve, a plot of absorbance versus concentration for each analyte. How should standard solutions be made up? From the listing of possible sources of interference and error in Table 14.2, it can be deduced that making up standard solutions to be as nearly identical in composition to the analytical solutions as possible may be the best way to minimize these effects. For example, neither matrix effects nor the amount of enhancement of sensitivity by solvents can be predicted with assurance. Ultrapure reagents and *solvents* should be used. This proves of considerable importance in the case of organic solvents. Similarly, chemical interference is possible as a result of the particular constituents in solution and some account must be taken of this situation.

Working Conditions. As is true of most photometric measurement systems (see Section 16.4), precision is best when absorbance falls in the range of 0.15 to 1.0. For high precision, concentrations should be adjusted accordingly. If an internal standard is used in a dual-channel instrument, the absorbance range can, of course, be extended with good precision.

If the concentration range of an element in a sample exceeds values that give optimum absorbance, several options are open. One is variation of the optical path length through the flame by rotating the burner (of the slot type) about its shaft. It

*E. E. Pickett and S. R. Korityohann, *Anal. Chem.*, **41**(14) 28A (1969).

should be possible in this way to vary absorbance by as much as a factor of 10. Another possibility is to use a less sensitive line of the element. Indeed, this option might yield a linear calibration curve at higher concentrations. A third strategy is to dilute the solution, though one should be sure that the diluent is free of traces of contaminants! Finally, where somewhat reduced precision is no problem, the concentration range can be extended even though a nonlinear calibration curve is obtained.

Standard Addition. In general, quantification by use of standard addition is more attractive than use of a calibration curve only if samples of relatively low concentration are to be assayed (see Section 11.4). As applied to atomic absorption measurements, a minimum of three solutions is needed: sample solution, sample solution after addition of a known amount of a standard solution, and a blank (solution) to which the same addition is made. It is important that the dilution of each solution be to the same final volume so that interfering substances are at the same overall concentration. In this method a reliable estimate of background absorption by use of the blank is especially important. If background is neglected, its nature should at least be known.

Background Correction. In general, correction for background should be made mainly in cases where concentrations are low or the other conditions cited in Section 14.4 apply. Necessary corrections may be made with the instrument at hand. In any event, the corrections must be determined by measurements at a wavelength close to that of the absorption line.

Trace Determinations. In addition to the use of preconcentration or extraction techniques (see Section 11.6) for enhancement of sensitivity, it is nearly always desirable to shift from use of a flame to an electrothermal atomizer.

In furnace determinations at the trace level background correction is especially important. For development of procedures it is useful to obtain the background correction as a separate signal. As emphasized in Section 14.4, temperature routines for flameless atomization procedures must be devised with care. If background correction can be plotted as a function of time for each trial temperature step, it should be possible to ascertain the best time to measure the analyte signal.

14.8 QUANTITATIVE FLAME EMISSION SPECTROMETRY

Flame emission spectrometry is also basically a quantitative technique applicable to more than 70 elements, including some nonmetals which are determined by indirect methods or by measurement of molecular emission. About 40 elements can be determined at the level of 1 ppm (1 μg mL^{-1}) and many at levels down to 1 ppb. Qualitative analysis is usually carried out with higher-temperature plasma spectrometric techniques, which offer greater versatility.

Excitation and Emission. As discussed earlier, local equilibrium may be attainable mainly in the center of the interzonal region of a premix flame; there the Boltzmann equation can be taken as defining the population of species in an excited state relative to its concentration in the ground state (see Section 13.3).

Much excitation in flames has been found to occur by straightforward exchange of energy during collision of a vibrationally excited molecule and an analyte atom. There is a good probability that a molecular vibration level will match an electronic level of the analyte atom and lead to energy transfer. Second, a chemiluminescent reaction may occur in which one "product" is a desired excited state of an atom. The analyte must be involved in a complex collision with other atoms that leads to formation of a molecule. For example, an excited potassium atom might be generated not far above the inner flame cone by the reaction $K + H + H = H_2 + K^*$. Shortly thereafter a photon would be emitted.

As discussed in Section 13.3, the intensity of an atomic emission can be shown to be just N_2E_{21}/τ, where N_2 is the number of excited atoms, E_{21} the energy of transition, and τ the lifetime in the excited state. Thus, the intensity of an emission line depends linearly on the transition energy and frequency (since $E = h\nu$) and exponentially on temperature [see Eq. (13.5)].

In flame emission spectrometry samples are also commonly presented for analysis as solutions. The complex set of processes in the nebulizer–burner that begins with formation of droplets of solution, solvent evaporation, and sample sublimation/atomization and ideally culminates in excitation of atoms. Yet other flame processes compete with excitation in a premix burner flame: the atoms formed may (a) associate into molecules, (b) be ionized, or even (c) become excited ions. All these processes result mainly from collisions with flame gas molecules. The discussion of excitation and equilibria in plasmas in Section 13.4 is helpful background.

How much of the spontaneous emission of excited analyte atoms can be captured by the spectrometer will depend on

1. the solid angle subtended by the collection optic, mirror, or lens;
2. the number of excited species undergoing nonradiative deactivation by collision; and
3. the extent to which emitted resonance radiation is absorbed by other analyte atoms in the ground state.

The best wavelength for determination of an element is in general its most sensitive emission line that is free of spectral interference from other elements in the sample and from flame species such as OH. For example, in the portion of a spectrum shown in Fig. 14.9, it would appear desirable to select a cobalt line that

---

Fig. 14.9 Flame background in the emission spectrum of a cobalt sample aspirated into an oxy-cyanogen flame. All lines are due to cobalt emission unless otherwise identified. A continuous OH band extending from 322 to 306 nm is superimposed on the emission lines of cobalt and elevates the background substantially. Note the impurity lines of copper and silver and two NH band emissions. Because of this interference cobalt is ordinarily determined at 345 or 241 nm. (Courtesy of Interscience Publ. Co.)

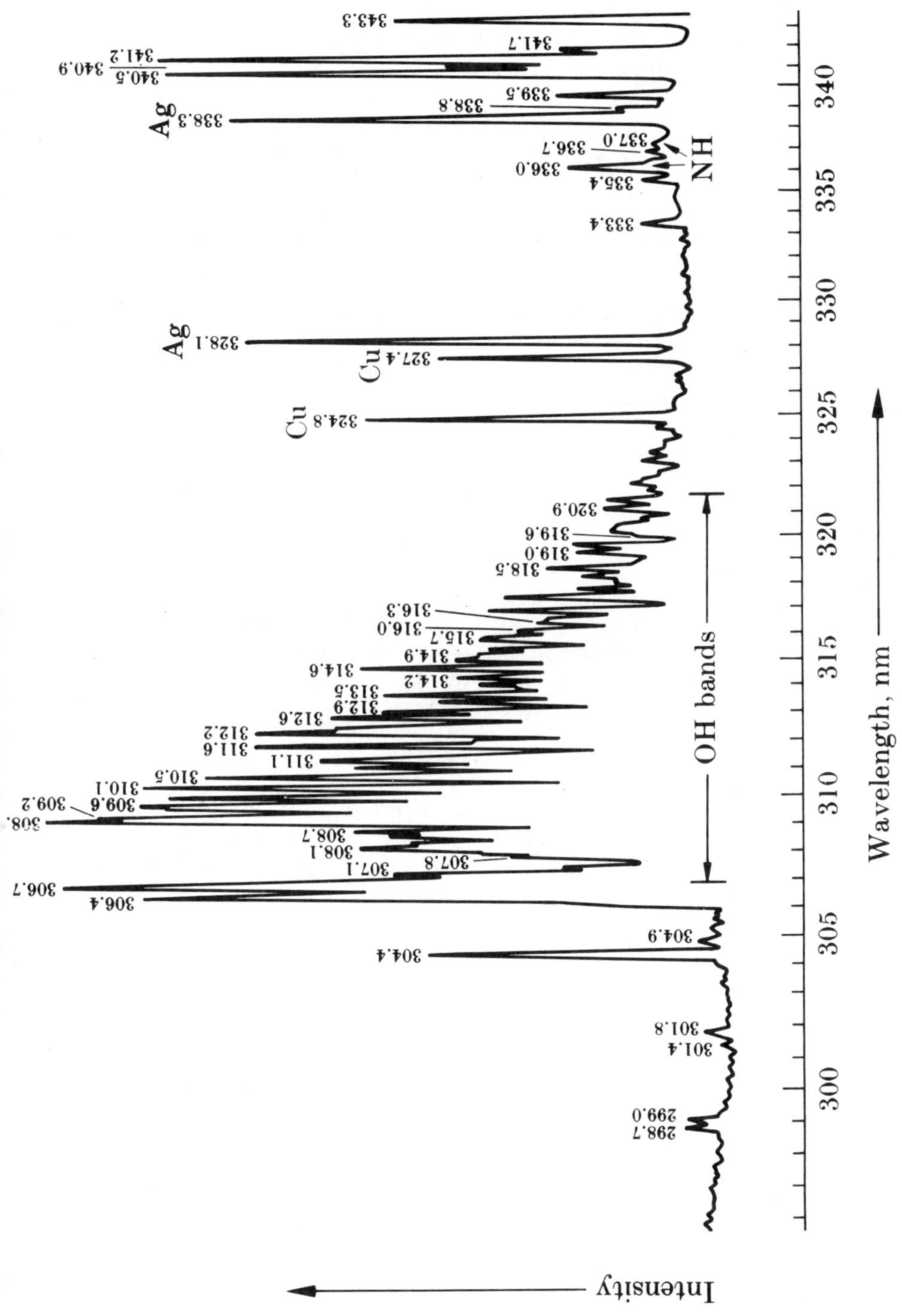
Intensity
Wavelength, nm
300
305
310
315
320
325
330
335
340
OH bands
NH
Ag
Ag
Cu
Cu
298.7
299.0
301.4
301.8
304.4
304.9
306.4
306.7
307.1
307.8
308.1
308.7
308.9
309.2
309.6
310.1
310.5
311.1
311.6
312.2
312.6
312.9
313.5
314.2
314.6
314.9
315.7
316.0
316.3
318.5
319.0
319.6
320.9
324.8
327.4
328.1
333.4
335.4
336.0
336.7
337.0
338.3
338.8
339.5
340.5
340.9
341.2
341.7
343.3

is free of the strong OH band spectrum as well as to avoid regions where the flame itself is radiating to lessen the possibility of loss of accuracy in determination.

The reliability of quantitative results by flame spectrometry is usually about ±2% for a series of measurements in which standards are run before and after a set of samples. If each sample is preceded and followed by a standard of closely similar composition, the reliability can be improved to ±0.5% to ±1%.

Background. Subtraction of spectral background is also desirable, especially for careful work. Ordinarily, the background intensity can be found (while analyte radiation is being observed) by measurement of intensity immediately on either side of the analytical line. If possible, the determination should be made at a distance (in nanometers) from the center of the line that is about twice the spectral slit width by averaging a scan of background over a nanometer interval. In Fig. 14.9 flame background is shown superimposed on a portion of the emission spectrum of a cobalt sample. Both uncorrected signal and background readings should be measured relative to the instrument baseline, which is observed on nebulization of pure solvent. Frequent checks of baseline are desirable.

Wavelength Modulation. An alternative to measuring background by the scanning procedure outlined above is the use of wavelength modulation. This procedure calls for repetitive scanning of a small wavelength interval on either side of the analytical line. In some instruments an oscillating quartz reflector plate located between the entrance slit of a monochromator and the collimating mirror serves as the modulator. As it oscillates, it introduces a limited modulation of wavelength. It is possible to select the second harmonic of the modulation frequency in processing and thereby obtain essentially the wavelength derivative of the signal. Often this measurement is more sensitive than a direct measurement of the signal.

Sample Preparation. Where interferents are present, often a *releasing agent* can be used to preferentially complex the interferent. Alternatively, it is valuable to add a chelating agent such as EDTA whose complex with the analyte will frustrate action by an interferent but will atomize readily in the flame. Suggestions for most types of interference, including ionization interference, will be found in Table 14.2. When chemical interferents cannot be eliminated satsifactorily, the method of standard additions is valuable (see Section 11.4).

If working curves are significantly nonlinear, as in Fig. 14.10, the probability is good that self-absorption is occurring. The answer is dilution of the solutions.

14.9 QUANTITATIVE ATOMIC FLUORESCENCE SPECTROMETRY

Atomic fluorescence has also been of continuing interest as a trace-level method, though the technique has had to compete with similar methods that are already well established. Its strength for this application arises from having a signal at low analyte concentrations that is proportional to the source intensity. Since it also requires a sampling module that provides atomization, the method is conveniently included here. Necessary fluorescence theory will be found in the next chapter.

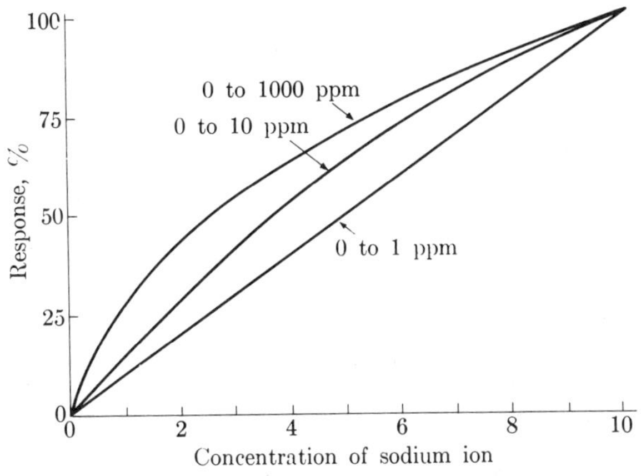

Fig. 14.10 Flame photometer calibration curves for sodium based on 589-nm doublet. Three concentration ranges are represented. The nonlinearity at high concentration is caused mainly by self-absorption of radiation by ground-state atoms.

Some types of *atomic* fluorescence that are possible are shown in Fig. 14.11. Of these the resonance transition of curve A to the well-populated ground state is of course the most common type and the most widely measured. In systems in which two higher electronic states are easily accessible, the direct-line and stepwise types of fluorescence shown as curves B and C are also possible. Theoretical interest in them is considerable because they offer the possibility of separating scattered radiation of the exciting (incident) wavelength from the fluorescent radiation. With laser irradiation, which will give rise to still more intense scattered

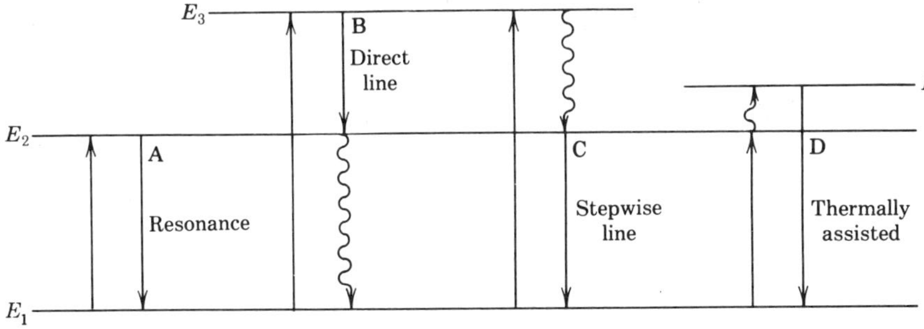

Fig. 14.11 Principal types of atomic fluorescence processes. Several occur in conjunction with a nonradiative step. (A) Regular (pure) fluorescence. Following absorption of a photon by a free atom, its full excitation energy is released radiatively. The absorbed and emitted wavelengths are equal. (B) Direct-line fluorescence. This type follows photon excitation of an atom to a still higher electron state. Fluorescence accompanies the downward transition from that state to an intermediate excited state. Subsequently the atom relaxes nonradiatively to the ground state. (C) Stepwise fluorescence. This process follows an initial collisional (nonradiative) relaxation to an intermediate electronic state. (D) Thermally assisted fluorescence. In this phenomenon a collision assists the absorption of a photon whose energy by itself is insufficient to produce excitation.

light, either of these types is especially advantageous since it will ensure that fluorescent radiation can be measured against reduced background.

The potential advantage of fluorescence methods over absorption methods can even be realized by a nondispersive spectrometer of the modular form shown in Fig. 14.12. Here spectral discrimination is provided because a single, narrow fluorescent line is emitted. In addition to low limits of detection, fluorescence offers virtual freedom from spectral interferences and a long dynamic range (4–5 orders of magnitude).

Instrumentation. Sources of high brightness that are especially attractive for atomic fluorescence are electrodeless discharge lamps, hollow cathode lamps, xenon arc lamps, and though expensive, tunable lasers. The last are often nitrogen-pumped tunable dye lasers coupled to a frequency doubler to give adequate radiation down to 217 nm. The conventional hollow-cathode lamp lacks sufficient intensity to make it useful for fluorescence, but the variety that can be pulsed at high current gives the needed intense output.

Figure 14.13 shows a schematic diagram of a nondispersive atomic fluorescence spectrometer based on the use of an inductively coupled plasma (ICP) torch of moderate temperature as the sampling module. Note that the source is a pulsed (for extra brightness) hollow-cathode lamp, and analyte fluorescence is detected by a photomultiplier tube. In this spectrometer, up to 12 source–detector combinations of the type shown in the figure can be assembled around a single ICP torch. If this is done, each hollow-cathode lamp is pulsed in turn and the signal received in its photomultiplier tube detected. Detection electronics are synchronized (gated) to achieve the necessary selectivity. Note that the purpose of the interference filter is to reduce the amount of background radiation falling on the photomultiplier tube so that it has no tendency over a period of time to show fatigue. Not only are measurements made rapidly, but as a result of the ICP torch the influence of interferents and background is minimized. Further, light scattering attributable to high salt concentrations seems minimal as well. The device is a general purpose atomic fluorescence spectrometer.

As was discussed for atomic absorption spectrometry in the opening section, the development of atomic fluorescence spectrometry as a general technique has required improvement in instrumentation. Scattering has made flames less attractive as atomizers for samples of high salt content. Unfortunately, background corrections are difficult. Nondispersive systems are usually shot noise limited.

Laser-Excited Fluorescence Spectrometry. It is possible to achieve saturation of the excited state with intense laser radiation. Saturation can be detected as a steady state in which the probability that an incident photon will stimulate fluorescence (a transition down-

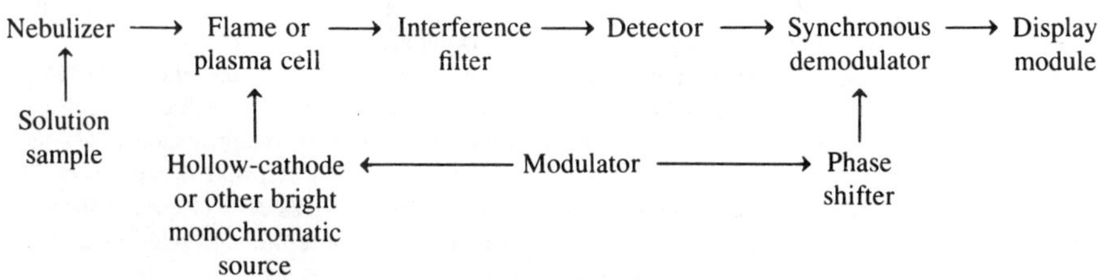

Fig. 14.12 Modular diagram of a nondispersive atomic fluorescence spectrometer.

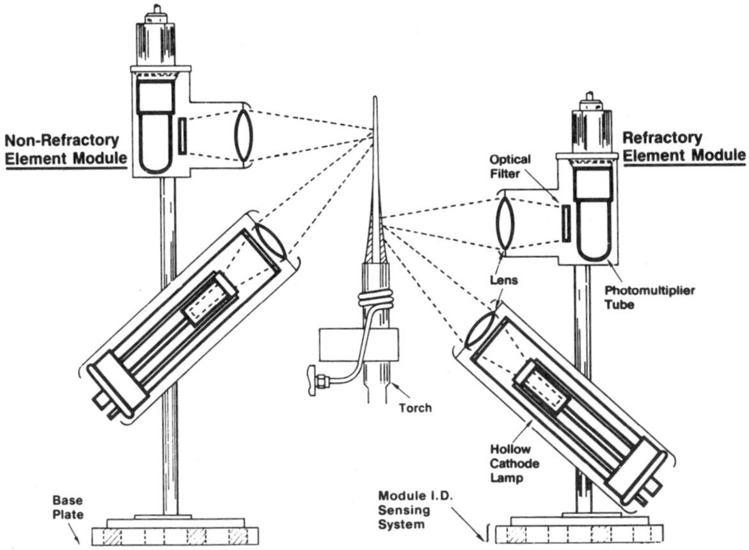

Fig. 14.13 Sketch of sets of characteristic modules for a multichannel nondispersive plasma atomic fluorescence spectrometer. Elemental lines emitted as each hollow cathode lamp is pulsed are focused on the inductively coupled plasma torch plume. The fluorescence of excited analyte atoms in the plasma is detected by the photomultiplier tube. Note that its lens collects a small solid angle of radiation at an angle about 45° to the exciting beam. An interference filter for each detector passes the chosen fluorescence wavelength and rejects most background that would otherwise contribute to fatigue of the photomultiplier tube. As many as a dozen elements can be determined by installing for each element an individual lamp–filter–detector set. As the spectrometer operates, each hollow cathode laser is pulsed in turn and the output of its photomultiplier tube integrated (averaged) over a chosen number of pulses. Torch power levels are low in this application (400–850 W) and argon coolant flow is smaller than usual. In the cooler plasma (3200–3800 K) that results noise is less by an order of magnitude. To reduce oxide formation when refractory elements are being determined, it is desirable to add a flow of propane. (Plasma/AFS Spectrometer, courtesy of Baird Corporation.)

ward) is equal to the probability that it will cause excitation (an upward transition). The solution is sometimes said to be "bleached." What is important is that saturated atomic fluorescence is relatively insensitive to many sources of fluctuation in intensity: unsteady laser output power and quenching processes (radiationless relaxation) are two.

Under conditions of saturation it is likely that extremely low detection limits for atomic species can be achieved (perhaps of the order of a few hundred or thousand atoms per cubic centimeter).

The best laser possibility at present is a nitrogen-pumped dye laser with frequency doubling. Pulse widths of 1 ns or less may be achieved. If coupled with suitable processing modules, time-resolved spectroscopy is possible. Perhaps the most important possibility is extremely low limits of detection.

14.10 COMPARISON OF ATOMIC METHODS

What are the relative strengths and weaknesses of the several elemental spectrometric methods considered in the last two chapters? It will be worthwhile to include X-ray fluorescence spectrometry in the comparison and to mention a quite new technique, laser-assisted flame ionization spectrometry (LAIS) at the end of this section. The set of methods compared is thus atomic emission, atomic absorption, atomic fluorescence, X-ray fluorescence, and LAIS. All but the fourth require that samples be dissociated into free, gaseous atoms, and all but the second that the atoms be excited. While it is necessary in emission spectroscopy to obtain excitation, for the other two techniques the need is for as large a concentration of *unexcited* atoms as possible.

Atomic emission and absorption methods are widely used, but atomic fluorescence is less commonly employed. For that reason most of the discussion will focus on absorption and emission methods. A summary of advantages and disadvantages of all the techniques is given in Table 14.3.

Emission spectroscopy enjoys the advantages of a greater dynamic concentration range, easier simultaneous determination of many elements, and applicability at all concentrations, high, moderate, and very low. In addition, most nonmetals can be determined, though for some it means measurement in the vacuum UV. The most significant disadvantage of atomic emission methods relative to the others is spectral interferences when line-rich elements are present.

Atomic absorption also has quite low limits of detection for many metals and can be used with more ease than the other methods since hollow cathode lamps provide selectivity for elements without careful wavelength setting. Sampling modules in this method operate at lower temperatures than for emission, and chemical interferences are somewhat more common. Both absorption (electrothermal sampling) and fluorescence are especially applicable at low analyte concentrations; indeed, fluorescence spectrometry is best as a trace method. Both instrument design problems and source factors limit the application of atomic absorption and fluorescence for simultaneous determination of several elements.

X-ray fluorescence spectrometry is another important atomic emission method (see Chapter 21) which involves electron transitions among inner shells. The simultaneous determination of elements is easily possible and of all the methods described in Table 14.3, sample preparation for X-ray fluorescence measurements is generally minimal. Spectral interferences are virtually unknown since X-ray spectra have few lines. Another virtue of the technique is that measurements with modern spectrometers are exceptionally fast. While matrix effects are often substantial, it is fortunately possible through available computer programs to make reliable quantitative calculations. A disadvantage relative to other elemental methods is the practical difficulty of determining elements of $Z < 9$ (fluorine), though new instrumentation is making such measurements easier. Finally, the method is not usable below concentrations of about 0.01% except for thin-film samples.

Laser-assisted ionization spectrometry calls for directing a laser beam at an atomic sampling module, usually a flame, to transfer sufficient energy to analyte atoms that they will be close to ionization. Through subsequent collisions they acquire additional energy and ionize. A flame ionization detector can register the increase in current.

Table 14.3 Comparison of Some Elemental Spectrometric Methods[a]

	Atomic Emission Methods			*Atomic Absorption Methods*			
	Plasma Sampling					*Atomic*	
	ICP	*DCP*	*Flame Sampling*	*Flame Sampling*	*Electrothermal Sampling*	*Fluorescence Methods*	*X-Ray Fluorescence*
				Performance Criteria			
Dynamic range	Wide		Wide	Limited	Limited	Wide	Low
Qualitative analysis	Excellent		Excellent	Poor	Poor	Poor	Very good
Elemental range	Excellent		Excellent	Excellent		Excellent	Very good
Trace analysis	Excellent		Excellent	Excellent		Excellent	Poor
Micro samples	Good		Good	Good	Excellent	Good	Good
Matrix interferences	Low		High	High	High	Low	High
Spectral interferences	High	Moderate	Low	Low	Low	Low	Low
				Costs			
Instrumentation	High	Moderate	Low	Low	Low	Moderate	Moderate
Maintenance	High	Moderate	Low	Low	Low	Moderate	Low
Single element	Moderate		Low	Low	Low	Moderate	Moderate
Sample preparation	Moderate		Low	Moderate	Moderate	Moderate	Low

[a]After M. L. Parsons, S. Major, and A. R. Foster, *Appl. Spectros.*, **37**, 411 (1983).

Li 0.001 A,G	Be 0.003 D,E											B 0.1 E	C 44 E	N * E	O * E	F * E
Na 0.004 A,D	Mg 0.0002 D											Al 0.01 D	Si 0.005 D	P 0.3 D	S 10 D,E	Cl * E
K 0.004 A,D	Ca 0.0001 E	Sc 0.4 E	Ti 0.03 E	V 0.06 D,E	Cr 0.004 D	Mn 0.0005 D	Fe 0.01 D	Co 0.008 D	Ni 0.05 D,E	Cu 0.005 D	Zn 0.0003 C,D	Ga 0.01 D	Ge 0.1 D	As 0.02 B,D	Se 0.02 B,C,D	Br * E
Rb 0.02 A	Sr 0.002 E	Y 0.04 E	Zr 0.06 E	Nb 0.2 E	Mo 0.02 D		Ru 30 B,E,F	Rh 0.1 D	Pd 0.05 D	Ag 0.001 D	Cd 0.0002 D	In 0.006 D,G	Sn 0.03 D	Sb 0.08 B,C,D	Te 0.002 B	I 3 D,E
Cs 0.02 A,D	Ba 0.01 D,E		Hf 10 E	Ta 5 E	W 0.8 E	Re 6 D,E,F	Os 0.4 E	Ir 0.5 D	Pt 0.2 D	Au 0.01 D	Hg 0.001 B,C	Tl 0.01 D	Pb 0.007 D	Bi 0.01 B,D		
		La 0.1 E	Ce 0.4 E	Pr 10 E	Nd 0.3 E		Sm 1 E	Eu 0.06 A,E	Gd 0.4 E	Tb 0.1 E	Dy 4 E,F	Ho 0.7 D	Er 0.3 D,E	Tm 0.2 E	Yb 0.01 D,E	Lu 0.1 E
			Th 3 E		U 1.5 E											

A. Flame Emission Spectrometry
B. Flame Atomic Absorption
C. Atomic Fluorescence Spectrometry
D. Electrothermal Atomic Absorption
E. ICP Emission Spectrometry
F. DCP Emission Spectrometry

Fig. 14.14 Best elemental detection limits in 1981 (ng mL^{-1}). The most favorable methods for detection are identified by letter in each box. Inductively coupled plasma (ICP) emission spectroscopy is listed as the best or second best method for 60 elements, electrothermal AA spectrometry for 55, and direct current plasma (DCP) emission spectrometry for 24. Spectra from elements indicated by an asterisk have been observed in ICP, but detection limits are not available in a comparable form (M. L. Parsons, S. Major, A. R. Forster, *Appl. Spectros.* **37,** 411, 1983; with permission of Plenum Publishing Corp.)

Trace Determinations. The limits of detection of elemental methods have fallen steadily with the refinement of atomic spectroscopic techniques. A comparison is made in Fig. 14.14. As is evident, at the time of compilation of data, detection levels for many elements are in the parts per billion (nanograms per milliliter) range. Recall that quantitative determinations call for concentration levels higher by *at least* a factor of 5, which are called the limits of determination (see Section 10.8). Further, in case an element is in a complex matrix, detection may be possible only at levels 20–100 times the ordinary limit of detection.

REFERENCES

Some general references for the spectrometric methods treated in this chapter are:

1. C. Th. J. Alkemade and R. Hermann, *Fundamentals of Analytical Flame Spectroscopy*. New York: Wiley, 1979.
2. G. F. Kirkbright and M. Sargent, *Atomic Absorption and Fluorescence Spectroscopy*. New York: Academic, 1975.
3. B. Welz, *Atomic Absorption Spectrometry*, 2nd English ed. New York: VCH, 1985.

Some sources of detailed practical information on methods for individual elements are:

4. W. J. Price, *Spectrochemical Analysis by Atomic Absorption*. Philadelphia: Heyden, 1979.
5. J. W. Robinson, *Atomic Absorption Spectroscopy*, 2nd ed. New York: Dekker, 1975.
6. M. Slavin, *Atomic Absorption Spectroscopy*, 2nd ed. (1st ed. under name of son, W. Slavin) New York: Wiley, 1978.
7. V. Sychra, V. Svoboda, and I. Rubeska, *Atomic Fluorescence Spectroscopy*. New York: Van Nostrand Reinhold, 1975.
8. J. C. Van Loon, *Analytical Atomic Absorption Spectroscopy*. New York: Academic, 1980.

References dealing with particular topics in flame atomic spectroscopy on an intermediate level are:

9. C. Th. J. Alkemade, T. Hollander, W. Snelleman, and P. J. Th. Zeegers, *Metal Vapors in Flames*. New York: Pergamon, 1982.
10. A. G. Gaydon, *The Spectroscopy of Flames*, 2nd ed. New York: Wiley, 1974.

Also of interest are:

11. S. D. Brown, Zeeman effect-based background correction in atomic absorption spectrometry, *Anal. Chem.*, **49,** 1269A (1977).
12. R. F. Browner and A. W. Boorn, Sample introduction: The Achilles' heel of atomic spectroscopy? *Anal. Chem.*, **56,** 786A (1984).
13. C. W. Fuller, *Electrothermal Atomization for Atomic Absorption Spectrometry*. London: The Chemical Society, 1980.
14. S. S. M. Hassan, *Organic Analysis Using Atomic Absorption Spectrometry*. New York: Wiley-Harwood, 1984.

15. W. B. Robbins and J. A. Caruso, Development of hydride generation methods for atomic spectroscopic analysis,'' *Anal. Chem.*, **51,** 889A (1979).
16. R. E. Sturgeon, Factors affecting atomization and measurement in graphite furnace atomic absorption spectrometry, *Anal. Chem.*, **49,** 1255A (1977).
17. J. D. Winefordner and T. J. Vickers, Calculation of the limit of detectability in atomic emission absorption flame spectrometry, *Anal. Chem.*, **36,** 1939, 1947 (1964).

See also references listed for Chapter 13.

EXERCISES

14.1 Assume a hollow-cathode lamp (HC) warms up in about 10 min. With the monochromator set to a desired emission line, sketch the detector signal as a function of time during warm-up in an operating (a) single-beam atomic absorption spectrometer and (b) double-beam atomic absorption spectrometer.

14.2 A sodium vapor discharge lamp relies on neon as a starter gas. When the lamp is first started, a strong neon spectrum is obtained, but when all the sodium is vaporized the neon spectrum is very weak. Explain briefly the cause of its suppression.

14.3 The spectrum of a titanium HC was observed at several operating currents. Up to a current of 50 mA all the lines gave sharp peaks. At higher currents the most intense line at 364.3 nm developed a cup-like crater in its peak. Explain.

14.4 Why is the formation of oxides a problem in flame atomic absorption and atomic emission spectrometry?

14.5 A two-element atomic absorption spectrometer can be defined as one that operates on wavelengths characteristic of two different elements. Consider how it might be designed if a variety of HCs and other modular units are available. (a) Choose one option from each set below and defend the selection on the basis of simplicity, cost, convenience, and processing. (b) Sketch one instrument based on the choices and describe any coupling arrangements you have included. Options: Set A—two single-element HCs or one dual-element HC; Set B—a single monochromator with a fast slewing arrangement or a pair of simpler monochromators; Set C—two detectors or a single detector (suggestion: consider modulation arrangements).

14.6 In Fig. 14.15 a schematic layout of the optical part of a double-beam atomic absorption spectrophotometer is given. (a) Draw a modular diagram. Indicate components associated with each module in the diagram. (b) What type of monochromator is used? Explain the pair of gratings G. (c) If the chopper were deleted and the current of the HC were modulated, what changes would need to be made in the instrument?

14.7 Should it be possible to perform *molecular* absorption flame photometry? For example, molecular sulfur has absorption bands at 320–261 nm, 230–180 nm, and 158–152 nm. The second listed is the most intense. For a determination of sulfur, suggest a source, type of wavelength isolation device, flame arrangements, and detector. [See K. Fuwa and B. L. Vallee, *Anal. Chem.*, **41,** 188 (1969).]

14.8 Why is it important in graphite furnace atomic absorption spectrometric determinations of a metal that the type of compound (e.g., nitrite) be the same in samples and standards?

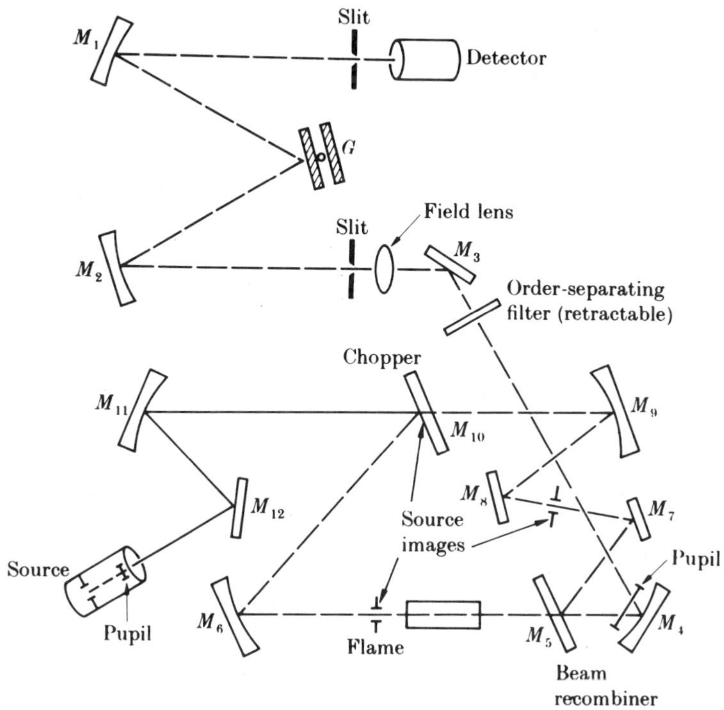

Fig. 14.15 Optical schematic of a double-beam atomic absorption spectrometer. M_{10} is a rotating sector mirror; M_5 is a semitransparent beam recombiner. (Courtesy of Perkin-Elmer Corp.)

14.9 What is the function of a flame or ionization buffer? Why can strontium act as a flame buffer for calcium?

14.10 In the atomic absorption spectrometric determination of zinc in 10% NaCl solutions it is found that light scattering occurs in the flame. (a) How will this scattering affect the instrument response? (b) What is the probable origin of the scattering? (c) To correct for the scattering, measurements are made at a zinc wavelength (213.9 nm) using both a zinc HC and a deuterium lamp. The standard addition method is used and the following data are recorded:

	Sample	First addition	Second addition
Added Zn ($\mu g\ mL^{-1}$)	0.00	0.1	0.2
Absorbance with			
zinc hollow-cathode lamp	0.10	0.27	0.44
hydrogen lamp	0.10	0.10	0.10

What concentration of Zn should be reported with and without the use of the hydrogen lamp?

14.11 The simple atomic absorption spectrometer shown below in Fig. 14.16 proved more sensitive for mercury determinations than a regular flame cell instrument. Mercury present

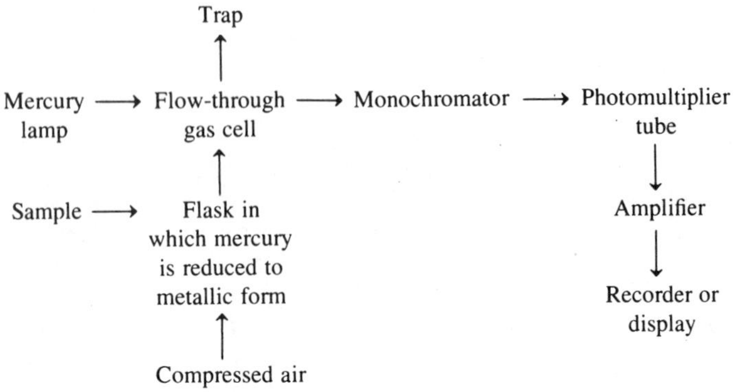

Fig. 14.16

in samples is reduced to the metallic form in aqueous solution, usually by pretreatment with tin(II) in 1–3 *M* H_2SO_4. With proper choice of conditions good accuracy is obtained. A quantity as small as 10 ng can be determined when the 253.7 nm mercury wavelength is used. (a) Show how this instrument differs from a flame cell device and justify the difference. Why might a flame cell lead to lower sensitivity? (b) State any assumptions that must be made in using the instrument of Fig. 14.16.

14.12 Atomic emission is observed from the excited state. At 2000 K only 9.9×10^{-6} of all sodium atoms in a sample are in the excited state. In view of the minimal fraction excited, why is flame emission such a common means of detecting sodium as compared to flame absorption?

14.13 Aluminum can be sensitively determined by flame emission photometry by complexing it with thenoyltrifluoracetone and extracting the chelate into methyl isobutyl ketone; this solution is aspirated into a flame and observed at 484 nm. An experimenter prepared a calibration curve for the system from the following data (in each case the first figure is the concentration of the standard solution in ppm and the second is the meter deflection): 20, 12; 45, 26; 90, 51; 120, 70; 150, 87. He found that a similar aliquot of an unknown gave a meter reading of 57. What was the aluminum concentration of the aliquot?

14.14 The atomic spectrometric methods, X-ray fluorescence (XRF), atomic emission (AES), and atomic absorption (AAS) are common elemental techniques. Recommend one of them for use in each of the following cases and support your choice. If sample preparation is necessary, describe it by a sentence or two at most. The cases are: (a) determination of beryllium in an ore; (b) analysis of a brass sample for traces of lead as well as the major metallic constituents; (c) elemental analysis of a painting suspected of being a forgery; (d) rapid analysis of river water for polluting metals.

ANSWERS

14.4 Oxides are molecules and give band spectra, not line spectra. *14.9* A flame buffer has an excitation energy smaller than that of the analyte and will return excited analyte atoms to the ground state by collisional exchange. Strontium is such an element for calcium. *14.10* (c) 0.06 μg Zn mL^{-1}; 0.0 μg Zn mL^{-1}. *14.12* Emission has the advantage of $I_0 = 0$ at the wavelengths for observation of sodium, that is, measurement in the absence of interference is against a dark background. An atomic absorption measurement by contrast begins with a high intensity source. *14.14* (a) Elements below $_8O$ are awkward to determine by XRF. Since a single element is involved, AAS would seem the method of choice. The ore would need to be ground and dissolved. (b) Probably either AAS or ICP-AES would be suitable. The sample would need to be dissolved. AES has the advantage of being able to handle high concentrations (of major elements) as well as lead at the trace level and would probably be the better choice. (c) XRF can deal with almost any specimen and as a nondestructive technique is the method of choice. (d) Preconcentration of analyte on an ion-exchange column would be helpful. Metals would be eluted for determination by AES. AAS would require many different hollow-cathode lamps.

Chapter 15

MOLECULAR LUMINESCENCE SPECTROMETRY

15.1 INTRODUCTION

Once the absorption of a quantum of radiation has excited an outer electron in a molecule to a higher level, emission processes called fluorescence or phosphorescence, or collectively, *luminescence*, are possible.* Fluorescence is the process of fast emission occurring about 10^{-8} s after photon absorption. Phosphorescence is a similar but slower and less common process which takes place 10^{-5} to 10 s after excitation. Measurement of the former is called *fluorometry* and of the latter, *phosphorimetry*. Both techniques provide highly sensitive, selective, and quantitative information about luminescent substances. Since fluorometry and phosphorimetry are both molecular emission methods, they are quite similar. The discussion of theory, measurement, and instrumentation in the following sections will apply to both methods unless stated otherwise.

Analytical applications begin with standard quantitative determinations. For these sensitive methods, trace-level measurements are common. As is true with spectrophotometric procedures, a great many nonfluorescing species such as metal ions (some lanthanide ions fluoresce) can be fluorometrically determined by complexing with a fluorescing ligand.

The interaction of small fluorescent molecules with biological species can also contribute secondary information about sites on biological membranes and conformation and binding of macromolecules such as proteins. The luminescence of small molecules is surprisingly sensitive to perturbations in structure. Thus, they are useful in probing the nature of absorption on a substrate, complexation with a metal ion, the character of the chemically bonded layer on a reversed-phase liquid chromatographic stationary phase, or phenomena connected with cell membranes.

Increasingly, they serve as labels in immunoassay procedures. Using time-resolved spectroscopy they also furnish information about themselves in excited states. Fluorometry and phosphorimetry thus have unusual breadth as well as sensitivity.

*The excitation energy for other forms of luminescence has a different origin. Thus, in *chemi-*, *bio-*, and *thermoluminescence* excited species are formed in chemical reactions, and in *triboluminescence* mechanical energy produces excitation.

15.2 LUMINESCENCE SPECTRA: ENERGY, INTENSITY, AND LIFETIME

To put luminescence spectroscopy in an energy context, in Fig. 15.1 a few of the many quantized transitions by which molecules can gain and lose energy are shown in Jablonski diagram. For molecules possible transitions are numerous in contrast to the few possible in atomic fluorescence: vibrational and rotational changes nearly always accompany electronic shifts in molecules. This discussion will be focused on molecular emission or relaxation processes by which a species can go from an excited electronic state to one of lower energy.

Some comments will aid the interpretation of Fig. 15.1. By definition in *singlet* electronic states all electron spins are paired, though in a singlet excited state one pair is divided between the excited and ground state. In a *triplet* state two spins are unpaired; while one spin remains in the S_0 level, the other is in the lowest triplet state T_1. Only the *singlet* ground state S_0 and the *first excited* singlet state S_1 are shown in Fig. 15.1 since other singlet states are ordinarily unimportant in luminescence.* Further, spin angular momentum is ordinarily conserved in spectral transitions and a transition that requires a change of spin such as $S_0 \rightarrow T_1$ is of low probability.

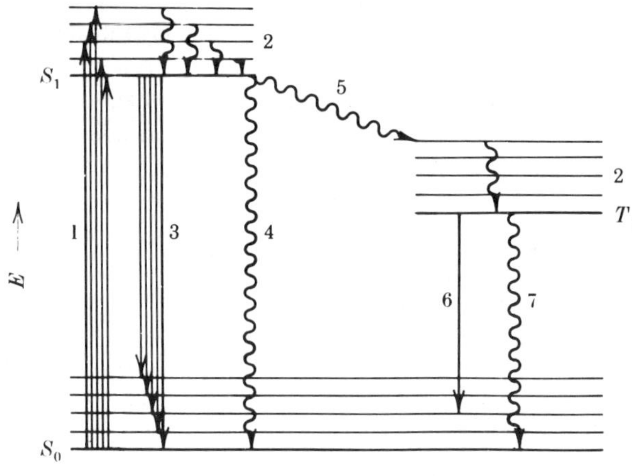

Fig. 15.1 Energy relationships among molecular absorption and relaxation processes. Shown are several vibrational levels in three electronic states, *singlet* states S_0 (ground) and S_1 (excited) and *triplet* state T_1. The processes shown are: 1, absorption; 2, vibrational relaxation (nonradiative); 3, fluorescence; 4 and 7, quenching (nonradiative conversion of electronic energy to heat); 5, intersystem crossing (nonradiative; change of spin); 6, phosphorescence. Phosphorescence may also be from the lowest vibrational level of T_1, to other vibrational levels of S_0.

*When a higher singlet state such as S_2 is excited, it usually loses much of its surplus energy *nonradiatively*. Initially, vibrational relaxation takes the species to the zero vibrational level of S_2. Then a nonradiative *internal conversion* that involves a shift to an *excited* vibrational level of S_1 is probable.

With many kinds of competing transitions possible, the changes that are dominant tend to be decided principally by their relative rates. Information about the time required for different kinds of transitions is given in Table 15.1. For example, since vibrational lifetimes are about 10^{-14} s, an electronically and vibrationally excited molecule nearly always will lose all its vibrational energy by collision during the average interval of 10^{-8} s in which the excited electronic state exists. For that reason fluorescence emission generally occurs from the bottom vibrational level of S_1. Ordinarily, the fluorescence of a species continues from 1 to 1000 ns after excitation is stopped.

Photodissociation. Two factors mitigate against the photodissociation of an excited molecule during excitation or fluorescence. First, a relatively small flux of exciting photons is actually incident: many are blocked by a monochromator or filter. Second, there is only a low probability that energy will be redistributed into a vibrational mode that favors dissociation of a molecule. Indeed, it is more probable in the usual liquid or solid system that high vibrational energy will be lost in collisions than will lead to dissociation. Nevertheless, the shorter the irradiating wavelength, the more likely dissociation or another photochemical reaction will be; for this reason exciting wavelengths shorter than 220 nm are avoided.

What processes lead up to phosphorescence? The nonradiative *intersystem crossing* from singlet state S_1 to triple state T_1 shown in Fig. 15.1, even though spin-forbidden, occurs with moderate probability in many molecules. Once a molecule is in a triplet state, the radiative *phosphorescence*, the transition $T_1 \rightarrow S_0$ becomes probable. There is also competition from nonradiative mechanisms bridging the same pair of states. From the figure, note that phosphorescence must also be spin-forbidden. For this reason, triplet-state lifetimes are much longer (10^{-5} to 10 s) than those for excited singlet states.

Phosphorescence is most likely to be observed if the triplet state molecule is at a low temperature, that is, in a rigid solution or glass in which the rate of collision-induced nonradiative transitions will be quite low. Phosphorescence is also enhanced by having an atom of high atomic number like I in a molecule. Such atoms

Table 15.1 Lifetimes for Molecular Luminescence Processes

States				
Initial		*Final*	*Transition Lifetime (s)*	*Process*
$h\nu + S_0$	$\longrightarrow$	S_n	10^{-15}	Absorption
S_n	$\rightsquigarrow$	S_1	10^{-14} to 10^{-11}	Internal conversion
S_1	$\longrightarrow$	$S_0 + h\nu$	10^{-9} to 10^{-7}	Fluorescence
S_1	$\rightsquigarrow$	T_n	10^{-8}	Intersystem crossing
S_1	$\rightsquigarrow$	S_0	10^{-7} to 10^{-5}	Internal conversion
T_1	$\longrightarrow$	$S_0 + h\nu$	10^{-5} to 10	Phosphorescence
T_1	$\rightsquigarrow$	S_0	10^{-3} to 10	Internal conversion

perturb electron spins and enhance state mixing, which will increase the rate of intersystem crossing.

Since fluoresc⌃nce involves two independent processes, absorption and fluorescence, it gives rise to both an *excitation* and an *emission* spectrum. A representative pair is pictured in Fig. 15.2. The fluorescence spectrum appears at longer wavelengths (sometimes called a Stokes shift) because of the energy lost in the excited state by nonradiative vibrational relaxation. As long as luminescence studies are carried out in condensed phases like solutions or frozen matrices the transitions broaden into bands.

Molecular Structure and Energy Levels. What kinds of molecules fluoresce? Most strongly fluorescent species have rigid, planar structures that lessen competing nonradiative transitions. For example, fluorescein is rigid and fluoresces, but the similar nonrigid molecule phenolphthalein does not. The only classes of organic substances that fluoresce reasonably efficiently are aromatic molecules and aliphatic molecules with extensive π-electron systems (alternating single and double bonds). Since both types have low-lying excited π^* singlet states, the most common electronic transition is $\pi \rightarrow \pi^*$. When an oxygen or nitrogen or other

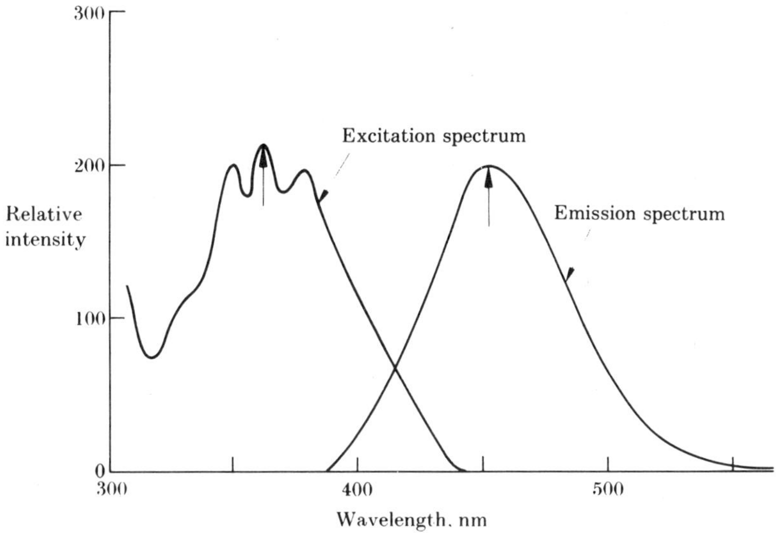

Fig. 15.2 A representative excitation (absorption) and emission (fluorescence) spectrum for a substance. Note the rough mirror-image relationship of the two spectra. This phenomenon arises because the excitation process usually begins from the ground vibrational state of S_0 and ends on a higher vibrational level of S_1. As a result, excitation requires an amount of energy equal to the basic electronic excitation *plus* a vibrational energy increase; conversely, emission yields an absolute energy equal to the basic electronic excitation energy *less* a vibrational energy change. The excitation spectrum was obtained by scanning the wavelength range of the excitation monochromator while observing fluorescence at 450 nm (see arrow), and the emission spectrum by exciting the species at 362 nm (see arrow) and scanning the range of the emission monochromator.

atom with a lone pair of electrons is *also present*, however, a lower-energy transition of $n \rightarrow \pi^*$ type occurs. The term *fluorophore* (or luminophore) describes a molecular group or entity that will give rise to fluorescence in a more complex structure. For example, the amino acid tryptophan and the small molecule chlorophyll are naturally occurring fluorophores.

Substituents strongly affect fluorescence. Thus, aniline fluoresces but nitrobenzene does not. Undoubtedly, fluorescence might also occur with aliphatic or saturated cyclic compounds, but both the high energy required for electronic excitation and the fact that a bonding electron would have to be involved ordinarily cause excitation to lead to photodissociation. These types of molecules are not observed to fluoresce under ordinary conditions.

A third class of molecules that would seem to meet the criterion are those with a nonbonding pair of valence electrons, for example, an amine with a lone electron pair on its nitrogen atom. Such electrons can be promoted without disruption of bonding. In general, a delocalized π-system must also be part of this type of molecule to ensure easy fluorescence.

Fluorescence Efficiency. The efficiency of fluorescence ϕ_f, also known as *quantum yield* or quantum efficiency, is defined as

$$\phi_f = \text{number of fluorescence photons/number of absorbed photons.}$$

The fluorescence efficiency is characteristic of a species and sets an upper limit on the sensitivity of measurement. Either rates or lifetimes of the several processes in competition can be measured to estimate the relative efficiency of fluorescence. For example, using the rate constants of some but not all the competing processes, ϕ_f can be defined as

$$\phi_f = k_f/(k_f + k_{ic} + k_{is}), \tag{15.1}$$

where subscripts identify fluorescence, internal conversion, and intersystem crossing processes, respectively. The one with the largest rate constant will mainly decide the fate of the molecular excitation energy.

The *luminescence lifetimes* of excited states basically are the reciprocals of decay rates, as was seen in Section 13.3. These lifetimes can generally be defined in terms of the first-order decay process by which a molecule relaxes from an excited state:

$$I_t = I_0 e^{t/\tau}, \tag{15.2}$$

where I_0 and I_t are the intensity of emission of a species during irradiation and at time t after such excitation is stopped, respectively. From Eq. (15.2) τ is seen to be the time required for luminescence to decay to $1/e$ of its initial value. Representative lifetimes in the singlet state are 1–20 ns (fluorescence lifetimes), and in the triplet state 0.001–10 s (phosphorescence lifetimes).

Example 15.1. Using Eq. (15.1), predict the effect of raising the temperature on molecular fluorescence.

An increase in temperature will increase the number of collisions, including those involving excited species. Since nonradiative processes depend on collisional transfer of energy, they will be favored. Experimentally, it is indeed found that ϕ_f decreases about 1% per degree increase in temperature.

Example 15.2 On the basis of lifetimes of excited states, which type of spectroscopy, fluorescence or phosphorescence, seems to offer the greater possibility for identifying species?

While the lifetime ranges given in Table 15.1 are approximate, the greater range over which triplet lifetimes occur should lead to better resolution of different components by phosphorimetry, if the method can be used. This possibility will be examined in Section 15.4.

Luminescence Intensity and Concentration. While molecules may be excited in many ways, only emission that follows the absorption of light photons is defined as fluorescence or phosphorescence. To define the relationship of these phenomena to species concentration, both absorption and emission processes must be taken into account. For absorption, Beer's law describes the relationship (see Section 16.2). The law states that the fraction of incident amount of power P_0 in an incident beam that is absorbed by a dissolved solute of molarity c and molar absorptivity ϵ (L mol^{-1} cm^{-1}) held in a cell of path length b (cm) is just $P_0(1 - e^{-\epsilon bc})$ at wavelength λ. Multiplying this factor by the fluorescence efficiency ϕ_f should give the intensity of fluorescence. The analytical signal produced in a measurement will be given by the expression

$$\text{signal} = P_0(1 - e^{-\epsilon bc})\,\phi_f. \qquad (15.3)$$

Clearly, factors relating to instrument modules must yet be added. In addition, quenching, which would result in an entirely nonradiative relaxation to S_0, is assumed minimal.

The exponential relation between fluorescence and species concentration just given can be usefully simplified by limiting the concentrations to quite dilute solutions. An advantage gained is that the dependence of measurements on quantum efficiency is minimized at low concentrations (below 10^{-4} M). In quite dilute solutions the exponential factor in Eq. (15.3) can be restricted to values such that $\epsilon bc \leq 0.05$. Then the factor $1 - e^{-\epsilon bc}$ can be approximated by ϵbc,* and Eq.

*A series expansion of $1 - e^{(-\epsilon bc)}$ gives

$$1 - e^{(-\epsilon bc)} = \epsilon bc - (\epsilon bc)^2/2! + (\epsilon bc)^3/3! - \cdots.$$

Inspection shows that when $\epsilon bc < 0.05$, retaining only the first term of the series will produce an error of less than 1%.

(15.3) can be rewritten in the approximate form

$$\text{signal} = P_0 \epsilon b c \phi_f. \quad (15.4)$$

How well does the approximation hold? Since typical values of ϵ are of the order 10^3–10^4, if $b = 1$ cm, the condition $\epsilon bc < 0.05$ requires that the analyte concentration be no greater than 10^{-5}–10^{-6} M. This limitation does not turn out to be serious, as will be discussed later.

For luminescence to occur in the condensed phases in which samples of interest are usually found, collisional processes that will lead to nonradiative relaxation must be minimized. The processes responsible for such losses are collectively described as *quenching* processes, that is, processes in which there is a loss of luminescence.

Quenching. There are several potent types of bimolecular quenching. *Collisional impurity quenching* leads to loss of fluorescence because of the formation of an *exciplex* or excited complex between the excited analytical species and a ground-state impurity molecule and subsequent nonradiative energy losses. Dissolved O_2 is the most common impurity quencher. *Energy transfer quenching* is a second type of quenching that is caused by an impurity whose first excited singlet state is at an energy below that of the excited singlet state of the analyte. In this case nonradiative energy transfer can occur (even though there is no collision), followed by a further radiationless loss by the impurity. The presence of longer wavelengths than normal in the solute fluorescence spectrum is indicative of the presence of energy transfer type quenching impurities. Aromatic substances are prime offenders in this category.

The third main type is *concentration quenching*. This kind of self-quenching with increasing concentration is responsible for the loss of fluorescence efficiency described above. In this case quenching occurs because of formation of an *excimer*, that is, a collisional complex between an excited analyte molecule and a second, identical, unexcited molecule. Subsequent radiationless processes degrade the extra energy to heat.

15.3 LUMINESCENCE SPECTROMETERS

A modular diagram of a generalized instrument for measuring fluorescence, a fluorometer, is shown in Fig. 15.3. While a fluorometer calls for the types of spectrometric modules described in Chapters 8 and 9, its design is distinctive in requiring a pair of wavelength-isolation modules, one for excitation and the other for emission, and also a right-angle bend in the signal channel at the sample module.

Design Criteria for Fluorometers. With this layout in view, it should be possible to complete the equation for the fluorescence analytical signal, Eq. (15.4), by adding factors for modular performance. The final expression will provide criteria for choosing modules and coupling that will enhance performance. Recall Eq.

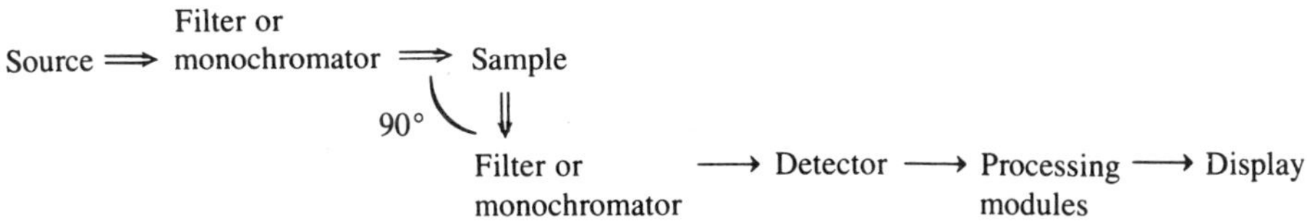

Fig. 15.3 Modular diagram of a generalized fluorometer.

(15.4) is restricted to concentrations that will give a linear curve. With the additions, Eq. (15.4) becomes

$$\text{signal} = P_0(\lambda)\,\epsilon b c \phi_L f_{ex}(\theta, \lambda)\, f_{em}(\theta, \lambda)\, f_{det}(\theta, \lambda). \qquad (15.5)$$

Here the terms $f_{ex}(\theta, \lambda)$ and $f_{em}(\theta, \lambda)$ define the transmission or throughput of the wavelength-isolation devices as they receive radiation of wavelength λ at solid angle θ, and $f_{det}(\theta, \lambda)$ defines detector response. Since optical filters provide greater throughput than monochromators, we shall need to consider both filter fluorometers and spectrofluorometers (with monochromators); both find wide use.

Sources. The linear dependence of Eq. (15.5) on P_0 makes clear that a high source intensity will directly enhance the analytical signal. Mercury and xenon arc lamps are commonly used. Implications with respect to lasers will be taken up below. For filter fluorometers, a low-pressure mercury arc is often used. High intensities are available in its emission lines and most samples absorb at one of the mercury wavelengths. In spectrofluorometers, however, a continuous source, and especially the xenon arc lamp, is preferred. At any wavelength in its range there is sufficient intensity for quantitative analysis and especially for obtaining excitation spectra and a two-dimensional excitation–emission matrix display, if they are desired.

Lasers, as sources of very high power, should permit much lower limits of detection. Yet in general the extra power of a laser cannot be translated into lower limits of detection. Luminescence measurements on solutions are blank-limited, that is, a blank itself (solvent, impurities, cell) exhibits so much Raman and Rayleigh scattering as to make this background the limiting factor rather than detector noise.* Lasers are preferred in applications such as remote sensing, for example, determination of pollutants in the atmosphere, some measurements at very low concentration, and in research where the special laser attributes of high intensity, spectral coherence, and the possibility of very short pulses have been especially valuable (see end of this section).

Light-collecting arrangements are necessary to couple sources efficiently to the wavelength-isolation module. Either mirrors or lenses are used, the former being more common. For example, an off-axis ellipsoidal mirror is especially effective

*F. E. Lytle, "Laser Excited Molecular Fluorescence of Solutions," *J. Chem. Ed.*, **59**, 915 (1982).

when the source is placed at its near focus and a monochromator entrance slit at its far focus (see Fig. 9.5).

Wavelength-Isolation Modules. This pair of modules performs two essential functions. First, they permit selection of the most favorable wavelengths for excitation and for emission. Second, they make it possible to minimize the intensity of scattered *exciting* radition that enters the detector along with the desired luminescence. In more sophisticated measurements they will also be essential to obtaining multidimensional luminescence information.

As suggested in Section 7.12, a substantial amount of light is scattered at right angles to an incident beam, precisely in the direction usually chosen for observation of luminescence. Fortunately, much of this stray light, which is at the exciting wavelength, can be blocked. Since most molecular fluorescence occurs at longer wavelengths, the two arms of a fluorometer can be tuned to different wavelengths to prevent the shorter wavelength exciting light from reaching the detector. For work near the limit of detection, however, scattered light must be reduced still more and a double monochromator is highly desirable in the emission arm.

Sample Module. Fused silica cells, preferably of square cross section for reproducibility of positioning, are ideal. Cells of quartz, which has some fluorescence, or Pyrex, which transmits only above 320 nm, see use also. Low-temperature quartz cells for phosphorimetry are inserted in liquid-nitrogen-cooled (77 K) dewars. The optical path is through an unsilvered wall. Room-temperature phosphorimetry is possible in some cases also.

For measurements at trace and ultratrace levels, specially purified solvents should be used to avoid quenching agents and impurities that may absorb either the exciting or luminescence wavelengths. Solutions should also be filtered to remove particles to minimize scattering.

The amount of *primary and secondary absorption* in a sample cell can also become troublesome. In the cell P_0 diminishes steadily with distance as exciting light is absorbed by the analyte or impurities (primary absorption). Clearly, the intensity of fluorescence must also diminish across the cell. Occasionally fluorescent light is itself absorbed (secondary absorption), leading to further loss of signal. Both of these effects, which are also called the inner-filter effect, must be kept small.

Example 15.3. How serious will primary *absorption effects* be in the optical configuration of Fig. 15.4? In that design only light emitted from fluorophores in the center of the sample cell will enter the emission monochromator. Primary absorption by sample, especially at higher analyte concentrations, will mean the fluorescence intensity measured will be substantially smaller than expected. Exercise 15.6 explores this situation. A similar result will be observed, of course, if the luminescence is itself absorbed in the cell. Even if the approximation that led to Eq. (15.4) holds, primary absorption especially can cause an appreciable negative departure of an analytical signal curve from linearity.

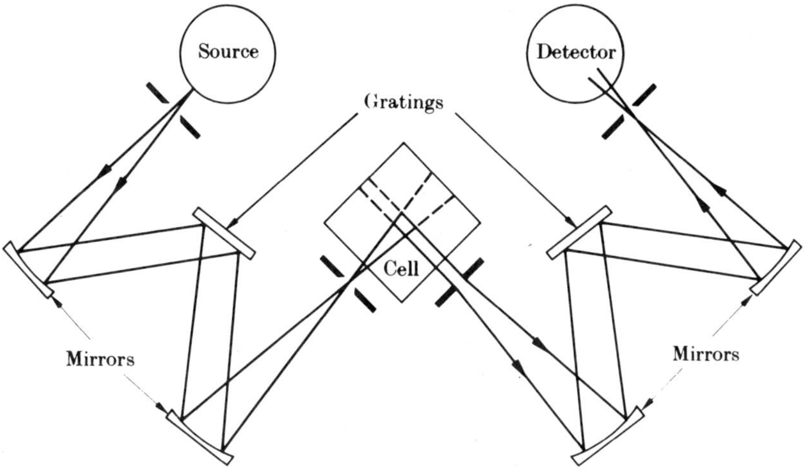

(a)

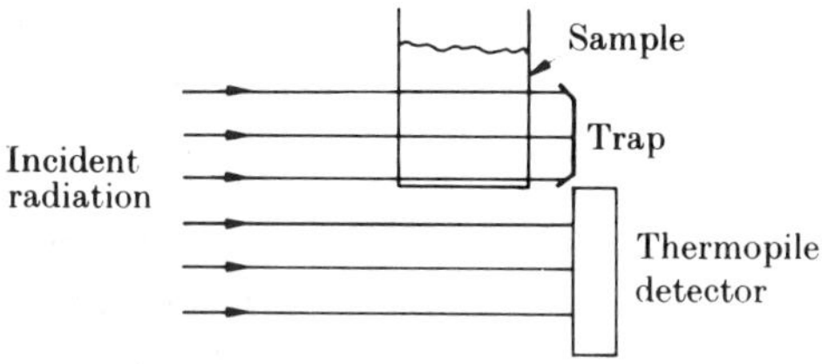

(b)

Fig. 15.4 An optical schematic of a spectrofluorometer using two Czerny–Turner monochromators. (a) Top view of optical modules. (b) Side view of sample cell showing an arrangement to monitor excitation intensity. The output of the wavelength-independent thermopile detector can be used to obtain excitation energy-corrected spectra.

Avoiding the Absorption Effect. Of several strategies that will reduce the effect, the simplest is dilution of the sample solution. Yet if any analyte is reduced to trace amounts in the process, this approach may be unattractive. The most workable answer is to observe luminescence from the sample cell surface. Fluorescence from the front of the sample cell is collected at an angle of 22° relative to the incident beam. Several commercial spectrofluorometers provide this feature. Another strategy is to choose an excitation frequency that is less strongly absorbed.

Filter Fluorometer. For routine quantitative analyses a filter fluorometer is, in general, preferred. With its broader bandpass, a larger aperture, and usually a photomultiplier tube detector, this type of fluorometer is capable of great sensitivity. A representative filter fluorometer is shown in modular form in Fig. 15.3. The filters may be sharp-cutoff types or at least interference filters with fairly narrow bandpass. When wide-band absorption filters are used, it is difficult to avoid some overlap in wavelengths passed by the two filters and scattered light can be a significant error. In solutions containing polymers and proteins, which are good scatterers, interference filters with their narrower bandpass are certainly a better choice.

This design nearly always calls for a partial double-beam arrangement to lessen effects of fluctuations and drift in source intensity and detector response. A variety of ingenious ways for obtaining the monitoring channel have been devised. It should be noted that a filter fluorometer can be quickly adapted for absorption photometric work in the visible by installing a detector in line with the sample and source.

"Reflectance" fluorometers, that is, instruments that rely on detection of fluorescence emitted 180° to the incident beam, are in use also. For example, paper and thin-layer chromatograms sprayed with ninhydrin or other selectively binding fluorophor can be monitored in this way. Various fluorescent solids (e.g., uranium ores) are also routinely assayed by such devices (see Exercise 15.16).

Fluorometers for remote monitoring of dangerous solutions often use a fiber-optic probe whose tip is shaped like a lens. Both exciting and emitted radiation travel the fiber optic. Another remote sensing arrangement is the laser-beam excitation of fluorescence of airborne pollutants. It also can be monitored by a filter fluorometer.

Spectrofluorometers. A representative fluorescence spectrometer or spectrofluorometer is shown in Fig. 15.4. For efficiency in the monochromator, the excitation grating is ordinarily blazed at from 250 to 300 nm, and the emission grating from 375 to 500 nm. Since absorption and fluorescence peaks are broad in most instances, the resolution of the monochromator need not be high.

The effectiveness with which scattered exciting radiation is excluded from luminescent radiation by an instrument will largely determine its S/N and limit of detection. Thus, for measurement of low analyte concentrations, spectrofluorometers with holographically ruled gratings are much preferred for their lower stray light. Indeed, a double monochromator in the emission (fluorescence) branch will reduce scattered light drastically. Modulation and lock-in amplifier detection are provided with many instruments and will extend the limit of detection still further (see Section 12.7).

Absolute Intensities. Where quantum efficiencies, other aspects of energetics, or simply accurate spectra are of interest, it is necessary to correct apparent intensities displayed by a spectrofluorometer. Basically, the wavelength dependence of each characteristic module must be measured and the resulting intensity data plotted or entered in memory. Then they can be used to correct apparent spectra by use of appropriate programs.

Example 15.4. What steps are necessary to *correct excitation spectra* from a particular spectrofluorometer? While exact corrections call for radiometric procedures, good corrections can be made more simply. According to Eq. (15.5) the light flux of the source/excitation–monochromator pair is the product $P_0(\lambda) f_{ex}(\theta, \lambda)$. To measure this quantity, it is feasible to use as the sample a "quantum counter," a dye whose fluorescence efficiency is essentially independent of wavelength. Its usable spectral range must be known.

For instance, Rhodamine 101 is useful as a quantum counter. A solution in ethylene glycol (5 g L^{-1}) is known to absorb virtually all incident radiation and have a high, nearly constant quantum yield.

With the dye solution in place, the emission monochromator is set to the wavelength of maximum response. The fluorometer output is recorded while the excitation monochromator is scanned through the spectral range of interest. The output of the spectrofluorometer is the needed excitation correction curve. Exercise 15.8 offers an alternative procedure.

To correct an observed excitation spectrum, the amplitude at each wavelength should be divided by the value in memory. Scaling will probably be necessary. Exercise 15.7 extends the discussion and deals with correction of emission spectra as well.

Example 15.5. How can one establish that spectral curves obtained in the laboratory are accurately corrected? Assume that the procedure of Example 15.4 is applied to a sample solution.

One should compare the corrected excitation spectrum with the absorption spectrum for the same substance; they should be identical. It is easy to make a comparison when the spectrofluorometer may also be used as a single-beam spectrophotometer (see Section 17.4).

To validate the correction of emission spectra it is best to determine the emission spectrum of a standard substance like quinine sulfate at a concentration reported in the literature.

Pulsed-Source Phosphorimeters. Today in phosphorescence spectrometry the time-resolution of spectral responses is commonly provided. This approach makes it feasible to resolve phosphorescence spectra of analytes with different lifetimes (recall that they spread over two to three orders of magnitude). This type of phosphorimeter also provides a way to separate these responses from background composed of Rayleigh scattering of the exciting radiation and fluorescence by ever present impurities and perhaps also the analytes.

For time-resolved phosphorimetry a pulsed source such as a xenon flash lamp that will give a burst of illumination with half-width no greater than 10 μs is adequate. It may be energized by short bursts of current from a discharging capacitor or a special electronic pulse generator, which should also trigger a counter to provide a delay interval sufficient to let scattering and fluorescence die out before the detector is activated. In Fig. 15.5 a time diagram of the flash lamp pulse and sample luminescence is shown. As the timed interval ends the circuit should trigger a second counter to activate the detector circuit for a precise interval. Note this interval in the figure. Where more than one luminescent species is present, if lifetimes are known and are sufficiently different, the delay interval can be adjusted to permit separate observation of the emission from each following successive flashes.

The observation or "gate" interval should also be adjustable. By keeping this interval small and varying the delay time, it is possible to obtain emission intensity

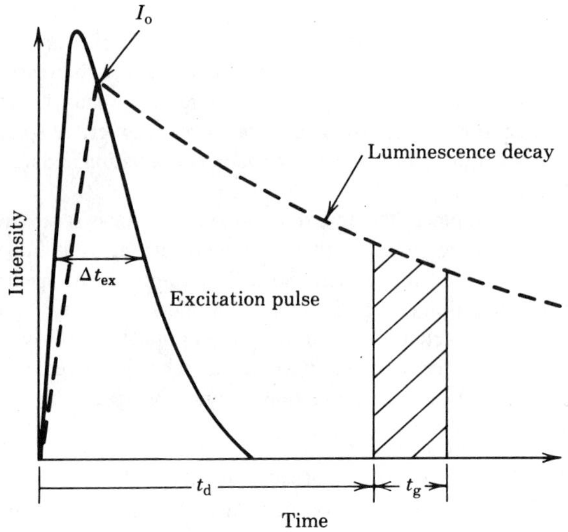

Fig. 15.5 Time dependence of an excitation pulse and the luminescence of a substance. Symbols are: Δt_{ex}, half-width of excitation pulse; I_0, maximum intensity of luminescence; t_d, the delay prior to observation of luminescence; and t_g, the observation or gate interval.

as a function of time and from that information to calculate the luminescence lifetime (see Exercise 15.19) if a single species is involved.

Time-Domain Spectrometers. Spectrometers capable of carrying out time-resolved luminescence spectroscopy permit resolution not only between phosphorescence and fluorescence spectra but also between Raman emission (and Rayleigh scattering) and luminescence emission. More important, they facilitate collection of data for accurate measurements of lifetimes of excited species. The advent of pulsed lasers (pulses of 70–100 ps can be achieved with a mode-locked laser and still shorter pulses are available if a dye laser is driven by a mode-locked laser) has made possible the short excitation pulses required for such measurements on many fluorescent species.

Pulsed-excitation methods are widely used for precise lifetime determinations. Two or as many as three different lifetimes can be determined in a system. Pulse sources are a fast flash lamp like the coaxial all-metal hydrogen thryatron* or a pulsed laser whose discharges can be precisely excited. Fast-rise-time photomultiplier tubes are also necessary to ensure the capability of operation at times as short as 1 ns. If relaxation times are longer than the average photomultiplier pulse, the decay time is measured by methods such as the one described for a phosphorimeter. For decays that occur more quickly than the average photomultiplier pulse length, a more complex procedure is needed. The *single-photon counting method* is especially valuable. It is beyond the scope of this text to describe these methods in detail and the reader is referred to references.†

*D. J. S. Birch and R. E. Imhof, *Rev. Sci. Instr.*, **52,** 1206 (1981).

†J. C. Wright in *Applications of Lasers to Chemical Problems*, T. R. Evans, Ed. New York: Wiley, 1982.

An alternative to pulsed excitation is phase modulation. In this approach the exciting radiation is modulated at any of several frequencies, usually sinusoidal, that can be as high as 250 MHz. Fluorescence from an analyte will be phase shifted by angle ϕ relative to the exciting light as defined by the expression

$$\tan \phi = 2\pi\nu\tau,$$

where τ is the lifetime of the excited state and ν the modulation frequency. Phase-sensitive detection will recover the phase angle, and the lifetime can be calculated. The name *phase-resolved fluorescence* spectrometry is applied to the method. A multicomponent approach in which data are taken for overdetermined matrices allows both analyte concentrations and lifetimes to be calculated.*

15.4 ANALYTICAL MEASUREMENTS

In dealing with analytical procedures, attention will be focused on quantitative approaches since luminescence measurements do not lend themselves to qualitative analysis. In the main this situation can be traced to the small number of species that luminesce, though it also reflects the lack of detail in regular luminescence spectra.

Fluorometry. Over what range of concentrations is fluorometry valuable for quantitative determinations? What degree of control of solution conditions is necessary to secure valid measurements? These important questions will be considered in sequence.

The relation between fluorescence and concentration involves, as was seen in developing Eq. (15.5), both the molecular system and the instrument system. Recall that as long as the product $\epsilon bc < 0.05$, Eq. (15.5) can be written essentially as given earlier

$$\text{signal} = P_0 \epsilon b c \phi_f \cdot \text{modular factors}. \tag{15.6}$$

In a 1-cm cell, for this product to have such a small value, the analyte concentration should be no greater than 10^{-5}–10^{-6} M if ϵ is of the order 10^3–10^4, which are representative values for molar absorptivity in the UV. Fortunately, concentrations can often be higher when quantification is by use of calibration curves, as will be seen below. Usually the dynamic range covers as many as four orders of magnitude, since for a given fluorometer, Eq. (15.6) reduces to

$$\text{signal} = Kc, \tag{15.7}$$

where K is a constant including the other terms.

Analyte concentrations would thus be kept low if only to ensure linearity of response. Another reason to do so, as noted earlier, is that efficiency ϕ_f holds constant only at low concentrations. A final reason for low concentrations is that primary and secondary absorption by the analyte and by impurities at higher con-

*L. B. McGown and F. V. Bright, *CRC Crit. Rev. Anal. Chem.*, **18,** 245 (1987).

centrations lessens the instrument response. Absorption by impurities is probably best handled by use of the method of standard addition since impurity concentration may vary from sample to sample.

Care must also be taken to minimize processes that will cause loss of excitation energy in solution, which are collectively termed *quenching* and were discussed in Section 15.2. Somehow these processes must be controlled and minimized to make quantitative luminescence measurements possible.

Dissolved O_2 is the most common impurity quencher and can be removed easily by bubbling pure N_2 through a solution or by alternate freezing and pumping of a solution. An indication that energy-transfer quenching, which involves impurities, is occurring is the presence of longer wavelengths than normal in the solute fluorescence spectrum. Starting with pure solvents will thus be essential for careful work. Attention also needs to be given to the possibility that samples may contain impurities, especially aromatics, that will quench part or all of the luminescence of the analytes. A preliminary separation is indicated if such quenching occurs. In a few instances dilution may also be effective in reducing such quenching if luminescence intensity is not lowered too greatly in the process. Finally, the type of quenching that occurs with increasing concentration will be evident from a calibration curve and is avoided by appropriate dilution.

All the regular analytical procedures for quantifying measurements, such as the use of a working curve or the standard addition method, are usable in fluorometry (see Sections 11.3 and 11.4). Under conditions outlined in Section 15.2 and extended below, concentrations of luminescent analytes can be reliably determined. High selectivity and precision and low limits of detection can be achieved. In practice, analytical or working curves (instrument readout versus c) can be extended upward in concentration until the fluorescence efficiency begins to fall (usually at about 10^{-4} M).

The scope of quantitative determinations of species is broad. A variety of crystalline salts, dye–metal complexes, and aromatic and other unsaturated organic compounds fluoresce. Thiamine and riboflavin belong to the last category and can be estimated directly. Most analyses, however, are based on the formation of a fluorescing species by combining an analyte with a fluorophore. For example, traces of uranium can be determined by fusing a sample with a KF–$NaKCO_3$ mixture to form a fluorescent salt complex. Complexation with a fluorophore is a more common mode of development of fluorescence; for example, aluminum may be analyzed as the fluorescent chelate formed with the dye Pontachrome blue-black.

Example 15.6. What will be the shape of a luminescence calibration curve at relatively high analyte concentrations?

If concentration is increased to a level where all entering light is absorbed, the exponential term $1 - e^{-\epsilon bc}$ in Eq. (15.3) will go to zero. The luminescence intensity will not be independent of concentration, however, because concentration quenching and other effects discussed in Section 15.2 will cause emission to decrease. As a result, most plots of luminescence are found to go through a maximum at higher concentrations and then decrease substantially.

Example 15.7. How can one determine the best wavelengths for excitation and emission for a single analyte? Assume that interfering substances are absent. Is knowing its luminescence spectrum sufficient?

Since fluorescence occurs after vibrational deactivation in the excited state, for a given analyte the emission spectrum is ordinarily independent of the excitation spectrum and the excitation wavelength may be any value that is absorbed that leads to emission. For phosphorescence the additional step of intersystem crossing provides comparable isolation from the excitation process. If the luminescence spectrum of an analyte is available, the optimum wavelength for a luminescence measurement will ordinarily be that of peak emission.

Especially for measurements in organic solvents, in addition to keeping concentrations below levels at which concentration quenching begins to affect precision and removing oxygen and other impurities that can lead to impurity quenching, pH and temperature should in most cases be regulated. The latter variables affect the point of equilibrium and alter fluorescence intensity. For instance, as simple a structural change as the addition or deletion of a proton is enough for many molecules to shift wavelengths of fluorescence or to destroy fluorescence. The optimum pH and temperature for a fluorescent species in a given solvent may be found experimentally by use of an optimization procedure.

Phosphorimetry. Solution samples are nearly always observed at 77 K in the glassy or vitreous state to slow down nonradiative processes. A thorough degassing of solutions is essential because O_2 seriously quenches the triplet state from which phosphorescence emission begins. Subsequently, they are frozen in a narrow quartz tube sealed to the lower end of a quartz dewar flask. Naturally, phosphorescence can also be observed from many solid samples including salts of most lanthanides and actinides.

Room-temperature phosphorescence was first extensively studied in matrices from thin-layer chromatography.* It was found possible to extend this system by adsorbing phosphorescent molecules on filter paper which could then be wrapped around a cylindrical drum. More recently solution "matrices" such as sodium lauryl sulfate micelles have proved appropriate supports as well as hollow-cylinder molecules such as cyclodextrin.†

Qualitative Analysis. Since a fluorescence spectrum has much less structure under ordinary conditions than an absorption spectrum, fluorometric methods tend to be used mainly for spot tests. Fluorometric methods are frequently used to locate, and in a preliminary way identify, species separated on thin-layer or paper chromatograms.

Probes in Biological Systems. Small, highly fluorescent molecules with moderate to good quantum efficiencies ($\phi_f > 0.1$) and solubility in polar media make good

*E. M. Schulman and C. Walling, *Science*, **178,** 53 (1972) and *J. Phys. Chem.*, **77**, 902 (1973).

†R. J. Hurtubise, Phosphorimetry, *Anal. Chem.*, **55,** 669A (1983).

probes, labeling agents, or tracers in biological systems: their values of λ_{max}, ϕ_f, τ, and degree of polarization of emission are quite sensitive to local molecular environment.

Example 15.8. Highly fluorescent molecules such as ANS (1-aniline-8-naphthalene sulfonate, λ_{max}. $\phi_f = 670$, $\tau = 16$ ns) make excellent probes in polar solutions. As ANS is placed in less polar solvents, the λ_{max} of its fluorescence spectrum shifts from 515 nm (in water) toward the blue and ϕ_f increases.

What information can be deduced by 1:1 binding of ANS with myoglobin if its λ_{max} shifts to 454 nm and its ϕ_f becomes 0.98? Further, ANS is displaced by the heme. These fluorescence data make it appear that binding is in a nonpolar region and the displacement of ANS by the heme points to the possibility it has bound in the heme pockets, which are of low polarity.

Example 15.9. Much can be learned about biological processes from lifetimes. For instance, though chlorophyll molecules are fluorescent, in daylight the photons they absorb are used for photosynthesis and not reradiated. Can the rate of internal transfer of excitation energy to centers that promote photosynthetic chemical reactions be estimated from what is known about chlorophyll fluorescence?

If 10^{-9} s is a representative lifetime in the excited singlet state, most energy transfers must occur in less than 1 ns. The lifetime of the process of internal transfer must have a lower limit of about 10^{-10} s.

Example 15.10. Further, changes in polarization yield information about biological changes. By irradiating a fluorescent probe molecule with linearly polarized exciting light while the probe is in a system of interest and then measuring the degree of polarization of emitted light, much can be learned about possible rotation of the probe in that system during the interval of its excitation. Both molecular absorption and emission are at a maximum when the vibration of the electric vector of light is aligned with the transition moment. Transition dipoles are oriented characteristically with respect to molecular structure. Processes in biochemistry that have been studied in this fashion include protein denaturation and aggregation and binding [3].

Analysis of Mixtures. The rather featureless character of most solution fluorescence spectra and their extensive overlap limits the possibility of determining the composition of mixtures without increasing the number of variables measured and using better computational methods. One strategy is to make measurements at a set of excitation and emission wavelengths and to seek solutions to the simultaneous equations for total fluorescence at each wavelength. The goal is to achieve sufficient selectivity to simplify the analysis. Standards must be run to determine sensitivity factors.

Several more powerful approaches have been developed. In *synchronous luminescence spectrometry* excitation and emission wavelengths are scanned simultaneously with a fixed $\Delta\lambda$ wavelength difference (e.g., 10 nm) between them. The method is straightforward instrumentally as long as a spectrofluorometer has a mechanism for interlocking the monochromator drives.

In Fig. 15.6 the manner in which synchronous scanning develops sharper spectral features is suggested. The narrow spectral window strongly affects the emission spectrometer output; only when the window includes a wavelength at which excitation is possible and another at which emission is possible will emission be observed. As shown in the figure, this method gives a much narrower curve with more sharply distinctive features.

Finally, *site-selection* (line narrowing) *luminescence spectrometric methods* that yield better resolution of fluorescent peaks have been developed and will be described below.

Total luminescence spectrometry is an even more powerful type of multidimensional fluorometry. Emission intensities are monitored while both excitation and emission wavelengths are varied. The data array obtained is usually called an *excitation–emission matrix* (EEM). Repeated scans are of course necessary to develop EEM displays. Spectrofluorometers in which stepper motors are used often have a program for the necessary raster-type scans and data acquisition. Older instruments can often be adapted simply.*

When the excitation wavelength is plotted on the y axis and emission wavelength along the x axis, *contour-mapping* may be used to represent the emission

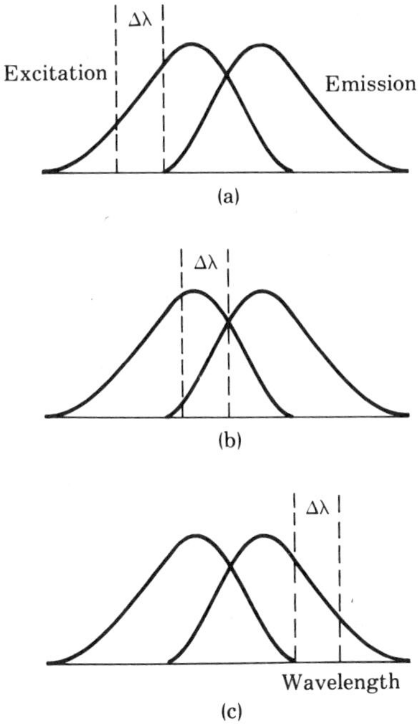

Fig. 15.6 Synchronous luminescence spectrometry. A constant difference $\Delta\lambda$ is maintained between the wavelengths of the excitation and emission monochromators. Emission can be observed only when the leading edge of $\Delta\lambda$ is within the emission band and the trailing edge within the excitation band. (a) No emission. (b) Emission. (c) No emission.

*E. R. Weiner and M. C. Goldberg, *Am. Lab.*, **14**(9), 91 (1982).

intensity. In the resulting display lines will indicate constant levels of intensity and peaks will be characteristic of components present. Alternatively, emission intensity can be represented on a vertical z axis and excitation and emission wavelengths plotted along y and x axes about 60° to each other. An example of this type of plot is given in Fig. 15.7. What would a contour map display of the data look like?

Site-Selection Spectrometry. Several modern luminescence procedures called site-selection or line-narrowing techniques have been developed for the analysis of mixtures. At room temperature luminescence bandwidths are wide because of thermal motion. At very low temperatures in the glassy or in the crystalline state, bandwidths are large because of microscopic flaws and molecular variability. In a glass, an analyte molecule can be said to occupy a solvent cage; since the cages differ considerably in form, a bandwidth for a particular fluorescence transmission is about 300 cm^{-1}.

Resolution of peaks for different species will require line narrowing. It is apparent that inhomogeneities need to be overcome and thermal effects reduced. With a tunable laser, for example, a N_2 laser-pumped dye laser, it is found possible once microscopic inhomogeneities are reduced to selectively excite to fluorescence species in those sites for which the laser frequency is appropriate. Since the laser bandwidth is narrow, only those species in sites attuned to the particular frequency give rise to fluorescence.

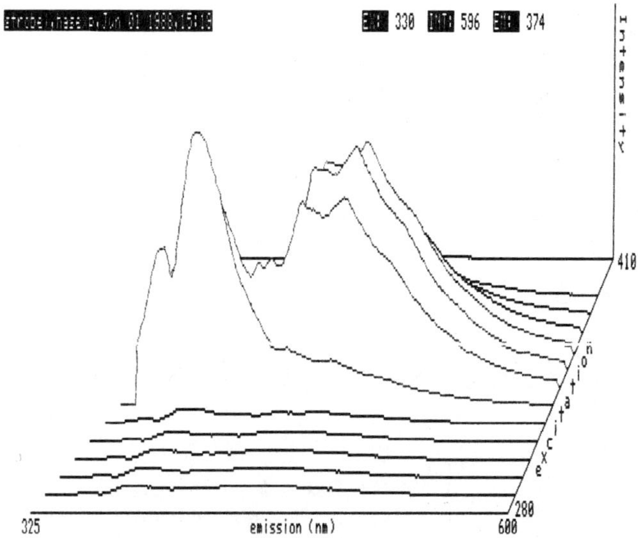

Fig. 15.7 A multidimensional luminescence plot (emission intensity × excitation wavelength × emission wavelength) of a mixture of 1,1-binaphthyl and fluoranthrene in cyclohexane at 25°C. Spectral slit widths were 10 and 1 nm for excitation and emission monochromators, respectively, and a Hg arc lamp was the excitation source. Each line is an emission spectrum at a particular excitation wavelength. Emission peaks are: 375 nm, 1,1-binaphthyl; 440 and 460 nm, fluoranthrene. These components can be spectrally resolved since the 313 nm Hg line excites both substances while the 365 nm line excites only binaphthyl. [Courtesy of Charles H. Lochmüller.]

Some valuable types of molecular *settings* (by type of analyte and molecular setting) are:

1. Ionic species dispersed in a solid CaF_2 matrix. Lanthanide ions in particular may be determined by coprecipitation in CaF_2, drying at about 500°C to ensure substitution in the CaF_2 lattice, followed by slow cooling to 13 K [J. C. Wright and F. J. Gustafson, *Anal. Chem.*, **50,** 1147A (1978) and **51,** 1762 (1979)].
2. Fluorescent molecular species dissolved in an appropriate solvent mixture and subsequently cooled to 4 K to form a clear glassy medium.
3. Luminescent molecules whose structure fits well into a frozen lattice. For example, polyaromatic hydrocarbons fit well in an *n*-heptane lattice. See Example 15.12 below.
4. Luminescent molecules vaporized and mixed with inert gas and subsequently crystallized in solid layers on a sapphire substrate cooled to 11–13 K. Matrix isolation spectrometry is a valuable technique.

Example 15.11. Spectra of a set of polycyclic aromatic hydrocarbons (PAHs) from solvent-refined coal were resolved in a glassy medium. A sample mixture was dissolved in an approximately 1:1:1 mixture of glycerol, H_2O, and dimethyl sulfoxide and cooled carefully to 4 K to form a clear glass. Though absorption bandwidths are inhomogeneously broadened to perhaps 300 cm^{-1}, with a narrow-line laser, absorption lines are no more than 1–3 cm^{-1} because of the site selection capability. An N_2 laser-pumped dye laser ensured the narrow linewidths for excitation. Triggering by the pulsed N_2 laser activated delay gates whose delay and width were varied to enhance resolution. In each analyte spectrum its $v = 0$ to $v = 0$ electronic transition was the most intense. Even substitutional isomers were spectroscopically distinguished [J. C. Brown, J. A. Duncanson, and G. J. Small, *Anal. Chem.*, **52,** 1711 (1980)].

Example 15.12. In laser-excited *Shpol'skii spectrometry* excellent selectivity for a mixture of benz[a]anthracene and 11 alkylated derivatives in a solid *n*-octane lattice was obtained [A. P. D'Silva and V. A. Fassel, *Anal. Chem.*, **56,** 985A (1984)]. In each case the 0–0 transition is the strongest. Both a XeCl and a KrF eximer laser-pumped dye laser were used for excitation of fluorescence. Approximately 10 ns pulses were produced. Two interchangeable detectors, a photomultiplier and a photodiode array, were used. Spectral positions of the 0–0 fluorescence lines for the alkylated PAHs ranged from 383 to 401 nm. Excitation wavelengths ranged from 373 to 389 nm. By use of an internal reference consisting of a deuterated analog of the analyte it is possible to carry out quantitative analyses.

Time-Resolved Fluorometry. With pulsed lasers that are tunable time-resolved measurements can be arranged. The initial pulse can trigger a delay circuit and a gate timer. The delay allows the excitation pulse to end and Rayleigh scattering to become minimal. When the gate opens briefly, a selected portion of the fluorescent spectrum is recorded. A limit of detection for rubrene in benzene of 1.8×10^{-13} *M* has been achieved by time-resolved observation of the fluorescence [G. R. Haugen and F. E. Lytle, *Anal. Chem.*, **53,** 1554 (1981)].

REFERENCES

Some well-written general volumes on an intermediate to advanced level are:

1. J. B. Birks, *Photophysics of Aromatic Molecules*. New York: Wiley-Interscience, 1970.
2. G. G. Guilbault (Ed.), *Practical Fluorescence, Theory, Methods, and Practice*. New York: Dekker, 1973.
3. J. R. Lakowicz, *Principles of Fluorescence Spectroscopy*. New York: Plenum, 1983.
4. S. G. Schulman, *Fluorescence and Phosphorescence Spectroscopy: Physicochemical Principles and Practice*. New York: Pergamon, 1977.
5. W. R. Seitz, Luminescence spectrometry (fluorimetry and phosphorimetry), in *Treatise on Analytical Chemistry*, 2nd ed., Part I, Vol. 7, P. J. Elving, E. J. Meehan, and I. M. Kolthoff (Eds.). New York: Wiley-Interscience, 1981.

A thorough, advanced treatment with special coverage of instrumentation will be found in:

6. C. A. Parker, *Photoluminescence of Solutions with Applications to Photochemistry and Analytical Chemistry*. New York: Elsevier, 1968.
7. J. D. Winefordner, S. G. Schulman, and T. C. O'Haver, *Luminescence Spectrometry in Analytical Chemistry*. New York: Wiley-Interscience, 1972.

A practical treatment of reliability in measurement is given in:

8. J. N. Miller (Ed.), *Standards in Fluorescence Spectrometry*. New York: Chapman and Hall, 1981.

Also of interest are:

9. L. J. Cline Love and L. A. Shaver, Time-Correlated single-photon technique: Fluorescence lifetime measurements, *Anal. Chem.*, **48,** 364A (1976).
10. A. P. D'Silva and V. A. Fassel, Laser-excited Shpol'skii spectrometry, *Anal. Chem.*, **56,** 985A (1984).
11. R. J. Hurtubise, Phosphorimetry, *Anal. Chem.*, **55,** 699A (1983).
12. H. H. Jaffe and A. L. Miller, The fates of electronic excitation energy, *J. Chem. Educ.*, **43,** 469 (1966).
13. W. J. McCarthy and J. D. Winefordner, The selection of optimum conditions for spectrochemical methods, Part II, Quantum efficiency and decay time of luminescent molecules, *J. Chem. Educ.*, **44,** 136 (1967).
14. L. B. McGown and F. V. Bright, Phase-resolved fluorescence spectroscopy, *Anal. Chem.*, **56,** 1400A (1984).
15. S. G. Schulman (Ed.), *Molecular Luminescence Spectrometry*, Methods and Applications: Part 1. New York: Wiley-Interscience, 1985.
16. I. M. Warner, G. Patoney, and M. P. Thomas, Multidimensional luminescence measurements, *Anal. Chem.*, **57,** 463A (1985).
17. E. L. Wehry (Ed.), *Modern Fluorescence Spectroscopy*. New York: Plenum, Vols. 1 and 2, 1976; Vols. 3 and 4, 1982.
18. J. C. Wright and F. J. Gustafson, Ultratrace inorganic ion determination by laser excited fluorescence, *Anal. Chem.*, **50,** 1147A (1978); **51,** 1762 (1979).

EXERCISES

15.1 Show that the term $P_0(1 - e^{-\epsilon bc})$ appearing in Eq. (15.3) expresses the intensity of incident radiation absorbed by a sample. [See Eq. (16.3).]

15.2 Assume a linear fluorometer using an intense source is to be designed. Indicate modules or types of components necessary if very-low-intensity fluorescence is to be quantitatively detected.

15.3 For a luminescent molecule, (a) how strongly is the *shape* (profile) of the emission spectrum (strongly, weakly, not at all) dependent on the excitation wavelength? (b) Is the *intensity* of the emission spectrum dependent on the excitation wavelength? (c) Is the shape (profile) of the excitation spectrum dependent on the monitored emission wavelength? Support the choices made.

15.4 Assume a nonfluorescing solute absorbs at the same wavelengths as a fluorescent analyte. (a) Briefly describe how a measured fluorescence intensity can be corrected. (b) Can the standard addition method be applied to compensate for the effect? If so, describe briefly.

15.5 Dilute solutions of pyrene in cyclohexane produce a fluorescence spectrum that obeys the "mirror-image" law and yields an emission maximum at about 390 nm. As the pyrene concentration is increased, however, the band at 290 nm diminishes and a new, structureless emission appears at 480 nm. Interestingly, the intensity of the 480 nm band is proportional to $[\text{pyrene}]^2$. Suggest an explanation.

15.6 A study is to be made of the primary absorption or inner filter effect. A fluorescent analyte is irradiated at λ_{max} in a 1-cm square cell at concentrations that give absorbances A of 0.05, 0.5, and 1. (a) Draw an *approximate* plot of P/P_0 versus distance into the cell. Assume that P_0 is the power incident on the cell face for a collimated beam and P the power *remaining at different distances into the cell.* (b) Consider that fluorescence is to be observed through a side wall of the cell 90° to the front wall. Considering only the power available for excitation at different distances, would it be best to observe fluorescence from portions of the solution 0–0.2, 0.4–0.6, or 0.8–1 cm into the cell? (c) Does the $A < 0.05$ upper limit on which Eq. (15.4) is based ensure a minimum absorption effect? Discuss.

15.7 Energy-corrected excitation spectra are to be measured. (a) Show by block diagram how to configure a spectrofluorometer with a xenon lamp source to determine its excitation correction curve by the procedure outlined in the Example 15.4. (b) Assume Rhodamine 101 is used as the quantum counter and that the acceptance angle of the monochromator includes the entire sample cell. Equation (15.3) applies to the measurement. What assumption must be made about the product ϵbc of the dye over the spectral range of interest for the correction curve to be valid? (c) Sketch an approximate excitation correction curve. What information does it give about the xenon source? If the excitation monochromator grating is blazed at 300 nm, what contribution has it made to the correction curve? Write an equation for the curve.

15.8 To obtain an emission correction curve for a spectrofluorometer, assume a diffuser plate is placed in its sample cell compartment and that its two monochromators are scanned in synchronism while the instrument output is recorded. (a) Sketch a rough emission correction curve. Explain how this correction curve would be used. (b) Write an equation for the emission correction curve. (c) If emission curves are to be obtained at several different excitation wavelengths, would the excitation correction curve be needed also?

15.9 An excitation spectrum is observed for a fluorescent species using a grating spectrofluorometer with a xenon lamp source. Several slit widths are employed in the excitation monochromator. As the slits are narrowed, increasing fine structure appears in the observed spectrum. Is the source of this "detail" likely to be the instrument or the species being excited? Suggest a procedure to check your answer.

15.10 Describe the effect on both the intensity of the fluorescence and the emission maximum for a 10^{-4} *M* solution of benzo[a]pyrene in pentane ($\lambda_{ex\,max} = 381$ nm, $\lambda_{em\,max} = 403$ nm) when the following conditions are changed. (a) Sulfuric acid is substituted for pentane. (b) The temperature is lowered from 25 to -25°C. (c) Concentration is doubled. (d) λ_{ex} is changed to 310 nm. (e) A 100-W Hg lamp source is used in place of a 150-W Xe lamp.

15.11 What is the concentration of anthracene at the high end of its dynamic range? Assume $\epsilon_{ex} = 7{,}400$ L M^{-1} cm^{-1} at 376 nm and a cell pathlength of 1 cm.

15.12 How successfully can scattered light be discriminated against in fluorometry by linear polarization of the excitation source output and use of a crossed polarizer between the emission monochromator and the detector? Does scattering alter the direction of polarization? [See C. S. Lim et al. *Anal. Chem. Acta*, **100,** 235 (1978).]

15.13 To compare substances as fluorescent probes for research studies, the product $\epsilon_{max}\,\phi_f$ has been used as a measure of relative fluorescence efficiency. (a) Evaluate the concept in terms of Eq. (15.4). (b) Rate the following substances as probes on the basis of their fluorescence sensitivity: (i) dansyl chloride (8-dimethylamino-3-chlorothionylnaphthalene), $\epsilon_{max} = 3.4 \times 10^3$, $\phi_f = 0.1$ and (ii) pyrene, $\epsilon_{max} = 40 \times 10^3$, $\phi_f = 0.98$. (c) What other properties of probes might affect their usefulness in particular systems?

15.14 Quinine sulfate undergoes a change in fluorescence wavelengths with change in pH (at pH 6 from blue to violet and at pH 10 from violet to colorless). Suggest an analytical use for this phenomenon.

15.15 Assume that radiation from a laser is polarized parallel to the *z* axis of a crystal or experimental coordinate frame. (a) Rewrite Eq. (7.23) using parallel and perpendicular subscripts and assuming the analyzer's polarization is parallel. (b) If the laser excites fluorescence in a motionless molecule, its fluorescence will have the same orientation. If, however, the molecule rotates far enough between excitation and fluorescent emission, the emission will be partially depolarized. Calculate the value of *P* for no rotation and for sufficient rotation that $P_\perp = 0.5P_\parallel$ on emission.

15.16 Uranium can be determined in solids at trace levels by fluorometry. Samples are ground, mixed with a flux, such as sodium carbonate or sodium fluoride–lithium fluoride, and melted. Suggest (a) a suitable shape into which the melt can be cast for measurement of fluorescence and (b) a compact design of instrument in which two exciting lamps (source) and detector are located on the same side of the sample. Indicate where baffles are used. Thermo Jarrell-Ash markets such an instrument.

15.17 It is found that a fluorescence emission profile varies with the excitation wavelength. What conclusion may one draw about the sample?

15.18 Show that by differentiating Eq. (15.2) and converting to logarithms to base 10 that one has

$$\log I = -2.303t/\tau + \log I_0.$$

Since the new equation is that of a straight line, obtain the equation for its slope. Show how the lifetime τ of the excited state of the luminescing species can be calculated.

ANSWERS

15.3. (a) No. Shape is a function of energy levels and is determined by the nature of the analyte. (b) Yes. Since molar absorptivity varies with wavelength, the emission intensity is wavelength dependent. *15.5* It has been hypothesized that pyrene forms an excimer, a dimer consisting of a molecule of pyrene in the excited state and another in the ground state. Delocalization of electrons over a molecular system twice as large accounts for the red shift. Dimer formation accounts for the second-order dependence on pyrene concentration. *15.6* (b) 0–0.2 cm best if relatively highly absorbing solutions are to be measured. (c) It appears a good limit. *15.7* (a) A solution of a ''quantum counter'' dye such as Rhodamine 101 should be placed in the sample cell. The emission monochromator wavelength should be set to give the peak intensity. *15.10* (a) Polar solvent effects cause emission to increase and λ_{max} to move to longer wavelength. (b) Emission increases and λ_{max} moves to shorter wavelength. (c) Intensity is increased but most likely not doubled. (d) Intensity is greatly diminished (ϵ is smaller). (e) Intensity will most likely diminish. *15.11* For a linear response $\epsilon bc \leq 0.05$, $c_{max} \approx 6.7 \times 10^{-6}$ *M*. *15.12* (b) (i) 340, (ii) 10,000, (iii) 6700. *15.13* (b) Pyrene is better. *15.14* Quinine sulfate could be used as a fluorescent indicator for pH titrations that would be especially useful for colored solutions. *15.15* (a) Clearly I', the maximum intensity must correspond to $I_{\parallel}$. Thus, Eq. (7.23) becomes $P = (I_{\parallel} - I_{\perp})/(I_{\parallel} + I_{\perp})$. (b) No rotation: $P = (1 - 0)/(1 + 0) = 1$; rotation sufficient to make $P_{\parallel} = 0.5P_{\perp}$, $P = (0.67 - 0.33)/(0.67 + 0.33) = 1/3$. *15.17* More than one luminescing component must be present.

Chapter 16

ABSORPTION SPECTROMETRY: SPECTROPHOTOMETRY

16.1 INTRODUCTION

For as long as color has been recognized as characteristic of particular materials under given conditions, it has been used as a means of identification. Qualitative analysis schemes, for example, have traditionally been committed to color tests such as observing the orange hue of lead chromate to confirm the presence of lead or chromium ions. Yet tests of this type are inherently limited in precision and range, for they rely on the human eye as a detector of radiant energy.

The perfection of other detectors of radiation in the optical spectral range, together with the general advance in instrumentation, has produced a vast extension of absorption techniques. In range, these now cover the optical spectrum (Table 16.2) from the far infrared through the vacuum ultraviolet. Since such techniques are concerned with the measurement of the intensity or power of radiation as a function of wavelength, the terms *spectrometric* or *spectrophotometric* may be used to identify them. In some spectral regions and for some analytes the absorption is so strong that concentrations are small as 0.01–0.001 ppm can be detected. The instrumentation of absorption spectrometers both of the dispersive and Fourier transform types will be considered in detail in Chapter 17.

Spectrophotometric techniques are based on the ability of substances to interact with characteristic frequencies of radiation. Since each isolated species of ion, atom, or molecule will exhibit an individual set of definite energy levels, it can absorb only the frequencies that correspond to excitation from one level to another. These matters will be considered in detail in Section 16.3. The absorption spectrum of an unknown substance may be measured in qualitative analyses to establish its identity. Quantitative procedures can be devised by relating the intensity of absorption to the number of the species of interest in the optical path. This chapter will be concerned with all of these topics.

Classic colorimetric methods, on the other hand, are distinguished by their dependence on the subjective perception of color originating in the human brain. By definition, color is restricted to the visible frequencies. It is always related to (a) some source of radiant energy, a fluorescent lamp, perhaps, (b) the chemical constitution of the material to which the color is ascribed, and (c) the eye of the individual who observes the color. As any of these varies, the "quality" of color will change. A distinction may also be drawn between the quality of color perceived by transmitted and reflected light. Generally, classic methods prove less reliable than instrumental methods as a result of fatigue, poor ability to establish intensities, and other characteristics of the average eye.

16.2 BEER'S LAW

Whenever a beam of broadly polychromatic radiation passes through a medium, for example, a liquid or gas, some loss of intensity occurs. First, reflection takes place at the phase boundaries as a result of refractive index differences between the medium and its surroundings (see Section 7.8). Second, scattering caused by inhomogeneities (in mixtures) or by thermal fluctuations in the bulk of the medium produces an additional small loss of power from the main beam (see Section 7.12). In general, neither of these is as significant in accounting for the intensity diminution, however, as the fact that the medium itself is not perfectly transparent but will absorb the radiant frequencies that promote energy changes within its molecules and ions. A schematic representation of the effect of reflection and scattering is given in Fig. 16.1.

A distinction should be made between the process by which the power level of the radiation is changed (e.g., absorption) and the quantitative measure of the effect. For easy identification let the suffix -ion refer to the process, and -ance to the measured value. For example, transmission, reflection, and absorption are occurring in Fig. 16.1 and lead to a measurable transmittance, reflectance, and absorbance. These terms are still to be defined. In general, the absorption is not directly measured but is derived from a determinable quantity, the radiant power *P* of the beam. *P* is simply the energy of the radiation reaching a given area per second.*

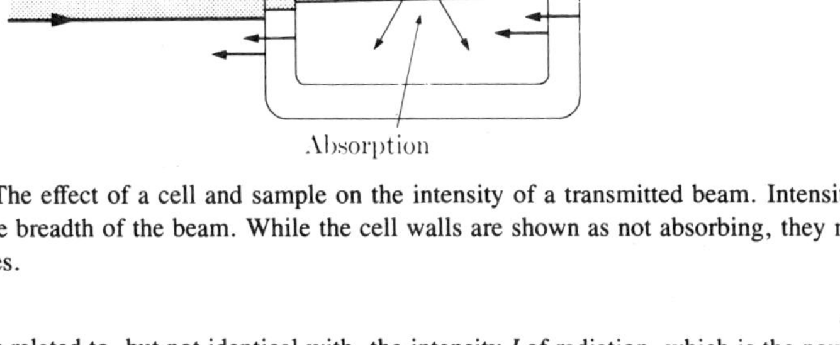

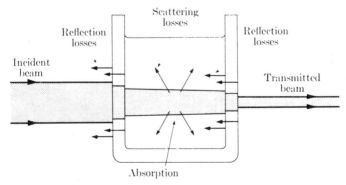

Fig. 16.1 The effect of a cell and sample on the intensity of a transmitted beam. Intensity is represented by the breadth of the beam. While the cell walls are shown as not absorbing, they may absorb in other cases.

*It is closely related to, but not identical with, the intensity *I* of radiation, which is the power per unit solid angle.

For each wavelength present in the beam, the absorbance depends on (a) the nature of the medium (i.e., its composition) and (b) the length of optical path in the medium. This dependence is expressed by Beer's law.* The assumptions made in obtaining the law are important:

1. The incident radiation is monochromatic.
2. The absorbing units (molecules and ions) act independently of one another regardless of number and kind.
3. The absorption is limited to a volume of uniform cross section.

We will consider the effect of these assumptions for practical work. Beer's law may be expressed as: The absorption of a medium is directly proportional to the number of absorption centers. In other words, each centimeter thickness of a solution obeying Beer's law will absorb an equal fraction of the incident power.

As light of power P passes through an infinitesimally thin cross section of the sample containing dn absorbers, it loses a fraction of its power dP/P. Beer's law connects these quantities by the equation $-dP/P = k\,dn$, where k is a proportionality constant. This differential equation can be integrated easily since the variables are separable. Carrying out the integration of dP/P between the limits of P_0 and P and of $k\,dn$ between 0 and n gives the equation:

$$\ln P/P_0 = -kn. \tag{16.1}$$

Here P_0 is the power incident on the sample, P is the power leaving the sample (see Fig. 16.1), and n is the number of absorbing centers of one kind in a volume of unit cross section. For the moment the cell holding the sample has been neglected. The equation predicts that the power of the emergent beam will drop off exponentially as the number of absorbers in the beam increases.

The *transmittance* T of the sample (i.e., the fraction of incident power transmitted), which is defined as $T = P/P_0$, makes its appearance now also. This quantity will have values from 0 to 1 or may be given as percent transmittance, 0–100% T.

Equation (16.1) is not operationally useful as it stands. First, the number of absorbers in the beam must be expressed in terms of concentrations. Assuming a rectangular sample cell, the total number N of absorbers in the beam will be the product $N = cN_0bS$, where c is the concentration of absorbers in moles per cubic centimeter, N_0 is Avogadro's number, b is the cell thickness in centimeters, and S is the cross-sectional area of the beam in square centimeters. Finally, as a matter

*Several scientists, Beer, Bouger, Lambert, Bunsen, and Roscoe, have contributed by their investigations and theorizing to the development of this relation. Often it is called the Beer–Lambert or the Bouger–Beer law. However, it has been shown by Liebhafsky and Pfeiffer [28] that Beer's original conception was sufficiently broad to include both concentration and length dependence. H. A. Liebhafsky and H. G. Pfeiffer, Beer's law in analytical chemistry, *J. Chem. Ed.*, **30,** 450 (1953). For that reason and for simplicity, the formal statement will be termed Beer's law.

of convenience, Eq. (16.1) may be shifted from the natural logarithmic base e to the Naperian base 10, designated by log, and the concentration changed to units of molarity. This modification can most easily be accommodated by substituting a new constant ϵ, called the *molar absorptivity*, for the constant k. With these changes Eq. (16.1) may be written

$$\log P/P_0 = -\epsilon bc \qquad \text{or} \qquad P = P_0\,10^{-\epsilon bc}. \tag{16.2}$$

Taking the reciprocal of the ratio P/P_0 removes the negative sign and gives

$$\log P_0/P = A = \epsilon bc. \tag{16.3}$$

This expression defines the absorbance A and is the simplest mathematical statement of Beer's law. Some other concentration units used in Eq. (16.3) are listed in Table 16.1.

In absorption photometry, both the transmittance and the absorbance figure prominently. The latter is more useful, however, because of its linear dependence on concentration and path length. The difference in behavior of T and A may be seen clearly in Fig. 16.2. For purposes of compound identification, the logarithm of A may be preferable to either T or A. (See Fig. 16.9 and accompanying discussion in the text.)

The important definitions and concepts of absorption photometry are summarized in Table 16.1. Other names and symbols that have been given these variables are noted in parentheses. Two typical calculations will provide illustrations of the interrelationship of the variables by use of Eqs. (16.2) and (16.3).

Example 16.1. In passing through 1 cm of a colored solution the incident power of a beam of a particular frequency is reduced 20%. (a) Develop an expression for absorbance in terms of transmittance. (b) What are the transmittance and absorbance of this solution? (c) If this solution is next placed in a cell with a 5-cm pathlength, what will the transmittance be?

Table 16.1 Terms and Symbols in Absorption Spectrometry

Name and Symbol	*Names No Longer Recommended*	*Definition*[a]
Absorbance, A	Optical density, extinction	$A = \log P_0/P$
Transmittance, T		$T = P/P_0$
Path length, b		
Absorptivity, a	Absorbancy index, extinction coefficient	$a = A/bc$, (c in g L^{-1})
Molar absorptivity, ϵ	Molar extinction coefficient, molar absorbancy index	$\epsilon = A/bc$, (c in mol L^{-1})

[a]Transmittance and absorbance can also be stated in terms of the intensity ratio I_0/I and $\log I_0/I$. Both absorbance and transmittance are dimensionless terms.

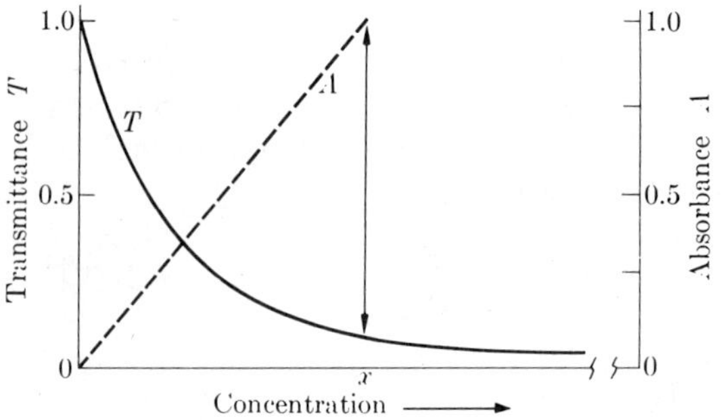

Fig. 16.2 The absorbance and transmittance of a solution at a given wavelength as a function of concentration. For example, at concentration x an arrow links the transmittance (0.10) and the absorbance (1.00). The path length b and other variables are constant.

Substituting the definition of transmittance ($T = P/P_0$) in Eq. (16.3) gives $\log(1/T) = A$ or $-\log T = A$.

The transmittance of the solution described is just $T = 0.80/1.00 = 0.80$. Its absorbance is $A = -\log T = -\log 0.80 = 0.096$.

To accommodate the change of path length, note that ϵc can be found for the 1-cm cell: $\epsilon c = A/b = 0.096/1$. Now the absorbance can be found for a 5-cm cell: $A = 0.096 \times 5 = 0.48$. Thus, antilog $0.48 = 0.33 = T$.

Note that for substances following Beer's law the molar absorptivity for a substance is independent of concentration and optical path length. It is a constant characterizing a substance at a particular wavelength when dissolved in a given solvent. Its occasional apparent dependence on temperature and other variables will be taken up in Section 16.4. Beer's law also extends to mixtures of different types of absorbers. As long as they act independently, the law holds. Each species has a different molar absorptivity, and the absorbances may be added. Equation (16.3) in the form applicable to homogeneous mixtures is

$$\log P_0/P = A_t = \epsilon_1 bc_1 + \epsilon_2 bc_2 + \cdots = b \sum_i \epsilon_i c_i, \qquad (16.4)$$

where solvent, path length b, and wavelength are presumed constant.

Concentration Limit on Validity. The restriction that absorption centers do not interact with themselves or other species causes Beer's law to be a limiting law applicable mainly in dilute solutions (concentrations $< 10^{-2}\,M$). The interference alters the charge distribution either in the absorbing or excited species or both and thus changes the energy needed for excitation. As a result, the position, shape, and height of the absorption region may be altered. Not all interactions are between neighboring molecules; through electrostatic forces, influences may be exerted at a relatively great distance. Many complex organic molecules, such as eosin and methylene blue, are particularly susceptible to interference and may follow Beer's law only below $10^{-5}\,M$ if certain simple salts are present.

A second limit on the validity of the law is that the index of refraction of the solution alters as the concentration changes. It may be shown that it is $\epsilon n/(n^2 + 2)^2$ rather than ϵ itself that is constant with concentration. Whenever the variation of $n/(n^2 + 2)^2$ exceeds the experimental error over a concentration region of interest, again Beer's law does not hold. In this instance a correction can be used at high concentrations; however, deviations larger than 0.01% from this source occur only at concentrations above 10^{-2} *M*.

16.3 ABSORPTION PROCESSES AND INTENSITIES

Beer's law relates absorption to concentration, but provides no clue as to the dependence of absorption on molecular structure. The connection is established through quantum-mechanical study of the characteristic "motions" that permit a molecule to interact with radiation. Since the energy of such motions, especially vibrations, is characteristic of particular molecules, absorption spectra are a sensitive structural tool. The manifold processes related to the absorption (and emission) of EM radiation are classified in Table 16.2. This section deals with the simpler aspects of molecular absorption; transition probabilities and intensities, line widths, and lifetimes of excited states were treated earlier in Section 13.3.

The energy of a molecule is usually characterizable as being distributed among motions of sufficiently different energy that each can be treated separately,* that is,

$$E_{\text{molecule}} = E_{\text{translation}} + E_{\text{rotation}} + E_{\text{vibration}} + E_{\text{electronic}} + E_{\text{nuclear orientation}}.$$

The absorption of a quantum of incident light may, depending upon its size and the particular molecule, simultaneously promote transitions in one or several categories of motion. Quantum-mechanical *selection rules* describe the combinations allowed. Neither translational nor nuclear orientation changes are treated here, the former because translational energies are essentially continuous and the latter because the topic is beyond the scope of this volume.

Rotational Transitions. For gaseous (i.e., isolated) molecules, energy changes associated with pure rotational shifts are observable, but only in the microwave and far IR range. In the easily accessible 4000–650 cm^{-1} IR range, only those rotational changes that accompany vibrational transitions are observed. Molecular rotational levels are spaced on the order of 10^{-2} J mol^{-1}, corresponding to radiation of a frequency of about 10 cm^{-1}.

It can be shown that the rotational energy of a molecule is generally expressible as a sum of terms, each proportional to the reciprocal of one of the molecular moments of inertia. These moments are taken around a set of internal, mutually orthogonal axes. For example, rotation of a rigid linear molecule lying along the

*At high levels of excitation, interactions among such "motions" are no longer negligible and correction terms must be added.

Table 16.2 Processes of Absorption of Radiation in Molecular Systems

Wavelength (nm)				*1000*	*100*	*10*	*1*	*0.1*
Wavenumber (cm^{-1})	*10*	*100*	*1000*	*10 000*	10^5	10^6		

	Microwave	Infrared (Far / Middle / Near)	Visible	Ultraviolet (Near / Far)	X-Ray (Soft / Hard)
Spectral regions	Microwave	Far, Middle, Near: Infrared	Visible	Near, Far: Ultraviolet	Soft, Hard: X-Ray
Characteristic process excited in region:	Molecular rotation in gases	Molecular rotations in gases, *inter*molecular vibrations in solids, liquids; Intramolecular vibrations in liquids, solids plus rotations in gases	Excitation of π, n, σ electrons as ν increases plus intramolecular vibrations and rotations		Electronic excitation of inner shell electrons in atoms
Characteristic spectrum	Pure rotational lines	Rotational–vibrational lines in gases and bands in liquids, solids	Electronic–vibrational–rotational lines in gases and bands in liquids and solids		Pure electronic lines
Energy of transition ($kJ\ mol^{-1}$)	0.1196	11.96	1196		1.196×10^5
Energy of transition ($eV\ mol^{-1}$)	1.24×10^{-3}	0.124	12.4		1240

z axis can occur around both x and y axes. Its two moments of inertia are equal, and the rotational energy is given by the equation

$$E_{\text{rotation}} = J(J + 1)(h^2/8\pi^2 I), \tag{16.5}$$

where J is the rotational quantum number ($J = 0, 1, 2, 3, \ldots$), h is Planck's constant, and I is the single moment of inertia. Exactly $2J + 1$ different orientations of a particular rotational axis in space are allowed. In the absence of an external field, these are of equal energy, giving each rotational energy level a $(2J + 1)$-fold degeneracy. The quantum-mechanical selection rule that applies to the general case of rotational transitions is $\Delta J = 0, \pm 1$.

Example 16.2. The microwave rotational spectrum of $^{12}C^{16}O$ consists of evenly spaced lines at 3.84 cm^{-1} ($J = 0$ to $J = 1$), 7.68 cm^{-1} ($J = 1$ to $J = 2$), 11.52 cm^{-1} ($J = 3$ to $J = 4$), 15.36 cm^{-1} ($J = 4$ to $J = 5$), and so on. The wavenumber spacing is $2B/hc$ where $B = h^2/8\pi^2 I$.

A further source of complexity in treating rotation may be noted. A molecule in a high rotational state is stretched, causing each moment of inertia to be larger. If a molecule is in a higher vibrational state, anharmonicity (see below) and an average lengthening of bonds leads to greater moments of inertia.

In liquids and solids rotational motions need not be treated as quantized. The reason is that molecular collisions or cooperative vibrations are frequent. Any rotation that occurs is satisfactorily treated by classical models. Actually, in crystalline solids free rotation is usually not possible.

Vibrational Transitions. In contrast to rotational changes, vibrational transitions persist through all the states of matter. Since chemical bonds are stretched or bent in molecular vibrations, much larger energies are involved. These are of the order of 10 kJ mol^{-1}, corresponding to frequencies of the order of 2000 cm^{-1}. Thus quanta in the middle IR range are sufficiently large to promote vibrational changes.

Each vibrational degree of freedom of a molecule or *normal mode* can be treated in first approximation as a separate harmonic oscillator. Its potential energy curve is parabolic, and during vibration atoms move equal distances on either side of equilibrium positions. The vibrational energy in each mode is given by the expression

$$E_{\text{vibrational}} = (V + \tfrac{1}{2})\, h\nu_0, \tag{16.6}$$

where V is the vibrational quantum number ($V = 0, 1, 2, 3, \ldots$) and ν_0 the frequency of vibration. There are no degeneracies, except as symmetry causes two or more modes to be identical in energy. At room temperature, most molecules are in the ground state ($V = 0$) and possess the so-called *zero point energy* $\frac{1}{2}h\nu_0$ predicted by Eq. (16.6). Thus, the vibrational transition commonly observed in molecular absorption spectra is of the $V = 0$ to $V = 1$ type.

Actually, molecular vibrations are slightly anharmonic, the degree of anharmonicity increasing with vibrational amplitude. The potential function is a distorted parabola. Although Eq. (16.6) predicts an even spacing of vibrational levels, the separation between them actually diminishes as V increases. Because of anharmonicity no selection rule really holds; any transition is allowed to a degree. For example, overtones of *diminished* intensity such as a $V = 0$ to $V = 2$ transition are observed and appear at a frequency somewhat less than twice the fundamental ($V = 0$ to $V = 1$). Correction terms can be added to Eq. (16.6) to compensate for anharmonicity.

In the quantum-mechanical treatment of molecular vibrations the formula for the frequency ν_0 is identical to that derived from a classical Hooke's law treatment. If we treat a molecule as a simple harmonic oscillator we obtain

$$\nu_0 = \frac{1}{2\pi}\sqrt{\frac{k}{\mu}} \tag{16.7}$$

where k is the bond force constant and μ the reduced mass. Of course the amplitudes of vibration are quantized as required by Eq. (16.6).

Example 16.3. What is the force constant k of the bond in carbon monoxide?

According to Eq. (16.7), for a diatomic molecule such as CO, both the reduced mass μ and molecular vibrational frequency ν_0 are needed. The latter can be found from tables, which show that the vibrational transition 0–1 for CO occurs at 2140 cm^{-1}. Its reduced mass [$\mu = m_1 m_2/(m_1 + m_2)$] is just

$$\mu = \left(\frac{12.0 \times 16.0}{12.0 + 16.0}\right) \text{g mol}^{-1} \times \frac{10^{-3}\ \text{kg g}^{-1}}{6.01 \times 10^{23}\ \text{atom mol}^{-1}} = 1.14 \times 10^{-26}\ \text{kg}.$$

Since $\nu_0 = c\sigma$, where σ is the wavenumber of the absorption,

$$\nu_0 = 2140\ \text{cm}^{-1} \times 3.0 \times 10^{10}\ \text{cm s}^{-1}.$$

Solution in Eq. (16.7) gives

$$\nu_0 = 6.42 \times 10^{13}\ \text{s}^{-1} = \frac{1}{6.28}\sqrt{\frac{k}{1.14 \times 10^{-26}\ \text{kg}}}$$

and

$$k = 18.4 \times 10^{2}\ \text{N m}^{-1}.$$

The selection rule that defines allowed transitions in the gas phase states that only vibrational transitions for which $\Delta V = \pm 1$ are allowed. Nearly always rotational changes accompany vibrational transitions for a gas molecule, and the selection rule $\Delta J = 0, \pm 1$ simultaneously applies. Accordingly, each vibrational absorption gives rise to a collection of lines or a band. Figure 16.3 shows a spectrum observed under conditions of both high and low resolution.

The so-called "P-branch" of such a spectrum includes all vibrational transitions for which the accompanying rotational shifts are $\Delta J = -1$ and is the "wing" at lowest frequencies. An absorption peak for which $\Delta J = 0$, the "Q-branch," appears for most molecules.* Note its presence as the strong absorption in the middle of the pattern in Fig. 16.3. Finally, transitions for which $\Delta J = +1$ comprise the highest frequency set, the "R-branch."

Infrared spectra become more complex the greater the number of atoms N in a molecule. Each of the $3N - 6$ degrees of vibrational freedom ($3N - 5$ for a linear molecule) gives rise to a fundamental frequency and an absorption pattern like that in Fig. 16.3. We approach the situation mathematically by establishing a set of internal coordinates that reduce complex vibrations to so-called *normal modes* of vibration that are mathematically more straightforward. While it is difficult to relate such modes to IR absorption peaks, assignments have been made for most simple molecules.

Exact motions of atoms during a vibration are hard to describe. All atoms participate but fortunately there is often a great deal of localization in one chemical bond or a set of bonds. For example, if a molecule has an "unassociated" N–H bond, a vibration that stretches the bond almost always absorbs IR in the 3550–3340 cm^{-1} range. The motion of other atoms in the molecule exercises a small perturbing force so that a wavenumber range rather than a specific value has to be given. Thus, vibrational spectra provide a very important basis for identification of chemical bonds in molecules (see Section 16.10).

Electronic Transitions. No other characteristic molecular "motion" has as great a charge displacement as an electronic transition. The resulting large dipole moment leads to high intensities for such transitions; even forbidden electronic transitions are commonly observed. For this reason and also because there is no general expression covering these transitions in terms of properties that are readily observable, electronic selection rules are seldom useful in the study of absorption electronic transitions. Characteristically, quanta of order of magnitude 20,000 cm^{-1} are absorbed, and most electronic transitions occur in the visible and UV from about 750 to 110 nm (15,400 to 91,000 cm^{-1}).

Typically an electronic transition in a molecule is accompanied by a change in vibrational and, in the gaseous state, rotational motion as well. In Fig. 16.4 a possible transition is shown for a diatomic molecule. The figure also provides a comparison with other characteristic transitions.

What vibrational change occurs during an electronic transition? We may partly resolve the question by applying the Franck–Condon principle, which states that

*For diatomic molecules, as predicted by quantum mechanics, the Q-branch is absent. It is also worth recalling that for homonuclear diatomics like N_2 and O_2 even the basic vibrational transitions cannot be brought about by IR radiation since there is no transition dipole moment.

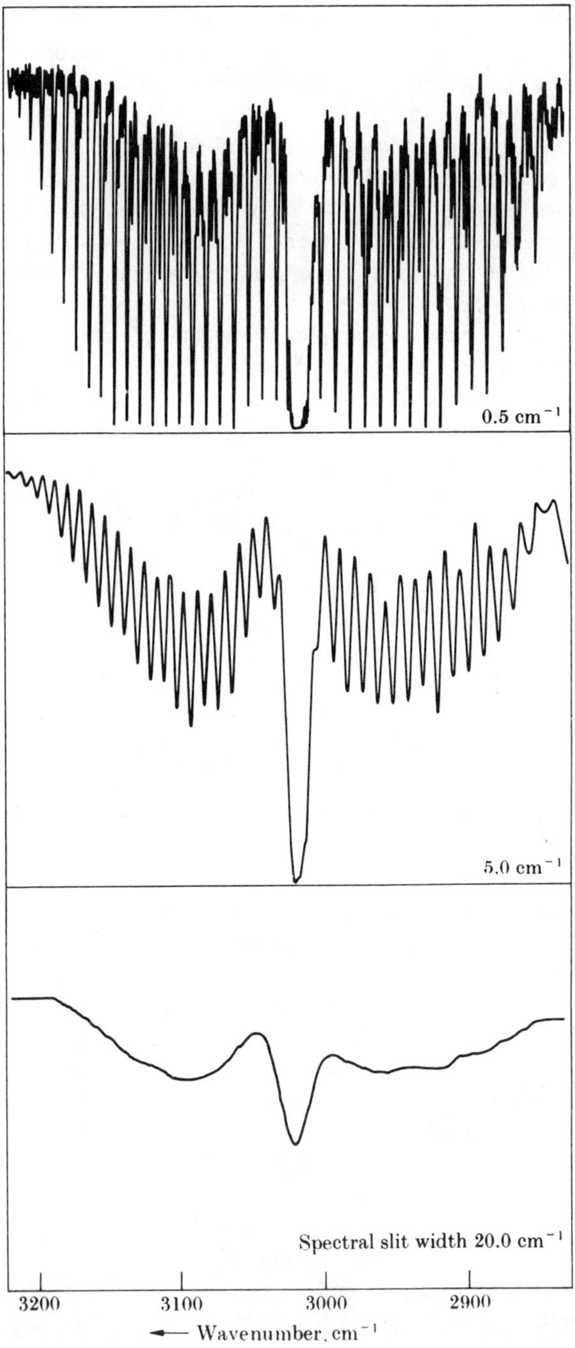

Fig. 16.3 The C–H stretch band of gaseous methane as observed at widely different resolutions. The instrument spectral slit width is noted underneath each trace. (Courtesy of Beckman Instruments.)

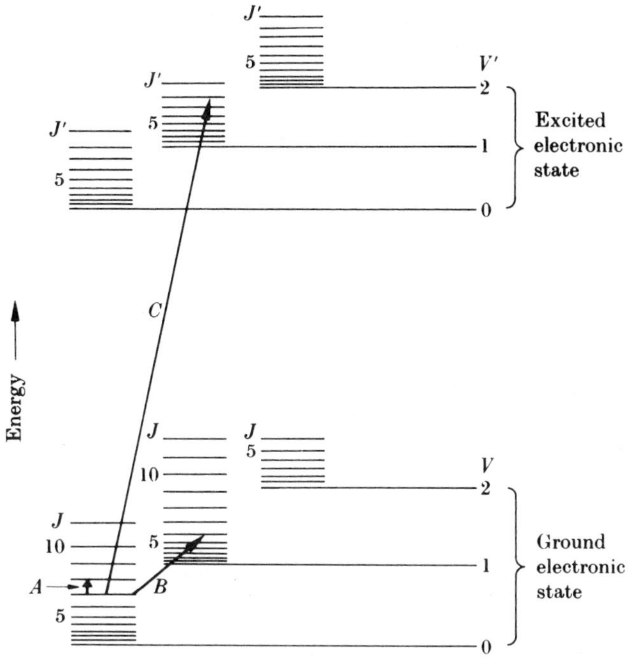

Fig. 16.4 Schematic illustration of energy levels in a diatomic molecule. Types of transitions and the spectral region in which they are observed are as follows: *A*, pure rotational (far IR); *B*, vibrational-rotational (middle and near IR); *C*, electronic-vibrational-rotational (VIS and UV).

nuclei in a molecule do not move during the short interval required for such excitation. The vibrational change also depends on the relative positions of ground and excited state potential energy curves (or surfaces for a polyatomic molecule).

Example 16.4. The Franck-Condon principle lends itself to prediction of the vibrational transitions likely to accompany electronic changes. On a step-by-step basis, how does one apply it?

To apply the principle, potential energy curves are sketched for the ground and lower excited electronic states of a species. If not known, a molecule is assumed to have an equilibrium bond length about 5% greater in the excited state than in the ground state. Vibrational states are added next, often by approximating a likely spacing with a simple Hooke's law calculation. On the basis of the Franck-Condon principle, one represents absorption or excitation of a molecule on the potential energy diagram by drawing a vertical line from the center of the ground vibrational state (a portion corresponding to the equilibrium bond length) to the first vibrational level it intersects in the excited state. Such transitions generally show that excited electronic species are also excited vibrationally.

Example 16.5. What effect does the coupling of vibrational and electronic transitions during excitation have on selection rules for molecules? Apply the answer to $n \rightarrow \pi^*$ molecular transitions, which should be forbidden since no electronic transition dipole is likely. Without coupling, only electronic transitions that result in an unsymmetrical movement of charge are probable (allowed).

Molecular orbital shapes (charge distributions) are distorted by mixing with other orbitals in a vibronic transition. Such mixing occurs naturally as suggested above. Evidence for such coupling is the appearance of $n \rightarrow \pi^*$ transition in many amino acids at about 290 nm (carbonyl absorption, $\epsilon = 20\ \text{cm}^{-1}\ M^{-1}$).

Classification of Electronic States. Molecular electronic states are classified mainly according to total electron spin or symmetry. The total spin or intrinsic angular momentum of a molecule is well described by the vector sum of the spins of its electrons, provided there is only weak coupling between spin and orbital angular momentum of electrons. In general this condition holds for simple molecules with light atoms. Given that the spin per electron is $+1/2$ or $-1/2$, for a molecule the possible value of total spin S for two electrons is 0 or ± 1. For a total of three electrons, $S = \pm 1/2$ or $\pm 3/2$; for four, $S = 0, \pm 1$, or ± 2.

Since many states of the same energy have the same total spin, the multiplicity, $2|S| + 1$, is frequently cited instead of the total spin. For example, any state in which all electrons are paired has $S = 0$ and a multiplicity of unity. Such a state is a *singlet*. If there is one unpaired electron the multiplicity is two, and the state is a *doublet*. Similarly, with two unpaired electrons with parallel spins, the state is a *triplet* since $S = \pm 1$ and $2|S| + 1 = 3$.

Though the electronic states of linear molecules can be further detailed in terms of angular momentum, for most polyatomic molecules states are classified principally in terms of symmetry. This is possible because the potential field in which electrons move has the same symmetry as the molecule.

Integrated Intensities. An absorption band may be characterized by its intensity and band half-width as well as its frequency. For quantitative work it is usually sufficient to know ϵ_{max}, the molar absorptivity at the absorption peak. For molecular studies, however, the integrated intensity is usually of greater interest. In particular it is closely related to the polar properties of molecules.

The absolute intensity of an absorption is defined by abs. intensity $= \int_{\nu 1}^{\nu 2} \alpha_\nu\, d\nu$, where α_ν, the absorption coefficient is integrated over the entire band. In most spectrophotometric work an *integrated* intensity is defined by

$$\int_{\nu 1}^{\nu 2} \epsilon_\nu\, d\nu = \frac{1}{cb} \int_{\nu 1}^{\nu 2} A_\nu\, d\nu, \qquad (16.8)$$

where the molar absorptivity (or the absorbance) is integrated. The reliability of the integration depends greatly on whether the spectral slit width (see Section 9.2) is appropriately narrow for the absorption peak. It should be noted that an integrated intensity can often be approximated as $\epsilon_{max} \cdot \Delta\nu_{1/2}$ where $\Delta\nu_{1/2}$ is the bandwidth at half-intensity. Representative integrated intensities in the IR are of the order of 0.11 $\text{mol}^{-1}\ \text{cm}^{-2}$ with the largest values appearing for OH and C=O vibrations.

In certain cases the integrated intensities of IR vibrational bands can be related more effectively to molecular properties than can molar absorptivities.* In any event better functional group indentification can be made if both intensity and frequency data are used.

Oscillator Strength. The absolute intensity of an electronic transition is often described in terms of its f number or *oscillator strength*. This term arises from the classical treatment of dispersion and is just the f given in Eq. (7.11). It can be regarded as the effective number of electrons set into oscillation in a given absorption. Further, f can be related to both the quantum-mechanical dipole strength D of an oscillator and the Einstein transition probability of absorption B_{mn} (see Section 8.3). It can be shown that

$$f = \left(\frac{8\pi^2 m_e g_n}{3he^2}\right) 2\pi\nu D,$$

where m_e is the mass of an electron, g_n is the degeneracy of the excited state, and ν the frequency of absorption. The second expression is

$$f = \left(\frac{m_e hc^2 2\pi\nu}{\pi e^2}\right) B_{mn}.$$

Finally, one has

$$f = \frac{2303 m_e c^2}{\pi N_0 e^2} F \int_{\nu_1}^{\nu_2} \epsilon \, d\nu,$$

where F is a factor near unity that corrects for the refractive index of the solution medium. This expression relates f to the usual integrated intensity of a band. Typical values of f for electronic transitions are 10^{-4} to 10^{-3}.

16.4 SOURCES OF ERROR

To obtain absorption spectrometric measurements with high values of S/N, it is mainly systematic error or bias that must be minimized. This favorable result will prevail whenever intense sources and sensitive detectors are available and analyte concentrations can be adjusted to give absorbances in the 0.1–1.0 range. For example, most measurements in the UV–VIS range will fall in this category.

In absorption measurements systematic error arises both from *general* and *instrumental* sources. The former will include faulty chemical procedures and apparent failures of Beer's law, and the latter bias arising from the type of modules and overall design chosen. Since general errors occur in all spectrophotometry, they will be taken up first. Instrumental sources of bias for spectrophotometers

*A. S. Wexler, Integrated intensities of absorption bands in infrared spectroscopy, *Appl. Spectr. Rev.*, **1**, 29 (1967).

based on monochromators will be taken up below in Chapter 17 for other types of spectrophotometer. Brief attention will also be given to precision, which is limiting (a) in measurements made near the limit of detection through its effect on the blank and (b) in measurements of high precision.

Beer's Law Errors. Probably the most generally useful indicator of error in quantitative spectrophotometry is whether Beer's law fits the data. True failures of the law in homogeneous systems (e.g., solutions) are unknown so long as there is no interference between absorbing species and no refractive index correction [1]. Both these factors were discussed in Section 16.2. The upper concentration limit for validity of the law ranges from 10^{-5} to 10^{-2} *M*. In dilute solution, in most cases of apparent failure it thus follows that one or more other errors must have occurred. The evidence for apparent failure will be the production of a nonlinear curve for an analyte when a series of absorbances at a given wavelength is plotted against concentration.

Whether the cause of nonlinearity should be traced to its source will probably depend on whether new analytical procedures, new reagents, or new instruments have just been introduced. In those cases introduction of bias is likely and efforts to reduce it will be important. If, after investigation, the source of deviation proves to be something that cannot easily be controlled, subsequent spectrophotometry can probably still be performed reliably on a relative basis, that is, by the comparison of absorbances of samples with those obtained under the same conditions with standards. Actually, most *quantitative* spectrometric analyses are performed in this way to minimize the influence of undetected errors. The calibration curves are usually obtained as plots of A or percent transmittance versus c and called analytical or *working curves*. Further attention will be given to these graphs in later sections.

In *qualitative* photometric studies, both the wavelength of an absorption peak and the integrated intensity (see the last section) are important. Wavelength and photometric calibration will be discussed below in connection with different types of spectrophotometers.

Chemical Error. Several chemical sources of error can be listed: uncontrolled pH, the presence of impurities, the changing of solvents and temperature variations. They may give rise either to positive or to negative Beer's law deviations. These variables influence absorption mainly through their effect on equilibria involving the dissolved species. Particular conditions for a given analysis should be understood and followed.

Example 16.6. Beer's law will not hold for sodium chromate in water unless a small amount of strong base is added, for the chromate tends to condense somewhat, depending on concentration, and the dichromate formed absorbs at different wavelengths. The equilibrium involved is

$$2CrO_4^{2-} + 2H_3O^+ \rightleftharpoons Cr_2O_7^{2-} + 3H_2O.$$

Adding a strong base will ensure that condensation is suppressed.

Example 16.7. If a particular solvent and solute interact, the absorption spectrum of the solute may be markedly altered when another solvent is used. Thus, acetic acid gives an absorption characteristic of the molecule in hexane, but in water the spectrum has many features attributable to the ionic species that result from dissociation.

$$CH_3COOH + H_2O \rightleftharpoons CH_3COO^- + H_3O^+$$

Example 16.8. When the absorbing species has been formed from the analyte by complexation with a chromophore, an excess of complexing agent must be present. The formation constant of the complex will determine the excess concentration needed. For example, copper (II) can be determined in aqueous solution by adding ethylenedianime (EDA):

$$Cu(H_2O)_4^{2+} + 2EDA \rightleftharpoons Cu(EDA)_2^{2+} + 4H_2O.$$

An insufficient excess of EDA will mean the presence of a little of the aquo complex.

Pressure Broadening in Gases. In the spectrophotometric analysis of gaseous mixtures some special sources of error enter. It is found, for example, that gaseous absorption curves vary not only as a function of concentration, but also with the total pressure. At a fixed concentration of a constituent, increasing pressure usually produces broadening of its absorption bands (see Section 13.3). An effectively larger absorbance will thus be detected at many wavelengths. At higher pressures the absorption is fortunately much less sensitive to further changes. If the total pressure can be maintained in that range and a narrow band of radiation used, Beer's law may be expected to hold.

Fluorescence. If a sample being analyzed spectrophotometrically fluoresces under the incident illumination, the fluorescent radiation will produce an error exactly like that induced by stray radiation. Since the fluorescence will be of longer wavelength than the incident radiation (see Section 15.2), it is not difficult to ensure minimum interference by keeping the spectral slit width narrow.

Monochromator-Based Spectrophotometers. Systematic error arising from the instrumentation of scanning spectrophotometers includes both static errors that occur at each wavelength and dynamic errors related to the incomplete response, for example, by the detector or AD converter at a particular speed of scanning. The latter are taken up in connection with the discussion of spectrophotometric instrumentation in Section 17.5.

Among static errors the proper choice of *spectral slit widths* is of major importance. In the accurate analysis (qualitative and quantitative) of a substance and especially of a mixture, if relatively sharply defined absorption peaks such as those of gaseous methane are involved, a very small value of S is needed. With too broad a value, a loss of optical resolution occurs that is evident as a broadening of bands and a severe loss of peak height. This loss of resolution was clearly shown in Fig. 16.3. By contrast, measurements based on the quite broad absorption band of bromthymol blue can be carried out with accuracy even when $S \approx 30$ nm, judging by Fig. 16.5.

To ensure sufficient optical resolution of absorption peaks a good rule of thumb is that $S \approx 0.1B_{1/2}$, where $B_{1/2}$ is the width of an absorption band at half height. Alternatively, an appropriate width can be chosen by scanning a spectrum at two different slit widths. If peaks are no higher or slimmer at the narrower width, resolution is adequate at the wider one and it should be used for the better S/N it affords. If peaks are sharper for the narrower slit, trials of still narrower widths are indicated.

Many apparent negative deviations from Beer's law may be traced to the use of instruments with spectral slit widths that are too large. The reason may be explained in terms of Fig. 16.6. Suppose the bandwidth isolated by the monochromator or filter is *ab*. At each concentration the instrument signal will translate into an effective absorbance that is between the extreme values at wavelengths *a* and *b*; it will not be the mean, however, because the detector response varies directly with the transmitted power and the absorbance varies logarithmically (and inversely) with the power.

For the case illustrated by curve B in Fig. 16.6, the high level of power at wavelengths near *a* will contribute more to the instrument response than will the much lower levels near wavelength *b*. The absorbance determined using radiation of bandwidth *ab* will thus be *smaller than the absorbance at the mean wavelength*. For the higher concentration (curve A) the difference in the values of the transmitted power at *a* and at *b* is greater than for curve B, and the difference between the measured and "mean" absorbance is larger. In general, the higher the concentration is, the greater the variation. An apparent negative deviation from Beer's law is observed. A similar deviation will occur for wavelength band *bc*, even though it is centered on a maximum, since the curve is asymmetrical. Clearly, the extent of the deviation will be determined by the exact wavelength dependence of the power in the beam, the transmitted power, and the sensitivity of the detector.

How seriously will working curves with negative deviations from Beer's law

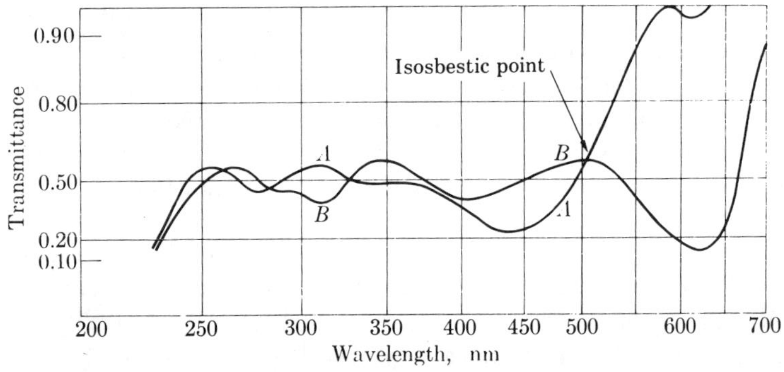

Fig. 16.5 Absorption spectrum of an aqueous bromthymol blue in solution. Curve *A*, pH 5.45. Curve *B*, pH 7.50. An isosbestic point is evident at about 500 nm.

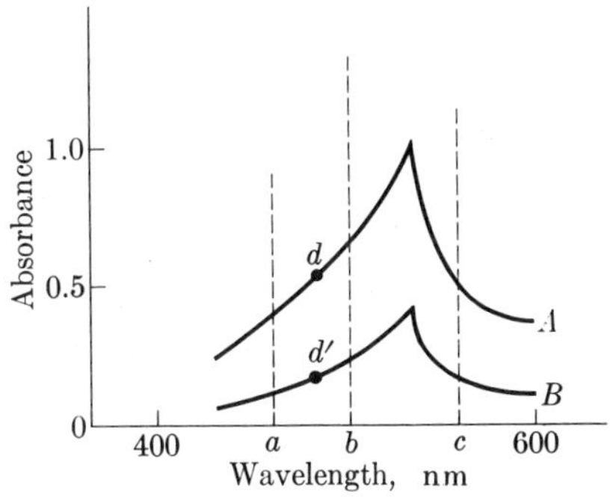

Fig. 16.6 Absorbance of a substance obeying Beer's law. Curve A, 3 concentration units; curve B, 1 concentration unit.

affect calculations? At high concentration, sensitivity will be less and the uncertainty in the calculated concentration greater. When bandwidth errors are likely to be large, it is desirable to dilute solutions and work at lower concentrations. For example, dilutions are often made in quantitative IR analyses where spectral slit widths are comparable to absorption band halfwidths.

Stray Radiation. In monochromator-based spectrometers some stray radiation adds to the regularly transmitted beam reaching the detector (see Section 9.7). How serious a systematic error is introduced will depend both on the intensity of stray radiation and the sample absorbance. At high absorbances the error will be greatest, and vice versa. As a result, Beer's law will appear to fail negatively.

Example 16.9. Assume the power S of the stray radiation in the wavelength range of interest is 1.5% of the power level P_0 transmitted by the reference cell. What absorbance error occurs for a sample with a transmittance of 10%?

If there were no stray radiation, the true absorbance A^0 would be

$$A^0 = \log(P_0/P) = \log(100/10) = 1.$$

With stray light S also present, both P_0 and P increase by the factor 1.5%. The measured absorbance is

$$A = \log(P_0 + S)/(P + S) = \log(100 + 1.5)/(10 + 1.5) = \log 8.83 = 0.95.$$

With stray light of this magnitude an apparent absorbance 5% smaller than the true absorbance is measured.

Example 16.10. Will modulation of a spectrophotometer beam and subsequent synchronous demodulation diminish the effect of stray light? Actually, all radiation entering the monochromator, including that which departs from the optical path and becomes stray light, is modulated at the same frequency. As a result, the detector and processing modules respond both to signal and to stray light at each wavelength.

Wavelength and Photometric Calibration. Lines of peaks from well-defined absorption or emission spectra, for example, the mercury vapor spectrum, various lasers, or a 4% solution of Ho_2O_3 in 10% $HClO_4$ (available as SRM 2034 from the National Bureau of Standards), serve for wavelength calibration. Tables of reference wavelengths are readily available [31, 32]. Similarly, solutions of known absorbance may be used to standardize instrument *photometric response* [1, 21, 31]. Initial checks and routine calibrations are highly desirable to minimize errors from loss of calibration.*

Precision of Measurement. It is possible to determine a range of optimum transmittance values for each type of spectrometric device and procedure. This information in turn will allow the most favorable concentration or cell path length for a measurement to be calculated.

Most spectrometers produce a response directly proportional to the power falling on the detector. In addition, the conventional operational procedure calls for adjusting slits or other beam intensity controls so that the transmittance of the reference is 1.00. In this situation Eq. (16.2) applies. Let T replace P/P_0 and solve for concentration. Rearranging Eq. (16.2) results in

$$c = -\log T/\epsilon b. \tag{16.9}$$

For a given absorbing species, wavelength and path length ϵ and b are constant and all the error in the determined concentration can be attributed to the uncertainty dT in the measured transmittance T. By taking the total differential of Eq. (16.9) and dividing through both sides by c one obtains

$$dc/c = -(1/c\epsilon b)(\log e)/T\,dT. \tag{16.10}$$

The term on the left, dc/c, is just the relative error in concentration, where dc represents a very small error in c. By making the simplest assumption about the error in transmittance measurements, namely, that it is constant regardless of the magnitude of T, dc/c as expressed by Eq. (16.10) has been graphed (curve A) as a function of transmittance and absorbance in Fig. 16.7.

The methods of Section 10.3 may be used to estimate the error in the concentration resulting from random error in the modules of an instrument. The range of absorbance values in which the propagation of error is smallest may be ascertained in a second operation. What sources of variance should be considered; should all modules be examined? While detailed treatments have been reported in the literature [see references in 1], for a general treatment it is more helpful to consider *limiting cases*. In Table 16.3 information about instrumental sources of variance has been organized in terms of limiting cases. Ways of improving precision are suggested as well. Once the range ensuring the best analytical results is known,

*The paper, L. S. Goldring, R. C. Hawes, G. H. Hare, A. O. Beckman, and M. E. Stickney, *Anal. Chem.*, **25**, 869 (1953), is an interesting study of such errors.

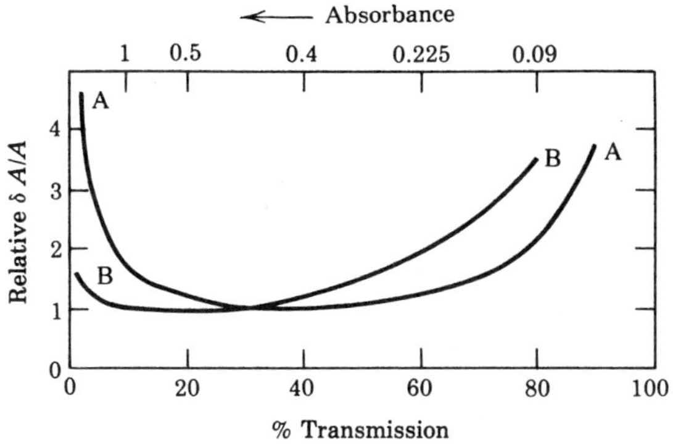

Fig. 16.7 Relative spectrophotometric precision as a function of transmission and absorbance. Curve A, constant uncertainty δp in beam power (or intensity); curve B, uncertainty proportional to $\sqrt{I}$. (After E.J. Meehan [1].)

Table 16.3 Origin of Instrumental Variance in Some Limiting Spectrophotometric Cases[a]

Major Modular Sources of Variance (Types of Noise Contributing)	*Optimum Transmittance and Absorbance*[b] T_{opt}	A_{opt}	*Instrumental Factor Leading to Variance (Appropriate Corrective Action)*
Case 1: Processing Modules[c]			
(Dark current shot noise Amplifier noise Readout noise)	0.33	0.48	Poor readout or processing precision (Arrange for scale expansion and/or increase bit specification of AD converter)
Case 2: Detector			
(Photocurrent shot noise)	0.11	0.96	Low light at detector (Increase throughput)
Case 3: Optical Source			
(Flicker noise in source)	(0)[d]	(∞)[d]	Unsteady source (Stabilize source)

[a] J. D. Ingle, Jr. and S. R. Crouch, *Anal. Chem.*, **44**, 1375 (1972). In their mathematical treatment they assumed stray light was negligible. For that reason and for mathematical reasons as well, the table offers no information useful for measurements at high absorbance.
[b] T_{opt} and A_{opt} are values at which S/N ratio is highest.
[c] Factors listed are independent of the photoelectron current.
[d] The level of stray radiation and dark current noise will instead set the limits.

either the concentration of samples or the cell path length should be adjusted accordingly. As would be expected, the range for optimum transmittance holds whether the data are obtained under single- or double-beam operation. Some examples will deal with the limiting cases.

Example 16.11. What conversion factor will translate instrumental variance to a variance in concentration for case 1 in Table 16.3?

When processing modules are the main source of variance in a spectrophotometer, it can be shown that $\Delta c/c = 2.72(\Delta P/P_0)$ [1]. The conversion factor 2.72 arises from the logarithmic dependence of concentration on beam power arising from Beer's law. On the basis of this equation, a modern spectrometer whose photometric precision is $\pm 1.0\% T$ should yield concentrations uncertain by $2.72 \times 1.0 = \pm 2.7\%$.

Example 16.12. A spectrometer with an analog (meter or chart recorder) readout that is linear in transmittance has a photometric precision of $\pm 1\%$. Even though it employs a photomultiplier tube detector, its precision is limited by its readout. What is the optimum value at which to measure sample absorbances? If better precision is needed, what type of instrument should be sought? Will an instrument of higher precision have the same optimum sample absorbance value? What precision can be hoped for with a well-designed instrument and precision manipulations?

To find an optimum absorbance value for limiting case 1, Eq. (16.10) may be differentiated and the derivative set to zero. One finds that $A_{\text{opt}} = 0.434$ when error is proportional to ΔA. Curve A in Fig. 16.7 shows the relative error for this situation.

According to Table 16.3, to obtain better precision the readout precision must be increased until it is no longer limiting. Substitution of an instrument with digital readout [instrumentation amplifier, 12-bit AD converter (0.25% precision) and a digital display device of good precision] for a meter readout will provide this improvement. The usual provision for scale expansion will ensure advantage can be taken of the full precision of this readout. With these changes instrument performance would fall under case 2 in the table. Its relative precision varies as $\sqrt{T}$ and follows curve B in Fig. 16.7. The region of optimum precision for measurement is now centered around $A = 1$.

With a well-designed instrument, a good protocol, and careful work, absorbances can in general be measured with a precision of $\pm 0.5\%$.

Example 16.13. When radiation reaching a detector is limited, it is often advantageous to use the method of photon counting in which current pulses resulting from reception of photons at its photocathode are in fact counted. It can be shown that in this situation the variance is now proportional to the square root of the total number of counts. Note that the collection of individual counts follows a Poisson distribution (see Section 21.8). As long variability in the arrival of photons is the major source of instrumental variance, it is advantageous to make measurements at higher absorbance values. The optimum measurement region lies around $A = 1$ $(T = 0.10)$ or higher.

Example 16.14. In spectrophotometry, how likely is random error in both the transmitted and incident power? Interestingly, in the earlier treatments of spectrometric precision the latter source of error was usually neglected.

A reasonable assumption is that the error in P and P_0 will be essentially equal and follow a normal distribution. The error in ln P/P_0 will then be given by

$$\ln(dP/P_0) = [(dP/P_0)^2 + (dP_0/P_0)^2]^{1/2}.$$

It can be shown that the optimum transmittance corresponds to the minimum in $[1 + \exp(\ln P/P_0)^2]^{1/2}$ $(\ln P_0/P)$, which occurs at $A = 0.48$ or $T = 0.33$. The most favorable region for transmittance measurements is the range 0.3–0.5. [D. Z. Robinson and R. Cole, *J. Opt. Soc. Am.*, **41,** 560 (1951) and N. T. Gridgeman, *Anal. Chem.*, **24,** 445 (1952).] Compare this result with the first limiting case listed in Table 16.3.

Example 16.15. In IR measurements the detector is usually the limiting source of noise and fluctuations in detector output appear as a constant error in transmittance. In spectral scanning can one expect that the optimum absorbance is 0.434?

In mixtures of analytes, absorbances range widely in value. When the maximum absorbance of any analyte is 0.86 and the average absorbance is 0.434, optimum results are secured. [D. E. Honigs, G. M. Hieftje, and T. Hirschfeld, *Appl. Spectrosc.*, **39,** 254 (1985).]

16.5 SAMPLING MODULE: SAMPLE CELL AND SAMPLE PREPARATION

Spectrophotometric samples are ordinarily examined in one of the following settings: in dilute solution; in a matrix with solid KBr (after grinding the sample with KBr, the mixture is pressed into a pellet); in a matrix with N_2 or Ar in the glassy state (after the volatilized sample is diluted with the gas, the mixture is rapidly frozen); or in the gaseous state. By selecting the analyte setting two important gains are secured. First, any *matrix effect*, that is, dependence of its absorption spectrum on the medium in which it is found, is minimized.* Second, ions or molecules of a species are dispersed sufficiently to behave essentially independently so that Beer's law is applicable.

Cells. For precision work, cells that are rectangular in cross section are ordinarily chosen to ensure that all rays traverse the same thickness of sample. Either pairs of matched cells or cells of known thickness b with appropriate windows or walls are readily available in the UV and through the NIR. Cells for the middle and far IR however, require constant attention, for they are either of the demountable type, and have a thickness that varies with sealing pressure, or have windows of relatively soft optical materials that erode as they are used. As a result, their thickness must be measured (see below). In work of lower precision,

*Absorption frequencies of a substance alter with its physical state, of course, as intramolecular forces vary. When an analyte is transferred from the gaseous to liquid state or from solvent to solvent, and only differences in van der Waals forces between analyte and matrix change, effects are small. If specific substance–solvent interactions like solvation or hydrogen-bonding are altered, either in the ground or excited state, frequency shifts are somewhat larger.

cylindrical cells such as test tubes offer no problem. In a gas cell mirrors are placed to fold a beam so that it traverses the cell several times and achieves a total path of 1–10 m.

The identical positioning of cells in a beam each time is also necessary for accurate work. Actually, rather than rely on exact positioning, it is preferable to use a cell that can be rinsed and refilled while remaining in place. If a cell must be removed from the sample compartment, it should be marked, if possible, to permit exact repositioning. If separate sample and reference cells are used, it is important that they be matched in transmission for full compensation of cell-wall reflection and absorption.

Cell Thickness. The cell dimension b needed for a Beer's law calculation can easily be determined interferometrically, especially for work in the IR. The cell of interest, filled with air, is placed in the sample beam of an IR spectrometer, the reference channel is left empty, and a regular spectral scan is made. The form of the curve that results is shown in Fig. 16.8. A small sinusoidal wave appears superposed on the transmission curve over a considerable range. The wave maxima and minima arise from constructive and destructive interference between the radiation that is reflected twice (or four times) internally and that which is directly transmitted. Since this case is analogous to that of the Fabry–Perot interference filter (see Section 7.10), Eq. (7.17) is also applicable here if radiation is vertically incident. After rearranging the equation and substitution of cell thickness b, we have

$$b = m\lambda/2n = m/2n\sigma,$$

where m is the interference order (an integer), λ and σ are the vacuum wavelength and wavenumber, and n is the refractive index of filling material (air). It is simplest to apply this equation to adjacent maxima since it can be shown that they represent orders m that differ by unity.

More commonly, to increase the precision of the estimate, the number of maxima is totaled over a long spectral region. Then the equation above becomes

$$b = \Delta m/2n\Delta\sigma, \tag{16.11}$$

where δm equals the number of maxima (and the increase in order) and $\Delta\sigma$ equals the wavenumber range over which maxima are tallied.

The equation also lends itself to determination of refractive indexes in the IR. To apply it several conditions must be met: (a) a cell of known thickness must be available to hold the sample; (b) the substance must have a transparent range in the IR at least 200–500 cm^{-1} long; and (c) $| n_{sample} - n_{window} | > 0.3$–$0.5$ (for adequate reflection). If a converging IR beam is used (true of most spectrophotometers), provision must be made for off-vertical incidence [9].

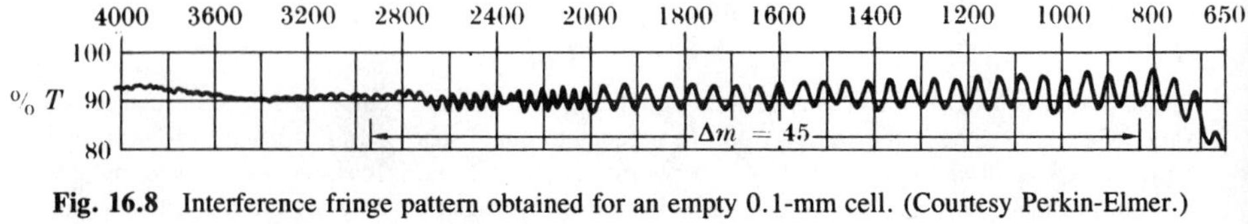

Fig. 16.8 Interference fringe pattern obtained for an empty 0.1-mm cell. (Courtesy Perkin-Elmer.)

Example 16.16. An air-filled demountable cell gives transmission maxima at 1000, 1050, 1100 cm^{-1}, ... What is its thickness? Assume $n = 1$ and essentially vertical incidence of light.

For the fringes at 1000 and 1050 cm^{-1},

$$2b = m_1/1000 \cdot 1 \quad \text{or} \quad 2000b = m_1$$

$$2b = (m_1 + 1)/1050 \cdot 1 \quad \text{or} \quad 2100b = m_1 + 1.$$

Subtracting the first equation from the second, $100b = 1$ and $b = 0.01$ cm. The value of m_1, surprisingly, is 200, a high order.

Solvents. For work in the visible and UV most species are soluble in at least one solvent that is transparent, and examination in dilute solution is common. In IR work, suspensions are used more often: both the dissolving ability and transparency of available solvents are limited.

Table 16.4 shows *in black* the ranges of transparency in the UV, VIS, and IR for several spectrophotometric solvents. The wavelength at which the absorbance of a 1-cm layer of the solvent reaches unity in the UV is noted at the left.

From the standpoint of transmission, all common solvents are suitable in the visible, many fewer are usable down to 210 nm, and very few transmit enough

Table 16.4

Solvent	*UV cutoff (nm)*
Acetone	330
Benzene	280
Carbon tetrachloride	265
Cyclohexane	210
Ethyl ether	220
Methanol	210
Tetrachloroethylene	290
2,2,4-Trimethylpentane	210

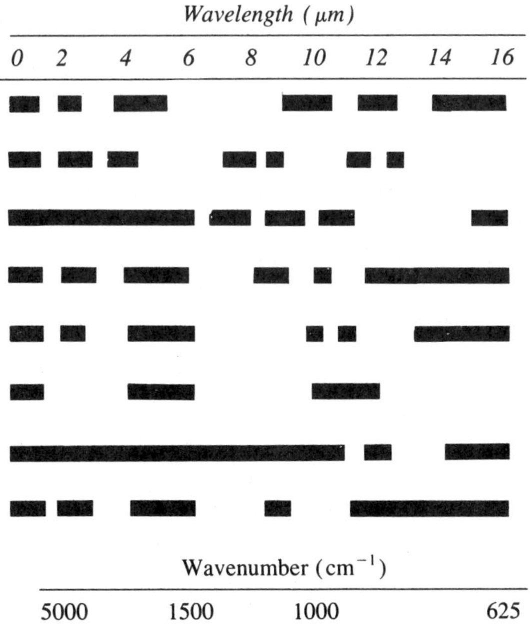

radiation below this wavelength. The useful solvents include water, most alcohols, ethers, and saturated hydrocarbons.

For the IR, few liquids have long regions of transparency, as Table 16.4 demonstrates. Carbon tetrachloride, tetrachloroethylene, and carbon disulfide are three of the best. Solvents useful in the IR range beyond the table (625–400 cm^{-1}) include carbon tetrachloride, *n*-heptane, and several others. Often, with polymers or other difficultly soluble materials, solvents must be employed that have limited windows, such as methyl ethyl ketone, dimethylformamide, or tetrahydrofuran. The volatility of solvents and the small quantities required make it desirable to prepare solutions for IR work by weighing and to transfer them to cells with care to avoid evaporation losses.

The amount of solvent absorption that can be tolerated in analytical procedures will depend ultimately on how seriously it limits the energy of the light beam. The greater the sensitivity of the detector and the more stable the electronics, the greater the tolerable absorption from this source. Cells as thin as 0.025 mm can be used to reduce such absorption.

Mulls and Pellets. Particularly in IR analysis it is desirable to examine solids as such. Both poor solubility in useful solvents and saving of time in sample handling are responsible. A few milligrams of solid sample suffices, since the final concentration in the suspension should be 0.01 to 0.5%. A widely used procedure is the preparation of a "mull" or suspension in a medium with few absorption bands and high refractive index, such as a viscous liquid like mineral oil (Nujol) or a solid like potassium bromide. The solid is first intimately ground into the suspending medium. In the case of liquid mulls, grinding must reduce the particle size to no more than one-fifth to one-tenth the wavelength of interest to lessen light-scattering losses. The requirement does not apply to solid suspensions, providing that sintering occurs during the pressing of the mixture into a disk. Liquid mulls are usually examined in a demountable liquid cell, KBr disks in a simple frame.

Alkali halide suspensions are compressed by a laboratory hydraulic press into thin pellets under pressure of 7×10^7 N m^{-2} to 14×10^7 N m^{-2} (10,000–20,000 lb $in.^{-2}$). The halide should be as dry as possible. A vacuum is often applied during the pressing to improve the clarity of the disks. The disk method offers the advantage of freedom from most extraneous absorption bands (sometimes weak O–H bands appear from water picked up during preparation) and is usable to the limit of transparency (KBr, about 250 cm^{-1}; CsI, about 200 cm^{-1}). Mineral oil mulls are especially satisfactory in the far IR, being virtually transparent throughout.

For quantitative work pellets and mulls must be prepared by careful weighing. Their thickness must also be known; alternatively, an internal standard such as KSCN may be used. Since mulls have an uncertain cell thickness, an internal standard is nearly always necessary with them.

Pyrolysis. Substances that are difficult to grind or dissolve in a suitable solvent, for example, carbon-filled rubber, are sometimes pyrolyzed before examination.

The process involves partial decomposition by heating. A sample is usually placed on an internal reflectance plate (see Section 16.8) and subjected to a controlled high temperature (up to 1300°C) for a selected period in an evacuated chamber. After pyrolysis its spectrum is interpretable in terms of molecular fragments present.

Complexing. Samples containing metals are usually dissolved. The resulting metal ions are then converted to a highly absorbing form by complexing with a chromogenic reagent such as an organic dye.

Extraction. This is a valuable approach to concentration of a trace constituent or to removal of a constituent from interfering absorbers. A liquid–liquid extraction is used. Often a reagent that will complex interfering species is added. By this means, the interferents are either prevented from reacting with the chromogenic agent used or kept from being extracted.

Matrix Isolation. This method permits simulation of gas-phase conditions for species given to strong matrix effects such as distinctive intermolecular interactions. The principle is to dilute the species with an inert gas such as argon and quickly freeze the mixture to the glassy state. Often the sample is diluted as it diffuses out of a Knudsen cell in which it has been volatilized by electrical heating. Usually sapphire plates (Al_2O_3), which are suitably transparent, are held at 15 K to collect the matrix.* In a solid argon matrix analyte interactions are weak and analyte spectra are greatly sharpened. A suitable matrix is one that is transparent and weakly scattering.

In recent years the availability of closed-cycle refrigerators, which can attain such temperatures without liquid cryogens, has made matrix isolation a much more accessible sampling technique. Not only does the method lend itself to examination of free radicals and chemical intermediates stable only in the gas phase (e.g., materials such as CN and FOO), but it is also ideal for trapping trace analytes eluted from gas chromatographs that are to be subsequently analyzed by Fourier transform IR spectrometry.

16.6 QUANTITATIVE PROCEDURES

This field represents the largest application of absorption spectrometry. Because of the high intensity of electronic transitions ($\epsilon \sim 10^3$–10^5) compared with vibrational transitions ($\epsilon \sim 10$–100), sensitivity is greatest for determinations that can be made in the visible or UV. At moderate to high concentrations all optical regions can yield accurate and precise quantitative results. The different optical regions are compared for their usefulness in Table 16.5.

From the table it is evident that in terms of *decreasing* convenience and ease of quantitative measurement, the regions may be ranked (a) visible, (b) near ultra-

*G. Mamantov, A. A. Garrison, and E. L. Wehry, *Appl. Spectrosc.*, **36**, 339 (1982).

Table 16.5 Relative Attractiveness of Molecular Absorption Spectrometry by Spectral Region

Region (Very rough molar absorptivity)	*Quantitative Analysis*		*Qualitative Analysis*	
	Single Analyte	*Multicomponent Analysis*	*Single Analyte*	*Multicomponent Analysis*
IR (10–100)	Good	Poor (Interferences)	Excellent	Fair
NIR (>100)	Very good	Good for some classes of substances	Limited	Fair
VIS (1,000)	Good[a]	Good[a]	Limited	Moderately good[a]
UV (10,000)	Good[a]	Good[a]	Limited	Moderately good[a]

[a]While the types of substances to which absorption spectrometry is applicable is limited, those species that absorb (either intrinsically or after reaction with a chromogenic reagent) can be very sensitively determined.

violet, and (c) infrared. Measurements in the visible and near UV tend to be preferred. Fortunately, many substances that do not absorb in the visible or near UV can be reacted with a chromogenic reagent to develop a characteristic and useful absorption in this region. Metal ions, nonmetals, and some classes of organic substances are included. Occasional interferences that arise can be minimized (see below). In favorable cases substances at the level of 0.1 ppm can be determined approximately. At the ppm level a precision of ±1% is common.

A very wide range of substances can be analyzed quantitatively in the IR. Virtually any substance with covalent bonds, whether molecular or ionic (e.g., a salt, one of whose ions is polyatomic), can be determined. While the IR region is not suited to trace analysis, determination carried out at the moderate concentration level (1-10%) or above can be quite reliable. Interferences are common and are dealt with by usual techniques (see below). Solids are nearly always analyzed by reflectance techniques (see Section 16.8).

Accuracies of quantitative spectrophotometric analyses are usually good. With a single-beam instrument a relative accuracy of ±0.5 to ±2% is realizable, depending on the reliability of the instrument and the care taken in analysis. With a double-beam optical null instrument the accuracy may be as great as ±1% if the optical wedge has a transmission that varies linearly along its length or ±0.5% for a ratio-recording instrument. Calibration or potentiometric compensation techniques can significantly improve accuracy. Indeed, with the latter an accuracy of ±0.3% is realizable.

Digital Recording. In quantitative work digital recording, especially when tied to computer calculations, offers advantages. The frequency of an absorption maximum can be located and the absorbance value observed automatically. It is straightforward to apply photometric and frequency corrections and to obtain integrated band intensities. In general, mathematical smoothing methods can be used to reduce noise while introducing minimal distortion and overlapping curves can be resolved.

Single Analyte. The simplest case is one in which the concentration of a single substance is sought. In devising a procedure, the absorption curves of all the constituents of the sample should be studied for two kinds of information, the wavelength and the concentration at which the analysis can best be done. A suitable concentration level will, of course, also depend on the instrument and cell.

Ideally the species of interest should absorb (a) in accordance with Beer's law or with little apparent deviation from it and (b) in a spectral region free of absorption by other constituents in the sample. Meeting the first condition requires preliminary trials to establish a suitable concentration range. If the second condition is not met, either a preliminary chemical operation must be performed or the sample must be treated as a mixture. Possible preliminary treatments of a species include (a) complexing to give an intensely absorbing species at a suitable wavelength, (b) separation from the sample by a method like precipitation or extraction, and (c) conversion to another oxidation state or other new form.

In inorganic analysis, complexing is probably the simplest and most effective means of minimizing absorption interference. Not only are the majority of inorganic ions colorless or weakly absorbing, but most organic agents are specific, that is, complex with one or at most a very few species of ions. Substances that would otherwise interfere can therefore usually be left in solution without ill effects. The reader is referred to the voluminous literature on the analysis of inorganic substances for details for different species.

Example 16.17. Nickel in steel can be determined photometrically by dissolving the steel in hot HCl solution, oxidizing with bromine water, and complexing the nickel with dimethyl glyoxime. A soluble red complex of unestablished identity is formed, and the other metals (Fe, Cr, Mn, etc.) do not interfere [15]. The analysis may be carried out with a spectrophotometer at about 350 nm or with a filter photometer by using a blue-green filter that is opaque above 430 nm.

Example 16.18. Substances whose maximum safe concentrations are specified in national and state legislation relating to water supplies are commonly determined by filter photometer or colorimeter after complexation. Their absorption curves are broad and filters with bandwidths (half maximum) of 30–35 nm give good results. Some examples and the concentration ranges over which they are usually determined for this purpose are: Al^{3+} over the range 0–1 mg L^{-1} at 525 nm (on complexation with aluminon indicator); Cd^{2+} over the range 0–70 μg L^{-1} at 525 nm (on complexation with dithizone); CN^- over the range 0–20 mg L^{-1} at 610 nm (on reaction with pyridine–pyrazolone indicator); and selenium at 420 nm over the range 0–1 mg L^{-1} (on reaction with diaminobenzedine hydrochloride and extraction into toluene).

If complexing fails to eliminate troublesome interferences, whether to attempt a chemical separation or a conversion procedure will have to be decided in each case. Precision will decrease with each step added, and it may prove advantageous to seek an entirely different quantitative method requiring fewer operations.

In organic analysis the high frequency of IR interferences dictates a variety of strategies. Where interferences are not severe, a direct determination, by differential procedure if necessary, is desirable. A shift of region is a good strategy. Clearly, substances like ketones that absorb both in the UV and IR can in all probability be determined directly in the UV. For some types of substances complexing with a chromogenic reagent develops useful absorption bands in the visible. For most organic mixtures, however, a combination of techniques is more fruitful. In these cases IR, NMR, UV, and/or mass spectrometry are used to identify the constituent as its concentration is quantitatively determined by gas or liquid chromatography or another separation technique.

What wavelength and slit width should be selected for an accurate and sensitive analysis of a species? In general, the wavelength chosen should be that of the maximum in the strongest absorption band that is free of interferences. There are two reasons. First, that wavelength will be the one at which there is the greatest change in absorption with concentration. Second, taking data at wavelengths on the side of an absorption band where absorbance is changing rapidly should ordinarily be avoided, since a slight inaccuracy in setting the monochromator wavelength may produce an appreciable error. Further, the spectral slit width or bandwidth of the instrument selected,* except in unusual cases, should be no wider than the value needed to obtain absorbances obeying Beer's law. Both resolving the selected absorption peak from neighboring absorptions and excluding regions of higher transmittance on either side of the peak are involved. In general, the broader the absorption peak of a species, the less stringent any of these requirements is.

Fortunately, many frequently performed photometric analyses are based on broad absorption regions and can be made with either a filter photometer or a spectrophotometer. If there is only a short spectral range free of interference, filter photometers are seldom suitable. In this case, the narrower bandpass of a spectrophotometer beam is indispensable.

Example 16.19. An analysis for manganese based on its oxidation to the permanganate form could be performed with a blue-green filter passing a broad band of from 440 to 560 nm according to the plots of Fig. 16.9. Ordinarily a much narrower bandpass filter would be used. A spectrophotometer would probably be set at about 520 nm for best precision. Unless interferences made it necessary to do so, it would not be set at 450 or 580 nm.

---

*Spectral slit widths are usually set by the slit program a manufacturer has prepared. In the case of filter photometers (Sections 17.3), the value is set by the filter bandpass. If the slit width of an instrument must be selected manually, data furnished about slit width versus spectral slit width over the wavelength range of interest should be consulted to decide on an appropriate value.

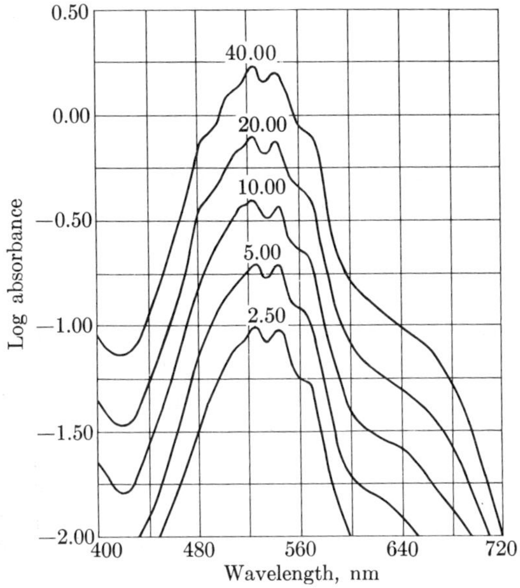

Fig. 16.9 Logarithmic plots of absorbance versus wavelength for aqueous solutions of potassium permanganate in a 1-cm cell. Concentrations given in parts per million.

Example 16.20. Naphthalene absorbs in about the same spectral region of UV as benzene does. The absorption curves are given in Fig. 16.10. If naphthalene in mixtures with benzene is to be determined by photometric analysis, a spectrophotometer is required. Only the naphthalene absorbs appreciably at 285 nm. The wavelength control can be set to that figure and the slit adjusted to give a spectral band width of from 1 to 2 nm.

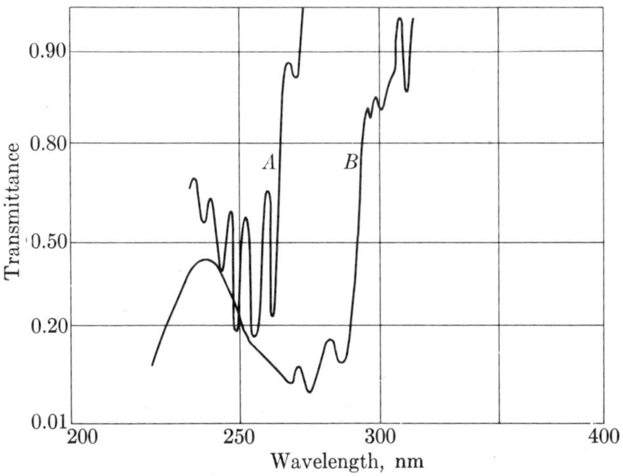

Fig. 16.10 Absorption spectrum of *A*, benzene, and *B*, naphthalene.

If the wavelength is on the side of an absorption band, special attention must be given to the reproducibility of the settings of wavelength and slit width on the instrument used. Further, a precise setting is nearly always called for in systems that have an *isosbestic* point. An example is shown in Fig. 16.5. Substances of biochemical interest and compounds useful as indicators exist in different forms at different pH's. If these forms are simply interrelated, there is a wavelength at which the molar absorptivity of a substance is the same in all its forms, a so-called isosbestic point. Determinations at this wavelength greatly simplify procedures.

It should also be noted that a correction for small amounts of extraneous absorption may be feasible. The conditions are that the absorbances be additive and that not more than one or two interfering substances be involved. The approximate concentrations and absorptivities (at the analytical wavelength) must be known for the interferents. The contribution of each to the measured absorbance may then be estimated and subtracted (see the discussion of mixtures below).

Example 16.21. In a copper determination an absorbance of 0.65 is measured. It is known that there is about a 10^{-4} M concentration of an interferent with molar absorptivity of approximately 100. The cell has a 1-cm path length. The correction is:

$$\epsilon bc = 100 \times 10^{-4} \times 1 = 0.01;$$

the absorbance due to copper alone is

$$0.65 - 0.01 = 0.64.$$

An uncertainty of 10% in the molar absorptivity of concentration of the interfering species will produce relatively little error in calculating the copper absorbance.

Finally, even if Beer's law appears to fail, photometric methods can still be used for a particular constituent. One then works entirely from a calibration curve.

Suspensions. Some absorption photometric procedures, like the analysis of nickel by complexing with dimethyl glyoxime, actually involve measurements on suspensions. To be useful in absorption photometry, the conditions of formation of these suspensions must be closely reproduced in each analysis and stable, very finely divided colloids produced. Since the power transmitted by this kind of sample will be reduced by both absorption and scattering use of calibration curves is necessary. Other examples of absorption photometric analyses involving colloids are the determination of acetylene by ammoniacal chloride and the determination of antimony as the sulfide.

Baseline Method. When the concentration of a single species in a mixture is to be determined, allowance must be made for the absorption of the other substances present. In such instances the baseline method can be useful.

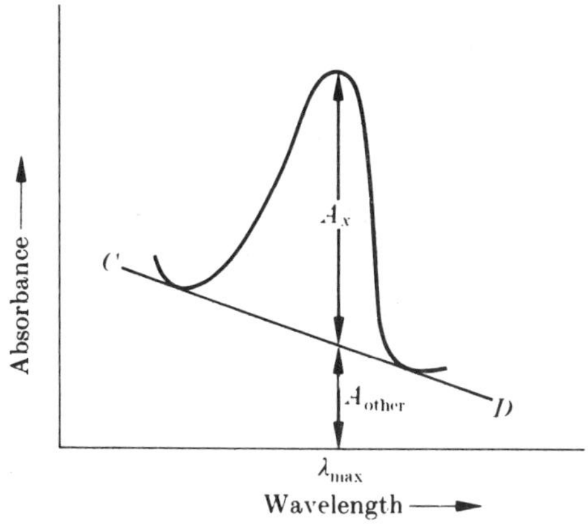

Fig. 16.11 Subtraction of background (baseline) of an analyte peak. It is assumed that baseline *CD* is the sum of the absorbances of all other components in this region. On this basis A_x, the absorbance of the analyte at λ_{max} equals $A_{total} - A_{baseline}$.

The procedure is illustrated in Fig. 16.11. A "baseline" *CD* is drawn tangent to the minima on either side of an analytical absorption peak. The peak absorbance is taken from the curve at λ_{max} and is just $A_{species} = A_{baseline}$. This method is valid if the spectra of the other substances change linearly with wavelength in the region of the absorption peak. Where the condition is not met, the usual Beer's law procedure for a multicomponent mixture to be described below can be carried out.

Multiple Analytes. Samples can be analyzed quantitatively for several components if the absorption curves of all the constituents are available. For example, petroleum refiners often make use of IR spectrophotometry for the direct determination of the composition of various hydrocarbon process streams. No preliminary separation is required provided there are not more than 8 or 10 constituents and the mixture is homogeneous. A primary requirement is that a spectrophotometer of high resolution and reliability be used.

The theoretical basis for a simultaneous photometric determination of several components is implicit in one of the assumptions of Beer's law, that each absorbing center interacts with radiation independently. In other words, each ion or molecule is assumed to absorb an amount of radiant energy that is unaffected by the presence of neighbors. Some severe difficulties are foreseeable in the case of polar mixtures; suffice it to say that such cases ought to be investigated thoroughly before photometric analysis is attempted.

If Beer's law holds for each constituent (or for all but one constituent [1]) under the conditions of the mixture, Eq. (16.4) can be applied. This equation states that at any particular wavelength λ,

$$A_\lambda = \log 1/T_\lambda = b \sum_i \epsilon_{i\lambda} c_i, \qquad (16.12)$$

where the subscripts i denote the several solutes. The manner in which the equation is used may be seen by noting that

1. all the solutes are in the same cell, and the cell thickness b is a known constant;
2. the molar absorptivities ϵ_i of the different solutes may be found by securing the individual absorption curves; and
3. the "unknowns" in the equation are the concentrations.

Since a single equation can be solved for only one unknown, a set of as many equations as there are absorbing constituents in the mixture is needed if the concentrations are to be found. For instance, analyzing for five constituents will necessitate measurements of A at five wavelengths. The optimum wavelengths will be those giving as widely different values of molar absorptivities as possible. The set of equations is then solvable for the concentrations by the usual methods applied to simultaneous equations.

Example 16.22. A two-component mixture is to be analyzed photometrically. The *molar absorptivities* of these substances are shown in Fig. 16.12. The measurements on the mixture are carried out at wavelengths λ_1 and λ_2, where the absorptivities differ most. The two equations

$$A_{\lambda_1} = (\epsilon_{11}c_1 + \epsilon_{21}c_2)\,b$$

$$A_{\lambda_2} = (\epsilon_{12}c_1 + \epsilon_{22}c_2)\,b$$

are obtained. On the right side the first subscript of terms refers to the substance, the second to the wavelength. The pair of simultaneous equations may be solved for c_1 and c_2.

Observations at different wavelengths must, of course, be made rapidly enough to avoid composition changes. It is essential that the spectral slit width be precisely that used in determining the absorption curves of each constituent. Otherwise, the accuracy of the determination will be low. If possible, the same spectrophotometer

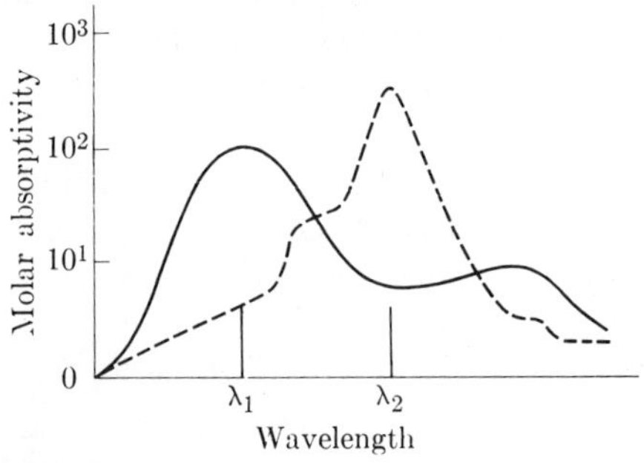

Fig. 16.12 Best wavelengths λ_1 and λ_2 for analysis of a two-component mixture.

should be used for all measurements. As suggested earlier, measurements on the steep slopes of absorption regions should be avoided where feasible.

Computer programs are available to solve sets of simultaneous equations. They are fed data consisting of the appropriate absorbances and molar absorptivities, and they yield composition reports almost immediately. They find a ready application in industry in the analysis of process streams.

Trace Analysis. Absorption spectrometry is widely used for trace analysis at the 1- to 0.1-ppm level and sometimes below. It is true that chemical and analytical steps requiring considerable skill and care are generally necessary prior to the actual spectrometric determination. Yet this disadvantage is offset by the fact that spectrophotometric measurements are simple, fast, specific, and suitable for automation.

In many trace determinations a preliminary separation by extraction (often after chelation), distillation, or a chromatographic procedure is necessary to isolate a trace constituent from one or more interferents present at much higher concentrations. Here efficiency is a key requirement. Further, a preconcentration step (e.g., by use of an ion-exchange resin column) is often essential in order to ensure either stoichiometric reaction with the chromogenic agent or to secure adequate absorption by the substance itself. During all these steps inadvertent introduction of interfering contaminants must be kept to a minimum. The sensitivity of spectrophotometric trace methods can be increased by preconcentration, use of a longer cell, and especially by finding more highly absorbing chromogenic agents.

As a result of continuing research good trace methods are presently known for most metals and the number of satisfactory nonmetal procedures is steadily growing. Most organic substances that absorb in the UV or VIS can be analyzed directly at the trace level. Those that absorb only in the IR can ordinarily be handled by combining preliminary concentration by wick stick with use of a KBr pellet.

Dual-Wavelength Spectrometry. Small changes in the concentration of a species are difficult to determine with any precision when it is present in a highly absorbing or strongly scattering (turbid) solution. Basically, the problem is one of measuring absorbance reliably to 1 part in 10,000. In such experimental situations a considerable advantage is gained by using a dual-wavelength procedure. Spectrophotometers that have a duochromator, that is, a device for isolating two wavelengths of interest when illuminated by a continuous source, are commercially available.

In dual-wavelength spectrometry, one wavelength is selected as that of an isosbestic point, a wavelength at which the absorbance of a system will be constant though the concentration of at least one species is changing. (In the usual more restricted sense, an isosbestic point is characteristic of a system in equilibrium such as an acid–base indicator.) The second wavelength is selected as that of a distinctive absorption peak of the analyte. The instrument is adjusted initially so that there is a balance between the signals produced at the two wavelengths. Departures from the balance may be very sensitively determined by a null procedure. Studies in biological systems are especially facilitated by such measurements. In Exercise 16.25 the application of Beer's law to the dual-wavelength method is examined.

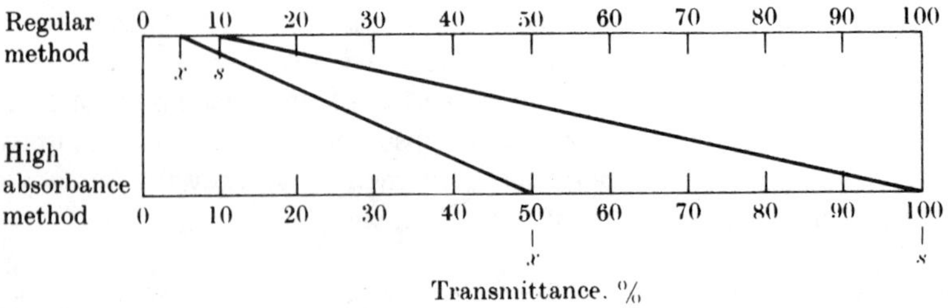

Fig. 16.13 The scale expansion achievable using the high-absorbance method. A sample x of 5% T is compared with a standard solution s of 10% T whose concentration is $[c_1]$. On the new scale the apparent transmittance of $[c_1]$ becomes 100% and that of the sample, 50%. The scale expansion factor is $100/10 = 10$.

Differential Spectrophotometry. In instruments that have neither digital readout nor provision for scale expansion, use of a differential spectrophotometric method may be attractive as a way to improve precision considerably. These procedures are not to be confused with derivative methods, which will be discussed in the next section. Differential methods are mainly useful with very concentrated *or* very dilute solution that cannot be brought with ease into the favorable absorbance range 0.2–1.2. They are basically useful for single-beam spectrometers and involve a scale-setting procedure.

With solutions of high absorbance the 100% T instrument setting is made with a standard solution almost as concentrated as expected analyte solution. In this case 0% T is still set with no light. The change is shown in Fig. 16.13 and can be simply interpreted as achieving good results by bringing about a scale expansion because more of the instrument scale is now used. In a high absorbance method the calibration curve obeys Beer's law in slightly modified form:

$$A'_x = \epsilon b(c_x - c_1).$$

where A'_x is the absorbance reading observed for the sample solution on the new scale, c_x is the analyte concentration in the sample solution, and c_1 is the concentration of standard. The calibration curve is *linear*.

Clearly the instrument being used must have low stray light. The precision of measurement may increase at most by a factor as great as the pseudo-scale-expansion factor. In practice, instrumental noise tends to make the augmentation smaller.

As applied to a *trace analysis*, the 0% setting is made with a standard solution of concentration c_1 that is somewhat more concentrated than the most concentrated sample. A regular blank is still used to establish 100% T. The calibration curve in this approach will be *nonlinear*.

16.7 DERIVATIVE SPECTROMETRY

When an absorption peak is largely obscured by an interfering peak, taking the first- or higher-order derivative of the spectral trace with respect to wavelength is often a useful way to extract quantitative information. Further, second- and higher-order derivatives often reveal vibronic fine structure in absorption and fluorescence spectra. Recall that condensed phase spectra in the UV–VIS like that of bromthymol blue in Fig. 16.5 are often featureless. Since differentiation also enhances noise (see Section 4.8), the method is seldom attractive outside the UV–VIS where initial S/N ratios are high.

Commonly, the derivative $dI/d\lambda$, or $dA/d\lambda$ is approximated by a difference ratio. Thus $dA/d\lambda$ is calculated as

$$(dA/d\lambda) \approx [A_{\lambda+\Delta\lambda/2} - A_{\lambda-\Delta\lambda/2}]/\Delta\lambda, \tag{16.13}$$

where the absorbance change ΔA is divided by the wavelength interval $\Delta\lambda$ over which it occurs. The first derivative will have units of nm^{-1}. The choice of a finite $\Delta\lambda$ leads to a loss of amplitude and to broadening of derivative features (see Example 16.24). It is straightforward to extend Eq. (16.13) to the calculation of higher derivatives.

Wavelength modulation is an alternative mode of obtaining derivatives. It can be accomplished in a dispersive spectrometer by oscillating a small refractor plate inserted in the light beam. By controlling its angle with respect to the beam and the rate of oscillation, the value of the wavelength interval $\Delta\lambda$ is set and an ac signal developed that can be extracted electronically as the derivative [22].

The effect of taking derivatives of higher order can be demonstrated simply. Consider the case of two substances in solution whose spectral bands A and B overlap and are essentially of the same shape. The higher the order n of the derivative, the more the bands are discriminated. It is possible to show that if the separate peaks have heights H_A and H_B and widths W_A and W_B, the derivative spectra of order n will show heights H_A^n and H_B^n given by the expression*

$$H_A^n/H_B^n = (H_A/H_B)(W_B/W_A)^n.$$

Heights H_A^n and H_B^n will be the vertical distances between a maximum and the subsequent minimum in each feature associated with the derivative curve. The equation is approximate unless the bands are of identical shape, but it still provides a useful estimate of the increase in discrimination obtainable by going to higher-order derivatives. The enhancement is shown well in Fig. 16.14. With higher-order derivatives repetitive scans and averaging are advisable to reduce the noise that also enters.

Example 16.23. In Fig. 16.14 the first derivative is taken of a UV–VIS peak that is known to result from superposition of an analyte peak on a substantially broader interfering peak. Customarily, as in this case, the derivative is of absorbance with respect to wavelength, $dA/d\lambda$. In the derivative spectrum, the vertical distance between the height of the maximum and the subsequent minimum for the analyte (see figure) is used as a measure of its concentration.

*J. E. Cahill, *Am. Laboratory*, **11**(11), 79 (1979).

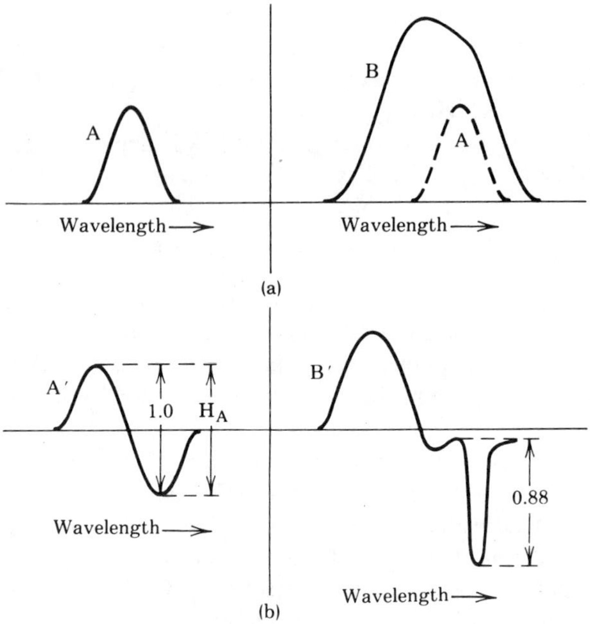

Fig. 16.14 Application of the first derivative approach to an absorption spectrum in which an analyte peak is overlapped by a large interfering peak. (a) Absorbance curves for analyte peak A and composite peak B. (b) A′, B′, first derivatives of the two absorbance curves. The vertical distance H_A (between the maximum and the following minimum) of each distinctive derivative's feature for the analyte is taken as the measure of its contribution. Note that H_A is nearly as large for the analyte in the composite as for analyte alone at the same concentration. (After T.C. O'Haver [22].)

Before differentiation of the composite peak no satisfactory measure of the contained analyte peak height can be obtained. When the peak height of the isolated analyte (upper left) is taken as 1.0, a method of tangents applied to the superposed peaks gives about 0.4 as the analyte contribution. In the first derivative curve, however, a clearly separated maximum and valley appear both for the interfering substance and the analyte (see the lower right curve in Fig. 16.14). The peak-to-valley vertical distance for the first derivative of the peak for the isolated analyte (at the lower left) is taken as 1.0. In the derivative curve for the composite peak note that the peak-to-peak height for the portion that can be identified with the analyte is 0.88 units and there is no ambiguity about the vertical distance. The usefulness of the vertical distance between the height of the derivative maximum and the subsequent minimum as a quantitative measure of the analyte concentration seems confirmed. For quantitative calculations a calibration curve based on these heights will of course have to be developed.

As a rule of thumb, this quantitative method is attractive whenever the overlapping band is at least twice the breadth of the analyte band [22].

Example 16.24. What wavelength increment $\Delta\lambda$ will give least distortion in derivatives approximated by a difference ratio?

In general, the *smaller* $\Delta\lambda$ is, the better the precision in amplitude and breadth of the derivative signal. Yet noise generated in taking the derivative will be smaller for *larger* $\Delta\lambda$ since it will be averaged over a larger interval. While $\Delta\lambda$'s for a first derivative will lead to some distortion in shape, the improved S/N ratio should more than make up for this loss. Nevertheless, the higher the derivative sought, the more important it will be that $\Delta\lambda$ be about the full width at half height of a feature of interest for precision and increased resolution of features.

Example 16.25. What factors affect the selection of the order of derivative in absorption spectrometry? Only in the case of a quite noisy spectrum or a simple curve overlap like that in Fig. 16.14 is the first derivative the best choice. Absorption peak positions are more easily identified in a second derivative where they correspond to minima. Often a second derivative provides a good compromise between solution and noise. Fourth-order spectra are valuable for developing fingerprint regions if the original S/N ratio is usually high.

16.8 REFLECTION METHODS

While transmission methods are the most widely used procedures for obtaining absorption spectra, reflection techniques are increasingly employed. If the sample is a film or a polished, flat solid, a regular or *specular reflection* technique is sometimes used. Measurements are made at the angle of reflection (equal to the angle of incidence) and a surface of known reflectance is employed as a reference. A spectrum can be obtained by wavelength scanning.

If the sample is a powder or a suspension, however, incident radiation is back-scattered (see Section 7.12) and the quantity measured is the *diffuse reflectance*. This useful quantity is often measured. In the visible and UV radiation diffusely reflected from the sample at many angles is collected by the walls of an integrating sphere, and a definite fraction is reflected to an exit slit and detector. A calibrated reflection standard such as magnesium oxide is generally used as a reference.

In the NIR *diffuse reflectance spectrometry* offers a way to obtain quantitative chemical information about complex samples. Some examples are finely ground coal, grains, polymers, and pharmaceuticals. NIR radiation excites harmonics and combinational overtones of vibrational transitions of molecules. Ellipsoidal collecting mirrors are substituted for integrating spheres to obtain good S/N ratios. Further, the spectrometer used is either a Fourier transform or filter instrument to ensure throughput as well as resolution.

To minimize unwanted scattering by a powdered sample, a proper, uniform particle size must be secured by grinding a sample sufficiently and sizing by use of standard sieves. For quantitative work relative measurements are required, that is, at each wavelength reflectance values of samples are compared with those of a standard such as powdered KCl, which has a high reflectance over a considerable wavelength range. Basic theory and the classical Kubelka–Munk equation for diffuse reflectance are introduced in a reference [7].

Example 16.26. NIR diffuse reflectance spectrometry can be used to obtain a multicomponent analysis of powdered milk for fat, lactose, total solids, and protein. Measurements must be made at least at four wavelengths. For such analyses instead of the Kubelka–Munk equation it has become customary to use more direct analytical expressions of the type:

$$\text{percent } x = Z + a \log 1/R^1 + b \log 1/R^2 + c \log 1/R^3 + \cdots , \qquad (16.14)$$

where R^i represents the relative reflectance at wavelength λ_i. Coefficients for each component are calculated at the measurement wavelengths using a regression equation derived from measurement of standards [10].

Where more detailed spectra are sought, however, the technique of *multiple internal reflectance spectrometry* that is based on the phenomenon of total internal reflection (see Section 7.11) is commonly used. The technique is also called attenuated total reflectance spectrometry. A standard spectrometer can be used; only the sample holder need be different (see Section 17.6). In this technique the sample is held tightly against a transparent plate of higher refractive index. Radiation is incident on the sample through the plate. When its angle of incidence on the plate–sample interface is greater than the critical angle, the radiation reflected is a spectrum characteristic of the sample that is nearly identical to a transmission spectrum [8, 9].

Total internal reflection is accompanied by penetration of the radiation only a fraction of a wavelength into the sample (the rarer medium). Indeed, the short depth of penetration is one of the most attractive features of internal reflectance spectroscopy. Even highly absorbing samples such as leather may be examined. If a transmission method were used instead, an absorption spectrum could not be determined because it would be impossible to cut a thin enough layer. On the other hand, if a sample is only weakly absorbing, *multiple* internal reflections can sometimes be used to secure sufficient signal.

Example 16.27. By means of internal reflectance spectrometry water and other agents on the surface of semiconductors have been detected (by IR measurement) at the level of 10^{-7} g. An internal reflection plate is used at an angle of incidence that permits hundreds of reflections at the semiconductor surface before the beam enters the spectrometer.

Several further advantages of internal reflectance spectrometry are

1. penetration of a sample is independent of its thickness;
2. absorption in a sample is independent of direction; and
3. interference and scattering phenomena do not occur in a sample.

As a result, two types of application are especially attractive: analysis of anisotropic substances; analysis of trace constituents in films whose small absorption might otherwise be obscured by interference fringes.

The chief decisions to be made in using the technique are the shape of the reflector plate and its composition. It is essential that the plate have a refractive index greater than the highest index of the sample to avoid serious distortion of absorption bands.* It is also desirable for purposes of strengthening absorption to use an angle of incidence on the sample somewhat greater (3–5°) than the average critical angle over the wavelength of the region of interest.

If the conditions outlined are met, a high-contrast spectrum can be obtained whose absorption bands are nearly identical to those in a transmission spectrum. Actually, the depth of penetration of a sample does increase slightly with wavelength, causing absorption bands at longer wavelengths to be relatively stronger. Further, for broad bands there is a noticeable widening on the long-wavelength side.

The limitations of the method are several. It is not particularly useful with weakly absorbing species such as gases, or for the determination of end points of titrations where there is a weak color change. Wherever surface properties of a material are different from its bulk characteristics, the technique is quite limited. Since the index of refraction of the plate and the angle of incidence must be chosen judiciously, it is certainly not a routine technique. Corrosive samples are difficult to use because they may attack the reflectance element. Perhaps the most serious limitation of the technique is that since contact problems may severely limit the reliability of intensities obtained it is difficult to use for quantitative measurements. If quantitative work is to be done, it is an advantage to work with polarized light [8].

16.9 DRY REAGENT PROCEDURES

The great majority of quantitative measurements in clinical chemistry are performed spectrophotometrically. Where at an earlier stage continuous flow methods were commonly used, today most analyses are done by what are called *discrete sample analysis*. In this approach a sample of blood serum or other body fluid is sampled by an automatic device that withdraws from a sample tube the precise aliquot necessary for each different analysis. The aliquot is transferred to an elaborate tiny *sampling module* designed for that analyte. In the module or cell the sample undergoes a reaction that develops a characteristic color and potential interferents are systematically taken care of. When the reaction process is complete after from 3 to 30 s, the system is examined either by transmission spectrophotometry or a diffuse reflectance procedure.† High selectivity is attained.

Here brief attention will be given to a different kind of measurement system in which a liquid sample such as blood serum and its reaction products move through

*Recall that the refractive index of a sample can be expected to increase in regions of absorption (see Section 7.7). If n_1(plate) $<$ n_2(sample) in an absorption region, a transmittance curve resembling a first derivative of an absorption peak is obtained.

†The approach has proved valuable also for clinical analyses based on use of ion-selective electrodes (direct potentiometry) and fluorescence spectrometry.

successive dry layers that contain appropriate reagents and barriers for a highly selective reaction for a particular component. A detailed illustration is given in Fig. 16.15. The final result is development of a distinctive color that may be determined by the diffuse reflectance methods discussed in the last section. The diffuse reflectance spectrometer may indeed have very simple characteristic modules consisting of a source, a suitable integrating sphere for collecting the diffuse reflection from the sampling unit, an interference or other type of filter if necessary, and a detector.

Determinations of reflectance are made relative to a known standard using the equation $(\%R) = (I_s/I_r)\,R_r$, where I_s, I_r, and R_r represent reflected intensities from the sample module and a reference and the percent reflectivity of the standard, respectively. In general an appropriate algorithm will provide for conversion of reflectance measurements to a function linear in concentration.

Advantages of the approach are simplicity, attainment of a high degree of precision in that the sampling module is sterile and highly stable, and its reactions are self-contained and highly reproducible. Further, most of the dry reagent devices

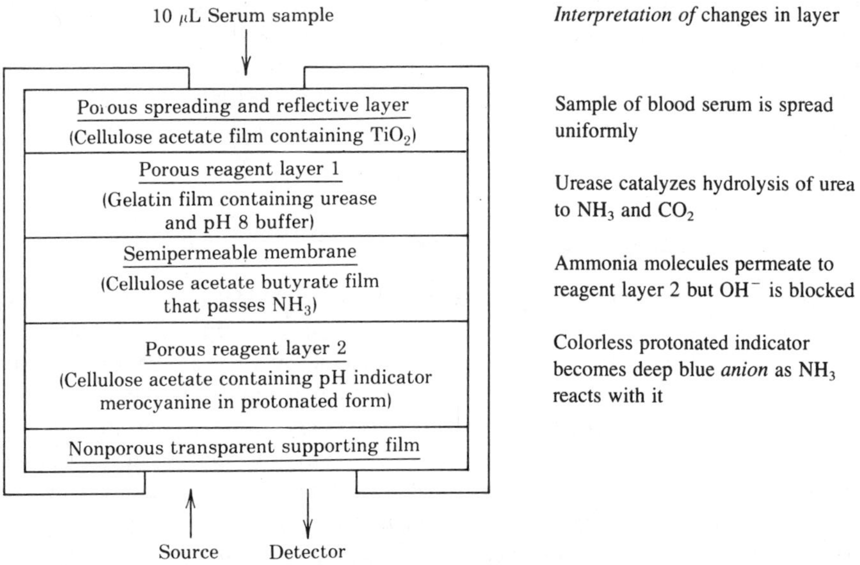

Fig. 16.15 Cross section of an Eastman Kodak Ektachem "slide" (about 1 cm^2 × 1.5 mm) for a diffuse reflectance analysis of blood urea nitrogen (BUN). A 10-μL sample of blood serum is deposited at the top. The serum and reaction products diffuse down through the layers. Chemical reactions and physical changes occurring in the four working layers are interpreted at the right. The bottom layer, which is a nonporous transparent film, supports the other layers. This film allows light from an optical source to enter the slide through the round aperture in the bottom, traverse all layers, be diffusely reflected by the finely dispersed TiO_2 particles in the top layer, traverse all layers once more, and emerge through the bottom aperture for detection. When no BUN is present, a baseline reflectance is obtained. When BUN is present, that reflectance is decreased by absorption by the merocyanine indicator in porous reagent layer 2 and is quantitatively detected. The logarithm of the ratio of reflectances is a measure of BUN concentration. (Courtesy of Eastman Kodak Clinical Products Division.)

require no more than about 10 μL of serum or 30 μL of whole blood. Biologically important species from small molecules to ions and enzymes can be determined over a considerable dynamic range. A final advantage is that algorithms specific for each test can be employed.

16.10 QUALITATIVE ANALYSIS

Spectrophotometry provides sufficient information for these kinds of qualitative analysis:

1. Pure substances can be identified and characterized as to structure.
2. Characteristic groupings of atoms can be identified.

Very few instrumental techniques provide as positive an identification of a pure substance as does absorption spectrometry. This is true because the *complete* absorption curve of a substance is unique, depending as it does on the kind, number, mass, and geometry of all the atoms in a molecule. Thus, not only does the absorption curve allow a substance to be distinguished from all others, but it also provides information about the kinds of atoms present and identifies the groupings in which they occur. Infrared and UV spectra are commonly cited as confirmatory evidence for postulated structural features. An illustration of the power of this method of analysis is provided by the comparison in Fig. 16.16 of the IR spectra of three similar substances. Even their small differences produce significant changes.

Spectrophotometry is much less well adapted to the study of impurities and gross mixtures. The basic difficulty is that the substances in mixtures present so many overlapping regions of absorption that the maxima and other shape details on which one relies to establish the identity of a particular compound are obscured. Ordinarily, in a qualitative analysis, mixtures are rather completely resolved before spectrophotometric study. For further information on dealing with mixtures, references should be consulted.

It should be emphasized that exploiting all the potentialities of a spectrometric analysis requires that the shape, height, and wavelength of absorption maxima be

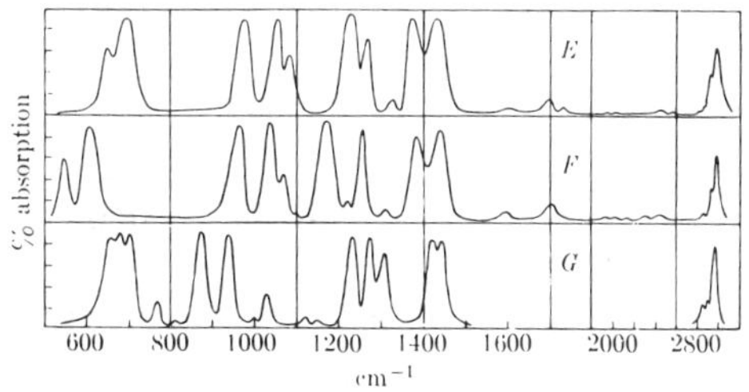

Fig. 16.16 Infrared absorption spectra of *E*, 1,1-dichloroethane; *F*, 1,1-dibromoethane; *G*, 1,2-dichloroethane. [Courtesy of the Faraday Division of the Royal Chemical Society; H.W. Thompson and P. Tarkington, *Trans. Faraday Soc.*, **42,** 432 (1946).]

known precisely. In most instances this requirement can be met by the use of an instrument of high resolution, that is, one with a good monochromator and a detector sufficiently sensitive (see Section 8.6) so that very narrow wavelength bands can be used.

Presentation of Data. An absorption curve is so dependent on instrumental and chemical variables that it will not stand by itself; it must be accompanied by certain data before its validity and accuracy are established. A thorough report should give or clearly imply (a) the type of instrument, (b) wavelength calibration points or references (c) spectral slit width or effective bandwidth, (d) the interval between points (manual instrument) or scanning speed (recording instrument), (e) the thickness of cells, and (f) all pertinent chemical information such as concentrations, pH, and temperature. In recognition of the necessity of reporting these data, appropriate blanks are placed on the charts for recording spectrophotometers.

Most qualitative studies require the examination of individual absorption curves as well as the intercomparison of different curves. Each of these processes is greatly facilitated by systematization in recording or plotting data. The routine devised should, of course, take into account both the available instrument(s) and the use to which the data will be put. Whether a spectrophotometer registers absorbance (optical density) or transmittance directly does not make much difference. Either variable gives a satisfactory representation of absorption when graphed as a function of wavelength or wavenumber (Fig. 16.16). For computer storage digital recording is desirable.

There are two graphical methods that are important because they lessen or remove completely any dependence on concentration. One involves plotting the logarithm of the absorbance, the other, the logarithm of the molar absorptivity. In Fig. 16.9 the character of the log A plot is demonstrated. Note that the *shape* of the absorption spectrum does not change with concentration though the curve "height" does. In other words, the slope of a log A curve depends only on the absorptivity of a substance. In the more conventional A vs. λ graphs the curves change both in slope and position as concentration is varied. The behavior of log A plot can be better understood by taking the logarithm of both sides of Eq. (16.3):

$$\log A = (\log \epsilon bc) = \log \epsilon + \log bc. \tag{16.15}$$

In this form the absorptivity ϵ, the only factor dependent on wavelength and the nature of a substance, is separated from the concentration-dependent term. Thus, the outline of a family of log A curves is specific for each substance. The height of a curve above the abscissa, however, is directly proportional to log bc.

If a library file of curves is being prepared, or if absorption data are primarily of research interest, it is probably worth the additional effort to calculate log ϵ and graph it. Although ϵ itself could be plotted, its range is ordinarily so great (perhaps from 0.1 to 1000 or more) that the logarithmic representation is advantageous. A straightforward and systematic computation procedure based on Eq. (16.15) can be devised.

Procedure. The IR and UV, in that order, are the best regions for qualitative analysis. Actually, the best evidence to establish structural features, and thus the identity, of a substance will come from the very important vibrational transitions appearing in the IR that are known as *group frequencies*. While a whole molecule vibrates as a unit, certain stretching or bending vibrations depend primarily on the mass of a small number of bonded atoms and on the force constants of the bonds joining them. In these cases other atoms have only a slight effect on the frequency of vibration. For example, the stretching of bonds between hydrogen and other atoms such as C—H and O—H occurs at about 2900–3600 cm^{-1}. Many group frequencies are listed in Table 16.6.

In general the smaller a group frequency, the more sensitive it is to the structure of a molecule as a whole. For example, the exact frequency of a group whose vibration appears below 1300 cm^{-1} is often useful in identifying the position of the group if it is bonded to an aromatic ring or details of chain structure if in an aliphatic substance. Since the entire molecule as well as the group determines

Table 16.6 Some Characteristic Group Frequencies[a]

	Approximate Position of Absorption Band	
Functional Group	*μm*	*cm^{-1}*
$-NH_2$ (primary amine)	2.85– 3.15	3180–3505
—OH (alcohols)	2.84– 3.22	3100–3520
$>$C—H (aliphatic)	3.3 - 3.6	2780–3050
$>$C—H (aromatic)	3.2 - 3.35	2995–3140
$-CH_2-$ (methylene)	3.42– 3.55	2810–2920
—C≡N (nitrile)	4.16– 4.55	2200–2405
—C≡C—	4.42– 4.65	2150–2270
$>$C=O (esters)	5.71– 5.81	1722–1752
$>$C=O (acids)	5.75– 6.01	1660–1740
$>$C=O (aldehydes and ketones)	5.76– 6.05	1650–1740
$-NH_2$ (amine)	6.08– 6.35	1575–1645
—C=C—	5.98– 6.17	1620–1670
$>$C—Cl	13.1 -15.2	660–765
$>$C—Br	15.6 -18.9	530–640

[a]Though not listed, inorganic ions such as SO_4^{2-} and NH_4^+ also exhibit characteristic group frequencies. These vibrations are found in the 4000–400 cm^{-1} region also.

group frequencies in this range, the lower frequency IR is often termed the fingerprint region.

Absorption in the near UV (400–200 nm) is found only for species with electron clouds that are somewhat loosely bound such as compounds with π-bonds. Thus, conjugate double-bond systems with their delocalized orbitals show especially strong UV absorption. Since electron levels are more nearly molecular in character than group frequency vibrations, less information can generally be obtained from the UV than the IR. A further disadvantage of the UV is the relative scarcity of absorption peaks; most substances show only one or two in the usual condensed phase sample.

Assuming that one has isolated an unknown substance in a fairly pure state, the best analytical approach is to determine its IR absorption from 4000 to 650 cm^{-1} (2.5–15 μm). Alternatively, the UV or the near IR can be scanned.* It must be emphasized that an analysis cannot be based on an examination of only a small spectral region (e.g., a range of 100 nm), because similar compounds may have nearly identical spectra over short ranges.

If an extensive file of curves (plotted on the same coordinates) is available, identification of the unknown may be a matter of intelligent comparison. For this operation, translucent charts and an illumination box (a glass-topped box lighted from inside) are advisable. If a known with the same absorption features (wavelengths of maxima, principally) is found, the identification may be assumed virtually conclusive. Identical UV features will indicate strongly that the substances are identical. To speed identification, the absorption maxima of all knowns can be compared by use of an appropriate computer program with the characteristic frequencies found for the unknown. If only a partial file or no file of curves is available, the unknown curve must be translated into the type of "groups," that are

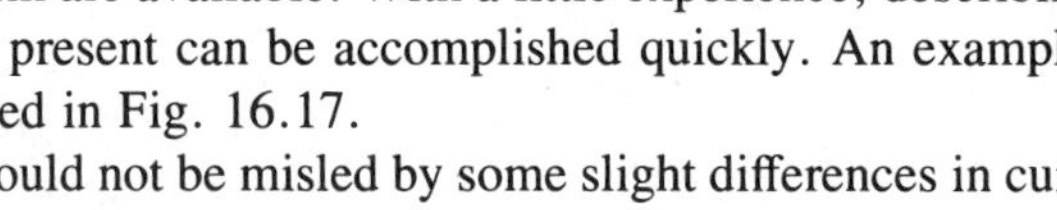
and so on,

present beyond reasonable doubt in the substance, for example, by use of a listing such as that in Table 16.6. Extensive compilations of this type of information in tabular form are available. With a little experience, describing which groupings of atoms are present can be accomplished quickly. An example of what is involved is illustrated in Fig. 16.17.

One should not be misled by some slight differences in curves in regions of high transmittance; these may result from impurities. Since their concentrations will be low, impurities will be unlikely to absorb enough radiation to show their presence except in regions where the substance being analyzed has a high transmittance. Impurities will ordinarily have little effect on the absorption peaks.

*In the near IR (14,000–5000 cm^{-1}) spectral absorptions are characteristic of overtones (harmonics) or combinations of fundamental vibration frequencies. Thus, they are helpful in confirming molecular identity but not generally useful in elucidating molecular structure.

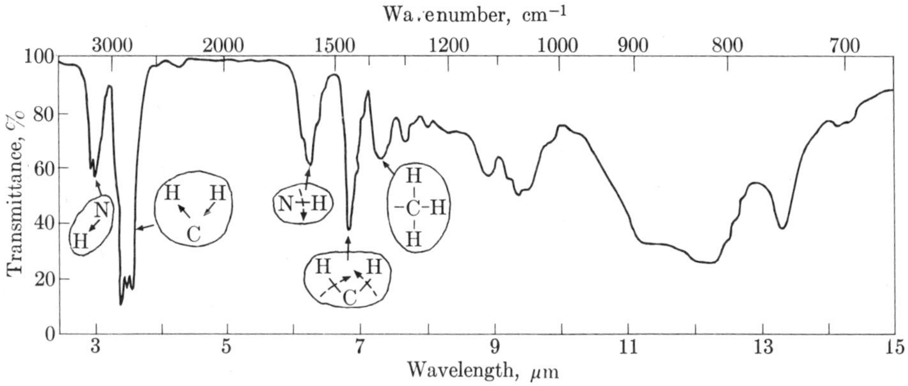

Fig. 16.17 Correlation of atomic groupings with absorption maxima in the 4000–650 cm^{-1} region for 3,3′-diamino-*N*-methyldipropylamine (0.01-mm cell path),

$$CH_3-N\begin{matrix} \diagup CH_2-CH_2-CH_2-NH_2 \\ \diagdown CH_2-CH_2-CH_2-NH_2 \end{matrix}$$

It is clear that functional group determinations in new compounds may also be made by spectrophotometric methods. Indeed the methods used are the same as those employed in identifying an unknown substance except that the search is limited to the wavelength regions that will reveal the groups of interest. Both the UV and IR are useful regions for this type of study.

Comparison of Qualitative Methods. Because of the increasing capabilities of modern analytical chemistry in identifying species (i.e., establishing their structure), an intercomparison of qualitative analytical techniques is useful. Each method has its areas of strength and its limitations. In Table 16.7 this type of comparison is made for the most important techniques of qualitative analysis. As noted, methods that furnish elemental analysis or serve mainly for functional group determinations are not included.

16.11 STRUCTURAL INVESTIGATIONS

One of the more important applications of spectrophotometry is obtaining information about characteristic energy levels and the structural parameters of molecules. The IR and UV regions provide a rich yield of such data. Bond stiffnesses, (i.e., force constants), are calculable from vibrational frequencies, moments of inertia, and internuclear distances from rotational frequencies (Section 16.3). From higher-order rotational transitions some notion of the distortion produced in a molecule by rotations may be secured. The excited molecular electronic states can be located and used to refine present mathematical molecular models. Bond dissociation energies are also accurately estimated from UV spectra.

Table 16.7 Comparison of Several Methods of Qualitative Analysis[a,b]

Technique	*Usefulness in Characterizing Type of Substance*	*Range of Usefulness*	*Finger-Print Type Identification*	*Deductive Type Indentification*	*Cost of Instrumentation*
Absorption spectrometry					
UV, VIS	Limited	Narrow	Fair at best	Poor	Low to high
IR	Good	Very wide	Very good	Fair	Low to high
Raman spectrometry	Limited	Narrow to moderate	Fair	Fair	Moderate to high
NMR spectrometry	Limited	Very wide	Excellent	Excellent	Moderate to high
Optical rotary dispersion	Very limited	Narrow	Fair to poor	Fair to poor	Moderate to high
Mass spectrometry	Good	Very wide	Excellent	Excellent	High
X-ray diffraction					
spectrometry	Limited	Narrow	Excellent	Excellent	High
single crystal powder	Limited	Narrow	Excellent	Poor	Moderate to high

[a]After S. Siggia, *Survey of Analytical Chemistry*, New York: McGraw-Hill, 1968, pp. 135–136.

[b]Qualitative analysis is used only in the sense of identification of a species or characterization of its structure. No methods are listed that lend themselves principally to elemental analysis, such as emission spectroscopy and neutron activation analysis, or to functional group determination, such as electroanalytical procedures.

Spatial aspects of molecules are also susceptible to IR analysis. Where rotational isomers exist, as in a molecule in which bulky substituents are attached to both sides of a carbon–carbon single bond, they can be detected. Conformational studies can be made of molecules such as cyclohexane derivatives. For example, an axial C—X bond vibration ordinarily is found at a lower frequency than an equatorial C—X vibration. As would be expected, tautomeric shifts such as keto–enol transformations can be followed. Structural isomerism, as in molecules that have *cis*- and *trans*-configurations, is also easy to examine.

Of equally great interest is the application of spectrophotometry to the investigation of intermolecular interactions. The degree of hydrogen bonding (either *inter*- or *intra*molecular), molecular association, and attraction of polar molecules to metal ions has been determined spectrophotometrically. These studies usually involve a measurement of the shift in (a) the frequency of a vibrational transition involving the O—H, N—H, or other bonds, as well as (b) intensities. Complexing studies on the other hand, have typically made use of spectrophotometry as a quantitative tool both for determination of formulas and information about bond character.

Studies of crystals with polarized IR radiation have yielded increasing information about the orientation of groups and the nature of vibrations. Either the depolarization ratio or the dichroic ratio $A_{\perp}/A_{\parallel}$ is determined.

REFERENCES

Some general references are:

1. E. J. Meehan, Fundamentals of spectrophotometry, in *Treatise on Analytical Chemistry*, 2nd ed., Part I, Vol. 7, P. J. Elving, E. J. Meehan, and I. M. Kolthoff (Eds.), New York: Wiley-Interscience, 1981.

Molecular spectroscopy is treated on an intermediate level in:

2. C. N. Banwell, *Fundamentals of Molecular Spectroscopy*, 2nd ed. New York: McGraw-Hill, 1983.
3. N. B. Colthup, L. N. Daly, and S. E. Wiberley, *Introduction to Infrared and Raman Spectroscopy*, 2nd ed. New York: Academic, 1975.

Absorption spectrometry is presented clearly with emphasis on instrumentation and practice in:

4. N. L. Alpert, W. E. Keiser, and H. A. Szymanski, *IR, Theory and Practice of Infrared Spectroscopy*, 2nd ed. New York: Plenum, 1970.
5. A. Knowles and C. Burgess (Eds.), *Practical Absorption Spectrometry.* New York: Chapman and Hall, 1984.
6. E. G. Brame and J. G. Grasselli (Eds.), *Infrared and Raman Spectroscopy*. New York: Dekker, 1977.

Reflection spectroscopy is treated lucidly in:

7. M. P. Fuller and P. R. Griffiths, Diffuse reflectance measurements by infrared Fourier transform spectrometry, *Anal. Chem.*, **50,** 1906 (1978).

8. N. J. Harrick, *Internal Reflection Spectroscopy*. New York: Interscience, 1967.
9. G. Kortum, *Reflectance Spectroscopy; Principles, Methods, Applications*. New York: Springer-Verlag, 1969.
10. D. L. Wetzel, Near-infrared reflectance analysis, *Anal. Chem.*, **55,** 1165A (1983).

Detailed practical information on visual and absorption photometric methods of analysis for elementary ions and compounds is given by:

11. F. Feigl, *Spot Tests in Organic Analysis*, 7th ed. New York: Elsevier, 1966.
12. F. Feigl and V. Anger, *Spot Tests in Inorganic Analysis*, 6th ed. New York: American Elsevier, 1972.
13. E. Jungreis, *Spot Test Analysis. Clinical, Environmental, Forensic, and Geochemical Applications*. New York: Wiley-Interscience, 1985.
14. Z. Marczenko, *Separation and Spectrophotometric Determination of Elements*. New York: Ellis Harwood (Halsted), 1986.
15. F. D. Snell, *Photometric and Fluorometric Methods of Analysis: Nonmetals*. New York: Wiley-Interscience, 1981.

Also of interest are:

16. G. D. Christian and J. B. Callis, *Trace Analysis: Spectroscopic Methods for Molecules*. New York: Wiley, 1986.
17. T. D. Harris, High-sensitivity spectrophotometry, *Anal. Chem.*, **54,** 741A (1982).
18. P. A. Jannson (Ed.), *Deconvolution—With Applications in Spectroscopy*. New York: Academic, 1984.
19. D. S. Kliger (Ed.), *Ultrasensitive Laser Spectroscopy*. New York: Academic, 1983.
20. S. H. Lin et al., *Multiphoton Spectroscopy of Molecules*. New York: Academic, 1984.
21. W. W. Meinke, Spectrophotometry in Standard reference materials for clinical measurements, *Anal. Chem.*, **43**(6), 28A (1971).
22. T. C. O'Haver, Derivative and wavelength modulation spectrometry, *Anal. Chem.*, **51,** 91A (1979).
23. J. A. Perry, Quantitative analysis by infrared spectrophotometry, *Appl. Spectr. Rev.*, **3,** 229 (1970).
24. B. Walter, Dry reagent chemistries in clinical analysis, *Anal. Chem.*, **55,** 498A (1983).
25. E. L. Wehry and G. Mamantov, Matrix isolation spectroscopy, *Anal. Chem.*, **51,** 643A (1979).
26. G. Svehla, Differential spectrophotometry, *Talanta*, **12,** 641 (1966).

Some volumes dealing comprehensively with applications of spectrophotometry in fields other than chemistry are:

27. F. S. Parker, *Applications of Infrared, Raman, and Resonance Raman Spectroscopy in Biochemistry*. New York: Plenum, 1983.

Compilations of absorption spectra, references to methodology for obtaining such spectra, are available in:

28. ASTM Committee E-13, *Manual on Practices in Molecular Spectroscopy*, 4th ed. Philadelphia: ASTM, 1980.

29. L. J. Bellamy, *Advances in Infrared Group Frequencies*. New York: Barnes and Noble, 1968.

30. L. J. Bellamy, *The Infra-Red Spectra of Complex Molecules*, 2nd ed. New York: Wiley, 1958.

31. C. Burgess and A. Knowles (Eds.), *Standards in Absorption Spectrometry. Techniques in Visible and Ultraviolet Spectrometry*. London: Chapman and Hall—Methuen, 1981, Vol. 1.

32. IUPAC, *Tables of Spectrophotometric Absorption Data of CPDS Used for the Colorimetric Determination of Elements*. London: Butterworth, 1963.

33. M. J. Kamlet (Ed.), Vol. 1, and H. E. Ungnade (Ed.), Vol. 2. *Organic Electronic Spectral Data*. New York: Interscience, 1960.

34. K. N. Rao, C. J. Humphreys, and D. H. Rank, *Wavelength Standards in the Infrared*. New York: Academic, 1966.

EXERCISES

16.1 (a) If a liquid absorbs strongly in the spectral region from 500 to 700 nm, what is its color? (b) If the absorption band of a solution is from 450 to 600 nm, what color does it have?

16.2 A solution of potassium permanganate in a 1-cm cell is found to have a transmittance of 60% at a particular wavelength. If the concentration is doubled, what will be (a) the percent transmittance? (b) the absorbance? (c) What concentration of the $KMnO_4$ must be used to give 60% transmittance in a cell 10 cm long?

16.3 The molar absorptivity ϵ of benzoic acid in methanol is about 1950 at 275 nm. What is the maximum concentration of benzoic acid in grams per milliliter that can be used in a 1-cm cell if the absorbance is not to exceed 1.3?

16.4 A 2.0×10^{-5} *M* solution of a substance gives 50% transmittance in a 1-cm cell at 230 nm. What is its molar absorptivity?

16.5 A 20-mL aliquot of river water is treated with an excess of potassium thiocyanate and diluted with distilled water to 50 mL. (Fe^{3+} ions form red $FeSCN^{2+}$ ions with $\epsilon_{580} = 7.00 \times 10^3$.) This solution gives an absorbance of 0.104 at 580 nm when measured in a 1-cm cell. (a) Give the dimensional *units* of the molar absorptivity ϵ. (b) Calculate the concentration of Fe^{3+} in the river water. (c) Later this result is challenged because the blank was made up in distilled water. The river water was found to have about 97% transmittance from 500 to 600 nm. What concentration of iron should have been reported?

16.6 A cadmium–zinc alloy is known to contain about 5% iron. It is proposed to analyze for the 1:1 complex with a 5-sulfoanthranilic acid ($\epsilon_{455} = 1306$). A 0.2-g sample is dissolved in dilute H_2SO_4, and pH adjusted, and complexing agent added. The final volume is 400 mL and $A_{455} = 0.637$ in a 1-cm cell. A blank containing only complexing agent at the supposed excess concentration gave $A_{455} = 0.015$. Calculate the concentration of iron in the alloy.

16.7 The manganese content of a steel sample is determined spectrophotometrically by oxidation to permanganate and measurement at 575 nm. The absorbance of the three standards and a blank was also determined at 575 nm. From the results tabulated below, calculate the percentage of manganese in the sample.

Solution	*Concentration*	*Absorbance*
Unknown	0.5000 g steel in 250 mL soln.	0.456
Blank	0.5000 g steel in 250 mL soln., but Mn unoxidized	0.020
Standard (1)	0.020 mg Mn mL^{-1}	0.600
Standard (2)	0.010 mg Mn mL^{-1}	0.301
Standard (3)	0.0040 mg Mn mL^{-1}	0.115

16.8 From the absorption spectra of two amino acids in 0.1*M* NaOH solution in Fig. 16.18 (a) select appropriate wavelengths for the simultaneous determination of their concentrations and justify the choice. (b) Using approximate numerical values from the figure write the pair of equations that must be solved to obtain c_A and c_B.

16.9 In spectrophotometry the terms stray light and scattered light are both used. How do they differ in meaning? Define each term. Give an example of each phenomenon.

16.10 It is found by a study on solutions of known absorbance that a given spectrophotometer has 0.2% stray radiation at 450 nm. (a) For a concentrated solution, an apparent transmittance of 2.3% is observed. What is its apparent and true absorbance? (b) What relative error in concentration will result if this apparent absorbance is used in a calculation without correction?

16.11 With a double-beam UV–VIS spectrophotometer it is found that the level of stray light at 400 nm is 0.1% of power P_0. (a) What relative error from stray light will result at that wavelength when $A = 1$? when $A = 3$? (b) With what relative precision can the concentration of a substance whose molar absorptivity is 3.50×10^3 L mol^{-1} cm^{-1} be measured in an approximately 7×10^{-3} *M* solution? (c) With a tungsten source in use in the spectrophotometer is it reasonable to expect the stray light level to be higher at 350 than at 400 nm?

16.12 The species HA has an absorption band centered at 500 nm in water. In this solvent the HA also dissociates somewhat according to the equation $HA \rightleftharpoons H^+ + A^-$. Neither H^+

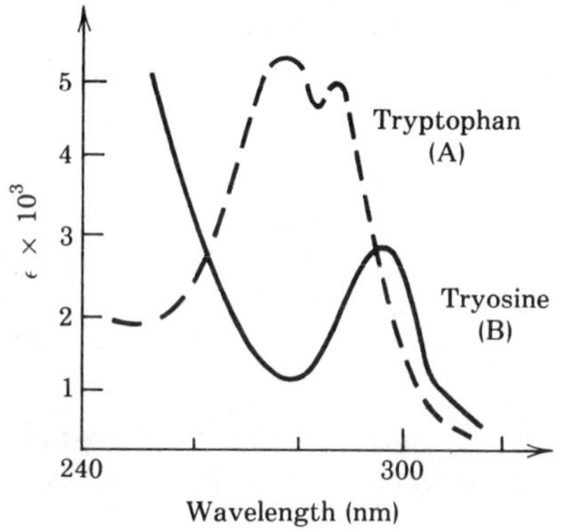

Fig. 16.18

nor A^- absorb in the visible. (a) Will a plot of the absorbance of HA solutions at 500 nm display a positive or negative deviation from Beer's law? (b) If a considerable excess of NaA, a soluble salt, is added and the measurements are repeated, will a deviation be observed? (c) Should the results obtained be reported as showing that the substance HA does not follow Beer's law? Explain.

16.13 Sketch the type of spectrophotometric titration curve that will be obtained if at the wavelength chosen for monitoring (a) both reactant and titrant absorb and the reactant absorbs more than the titrant, (b) reactant and product absorb and the reactant absorbs more than the product, (c) only the reactant absorbs.

16.14 A differential spectrophotometric procedure is used to determine a metal A in an alloy. After dilution and complexing with a chromogenic ligand, the transmittance at the absorption maximum for complexed A is found to be less than 10%. A 1-cm cell is used. A 1.35×10^{-3} M solution of A similarly complexed and diluted is accordingly used to set the 100%T reading while the 0%T setting is made with no light. Under these conditions, the sample transmittance is 55%T. The molar absorptivity of the complex is 3.87×10^3 L mol^{-1} cm^{-1}. What is the concentration of A in the solution?

16.15 The thickness of a polystyrene film (refractive index 1.6 in the IR) is to be determined by use of an IR spectrophotometer. A regular spectrum is obtained with the film in the sample beam and air in the reference beam. Between 2000 and 2800 cm^{-1}, where polystyrene is transparent, the baseline shows 30 small interference maxima. How thick is the film?

16.16 (a) For ethylene, identify molecular orbitals by conventional notation and state their electron occupancy in the ground state. (b) What electronic transitions in ethylene are open to the most easily excited electron(s): $n \rightarrow \pi^*$, $\sigma \rightarrow \pi^*$, or $\pi \rightarrow \pi^*$? (c) Will the same type of transition be the easiest to excite in 3-octene? in cyclopentene? (d) The longest wavelength UV–VIS absorption of these olefins is at about 185 nm. Is this result consistent with the answers to (b) and (c)?

16.17 What is the wavelength and wavenumber of the absorption maximum for the fundamental vibrational band for HBr? Assume a force constant equal to 0.5×10^3 N m^{-1}.

16.18 Which bond will give more closely spaced vibrational levels to a molecule

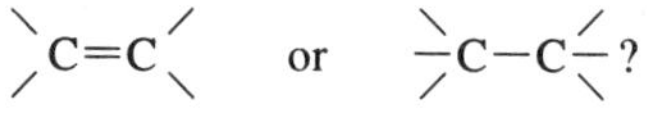

Explain.

16.19 At about what wavenumbers in the IR should the molecule 4-hydroxybenzaldehyde absorb? Indicate the characteristic vibration responsible for each absorption.

16.20 If a colorless species is to be determined by absorption photometry in the visible range, it must be complexed with an appropriate chromogenic reagent. Suggest several properties desirable in such a complex if it is to be used in quantitative analysis.

16.21 Describe a spectrophotometric procedure that can be used to determine the pK_a of an indicator such as phenolphthalein.

16.22 A source of difficulty in IR spectrometers in that a sample or any instrument part that is above room temperature is also a net IR emitter. Assume the IR absorption spectrum of a polystyrene film is to be measured in a double-beam spectrophotometer with air in the other path. (a) What effect will the gradual heating of the film by the radiation beam have on the wavelength of the absorption peaks? on absorbance at the peaks? (b) If the instrument is at 298 K, the film is heated to 325 K, and the real source is at 1500 K and both "emitters" behave as blackbody sources, at what wavelength will film emission generate the larger error, 2000 or 700 cm^{-1}?

16.23 It is desired to analyze heavy water that is more than 99% D_2O for isotopic purity. Differential IR spectrometry is chosen as a method that is suitably accurate. Since small concentrations of ^1H appear almost entirely in HDO molecules, the analysis may be made at 5980 cm^{-1} (1.67 μm), a combination vibrational wavelength of HDO. Describe roughly how such an analysis might be arranged, mentioning choice of cell (window) and a basis for setting 100%T and 0%T.

16.24 Which technique, spectrophotometry or potentiometry, would offer better precision as an end-point detection technique in the following titrations? State any assumptions made in each case: (a) a dilute organic analyte in aqueous solution, (b) a colored analyte in nonaqueous solution; (c) an aqueous HCl solution; (d) a concentrated, aqueous $ZnSO_4$ solution.

16.25 The dual-wavelength photometric procedure described in Section 16.6 is to be applied to a kinetic study of an oxidation of cytochrome *c*. At wavelength λ_1 and λ_2, let ϵ_{R1} and ϵ_{R2} be the molar absorptivities of reduced form, ϵ_{O1} and ϵ_{O2} be the molar absorptivities for the oxidized form, and A_1 and A_2 be the total absorbances. Let C_R and C_O be the concentration of reduced and oxidized forms at a given time. Assume λ_1 is an isosbestic point and that $C_R + C_O$ = constant. Derive an expression that relates the change in concentration of the oxidized form, ΔC_O, to the other parameter.

16.26 While Beer's law is theoretically valid only if monochromatic radiation is used, the strictness of this requirement depends on the degree to which the absorptivity (a or ϵ) of a sample is constant over the wavelength band corresponding to the spectral slit width. In a determination what is actually measured is not the power transmitted at one wavelength but an intensity $\int I_0(\nu)\, d\nu$ where the integration is carried out across the spectral slit width. Thus an apparent absorbance A' is determined for a sample where A' is given by

$$A' = \log\left[\int I_0(\nu)\, d\nu\right] \Big/ \int I_0(\nu) \times 10^{-\epsilon(\nu)bc}\, d\nu.$$

(a) Show that this expression follows from Beer's law. (b) Show that if $\epsilon(\nu)$ = constant over the spectral band passed by the slit, Beer's law will hold. [See W. E. Wentworth, *J. Chem. Ed.*, **43,** 262 (1966).]

ANSWERS

16.1 (a) Blue; (b) purple. *16.2* (a) 35.9%; (b) 0.44; (c) 0.1 original. *16.3* 8.1×10^{-5} g mL^{-1}. *16.5* (a) L mol^{-1} cm^{-1}; (b) 3.7×10^{-5} *M*; (c) 3.2×10^{-5} *M*. *16.6* 5.33%. *16.7* 0.72%. *16.10* (a) Apparent, 1.638; true, 1.688; (b) 1.8%. *16.14* 1.42×10^{-3} *M*. *16.15* 0.16 mm. *16.17* 3.45 μm, 2900 cm^{-1}. *16.22* (b) at 700 cm^{-1}. *16.25* $\Delta C_0 = \Delta A_2/(\epsilon_{O2} - \epsilon_{R2})$.

Chapter 17

INSTRUMENTATION FOR ABSORPTION SPECTROMETERS

17.1 INTRODUCTION

The basic absorption spectrometer or spectrophotometer used in the kinds of measurements described in Chapter 16 is shown in modular form in Fig. 17.1. The diagram pictures a *single-beam* or single-channel instrument that requires separate observation of light transmitted by a blank, or reference, and a sample. In the majority of *absorption* spectrometers a monochromator serves as the wavelength isolating module; they are wavelength-dispersive devices. In this chapter single- and double-channel dispersive spectrometers, colorimeters or filter photometers, and Fourier transform spectrometers will be discussed. Modules will also be described to adapt any of these spectrometers for (a) reflection and (b) photothermal or photoacoustic measurements.

Much background that is essential to understanding this instrumentation was developed earlier in the book. Chapter 7 presented many principles of physical optics, Chapter 8 discussed the theory and design of optical sources and detectors, and Chapter 9 presented monochromator and polychromator theory and design. In addition, Chapters 2–5 offered insights into the electronics of instrumentation, data acquisition, and processing and Chapters 6 and 12, background on computational and S/N aspects of processing.

17.2 DESIGN CRITERIA FOR DISPERSIVE SPECTROMETERS

How should one go about selecting the best modules to make up dispersive spectrophotometers for different applications? While most weight must be given to the performance of different modules, basic design considerations having to do with the order of modules in the signal channel, effective modes of coupling, and the choice of a double- or single-beam instrument layout cannot be neglected.

To apply these criteria, consider the static operation of the instrument shown in Fig. 17.1. Later (see Section 17.5) dynamic aspects will be discussed. The amplitude of the detector signal with a sample in place can be related to modular performance by the general equation

$$\text{signal from a sample} = [p(\lambda)\, m(\theta, \lambda)\, l(\theta, \lambda)]\, 10^{-\epsilon bc} f(\theta, \lambda). \quad (17.1)$$

Continuous source → Wavelength-isolation device → Sample module → Detector → Amplifier → Processing modules → Readout

Sample module ↕ Reference

Fig. 17.1 Modular diagram of a single-beam absorption spectrometer. In simple instruments the signal-processing stage that appears between amplifier and readout is omitted. The order of the wavelength-isolation module and sampling module are sometimes interchanged.

Each factor will be identified in turn. It is helpful to view this entire equation as an operational adaptation of Eq. (16.2), $P = P_0\, 10^{-\epsilon bc}$, to the modular diagram of Fig. 17.1. In Eq. (17.1):

1. $p(\lambda)$ is the power furnished by the optical source as a function of wavelength, $m(\theta, \lambda)$ is the efficiency of the wavelength isolator as a function of wavelength and the solid angle θ subtended by the source at its entrance aperture and $l(\theta, \lambda)$ is the transmission of a blank. Thus this collection of factors in the brackets makes up the term P_0. Operationally, it is the power transmitted by a blank.
2. The term $10^{-\epsilon bc}$ defines the factor by which a sample decreases P_0 at wavelength λ.

Finally, in Eq. (17.1) $f(\theta, \lambda)$ expresses the efficiency of the detector as a function of both the solid angle of radiation it receives and the wavelength.

Example 17.1. Does Eq. (17.1) imply that the sensitivity of measurement will increase if the source intensity is increased in spectrophotometry? What level of intensity will ensure an adequate S/N?

To respond to these queries we need to examine both Eqs. (17.1) and (16.2). Equation (17.1) predicts that sample transmission P is proportional to P_0. With a more intense source both power levels will increase; it is not P that is proportional to analyte concentration, but the ratio P/P_0, which will remain unchanged with a source intensity increase. As a result, the sensitivity of measurement will be unchanged.

On the other hand, if S/N for a measurement is initially low, use of a brighter source will increase the signal relative to noise and better precision will be attained.

Coupling of Characteristic Modules. Good coupling can be defined as an efficient transfer of optical energy (*throughput*, luminosity, or etendue) through the optical modules of a spectrophotometer. For optimum throughput the beam leaving one module must just fill the *entrance aperture* of the next, as implied in Eq. (17.1). For example, for efficient operation of a grating, mirror, sample cell, detector, or other component, its full working area must be illuminated. There will also be places along the instrument channel at which the beam should be reimaged (see Section 9.3) so that power density in the beam is uniform to permit operations such as diffraction to be carried out with precision.

Good coupling will ensure that beam intensities will be as high as possible for a given source even under conditions of high absorbance. This requirement translates into maximizing throughput by adding to the modules shown in Fig. 17.1 a variety of front-surface mirrors, masks, and lenses. They can be placed between modules to redirect a beam or alter its diameter. Good coupling calls for keeping the beam from being unnecessarily broadened, absorbed, or distorted. Commonly mirrors, lenses, and other optical elements are coated with dielectric layers of appropriate thickness to maximize throughput, that is, reflectance for mirrors and transmission for lenses, at least in important wavelength ranges. Designing mirrors and lenses to minimize aberrations such as astigmatism and coma that distort images such as the rectangular pattern of a slit calls for careful design of each element and of the layout.

The high power densities of laser beams impose still other requirements. For example, pencil-width laser beams are always expanded in diameter before they fall on a diffraction grating.

Coupling must also minimize noise. For example, coupling should include baffles to reduce the possibility of radiation that has left the optical path from reentering it as stray radiation. To remove undesired orders of spectra or spectral regions may also call for insertion of cutoff filters or dispersing prisms, or very occasionally polarizing devices, into the optical beam. The efficient transfer of desired wavelengths and diminishing the intensity of others may call for changing the sequence in which modules appear. Some examples will illustrate decisions about coupling, energy transfer, and imaging. The reader is also referred to Chapter 9 for a more detailed examination of the subject.

Example 17.2. When is an optical beam of uniform power density or cross section needed in a spectrometer? How does one secure this behavior?

To develop a uniform optical cross section requires that a source be used that has at least one region of uniform irradiance (see Section 8.2). An image of that area in a source can be captured and reformed whenever a beam of uniform cross section is needed in an instrument. In general, an image can be reformed many times along the path of a beam.

Some places where uniform beam power is advantageous are (a) at a monochromator entrance slit, so that inside the monochromator the beam spreading from the slit will evenly illuminate the diffraction grating, (b) at an optical wedge in an optical null system to ensure that beam attenuation will be smooth as it is driven into a beam, (c) at the responsive surface of a detector to ensure reproducible output, and (d) at the center of the sample module to permit a microcell to be used if needed.

Example 17.3. For solution measurements of moderate precision in the visible, will a set of interference filters or a monochromator make a more appropriate wavelength isolation module? Will a need for versatility alter the choice?

A set of filters is usually more advantageous. The key factor is that most analytes have broad (electronic) absorption bands in the visible and filters with 12–30 nm bandwidths will usually ensure adequate Beer's law performance. By virtue of the high filter throughput simpler instruments can be used. They will offer the virtues of design simplicity, ruggedness, lower cost, and lower maintenance. For routine measurements with a limited set of analytes, filter instruments are a good choice.

For spectral versatility, however, one would be likely to choose a monochromator with 2-nm resolution or better. It would allow measurement of analytes with absorption bands of any wavelength and different bandwidths.

Example 17.4. In most UV–VIS spectrophotometers the sample module follows the wavelength isolation device whereas it is placed before it in IR spectrophotometers. Why is the order of modules important? How important is it?

Look again at Fig. 17.1. In principle the relative position of these two modules should not affect measurements. Their order will, however, influence the intensity and wavelengths of light that enter the monochromator. The sample cell will always absorb and scatter some of the radiation falling on it. If that cell is first, the monochromator will receive much less light and thus generate less stray light. If sample solutions have high absorbance, the amount of light entering the monochromator will be dramatically less and measurements will be more accurate.

Now let us look at dispersive spectrometers, one to be used in the IR and another for the UV–VIS range. In the IR the usual source is a heated solid whose broad emission region peaks at in the NIR (5000–10,000 cm^{-1}). The analytical part of the spectrum, however, is the mid-IR (4000–250 cm^{-1}), where intensities are perhaps two orders of magnitude smaller. Placing a monochromator next to the source would permit its intense 5000–10,000 cm^{-1} light to enter and contribute strongly to the stray light level at its exit slit. An appreciable fraction of this NIR stray light may be transmitted by a sample as well as the mid-IR radiation of interest, causing a spuriously large detector response at mid-IR wavelengths. But if the sample cell were next to the source, the cell would absorb or scatter out of the beam a substantial amount of the NIR radiation that would otherwise enter the monochromator. This arrangement reduces greatly the generation of stray light. Nevertheless it poses a small trade-off, in that samples in this position are strongly heated.

Why does the sample cell *follow the monochromator* in UV-VIS instruments? In these instruments emission at all wavelengths by sources (halogen–tungsten in the VIS and deuterium in the UV) is much stronger. The peak emission of the lamp will often fall *within* the spectral range of measurement. By having the full intensity enter the monochromator, that energy will be available for good S/N and some stray light can be tolerated. At the blue end of the spectrum a tungsten lamp will cause considerable stray light; it can be reduced, however, by dropping a broad-band blue filter into the beam.

Example 17.5. Since scattering has been mentioned, it is worth noting that if the absorption spectrum of a strongly scattering sample is to be examined, for example, a solution containing macromolecules, the sample cell should be placed as close as possible to the detector to avoid the loss of higher frequency light, which is much more efficiently scattered.

Example 17.6. For measurements in the far or vacuum UV (wavelengths shorter than 185 nm) two coupling problems exist in the instrument shown in Fig. 17.1: (a) the principal gases in air, oxygen and nitrogen, absorb strongly and (b) photomultiplier tubes respond poorly because of the low emissivity of photocathode surfaces. Evacuating the optical part of a spectrometer will take care of the first problem. An ingenious way to overcome the second is to seal to the photomultiplier entrance window a *coupling plate* coated with a *fluorophor* of high quantum efficiency such as sodium salicylate. Ultraviolet light will simply excite the fluorophor to emit at the *longer wavelengths* that elicit a good photomultiplier response.

17.3 COLORIMETERS

For routine quantitative determinations in the UV-VIS an absorption instrument can be designed with a set of filters (see Section 7. 10) as the wavelength isolation device. These instruments are commonly called colorimeters.* They offer the advantages of simple construction, minimal maintenance, ruggedness, good throughput, low cost, and usually portability. With their high throughput an amplifier is seldom necessary.

The simplest arrangement is the single-beam filter photometer shown schematically in Fig. 17.2. Since it has a meter readout and a scale linear in percent transmittance, under ordinary conditions an accuracy of about ±2% in transmittance is obtained.

To minimize error in the readout module, source, and other modules, a split-beam arrangement and a ratiometric or null-balance readout may be added. This type of design is illustrated in Fig. 17.3. In this design, the voltage drop generated by the reference beam is reduced by variable voltage divider R_1 until it matches the voltage drop generated by the sample beam. The accuracy of a measurement may be as good ±0.5%*T* with electrical ratioing.

Example 17.7. Colorimeters are widely used for the water and waste-water testing required by such governmental regulatory agencies as the EPA. Since colorimeters used filters for

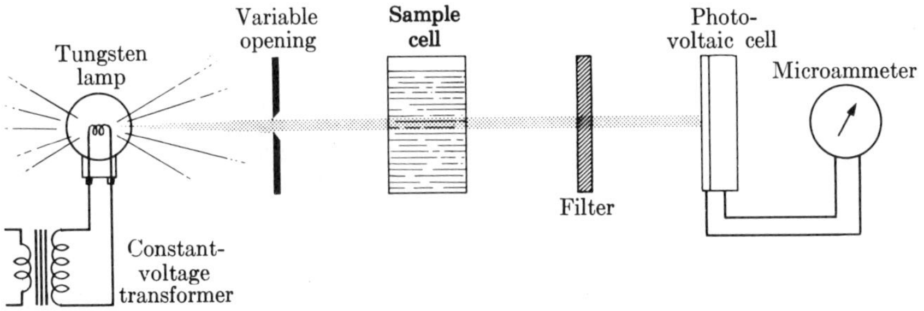

Fig. 17.2 Schematic of a simple filter photometer. Several filters are mounted on a filter wheel, allowing easy selection. To make a measurement two calibration adjustments must be made: (a) with no light falling on the detector, the meter must be adjusted to read 0%*T* and (b) with a blank replacing the sample shown in the beam, a variable opening (intensity control) must be adjusted to cause the meter to read 100%*T*. Frequent resettings are desirable.

*Earlier the name colorimeter identified *only* devices for which one's eye was the detector because color is by definition a visual phenomenon. Usage today also designates *filter photometers* as colorimeters.

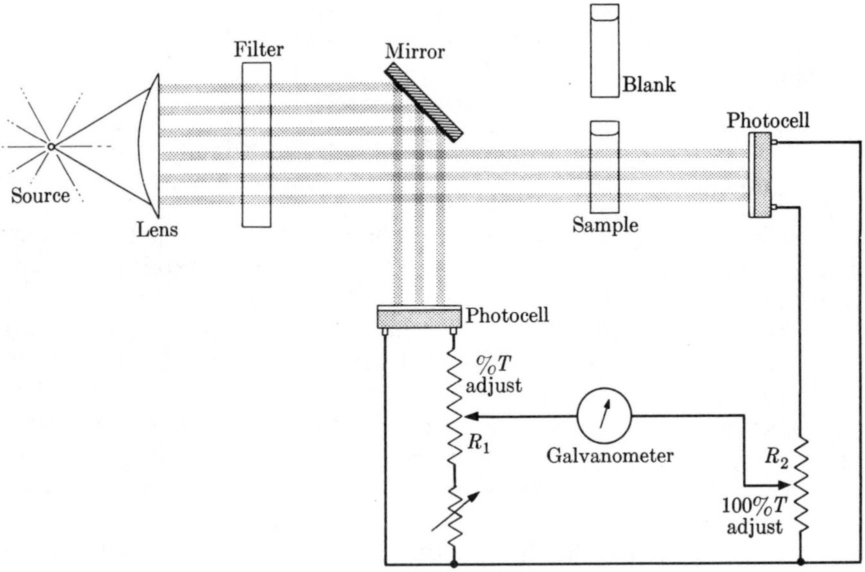

Fig. 17.3 Schematic diagram of a simple split-beam, single-channel, null-balance filter photometer. A portion of the incident beam is deflected to a second photocell, and the currents produced by the two cells are brought into electrical balance. With a blank in the measuring beam and R_1 wiper at the top of the slidewire, an initial balance is obtained by adjusting the 100%T wiper. Then the blank is replaced by the sample and the %T measurement made by adjusting R_1 downward.

wavelength isolation, it would seem that the overlap of analyte bands with those of other substances would seriously limit this application. How can this problem be dealt with?

Fortunately, for a variety of analytes selectivity or specificity can be secured by selective complexation (see Section 16.6). Remaining interferences can often be blocked by selective chelation or masking.

17.4 THE SPECTROPHOTOMETER

The incorporation of a monochromator as the wavelength isolation device distinguishes a spectrophotometer from simpler filter devices. The better the monochromator, the more versatile the spectrophotometer. Many advantages are gained:

1. operability over the extended wavelength range common to the optical sources and monochromator used;
2. greater wavelength accuracy traceable to the precision of monochromator setting; and
3. good wavelength resolution.

Accompanying disadvantages are mainly

1. reduced throughput,
2. less ruggedness, and
3. more cost and maintenance.

In general, all other modules are also upgraded. The most striking change is usually the introduction of a microcomputer in place of traditional processing modules. Its incorporation will permit programmatic control of module parameters such as source current and monochromator wavelength setting and slit width. In addition, programs usually will have been put in ROM, which allow the calculation of absorbance, the storage of calibration data for particular analytes, and the calculation of their concentrations. Computer control also makes possible correction for some kinds of errors such as stray light and baseline variation as long as these quantities have reproducible, steady values as a function of wavelength. Further, because of the inherent precision possible with digital readout, precision in absorbance measurements is no longer significantly better for values centered on 0.43, but is nearly optimum from about 0.2 to 1.1 (see Section 16.4). As a result, a wider range of analyte concentrations can be measured with precision.

Single-Beam Spectrophotometers. Many inexpensive, manually operated single-beam instruments with meter readout and other modules of comparable precision are particularly well adapted to quantitative analysis. They are little used for qualitative analysis, which often requires spectral scanning. Such instruments have moderate resolution and sensitivity, stray light levels of about 0.3%, and an optimum precision of $\pm 2\%$ of full-scale deflection. With their simple design they are sturdy, require little maintenance, and are relatively inexpensive. The optical layout of a widely used instrument of this type is shown in Fig. 17.4.

Precision single-beam spectrophotometers include one or more microprocessors and optical modules of high quality to ensure stable operation. For example, UV-VIS instruments commonly have tungsten–halogen/deuterium sources and use silica-coated toroidal mirrors in their monochromators. To minimize stray light filters may be inserted between the monochromator and sample cell to block unwanted radiation of wavelength well away from the region of measurement. Measurements of stray light have become refined.* To compensate for drift and most source intensity fluctuations they may use a split-beam design. One way to complete such a design is with two detectors, as shown in Fig. 17.4. Alternatively, the source output can be chopped and optics added to permit a single detector to respond in turn to radiation transmitted by the sample and that coming directly from the source. Microcomputer programs can provide control of modules which will ensure stability and freedom from drift.

With quality components, including high-intensity sources and microcomputer control, a single-beam UV–VIS instrument offers high S/N. As a result, fast wavelength scanning can often be arranged with the dividend that drift can be minimized during measurements to make it feasible to record the transmittance of a blank at sampled wavelengths in the memory of the computer as the baseline for a subsequent sample scan. Then the P_0/P of a sample can be taken point by point with high precision over the same wavelength range. As a consequence of such developments, a modern single-beam instrument often equals or at least approaches the effectiveness of a double-beam spectrophotometer. Note that with

*W. Kaye, *Anal. Chem.* **53**, 2201 (1981).

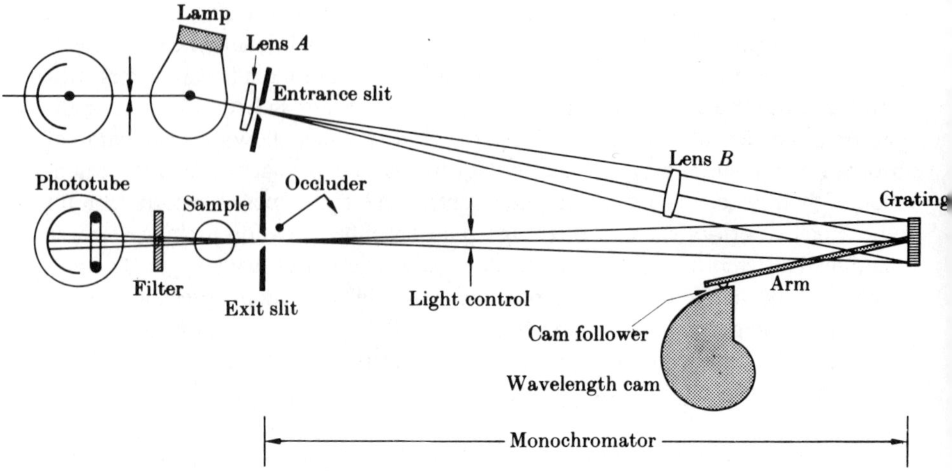

Fig. 17.4 Optical schematic of a manual single-channel grating spectrophotometer, the Spectronic 20. The regulated model is shown in which the output of the tungsten lamp is stabilized by having the upper phototube monitor the lamp intensity as part of a negative feedback system. For example, when lamp intensity increases, the monitoring signal causes the lamp power supply to diminish proportionately the power it furnishes. Within the monochromator, a field lens focuses light from the entrance slit on the objective lens. This lens in turn provides a slit image at the exit slit. The filter in front of the measuring (lower) phototube detector passes only the visible range and blocks unwanted orders of spectra from the grating. The spectral slit width is 20 nm and range is basically limited to the visible (350–650 nm), but can be extended up to 900 nm in the near IR by the use of a red-sensitive phototube and a filter to reduce stray radiation. (Courtesy of Milton Roy Analytical Products Division, formerly a division of Bausch and Lomb, Inc.)

digital readout precision is limited basically by the optical modules and not by the processing modules, which may be as precise as the AD converter in the channel.

It is generally in the UV–VIS range that high precision single-beam dispersive spectrophotometers are available. In the IR, similar precision can be obtained at the same speed only with a Fourier transform spectrophotometer (see Section 17.8).

Spectrophotometers incorporating a *polychromator* (see Section 9.9) instead of the usual monochromator and one-channel detector are inherently much faster because of the electronic scanning (0.1 s per scan) possible with a multichannel detector. A schematic design of such an instrument is shown in Fig. 17.5. (Recall that its dispersive element, usually a diffraction grating, is stationary.) It will usually be the instrument of choice for measurements that call for rapid scanning, precision simultaneous multicomponent analyses, and for kinetic measurements.

Double-Beam Spectrophotometers. A double-channel design offers, with respect to a single-channel design, the advantage of compensation both for drift and for all fluctuations occurring somewhat more slowly than the modulation rate. Thus, it compensates for the effect of gradual alterations in thermal, mechanical, electronic, and atmospheric factors on the performance of the source, monochromator, detector, and amplifier. For instance, much of the slower flicker in lamp intensity

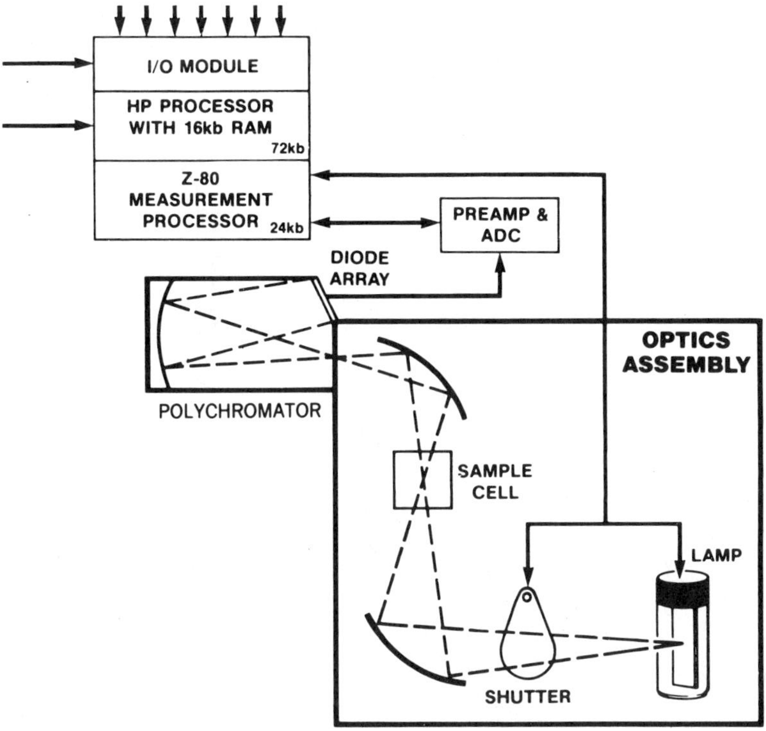

Fig. 17.5 Optical schematic of a single-channel diode array UV–VIS diode array spectrophotometer, the Hewlett-Packard 5481. Its polychromator employs a concave holographic grating and 328-element diode array detector and covers the range from 190 to 820 nm. Note that a total of only three reflecting surfaces is employed, guaranteeing high throughput. (Courtesy of Hewlett-Packard Co.)

will be compensated. In addition, double-channel design provides the possibility of easy differential measurements and allows the use of somewhat less expensive modules. Some limitations of double-beam design are its higher cost, greater mechanical and optical complexity, and somewhat greater maintenance requirements.

For a double-beam spectrophotometer, the most favorable arrangement is to avoid the difficulties of module matching and the cost that entirely separate channels would entail. Instead a single module, the sample module, is doubled and a new mode of coupling is introduced. The optical beam is simply split before it enters the sample area and the two beams are rejoined immediately following the sample area. Beam splitting is usually combined with modulation, often by use of a rotating sector mirror that has alternating mirror and open sectors. The modulation facilitates signal processing, as will be discussed below. As the sector mirror turns, open sectors pass the beam along one channel and mirror sectors reflect it to the second channel. One channel is permanently assigned to accommodate a reference or blank and the other, the sample. Each channel has pulses of light nearly as bright as the original beam. A precision spectrophotometer with this arrangement is shown schematically in Fig. 17.6.

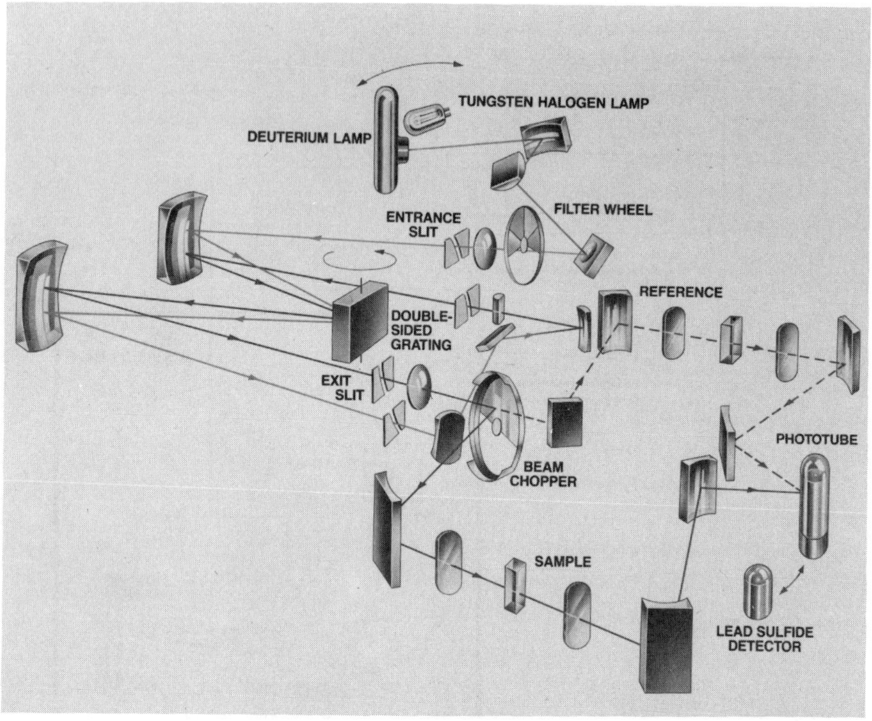

Fig. 17.6 Optical schematic of a double-channel UV–VIS–near IR spectrophotometer, the Varian 2300. A double-pass Czerny–Turner monochromator with two entrance and exit slits is used to double the dispersion ordinarily available. The diffraction grating is double-sided also to secure an extended range. (Courtesy of Varian, Inc.)

Analog Balancing. Extensive lines of spectrophotometers that used analog methods to obtain the ratio P/P_0 were developed prior to the general advent of microcomputer-controlled instruments. The latter of course rely on digital methods. The analog methods are of two types, identified as *optical* or *electrical null*. They provide a good photometric accuracy ($\pm 1\% T$ or better) and instruments based on these approaches are still available. The optical null type will be described more fully as a good example of an analog processing module. It secures an optical null, that is, matching of beams, by means of an ingenious feedback loop that includes a servomotor. Remember that light is directed through both the sample and reference by a rotating sector mirror or other arrangement. The beams thus consist optically of square pulses of light. As long as equal amplitude pulses reach the detector from each channel, the detector has a steady or a dc output. Whenever sample beam power P begins to deviate from reference beam power P_0 during a spectral scan, the detector output consists of an ac signal superposed on the dc base. Both the output pulses from the detector as it receives first sample and then reference pulses and the superposed ac output pattern are shown schematically in Fig. 17.7. The appearance of ac in the output automatically initiates a process of rebalancing the beams. To accomplish this goal, the ac output is amplified and directed to a servomotor, which in turn moves a wedge of steadily

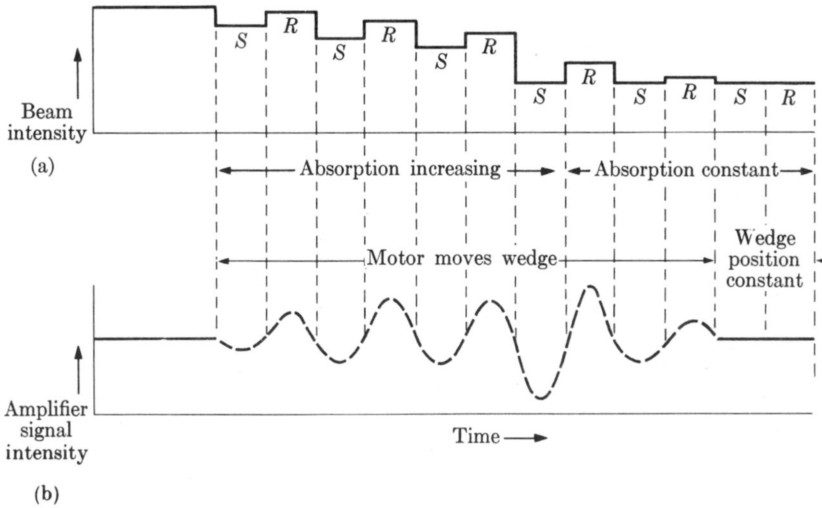

Fig. 17.7 Signals leaving the detector of an optical null spectrometer during a wavelength scan. As an absorption peak is approached, absorption at first increases rapidly. (a) Intensity of beams reaching the detector: S, sample beam power, R, reference beam power. (b) Amplifier signal to servomotor driving the optical wedge. With G as the amplifier gain, the ac signal is $G(R - S)$; it is shown as a broken line. It drives the motor and moves the wedge at a speed determined by ac signal amplitude, in a direction determined by its phase. The wedge is driven just far enough to reduce the intensity of the reference beam to that of sample beam.

changing transparency into, or out of, the reference beam. The servomotor is wired to drive the wedge *in the direction* that will reduce the difference signal $(P_0 - P)$. From the description given it is evident that the wedge is no longer moved when the power level of the reference beam again equals that of the sample beam. The final position of the wedge after each rebalancing is displayed as P/P_0.

In the electrical nulling procedure the detector output P_0 develops a stable voltage across a sidewire resistance. A fraction of this voltage is tapped off by a wiper (moving contact) and directed to one input of a difference amplifier. Its second input is simply a corresponding voltage version of sample beam power P. In this case a null balance is obtained by moving the wiper until the differential amplifier output is zero. The relative position of the wiper then corresponds to the ratio P/P_0. Usually somewhat greater precision is secured by electrical ratioing, especially with strongly absorbing samples.

The disadvantages of these analog arrangements are their moving parts, which wear with time and have sufficient inertia to slow spectral scanning considerably. Further, the optical null balancing arrangement is relatively inaccurate at low values of transmission ($T < 2\%$) since the energy throughput of both beams is severely reduced.

In modern precision spectrophotometers program-controlled processing provides P/P_0, mechanical wear of moving parts is minimized, speed is enhanced, and reliability of measurement is high. Detector signals are amplified, digitized, and processed in real time. Usually they are also entered in memory for further processing and subsequent study. Processing routines are selected by the user (see

Section 6.6). These procedures realize the inherent gain in accuracy, speed, reliability, and flexibility in calculation possible by processing digital signals. Further, tradeoffs among adjustable parameters of characteristic modules such as slit width and source intensity, which will be taken up in the following section, can be established automatically by program once the user has entered information about the type of measurement to be made.

Further, by including a reading of detector dark current in each cycle of modulation, sample and reference signals can be precisely corrected for background before their ratio is taken, though this is seldom done except in high-precision instruments. Otherwise, the correction can be made by consulting data in memory as to background observed at a previous time under the same conditions. In some instruments the reference signal is held constant by a feedback loop that utilizes a difference signal to control either monochromator slit width or the first dynode voltage of the detector (if a photomultiplier detector is used).

Curve Reconstruction. With a digitally recorded spectrum the data available consist of a series of points giving absorbance or $\%T$ versus wavelength. How can the curve be reconstructed for display?

It is straightforward to provide a program that will interpolate between points. Naturally only the data points will be accurate. Alternatively, electronic filters may be used to smooth the data.

Diagnostic Routines. It is important to have in the programs that are routinely available to the spectrophotometer microcomputer a set of instrument diagnostic routines. Since the user has little sense of the instrument adjustments being made and of the magnitudes of signals during measurements, it is essential that the manufacturer provide programs that will

1. make initial tests such as a check of memory and appropriate levels of power,
2. monitor the energy in the reference beam to ensure that it is above the minimum acceptable S/N and that other variables are within appropriate ranges,
3. ensure that user-entered commands are within acceptable ranges,
4. monitor performance factors, for example, the wavelength at a reference point, to be sure that they have correct values, and
5. make diagnostic checks when there are component failures or when operation seems to be impaired.

17.5 TRADE-OFFS AMONG ADJUSTABLE PARAMETERS

In precision scanning monochromators or spectrometers there are important tradeoffs to be made among instrumental variables. In the list are quantities that relate to both characteristic and processing modules, slit width, scanning speed, amplifier gain, system time constant, signal sampling interval, and conversion rate of the

AD converter. The nature of a trade-off is, of course, that by changing a variable we improve one aspect of performance while diminishing another. That being true, by what criteria can the user, a programmer, or manufacturer, decide on appropriate trade-offs for a given instrument system to permit data of required precision to be obtained as quickly as possible? (Where the manufacturer has set these parameters, we will see the choices reflected in instrument specifications.)

Slit Width. The central role of the slit in both resolution and throughput makes its width the variable to consider first. In general, the entrance slit must ensure a resolution sufficient to

1. resolve adjacent peaks and
2. render accurately peak heights and bandwidths (widths at half maximum height).

For a particular spectrometer the choice of slit width will depend mainly on the bandwidths of analyte peaks. To develop a general approach to rendering an individual peak accurately, assume that it is of Gaussian shape. It can be shown that, under these conditions if the ratio $S/B_{1/2} \approx 0.1$, where S is the spectral slit width and $B_{1/2}$ the peak bandwidth (FWHM), the spectrometer will report the peak height (and breadth) with no more than 0.5% error. The reliability of measurement will *decrease* as the slit is widened. When $S/B_{1/2} = 0.5$, the height will be "read" as 90% of the true value and when the ratio rises to 1.0, the apparent peak height will be about 70% of the true value. An error of the same magnitude will occur if concentration is calculated from peak height; this error may, however, be cut in half by basing the calculation on the integrated peak intensity (see Section 16.3).

If better resolution is needed than a particular monochromator–spectrometer can give, a spectrometer capable of higher resolution should be sought. Its monochromator will have wide gratings ruled as finely as feasible for the spectral region. To take advantage of the dispersion thus provided, quality optics of long focal length (usually 0.5 m minimum) will be required.

A decision to reduce slit width also should take into account whether the resulting throughput will give sufficient S/N for photometric accuracy. Fortunately, the trade-off between resolution and energy need seldom be examined except when beam energy is limited. For UV–VIS spectrometers the question will arise infrequently. In the IR, where sources are of low energy, the question is ever present (see Example 17.11).

Example 17.8. It is desired to obtain a precise measurement of a broad UV absorption peak whose natural bandwidth is 21 nm. What slit setting should be used with an instrument whose reciprocal linear dispersion is 2 nm mm^{-1}?

First, using the criterion for adequate resolution given above, the spectral slit width should be $0.1 \times 21 = 2.1$ nm. Second, recall that spectral slit width S is a triangular function of wavelength for a high-resolution monochromator and that it is related to the geometrical slit width W by the expression $S = WD^{-1}$, where D^{-1} is the linear reciprocal dispersion (see Fig. 9.3). While the expression will be only approximate for a low-resolution monochromator, it permits a value to be estimated. One has $W = S/D^{-1} = 2.1/2 \approx 1$ mm: a slit opening of 1 mm (or less) can be used. Unless the source energy is quite limited (unlikely in the UV) or the solvent is strongly absorbing, there should be no problem in obtaining the measurement at a good S/N ratio.

Example 17.9. At what slit width should a spectrometer be set if little information is available about a system to be studied?

In this case it is best to scan its spectrum with several different slit widths if possible. The best choice will be the *widest* width that still gives maximum peak heights and as deep valleys between peaks as possible. For a system with complex spectra and close-spaced, narrow peaks, a better way to estimate peak parameters may be to differentiate such spectra (see Section 16.7).

Example 17.10. What is the energy "cost" of doubling spectrometric resolution? Can the loss be overcome by increasing slit height?

In Section 9.3 the relationship of geometrical slit width W and spectral slit width S to throughput was shown to be: throughput $\propto W^2h \propto S^2h$, where h is the slit height. If S must be halved to double resolution, this change will reduce throughput by a factor of 4.

According to this throughput expression, increasing the slit height will improve throughput linearly. Recall also that in Section 9.3 it was stated that resolution will decrease as height increases much beyond 5 mm, unless the monochromator has curved entrance slits.

Example 17.11. What rules of thumb can be developed to guide the trade-off of throughput and resolution in IR spectrophotometry?

1. In a survey scan, resolution can be traded off for energy to permit a fast scan since only the presence or absence of peaks is of interest.

2. In quantitative work the decision must ordinarily be made in favor of energy since a good value of S/N is essential. Note that the value of S/N can often be maintained at an adequate level without serious loss of resolution simply by reducing noise (see below).

3. Two factors may make it impossible to sacrifice resolution in quantitative work: spectral interference and fine detail in analyte peaks in an absorption spectrum. When the signal energy is reduced, the best strategy is to lower noise by a larger factor by going to a longer response time (see below) and slower scan.

If the process of obtaining adequate resolution has reduced photometric accuracy below a tolerable level, at least three instrumental means to improve S/N are available even at fixed slit width. First, it may be worthwhile to increase source intensity. Either a brighter source can be used, if available, or current can be increased.*

*Increasing brightness (at the cost of lifetime) is likely to be an attractive trade-off for VIS–UV sources. For IR sources it is less so since brightness at wavelengths far from the intensity maximum increases only linearly with absolute temperature.

It may also be feasible to substitute: (a) a monochromator of higher efficiency, that is greater throughput but comparable resolution; (b) a double monochromator—since its dispersion is also usually doubled, a slit twice as wide will give a beam with the same resolution and four times the energy; or (c) a Fourier transform spectrometer with its higher throughput and simultaneous observation of all wavelengths (see Section 17.8).

Third, the time constant of the processing electronics can be increased as a means of increasing S/N by diminishing white noise. Alternatively, signal averaging can be added (see Section 12.8).

Compensation may also be incorporated for the often drastic energy decreases produced by solvent absorption or cell window absorption, for example, in multiple internal reflection measurements. Usually this operation depends on monitoring the reference beam intensity to keep its level steady. If its intensity is falling, a signal can be directed either to a slit servo system that can open the slit appropriately or to the power supply of a photomultiplier detector to increase the first dynode voltage and thereby the gain. Since these changes will also increase the noise level, the maintenance of a proper S/N clearly requires selecting a suitable slit width at the beginning.

Signal Sampling Interval. In acquiring information from time-domain signals such as exist in a Fourier transform spectrophotometer, the Nyquist theorem sets sampling at a rate twice that of the highest frequency of interest as a lower limit. What guide exists for sampling in the *frequency* domain in which dispersive spectrometers operate? To obtain enough data points to define peak shapes, a rule of thumb is to obtain eight points per peak. The user will of course need to ascertain the narrowness of *analyte* peaks.

Example 17.12. In a UV–VIS spectrophotometer with a 512-element photodiode array polychromator, it is possible to select the spectral range to be registered. For a sample in which the narrowest bandwidth is expected to be 10 nm, and the total range to be observed is 400–700 nm, what spectral range should be observed by the diode array at one time?

It appears the entire 300-nm range can be registered. Since each diode will record 300/512 $\approx$ 0.6 nm, about 15 diodes will provide data even for the narrowest peak.

Scanning Speed. Spectra can be accurate only if both the "static" sources of error already considered and the "dynamic" errors that arise from the finite response time of modules and data acquisition rate are minimized. Usually the fastest scanning speed is sought that will permit the spectral resolution and accuracy already established by the choice of slit width and monochromator design to be maintained. The dynamic factors which must now be considered are the time constant of the electronics, maximum data acquisition rate, and the wavelength scanning speed.

Any system requires a certain time interval to respond fully to a signal. In analog terms, if its response to a step signal is exponential, the response can be characterized by a *time constant* τ, the interval required for a 63% $(1 - 1/e)$ response. In an interval equal to 4τ a 98% response will be achieved. We define the *maximum scanning speed* S_{max} that will allow the inherent optical resolution to be realized as

$$S_{max} = S/4\tau,$$

where S is the spectral slit width.

Most spectrometers provide several scanning speeds. Surveys can be made at fast speeds and careful studies at slow ones. Some instruments feature a time-saving arrangement that does not sacrifice resolution or photometric accuracy known as a speed suppression system. It allows fast scanning in spectral regions where there is little absorption but greatly slows the scan whenever absorption begins to occur. Such an arrangement is advantageous: it gives quality spectra, conserves time, and yet does not greatly increase mechanical complexity.

Example 7.13. An IR spectrometer whose band pass or spectral slit width is 1 cm^{-1} and whose time constant is 0.1 s is available. What is the maximum scanning speed that will allow the inherent resolution to be realized? How may this speed be validated as adequate?

According to the equation given above the speed is $1\ cm^{-1}/4 \times 0.1\ s = 2.5\ cm^{-1}\ s^{-1}$. If one is to scan a spectrum from 4000 to 2000 cm^{-1} (2.5 to 5 μm), the time required will be 2000/2.5 = 800 s, or about 13 min.

A straightforward check to ensure that dynamic error has been minimized is to compare a spectrum run at the desired speed with one run at one-fourth that speed. If peak heights differ by less than 0.02 absorbance units, the desired speed is satisfactory.

Example 17.14. An older, versatile IR spectrometer that has been used for spectral surveys (fast scan) is to be reprogrammed for quantitative analysis. In the survey mode it is known to have a value of S/N of about 300, a scan time of 15 min for the range 4000–200 cm^{-1}, and an attenuator response time of 3 s. What adjustments should be made to shift to conditions of high quantitative accuracy?

First, resolution may be traded off for energy. Doubling its S/N and beam energy will be desirable. The slit width should be increased $(600/300)^{1/2} \approx 1.4$ times. Now the amplifier gain can be reduced by about a factor of 2; if this change is not made, the recorded response will be too strong.

Second, the desired increase in accuracy can be assured by halving the scan speed. Note that recording accuracy could also have been increased by halving the response time. The latter adjustment, however, would have increased photometric noise and offset the benefit of opening the slits wider.

Finally, since slits have been opened somewhat, as broad a band as possible should be selected for the analysis to keep the ratio $S/B^{1/2}$ at about 0.1–0.2 if possible.

Scale Expansion. In either regular or differential analysis it is often necessary to examine details of spectra that are too small for the regular photometric scale. For this reason ordinate expansions up to 10 times or more are provided. Often abscissa expansions are also available. Provisions for accurate expansion will depend not only on the quality of electronic circuits, but especially on the use of an AD converter of at least 14-bit capacity.

17.6 REFLECTION ASSEMBLIES

In most instances reflection spectrometry is carried out with a standard spectrophotometer by substituting an appropriate reflection assembly for the regular sample cell. Three reflection modes are in use, *total internal reflection*, *diffuse reflection* (i.e., reflection from an irregular surface), and *external* or *specular reflection*.

Some representative arrangements for *multiple internal reflection* (MIR) are shown in Figs. 17.8 and 17.9. The internal reflection mode is ideally suited for measurements on solids, viscous liquids, and solutions in spectral regions where they absorb strongly. Liquid or solution samples are often placed in a cell that surrounds an *internal reflection rod*, as shown, whereas all types of samples can be examined if brought into *intimate contact* with one or both sides of a flat *internal reflection plate*. The advantage offered by internal reflection as a mode is that it allows light to penetrate only a fraction of a wavelength into a sample (see Section 7.11). For instance, even though water is highly absorbing in the IR, this mode makes possible IR absorption measurements on aqueous solutions. Note that the design in Fig. 17.8 provides some five internal reflections, thereby increasing the light path *within* the sample.

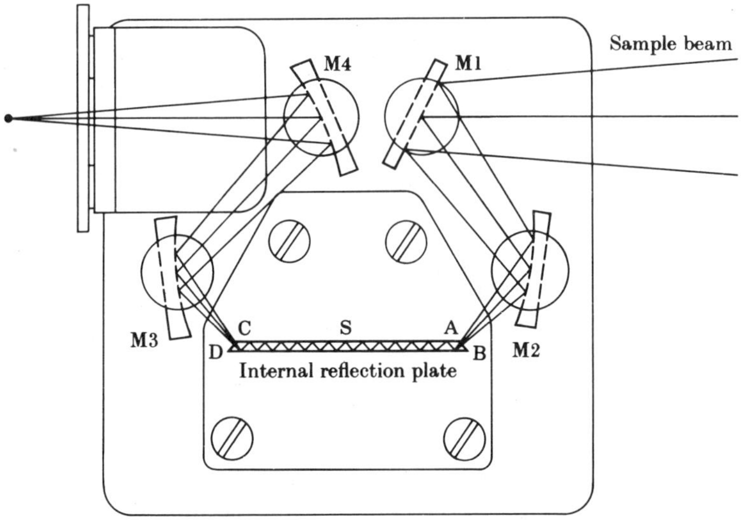

Fig. 17.8 Optical schematic of multiple internal reflection (MIR) attachment. A sample S is pressed against one or both sides of the plate by a holder (neither are shown). AB is the entrance aperture of the internal reflection plate, which is a fixed-angle type, and CD is the exit aperture. The entire plate is filled by light. (Model 9 Single Beam MIR Attachment, courtesy of Foxboro/Analabs.)

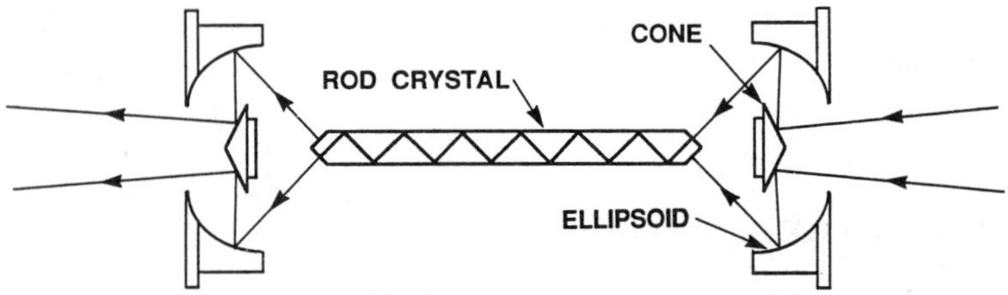

Fig. 17.9 A cylindrical rod for spectrophotometric examination of aqueous and other solutions by multiple internal reflection. Together with a glass cell (not shown) that surrounds the rod with a solution sample, it comprises a Circle cell. The optics provide that most of the beam is directed into and out of the rod. (Courtesy of Spectra-Tech.)

Because of the many reflections involved, however, an MIR cell also reduces the beam energy substantially. For that reason good internal reflectance assemblies have a dual mirror system to focus the beam down sufficiently to permit all of it to enter the flat, beveled face of a reflection plate or rod and also ensure that all the leaving beam reaches the next module. Usually a simple attenuator is added to the reference beam in a double-beam spectrophotometer to compensate.

Figure 17.10 shows a reflection assembly for *diffuse reflectance*. The crucial aspect is its pair of large ellipsoidal mirrors for efficient collection of diffusely reflected radiation. Indeed, the solid angle of collection is about 20% of the possible total of 4π. The optical beam of the spectrometer should not be defocused or diffused. Powdered samples are widely examined by DRIFT (an acronym based on diffuse reflectance IR Fourier transform) spectrometry. Diffuse reflectance is attractive for weakly absorbing samples since it has been found* to be of greater sensitivity and to produce a greater S/N than photoacoustic spectrometry (see Section 17.8).

External Reflection. The assemblies discussed are easily adaptable for *external reflectance* measurements. This mode is much less common than the others since it is useful only for samples having highly reflective surfaces or for molecular layers deposited on such surfaces. For the unit shown in Fig. 17.8 a single reflectance measurement can be made from a flat specimen installed in place of the internal reflection plate by adjusting mirror angles. Multiple external reflectance measurements (from 3 to 100 reflections) require adding a flat mirror parallel to the specimen and adjusting their separation. The assembly of Fig. 17.10 can also be adapted by tilting the sample forward. With little other modification it is possible to obtain at least two reflections.

Internal Reflection Plates and Rods. The choice of a material for an internal reflection plate or rod depends on the spectral range in which it will be used, relative physical properties desired (%T, hardness, and solubility) and the types of samples to be studied. It must, of course, have a higher refractive index than samples over the spectral range of interest.

*J. W. Childers and R. A. Palmer, *Am. Lab.*, **18**(3), 22 (1986).

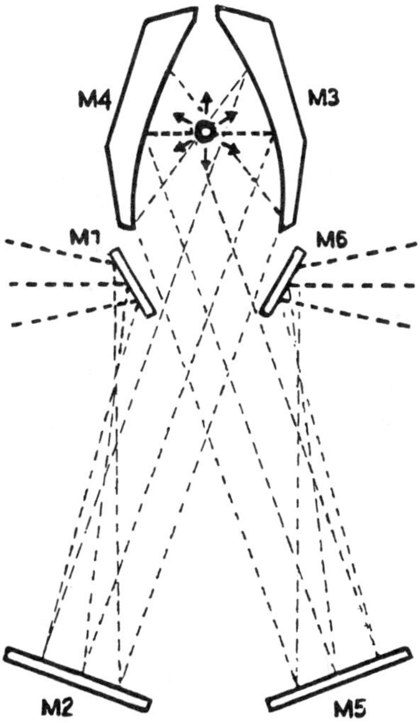

Fig. 17.10 A diffuse reflectance attachment. As the beam comes in from the left, it is reflected off the upper right elliptical mirror onto the top of a specimen. Mirrors are adjusted to minimize specular reflection. The upper left mirror receives the diffusely reflected light and reflects it to the exit mirrors. Where desired, mirrors can be rotated to capture the specular reflection of a sample. (Courtesy of Harrick Scientific Co.)

Given that the typical path length in a reflection plate is as long as 70 mm, it should also be of high purity to minimize absorption by impurities. Properties of several commonly used substances (for IR studies) are listed in Table 17.1. KRS-5, thallium bromide–iodide, is one one of the best substances for plates for the IR. Its main disadvantages are poor chemical resistance and softness. For UV, VIS, or NIR spectrometric work both aluminum and magnesium oxides are ideal.

TABLE 17.1 Properties of Some Crystalline Materials Useful for Internal Reflection Plates[a]

Substance	*Refractive Index (Mean)*	*Range of Transparency (μm)*	*Hardness*	*Chemical Resistance*
KRS-5	2.5	0.6–40	Soft	Poor
Germanium	4.0	2–12	Brittle	Inert
Silicon	3.4	1–100	Hard	Inert
Zinc sulfide	2.2	0.1–14	Hard	Inert
Zinc selenide	2.4	0.5–20	Soft	Relatively inert
Quartz, SiO_2	1.5	0.2–2.8; 40–333	Hard	Inert
Sapphire, Al_2O_3	1.75	0.15–5	Hard	Inert

[a]Polycrystalline materials such as the Irtran series are useful as windows but not as internal reflection materials. They scatter light strongly at shorter wavelengths and thus have very low transmission for 70-mm path lengths.

17.7 THE FOURIER TRANSFORM IN SPECTROMETRY

In addition to the dispersive spectrometric methods treated in detail up to this point, Fourier transform methods have gained an increasingly important position in absorption measurements.* The earliest and still the most widely used application is certainly in IR spectroscopy. More recently Fourier transform methods have also been applied to pulsed NMR spectrometry and ion cyclotron mass spectrometry. All Fourier transform techniques are distinctive in that they enable the spectrometer to operate in the *time domain*. In this domain measurements are highly efficient since all frequencies are detected simultaneously. The analytical signal obtained is, of course, complex and must be Fourier transformed to the frequency domain to obtain a normal spectrum. The development of a fast transform algorithm and the great decrease in computation costs have made this approach attractive.

What will be important to know about Fourier transform processing? The most important condition for a successful transformation is that signals combine *linearly*, as electromagnetic signals do. They superpose as

$$f(t) = \sum_{i=0}^{N-1} A_i \cos 2\pi\nu_i t,$$

where A_i is the amplitude of frequency component ν_i. A more general and graphical treatment of superposition that includes spatial and phase considerations of sinusoidal waves is given in Section 7.3.

In the development of Fourier transform theory it is necessary to let the number of frequencies be very large. One type of time → frequency transform that results is

$$H(\nu) = \left(\frac{1}{2\pi}\right) \int_{-\infty}^{+\infty} h(t) \cos 2\pi\nu t \, dt, \tag{17.2a}$$

where amplitudes $h(t)$ are for time functions. By integrating over infinite time, the frequency (or wavelength) spectrum $H(\nu)$ is secured. Actually, an *inverse transform* that takes a frequency domain signal to the time domain also exists:

$$h(t) = \int_{-\infty}^{+\infty} H(\nu) \cos 2\pi\nu t \, d\nu. \tag{17.2b}$$

If only cosine terms are incorporated, as in the equations shown above, the transforms can treat only functions of *even symmetry*. Recall that the cosine function falls symmetrically from unity at the origin to zero within 90° on either side.

*Hadamard and Hilbert transform methods, which are alternatives to Fourier transform techniques, also find applications in chemical measurements [6].

Conversely, a sine transform in which sine functions replace cosines will treat only functions of odd symmetry.

Fourier Transforms and Exponential Notation. Most Fourier transforms utilize both sine and cosine terms to permit the expression of functions of arbitrary symmetry. The essential element of orthogonality between sine and cosine can be preserved if the sum of terms, $\cos x + i \sin x$, where $i = \sqrt{-1}$, is taken at each angular frequency x and integrated over all frequencies. Finally, advantage is taken of the identity $\cos x + i \sin x = e^{ix}$ to substitute e^{ix} for the sum.* This notation provides a simple way to combine needed orthogonality, symmetry, and phase relations.

Accordingly, the pair of Fourier transforms for a complex signal can be given in the general form†

$$H(\nu) = \int_{-\infty}^{+\infty} h(t) \exp(-2\pi i\nu\, t)\, dt \tag{17.3a}$$

$$h(t) = \int_{-\infty}^{+\infty} H(\nu) \exp(2\pi i\nu\, t)\, d\nu. \tag{17.3b}$$

The relationship between transform $H(\nu)$ and its inverse $h(t)$ may be symbolized by $H(\nu) \leftrightarrow h(t)$.

Discrete Fourier Transform. There is now a need to simplify the transform to permit necessary calculations to be done in finite time by available computers. Information theory provides the steps. First, the range of frequencies is limited to those in the range of interest. In particular, an upper frequency limit ν_m is set. Second, the rate of sampling of the time domain signals is set to be twice the Nyquist frequency, as discussed in Section 12.5. At this frequency, the sampling interval Δt will be related to the maximum frequency by the expression $\Delta t = 1/2\nu_m$. As a consequence, the integral of Eq. (17.3a) can be approximated by the summation

$$H(\nu) = \sum_{n=0}^{N} h_n \exp(-i2\pi\nu n\Delta t), \tag{17.4}$$

where h_n is a series defined by the sampling interval and total number of samples N. Finally, use of the Cooley–Tuckey fast Fourier transform algorithm with subsequent improvements will greatly shorten the computation involved. References will provide details of its application.

*The validity of this expression can be demonstrated by substituting in it the Taylor expansions for $\sin x$, $\cos x$, and e^x. They are $\sin x = x - (x^3/3!) + (x^5/5!) + \cdots$, $\cos x = 1 - (x^2/2!) + (x^4/4!) + \cdots$, and $\exp(x) = 1 + x + (x^2/2!) + (x^3/3!) + \cdots$.

†The negative sign is necessary in (17.3a) to ensure that the original function $h(t)$ is regained after two integrations.

17.8 FOURIER TRANSFORM ABSORPTION SPECTROMETERS

In the common types of Fourier transform absorption spectrometer a two-beam, variable-pathlength *interferometer* is used in lieu of the wavelength isolation device shown in Fig. 17.1. A modular layout of the new instrument is given in Fig. 17.11. In the Fourier instrument the signal leaving the amplifier is said to be in the *time domain* because it contains all wavelengths being examined at all times. A customary wavelength spectrum is extracted from this signal by taking its Fourier transform.

The importance of this type of spectrometer derives from several design features:

1. All wavelengths are present simultaneously (*Fellgett* or *multiplex advantage*).
2. Optical beams are ordinarily wide (*throughput* or *Jacquinot advantage*).
3. Mirror travel is nearly always monitored by a laser (*precision wavelength advantage*).

Each feature makes possible an enhancement of performance. How these advantages are used to secure faster scanning, better resolution, or higher precision will be examined at the end of the section.

Interferometer. Most Fourier transform spectrometers are built around a Michelson interferometer.* Its design is shown schematically in Fig. 17.12. A *beamsplitter*, a thin, semireflecting film, divides the collimated light from the source into two beams whose relative length can be varied. One beam is returned to the right-hand surface of the splitter by movable mirror M_1 and the second beam is returned to the same splitter plane by fixed mirror M_2. Even though the incident light is incoherent, splitting it into two beams ensures that they will recombine as if coherent. As a result, they undergo interference at the splitter, and as M_1 moves, the wavelengths interfere in differing amounts, making it possible to identify them later. The combined beam A passes through the sample cell to an appropriate detector.

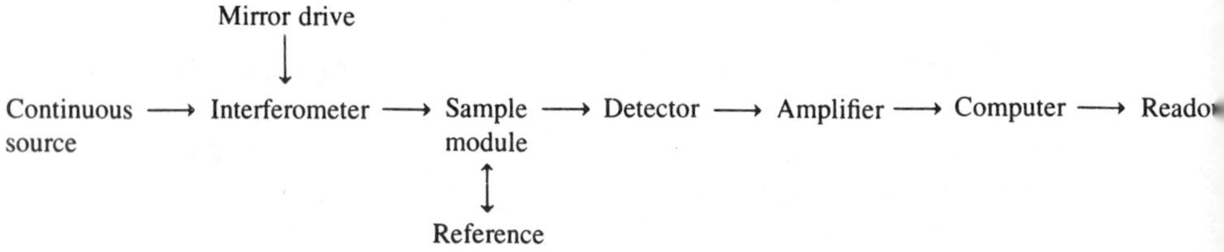

Fig. 17.11 Modular diagram of a Fourier transform spectrometer.

*A very few Fourier transform spectrometers based on a Fabry–Perot interferometer (Section 7.10) have been developed commercially.

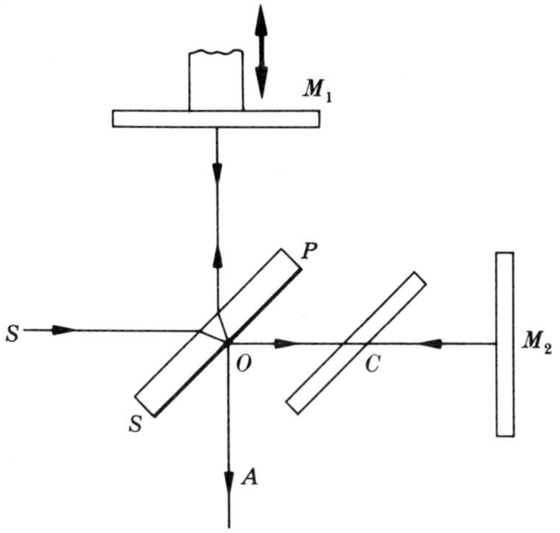

Fig. 17.12 Optical schematic of a Michelson interferometer. Essentially parallel light is directed from the source S toward the beam splitter SP. Beam width is limited by the diameter of mirrors and beam splitter as well as any diaphragm (not shown) placed in the path. The incident beam is divided into two rays by beam-splitter layer SP, which is a semireflecting film. The rays are reflected by planar mirrors M_1 and M_2 and recombined at O, where they interfere. Beam A is the analytical beam. If the plate backing the beam splitter is optically "thick," a compensator place C is affixed to equalize optical path lengths. In spectrometric scanning, mirror M_1 is moved vertically.

To examine the nature of the interference, consider the change in beam path lengths as mirror M_1 moves upward. As it does, the optical path OM_1O increases in length relative to the path OM_2O, a difference symbolized by δ and called the *retardation* so that $\delta = OM_1O - OM_2O$. When the two beams recombine at the beamsplitter surface to form beam A, the interference observed for wavelength λ will be constructive when δ is an integral number of wavelengths ($\delta = m\lambda$). Similarly, when $\delta = (m + \frac{1}{2})\lambda$, there will be destructive interference. For instance, for monochromatic light the interference will vary sinusoidally as mirror M_1 moves at constant velocity and yield alternating light and dark bands called *interference fringes*. With polychromatic light, the plot of beam intensity versus mirror position or δ is called an *interferogram*.

Other types of two-beam interferometers are also available. An interferometer based on *refractive scanning* is obtained by arranging that a wedge-shaped prism be moved in and out of the variable beam. It offers the advantage of minimal constraint on the solid angle of the beam and a better use of space.* Still another design employs rotating refractor plates.

Development of an Interferogram. In order to extract chemical information effectively it will be essential to understand how an interferogram is developed by an interferometer. In this discussion wavenumber σ ($\sigma = 1/\lambda$) will be used instead

*P. R. Griffiths, *Science*, **222**, 299 (1983); W. M. Doyle, B. C. McIntosh, and W. L. Clarke, *Appl. Spectrosc.* **34**, 599 (1980).

of wavelength to simplify the handling of notation. This examination may also expose some ways to minimize error in the development. First, consider the effect of the beamsplitter on an interferogram. Ideally, for the Michelson interferometer the beamsplitter will have at wavenumber σ a reflectance R that equals its transmittance T, that is, $R_\sigma = T_\sigma = 0.5$. Then, as retardation δ is varied, the intensity of radiation leaving the interferometer as beam A is given by the expression $I_A = 0.5 I_T(1 + \cos 2\pi\sigma\delta)$, where I_T is the initial intensity incident on the beamsplitter from the source. In the common case, however, $R \neq T$ and the product $2R_\sigma T_\sigma I_T$ is substituted for $0.5 I_T$, giving the new expression $I_A = 2R_\sigma T_\sigma I_T\,(1 + \cos 2\pi\sigma\delta)$. It is the ac component of intensity represented in this equation by the term, $2R_\sigma T_\sigma I_T \cos 2\pi\sigma\delta$, plotted as a function of δ that constitutes an *interferogram*. As noted above, for the interferometer in Fig. 17.12, if the source is monochromatic and the sample nonabsorbing, the interferogram produced will be a simple sinusoidal curve like that in Fig. 17.15*a*.*

When a sample is placed in beam A, its absorption should ideally follow Beer's law for each analyte and will add a multiplicative term 10^{-A}, where A is the absorbance at each wavenumber. Now let $B(\nu)$ represent the product of all terms multiplying the cosine term. On being detected and amplified according to the response function $G(\nu)$ of detector and amplifier, the analytical signal can be expressed for this sample, wavenumber, and retardation as

$$x_{\sigma,\delta} = 2G(\sigma)\, R_\sigma T_\sigma I_T\, 10^{-A} \cos 2\pi\sigma\delta = B(\sigma) \cos 2\pi\sigma\delta. \qquad (17.5)$$

If a continuous source is now substituted, all its wavelengths will be present simultaneously and each will independently undergo interference and absorption by the sample at each retardation. The result will be a complex interferogram like that in Fig. 17.15e.

Interferometer Performance. Some mechanical aspects that affect the quality of measurement also deserve attention. First, we consider collimation.† If a spreading beam is sketched in Fig. 17.12, it becomes evident that the path length is greater for off-axis rays than for paraxial rays. Good collimation will lead to clean-cut interference patterns. Thus, the greater the divergence, the poorer the spectrometer resolution. A simple strategy to keep resolution in an acceptable range is to limit the solid angle of the beam at the cost of throughput. If the highest wavenumber of interest in a spectrum is σ_{max} and the resolution desired is $\delta\sigma$, it can be shown that the maximum solid angle of divergence should be given by $\Omega_{max} = 2\pi\delta\sigma/\sigma_{max}$. This angle can easily be reduced by inserting a diaphragm or variable aperture at a focal point. Attention to beam divergence proves especially important when measurements are taken at low resolution or when the frequency range is small.

*Often small phase lags caused by wavelength dispersion in the beam splitter or filters in the amplifier must be included in the argument of the cosine [5].

† Perfect collimation requires an infinitely small source, so that with all finite sources there is some beam divergence.

Second, the *precision of the mirror drive* must be closely controlled. Again resolution is at stake. In the far IR ($\sigma < 200$ cm^{-1}), no more than a crude drive is ordinarily required; for example, it is possible to mount the mirror on the end of a piston luricated by grease. In the middle IR range, however, to ensure precise mirror motion, air bearings are in common use. This arrangement provides high to moderate resolution (0.05–10 cm^{-1}). For still higher resolution and in the NIR and VIS–UV reflecting corner cubes, sometimes called retroreflectors, usually replace the mirrors since they can compensate for mirror tilt.* Fortunately, in the NIR and UV–VIS a less expensive ball-slide bearing provides suficient support for the moving reflector and also allows a long mirror path for better resolution, as will be discussed below. In general, constant speed is also important for good resolution.

Finally, a valuable but perhaps unexpected dynamic result is that mirror movement *modulates all wavelengths (frequencies)* in the beam. If the mirror has a constant velocity of V cm s^{-1}, an increase in path length by δ cm takes $t = \delta/2V$ s. (Recall that the retardation and path difference are double the actual mirror travel.) Solving for δ and substituting the value in the cosine term of Eq. (17.5) gives cos $2\pi 2V\sigma t$. The product $2V\sigma$ is just the *modulation frequency* sought. For example, a mirror velocity of 0.3 cm s^{-1} will modulate 1000 cm^{-1} light at a frequency equal to 2 times 0.3 cm s^{-1} times 1000 cm^{-1} = 600 Hz. These modulation frequencies contain the spectral information of interest in the interferogram and must be detected along with amplitudes and phases.

Detection. At all frequencies greater than the far IR, a pyroelectric detector is commonly used to accommodate the high modulation rates (in the kilohertz range) that are produced by rapid-scanning interferometers. From the data in Table 8.4, note the relative superiority of a pyroelectric detector such as deuterated triglycine sulfate (DGTS) over other thermal detectors for this application. For higher sensitivity or quite rapid scanning, a photoconductive detector such as a mercury cadmium telluride (MCT) detector operated at 77 K is valuable. It gives good sensitivity at modulation frequencies up to 20 kHz, but is limited in spectral range as other quantum detectors are.

Timing Diagrams for Signal Sampling. To develop an interferogram it is necessary to synchronize sampling of the amplified detector signal with mirror *movement*. For this purpose a laser interferometer is generally added as an internal reference. As shown in the schematic optical diagram of Fig. 17.13 it is an integral part of the main assembly. As the interferometer mirror moves, the monochromatic laser emission generates a series of interference fringes that its detector reports as a sinusoidal signal as shown in Fig. 17.14. These fringes provide accurate information about the position of mirror M_1 and allow sampling of interferograms at precise retardation intervals. It should be emphasized that by synchroniz-

*A corner cube consists of a pair of planar front-surfaced mirrors cemented at a 90° angle to each other. This device consistently reflects an incoming ray 180° as a Porro prism does (see Fig. 7.15b) over a considerable range of angles of incidence.

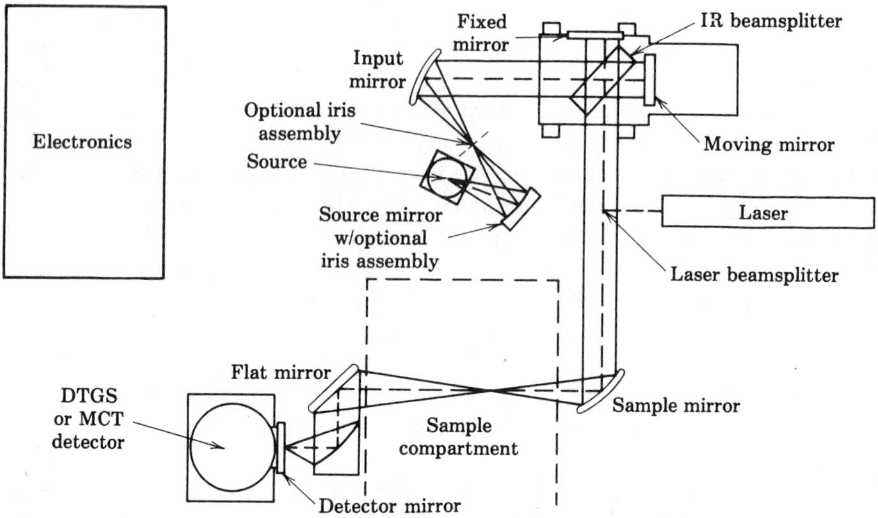

Fig. 17.13 Optical schematic of a Fourier transform infrared spectrometer (FTIR). Note that a laser reference interferometer is present as a part of the basic FTIR. To accommodate it, a small portion of the IR beamsplitter is modified for the laser wavelength and a photodiode detector (not shown) is added. (Basic layout for 500 and 700 series FTIRs, courtesy of Nicolet Instrument Corp.)

ing mirror movement and sampling, spectral frequency can also be established precisely (to about 0.02 cm^{-1}).

By splitting the laser beam into two beams, retarding one 45° and then recombining, it is possible to identify the point of zero path or retardation difference. Sampling in a subsequent mirror scan can then be initiated at a suitable point, either at the zero path difference as is commonly done, or sufficiently before the point so that a "double-sided" collection of data can be made. Actually, the collection of data around the zero retardation point allows correction of phase shifts introduced by any filtering of the interferogram to minimize de-

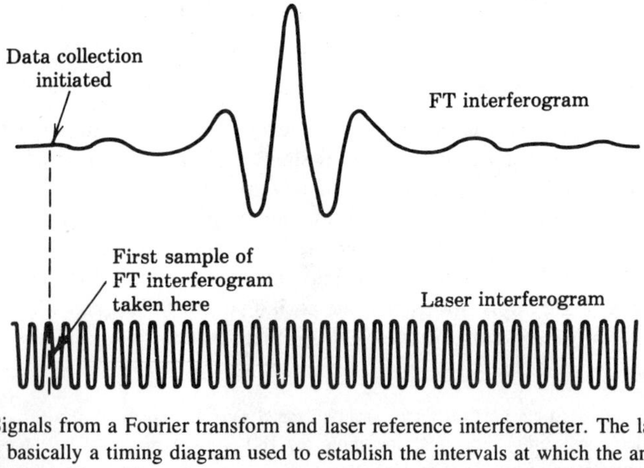

Fig. 17.14 Signals from a Fourier transform and laser reference interferometer. The laser fringe interferogram is basically a timing diagram used to establish the intervals at which the analytical interferogram is sampled. The taking of each sample can be related to the zero crossing of each falling or rising edge of a laser fringe.

tector noise. For example, one source of shift is the active filter that limits the highest wavenumber passed; other elements of the electronic system contribute additional phase shift.

Data Acquisition. It is helpful to state the spectrometer variables that must be set by the user to ensure recording an interferogram that will give an accurate wavelength or wavenumber spectrum on Fourier transformation. They are (a) maximum retardation Δ_{max} and (b) mirror velocity V. In addition, values of two processing variables must be chosen by the manufacturer or user: (c) the signal sampling interval and (d) the maximum spectral frequency or wavenumber σ_{max}. How should reasonable values be selected?

Choosing the interferometer retardation Δ_{max} basically defines the spectrometer resolution $\delta\sigma$ as has already been suggested. The connection with resolution can be shown from information theory to yield the expression $\delta\sigma = 1/\Delta_{max}\ \text{cm}^{-1}$. For instance, if 4 cm^{-1} resolution is desired in an IR spectrum, the retardation need only be $\Delta_{max} = 1/\delta\sigma = 0.25$ cm and the mirror movement half that distance, 0.125 cm.

The maximum retardation also affects the S/N of a spectrum, which will depend both on the number of resolution elements M and the S/N of the source. Here the connection with the Fellgett advantage is made. For instance, if a 500–4000 cm^{-1} scan is to be taken with a retardation giving a resolution of 1 cm^{-1}, $M = (4000–500)/1 = 3500$. It can be shown that

$$(\text{S/N})_{\text{spectrum}} = M^{1/2}\,(\text{S/N})_{\text{source}}.$$

Quantitative accuracy will increase as S/N increases.

The choice of mirror velocity will depend greatly on the scanning speed desired and the dynamic range of the available AD converter. The greater the velocity, the higher the modulation frequency, and the stronger the demand on the ADC. For example, for an IR source at least a 14–15 bit AD converter is desirable. Remember that a significant amplitude must be digitized at each point since transmission measurements are being made. Further, the more complex the spectrum to be observed, the more rapidly amplitude will change and the greater the demand on the ADC. Clearly, the high wavenumbers in the UV–VIS (12,000–50,000) will yield higher rates of modulation and will not only call for high-speed data acquisition, but also abundant memory.

As noted in Section 12.5, the sampling interval must be sufficiently short to satisfy the Nyquist criterion: it must be at least twice the largest modulation frequency of interest. An illustration is given in Example 17.18. By contrast, establishing a high frequency or wavenumber limit, usually by adding an optical cutoff filter, has one main effect, to minimize generation of spurious frequencies on sampling, an effect called *aliasing* or folding (see Section 12.5). If a wavenumber $\sigma > \sigma_{max}$ is not blocked, it will be folded back into the output to appear as a spurious peak of wavenumber lower than σ_{max} by the amount $\sigma - \sigma_{max}$. Certainly, σ_{max} should be no smaller than the Nyquist frequency.

Example 17.15. In the far IR, what will be the number of resolution elements M obtained in scanning from 20 to 400 cm^{-1} when $\Delta_{max} = 1$ cm? If a slow scan, perhaps 2.5 $\mu m\ s^{-1}$, is used, what will be the total scan time and the modulation rate?

Since $\delta\sigma = 1/\Delta_{max} = 1\ cm^{-1}$, we have for the number of resolution intervals $M = (400 - 20)/1 = 380$. The total scan time will be 0.5 cm/2.5 $\mu m\ s^{-1}$ = 2 times 10^3 s or about 30 min. For a σ_{max} of 400 cm^{-1}, the modulation rate will be no greater than 2 times 2.5 $\mu m\ s^{-1} \times 400\ cm^{-1} = 0.2$ Hz.

Example 17.16. What number of data points will need to be taken per second for absorption spectra in the middle IR (characteristically 400–4000 cm^{-1}) with a mirror movement of the order of millimeters per second and a total retardation of 1 cm? Will an ADC of high dynamic range be desirable?

The scan will require 3–4 s and the modulation rate will be as great as 1.2 kHz. The total data points will be $(4000 - 400)/1 = 3600$. Since a continuous source is used, there will be a signal of reasonable intensity at every sampling point. These signal intensities together with the rapid rate of sampling required by fast mirror movement will require a high data acquisition rate.

There will be so many data to be recorded that care must be taken to avoid exceeding the dynamic range of the AD converter. While the mirror scan rate can be decreased, it may cause the S/N ratio to be limited by *digitization noise* rather than by another source such as detector noise. If this occurs, it is to be expected that signal-averaging techniques will be necessary to increase S/N further. In general, it is *interferograms* that are signal-averaged, rather than the final spectra.

Example 17.17. An FTIR spectrometer has been used on many spacecraft (such as the Voyager pair) to determine the composition of the atmosphere of other planets.* In this case since the spectrometers could be kept at 200 K by insulation and devices that radiated most of their heat into deep space, they were designed to record IR *emission* spectra. What data acquisition rates were appropriate, for example, in the Voyager spacecraft that sampled the atmosphere of Jupiter, Saturn, and Uranus in 1979, 1981, and 1986, respectively, and will have an opportunity to obtain similar information from Neptune in 1989 before leaving the solar system?

Two factors helped keep the acquisition rate down. One was that emission intensities were small at low planetary temperatures, giving fewer bits of data. Further, slow scanning was tolerable since data could be stored and transmitted later. Their interferograms were acquired in 45.6 s at a rate of 80 words s^{-1}. Since the S/N of each was low, hundreds of interferograms were averaged to improve the value.

Example 17.18. It is desired to obtain an IR spectrum from 400 to 4000 cm^{-1} with a resolution of 0.2 cm^{-1} by use of a Fourier transform spectrometer. Assume the mirror velocity has been set at 0.1 cm s^{-1}. What other settings must be chosen for the run?

First, the maximum mirror travel Δ_{max} must be set. The desired resolution defines the maximum path difference Δ_{max}:

$$1/\Delta_{max} = \delta\sigma = 0.2\ cm^{-1} \text{ and } \Delta_{max} = 5\ cm.$$

The mirror must be driven half this distance or 2.5 cm.

*S. A. Borman, *Anal. Chem.*, **53,** 1544A (1981).

Second, the maximum modulation frequency will be $2V\sigma$ = 2 times 0.1 times 4000 = 800 Hz. The Nyquist frequency will be twice that, necessitating sampling at a rate of 800 times 2 = 1.6 kHz or a point every 0.625 ms or 1/1600 s^{-1}. Given the rate and distance of mirror travel the number of data points taken will be 2.5 cm/0.1 cm s^{-1} = 25 s; 25 s/6.25 $\times$ 10^{-4} s = 40,000.

Resolution and Apodization. Resolution is also influenced in a more subtle way by imposing the wavenumber cutoff σ_{max}. This limit adversely affects the Fourier transform process. The shape of peaks is affected; each peak now appears with side lobes of modest amplitude, as in Fig. 17.15*b*. Note in the figure that the lobes diminish in amplitude and number as maximum retardation is increased. It is not difficult to see that these lobes will reduce resolution for close-lying peaks and that the baseline will show ripple.

Mathematically, limiting the frequency is equivalent to *boxcar truncation*, allowing a time–domain boxcar function to *convolute* the regular spectral time-domain function $h(t)$. A boxcar function $b(t)$ has values $b(t) = 1$ for $t \leq T$ and $b(t) = 0$ for $t > T$. This function is included in Eqs. (17.3a) and (17.3b) and is also transformed into the frequency domain as a function $B(\sigma)$. It may be shown that $B(\sigma) = (2T \sin 2\pi\sigma T)/2\pi\sigma T$, a term usually called a $\sin x/x$ or *sinc* function.

To enhance resolution the amplitude of the side lobes can be reduced by applying the processing technique of *apodization*. The instrument signal is simply convoluted, that is, multiplied by a suitable weighting function, before the transformation is carried out. The spectrum obtained on Fourier transformation is then simply $G'(\sigma) = \int_{-\infty}^{\infty} g(x)a(x) \exp(2\pi i\sigma\delta)\, d\delta$. A widely used apodizing function is a simple *triangular function* $a(x)$, where $a(x) = 1 - |x|/L$.

Triangular apodization drastically diminishes side lobes, and even with the trade-off of a broadening of the main peak, as shown in Table 17.2, is regarded as improving resolution. Other apodizing functions studied give about the same enhancement. To illustrate the two types of peaks, boxcar truncated and triangularly apodized peaks are compared in Table 17.2 and Fig 17.15*d*.

Photoacoustic Detection. Many FTIR spectrometers feature optional photoacoustic detector modules for measurements on opaque materials, solids of irregular shape, powders, and other special samples that are to be examined without dissolution. These detectors sense the *absorption* of a modulated FTIR beam by a sample.

Many important variables for the measurement as well as the physical layout of the cell and detector are depicted in Fig. 17.16. As absorption occurs, the fact that each frequency has been modulated by the moving interferometer mirror becomes evident in the sinusoidally varying heat waves that are produced in the sample. The waves traveling toward the sample surface transfer their energy to the gas atmosphere next to the sample. Its resulting thermal expansion and contraction constitutes a sound wave whose amplitude and frequency can be sensitively detected by a microphone. While the transmitting medium surrounding the sample and the detector is usually a gas, it may also be a liquid. The medium is selected for its minimal absorbance and its coupling ability.

At the detector the usual interferogram is generated. To minimize the effect of gas absorption, wavelength-dependent detector response, and many other sources of error, the

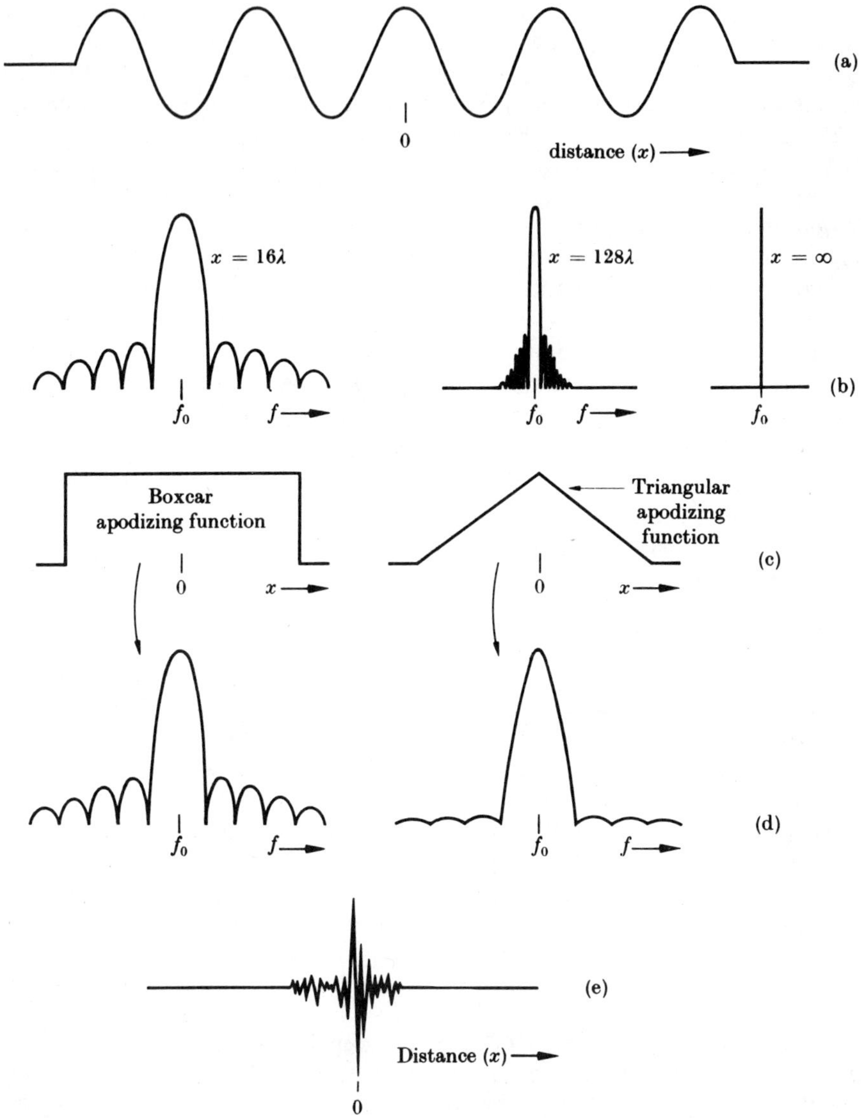

Fig. 17.15 Interferograms, apodizing functions, and Fourier transforms of interferograms. (a) Cosine wave interferogram (a fringe pattern) obtained with monochromatic light of frequency f_0. (b) Fourier transforms of monochromatic light interferograms for the different retardations shown in the figure. Retardations are given in terms of multiples of the wavelength of the light. (c) Graphical representation of the boxcar and triangular functions that are convoluted with spectral interferograms during Fourier transformation. (d) Fourier transforms of the interferogram obtained with retardation $\delta = 16\lambda_0$ of an isolated spectral peak of frequency f_0. On the left the interferogram has been boxcar truncated by imposition of a maximum frequency limit σ_{max}; on the right, subjected to triangular apodization. As is evident, triangular apodization diminishes side lobes but also broadens the central peak. (e) An interferogram of a multifrequency spectrum. The large central maximum contains most of the information about the background intensity and the wings define spectral peaks.

Table 17.2 Resolution in Fourier Transform Spectrometry for Maximum Retardation of Δ_{max}

Type of Apodization	Definitions of Resolution[a]		
	Peak Width at Half Height	*Baseline Peak Separation*	*Rayleigh Criterion*[b]
Boxcar	0.60	1	0.73
Triangular	0.88	2	1

[a]Resolutions are calculated as multiples of the retardation Δ_{max}. For example, for separation of adjacent peaks at the baseline, the distance from the origin to the first zero values on either side of the peak equals $1/\Delta_{max}$ with boxcar apodization and $2/\Delta_{max}$ with triangular apodization.
[b]A 20% valley between peaks.

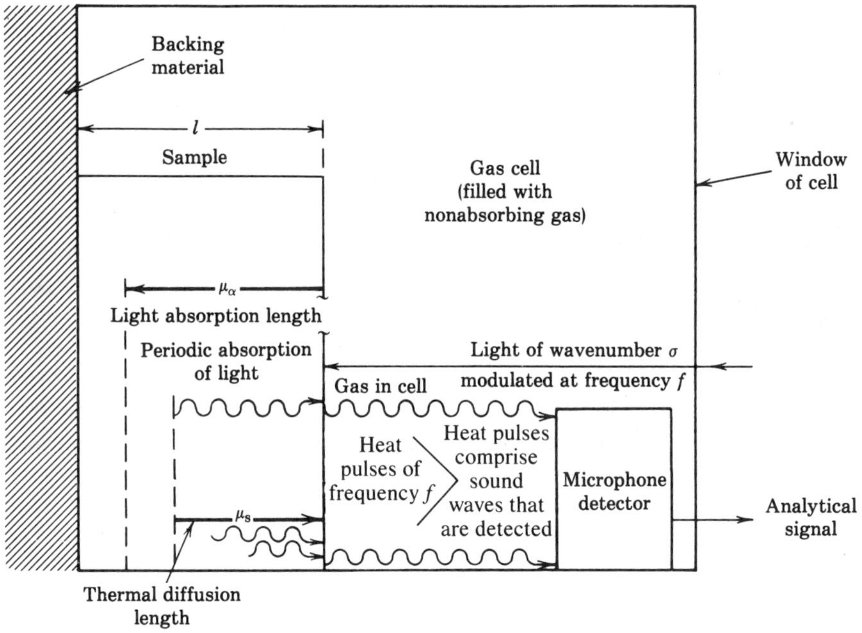

Fig. 17.16 Block diagram of a photothermal (photoacoustic) gas-coupled detector. A sample is placed inside the cell against its back edge (opposite the window through which radiation enters). The gas is assumed to be nonabsorbing. Radiation is absorbed by analytes in a sample in characteristic ways and the extra energy is released as heat. Three distances *in the sample* are of interest: its actual thickness l; its absorption length μ_α, which is the greatest distance irradiating light can penetrate; and its thermal length μ_s, the greatest distance inside the sample from which the heat released can escape to the front surface. Only heat waves that escape to this surface can generate sound waves in the surrounding gas and be sensed by the microphone detector. For a successful spectral measurement, it is necessary that $\mu_s < \mu_\alpha$. In an FTIR spectrometer, the analytical signal from the detector is of course an interferogram; it must subsequently be Fourier-transformed to obtain the desired spectrum. (A photothermal detector can also be used with a dispersive spectrometer in any optical spectral region. In this case it is necessary to modulate the irradiating light at a selected audiofrequency. As a wavelength scan is made, the microphone signal is demodulated and plotted versus wavelength to give the spectrum.) Piezoelectric detectors placed against the side or back of a sample can also be used.

interferogram is ratioed to the photoacoustic interferogram of a thermally saturating substance such as carbon black. Then the ratio is Fourier transformed to convert the sample spectrum into the wavelength (frequency) domain.

Photoacoustic detection can also be used in the UV–VIS with a dispersive spectrometer by modulating the optical beam in the audiofrequency range and use of phase-sensitive detection. In this situation one can also select the modulation frequency that best accommodates the thickness of the sample and its rate of heat conduction.

With photoacoustic detection spectra can be obtained for samples that are quite difficult to handle. Very little sample preparation is required. Even essentially opaque or highly scattering samples can be examined as they are. References should be consulted for the theory and details of measurement [9, 10]. The photoacoustic detector complements the diffuse reflectance procedures described in Section 17.6.

Comparison of FT-IR and Dispersive Spectrometers. As noted before, Fellgett's advantage of having all wavelengths simultaneously incident on the detector is one of the striking advantages of FTIR over dispersive IR spectrometers that have a single-channel detector. In the latter, a particular wavelength is incident on the detector only for a brief interval during observation of a spectrum. It is common to describe the advantage by talking about resolution intervals M. In measurements made in the same total time, the FTIR spectrometer theoretically yields an advantage of $M^{1/2}$ in S/N over the dispersive spectrometer. Conversely, if equal values of S/N are satisfactory, measurements can be made M times more quickly on a Fourier instrument. In these comparisons it is assumed that measurements are taken at equal resolution with the same detector and the same optical throughput.

Since each wavelength will be modulated and detected at a different frequency, it will be subjected to *noise* only over its own narrow bandwidth. As a result, there is really no counterpart in Fourier transform measurement to stray light. The freedom from stray light reinforces the Fellgett advantage in FTIR for signals of low S/N. In a dispersive spectrometer all wavelengths are modulated at the same frequency. Those reaching the detector by routes other than the optical path, that is, as stray light, are registered as though they were signals.

Actually, it is clear that the throughput is substantially greater with an interferometer, a gain known as the Jacquinot advantage. While it is beyond the scope of the text to obtain a precise evaluation of this gain, it may be noted that the necessity of using a faster detector substantially offsets the gain.

The major kinds of measurements in which FT spectrometers are used with advantage are those in which

1. energy is quite low, such as in the middle and far IR or in astronomical measurements;
2. samples are highly absorbing or scattering;
3. samples contain transient species;
4. the highly reproducible interval of a laser is used for referencing; or
5. high resolution is required.

17.9 FTIR SPECTROMETERS WITH CHROMATOGRAPHIC SAMPLE CHANNEL

With the lower limits of detection possible with the Fourier transform spectrometers just discussed, even weak IR spectral transitions ($\epsilon \leq 10^2$) become attractive for quantitative analysis. Two difficulties impeded the use of this approach for mixtures for some time, the frequency of spectral interference and the instrumental problem of low source intensities. An attractive solution was to remove interferences by a preliminary quantitative chromatographic separation and then to employ an FTIR spectrometer to identify components and quantify their measurement. Combining these operations can often be accomplished in a single instrument, an FTIR spectrometer with a chromatographic *sample channel* (see Section 11.2). Fortunately, continued research has provided steadily improved ways to interface sample channels and spectrometers. At present success has been found mainly with gas chromatographic channels by virtue of the fact that their mobile phases (mainly He and H_2) are transparent in the IR.* Quantitative determinations can be based on a calibration curve approach involving the use of Gram–Schmidt reconstructions of the curve rather than a fast Fourier transform.†

Coupling Requirements. To couple a gas chromatographic channel to an IR spectrometer there must be (a) a temporal match between peak elution times and spectrometer scan times and (b) a correlation between minimum concentrations of components emerging from a column and spectrometer limits of detection. For example, a wide bore, wall-coated open tubular (capillary) column (WCOT) can be coupled effectively to an FTIR spectrometer. Temporal requirements match well: about 3–15 s is required for complete elution of a peak and only about 1.5 s for an FTIR scan of modest resolution (8 cm^{-1}) by a spectrometer with a fast, sensitive detector. In the event that scans of the significant spectral range take less time than component elution, it is possible to make several scans and average them to reduce noise.

Very seldom is a dispersive IR spectrometer coupled to a chromatographic channel. Since its scans take a longer time, coupling can best be done using a stopped-flow device in which samples are first collected and spectra are taken later. Detection of trace components is problematical.

If chromatographic samples are too small to ensure that component concentrations will exceed spectrometer limits of detection, column capacities and loading can be varied somewhat to alter analyte sample size. If emphasis is on speed and resolution, a capillary rather than a packed column may be attractive. While an open tubular column delivers minute amounts of components, many at the nanogram level, with an FTIR spectrometer the limit of detection with a room-temperature pyroelectric detector is low and is even lower when using an MCT photoconductive detector, which is customary with GC–FTIR combinations. Nevertheless, because the latter detector operates at 77K, both initial investment and op-

*Mobile phases used in liquid chromatography (e.g., water and other liquids) absorb in broad regions of the IR. Modes of solvent removal and concentration of analyte continue to be investigated.

†D. T. Sparks, R. B. Lam, and T. L. Isenhour, *Anal. Chem.*, **54,** 1922 (1982).

erating costs are substantial. For mixtures with many components, a greater amount of sample is required and columns of greater capacity become attractive.

GC–FTIR Interfaces. A *gold-plated light pipe* is a common GC–FTIR coupling device. A sketch of a suitable pipe is shown in Fig. 17.17. Eluant enters it from the chromatographic column through a small diameter stainless-steel tubing column that ensures minimum dead volume. It leaves by the far end, often to continue to a GC detector such as a flame ionization detector. In an average light pipe about 25% of the IR beam traverses its length. Whether components can pass through without being band broadened can be ascertained by observing the response of a GC detector to the eluant from the light pipe.*

A *matrix-isolation* interface that provides more efficient *collection* of components has been developed recently. It avoids the need to correlate flow rate with peak width and offers quite low limits of detection. With argon as the matrix gas (added at a low level to the carrier) the column eluate can be sprayed on a slowly moving cryogenic surface to collect components. Infrared spectra are taken continuously after a standard, short time lag. Its cost is high.

Spectral Searching. To identify analytes from spectra, some mode of spectral searching is desirable. One appropriate technique is examination of the spectrum of each component for types of functional groups that should be present. In any case, the final step is library scanning of related IR spectra, seeking a match to complete the identification of an analyte.

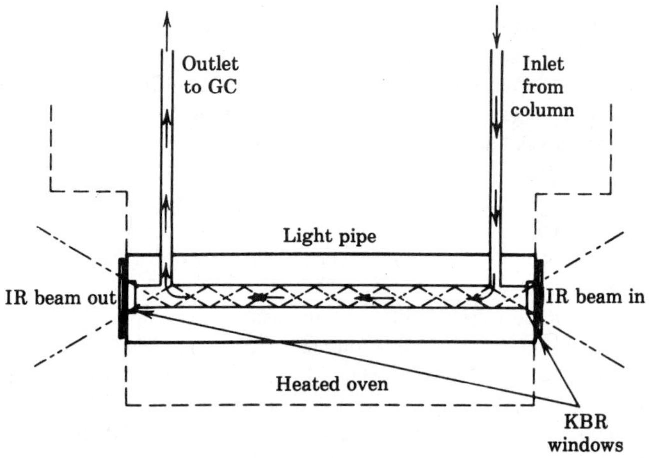

Fig. 17.17 A representative IR light pipe. Eluate from a gas chromatograph flows the length of the cell (dimensions: about 1 mm × 50 cm with 0.2-1-mL volume). The reflective coating of gold on the inside ensures that the IR beam entering through the KBr window is reflected back and forth through the eluate many times. (Courtesy of Digilab Div. of Bio Rad.)

*See P. R. Griffiths, J. A. deHaseth, and L. V. Azarraga, *Anal. Chem.*, **55,** 1361A (1983) for a good discussion of many aspects of GC–FTIR.

17.10 SPECTROPHOTOMETRIC PROCESS ANALYZERS

In the continuous analysis of process streams or plant effluents, filter photometers and nondispersive photometers are most often employed. Their use is most attractive when information needed for control can be secured by monitoring a single analyte which has at least one absorption band in the UV–VIS, NIR, or IR that is not overlapped seriously by bands of other components. Their ruggedness, high throughput and sensitivity, and low cost are strong advantages.

More recently Fourier transform IR analyzers have appeared on process streams. They feature room-temperature detectors, assembly-language fast Fourier transform programs, long-stroke mirrors using ball-slide bearing corner-cube reflectors for high resolution, and an appropriate optical bench for relative freedom from vibration and noise.

Sample cells are adapted to the type of process stream. Cells for gases provide long path lengths. Solution cells have much shorter optical paths. Where solvents are absorbing, cylindrical multiple internal reflection cells (see Fig. 17.9) offer advantages because of their short penetration depth. Further, they are valuable for process streams that are viscous or may contain high concentrations of dissolved solids. In a cylindrical cell radiation from a source enters one end and is multiply reflected internally until it passes out the other end and is directed to a IR detector.

Filter Photometers. With a filter photometer process analyzer, absorption is usually determined at two different wavelengths. A modular diagram of a split-beam design that lends itself well to this approach is shown in Fig. 17.18 and is interpreted in the legend. Since the analytical signal is taken relative to a light intensity at which the sample is transparent, the signal is relatively unaffected by changes in source intensity and drift in the other modules. For example, accumulating dirt should affect both beams similarly.

In a dual-beam instrument employing both reference and sample cells, however, dirt in either cell will give rise to error. In this design it is important to provide for periodic standardization using an appropriate reference. For initial establish-

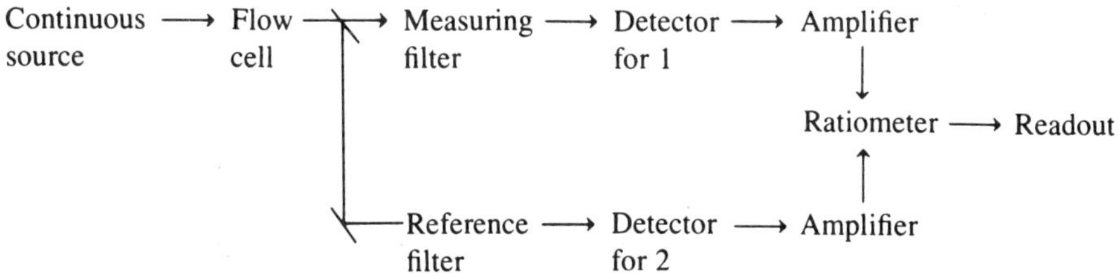

Fig. 17.18 A modular diagram of a process analyzer based on a split-beam filter photometer. The beam splitter is a half-silvered mirror or a variable splitter. The process stream is continuously sampled to keep the sample cell full. The measuring filter passes mainly wavelengths absorbed by the analyte. The reference filter passes wavelengths that are little absorbed by representative samples. Thus, analyte transmittance is measured by taking the ratio of detector outputs.

ment of parameters it will be important that the cell path length be variable to permit selection of a length that will give 0.2–1 absorbance units for good measurement sensitivity. Calibration will be with standards.

IR Nondispersive Photometers. Nondispersive IR photometers, that is, intensity measuring devices without any type of wavelength isolating module, are commonly used as process analyzers. As shown in Fig. 17.19, most are used in a dual-beam design. In this design, if the transmission of light through the cells is efficient, the total power of the beam is available for transmission by a cell. Transmission by a sample can be compared with that of a reference cell filled with the analyte. The transmitted radiation in each beam is then absorbed by a pair of detectors forming arms of a Wheatstone bridge (see Section 3.4). The point of balance is indicative of the concentration of the analyte in the sample. Again, low cost, ruggedness, and generally good throughput are advantages.

The slightly different Luft-type nondispersive IR analyzer is also in common use. Sample and reference cells are again placed side by side. Two heated nichrome filaments are used as IR sources and their output is chopped so that light is alternately passed through the sample cell and the reference cell. Transmitted radiation falls on a single Golay-type detector (see Section 8.6). Since its diaphragm is one plate of a condenser microphone, when the transmitted beams are of different intensity, the gas volume alternately expands and contracts, causing the diaphragm microphone to produce an ac signal. The Luft-type analyzer operates well in the 4000–600-cm^{-1} region and has good sensitivity. Nevertheless, optical alignment is critical and corrosion or dirt in the sample cell leads to an imbalance that cannot be compensated for by the reference.

Dual-beam nondispersive analzers have a response time of 3–5 s, low noise levels, and drift of less than 1% of full scale per day; repeatability is within ±1%.

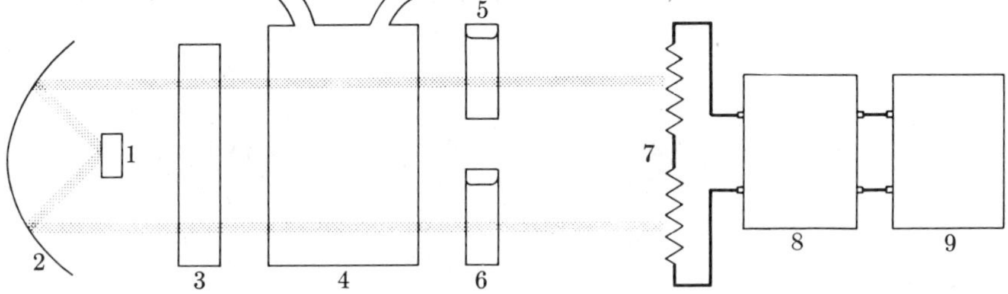

Fig. 17.19 A representative nondispersive IR analyzer. Components are as follows: 1, source; 2, spherical focusing mirror; 4, sample cell; 5, reference cell (filter cell); 8, bridge incorporating bolometers, 7, as arms; and 9, display. Two additional cells, 3 and 6, are available to reduce the effect of spectral interference. If substances whose absorption bands may overlap the chief analyte bands are present, a variation in their concentration will appear as a *change* in analyte concentration. This effect can be largely compensated for by adding appropriate concentrations of the interferents to cell 3 so that the energy these are capable of adsorbing is removed ahead of time in reproducible fashion. In severe cases, interferents can also be added to cell 6, through S/N will be seriously reduced where they absorb.

Most sample cells are plated with gold or another precious metal to minimize corrosion.

Remote Sensing. Process streams can be analyzed spectrophotometrically at a considerable distance from an instrument by use of a *fiber optics* connection. This arrangement is also adaptable to *in situ* measurements on many a biological or chemical system that cannot be sampled without upsetting its dynamics. Ordinarily a pair of fiber optics, one to transmit light from a source to the probe point and the other to return the transmitted radiation from the system, gives best results. Path lengths at the measurement site are easily varied from 1 to 40 mm. Standard probes can withstand temperatures to 350°C and pressures up to 500–1,000 bar. Advantages are not only minimal invasion of a system and avoidance of duplicating instrumentation but also better S/N in many cases because more light can be sent and returned from a system.

Applications include *in situ* monitoring of process streams that may be dangerous to approach closely; study of the dynamics of catalysis under extreme conditions; measurement of reaction kinetics and identification of reaction intermediates; and analysis of samples of plant and animal tissue.

REFERENCES

Discussions mainly restricted to instrumentation and components are available in:

1. W. Kaye, D. Barber, and R. Marasco, Design of a microcomputer-controlled UV–VIS spectrometer, *Anal. Chem.* **52,** 437A (1980).
2. L. T. Skeggs, Jr. and H. Hochstrasser, Multiple automatic sequential analysis, *Clin. Chem.*, **10,** 918 (1964).
3. J. V. White and M. O. Liston, Construction of a double beam recording IR spectrophotometer, *J. Opt. Soc. Am.*, **40,** 29 (1950).

Fourier transform spectrometery is discussed in:

4. P. R. Griffiths and J. A. deHaseth, *Fourier Transform Infrared Spectroscopy.* New York: Wiley, 1986.
5. P. R. Griffiths (Ed.), *Transform Techniques in Chemistry.* New York: Plenum, 1978.
6. A. G. Marshall (Ed.), *Fourier, Hadamard, and Hilbert Transforms in Chemistry.* New York: Plenum, 1982.

Other references of interest are:

7. J. G. Grasselli, P. R. Griffiths, and R. W. Hannah, Criteria for presentation of spectra from computerized IR spectrometers, *Appl. Spectrosc.*, **36,** 87 (1982).
8. T. D. Harris, High sensitivity spectrophotometry, *Anal. Chem.*, **54,** 741A (1982).
9. M. D. Morris and K. Peck, Photothermal effects in chemical analysis, *Anal. Chem.*, **58,** 811A (1986).
10. A. Rosencwaig, *Photoacoustics and Photoacoustic Spectroscopy.* New York: Wiley-Interscience, 1981.

The titles listed for Chapters 8, 9, and 16 should also be consulted.

EXERCISES

17.1 Would amplification be desirable or appropriate in a filter photometer? Support your answer with several reasons.

17.2 Why do colorimeters with broad bandpass filters give results of low sensitivity and accuracy when used as wavelength isolation devices in the determination of species with narrow absorption bands? What is the connection between photometric accuracy and instrument sensitivity?

17.3 What reasons can you give for better precision for quantitative spectrophotometric measurements in the UV–VIS compared with the IR region?

17.4 Suggest reasons why spectrophotometric precision is poorer (a) when the absorbance is near zero; (b) when it is greater than three. Be specific.

17.5 For quantitative measurements both a single-beam UV–VIS spectrophotometer of the split-beam type (about 10% of the beam intensity is diverted by a beam splitter to an alternate path, chopped, and then returned to a common detector) and a double-beam instrument are available. Since only the latter accommodates a reference cell, the split-beam spectrophotometer requires two spectral scans, one with a reference cell to record a baseline in memory, and a second with the sample cell. Discuss the comparative effectiveness of the two types in: (a) measurement of UV–VIS absorption curves of analytes; (b) quantitative measurements of analytes with narrow absorption peaks; (c) S/N ratio; (d) ability to compensate for short-term fluctuations in source, detector, and amplifier; and (e) ability to compensate for solvent absorption.

17.6 Either a tungsten–halogen or a deuterium lamp can be used as a source at 340 nm. Which source is the better choice for accurate spectrophotometry? Does one need to know their relative intensity at 340 nm to answer? Explain.

17.7 How do data acquisition requirements (number of bits and speed of the AD converter and avaliable memory, for atomic emission spectrometry compare with those for IR absorption spectrometry? Consider average amplitudes available at each sampling of the analytical signal.

17.8 Does the method of inserting a substance that is opaque at wavelength λ_1 between monochromator and detector (see Section 9.7) actually give the level of spectrophotometer stray light at λ_1? Explain. Show how a series of different concentrations of an analyte absorbing at λ_1 can be used to measure stray light at that wavelength. Develop a curve or an equation that will yield the quantity. [A. H. Mehler, *Science*, **120,** 1043 (1954).]

17.9 Would you expect the stray radiation in an IR single-prism spectrophotometer to increase or decrease with increasing wavelength in the range from 5 to 15 μm? Assume that the instrument is operated so as to provide constant power at the detector. What steps may be taken to minimize the stray radiation?

17.10 A "double-pass" Littrow prism monochromator is to be fabricated to achieve enhanced resolution. In this design, the dispersed beam leaving the Littrow prism is reflected back for a second pass through it and further dispersion before being directed to the exit slit. Where should a beam chopper be placed for maximum effectiveness in reducing stray light? Consider that the chopper could be located before the first or second pass or after the second, and that the detector output will be selectively amplified at the modulation frequency.

17.11 Assume a laser whose emission can be tuned to cover the visible range is available. (a) Consider briefly how an absorption spectrophotometer might be designed to use its distinctive properties. (b) Would it have any advantages with microsamples? [See M.J. Houle and K. Grossaint, *Anal. Chem.*, **38,** 768 (1966)].

17.12 A helium rf glow discharge will excite dissolved metals in samples as small as 10 nL. A drop of solution is deposited on a heated iridium filament and vaporized into a helium atmosphere. The glow discharge excites the metallic vapor species to emission. (a) Show by block diagram the types of optical modules needed for quantitative determination of metals like sodium. (b) Suggest by diagram or in words the essentials of the associated electronics. Discuss the relative merits of an instantaneous versus an integrative type of observation. [See also G. V. Vurek, *Anal. Chem.*, **39,** 1599 (1967) and W. W. Harrison et al. ibid., **58,** 341A (1986)].

17.13 A simple absorption photometer for detecting mercury vapor can be built around a low-pressure mercury lamp that furnishes principally resonance radiation of 253.7-nm wavelength. No wavelength isolator is needed. (a) Draw a block diagram for the device and classify it as to type of photometer. (b) Explain its operation. (c) Will any atomic or molecular interferences be likely? Will sensitivity be as great for any interfering species as for mercury?

17.14 A program for the slit of a dispersive IR spectrophotometer is to be developed to ensure a good S/N at longer wavelengths. Assume that a Globar source at 1500 K, a grating monochromator, and a thermal detector are used. (a) What is the wavelength (wavenumber) of maximum emission? (b) The instrument is to scan from 4000 to 400 cm^{-1}. Take the source energy as unity at 4000 cm^{-1}. What is its relative value at 3000, 1500, and 500 cm^{-1}? (c) If a 0.2-mm slit width is adequate at 4000 cm^{-1}, what should be its width at 1500 and 500 cm^{-1} to give a constant S/N? (c) With adjusted widths, what are the relative *spectral* slit widths at 1500 and 500 cm^{-1}?

17.15 If a hot sample is examined in an IR spectrometer its blackbody radiation will add to the transmitted beam intensity. (a) Will this spurious effect be minimized by chopping the beam after the sample? Before the sample? (b) Suggest a procedure or instrument design for IR spectrometry of molten salts that will minimize this problem.

17.16 In internal reflectance spectroscopy, why is perpendicular incidence on the face of an internal reflection plate sought for the beam entering or leaving it?

17.17 Consider that the nondispersive IR analyzer shown in Fig. 17.19 is to be used for analysis of CO_2 in air over the concentration range 0–10%. (a) Indicate the gas or gas mixture that should be placed in the various cells. (b) Explain the operation of the device in this analysis.

17.18 Draw transmission curves to show how a nondispersive spectrometer is sensitized to CO in CO–CO_2 mixtures by filling the interference cell with CO_2. Carbon dioxide has an absorption band at 2300 cm^{-1}, CO has one at 2600 cm^{-1}. Draw a rough transmission wavelength curve for the reference path and for the sample path.

17.19 Several types of malfunctions possible with a dispersive spectrophotometer are listed below. Some can be corrected by calibration, some are compensated for by the differential character of double-beam operation, and others require an appropriate instrument adjustment. For each malfunction listed, suggest (a) how it would be discovered from an inspection of absorption curves and (b) whether compensation will usually take care of it, or whether calibration or adjustment of some type is necessary.

(i) Speed of scan is too high.
(ii) Gain of amplifier is set too high.
(iii) Mechanical beam attenuator (in optical null instruments) is nonlinear or not functioning properly.
(iv) Good optical alignment is lost.
(v) Stray light levels are high over some regions of wavelength.
(vi) Nulling system is not balancing correctly for zero (baseline) signal.
(vii) Monochromator mirror is fogged.

17.20 Spectrophotometers are subject to both systematic and random error. Indicate the error category into which each of the following errors falls and suggest a design strategy by which it might be minimized. *Errors*: (a) flicker in source intensity; (b) amplifier unsteadiness; (c) high detector dark current; (d) a dirty cell window; (e) scattering of light by a sample solution; (f) aging of light source; (g) a change in monochromator alignment during warm-up; (i) substantial detector shot noise; (j) high monochromator stray light.

17.21 Consider that an emission spectrum is to be recorded first by a dispersive spectrometer and then by a Fourier transform spectrometer. (a) Is the *incident* radiation in the time domain or the frequency domain? How can it be represented mathematically? (b) Can it be said that a dispersive spectrometer takes the Fourier transform of the incident radiation? Discuss briefly. (c) Does the optical system plus detector of the Fourier transform spectrometer take the Fourier transform of the incident radiation?

ANSWERS:

17.3 In the UV–VIS molar absorptivities are much larger, sources are brighter and detectors with high internal gain are available. *17.4* (a) When $A \approx 0$, P_0 and P are nearly equal and a small change in P relative to P_0 must be measured. Noise will affect both unless a long time constant is used. (b) When $A > 3$, there is so little energy in P that it is hard to distinguish signal from noise.

Chapter 18

RAMAN SPECTROMETRY

18.1 INTRODUCTION

Raman spectroscopy is an important mode of gaining information about molecules based upon the scattering of monochromatic light. While most light scattered by a sample is essentially unchanged in wavelength, to a minute extent incident light interacts with characteristic molecular motions in a process called Raman scattering to give new wavelengths called Raman wavelengths. Each of these is shifted from the incident wavelength by an amount that corresponds to the energy involved in a change in vibration or another characteristic motion for an analyte. The collection of such lines for a particular species is simply its Raman spectrum.

Raman spectroscopy is especially valuable for the study of vibrational motion, but has wide application for qualitative and quantitative measurements of molecular species under a variety of conditions. For example, normal or spontaneous Raman spectroscopy complements IR spectroscopy in the study of molecular structure and conformations. Perhaps surprisingly, biological applications of resonance Raman spectroscopy have been especially valuable.

The development of new Raman methods and spectrometers can largely be traced to the steady improvement of lasers. Resonance Raman spectrometry and the stimulated or coherent Raman techniques are good examples of such developments. Since the field has been growing rapidly, special attention will be given to more advanced techniques.

18.2 RAMAN SPECTROSCOPY

Raman scattering can be explained most easily by using the classical model for interaction of EM radiation and matter (see Section 7.7). Only objects whose major dimension is less than $1\text{–}1\frac{1}{2}$ wavelengths will scatter light; others will specularly or diffusely reflect it. In this model incident radiation induces an oscillating dipole in small objects like molecules. Unless the incident frequency coincides with that of a characteristic molecular transition, the oscillating dipole reradiates virtually all the incident energy immediately as a secondary wave of the same frequency and the result is *Rayleigh scattering*. But the incident radiation can also interact to a small degree with molecular motions such as vibrations with the result that some oscillating dipoles radiate at new wavelengths and give *Raman scattering*.

A more precise way to describe the latter type of interaction is to say that the scattered light has been weakly modulated by at least some of the characteristic frequencies of the molecule.* The amplitude modulation that occurs gives rise to sum and difference frequencies, $\omega_i - \omega_v$ and $\omega_i + \omega_v$, where angular frequencies ω are used ($\omega = 2\pi f$). Here ω_i is the incident frequency and ω_v corresponds to a characteristic molecular frequency. Because the amplitude of the new frequencies is quite small, only with an intense monochromatic source like a laser is Raman spectroscopy routinely feasible.

Spontaneous Raman. The existence of sum and difference frequencies, $\omega_i \pm \omega_v$, is the basis for normal or spontaneous Raman spectroscopy. A representative set of transitions is shown in Fig. 18.1. The upper levels shown as broken lines represent distorted or polarized forms of the molecules (sometimes they are called virtual states), rather than actual excited states. For that reason, the delay in scattering is only the usual 10^{-14} s required for interaction of radiation and matter. Indeed, an important difference between Raman scattering and fluorescence emission is the much longer delay characteristic of the latter. When a photon is absorbed, as in fluorescence, the finite lifetime of the excited state causes emission to be delayed on the order of 10^{-8} s. The delay provides a basis for distinguishing or separating these spectra (see Section 18.4).

At room temperature Raman transitions tend to originate from the lowest vibrational level of the ground electronic state, since most molecules are in this level. These more intense changes are called *Stokes lines* and are of smaller frequency ($\omega_i - \omega_v$), as shown in Fig. 18.1. Their formation means that energy has been subtracted from the incident light. Similarly, their wavelengths are longer than those of the exciting radiation. As shown in the figure, vibrational changes may also orignate from an excited vibrational level, whose population is smaller. These *anti-Stokes lines* appear at shorter wavelengths, and are of lower intensity than Stokes lines. While a Raman spectrum comprises both the longer and shorter wave-

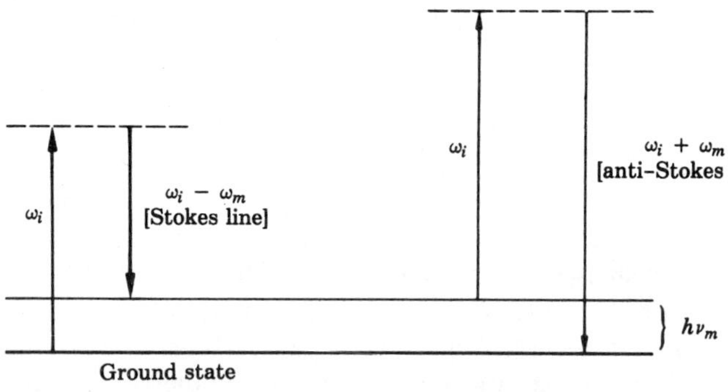

Fig. 18.1 Schematic representation of the quantum-mechanical origin of a Stokes and anti-Stokes pair of Raman lines. Monochromatic light of frequency ω_i is incident on a molecule. In the light scattered by the molecule three frequencies appear: ω_i and Raman lines $\omega_i - \omega_m$ and $\omega_i + \omega_m$.

*Most Raman frequencies represent vibrational transitions. Raman rotational changes are observed mainly in gaseous samples since in the usual liquid or soild sample rotation is not possible. In specialized cases, Raman electronic spectra may also be observed.

lengths, usually only the former, the Stokes lines, are measured. Not only are they more intense, but since the lines occur in pairs, only one need be measured to establish the main information of interest, the *difference* in wavelength from the exciting line. Intensities and polarizations are also of interest, as will be seen below.

Example 18.1. Ordinarily, Raman transitions are given in wavenumbers. What is the wavenumber of a Stokes line that appears at 680 nm when the exciting light is the 632.8-nm wavelength of a helium–neon laser? What will be the wavenumber and wavelength of the corresponding anti-Stokes transition?

The Stokes line will have a wavenumber of

$$(1/632.8 \times 10^{-7}\text{cm} - 1/680.0 \times 10^{-7}\text{ cm}) = (15{,}802 - 14{,}706)\text{ cm}^{-1} = 1096\text{ cm}^{-1}.$$

As explained, the anti-Stokes line will be of identical wavenumber, but the line will appear in the measured spectrum at $1/(14706 + 1096) = 500$ nm.

For normal Raman vibrational spectra to be observed, two general energy conditions are necessary. First, an incident photon must have an energy substantially larger than that required to excite a vibration. Second, the incident photon must have less energy than is needed to produce electronic excitation. Both conditions are satisfied by wavelengths in the visible region.

Raman scattering is described more precisely as *second-order interaction* between the electric vector of the incoming radiation and the polarizability α of the molecule when its polarizability changes during a molecular transition.* An oscillating dipole is created by interaction. It has a moment μ defined by the expression

$$\mu = \alpha E = \alpha E_0 \cos \omega_i t, \tag{18.1}$$

where E_0 is the amplitude factor of the electrical vector of the wave and t is the time. For a molecule that is spherically symmetrical, or for an isotropic medium, α is a constant, that is, it does not vary with direction. More commonly, however, molecules are asymmetrical and their polarizability must be represented by a tensor.† The oscillating polarization (moment) is responsible for reradiation.

How should one describe alterations in polarizability that lead to Raman scattering, for example, during characteristic vibrations of frequency ω_v? It is useful to suggest how a vibration affects polarizability by expressing α in a series expansion as a function of one normal vibrational coordinate X:

$$\alpha = \alpha_0 + (\partial\alpha/\partial X)\,X + \cdots, \tag{18.2}$$

*Recall that an interaction which leads to *absorption* of energy is possible only if a dipole is created during the interaction. Thus IR, VIS, and UV absorption spectrometry are possible when molecular vibrations, rotations, or electronic changes yield an asymmetric charge distribution. Such interactions are sometimes called first-order interactions.

†The magnitudes of the nine tensor coefficients are usually expressed as a polarizability ellipsoid.

where α_0 is the polarizability at the equilibrium position of atoms in the molecule and $\partial \alpha / \partial X$ represents the change in polarizability for small vibrational amplitudes along the selected normal coordinate. Substituting this expression into Eq. (18.1) gives as a first term an oscillating dipole moment $\alpha_0 E_0 \cos \omega_i t$ that is responsible for *Rayleigh scattering*: it is independent of the vibration.

It is the second term in Eq. (18.2) that is connected with Raman scattering. For a harmonic vibration of frequency ω_v, X takes the form $X_0 \cos \omega_v t$, where X_0 is the equilibrium value of the normal coordinate. Substitution of this expression into Eq. (18.1) yields the equation of a complex oscillating dipole moment:

$$\mu' = (\partial \alpha / \partial X) X_0 E_0 \cos \omega_i t \cos \omega_v t. \tag{18.3}$$

Let us focus on the cosine product in the expression. By use of trigonometric identities this product can be reduced to cosines of sum and difference frequencies $\cos(\omega_i + \omega_v)t$ and $\cos(\omega_i - \omega_v)t$. These are precisely the frequencies that are the anti-Stokes and Stokes Raman lines, respectively.

In addition to a polarizability change, the Raman selection rule for a vibrational change in $\Delta v = \pm 1$. Actually, the high electric field imposed by lasers will also allow nonlinear changes in polarizability that are of third and higher order, as will be examined below.

How should rotation affect α? It will also vary during this motion. Twice per rotation α will have the same value, and Eq. (18.3) will apply if $2\omega_R$ is substituted for ω_v. The selection rule for rotational Raman changes is $\Delta J = 0, \pm 2$.

It is of interest to note that *stimulated* Raman scattering is also possible. If a Raman-active species is placed within the optical cavity of a pulsed laser, lasing will occur at the one or two most intense Stokes and anti-Stokes frequencies as well as at the characteristic laser frequency.

Comparison of Raman and IR Spectrometry. Since Raman and IR spectra arise from the same molecular transitions, a comparison is of interest. For the most part, frequencies found in a Raman spectrum are also those observed in an IR spectrum. Yet as mentioned earlier, Raman spectroscopy complements IR spectroscopy. One way it does is to provide information about vibrational transitions missing in IR spectra. Two examples will illustrate this point.

Example 18.2. The degree of symmetry of a molecular species bears in a central fashion on the frequencies that can be excited in Raman scattering. What frequencies ought to appear in the Raman spectrum of a molecule that has no symmetry elements or has several symmetry elements? Further, what lines will appear only in a Raman spectrum?

If a molecule lacks any symmetry element, each normal vibration will alter its polarizability and give a Raman line. All vibrations are said to be *Raman-active*. In a symmetrical molecule, however, not all vibrations will alter polarizability. Any that do not fail to appear in a Raman spectrum and are termed *Raman-inactive*. Lines that appear *only* in the Raman spectrum of a molecule result from normal vibrations that do not develop an electric dipole but are accompanied by a change in polarizability. Given the great number of combinations of dipole moment and polarizability for different vibrations in most substances, full information about their molecular vibrations and structure requires examination of both Raman and IR spectra.

Example 18.3 Carbon dioxide O=C=O is a familiar nonpolar linear molecule with a center of symmetry. How many of its four normal modes of vibration are Raman-active?

Its symmetrical stretching vibration, designated ν_1 in which oxygen atoms move away from the carbon in phase, a "breathing" mode, is not excited in the IR. No transition dipole moment exists. This mode is Raman-active, however, since the polarizability changes.

The other normal vibrational modes identified as ν_{2a}, ν_{2b}, and ν_3,

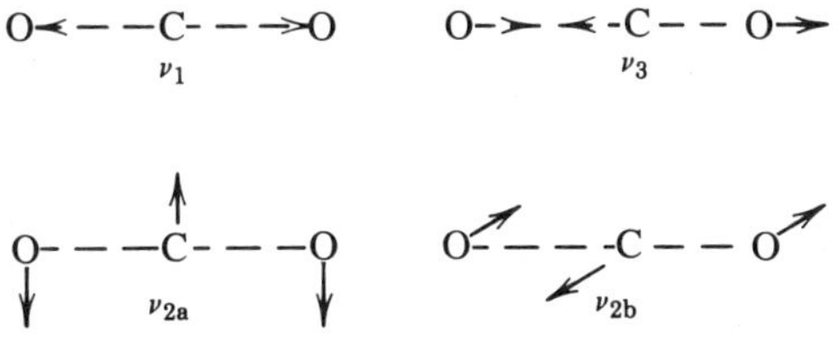

however, clearly create asymmetry in the molecule and are IR-active. Yet *all* prove Raman-inactive. In the ν_3 vibration it appears that the increased polarizability resulting from stretching one bond is offset by the loss on shortening the other bond. In the bending modes ν_{2a} and ν_{2b} there are similar offsetting changes.

Actually, Raman spectra often yield more information than IR spectra. Since absorption and fluorescence spectra arise from a transition dipole, they allow at most a calculation of the three orthogonal dipole components of a molecule. By contrast, Raman spectra ideally yield the nine components of the molecular polarizability tensor and thus offer the possibility of greater molecular characterization.

A second area of complementarity relates potentially to analytical applications. For organic substances Raman spectra tend to characterize the skeletal or hydrocarbon portion. By contrast, IR absorption provides most information about the polar, functional groups in a molecule. Both kinds of information are essential in most analyses (see Section 18.4).

Resonance Raman. With tunable lasers it became possible to develop a resonance form of Raman spectrometry. In this method the wavelength of scattering is slowly increased toward or into the UV until some Raman lines of a species are suddenly enhanced. Enhancement generally occurs at wavelengths corresponding to electronic transitions that also involve excitation of a vibrational mode of the chromophore, that is, the part of a molecule undergoing the electronic excitation, which is coupled with a similar vibration in the excited state. Enhancement may be as great as 10^6 or more.* Equally important, increased selectivity becomes

*At slightly longer wavelengths a somewhat smaller enhancement, called a preresonance effect, often occurs.

possible since only certain vibrational modes are enhanced. Since a great deal of energy may be absorbed in this technique, care must be taken to avoid damage to the sample.

With the coherence, high intensity, and monochromaticity of laser radiation it becomes possible to develop new possibilities in Raman spectrometry. Three new techniques will be briefly described.

Coherent Anti-Stokes Raman (CARS). By simultaneous irradiation of a sample by a laser of fixed-frequency ω_p *and* a second, tunable laser of frequency ω_s, it is possible to observe anti-Stokes spectra attributable to *third-order* Raman scattering. The nature of the complex transition involved is shown in Fig. 18.2a. In terms of the three cosine terms in an equation analogous to Eq. (18.3), the anti-Stokes frequency of scattering will equal $2\omega_p - \omega_s$, as shown in the figure. For a particular analyte, appropriate tuning of the laser of frequency ω_s should yield stimulated anti-Stokes Raman scattering at each characteristic frequency. Note that the intense laser emission also guarantees a large population in the vibrational state $\omega = 1$ of the ground electronic state. The continued irradiation by frequency ω_p will promote large numbers of the excited vibrational species to the virtual state from which anti-Stokes scattering can occur.

From a quantum mechanical point of view, Raman scattering involves the mxing of states. Stimulated anti-Stokes scattering can be described as four-wave mixing of ω_p, ω_s, ω_p, and ω_v. Clearly, while stimulated Stokes scattering might also be observed, it would be simply an addition to the output of the dye laser. In Table 18.1 a comparison is made of

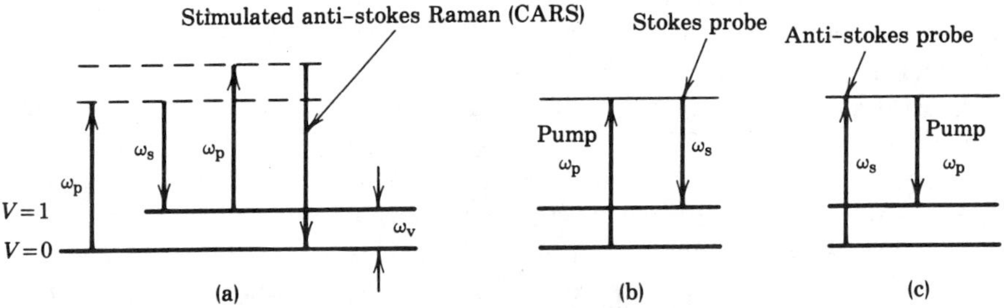

Fig. 18.2 Energy relationships for some coherent types of Raman spectroscopy. In each technique the output of a pair of pulsed lasers furnishing angular frequencies ω_s and ω_p is directed collinearly and simultaneously on the sample. One laser is of fixed frequency and the other is a dye laser or other tunable laser. (a) Energy levels and laser frequencies for stimulated coherent anti-Stokes Raman spectrometry (CARS). As ω_s is varied by tuning, the algebraic sum $2\omega_p - \omega_s$ yields anti-Stokes Raman lines. (b) In *Raman-gain* spectroscopy stimulated Stokes scattering is observed. It is detected as an *increase in the intensity* of the probe laser, which in this case is a lower-power tunable dye laser, whenever $\omega_p - \omega_s$ equals a characteristic molecular frequency. Photons coherently transferred from pump to probe beam are detected by modulating the pump beam and monitoring the probe output at the modulation frequency. (c) In *inverse Raman* or *Raman loss* spectroscopy stimulated Stokes scattering is observed. The fixed-frequency pump is of lower frequency and a tunable dye laser operates as the higher-frequency probe. The recurrence of Raman scattering whenever $\omega_p - \omega_s$ equals a characteristic molecular frequency is monitored as an *intensity loss*. In this case photons lost from the probe beam to generate Raman Stokes scattering are detected by modulation of the pump beam and detection of the probe output at the modulation frequency.

Table 18.1 Comparison of Intensity of Spectral Peaks in Normal and Stimulated Anti-Stokes Raman Spectroscopy[a]

	Spectral Power	
Experimental variable	*Normal Raman*	*(Stimulated Coherent Anti-Stokes Raman (CARS)*
Number density (cm^{-3})	N	N^2
Raman cross section, ($cm^2\ sr^{-1}$)	$(d\sigma/d\Omega)$	$(d\sigma/d\Omega)^2$
Power of laser beam	P	$P(\omega_p)^2$
Line width (cm^{-1})	Γ	Γ^2
Excitation frequency (s^{-1})	ω^4	ω^4

[a]After R. F. Begley, A. B. Harvey, R. L. Beyer, and B. S. Hudson, *Am. Lab.*, **6**(11), 11 (1974).

the dependence of peak intensity on main variables for normal and CARS spectroscopy. By virtue of the quadratic dependence on both the number density of the active species and the power of the exciting laser beam, CARS peaks are about 10^5 more intense than are normal Raman peaks. With high-peak-power lasers (15–50 kW) a 1% conversion is achievable.

Coherent anti-Stokes Raman has several marked advantages: (a) high efficiency; (b) a spectrum displaced toward higher frequencies that largely avoids the region of luminescence; and (c) spectral lines that are also separable by correlation procedures from fluorescence emission, which is not coherent with the exciting radiation. In addition, minimal resolution is needed in the monochromator in some instances since line widths are established by the lasing process.

The method is particularly well suited to examination of gaseous systems since its intensity is as much as 10^6 times greater than that for spontaneous Raman. In addition, neither the nonresonant part of the third-order polarizability nor optical background affects the spectra. CARS has been widely applied in the study of flames and combustion processes.

The disadvantages of the method are several. Efficiency falls as analyte concentration drops, giving ultimately a linear dependence on concentration. Because of the dependence of signal strength on the square of both the Raman cross section and number of scattering molecules, it is difficult to correct for perturbations caused by molecular intractions. Raman scattering by the solvent also unfortunately tends to dominate at low concentrations. Finally, considerable cost is involved in the sophisticated instrumentation and many dyes are needed if an entire vibrational spectrum is to be observed.

Other Types of Raman Spectrometry. Another type of stimulated Raman scattering is called *Raman gain spectroscopy.* Two lasers are used, one furnishing a "pump beam" of intensity I_p and frequency ω_p and the other a "signal beam" of intensity I_s and smaller frequency ω_s. One of the lasers must be tunable and they must be precisely collinearly directed toward the sample. As one laser is tuned, whenever the difference $\omega_p - \omega_s$ equals a Raman frequency, the phenomenon of Raman gain occurs: each photon lost from the pump beam will give rise coherently to a scattered photon of the Raman frequency. The Raman gain process is illustrated in Fig. 18.2b. Expressed differently, in this technique the stimulating laser causes a constant transfer of photons to the scattered beam. The pump beam is modulated and the measurement itself consists of monitoring the intensity of Raman scattering at the modulation frequency.

By interchanging the two beams so that excitation is by the probe beam, *inverse Raman* or *Raman loss* spectrometry can be carried out (see Fig. 18.2c). In this case it is the loss of a photon from the exciting beam that is coherent with the emission of a scattered photon.

In both methods intensities are dependent on a single frequency and vary linearly with analyte concentration. Nevertheless, the signal is coherent because photons in the monitored beam are at the frequency of the probe laser and are in phase with the stimulating beam. References should be consulted for further information about these and still other Raman methods.

18.3 RAMAN SPECTROMETERS

The modules of a basic Raman spectrometer are displayed in Fig. 18.3. While the pattern is familiar, the special conditions encountered in Raman measurements impose severe requirements on modules:

1. Spectral peaks in normal Raman spectrometry are commonly 10^{-5} to 10^{-7} times weaker than the incident radiation.
2. Some Raman lines are closely adjacent to the exciting wavelength.
3. Raman peaks must also be observed against a background of stray light originating from intense Rayleigh scattering of the exciting radiation.
4. Often Raman peaks must also be separated from the fluorescence of analytes or sample impurities.
5. Today there are a variety of useful Raman methods, each with different requirements.

Given these conditions, the number of Raman methods, and the variety of experimental conditions encountered, it is common for Raman spectrometers to be assembled from appropriate modules. Let us look at useful modules.

Sources and Cells. To give sufficient analytical signal it will be essential to employ a high brightness monochromatic source. Continuous-wave (cw) lasers as well as pulsed lasers are in wide use (see Section 8.5); very occasionally a mercury arc-filter source is employed. Dye lasers will provide needed tuning capability though additional frequency doubling optics will be required for work in the UV.

Two widely used methods of sample irradiation are shown in Fig. 18.4. In the

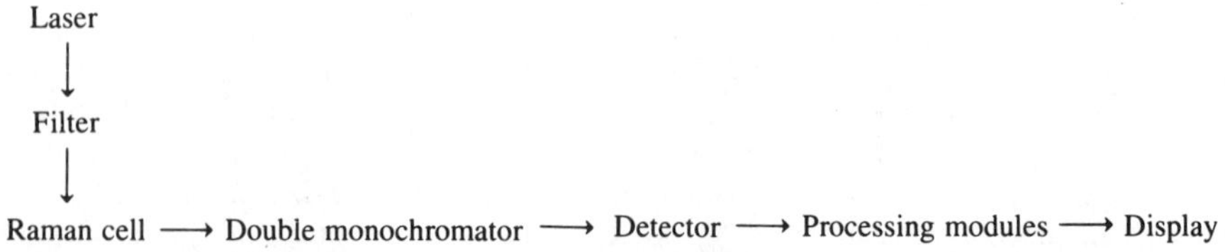

Fig. 18.3 Modular diagram of a Raman spectrometer. Raman scattering is detected at 90° relative to the direction of irradiation to lessen stray light problems. A filter is chosen that passes the incident frequency and excludes frequencies such as the nonlasing lines of the laser.

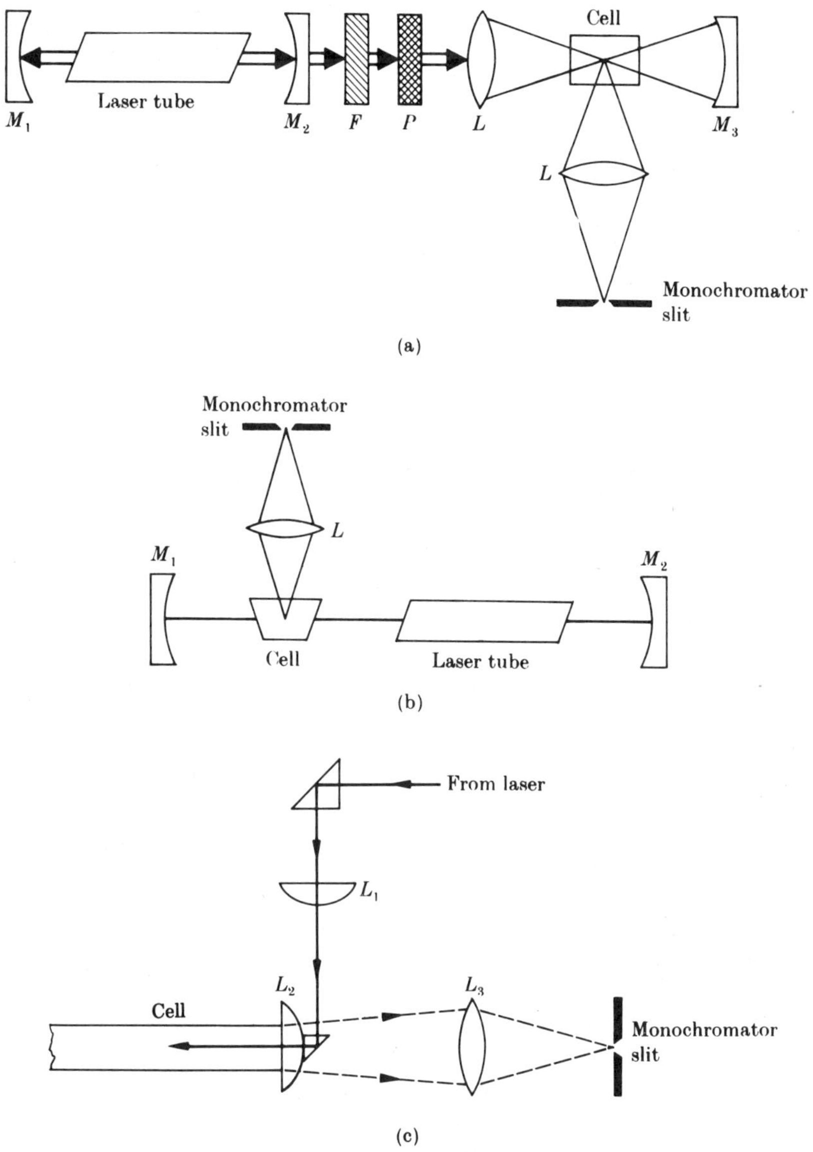

Fig. 18.4 Methods of irradiating a Raman cell by use of a laser. In (a) the cell is external to the laser cavity, which extends from resonating mirror M_1 to M_2, which is partially transmitting. F is a filter to exclude nonlasing frequencies, P is a half-wave plate to rotate the plane of polarization, if desired. In (b) the Raman cell is within the laser resonant cavity to take advantage of the much higher intensity inside. Some of the scattered light, which contains the Raman frequencies, is collected at 90° to the incident light by lens L and focused on the slit of a monochromator (not shown). Components F and P may also be used in (b). In (c) a tubular microcell (about 0.7 mm i.d.) is shown. Scattered light collected by hemispherical lens L_3 is directed toward the monochromator slit. Thus, it is viewed at an angle 180° to the irradiating beam. Note the two total internal reflection right-angle prisms used to bend the laser beam with minimum loss.

first, the effective source intensity is enhanced by placing a reflector behind the Raman cell and in the second by placing the cell within the laser cavity. For gaseous systems, multiple-pass cells are common, and for minute samples microcells find wide use. Alternatively, in a laser Raman microprobe, the optics of a microscope are used to focus a laser beam to a spot of 1–2-μm diameter, as will be discussed below.

Monochromator. To resolve Raman lines and minimize stray light for samples with moderate analyte concentrations, at minimum a monochromator with a holographic grating is needed. In a general purpose spectrometer, however, it will be preferable to use a *double monochromator* (with holographic gratings). Its efficient exclusion of stray radiation will permit even weak Raman lines close to an exciting wavelength to be detected.

If high resolution is sought, a double monochromator coupled to a polychromator is attractive. Recall that the latter module is both a monochromator and array detector. The double monochromator can be designed so that the dispersion in the second unit reverses that in the first, a process called subtractive dispersion. Such a unit provides high throughput and high rejection of stray radiation. The polychromator slits can be set for the high resolution with appreciation that the multiplex advantage of its multichannel detector will ensure good sensitivity even for low-intensity Raman lines.

Attention will need to be given to the degree of polarization that monochromators introduce (see Section 9.6) if polarization effects are to be studied. Polarization–intensity corrections may be applied by computer program using data in memory for the monochromator, or manually. Otherwise a weak Raman peak may go undetected. The correction is less if the instrument beam is depolarized prior to entrance into the monochromator. In the diagrams of Fig. 18.4, a crystalline quartz polarizing scrambler wedge might be inserted for this purpose.

Detector. Commonly the detector is a photomultiplier tube (preferably one with a Ga–As photocathode, which ensures a nearly flat response through the visible range), a diode array, or an intensified diode array. The detection mode is usually that of photon counting. With modern photon counters a wide dynamic range (several orders of magnitude) can be registered.

Types of Spectrometers. Several designs of Raman spectrometer are described in terms of their modules in Table 18.2. Manufacturers provide a range of modules. The simplest spectrometers are those suitable for normal or spontaneous Raman measurements. Modifications to form a resonance Raman spectrometer are straightforward.

To secure greater incident power, pulsed lasers are often incorporated. To avoid sample damage some way must be provided to ensure that a laser beam does not fall too long on one area of a sample. Use of sample cells capable of being spun proves especially effective.

Table 18.2 Some Choices of Characteristic Modules for Different Types of Raman Spectrometers

	Technique			
Module Designation	*Normal Raman*	*Resonance Raman*	*Time-Resolved Raman*	*Stimulated Coherent Anti-Stokes Raman (CARS)*
Source	Continuous-wave (cw) laser; usually with output in VIS	Tunable dye laser (cw or pulsed, VIS or UV)	Synchronous-pumped dye laser with frequency doubler, cavity-dumping device	Pulsed laser *and* tunable dye laser
Sample cell	Glass cell			
Wavelength isolation module	Double monochromator; holographic grating monochromator	Double monochromator to minimize stray light followed by polychromator with diode array, or SIT vidicon detector in photon-counting mode	Double monochromator to minimize stray light followed by polychromator with diode array, or SIT vidicon detector in photon-counting mode	Double monochromator to minimize stray light followed by polychromator with diode array, or SIT vidicon detector in photon-counting mode
Detector	Photomultiplier tube with constant response across VIS, or diode array used in photon-counting mode			
Double channel or ratioing provision	Use split cell; spin and use optical encoder to label detector output for analytical signal processing			Use ratiometer to minimize effect of fluctuation in dye laser output by relating Raman wavelength to laser output or reference channel output

Raman Microprobe. Any type of spectrometer can be coupled to a microscope to permit both microscopic examination of a specimen and determination of the Raman spectrum of selected regions as small as 1–2 μm in diameter. Such a device permits identification of the molecules and complex ions that comprise the sample.

Other Spectrometers. For a time-resolved spectrometer, not only mode-locked, tunable, pulsed lasers, but sophisticated gating of the detector will be required. Fast photomultipliers must also be used.

For a spectrometer capable of employing coherent Raman techniques such as CARS, two different lasers will be required as well as special control arrangements. Good collinearity of the exciting laser beams is essential as they pass through a sample. In techniques such as Raman gain spectrometry, since the intensity of a probe laser is monitored, it is essential that it be a low-noise type to ensure precision. In such methods, the output of the pump laser is modulated by an electrooptical modulator. Then changes in probe laser intensity attributable to Raman lasing can be detected by use of a lock-in amplifier detector keyed to the modulating frequency.

Some designs of spectrometer can successfully reject luminescence and minimize the troublesome overlap of fluorescence with Raman spectra. For example, a time-resolved spectrometer will reject most fluorescence if its detector is gated to respond only during the 10 ns required to receive the laser pulse and detect the Raman scattering. Another approach is to irradiate at 210–230 mn in the UV.* In this case, since fluorescence wavelengths are established by molecular parameters and usually appear above 300 nm, a region at least 70 nm wide will be open for Raman lines.

Alternatively, a separation of luminescence can often be effected by use of software. Note that the overall signal is in the time domain. Further, this signal will generally be a linear combination of the detector responses to light of all types including stray light and fluorescence. The latter contributions are very broad, that is, of very low frequency, while Raman lines are narrow. By Fourier transforming the total signal into the frequency domain and blocking the lower frequencies, the background can be minimized. On transforming the signal back to the time domain, stray light and fluorescence will have largely been removed.†

18.4 ANALYTICAL MEASUREMENTS

The advantages of laser Raman spectrometric measurements are several:

1. Studies of many vibrational motions of molecules can be made within the visible range *and* in aqueous solution. For example, glass cells can be used for solutions and mulling procedures are not required for solids. It is also unnecessary to change gratings and other spectrometer components to cover the entire fundamental vibrational region.
2. Analyte selectivity is good since most species give quite narrow Raman peaks.

*S. A. Asher, C. R. Johnson, and J. Martaugh, *Rev. Sci. Instr.*, **54,** 1657 (1983).

†G. Horlick, *Anal. Chem.*, **45,** 319 (1973).

3. Intensities of spontaneous Raman lines increase linearly with concentration.
4. Depolarization studies (measurement of the intensity ratio $I_{\parallel}/I_{\perp}$) are especially easily made. (A laser gives 99.9% linearly polarized incident radiation.)
5. Wavelengths are easily established. (There are many excellent emission standards in the visible and UV.)

On the other hand, there are certain relative disadvantages.

1. Laser excitation may induce fluorescence, in an analyte or in impurities, that will tend to mask weak Raman lines.
2. Precision instrumentation is required to isolate Raman lines because of their low intensity relative to the incident radiation. Further, stimulated or coherent Raman methods call for instrumentation of high sophistication and cost.
3. Spectra spread over many micrometers in the IR are compressed into a small region of the visible spectrum.

A range of 20–50 nm on either side of the exciting wavelength is sufficient to cover most Raman transitions. Figure 18.5 shows the Raman spectrum of *p*-diox-

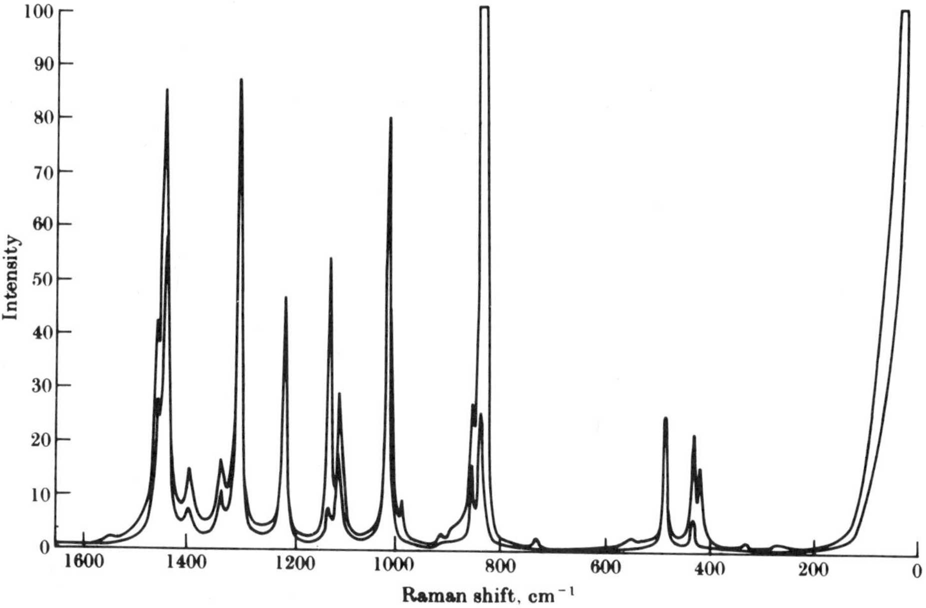

Fig. 18.5 A portion of the Raman spectrum of *p*-dioxane obtained with a Cary Model 82 Raman Spectrophotometer. The sample was run neat in a multipass cell with excitation at 514.5 nm, a 3-cm^{-1} spectral slit width, and a scan rate of 3 cm^{-1} s^{-1}. The upper curve represents the intensity of the Raman emission component whose electric vector is vibrating parallel to the cell axis, and the lower curve that whose electric vector is vibrating perpendicular to it. The depolarization ratio of a Raman line is then $I_{\parallel}/I_{\perp}$. (Courtesy of Varian Instruments.)

ane, obtained with an instrument of high resolution. Several applications are described at the end of this section.

For some of the reasons just cited, Raman spectroscopy is used only occasionally in qualitative analysis. Nevertheless, the advent of the Raman microprobe and the growth of library collections of Raman spectra and tables of characteristic group frequencies have considerably increased its use in identification. In comparison to IR spectrometry, Raman is especially valuable for identifying symmetrical bonds, such as C = C bonds, and essentially nonpolar bonds, such as carbon–sulfur links in molecules.

How should an excitation wavelength be selected? For a particular sample and type of spectrometry, as short a wavelength as feasible (just longer than that at which photodecomposition begins) is desirable. Like other phenomena originating in scattering, Raman lines increase in intensity with the fourth power of the exciting frequency. For normal Raman spectrometry, the wavelength also should be longer than those absorbed for electronic transitions. Preferably, it should not be one that gives rise to luminescence or energy will be lost, and the resulting luminescence may well obscure Raman lines unless measures described earlier are used to avoid this interference.

Quantitative Analysis. In most quantitative studies, it is necessary to reference the intensity of Raman bands to those of an added internal standard. When a rotating cylindrical split cell is used, one half may contain a reference, removing the need for an internal standard. In general, calibration curves are developed with standards, or standard addition techniques are used.

Spontaneous Raman scattering is directly proportional to analyte concentration. The power of resulting Raman lines is given by the expressions

$$P = cP_0\,\nu_s^4\,\alpha^2, \qquad \text{or} \qquad P = N_1(d\sigma/d\Omega)\,l\,P_0\,d\Omega. \tag{18.4}$$

In the first expression c is a constant, P_0 is the power in the incident beam, and ν_s is the laser frequency. In the second expression N_1 is the analyte number density, which is basically a concentration, $d\sigma/d\Omega$ is the differential Raman scattering cross section (usually 10^{-30}–10^{-31} cm^2 sr^{-1}), l is the sample path length along which radiation is being collected, and $d\Omega$ is the solid angle of collection. The second expression holds for a given frequency. The direct dependence on N_1 ensures for Raman methods a large dynamic range and makes clear the linear dependence of response on analyte concentration.

For trace-level measurements, resonance Raman is of course especially attractive. With the careful tuning possible with a dye laser, sensitivity and selectivity are high. The consequent signal enhancement lessens the importance of stray light. It should be noted from the first expression in Eq. (18.4) that the greater intensity from resonance Raman must be associated with enhanced polarizability.

Sample Preparation. It is fortunate that in Raman spectrometry sample preparation is comparatively simple. Aqueous solutions are useful since the Raman scattering of water is so weak that it produces minimal background.* Filtration of sample solutions through a microporous filter to lessen particulate scattering is a prerequisite to all subsequent steps. For biological samples, luminescence may also drastically lower the Raman S/N. Offending species—if not the analyte itself—may sometimes be selectively adsorbed on activated charcoal or selectively photolyzed. With solids, little or no sample handling is required. For example, plastics can be examined directly in the form of sheets and pipes.

Example 18.4. For inorganic substances a Raman microprobe offers a ready means of identifying species by reflectance Raman spectrometry. For instance, the nature of the changes undergone by lead oxides during the discharge–recharge cycles of a typical lead–sulfuric acid storage battery can be discerned by examining oxide granules on a plate under the microprobe. Spectra are substantially different for litharge (PbO), massicot (amorphous Pb_2O_3), and other oxides. Knowing the compounds present can lead to battery improvement.

Example 18.5. Structural changes in cell membranes can be studied by Raman spectroscopy. The typical membrane is found to be a lipid bilayer in which proteins are immersed. It is known that the absorption of heat produces a "melting" of the hydrocarbon chains and that as the membrane goes through such a phase transition, chains move from mainly a trans conformation to one in which many are in the gauche form. Membrane transport properties are greatly affected. Raman spectroscopy can give information about the nature of the melting process at the molecular level because of its high sensitivity to changes in the C–H and C–C stretching frequencies [4].

Example 18.6. Raman scattering techniques lend themselves to the study of fast reactions. To investigate such reactions in solution, the medium is allowed to flow rapidly as a free jet (modules that form jets 1 cm broad and 0.1 mm thick are available) under a pulsed laser beam that has been focused down to about a 1-μm spot. If a synchronously pumped, mode-locked Ar^+ ion laser is used, it will release about 5 nJ of energy into the sample in each 30-ps pulse, and the pulse repetition rate will be high ($\sim$ 0.8 MHz). In measurements, Raman scattered radiation is collected by a fast multichannel detector. Further, the response is integrated over many pulses to lower the noise level.

Systems studied by this procedure have been the mechanism of photodissociation of CO from carbon monoxyhemoglobin and changes in the conformational structures of oxyhemoglobin and related species. [J. Terner et al., *Proc. Natl. Acad. Sci. USA*, **78,** 1313 (1981); also R. E. Hester, *The Spex Speaker*, **27**(2), August 1982.]

Example 18.7. The characterization of many aspects and uses of polymers has been greatly benefited by Raman spectrometry. Not only is little or no sample handling required, but measurements can be made even on composite polymers that contain "fillers," such as clay. Such fillers are usually poor Raman scatterers but strongly absorb IR radiation, making that technique unusable without prior removal of the filler.

*By contrast, in IR spectroscopy, water absorbs over a long spectral range.

Raman spectroscopy permits the identification of polymer functional groups, end groups, structure, orientation of chains, and conformation. Importantly, it is also possible to study changes that occur in polymers as they are exposed to mechanical stress and different environments such as bright sunshine. Over time these vibrational Raman spectroscopic studies seem likely to lead to a better understanding of the relationship between polymer design and specific physical properties that are of interest.

REFERENCES

1. S. A. Asher, C. R. Johnson, and J. Murtaugh, Development of a new UV resonance Raman spectrometer for the 217–400 nm spectral region, *Rev. Sci. Instr.*, **54,** 1657 (1983).
2. F. R. Dollish, W. G. Fateley, and F. F. Bentley, *Characteristic Raman Frequencies of Organic Compounds.* New York: Wiley-Interscience, 1971.
3. S. K. Freeman, *Applications of Laser Raman Spectroscopy.* New York: Wiley, 1974.
4. J. G. Grasselli, M. K. Snavely, and B. J. Bulkin, *Chemical Applications of Raman Spectroscopy.* New York: Wiley-Interscience, 1981.
5. A. B. Harvey (Ed.), *Chemical Applications for Nonlinear Raman Spectroscopy.* New York: Academic, 1981.
6. K. Nakamoto, *Infrared and Raman Spectra of Inorganic and Coordination Compounds*, 3rd ed. New York: Wiley, 1978.
7. M. D. Levenson, *Introduction to Nonlinear Spectroscopy.* New York: Academic Press, 1982.
8. D. A. Long, *Raman Spectroscopy.* New York: McGraw-Hill, 1977.
9. B. P. Straughan and S. Walker (Eds.), *Spectroscopy*, 2nd ed., Vol. 2. New York: Wiley, 1976.
10. D. P. Strommen and K. Nakamoto, *Laboratory Raman Spectroscopy.* New York: Wiley-Interscience, 1984.

Applications to biology are thoughtfully presented in:

11. P. R. Carey, *Biochemical Applications of Raman and Resonance Raman Spectroscopy.* New York: Academic, 1982.
12. A. T. Tu, *Raman Spectroscopy in Biology.* New York: Wiley, 1982.

EXERCISES

18.1 Should the nitrogen molecule absorb in the IR? Should it be Raman-active? Why?

18.2 Which of the following vibrational modes in the molecules given will be IR-active, Raman-active, or both: (a) symmetric stretch of O_2, (b) asymmetric stretch of CO_2, (c) bending mode in H_2O, (d) twisting mode in C_2H_4?

18.3 The IR spectrum of carbon tetrachloride contains one fundamental absorption peak at 790 cm^{-1}. The Raman spectrum of CCl_4 shows fundamental vibration peaks at wavenumber shifts of 218, 314, 459, and 790 cm^{-1}. Explain the absence of the first three peaks in the IR spectrum.

18.4 In the Raman spectrum of a compound obtained with He–Ne laser excitation (see Table 8.2) lines are observed at 650.0, 661.5, and 670.3 nm. Calculate the wavenumber shift ($\Delta\bar{\nu}$) of each line.

18.5 Should a He–Ne laser or an argon–ion laser be used to obtain a Raman spectrum of a substance that absorbs strongly in the blue-green region of the spectrum?

18.6 Raman spectra are frequently rerun using a different exciting frequency to identify possible spurious spectral lines. The Raman spectrum of one compound excited with the 488.0-nm Ar^+ laser line has peaks at wavenumber shifts of 129, 378, 529, 1056, and 3301 cm^{-1}. The same compound excited with the 514.5-nm Ar^+ laser line has peaks at 129, 378, 529, and 3301 cm^{-1}. Identify the source of the spurious 1056-cm^{-1} peak in the former spectrum.

18.7 Suggest two or three ways in which one might distinguish whether a peak obtained on laser excitation is caused by fluorescence or Raman scattering.

18.8 It is desired to calibrate the wavelength scale of a laser Raman spectrometer. Suggest at least two basically different ways to do so.

18.9 The He–Cd laser that emits at 441.6 nm is also a useful source in Raman spectroscopy. What is the ratio of its scattering power to that of the He–Ne laser characterized in Table 8.2?

18.10 Consider that an argon–ion laser source, double-grating monochromator, and photomultiplier detector are to be employed as elements of a Raman spectrometer. Would it be preferable to use gratings blazed at 500 (first-order) or at 400 nm? Consider the response of monochromator and detector as well as any other relevant factors.

18.11 Draw a block diagram of a laser Raman spectrometer designed to operate with phase-sensitive detection. Draw a second block diagram for one with photon counting detection.

18.12 Why is Raman spectroscopy considered to be superior to IR spectroscopy for the analysis of many organosulfur compounds such as sulfides, disulfides, and thiophenes?

18.13 Rate the relative ability of Raman spectroscopy to handle the following sample types compared with UV–VIS, IR, and NMR methods: (a) gases, (b) neat liquids, (c) aqueous solutions, (d) powders, (e) surfaces.

18.14 What features of Raman spectroscopy make it a better choice than IR spectroscopy for the *in vivo* study of biological materials such as chlorophylls and polypeptides?

ANSWERS

18.1 No; yes. *18.2* (a) Raman-active, (b) IR-active, (c) both, (d) neither. *18.6* It is due to the 514.5-nm plasma emission line of the Ar^+ laser: 1/488.0 nm = 20,492 cm^{-1}, 20,492–1,056 cm^{-1} = 19,436 cm^{-1} = 1/514.5 nm. *18.7* Change the excitation frequency. If the $\nu_{peak} - \nu_x$ difference remains constant, it is caused by Raman scattering. Or use time-resolved methods, which will separate Raman (fast emission) from fluorescence (slow emission). *18.9* 4.2. *18.12* Sulfur is a highly polarizable substituent atom, so it produces a high-intensity Raman signal. Sulfur–carbon bonds do not produce an especially strong dipole, however, so the corresponding IR bands are weak. *18.13* Responses follow the order: Raman, UV–VIS, NMR, IR: (a) difficult, simple, very difficult, simple; (b) very simple, very simple, simple, very simple; (c) very simple, very simple, simple, difficult; (d) very simple, difficult, difficult; (e) simple, difficult, impossible, difficult. *18.14* An aqueous, ionic environment that most commonly provides the solvent media in *in vivo* studies poses no problems for Raman spectroscopy, but defeats IR measurements. Also, the additional information that can be gained from depolarization values can offer a sensitive probe of conformation and higher-order structure in biological materials.

Chapter 19

LIGHT SCATTERING AND REFRACTOMETRY; CHIROPTICAL METHODS

19.1 INTRODUCTION

Measurable interactions of EM radiation and matter occur over the whole range of known frequencies. Those classifiable as emission or absorption we have already found to apply to a wide variety of analytical and measurement situations. Now we turn to three minor optical methods: the essentially nonabsorptive interactions of scattering and refraction and the complex phenomenon of optical activity, which is nonabsorptive over most of the spectrum. Each phenomenon has been discussed from a theoretical but qualitative point of view in Chapter 7 and must now be related to appropriate instrumentation and analytical procedures. While many of these interactions are unquantized and thus nonspecific for a given molecule or atom, the magnitude and geometrical aspect of interaction depends on the physical properties of particular substances as well as the conditions of measurement. In general, these techniques will prove valuable in analysis to the degree that they can be used

1. to confirm identity or provide information favoring just one of alternative geometric or stereoisometric structures of pure substances;
2. to analyze certain mixtures when the constituents are known; or
3. to determine molecular weights and obtain shape and size data for polymers, macromolecules, and colloidal aggregates.

19.2 LIGHT-SCATTERING METHODS

The scattering of light is a sufficiently general phenomenon that it can be observed in gases, liquids, and solids. From a classical perspective the alternating electrical component of radiation sets atomic and molecular electron clouds and nuclei in oscillation. Nearly simultaneously, these atoms and molecules reradiate the light. The phenomenon of Rayleigh scattering is described in Section 7.12 and that background will be assumed familiar in the following discussion. Scattering will *fail* to occur from a region (a) of dimensions about the same as the incident wavelength and (b) momentarily uniform in refractive index.

Since no appreciable change in incident frequency occurs from Rayleigh scattering, that process can be described as elastic. In practice, minute Doppler shifts in frequency ($\Delta\nu/\nu \approx 10^{-12}$) are caused by the modulating effect of internal and translational motions of molecules.

Quasi-Elastic or Dynamic Light Scattering. Quasi-elastic scattering measurements on bulk media, solutions, and suspensions can be carried out by autocorrelation procedures. These methods permit the estimation of particle size in suspensions and the study of Brownian motion in fluids. Included in this category are bulk light-scattering phenomena such as *Brillouin scattering*, which is attributed to the effect of thermally excited sound waves, and Rayleigh scattering, which occurs because of local density changes attributable to random temperature fluctuations [1].

Scattering occurs only if the major dimension of the scattering molecule, colloidal aggregate, or suspended particle is less than $3/2\lambda$. Larger particles will reflect the radiation. For instance, in the UV–VIS spectral region scattering particles will range from 1 nm to 1 μm in greatest dimension. The scattering of suspended particles is often termed the Tyndall effect.

To describe scattering consider that each interaction between radiation and matter occurs at a *scattering center*. The center may be a polymer molecule, a biological macromolecule, a suspended dust particle or a colloidal aggregate whose refractive index is different from that of the bulk medium. It may also be a region in which the density of molecules making up the medium is momentarily different. Each center can be characterized by a complex refractive index ratio n_s/n_m, where s identifies the scattering center and m the medium. Recall that this ratio reduces to the ordinary refractive index ratio if no absorption occurs since the absorptive part of the index is imaginary. A second parameter, α, denotes the size of the center relative to the wavelength of light ($\alpha = 2\pi r/\lambda$), and the last parameter is the angle of observation θ of a scattered ray relative to the incident ray.

The dependence of scattering on these parameters is best described through use of the model in which the transverse electrical vibrations of incident light I_0 are resolved into two mutually perpendicular components. The magnitude of parameter α determines the phase distribution of these two components of scattered radiation. In general, when $\alpha \ll \lambda$, the pattern or envelope of scattering is symmetrical about the axis of the incident radiation. For large centers, however, scattering is substantially in the forward direction (see Fig. 7.27). The intensity of scattering I_s also has a strong wavelength dependence. As long as $\alpha \ll \lambda$ (i.e., for Rayleigh scatters), $I_s \propto \lambda^{-4}$ and shorter wavelengths (blue) are preferentially scattered. Both the scattering pattern and the absolute or relative intensity at given angles are measureable and are the basis for analytical determinations.

Example 19.1 Dust particles and water droplets in the sky are essentially invisible yet are effective scatterers of light. Explain their role in producing the blueness of a clear sky and the whiteness of clouds on a sunny day.

Since dust particles and water droplets usually have small values of α, they scatter blue wavelengths of sunlight especially effectively and establish that color for the daytime sky. As water droplets grow in size, however, they tend to scatter all wavelengths nearly equally.

Thus, masses of water droplets such as clouds are white. They may also be bright since larger particles scatter more radiation overall and do so basically in the forward direction toward an observer.

To measure concentrations of suspended materials, turbidimetric or nephelometric measurements of transmitted or forward-scattered light are commonly used. Optimum conditions of measurement will be described in the next section.

If a measurement of the molecular weight and shape of synthetic polymers or biological macromolecules is sought, one needs to determine their scattering at several concentrations and angles. If there is a range of particle sizes, measurements yield an average molecular weight and approximate values of size and shape. For collections of both small and very large particles, molecular weight information obtained by light scattering supplements that available from the more standard techniques such as osmometry. In the intermediate size region (dimensions from 0.03 to 0.10 μm), light scattering is invaluable, for none of the standard approaches gives good information. Naturally, results will be most meaningful for monodisperse particles in which scatterers are of the same size. Discussion of the theory and instrumentation of light-scattering methods for measurement of molecular parameters is beyond the scope of the book and the reader is referred to volumes such as those by Kerker [5], Chu [3], and Berne and Pecora [1] for study.

19.3 MEASUREMENT OF TURBIDITY AND TURBIDIMETERS

The name *turbidimetry* designates the measurement of concentrations of suspended particulate matter of refractive index different from that of the fluid by transmitted or forward-scattered light. The term *nephelometry* is applied when scattering is observed at an angle to the irradiating beam. Here turbidimetry will be used for all simple scattering measurements and when they are made at angles other than zero, the angle will be stated.

Turbidity. When a beam of radiation of intensity I_0* passes through a nonabsorbing medium that scatters light, the transmitted intensity I is given by the expression

$$I = I_0 e^{-\tau b}, \qquad (19.1)$$

where τ is the turbidity and b the path length in the medium. The quantities τ and I vary with concentration, molecular weight, and so on. When b is in centimeters, τ is in reciprocal centimeters. Frequently Eq. (19.1) is rearranged to define directly

$$\tau = (2.303/b) \log I_0/I \qquad (19.2)$$

Note the resemblance of this equation to Beer's law, Eq. (16.3).

*Strictly, the irradiance, instead of the intensity, defines the turbidity. Irradiance has units joules per centimeter squared per second, and I has units joule per staradian per second, where Ω is the solid angle.

At low concentrations measurement sensitivity is poor if the transmitted intensity (0°) is measured. The problem is simply that small deviations from high-incident intensities must be detected. In practice, to obtain good sensitivity, scattered light is usually observed at angles from 75 to 135° relative to the incident beam so that it can be detected relative to a fairly dark background. Even measurement at large angles becomes unreliable at high particulate concentrations as a result of interparticle interference and direct transmission is again used. As is evident, flexibility in setting the angle is desirable for measurements over a wide concentration range.

When we determine scattered intensities, however, it is ordinarily preferable to work directly with them instead of converting to turbidities. The transformation requires an integration of the intensity over 360°.

Turbidimeter. In a single-channel turbidimeter it is desirable to minimize (a) instrument drift, (b) the effect of multiple-particle scattering at high concentrations, and (c) stray light. It is also desirable to (d) measure scattering independently of absorption and (e) achieve a large dynamic range. It is valuable to use a ratio method to compensate for most drift in the source, detector, and amplifier as well as for gradual changes in the optical path such as accumulation of dust on optics. By minimizing drift one can also ensure that calibration (see below) will be needed only infrequently.

A simplified optical schematic of a versatile turbidimeter is shown in Fig. 19.1. Good baffling and careful placement of the 90° detector minimize both the effect of multiple scattering and stray light. By taking the ratio of 90° scattering to transmitted light one compensates for most drift and also for absorption in colored samples. The silicon photodiode detectors are operated in the photovoltaic mode to furnish a current signal and are followed by current-to-voltage operational amplifier circuits to effect necessary amplification. Finally, adjustment of the amount of the forward-scattered signal added to the transmitted light signal will effect the best linearity and longest dynamic range.

Calibration. The primary standard for turbidimetry is an aqueous suspension of *formazin*, an insoluble polymer formed by condensation of hydrazine sulfate $[N_2H_4 \cdot H_2SO_4]$ and hexamethylenetetramine $[(CH_2)_6N_4]$. Appropriate amounts of these substances are dissolved in ultrapure water and allowed to stand one day at 25°C to develop a standard formazin suspension. Secondary standards consisting of very fine metal oxide particles suspended in a silicon polymer gel are also available. The gel prevents coagulation over time. Indeed, turbidity is now generally reported in turbidity units (TU) based on formazin.

Example 19.2 Perhaps the best known *chemical turbidimetric analysis* involves the precipitation of barium sulfate under conditions that yield a stable monodisperse suspension. Either sulfur or "sulfate" may be determined by this method. How can appropriate control of precipitation be exercised?

The precise duplication of precipitation conditions calls for close control of temperature, amounts of concentrations of samples and reagents, rate of stirring, and length of mixing.

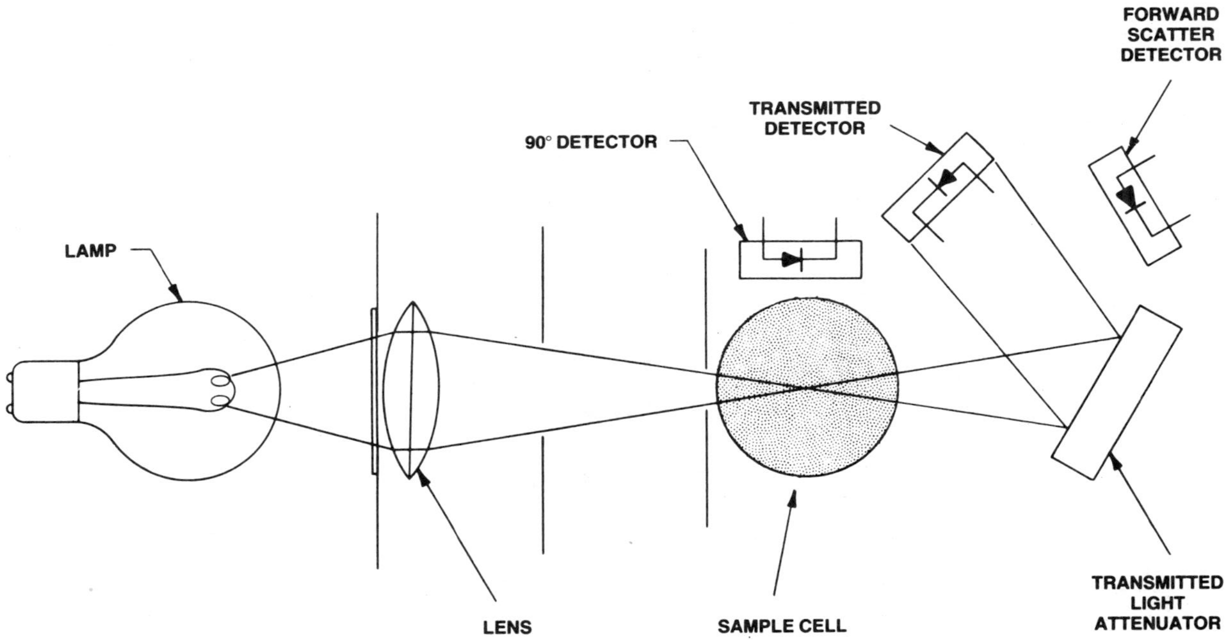

Fig. 19.1 Simplified optical schematic of the Hach ratio turbidimeter. This top view does not show the elaborate baffling that minimizes stray light. The tungsten lamp output peaks at about 1040 nm (color temperature ≈ 2800 K) and the photodiode detector response peaks just over 800 nm. The filter brings the peak instrument response down to about 575 nm. Note the three photodiode detectors. The attenuator, by reflecting no more than 4% of the intense transmitted beam, ensures that the detector of transmitted light does not saturate. The turbidity values reported is the ratio of detector responses $(I_{90} - I_0)/(I_{\text{forward}} - I_{\text{transmitted}})$. (Courtesy of Hach Company.)

Usually a stabilizer, such as gelatin, is added to protect the colloid, that is, to prevent coagulation beyond a certain size. Some sodium benzoate is often used as a preservative.

Example 19.3. The concentration of smokes, fogs, aerosols, and so on, may be ascertained quite easily by light-scattering measurements. Samples are continuously drawn through the scattering cell. In this case calibration curves are obtained with an aerosol of known particle size and physical properties.

Automatic Turbidimeters. A variety of automatic devices have been developed for plant stream monitoring. For example, the ''makeup'' of acceptably pure boiler feed water can be monitored by a recording turbidimeter. In particular it allows estimation of the amount of suspended matter such as silica in liquids.

The idea may also be applied to the analysis of a pair of mutually immiscible liquids if the suspension is reasonably stable. Control of the homogenization quality of milk and other emulsions can be carried out with an automatic instrument by use of calibration curves that include fat concentration. [W. D. Pandolfe and S. F. Masucci, *Am. Lab.*, **16**(8), 40 (1984)].

19.4 REFRACTOMETERS AND ANALYTICAL MEASUREMENTS

Although refraction is a nonspecific property, at a given temperature and wavelength few substances have identical refractive indices. For example, in liquid chromatography, refractive index detectors see considerable use in monitoring the elution of components (see Section 26.3). Measurements can be made simply, rapidly, and nondestructively. Refraction originates in optical polarizability; the greater the polarizability of molecules, atoms, or ions, the higher their refractive index is. Familiarity with the background material on refraction in Sections 7.5–7.7 will be assumed in the discussion that follows.

Refractometers. Instruments based on the *critical angle principle* are in common use. They provide sufficient precision of measurement for most liquid and solid samples, cover a broad range of refractive index, and are easily operated. The critical angle is defined as the angle of refraction in a medium when radiation is incident on the boundary of a medium at grazing or 90° incidence. In this case, Snell's law [Eq. (7.8)] simplifies to

$$n_2/n_1 = 1/\sin \phi_c, \tag{19.3}$$

where ϕ_c is the critical angle at the wavelength used. In making an observation, medium 1 is usually the sample and medium 2 is a prism of known index.

Figure 19.2a shows the most widely used critical-angle type of refractometer, the *Abbe*. Since in this design the critical angle is formed in prism R, that angle itself is immeasurable. What is determinable is the angle of emergence α of the critical ray (Fig. 19.2b). Note that the design limits measurements to substances whose n is less than that of the refracting prism; otherwise total reflection would

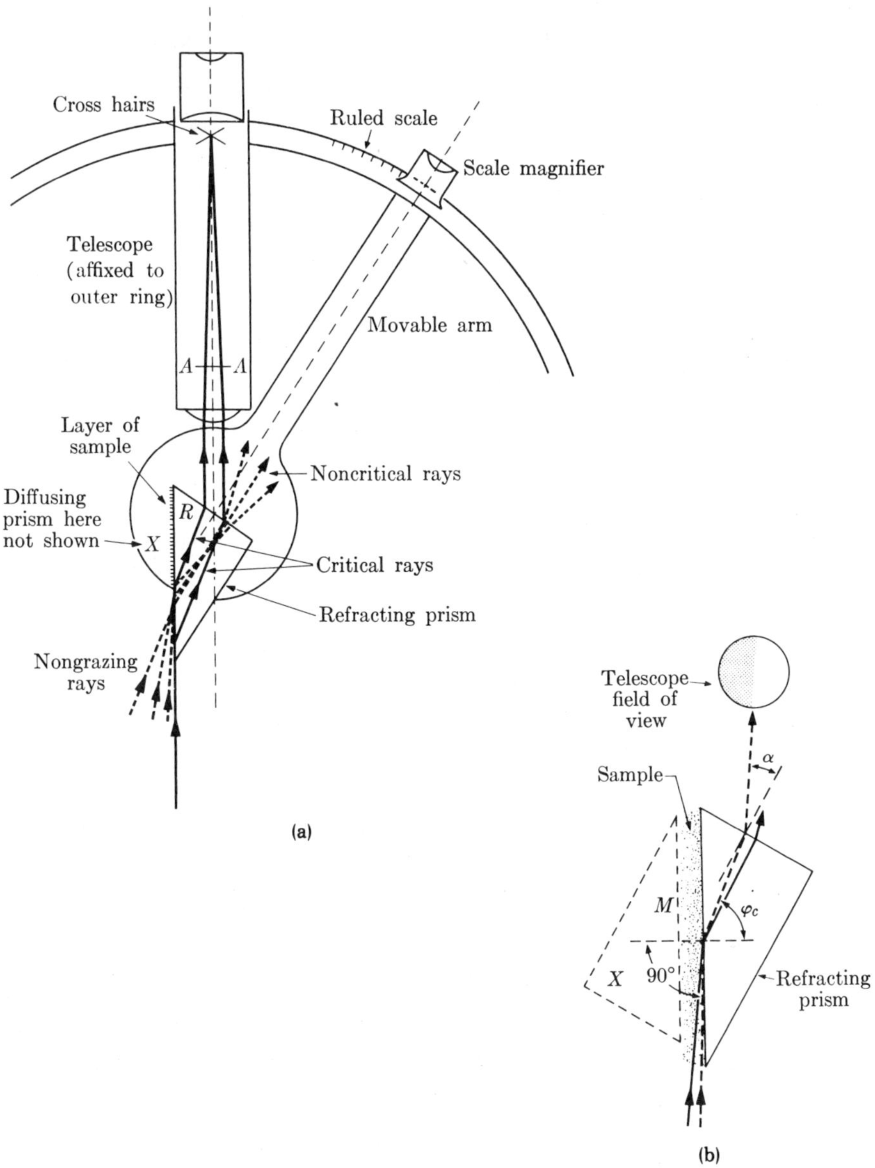

Fig. 19.2 Schematic diagram of an Abbe refractometer. (a) General view. As shown, the Abbe is intended for use with monochromatic light. Usually a sodium lamp, 99% of whose light is the intense sodium doublet is used. The prisms R and X are partially hollowed to allow circulation of thermostated fluid. Light enters from below. The angle α may be found by "scanning" the upper edge of the refracting prism R with the telescope until the sharp edge of the light-dark boundary is centered on the cross hairs in the telescope ocular. (b) Detailed schematic of auxiliary and refracting prisms. Here the critical ray is shown as a broken line. The X prism is used only with liquid samples and is an auxiliary prism with a matte surface M that diffuses light. The thickness of the layer of sample is greatly exaggerated.

result. With the usual prism in an Abbe we may determine indices over the considerable range of 1.30–1.70. Only a few substances have n's outside these limits.

Example 19.4. What sources of error are likely in an Abbe refractometer? How may they be reduced? How may an Abbe be calibrated?

Probably the most serious source of *instrumental error* in the Abbe is the cutting off of near-grazing rays, those incident on the refracting prism at very small angles. Figure 19.2b shows how the standard Abbe refracts only the radiation diffused from the rough or matte surface of prism X into the liquid. Unfortunately, this radiation is poor in grazing rays. As a result, the true critical ray boundary may be difficult to perceive, and instead we may select an apparent boundary somewhere within the shaded region, which is often of the order of 1×10^{-4} in n. A precision Abbe refractomer reduces this error by employing a smooth-surfaced diffusing prism X.

Uncertainty in temperature (constancy to $\pm 0.2°C$ is desirable) will lead to a variation of about 4×10^{-4} units in n per °C for most liquids. Another source of uncertainty occurs when measurements are made with a continuous source like a tungsten lamp instead of a sodium lamp, mercury arc (critical angle boundaries for the separate intense mercury lines can be seen), or other reasonably monochromatic lamp. Because dispersion of the critical ray over a range of angles occurs, the uncertainty in n will be about 1×10^{-3}. With compensation for dispersion, even a tungsten lamp allows a precision of 1×10^{-4} in n.

For *calibration* of a refractometer, common pure materials, after drying and distillation are ideal. Water, chloroform, and toluene (with n_D values 1.33–1.49) will cover the ordinary range of n. Since procedural errors can also influence measurements, liquids ought to be used for calibration if most measurements are to be made on liquids. Preserving the calibration of a refractometer is in general a matter of ensuring that the refracting prism or the scale are not somehow jarred from their original alignment. The glass test piece supplied with many Abbe refractometers is also useful in checking the calibration stability. If needed, it is possible to reposition the telescope using a standard liquid.

A second widely used type of instrument, an *image displacement refractometer* can be constructed from any prism spectroscope or spectrometer (see Fig. 9.1). The angular displacement of the optical image of its rectangular entrance slit is a measure of the n for the prism material at each incident wavelength. If we are to examine a solid, we fashion it into a prism; liquids can be placed into a hollow prism-shaped vessel. An optimum precision and accuracy of $\pm 1 \times 10^{-6}$ in n is possible with a spectrometric refractometer with monochromatic light, careful instrument manipulation, and close temperature control ($\pm 0.002°C$ for liquids). Yet such an instrument is ill-suited for routine work.

A more attractive approach is a *differential refractometer* of the image displacement type that will respond to the *difference* between the n of a sample and that of a reference. Such a design with two hollow prisms set in opposition, one for the sample and the other for the reference, as illustrated in Fig. 26.11a provides that the displacement of a light beam by the sample prism will be substantially offset by that of the reference. Its precision will be about 3×10^{-6} in n. This the type mainly used with liquid chromatographs.

Schlieren Techniques. These refractive index methods properly fall within the category of image displacement refractometry. They are conventionally used not in measuring refractive indices, but in locating concentration gradients by means of their accompanying refractive index gradients. For example, in protein solutions we may obtain a quantitative estimate of concentration by Schlieren scanning of the different layers set up by electrophoresis. Under an electrical field, most of the components of the mixture will migrate in a cell at characteristic rates. As in chromatographic procedures (Chapter 24), each substance tends to occupy a definite zone at the edge of which there is a detectable refractive index gradient.

Automatic Refractometers. Automated refractometers are usually differential image displacement instruments. Many use a split photocell detector or dual phototubes. When there is a change in n, light from the slit shifts principally to one photocell and an unbalance signal appears. This signal is amplified and fed to a servomotor, which acts to restore balance by inserting an optical wedge or prism or by turning the sample prism table. Either operation provides a readout. In other words, a reading is secured by this "nulling" operation. Figure 19.3 gives a schematic drawing of a process-stream differential instrument. The precision of these devices varies from 2×10^{-6} to 1×10^{-4} in n.

Automatic differential refractometers are finding increasing application in monitoring chemical streams such as distillate from an industrial fractionating column or the effluent

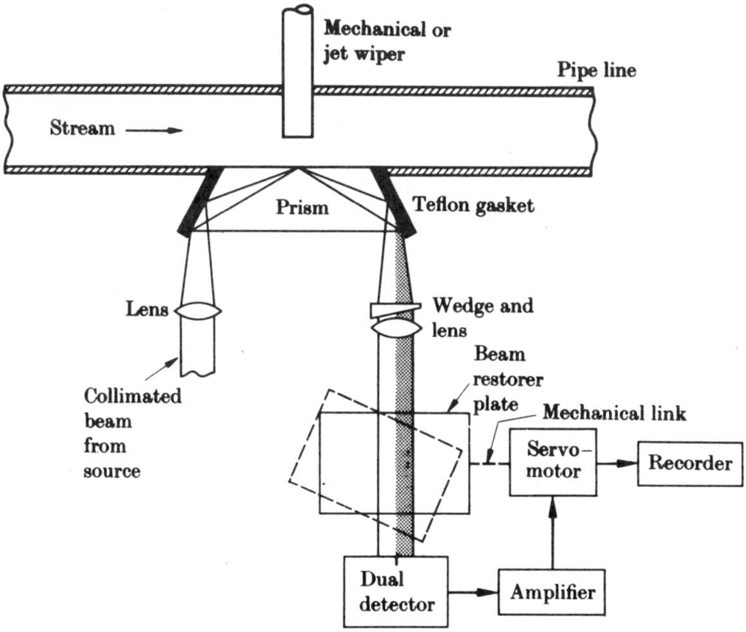

Fig. 19.3 Simplified schematic of an automatic, critical-angle refractometer, the Anacon Inline Refractometer. When there is a change in refractive index of the stream, the beam restorer plate is used to return the beam to its original position. Since the instrument relies on total internal reflection with its minute penetration into a sample, the refractometer can be used to monitor the index of highly opaque liquids or suspensions. (Courtesy of Anacon Inc.)

from an HPLC column. Illustrations of the latter application may be found in Fig. 26.11. In industrial fractionation we may set an instrument to register deviations from the desired value of n. By connecting the refractometer to an appropriate control unit that will alter an operating variable like the take-off rate, we can usually secure constancy of composition of the output.

Interferometer. By utilizing interference techniques, we can extend the reliability of differential refractive index measurements into the seventh decimal place for liquids and the eighth for gases. Each extension in precision further restricts the difference between the n of the standard and that of the sample. For liquids, a precision of $\pm 2 \times 10^{-7}$ in n is realizable, and for gases in a long cell, about $\pm 3 \times 10^{-8}$.

In the typical interferometer, the sample and the standard are placed in matched cells of equal length. Rays from a slit illuminated by white or monochromatic light are collimated, directed through the cells and onto a slit at the end of each cell, and then suitably merged by a lens arrangement. A pattern of interference bands or fringes results. The relative position of the pattern depends on the difference in the two optical path lengths (see Section 7.17). In turn, these path lengths are directly related to the refractive indices of the contained gases or liquids. The eye is the sensing device. To avoid counting bands or fringes, we set up a fixed band pattern below the desired pattern, and a variable glass compensator in the sample path is pivoted until the two patterns are in alignment. The Δn is obtained from the reading of the compensator micrometer screw.

When interferometer measurements are to be made in the IR, it is convenient to count interference fringes in a regular sealed cell (for liquids). The procedure has been discussed in Section 16.6.

Analytical Measurements. The index of refraction is broadly useful for its corroborative value in *qualitative analysis*. For a substance in a reasonably pure state, determining n immediately narrows the possibilities that must be examined to identify the compound. Extensive tables listing compounds according to refractive index will indicate possible formulas. For instance, if a considerable number of unknowns are to be identified by IR spectrophotometry, this approach would reduce spectral searching.

A *quantitative measurement* of the composition of simple, homogeneous liquid or gaseous mixtures can be determined reliably by refractive index measurements as long as the refractive index–composition curve is nearly linear for the analytes. Often that condition is met over small composition ranges.

Example 19.5. The pentane–hexane system gives a linear plot of refractive index against mole fraction with n_D values at 25°C ranging from 1.3530 to 1.3732. What type of refractometer should be used to determine composition in pentane–hexane mixtures?

For this mixture $\Delta n = 0.0202$ is the available range. Composition can be determined to *about 2 mole%* if an Abbe is used since its precision is ± 0.0001. With a differential image displacement refractometer of precision 3×10^{-6}, however, a composition difference as little as $(3 \times 10^{-6}/20200 \times 10^{-6}) \times 100 = 0.015\%$ mole% could be determined under optimum conditions.

19.5 OPTICAL ACTIVITY AND CHIROPTICAL METHODS

In the final portion of this chapter we take up two closely interrelated methods based on the optical activity or chirality of substances, *circular dichroism* (CD) and *optical rotatory dispersion* (ORD). Measuring these properties as a function of wavelength provides stereochemical information about substances that is unusual in its precision and specificity, especially for CD. The name chiroptical methods has been suggested for the pair.

Substances are described as optically active or having chirality when they rotate the direction of vibration of linearly polarized light or interact differently with left- and right-circularly polarized light. Optical activity has its origin in structural asymmetry such as exists in any substance that does not have either a plane or a center of symmetry. The asymmetry may be

1. inherent in the structure of a molecule, which is often called a *chiral molecule* (e.g., dextrose).
2. peculiar to the crystalline form of the substance, as in quartz, and not appear at all in the liquid and gaseous states; and
3. at least in part the result of a particular conformation, as in the helical form of a polypeptide.

In the first case optical activity is evidenced in all physical states and in solution; in the second, it is displayed only by the crystal. Figure 19.4 shows three kinds of asymmetrical molecules, one of which has a plane of symmetry and is thus optically inactive.

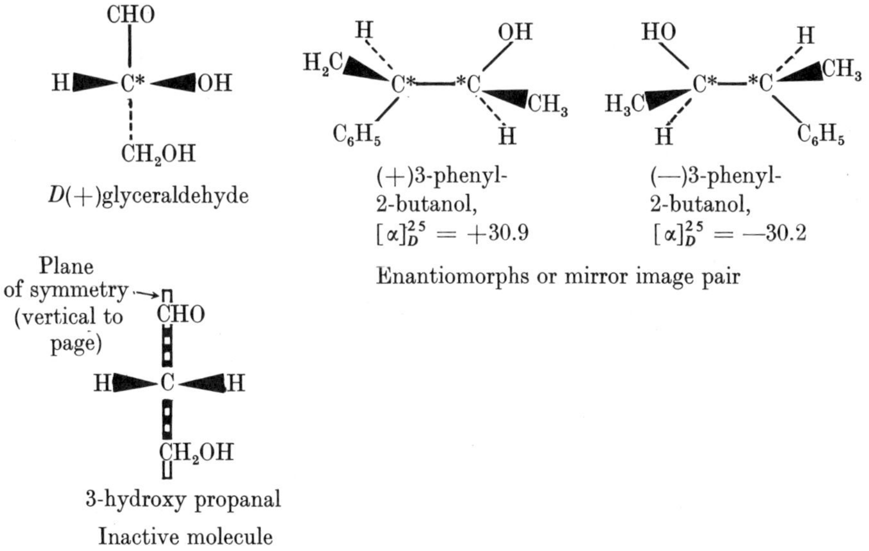

Fig. 19.4 Optical rotatory power as a function of structure. A heavy bar designates a bond coming out of page, a broken line one going back of page. The asymmetric carbons are starred.

The interaction of substances with linearly polarized light has been a straightforward measurement since the time of Louis Pasteur. A sample alternately irradiated by left- and right-handed circularly polarized light can display more complex behavior. It may

1. absorb the polarizations to a different extent, or
2. reduce the velocity of each to a different degree.

The first interaction can be observed mainly during electronic transitions in the UV and is measured as a differential absorption (CD), while the second appears as a rotation of the direction of vibration of linearly polarized light (ORD) in both the visible and UV. The connection between these properties can be expressed theoretically by Kramers–Kronig relations [2].

In earlier years available instrumentation made it simpler to measure ORD throughout the visible range; with advances in methods of generating circularly polarized radiation and determining its differential absorption, circular dichroism has become as easy to determine. The brief theoretical treatment given these effects in Section 7.18 will be assumed familiar in the following discussion.

Variables Influencing Optical Rotation. Optical rotation varies mainly with wavelength, optical path length, temperature, and density, and in the case of solutions, the concentration of optically active analyte. There is also a marked dependence of such activity on the magnetic field strength; such fields must be small even for measurements of ordinary sensitivity unless the field effect itself is of interest. When an appreciable magnetic field is imposed, even optically inactive substances exhibit optical rotation, a phenomenon termed the *Faraday* or *magnetic rotatory effect*. If a substance with optical rotatory power is also anisotropic, ordinary double refraction is superimposed on the optical rotatory effect (see Section 7.18).

Origin of Optical Activity. In a molecule with optical activity both magnetic and electrical changes occur in those electronic transitions that are "active." For electronic excitation by light, a transition dipole moment μ must exist. What distinguishes an active transition is that the moment has a circular component, that is, a rotating current. It gives rise to a finite magnetic moment **m** perpendicular to the plane of the circular current. It is because the vector product $\mu \times \mathbf{m}$ corresponds to a helical displacement of charge that left- and right-circularly polarized rays interact differently with such molecules, even in solution. The magnitude of ORD and CD is proportional to that vector product.

Example 19.6. Because of the connection of optical activity with an electronic transition, will only asymmetric chromophores have active transitions? Can symmetric chromophores ever generate optical activity?

An intrinsically optically active chromophore, the amide bond $-(C{=}O){-}N{-}$ (absorption peak at about 240 nm), is found in proteins and peptides. Carbohydrates and polysaccharides also have asymmetric chromophores but their absorption bonds are less accessible (λ < 200 nm). The active electronic transition for substances containing this chromophore may be $n \rightarrow \pi^*$.

Symmetric chromophores that exist in an asymmetric environment can also contribute to optical activity. For instance, the bases found in DNA are optically inactive, but when they are built into nucleosides optical activity is induced by the environment. Their ORD and CD peaks are often said to result from an *extrinsic* Cotton effect.

Defined Parameters. Measurements of the angular rotation of a substance are usually defined in terms of *specific rotation* $[\alpha]_\lambda$ where $[\alpha]_\lambda = 100\alpha/lc$, where α is the observed rotation, l is the path length in decimeters (1 dm = 10 cm), and c is solute weight in 100-mL solution. To designate the direction of rotation of the polarization plane and thus the sign of α or $[\alpha]_\lambda$ the following convention is used: when the polarization plane is rotated clockwise as one faces the emerging beam, the substance is said to be dextrorotatory (+); when the rotation is counterclockwise, it is levorotatory (−).

Results may also be expressed as the *molar rotation* $[\phi]$, which is defined in terms of the specific rotation as

$$[\phi]_\lambda = [\alpha]_\lambda M/100, \qquad (19.4)$$

where M is the analyte molecular weight. The use of $[\phi]$ facilitates both comparisons between the rotatory behavior of related compounds and the addition and subtraction of data. The units of $[\phi]$ are degrees centimeters squared per decimole.

For circular dichroism the differential absorption is observed directly as $\Delta A = cb\Delta\epsilon$, where b is the path length and $\Delta\epsilon = \epsilon_L - \epsilon_R$. Often the *molar ellipticity* $[\theta]$ is used where

$$[\theta]_\lambda = \theta_\lambda m/100Lc. \qquad (19.5)$$

Here θ, the observed ellipticity in degrees is defined by $\tan^{-1}\theta = b/a$ in which b is the minor axis and a is the major axis of the polarization ellipse. The conversion of $[\theta]$ to $\Delta\epsilon$ can be made using the expression

$$[\theta]_\lambda = 3300\Delta\epsilon_\lambda \text{ degree cm}^2 \text{ dmol}^{-1}.$$

Analytical Measurements. Organic mixtures containing one or more optically active substances can often be analyzed quantitatively by *polarimetry*, the technique of observing the rotation of plane of polarization of linearly polarized light. For example, the sugar content of foodstuffs is routinely determined by this method.

Polarimetric measurements are often of the optical rotation of a sample at a given wavelength, for example, at 589.3 nm (sodium D-lines). Such determinations provide a way to complete the identification of a substance after other data have been used to narrow the possibilities.

For most chiral substances specific rotation $[\alpha]_\lambda$ varies strongly with wave-

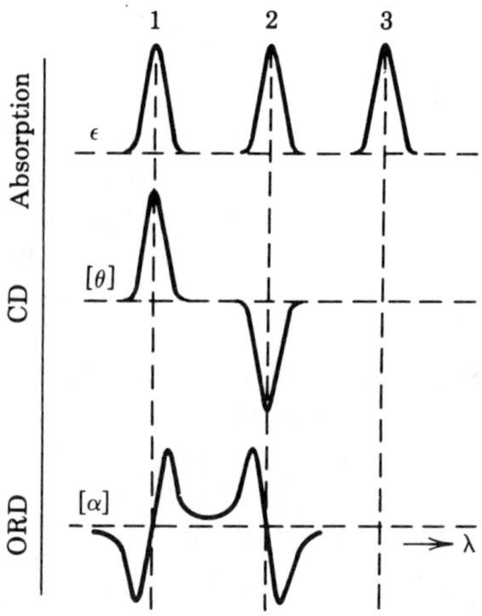

Fig. 19.5 A comparison of representative CD and ORD curves for a substance with two electronic transitions that are optically active (1, 2) and one that is inactive (3). For absorption 1, the CD band is positive since $n_L > n_R$, and for absorption 2, the CD band is negative since n_R is larger.

length, that is the substances display optical rotatory dispersion. A strong advantage of ORD is that these spectral measurements with a spectropolarimeter may be made in the visible region where there are no absorption peaks and S/N values are higher. As a result, both conformational analyses and the determination of the absolute conformation of optically active substances are ordinarily carried out by ORD determinations. These measurements contribute importantly to the study of the structure of complex molecules in other ways as well.

For a wide variety of complex molecules, however, CD spectra prove sufficiently distinctive to serve as a good basis for identification. In Fig. 19.5 these types of spectra are contrasted. As shown, the simpler shape of CD bands and the fact that some are positive and some negative makes them inherently easier to identify and resolve than ORD curves.

19.6 POLARIMETERS AND SPECTROPOLARIMETERS

Figure 19.6 gives a modular diagram of a generalized *polarimeter*, a device for measuring the optical rotation of a sample. The polarizer → sample cell → analyzer sequence of modules described in Section 7.14 is the heart of this device. Some common choices for its modules are now described.

Source. Sodium and mercury arc lamps are widely used monochromatic sources for polarimetry in the visible. In routine studies the sodium lamp output is seldom filtered, since 99% of its intensity is in the D-line doublet. A continuous source is used only with a saccharimeter, a device for measuring sugar concentration. In this instrument the dispersion produced by sugar is just compensated for by that of movable quart wedges. A xenon lamp and a Globar or Nernst glower, with a series of filters or monochromator, provide entirely adequate sources in the UV and IR, respectively.

Polarizer and Analyzer. Calcite prism polarizers are widely used as polarizer and analyzer in the visible range. They should not be used with a strongly convergent beam since

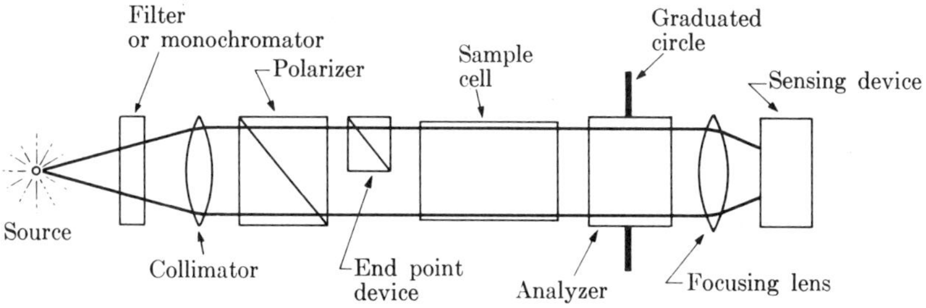

Fig. 19.6 A schematic diagram of a polarimeter.

the angular field of calcite prisms is less than 30°. These polarizers can also be used to about 230 nm in the UV if an air gap is substituted for the Canada balsam layer. At still shorter wavelengths calcite itself begins to absorb. Reflection polarizers (stacks of 3–6 selenium films or silver chloride plates placed at the polarizing angle) are commonly used in the IR.

Cells. Polarimeter tubes or cells are usually 10 or 20 cm long. Their internal diameters must be great enough that internal reflection and spurious polarization effects are unlikely. Considerable error can arise from strain in cell windows, lack of reproducible cell positioning, and temperature changes. For cells fused windows, annealed if necessary, are usually strain-free but screw-on cap windows often generate strain birefringence and apparent changes in sample rotation. Many cells are jacketed for thermostating to 0.1 °C to minimize error. Perhaps the best answer to cell position errors that lead to depolarization effects is to choose a stationary cell filled and emptied by syringe.

Detectors. The eye is quite satisfactory in the visible. Photoelectric and photoconductive detectors are widely used in the visible and UV. The latter also serves in the IR.

End-Point Devices. Recall that the analyzer of a polarizer–analyzer pair is ordinarily rotated until minimum transmission is secured. For a precision visual determination of this point, a null type of procedure, often with the Lippich *half-shadow end-point device*, is common. As shown in Fig. 19.7a,b, this design uses a small Nicol prism, turned slightly to alter the plane of polarization of *half the beam* emerging from the polarizer. The Lippich prism is also tipped slightly toward the polarizer to obtain a sharp boundary between the two optical fields so that the two portions of the beam differ in plane of polarization by about 1–7°. The condition of minimum transmission is found by rotating the analyzer until one sees fields of matched intensity (setting 2 in Fig. 19.7c). Recall that great visual sensitivity is obtained by matching intensities. Nevertheless, it is apparent that the angular setting observed with an end-point device must differ from the true rotation of the vibrational plane in the sample. Fortunately, the difference is some constant value called the *half-shadow angle*. Thus when absolute angular rotations are desired, this difference is taken into account in the calibration.

An advantage of the Lippich device is that measurements can be made even with turbid or highly absorbing samples by turning the small Lippich prism further to increase the half-shadow angle. In this way more light will be available at the end point, though some accuracy will be sacrificed.

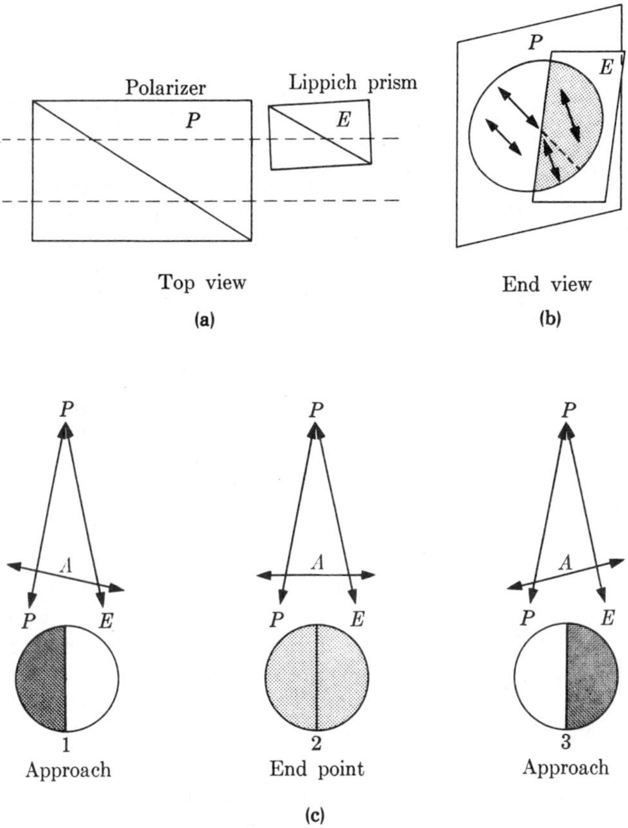

Fig. 19.7 Lippich prism end-point design. (a) Top view of polarizer and Lippich prism, a small Nicol prism, which covers half the field of the polarizer and may be rotated with respect to it. (b) End view of these parts. (c) The views seen in the eyepiece at different times. Arrow A represents the vibration favored by the analyzer or sample (neither is shown), while P and E designate the directions passed by the polarizer and Lippich prism, respectively. As the analyzer is turned to change its degree of "crossing" with respect to the incoming vibrations, first one half of the field and then the other becomes darker. The half-shadow point is reached at setting 2, when the fields are of equal intensity.

Calibration. To calibrate a polarimeter it is often simplest to use a known concentration of an optically active compound as a standard. For precise work, it is preferable to calibrate with a plate of crystalline quartz that is free of flaws and constant in thickness to better than 100 nm.

Automatic Polarimeters. An optical nulling procedure also proves an effective approach to the design of a simple, automatic, sensitive polarimeter. Figure 19.8 shows one type of design in modular form. In order to secure automatic operation and improve S/N, the radiation beam is modulated by oscillating the polarizer through a small angle about its axis (no more than $\pm 1°$) perhaps a dozen times a

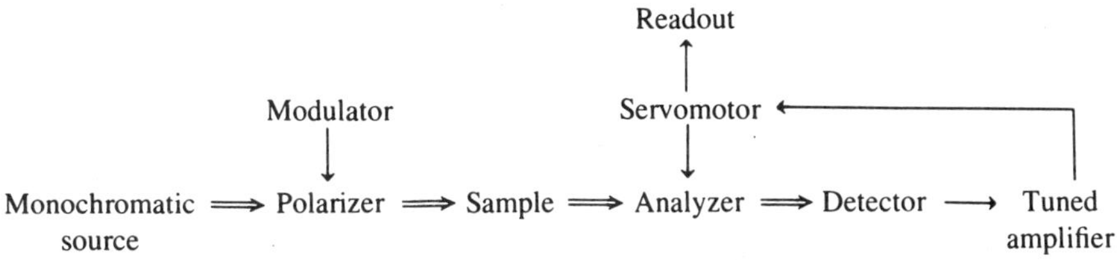

Fig. 19.8 Modular diagram of an automatic polarimeter.

second. At the equilibrium position the analyzer is crossed with respect to the polarizer.

When no sample is present, or when an optically inactive sample is placed in the cell, the detector receives pulses of equal intensity as the polarizer uncrosses in either direction whose frequency is twice that of the modulation frequency. If an optically active sample is placed in the beam, a component of the basic modulation frequency is also present. An amplifier tuned to that frequency sends energy to the servomotor. We can easily arrange phasing so that it turns the analyzer until a "null" is again attained. The readout is the rotation produced by the servomotor. Reproducibility is of the order of millidegrees.

Since a large light flux would create substantial shot noise at the detector, large modulating angles are seldom used. Yet there must be sufficient intensity on either side of the crossed orientation to maintain a reasonable S/N even if a sample absorbs at the wavelength(s) used. It is usual to employ discrete sources in conjunction with a set of filters to permit measurements at several frequencies. If we are examining highly absorbing solutions, a very intense soure (e.g., a laser) can be used with a microcell.

Spectropolarimeters. Polarimeters that operate over a wavelength range are called spectropolarimeters. Most are capable of determining both circular differential absorption (circular dichroism or CD) and optical rotation (optical rotatory dispersion or ORD). We can construct a spectropolarimeter *to determine ORD curves* from an automatic polarimeter by adding a suitable scanning monochromator. It is customary to develop a CD instrument around a double monochromator. These determinations are made in absorption regions where the energy of the transmitted beam is very small and stray light must be drastically minimized. Figure 19.9 shows a spectropolarimeter that is used mainly in the CD mode. It features a piezoelectric cell to develop modulated left- and right-circularly polarized light. If no sample is present, the circularly polarized components are incident on the detector in equal intensity and its output is unmodulated dc. If we insert an optically active sample and examine it in a region of a chiral transition, unequal absorption is evident and we obtain an ac output whose phase corresponds to the sign of the circular dichroism and whose magnitude is proportional to the amplitude of the CD. Taking the ratio of ac to dc components and amplifying secures a readout equal to differential absorption.

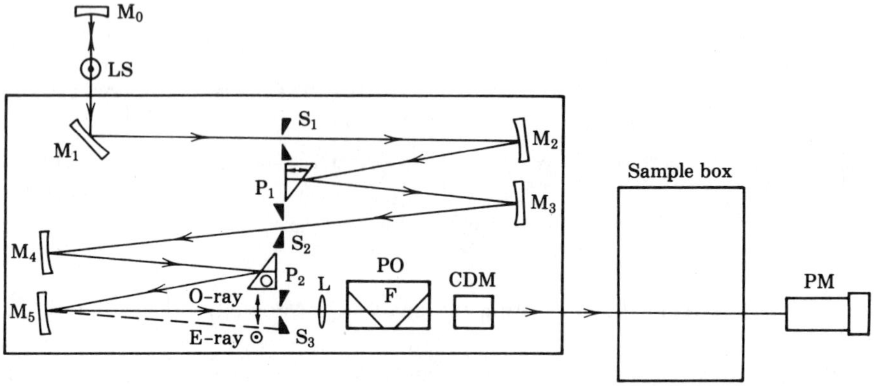

Fig. 19.9 Optical schematic of the Jasco J-600 spectropolarimeter. In principle, the layout is the same for the J-500 model. A double monochromator using natural quartz prisms ensures minimal stray light. Symbol identifications are as follows: M, mirrors; LS, light source; S, slits; P, prisms (optic axis horizontal in P_1, vertical in P_2); L, lens; F, filter; CDM, 50-kHz piezoelectric modulator; PM, photomultiplier. The linearly polarized light emerging from the monochromator (note double-headed arrow at slit S_3) is modulated to left- and right-circularly polarized light by the CD modulator. (Courtesy of Jasco Inc.)

REFERENCES

General volumes dealing with optical activity and light scattering are:

1. B. J. Berne and R. Pecora, *Dynamic Light Scattering.* New York: Wiley-Interscience, 1976.
2. E. Charney, *The Molecular Basis of Optical Activity.* New York: Wiley, 1979.
3. B. Chu, *Laser Light Scattering.* New York: Academic, 1974.
4. P. Crabbe, *ORD and CD in Chemistry and Biochemistry.* New York: Academic, 1972. (A clearly written introductory volume, requiring minimum theoretical background, that emphasizes applications of these methods to structural and stereochemical problems.)
5. M. Kerker, *The Scattering of Light and Other Electromagnetic Radiation.* New York: Academic, 1969. (Provides a careful treatment of scattering by particles of classical geometry.)
6. S. F. Mason, *Molecular Optical Activity and the Chiral Discriminations.* Cambridge: Cambridge University Press, 1982.

Also of interest are:

7. I. A. Campbell and R. A. Dwek, *Biological Spectroscopy.* Menlo Park, CA: Benjamin/Cummings, 1984.
8. M. B. Huglin (Ed.), *Light Scattering from Polymer Solutions.* New York: Academic, 1972. (An extensive reference on scattering by synthetic polymers.)
9. G. D. J. Phillies and F. W. Billmeyer, Jr., Elastic and quasielectric light scattering by solutions and suspensions, in *Treatise on Analytical Chemistry*, 2nd ed., Part I, Vol. 8, I. M. Kolthoff and P. J. Elving (Eds.). New York: Wiley-Interscience, 1986.

EXERCISES

19.1 To measure the optical rotation of a sample with a polarimeter must the transmitted intensity be measured when the solution cell is in the instrument and the analyzer and polarizer are set to "extinction"? Explain.

19.2 The absorbance of a cloudy, colored solution is measured in standard fashion in a 1-cm cell as 0.15. (a) How will correction for scattering affect this measurement? (b) Later the turbidity of the solution is determined to be 0.012. What is the true absorbance of the solution?

19.3 What will be the relative scattering effectiveness of two particles (a) when one is 100 times the volume of the other, and (b) when both are spherical and one is 100 times the diameter of the other? Assume that the molecular weight of a particle is roughly proportional to its volume.

19.4 It is ordinarily advantageous to make angular measurements of the intensity of scattered light with vertically polarized light rather than with unpolarized light. One reason is that the volume correction for the former is $\sin \theta$, while that for the latter is $1/(1 + \cos^2\theta)$. If the intensity reading is more accurate than the observation of the angle, show that the $\sin \theta$ correction is preferable. [Hint: calculate the change in each correction factor between $\theta = 45°$ and $\theta = 30°$.]

19.5 A sample of wet butanol is submitted for analysis. The refractive indices n_D at 25°C of butanol and water are 1.37226 and 1.33252, respectively. In order to estimate the minimum amount of water detectable, assume a linear relationship between the n of the mixture and the percentage composition. Using in turn a standard Abbe, a differential (image displacement) refractometer, and an interferometer with a long liquid cell, what amounts of water would be detectable by refractometry?

19.6 Why is it necessary to correct the Abbe refractive index scale when wavelengths other than the D-line are used? Would you expect the correction to be greater for the mercury 546.1- or 404.7-nm line?

19.7 A hollow 60° prism of Pyrex is used to hold liquid samples for refractive index measurements with the precision spectrometer. The prism is 4 cm to a side and has 2-mm walls uniformly thick. What correction, if any, will have to be made because the radiation traverses the glass walls as well as the sample?

19.8 The refractive index of a benzene–carbon tetrachloride mixture varies linearly with mole fraction. At 25°C, a length of Pyrex tubing is found to be invisible in a mixture whose mole fraction of benzene is 0.345. What is the refractive index of Pyrex glass? At 25°C, the refractive indices n_D of carbon tetrachloride and benzene are 1.4576 and 1.4979, respectively.

19.9 Since the refractive index of gaseous mixtures varies little with composition, how is the necessary sensitivity of measurement secured? Consider that the percentage of CO_2 in dry air is to be determined. At standard conditions the refractive indices n_D for air and CO_2 are 1.000292 and 1.000450, respectively.

19.10 A refractometer based on the use of fiber optics was developed some years ago. It uses a single uncoated rigid "fiber" of about 1 mm diameter and 25 mm length. Monochromatic light is focused onto a spot at one end of the rod and a photocell is placed at the far end. The rod is positioned in the middle of a liquid cell with opaque walls, that is, each liquid sample surrounds the rod. (a) Sketch the device and show the paths of reflected and refracted rays in the fiber. (b) Deduce whether the intensity of radiation reaching the detector increases or decreases with increasing *n* of a sample. (c) For a given sample, predict whether a fiber of closely similar *n* or greatly different *n* will give a more accurate measurement.

19.11 The specific rotation of dextrose is +52.5° in water solution at 20°C and 589.3 nm. (a) What will be the rotation of a solution containing 0.0055 g in 100 mL of solution if a 20-cm sample cell is used? (b) Can so dilute a solution be analyzed with 1% accuracy by a polarimeter whose precision is 0.1°? 0.05°?

19.12 The Dupont 420 Process Polarimeter is an automatic version of the polarimeter discussed in Section 19.6. A linearly polarized beam is sent through the sample cell (a flow cell) and then split into two beams. On the basis of this information, draw a schematic of a possible optical layout that will allow automatic operation.

ANSWERS

19.2 0.145, *19.3* (a) The larger particle is a 100 times better scatterer; (b) the larger particle is a 10^6 times better scatterer, *19.5* 0.3%, 0.01%, 0.0007%, *19.8* 1.472, *19.9* The mixture can be analyzed by an interferometer. An interferometer with 100-cm cells will allow differences in *n* as small as 2×10^{-8} to be observed. Accordingly, the concentration of CO_2 may be estimated to about $(2 \times 10^{-8}/15800 \times 10^{-8}) \times 100 = 0.014\%$. It is assumed that *n* varies linearly with composition at constant total pressure.

ENERGETIC PARTICLE AND X-RAY METHODS

Chapter 20

MASS SPECTROMETRY

20.1 INTRODUCTION

This division of the book opens with the first of the ''particle'' methods, mass spectrometry. In this major technique, analytes are converted to gaseous ions, the ions separated by mass-to-charge ratio m/e. For an analyte the display of relative currents or ion abundances versus m/e is its *mass spectrum*.

The power of mass spectrometry derives from its high specificity and unusually low limits of detection. Selectivity or specificity arises because the ions from a given analyte are a few of the multitude of characteristic values of mass and abundance that are possible. Low limits of detection can be traced to the high sensitivity of the electron multiplier detector used in mass spectrometers; this detector can respond to individual ions. Further, measurements are fast: from ion formation to ion analysis the elapsed time is 10^{-5}–10^{-2} s. As a result of these strengths, applications of mass spectrometry extend from elemental analysis and the measurement of isotope ratios to the analysis of complex organic and biological mixtures and the elucidation of the structure of large molecules. The special importance of mass spectrometry in biochemistry, pharmacology, and other fields related to biology should be noted.

Because the minutest of samples is required, mass spectrometry is also well suited as an ultratrace *and* micro technique. Samples as small as 100 pg can under favorable conditions produce a satisfactory qualitative and quantitative analysis as well as yield a great deal of structural information about a molecule. Quantitative analysis by mass spectrometry is a well-advanced technique.

In the past, mass spectrometric analysis was limited to substances that could be volatilized thermally without degradation. Accordingly, many compounds, including peptides, nucleotides, and other polar or high-molecular-weight substances were uncharacterizable. Sample techniques introduced since 1981 have allowed mass spectrometry to be applied to many types of samples previously untouched. The development of advanced magnet systems has kept pace with the new sampling techniques.

Terminology. In mass spectrometry analyte ion signals are displayed as a function of mass-to-charge ratios m/ze, where $z = 1, 2, 3$ since multiple charges per ion are possible though of low probability. How are these ratios reported since in the SI system m is in kilograms and e, the basic unit of charge, equals 1.60 ×

10^{-19} C? It is conventional to *report* m/ze in terms of the simpler quantity m/z with m in atomic mass units (amu). For example, while the most common trifluoromethyl ion $^{12}CF_3^+$ has an m/ze ratio of 7.1509×10^{-7} kg C^{-1}, its m/z value is simply 68.9952 amu.

20.2 ION SOURCES FOR GASES: IONIZATION AND FRAGMENTATION

In Fig. 20.1 the characteristic modules that make up a mass spectrometer are shown. As the diagram implies, it is in the first module, the *ion source*, that a nongaseous sample must be at least partly (a) volatilized and (b) converted to gaseous ions. The vapor of a volatile sample can of course be introduced (in minute amount) directly into the ion source. While the introduction of a gas is ordinarily a simple process, a discussion of the steps of sampling will be deferred to Section 20.7 when more complex aspects can also be conveniently taken up. We shall begin with a gas in the ion source.

In an ion source gaseous molecules are commonly ionized by bombardment with electrons or reagent ions, processes that usually also lead to molecular fragmentation. Here principles that underlie these processes as well as customary types of ion sources will be explored. For example, an important theoretical question is whether fragmentation can be controlled so that fragments will be generated in reproducible fashion and thus will reveal information about the structure of the parent molecules.

Ion-Source Pressures. The control of partial pressures in an ion source will establish probabilities of molecular collisions and thus also ion lifetimes. To ensure high probabilities of ionization, the pressure in many ion sources is held at about 1×10^{-1}–1×10^{-2} Pa (10^{-3}–10^{-4} torr). Somewhat higher pressures will be needed in chemical ionization, as will be discussed later in the section.

Between the ion source and detector pressures are maintained in the range of 10^{-4}–10^{-5} Pa (10^{-6}–10^{-7} torr) to keep the collision of analyte ions with residual gas or other analyte ions minimal.* In fact, pressures below 10^{-5} Pa (10^{-7} torr) in intermediate modules are also *unattractive* as they result in too few ions of some species reaching the detector.

Sample
↓
Ion source (ionization module) ⟶ Mass analyzer ⟶ Detector

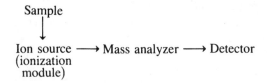

Fig. 20.1 Characteristic modules of a mass spectrometer.

*Clearly, reactions between gaseous ions and molecules are of interest in their own right as are molecular dynamics and kinetics in the gas phase. Much research in these areas is performed by specially adapted mass spectrometers that operate at higher pressures in particular parts of their systems.

Electron-Impact Ionization. The most common mode of ionization of gaseous substances is by electron impact (EI) in a source such as that shown schematically in Fig. 20.2. Assume that gaseous molecules from a sample have been introduced. Electrons emitted from a metal (often a rhenium alloy filament heated by current to a red or white heat) are accelerated in a stream (apertures help define its cross section) toward an anode 70–100 V more positive. Electron energies thus reach 70–100 eV ($E = eV$, the product of electron charge e and potential difference V).

To put these energies in perspective, the energy of a mole of 70 eV electrons is 6700 kJ or 1600 kcal. Energies that large are used since they guarantee that the rate of production of ions will be essentially independent of electron energy: the ionization process will be saturated. Actually, only a fraction of molecules is ionized (perhaps 1%) and the whole transfer process is inefficient. One example is that even though ionization energies tend to be from 7 to 13 eV, the excess energy left in most ions is in the range 2–8 eV (200–775 kJ mol^{-1}).

Ions are formed from molecules M on electron impact by a reaction such as

$$M + e = M\cdot^{+} + 2e.$$

The ion $M\cdot^{+}$ formed is called a *molecular ion* and the symbol identifies a species with an unpaired electron, an ion radical. As implied, positive ions are usually formed. Doubly charged ions are formed only occasionally (they are more probable for aromatic species compounds with several rings), and triply charged ions are rare. Similarly, the capture of a 70-eV electron to give a negative ion M^- is less probable than the formation of a positive ion by a factor of about 100. Negative ions can be formed, however, by capture of lower-energy electrons or from

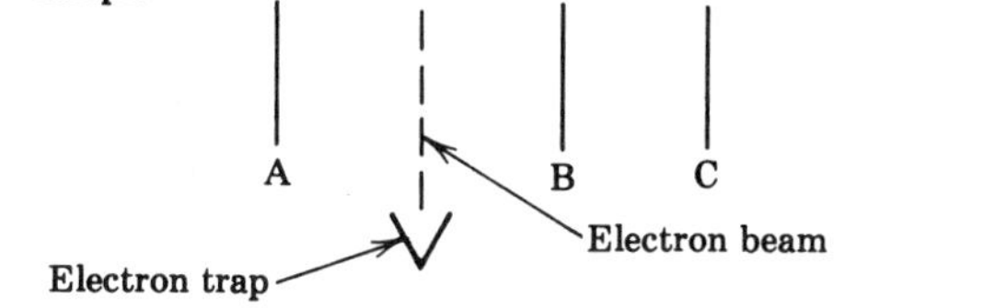

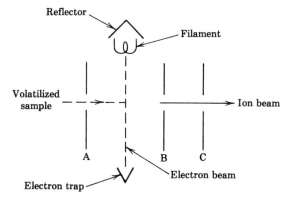

Fig. 20.2 Simplified schematic of an electron-impact ion source and its coupling to a mass analyzer. Electrons emitted by the heated filament are accelerated toward the anode so that their energy is at about 70 eV on impact with the sample. Pressures are maintained at 10^{-4}–10^{-5} Pa. Molecules enter through the slit in plate A. It is made slightly positive (given a repeller potential) relative to plate B to ensure that positive ions formed in the source are extracted only through the exit slit. To extract ions plate C is strongly negatively charged (a few kilovolts) relative to plate B. Ions leaving are accelerated to a high uniform velocity in preparation for entering the next stage.

gas-phase chemical reactions. Negative-ion mass spectrometry has some special applications that will be mentioned later in the chapter, but unless otherwise stated, discussion will deal with positive-ion mass spectrometry.

In general enough excess energy is transferred to gaseous molecules during ionization so that they subsequently undergo *unimolecular* decomposition. The fragmentation of isolated ions is of considerable interest for the information it yields about the species in a sample.

How does excess energy lead to fragmentation? Since collisions with other gaseous species are unlikely, the internal energy of newly formed ions must be handled internally. It may initially lead to electronic excitation but eventually dissociation resulting in formation of neutral or ionized fragments is likely:

$$\mathrm{M}\cdot^{+} = \mathrm{D}\cdot^{+} + \mathrm{E} \quad \text{or} \quad \mathrm{M}\cdot^{+} = \mathrm{F}^{+} + \mathrm{G}.$$

In these reactions the interrelation of species is suggested by their names: $\mathrm{M}\cdot^{+}$ is a precursor or parent ion and $\mathrm{D}\cdot^{+}$ and F^{+} are fragment or daughter ions. We shall ignore the neutral species; they can only be measured indirectly by a mass spectrometer. A mass spectrum for a particular species from a sample will thus usually be made up of parent and daughter ions, all having been formed in the ion source.

No species is more important than the *molecular ion* since its presence allows the molecular weight of an analyte to be determined precisely. Yet, it is difficult to predict with electron-impact ionization for an unknown compound whether a molecular ion will survive to appear in a spectrum. Fortunately, chemical ionization (see below) and some of the sophisticated modes of sampling to be discussed in Section 20.7 overcome this problem.

Quasi-Equilibrium Theory. This theory has been the most useful in the attempt to understand unimolecular fragmentation and interpret mass spectra. It derives from quasi-equilibrium theory. Its fundamental assumptions are: (a) that the unimolecular fragmentation reactions of the energetically excited molecular ion are in competition with each other and that many are consecutive in character; (b) that the energy of excitation is distributed in random fashion among the different vibrational modes of the ion at a rate that is fast relative to dissociation or fragmentation.* The second assumption leads to the conclusion that fragmentation will be likely to occur when a sufficient amount of energy (number of quanta) has accumulated in the bond most critical for a particular dissociation.

From the theory the simplest equation developed for reaction rates of dissociation reactions is

$$k = \nu[(E - E_0)/E]^{s-1}, \qquad (20.1)$$

*Randomization that includes rotational modes is improbable because of differences in angular momentum between vibrational and rotational modes. It may, however, include internal conversions that result in transfer of a species from the ground electronic state to an excited electronic state if the total of vibrational and electronic energy is nearly equal in the two states.

where E is the internal energy of the ion, E_0 its activation energy for a particular dissociation, ν the "frequency factor," and s the number of effective molecular vibrations. The frequency factor ν basically relates to the vibrational frequency of bonds and would be expected to be about 10^{14} s^{-1}. The crucial factor appears to be the amount by which the internal energy exceeds the activation energy for a particular dissociation. For $E \gg E_0$, dissociation might occur in one vibration. In general E will be of the same order of magnitude as E_0 and random processes will govern whether enough quanta accumulate at some time in a given bond and produce dissociation. With regard to the last factor, it is found empirically that no more than one-half to one-third of the $3N - 6$ allowed modes is effective.

Data of two sorts may be obtained for rough testing of the theory. Ion currents I_0 obtained after electron-impact ionization of a given species can be plotted against electron acceleration potential V (since it is proportional to the excitation energy eV). By judicious extrapolation to $I = 0$, one can estimate the ionization potential. Further, *appearance potentials*, that is, acceleration voltages necessary for daughter species to appear during fragmentation, can be obtained in similar fashion. Is it justifiable to substitute these values for E_0 in Eq. (20.1)? In the absence of other data, the substitution does permit an approximate calculation. Clearly, a knowledge of the transition state involved in the ionization process would be required for more precision.

In Fig. 20.3 potential energy curves for a diatomic system are used to show the sort of redistribution of energy that may occur with a polyatomic species. The

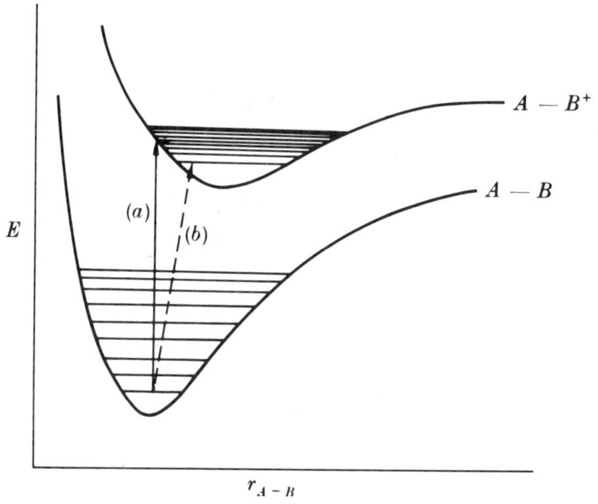

Fig. 20.3 Potential energy curves for the ground electronic state of molecule A–B and its singly charged positive ion. The solid vertical line (a) represents the ionization energy measured in electron-impact ionization, which is considered to occur so rapidly that atomic nuclei have insufficient time to move (Franck–Condon principle). As a result the transition probability measured is from the heavily populated ground vibrational level to an excited vibrational level for the ion. The broken line (b) is the transition measured spectroscopically. The latter ionization energy is smaller since the final state for the ion is also a ground vibrational state.

vertical transition labeled (a) is believed to occur on electron impact. A bombarding electron will move so rapidly that it will remain in the vicinity of a sample molecule very briefly. For that reason, ionization will occur before any movement of the nuclei can take place (Franck–Condon principle). The smaller transition measured spectroscopically is shown by the broken-line transition.

In Table 20.1 some first ionization energies are given for a few simple organic compounds. Note that the values are far below the usual 50–70 eV electron-impact energies employed.

Application of Quasi-Equilibrium Theory (QET). To apply QET theory one can assume an appropriate set of consecutive unimolecular decompositions for the molecular ion. These would be taken to be in competition, some occurring more rapidly, others more slowly. Ions of lower mass would be produced entirely by fragmentation where sufficient energy had been imparted to the molecular ion.

To apply the theory to a particular substance, its initial ionization potential and appearance potentials of fragments would be needed. Values for the constants in the equation can be found as follows. First, assume that appearance potentials can be substituted for E_0. A reasonable assumption for the frequency factor would be $10^9\ s^{-1}$ for a complicated dissociation (and even smaller values if some rearrangement of atoms is also required) up to values $10^{13}\ s^{-1}$ for quite simple types of dissociation. A routine assumption that only one-third of the vibrational modes are "effective" would complete the selection.

It is also important to remember that the molecular ion is likely to remain in the ion chamber no more than 10^{-5}–10^{-6} s after ionization. At any later time positive ions would be traveling through the mass analyzer. The residence time has an important bearing on experimental values of appearance potentials. For a particular species of ion to appear in a source and be determined in the analyzer, its rate constant must be at least $10^6\ s^{-1}$. Thus, measured appearance potentials include an amount of energy called the kinetic shift that ensures such a rate.

For many molecules the mass spectrum calculated in this fashion correlates reasonably

Table 20.1 First Ionization Energies of Some Simple Organic Compounds[a]

	Ionization Energy (eV)	
Compound	*Electron Impact*	*Photoelectron Spectrometry*[b]
$CH_3CH_2CH_2Cl$	10.82	—
$CH_3CH_2CH_2OH$	10.17	—
C_6H_5Cl	9.60	9.31
C_6H_5OH	9.16	8.75
$C_6H_5N(CH_3)_2$	7.95	7.51

[a]Second ionization energies, I_2, are substantially larger. Only for large aromatic molecules is the formation of doubly charged ions moderately probable. For example, for pentacene, $C_{22}H_{14}$, $I_2 = 19.6$ eV.

[b]The data were taken by ultraviolet photoelectron spectrometry (UPS) in which a sample is irradiated with the He(I) line (21.21 eV) and photoelectrons are ejected. (Since their kinetic energy is given by $E_{KE} = 21.21 - I$, the ionization energy can be calculated. Measurements are of good precision. See Section 22.6.)

well with experiment. Where there is disagreement between predicted and observed spectra, usually there has been a failure to allow for some rearrangement of the molecular ion.

Extension. The Rice–Ramsperger–Kassel–Marcus (RRKM) equation (RK) has been developed from a more rigorous treatment of unimolecular reactions:

$$K(E) = (Z^{**}/hZ^{*}) \sum P^{**} (E - E_a)/\rho^{*}(E), \tag{20.2}$$

where h is Planck's constant, Z is the partition function, ** and * refer to the activated complex and the molecule in question, respectively, $P(E - E_a)$ is the number of energy states in the molecular range $E - E_a$, and $\rho(E)$ is the density of states. This expression is more difficult to apply than Eq. (20.1) but it does take advantage of modern quantum mechanical insights. For example, it is evident that the number of states involved in going from the molecule to the transition state (activated complex) must be estimated rather than a frequency factor as in Eq. (20.1).

Chemical Ionization. The chemical mode of ionization (CI), is "softer" (less energy is transferred than with EI) and produces less fragmentation. A simple molecule such as methane, isobutane, or ammonia is introduced into a closed ion source which is at about 100 Pa (1 torr). This substance serves as a *reagent* to produce analyte ionization.

How chemical ionization occurs can be illustrated by taking methane as an example. This gas is first ionized by electron impact, quickly yielding species like CH_4^+ and CH_3^+ that react to form other species such as CH_5^+ and $C_2H_5^+$ as follows: (a) $CH_4^+ + CH_4 \rightarrow CH_5^+ + CH_3$ and (b) $CH_3^+ + CH_4 \rightarrow C_2H_5^+ + H_2$. A steady state is achieved in this reagent plasma, and the sample is introduced at about 0.1 Pa. Reagent ions mainly behave as *strong acids*, protonating an analyte species such as AH:

$$AH + CH_5^+ \rightarrow AH_2^+ + CH_4$$

or facilitate a hydride–ion transfer $C_2H_5^+ + AH \rightarrow A^+ + C_2H_6$. Both AH_2^+ and A^+ are even-electron ions with some excess energy; their fragmentation will commonly be less extensive than with EI.

A chemical-ionization module permits analysis of many mixtures that would be difficult to analyze by electron impact.* Clearly, CI sources should also be valuable for the study of ion–molecule reactions in their own right. Products produced in this kind of source are often identified and afterward studied by ion–cyclotron resonance mass spectrometry (see Section 20.4).

Negative ions are also generated in a chemical ionization plasma when an appropriate reagent gas is chosen. For instance, when methane is the reagent and

*If detection limits of a class of analytes are low, a reagent can be chosen that will selectively analyze them. For example, water vapor might be used for selective ionization of alcohols, ketones, esters, and amines.

CH_4^+ is formed by electron impact, lower-energy electrons are by-products. An analyte has a resonable probability of capturing these electrons if it has sufficient electron affinity. To convert from positive to negative ion measurements in mass spectrometry the main change required is reversal of electrical polarities at many points in a spectrometer (see Section 20.6).

What are the principal differences between chemical ionization and electron impact ionization? First, in CI the reagent gas produces an ion that protonates the analyte, that is, causes M or AH to form $[M + H]^+$ or AH_2^+, respectively. These ions, in which all electrons are paired, are inherently more stable than the radical species that result from EI ionization. Second, the pattern of fragmentation in CI is essentially complementary to that produced by EI. Third, the degree of fragmentation in CI is smaller, giving a spectrum that can be interpreted more easily.

In most cases dual ion sources featuring both EI and CI are provided. An outside lever or switch permits the user to open or close the ion sources to provide the lower or higher pressure required by EI or CI, respectively, without breaking the spectrometer vacuum. What must be altered to effect the change? Ordinarily, in the change from the EI to CI modes both the (a) size of the electron entrance aperture and (b) width of the ion exit slit are greatly reduced. The first change reduces the intensity of the electron beam while the latter ensures that analyte molecules will remain accessible to ionization as long as possible and not be drawn into the vacuum system.

Other Methods. Can ions be formed from solids of minimal volatility in such a way that they characterize the sample? Methods such as laser desorption-ionization, field ionization-desorption, electrical spark ionization, and secondary ion mass spectrometry (especially in its fast atom bombardment version) do just that. These techniques combine volatilization and ionization and will be treated in Section 20.7. Another technique, *photoionization*, serves mainly as a means of measuring ionization and appearance potentials, but is also the basis of both multiphoton ionization (MPI) and photoelectron spectrometry (UPS), an important technique itself (see Section 22.6).

Extraction of Ions from the Source. It is worthwhile to look ahead. How should an ion source be coupled to the next module, which is nearly always a mass analyzer? Many types of analyzer require an essentially monoenergetic beam of ions. Will this requirement be satisfied by the sort of ion source shown in Fig. 20.2?

Note that ions leave this source through the apertures in plates B and C; these plates can be regarded as coupling the source to mass analyzer. It is only necessary to charge them appropriately (for the usual positive ions, to apply a strong, steady negative dc potential between plate C and B). The plates provide for (a) extracting the ions, (b) forming them into a beam, and (c) accelerating them toward the entrance aperture of the mass analyzer.

How uniform an ion kinetic energy is achieved? If ions of charge e are accelerated through a potential difference V between plates B and C, all will receive the same energy eV and will emerge with essentially equal kinetic energies:

$$\frac{1}{2}mv^2 = eV \tag{20.3}$$

where m and v represent the mass and final velocity of each mass of ion. Equal kinetic energies imply velocities that vary systematically with mass (see Exercise 20.5). The residual variations in kinetic energy will be explored when resolution is discussed in the following section.

Alternatively, where mass analysis is quite weakly dependent on kinetic energy as for a quadrupole analyzer, the coupling device need be no more than an ion lens that forms a beam (see Section 20.4).

20.3 SECTOR MASS ANALYZERS

The heart of a mass spectrometer is certainly its mass analyzer that separates ions by their values of m/z. This module substantially determines the resolution of species achieved and by its throughput or transmission of ions contributes a great deal to the instrument S/N. Its scanning speed will also set the minimum time required to obtain mass spectra for a system. These quality criteria are defined in Table 20.2. In examining types of mass analyzers these criteria will help to establish their advantages and limitations.

The widely used sector mass analyzers separate ions that have been formed into a well defined beam of uniform kinetic energy by application of either a magnetic or electric field perpendicular to the direction of ion motion. The name sector reflects that fact that these devices have longitudinal cross sections in the shape of a sector, a geometric figure whose sides are the radii of a circle and whose curved inner and outer edges are the arcs of a circle (see Figs. 20.4 and 20.5).

The Ion Beam. As discussed in the last section, positive ions are extracted from an ion source by a potential difference applied between extraction plates. For sector mass analyzers the ions are shaped into a flat beam by slits as they are accelerated. Because ions are extracted from different places in the source their initial energies

Table 20.2 Some Quality Criteria for Mass Analyzers

Resolution R is the capability of an analyzer to present as separate two nearly equal m/z peaks. It is defined by the expression $R = m/\Delta m$, where m is the average mass of just-resolved adjacent mass peaks and Δm their separation.

Transmission measures the fraction of ions of given m/z that traverses a mass analyzer on the average.

Scanning speed is roughly the reciprocal of response time. Its upper limit is set by how rapidly the scanning field (electric or magnetic) can be changed without loss of mass resolution.

differ slightly ($\Delta E = 0.2$–5 eV, depending on source design). After extraction and the accompanying acceleration necessary for a sector analyzer, however, the beam will be nearly homogeneous in energy. For example, an initial range in kinetic energy of 2 eV will amount to only a 0.2% difference when the average energy is increased to 1 keV. When sector analyzers are used, acceleration voltages are commonly 1–8 kV. The highest resolution instruments use > 8 kV.

Magnetic Sector Analyzers. A schematic of the earliest magnetic sector is shown in Fig. 20.4. Ions enter the sector through slit S_1 and exit through slit S_2. Modern magnetic sectors differ in geometry from this one mainly in that ion beams are bent through only a 60–90° arc (see Fig. 20.8). Note that the sector radius r can be defined by projecting the perpendicular entrance and exit faces to their point of intersection.

This sector performs a mass analysis by applying a steady magnetic field B perpendicular to the ion beam. This field subjects the ions of steady velocity v and charge e to a torque of magnitude Bev that causes the ions to move in a circular path of radius r. Since the torque is exactly opposed by a centrifugal force, we have the equation

$$Bev = mv^2/r. \tag{20.4}$$

Rearrangement of terms defines the value of m/e related to a given radius: $m/e = Br/v$. Note that the ion velocity should be just the steady value established by the earlier acceleration stage. Substituting for v from Eq. (20.3) we obtain

$$m/e = B^2r^2/2V \quad \text{or} \quad m/ze = B^2r^2/2V, \tag{20.5}$$

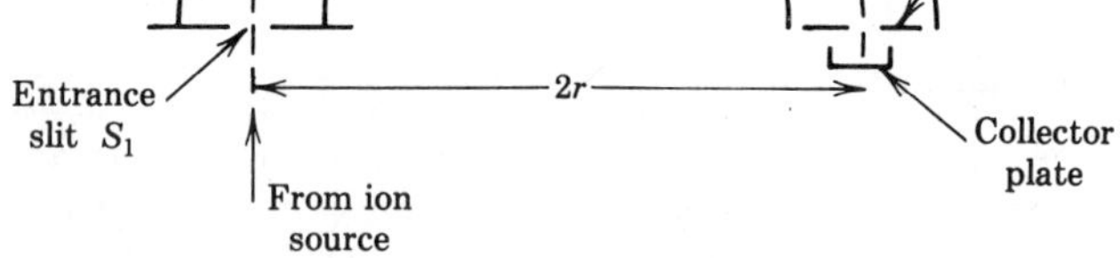

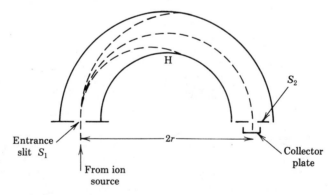

Fig. 20.4 Simplified schematic of the 180° Dempster magnetic mass analyzer. A magnetic field (not shown) imposed at right angles to the plane in which ions move acts to bring the divergent ion beam entering at slit S_1 with velocity v to a focus at the exit slit S_2. It is the steady perpendicular magnetic field of strength B that subjects the ions to a torque Bev and causes ions of a particular m/e ratio to move in a circular path of radius r. Note the Faraday cup detector at S_2. This analyzer is similar in design to analyzers used in electron spectrometry (see Section 22.5).

where the latter equation applies to an ion of charge ze. As Eq. (20.5) shows, for given values of magnetic field strength B and accelerating potential V, only ions with a particular mass-to-charge ratio (m/ze) fall into the circular path that allows them to pass through the exit slit of the analyzer in Fig. 20.4 and reach the collector or detector; others strike the walls of the analyzer. As discussed in Section 20.1, the ratio is always reported as m/z (i.e., as amu per unit charge). Another strength of a magnetic sector is its directional focusing: even though ions diverge after entering the sector, they are brought to a focus again at its exit slit.

To scan the masses of ions from a sample either the magnetic field B or the electrostatic potential difference V may be varied. During a scan ions of different mass will in turn be brought to the exit slit and a mass spectrum produced. According to Eq. (20.5):

1. If accelerating potential V is decreased while field B is held constant, the scan will be toward increasing values of m/e.
2. If magnetic field B is increased while V is held constant, ions will appear at the exit slit in order of increasing m/e.

Today most scanning is done by increasing the magnetic field strength.* Nevertheless for a given range of field B, the upper limit of the mass range will be inversely proportional to V. For example, if for $V = 4$ kV the highest mass that is resolvable for singly charged ions is 700 amu, that limit can be increased to 2800 amu by reducing V to 1 kV.

Example 20.1. For precision measurements of ion masses, which variable must be controlled more carefully, the magnetic field strength B or the accelerating potential V? What impact does this have on mass spectrometer design?

The first query can be answered by partial differentiation of Eq. (20.5) with respect to each variable in turn. Note this approach is also taken in tracing the propagation of error (see Section 10.3). Partial differentiation gives, *on a relative basis*, after substitution of finite increments for infinitesmals,

$$\Delta m/m = 2\Delta B/B - \Delta V/V.$$

From the coefficients of the ΔB and ΔV terms, two and one, respectively, it appears that a given instability in magnetic field will lead to twice as great an error in mass at a particular value of m/ze as will an equal instability in electrostatic potential. For this reason special care is taken in designing magnets for sector mass spectrometers.

Magnetic Scanning. To scan over a range of masses with a magnetic sector mass analyzer the current in the winding of an electromagnet is changed. A straightforward approach

*Scanning by varying V has the disadvantage that some of this voltage "leaks back" into the ion source, leading to altered source conditions. The result is that the degree of separation for increments of the m/e ratio changes and some information is lost in the region of the spectrum where measurement sensitivity is low anyway.

is to vary the magnet current linearly. If the current is small and is increased at a moderate rate, magnetic field B also increases linearly. Yet according to Eq. (20.5) this will lead to a quadratic mass scan. A second possibility is to increase current exponentially and settle for an exponential scan of magnetic field. This approach is attractive since it will lead to observing each mass peak for an equal interval and will ensure that data acquisition will occur at the same rate across a spectrum.

Three sources of error in magnetic scanning must be kept in view. While field B increases linearly with current at small currents, the rate of increase gradually falls to zero as the magnet saturates. Second, *hysteresis* effects caused by residual magnetism cause field values to differ for a given current depending on whether the current is being increased or decreased. Finally, fast scanning of a mass spectrometer has always been difficult because of the large inductance of electromagnets. Before addressing these problems, what modes of scanning are ideally available?

The first two difficulties with scanning have been overcome by relying on the field control of B. The output of a Hall probe detector* inserted between magnet pole pieces is continuously compared with a programmed value and the difference signal used in a feedback circuit to correct the field. The last difficulty has been reduced by development of sophisticated magnets with laminated cores, a change which greatly reduces inductance and also reduces eddy currents and heating.

Resolution. What factors will limit the spectral resolution obtainable if a magnetic sector is used as the mass analyzer? Remember that the analytical signal will be the size of ion current for each value of m/e. Resolution will of course be affected by the degree of homogenity of magnetic field over the length of the sector and the constancy of B. More importantly, it will depend critically on S_1 and S_2, the entrance and exit slit widths of the magnetic sector (see Fig. 20.4). For the moment let us assume that the range of velocities of ions produced at the exit of the ion source is insignificant.

From a consideration of the instrumental factors noted, Dempster developed the following expression for resolution:

$$R = m/\Delta m = r/(S_1 + S_2 + \beta r), \tag{20.6}$$

where all terms have been defined except the proportionality constant β. Ion optical aberrations cause the appearance of the last term in the denominator ("chromatic aberration" caused by the range of ion velocities and scattering resulting from ion–molecule collisions along the flight path). These aberrations lead directly to peak broadening and a lower resolution.

It appears from the equation that reducing slit widths would be a straightforward way to improve resolution. In most modern spectrometers this step has been taken. The resulting loss of ion current is usually offset by providing a highly sensitive electron multiplier detector (see Sections 20.5 and 22.5).

*In a Hall probe the output signal is the transverse electrical potential developed between the sides of a current-carrying conductor placed so that the current flow is perpendicular to the magnetic field to be monitored.

How can the range of ion velocities from an ion source be reduced? To minimize this source of broadening, an electrostatic sector is ordinarily inserted just after plate C. As will now be seen, it acts an energy filter, passing through its exit slit only ions of a given value of kinetic energy.

Electrostatic Sector Analyzer. As shown in Fig. 20.5, this analyzer consists of a pair of coaxial metal plates that are sectors of a cylinder between which a uniform potential V' is maintained. Since the plates are separated by distance d, the strength of electric field E between them is $E = V'/d$. This type of analyzer was first inserted as a kinetic energy filter between the extraction plates of the ion source and the magnetic sector to improve the mass resolution achievable.

It can be shown that an electrostatic sector also provides directional focusing at the exit slit. In a mass spectrometer with both electrostatic and magnetic sectors, ions that reach the final detector are focused twice. Such *double-focusing* combinations can provide very high resolution (up to 150,000).

For an ion to traverse the sector, the governing condition is equality of electrostatic and centrifugal forces on it during passage: $eE = mv^2/r_e$ where r_e is the radius of the sector. By rearranging this equation one can put it in the form of ion kinetic energy. Since that energy was determined by the acceleration potential V and remains unchanged in the absence of collisions, we can obtain by use of Eq. (20.3) the expression

$$mv^2/2 = r_e eE/2 = eV \quad \text{or} \quad r_e = 2V/E. \tag{20.7}$$

The appearance of acceleration potential V in Eq. (20.7) indicates clearly that only ions of a particular kinetic energy travel the circular path of radius r_e that will allow them to emerge through the exit slit. Slits of this sector can be narrowed to define more precisely the energy band that will pass. (See Example 20.6.)

Mass analysis is also possible with an electrostatic sector. For this purpose both V and V' are varied to bring ions of different kinetic energy, and thus, different

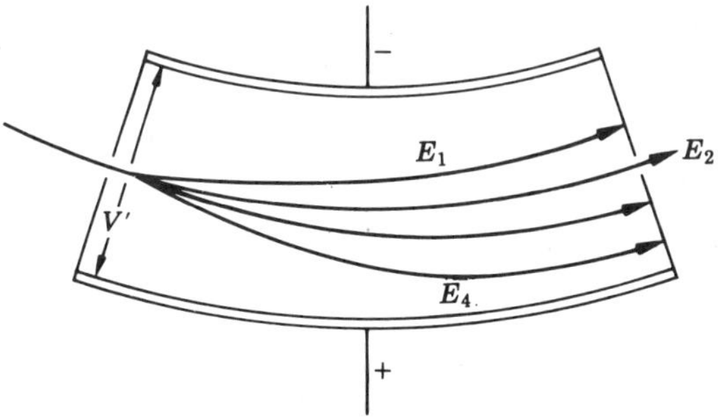

Fig. 20.5 Electrostatic sector to filter ions by energy. A radial electrical field is set up between the cylindrical metal plates by applying a potential V' between them. Note the outer plate is positive. In this field only positive ions of kinetic energy E_2 are able to enter the slit at the left and leave through the exit slit. They travel in a circular trajectory. The paths of ions with other kinetic energies cause them to be spread out along the exit plane. Ions of energy E_2, regardless of mass, leave the sector.

m/e ratio to the exit slit. It has mainly been with the development of tandem and multiple analyzer mass spectrometers than the mass-sorting properties of electrostatic sectors have been exploited (see Section 20.8).

20.4 QUADRUPOLE AND OTHER DYNAMIC MASS ANALYZERS

In addition to the mass analyzers of the sector type just discussed, several kinds of *dynamic analyzer* rely either on time-varying electric fields or on time-dependent ion movement to effect mass separations. They include the quadrupole, Fourier transform (ion cyclotron), and time-of-flight types. Each has strengths and limitations and finds special areas of application.

Quadrupole Mass Analyzers. The most widely used mass analyzer today is the quadrupole analyzer or mass filter, as it is often called. As shown in Fig. 20.6, it consists of four parallel rod-shaped electrodes arranged at the apices of a diamond.* Two necessary design conditions are that the rods must be *precisely* machined and aligned so that electric fields with defined gradients are set up, and the usual high vacuum of 10^{-5}–10^{-6} Pa (10^{-7}–10^{-8} torr) must be maintained in the analyzer to prevent collisions that would change the direction of ions or the focus of the beam.

With a quadrupole analyzer kinetic energy differences are inconsequential and ions are extracted from an ion source and focused into a beam by an ion lens at low kinetic energies, 0 to 20 eV. The ion beam moves along the axis of the rod array toward a detector at the far end.

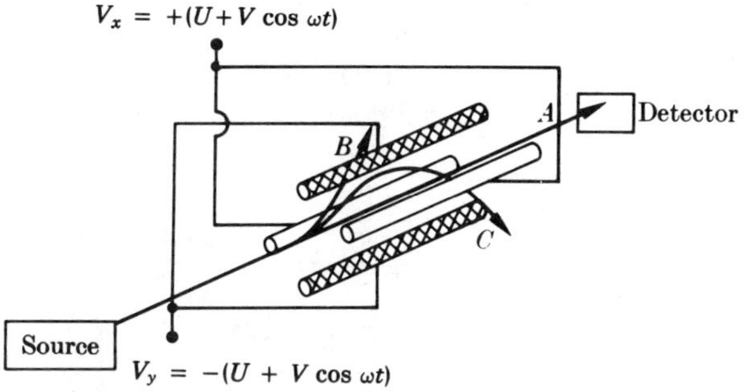

Fig. 20.6 Schematic diagram of the quadrupole mass filter. Ions B and C have improper m/z ratio and are thrown against the rods; A has the correct m/z ratio and is detected at the end of the analyzer. While the names positive pair and negative pair are used for rods along the x and y axis, respectively, their polarities are of the opposite sign twice during each period as the rf amplitude exceeds the dc potential. Ions in fact follow oscillatory paths through rods.

*A theoretical treatment of quadrupole analyzers suggests that the rods should be of hyperbolic cross section, rather than circular as shown, to establish between them the desired axially symmetric electric field. In practice, the loss in field symmetry with cylindrical rods is not severe and they are in common use.

The principle of a quadrupole filter is that application of an appropriately varying electrical field to the rods will cause all ions to impact on the rods during transit except those of a particular value of m/z. To accomplish this result, both a dc potential of amplitude U and a radiofrequency (rf) potential $V_{rf} \cos \omega t$, where V_{rf} is the amplitude constant, is applied between the pairs of opposite electrodes. As shown in Fig. 20.6, the potentials on the x and y electrode pairs must be $V_x = +(U + V_{rf} \cos \omega t)$ and $V_y = -(U + V_{rf} \cos \omega t)$. Ions moving along the z axis through the resulting electric field experience a combination of forces that cause them to undergo oscillations. For any ion oscillatory trajectory to be stable, the dc potential must be less than one-sixth the maximum amplitude of the rf potential: $U < V_{rf}/6$. Except for ions of a single value of m/z their oscillations tend to increase indefinitely in amplitude causing them to strike a rod. For a given frequency and set of voltages only ions in a narrow range of values of m/z are able to traverse the analyzer. Representative maximum values for U and V_{rf} are 500 V and 3 kV and the rf frequency is commonly between 0.5 and 3 MHz.

How is *scanning* accomplished? Theory shows that a linear mass scan with a quadrupole analyzer is straightforward if at a given rf frequency the ratio U/V_{rf} is kept constant while both the dc and rf potentials are increased linearly. Alternatively, the rf frequency can be increased while keeping U/V constant.

Since quadrupole mass filters have both low inductance and capacitance, they can be scanned quite rapidly from lowest to highest masses (in a few milliseconds). Their high scanning speed is responsible for their inclusion in many new designs of mass spectrometer. For instance, this advantage makes quadrupole filters especially attractive mass analyzers for high resolution gas chromatography (see Example 20.4). They can also mass analyze ions of *either* polarity without change, since ions of both charges will perform the requisite oscillatory motions.

Resolution can be changed by altering the dc/rf voltage ratio while maintaining a stable voltage ratio that does not exceed the maximum value. Yet resolution also depends in a complicated way on operating parameters. For example, the length of rods affects resolution since it determines the number of oscillations to which ions of a given m/z will be subjected. Further, if a resolution greater than 100 is needed, a trade-off must be worked out between increasing the rf frequency and use of longer rods. In practice, only low-resolution quadrupole spectrometers are presently available.

From a mathematical analysis of the equations of motion for ions it can be shown that for the entire range of stable ion paths there will be a fixed mass window Δm centered around the condition for maximum transmission through the analyzer. Thus all peaks in a quadrupole spectrum will have the *same width regardless of ion mass*.

The advantages of a quadrupole mass analyzer are low cost, simplicity, capability of rapid scanning, tolerance of "high" pressures, and ease of control by computer. The analyzer also has high transmission since paths do not depend on ion kinetic energy. The disadvantages of the analyzer relate mainly to the rods. They are difficult to machine precisely and to align well. The rods must also be cleaned periodically to remove deposits; layers that build up are normally nonconducting and become charged as ions impact. Such charges lead to variation in the quadrupole electric field and loss in analyzer resolution and sensitivity. Sometimes a quadrupole unit with short rods is installed as a prefilter to trap most deposits.

Sets of quadrupole filters with a resolution of about 250 and a mass range up to 750 amu appeared in the 1970s. By the mid 1980s quadrupole analyzers of higher resolution and upper mass limits of 2000 amu became available.

Example 20.2. If a quadrupole filter is operated to give a *resolution* of 100 at 100 amu, what resolution will it have at 1000 amu? How may the dependence of quadrupole resolution on mass be easily stated?

Since peak width is invariant with ion mass for this type of analyzer, the minimum separation of peaks Δm required for 10 or 50% valley resolution should be the same at both 100 and 1000 amu. From the definition $R = m/\Delta m$, however, resolution at 1000 amu will have increased to 1000 because the numerator will increase by a factor of 10 while the denominator is constant. This type of dependence of resolution can be described by a statement that the quadrupole has *unit mass resolution*. Its resolution may also be stated as 1M, where M is ion mass in daltons. For instance, at mass 500 the resolution will be 500. Clearly, having resolution improve with mass will greatly benefit spectral displays at higher m/z.

Example 20.3. How might a quadrupole analyzer with a resolution of 100 or less be effective as a *leak detector* for a vacuum system? Or as a pressure gauge?

Leaks in a vacuum system are commonly detected by connecting the system to a quadrupole analyzer (coupled to an appropriate detector and a display module). While the system is hooked to the quadrupole leak detector, it is evacuated and a stream of a low-molecular-weight gas such as helium is directed to each valve and seal that is to be checked for leaks. Where there is a leak, helium is drawn in and there in an output from the detector at $m/z = 4$.

A similar quadrupole arrangement can function as a pressure gauge. It is simple to calibrate its output in terms of partial pressure at values of m/z appropriate to the gas(es) in the system.

Example 20.4. Will a quadrupole mass filter provide sufficient speed and resolution as a gas chromatograph detector? (Chromatograph and mass filter detectors are commonly coupled through an interface that separates the carrier gas—usually He or H_2 at 1 atm—from the slower traveling analyte molecules that are eluting from the column. The momentum of the analytes is sufficiently greater to carry them into the ion source.) Will a quadrupole filter offer any advantages for this GC application?

Actually, a quadrupole mass analyzer is a common gas chromatograph detector (in GC/MS). As long as the chromatographic column separates components well, their identification seldom requires a mass spectrometer of high resolution. A fast scanning quadrupole mass analyzer accommodates better to the rapid appearance of analyte peaks than do many magnetic sector analyzers. Where *high* resolution is required, the latter type of analyzer would be desirable.

Ion Cyclotron Resonance Mass Analyzer. As discussed in the last section an ion that enters a steady transverse magnetic field with constant velocity will be caused to travel in a circular orbit. It can be shown that the angular frequency ω of circular motion for each ion depends on its value of m/e as well as the magnetic field strength B according to the expression

$$\omega = v/r = eB/m. \tag{20.8}$$

If an rf field of frequency ω is applied, an ion of the same frequency is in resonance with the field, absorbs energy from it, and expands its orbit steadily.

To determine a mass spectrum, sample ions are placed in crossed magnetic and rf fields, and the magnetic field is increased at a given rf frequency. Ions of different mass are then brought into resonance and the absorption of energy at different rf frequencies is monitored by a sensitive device like a marginal oscillator to obtain the mass spectrum. A schematic diagram of this analyzer is shown in Fig. 20.7. Note that ions enter from the source and that one kind, A, is experiencing resonance and absorbing energy from the rf field. This analyzer has often been used for the study of ion–molecule reactions.

Yet the resolution of the ion cyclotron resonance analyzer is limited, and further, is a linear function of detection time. In part it was the attempt to reduce the detection time that provided the impetus for the successful development of the Fourier transform mass spectrometer of the ion cyclotron resonance type. It is discussed in Section 20.6.

Time-of-Flight Analyzer. In this analyzer ions are extracted from the ion source in batches. Brief acceleration–extraction pulses V of about 100 volts are imposed

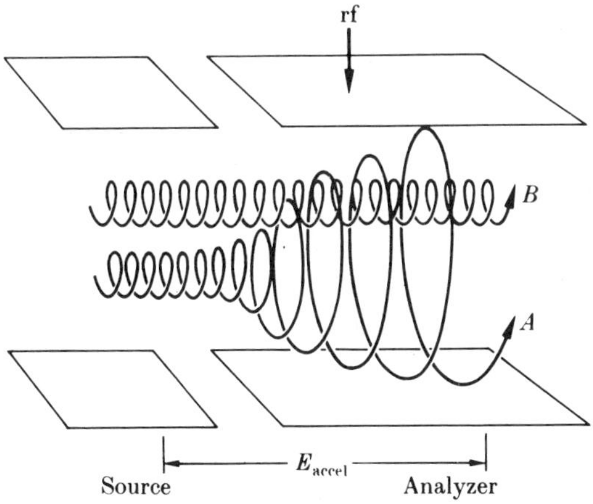

Fig. 20.7 Principle of the ion cyclotron resonance spectrometer. Ion A has the proper m/z, and its frequency corresponds to the rf frequency; it absorbs energy and describes an orbit of increasing radius while maintaining the same frequency. Ion B has a different m/z and does not absorb energy. Ions of the same m/z may have different radii of orbit, but must have the same frequency.

to establish a batch and achieve uniform kinetic energy. Each batch of ions is allowed to drift without collision down a long, evacuated, field-free tube. For ions of each m/z represented in the batch to reach a detector d cm away from the point of extraction the time of flight can be calculated from $eV = 1/2\ mv^2$ to be

$$t = dv = d(1/2V)^{1/2}\ (m/e)^{1/2}, \tag{20.9}$$

where v is the ion velocity. Equation (20.9) confirms that ions in each batch will become separated according to their values of m/z.

In a conventional arrangement with a drift tube about 2 m long and an acceleration pulse of about 100 V, the interval between the time of arrival at the detector of ions of m/z 300 and 301 will, however, be only tens of nanoseconds! To resolve such peaks, it will be necessary (a) that the duration of the initial accelerating pulse be shorter than the interval between arrival times and (b) that pulse resolution not be diminished in the detector and processing modules.

Since earlier detectors had rise times too long to resolve close-lying time-of-flight peaks, they were electronically gated to respond to ions of only a single m/z in each batch of ions. To cover the desired mass range, the interval of detection was advanced after each pulse for detection of the next m/z value; a scan over a mass range thus required a great many pulses and involved a relatively long measurement. While as a result the use of time-of-flight analyzers has been limited, the recent development of a channel electron multiplier of very short rise time ($<$ 1 ns) and narrow pulse width ($\sim$ 1 ns) seems to offer a solution. This detector should permit the detection of ions of all values of m/z in a single source pulses of sample ions. In addition, new methods allow production of pulses with very fast rise times ($<$ 100 ns). Further, post-acceleration methods have ensured sufficient momentum for heavy ions to be effectively detected by the first stage of the electron multiplier detectors.

20.5 DETECTORS

In a mass spectrometer the ion current at all or at particular values of m/z is measured, in most cases as ions leave the (last) mass analyzer. In a complex spectrometer mode such as MS/MS (see Section 20.8) the ion current may also be determined at other points, for example, after the first analyzer. In any spectrometer at least one detector (sometimes called a collector) will be needed. A detector must also be supported by electronic circuits of high gain, long dynamic range (three to five decades), fast response, and minimal noise. Electronics with these characteristics will provide for sensitive determination of the characteristically minute ion currents that change suddenly through a wide range as a scan is made.

The elegantly simple *faraday cup* type of detector is commonly used in lower-resolution instruments. It captures all ions of a particular m/z by letting them impact on a sloping plate electrode deep in a metal cup. Since their impact produces secondary electrons, the cup entrance usually contains a negatively charged screen or plate to repel them back into the cup where they will be recaptured, avoiding any loss in signal. The detector output is amplified by a high-input impedance operational amplifier. Its circuit also provides extensive negative feedback to increase stability and reduce the effective input capacitance, thereby shortening the circuit time constant and reducing response time. The faraday detector is inexpensive, is mechanically and electrically simple, and offers a response that is independent of ion mass or other variables. Its main disadvantage is lower sensitivity because it provides no internal amplification.

Most mass spectrometers today employ an *electron multiplier detector* that provides considerable internal amplification and thus high sensitivity and lower limits of detection. To see the importance of their internal gain of 10^4–10^6 one has only to calculate that the arrival of 6000 singly charged ions per second will produce a current of just 10^{-15} A. The types of electron multiplier in general use are the box-and-grid, Venetian blind, and continuous dynode varieties. It is essential that ions impact on the initial cathode with sufficient energy. Higher energies are particularly important for more massive ions ($m/z > 2000$). When the first two kinds are used with sector mass analyzers, incoming ions are often further accelerated (they enter the analyzer with 3–8 keV kinetic energies) toward a metal grid held at a high potential. Most of them pass through the grid, strike a Cu–Be cathode, and eject electrons, which are accelerated down a series of dynodes at successively higher potentials, with each impact ejecting a large burst of secondary electrons. The devices are nearly identical to photomultiplier tubes. Periodic cleaning of the Cu/Be surface renews performance as deposited material builds up.

In the continuous dynode multiplier a 1.8–2 keV potential difference is applied across a long trumpet-shaped channel. Ions impact at its entrance to eject electrons, which subsequently skip along its surface, giving rise to further secondary emission bursts with each impact. The greatly amplified flow of charge, now as an electron flow, is collected at its end. A full description will be found in Section 22.5. When a continuous dynode or channel electron multiplier is used with a quadrupole mass analyzer, usually external entrance electrodes are added. These electrodes are necessary to accelerate the quite low energy ions (2–20 eV) separated by quadrupole mass filters sufficiently so that their impact will induce secondary emission. Two accelerating electrodes are needed since quadrupole mass filters can analyze ions of either charge. To accelerate positive ions and deflect them into the detector one electrode is charged negatively and *vice versa*.

For higher ion currents the detector electronics are designed in the *dc mode*: the detector output goes to a high input impedance operational amplifier and its output current is averaged for a time corresponding to the circuit time constant. For small ion currents, however, a *pulse counting* or individual ion counting mode, which is analogous to photon counting, is more attractive. Here the detector output goes to a pulse amplifier, which is linked to a voltage discriminator circuit for rejection of lower amplitude noise pulses.

Advantages of multiplier detectors are stability, relative simplicity, and good internal amplification. A disadvantage is that efficiency of secondary electron ejection by an ion decreases as its mass increases. For higher mass ions ($m/z > 2500$) provision is made for extra ion acceleration before they enter the multiplier.

To determine precise *mass ratios of ions*, for example, to establish isotope abundance ratios, a single detector is seldom satisfactory. Small fluctuations in ion source conditions limit accuracy. To avoid this problem, a double beam or dual collection system is frequently employed. An instrument modified in this way is called an isotope ratio mass spectrometer or an isotope ratiometer. Its dual slit-and-detector units may be placed in appropriate positions for comparison of masses of interest.

Other Detectors. *Mattauch–Herzog* mass spectrometers bring all masses of ions to a focus along a plane, permitting use of a multichannel detector. *Photographic emulsions* on glass plates were the earliest such detectors. Ions captured by the emulsion leave energy deposits that can be translated by chemical development into plate darkening proportional to the integrated ion signal. Calibration and making some allowance for varying m/z ratios and energies is necessary. Emulsions available for positive-ion detection include Ilford Q plates, Kodak SWR, and Schumann plates. The relatively low limits of detection and moderate contrast of the Ilford QII plates cause them to be commonly used. Sensitivity is roughly dependent on $(m/z)^{-1/2}$ and energy, giving an overall relationship of sensitivity and the square root of ion mass.

A scintillation or *fluorescence* detector is also serviceable. A crystalline phosphor is dispersed on a thin aluminum sheet that can be charged to either the accelerating potential of the mass spectrometer (for stable ions) or to a level of about 50 V below that (for detection of metastable ions). Since the number of photons released per ion is moderate, the output must be passed through a window to a photomultiplier outside the spectrometer chamber. To make this detector especially useful for metastable ions, it is arranged that the ion beam pass through a slit in an enhancer plate just before striking the scintillation plate. Metastable ions will be repelled by the scintillator, strike the enhancer, and release secondary electrons. These in turn will be attracted to the scintillator and produce photons. The main advantage of the detector is its ready response to metastable ions.

20.6 MASS SPECTROMETERS

A great variety of mass spectrometers has been developed commercially for different applications. Their design has required appropriate choices among many types of ion sources, mass analyzers, other modules, and their coupling. Some choices enhance resolution, others mass range, limit of detection, speed, accuracy, and versatility. In describing the main designs we will draw heavily on the sections on specific modules that immediately precede this one. Description of the more complex hyphenated mass spectrometers, which are designated by acronyms such as GC/MS and MS/MS, will be postponed to the next section.

Single-Focusing Spectrometers. In a single-focusing instrument the modular layout is exactly as shown in Fig. 20.1: ion source → mass analyzer → detector. A good mass spectrometer of this classic design will have (a) both an electron-impact and chemical ionization ion source—providing versatility and a narrow range of ion kinetic energies, (b) a several-kilovolt ion accelerator–coupler, (c) a magnetic sector mass analyzer with narrow slits for good resolution, followed (d) by an electron-multiplier detector. Such an instrument can scan masses from 0 to 600 amu with resolutions up to 1800. Today common magnetic sector angles are 60° or 90° (see Fig. 20.8). The mass range can be increased by lowering the accelerating voltage, though this will also mean a small loss of resolution. For high scanning speed newer spectrometers employ a laminated core magnet. Single-focusing *magnetic sector mass spectrometers* are relatively inexpensive and have considerable versatility.

Alternatively, by designing around a quadrupole mass filter, a *quadrupole spectrometer* is obtained that costs less and can scan faster with better sensitivity be-

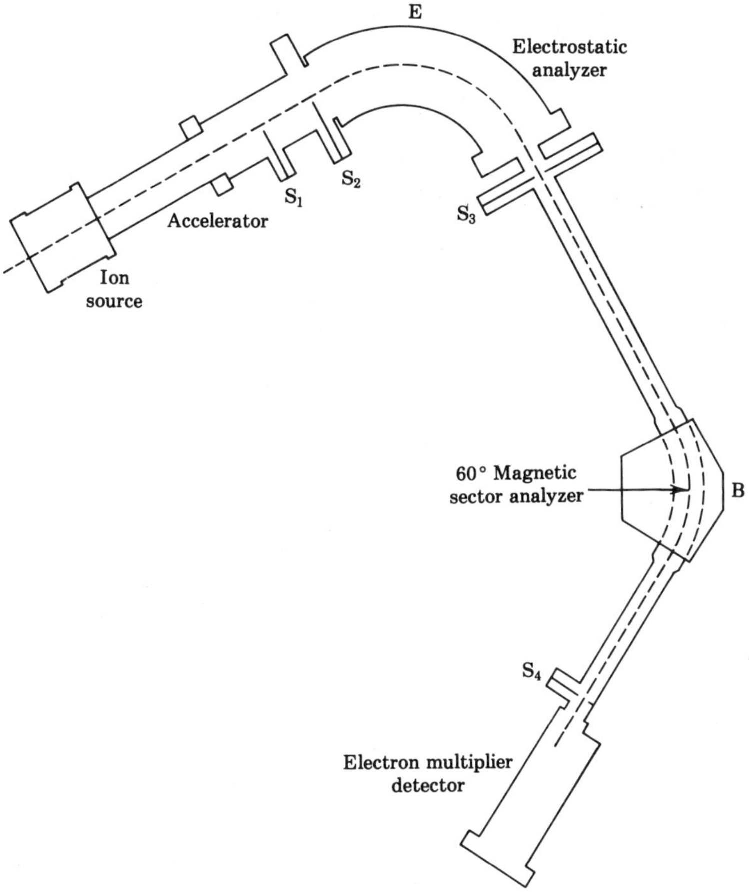

Fig. 20.8 Schematic of characteristic modules of a mass spectrometer of Nier–Johnson geometry. The instrument uses a 90° electrostatic sector and 60° magnetic sector to achieve double focusing of a high order. All masses are recorded for a particular radius r_m. To achieve velocity focusing in the electrostatic sector the ratio r_m/r_c must be 0.81. Four slits are employed. To secure high resolution, slit widths are made narrow, 12.5–1 μm.

cause higher ion transmission is secured. Its resolution is lower, however. A simple electrostatic extraction lens is sufficient to couple ion source and mass analyzer instead of an ion accelerator. This type of spectrometer is attractive for many applications.

Double-Focusing Spectrometers. As mentioned earlier, the greatest contribution to higher resolution in an instrument with a magnetic sector analyzer is addition of an electrostatic sector. It will ensure that ions of uniform energy enter the mass analyzer. A widely used type of spectrometer incorporating these two sectors, the Nier–Johnson design, is shown in Fig. 20.8. Ions are focused at the exit slits of both sectors. With large sector radii (30–40 cm) and slit widths narrowed to 2 μm, resolution can be increased to 1–2 times 10^5. The newest Neir–Johnson high-resolution instruments have provision for rapid scanning (1–3 s) of a normal sample over the mass range m/z = 10–2000. The Mattauch–Herzog double-focusing design, which is quite different, is described below.

Example 20.5. How are voltages and magnetic field strength varied in a double-focusing mass spectrometer when a mass scan is made? Is it sufficient to vary the strength of the magnetic field systematically?

As described earlier, the accelerating voltage must first be set sufficiently low to permit the desired upper limit on the mass scan, which will be defined by Eq. (20.5). In concert with that change, the electric sector voltage must be reset to accommodate the new average ion kinetic energy. Varying the magnetic field strength will then accomplish the mass scan.

Example 20.6. How important are sector size and slit width in determining the resolution of a double-focusing spectrometer?

They are of central importance. Indeed they affect resolution in the manner one would intuitively expect. With the Nier–Johnson design and sector radii r_e = 18.8 cm and r_m = 15.2 cm and slit widths S_1–S_4 = 12.5 μm a resolution of 14,000 can be secured. If sector radii are increased to r_e = 50.3 cm and r_m = 40.6 cm and slits are narrowed to 1 μm, resolution increases to 75,000.

Example 20.7. How feasible is a double-beam mass spectrometer? Would its possibilities for referencing a signal from the sample channel against one from the reference channel be attractive?

At least one double-beam spectrometer has been developed using Nier–Johnson geometry. This device uses a pair of ion sources and a pair of detectors, but directs the two ion beams through the same electrostatic and magnetic analyzers. By having sectors sufficiently wide and separate slits for each channel, stability is increased and the cross talk between the beams is kept to about 5%. In this design it is possible to compare an unknown compound in one channel directly with that of a known sample or a reference material such as perfluorokerosene in the other channel.

Vacuum Systems. Brief attention to spectrometer vacuum systems is also needed. A high vacuum must be maintained in several modules of a mass spectrometer. Chief among the many reasons already mentioned is the need to minimize *unexpected* ion–molecule collisions and changes in ions induced by such collisons after the ion source. In addition, operation at kilovolt potentials requires a high vacuum to prevent electrical arcing. A dividend of maintaining lower pressures is that contamination of the ion source and mass analyzer(s) is reduced.

To secure low pressures in a mass spectrometer roughing or fore pumps first lower the internal pressure to about 1 Pa (10^{-2} torr). Diffusion pumps that employ jets of organic liquids of low vapor pressure further reduce the internal pressure to about 10^{-3} Pa or ultimately to about 10^{-6} Pa (10^{-8} torr). Alternatively, turbomolecular pumps that employ high-speed bladed rotors (rpm 40,000 or higher) to push gas molecules out of a system are used. Sometimes cryopumps that remove gases by condensation are added. The more flexible and sophisticated the mass spectrometer, the more intricate its pumping system. Differential pumping of modules is common. For example, in a chemical ionization module collisions are necessary and pressures are expected to be 10^2–10^4 times higher than in a mass analyzer or detector where they must be minimized.

Mattauch-Herzog Geometry. In this type of spectrometer ions of all values of m/z are brought to a focus along a plane. Its layout is shown in Fig. 20.9. The attractiveness of the design lies in the possibility of use of a multichannel detector such as a photographic plate, a microchannel plate, or a linear diode array. Simultaneous dectection of all peaks is a distinct advantage with a small sample or a sample whose composition changes rapidly. While photographic plates are seldom employed today because of the slowness of development and densitometer scanning, modern multichannel detectors avoid this limitation. This type of spectrometer was especially popular at an earlier time for elemental analysis of inorganic samples with spark-source excitation.

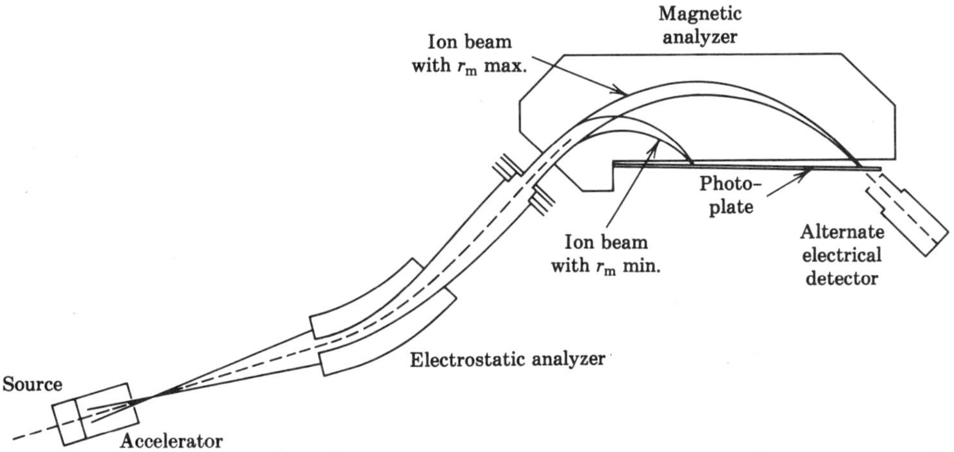

Fig. 20.9 Schematic of characteristic modules of a mass spectrometer of Mattauch–Herzog geometry. With this instrument an entire spectrum can be detected simultaneously by a multichannel detector, usually a photoplate or microchannel plate. The mass scale is quadratic since r_m is proportional to $\sqrt{\text{mass}}$ for constant ion kinetic energy. Known mass lines are used to calibrate the multichannel detector or the output of the alternate, movable electrical detector.

Fourier Transform Spectrometer. The heart of the Fourier transform mass spectrometer is an ion cyclotron analyzer much like the one illustrated in Fig. 20.7. The Fourier transform device differs in that all ions from a sample are introduced into the analyzer nearly simultaneously and then excited. The exciting radiofrequency signal may be generated by a broad- or narrow-band generator whose output can be swept quickly (in a 2–5 ms "chirp") through the requisite frequency range. The time domain signal thus generated must be Fourier-transformed to yield ion cyclotron resonance frequencies. It is Eq. (20.8) $\omega = 2\pi f = B/(m/e)$ and standardization of frequency against known-mass ions that provide the m/z values in the spectrum. Note that frequency is inversely proportional to m/e. For example, for a magnetic field of 1 T, an m/z 500 ion has a cyclotron frequency of 30.7 kHz, and a m/z 250 ion, a frequency twice that.

What must occur in the ion cyclotron chamber? How can the response of all ions be followed? What changes in instrumentation will make this possible? A single-cell arrangement that has been used is shown in Fig. 20.10. In this cell the several steps in measurement are separated *in time*. After an initial quench pulse (see below) the sequence of operations is as follows:

1. A sample (or for CI, a sample plus reagent gas) is admitted.
2. A short electron pulse is generated to cause EI (or CI) ionization of the sample.
3. Ions are trapped by a strong, steady magnetic field that induces ion cyclotron motion.

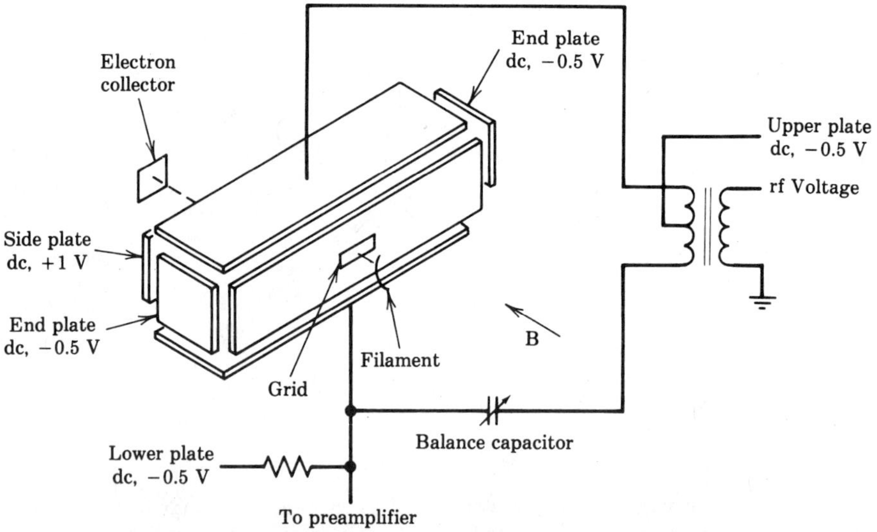

Fig. 20.10 Schematic drawing of a one-region ICR cell for a Fourier transform mass spectrometer. It stores gaseous positive ions by the combined effect of the strong homogeneous magnetic field B and an electrostatic field created by small dc voltages on the six metal plates.

4. As soon as the electron pulse ends (or after an appropriate interval in the CI mode) a fast radiofrequency sweep (about 1 ms) coherently excites all ion masses (believed present) to larger cyclotron orbits.
5. Following the rf sweep, the decay of cyclotron motion induces *image currents* in a receiver circuit in the cell walls. This time-dependent signal is observed for milliseconds to seconds, the larger time being possible at lower cell pressures.
6. At the end of the measurement cycle a quench pulse expels ions.

Subsequently, the complex decay signal is amplified, phase-detected, filtered, digitized, and stored in computer memory. The final step is taking its Fourier transform to shift information from the time to frequency domain and display of the mass spectrum. To obtain the entire frequency spectrum the signal bandwidth for data acquisition may be required to be as much as 1 MHz, which means that a large number of bits will be recorded in memory. For example, if $B = 1$ T (1000 G) and the mass range is 15–1200 amu, Eq. (20.8) shows that the angular frequency extends from 12.5 kHz to 1 MHz.

As is true of other Fourier spectrometric procedures, the Fellgett or multiplex advantage of simultaneous observation of all masses of ions allows either low resolution and very fast scanning or the realization of much higher resolution with a data acquisition time as long perhaps as 1 s. Usually the latter choice is made; the Fellgett advantage is used to increase the resolution as much as possible. Data showing resolutions of 700,000 at m/z values of 80 using magnetic field strength of 2 T have been obtained. As the value of m/z increases, resolution decreases. The higher the magnetic field, however, the higher the resolution.

Fourier transform mass analyzers of the ion-cyclotron type have the capability of very rapid scanning. The ionization, excitation, and data collection can occur with sufficient speed that scan repetition may approach 10 Hz. The time-of-flight mass analyzer has a potential to produce a complete spectrum at very high rates (where the heaviest ion is the mass 800 amu, a repetition rate of about 11 kHz is possible). Yet time-of-flight spectra are usually obtained by sampling only one mass for each pulse of the source. In this case with an adequate S/N ratio an acquisition rate of about 8–10 Hz can be attained.

The Fourier transform mass spectrometer (FTMS) is able to offer the different aspects of instrumentation in time rather than in space. Since all masses can be scanned simultaneously, the device lends itself to development of time profiles on collision-induced reaction.

As mentioned for ion-cyclotron mass analyzers, FTMS systems require very high vacuum or resolution may be lost. Thus, molecular sampling techniques such as fast atom bombardment, which will be discussed in the next section, are not adaptable to the single-cell FTMS designs. Adaptation of these alternative sampling systems is still under development at present.

Processing Modules. What should be said about the many systems for data acquisition that are part of mass spectrometer systems? It is characteristic of design that many data systems are available and that one with appropriate capacity and other properties is coupled to an assembly of characteristic modules. Fast *sampling rates* are essential in ranges from 20 to 250 kHz. These rates must, of course, be coordinated with the number of peaks anticipated for a spectrum, the resolution required, and the actual scan time allowed per decade of m/z values. For a conventional scanning rate of 2 s per decade of mass, a 20-kHz rate of sampling will probably be adequate to provide a resolution of about 1500. If the time per decade is decreased to 0.3–0.1 s, sampling must move up toward the top of the range suggested above to maintain resolution. If high resolution is sought, the higher sampling rates are required. An AD converter of 12-bit quality will give an amplitude precision of 0.03%, but 16-bit converters are common where high quality is desired, and the best system can use 18- to 20-bit converters. Very high storage capacity is important and disks that will take care of 25 Mbytes or more are usually provided.

For effective processing it is important to choose between continuous spectral scans and measurements of selected ions. Most modern spectrometers provide both modes. For most samples data points will not occur for all values of m/z and an exploratory run on a new sample will locate such regions as well as allow one to identify the significant peaks to be monitored. The gain if one can concentrate on selected peaks is not only in the processing efficiency but also in measurement precision, sensitivity, and in lower limits of detection.

Selected-Ion Monitoring. Selecting from 1 to n peaks to monitor is especially attractive for routine measurements of from 3 to 10 analytes in samples. For this method, from the instrumental side it is only necessary to select from 1 to n voltages at which selected ions will reach the detector. With a quadrupole mass analyzer, voltages can be changed rapidly to monitor selected ions of m/z of 50–500 or more. When a magnetic sector mass spectrometer is used, a limiting factor is the magnetic reluctance or "inertia" of the magnet itself, which causes response times to change operating currents to be relatively long. The recent development of laminated, grain-oriented magnets has allowed faster scanning rates and has made this technique routine for magnetic instruments. In Fig. 20.11 selected monitoring of four ion masses is shown. Note at the bottom the timing diagram controlling the data collection time for each peak. In this instance the intensity of the magnetic field was switched to refocus the mass analyzer sequentially on the selected ions.

Full Scanning of Spectra. With the advent of possibilities for fast scanning (1 s or less per decade) and large memories, continuous scanning over a wide range of masses has become possible. Even where the desired mode is monitoring selected ions, an initial survey scan is likely to be required. Slower scans to obtain complete spectra are, of course, straightforward. Where high resolution data ($R > 10{,}000$) are sought, somewhat slower scans would be advisable. Often a mixture of a standard such as perfluorinated kerosene (PFK) and a compound of interest are used. Computer programs controlling the data acquisition process, such as *thresholding*

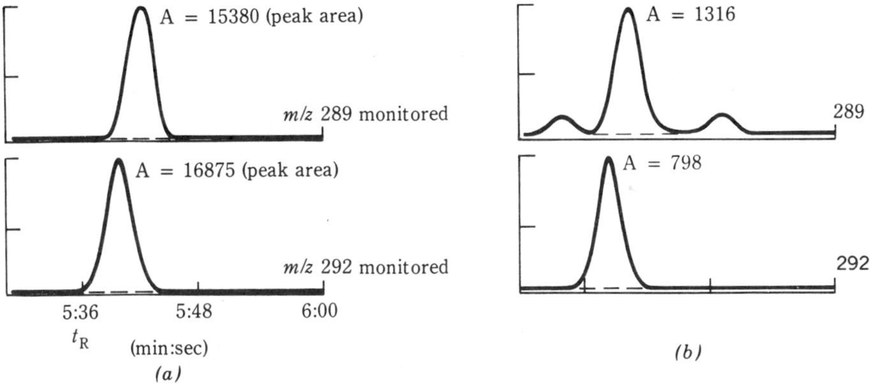

Fig. 20.11 Application of selected ion monitoring of methyl malonic acid (MMA) as a test for vitamin B_{12} deficiency in patients. (The acid is a required cofactor for the enzyme that converts MMA to succinic acid. Urine samples are extracted with a diethyl ether-ethyl acetate mixture, the extract dried and derivatized to obtain t-butyldimethylsilyl ester derivatives.) A stable isotope analog $^2H_3$ methyl malonic acid is used as an internal standard. Derivatized samples are analyzed by GC/MS and the $(M\text{-}57)^+$ ions characteristic of fragmentation of derivatized MMA (m/z 289) and internal standard MMA (m/z 292) are selectively monitored. The MMA derivative is well resolved from isobars. (a) Ion profiles (response versus retention time) for 1 : 1 mixture of derivatized standard MMA and internal standard. The ratio of peak areas of m/z 289/292 equals 0.91 and is one point on a standard curve that permits the absolute concentration of MMA in human urine to be determined. (b) Scaled response curves for urine sample with internal standard added. (Courtesy of Drs. David S. Millington and Daniel L. Norwood.)

and *peak acquisition*, are discussed in Chapter 12 in some detail; suffice it here to note that thresholding insures that processing of data occurs only when four successive data samples indicate that a single amplitude at the detector is greater than a preset figure. Peak acquisition implies that once the thresholding requirement has been met, data are sampled until there is again a return to the baseline, following which the peak maximum position in terms of m/z ratio and amplitude are determined by a program. If the standard and a compound are both involved with the mass spectrum, it is possible to observe the early values of PFK peaks and arrange to convert subsequent peaks to masses.

Where a photoplate detector is employed, data acquisition takes the form of using a densitometer scan of the plate. While this method proves to give highly accurate data, it is, of course, very much slower and less convenient.

For a mass calibration under low-resolution conditions, it is usually sufficient to calibrate against a standard once a day. If the instrument is not detuned or the settings changed, the reproducibility of a scan is very good. Indeed quadrupole spectrometers frequently will maintain the calibration over a period of days when routine measurements are made.

Display Devices. Today most mass spectral data are digitized and stored on magnetic disk or in the memory of a computer for processing. Spectra can be displayed on a printer–plotter as needed. Interactive cathode ray tube displays are generally used to manipulate the data and choose the form of spectrum to be printed. One design of spectrometer presents spectra photographically. In some other spectrometers of earlier design a fast recording oscillograph display is also used. The output of the detector amplifier is directed to several mirror galvanometers with different sensitivities and several traces of a spectrum are made simultaneously. With four or five galvanometers, intensities differing by a factor of 10,000 can be read from the display.

20.7 ION SAMPLING METHODS; ION SOURCES FOR INVOLATILES

The two most common processes of producing of gaseous ions for mass spectrometry, electron impact and chemical ionization, were discussed in Section 20.2 for volatile samples. Now those processes need reexamination to emphasize their role in *sampling* as well as development of ions. This additional exploration will set the stage for description of the host of ion-sampling processes developed for unstable and/or involatile substances in recent years. They have greatly extended the range of samples that can be mass analyzed. A comprehensive listing of current ion-sampling modes will be found in Table 20.3.

Pyrolysis. Once the only method of applying mass spectrometry to unstable or nonvolatile organic and biological materials was to *pyrolyze* them, that is, subject them to thermal fragmentation and mass-analyze the fragments generated. Today this method continues to be quite useful for some classes of materials. The volume by H. L. C. Meuzelaar et al., *Pyrolysis Mass Spectrometry of Recent and Fossil Biomaterials*, New York: Elsevier, 1982, is a useful account. For example, coals are often classified by the masses of their pyrolytic fragments. In many other cases, however, laser desorption/MS has largely replaced pyrolysis/MS.

Molecular Sampling of Volatile Materials. The introduction of volatilizable samples into an ion source is straightforward. Most low-molecular-weight samples are admitted directly as gases. Others need to be heated to 325–450 K. The sample vapor is usually stored in a glass bulb (reservoir) and "leaked" into the ion source through an aperture sufficiently large to avoid fractionation of analytes caused by different effusion rates. Both the reservoir and its inlet must be kept warm enough to maintain a pressure of 0.3–10 kPa (2–80 torr).

An unstable or low volatility sample may ordinarily be introduced by coating it on a *direct-insertion probe*. The probe tip may have been silanized or otherwise chemically treated to enhance the release of molecular ions. The probe with adhering sample is inserted into the ion source until the sample is 1–3 mm from the electron beam in an EI source or centered in a CI source, and the tip heated briefly to speed "desorption" of analyte.* Once volatilized, analyte ionization proceeds

*R. J. Cotter, *Anal. Chem.*, **52,** 1589A (1980).

Table 20.3 Ion-Sampling Methods in Mass Spectrometry

Mode of Volatilization	*Steps in Sampling*	*Mode(s) of Ionization*
	A. Thermal Volatilization	
Normal evaporation or sublimation at accessible temperatures	Store sample vapor in thermostated reservoir and admit it from there to ion source through a controlled leak	Electron impact, chemical ionization, or field ionization
	B. Volatilization of Atoms/Ions in Ion Source	
Atomic sputtering by ions (SIMS)	Prepare flat surface on sample and expose to bombardment by beam of Ar^+ or other ions of keV energy	Sputtering by energetic ions
Plasma heating (DC or inductively coupled AC) (ICP–MS)	Dissolve, nebulize to aerosol, and introduce into Ar plasma torch	Radiative and collisional heating
High voltage spark (spark source MS)	Powder, add conductor and binder, if necessary, and form into electrode for sparking	Electrical spark
	C. Volatilization of Molecules/Molecular Ions in Ion Source	
Fast atom or ion bombardment (liquid SIMS)	Dissolve sample in drop of glycerol and place on insertion probe; bombard with atoms or ions of keV energy	Sputtering by energetic, heavy ions or atoms
Thermal desorption	Coat sample on insertion probe; sublime with resistive heating	Electron impact, chemical ionization
Laser desorption	Deposit sample on metal or mesh; expose to finely focused and pulsed laser beam	Multiphoton ablation (of protonated ions); thermal desorption (often of cationized molecules)
Field desorption	Deposit sample on multiple fine carbon needles on electrode and subject to high electric potential	Very strong electric field
^{252}Plasma Desorption	Deposit sample on foil and bombard with fission fragments of MeV energy	Bombardment by highly energetic ions

normally by the EI or CI mode.

Molecular sampling of involatile materials. Over the years mass spectrometrists have worked steadily and successfully to extend their techniques to substances with molecular weights of 10,000 and higher. Their newer methods permit the addition of energy to large molecules in the small amounts needed to (a) overcome intermolecular forces and (b) facilitate gentle ionization that will preserve the integrity of these large ions. Sampling and ionization are almost a single process. Sometimes the molecular ions formed are protonated, sometimes cationized, for example, released as $(M + Na)^+$. The cation attached is often from an impurity. The range of molecular ion sampling modes currently in use is suggested in Table 20.3. While that listing is inclusive—elemental or atomic sampling for isotopic and elemental analysis is included as well as the molecular sampling now being discussed—it is not exhaustive. First, molecular methods will be described.

The sputtering or desorption technique called *fast atom bombardment* (FAB), which is really a liquid or molecular form of secondary ion mass spectrometry (SIMS), is especially attractive in yielding molecular ions in appreciable concentrations. The two qualities that distinguish this method from straight ion bombardment are

1. the sample is dissolved in a low-volatility, viscous solvent such as glycerol, and
2. energetic atoms, for example, 4–8-kV Xe or Ar atoms, or sometimes energetic ions, are used.

Molecules of analyte at the surface of the glycerol solution apparently get caught up in the collision chain (see Section 22.3) that results from the impact of energetic atoms; as a result, they appear in the plasma essentially whole. Importantly, for a fraction of sputtered molecules ionization occurs as a natural consequence of the collision chain disruption. The bombarding atoms are found to penetrate several layers and distribute their kinetic energy broadly. The energy is not focused into one molecule, and fragmentation such as occurs in EI does not occur. Many compounds give spectra that are more comparable to CI spectra. Molecular ions are generally observed even for very labile compounds. A definite advantage of having the sample in a solution is that analyte at the surface is constantly replenished by diffusion from the interior, ensuring good sensitivity in measurement. Thus the choice of solvent (matrix) is important.

What is the upper limit on molecular size for this ion-sampling procedure? If insolubility is a problem, addition of up to 1% of a solubilizing agent such as Triton X100 often helps. (Naturally, both solvent and solubilizing agent will contribute peaks to the mass spectrum.) In general, good results have been obtained up to molecular weights of about 9600 amu. Species covering the complete range from hydrocarbons to polar materials such as oligopeptides and oligosaccharides have been examined.

The necessary fast atom beam can be obtained from an ion gun in which charge exchange is arranged at its exit between accelerated ions and neutral counterparts. Accelerated atoms emerge, usually with little loss of momentum. Any ions remaining can be deflected from the beam by a charged electrode. It has been found that a sputtering angle about 60° to the normal to the solution surface gives optimum ion production.

Laser Desorption. Light pulses from a laser represent still another way to release ions from a sample. Laser irradiation of the smooth surface of a sample at about a −90° angle will produce a cloud of atoms and ions. Ions can be extracted into a spectrometer at about a +90° angle. Often such a system is part of a *laser microprobe.*

The laser beam can be focused down to about a 1-μm spot. The ensuing multiphoton interaction with the sample appears to overcome binding forces so that all structure is lost in a sample and ions are released. [R.J. Cotter and J. C. Tabet, *Am. Lab.*, **16**(4), 86 (1984).]

Since ion production is also pulsed, a time-of-flight or a Fourier transform mass spectrometer that can handle ions of all m/z ratios at one time is best adapted to the analysis. In a laser microprobe spectrometer the user ordinarily first examines a sample carefully by an optical microscope and then directs the laser beam to regions of interest.

Though very thin samples (thickness $< 0.3\ \mu$m) can be used, a sufficient amount of sample can be volatilized by a pulse to obtain very low absolute detection limits (down to 10^{-19} g for species such as lithium and other alkali atoms). The absorption characteristics of the sample partly determine the response as well. The probe has good capability for molecular species and generates molecular ions from many high-molecular-weight (<800 amu) materials. Mass resolution is ordinarily of the order of 200–800. For many samples detection limits appear comparable with those attainable by electron microprobe X-ray spectrometry.

In general, the method provides good qualitative information. It furnishes only semiquantitative data on concentrations unless considerable care is taken in controlling laser parameters such as power density, pulse repetition rate, and scan area and in cleaning the sample surface. With a laser probe it is also difficult to define the analytical volume. Chemical interferences arise, apparently as a result of matrix effects having to do with mechanisms of laser absorption. Suitable standards are also a problem.

Californium-252 Plasma Desorption. Fission products from a ^{252}Cf nucleus provide high energy (75-MeV) sputtering, which is valuable in bringing about the release of sample ions.* Their impact on a sample generates so much electronic excitation that the collision chain mentioned earlier seems to become a *localized plasma.* Both molecular ions and fragment ions from the sample are ejected from the plasma.

Measurement begins when a thin layer of ^{252}Cf is placed behind a foil holding a thin layer of sample. When an atom of this highly unstable isotope undergoes fission, it yields a *pair of ions*, each with a mass of at least 100 amu, charge as great as 20^+, and kinetic energy as large as 75 MeV. The fission ions from ^{252}Cf travel in opposite directions: one is captured by a detector on the left and the other impacts through the foil on the sample.

Repeated fissions will occur at microsecond intervals, so a mass spectrum is best obtained by a time-of-flight method. Timing is done by a digital clock, which is started when

*R. D. Macfarlane, *Anal. Chem.*, **55,** 1247A (1983).

the particle detector on the left receives its fission fragment. Molecular and fragment ions from the sample are accelerated toward a charged grid where they are captured by a particle detector. The ions ejected from the plasma can be mass analyzed. Their time of arrival at the detector is measured precisely by the digital clock. A precise m/z time scale is also established for the clock by recording the time of arrival of a known, simple inorganic species such as Na^+ ion. Ion flight times are recorded in channels of computer memory. Within 3 s a mass range of m/z up to 10,000 can be analyzed.

If it is desired to detect light ions, such as from air, the acceleration voltage is increased. By design, path lengths from sample to detector are no longer than 10–50 cm to minimize the possibility of unimolecular ion fragmentation before detection. Resolution ranges from 500 to about 4000. Large biomolecules were analyzed in this way prior to the introduction of fast atom bombardment.

Field Desorption. Molecular ions can also be released from samples deposited on a conducting surface by subjecting them to an intense electric field under high vacuum.* Point electrodes are required to attain the very high field strength necessary (10^7–10^8 V cm^{-1}). Large numbers of points can be attained within a small area by growing carbon needles (10^1–10^2 nm radius) on a fine tungsten wire by placing it in benzonitrile at high temperature. In such strong fields ionization appears to occur by (quantum mechanical) tunneling. The process is not sharply defined in time as in electron impact ionization. Positive molecular (or fragment) ions are released. There is very little fragmentation since (a) there are no electron collisions and (b) ions move into the mass analyzer in about 10^{-6}–10^{-7} s.

With the array of microneedles a relatively large mass of sample can be subjected to high fields. Once a sample has been deposited it is heated to assist desorption by passage of current through the wire or directing a laser pulse toward it.

If the sharply pointed electrode is a metal and has not been coated with sample, positive metal ions are released. This effect is known as *field ionization*. This process is sometimes used as a spectrometer ion source by introducing a gaseous sample that will deposit on the tip of the electrode and undergo desorption and ionization.

Atomic Sampling. Ion sampling of metallic and inorganic solids basically yields elemental ions such as Sn^+, Fe^+, and O_2^+. In the type of ion source called a *spark source*, a pulsed (commonly 50–100 pulses s^{-1} of 50–100 μs duration), high voltage (20–60 kV peak-to-peak) radiofrequency spark is used to vaporize and ionize a sample. Such sources are useful for materials that can be fashioned into an electrode, much as is done in arc and spark atomic emission spectrometry (see Section 13.4). A spark source simply replaces the usual ion source. Nonconductors and inorganic solids can be powdered, mixed with a conductor of known composition, thoroughly blended, and compressed into electrodes. Samples existing as powders are treated similarly.

In recent years plasma torches, and especially the *inductively coupled plasma torch* (ICP) have provided a valuable alternative to the spark source. Samples are first dissolved, sometimes after fusion with a substance like sodium borate, to give a soluble glass. Then the solution is nebulized to an aerosol and introduced into

*W. D. Reynolds, *Anal. Chem.*, **51,** 283A (1979).

the 5000–6000-K argon plasma of a torch (see Section 13.4). Excitation of samples produces an abundance of monovalent ions since most elements have much smaller first than second ionization energies. More than 50 elements are singly ionized with an efficiency of 90% or higher.

Extraction of the ionic portion of the plasma into a mass analyzer generally involves placing a pair of nested conical water-cooled electrodes that have small central apertures into an atmosphere-pressure plasma. The electrodes are strongly charged to attract elemental ions. As the ions accelerate, they pass through the orifices in the plates into the mass spectrometer. Strong pumping after each electrode finally lowers the pressure to the usual mass spectrometer value. The limits of detection of atomic species by this mode of mass spectrometry are from 20 to 1000 times better than by regular atomic emission spectrometry. Isotopic analyses can be made routinely also.

The valuable technique of *secondary ion mass spectrometry* (SIMS) employs ion sputtering to atomize and ionize the surface layers of a sample. The ionized portion of the material is capable of extraction into a mass spectrometer as discussed above. An ion gun bombards a selected area with a focused beam of 1–10 keV Ar^+ or Xe^+ ions. Sputtering is continued until there are enough gaseous ions for a quantitative elemental determination of the analytes, even at the ultratrace level. Sometimes one finds that a large crater must be sputtered in a sample. A full presentation of SIMS will be found in Section 22.9.

Actually, for release of ions, bombardment with energetic atoms also proves effective. Efficiency in sputtering proves mainly dependent on the transfer of momentum from bombarding species rather than their charge.

There are also low-energy elemental ion-sampling processes such as the hollow-cathode generation of ions and the closely related method of *glow discharge*.* These methods rely on low-energy noble gas ions to sputter and ionize sample atoms. A potential of a few hundred volts dc is applied between two electrodes in a rare gas at $10–10^3$ Pa (0.1–10 torr) to accelerate positive ions for vigorous sputtering. The sample may be attached to a cathode by a spring clip, or placed on or within it as a foil or solid disk. Because of the gas flow that accompanies the process of ion extraction for this type of source, differential pumping is required.

20.8 SEPARATIVE-CHANNEL MASS SPECTROMETERS: CHROMATOGRAPHIC AND MS/MS TYPES

The remarkable powers of mass spectrometry can be extended to complex organic mixtures and other difficult samples by adding a *separative sample channel* to a mass spectrometer. As discussed in Sections 1.5, 1.8, and 11.2, such a channel ideally allows separation of the components of a mixture quantitatively, in reproducible fashion, and with high resolution. At present the main types of separative channel in general use with mass spectrometers are based on either chromatographs—gas, liquid, or supercritical fluid—or mass spectrometers. These "hy-

*W. W. Harrison et al., *Anal. Chem.*, **58,** 341A (1986).

phenated'' or tandem instuments are usually identified by the acronyms GC/MS, LC/MS, SCF/MS, and MS/MS.

The introduction of separated analytes into a mass spectrometer greatly increases the specificity of measurement. Since the chromatographic aspects of separation will be discussed in Chapters 24–26, here the focus will be on (a) the interfacial device by which a separative channel is coupled to a mass spectrometer and (b) the strengths and limitations of hyphenated or tandem mass spectrometers.

Chromatographic Separative Channels. Since gas chromatography was the first chromatographic technique to develop in sophisticated fashion, GC/MS was probably the first hyphenated instrument. Later as high performance liquid chromatography came of age, developing LC/MS interfaces also received priority, and SCF/MS has been the newest hyphenated methodology to be undertaken.

What sort of interface best couples a gas chromatographic column and a mass spectrometer? The coupling must ideally ensure that (a) the carrier gas at about 1 bar (atm) pressure is removed while (b) the eluting analytes move into the spectrometer. For packed gas chromatographic columns, the best interface appears to be the *jet separator* shown schematically in Fig. 20.12. In it the carrier gas molecules (He or H_2) in the eluting jet diffuse rapidly into the vacuum around the nozzle and are pumped out of the connecting chamber while the higher momentum of the slower moving analyte molecules carries them into the aperture of the spectrometer ion source.

For open tubular or capillary columns, as shown in Fig. 20.12, *direct coupling* to the mass spectrometer is feasible because of the much smaller gas flow. The carrier gas going into the ion source will be evacuated satisfactorily as long as a high-capacity vacuum pump is used. With a fused silica capillary the column can

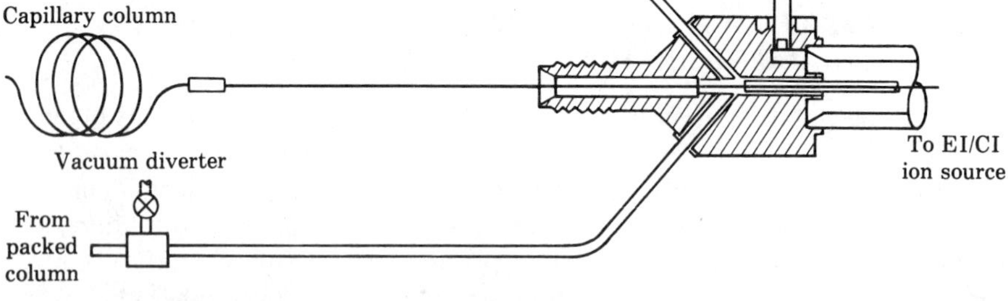

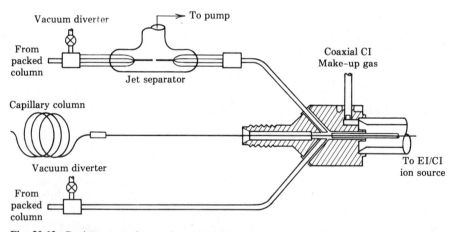

Fig. 20.12 Devices to couple gas chromatographic sample channels to a mass spectrometer. Only true capillary columns are ordinarily coupled directly. The two packed column interfaces have splitting devices (vacuum diverters) to reduce gas flow into the spectrometer substantially. (After G. M. Message, *Practical Aspects of GC/MS*. New York: Wiley, 1984.)

be "threaded into" a mass spectrometer to within 1–2 cm of the ion source to ensure that component resolution is not degraded. Direct coupling appears feasible for SCF/MS coupling.

Temporal Compatibility. For effective coupling it is important that the response time of the mass spectrometer be sufficiently short to handle the output of the separative sample channel. The time scale on which components emerge from a packed-column chromatographic channel is usually of the order of minutes with peaks having half-widths of 10–25 s. Mass spectrometers can admit eluant continuously since they require no more than microseconds to ionize and mass analyze a component and can scan a mass range of 50–500 m/z in a time as short as 1 s. Indeed many mass scans can be made during the elution of GC peaks, allowing several mass spectra to be averaged to achieve better limits of detection and precision.

With open tubular gas chromatographic columns temporal peak widths are often less than a second and ensuring temporal compatibility between sample channel and mass spectrometer becomes more difficult. For example, high-resolution gas capillary columns can produce chromatographic peaks less than 0.1 s in breadth. Commercially available columns routinely deliver peaks from 1 to 5 s in width. A mass scan from m/z 60 to 600 takes at least 0.1 s at present, taxing the limits of most modern GC/MS systems. In any event, it is important that mass spectral scans be sufficiently fast to (a) minimize changes in sample concentration during a scan and (b) allow reconstruction of the emerging chromatograms (see below). For fast mass scans few ions of a given m/z ratio will be collected, and the randomness of ion arrival (Poisson statistics) will be limiting. There will be error both in terms of the useful range and the apparent concentration. Spectrometric precision will be proportional to $\sqrt{N}$, where N represents the number of ions detected per second. [J. F. Holland et al., *Anal. Chem.*, **55,** 997A (1983).]

In this situation selected ion monitoring in which one to three representative m/z peak are observed for each analyte offers a possibility of higher speed if routine samples are being analyzed. Much less time is required to step the mass spectrometer between significant peaks, allowing more time spent at peaks with a considerable increase in signal and thus precision.

The temporal demand on the mass analyzer is also of concern. Comparable rates of scanning (up to 0.1 s per decade) can be achieved by quadrupole instruments and magnetic sector instruments with laminated inhomogenous field magnets. Yet with magnetic sector analyzers, faster scanning means significant losses in resolution. At present a reasonable rate for repeating scans for successive components entering a magnetic sector is of the order of 3–4 Hz. A faster scan will simply produce distortion in the spectrum. In all considerations of scanning speed, trade-offs between sensitivity, resolution, and S/N are involved.

Example 20.8 Will a quadrupole mass spectrometer provide sufficient resolution for a GC/MS? Will a quadrupole filter offer any advantages as a mass analyzer for this application?

As long as a chromatographic channel separates components well, their identification seldom requires a mass spectrometer of high resolution. A quadrupole spectrometer provides sufficient resolution for the majority of components separated by GC. A quadrupole analyzer also offers the fast scanning required if an open tubular column is used. With complex organic mixtures, however, the higher resolution of a double-focusing mass spectrometer (see Section 20.6) becomes attractive.

Coupling Devices for a Liquid Chromatographic Sample Channel. In the *moving-belt interface* for LC / MS coupling, eluant from the liquid chromatographic column is transferred onto a moving ribbon. This ribbon passes through a heated and evacuated zone to remove solvent and then to a zone of flash heating in which solute is also vaporized. As each solute volatizes, it moves into the spectrometer ion source, which is at a much lower pressure. Problems arise in analyte transfer when inorganic buffers have been used in the LC or when the solute is too polar or labile to vaporize.

The widely used *thermospray coupler* expels the column eluant (flow rate 1 mL min^{-1}) through a heated nozzle with a fine point. The transfer line from the column is heated to vaporize the solvent partially, even as it forms into a jet of fine particles (an aerosol). As the hot aerosol emerges into the intermediate vacuum chamber, volatilization continues and is largely complete as the spray nears the spectrometer aperture.

During the evaporation, ionization of analyte molecules occurs as long as the eluant is at least 10% water; the solvent must be polar. Ionization is aided by (a) the strong electrical fields that build up on the surfaces of minute aerosol particles and (b) the presence of electrolytes such as ammonium acetate. Indeed this salt is often added to the original sample at a concentration of at least 10^{-3} *M*. In addition, a filament that can be electrically charged is sometimes included in the interface to aid ion formation.

Simple Mass Spectrometers as GC Detectors. Simple, efficient mass analyzers such as the *ion trap* (mass spectrometer) have been designed specifically as gas chromatographic detectors. For an open tubular or capillary GC, direct coupling permits eluting components to enter the ion trap. These species are ionized in the trap by electron impact and the ions are held in its ion storage region. For a set interval, ions of a particular m/z are extracted (selected ion scanning) and directed to an electron multiplier detector. Low limits of detection are possible by careful choice of ions specific for particular analytes and by holding the extraction conditions a set time for each ion selected. In this device the relatively long periods between scans are almost eliminated and a mass scan can be completed in 1 s. [G. C. Stafford, Jr., et al., *Am. Lab.*, **15**(6), 51 (1983).] Quadrupole mass filters are also employed as GC detectors.

Tandem Mass Spectrometers. As ion sampling has been increasingly extended to involatile materials, as discussed in the last section, the potentialities of a mass spectrometer as a separation channel have become apparent. It can isolate specific single ions, such as molecular ions, for each analyte. The result has been the development of the tandem mass spectrometer, which is commonly identified by the acronym MS/MS. The MS/MS spectrometer, even more than GC/MS and LC/MS spectrometers, has the potential to analyze complex mixtures. In addition, for large molecules it provides a means of structure elucidation, a process often difficult or impossible with a regular mass spectrometer.

A modular diagram of the characteristic modules of a MS/MS spectrometer is shown in Fig. 20.13. The separative channel, MS-I, consists of an ion source followed by a mass analyzer. A CI ion source is a good choice for MS-I since its soft ionization will be more likely than not to produce analyte molecular ions. At a given point in a scan of this analyzer, it will produce an analyte ion of particular

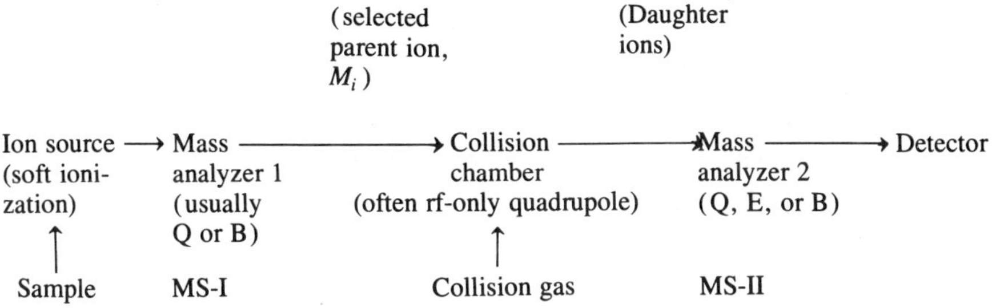

Fig. 20.13 Characteristic modules of a tandem mass spectrometer (MS/MS).

m/z at its exit slit. Thus the separation channel selects a *parent ion* representative of one analyte at a time introduce to MS-II. This ion has an opportunity to undergo fragmentation in a gas reaction cell following MS-I. It contains either a reagent gas or an inert collision gas to allow soft fragmentation of a particular species of ion from MS-I by *collisionally activated dissociation.* Characteristic product ions of many masses are produced and directed by an ion lens into the second mass analyzer, MS-II for resolution and identification. When these fragment ions are resolved by MS-II, the identity of the parent ion should become clear. The two stages of the tandem instrument generate a vastly greater amount of chemical information about a sample than a regular mass spectrometer.

Each component of the mixture should be identifiable in turn. To identify the other analytes it is only necessary to scan the output of MS-I sufficiently slowly to allow a complete scan of the output of MS-II for each parent ion that enters.

Other Representative Measurements. In a tandem spectrometer, mass analyzers are independently operated; as a result, other types of mass measurements are available. Two of these are:

1. Screening for functional groups or ion structure (MS-I is set to scan the mass range and MS-II to select a particular daughter ion). The spectrum generated in this mode of scanning displays all parent ions that fragment to the selected daughter ion.

2. Neutral loss spectrum (scans for both MS-I and MS-II are coordinated so that a chosen mass difference is maintained). Further insight is obtained about functional groups.

Types of MS/MS Spectrometers. The best choice for MS-I will be a high resolution mass analyzer. Good resolution is important for it. Since somewhat simpler ions are to be resolved by MS-II, it may be a device of lower resolution. Either a magnetic sector or a double-focusing combination with a quadrupole device will be suitable. Often an electrostatic sector is used for MS-II and mass analysis is accomplished by scanning the sector field strength E.*

*The acronym MIKE or MIKES is sometimes used to identify mass analysis with an electric sector. It stands for mass analysis by ion kinetic energy.

Tandem spectrometers with as many as four mass analyzers have been used. To describe their design it is useful to label electrostatic sectors E and magnetic ones B. For example, sometimes two double-focusing instruments have been hooked together to provide a large mass range for both parent and daughter ions. With this combination it is possible to have a conventional EB EB pattern or a BE EB instrument. In both designs a reaction region is sandwiched between the initial pair of analyzers and the final pair.

In addition, it is possible to use a set of three quadrupoles as an MS/MS arrangement. MS-I serves as the initial mass separator to select an ion, the second contains reagent gas and through chemical ionization provides soft fragmentation of the selected ion. It is set up as an rf-only quadrupole (transmits all ions). Mass analyzer MS-II resolves the daughter ions. These quadrupole spectrometers are called triple-stage quadrupoles (TSQ). Because of their versatility and ease of control with modern data systems, they have gained wide acceptance. Their cost is about 60% of that of MS/MS instruments with magnetic sector analyzers. Naturally, mixing sector and quadrupole analyzers is of interest as well and the reader is referred to the references for further information.

20.9 MASS MEASUREMENT AND SPECTRAL INTERPRETATION

The interpretation of mass spectra of substances will yield their identities. For simple, pure compounds, ion masses need in general be known only with a precision of 1 amu to carry out an identification. Example 20.14 furnishes a good illustration of such an application for a low-molecular-weight (m < 200 amu) compound. More precision is needed, however, to identify the components of a mixture. Pattern recognition programs (see Section 11.9) can help to establish the identities of components (with perhaps 90% probability), but a more powerful approach is the use of a tandem mass spectrometer, as discussed in the last section.

In addition to identification, mass spectra can be interpreted to obtain

1. precise isotopic masses and isotopic ratios;
2. molecular weights, and if high resolution is used, elemental composition and empirical formulas; and
3. structure elucidation.

Given the uniqueness of elemental isotopic abundances and mass defects the precise measurement (± 0.00001 amu) of the mass of an ion (m < 1000 amu) will in general be sufficient to identify it. As noted in Section 20.2, such data are invaluable in research on the dynamics and kinetics of ion–molecular gas phase reactions and many other aspects of gas behavior.

Mass Measurements. The classical method of accurate measurements of mass with magnetic sector mass spectrometers is by *peak matching* under conditions such that the knowledge of the elemental composition of a reference ion allows calculation of the exact mass of an ion from the analyte. For example, one peak might be a molecular ion of mass m and the other a reference ion (see discussion below) of mass km. A measurement is based on the observation that in a magnetic mass analyzer at constant magnetic field, an ion of mass *km* will travel a path identical to that of one of mass *m* if the accelerating voltage (prior to the magnetic sector) is changed by a factor of $1/k$. Good spectrometers provide for measurement of the voltage ratio necessary to cause the two ions to travel the same path with very high accuracy. Fortunately, high resolution is not a requirement for precision in mass measurement. Instead, the measurement depends on the product of sensitivity and resolution. These measurements can be made with parts per million accuracy at a resolution of about 3000. If approximate ion masses are sought, measurements reliable to 0.1 amu are desirable.

Example 20.9. It is desired to use peak matching to determine the mass of an unknown ion by comparison with that of a reference ion. Take the mass of the reference ion as M, and that of the unknown ion as $M + \Delta M$. Following the method outlined in the last section, it is only necessary to change the accelerating voltage for the unknown ion until its peak falls at the same point in the display as that of the reference ion. If the change in voltage is ΔV, the following expression allows the unknown mass to be calculated:

$$\Delta V/V = \Delta M/M.$$

Most spectrometers are capable of making this type of measurement. Other methods of matching can also be used.

Reference Compounds. Perfluorotributylamine (PC 43 or PFTBA) is a widely used mass reference whose many useful peaks range from m/z 50 to 614, the principal one being the $^{12}CF_3^+$ peak at 69. Other fluorinated compounds that serve as references are perfluorokerosene, which is less volatile than PFTBA, and tris(perfluoroheptyl)-S-triazine. The first two are used for GC/MS calibration. They are introduced into the mass spectrometer from a small bulb connected to one of the inlets of the spectrometer. The triazine is attractive for higher mass compounds because it furnishes significant peaks at mass intervals of about 13–10 up to m/z 1185.

These compounds are attractive for mass calibration because fluorine has a single isotope and the mass defects of these compounds are sufficiently different from those of common organic ions containing C, H, N, and O that analyte and reference ions of the same mass number can be resolved and measured easily on a high-resolution instrument.

Example 20.10. How valuable is the precise determination of molecular ion mass (mass well below 1000 amu) in obtaining empirical formulas? Apply the method to the isobaric (like mass) ions CO^+ and N_2^+, using Table 20.4, which lists exact masses of the major isotopes found in organic compounds.

According to Table 20.4, $^{12}C^{16}O$ has an exact mass of 27.994914 amu and $^{14}N_2$ one of 28.006148 amu. A mass spectrometer with resolution of more than 3000 will differentiate CO and N_2; if its precision is at least 10 ppm it will permit one to identify the ions unambiguously as those of the molecular ions. Conversely, if a precision mass spectrometer yields ions with these masses, the empirical formulas of the ions will be evident immediately.

Today every attempt is made to learn precise masses of molecular ions. A precision of

Table 20.4 Mass and Approximate Natural Abundance of Some Common Nuclides

Isotope	*Mass*	*Natural Abundance (%)*	*Isotope*	*Mass*	*Natural Abundance (%)*
^{1}H	1.007825	99.985	^{29}Si	28.976497	4.700
^{2}H	2.014102	0.015	^{30}Si	29.973722	3.090
^{12}C	12.000000	98.893	^{31}P	30.973763	100.000
^{13}C	13.003354	1.107	^{32}S	31.972073	95.000
^{14}N	14.003074	99.634	^{33}S	32.971459	0.760
^{15}N	15.000108	0.366	^{34}S	33.967870	4.220
^{16}O	15.994915	99.759	^{35}Cl	34.968854	75.77
^{17}O	16.999133	0.037	^{37}Cl	36.9659030	24.23
^{18}O	17.999160	0.204	^{79}Br	78.918332	50.69
^{19}F	18.998405	100.000	^{81}Br	80.916292	49.31
^{28}Si	27.976927	92.210	^{127}I	126.904476	100.000

1 ppm is not difficult to achieve and precision of about 1 ppb is available with special spectrometers. Critical fragment ions also deserve this kind of attention. Lengthy tables have been developed that give exact masses of a great variety of compounds based on standard isotopic abundances. Thus, if the mass of an ion is found to be 131.0860, its empirical formula is almost certainly $C_{10}H_{11}$ and if 131.0497, it is C_9H_7O.

Example 20.11. Why is the precise determination of mass less valuable above m/z 1000?

It has been estimated that a measurement to ± 10 ppm for a mass of 1000 will restrict the field of empirical formulas to about 1000 possibilities. Determination of ± 1 ppm will narrow that field to about 100 formulas. If in addition to knowing a precise mass (± 1 ppm) for a molecular ion one also knows m/z values for four daughter ions, however, the number of possible formulas drops to about 10.

Example 20.12. Can mass spectrometry be used to make faster carbon-14 dating measurements? Will the greater precision of mass spectrometry permit extension of measurements to still earlier eras? Recall that the customary ^{14}C measurement requires counting disintegrations until the statistical error is small (sometimes more than 24 h).

For a mass spectrometric measurement the $^{14}C/^{12}C$ isotope ratio must be obtained. Sample preparation steps are necessary. One satisfactory procedure calls for all the carbon from a sample to be converted to lithium carbide, then to methane, and finally back to carbon, which is deposited on a tantalum wire. After carbon ions are formed, high acceleration (up to 3 MV) ensures sufficiently precise quantitative resolution of ^{12}C, ^{13}C, and ^{14}C to date artifacts at ages up to 60–100 thousand years. This process is often called accelerator mass spectrometry (AMS).

Example 20.13. Can selected ion monitoring be useful in precision mass measurement and deduction of empirical formulas of analytes by gas chromatography/mass spectrometry?

Selected ion monitoring is nearly always performed with computer assistance. To measure ion masses precisely it is valuable to enhance S/N by stepping a mass spectrometer between analyte peaks so that measurement time is mainly at values of m/z where there is an ion of interest. In addition, it has been shown that exact masses can be best obtained for chromatographic peaks by recording and averaging peak profile data [Y. Tondeur et al., *Anal. Chem.*, **56,** 373 (1984)].

Dynamic Aspects. There is a small but definite probability that an ion of mass m_1 with excess energy will dissociate somewhere along the flight path between accelerator and detector rather than in the ion source, as implied in Section 20.2. If the dissociation $m_1 \rightarrow m_2$ occurs *in the source* (residence time about 1–5 μs), the ion will be accelerated and mass-analyzed as the daughter ion m_2. If dissociation occurs between the source and detector (travel time very roughly 15 μs), however, the ion is called a *metastable ion* and may often be detected. Since its correct identification can provide valuable information about gas-phase reactions, many spectrometers are designed for such detection.

Identification is often difficult since m_2 is actually the ion to be detected, and it leaves an identifiable trace in the mass spectrum only if ion m_1 dissociates in a region free of either magnetic or electrostatic fields. Note that there is a *field-free region* between the accelerator and electrostatic sector, and in a double-focusing instrument, a second such region between electrostatic and magnetic sectors. The probability of observing m_2 will depend on the lifetime of the metastable ion m_1 and its relative time in a field-free region. Such residence times will in turn depend on the acceleration voltage V, mass m_1, and length of these regions. Sector spectrometers designed for metastable ion measurements allow the parameters V, E, and B to be varied to facilitate the measurement.

Structure Elucidation. To elucidate molecular structure, the mass and abundance of fragmentation products from analyte molecules are measured. The connections between fragment ions and precursor ions are deduced by studying the energetics and kinetics of these reactions. Ion cyclotron resonance spectrometry lends itself well to such investigations. Since the activation energy of an ion–molecule reaction is mutually zero or small, degradation under conditions that ions acquire low energies can be followed. In other words, ion–molecule studies complement the free-composition studies ordinarily made by fragmentation.

Another useful approach is the study of metastable ion characteristics. The reader is referred to some of the references at the end of the chapter for further information.

Searching of Spectral Libraries. In nearly all cases software available with mass spectrometers provides the possibility of spectral searches. As discussed in Section 11.8, strategy is perhaps the most important aspect of computer searching. If instead it becomes necessary to search a large library of full mass spectra for a match with an unknown spectrum, the cost in time and money will be considerable.

In the *Biemann* method, a strategy of abbreviation is employed (24). For every 14 mass units beginning with m/z 9, only the two most intense peaks are noted. The library spectra are similarly abbreviated and a comparison is made. The degree of agreement between spectra is calculated by employing a similarity index S which is defined as S = (average weighted ratio of amplitudes of all mass numbers matched)/(1 + fraction of unmatched spectra). An index of unity indicates a perfect fit, and one of zero, totally dissimilar spectra. Since spectra may be distorted by variations in speed of scanning, provision is made for taking ratios *R* of unknown peaks and of each peak in the matching spectrum in such a way that the value of *R* is always less than 1. Note the term ratio in the numerator of the similarity index. It is qualified by being weighted as well: weighting factors are 12 (peak amplitudes > 10% of base peak), 4 (amplitudes, 1–10%), and 1 (amplitudes < 1%). Finally, the average of the weighted ratio is calculated by adding all weighted ratios and dividing by the weighting number.

Another strategy known by the acronym *STIRS* was developed by McLafferty and coworkers to utilize information about mass spectral fragmentation processes [25]. Sometimes reverse searches are used. Still another procedure known as the *Clerk* system arranges that the 16 most heavily weighted features of the unknown spectrum are compared with library spectra [26]. Dissimilar spectra are then rejected rapidly.

In any event, it is an advantage to have sublibraries of compounds that may contain likely candidates for the components expected for a given set of samples. Thus when samples are being analyzed for drugs, libraries of spectra of steroids and a range of drugs are useful.

Example 20.14. Identification of an Unknown. The spectrum of an unknown was obtained in a single-focusing instrument. To introduce the sample the inlet and ion source were maintained at 200°C. High-resolution information was secured on the unknown with a separate instrument whose mass scale was calibrated with a known compound. The spectrum produced is given in the form of a bar graph in Fig. 20.14. We take the intensity of

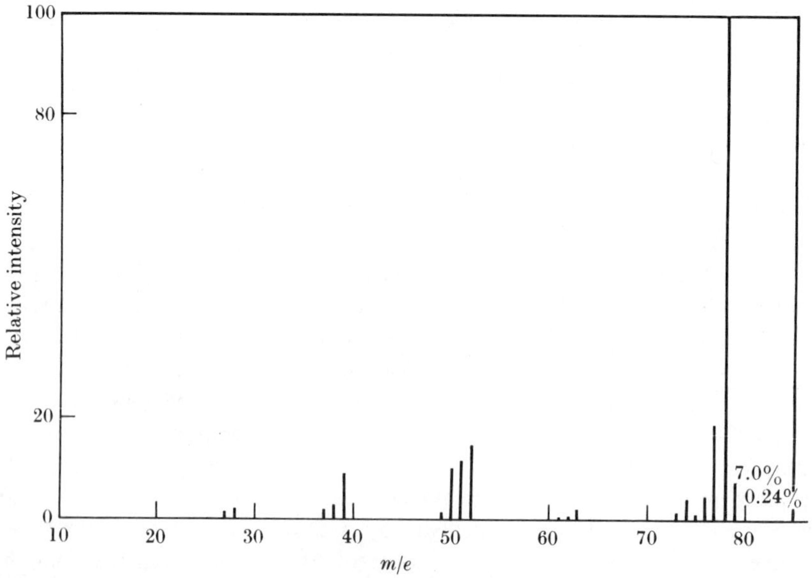

Fig. 20.14 The mass spectrum of an unknown liquid.

the largest peak as 100. How can we identify the substance by employing typical methods to interpret the spectrum?

ANSWER

As a first guess, we assume reasonably that the intense peak at the high-mass end of the spectrum (at $m/z = 78$) corresponds to a singly-charged molecular ion M of the unknown compound. We assume further that the small peaks just above this peak in m/z ratio are isotope peaks, that is, peaks due to the expected content of naturally occurring minor isotopes in the compound. Since there is no large peak at mass $M + 2$, it is obvious that the compound does not contain Cl or Br (see Table 20.4). We assume then that the $M + 1$ peak is the result of contributions of minor isotopes of C and N, and that the $M + 2$ peak is due to contributions for isotopes of O and C (principally, for the latter, contributions of molecules containing two ^{13}C atoms rather than one ^{14}C atom).

Probability laws derived from the binomial theorem state that the intensity of the $M + 1$ peak relative to the intensity of the M peak is given by

$$I_{M+1}/I_M = 0.011(\#\mathrm{C}) + 0.0037(\#\mathrm{N}), \qquad (20.10)$$

where (#C) and (#N) are the number of atoms of carbon and nitrogen in the compound. Assuming that the nitrogen content is low,* we neglect it at the start and calculate

$$7.0/100 = 0.070 = 0.011(\#\mathrm{C}).$$

The compound apparently contains six carbon atoms; the slight amount of excess intensity over that calculated for six carbons may be a small amount of background material in the instrument, the result of a slight fluctuation in source conditions, or possibly a small amount of protonated species resulting from an ion–molecule reaction in the source. The contribution to the $M + 1$ peak from ^2H is not significant.

We calculate the $M + 2$ peak from the formula

$$I_{M+2}/I_M = \tfrac{1}{2}[0.011(\#\mathrm{C})]^2 + 0.0020(\#\mathrm{O}). \qquad (20.11)$$

If we assume that the six carbons alone produce the $M + 2$ peak, we calculate an expected intensity ratio of 0.0022; the observed value is close to this. Carbon atoms account for 72 of the 78 mass units present; the remainder is logically hydrogen, and so the empirical formula of the ion is C_6H_6. A reasonable compound fitting these requirements is benzene, but the spectrum of hexa-1,3-dien-5-yne is similar.

*Since the molecular weight, 78, is even, the number of nitrogen atoms present must be even. This follows from the coincidence that among all the common elements in organic compounds, only N has a principal isotope of even weight and an odd valence. In all other cases, both are even or both are odd.

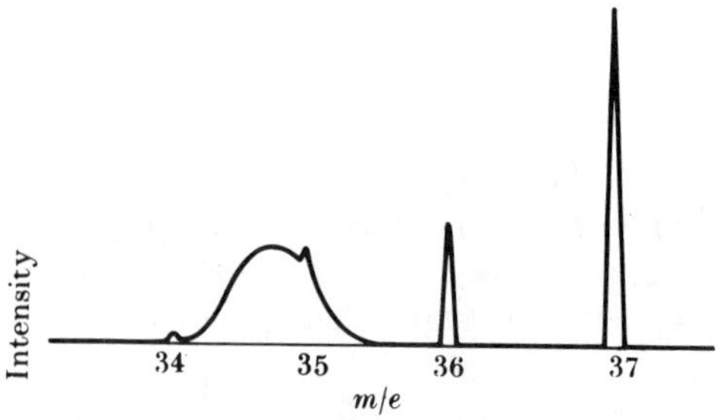

Fig. 20.15 A metastable ion peak at m/e 34.7, with a few low background peaks at integral masses that are useful in interpreting the spectrum.

In addition to the tabulated peaks, there is a broad, triangular peak in the spectrum centered at m/e 34.7 (Fig. 20.15). This is a metastable peak. It is produced by a metastable ion, one whose lifetime is sufficiently long for it not to decompose in the source, but instead to break apart as it passes through the analyzer of the instrument; its lifetime is thus about 10^{-5} to 10^{-6} s. It is accelerated at the heavier mass, that of the precursor ion as it existed in the source, but is deflected at the lighter mass of the final fragment. As this sort of fragment ion enters the magnetic field, it is moving more slowly than normal ions of its mass because it has roughly only the velocity which its precursor had. Hence it is deflected by the magnetic field into an orbit of smaller radius than the orbit of normal fragment ions of its mass. To a first approximation, the radius corresponds to that of an ion of apparent mass m^*, where

$$m^* = \frac{(m_{\text{fragment}})^2}{m_{\text{precursor}}}. \tag{20.12}$$

The broadening is a result of the possibility of formation of the fragment ion in the fringe of the accelerating field or the fringe of the magnetic field.

We have one equation and two unknowns for determining the nature of m/e 34.7. Though with double-focusing instruments there are alternate ways of studying metastable ions to identify them, we can make use of the general observations that the fragment ion formed in a metastable decomposition is almost always indicated by an intense normal peak, and that the precursor ion is usually indicated by an intense normal peak. While there may be an infinite number of solutions to Eq. (20.12), the most likely one is that for which m_{fragment} and $m_{\text{precursor}}$ correspond to large peaks in the spectrum. Trying various combinations of masses suggested by Fig. 20.14, we note that $52^2/78 = 34.7$; it appears that the decomposition $78^+ \rightarrow 52^+ + 26$ explains the metastable peak. Such a fragmentation supports the hypothesis that the unknown is benzene. For further discussion of this spectrum and of qualitative analysis by mass spectrometry, see the references at the end of the chapter.

20.10 QUANTITATIVE MASS SPECTROMETRY

To determine the composition of mixtures a range of approaches is available. A multicomponent analysis can be made by solving a set of simultaneous equations. A variety of software is available to assist in such determinations. Necessary analytical data are simply taken from conventional scans. Alternatively, a tandem mass spectrometer that incorporates a separative sample channel, as discussed in Section 20.8, can be used.

If a sample is thermally stable at a temperature at which it has at least modest vapor pressure, it can be introduced easily into an ion source. Often it is advantageous to develop ions from a sample by chemical ionization. With this mode of ionization, there is generally a much higher S/N and an abundance of molecular ions. Both contribute to higher quantitative accuracy. Perhaps a more fundamental choice is between sampling of a mixture by any of the available techniques and choice of a GC/MS or MS/MS approach, both of which offer significant advantages.

Criteria that must be met for a reliable quantitative analysis include

1. reproducible mass spectra must be obtained for both standards and unknown;
2. pressures must be sufficiently low so that the spectral intensities of peaks of common mass number from different components add; and
3. ion abundances must be proportional to partial pressures.

For most mixtures it is important that the spectra of components be substantially different. For example, it is difficult to analyze a mixture of cis and trans isomers because of similar mass spectra.

Calibration for Measurement of Ion Abundances. Decafluorotriphenylphosphine (DFTPP) is widely used for calibrating or tuning a mass spectrometer or a GC/MS spectrometer for ion abundance measurements [see J. W. Eichleberger, L. E. Harris, and W. L. Budde, *Anal. Chem.*, **47**, 995 (1975)]. Indeed its fragmentation pattern has been adopted as a standard for measurements of priority pollutants in the USA. It is attractive as a standard since it is available in pure form as a crystalline solid, soluble in a variety of solvents, stable, has a molecular ion at about m/z 450 (where many pollutants have peaks) and gives a spectrum with relatively abundant ions at about 75-amu intervals from m/z 51 to 275.

Example 20.15. How do sensitivity and fragmentation efficiency enter into ion abundance measurements for analytes in a gaseous mixture? Assume the mixture is volatilized completely and without sample fractionation and that electron impact ionization is used.

Sensitivity S_i for each analyte i will by definition be given by $S_i = \Delta I_i/\Delta C_i$. What factors will affect the ratio? A major factor will certainly be the cross section of the analyte molecule for simple ionization, which will determine the probability of molecular ion formation. Another will be its relative tendency to fragment. That both factors will vary with electron energy and ion source pressure suggests the importance of ion abundance calibration before carrying out a quantitative analysis.

By obtaining mass spectra of standards under conditions also used for unknowns, it is possible to obtain a set of linear simultaneous equations:

$$I_1 = a_{11}p_1 + a_{12}p_2 + \cdots + a_{1n}p_n,$$

$$I_2 = a_{21}p_1 + a_{22}p_2 + \cdots + a_{2n}p_n,$$

$$\vdots$$

$$I_m = a_{m1}p_1 + a_{m2}p_2 + \cdots + a_{mn}p_n, \qquad (20.13)$$

where I_i represents the measured peak height at mass i in the spectrum of a mixture, p_n is the partial pressure of component n, and a_{ij} is the peak height of mass i when the component j is present at unit pressure. As with all sets of simultaneous equations, it is essential to have at least as many equations as there are independent unknowns.

Example 20.16. The ruggedness, simplicity, and speed of a low-resolution quadrupole mass spectrometer makes it potentially attractive in process gas analysis. The approach of choice is *selected ion monitoring* (see Section 20.6), but the problem of overlap of isobaric (equal mass) fragment ions and molecular ions must be resolved for this method to work. Consider the use of such a spectrometer to monitor the O_2/CO_2 ratio in the headspace atmosphere in a fermentor used in drug production. How can N_2, O_2, Ar, and CO_2 be monitored in the headspace gas?

A mass spectrum taken during calibration runs show that m/z 28, 32, 40, and 44 are the peaks with highest signals for the gases listed. Since the CO^+ fragment from CO_2 also falls at m/z 28, this interference must be dealt with. Fortunately, the 44/28 m/z ratio for CO_2 is stable at 1:0.1100. To calculate the percentage of N_2 the value $m_{28} - 0.11 \times m_{44}$ is calculated and multiplied by the N_2 calibration factor by computer.

Example 20.17. How can the set of simultaneous equations for quantitative analysis, Eqs. (20.13), be solved by matrix algebra? What spectral peaks should be used? What standards are appropriate? What sources of error should be watched?

The set of equations can be represented as $\mathbf{I} = \mathbf{AP}$ and their solution as $P = \mathbf{A}^{-1}\mathbf{I}$, when $\mathbf{A}^{-1}$ is the inverse matrix. Assuming ideal gas behavior, the concentrations of components i can then be found from the equation $C_i = p_iRT/V$, where R, T, and V are constants for the gaseous system.

Suitable standards to employ to obtain values of a_{ij} are either pure compounds or known mixtures. Unfortunately, partial pressures are difficult to measure accurately. As a result, when standards are run, it is difficult to obtain coefficients a_{ij}. Measuring more intensities than the minimum serves as a check on the reliability of the partial pressures obtained.

Peaks to include in the analysis are selected so that each component is represented by at least one peak that it yields with high sensitivity.

Perhaps the most serious source of error will be the presence of peaks from unexpected substances. This matrix approach works best for routine analyses following good qualitative analysis. Background may also affect accuracy. Both types of interference will tend to be smaller when higher resolution is employed. If samples are admitted from a reservoir system, there may be a slight amount of fractionation of components as the sample is admitted to a mass spectrometer. Ordinarily this systematic error can be neglected, but it is important to know its probable magnitude.

Perhaps the major difficulties in quantification for a GC/MS system lie in the area of the dynamics of the instrumentation. For example, the shorter the scan time per decade and the higher the resolution, the greater must be the frequency range to which the amplifier and data system are responsive. *Selected-ion monitoring* eases these requirements for samples with only a few components by allowing a single characteristic peak of each component to be monitored. Finally, a tandem MS/MS system offers good possibilities for quantification at rates with which the system can cope. Some of the references at the end of the chapter will be useful for further information.

Statistical Assessment of Error. Today precision of operation of mass spectrometers with processing modules incorporating one or more microprocessors is high. Naturally, suitable calibrations are required for characteristic modules from time to time. As a result, most error in measurement is likely to arise in sampling and the many other preparative steps, such as dilution, that are part of a measurement. A helpful list was given in Fig. 10.5. To ascertain whether this is the case, regression analyses or analyses of variance will be very helpful (see Sections 10.4 and 10.5).

REFERENCES

Some clearly written introductory volumes are:

1. J. R. Chapman, *Practical Organic Mass Spectrometry.* New York: Wiley, 1985.
2. J. R. Majer, *The Mass Spectrometer.* London: Wykeham, 1977.
3. G. M. Message, *Practical Aspects of Gas Chromatography/Mass Spectrometry.* New York: Wiley, 1984.
4. B. S. Middleditch (Ed.), *Practical Mass Spectrometry: A Contemporary Introduction.* New York: Plenum, 1979.
5. J. T. Watson, *Introduction to Mass Spectrometry*, 2nd ed. New York: Raven, 1985.

Well-presented general volumes on an intermediate to advanced level are:

6. I. Howe, D. H. Williams, and R. D. Bowen, *Mass Spectrometry: Principles and Applications.* New York: McGraw-Hill, 1981.
7. K. Levsen, *Fundamental Aspects of Organic Mass Spectrometry.* New York: Verlag Chemie, 1978.

Clear, more specialized volumes are:

8. H. D. Beckey, *Principles of Field Ionization and Field Desorption Mass Spectrometry.* New York: Pergamon, 1977.

9. M. V. Buchanan (Ed.), *Fourier Transform Mass Spectrometry: Evolution, Innovation, and Applications.* Washington, DC: American Chemical Society, 1987.
10. J. R. Chapman, *Computers in Mass Spectrometry.* New York: Academic, 1978.
11. P. H. Dawson (Ed.), *Quadrupole Mass Spectrometry and Its Applications.* New York: Elsevier, 1976.
12. T. A. Lehman and M. M. Bursey, *Ion Cyclotron Resonance Spectrometry.* New York: Wiley-Interscience, 1976.
13. P. A. Lyon (Ed.), *Desorption Mass Spectrometry.* Washington, DC: American Chemical Society, 1985.
14. F. W. McLafferty (Ed.), *Tandem Mass Spectrometry.* New York: Wiley, 1983.
15. H. L. C. Meuzelaar et al., *Pryolysis Mass Spectrometry of Recent and Fossil Biomaterials.* Amsterdam: Elsevier, 1982.

A volume keyed to the interpretation of spectra is:

16. F. W. McLafferty, *Interpretation of Mass Spectra*, 3rd ed. Mill Valley, CA: University Science Books, 1980.

Also of interest are:

17. J. W. Eichelberger, L. E. Harris, and W. L. Budda, Reference compound to calibrate ion abundance measurements in gas chromatography–mass spectrometry systems, *Anal. Chem.*, **47,** 995 (1975).
18. W. W. Harrison, K. R. Hess, R. K. Marcus, and F. L. King, Glow discharge mass spectrometry, *Anal. Chem.*, **58,** 341A (1986).
19. J. F. Holland et al., Mass spectrometry on the chromatographic time scale: Realistic experimentations, *Anal. Chem.*, **55,** 997A (1983).
20. R. D. Macfarlane, Californium-252 plasma desorption mass spectrometry, *Anal. Chem.*, **55,** 1247A (1983).
21. C. L. Wilkins, Linked gas chromatography/infrared/mass spectrometry, *Anal. Chem.*, **59,** 571A (1987).
22. C. L. Wilkins and M. L. Gross, Fourier transform mass spectrometry for analysis, *Anal. Chem.*, **53,** 1661A (1981).

Tables of data will be found in:

23. J. H. Beynon and A. E. Williams, *Mass and Abundance Tables for Use in Mass Spectrometry.* New York: Elsevier, 1963.

Articles describing some major methods of mass spectral searching are:

24. H. S. Hertz, R. Hites, and K. Biemann, *Anal. Chem.*, **43,** 681 (1971); H. S. Hertz, D. A. Evans, and K. Biemann, *Org. Mass Spectrom.*, **4,** 453 (1970).
25. K.-S. Kwok, R. Venkataraghavan, and F. W. McLafferty, *J. Am. Chem. Soc.*, **95,** 4185 (1973).
26. J. T. Clerc, F. Erni, C. Jost, T. Meili, P. Nageli, and R. Schwarzenbach, *Z. Anal. Chem.*, **264,** 192 (1973).

EXERCISES

20.1 What resolution is required to separate the molecular ion of benzene at m/z 79 from the molecular ion of pyridine in a mixture of benzene and pyridine?

20.2 One of the ions found in the spectrum of a fluorocarbon standard for high-resolution work is $C_2F_5^+$. How suitable a choice would this be for peak-matching the $C_9H_9^+$ ion? Is it more useful than $C_8H_5O^+$?

20.3 The ionization potential of the methyl radical was found by a photoionization technique to be 9.82 ± 0.04 V. The ionization potential of methane was determined as 12.70 ± 0.01 V and the appearance potential of CH_3^+ in the spectrum of methane was 14.25 ± 0.02 V. Calculate the strength of the C–H bond broken to produce $\cdot CH_3^+$ from methane. How does this value compare with the bond strength of the C–H bond quoted in tabulations of data?

If the ionization potential of ethane is 11.52 ± 0.01 V and the appearance potential of CH_3^+ in the spectrum of ethane is 13.6 ± 0.01 V (obtained by another method than photoionization), calculate the C–C bond strength in ethane and compare it with values quoted in other tabulations.

20.4 An ion of unit charge and mass leaving the ion source and entering the accelerating potential of V volts has an initial velocity v_0. Develop an expression for the final velocity it will have as it leaves the accelerator to enter a mass analyzer.

20.5 Two ions of masses m_1 and m_2 and negligible initial velocity are accelerated through a potential difference of V volts. They enter a magnetic sector mass analyzer. (a) What are their kinetic energies on leaving the accelerator? (b) Assume the analyzer is followed by a detector and processing modules. Further, assume that fast scanning of the spectrum is sought. Should scanning be from low to high mass or vice versa? Support your answer.

20.6 Show by rearrangement of Eq. (20.4) that a magnetic sector can also be considered a momentum analyzer.

20.7 It is proposed to use a standard double-focusing mass spectrometer as a MS/MS tandem instrument by scanning the accelerating potential V to select analyte (parent) ions. The second MS state is therefore the electric sector–magnetic sector (EB) combination. (a) Comment on the usefulness of this design. If the field-free region between accelerator and electrical sector E is lengthened to permit some fragmentation to occur, would it be more useful? (b) Would the resolution of the accelerator as a mass analyzer be poor or good? Explain. (c) In practice, spurious peaks (not traceable to the expected analyte ion) appear in the output of MS II, the EB double-focusing unit. What may account for this phenomenon? [See D. H. Russell et al., *Am. Lab.*, **12**(3), 50 (1980); A. P. Bruins et al., *Int. J. Mass Spectrom. Ion Phys.*, **26,** 395 (1978).]

20.8 The major peaks of the mass spectrum of an unknown halogen-containing compound introduced through a heated inlet at 150°C are listed below. Identify the compound; no high-resolution data are needed. Postulate plausible structures for the ions.

m/z:	37	38	50	51	56	73	74	75	76	77	78	112	113	114	115
Intensity:	2.0	5.7	9.6	12	3.9	2.1	4.3	4.6	3.4	45	3.0	100	6.7	33	2.2

20.9 The spectrum of a common natural product introduced as a liquid into the heated inlet system maintained at 185°C is given below. What is the compound?

m/z:	14	15	26	27	28	29	30	31	32	42	43	45	46	47
Intensity:	2.3	4.7	5.2	18	4.5	15	5.3	100	1.1	3.5	11	50	23	0.50

20.10 The principal peaks in the spectrum of toluene have m/z ratios of 39, 50, 51, 63, 65, 89, 91, and 92. There is a broad, triangular peak centered at m/z 46.5. Interpret this peak.

Table 20.5 Spectra of C_6H_{12} Isomers

m/z	*1-hexene*	*2-hexene*	*4-me-1-pent*	*2-me-2-pent*
15	7.8	5.0	7.2	7.8
26	6.8	4.6	2.3	2.2
27	68	36	34	23
28	13	7.0	3.9	3.4
29	28	27	6.2	9.2
39	49	29	33	29
40	8.7	4.9	5.7	5.8
41	100	41	72	100
42	75	48	32	9.3
43	60	12	100	5.5
53	7.5	7.8	3.4	8.0
54	6.0	7.3	1.2	1.8
55	59	100	7.9	12
56	86	23	43	7.9
57	3.8	0.93	2.0	0.48
69	19	18	13	92
84	28	29	11	30
85	1.8	1.9	0.7	2.0
Sensitivity of 100% peak, div μm^{-1}	20	29	30	32

20.11 A mixture of methyl chloride and ethyl chloride was studied by ion cyclotron resonance. Peaks at m/e 79 and 81 in an approximate ratio of 3 : 1 were observed, and were found to be coupled to the ions of m/e 51 and 53 (ratio again about 3 : 1) and to m/e 65 and 67 (ratio 3 : 1) in a way suggested by the following. If the ion of m/e 81 was monitored while those of m/e 65 and 67 were irradiated at their resonant frequencies, the ratio of the double resonance signals was 3 : 1. If the same ion was observed while those of m/e 51 and 53 were irradiated, the ratio of the double resonance signals was 0 : 1. The peaks at m/e 79 and 81 were thought to correspond to the structure $CH_3ClC_2H_5^+$. Comment on the origin of the Cl atom in this ion.

20.12 A mixture of 1-hexene, 2-hexene, 4-methyl-1-pentene, and 2-methyl-2-pentene was admitted to the source of a mass spectrometer and intensities of ions found for the mixture as follows:

m/z:	15	26	27	28	29	39	40	41	42	43	53	54	55	56	57	69	84	85
Intensity:	8.6	4.9	50	8.6	21	44	7.9	100	50	56	8.0	4.7	48	53	2.4	45	29	1.9

The 100% peak was 2200 units high on the recorder. Using the tabulation of spectra of pure components (Table 20.5), determine the partial pressures of each component of the mixture.

ANSWERS

20.4 ($v = v_0 + 2eV/m$); *20.5* (a) $\frac{1}{2}m_1V_1^2 = \frac{1}{2}m_2V_2^2 = eV$; 20.6 m$v$ = Bev.

Chapter 21

X-RAY FLUORESCENCE SPECTROMETRY

21.1 Introduction

The X-ray region of the spectrum, lying "beyond" the optical range of everyday experience, gives rise to electromagnetic phenomena that are not measurable with instrumentation and methods used in the optical range, but require the development of new instrumentation and techniques. The kinds of wave phenomena examined earlier in the optical region, such as refraction, polarization, diffraction, and scattering, of course hold for X rays. In the X-ray region, however, the higher frequencies (10^{17}–10^{21} Hz) and shorter wavelengths [10 nm (100 Å) to 1 pm (0.01 Å)] affect the properties of matter and as a result give X radiation distinctive behavior. In particular, refractive indices of substances are nearly *unity* at X-ray frequencies (see Fig. 7.10 and Section 7.7). As a result neither mirrors nor lenses can really be fashioned for this region, and it is difficult to form and direct beams of X rays. Because a small amount of reflection is possible at grazing incidence some shaping is possible. In addition, diffraction from crystals or gratings can be used to disperse X rays angularly.

Early research elucidated the relationship of X-ray spectra to inner electronic transitions in atoms. Table 16.2 provides a comparison of these transitions to other electronic transitions.

Moseley's Law. Moseley's pioneering X-ray studies established a basis for determining the atomic number of any atom or element beyond lithium by examination of its X-ray spectrum. He demonstrated that X-ray frequencies were correctly predicted by an expression of the form $\sqrt{\nu} = K(Z - \sigma)$, where Z is the atomic number and K and σ are constants that vary with the spectral series. In particular, σ is a nuclear screening constant and has the value unity for an electron falling from level $n = 2$ to $n = 1$.

Since the spectra obtained are nearly independent of the chemical state of atoms, Moseley also foresaw that quantitative elemental determinations by X-ray emission spectrometry would be quite attractive in all kinds of samples. [H. G. J. Moseley, *Phil Mag.*, **26,** 1024 (1913); **27,** 103 (1914).

X-ray fluorescence spectrometry is widely used as a quantitative elemental technique and less often for qualitative analysis. Measurements call for the determination of both the wavelength and intensity of X-ray spectral peaks. These methods are applicable to all elements down to beryllium ($Z = 4$) though standard instruments are usually limited to analyses of elements down to $_{9}F$ and sometimes only down to $_{11}Na$.

The major advantages of X-ray spectrometric measurements are:

1. The spectra are simple.
2. The spectra (number of lines and relative intensities) are independent of excitation conditions. Further, X-ray spectra are essentially independent of the chemical state of an analyte (at most small corrections are needed).
3. Minimal sample preparation is needed.
4. Methods are nondestructive and are applicable over wide ranges of concentration without special attention.
5. Precision and accuracy are good.

Some disadvantages of X-ray emission spectrometry must also be logged.

1. X-ray penetration of samples is minimal: only atoms in the top 0.01–0.1 mm layer really contribute to a measurement.
2. Interelement effects are usually substantial and generally require computer correction.
3. Limits of detection are only modest.
4. Spectrometers are somewhat expensive.

An alternative title for this chapter would be X-ray *emission* spectrometry, since excitation of X-ray spectra by electron, photon, and other particle bombardment is increasingly common.

21.2 X RAYS, ELECTRONS, AND ATOMS

To begin to elucidate the relationship among X rays, electrons, and atoms, it is instructive to examine the spectrum generated by an X-ray tube. When energetic, electrically charged particles are allowed to bombard a target, X rays are produced. If the particles are electrons and the target is molybdenum, X rays are generated with spectral patterns like those shown in Fig. 21.1. In the tube, a stream of electrons emitted thermionically from a heated cathode is given high kinetic energy by acceleration through a potential difference maintained between the cathode and a positively charged metal target (anode). X-ray emission occurs where the beam strikes the anode. From Fig. 21.1 it is evident that the intensity and character of X-ray emission change appreciably as the accelerating potential is increased. It appears that at least two different processes may be operating. At low acceleration voltages V_0(V_0 < 20 kV), a broad band of continuous X radiation is obtained. But at higher potentials sharp peaks begin to be superimposed on the envelope of continuous radiation. While most of the kinetic energy of the bombarding electrons is converted into heat in the X-ray tube, it is one of the most useful devices available for X-ray production and will be discussed in detail in Section 21.4.

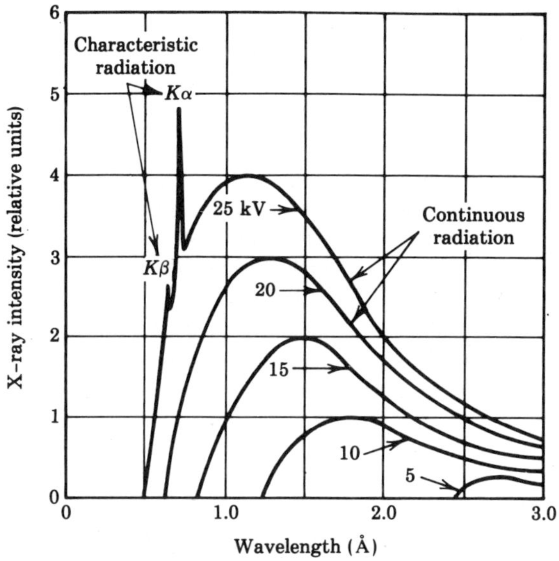

Fig. 21.1 Spectrum produced at different accelerating voltages by an X-ray tube with a molybdenum target. Neither wavelengths nor the widths of characteristic lines are to scale. The wide range of wavelengths emitted in the continuum is brought about by interaction of incoming "monochromatic" electrons with molybdenum electrons that are widely different in energy in the surface layer. (After B. D. Cullity, *Elements of X-Ray Diffraction*. Reading, MA: Addison-Wesley, 1956.)

Continuous Spectrum. The continuous spectrum or *continuum* so evident in the curves of Fig. 21.1 can be shown to result from deceleration of electrons by atoms in a target. Indeed, the original German name, bremsstrahlung, meaning "braking radiation," is also commonly applied. Since the decleration of electrons is incremental, it gives rise to a broad, continuous band of X-ray wavelengths.

Properties of the Continuum. The existence of a *short-wave cutoff* λ_{min} for the X-ray continuum is evident in Fig. 21.1. The gradual shift of the cutoff as V_0 is increased implies a dependence on energy. Indeed, It is possible to show experimentally that the minimum wavelength λ_{min} (or maximum frequency ν_{max}) fits the equation

$$eV_0 = \tfrac{1}{2}mv^2 = h\nu_{max} = hc/\lambda_{min}, \tag{21.1}$$

where e and m are the charge and relativistic mass of an electron. At this limit, the most energetic bombarding electrons have their entire kinetic energy converted into the energy of a photon. The cutoff λ_{min} is often called the *Duane–Hunt limit* and can be calculated from Eq. (21.1). For wavelengths in nanometers and voltages in kilovolts, rearrangement of the equation yields the expression $\lambda_{min} = 1.24/V_0$.

The distribution of bremsstrahlung intensities is always characterized by a maximum and a gradual drop-off of intensity toward longer wavelengths. In general, the wavelength of maximum intensity is 1.5–2 times λ_{min}.

The intensity of X rays produced at different wavelengths in the continuum can be calculated by the equation formulated by Kramers for an infinitely thick target of an element of atomic number Z:

$$I(\lambda)d = kiZ(\lambda/\lambda_{\text{min}} - 1)\,d\lambda/\lambda^2, \qquad (21.2)$$

where k is a constant, i is the tube current, and λ_{min} is calculated from Eq. (21.1) by substituting the value of V_0. From this expression it is clear that if continuous radiation of substantial intensity is desired, one should use an X ray tube with a target of high atomic number, perhaps tungsten ($Z = 74$), at high voltages and currents.

Equation (21.2) can be refined by (a) correcting for absorption by the X-ray tube window at longer wavelengths, (b) correcting for self-absorption by the target, especially at long wavelengths, and (c) inclusion of a more complex dependence on the applied potential V_0.

Characteristic Spectra. Electron bombardment also produces X rays characteristic of elements. For instance, two peaks labeled Kα and Kβ in the top curve in Fig. 21.2 are characteristic of molybdenum. Since they were absent at lower accelerating voltages, it appears that characteristic wavelengths are generated only at potentials $V_0 > V_K$, where V_K is a *critical excitation voltage*. For example, for molybdenum K X rays, $V_K = 20.01$ kV. When the accelerating voltage in the molybdenum tube is increased to 35 kV, as in Fig. 21.2, the increase in the height of Kα and Kβ peaks is striking. Note that the spectral distribution of the continuous radiation changes as one would predict from Eq. (21.2).

Other series of characteristic X rays designated L, M, N, . . . related to initial vacancies in those shells (a) will appear at progressively longer wavelengths since they involve smaller energies and (b) have smaller intensities because of competition from the Auger effect (see below). For instance, molybdenum L wavelengths appear at about 0.5 nm (5 Å). Each X-ray line represents a true quantum transition and has a characteristically narrow width (about 1×10^{-4} nm). Doppler broaden-

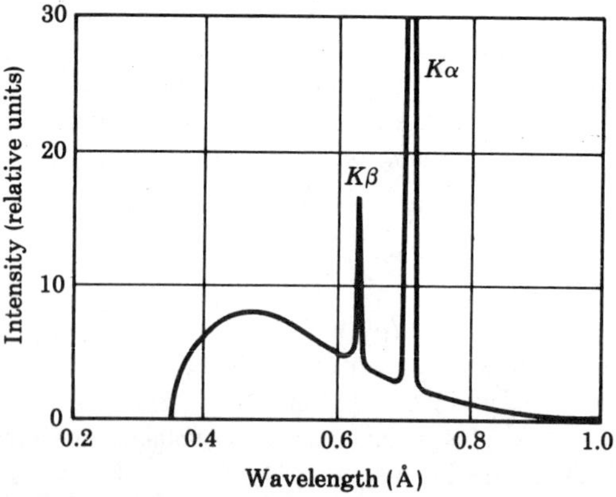

Fig. 21.2 X-ray spectra produced by 35-kV electrons bombarding molybdenum. The wavelength of maximum emission in the continuum has shifted to wavelengths shorter than those of Kα and Kβ as the accelerating voltage has been increased from 25 kV. Note that the Kα and Kβ peaks are also much more intense. (After B. D. Cullity, ibid.)

ing is a minimal factor in X-ray production as long as emission occurs from "immobilized" atoms in a solid.

Absorption of energy by an atom to produce the excited state must, of course, precede emission. The process is more dramatic for X rays than for optical spectra, however, because (a) the initial displacement of an inner-shell electron virtually always leads to ionization and (b) filling the vacancy may involve any electron of lower binding energy. One way to picture this transfer of energy is suggested in Fig. 21.3a for a sodium atom. First, an energetic *primary* electron ejects a K-shell electron.* With a vacancy in the K shell, the ion is unstable. As suggested, the

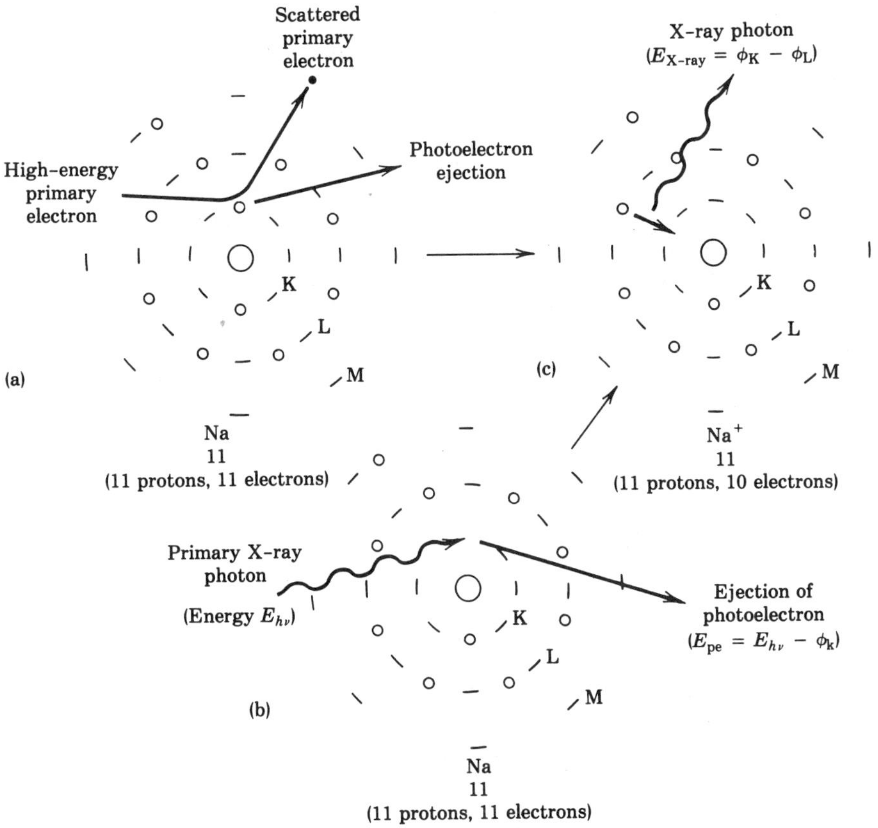

Fig. 21.3 Schematics of two modes of excitation of a sodium atom ($Z = 11$) to X-ray emission. (a) Electron bombardment causes ejection of a K or 1 s ($^2S_{1/2}$) electron. The ion formed is in an excited state because of the vacancy in the K shell and the incident electron has been scattered. (b) Alternatively, photon bombardment causes a K electron to be ejected. In this *photoelectric process* the entire energy ($E = h\nu$) is consumed in supplying kinetic energy to the photoelectron and overcoming the binding energy ϕ_K of the electron. (c) Both processes can lead to Kα X-ray emission if an L_{III} ($^2P_{3/2}$) or L_{II} ($^2P_{1/2}$) electron fills the vacant level. Other processes may also occur.

*It is also possible that excitation will simply promote an electron to a conduction band, which also effectively removes it from atomic influence.

most probable mechanism of relaxation is the filling of the vacancy by an electron of lower binding energy, for example, one from the L shell. Since a stable state has still not been achieved, other events will occur. A generally likely possibility is shown in Fig. 21.3c, the simultaneous emission of an X-ray photon of an energy equal to the transition energy $\phi_K - \phi_L$, where ϕ_K and ϕ_L are *binding energies* in the respective shells.

Another mode to achieve the identical excitation is shown in Fig. 21.3b: incidence of an X-ray photon of energy greater than the binding energy of a K-shell electron. The entire energy of the incident photon is absorbed by the interaction, the ejected electron is called a *photoelectron*, and the process is an example of the *photoelectric effect*. The photoelectron leaves with the relatively small kinetic energy $E - \phi_K$, where E is the energy of the incident photon. Because emission was caused by photon absorption, the process is described as *X-ray fluorescence*.

Example 21.1. Many of the K and L X-ray lines for gold ($Z = 79$) originating from the L and M shells are shown in Fig. 21.4. How do so many different electronic levels arise in these shells? Do X-ray transitions follow regular electronic selection rules?

Figure 21.4 illustrates the origin of some common but more complex X-ray lines for a heavy metal. Some general comments about the figure will provide an answer to the second question. First, note that the initial or final state of electrons is indicated at the left of the

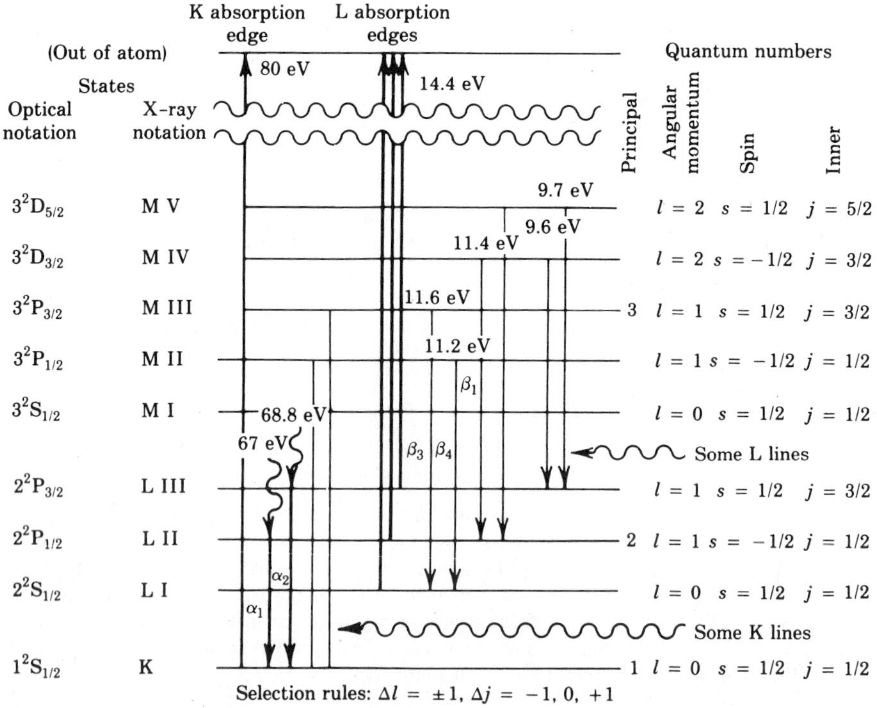

Fig. 21.4 Electronic transitions responsible for some of the common X-ray lines of gold ($Z = 79$).

diagram in both optical and X-ray notation. Second, excitation leading to photoelectron production from K and L shells is represented as heavy lines. These vertical lines from inner levels to the "out-of-atom" or zero-potential energy level represent electron *binding energies* and thus also designate the *critical excitation energies* for initiation of K, L, . . . X-ray production. Third, M lines are omitted since excitation of M-shell electrons will be less common and thus, less important, in analytical measurements. Finally, the selection rules identifying allowed or probable transitions are summarized at the bottom of Fig. 21.4 and are seen to follow regular patterns by comparison with Section 13.2.

In the figure, three discrete electronic levels are shown for the L shell and five for the M shell. Their origin can be traced to connections between quantum numbers and electronic energies. Here quantum number n is, of course, the principal quantum number, l is the angular momentum quantum number, and s is the spin quantum number. The so-called inner quantum number j reflects the combined effect on energy of angular momentum and electron spin and results from the coupling of l and s. For each electron the unique combinations obtained are $j = l + s$ or $j = l - s$. Reference to Table 13.1 and the surrounding discussion will clarify this notation.

The energy states listed are the levels that can exist in each shell for gold; some levels are, of course, degenerate and contain more than one orbital. In the K shell, since $l = 0$, a single level exists. In the L shell, three levels are possible since l can equal zero or unity. These levels, designated by Roman numeral subscripts, are: $L_I(l = 0, j = \frac{1}{2})$, $L_{II}(l = 1, j = \frac{1}{2})$, and $L_{III}(l = 1, j = \frac{3}{2})$. Similarly, in the third or M shell it is the occurrence of three values for l that causes five electronic levels to exist, as the reader can readily ascertain.

Primary versus Secondary Excitation. By convention, X rays excited by the impact of electrons with energies greater than critical excitation energies are termed *primary* X rays. An equally energetic beam of accelerated protons and other particles can be used instead and an identical X-ray spectrum will be produced. Lines characteristic of the elements present will appear superimposed on a strong continuum background.

When X-ray photons are used for the excitation, however, the resulting radiation is called *secondary* or fluorescent X radiation. With photon excitation no continuum is generated since quanta are completely absorbed. Will there be any background? Because of scattering, the characteristic lines will still appear against some background, as will be discussed in the following section. Nevertheless, the strong continuum background will be absent and more sensitive measurements can be made.

Example 21.2. Do any instruments produce X rays while carrying out a quite different primary function? Might useful information be extracted by adding an X-ray detector and analyzer to such instruments?

Actually, *primary* X-ray excitation occurs in electron microprobes, scanning electron microscopes, and analytical transmission electron microscopes while they are performing their main function. In most research instruments appropriate modules are added to probes and microscopes to obtain X-ray emission spectra of elements in the region being observed. An in situ microanalysis is possible since the exciting electron beam is focused to a spot that is usually smaller than 1 μm^2. Information is obtained about the elements present and their quantity and distribution.

Fluorescence Yield. Ionization from an inner shell leaves a vacancy in an atom that is immediately filled by the descent of an electron from a shell farther out. While the energy released in the process often appears as an X ray, an alternative path is for energy to be redistributed in the atom as a part of the Auger (pronounced oh-jay') effect. In this case a second ionization occurs and energy that would otherwise have been released from the atom as an X-ray photon now goes toward ejecting an electron from an L or M shell. Any excess energy that remains appears as kinetic energy of the newly ejected electron. For instance, in Fig. 21.3b there was a good probability (over 0.5) that an M electron would have been ejected from the excited sodium ion nearly at the same time an L electron moved into the K-shell vacancy. This *Auger electron* would have had a kinetic energy $E_{ae} = \phi_K - \phi_L - \phi_M$.

How seriously is the intensity of X-ray transitions reduced by the Auger effect? Part of the answer is given in Fig. 21.5, in which fluorescence yields for K and L X-rays for all elements is plotted versus Z. By definition, the *fluorescence* or *emission yield* ω is the ratio of X-ray photons emitted from a given shell to the number of vacancies initially created in the shell. Since Auger electron production is the only competing reaction, the ratio of Auger electrons to vacancies equals $1 - \omega$. From the figure it is clear that at low atomic number Auger processes dominate, but the X-ray yield ω_K increases steadily with atomic number and surpasses 50% for elements with $Z > 30$. In general, whenever the energy required to ionize a second electron from a higher shell is just a little smaller than that of the X-ray photon that could be emitted, the energy is transferred without emission and an Auger electron appears. This factor is mainly responsible for the reduced intensity

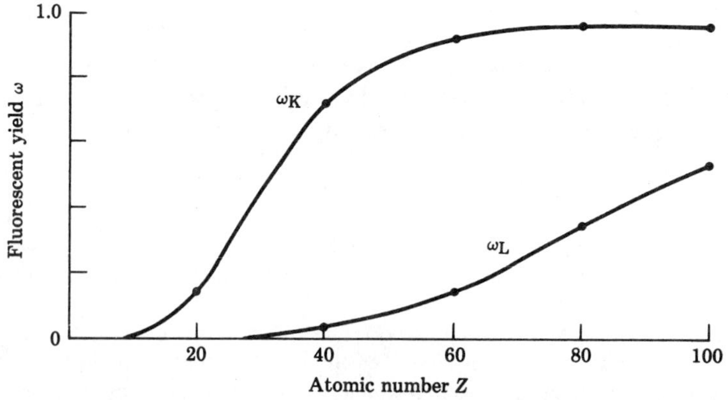

Fig. 21.5 X-ray fluorescence yield ω for excitation in the K and L shells over most of the periodic table.

of X-ray emission from L, M, and N shells. As is evident in Fig. 21.5, ω_L never reaches 0.5 and fluorescence yields are still smaller for M and N shells.

Interestingly, kinetic energy spectra of Auger electrons may also be obtained. Escape of electrons from condensed phases is difficult, however, and Auger electron spectroscopy provides mainly elemental information about surfaces. It will be discussed in the following chapter.

Satellite Lines. The possibility of ionization of an atom in more than one orbital at the same time gives rise to the phenomenon that energy connected with a double electron jump is emitted as a single photon. This is an inverse analog of the multiphoton transitions of interest in laser spectrometry. These double transitions are identified by notations such as LL–LK. The lines are not as intense, are usually called satellite, or nondiagram lines, and can be characterized as lines that are not described by the usual selection rules for transitions.

21.3 X-RAY ABSORPTION AND SCATTERING IN CONDENSED PHASES

How transparent are substances to X rays? If an analyte atom emitting X-ray photons is some distance inside a solid sample, what is the likelihood that the photons will be absorbed by unexcited analyte atoms before emerging? Or that the photons will be scattered out of the path to the surface as they interact with electrons? Because these absorption and scattering phenomena are strong for materials ordinarily examined in the condensed phase X-ray photons emitted below the top 0.1-mm layer of atoms are unlikely to emerge from a specimen. Results will also be applicable to gases if the mass attenuation coefficients described below are used.

It is easiest to categorize these effects by developing the X-ray version of Beer's law. When a monochromatic beam of parallel X-rays of unit cross section and intensity I_0 falls on a homogeneous sample of thickness dx, a certain intensity I will be transmitted. The difference $(I_0 - I)$ represents photons either absorbed or scattered. The latter term includes the special phenomenon of diffraction in which constructive interference favors Bragg's law behavior in crystals. These losses will be proportional to the initial intensity I_0, sample thickness dx, mass of substance dm, numbers of atoms dn of each kind, and moles d (mol) of each kind in the path. One obtains two important relationships:

$$dI_0 = -I_0 \mu \, dx, \tag{21.3a}$$

$$dI_0 = -I_0 \mu_m \, dm, \tag{21.3b}$$

where μ is the *linear* attenuation or absorption coefficient giving attenuation per unit area per centimeter (units, cm^{-1}) and μ_m the mass attenuation or absorption coefficient giving attenuation per unit area per gram (units, $cm^2\ g^{-1}$). From the units it is clear that $\mu_m = \mu/\rho$, where ρ is density. As will be seen shortly, the adjective attenuation better describes these losses than absorption and will be mainly used hereafter.

Actually, the *mass attenuation coefficient* μ_m sees widest use. Its value for an element is essentially independent of its chemical or physical state. For example, for tin the same value of μ_m applies to the element when it is present in tin metal, $SnCl_2$ or Na_2SnO_3, and regardless of whether these substances are in the liquid, solid, or gaseous state.

For multicomponent samples, it is customary to use mass attenuation coefficients rather than linear attenuation coefficients, and for convenience, to represent the mass attenuation coefficient for an element i simply as μ_i. One can calculate the attenuation coefficient of such a sample by adding the individual elemental contributions $\mu_i w_i$, where w_i is the weight fraction of element i present. One obtains for the sample the expession:

$$\mu_{\text{sample}} = \mu_A w_A + \mu_B w_B + \cdots = \sum \mu_i w_i.$$

This equation holds both for monochromatic radiation and for polychromatic radiation if a satisfactory effective wavelength for the latter exists [1, 2]. Integration of Eq. (21.3a) gives $I = I_0 \exp(-\mu x)$. If now μ_i is substituted for μ, one obtains

$$I = I_0 \exp(-\mu_i \rho x). \tag{21.4}$$

Note that Eq. (21.4) is Beer's law applied to X-ray absorption.

What general dependence can be expected for Eq. (21.4) with respect to atomic number and X-ray wavelength? Perhaps the question seems strange since there is no *general* dependence on such variables in the optical region. For X rays the answer can be found more easily if absorption and scattering can be separated as phenomena. Fortunately, the mass attenuation coefficient can generally be formulated as the sum of two components,

$$\mu_i = \tau_i + \sigma_i, \tag{21.5}$$

where τ_i is the photoelectric absorption coefficient and σ_i the scatter coefficient. Photoelectric absorption is by far the more important effect for most elements, for example, for elements of $Z > 10$, $\tau_i \geq 10\sigma_i$.

Photoelectric Absorption. In Fig. 21.6 X-ray mass attenuation coefficients for two elements, titanium and aluminum, are plotted against wavelength. Some thought suggests that the abrupt changes would correspond to the *critical excitation energies*, in this instance for K electrons, that were discussed in the last section. They are now appropriately called *absorption edges* (evidence shows that scattering in general changes smoothly with wavelength). For instance, it is at precisely the wavelength of the edge that quanta first become sufficiently energetic to eject a K electron from a titanium atom, a process which was earlier seen as a first step in the production of a titanium K X ray.

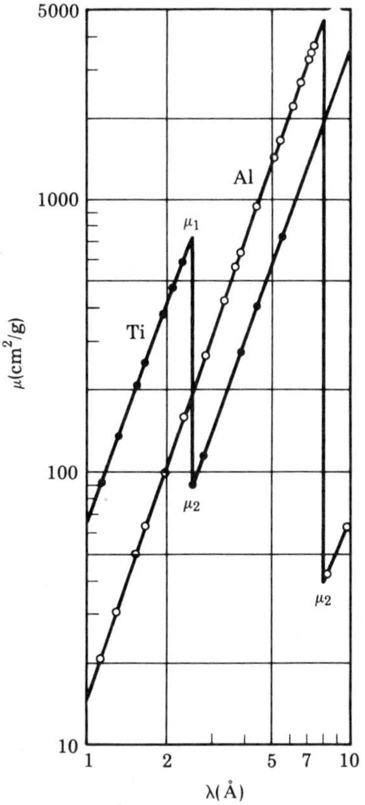

Fig. 21.6 Mass attenuation (absorption/coefficients as a function of wavelength (log/log plots) for two metals. Solid circles are for titanium ($Z = 22$) and open circles for aluminum ($Z = 13$). Both metals have a K *absorption edge* (vertical lines μ_2 to μ_1) in the wavelength region examined. On either side of the edges the data are fitted to the equation $\tau/\rho = N_0 A_i^{-1} C_i Z^4 \lambda^n$, where τ is the photoelectric absorption coefficient, ρ is the density, C_i is a constant varying slowly with atomic number and wavelength, and values of n range between 2.5 and 3. It is necessary for application of the equation that the area of the sample somewhat exceed that of the cross section of the X-ray beam.

The change in absorption at an edge is startling: as seen in Fig. 21.6, on moving across an edge toward shorter wavelength, the attenuation coefficient for titanium rises perpendicularly from about 90 to 800 and that for aluminum rises from about 30 to 4600. At each edge, the attenuation coefficient is the higher figure μ_1. Note also that attenuation generally decreases with *decreasing* wavelength; while it rises abruptly at an absorption edge, attenuation continues its relentless downward trend as wavelength decreases further.

Absorption Edges for L and M Electrons. Because there are several electronic levels in L and M shells in contrast to one in a K shell, what complications will be introduced? From curves like those in Fig. 21.6, what wavelengths should be chosen for exciting particular X-ray lines?

Since the earlier discussion indicated that there are actually three energy levels in an L shell, it can be assumed this shell should have three edges. Actually two edges, L_{III} and L_{II}, will be close together and appear as "saw teeth," while the third at the longest wavelength, L_I, will be a steep rise that marks the onset of absorption from the lowest energy level in the L shell. Similarly, it may be assumed that as many as four small edges and one steep rise will characterize M-absorption edges. These edges will lie quite close together and be difficult to resolve except for elements of quite high atomic number.

Example 21.3. What wavelength X rays will be most efficient in exciting Kα X rays for a given element? According to Fig. 21.6, what wavelengths will most effectively excite Kα X rays of titanium?

In every case the wavelengths best for exciting particular lines will be those just *shorter* than the related absorption edges. At those wavelengths the mass attenuation coefficients are highest. For the Kα line of titanium, practical exciting wavelengths should be just shorter than the K-absorption edge at 0.23 nm (2.3 Å). While titanium L and M absorptions have not fallen to zero at this wavelength, they are relatively weak and little energy should go to those transitions.

Example 21.4. Can the mass attenuation coefficient of a compound at a particular wavelength be found from the coefficients of the constituent elements? Is it the same for an equal mass of a substance at a given wavelength regardless of its state?

As long as elemental wavelengths are independent of bonding, it should be possible to calculate the attenuation coefficient of a compound μ_c at a given wavelength by multiplying the weight fraction w_i of each element in the substance by its mass attenuation coefficient μ_i at that wavelength and adding all elemental contributions: $\mu_c = \Sigma\, w_i \mu_i$. The process may be repeated at other wavelengths. Since the mass attenuation coefficient of an element is independent of physical state, the absorption curve obtained for a substance should also be.

Example 21.5. Contrast X-ray and optical atomic absorption spectra. Explain the marked difference. Are emission spectra, whether for optical or X-ray emission, more alike than absorption spectra? When do X rays reveal anything about the chemical state of an atom?

First, an optical atomic absorption has a narrow bandwidth and accompanies a discrete quantum process in which an outer-shell electron is excited to an unoccupied higher electronic level without ionization of the species. X-ray absorption causes a photoelectric process by exciting an inner-shell electron sufficiently to ionize it. In this case the energy of the primary photon in excess of what is required for ionization appears as kinetic energy of the ejected photoelectron. Thus, X-ray atomic absorption does not occur at a certain wavelength but changes gradually with wavelength.

Second, X-ray emission is more like optical atomic emission: in both processes an electron falls from a higher to a lower quantum level. Assume for the moment that a gaseous atom is involved for each process. For the optical process, however, emission energies (wavelengths) vary periodically with Z because the terminus of the transition is an outer level. In the X-ray case an electron falls to a core shell and energies and wavelengths vary *monotonically* with Z (actually, according to the Moseley law).

Finally, assume the atom is Cr in a Na_2CrO_4 crystal. The lower terminus in X-ray emission is likely to be a K shell in which the electron is tightly held. The extremely high binding energy of the lower level largely determines the energy of transition, an energy which will be virtually the same as for a $Cr_{(g)}$ atom. It would be expected that X rays whose lower levels are further from the nucleus (M levels or higher) would show some dependence on bonding. By contrast, optical emission energies would be strongly affected by the chemical state.

Example 21.6. Can one determine the concentration of an element in a solid by *X-ray absorptiometry*?

To make such a determination it is best to select a wavelength just shorter than an absorption edge. A very thin specimen would be needed to be sure that interelement effects are minimal. Note that in Fig. 21.6 an equation has been fitted to the data; it is easily possible to interpolate between tabulated absorption coefficients for the analyte to obtain values at the particular wavelength used. With that value, substitutions can be made in Eq. (21.4) and the density ρ of analyte atoms, which is actually a mass concentration, can be calculated.

Scattering. The background observed in X-ray spectra is attributable mainly to scattering. It is worth noting that scattering represents the only effective way in which X-rays can be polarized (see Section 7.14). Much of it is *coherent scattering* in which photons are scattered without a change of energy or wavelength. Basically, it is Rayleigh scattering, though of much higher magnitude because of the short wavelengths in the X-ray region. The rest is inelastic or Compton scattering at wavelengths *longer* than the exciting radiation and will be described below. The contribution of both types of scattering to the mass attenuation coefficient is usually described by splitting the linear scatter coefficient σ_i in Eq. (21.5) into coherent and incoherent terms: $\sigma_i = \sigma_{COH} + \sigma_{INC}$. An idea of the relative values of σ_{COH}, σ_{INC} (Compton scattering), τ_i and μ for a heavy atom can be gained from Fig. 21.7.

Scattering Theory. J. J. Thomson showed that electrons are responsible for X-ray scattering. The changing electrical field of the waves causes the electrons to oscillate around an equilibrium position. As a result of the acceleration they experience, the electrons radiate

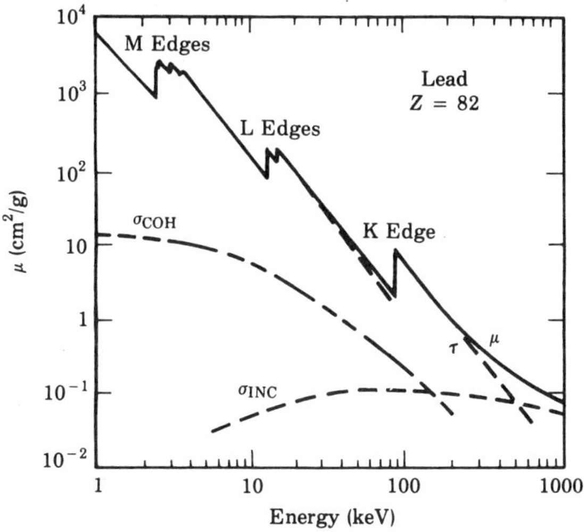

Fig. 21.7 Comparison of the size of the several different attenuation coefficients for lead as a function of energy. The symbols used are μ_m, mass attenuation coefficient, τ, the photoelectric absorption coefficient, σ_{COH}, the coherent or Rayleigh scattering coefficient, and σ_{INC}, incoherent or Compton scattering coefficient. The effectiveness of lead as an absorber of energetic radiation is clearly shown by the domination of photoelectric absorption at low energies and Compton scattering at high energies over Rayleigh scattering.

a linearly polarized scattered wave of the same frequency in all directions. Indeed, even in the optical region, scattering of light by molecules, atoms, or particles is usually viewed as traceable to the interaction of the electromagnetic waves with single electrons.

When a *free electron* is irradiated by an unpolarized X-ray beam with an energy flux I_0 (per square centimeter), the amount of energy I_e scattered (radiated) per unit solid angle per second is

$$I_e = 7.90 \times 10^{-26} I_0(1 + \cos^2 \psi)/2, \tag{21.6a}$$

where ψ is the scattering angle taken relative to the direction of the beam. For example, $\psi = 0$ for radiation scattered in the direction of the beam. Equation (21.6a) is known as Thomson's law. The magnitude of I_e serves as a convenient reference in scattering calculations, as will be seen below.

For a *bound electron* in a spherically symmetrical atomic cloud of electrons the amplitude of coherent scattering, f_e, taken relative to I_e, can be shown to be

$$f_{atom} = \sum_{n=1}^{Z} \int_0^{\infty} 4\pi r^2 \rho(r) \frac{\sin kr}{kr}\, dr. \tag{21.6c}$$

where r is the distance of the electron from the center of the cloud, $\rho(r)$ is the radial electron density distribution, and $k = (4\pi \sin \theta)/\lambda$. Here $\theta = \psi/2$ and λ is the X-ray wavelength. For an atom with Z electrons Eq. (21.6b) must be summed over all n electrons to obtain the scattering by an entire atom:

$$f_{atom} = \sum_{n=1}^{Z} \int_0^{\infty} 4\pi r^2 \rho(r) \frac{\sin kr}{kr}\, dr. \tag{21.6c}$$

The radiation coherently scattered by an atom equals $f^2_{atom} I_e$. Further, as $\sin \theta/\lambda$ approaches zero (at low angles of scattering) or for low X-ray energies as $\lambda_{X\text{-ray}}$ becomes large (for example, for an element of low Z), $f^2_{atom} \rightarrow Z^2$ and $I_{COH} \rightarrow Z^2 I_e$. In these cases, coherent scattering dominates.

For an atom the intensity for incoherently scattered X rays is the sum of contributions from individual electrons since there is no interference between their scattered waves. It is given by $\Sigma_{n=1}^{Z} (1 - f_e^2) I_e$.

Compton scattering is by contrast an incoherent type of scattering in which an energetic X-ray photon interacts with an outer shell electron and causes its ejection as a "recoil electron." As a result, the X ray leaves the scene at an angle ϕ relative to its original direction and with less energy and a longer wavelength. This process is illustrated in Fig. 21.8a.* As shown in the last figure, this type of scattering is stronger for more energetic X rays. It is also stronger for organic substances and other samples containing elements of low atomic number.

*The recoil electron has a direction and velocity that ensures conservation of energy and momentum. The change in wavelength is given by the equation $\Delta\lambda = (h/m_e c)(1 - \cos \phi)$, where m_e is the electron rest mass.

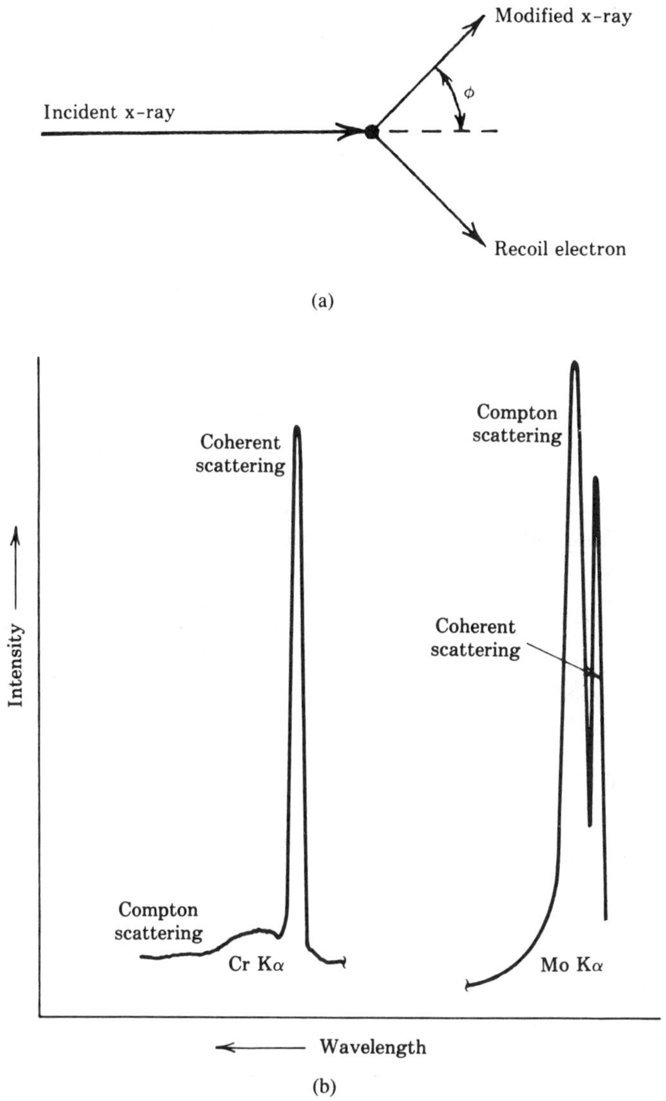

Fig. 21.8 A comparison of intensities of coherent and Compton (incoherent) scattering from a strongly scattering sample. (a) The origin of Compton scattering. (b) Relative intensities of coherent and Compton scattering by Kα X rays from $_{24}Cr$ (0.23 nm) and $_{42}Mo$ (0.071 nm).

In Fig. 21.8b the intensities of coherent or Thomson scattering intensities and Compton scattering are compared for different X-ray energies. It is evident that Kα rays of $_{24}Cr$ produce relatively little Compton scattering compared to such rays from $_{42}Mo$. Similarly, in X-ray emission spectra shorter wavelength lines show stronger Compton scattered background on their longer wavelength side than do longer lines.

21.4 X-RAY SOURCES

This section is the first of four dealing with instrumentation. Here several sources, including the regular or Coolidge X-ray tube, which was introduced earlier in the chapter, the pulsed X-ray tube, and radioisotope sources are treated. Particle accelerator and synchrotron sources fall outside the scope of the text, and may be studied through references.

In Fig. 21.9 a representative X-ray tube is shown schematically. It is energized by a high-voltage power supply that will in all modern instruments be a regulated, constant-voltage supply with an output of at least 50 mA at potentials ranging from 0.5 to 50 kV or higher. The head of the evacuated tube (toward the left), contains the target (tube anode), which is usually aluminum, magnesium, molybdenum, chromium, rhodium, tungsten, silver, or gold. As accelerated electrons strike its surface, X rays are emitted. The head is made of heavy metal to shield the external area from X-rays.

The tube must also form X rays into a beam. Fortunately, the exciting electron beam can be focused so that rays originate from a small spot on the anode (see Fig. 21.9). The resulting diverging beam of X rays is limited by the beryllium window to a conical beam of solid angle of at most 20°.

Though this type of tube is in wide use, in recent years end-window tubes in which the target is perpendicular to the tube axis (and parallel to the window) have also become common. In these, electrons follow a circular path from filaments placed at either side of the target. These tubes offer higher energies at long wave-

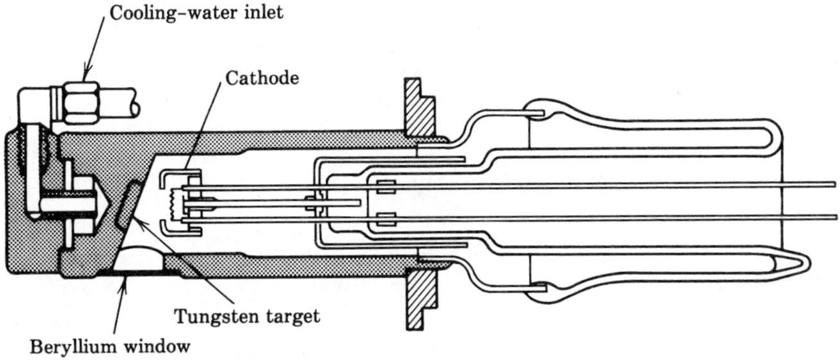

Fig. 21.9 Schematic diagram of an X-ray tube. X rays are generated in a target (in this tube, a tungsten insert in a copper anode) when it is bombarded by an electron beam. This beam is generated by thermionic emission from the cathode and accelerated through the large potential difference between it and the anode. The latter is kept at ground potential for reasons of safety while the cathode is made strongly negative. A geat deal of heat is dissipated at the anode and carried away by circulating water. The X-ray beam emerges through the beryllium window placed well to the side. It is important for the window to be out of the line of strongly scattered electrons as well as materials sputtered from the target so that it will suffer minimum thermal stress and plated layers, which would reduce its transmission. Air is removed from inside the tube to reduce absorption of long wavelength X rays. The beam is brought to a focus on the cathode by making the cylindrical electrode around the cathode sufficiently more negative than the cathode.

lengths and are thus useful for measurements of elements below $_{11}Na$. An ingenious design calls for a transmission anode in which the target metal is sputtered on the inner surface of a beryllium end window. A low-power electron beam generates X rays. This type of tube can be placed close to a sample, yielding efficient coupling.

Choice of Target. The atomic number of the target element is of some importance since an X-ray tube must provide wavelengths sufficiently short to excite K or L lines in a range of elements. Yet the use of anodes of high Z will fail to take advantage of the fact that analyte atoms best absorb X rays that are just shorter in wavelength than their relevant absorption edges. To arrange this condition for many different analytes would require a range of targets and a full shelf of tubes. A simpler answer is use of a dual-target tube. For example, tubes with tungsten and chromium targets within a single envelope are available. Their geometry permits the electron beam to be directed to either target by shifting the attracting potential to it. In such a tube tungsten, $_{74}W$, will provide characteristic lines to excite either $K\alpha$ or $L\alpha$ lines of high Z elements and chromium, $_{24}Cr$, will take care of elements below $_{22}Ti$.*

The power of X-ray tubes ranges widely (about 100 W to 4 kW). Low power tubes see extensive use in energy-dispersive spectrometers (see Section 21.7). High power X-ray tubes, on the other hand, are widely used for excitation of analytes of high atomic number in which K lines must be used rather than L or M lines. Naturally, a 50-kV X-ray tube will consume less power and require less cooling and general attention than a 100-kV tube.

Special attention must be given to the power supplies that provide both constant potential high voltages (10–100 kV) and good control of X-ray tube current (5 mA–100 mA for a wavelength-dispersive spectrometer and 1 μA–5 mA for an energy-dispersive spectrometer). To minimize drift in tube output requires that the influence of changes in ambient temperature, line voltage, and other factors be made as small as possible.

Monochromatic X radiation can be obtained by filtering. Usually it is secured by (a) using intense continuum radiation from a high power X-ray tube to excite characteristic wavelengths from an outside target of a selected metal and (b) filtering out continuum radiation scattered from the outside target by a thin metal filter of the same metal. Monochromatic sources will be described in more detail in Section 21.11 in connection with trace-level measurements in which they are valuable. The radioisotope sources now to be described are also essentially monochromatic in output.

Radioisotope Sources. While X-ray tubes offer high power and versatility, they use large amounts of power and require power supplies. By contrast, radioisotope

*Anodes consisting of layers of different elements have also been developed. The simplest has a thin layer of a lighter element plated on a base of a heavy element. The tube is operated at a low voltage to excite the element on the surface and at high voltage to energize electrons to excite the element in the base layer.

sources are light, small, and consume no power. In particular uses, these are important advantages. In addition, many radioisotope X-ray sources are monochromatic. While such sources are of low power, precisely for that reason, they can be placed close to both the sample and detector (see Section 21.7), which is often a favorable geometry for reducing background. Spectrometers with such sources are also easily portable and can be used, for example, in routine process control, environmental or mineralogical studies in the field, and in Viking and Venera space probes.* The main disadvantages of radioisotope sources are relatively low power and gradual loss of power over time with radioactive decay (the aspect of drift can be corrected for simply).

Radioisotope Sources. Only a relatively small number of isotopes qualify as X-ray sources. In radioisotopes such as cobalt-57, americium-241, and curium-244 X rays are yielded by a phenomenon termed internal conversion. An X ray is produced after a γ ray emitted from the nucleus excites a K-shell electron to ionization. On the other hand, in the widely used sources ^{55}Fe (5.9- and 6.4-kV X rays) and ^{109}Cd (22.2–25.5 kV X rays) X rays result from orbital-electron capture. A K or L electron is captured by the nucleus, causing Z to decrease by one; the X ray produced is characteristic of the new element. Other nuclear events also occur with some of the radionuclides. In addition, some α and β emitters are used *to generate* X rays by energetic particle bombardment: ^{241}Am, an α emitter generates X rays from 14 to 21 kV and ^{3}H and ^{147}Pm, β emitters generate brehmsstrahlung up to 18.6 and 22.5 kV, respectively.

21.5 DETECTORS

How does one determine the energies of wavelengths of X-ray photons from an analyte and also their number? The interaction of X rays with matter by photoelectric absorption, discussed earlier, provides an attractive mechanism for detection since ideally the full energy of each photon is transferred to the system. (Energetic particles such as beta particles also interact to produce extensive ionization and may be similarly detected. This section will be expanded to serve their detection also.) The main types of detector are designed around the different ways to measure that transfer. An X-ray photon or energetic particle is measured

1. in a gas-filled detector as a current pulse by collection of the ion–electron pairs it forms;
2. in a semiconductor detector as a current pulse by collection of the electron–hole pairs it forms;
3. in a scintillation detector as light pulses it releases from a phosphor; and
4. in a photographic detector as a silver deposit by subsequent chemical development of the latent silver image it forms.

*B. C. Clark, III, et al., *J. Geophys. Res.*, **82,** 4577 (1977); Yu. A. Surkov et al., *Anal. Chem.*, **54,** 957A (1982).

These detectors are usually operated in digital fashion as *photon counters*. This mode has all the advantages described earlier for photon counting (see Section 8.7), including a higher S/N ratio in measurements and lower limits of detection. Especially, digital detection permits operation at the low levels of X radiation at which pulse-height selection (analysis) and energy discrimination become feasible. The electronic circuits necessary to present the output as clearly countable pulses will be taken up toward the end of this section. In a very few cases where the number of photons is large, the output will be the amplitude of a smoothed analog signal.

Gas proportional detectors, Si(Li) solid-state detectors, and scintillation detectors are the most widely used types. They are also valuable for detection of γ rays and the energetic particles ejected by radioactive atoms. For that reason, Section 23.4 should be read along with the present one for completeness. For reasons of economy, however, reference hereafter will mainly be to photons.

Gas-Filled Detectors. The common gas-filled detector has a cylindrical outer cathode and a central (axial) anode. Figure 21.10a shows graphically the field set up in such a detector and Fig. 21.10b a simplified electronic circuit. The electrode geometry takes advantage of the fact that the field strength of a conductor of circular cross section is inversely proportional to its radius and will be large at the central wire anode. Such detectors are often filled with an inert gas such as argon. When an X-ray photon enters through the window (not shown), it travels linearly to its point of interaction with a gas molecule. There it ejects a photoelectron from an outer shell of the argon and leaves a positive ion behind. After the ionization event the energy remaining appears mainly as the kinetic energy of the photoelectron. That electron in turn loses energy by ionizing new gas molecules until it is brought to rest. By applying a potential difference the collection of electrons and positive ions formed can be separated and collected at the electrodes.

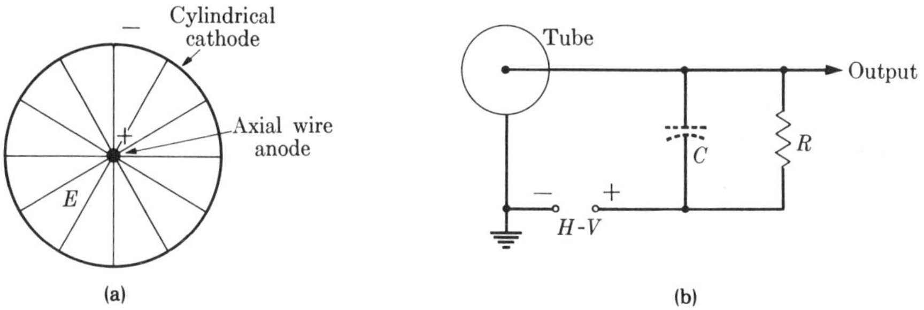

Fig. 21.10 A simple gas detector and its circuit. (a) Schematic cross section of the detector tube. Side entrance windows are customary. A central wire (0.1–1 mm diameter) seves as the anode and an outer cylinder (1–2.5 cm) as cathode. Lines of force are drawn to show the electrical field. As a result of the high field near the anode, perhaps 1–10 kV cm^{-1}, electrons are greatly accelerated and are collected rapidly at the anode. Since the radius of the cathode is so much greater, positive ions experience a small and diminishing electric field as they approach the cathode and are collected more slowly. (b) Simplified schematic of detector circuit. R is the order of 10^9 Ω and the distributed capacitance C is no more than a few picofarads.

Ideally, all X rays or radioactive particles that penetrate the *active volume* (between the electrodes) create a full cloud of charge. The electrons are swept to the central wire in about 1–2 μs while positive ions require 100–500 μs to reach the cathode.

To picture the effect of voltage on the detector, assume that photons of a single energy are arriving. If the potential across the electrodes is now gradually increased, a current–voltage curve like that in Fig. 21.11 is obtained. From 0 to 350 V, the total change collected per photon becomes steadily larger and then remains constant. To interpret these results, we recall that ion pairs tend to recombine on collision. As the electric field is increased, the fraction of electrons and ions existing long enough to reach the electrodes grows. When the electric field is sufficiently large to collect essentially all electrons and ions formed by the entering radiation, the curve flattens. A condition of *saturation* exists. This is the region of *ionization chamber* behavior.

With further increase in voltage, the current rises again. Now electrons produced by an X-ray photon are strongly accelerated in the high field *near the anode* and produce additional ion pairs by collision. *Gas amplification* is occurring. In this middle-voltage region the amplification fortunately spreads only within a small sheath around the path of each accelerated particle and each avalanche of ioniza-

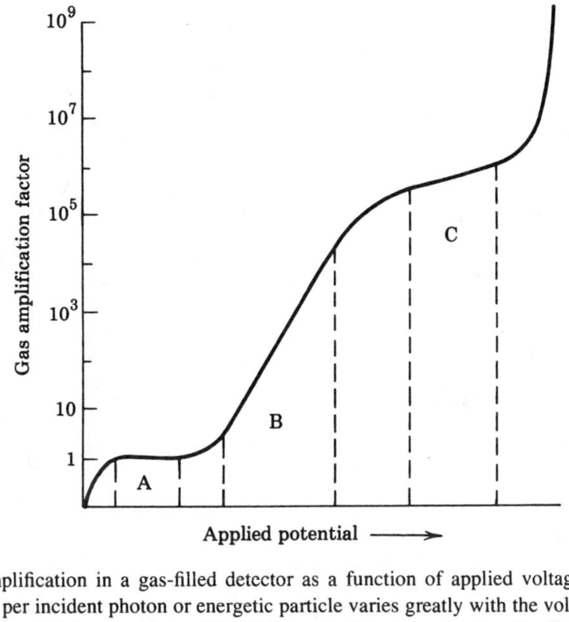

Fig. 21.11 Amplification in a gas-filled detector as a function of applied voltage. The amount of charge collected per incident photon or energetic particle varies greatly with the voltage applied. Until application of a potential as much as 20–200 V, recombination of some fraction of electrons and positive ions from ionizing events occurs. Through region A amplification of the original charge is unity and charge pulses collected are of miniscule size. By increasing the potential difference to 500–700 V, however, region B of proportional amplification is reached and amplification of the original charge by factors of from 10^2 to 10^5 times is secured. In the Geiger region C amplification is so large that all collected pulses are essentially of equal amplitude.

tion proceeds essentially independently. Here the total charge collected per photon or particle is proportional to its initial energy. This is the region of *proportional counting*.

As progressively large voltages are applied, the slope of the curve decreases (see Fig. 21.11) and the *Geiger region* is entered (see Section 23.4). Finally, the potential difference is so great as to sustain a continuous discharge. For X-ray photon detection only the proportional counter region in which there is attractive linear gas amplification of 10^3–10^6 is commonly used.

Example 21.7. A β-emitting radioisotope, on the known basis of its disintegration rate, is predicted to give off about 50 β particles per second. All enter the active volume of a simple gas-filled detector. What size current pulse on the average will each β particle yield if the detector operates in the ionization chamber region?

If particle energies are in the range of 0.05–1.0 MeV, about 100 primary ion pairs will be formed per photon per centimeter of travel. Assuming a 5-cm active path in the detector, the number of electrons collected per photon per second is $50 \times 100 \times 5 = 2.5 \times 10^4$. The current in the external circuit is $2.5 \times 10^4 \times 1.6 \times 10^{-19}$ C $= 4 \times 10^{-15}$ A. Before either the individual pulses or the average current can be registered, the signal must be considerably amplified.

Example 21.8. In a gas detector, how strict a proportionality between the number of ion pairs and the energy of the incident photon is observed?

Complexities of the interaction of the photoelectron and gas molecules show that a calculation of ion pairs produced by the photoelectron can only give an average figure for the number of ion pairs; in fact, the number of electron–ion pairs formed by photons or energetic particles follows a Gaussian distribution.

Example 21.9. In an X-ray spectrometric measurement, 17.4-keV Mo Kα photons enter the active volume of a gas detector. What number of initial ion pairs does each produce if it is totally absorbed? What amount of charge is collected when it operates in the proportional region with a gas-amplification factor of 1.0×10^4?

To estimate the production of ion pairs, an effective ionization energy must be known for the gas. Assume it is found to be 30 eV. An averge of $17{,}400/30 = 580$ ion pairs are formed per photon. The charge pulse registered at the anode will be $580 \times 1.6 \times 10^{-19}$ C $\times 1.0 \times 10^4 = 9.3 \times 10^{-13}$ C.

Proportional Detectors. In this detector an inert gas, usually argon, serves as the detector gas making up 90% of a mixture. Neon or helium are also good choices where X rays from second-row elements are to be detected.The other 10% is a quench gas, perhaps methane. Its role is the termination of spurious ion pairs.*

*One process that produces such ion pairs is initiated by cations. As they approach the cathode, they tend to attract electrons from it, which leads to production of UV photons and further spurious ionization. A gas like methane with a low ionization potential will supply electrons more easily and circumvent the process.

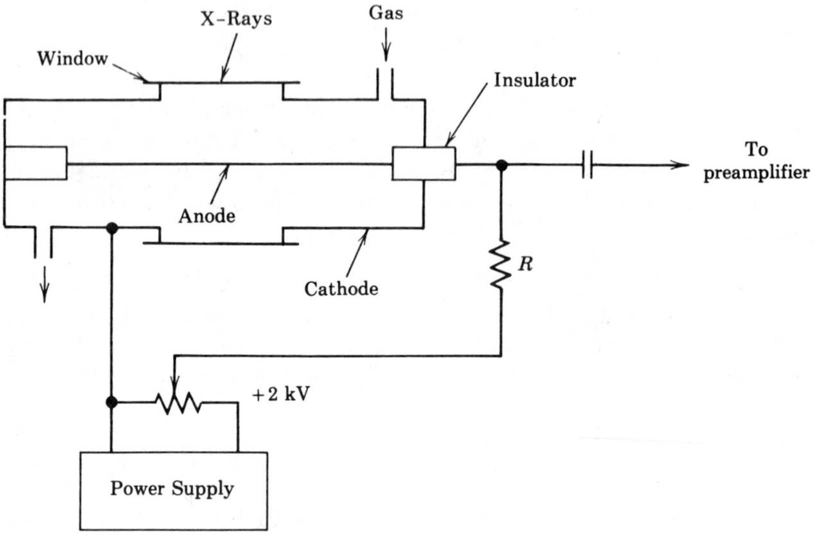

Fig. 21.12 Schematic diagram of a proportional (gas-flow) detector.

In Fig. 21.12 a type of proportional detector or counter called a *flow detector* and the associated electrical circuit are shown schematically. It has very thin windows—0.025-mm Mylar film or still thinner polypropylene supported by a grid of 200-mesh nickel or nylon. An advantage of the thin plastic windows is their good transmission of longer wavelength X rays. Whatever the window material, it is important that it have a conductive coating to ensure a more uniform electric field. In the flow detector gas moves at a slow rate through the active region and also diffuses out through the windows. Different flow rates are usually tried until one that gives the best detector response is found. Finally, for any proportional counter it is important that the anode wire be kept clean lest the electrical field gradient be reduced or made irregular. Such alteration is detected ordinarily by a reduced pulse height or sometimes by pulse broadening.

The two-window arrangement shown ensures that X rays that fail to be absorbed in the gas will leave the active volume. An alternative end-window arrangement may be used, but it is subject to difficulties with respect to uniformity of electrical field. The main characteristics of gas proportional detectors are compared with those of other detectors of energetic photons and particles in Table 21.1.

Sealed proportional detectors with windows of 0.5–1 mm beryllium or other suitable material are also made. They are efficient for intermediate X-ray wavelengths and are often filled with krypton.

Escape Peaks. While a Guassian distribution of pulse heights is observed for a given X-ray transition, some peaks it causes appear with much lower amplitude. They can be traced to a short-circuiting of the process of forming the charge cloud in a detector. In this case an analyte X-ray photon generates a *secondary X-ray photon* from the detector gas

Table 21.1 Characteristics of Some X-Ray Photon and Energetic Particle Detectors

	Properties						
Type of Detector	*Window Material and Thickness*	*Internal Gas; Maximum Gain*	*Wave-length Range (nm)*	*Dead Time (μs)*	*Max. Useful Count Rate (s^{-1})*	*Back-ground Count Rate s^{-1}*	*Resolution For Fe Kα (%)*
Flow proportional	Mylar (metallized) 0.4, 2, 6μm	Ar or Ne, CH_4; 10^6	0.07–1.8	0.5	5×10^4	0.2	15
Sealed proportional	Mica 10 μm	Xe or Kr, CH_4; 10^6	0.05–0.4	0.5	5×10^4	0.5	12
Geiger	Mica 10 μm	Ar, Br_2; 10^9	0.05–0.4	200	2×10^3	2	—
NaI(Tl) scintillation	Be 0.2 mm	— 10^6	0.01–0.3	0.2	10^5	10	50
Si(Li)	Be, Au 25 μm	—	0.04–1	2×10^4		—	5

instead of an energetic *photoelectron*. A two-step process then tends to occur: (a) the secondary X-ray passes out of the detector with no further interaction and (b) the ion pairs that are produced by the weak photoelectron that accompanied the secondary photon are collected as a relatively small *escape peak*.

If possible, an escape peak should be measured along with the main peak. Both originate from the photon being detected. Sometimes the escape peak is too low in amplitude to pass an imposed energy threshold set to reject peaks that are essentially indistinguishable from noise and is lost. If pulse-height analyzers (see Section 21.6) are being used, they can be set separately for main and escape peaks. If only the main peak can be measured, the overall signal or count will be spuriously low. In this case a different detector gas should be tried. Fortunately, escape peaks are seldom a serious problem in analytical measurements.

Scintillation Detectors. For X-ray photons the most common scintillation detector is the thallium-doped sodium iodide crystal, designated NaI(Tl). A large, single crystal that is a few millimeters thick will absorb all but the shortest wavelength X rays. Further, it is transparent to the 410-nm (fluorescent) photons generated. Napthalene is also a satisfactory phosphor, but has too small a mass absorption coefficient to be effective except with long wavelength X rays.

The scintillation crystal needs to be optically coupled to a transducer that will furnish an electrical signal. A photomultiplier tube serves well. It is fortunate that it also provides amplification, for use of this two-stage detector is inefficient: it yields one output pulse for about 40 eV particle energy compared with the roughly 3 eV excitation energy of NaI(Tl). There is also a loss in peak resolution. The scintillator crystal is covered with aluminum foil for good reflectivity except on the side toward the photomultiplier tube, where it is sealed to an appropriate intermediate refractive index layer to ensure good optical coupling. Outside the aluminum foil is a light-tight housing containing a thin beryllium window to admit X rays. The decay time following a scintillation burst is 0.25 μs. Note the generally favorable properties of scintillation detectors listed in Table 21.1. Escape peaks are seldom a problem.

Solid-State Diode Detectors. The development of solid-state diodes based on pure silicon and germanium single crystals for detection of X rays, gamma rays, and other radioactive emanations has provided a solid-state analog of the gas-ionization detector. In these detectors also, the arrival of photons or particles can be registered as events, the average energies determined by conversion to charged "particles," and their arrival time precisely noted. The possibility of the achievement of a high degree of purity and carefully controlled doping with impurities makes possible the development of appropriate resistivities and electrical junctions.

For X-ray detection the silicon detector is commonly used. Its band gap of about 1 eV ensures that high resistivity can be secured in pure silicon. A lithium-drifted silicon detector Si(Li), is shown schematically in Fig. 21.13a. Purified *p*-type silicon is cut into a disk roughly 1 cm dia by 3–5 mm. Lithium, an *n*-type dopant, is then diffused in from the bottom (at 350–450°C under an electrical gradient) to compensate the charge carriers in the *p*-type silicon and provide a wider "intrinsic" region of high resistance. This central region is the active volume in which detection occurs. These diodes are continuously refrigerated at 77 K with liquid nitrogen (an appropriate cryostat is attached) to prevent further diffusion of lithium.

For operation, the Si(Li) detector is reverse-biased. The right electrode is grounded and from −800 to −1000 V is applied to the left electrode. Under this biasing majority charge carriers that do remain in the "active region" will be attracted toward the electrodes, leaving the region still more of a *depletion layer*. X rays enter through the left-hand gold layer. In the figure an incoming X-ray photon is shown at the point of producing the first electron–hole pair. Pair production continues as the original photoelectron loses more and more energy in collison with new silicon atoms.

Charge collection is efficient. Recall that the density of the solid will be roughly 10^3 times that of a gas, resulting in a far higher possibility of dissipation of all kinetic energy, even in a layer no more than a few millimeters thick. A further advantage of the solid-state device is that holes and electrons have mobilities of the same order of magnitude. As a result, the charge collection time is short and typically in the range of 25 to 100 ns. The amount of charge per photon is found by dividing the photon energy by the average energy required to produce an electron–hole pair, 3.8 eV.

To ensure uniform electric fields and improve charge collection in Si(Li) detectors, diodes have been developed with other than disk-type shapes. One possibility is the diode with the annular grove, shown in Fig. 21.13b.

Trapping centers throughout the diodes may also serve to diminish charge collection, causing a certain tailing as pulses appear. Fortunately, noise, which in this

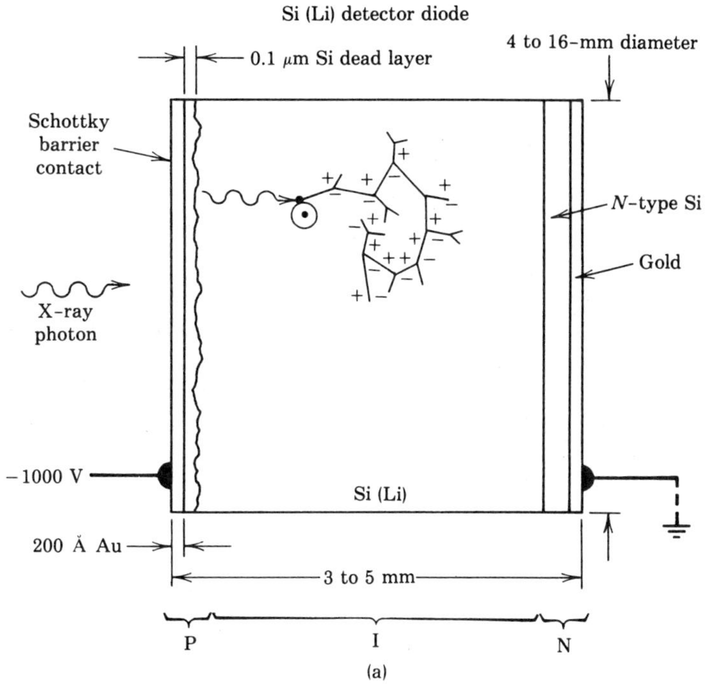

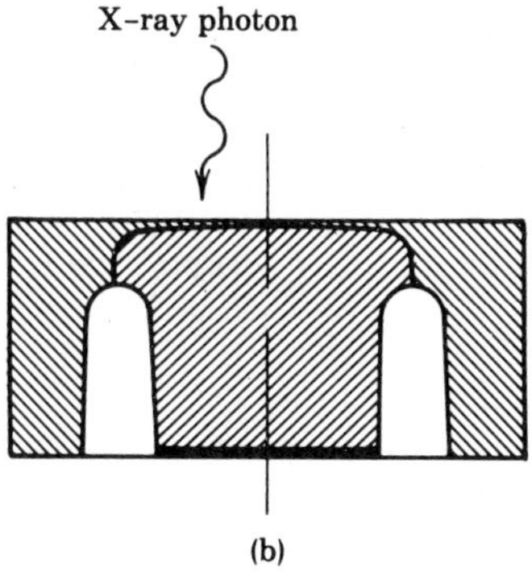

Fig. 21.13 Schematic diagram of a Si(Li) diode detector based on *p*-type silicon. The top gold layer serves as the X-ray window. Detection occurs because electron-hole pairs are produced by entering X-ray photons and collected as charge pulses. (a) Representation of mechanical and electrical aspects of the detector. The 20-nm gold layers deposited at the sides serve as electrodes. At the left it also sets up a Schottky barrier contact and provides roughly in the first 100 nm of silicon a dead region of *p*-character. Lithium diffused in from the right dopes the entire interior of the diode. As an *n*-type interstitial dopant, it pairs with *p*-type impurities in the central part of the diode and compensates the silicon to yield a large volume in which so few charge carriers exist that it has in effect been returned to an intrinsic or *i*-state. This high resistance region is the *active* volume in which energetic photons or particles dissipate their energy in electron-hole pairs that are collectable as a charge pulse. *n*-type doping of the right layer of diode assures an *ohmic* junction when the Au layer is deposited. The overall diode is thus *p-i-n* in type. Not shown is the large cryostat with capacity sufficient for 25 L of liquid nitrogen that maintains the detector indefinitely at 77 K. (b) Cross-section of a representative diode. The deep annular groove both limits the active region and ensures a more uniform electric field which improves charge collection substantially. Commonly, a metal mask blocks access by photons to some of the peripheral volume shown in (a). (Courtesy of EGG ORTEC.)

case is shot-effect noise in the small leakage current the diode conducts under the conditions of reverse biasing, is small at 77 K. Increasing the bias voltage further would improve the rate of charge collection but would also increase the leakage current and noise. Finally, it should be noted that Ge(Li) diodes are also available for detection of X rays of higher energy. Alternatively, thicker Si(Li) diodes may serve that purpose.

Pulse Shaping. Each of the detectors just described is capable of transducing the energy of an X-ray photon into a charge pulse. Most pulses, however, are poorly shaped: they are low in amplitude, noisy, and tail too much. To prepare these pulses for later processing to extract the desired chemical information, two electronic circuits must be added to the detector module. The first, a preamplifier designed to have very low noise, provides initial amplification and is placed very close to the detector to ensure that its leads add little noise. It is followed by a pulse-shaping amplifier to complete the amplification and reduce tailing by clipping the pulse shortly after its fast initial rise corresponding to electron collection. Recall that electron bursts occur essentially separately around the anode. It is thus possible to proceed with the pulse collection for a second electron avalanche without waiting to collect the earlier set of positive ions, a process that would require more than 100 μs.

Preamplifiers. A representative preamplifier suitable for use with proportional counters and scintillation detectors is shown in Fig. 21.14. Note the detector at the left of the circuit. Its high voltage input includes an *RC* filter (R_1 and C_1) to minimize ripple as well as interference from the high-voltage power supply that might affect the detector or the preamplifier. Capacitor C_2 blocks high voltage from the preamplifier and should be charged fully

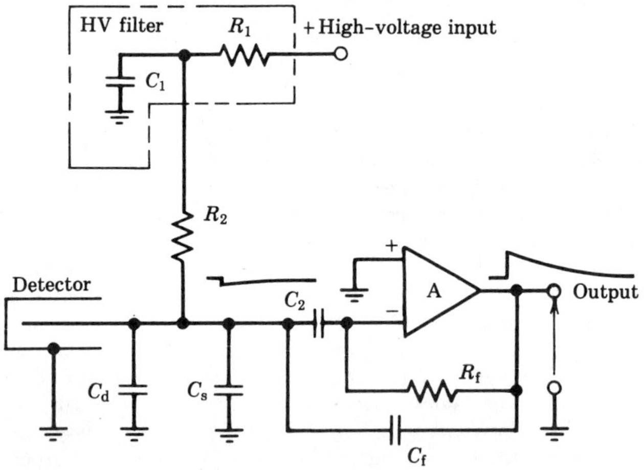

Fig. 21.14 Schematic of a charge-sensitive preamplifier commonly used with proportional (gas-filled) and scintillation detectors. (Courtesy of EGG ORTEC.)

by the steady high-bias voltage of the detector. That being so, output charge pulses from the detector will tend to collect on C_f, especially since C_f is made substantially smaller than the detector capacitance and circuit stray capacitance, $C_d + C_s$. As each charge pulse Q collects on C_f it is converted to a voltage. The op-amp will both buffer the detector from the following circuit and provide a high-amplitude output voltage pulse described approximately by the expression

$$V_0 = (Q/C_f)\exp(-t/R_f C_f).$$

Because the pulse emerging from the preamplifier tails strongly, a shaping amplifier becomes essential. It both further amplifies and reduces the duration of pulses, permitting a higher counting rate to be achieved in the detector circuit. It is beyond the scope of this text to discuss the pulse-shaping amplifiers in common use, which are either of the delay-line clipped variety or the semi-Gaussian amplifier type [12].

A preamplifier suitable for a solid-state Si(Li) diode detector is shown schematically in Fig. 21.15. Again the diode high-voltage bias is applied through a filtering network designed to remove ripple or transient noise. The FET in the preamplifier is placed as close to the detector as possible. With the strong reverse bias on the detector, electron–hole pairs separate under the electric field, holes drifting to the left and electrons to the right. The collected electrons appear at the gate of the FET and charge C_f.

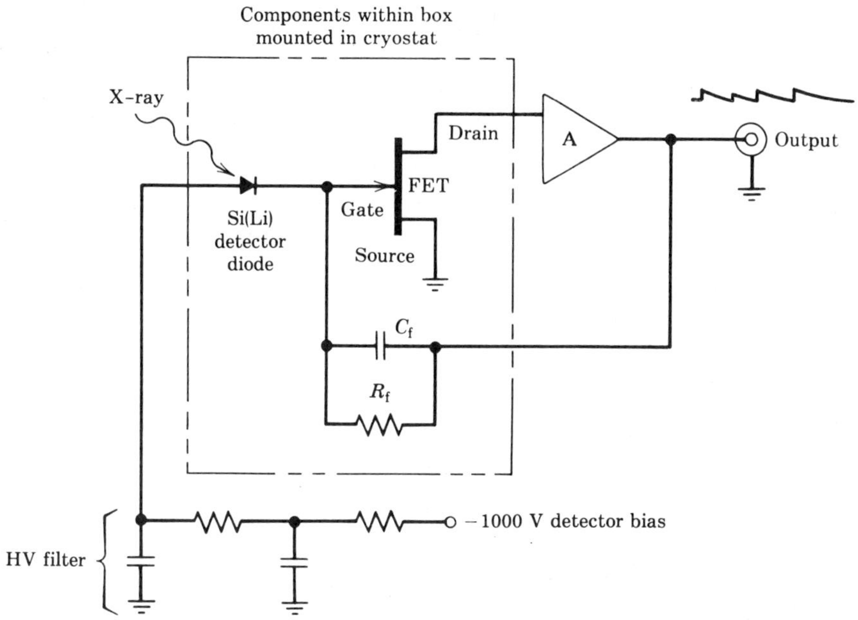

Fig. 21.15 Schematic diagram of a resistive-feedback preamplifier commonly used with a solid state Si(Li) detector. To minimize noise the detector and field effect transistor (FET), which is mounted close to the detector, are kept at 77 K. Components within the broken line box are inside the detector cryostat though C_f and resistor R_f in the feedback network are kept at 150 K by a small heater. The op-amp in the inverting configuration (its noninverting terminal is grounded) provides that the detector output charge capacitor C_f by driving the FET gate and the inverting terminal to ground. (Courtesy of EGG ORTEC.)

Almost all the output of the op-amp is returned to the gate of the FET by negative feedback through the network consisting of the parallel combination of C_f and R_f. In this way extraordinary stability is built into the system. The noninverting (+) input of the amplifier is grounded, forcing the op-amp to drive the other input and the gate of the FET essentially to ground potential. As a result, each pulse of charge collected from the detector is forced to charge feedback capacitor C_f to the potential available, giving a maximum preamp output voltage $V_0 = Q/C_f = (E/\epsilon)(e/C_f)$. Here E/ϵ is just the energy of the X-ray photon divided by the average energy to form an electron–hole pair, thus the number of electrons that will be collected, and e the charge per electron.

In the intervals between pulses, C_f will gradually discharge through R_f. When a pulse is received, the amplifier output voltage will, as shown, rise steeply to a maximum value as C_f charges and then fall exponentially to a base line at a rate controlled by time constant $R_f C_f$. Clearly, the smaller R_f, the shorter the time that must be allocated to a given pulse and the larger the counting rate that may be realized. Offsetting this behavior is a substantial increase in noise as R_f is reduced. In practice a trade-off is worked out. Since pulses of all amplitudes, that is, all energies, must be discriminated correctly by a multichannel analyzer, a pulse-shaping system that will guarantee a more easily recognizable and measurable shape of pulse and low noise is important. Its time constant is long (4–16 μs) as a result. As with the proportional and scintillation detector, a semi-Gaussian amplifier is commonly used [12]. Because operation is usually also at about the maximum number of pulses the shaping amplifier can accommodate, this amplifier usually establishes for the detector a maximum count rate, which is smaller than is feasible with the other gas-filled and scintillation detectors (see Table 21.1).

21.6 WAVELENGTH-DISPERSIVE SPECTROMETERS

Having dealt at length with X-ray sources and detectors as well as related electronics, we examine now the major X-ray spectrometric systems. The earliest and still most widely used ones are wavelength-dispersive spectrometers and employ Bragg's law of diffraction to disperse X-rays angularly in space. The well-known Bragg equation is followed: $m\lambda = 2d \sin\theta$, where m is the diffraction order, d the interplanar spacing of atoms in a well-formed crystal, λ the wavelength of incident parallel radiation, and θ the angle of incidence (and diffraction) of the X-ray beam relative to the normal to the crystal surface. What arrangement of components best facilitates precision-angular dispersion of X rays?

Dispersing Module. The collection of several components associated with dispersing X rays, the dispersing module, appears in Fig. 21.16 just after the sample holder. With a flat analyzer crystal, one needs primary and secondary collimators to ensure parallel rays and a goniometer. The last device is a table with precision drives for moving arms (not shown in the figure), one holding the crystal, and a second holding the secondary collimator and the detector, even though the latter is not part of the module. Sometimes this module is called a crystal analyzer or monochromator.

To see how one scans a spectrum, consider the "optical schematic" shown in Fig. 21.16. To scan, it will be necessary to increase θ, the angle of incidence of

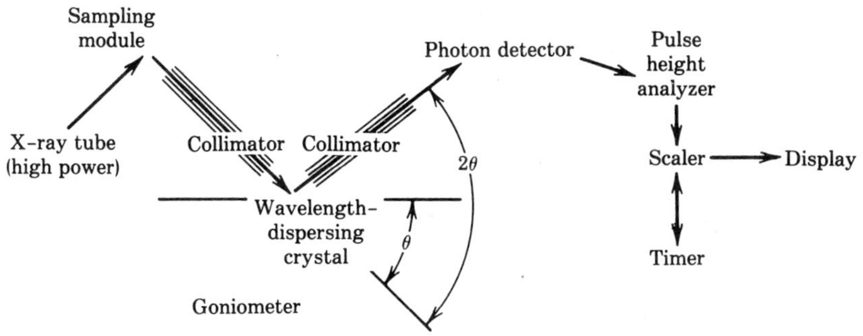

Fig. 21.16 Modular diagram for a wavelength-dispersive X-ray spectrometer. The readout will be the number of pulses per second versus the angle of dispersion or wavelength.

primary X rays on the crystal, from nearly zero degrees by pivoting the crystal through the angle θ about an axis through the center of its front face and perpendicular to the paper. Simultaneously, the angle θ the detector makes with the crystal surface must be equally increased. Regardless of the type of drive—gear train, dc motor, or other device—today it is usually software controlled.

A common type of collimator is a set of precisely parallel flat plates of heavy metal such as tantalum. For fine collimation the spacing may be 0.015 mm, for coarser collimation and higher throughput, 0.45 mm. The Soller type of collimator consists of a bundle of straight, hollow metal tubes. Sometimes their interiors are roughened to minimize internal scattering and reflection.

Some losses in intensity and resolution are inherent in the use of collimators. The first may be traced to absorption of X rays that do not enter by paths nearly parallel to the walls and the second to a small but finite divergence of rays in the collimator. Collimators with wider spacing or greater apertures require use of analyzer crystals and detectors with larger surfaces. Actually, if a Rowland circle spectrometer based on a curved crystal is used, collimators can be eliminated as will be discussed below.

Table 21.2 lists a variety of crystals that have seen use in dispersive devices together with wavelength ranges appropriate to each. A practical angular range for θ that will avoid mechanical problems in the movement of crystal and detector plus its collimator is from 15° to perhaps 75°. Some important crystal properties are also listed in the table.

Both the dispersion and the wavelength range a crystal provides will be important. How the first is affected by the properties of a crystal can be learned by differentiating the Bragg law to obtain the expression for angular dispersion: $d\theta/d\lambda = m/2d \cos\theta$. Clearly, crystals with the smallest values of d, or $2d$, will offer the highest dispersion. For example, LiF offers a dispersion of 0.27 rad/Å, whereas potassium hydrogen phthalate (KAP) offers a dispersion of only 0.044 rad/Å. From the Bragg law and Table 21.2, it is clear that the larger the wavelengths to be observed, the greater must be the crystal spacing.

For high *resolution*, the dispersing module should have a crystal with the largest

Table 21.2 Properties of Common Crystals for X-ray Spectrometers

Crystal	*Reflecting Plane (Miller Indices)*	*Lattice Spacing d(nm)*[a]	*Primary Useful Wavelength Range (nm)*[a]	*Diffraction Power*	*Resolution*
Lithium fluoride, LiF	(220)	0.1424	0.025–0.272	High	Very high
	(200)	0.2014	0.035–0.384	High	Very high
Topaz (hydrated aluminum fluorosilicte	(303)	0.1355	0.024–0.259	Medium	Very high
Silicon, Si	(111)	0.3138	0.055–0.598	Medium	Very high
Germanium Ge	(111)	0.3136	0.055–0.598	Medium	Very high
Pentaerythritol, PET	(002)	0.4371	0.076–0.834	Medium	Very high
Ethylene diamine *d*-tartrate, EDDT	(020)	0.440	0.077–0.840	Medium	Low
Gypsum, $CaSO_4 \cdot 2H_2O$	(020)	0.7595	0.132–1.45	Medium	Medium
Potassium acid phthalate, KAP	$(10\bar{1}0)$[b]	1.332	0.232–2.54[c]	Medium	Medium
Thallium acid phthalate, TAP	$(10\bar{1}0)$[b]	1.295	0.226–2.47[c]	Medium	Medium

[a] 1 nm = 10 Å.
[b] *hkil* indices.
[c] Valuable for Kα wavelengths of O, F, Na, and Mg.

value of $2d$ that will still give access to wavelengths desired (see Table 21.2) and collimators with the smallest divergence that will still give adequate S/N. Resolution will, however, be only fair at or with (a) small values of θ, (b) dispersing crystals that provide minimal dispersion, (c) collimators that have a large divergence, and (d) crystals that have defects (see below).

Rocking Curve of Crystal. Most crystals give poorer than theoretical resolution because of defects. An average crystal behaves as though its planes of atoms extend only a few hundred angstroms before they change orientation slightly. Consequently, most crystals diffract monochromatic radiation over an angular range $\Delta\theta$ rather than at a given angle. It is unnecessary to assess such imperfections directly, since an equivalent broadening of a peak may be produced by rocking a nearly perfect crystal through the same angle. Thus, the term *rocking curve of a crystal* commonly describes this behavior. Because of this behavior the integrated intensity of an X-ray peak is usually a more accurate measure of elemental concentration than peak intensity.

Rowland Circle Spectrometers. While planar analyzer crystals are common in wavelength-dispersive spectrometers, a considerable amount of attention has been given to making curved crystal "optics." Where available, their use makes possible the development of spectrometers based on the Rowland circle (see Section

13.6). Since collimators are eliminated in the curved-crystal spectrometer design there is some improvement in throughput. Though slits are necessary, they may have fairly large solid angles of acceptance without serious loss of resolution.

The original impetus for the development of curved-crystal instruments was to accommodate the diverging X-ray beam produced in electron-probe devices and electron microscopes that are equipped for concurrent measurement of X-ray spectra. Both devices produce a diverging X-ray beam as seen earlier in an X-ray tube, because a point area of the sample is bombarded by the focused electron beam. Another important application of the Rowland circle design is the development of multichannel X-ray spectrometers.

In a Rowland circle spectrometer the entrance slit S and its image lie on a circle of radius $R/2$. Since the analyzer crystal also lies on the circle and acts as a concave focusing device, its radius must be R, as discussed in Section 13.6. To make such a crystal one usually bends a suitably soft crystal around a cylindrical form. This operation requires careful heating of a crystal to a temperature somewhere below its melting point before bending. The resulting *Johann* geometry spectrometer is shown in Fig. 21.17. After bending, the planes of atoms follow a cylindrical path. Thin crystals of mica, gypsum, and quartz can be bent as well as those of some salts. The curvature guarantees that the Bragg condition of equal angles of incidence and diffraction is realized for a range of angles of incidence. A disadvantage is that there is some defocusing with this geometry; for Bragg's law to be obeyed generally, the surface of the crystal would have to be tangent everywhere to the Rowland circle, thus, with radius $R/2$.

A solution to this dilemma has been developed for some crystals. After they have been bent cylindrically to a radius of R, their front surface is ground to a

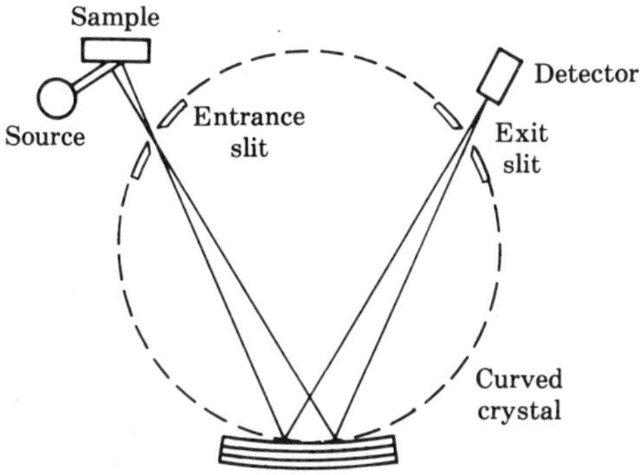

Fig. 21.17 Rowland circle spectrometer with a crystal cylindrically bent to an arc of a circle of radius R. In this *Johann geometry* design those X-rays emitted by the sample that pass through the rectangular entrance slit and are diffracted by the crystal ideally form a rectangular image of the entrance slit at the position of the exit slit. Some defocusing unavoidably occurs for rays that are diffracted from the ends of the crystal. Only if the crystal were bent to the radius of the Rowland circle, which is $R/2$, would its surface be tangent along its entire length and would extreme rays of the same wavelength focus to the same rectangular image. The length of crystal has been exaggerated to show the planes of atoms and the separation of crystal and circle toward the ends.

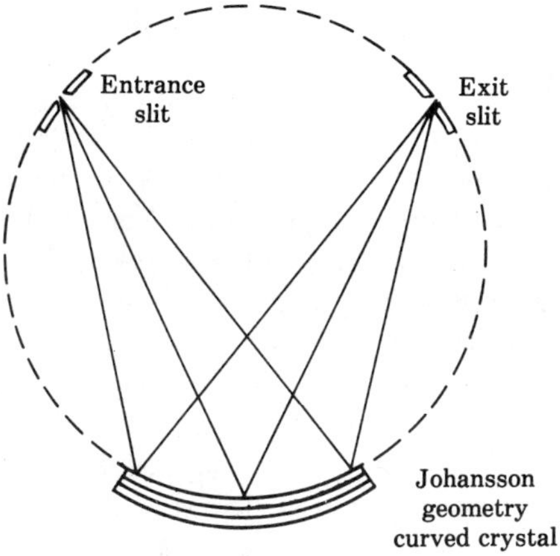

Fig. 21.18 A simplified version of a Rowland circle spectrometer for a curved *Johansson geometry* crystal. This crystal is first bent around a cylinder of radius R and then its surface is ground to the radius of the focal circle ($R/2$). Its surface is tangent at all points and X rays of a given wavelength that are incident toward the ends of the crystal are focused to the same image at the exit slit.

radius of $R/2$, giving a surface tangent to the focal circle at all points. This *Johansson* geometry is shown in Fig. 21.18. While the technique minimizes defocusing, unfortunately many crystals cannot be ground.*

Sources, Sample, Modules, and Detectors. Usually an X-ray tube with an appropriate target serves as the source for a spectrometer. Somewhat miniaturized wavelength-dispersive spectrometers are also being attached to scanning electron microscopes and electron probe devices. In standard wavelength-dispersive spectrometers, higher power X-ray tubes are commonly used because necessary collimation leads to serious intensity losses. As a result, large amounts of X-ray continuum are generated and a great deal of background is inevitable. Limits of detection are fairly high unless shielding is very good, and unwanted scattering and spectral background are minimized.

A compartment is ordinarily provided to hold a sample or specimen. It is important for reasons of S/N that the sample be held at the point where the X-ray beam from the source crosses the solid angle aperture of the dispersing module, in this case entrance aperture of the primary collimator (see Fig. 21.16).

Proportional flow counters, sealed proportional detectors, and at shorter wavelengths, scintillation counters, are commonly used in these spectrometers. Indeed, a scintillation counter is often placed behind a proportional counter. In Fig. 21.19 the detector and representative processing modules of a spectrometer are shown. Following the pulse-shaping amplifier that prepares detector pulses for information extraction, the pulses move to a single-channel pulse height analyzer or discriminator and are counted. It is the digital value of the count that is stored in memory and displayed.

*An alternative is to use a logarithmically curved crystal. The crystals are made by bending crystals or cementing minute oriented crystals to a shape in the form of a logarithmic spiral. Focusing quality is not as high as that of the Johansson crystal, but can be improved by use of a wider detector slit.

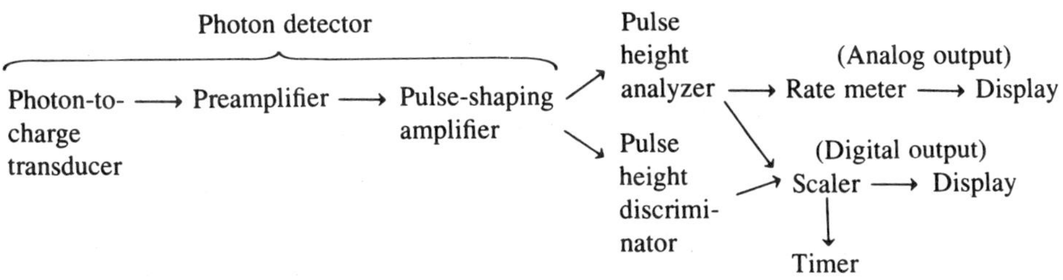

Fig. 21.19 A representative photon detector and set of processing modules for a wavelength-dispersive spectrometer. The X-ray photon detector consists of a photon-to-charge transducer, for example, a scintillation counter and two electronic circuits, a preamplifier producing a voltage pulse and a pulse-shaping amplifier.

Resolution. What module(s) mainly establish the resolution of a wavelength-dispersive spectrometer? It is useful to define resolution B as the full width at half maximum of an X-ray peak. This perspective allows definition of resolution in terms of the standard deviation of a Gaussian curve, which analyte lines approximate in shape. Because of the angular dispersion effected in this spectrometer, there will be minimum contribution from the detector since it can only report intensity versus angle. One strong contribution will be the full width at half maximum of the "rocking curve for the analyzer crystal," W, and a second contribution B_C will express the spread created by the collimator. For the overall resolution one has $B = \sqrt{W^2 + B_C^2}$. In other words, the rocking curve for the crystal convolutes the broadening introduced by the collimator. With a LiF (200) analyzer crystal the best resolution realizable is about 13 eV for 6-keV photons and 250 eV for 25.3-keV photons.

Pulse Height Analyzers. An *integral mode* pulse height analyzer or *discriminator* provides a basic voltage threshold above which all pulses will be counted. At the output of the pulse-shaping amplifier, low-frequency noise may be about 1 V while signal pulses will range up to 10 V or higher. In this case, it would be appropriate to set the analyzer to pass all pulses with heights greater than about 1.5 V. All pulses above the threshold are scaled and counted for a known interval (see Fig. 21.19) to obtain an intensity reading. Alternatively, as indicated in the figure, they may be registered in an analog mode by a rate meter.

Errors may arise. A pileup of two pulses in the detector will give a large composite pulse that will register as a single count though it should show as two. A higher-order peak ($m = 2$ or perhaps 3 in the Bragg equation) that inadvertently overlaps an analyte line will register spuriously. Since first-order X-ray lines are generally more intense than higher-order versions, one can in fact set the pulse-height threshold a bit higher to reject all pulses of higher-order peaks. With such discrimination against background and higher-order peaks, one may obtain the equivalent of a first-order spectrum even with a wavelength-dispersive spectrometer. Finally, for each wavelength to be measured, amplitudes will be different for background and for peaks. A pulse-height selector will need to have different settings for the baseline and peaks if background corrections are deemed necessary.

A *window mode* or *differential mode* is often available with an analyzer to convert it to a *single-channel pulse height analyzer*. For this mode, an analyzer incorporates a second discriminator circuit. Let V_L be the lower threshold and V_U be the level set an increment of voltage higher. Let a logic circuit test the response of the upper level circuit whenever the first level circuit produces a pulse. If the upper level also forms a pulse, that pulse is rejected since its amplitude exceeds both thresholds. Whenever a pulse falls short of V_U but exceeds V_L, it is passed on since its amplitude lies within the *window* $V_U - V_L$ provided by the pair of analyzers. For example, a single-channel analyzer can be used in a wavelength-dispersive spectrometer to identify the weaker pulses corresponding to a higher order of diffraction.

Multichannel Spectrometers. Many wavelength channels can be measured simultaneously with a wavelength-dispersive spectrometer by placing an analyzer crystal–detector pair at appropriate locations around the circumference of a goniometer. Each crystal must be at an angle θ relative to the sample holder that corresponds to a desired analyte line. The aperture of the accompanying detector must be tilted at twice that angle with respect to the sample holder. From 10–30 channels may be established for as many different elements.

Diffractometers. The *goniometer* is also a versatile device that permits measurement of angles with precision. While it is the heart of a wavelength-dispersive spectrometer when a crystal of known d is used, it can be used with unknown crystals as an *X-ray diffractometer*. In this application, X rays of known λ from a monochromatic source are directed toward a variety of crystals of unknown structure placed in the position of the crystal analyzer in Fig. 21.16. This lattice spacing is determined. It is beyond the scope of this text to deal with determination of crystal structures by more complex methods by involving X-ray diffractometry though this is an important analytical procedure as well.

21.7 ENERGY-DISPERSIVE SPECTROMETERS

X-ray spectra can also be presented as a function of photon energy since $E = hc/\lambda$. Yet the general development of energy-dispersive spectrometers had to await the advent of a proportional detector (output peaks proportional to energy of photons) with sufficiently good resolution. The development of solid-state detectors, and chiefly the Si(Li) detector with its resolution of 150 eV at 6 keV, made energy dispersion practical. Even though all wavelengths fall simultaneously on the detector, it can resolve the spectra of most samples since usually there are only a few spectral peaks per element.

Instrumentation. A modular version of an energy-dispersive spectrometer is shown in Fig. 21.20. Two of the clusters of modules need general attention. The

Sampling module

↗ ↘

X-ray tube (low power) → Si(Li) detector → Processing modules → Multichannel pulse height analyzer → Display

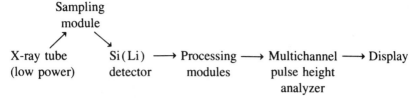

Fig. 21.20 Modular diagram of an energy-dispersive X-ray spectrometer.

first is the source–sample module–detector set. The ability of the detector to respond sufficiently rapidly to a barrage of the different energies of X-ray photons from a sample will be sorely taxed unless a low power source, for example, radioisotope emitter or a 100-W X-ray tube, is employed.

What will guarantee a satisfactory S/N at the detector output? Clearly, optimizing this ratio must begin with securing an appropriately large signal for each analyte. If an X-ray tube source is used, its current needs to be adjustable in small increments over perhaps two orders of magnitude to secure a high count rate from the detector without overloading it or the pulse-processing system. As suggested earlier, tube voltages need to be adjustable to ensure efficiency in excitation of particular analyte lines. In some cases a set of voltages, one at a time, may need to be applied for efficient excitation of particular samples.

Figure 21.21 shows configurations for both primary and secondary excitation of samples. By choosing one or the other, one has control over the distribution and

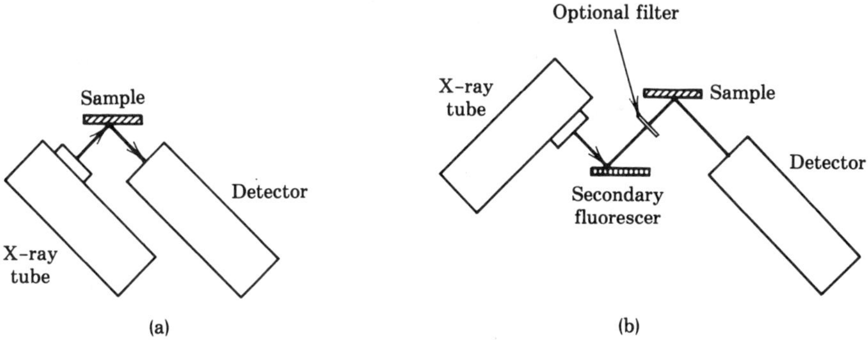

Fig. 21.21 Modular configurations in an energy-dispersive instrument for primary and secondry excitation of samples. (a) For primary excitation, the X-ray tube is sometimes a bremsstrahlung tube (with V_0 = about 10 kV) that provides polychromatic radiation. Alternatively, it may be a transmission target tube in which 30 kV bremsstrahlung are passed through a thin filter of 0.05 mm Mo to give mainly molybdenum Kα and Kβ X rays. (b) Secondary excitation is used to furnish nearly monochromatic exciting radiation to the sample. Bremsstrahlung or continuum radiation excites characteristic lines of the secondary (external) target, also called a *secondary fluorescer*. The continuum reaching the sample can be further reduced by insertion of a thin (60–250 μm) flat filter of the secondary fluorescer metal at the optional filter position. This filter will pass characteristic wavelengths, which will fall in the low absorption notch just above the absorption edge for this metal, and absorb bremsstrahlung. Actually, the absorbed brehmsstrahlung will generate additional characteristic photons. The name *regenerative filter* has been suggested to describe this device.

level of exciting radiation. With the direct (primary) excitation shown in Fig. 21.21a, one can use the high power available in the brehmsstrahlung emission. Continuum radiation provides efficient excitation of different elements. The trade-off it poses is greatly increased background from scattered radiation. That intensity is sufficient to obscure emission of X rays from trace constituents and in many cases also to overload the detector. An alternative approach, if it is known that samples will contain wide concentration ranges of elements, is control of excitation energies by the computer at appropriate points during the taking of the spectrum.

By contrast, by use of the secondary mode of excitation displayed in Fig. 21.21b, essentially monochromatic radiation can be directed toward a sample. The bremsstrahlung radiation irradiates a pure elemental target to obtain its characteristic radiation which is then perhaps directed through a very thin layer of the target metal to reduce scattered intensity. While considerably greater amounts of X-ray tube energy are required for this approach, it is effective in trace determinations. Note that one may select the wavelength at will by changing targets.

A grided X-ray tube that can be pulsed offers protection against pulse pileup in the detector by controlling the rate of emission. Application of a negative voltage to the grid will halt the electron beam and the appearance of new X-ray photons. If that provision is activated when a photon is being detected, there can be more efficient use of detector time. Data collection rates can be increased by a factor of 2 or 3 over dc excitation.

The second set of modules needing attention is the detector–multichannel pulse height analyzer–diplay combination shown in Fig. 21.22. In the photon detector circuits the inevitable trade-off between resolution and rate must be dealt with in the selection of the time constant for the pulse-shaping amplifier. Choice of a longer time constant will ensure that the best resolution possible is attained, though at the cost of speed. A range of possibilities is suggested in Fig. 21.23. Setting the processing system to a ''high'' processing rate will ensure a fast throughput of pulses, but will reduce resolution about 25% below that realized when the ''low'' range is chosen.

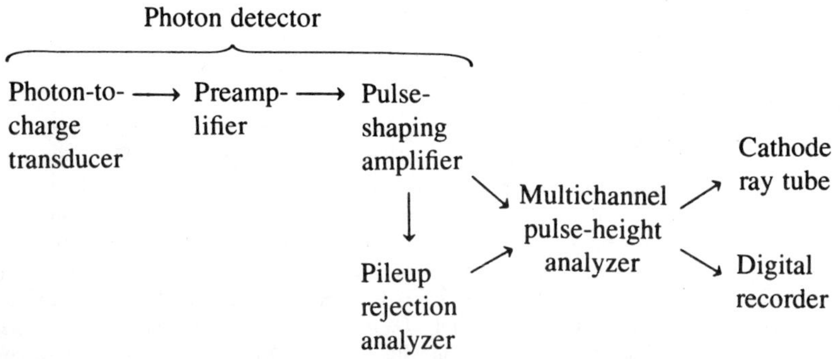

Fig. 21.22 A representative set of processing modules for an energy-dispersive spectrometer. The detector is included to establish the way in which this part of the signal channel connects with the one shown in Fig. 21.19.

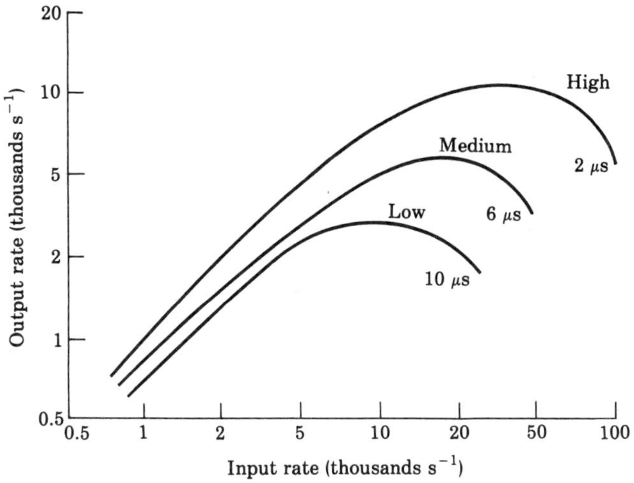

Fig. 21.23 Throughput of output signal as a function of filtering provided in the amplifier and signal processor modules. Approximate circuit time constants are shown. When a shorter time constant is selected, throughput is higher. The trade-off is that more noise is also passed and resolution is lower.

What underlying phenomena are affecting the efficiency of counting? The first is the randomness of emission of X-ray photons, which is characteristic of relaxation from an excited state. As a result, pulses "pile up" on each other at high processing rates. Modules designed to minimize pulse "pileup" often follow the detector module, as described below. The other possible error is that photons may arrive so fast as to saturate the detector. Either the development of a still faster detector or use of lower power sources, which is the common solution, take care of this possible error. Error is kept at a minimum only by keeping X-ray intensities low.

Pileup Rejection. The train of modules outlined in Fig. 21.22 calls for a pileup rejection amplifier to appear *in parallel* with the pulse-shaping device. Pulse pileup is said to occur whenever a second photon reaches a detector during its deadtime interval. When such a coincidence occurs, the additional ion pairs formed by the second photon are added to those produced by the first, leading to a pulse of spuriously high amplitude (energy).

An alternative approach is to use a gate to close the channel to the pulse-shaping amplifier for an interval known to be just longer than the deadtime of the detector and pulse-shaping amplifier. The gate is activated by a very fast rise-time amplifier in a pileup rejection circuit. Each detector output pulse activates both amplifiers. The blocking signal from the gate will cease after any single pulse has been shaped by the pulse-shaping amplifier. If while the linear gate providing this signal is closed, the fast amplifier responds to a second photon which reaches the detector, it reactivates the linear gate and lengthens it slightly by a time that would correspond to shaping the second pulse. The possibility of processing two photons within a given operation in the detector is thus avoided.

For accurate quantitative analysis it is important to correct for the loss of counts represented by the upper ends of the curves in Fig. 21.23. By keeping an accurate account of the intervals during which the detector is unable to accept another pulse an appropriate correction can be made.

Resolution. As implied earlier, in an energy-dispersive spectrometer resolution is defined mainly by the detector. The main contribution to the full width at half maximum of an X-ray photon peak is the statistical one given by the expression

$$2.35\sqrt{F\epsilon E},$$

where F is the Fano factor, about 0.13 for Si(Li), ϵ is the electron–hole production energy 3.8 eV, and E is the X-ray photon energy in eV. For a 6-keV photon the formula gives about 125 eV. In addition, noise is the detector and preamplifier contributes about 80 eV, giving a total resolution B of about $\sqrt{125^2 + 80^2} = 150$ eV for 6-keV photons.

Multichannel Analyzer. This module consists of a peak height circuit, an AD converter, a memory, and a display unit. Pulses leaving the pulse-shaping amplifier yield information about their peak height, which is then converted to a digital output by a Wilkinson-type AD converter.* In this unit the peak input pulse charges a capacitor. The capacitor is then rapidly discharged by a constant current. To avoid translating noise to an output signal, only the height of pulse above some discriminator threshold is actually transferred to the capacitor. During the period of discharge a fast counter operates to report the length of the interval. This number is labeled an address in memory. On completion of the discharge the count in the memory cell addressed is incremented by 1. The collection of counts in the channels of the multichannel analyzer will be a histogram of the X-ray spectrum measured.

Nearly always display of spectra will be on a cathode ray tube. Often color graphics simplify the interpretation of spectra. Since energy-dispersive spectrometers are widely used with scanning electron microscopes, display will be treated more fully in Section 22.2.

21.8 ERRORS IN COUNTING PHOTONS AND PARTICLES

In measuring the intensity of X-ray emission by excited analytes or particle emission by radioactive samples, we must allow for characteristic errors that include absorption and scattering by the sample itself, counting difficulties arising from randomness in the arrival of photons or particles in the detector, and other detector errors. Here attention will be given solely to the second type, which are called counting errors, and to other detector errors. Ways to control or allow for the first type of error, which may be termed sample errors, will be taken up in Section 21.11.

*R. W. Schumann and J. D. McMahon, *Rev. Sci. Instr.*, **27**, 675 (1956).

Statistical Uncertainty. Since the radioactivity of unstable atoms and the emission of electronically excited atoms are random processes, measurements of those phenomena (counting rates) are subject to statistical fluctuation. The rates typically become more precisely established with higher numbers of recorded counts, for the variations average out as they would in most types of measurements. It can be shown that a series of replicate measurements of X-ray emission or atom radioactivity* should fall strictly into a Poisson or binomial distribution. From the characteristics of the ideal distribution, we can find the probability of obtaining a certain number of nuclear disintegrations or electronic emissions in time t for a sample containing N_A analyte atoms and the statistical uncertainty in the measured level of activity. If the number of counts is reasonably large ($N > 100$), it can be shown that the standard deviation σ_N in the total count is

$$\sigma_N = \sqrt{N}, \tag{21.7}$$

and that the standard deviation in the intensity I or counting rate r is $\sigma_I = \sqrt{N}/t$.

Clearly, the precision of both the total count and the rate may be improved indefinitely by increasing the time of observation. For example, there will be approximately a 10% uncertainty if 100 counts are recorded, but only 1% uncertainty if 10,000 counts are noted.

A correction for background should often be made also. Further mention of this point will be made in the discussion of the ratio method. In Section 10.2 we saw that the standard deviation of a quantity R, which is defined by the expression $R = A - B$, is related to the standard deviations of A and B by the expression

$$\sigma_R = \sqrt{\sigma_A^2 + \sigma_B^2} = \sqrt{N_A + N_B}. \tag{21.8}$$

Let the subscripts R, A, and B stand for the net, analyte, and background rates or intensities, respectively. In practice, experiments where an accuracy of 1% or better is required are designed so that the standard deviation of the total counting rate A is several times the standard deviation of the background rate B. Inspection of Eq. (21.8) shows that it is not necessary that the background level be known with high precision if this ratio is maintained. In the interest of time, some such arrangement is almost essential, for an unduly long period would be required to obtain as many counts for background as for an analyte.

Fortunately, the Poisson distribution can be well approximated by a Gaussian distribution when N is large. One can thus use statistical procedures related to a normal error distribution, which is Gaussian, to express confidence limits. For instance, stating the confidence interval in a total accumulated count for photon or energetic particle data as one standard deviation, $N \pm \sigma$, will provide a confidence level of 68%. Similarly, stating it as $N \pm 2\sigma$, will imply that a second measure-

*By imposing the restrictions that (a) a large number of active atoms are observed for (b) a time short compared with their half-life and that (c) a large number ($N > 100$) of counts are taken, the Poisson and Gaussian distribution laws may also be used to predict probabilities of disintegration.

ment will fall within the limits 95% of the time. It is reasonable to assume the mean will lie within the 95% confidence limits. For example, if $N = 10{,}000$, the mean N_0 with reasonable probability will be between $N \pm 2\sqrt{N}$ or 9,800 and 10,200.

Often a counting rate or intensity I is registered instead of a total count. Intensity has the dimensions of counts per second and it is related to an accumulated count by the expression $N = It$, where t, the elapsed time, is assumed measured with insignificant error. The relative standard deviation will be simply $\sigma_I/I = \sigma_N/N$. From this relationship it may be shown that the standard deviation of the intensity or rate is just $\sigma_I = \sqrt{I/t}$.

Example 21.10. Assume that 400 counts have been obtained from a sample over a 10-s period. What is the standard deviation in the accumulated count and the percent relative standard deviation? What confidence limits are appropriate for each at the 67% level? What is the standard deviation σ_I?

The standard deviation in the count is $\sqrt{400} = 20$, and the percent relative standard deviation is $(20/400) \times 100 = 5\%$. Also, the counting rate is 40 s^{-1} and standard deviation in the rate $\sqrt{40/10} = 2\ s^{-1}$. The results may be expressed with confidence limits at the 67% level as follows: total counts 400 ± 20; counting rate, 40 ± 2 counts per second.

Example 21.11. In a 10-min count on a sample, a total of 1600 disintegrations is registered. The background is counted for 15 min and totals 450 counts. What is the standard deviation for the corrected analyte counting rate and 67% confidence limits for it?

The standard deviation for the total analyte count is $\sqrt{1600} = 40$. For the background it is $\sqrt{450} = 21.2$. The net counting rate is

$$\frac{1600}{10} - \frac{450}{15} = 160 - 30 = 130.$$

The standard deviation in the analyte intensity or rate is $40/10 = 4$ counts per minute; that of the background is $22/15 = 1.41$, and the standard deviation in the net counting rate is $\sqrt{4^2 + (1.41)^2} = 4.2$. The net counting rate is then 130 ± 4.2 counts per minute.

Ratio Method. In automatic X-ray spectrometers, it is common to take a ratio R of numbers of counts of analyte in a sample X to the ratio of its counts in a standard S. A counting time t_S is set with the standard and represents the interval required to reach the desired precision in counting: $t_S = N_S/I_S$. An advantage of the ratio method is that as long as the ratio of analyte to background counts is greater than 10/1, correction for background can in general be ignored. The sample is also counted for t_S. The counts ratio for the analyte is $(N_A - N_B)_X/(N_A - N_B)_S$. Sums of standard deviations will be involved in expressing the relative standard deviation of background-corrected ratios:

$$\sigma_R = \left[(N_A + N_B)_X/(N_A - N_B)_X^2 + (N_A + N_B)_S/(N_A - N_B)_S^2\right]^{1/2}. \quad (21.9)$$

When replicate sets of measurements are made, mean values and standard errors may be worked out (see Section 10.2). If no background corrections are made, Eq. (21.9) reduces to $\sigma_R = \sqrt{N_S + N_X}$, Eq. (21.8).

Counting Strategies. At least three different approaches to counting X-ray emission have been outlined and analyzed; fixed total count, fixed time of counting, and optimal fixed time. They are illustrated in Example 21.12. Surveying the scene briefly will illuminate the kinds of considerations that are important. In general, analyte peak intensities will be from 10 to 100 or more times the intensity of the background. While it will be important to count a sufficiently large number of photons to diminish the relative standard deviation of the analyte peak intensity, it becomes wasteful of time, in general, to diminish the relative standard deviation of the background count to the same degree. In other words, a 5% error in a variable that is consistently 10 times smaller than the main variable will be roughly equivalent to a 0.5% error in the larger quantity. Counting sufficiently long to bring both analyte intensity and background intensity to the same total number of counts is seldom done; more common is selection of an appropriate interval for accumulation of counts for an analyte peak and then applying it also to counting background. It can be shown that the best times in the optimal fixed method are relative, that is, $t_A/t_B = \sqrt{I_A/I_B}$. The smaller the background intensity, the shorter the time of counting background. The optimal fixed-time method has application mainly when differences between two intensities must be accurately known [7].

Example 21.12. Consider the relative errors involved in an X-ray measurement where an analyte peak giving an intensity $I_A = 100\ \text{s}^{-1}$, with background at the level $I_B = 5\ \text{s}^{-1}$, and a total counting time of $t = 100$ s. (a) What standard deviation and relative standard deviation will be obtained by the fixed-count strategy? (b) by the fixed-time strategies based on counting both background and analyte for the same total time? (c) What values will be secured if background is counted for only the length of time chosen for counting the analyte?

(a) For the *fixed-count* method, both analyte peak and background will be counted until the same total count is attained. Thus, if I_A and I_B are the counting rates for analyte and background, respectively, the number of counts for each will be $I_A t_A = I_B t_B$. For convenience let the total time be 100 s so that $t_A + t_B = 100$. Solving for t_B in the second equation and substituting in the first, $100\, t_A = 5(100 - t_A)$. Thus $t_A = 4.76$ s and $t_B = 95.24$ s.

The standard deviation for the corrected intensity or rate (total less background) can be found from the expression

$$\sigma_{\text{I}} = \sqrt{1/t} \cdot \sqrt{I_A + I_B} \cdot \sqrt{I_A/I_B + I_B/I_A}. \tag{21.10}$$

On substitution we have $\sigma_{\text{I}} = \sqrt{1/100} \cdot \sqrt{105} \cdot \sqrt{20.05} = 1.47$.

(b) For the *fixed-time* strategy based on counting for the same total time, $t_A = t_B$ and $t_A + t_B = 100$ s. The standard deviation for the corrected rate in this case can be calculated from the equation

$$\sigma_{\text{I}} = \sqrt{2/t} \cdot \sqrt{I_A + I_B}. \tag{21.11}$$

By substitution we obtain $\sigma_{\text{I}} = \sqrt{2/100} \cdot \sqrt{100 + 5} = 1.45$.

(c) Finally, for an optimal fixed-time strategy, let $t_A + t_B = 100$ s $= t$. It can be shown that optimum times stand in the ratio $t_A/t_B = \sqrt{I_A/I_B}$. Now find t_A by substitution into this equation from the time equation

$$t_A/(100 - t_A) = \sqrt{I_A/I_B} = \sqrt{20} = 4.47 \quad \text{and} \quad t_A = 81.7 \text{ s} \quad \text{and} \quad t_B = 18.3 \text{ s}.$$

The standard deviation for the corrected rate can be found from the equation: $\sigma_I = \sqrt{1/t}\,(\sqrt{I_A} + \sqrt{I_B})$. On substitution we have

$$\sigma_I = \sqrt{1/100}\,(\sqrt{100} + \sqrt{5}) = 1.22.$$

Coincidence Loss. We may trace uncertainty in detection either to coincidence loss or to poor counter efficiency once a particle has entered a properly operating detector. Associated with any counter and its electronic circuitry is a certain pulse resolving time. For example, the Geiger tube has a resolving time of about 200 μs, whereas a scintillation counter with a finer scaler may show a total resolving time of 1 μs. Since the resolving time represents an interval during which a counter and scaler fail to respond to new particles that enter the detector, there will be some coincidence error in general. If the recovery time is τ, the total time during which the unit is insensitive is $R\tau$, where R is the observed counting rate. The number of lost counts per unit of time is just $R' - R\tau$, where R' is the rate we would observe if there were no coincidence loss. To find the true count, note that the number of lost counts is also given by $R'R\tau$. Since the rate at which the count should be observed is R', that rate times the insensitive time $R\tau$ will also express the loss. Using these statements, we can calculate the true rate from Eq. (21.12):

$$R' = R/(1 - R\tau) \tag{21.12}$$

without serious error as long as $R\tau \leq 0.1$.

Example 21.13. A photon detector based on the use of a solid-state Si(Li) detector and pulse-shaping amplifier has a time constant of 2 μs. At a registered rate of 10,000 counts s^{-1}, what is the calculated true rate? How reliable is this estimate? What is the apparent error?

The true counting rate based on Eq. (21.12) is

$$R' = 10{,}000/(1 - 10{,}000 \times 2 \times 10^{-6}) = 12{,}500.$$

The percentage error, $100 \times (12{,}500 - 10{,}000)/10{,}000 = 25\%$. Since the correction is greater than 10%, the corrected rate can be taken only as an estimate.

Counter Efficiency Error. Finally, the possibility that error arises from poor or variable counter efficiency must be suggested. Not all particles that enter a detector give up all their energy as is desired, even when the detector is capable of responding. Usually poor transfer of energy occurs when radiation traverses a short path in the detector, in terms of the absorbing power of the sensitive volume for that kind of radiation and energy. In pulse-height determinations, this type of behavior produces spurious results. Note that in simple activity determinations we must know the detector efficiency to estimate the absolute rate, but we need only know it to be constant to obtain reliable relative measurements. Table 23.1 lists the efficiencies of some representative detectors.

21.9 SAMPLE PREPARATION

How can one select the most appropriate analytical X-ray fluorescence procedure(s) for the samples at hand? General aspects of preparing samples or specimens should claim our attention first; we will take up qualitative and quantitative procedures in the next two sections. The amount of preparation of samples needed will certainly depend on whether the requirements of analysis suggest that qualitative or semiquantitative results will provide sufficient information or whether only quantitative accuracy will suffice.

Sample Preparation. Since X-ray spectrometric procedures are nondestructive in most instances, it is possible to accommodate a variety of forms of specimens. In energy-dispersive spectrometry, it is common for a sample cup (see below) or other holder to be provided for specimens. Wavelength-dispersive spectrometers are more versatile with regard to the form of specimen.

Reproducible sample preparation methods are essential for good X-ray analysis. Samples must give specimens that have properties similar to available standards in terms of matrix, density, and particle size.

Homogeneous samples: powders, liquids, solutions, and bulk solids, can often be examined with little additional attention. At most, a standard type of pelletizing or preparation of a surface may be necessary. It is essential that all samples and standards provide the same area for exposure to the X-ray beam. When standard sample cups are employed, it is only necessary that samples be taken that will fill the cup. For instance, a diamond tool can be used to cut a sample of appropriate size from most solids.

For solid specimens surface preparation is unneeded if the surface is smooth and free of contamination, for example, a chemically clean piece of glass plate. Other surfaces must be ground and polished by methods that introduce minimum surface contamination. It is helpful to monitor polishing by comparing X-ray intensities at different stages. A rough surface will cause the paths taken by X rays to differ in length from place to place along the surface and give erratic results. With irreducible scratches or furrows care must be taken that they are kept minimal and are oriented parallel to the plane of incident radiation. Further, nonuniform particle size may influence peak intensities through its effect on the density of material in the X-ray beam. In general, roughness has a greater effect as analyte wavelengths and mass attenuation coefficients increase. Maximum peak-to-valley roughness tolerable for elements of $Z > 25$ is about 50 μm and for elements of $Z < 12$, about 10 μm. Longer wavelengths penetrate about a micrometer.

Powders are most appropriately pressed into pellets without a binder, but can also be examined in slurry form in a suspension.

Liquids are especially easily presented. A cover to the sample cup must be provided in the event of volatility or examination under vacuum conditions. The windows of cells for liquids should transmit the exciting and the fluorescent X rays. Sometimes a film of Mylar (polyethylene terephthalate) or fully oriented polypropylene can be used. Depth of sample should be controlled carefully if the sample is less than "infinite thickness."

Heterogeneous samples require substantial pretreatment. Some will need grinding and mixing to achieve homogeneity and others a matrix dilution treatment to reduce absorption-enhancement effects. It may also be that particle size is a problem requiring attention. Where pretreatment is involved, preparation of appropriate surfaces may challenge an operator. This will be especially true for biological samples that require microtoming and for other samples that have a mixture of soft and hard components. For heterogeneous powders, fusion with potassium pyrosulfate ($K_2S_2O_7$) or a tetraborate ($Na_2B_4O_7$ or $Li_2B_4O_7$) will homogenize the material sufficiently so that it can be ground, or in favorable cases, used directly as a pellet specimen.

Finally, if only very small quantities of material are available, it may be most appropriate to dissolve the sample in a suitable solvent and to evaporate the material onto filter paper disks. Such preconcentration methods may be necessary to examine particulates from air or other gas samples, by concentrating a dissolved material on an ion-exchange resin or ion-exchange filter paper. Samples that exist as films may also be examined, though a layer of about 1 μm is a lower limit.

21.10 QUALITATIVE ANALYSIS

In examining a sample by X-ray emission spectrometry it often suffices to identify the elements present and estimate concentrations roughly from peak intensities. Qualitative analysis by X-ray emission is attractive because it is a nondestructive technique that at most requires a sample that fits the sample holder of the spectrometer(s) available and can be irradiated. Minimal sample preparation is required over a wide range of elemental concentrations. For cases in which only the surface layer is analyzed, that may be a disadvantage. Other disadvantages are the possibility of spectral overlap when many analytes are present and moderately high limits of detection.

With an energy-dispersive spectrometer a spectrum may be obtained in a few minutes. Different X-ray tube conditions may be needed to cover the desired energy range sensitively, however, requiring more time. With a wavelength-dispersive spectometer a spectral scan will require substantially more time. If a 0.02–2 nm (0.2–20 Å) wavelength scan is to be made, several crystals and a pair of detectors, perhaps one in front of the other, will be needed. Spectra will be displayed as X-ray intensity versus energy for the first type of spectrometer and as a function of the 2θ position of the detector for a wavelength-dispersive instrument.

While the necessary intensities can be taken from peak heights, a more accurate estimate of concentrations can be made if the spectrum of one or more pure analytes can be obtained at the same time. With this information, the analyte weight fraction w_A can be calculated approximately using the expression $I_{A,X}/I_{A,A}$, where values are intensities for the sample and pure analyte, respectively, at the analyte wavelength. No correction is made for background.

Measurement Conditions. If little is known about a sample, it is advantageous to use an X-ray tube with a high atomic number target such as tungsten. If samples are believed to contain only elements below vanadium, a chromium anode tube is preferable.

Identification of Peaks. Identification of the elements present will depend on matching spectral peaks to literature values. Tables of X-ray peak wavelengths are available [16]. Where spectrometers are routinely employed for qualitative studies, organized procedures are developed for identifying lines. In cases where a chart spectrum is displayed, it is possible to develop transparent overlay scales that show positions of most characteristic X-ray lines. With in-line microcomputers characteristic peak positions can be stored in memory and made available for comparison with observed peaks.

In general, all peaks should be identified. Some relating to elements expected in the sample will be easily established. Both coherent and incoherent scattering will produce spurious peaks from continuum and characteristic radiation from an X-ray tube. The tube may also yield "contamination lines" from its filament (W), anode, and window seal (Ca, Fe, Ni). In a wavelength-dispersive spectrometer the crystal may also contribute higher-order peaks and its own X-ray lines.

As in any spectral analysis clusters of lines offer a way to establish that a given element is present with certainty. For instance, if Kα lines of an element are excited, one should seek both the Kα and Kβ lines. If L lines are used, an approach attractive for $Z > 40$, it is helpful once an Lα line is found to seek Lβ and Lγ lines for confirmation.

21.11 QUANTITATIVE MEASUREMENTS

In quantitative elemental or atomic measurements an analytical precision of the order of $\pm 0.2\%$ will be commonly obtainable by X-ray emission spectrometry except on the micro or trace level. Other advantages and disadvantages of the technique were summarized in Section 21.1. To achieve this high level of precision and a comparable level of accuracy, attention must be given to many factors, including systematic spectrometer errors, the counting protocol, excitation wavelengths and intensities, the manner of extraction of information from X-ray peaks, the mode of quantification of the spectrometric process, specimen preparation, and minimization of interelement effects.

This section deals with the last three topics listed. Included is a discussion of some general aspects not yet mentioned: reasons for choosing monochromatic versus polychromatic excitation, a basis for selection of analytical lines, concerns in trace-level measurements, and some ways to optimize the determination of low-atomic-number elements.

Peak Intensities. To establish concentration reliably either integrated peak intensities or carefully measured peak values, both corrected for background if necessary, are needed. In energy-dispersive spectrometry there is relatively little difficulty in obtaining integrated intensities since all peaks are determined simultaneously and measurement times are inherently short. In wavelength-dispersive spectrometry, however, scanning takes appreciable time since total counts must be large enough to reduce random error and it is customary to set a spectrometer at the 2θ angles of the several desired X-ray peaks and simply measure peak intensities. Fortunately, wavelength setting errors are minimal at peaks, and evidence shows that peak intensity values give minimal error in calculating concentrations except perhaps for the K lines of elements of low Z.

Background Corrections. Since correcting for background will introduce both systematic and random error, experimental steps should always be taken in quantitative work to keep background small. For trace analyses background reduction will be crucial. Standard methods are (a) use of good geometry, (b) adding baffling to reduce the amount of extraneous radiation that can reach the detector, and (c) minimizing extraneous X rays produced by the X-ray tube or other source. Some of the sources of background intensity I_{B} are suggested by

$$I_{\mathrm{B}} = I_{\mathrm{sc,prim}} + I_{\mathrm{sc,C}} + I_{\mathrm{em,C}}.$$

where the terms on the right denote (a) scattering of primary X rays in the X-ray tube, (b) scattering of secondary X rays by the sample, crystal (in a wavelength-dispersive spectrometer), or other instrument parts, and (c) emission of X rays by the crystal, detector, and other components.

Having taken steps to minimize background, one should also establish whether a concerted effort to measure the remaining background is important before instituting elaborate procedures. For either major or minor constituents there may be a modest gain at best in making corrections. A rule of thumb is that correction may be neglected if the total counts $N_A > 10\ N_B$. A preliminary comparison of peak/background ratios in an analysis will suggest when more than routine methods are advisable for dealing with background.

How should one determine the background level? If a number of similar samples are being analyzed, it may be feasible to measure background just before and just after absorption peaks and average readings. Alternatively, measurement at intervals between peaks may be appropriate. Or, a plot of the peak intensity of a characteristic analyte line versus concentration may be extrapolated to zero analyte concentration to obtain an intercept proportional to the background intensity. Where many elements are present in a sample, background I_B may also be estimated at the wavelength of the analyte peak for element i as $(I_B)_i = w_i I_i + \Sigma\, w_j I_j$, where j refers to other elements known to be present and I_j is an intensity from a pure specimen of j measured at the wavelength of the analyte peak.

One-Component Calibration Curves. Many modes of quantification are used in X-ray emission spectrometry. All must deal with the significant level of interelement effects associated with liquid, and especially solid, samples or specimens. Some ways to minimize or control these effects are summarized in Table 21.3. Many of these procedures lend themselves to the development of sets of one-component calibration curves (see Section 11.3) for analytes.

The use of *external standards* is a common procedure for establishing analytical curves. Yet standards are acceptable only if they have matrices similar to the sample matrices, for instance, a binary standard alloy for a binary alloy and analyte and nonanalyte concentrations close to those in samples. As indicated in Table 21.3, these standards will compensate for interelement, particle size, and sample density effects and even long-term drift in instrument modules.

A valuable, broadly usable alternative is the *matrix-dilution method*. It involves adding to each sample relatively large quantities of a diluent whose absorption-enhancement effects are minimal to "swamp out" the interelement properties of the original matrix. To be effective, the addition must reduce the variation in matrix effects with analyte concentration to about $\pm 5\%$.

There is no skirting the fact that dilution in this procedure means complete mixing. For geological samples for which dilution methods are especially effective, for mixing one usually relies either on fusion of the sample with the diluent or dissolution of both in an appropriate acid or aqueous solution. Since dilution will reduce measurement sensitivity, the method is unattractive when trace analytes must be determined.

Table 21.3 Effectiveness of Different Methods in Minimizing Systematic Error[a]

	Type of Error					
Method	*Absorption*	*Enhancement*	*Surface*	*Particle Size*	*Density*	*Long-Term Instrumental*
Calibration curve						
External standard[b]	C	C	—	C	C	C
Internal standard	C	C	—	—	—	C
Standard addition[c]	C	C	—	—	—	C
Thin-film[c]	C	C	—	C	C	—
Matrix dilution						
Mix with low absorber	C	C	C	C	C	—
Mix with high absorber	C	C	—	—	—	—
Fusion with low absorber	C	C	C	C	C	—
Solution	C	C	C	C	C	—
Mathematical corrections	C	C	—	—	—	—
X-Ray scattering[d]	C	—	—	C	C	C

[a] Adapted from R. Jenkins and J. L. De Vries, *Practical X-ray Spectrometry*. New York: Springer-Verlag, 1967. Methods that correct for an error are labeled C.

[b] Standards should be chemically similar and used over a limited concentration range.

[c] When used with trace-level analytes.

[d] For an account of standardization by use of scattered X rays, see E. P. Bertin, *Principles and Practice of X-ray Spectrometric Analysis*, 2nd ed. New York: Plenum, 1975.

Internal standards are also effective. A substance whose excitation and interelement properties are quite similar to those of an analyte in the matrix of interest makes an ideal standard. One must, however, be sure that the absorption edges of the standard will not interfere with analyte wavelengths. This approach compensates well for variability in most effects, but is difficult to use with solid samples and of course impossible if a nondestructive examination of samples is needed.

Multiple Regression Procedures. Fortunately, interelement effects are systematic by element and, thus, also correctable by multiple regression procedures. Several empirical computational procedures have been developed in X-ray spectrometry to relate the intensity of analyte lines to concentrations. One of the best is the use of empirical factors called *influence coefficients*.* The basic assumption of the approach is that the intensity of X-ray emission at a particular wavelength will in some way be affected by each element present. If attention is focused on a particular analyte i, the dependence of its concentration on its measured relative intensity $I_{r,i}$ when present in a matrix of interferents j can be expressed by

$$c_i/I_{r,i} = \alpha_{i,0} + \sum_{j=1}^{n} \alpha_{i,j} c_j, \tag{21.13}$$

where $\alpha_{i,j}$ is the *influence coefficient* representing the effect of element j on i. Uniquely, $\alpha_{ii} + \alpha_{i0} = 1$. Furthermore, $\alpha_{i,0}$ is the intercept of a plot of $c_i/I_{r,i}$ versus the concentration of a matrix element. Values of $\alpha_{i,j}$ are obtained by the use of reference samples of known composition. They are straightforward to solve for because the equations are linear in the α's since concentrations c_i and intensities $I_{r,i}$ are known for the reference samples. Measurements are made on a series of such samples and the set of simultaneous equations based on Eq. (21.13) are solved for the analyte concentrations. Note that the equations are linear for an unknown given the fact that values of $I_{r,i}$ are measured. A strong advantage of the influence coefficient approach is that it is based on nonlinear equations which are not forced to go through the origin and thus offers realistic analytical curves.

An example of an interactive computational procedure for the simultaneous determination of many elements will be taken up in the following chapter in discussing electron-beam X-ray spectrometry. A more complete discussion of this approach and of regression procedures developed in X-ray spectrometry may be found in many of the references.

Polychromatic Excitation. The full power of an X-ray tube can, of course, be used if polychromatic excitation is appropriate for a sample. At the outset it becomes important to relate absorption edges of elements to the X-ray wavelength available. The connection is illustrated in Fig. 21.24 by including both K absorption edges and plots of λ_{lim} of their continuum emission for many elements of

*R. Jenkins, *Pure Appl. Chem.*, **52,** 2541 (1980).

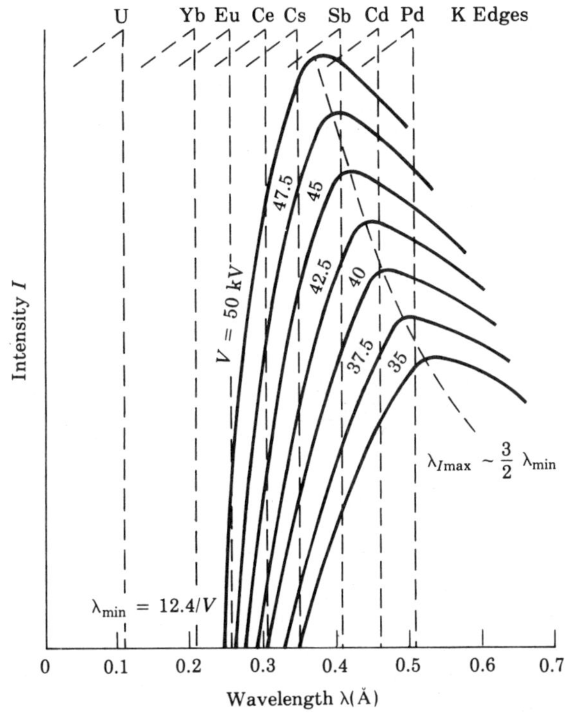

Fig. 21.24 Excitation of K lines of elements $_{46}$Pd through $_{92}$U by continuum generated by an X-ray tube as a function of X-ray tube operating potential. Other elements will fit into the sequence according to their atomic number. A similar curve would be set up for L lines and L absorption edges. K absorption edges are noted at the top and Duane–Hunt short-wavelength limits λ_{min} for the different voltages at the lower left for each curve.

higher Z versus excitation potential. Attention must be given to the short wavelength limit in deciding whether sufficient excitation energy is available for all the elements that are to be determined in a sample. For instance, it is clear from the figure that at a tube voltage of 35 kV continuum emission will excite cesium to X-ray fluorescence but not cerium. This figure makes clear why L lines are often sought for elements of higher atomic number.

Selection of Analytical Wavelengths. Since interelement effects are specially significant, analytical wavelengths should be selected to minimize such influences. Two examples will illustrate how this approach is applied.

Example 21.14. What strategies can be employed to reduce absorption-enhancement effects? Consider the absorption curves shown in Fig. 21.25 for a number of metals which are commonly found in high alloys steels, $_{24}$Cr, $_{25}$Mn, $_{26}$Fe, and $_{28}$Ni. Note that the diagram includes their Kα and Kβ emission lines at the bottom. How much can interelement effects between iron and manganese and between iron and chromium be reduced?

Often by a judicious choice of analytical line, absorption by other elements can be re-

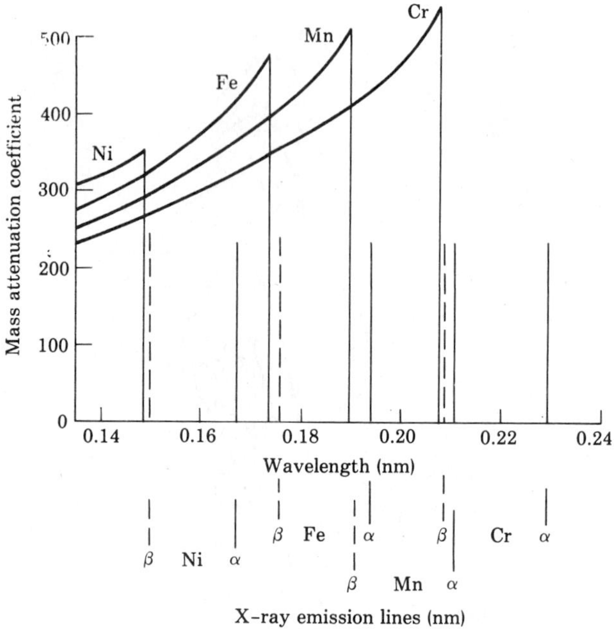

Fig. 21.25 Adsorption curves and X-ray lines (Kα and Kβ) of some metals in high-alloy steels. Their K absorption edges appear as vertical steps just to the left of their emission lines. No attempt has been made to resolve the separate Kα and Kβ lines for each element.

duced. In Fig. 21.25 the Kα line of iron appears on the long wavelength side of the manganese absorption edge. In a steel with more than minimal manganese, using this wavelength as the analytical line for iron would ensure that it could emerge with minimal loss from absorption. On the other hand, a poor strategy would be to select the Kβ line of iron since it falls on the short wavelength side of the manganese edge and would almost certainly be strongly absorbed. The validity of the first procedure is demonstrated in Fig. 21.26 by the essentially linear plot secured when the intensity of the Kα iron line is graphed against iron concentration in an iron–manganese sample.

On the other hand, from an inspection of Fig. 21.25 the Kα line would appear to give poorer results in iron determinations in binary iron–chromium alloys. Since the Kα line falls on the short wavelength side of the chromium absorption edge, it will be strongly absorbed in such mixtures. Again there is confirmation in the concave iron calibration curve in Fe–Cr alloys in Fig. 21.26. See Exercise 21.6 for discussion of a better strategy.

Example 21.15. Does the finding that the iron analytical curve in an iron–nickel mixture is convex mean that iron emission is enhanced (see Fig. 21.26)?

This is a clear case of enhancement. Evidence is supplied by returning to Fig. 21.25. In this case nickel Kα radiation lies just to the high-energy side of the iron absorption edge. As a result, the Ni line will be strongly absorbed by iron leading to extra emission from that element. There seems no doubt that this effect is the origin of the convex shape of the curve for iron Kα radiation versus composition. A correction must be made.

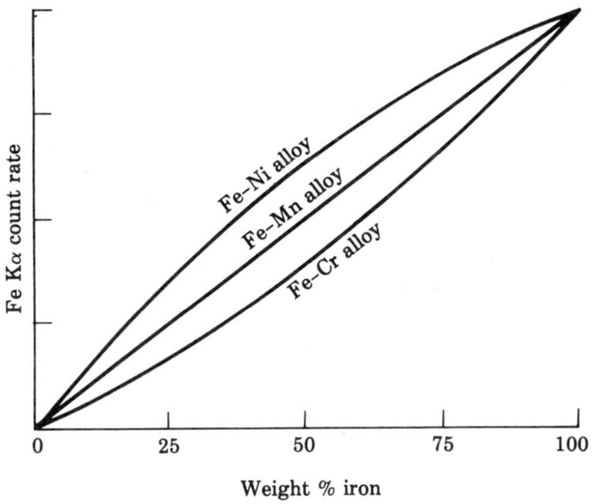

Fig. 21.26 Intensity of Kα line of iron in binary alloys with indicated elements plotted as a function of ion concentration.

Trace-Level Measurements. For trace-level studies the preparation of samples as *thin films* is especially valuable. By definition, a film is thin when it provides no interelement effects. Then all atoms will interact essentially independently with the X-ray beam and analyte line intensity will depend linearly on concentration over several orders of magnitude. Where samples are not initially available as thin films, it is often possible to grind them and disperse the powder on a support such as Mylar or filter paper. A secondary fluorescer or radiator may be used to advantage as a support for the film. Standards for comparison can be prepared by similar procedures.

For trace elements background from all sources becomes a limiting factor in measurement. In this case elemental peaks are difficult to quantify against the large statistical fluctuations in background. Continuum radiation from an X-ray tube, if not minimized, is a problem, especially if it is rather efficiently scattered into the detector. Often the answer is to achieve a more nearly *monochromatic source* by letting a secondary X-ray fluorescer be the true source. A suitable spectrometer configuration was discussed in connection with Fig. 21.21. Since a great deal of power is lost in this process, the approach works best with a lower-power spectrometer such as an energy-dispersive spectrometer. The extra filter of the target metal should also be inserted between the fluorescer and the sample. (See Fig. 12.21b and also Exercise 21.9 for further insight.)

Greater background reduction may also be secured by *ion-induced excitation* of X rays. An ion accelerator is needed. This mode of excitation eliminates a great deal of Compton scattering and other background. Proton-induced emission of X-ray fluorescence (PIXE) in thin-film samples is often used for trace measurements (see below).

Still another approach to background reduction is the use of a *total internal* *reflection* X-ray spectrometer to examine thin films.* Scattering and matrix effects are minimal since penetration under such reflection is only a few nanometers and is restricted to the film. Detection limits in the picogram range are achieved for many elements.

Finally, if only very small quantities of material are available it may be most appropriate to dissolve the sample in a suitable solvent and to evaporate the material onto specially purified filter paper disks or preconcentrate it on an ion-exchange resin or filter paper. Samples that exist as films may also be examined, though a layer of about 1 μm is a lower limit.

Measurement of Low-Z Elements. For elements of low atomic number ($Z < 18$), and especially for elements from $_4$Be through $_9$F, measurement sensitivity is quite low. One cause, as Fig. 21.5 shows, is that fluorescent yields are less than 0.1. In addition, spectrometers are inefficient at the long wavelengths characteristic of this region. Improvements have nevertheless been made in sources (layered targets and end-window tubes) and diffracting crystals (sputtered layer microcrystals). Finally, samples containing these elements require special care in polishing.

In this elemental range X-ray wavelengths are often slightly dependent on the state of bonding or oxidation, that is, on *chemical state*. The influence of bonding may also lead to peak broadening, especially at very low atomic numbers. The shape of the line may sometimes be used to estimate the specific contributions of different types of bonding to the spectrum. For instance, in alloys it is sometimes possible to examine X-ray spectra and determine which atom is a donor of electrons and which is an acceptor. For such applications measurements relative to standards will be essential. When quantitative results are the main objective, chemical effects can frequently be minimized by reducing the resolution of a spectrometer deliberately and then determining a reliable integrated intensity.

Information about bond lengths and bonding for some systems is also obtainable by measurement of the fine structure of X-ray absorption edges using a method called *extended X-ray absorption fine structure spectrometry* (EXAFS).† This technique furnishes information about the numbers and types of neighboring atoms as well as their deviations from equilibrium positions. Short-range order and local structure around particular atoms in a material can be deduced. For example, it is useful in the study of metal-on-support catalysts and the environment of metals in metallo-enzymes. For sufficient S/N either synchrotron radiation facilities or a rotating anode X-ray tube and a precision Johansson–Rowland circle-dispersing module (see Section 21.6) are required.

In measurements on light elements special attention must be given to *signal absorption by modules* of the instrument. Longer X-ray wavelengths from such elements are especially susceptible to attentuation by air in the beam path, the detector window, or the medium in the detector. As noted above, a helium atmo-

*W. Michaelis, J. Knoth, A. Prange, and H. Schwenke, *Adv. X-Ray Anal.*, **28,** 75 (1985).

†D. C. Koningsberger and R. Prins, *Trends Anal. Chem.*, **1,** 16 (1981) (a review article); B. K. Teo and D. C. Joy (Eds.), *EXAFS Spectroscopic Techniques and Applications.* New York: Plenum, 1981.

sphere, or preferably a vacuum, should be substituted for air in such cases to avoid severe attenuation. Detector windows may be of Mylar up to 0.3 μm, or very thin layers of beryllium or aluminum may be substituted. Almost all detector materials will tend to absorb longer wavelengths and appropriate allowance must be made for this possibility.

Electron Excitation. For light elements in particular, electron excitation has many advantages. It offers greater efficiency, eliminates or reduces the possibility of the absorption-enhancement effect that usually must be allowed for, is better suited to thin-film samples, and allows for the examination of minute areas (down to 1 μm^2). Nevertheless, electron excitation produces substantial background (the X-ray continuum), gives higher limits of detection as Z increases, requires that samples be in a highly evacuated chamber, is more likely to cause deterioration in specimens, cannot be used with liquid samples, and is especially troublesome with samples whose surface is not electrically conductive. Sometimes a thin coating of a conductor can alleviate the last problem. More complete attention will be given to electron excitation of X-ray spectra in the next chapter.

Absorption Measurements. The striking character of X-ray absorption edges (see Fig. 21.6) suggests that in some instances where the thickness of manufactured sheeting is to be controlled or a single element determined, a simple absorption procedure using monochromatic X radiation may suffice (see discussion below). It should be possible to obtain *monochromatic X radiation* for the measurements from one of several sources: (a) a simplified wavelength-dispersive spectrometer; (b) any source in which a characteristic emission line dominates the background; (c) an appropriate radioisotope source; or (d) a filtered X-ray beam. Two measurements are required. For instance, to determine film thickness, both the transmitted intensity and the incident intensity must be measured. It is also possible to carry out such absorptiometric measurements with polychromatic X-ray beams [4].

21.12 RADIATION HEALTH HAZARDS

Since X radiation ionizes matter with which it interacts it is important that users of X-ray spectrometers be knowledgeable about ways to minimize their exposure. Manufacturers seek to provide adequate shielding but cannot guarantee good practice on the part of users. Accordingly, it is important that radiation warning signs be clearly displayed at spectrometers, and if possible, the instruments be located in a separate room. These rooms should be equipped with permanently installed radiation monitors.

X-ray spectroscopists usually wear a type of dosimeter such as the film badge. Ordinarily its two pieces of film lie behind several absorbers with the badge worn on the left breast on the outside of all clothing. The film is renewed on a regular basis, for instance, every 2 weeks, and the previous film developed to provide an indication of any above-background exposure. The monitor is to measure the actual energy deposited, that is, the amount of ionization produced in the absorbing material.

The units in which radiation is measured have evolved over the years and several are in current use. The roentgen is defined as the radiation flux that will produce 2.08×10^9 ion pairs in 1 cm^3 of dry air at STP and amounts to an energy deposit of about 8.4×10^{-6} J. An alternate unit is the roentgen-absorbed-dose (rad), which is defined as 1×10^{-5} J g^{-1} deposited or converted to ions. Since living tissue has a slightly different absorption from many other substances, a unit called the roentgen-equivalent-man (rem) is the number of rads corrected for relative biological effectiveness. For practical purposes the rad and rem may be considered equal.

As suggested above the film badge basically measures exposure over time. Indeed, the safe dosage is defined for spectroscopists as up to 5 rem/yr for whole-body exposure (head, trunk, and so forth) and 75 rem/yr to hands and forearms (i.e., extremeties). An exposure of 5 rem/yr or 3 rem for any quarter year is regarded as acceptable for safe operation of X-ray spectrometers. References should be consulted for more details.

REFERENCES

Clearly written volumes in the introductory to intermediate level are:

1. E. P. Bertin, *Introduction to X-Ray Spectrometric Analysis*. New York: Plenum, 1978.
2. R. Jenkins, *An Introduction to X-Ray Spectrometry*. London: Heyden, 1974.
3. R. Jenkins and J. L. de Vries, *Practical X-Ray Spectrometry*. New York: Springer-Verlag, 1967.
4. H. A. Liebhafsky, H. G. Pfeiffer, E. H. Winslow, and P. D. Zemany, *X-Rays, Electrons, and Analytical Chemistry*. New York: Wiley-Interscience, 1972.
5. B. K. Teo, *EXAFS: Basic Principles and Data Analysis*. New York: Springer-Verlag, 1986.
6. R. Jenkins and B. de Vries, *Worked Examples in X-Ray Spectrometry*. New York: Springer-Verlag, 1970.

Well-written, more specialized volumes are:

7. E. P. Berlin, *Principles and Practice of X-Ray Spectrometric Analysis*, 2nd ed. New York: Plenum, 1975.
8. L. S. Birks, *Electron Probe Microanalysis*, 2nd ed. New York: Wiley-Interscience, 1971.
9. J. I. Goldstein, D. E. Newbury, P. Echlin, D. C. Joy, Ch. Fiori, and E. Lifshin, *Scanning Electron Microscopy and X-Ray Microanalysis*. New York: Plenum, 1981.
10. K. F. J. Heinrich, *Electron-Beam X-Ray Microanalysis*. New York: Van Nostrand–Reinhold, 1981.
11. D. W. L. Hukins, *X-Ray Diffraction by Disordered and Ordered Systems*. New York: Pergamon, 1981.
12. R. Jenkins, R. W. Gould, and D. Gedcke, *Quantitative X-Ray Spectrometry*. New York: Dekker, 1981.
13. C. Kunz (Ed.), *Synchrotron Radiation: Techniques and Applications*. New York: Springer-Verlag, 1979.

14. P. Luger, *Modern X-Ray Analysis on Single Crystals*. Berlin: Walter de Gruyter, 1980.
15. R. Tertian and F. Claisse, *Principles of Quantitative X-Ray Fluorescence Analysis*. London: Heyden, 1982.

Tables of X-ray wavelengths are presented in most texts on X-ray spectroscopy and also in:

16. E. W. White and G. G. Johnson, Jr., *X-Ray Emission and Absorption Wavelengths and Two Theta Tables*, 2nd ed., ASTM D537A, Philadelphia: ASTM, 1970.

Also of interest are:

17. W. Michaelis, J. Knoth, A. Prange, and H. Schwenke, Trace analytical capabilities of total-reflection X-ray fluorescence analysis, *Adv. X-Ray Anal.*, **28,** 75 (1985).

EXERCISES

21.1 What is the short wavelength (Duane–Hunt) limit to the continuum produced by an X-ray tube with a molybdenum target that is operated at 50 kV?

21.2 Calculate the minimum X-ray tube voltage required to excite the K lines of selenium. The absorption edge is (0.098) nm.

21.3 What thickness aluminum and beryllium windows are possible for an X-ray tube with a silver target if no more than 5% of the intensity of its Kα line at 0.056 nm is to be absorbed by them? The mass absorption (attenuation) coefficients for the metals at the silver Kα wavelength are about 2.7 and 0.22, respectively.

21.4 In preparation for analysis of copper samples with a wavelength-dispersive spectrometer spectra are observed with both a lithium fluoride (d = 0.2013 nm, 2.013 Å) and potassium hydrogen phthalate (KAP, d = 0.1332 nm) crystal. At what angles will the Kα (0.154 nm, 1.542 Å) and Kβ (0.1392 nm, 1.392 Å) lines appear with the two crystals?

21.5 Why is little if any sample preparation necessary in X-ray spectrometry?

21.6 Why is the number of spectral lines smaller in X-ray fluorescence spectrometry than in most optical emission spectrometry? Is this aspect an advantage or disadvantage in most cases?

21.7 Compare an X-ray and an optical absorption curve for chromium in the form of $Na_2Cr_2O_7$. Assume the chromium X-ray absorption edge occurs at 0.2 nm (2 Å) and its mass attenuation coefficients have values $\mu_2 \simeq 50$ and $\mu_1 \simeq 500$. Optical absorption involves the lowest lying electronic transition upward from a $3d^5$ state that has a maximum molar absorptivity of 3×10^3 at 425 nm. Draw a rough plot of (a) the X-ray absorption curve μ_m versus λ between 0.05 and 1.0 nm (0.5–10 Å) and (b) the electronic absorption curve (ϵ versus λ) between 300 and 800 nm. (c) What are the excited states in the two cases (identify possible electron configuration of atom and show any changes). (d) Sketch rough emission curves for the two cases (I versus λ) over the wavelength range used before.

21.8 Assume an alloy of several metals is to be analyzed qualitatively and quantitatively and that both an atomic emission (AES) and X-ray fluorescence (XRF) spectrometer are available. (a) Which type of spectrometer should be used? (b) What advantages and disadvantages does each type have?

21.9 Estimate from the curves in Fig. 21.25 to the nearest 0.05 nm the wavelengths at which absorption edges and Kα emission occurs. Tabulate the data roughly. Predict to your own satisfaction from the data the wavelength of (a) cobalt Kα emission, (b) the cobalt K absorption edge, and (c) the Kα emission wavelength for chromium. (d) What basis did you use for your prediction? (e) State specifically for each element whether its emission (Cr through Fe including Co) can be used to excite each of the other elements in the series.

21.10 Explain why the major factors determining the intensity of X-ray fluorescence lines are the fluorescent yield or efficiency ω and what has been called the excitation overvoltages, $V - V_{\phi}$.

21.11 According to Example 21.15 the choice of the Kα iron line as a way to determine the composition of iron–chromium mixtures does not work out well. Would a better result be secured by employing the iron Kβ line or the chromium Kβ line?

21.12 A secondary fluorescer is to be used as a source for trace element analysis. A choice must be made between selecting (a) a set of secondary fluorescers, each of which produces a characteristic wavelength just shorter than the absorption edge of a trace analyte and (b) a silver slab as secondary fluorescer. Under what conditions might (a) be more attractive than (b)?

21.13 By use of selective excitation it is possible to set up nondispersive X-ray spectrometry. A specimen of a high alloy steel containing nickel and chromium is to be analyzed for iron, chromium, and nickel. Bars of the pure metals are available to use as secondary fluorescer sources. Data that will be needed are:

Metal	$\lambda_{K_{abs}}$	$\lambda_{K\alpha\,line}$
$_{24}Cr$	0.2070	0.2291
$_{26}Fe$	0.1743	0.1937
$_{28}Ni$	0.1488	0.1659

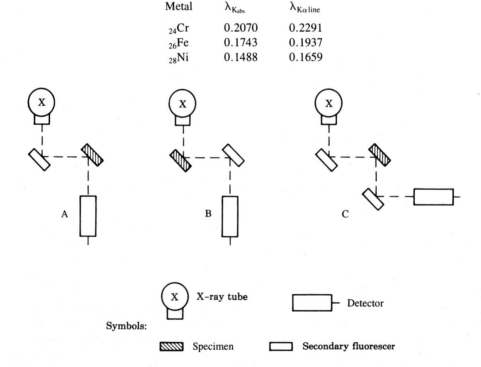

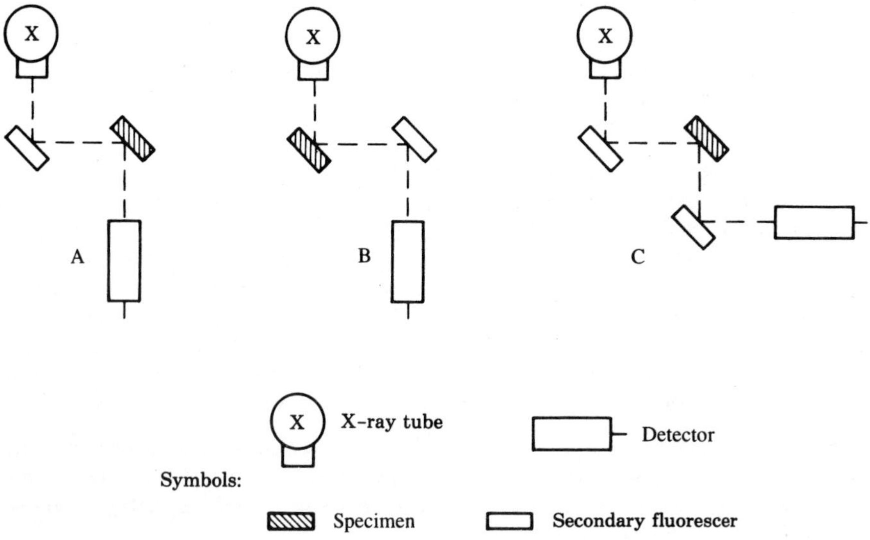

Consider the data and arrangements of modules labeled A, B, and C. One arrangement is ideal for each of the elements. Report the element to be measured with each configuration, and with each arrangement the metal(s) chosen as the secondary fluorescer(s). Briefly justify the choices. [Suggested by L. S. Birks, *X-Ray Spectrochemical Analysis,* 2nd ed. New York: Wiley-Interscience, 1969, p. 63.]

21.14 In Section 21.10 use of a secondary fluorescer is proposed as a way to develop a more nearly monochromatic X radiation for a particular analyte present at the trace level. Assume the Kα line of the analyte is to be excited. (a) Draw a plot showing this line and the related absorption edge. (b) Add to the plot the wavelength for the proposed monochromatic, secondary fluorescer. If the trace analyte is cobalt, what element would be appropriate as a secondary fluorescer? Explain. (c) Finally, show the absorption edge proposed for the filter that is to block the continuum of the secondary fluorescer.

ANSWERS

21.8 (a) Use AES with ICP excitation for qualitative analysis if one or more components is believed to be at the trace level. Use XRF for quantitative analysis of minor and major components where precision better than $\pm 1\%$ is required. *21.13* A: for measurement of Cr, use Fe as secondary fluorescer. B: for measurement of Ni, secondary fluorescer may be Cr. C: for measurement of Fe; initial and second secondary fluorescers are Ni and Cr, respectively.

Chapter 22

SURFACE SPECTROMETRIC TECHNIQUES

22.1 INTRODUCTION

With increasing maturity and sophistication of theory, instrumentation, and technology, the study of surfaces of solids and interfaces has become an important area of chemical measurement and analysis. In this chapter several spectrometric methods that provide information about surfaces will be taken up. The rationale for treating these techniques in a single chapter is that the chemical information they provide is often complementary and a single technique seldom provides enough data to characterize a surface clearly.

What we seek in applying these methods to a surface is as full an account as possible of

1. its microtopological features such as crystal or grain boundaries and irregularities like valleys and pores;
2. a point-to-point map of elemental composition with high lateral resolution (x and y directions) and, when desired, also a depth dimension (z direction);
3. a point-to-point map of compounds, functional groups, and elemental oxidation states with high lateral resolution that is correlated dimensionally with the elemental map.

At present *qualitative* analyses of surfaces are highly accurate and quantitative measurements are becoming more reliable. Challenges in quantitative measurement that must be resolved for better reliability include finding suitable standards for quantifying results, matrix problems, the inherently low surface concentrations of species, and the complexity of surface instrumentation and procedures. These aspects will be taken up as individual methods are discussed.

Especially important in the development of modern surface studies was establishing reliable methods for measurement of elemental compositions. Indeed, until such compositions could be found on a microscopic basis, there was no sure way of knowing whether a given surface was "clean" or "dirty," making it difficult to interpret such data as were obtained. The new spectrometric methods provide not only an assessment of cleanliness but allow a host of physical and chemical phenomena related to surfaces to be illuminated. These include both theory and practice related to heterogeneous catalysis, nucleation, sintering, adhesion, metal failure, corrosion, energy conduction, and energy conversion. In addition, all aspects of design, fabrication, and testing of patterned layers in solids such as very large scale integrated circuit semiconductor devices, computer memories, and photographic records have benefited.

In this chapter attention will be quite deliberately focused on solid surfaces. They have reasonable permanence in the absence of rapid physical or chemical processes affecting them and represent a large fraction of the surfaces of chemical interest. With such surface information it should be possible to develop a coherent account of phenomena at interfaces between phases.

Here, surfaces are taken to extend from the outside of a specimen down to, in some cases, as many as the first thousand atomic layers. Clearly, some "peeling back" of layers by sputtering or other techniques will be necessary for layer-by-layer depth profiles.

Finally, attention will be given to interpretations of surface phenomena based on useful models. While ideal crystals have been well described theoretically, actual solid surfaces with all their complexity pose difficult problems. As a result, approximate theoretically based models have been necessarily substituted in most cases for more theoretical approaches. There has, nevertheless, been a great deal of development in methods and in experimental procedures.

Preparation of Surfaces for Analysis. To obtain good information surfaces should of course be prepared as carefully as possible. Are general guidelines available? How clean should surfaces be (assuming that adsorbed species are not of primary interest)? Must surfaces be essentially flat?

Answers to such questions are often evident as soon as the object of study is known. For example, study of the corrosion of a metal surface can begin with examination of the surface of cross sections of samples of newly prepared and aged metal. On the other hand for heterogeneous catalysis one examines the surface of a catalyst under different conditions.

A *clean surface* may be defined operationally as one whose impurity content is below the detection limit of the methods to be used for examination. Since much of the interest in surfaces arises from the high reactivity of clean surfaces, it can be anticipated that once a surface is clean, gas must be rigorously excluded from the system. A good rule of thumb is that a specimen should be under a vacuum high enough so that the time required for its complete coverage by a monolayer of gas exceeds the time required for careful measurement.

Cleaning a Surface. Existing surfaces may be cleaned by one or more of the following procedures:

1. heating to high temperature to desorb materials that have accumulated;
2. etching (sputtering) with noble gas ions;
3. treatment with a specific reactant to remove an impurity.
   In some instances, however, it is worthwhile to prepare a new surface by
4. deposition of a thin film of the substance or
5. cleavage of a specimen under conditions of high vacuum.

The following series of steps for cleaning a specimen is recommended: (a) polishing, (b) ultrasonic washing in a sequence of solvents, (c) heating to 150°C in vacuum, preferably in the chamber and at the vacuum to be used experimentally, and (d) bombarding with a diffuse beam of very low energy Ar^+ ions (0.1–1 eV at a current density of 1 mA cm^{-2}) to remove nonbonded impurities or directly with 300–500 eV Ar^+ ions at 20 $\mu A\ cm^{-2}$ to remove an oxide layer. (One should omit any inappropriate step.)

Example 22.1. Approximately what level of vacuum must be attained in a system to ensure a clean surface for one second? one hour?

From kinetic-molecular theory, it can be shown that the number of gas molecules striking one square centimeter of surface per second is given by $Nv/4$, where N is the number of gas molecules per cubic centimeter and v is the root-mean-square molecular velocity. The amount of a surface covered may be put in terms of the number of molecules required for a standard monolayer, about 10^{15} cm^{-2}. By assuming ideal gas behavior, an average molecular weight of 28, and a temperature of 300 K, one obtains for the number of monolayers deposited per second the product $10^6 p$, where p is the pressure in torr (1 torr $\approx$ 100 Pa). By adding to the expression a sticking coefficient s, (i.e., the probability that a molecule striking the surface will be adsorbed), one obtains for the time of coverage t_c by a monolayer the expression $t_c = 10^{-6}/sp$.

Thus, if all collisions result in sticking, a standard high vacuum (10^{-4} Pa or 10^{-6} torr) will each second offer a sufficient number of molecules to establish a monolayer! A clean surface can hardly be assumed in such a system. For 1 h of measurement time when s = 1, the level of vacuum needed would need to be about 10^{-8} Pa or 10^{-10} torr.

Probing a Surface Analytically. The determination of surface microstructure and composition point-by-point across—and a few atomic layers into—a specimen can be thought of as a process of probing. In Fig. 22.1, three kinds of probing or exciting beams are shown: photon, electron, and ion. The most important spectrometric methods used principally for surface studies are listed by name and acronym in Table 22.1. To keep the discussion of methods clear, incident particles will be termed *primary*, and those leaving *secondary*. We can select the type of probing particle and average particle energy, beam intensity (number of particles/area), and angle of incidence. For the secondary beam we can measure the angle of

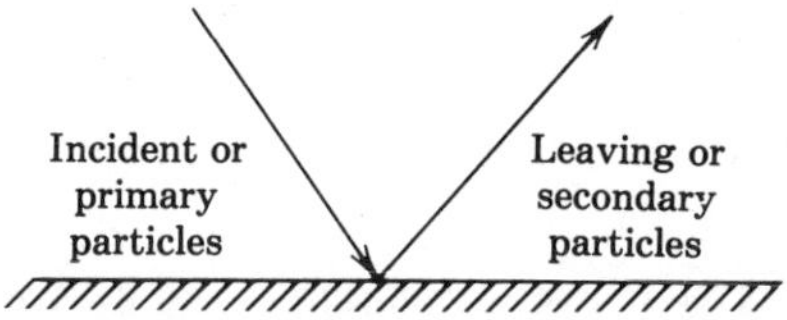

Fig. 22.1 Simplified representation of spectrometric surface measurements listed in Table 22.1. Exposure is to (a) IR, VIS, UV photons in reflectance spectrometry and VIS and UV photons in Raman spectrometry; (b) X-ray photons in XRF and XPS; (c) electrons in XRF, AES, and XPS; (d) ions in SIMS, ISS, and RBS. Particles leaving a surface are (a) unabsorbed IR, VIS, and UV photons in reflectance spectrometry and inelastically scattered photons in Raman spectrometry; (b) emitted X-ray photons in XRF; (c) emitted electrons in AES and XPS; (d) ions in SIMS, ISS, and RBS.

Table 22.1 Acronyms and Names for Some Surface Spectroscopic Methods

Acronym	*Name of Technique*[a]
AES	Auger electron spectroscopy
APS	Appearance potential spectroscopy
ELS, EELS, (HREELS)	Loss spectroscopy, electron energy loss spectroscopy, (high resolution electron energy loss spectroscopy)
EXAFS	Extended X-ray absorption fine structure
ISS (LEIS)	Ion scattering spectroscopy (low-energy ion scattering)
LEED	Low-energy electron diffraction
RBS (HEIS)	Rutherford backscattering spectroscopy (high-energy ion backscattering spectroscopy)
SAM	Scanning Auger microscopy
SEM	Scanning electron microscopy
SERS	Surface enhanced Raman spectroscopy
SIMS	Secondary ion mass spectrometry
STEM	Scanning transmission electron microscopy
UPS	Ultraviolet photoelectron spectroscopy
XPS or ESCA[b]	X-ray photoelectron spectroscopy
XRF	X-ray fluorescence spectroscopy

[a]Nomenclature recommended by IUPAC.
[b]Acronym for electron spectroscopy for chemical analysis, an unsystematic name.

ejection and particle energy (or mass), and if needed, the kind of particle. From surface spectrometric measurements we obtain information about the kinds of atoms and their location and oxidation state. Deductions can be often be made about chemical structure in the surface layers.

The nature of the probing process can be explored usefully in the study of Table 22.2, which briefly characterizes the main surface techniques in wide use. These methods range from general techniques such as IR, Raman, and X-ray fluorescence spectrometry to methods that are intrinsically surface procedures such as X-ray photoelectron spectrometry (XPS or ESCA) and ion-scattering spectrometry (ISS). Scanning electron microscopy is included, first, because it offers an excellent way to observe the microstructure of a surface, and second, because today's scanning electron microscopes are frequently equipped to operate also as electron microprobes yielding x ray and sometimes Auger electron elemental maps and analyses of sample surfaces. All these methods are discussed in later sections of this chapter; a few other methods will also be described briefly.

As Fig. 22.1 suggests, the optical spectrometric methods in general can be used to probe surfaces by being applied in a reflectance or emission mode, or alternatively, with ultrathin films of a sample, in a transmission mode. For example,

Table 22.2 Some Surface Spectrometric Techniques

	IR Spectrometry[a]	*Raman Spectrometry*[a]	*X-Ray Fluorescence Spectrometry*[b]	*Scanning Electron Microscopy*	*Auger Electron Spectrometry (AES)*	*X-Ray Photoelectron Spectrometry (XPS)*[c]	*Secondary Ion Mass Spectrometry (SIMS)*	*Ion Scattering Spectrometry (ISS)*
Probing beam	Photons of many frequencies	Photons of one frequency	X-ray photons	Electrons	Electrons	Monochromatic X-ray photons	Ions	Monoenergetic ions
Particles detected	Photons of same frequency	Frequency-shifted photons	Characteristic X-ray photons	Secondary electrons	Auger electrons	Photoelectrons	Sputtered ions	Scattered ions
Analyzer	Monochromator	Monochromator	Si (Li) diode, multichannel analyzer	—	Cylindrical mirror analyzer (CMA) or other electrostatic analyzer	CMA or other electrostatic analyzer	Mass analyzer	CMA or other electrostatic analyzer
Spectrometer range	IR region	VIS region	X-ray region	—	50–2500 eV	100–1500 eV	0–500 amu	10–1000 eV
Sampling depth (nm)	See footnote *a*			—	0.5–2	0.5–2.5[d] 2–10[e]	0.5–1 (static) 3–4 (dynamic)	Top atomic layer

[a]IR and Raman reflectance spectroscopies are widely used to study surfaces that have been modified by adsorption or chemical reaction. These spectroscopic techniques were discussed in Chapters 16 and 18, respectively.
[b]Discussed in Chapter 21.
[c]The earlier name, electron spectroscopy for chemical analysis, and acronym ESCA are also used.
[d]Metals, metal oxides.
[e]Organic substances, organic polymers.

focused IR polychromatic beams under conditions of attenuated total reflectance (see Sections 7.11 and 17.6) allow determination of the IR spectrum of a surface layer of a thickness that is a fraction of a wavelength (150–2000 nm, depending on the spectral region). Determination by photothermal or photoacoustic detection is also possible. This type of approach is also valuable with solutions at electrode surfaces. Adsorbed monolayers on a surface can be also be studied by direct reflectance. Surface application IR measurements have been enhanced by the use of Fourier transform procedures and photoacoustic detection (see Section 17.9). Raman spectrometry is often used for surface studies (see Chapter 18). Whether probing beams of radiation will be destructive of specimens under examination will depend on radiation frequency and intensity as well as fineness of focus.

X-ray techniques permit examination of surfaces at depths up to the escape depth of X-ray photons. With this technique, elemental compositions for elements of atomic number is $Z > 11$ become quite feasible (see Section 21.2). The determination of X-ray absorption fine structure (EXAFS), a more specialized technique which often relies on a beam of synchrotron generated X rays, is a valuable method for determination of chemical structure at the surface.

An older series of physical techniques that includes measurement of gas adsorption isotherms (to learn surface areas and distribution of pore sizes), ellipsometry (to estimate thickness of surface layers), determination of photoelectric work function (to characterize electronic levels), and optical microscopic observations (to elucidate larger aspects of surface topography) remain useful. In most cases they provide information that complements the data furnished by the spectrometric techniques to be discussed here, and in some instances these older methods furnish information easily obtainable in no other way. Since these methods fall outside the scope of the present work the reader should consult references for information.

22.2 ELECTRON MICROSCOPES: DIRECT AND INDIRECT IMAGING

It is always valuable to begin analysis of a surface by obtaining a physical picture of its topology on a microscopic scale. An optical microscope will reveal a great deal about most specimens, especially if their surfaces are smooth. For a resolution of features of chemical interest, however, an electron microscope (EM) usually of either the scanning (reflection) variety (SEM) or scanning transmission variety (STEM) will be essential. Only with information about a surface that includes details of grain or crystal boundaries, lattice imperfections, foreign inclusions, and other microphysical data can elemental and species composition be interpreted as more than average values.

As is well known, electron microscopes can provide much higher resolution than light microscopes. Rayleigh showed that diffraction effects limit the resolution of features in an image to those approximately equal to the wavelength of the radiation employed. Resolution is thus of the order of 500 nm for visible light. Moving particles like electrons and ions also display wave properties: the de Broglie relation defines for particles of mass m and velocity v a wavelength λ equal to h/mv. For electrons accelerated through a potential difference V (in volts), the expression becomes $\lambda = 1.25/V^{1/2}$ nm. For example, a limiting resolution of 0.04 nm should be possible with a 10-kV electron microscope. Practical resolution attainable is much less as a result of electron optical aberrations as will be discussed below.

Imaging. Electron microscopes form greatly enlarged images of a surface. If they form images by use of electron optics their operation can be called *direct imaging*; if instead the images are reconstructed from intensity data obtained by a point-to-point sampling of the surface, they provide indirect imaging as will be described below. For example, cathode ray tube displays provide indirect images, as described below. Each approach has advantages. Clearly, the time required for direct imaging is shorter and both resolution and S/N ratio can be greater because of the multiplex effect. On the other hand, the atomic processes that underlie indirect imaging may yield more chemical information from a specimen.

How are most *indirect images* formed? A finely focused electron beam or other probing beam is scanned linearly across the surface of the specimen in a *raster pattern*. The deflecting voltages that position the electron beam are varied by a raster program so that the beam sequentially (a) moves across a surface in a straight line, (b) is returned rapidly to the starting edge of the pattern, and (c) is shifted downward by a standard increment for each succeeding scan. Present standard television rasters have 530 lines. Usually a cathode ray tube (CRT) forms an image from secondary electrons emitted or scattered from each point of a specimen. To form the image the electron beam in the CRT is driven by (a) a raster identical to that employed in scanning the specimen and (b) its intensity at each point is adjusted at each instant in accord with the detector output from the corresponding point of the specimen surface. By selecting the type of detector and its processing, different secondary particles can be used for image formation.

Scanning Electron Microscope (SEM). In this device, a finely focused beam of electrons is used to scan the surface of a specimen in a rectangular raster. Its electron beam actually yields a variety of products on interaction with a specimen: reflected or backscattered primary electrons, slowly moving (secondary) ejected electrons, Auger electrons, and X rays. Any or all of these products of probing can be collected, detected, processed, and displayed. At minimum, a good fraction of either reflected or secondary electrons produced at each point along the surface of a specimen is displayed. If secondary electrons provide the intensity data, they are collected by a detector placed at a low angle to the primary beam. The process of formation of an image or picture is described below.

A schematic diagram for a representative scanning electron microscope is shown in Fig. 22.2. Note the electron gun at the top of the figure. (See Section 22.5 for more details about the gun.) Potentials used on the electrodes of the gun cause the emitted stream of electrons to converge to a circular crossover point (d_0 in the figure) about 10–50 μm in diameter. Subsequent magnetic lenses develop an image

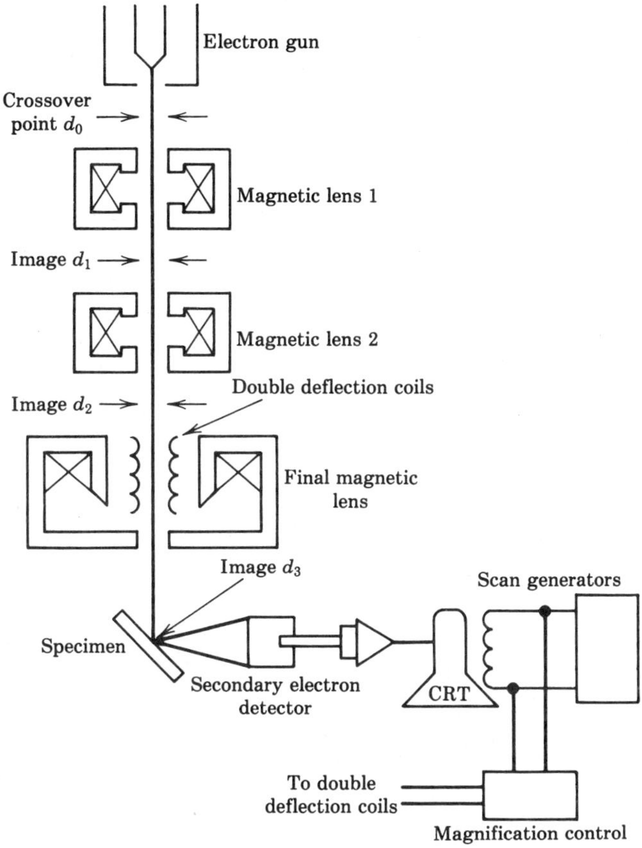

Fig. 22.2 Schematic of electron optics of a scanning electron microscope. The electron gun at the top generates an electron beam of fairly uniform intensity at its crossover point of diameter d_0. The lens system forms a final image of this region of diameter d_3 at the specimen surface. The scan generator deflects the electron beam in the cathode ray tube (CRT) in synchronism with the scanning microscope beam following a raster pattern. The result is that an indirect image of the specimen surface is generated in the CRT. The focus of the scanning beam is made steadily finer than d_0 by passing the electron beam through the succession of magnetic lenses. Thereby magnification is increased since the CRT dimensions remain constant. Each lens exists as magnetic lines of force between apertures in a magnet's iron core. Components not shown include aperture plates that block electrons traveling too far from the microscope axis and a device to minimize astigmatism. (After Goldstein *et al.* [9].)

or optical reproduction of the crossover point and ultimately reduce its diameter to 5–200 nm at the specimen surface. The final beam diameter fixes the ultimate resolution of the microscope. Electrostatic lenses might also have been used in the microscope. The collection of lenses and apertures is termed the electron optics of the device.

Example 22.2 How may useful indirect images of integrated circuit semiconductor devices be developed?

One possibility is to detect Auger electrons representing the silicon LMM peak at about 87 eV. In this case a silicon image of the device is displayed. It is brightest where silicon is most concentrated and dark where no silicon atoms are present, perhaps because of deposition of a gold conductor.

Aberrations. As electrostatic and electromagnetic lenses refocus images, aberrations in their performance lead to distortion. Electron lenses suffer all the aberrations known in light optics (see Section 7.20). Unfortunately, some distortions, such as spherical aberration, even in theory, cannot be corrected as well in electron lenses. At present, the optimum resolution realizable is ordinarily a few angstroms.

Spherical aberrations occur because electrons moving at some distance from the microscope axis are focused more strongly than those traveling nearer its axis. For instance, axial electrons are focused farthest along the microscope axis. This type of aberration is mainly responsible for the inability of electron optics to approach theoretical resolution more closely. In light optics spherical aberration can be overcome by using parabolic lens surfaces or by appropriately combining surfaces of positive and negative curvature, but this type of correction is inaccessible in electron optics.

The extent of spherical distortion is measured in terms of the size of the circular spot—often termed the circle of least confusion—that appears as the image of a specimen point. The diameter d_s of the circle or disk is defined by $d_s = 1/2\ C_s\ \alpha^3$, where α is the maximum angle of divergence of rays at the image plane and C_s is usually about 2 cm for an SEM. By narrowing the lens aperture, α and thus, spherical aberration, is decreased, but this gain is traded off for intensity.

In addition, some *chromatic aberration* will occur because of differences in velocities of electrons in the beam (energy range about ± 0.5 eV). Fluctuations in the accelerating potential in the electron gun and in intensities of magnetic fields of lenses also lead to distortion. Stabilization of both to 1 part in 10^6 per minute will keep this source of noise to a minimum. Finally, *astigmatism* also can affect the electron beam. Correction may be introduced in the objective lens to provide for this type of distortion.

Charging Effects. Ways must often be found to eliminate *charging effects* as the SEM electron beam sweeps across a specimen surface. If specimen electrical conductivity is low, local potentials developed at normal electron beam energies may be sufficient to cause dielectric breakdown. The resulting current surges will give rise to spurious detail, artifacts, in an image. One option is to use a low energy beam. Another is to coat specimens with a thin film of a good conductor such as copper, aluminum, or gold by sputtering the metal on in a high vacuum. With biological and many organic specimens a precoating of graphite is often applied at low vacuum and the metal sputtered on in a high vacuum. The precoating ensures an even deposition of metal.

Since electron beams are strongly absorbed or scattered by molecules in air, a high vacuum is essential in an electron microscope. If such an instrument also has surface analytical capability, it is likely to feature differential pumping of the compartment into which specimens and detectors are introduced. This region calls for an ultrahigh vacuum.

In most scanning electron microscopes, secondary electrons ejected from the specimen are used to form an indirect image. The double deflection coil shown in Fig. 22.2 provides for deflection in both x and y directions across the sample. The current and beam energy furnished by the electron gun may be varied to alter image brightness and contrast.

As pictured in Fig. 22.2 the secondary electrons emerging from a spot on the specimen usually interact with a scintillator and are detected as light bursts by an attached photomultiplier tube. As already noted, the photomultiplier tube output modulates the intensity of the CRT electron beam as it is scanned across the face of the tube. It is possible to achieve the magnification desired—within the limits that resolution permits—by altering the relative distance of scanning in the CRT and SEM.

As desired, other detectors and processing modules can be incorporated to take advantage of other secondary emission from a specimen. Perhaps an energy-dispersive X-ray spectrometric assembly is the most common of these.

Analytical Electron Microscope. An instrument that provides several modes of extraction of chemical information is the modern scanning *transmission* electron microscope (STEM); often it is called an analytical electron microscope. Its electron beam can be varied from a diameter of less than 20 nm to 10 μm at energies as great as 100 keV. While thin (up to 100 nm) specimen must be used to ensure adequate transmission, resolution is automatically improved because of the thinness. Some bombarding electrons will be backscattered strongly by atoms, others will produce secondary electrons, and still others will be absorbed and produce X rays (or Auger electrons). To capture most of the information available from these secondary results, usually both electron and X-ray detectors are placed above a specimen. Their outputs, after processing by modules that are not shown, will yield a distinctive type of chemical information about each area of the specimen scanned or they can simply be used to furnish distinctive kinds of images of the specimen surface.

Below the specimen, in addition to lenses to project a direct image of the specimen on a fluorescent viewing screen, an *electron energy loss spectrometer* is a useful addition. As its name suggests, the spectrometer measures the kinetic energy spectrum of electrons that have undergone mildly inelastic collisions with atoms in the thin specimen. This type of spectrometry will be discussed in Section 22.3. A particular advantage of energy loss spectrometry is its suitability for determination of elements of low atomic number ($Z < 10$) such as C, N, and O for which conventional X-ray methods are quite limited.

Elemental Mapping. To determine elements and their surface concentrations in a specimen only a few modules will need to be added to a scanning electron microscope. Usually an energy-dispersive X-ray detector such as the Si(Li) module shown in Fig. 21.13 is employed and an X-ray wavelength characteristic of an element is detected. Usually it is possible to select an analyte X-ray frequency relatively free of spectral interference by carrying out a preliminary qualitative analysis. Other useful X-ray modules that might be employed, such as a wavelength-dispersive spectrometer, were also described at length in the last chapter. In the indirect X-ray image of the surface developed intensity will be proportional to the local concentration of the analyte. If an Auger spectrometric array is substituted as suggested in Example 22.2, an Auger electron elemental map can be registered in similar fashion.

For an SEM with an energy dispersive X-ray spectrometer attached, the elemental concentration for the top 1000 nm will be represented since it is over this depth that X rays will usually have sufficient energy to escape from the surface of the specimen. For a transmission microscope concentrations in the entire thickness of the specimen will be shown. Factors relating to the precision with which elemental compositions are known by X-ray measurement have been dealt with in Section 21.11. If an Auger spectrometric setup has been attached, the composition of a much thinner layer will be represented since electron escape depths are smaller.

22.3 IONIZATION AND EXCITATION PROCESSES IN ELECTRON SPECTROSCOPIES

Both energetic electrons and photons interact with matter in ways that cause scattering and ejection of electrons. As just discussed, these interactions can be used to develop images of specimen surfaces. In addition, their quantum interactions with core-level electrons form the basis for electron spectrometric methods. It is appropriate to lay groundwork for these methods by extending the discussion of electron interaction with atoms in solids that was begun in Sections 21.2 and 21.3. That treatment will be reviewed and refocused first on Auger electron production and then on the photoelectric ejection process. We should, in principle, be able to learn from these spectrometric methods much about atomic energy levels, chemical bonds, and species present in samples since characteristic electronic orbitals are the initial and final states in all these processes.

Electron Interaction with Matter. Some electron interactions with atoms are elastic in character and give rise to Rayleigh scattering, that is, scattering without loss of energy or change in frequency. Other collisions are inelastic and produce an X-ray continuum (bremsstrahlung), characteristic X rays, extended X-ray absorption fine structure, or Auger electrons.

Some appreciation of the experimental side of this complex range of interactions can be obtained from the electron distribution curve shown in Fig. 22.3. This plot reports the numbers of secondary electrons of different energies "collected" at low resolution obtained from a target on bombardment by monoenergetic electrons. The curve is interpreted in the legend of the figure since different types of interaction are occurring in different energy regions.

In Fig. 22.3 it is of interest that some width is contributed to peak C by interaction of electrons with both plasmons and phonons. *Plasmons* are oscillations of conduction electrons in a solid with respect to its ionic (atomic) cores. Plasmon excitation is a highly probable electron scattering process. For example, in a metal such as aluminum, as much as 15 eV of energy may be transferred from a bom-

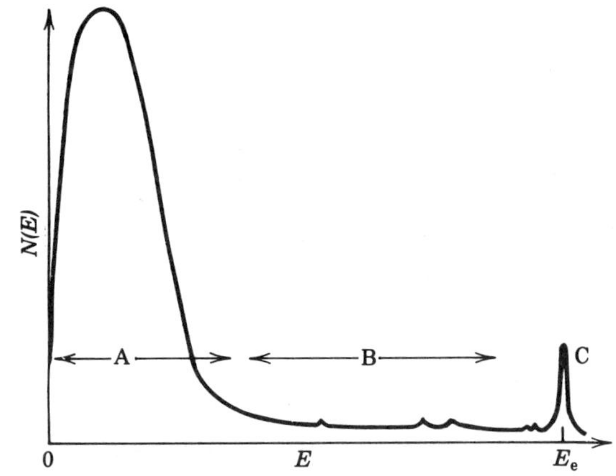

Fig. 22.3 Energy distribution $N(E)$ versus E of secondary electrons from a target bombarded by a beam of primary electrons of energy E_e. In region A many low-energy electrons are ejected by each primary electron as it undergoes a succession of inelastic collisions. Such electrons provide little information about surface characteristics. In region B the small peaks primarily represent emitted Auger electrons of characteristic energies. In region C the substantial peak represents elastically backscattered primary electrons. (After Ertl and Kueppers [15].)

barding electron to plasmon excitation. Electrons also interact weakly (with an energy of about 10 meV) with lattice vibrations, which are called *phonons*.

Electron Energy Loss Spectrometry (EELS). If the highest energy range of Fig. 22.3, peak C, is examined under closely defined experimental conditions such as those provided by a (transmission) analytical electron microscope (see Section 22.2) much can be learned. It is only necessary to subject the electrons which have passed through an ultrathin specimen to an energy analysis. By comparing their energies with E_e, the energy of the monoenergetic incident electrons, the energy loss can be measured and a plot of $N(E)$ versus E_{loss} obtained. Such a curve is shown in Fig. 22.4 for a beryllium specimen. As would be expected, this curve peaks at zero energy loss for elastically scattered electrons. Another peak corresponding to small loss as the first plasmon is excited appears at about 15 eV. Energy loss measurements can also be made by reflection from a surface as will be seen below. Other losses are related to the electronic and bonding state of the solid specimen (note the abrupt K X-ray absorption edge for beryllium at 115 eV in Fig. 22.4) and to local atomic environment (spectral features that are wavelike in appearance and begin just beyond the top of an absorption edge). The latter phenomenon has been briefly discussed in Section 21.11 as extended X-ray absorption fine structure (EXAFS) spectroscopy.

Vibrational spectroscopy of adsorbed molecular species and of surfaces can also be carried out with high-resolution EEL spectrometers. These instruments are highly sophisticated since minute energy losses are to be measured (10–100 meV). They usually feature one dispersive energy analyzer to develop the monoenergetic electron beam that is to be imposed on the surface followed by a second high-quality analyzer to measure the energies of electrons after they have been inelastically reflected (after loss of kinetic energy to excitation of a molecular vibration).

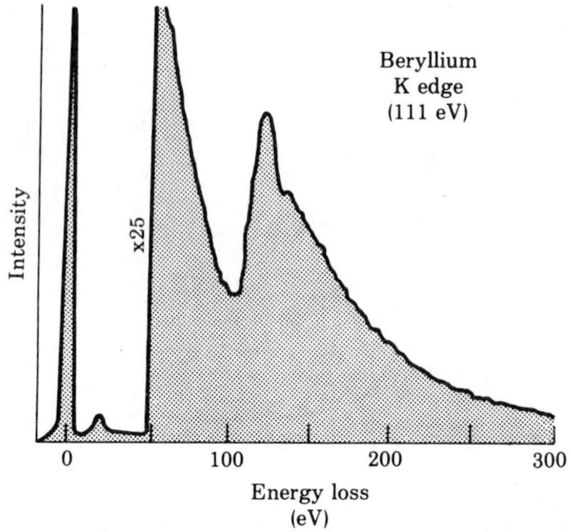

Fig. 22.4 Partial electron energy loss spectrum for a beryllium specimen using incident electrons of 1 keV energy. Note the 25X ordinate scale expansion at about 50 eV. [N. J. Zaluzec, *Thin Solid Films*, **72,** 184 (1980).]

To define the microscopic chemical and physical features of a surface calls for detection of secondary particles produced by a probing beam. Yet as the beam moves across a surface the precision with which its location is known depends on factors such as the beam diameter, particle energy, and the size of the *interaction volume* of particles and sample. From a representative answer for an electron beam, an estimate can be made of the degree to which the fineness of focus is actually responsible for the resolution obtainable in an SEM image and in an X-ray map that may be constructed of the specimen surface.

Experimental data about the interaction volume can be obtained, for example, by scanning an electron beam across a polymethylmethacrylate sample. If the polymer is next treated with a solvent that will etch it wherever electrons have broken bonds, the interaction volume will be shown. Etching for different intervals shows that the interaction volume is basically pear-shaped and has a greater depth than width. Where the incident beam strikes a specimen the active volume has a cross section approximately the diameter of the incident beam. Computer simulations of electron trajectories in solids assuming a Monte Carlo pattern of interaction confirm the pear-shaped interaction volume. They also show that the volume becomes deeper as the atomic number increases.

What fraction of secondary (ejected) electrons actually break free from the interaction volume? Electrons released at considerable depths in a specimen are generally reabsorbed. The average greatest depth from which electrons are released is termed the *escape depth*. This depth as would be expected is of the order of a few atomic layers and is related both to the matrix and to electron kinetic energies. For electrons with kinetic energies of 10–1200 eV in metals, escape depths average from 0.6 to 1.2 nm. For most substances depths are of the order of 1–3 nm. It is precisely this small escape depth that makes electron spectrometric techniques attractive as surface methods.

The Auger Process. Secondary electrons with discrete kinetic energy are produced from substances by the multistage Auger process. As was described in Section 21.2, the initial step is identical to that in emission of a characteristic X ray: transfer of energy to a core electron shell of an atom (e.g., K or L), and ejection of an electron. Next (a) an electron from a shell beyond the core shell falls into the hole that has been formed and (b) another electron called the Auger electron is ejected with characteristic kinetic energy. A representative Auger process is shown schematically in Fig. 22.5. Three electrons in total have been involved in the process and the atom has been left in a doubly ionized state. The kinetic energy E of the Auger electron can be calculated from the expression:

$$E = E_{\mathrm{K}} - E_{\mathrm{L(II)}} - E_{\mathrm{L(III)}}, \tag{22.1}$$

where X-ray notation has been used. Not all the energy $E_{\mathrm{K}} - E_{\mathrm{L(II)}}$ released in the downward transition is used up in the subsequent ionization (energy $E_{\mathrm{L(III)}}$ is required) and many Auger electrons can escape from the surface. Each type of Auger electron is identified by indicating all orbitals affected. For the process shown in Fig. 22.5, the Auger electron is designated KL_2L_3. Since the radiationless energy transfer between levels* is entirely an internal one, it is only necessary to determine the kinetic energy of the Auger electron to identify the element involved.

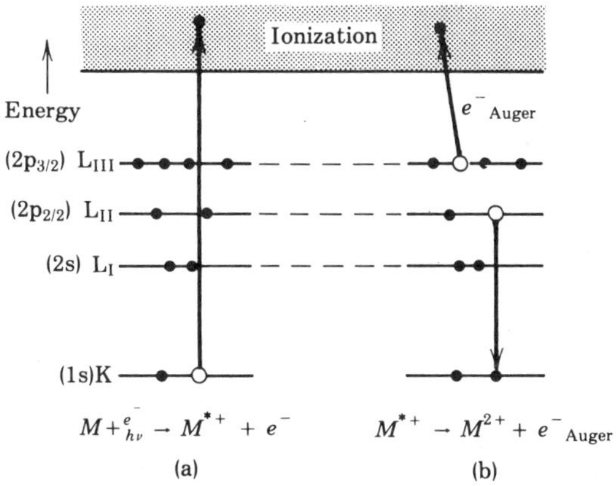

Fig. 22.5 The Auger electron process for a ground-state atom M that has two or more complete electron shells, which are shown as initially fully occupied. Both X-ray and optical notation are given for levels. (a) Step one: excitation (by photon or electron) of an electron from a K level to the ionization band, leaving a vacancy (hole) in the K level. (b) Step two: the core vacancy is filled by an L_{II} electron. The energy released in this radiationless process is transferred internally to an L_{III} electron, which leaves the atom as an Auger electron.

*It can be shown that the process is a radiationless one since it is not precisely governed by selection rules related to dipole transitions.

Each element can yield a number of Auger transitions. Note that the energy of bombarding electrons must exceed the X-ray absorption edge (see Section 21.2) for the core electron that is to be ejected. Not all possible Auger emissions will be intense; the processes by which they occur always compete with those that lead to formation of X-ray photons. While this competition favors the production of KLL Auger electrons only in lighter elements ($Z < 35$), the emission of LMM and MNN Auger electrons is more probable for virtually all elements.

Some additional considerations are involved if one wishes to correlate peaks in an Auger energy distribution with theoretical estimates of intensities, energies, and numbers of peaks for Auger transitions in a given sample. Three cases will be dealt with briefly.

For substances present as solids or films one must deal with energy bands rather than discrete orbitals at least for valence electrons. Levels through the Fermi level E_f are filled. Between this level and the surface of the solid a potential energy barrier termed "the work function" of the substance exists. In Fig. 22.6a, these levels are shown graphically. With respect to the vacuum level $E_{vac,sam}$ the measured Auger electron energy should be $E = E_K - E_l - (E_u + e\phi)$, where ϕ is the work function of the *sample*. As a result of contact potentials between sample and sample holder the work function of the sample will be altered; in this event the work function is best referred to the spectrometer ϕ_{sp}. This possibility is illustrated in Fig. 22.6b. Note that band energy details have been omitted in the right-hand part of the diagram.

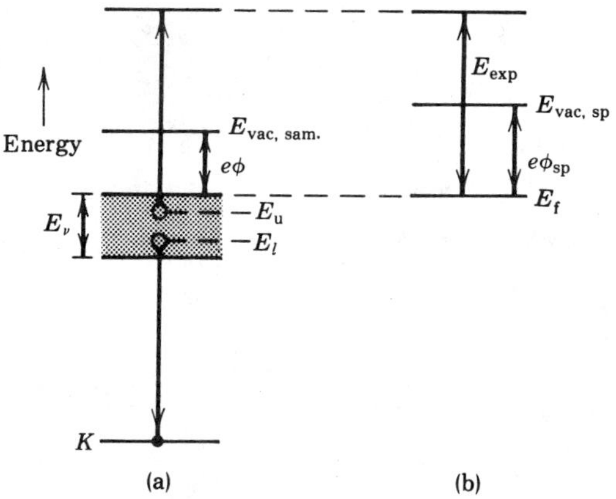

Fig. 22.6 Energy level diagram for measurement of electron kinetic energy in Auger emission from the valence band of a solid. Part of the energy to be overcome in ionization is the work function of the solid $e\phi$. (a) Electron transitions in Auger emission from a hypothetical isolated solid. An electron falls from valence band level E_l to fill the vacancy that has been created in E_K. The Auger electron leaves the atom (is ionized) from E_u. (b) If, as is customary, the specimen has been electrically connected to the spectrometer to reduce charging effects during electron-beam bombardment, the work function is that of the spectrometer itself ϕ_{sp}, and the kinetic energy of the Auger electron will be E_{exp}. Since Auger transitions originating in the valence band are being measured, line widths will be broadened by $2E_v$ because of uncertainty. (After Ertl and Kueppers [15].)

When working with levels in a valence band of width E_v an uncertainty of about 2 E_v is likely to be reflected in a broader spectral peak. This uncertainty is not easily removed since corrections are dependent upon knowing something about transition probabilities and the density of particular energy states. If one assumes, however, that the valence levels involved in an Auger transition, for example, E_u and E_l in Fig. 22.6, coincide with the Fermi level of the metal, the error is likely to be no larger than about 10 eV.

Chemical bonding and intermolecular forces may affect all three atomic levels on which the basic Auger transition depends and thus influence Auger transitions. If chemisorption has occurred, levels will be further shifted. Studies of such effects are undoubtedly best made by electron spectroscopy, a method which permits a more direct approach. If these questions must be pursued by Auger spectrometry, it will be important to have as high resolution in the spectrometer as possible.

Spin–Orbital Coupling. At the energies of interest for Auger spectrometry, which are usually smaller than about 2 keV, correlation forces between electrons dominate the scene. As a result, the usual *j-j* coupling between electron spins and angular momentum becomes less important. Instead Russell–Saunders (L-S) coupling dominates, and an unexpectedly large number of final states is possible. To illustrate the manner in which transitions are found to increase, consider Auger transitions in which the final state is of the KLL type. There should be six final states, given the three L states. With L-S coupling, however, two final states are possible for transitions identified by KL_1L_2, KL_1L_3, or KL_3L_3. For example, for the transition identified as KL_1L_2, singlet and a triplet final states are possible. In the gas phase a total of nine different transitions ending in a KLL state may be observed experimentally.

The Photoelectron Process. Illumination of molecular samples by UV or X radiation also leads to ejection of electrons. To interpret this phenomenon it is only necessary to extend Einstein's analysis of the photoelectric effect on metals. For any chemical species, for example, it is in principle possible for an incident photon of sufficient energy to ionize an electron from any of its energy levels. For a gaseous molecule the photoelectron ejected should have a kinetic energy E equal to the difference between the energy of the incident photon, which is completely absorbed, and the binding energy or ionization energy E_{Bi} of a particular molecular electronic level i that is affected:

$$E = h\nu - E_{Bi} \qquad (22.2)$$

Since momentum as well as energy is conserved, the electron will carry off almost all of the energy while the relatively massive molecular ion formed will acquire very little. If a reasonably large number of molecules of a given species is present, it is likely that photoelectrons appearing will characterize all energy levels in the ground state.

For gaseous molecules it is also possible that the ion produced will be in an excited rotational or vibrational state. Accordingly, the Einstein photoelectric law may be rewritten for isolated molecules interacting with photons* as:

$$E_i = h\nu - E_{Bi} - E_{vib} - E_{rot}. \qquad (22.3)$$

Comparison of Electron Spectroscopies and Radiation Spectroscopies. Finally, it will be helpful to compare electron spectroscopies with the classical absorption and emission spectroscopic techniques covered in Chapters 14–19. In classical spectroscopy transitions between definite quantum levels are observed. It is sufficient in measurement to use large numbers of photons of known energy and determine the number emitted or absorbed, that is, to measure initial and final beam intensities. In electron spectrometric processes, transitions are from discrete or stationary quantum states to the ionized state, which is part of the energy continuum. What is known is that some incident photons are *annihilated*. For such photons all the energy not used in an ionization process goes to provide kinetic energy to photoelectrons that are released. For measurements in electron spectrometry the energy of the incident photons or electrons must be known and the kinetic energy of photoelectrons or Auger electrons produced must be measured. In an XPS spectrum, for example, separate peaks should ideally appear for photoelectrons from each core atomic orbital and from each molecular orbital of species present. They are plotted versus $h\nu - E$ [see Eq. (22.2)] to give the spectrum.

22.4 ION SPUTTERING; DEPTH PROFILING

Having examined briefly some ways in which electron and photon beams can probe surfaces we now need to describe how ion beams can be used not only as surface probes but also for depth profiling in a sample.

The Sputtering Process. The considerable exchange of kinetic energy, momentum, and charge between a beam of bombarding ions and the atoms in a sample leads to turmoil. If the incident ions are sufficiently energetic, there is massive displacement of sample atoms from equilibrium positions down through several layers leading to the ejection of some atoms and ions. Two interrelated models are helpful in interpreting this complex phenomenon. One model developed mainly by Werner† proposes that a cascade of binary collisions occurs within the solid and a second model developed by Andersen and Hinthorne‡ describes phenomena at the surface thermodynamically.

*At the high-light intensities furnished by lasers there is some probability that two photons will collide simultaneously with a molecule and cause a two-photon transition. Further, if molecular concentration is greatly increased, as in a condensed state, it will also become probable that some photon collisions will coincide with molecular collisions, causing a slight modification in energy of transition. Equation (22.3) will need to be rewritten if either of these conditions obtains.

†H. W. Werner in *Electron and Ion Spectroscopy of Solids*, L. Fiermans, J. Vennick, and W. Dekeyser (Eds.). New York: Plenum (1978).

‡C. A. Anderson and J. R. Hinthorne, *Anal. Chem.*, **45,** 1421 (1973).

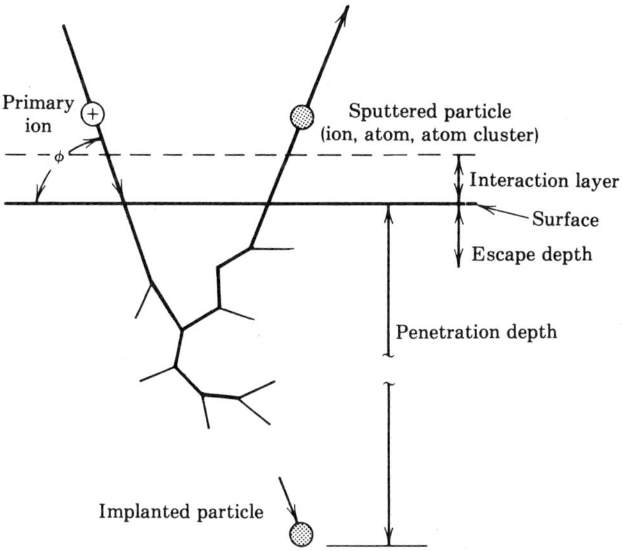

Fig. 22.7 The cascade model of the origin of atomic sputtering at a surface. The variety of interactions postulated to occur when a beam of moderately energetic (5–10 keV) ions bombards a solid are sketched for a single ion. Its charge is probably neutralized by a field-emitted electron just before impact, and once the atom is in the solid, it is slowed by successive binary collisions that yield "recoil" atoms. A cascade of moving atoms (not shown) results.

In Fig. 22.7, the cascade model is depicted schematically. The incidence of a single primary ion is shown as leading to a series of binary collisions. Although just a few collisions are shown, later impacts by recoil atoms must be also be imagined. If the solid has a single kind of atom of mass M, and the bombarding ions are of mass m, it can be shown that energy is transferred from incident ions to the solid in proportion to the factor $mM/(m + M)^2$. The ejection of atoms occurs when three conditions are met in a binary collision. First, the velocity vector of the atom struck must be pointing outside the specimen surface. Second, that atom must be at least as close to the surface as the *escape depth*, and third, the energy transferred to it must be sufficient to overcome the binding energy. It is of interest also to note the eventual fate of the primary ion: it comes to rest in the solid as an impurity atom. As such atoms accumulate they alter sample composition and local structure.

In what energetic state and with what charge are atoms ejected? Do the relative abundances of newly formed gaseous species correspond to proportions in the sample? Experiment shows that uncharged atoms emerge in abundance; singly charged ions are a thousand times less common than neutral species. Ions with a double or higher charge are a hundred times yet more scarce. There is also a small chance that molecular or cluster ions will appear and probabilities will depend on the dissociation energy involved in their release.

To account for the variety of species ejected and their relative abundance, it is helpful to examine the local thermal equilibrium model of Andersen and Hinthorne. They assume that under the conditions of ion bombardment a plasma exists in the surface layer (or selvedge). The plasma appears to extend about an atomic layer above the original surface as well as down into it perhaps 3–5 layers. This model assumes equilibrium between ions, atoms, and electrons in the plasma. Experimental evidence bears out the fact that at the surface atoms exchange charge and gain or lose excitation energy with ease. Accordingly, the model examines a series of equilibria including $M^0 = M^+ + e^-$ and $M^- = M^0 + e^-$ and applies the Saha–Eggert ionization equation to predict sputtered ion intensities. Their model fits a variety of experimentally observed sputtering situations. Temperatures derived in their systems were surprisingly high, from 5,000 to 15,000 K, and were interpreted as ionization temperatures which hold under local equilibrium conditions.* Other theoretical support for the model has developed.

It is apparent that any integrated theory of the ion-solid interaction must define probabilities for phenomena resulting not only from atom-atom collisions but the variety of other interactions in which electrons and photons are exchanged between atoms and clusters of atoms. Since sputtering is the process that enables secondary ion mass spectrometry (SIMS, see Section 22.9) to be carried out, further development of theory will be especially important to understanding this technique and in enhancing its applicability.

Depth Profiling of Samples. Sputtering is essentially the standard depth-profiling technique. Obtaining chemical information at surfaces and at greater and greater depths allows both surface and subsurface layers to be characterized. With such layers the questions begging to be answered are how far they extend in the z direction and whether their compositions are uniform. An alternative method for characterizing shallow surface layers will be described below.

Periods of sputtering must usually be followed by appropriate spectrometric measurement as depth profiling is carried out. For example, this sequence will be followed for X-ray photoelectron spectrometry. Perhaps the single exception occurs in examining a specimen by SIMS; in this case depth profiling and sampling will occur simultaneously because the technique itself employs ion bombardment.

What are the best conditions for sputtering? An appropriate protocol is suggested in Fig. 22.8. First, the sputtering ion beam is defocused somewhat so that atoms are sputtered out uniformly over a considerable area. Alternatively one can sputter in a raster pattern over the region to be excavated, a process sometimes termed "ion milling." In any event, after sputtering the crater bottom should be flat. Second, sampling should be from the central (i.e., flat) portion. In the figure several errors that can occur in depth profiling are also illustrated and explained.

Careful *calibration* of the sputtering rate is necessary. A common way to calibrate is to measure the time required to uncover during steady sputtering a well-defined layer that is a known distance below the surface. Such a layer may be

*C. A. Anderson in *Secondary Ion Mass Spectrometry*, K. F. Heinrich and D. E. Newbury (Eds.). NBS Special Publication 427. Washington, DC: U.S. Government Printing Office, 1975.

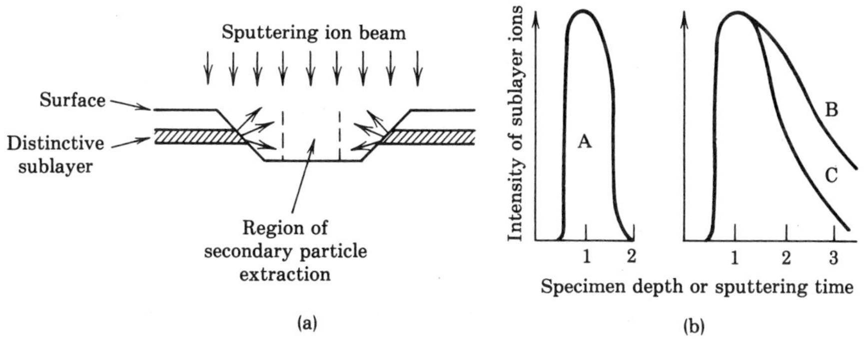

Fig. 22.8 A schematic diagram showing appropriate conditions for depth profiling of composition by sputtering. (a) Cross-sectional view of the sputtering of a specimen with a distinctive (note hatching) sublayer. Sputtering is well advanced. Note that the diameter of the ion sputtering beam is large compared with the width of the region over which secondary particles are to be extracted for analysis. (b) Counts per minute of sublayer species versus depth in sample or time of sputtering (arbitrary units). Curve A, ideal profile obtained when appropriate sputtering conditions are maintained. Curve B, profile observed when secondary ions are extracted over the entire width of the sputtering beam instead of the flat valley alone shows severe tailing. Now the sublayer contributes analyte continuously over a certain interval of sputtering as the crater grows deeper. Curve C, profile obtained when sputtered material collects on the edges of the crater and is resputtered and measured as the crater is deepened.

located with precision by careful implantation of atoms such as suggested in the discussion of the cascade model of sputtering. Periodic calibration of sputtering rates is desirable since the matrix changes that occur on going from one layer to another are likely to alter the rate even though ion beam intensity is held constant.

Resolution in depth profiling is of the order of 2–4 nm. The *absolute resolution* Δz, by definition, is the thickness of sample that must be sputtered out for an analyte signal to vary from 5 to 95% of the total change that marks the beginning of a well-defined layer. The term "relative depth resolution," $\Delta z/z$, is also in use. Depth resolution is reduced by poor sputtering techniques, surface roughness, ion beam inhomogeneity, and atomic mixing during profiling.

Varying the take-off angle, the angle at which ejected particles are accepted by the module measuring their energies, is a basically nondestructive technique of depth profiling for shallow surface layers (up to 10 nm) that is useful with electron spectroscopies. The geometry of the method is shown in Fig. 22.9. The length of take-off arrow in each sketch represents the average electron escape depth. By comparing sketches (a) and (b) it is evident that secondary electrons collected for a take-off angle of 20° are much more likely to have originated from atoms in the surface layer while those taken at 60° are more likely to have come from atoms in the bulk of the sample. Example 22.6 demonstrates this procedure. While this method is commonly used with X-ray photoelectron spectrometry, it is also important because it provides independent evidence of the composition profile with depth.

The signal intensity in the variable take-off angle method as related to emission

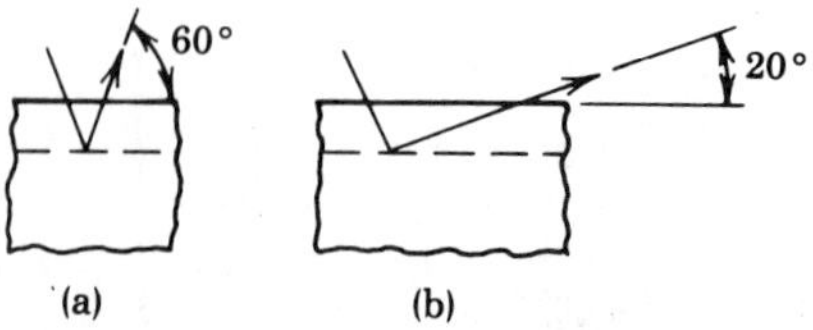

Fig. 22.9 Effect of variation of take-off angle (of secondary particle beam) on depth of specimen sampled. The length of the arrow in each sketch indicates the relative contribution of surface layer atoms. (a) More than half of the electrons that emerge at an angle of 60° will be from the bulk rather than the surface film. (b) At 20° nearly all of the electrons will emerge from the surface layer. By use of low take-off angles and variation of the angle it is possible to establish relative concentrations in about the top 10-nm layer.

from species in the surface layer will be given by the expression

$$I = K \exp(-d/\sin\theta),$$

where d is the thickness of the surface layer and θ is the take-off angle.

22.5 MODULES FOR PARTICLE SPECTROMETERS

Since the optical types of surface spectrometers rely on standard modules described earlier, only basic modules needed for X-ray photoelectron spectrometers and other *particle* instruments need be described here. Figure 22.10 provides a modular view of a single-channel surface spectrometer. As may be inferred from the figure, the geometry around a sample can be varied. Appropriate angles of incidence of probe and take-off beams for secondary electrons or ions must be optimized in terms of the type of measurement sought. All particle spectrometers will also require provision of a high or even an ultrahigh vacuum and magnetic shielding for many modules.

Sources. Characteristics of the X-ray and UV sources commonly used in photoelectron spectrometers are listed later in Table 22.3. For most other surface spectrometers, however, particle sources such as electron and ion guns are required. Electron microscopes, some types of electron spectrometers, and electron microprobes also need *electron guns*.

A standard electron gun is shown schematically in Fig. 22.11. In general, a

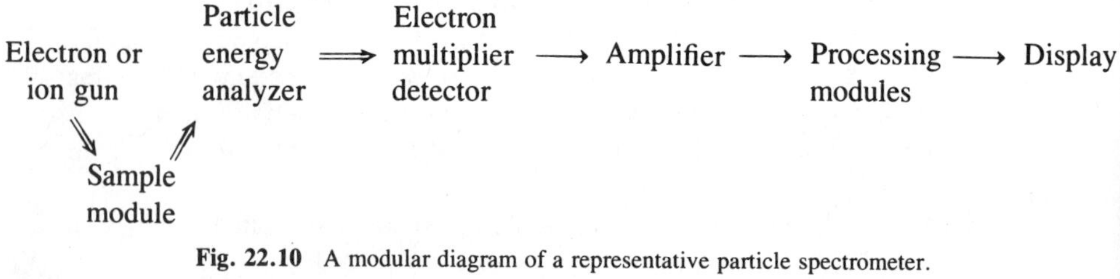

Fig. 22.10 A modular diagram of a representative particle spectrometer.

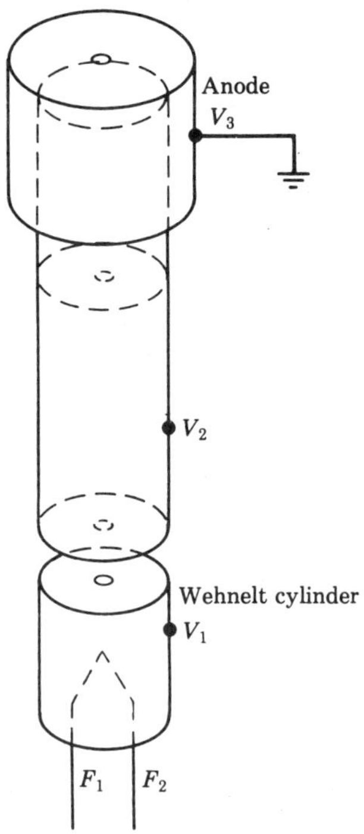

Fig. 22.11 Schematic diagram of an electron gun. It consists of a heated tungsten filament F_1/F_2 (cathode) that serves as the electron source; an electrode termed the Wehnelt cylinder that is charged to a sufficiently negative potential V_1 relative to the cathode that it constricts electrons into a beam of smallest diameter d_0 just beyond its exit aperture; a focusing electrode at potential V_2; and an anode at ground potential V_3 to provide the positive potential (relative to the Wehnelt cylinder) to accelerate the electrons. The strong positive (relative) potential of the cathode causes an electron cloud (space charge) to form around the filament, thereby stabilizing its operation since its emission saturates. Often the potential of the cathode is established by biasing it through a variable resistor connected to the high voltage that supplies the Wehnelt cylinder. This design ensures that both filament and anode are at ground potential.

hot-filament cathode generates electrons, in this case a 0.1-mm diameter tungsten wire that is bent to a V-shaped tip of effective area about 100 × 500 μm and heated resistively to about 2700 K.

Where a more intense beam is needed, at least two other electron sources can be substituted for the tungsten filament. One is a rod of *lanthanum hexaboride*. Because of its smaller work function relative to tungsten, LaB_6 will give a current at least five times greater than tungsten at a lower temperature. The boride is highly reactive, however, and a good lifetime is secured only by having a quite high vacuum in the gun, perhaps 100 nPa.

A second option is incorporation of a *field emission source*. In this device, a rod (usually tungsten) sharpended to a point no more than 100 nm in diameter serves as the electron source. When a strong negative potential is applied, the electrical field at the point is so great that electrons can "tunnel" through the surface potential barrier. At the same operating potentials a brightness hundreds of times larger than that of a thermionic source is secured. This source also calls for an exceptionally good vacuum.

The stream of electrons from the cathode is focused into a beam by electron optics exactly as in an electron microscope. In Fig. 22.11 the needed electrostatic fields (lenses) are achieved by holding the Wehnelt cylinder at −100 to −1000 V relative to the cathode as ground, and grounding the anode (V_3). The shorter the distance between the filament and Wehnelt electrode, the higher must be the filament temperature and the Wehnelt potential. Further, as those values are imposed, the greater will be the beam current.

In the gun an intermediate electrode is varied in potential to ensure good focusing at its anode aperture. As electrons are accelerated to maximum kinetic energy ($1/2\, mv^2 = eV_3$) on passing through the anode aperture, the beam is focused again. Deflection plates can be added following the anode to position the beam precisely.

Ion Guns. A variety of *ion guns* is also available for surface spectrometers. Their beams, usually of positive noble gas ions, are valuable for *etching* (i.e., cleaning a specimen surface) and for *sputtering* (i.e., removal of layers of atoms for depth profiling) as described in the last section.

In a representative type ions are formed by electron bombardment of a gas (at about 0.1 Pa or 10^{-3} torr), giving at room temperature ions with a small spread in kinetic energies. These gaseous ions are extracted into a focusing set of electrostatic fields and accelerated, much as in an electron gun. Final beam energies may range from 0.3 to 20 keV. Deflection plates can be added to position an ion beam precisely.

A more elaborate ion source, the *duoplasmatron*, is shown in Fig. 22.12. It can supply intense beams of ions of either charge. In the device a plasma consisting of ions, electrons, and molecules is generated at low pressure by a hot cathode arc and dual electrical and magnetic fields then constrict the plasma to a high density. Ions commonly furnished by a duoplasmatron are O^-, Ar^+, and F^-. The source is highly efficient and provides high ion density at the plasma boundary at the aperture through which ions are extracted. Electrostatic lenses and a stage of acceleration are added to obtain desired energies. Addition of a stage of mass analysis will even secure a monoisotopic beam.

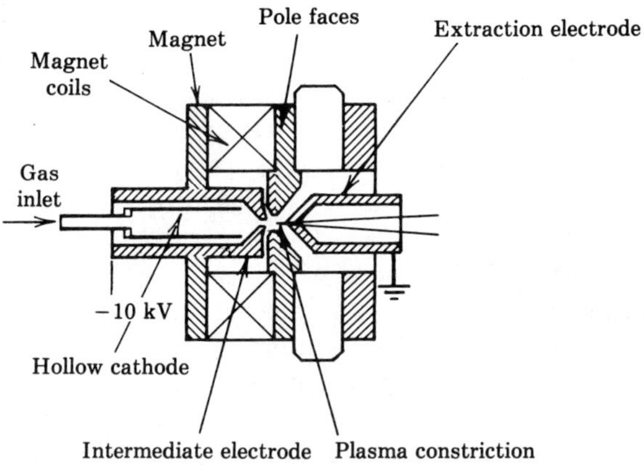

Fig. 22.12 Schematic of a duoplasmatron ion source. This source has exceptionally high brightness. The gas which is to furnish ions is introduced at low pressure into the hollow cathode and ionized by a 10-kV dc field. The plasma formed is constricted by the shape of the intermediate electrode. A second "squeezing" occurs because of the inhomogeneous magnetic field between its pole pieces, giving a high concentration of ions. Either positively or negatively charged ions may be drawn from the aperture of the source by placing an appropriate charge on the extraction electrode. The average energy of ions leaving the duoplasmatron is about 20 eV and the energy spread is small.

Particle Energy Analyzers. These devices sort the electrons and ions produced as a specimen is probed. Commonly they employ electrostatic acceleration and deflection to analyze particles according to kinetic energy.* Precision electrostatic analyzers require good magnetic shielding, for example, an enclosing cylinder of mu-metal, because even small magnetic fields deflect charged particles.

In principle, particle energy analyzers can be adapted for all kinds of charged particles. Most disperse particles angularly according to energy. In one type concentric metal sheets are formed into a cylindrical sector and in another into concentric hemispheres.

The *concentric hemispherical analyzer*, which has been developed to a high degree of perfection, is illustrated in cross section in Fig. 22.13. In this device a focused electron beam presented at an entrance slit will be refocused at the exit slit. Note that the voltage divider employed to supply a potential to the two plates is balanced with respect to the beam potential V_e (if electron energy is E_e, $E_e = eV_e$), that is, the outer plate is as much negative with respect to the electron beam as the inner one is positive. A double-focusing capability is provided by this arrangement, that is, the analyzer focuses both vertically and horizontally. For this reason a well-defined beam appears at the exit aperture. Electrons appearing at the entrance are focused at the exit aperture according to the relationship

$$V = E_e(R_1/R_2 - R_2/R_1),$$

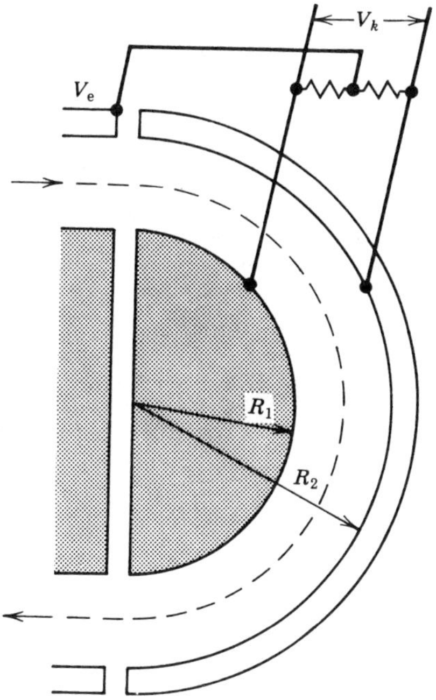

Fig. 22.13 Cross section of a concentric hemispherical particle energy analyzer. Guard rings are held at V_e, the potential of the electron beam. The potential difference across the hemispheres is fixed and the voltage divider arranges that their voltages are balanced.

*Magnetic deflection designs have proved more complex and less effective, in part because providing necessary magnetic shielding to other parts of an instrument is difficult.

where R_1 and R_2 are the radii of curvature of the inner and outer plates, respectively. Interestingly, resolution in this type of analyzer tends to be a particular percentage of the electron energy. For that reason and also to ensure particles are of optimum energy for the fixed radius hemisphere, the device inevitably incorporates a retarding grid at its entrance to ensure that electrons enter at relatively low energy (see Exercise 22.1).

In Fig. 22.14, a second type of dispersive electrostatic analyzer, the widely used *cylindrical mirror analyzer* (CMA) is shown schematically. In terms of the beam of secondary particles that reaches the detector, the CMA has one focus at the surface of the specimen and a second at the detector. While the sample and the inner concentric cylinder are grounded, the scanning potential is negative and is applied to the external cylinder. A magnetic shield of mu-metal (not shown) ordinarily surrounds the whole. Particles emitted from the surface of the specimen over a reasonable solid angle $\Delta\theta$ are passed through the fixed slit in the inner cylinder. As the particles experience the retarding field of the outer cylinder, if they have kinetic energies in a narrow range, they are reflected back through to the second set of slits in the inner cylinder and come to a focus at the first dynode of an electron multiplier or other collector/detector.

Focusing in the radial direction is secured for a particular angular divergence $\Delta\theta$ of the electron beam passing through the slit as long as the combination of values of V_p and electron accelerating potential V_e is suitable for the collecting angle θ fixed by the design. In addition, if this angle can be made a little smaller than 43°, it is also possible to secure focusing in the lateral direction with resulting improvement in resolution. Other options for increasing resolution are to provide a retarding grid input stage to the energy analyzer, to restrict entrance of high-

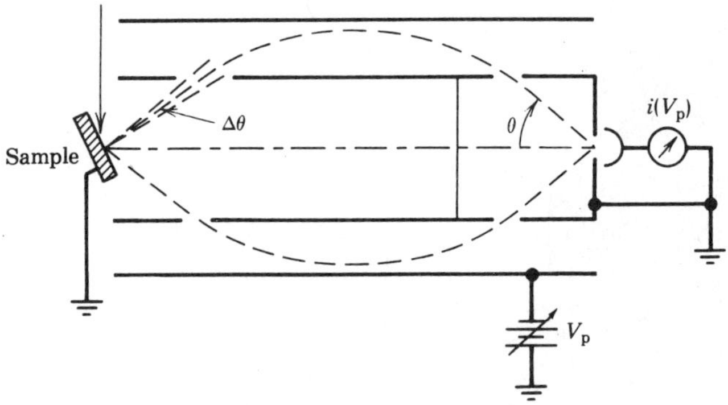

Fig. 22.14 Schematic cross section of a cylindrical mirror analyzer (CMA), detector, and readout. A primary particle beam is shown incident on a sample at about 60°. Ejected secondary particles are extracted through a series of apertures (not shown) into the CMA. The beam spread permitted by the cylindrical entrance slit is $\Delta\theta$. The inner cylinder is grounded and a potential V_p is applied to the outer cylinder where $V_p < V_e$, the accelerating potential that fixes the average energy of entering particles. Usually $V_e/V_p \approx 2$. In many applications a channel multiplier plate would be inserted before the detector to ensure a better S/N ratio.

energy species as well as bring all species within a more acceptable range, and to use a pair of CMAs in series.

With the CMA several quality parameters must be examined—sensitivity, resolution, and transmission. With a CMA in a multimode type of instrument (see Section 22.10), it becomes important to be able to trade off one quality parameter with another. Fortunately the resolution $\Delta E/E$ for a particular CMA is constant over a wide energy range. A retarding grid input stage is especially valuable for high-energy resolution, but could be traded off where a wide energy spectrum is to be covered to ensure more constant resolution across the spectrum. The resolution or width of energy band ΔE passed can also be adjusted by varying the pass energy of the analyzer or by altering the size of aperture between a pair of CMAs and between the output of the CMA and an electron multiplier detector. Shot noise is low since particles over only a narrow range of energies impinge at one time.

Where highest sensitivity is important, high transmission of particles is crucial, and large apertures between stages and in front of the detector are valuable. It also becomes important that the sample be placed as close to the entrance of the CMA as possible. This feature may be maximized by arranging that the electron gun be coaxial with the cylindrical mirror analyzer. An arrangement featuring this design is shown schematically in Fig. 22.22.

Electron Multiplier Detectors. A highly sensitive particle detector is desirable following a dispersive energy analyzer if measurements are to be made at trace levels. The usual device, termed a *channel electron multiplier*, is a hollow glass cylinder of about 1 mm i. d. coated inside with a special heavily lead-doped glass that has favorable secondary emission. The glass behaves as a continuous strip resistor; thus, the voltage applied across its length appears as a gradual voltage drop. In Fig. 22.15 the essential multiplier action is depicted schematically. Effi-

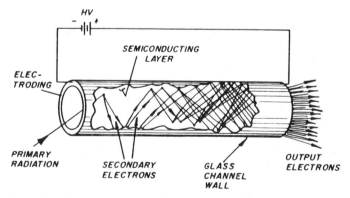

Fig. 22.15 Schematic of the multiplier action of an electron multiplier detector. A channel-type device in which the electron-multiplying lead-doped glass is used to coat the inside of the channel. A sufficient electrical potential difference is applied between the entrance and exit of the electron multiplier so that secondary electrons emitted from one part of the multiplier are accelerated and impact farther down the multiplier channel with sufficient energy to eject additional secondary electrons. Following the exit aperture a collector is required. (Courtesy of Galileo Electro-Optics Corp.)

ciencies in terms of secondary emission are in the neighborhood of 100% for entering electrons and ions with kinetic energies of 10 keV or more but fall off to 10% and lower for lower energy particles and photons of wavelengths longer than 80 nm.

Clearly, this multiplier–detector immediately transduces incident particles and photon signals to an electron signal. For an incoming particle to be detected it is only necessary that it have sufficient energy to eject a single photoelectron on impact. Subsequently this electron is accelerated and attracted back to the surface further down the tube, where it causes secondary electrons to be emitted. Successive collisions lead to growing numbers of secondary electrons and a gain as high as 10^5–10^8 can be secured. This detector mimics a photomultiplier tube.

Two conditions essential for gains of this magnitude are (a) operation under sufficiently high vacuum that residual gas will not impede electrons and (b) curvature in the channel, or at least a change in direction, to avoid *ion feedback*. The latter phenomenon can be traced to ionization of residual gas. Since the positive ions formed travel in the direction opposite to the electron stream, they not only reduce the electron current but their bombardment may damage the photoemissive surface. Ion travel is minimized by introducing curvature in the channel. The gain of this type of multiplier is a function of the length-to-diameter ratio: representative ratios are 20–60.

What other electrical characteristics of electron multipliers are important? Both bias and dark currents will exist. A *bias current* results from the application of a potential difference of 2–3 kV between the ends of a channel multiplier. With a representative resistance along the channel of 500 MΩ, the dc bias current will be about 5 μA. A schematic circuit diagram for a multiplier which is to be operated in a pulse-counting mode is given in Fig. 22.16. *Dark currents*, which are associated with secondary electron emission in the absence of incoming particles or photons, are extremely low (about 5 pA). Thus, S/N ratios are excellent. How large an output current can be tolerated? It is usually limited to about 10% of the bias current. Nonlinearity in gain as well as shorter lifetime will result from higher output currents.

Microchannel Plates. Coated microscopic hollow glass cylinders can also be assembled into multiplier arrays called microchannel plates. Plates have from 18 to 75 microchannels mm^{-2}. Even thin plates have high gain since the L/D ratio determines the gain. Good spatial resolution is preserved since channels are straight. Gain is limited nominally to 10^4 to avoid ion feedback.*

For a surface spectrometer that employs a hemispherical particle energy analyzer parallel detection of electrons of many energies can be achieved. The exit slit must be replaced by a *multichannel electron multiplier–detector* such as a microchannel plate coupled to an anode plate collector. In the hemispherical analyzer

*A pair of microchannel arrays, back-to-back, but with the second with channels at an angle of perhaps 8° with respect to the first, avoids the ion feedback problem. In such "chevron" arrays an overall gain of about 10^7 is realized.

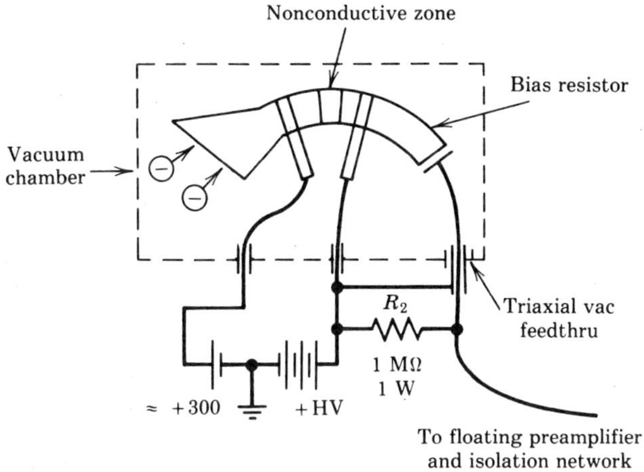

Fig. 22.16 Schematic diagram of the electrical circuit for a channel electron multiplier detector. A small positive voltage is applied to the extractor horn to attract particles of negative charge into the device and a high positive voltage is applied through resistor R_2 to the end of the coiled channel. A value of resistor is chosen so that the anode plate at the end of the detector is at least 100 V more positive than the end of the channel. Since the pulse signal from the multiplier is at a high voltage, it is passed to the processing electronics through a capacitor which blocks dc but provides capacitive coupling for the signal. The multiplier and detector must of course be in a high vacuum chamber. (Courtesy of Galileo Electro-Optics Corp.)

high-energy electrons sweep through a path of larger radius than do those of lower energy, causing the electrons to enter different channels of the detector. To sense the outputs of the many channels of the microchannel plate and identify their location, a position–sensitive anode is abutted against the plate. That analysis is performed by measuring the current from electrodes at the four corners of the plate and interpreting the results as a function of time by a computer program. This multichannel electron multiplier–detector provides much faster detection, a special advantage when high-energy bombardment of a sample will quickly lead to surface degredation. Alternatively, by trade-off of speed for sensitivity S/N for the measurements can be improved.

22.6 PHOTOELECTRON (XPS AND UPS) SPECTROMETRY

This discussion of photoelectron spectrometry is the first of several presentation of electron or ion spectrometric surface methods. Photoelectron spectrometry calls for directing a beam of monoenergetic photons (X ray or UV) toward a sample and analyzing the kinetic energy of ejected photoelectrons, sometimes as a function of take-off angle. In X-ray photoelectron spectrometry (XPS)* knowing electron energies will allow identification of elements and oxidation states of atoms.

*The technique was called electron spectroscopy for chemical analysis by Siegbahn, whose research group developed the method extensively. The euphonious acronym ESCA is still in use.

The wide usefulness of XPS as a surface technique derives from its ability to provide information about chemical state and the spatial distribution of species in heterogeneous systems such as catalysts. Both XPS and UPS (UV photoelectron spectrometry) can be used for theoretical characterization of molecular states and assessment of the effectiveness of calculations of orbital energies.

Of the two methods, XPS will be given almost exclusive attention here; UPS, which is instrumentally essentially identical, will be briefly examined at the end of the section. Some nonsurface applications of UPS will also receive brief consideration. The relative advantage of XPS for surface analyses arises from the roughly hundredfold greater energy of X-ray photons. As a result, these photons interact with many orbital levels of most elements and the photoelectrons ejected can escape from greater depths. The theoretical background of photoelectron spectrometries was developed in Section 22.3.

X-ray photoelectron spectrometry is the least destructive particle surface technique. Recall that an X-ray beam is a fairly mild probe for most samples. For this reason and others, XPS offers one of the best approaches to surface analysis of organic materials such as polymers. On the basis of the escape depth of emitted photoelectrons, the technique provides information about the top 2 to 10 nm (20–100 Å) for organic materials and the top 0.5 to 2.5 nm (5 to 25 Å) for metals and metal oxides.

Characteristic Modules. To assemble a suitable X-ray photoelectron spectrometer we draw on the modules that were presented in the previous section and Section 21.4. In Fig. 22.17, the generalized diagram for this type of spectrometer is repeated as a basis for discussion.

Conventional X-ray tubes of at least 15-kV power (Section 21.4) that provide either Mg $K\alpha_{1,2}$ or Al $K\alpha_{1,2}$ are appropriate sources for this type of spectrometer and are listed in Table 22.3. It is possible to incorporate both anodes in a single tube. Where still higher energy X-ray photons are required, a tube with a Cu anode furnishing the Cu $K\alpha$ doublet at 8048 eV is often used. Clearly, a sample should be irradiated with essentially only the $K\alpha_{1,2}$ doublet of one element to make possible more precise calculations of binding energies using Eq. 22.2. A minimum width or uncertainty in measured photoelectron peak energy will arise because of the intrinsic X-ray line widths (Table 22.3) and will introduce uncertainty in calculated binding energies. With elements with still higher Z, the separation of the lines in the doublet becomes appreciable, giving rise to unexpected detail and poorer resolution in photoelectron spectra.

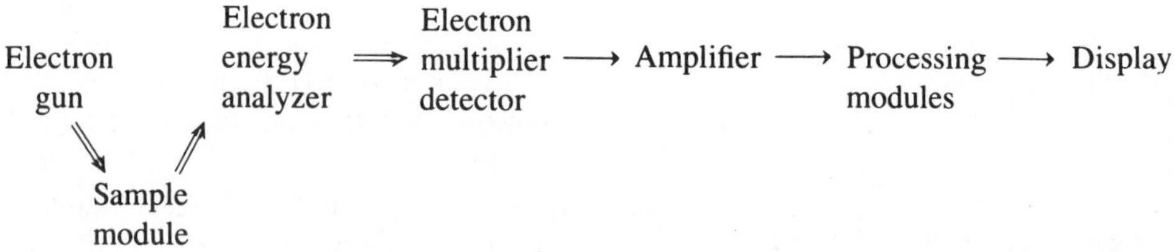

Fig. 22.17 Modular diagram of an X-ray photoelectron spectrometer.

Table 22.3 Common Sources in Photoelectron Spectroscopy[a]

X-Ray Photoelectron Spectroscopy (XPS)			*UV Photoelectron Spectroscopy (UPS)*		
Nature of Source	*X-Ray Line*[b]	*Energy (eV)*[c] *(wavelength, nm)*	*Nature of Source*	*Emission Line*	*Energy (eV)*[d]
X-ray tube	Mg $K\alpha_{1,2}$	1253.6 (0.989)	Gas-flow discharge lamp[e,f]	He I[a]	21.2175
	Al $K\alpha_{1,2}$	1486.6 (0.834)		He I	23.0865
	Cu $K\alpha_{1,2}$	8048 (0.01542)		He II	40.8136
				He II	48.3702

[a]Synchrotron radiation, though it is accessible only from accelerators, is an attractive source in the energy range between UV and X-ray sources.
[b]The designations $K\alpha_{1,2}$ indicate uresolved doublets.
[c]Linewidths are: Mg, 0.7 eV; Al, 0.85 eV; and Cu, about 2 eV or larger.
[d]Linewidths are less than 0.01 eV.
[e]Gases used besides helium are hydrogen, argon, and neon. In the 3 to 80-eV range, noble gas continuous sources may be used with a monochromator, but intensity is less.
[f]DC or microwave energized; pressure 10–100 Pa (0.1–1 torr) with discharge arranged to be within capillary to limit helium flow. Lamp design determines whether emission is mainly He I (atomic), or He II (singly-charged ions).

The source–specimen–detector sequence should be compact to ensure a high flux of X-ray photons at the sample surface and the collection of as large a solid angle of emitted photoelectrons as possible. Further, it is often desirable to position the X-ray tube so that its anode makes an angle of about 10° with respect of the surface of the specimen and thus approximates a slit source so that a smaller area of the specimen is irradiated. The detector take-off angle relative to the sample surface may be fixed, but if variable, will offer the possibility of shallow depth profiling.

An ion gun is usually provided as well. It will serve both to clean specimen surfaces and to provide substantial depth profiling in determining the composition of a specimen.

Of the four main surface spectrometric methods, XPS provides the poorest lateral resolution, about 1 mm. A cylindrically curved quartz mirror can be inserted in some XPS spectrometers to improve resolution by taking advantage of reflection at the Bragg angle (see Section 21.6) to focus monochromatic X-rays to a spot about 100 μm across. Alternatively, masks are sometimes used to limit the area of sample from which secondary electrons are extracted with about the same improvement in lateral resolution.

Both types of nondispersive particle energy analyzers have been incorporated in spectrometers. Cylindrical mirror analyzers are most often used because of their good resolution, simplicity, and large transmission factor. Commonly the detector is a channel electron multiplier. With at least some types of energy analyzers a microchannel (electron multiplier) plate is located where an exit slit would be and is followed by some type of phosphor-coated surface to display the output. In this multichannel detector scanning is unnecessary and particle energies are a function of position on the detector array.

In an XPS spectrometer an oil-diffusion pump backed by a mechanical fore-pump will provide sufficient vacuum. Yet ultrahigh vacuum capability will be necessary if surfaces are to be examined free of absorbed layers. Proper cryogenic baffling is important to prevent the possibility of diffusion pump contamination of a specimen surface. An ion, turbomolecular, or sorption pump is desirable.

Processing modules that follow the detector may be either of an analog or digital type. Since photoelectron events are randomly spaced in time, it is more common to have pulse capacity in the detector and other modules (see Section 21.5). Digital systems have the advantage of better precision and lead either to employing a multichannel pulse analyzer or a computer to acquire and process data.

Measurement and Analysis. Specimens for surface study are solids with regularly defined surfaces. For spectroscopic information alone, liquid and gaseous samples can also be examined. Many solid specimens are examined as received, but cleaning by ion etching is often used to remove an oxide or other surface layer. Powdered samples may be pelletized and compressed into a suitable sample compartment; they may also be spread on a double-sided adhesive tape.

To reduce *charging effects* electrical contact is usually made between a conducting sample and spectrometer so that the steady emission of electrons when only photons are incident is compensated. The charging problem can be serious with nonconducting and poorly conducting samples; when charge develops on a sample, spuriously high energy values are measured for photoelectron peaks. Some build-up of charge is relieved by the incidence of scattered electrons from the X-ray source. Usually an XPS spectrometer has an extra tungsten filament that can be heated to flood a specimen with electrons to offset specimen charging.

Having too many electrons fall on a specimen, as well as too few, will lead to spurious energy readings. Since a spectrometer must be calibrated in any event, while setting the intensity of the flooding beam it is good practice to monitor the position of a known XPS peak. When the peak appears at its standard position, flooding is of correct intensity. Often the C_{1s} peak (energy = 285 eV) is used as a reference since adventitious carbon from the atmosphere usually provides the line even when samples technically are carbon-free. If gold has been sputter-deposited on a sample, its $4f_{7/2}$ peak should be used as a reference. The gold will be likely to be in electrostatic equilibrium with the sample even though a precise charge balance has not been attained.

The referencing of energies is made clearer by Fig. 22.18. As can be inferred from the figure, for solids binding energies E_B are measured relative to the spectrometer and are given by the expression

$$E_B = h\nu - E_{exp} - e\phi_{sp} \tag{22.4}$$

where ϕ_{sp} is the work function of the spectrometer. The last factor arises whenever the spectrometer is electrically connected to the sample. By referencing to a known

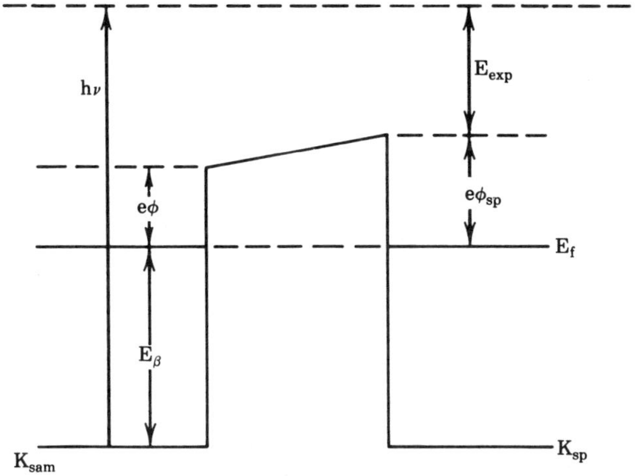

Fig. 22.18 Energy levels involved in XPS spectra for conducting solids.

elemental binding energy, one essentially calibrates measurements with the spectrometer.

In a representative XPS spectrum for a solid one obtains an electron distribution curve with kinetic energy peaks corresponding to discrete kinetic energies of bonding for particular atoms in the sample. Peaks appear against a strong background of secondary electrons that have been generated by collisions of photoelectrons with sample atoms; usually this background increases with energy as well. For that reason quantitative measurements will be difficult since background subtraction is likely to be arbitrary.

Photoelectron spectrometry is excellent for *qualitative analysis*. Spectral interferences are minimal since XPS peaks for core electrons tend to be fairly widely separated as illustrated in Fig. 22.19 by the XPS spectral traces for K shells. (Spectral transitions for outer shell electrons tend to be featureless since many factors contribute to their broadening.) Thus, an XPS elemental analysis based on core electron peaks is usually straightforward. Further, because peaks undergo chemical shifts, qualitative analysis leading to identification of compounds and of oxidation states of elements (speciation) is often possible. *Chemical shifts* result from the effect of changes in chemical environment of atoms on electron-binding energies. Fortunately, such shifts are most marked for the most common atoms: core orbital energies of second- and third-period elements are especially sensitive to their environment as would be expected. Where precise peak energies are sought, sputter-deposition of gold on a sample is advantageous to provide its $4f_{7/2}$ line as a reference as noted above.

Example 22.3 According to Eq. (22.2) binding energies E_{Bi} of electrons in the atomic and molecular orbitals^i of a substance can be calculated from its XPS spectrum. One secures the values E_{Bi} simply by subtraction of measured electron kinetic energies E from the energy of incident photons: $E_{Bi} = h\nu - E$. Figure 22.20 gives the XPS spectrum secured for

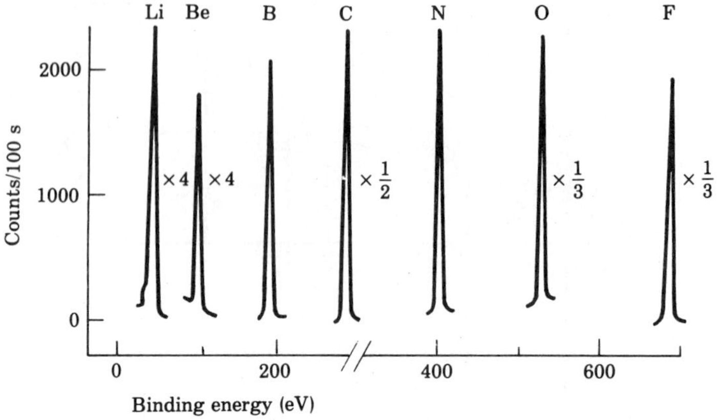

Fig. 22.19 X-ray photoelectron spectral lines for 1s core electrons (K shell) of second period elements. Note that intensities are registered in terms of counts per 100 s. Photons from Al Kα X radiation excited photoelectron emission. Binding energies have been calculated from the expressions 1486.6 − E_{EK}, where the last quantity is the kinetic energy of the photoelectron ejected from the designated orbital. Heights of peaks have been adjusted to comparable values. (W. C. Price [p. 151 of [4], Courtesy of Academic Press.)

gaseous carbon monoxide with a photoelectron spectrometer on bombardment by monoenergetic X-ray photons. Can the peaks be assigned on the basis of simple molecular orbital theory?

In the inset of Fig. 22.20 a standard energy plot of core-level orbitals and MOs is shown for CO. Its 14 electrons are shown as x's in the 1s carbon and oxygen orbitals and the σ and π MOs formed on overlap of their 2s and 2p orbitals. Since ionization energy is plotted from right to left along the abscissa, peaks to left of the spectrum such as the O 1s and C 1s are most negative in energy. Note the abrupt change in scale as the transition to MOs is made. In this case it is straightforward to assign XPS peaks of decreasing energy to atomic orbitals and MOs in order of their decreasing orbital potential energy. Corroboration of the assignment should of course be sought.

Example 22.4. How can one estimate *electron binding energies* in atomic orbitals in order to interpret XPS spectra reliably? The *equivalent cores approximation* offers a good approach. It states that an atomic core with an electron vacancy (hole) is chemically equivalent to a full core for an element with one more proton. Show how the idea can be used to estimate the binding (ionization) energy of the 2s orbital in sodium.

The ground configuration of a sodium atom is of course $1s^2 2s^2 2p^6 3s^1$. Hess' law can be applied to the following series of ionizations of known energy to give the desired direct ionization from the 2s shell:

$$\mathrm{Na}\ (2s^2 2p^6 3s^1) = \mathrm{Na}^{2+}(2s^2 2p^5) + 2e^- \qquad 52.43\ \mathrm{eV}$$

$$\mathrm{Na}^{2+}(2s^2 2p^5) = \mathrm{Na}^{2+}(2s^1 2p^6) \qquad 32.79\ \mathrm{eV}$$

$$\mathrm{Na}^{2+}(2s^1 2p^6) + e^- = \mathrm{Na}^{+}(2s^1 2p^6 3s^1) \qquad -15.03\ \mathrm{eV\ (estimated)}$$

$$\mathrm{Na} = \mathrm{Na}^{+}(2s^1 2p^6 3s^1) \qquad 70.2\ \mathrm{eV}$$

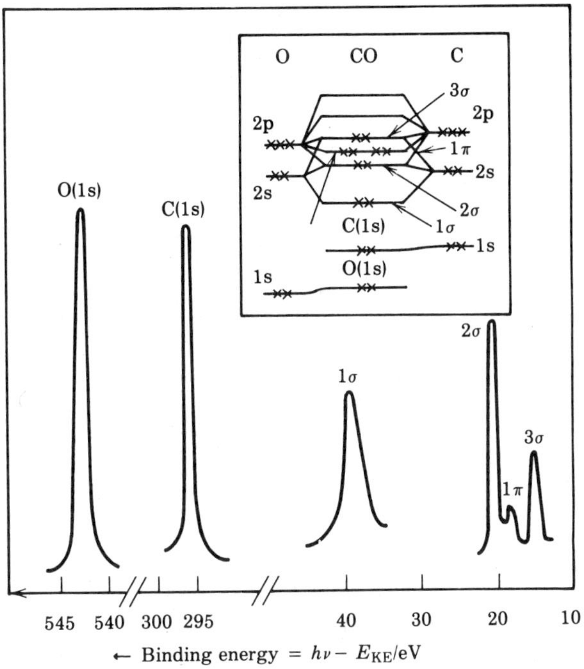

Fig. 22.20 An X-ray photoelectron spectrum of gaseous carbon monoxide. Note the change in scale along the abscissa. The binding energy in eV displayed along the abscissa is calculated by taking the difference $h\nu_{X\text{-ray}} - E_{KE}$ for electrons ionized from different orbitals. A representative molecular orbital (MO) energy diagram is given in the inset to show the correlation between spectral peaks and orbitals. Note the different scales for peaks at the left corresponding to ionization from atomic core levels of O and C and from σ and π MOs of the molecule. Under the conditions of measurement it was not possible to resolve peaks corresponding to particular vibrational transitions. (K. Siegbahn *et al*,. *ESCA Applied to Free Molecules.* Amsterdam: North-Holland, 1969; Courtesy of Elsevier Science Publishing Co.)

The equivalent cores approximation allows us to substitute the negative of the ionization energy of Mg^+ for that of the third reaction. The binding energy estimated for a 2s electron, 70.2 eV, agrees well with the experimental value 71.1 eV.

Actually the equivalent cores approximation has proved useful in organization of photoelectron ionization values in general. In addition, it has found application in predicting chemical shifts of elements in compounds (W. L. Jolly in [4]).

Some of the possibilities of XPS for qualitative analysis can be deduced by considering carefully the carbon and nitrogen photoelectron spectra shown in Fig. 22.21 for three compounds. Example 22.5 below discusses a catalytic application that involves oxidation states.

In *quantitative analysis* XPS offers all the challenges characteristic of X-ray determinations. Nevertheless, even without calibration, compositions can usually be estimated within 30% (relative). With proper calibration, relative standard deviations are ordinarily ±5% or better. Reference should be made to Chapter 12

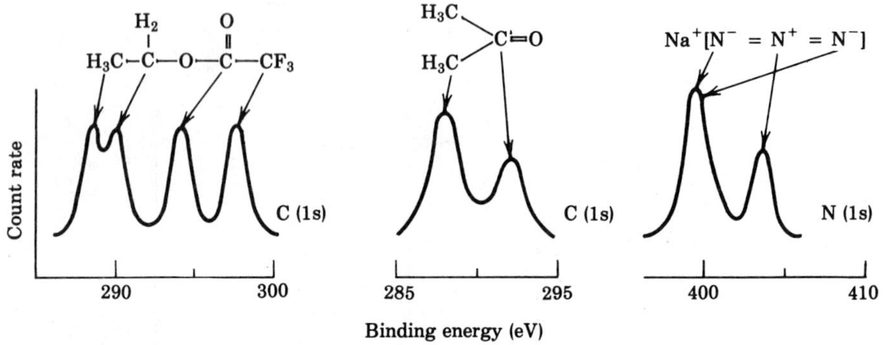

Fig. 22.21 Photoelectron spectra (XPS) for C_{1s} and N_{1s} in the designated compounds. A much larger scale has been used on the abscissa than in the previous figure. The "chemical shift effect" for electrons in core atom orbitals is seen to be a sufficient basis to permit assignment of peaks to atoms in particular chemical environments (W. C. Price [p. 151 of [4]; Courtesy of Academic Press.)

for examination of techniques for dealing with background and noise. Detection limits for species are high, about 0.1 atom percent.

The dependence of photoelectron signal intensity I_s on instrumental and sample parameters is given approximately by the expression

$$I_s = I_p \rho \lambda(e) (d\sigma/d\Omega) A \phi T,$$

where I_p is the photon flux in the primary beam, ρ is the emitting atom density, $\lambda(e)$ is the electron escape depth, A is the area sampled by the electron energy analyzer whose transmission factor is T, ϕ is a factor defining angular dependence, and $d\sigma/d\Omega$ is the differential cross section for photoelectrons from the given subshell. The S/N ratio can be enhanced by increasing any of the instrumental factors, though trade-offs will need to be balanced.

Resolution is limited in XPS and UPS by factors such as

1. the natural line width of the probing X radiation,
2. the natural width of photoelectron peaks,
3. the resolution of the electron energy analyzer, and
4. specimen charging effects.

Several of these factors have been discussed. If the primary X-ray line width is broad, it is often possible to improve resolution somewhat by mathematical deconvolution to remove the contribution of the X-ray line width. For example, the half width (FWHM) of the $3d_{5/2}$ Ag peak at 368 eV in an XPS spectrum can be reduced from about 0.9 eV to 0.6 eV in this way. As this moderate improvement suggests, the natural line width of photoelectron peaks themselves is usually the factor limiting resolution; this width is seldom less than 1.5 eV.

If peak overlap is severe, use of conventional curve deconvolution techniques with an analog or digital computer often permits extraction of more spectral information. However, parameters such as peak widths and shapes must be known from reference spectra for its reliable application. Further, if peaks are weak or noisy, deconvolution may well introduce error and caution must be used in interpreting results.

Auger electrons are also produced by X-ray photons (see Section 22.3). Indeed, in XPS spectra Auger peaks nearly always appear. It is of interest that they are more pronounced in this case since background caused by inelastically scattered electrons is low. As a result, Auger peaks are easily identified without differentiation of spectra and contribute additional useful chemical information.

Depth profiling of composition can be accomplished in XPS both by sputtering and variation of the photoelectron take-off angle as discussed in Section 22.4. Examples 22.6 and 22.7 illustrate applications.

Example 22.5. The oxidation state of molybdenum in a Mo/Al_2O_3 catalyst can be deduced by inspection of the $3d_{3/2}$ peak of Mo in XPS spectra of such catalysts. What sort of spectrum can be expected? How can it be used for at least a semiquantitative analysis?

A change of oxidation state of a particular atom can be expected to affect the electron probability density around it. The $3d_{3/2}$ peak is found at 236 eV in the +6 state, at 233 eV in the +5 state, and at about 230 eV in the +4 state. Though the peaks overlap, their areas can be estimated by deconvolution. The relative area of each peak in a Mo spectrum will be proportional to the number density of Mo in that oxidation state. By monitoring the Mo XPS spectrum while a catalytic process takes place and correlating peak areas with other XPS data, it becomes possible to hypothesize about the Mo species present and the role of such species in catalyst performance [T. A. Patterson, J. C. Carver, D. E. Leyden, and D. M. Hercules, *J. Phys. Chem.*, **80,** 1700 (1976).]

Example 22.6. In general a polymer is chosen for applications such as packaging on the basis of bulk properties. Yet its surface must often have different properties. For instance, it may be important to modify a polymer surface to increase wetting or adhesion. One standard way is to subject it to a mild electrical discharge in the presence of air or oxygen. But how can one be sure the desired changes such as chemical incorporation of oxygen have been effected and that only the surface has been changed?

X-ray photoelectron spectrometry allows the surface of a polymer to be distinguished chemically from its bulk. Dilks reported on the effectiveness of electrical discharge in surface oxidation of hydrocarbon films under different conditions [A. Dilks, *Anal. Chem.*, **53,** 802A (1981)]. Two studies were conducted on polyethylene and another on polystyrene. Since it was known that oxygen would be incorporated, the decision was made to examine the effects of an electrical discharge by determining the relative intensity of XPS peaks for oxygen (1s peak, 534 eV) and carbon (1s peak, 285 eV). The method of varying the take-off angle method was used to determine the depth of oxidation. The ratio of intensities of oxygen and carbon peaks was determined at 65° and at 30° relative to the surface. Results were as follows:

Sample type and treatment	*Ratio of peak intensities O_{1s}/C_{1s} at take-off angle indicated*	
	65°	35°
Polyethylene		
Oxidized by corona discharge in air	0.24	0.24
Deduction: Surface was homogeneously modified to a depth of at least 5 nm.		
Polyethylene		
Oxidized by microwave plasma in O_2	0.31	0.24
Deduction: Oxidation of surface greatest at top and gradually decreased with increasing depth down to perhaps 10 nm.		
Polystyrene		
Oxidized in rf plasma in low-pressure O_2	0.66	0.95
Deduction: Oxidation was confined to the outermost layer.		

Dilks contrasted the latter two cases. Since apparent penetration of oxygen was greatest for polyethylene, it appeared that an energetic species such as singlet O_2, which has a mean free path of about that magnitude, was involved in the oxidation. On the other hand, the short depth of oxidation in the last case suggests that its oxidation was dominated by atomic oxygen which has a very short mean free path.

Example 22.7. The thickness of a polymer film on a platinum electrode was found by XPS by variation of the take-off angle while observing the Pt_{4f} photoelectron peak. To cancel out dependence on as many variables as possible, the ratio of photoelectron peak intensity from platinum, "with film" to that from platinum "without film," $I_{Pt,\,substrate}/I_{Pt,\,\infty}$ was taken. The latter signal was, as denoted, equivalent to the signal from an infinitely thick platinum layer. The escape depth of platinum photoelectrons through the film $\lambda_{subs,\,film}$ was estimated. Photoelectron peak intensities at different take-off angles were substituted in the equation

$$I_{Pt,\,substrate}/I_{Pt,\,\infty} = \exp\left\{-d/[\lambda_{subs,\,film}\sin\theta]\right\}$$

to calculate film thickness d. From the study the thickness of one vinylferrocene film was estimated to be 10.7 nm. [M. Umana, R. W. Murray, et al., *Anal. Chem.*, **53,** 1170 (1981).]

UV Photoelectron Spectrometry (UPS). This technique sees much application in basic research within both organic and inorganic chemistry. In conjunction with XPS as necessary, UPS has been employed to deduce molecular orbital binding energies and work out peak assignments of many organic and inorganic species. Molecules and salts that have vapor pressures of at least $10–10^{-1}$ Pa ($10^{-1}–10^{-3}$

torr) at temperatures below 450 K are examined by UPS in the gaseous state. Since the usual helium sources are windowless because of the opacity of most substances in the 30 to 60 nm emission range, differential pumping between the source–sample area and electron energy analyzer, which operates at the much lower pressure of 10^{-5}–10^{-6} torr, is necessary. Electrostatic deflection energy analyzers are used and routine resolution is about 0.020 eV.

Example 22.8. Would examination of carbon monoxide by both XPS and UPS be desirable to obtain optimum spectra?

The answer can be found in Fig. 22.20, which provides a complete XPS spectrum for CO. Resolution of the core 1s peaks is good, but only with the greatly enlarged energy scale provided by moving from X-ray to HeIα excitation of UPS can the σ and π molecular orbital level peaks be resolved. Probably individual vibrational levels associated with the MOs would also be resolved as suggested by Eq. (22.3).

Both UPS and XPS studies relating to electronic structure and orbital energies have yielded very productive results. Photoelectron spectrometry has provided valuable data for checking calculations of orbital energies, bond lengths, dipole moments, and heats of formation as well as helping assign the proper experimental ionization potentials to the correct MOs.

22.7 AUGER ELECTRON SPECTROMETRY (AES)

In most Auger electron spectrometry electron beam irradiation is used to create the core-level vacancy that initiates the process. As described in Section 22.3, the Auger process is a two-electron decay—one electron filling the vacancy and the second, the Auger electron, being emitted. It is the kinetic energy of this ejected electron that provides analytical information about the atom. Auger spectrometry is useful for all states of matter, but it is particularly attractive as a surface method because of its high sensitivity, applicability to all elements except hydrogen and helium, and fast response. Relevant theory was presented in Section 22.3 and to some extent in Section 21.2.

Sensitivity is inherent in AES: the ionization of core electrons in atoms by electron impact is not only a quite probable process (cross sections in the range of 10^{-20} cm^2), but also occurs with reasonable independence of electron energy.*

Auger Spectrometers. For Auger capability in a surface spectrometer, it is desirable to design around a cylindrical mirror analyzer (CMA) as the particle energy analyzer. This module and others characteristically required for Auger measurements have been discussed in Section 22.5.

*X-Ray excitation of core electrons in atoms is actually often more probable. That advantage is offset, however, by the disadvantage that a normal X-ray tube provides much lower flux of photons (about 10^6 s^{-1}) than an electron gun does electrons (about 6×10^{12} s^{-1}).

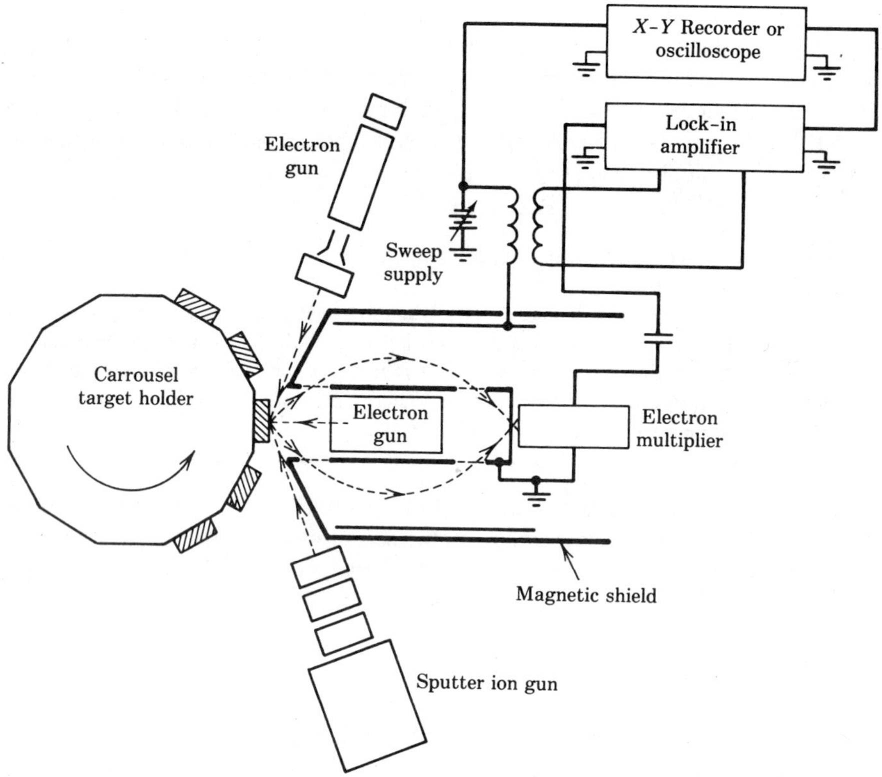

Fig. 22.22 Schematic of a representative Auger electron spectrometer. The particle energy analyzer is a cylindrical mirror analyzer. A dc sweep potential is supplied to the outer electrode to scan the Auger electron energies. Auger peaks correspond to low amplitude audiofrequency ac are superimposed on the dc sweep. The Auger signal emerges from the CMA as an ac signal. Note the channel electron multiplier that follows the CMA. The multiplier, in addition to greatly amplifying the CMA output, presents it superimposed on a dc bias voltage of from 500 to 3000 V. The capacitor between the channel electron multiplier and the readout blocks this dc while coupling the Auger signal to the processing train of modules. A lock-in amplifier modulates the ac Auger signal and also provides the synchronous demodulation to develop an output proportional to $dN(E)/dE$ [1]. (Courtesy of Elsevier Science Publishing Co.)

A schematic of a versatile Auger spectrometer is shown in Fig. 22.22. Note the alternate positions shown for the electron gun. Having it coaxial with the CMA is preferable both when a specimen surface is to be scanned to develop an Auger elemental map and whenever compactness is desired. Alternatively, employing an electron gun external to the CMA allows use of a low angle of electron beam bombardment. In this way a higher intensity signal can often be secured. Since primary electrons will now travel farther in the surface layer, more Auger electrons will be likely to escape. Further, by measurement with first one gun and then the other, some depth profiling of composition is possible (see below). The ion gun may be employed for surface etching or depth profiling of composition.

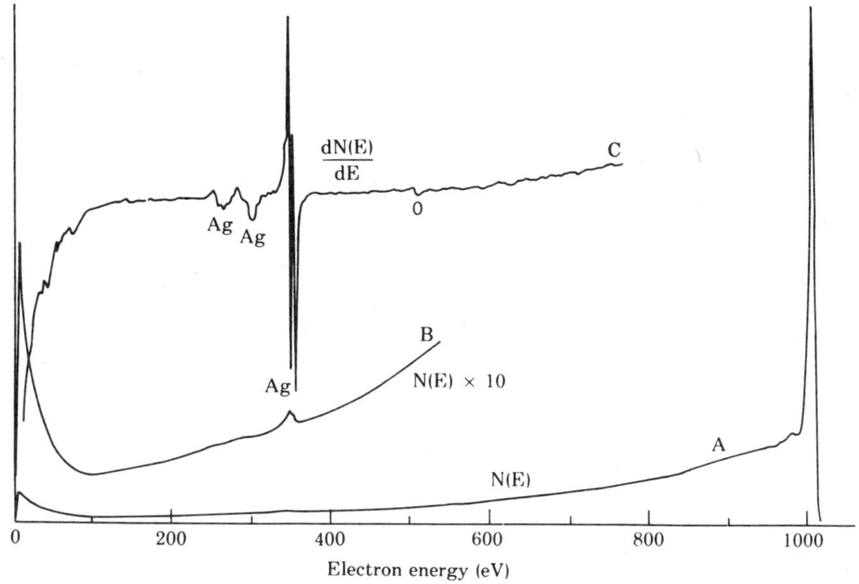

Fig. 22.23 Auger spectrum of a silver sample obtained with a 1-keV electron beam. Curve A shows the normal energy distribution $N(E)$. In curve B, curve A is shown with a 10X vertical scale expansion. Curve C displays the first derivative of curve A and presents Auger peaks clearly.

How should Auger spectra be displayed? Three possible modes are shown in Fig. 22.23. The simplest plot is one in which $N(E)$ is graphed as a function of E; it displays Auger peaks as small pips because of the high-intensity background from backscattered electrons. A second approach is to magnify the first curve by a factor of 10 (see the middle curve). This display seems little better. Instead the mode most often used is to graph the first derivative $dN(E)/dE$ as shown in the top curve and thereby improve the S/N or contrast between Auger peaks and background.*

The spectrometer design in Fig. 22.22 offers several signal processing advantages. It yields a first-derivative Auger spectrum directly (by ac modulation of the output of the CMA electron energy analyzer), efficient transmission of the information-carrying pulses from the high-voltage output of the electron multiplier to the amplifier, and noise reduction by phase-sensitive detection of the signal in the lock-in amplifier The amplitude of the *modulating* ac is kept low (less than 5 V) to avoid modulation error in the first-derivative output. The ac (of perhaps 30 kHz) is supplied by the lock-in amplifier and mixed with the dc sweep voltage applied

*By convention in a first-derivative spectrum for qualitative analysis, the energy of Auger peaks is reported as the point of maximum slope on the high-energy side of the regular peak. This point corresponds to the most negative point in the first derivative. For quantitative analysis, it is customary to use peak-to-peak height. (Y. E. Strausser in *Applied Surface Analysis*, T. L. Barr and L. E. Davis (Eds.). Philadelphia, PA: ASTM, 1980.]

to the CMA. The derivative is taken since the amplifier is tuned to accept the in-phase ac signal.

Measurement and Analysis. Elemental *qualitative* analysis is a strength of Auger electron spectrometry since peaks can be measured with somewhat higher sensitivity than is available in XPS. For this purpose the most intense Auger transitions, those whose final state fall in the families KLL, LMM, MNN . . . , should be used. It is found that Auger KLL spectra are readily obtained for lighter elements ($Z < 35$) and that LMM transitions are easily observed for *all* elements that give Auger spectra. Peak overlap is only occasionally a problem since natural peak widths are only of the order of about 3–10 eV.

If precise peak positions are sought, deposition of a thin gold layer on a sample by evaporation or sputter-deposition is advantageous. As in XPS, gold is valuable as a reference since the metal will usually be in electrostatic equilibrium with a sample.

How precise is AES as a quantitative method? In part the answer depends on whether a high S/N can be measured. It can be shown that the Auger portion of the readout for an *undifferentiated* peak is given by the equation

$$I_{\text{total}} - I_{\text{background}} = I_{\text{A}} = \beta[cI_{\text{p}}]\, Tf(\Delta E/\gamma),$$

where β is the Auger yield (see below), c is analyte concentration, T is the instrument transmission factor, I_p is the primary current, ΔE is the width of the energy passed by the analyzer (its resolution or energy window), γ is the natural peak width. Ideally, $T = 1$, and if $\Delta E >> \gamma, f(\Delta E/\gamma) = 1$. Noise will be represented by shot noise because of the discrete events occurring as electrons bombard the sample and as scattering occurs. It can be shown that $S/N \approx (I_p \tau_p)^{1/2}$, where τ_p is the sampling time. Clearly, increasing the primary beam current and sampling time will enhance S/N. To lessen noise the energy analyzer "window" is usually set to be close to the width of a typical Auger peak. In that way, noise on either side of the window is excluded and a maximum S/N ratio of about 1×10^4 is theoretically possible. This would correspond to an AES detection limit of about 100 ppm. Representative detection limits are about 1000 ppm since resolution of a CMA seldom matches Auger widths and many sources of uncertainty affect performance.

Use of digital data acquisition and processing will permit a considerable additional gain in S/N ratio. Signals can be acquired directly (without modulation) from a spectrometer by following its electron multiplier stage (see Fig. 22.22) with a high-voltage isolation amplifier. The $N(E)$ signal can be multiplied by E, the electron energy, and then stored on disk. Data can be processed as desired. Processing *may* include smoothing, deconvolution of overlapping peaks, sometimes background subtraction, and desired calculations.

The bounds of reliability of *quantitative* results from AES are the same as for XPS. Elemental compositions can be estimated to $\pm 30\%$ on an absolute basis and to $\pm 5\%$ by use of reference standards as nearly like the unknowns as possible.

The use of *sensitivity factors*, while less accurate, provides a simple and rapid alternative for a wide range of elements and compounds except in cases where so-called matrix effects are likely to be important, where their use leads to uncertain results. In this approach an unknown concentration is found from the equation $C_x = (I_x/S_x)/\Sigma_i(I_i/S_i)$ where S_i is the sensitivity factor for element i. A good example of application of Auger electron spectrometry follows.

Example 22.9. A glass window for a He–Ne laser was given a low-reflectance coating optimized to pass the laser radiation at 632.8 nm. First a thin layer of TiO_2 was applied and then a MgF_2 coating. When the window was later baked at 500°C to remove impurities, it was found to have become essentially opaque at the critical wavelength! How can depth profiling with AES be used to learn what happened to the window on baking?

To explore what had happened to the surface, AES depth/composition profiles were taken of both new and baked windows. Four major elements (Ti, O, Mg, and F) were monitored during depth profiling through the coating layers.

For the unbaked sample it was found that the Mg and F concentrations were initially high and paralleled each other to a depth of about 140 nm and then fell precipitately. At about that depth the concentration of O and Ti rose substantially.

A quite different result was obtained from the baked sample. For it the concentration of Ti remained virtually constant with depth while the oxygen concentration fell off gradually. The fluorine concentration began essentially at zero and rose to a maximum at a depth of about 140 nm; the concentration of Mg fell off more gradually than before.

A comparison of baked and unbaked samples showed a reversal of elemental concentrations in the top layers of the baked sample relative to the unbaked one. On baking the outer layer appeared to have become mainly MgO. Thus, one can hypothesize that the MgF_2 layer had basically diffused into the bulk of the window. As a consequence the surface no longer had a refractive index appropriate to high transmission at the laser wavelength. To confirm the hypothesis the baked window was also examined by XPS; this analysis indeed showed that the outer layer was mainly MgO. [Perkin–Elmer Corporation, PHI Data Sheet 1051.]

Raster scanning of the exciting electron beam, of course, makes it possible to obtain Auger elemental images of the surface. A resolution of 0.5 μm or better can be obtained. Often Auger spectrometers with scanning capability are called scanning Auger microscopes (SAM) though Auger element maps give only relative indications of concentration.

Depth profiling is possible by simultaneous or intermittent use of the ion gun (see Section 22.4). Surface contamination is greatly reduced if depth profiling is done at the same time that spectrometric measurements are taken. Depth resolution can approach about 3% of the sputtered depth. Auger spectrometers are available that can monitor several elements simultaneously by use of multiplexing.

22.8 ION SCATTERING SPECTROMETRY (ISS and RBS)

In ion scattering spectrometry a monoenergetic, monoisotopic, well-collimated ion beam with energy 200–3000 eV is backscattered off the surface of a specimen.

The incident beam is 1–2 mm in diameter with a current of from 10 to 200 nA. In most cases the singly charged noble gas (He, Ne, or Ar) ions engage in a binary elastic collision with an atom or ion in the top layer of surface. From the energies of the scattered ions, information is obtained about the mass and location of atoms in the surface. The technique is unusually sensitive for the surface layer alone. Mass identification by measurement of energy of the scattered ions is straightforward. For ions incident with energy E_0, the energy E after backscattering can be shown to be given by the expression

$$E/E_0 = M_1M_2/(M_1 + M_2)^2\left\{\cos\theta + \sqrt{(M_2/M_1)^2 - \sin^2\theta}\right\}^2,$$

where M_1 is the mass of incident ions, M_2 is the mass of target atoms, and θ is the scattering angle (180° − the angle between incident and scattered beams). When $\theta = 90°$, the expression reduces to

$$E/E_0 = (M_2 - M_1)/(M_2 + M_1).$$

Each elemental species present in a surface is represented by an approximately Gaussian peak in the distribution curve for scattered ions.

The intensity of the backscattered beam in terms of species i in a surface essentially depends linearly on the interaction potential between an incoming ion and a surface atom of species i, the surface density N_i of this species, and the probability P_i that the incoming ion can leave the surface without undergoing neutralization. Probability P_i is of the order of 10^{-2}.

Characteristic Modules. From the modules described in Section 22.5 an ion-scattering spectrometer can be assembled by selection of an ion gun furnished with a noble gas containing a single isotope, a quadrupole mass filter (see Section 20.4), a magnetically shielded scattering chamber, an electrostatic energy analyzer such as a CMA, an electron multiplier of the channel or dynode type, a detector, and appropriate processing modules. Clearly, attention must be given to vacuum requirements. The pressure in the ion gun region needs to be 10^{-3} Pa (10^{-5} torr) or slightly higher for an adequate ion flux, but in the scattering chamber the pressure should be lower than 10^{-6} Pa (10^{-8} torr) if possible. As a result of the high sensitivity to the surface layer, the possibility that foreign molecules may contaminate it must be kept low through having a very good vacuum. Indeed, an initial bakeout of all of the elements of this region of the spectrometer is also usually required.

Most spectrometers employ a cylindrical mirror analyzer to determine the energy distribution curve. Its large solid angle of acceptance insures a good S/N ratio. The ion gun may either be placed outside the analyzer to give scattering at an oblique angle with respect to the specimen or aligned coaxially with the analyzer.

Measurements and Analyses. It should be noted that many of the incident ions are neutralized in the process of interaction with the specimen surface. Indeed the *number* of scattered ions is only a fraction of the total yield from this scattering process. Most observers interpret the great sensitivity of ISS measurements to the surface layer as originating in this phenomenon. Neutralization occurs generally because an ion that penetrates more than one layer or two will tend to be neutralized. Clearly, the energy analyzers will "reject" neutral species.

Similarly, ISS is insensitive to subleties such as electronic energy levels of atoms or electron density changes due to chemical bonding. It is the location and masses of the atoms in the surface that mainly determine scattering. Heavy beam currents can be employed without danger of pileup at the detector since so many ions are neutralized during scattering and each increment of energy is scanned in turn during a measurement. Nevertheless, care must be taken to insure that the beam current is not large enough to cause surface sputtering and thus alter surface composition. When depth profiling is sought, it is only necessary to increase the beam current density.

Ion scattering spectrometry is especially valuable to follow surface effects after treatments such as annealing, sputtering, and adsorption of a layer of interest. Interaction of surfaces with adsorbed layers can be followed by isotopic labeling of the surface layer and monitoring of ion scattering while the surface is heated. Since ion bombardment is known to cause desorption from a sample, how can one guarantee that ISS will give a "static" view of an adsorbed layer? Experiment shows that if the current density of a primary He^+ beam is kept at 10^{-8} A cm^{-2} only about 10^{-4} of a monolayer is removed during the time required to obtain 10^4 counts in an oxygen peak.

The relative strength of signal from different elements has been used often to yield information about the orientation of adsorbed molecules. Typical ISS spectra are taken by plotting ion signal in counts per second versus ion energy E/E_0.

Example 22.10. The adsorption of carbon dioxide molecules on tungsten was followed by ISS to secure information about orientation of the molecule. At room temperature it was found that the oxygen/carbon ISS intensity ratio was roughly 17, but after heating tungsten to about 700 K the ratio was only 3.7. How can these data be interpreted?

These data imply that at room temperature CO is adsorbed with the carbon atom directed toward the metal and the oxygen pointing away. On heating to 700 K it appears that the CO molecules are lying flat or might even be dissociated. [W. Heiland, W. Englert, E. Taglauer, *J. Vacuum Sci. Tech.*, **15**, *419 (1978)*]

Rutherford Backscattering Spectrometry (RBS). It is of interest that there has also been extensive experimental development of ion backscattering at comparatively high energies (0.1–3 MeV). Sampling depths of a few hundred nanometers can be analyzed at 2 MeV with a depth resolution of about 20 nm. Usually He^+ or H^+ beams are employed.

Bombarding ions at these energies accomplish the sampling without the need of removal of surface layers. Furthermore, energies are so large that incident ions interact with *nuclei* of atoms according to the Rutherford scattering law (simple coulomb interactions). A representative application is examination of films on surfaces.

In high-energy scattering measurements silicon surface barrier semiconductor detectors are used. In analogy to Si(Li) detectors, the barrier detectors produce pulses proportional to the kinetic energy of the incoming ions and *neutral atoms*. Multichannel analyzers provide a distribution curve. Clearly, beam currents must be kept sufficiently low to limit pulse pile-up problems (see Section 21.7).

22.9 SECONDARY ION MASS SPECTROMETRY (SIMS)

This true *mass* (not energy) technique is one of the most versatile available for the analysis of solid surfaces. In SIMS, a focused beam of ions of several keV energy is directed at a specimen. Particles sputtered from its surface include atoms (ground or excited state), ions (positive or negative and either in a ground or excited state), X-ray photons, and Auger electrons. From this cloud of particles, secondary ions are "extracted" from as near the point of generation as possible and separated according to their mass-to-charge (m/z) ratios. After detection, abundance or intensity data is processed and read out as a plot of intensity against m/z. A related technique, valuable with involatile organic and biological substances, molecular or liquid SIMS, was described in Section 20.7.

Example 22.11. What kinds and intensities of *positive* ions are ejected when a 5 kV Ar^+ ion beam bombards an aluminum specimen at low intensity in a SIMS determination? Positive ions ejected and the secondary current in amperes (A) or (abundance) attributable to each are as follows:

10^{-10} A:	only Al^+;
10^{-12}–10^{-10} A:	Al^{2+}, Na^+, K^+, Al_2^+;
10^{-14}–10^{-12} A:	H^+, Al^{3+}, Al_3^+, Al_4^+;
10^{-14} A:	Al_2O^+, Al_2OH^+.

Example 22.12. What kinds and abundance of *negative* ions are ejected in the same experiment? Negative ions appearing and their secondary currents (abundances) are as follows:

10^{-10} A:	none;
10^{-12}–10^{-10} A:	O^-, C_2^-;
10^{-14}–10^{-12} A:	C^-, Al^-, Cl^-, AlO^-, AlO_2^-.

The background information on ion sputtering in Section 22.4 now becomes important. In that section it was established (see Fig. 22.7 in particular) that ion sputtering reveals information about a layer extending down as far as the *escape depth*. For example, 5-keV Ar^+ ions show an average escape depth of 0.6 nm (6 A) in a silicon sample. In general, on ion bombardment only a few very energetic species escape from depths much in excess of the average depth.* Matrix

*At least two other modes of sputtering, use of electrical discharges such as a spark or arc, and use of a laser beam, are also available. Unfortunately, each has disadvantages relative to ion sputtering that preclude general use. With electrical discharges the position or the area of the zone sputtered can seldom be predicted with any certainty. With a laser beam, it is difficult to control depth of sputtering. Both modes are used in particular applications.

effects strongly influence sputtering yields of different species, as will be seen below. The relative yield of positive and negative ions by aluminum in Examples 22.11 and 22.12 can be generalized. Similarly, the much greater abundance of singly charged ions evident in the examples is characteristic of all samples. Electropositive elements show up more sensitively in positive SIMS and electronegative ones in negative SIMS. It should also be apparent from the examples that a SIMS determination in no sense provides a mass spectrum of a surface! Finally, molecules adsorbed at the surface can also be sputtered and sensitively determined by the technique, as will be discussed later.

Several types of surface analysis can be effectively done with SIMS. These include point-by-point

1. determination of elemental composition;
2. determination of the identity and concentration of adsorbed species;
3. determination of elemental composition as a function of depth; and
4. trace analysis.

In cases 1 and 2 above, analysis of the surface layer of a specimen alone will be sufficient. But in the types of analysis listed in cases 3 and 4, it will be important to be able to sputter large craters in a sample. These different requirements have over the years led to the development of two kinds of secondary ion mass spectrometry. In *static* SIMS low-energy ions (0.5–2 keV) in a wide beam (perhaps 1 mm in diameter) with low current density (10^{-9} A cm^{-2}) are used. Since minimal sample erosion (about 0.1 nm h^{-1}) occurs during a measurement, a "true" surface sample is taken. In *dynamic* SIMS energetic ions (5–10 keV) are used in relatively intense beams of high current density (10^{-6}–10^{-3} A cm^{-2}) and a crater is eroded in a specimen at the rate of 30–10,000 nm h^{-1}. Each will be examined later.

Characteristic Modules. In Fig. 22.24 a modular diagram of a secondary ion mass spectrometer is given. It should be noted that angles of ion bombardment and take-off are unspecified in the figure since they will be determined by the needs of a given analysis. But it is now appropriate to consider the modules needed.

In an ion gun a noble gas or other type of gas supplied at about 10^{-2} Pa (10^{-4} torr) is ionized by electron impact. The pressure in the sample module, however, must be much lower, 10^{-6}–10^{-8} Pa. An economical and practical solution is to provide *differential pumping* between the two regions. Yet pressures at the lower

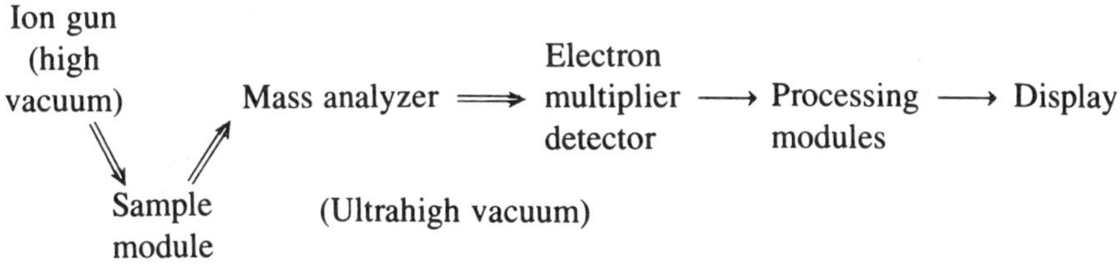

Fig. 22.24 Modular diagram of a secondary ion mass spectrometer.

end of the range can be approached only if components can be outgased by baking. In addition, measures should be taken to keep the sample chamber free of organic species when under vacuum. Such compounds tend to condense on a specimen and char under ion bombardment, leaving a carbon layer that causes uneven sputtering. Spurious peaks are also added to the spectrum! For this reason, an ion getter pump, as well as cryogenic baffling, is desirable.

Many ion guns also incorporate electrostatic deflection plates. These make possible precise placement of the beam at points of interest, such as on either side of a crystal (grain) boundary in a specimen. More importantly, deflection plates permit beam scanning. Ion guns to be used for elemental depth profiling usually require a more intense electron source and have ionizable gas furnished at higher pressure. Finally, for trace analysis, and certainly for higher mass resolution, it may be important to add a mass analyzer module to an ion gun. Good resolution will depend on removal of neutral species, which are of lower energy, and impurity ions, which will have a mass different from that of beam ions.

Once secondary ions appear, how efficiently can they be extracted from the cloud of particles sputtered out? Usually ion extraction is accomplished by placing an electrically charged conical shell of large entrance aperture near the specimen surface where the ion beam impacts. Secondary ions drawn into it are passed on to the mass analyzer. The higher its transmission, the more sensitive the measurements. In general, transmission and resolution must be traded off. A good value for a transmission factor is 5–10%.

Mass analyzers are of the several kinds described in Chapter 20, such as the single-focusing or double-focusing sector layouts and quadrupole types. The last kind has a large entrance aperture and is of simple, rugged design. As a result, it is widely used. Many instruments feature a single magnetic sector as mass analyzer (resolution 200–300). For higher resolution a double-focusing arrangement is required. Ordinarily the first stage is an electrostatic sector that sorts ions by velocity (energy) and the second, a magnetic sector. With this design resolution rises to 1000–5000. If mass defect differences are to be used to distinguish molecular or cluster ions of the same mass number but different composition, it is important to have a resolution of several thousand at minimum.

Processing modules appropriate to the type of readout will complete the secondary ion mass spectrometer.

Ion Microprobe and Ion Microscope. These special designs of SIMS deserve attention as well because of their versatility and sophistication. They provide scanning of a specimen surface in a raster pattern by a focused ion beam of 2–300 μm diameter. Their mass analyzer is held at a particular set of voltages and current to obtain the readout of a particular secondary ion. The device develops an indirect image in standard manner (Section 22.2). Naturally, a mass image can be developed for each isotopic species of interest. Alternatively, its ion probe can be directed toward an area of interest and present a mass spectrum of all species in that area. If a quantitative determination is of interest, areas of spectral peaks can be determined and appropriate calculations made.

Measurements and Analyses. At the outset values of ion-beam parameters that are appropriate to the type of SIMS measurement must be selected. Choosing values is often difficult because of the close interdependence of variables. In Fig. 22.25 the interrelationship of several factors such as sensitivity, sample erosion rate, ion beam current I_p, beam diameter d, and beam current density j_p is shown graphically. Note that sensitivity is directly proportional to the number of atomic layers removed from a sample per second. Let us see how the figure may be used.

Example 22.13. In a static SIMS analysis, sputtering should occur at a rate no higher than 10^{-4} monolayers s^{-1}. If a primary ion beam of 1-mm diameter is used, Fig. 22.25 suggests that a sensitivity of 100 ppm should be realizable. How can sensitivity be increased? Since the sputtering rate must remain minimal, a larger beam diameter and thus a larger current can be used if practicable. Use of a 1-cm diameter beam will increase sensitivity nearly 2 orders of magnitude.

Example 22.14. From Fig. 22.25 it is clear that to profile composition by depth or to perform a trace analysis calls for a similar choice of magnitudes of basic parameters. Re-

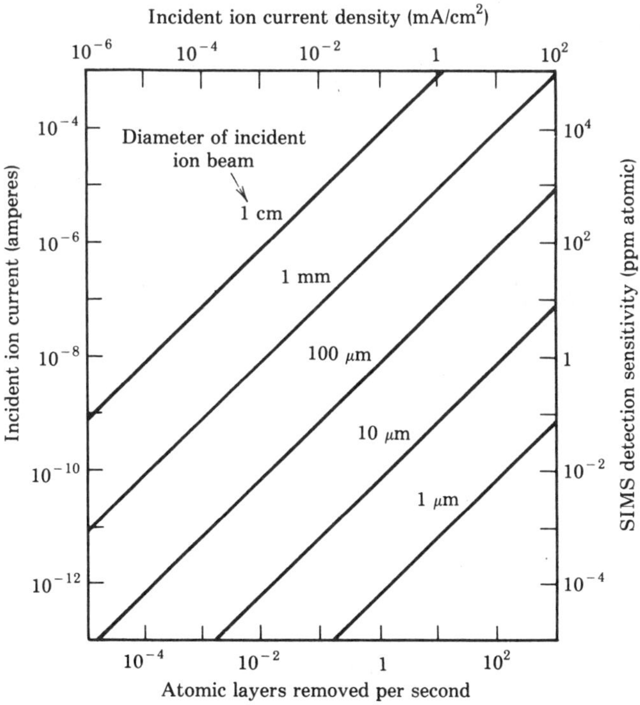

Fig. 22.25 Idealized relationships between several variables important in secondary ion mass spectrometry. The surface erosion rate for removing atomic layers is representative of the yield from a typical matrix. The detection sensitivity represents sensitivities found experimentally for a representative element-matrix combination [1]. (Courtesy of Elsevier Science Publishing Co.)

moving one atomic layer per second will give a sensitivity of 1 ppb for some elements if the primary ion current is 10^{-5} A and beam diameter is 1 mm.

Qualitative and at least semiquantitative analysis will be implicit in all SIMS measurements. Qualitative analysis in mass spectrometry was thoroughly discussed in Chapter 20, but attention needs to be given to differing elemental sensitivities and to so-called molecular SIMS. In terms of elemental quantitative analysis, it is important to be able to determine both positive and negative ions. As would be expected from usual ionization patterns, elements on the left side of the periodic table tend to form positive ions and those on the right, negative ions. Sensitivities vary dramatically from element to element and matrix to matrix. For this reason substitution of O^- or Cs^+ ion beams for the standard Ar ion beam may be valuable, as will be discussed below.

In *molecular secondary ion mass spectrometry* samples are deposited on a metal disk after it has been abrasively cleaned. The sample may be a solution or solid and must, of course, be pumped down before placement in the sample chamber. The method is particularly attractive because it provides a range of ionization mechanisms as will be seen in the following example.

Example 22.15. On bombardment of organic molecules by a low-intensity ion beam (5×10^{-8} A), some modes of ionization are:

(1) *Direct desorption* of existing cations or anions from the surface; for example, the desorption of methylene blue cation of mass 284 amu from a surface.

(2) *Cationization* or *anionization:* reaction of a sample molecule M with a cation or anion present at the surface followed by desorption; for example, the appearance of $(\text{Ag-1,10 phenanthroline})^+$ when phenanthroline is on a silver surface;

(3) *Electron ionization* of a sample molecule M at the surface to form M^+ or M^-; for example, the appearance of $(\text{butyrylcholine})^+$ from the parent molecule. Efficiency of these three processes decreases from top to bottom.

Previously intractable species as well as organic molecules of molecular weight up to nearly 1000 have proved susceptible to SIMS analysis. Fragmentation of ions also occurs in the gas phase according to mechanisms already worked out.

In approaching the challenges of quantification in SIMS the sputtering process must be described more quantitatively. The basic sputtering yield S is defined by the expression:

$$S = [(M^+ + (M^*) + (M^\circ) + (M^-) + (\text{molecules})]/(A^+),$$

where quantities in parentheses are concentrations of the indicated species and A^+ represents the bombarding ion species. The yield of a particular ionic species (e.g., M^+) is defined as $S_M^+ = (M^+)/(A^+)$. The quantity detected will be the secondary ion current I_s, which is given by $I_s = I_p S \beta^+ cf$, where I_p is the primary ion beam current, β^+ is the degree of ionization of species M^+, c is the concentration of the analyte, and f is the transmission factor of the mass analyzer and detector.

Because the degree of ionization for different elements can vary by five orders of magnitude, there are difficult problems in quantification in SIMS. Matrix problems add further complexity. Yet in recent years quantification has greatly improved. However the secondary ion current I_s of a species is measured, the average background current intensity must be subtracted, that is, $I_s = I_{peak} - I_b$. One comparative method that gives good results compares m/z analyte intensities for new samples with results from similar old samples. Another approach is to use *sensitivity factors*, that is, factors derived from reference samples whose composition is similar to that of the unknown. For a species i, one has the expression $c_i = I_i/\rho_i$ where I_i is the measured ion current for the species and ρ_i is its sensitivity factor. The method is tedious but gives accuracies of $\pm 5\%$.

Another approach corrects for matrix dependence of the yield of the particular ion species by means of parameters termed temperature and electron density (see Section 22.4). Ion currents are determined for two elements present in a sample of known concentration to determine the parameters.

Other Primary Ions. Varying the nature of the bombarding ion and increasing the beam accelerating voltage (to 15 keV) have proved valuable in enhancing secondary ion yields. In addition to argon ions, oxygen and cesium ions are often used. Oxygen will give either O_2^+ or O^- ions (from a duoplasmatron by varying the sign of the extraction voltage), and mass filtering of the ion beam will allow isolation of a desired isotope. Cesium, which has a single natural isotope, is supplied by a source that vaporizes the low-melting metal and ionizes and accelerates the Cs^+ ions produced. With Cs^+ bombardment high-sensitivity SIMS analyses are possible on the basis of negative ion yield (M^-) of a variety of elements detected as positive or negative ions with lower sensitivity by argon or oxygen analysis: carbon, hydrogen, the noble metals, and Groups 15(VA), 16(VIA), and 17(VIIA) [H. A. Storms, K. F. Brown, and J. D. Stein, *Anal. Chem.*, **49,** 2023 (1977)].

The *limit of detection* for an element, c_L, can be shown to be $c_L = I_L/fS_M^+I_p$. If trace analyses are to be performed frequently, it is advantageous to consider the possibility of adding an electron multiplier detector. While I_p might also be increased, the erosion rate may be higher than would be feasible in analysis of a thin film. In favorable cases, as little as 10^{14} atoms cm^{-3} of some elements can be detected.

Statistical fluctuations in measured ion currents, erosion rates, and in other parameters will limit the precision of quantitative analysis quite apart from any considerations having to do with matrix problems.

In *depth profiling* it is important to maintain both I_p and j_p constant. Often a wide stationary ion beam is used to sputter out a crater. For good depth resolution extraction of secondary ions is arranged to be primarily from the central area of the sputtered crater. A full discussion of depth profiling has been given in Section 22.4.

22.10 COMPARISON OF METHODS

After examining several different surface spectrometric techniques some guidelines are needed to identify the technique or set of techniques most appropriate to answer the questions about a particular sample. Three guidelines suggest themselves. One should

1. Weigh the relative strengths and weaknesses of each method.
2. Consider applying at least two different methods to take advantage of their complementary and independent nature.
3. Weigh the relative merits of different surface spectrometric instrumentation, which increasingly means choosing a surface analytical system.

The last and first of these aspects will now be addressed briefly. We shall look at the organizing principles used in designing some surface analysis systems and then summarize the main strengths and limitations of the techniques that have been described.

While spectrometers dealing with individual techniques are widely used, it has become attractive to combine several capabilities for surface study in an integrated system. This approach is attractive because there has been increasing demand to examine samples by several techniques to obtain a composite perspective on the character of their surface. Note two other advantages: such a system ensures both greater utilization of the necessary but costly high vacuum system and particle energy analyzer and the possibility of addition of still other surface capabilities less expensively.

Several systems have appeared: the augmented scanning electron microscope, the generalized surface analysis system, the electron probe microanalyzer, and the ion probe microanalyzer. The organizational principle informining each is worth describing briefly.

The rationale for adding spectrometric capability to a *scanning electron microscope* is that many kinds of interaction go on between an electron beam and specimen beyond that required for a microscope image. For example, it is not difficult to add a solid-state energy-dispersive detector to measure X-ray fluorescence and a particle energy analyzer to measure the electron energy loss characteristics of a sample. Adding the processing modules is of course required as well. Occasionally, capability for low-energy electron diffraction (LEED) is also provided. More often, Auger electron and/or secondary ion mass spectrometric capability are added. For the last method an ion gun and quadrupole mass analyzer are the key elements. It is attractive to be able to correlate a SIMS elemental analysis with its good lateral resolution with a microscope image.

Though not as versatile, the *electron microprobe* just discussed and the *ion microprobe* offer scanning and elemental mapping capability and qualifies them for inclusion as surface spectrometric systems. They provide intense particle beams and accompanying high elemental sensitivity.

Several characteristics of an electron probe microanalyzer are compared with

Table 22.4 Characteristics of Scanning Electron Microscopes and Microprobe[a]

Characteristic	*Scanning Electron Microscope*	*Electron Microprobe*	*Analytical Electron Microscope*
Electron image resolution	7–10 nm	7–10 nm	0.2 nm (line), TEM 2 nm (STEM)
Elemental analysis technique(s)	X-ray	X-ray	X-ray, energy loss
Resolution of X-ray map	~1 μm	~1 μm	~50 nm
Minimum detectability limit (wt%)	~0.1%	~0.1%	~0.1%
Minimum detectable mass (g)	10^{-14}	10^{-15}	10^{-19}
Accuracy	—	±1% (rel.)	±5–10% (rel.)

[a]After D. B. Williams and J. I. Goldstein, *Am. Lab.*, December 1981.

those of an analytical electron microscope in Table 22.4. Note that resolution is much higher for a microscope than for a microprobe, but that the elemental analysis provided by an electron probe microanalyzer (by X-ray fluorescence) is surprisingly accurate (±1% relative). Such analyzers incorporate good computational ability to obtain high-quality quantitative analysis.

The collection of techniques called surface analysis systems provide the most important techniques in a single instrument. In spite of cost, their versatility and comprehensiveness in garnering information about a specimen strongly commends their use. The particular advantages are that a costly high-vacuum system and a first-class particle energy analyzer are available to all the methods. Nevertheless, their versatility is sometimes gained at the expense of being able to optimize the arrangement for any single surface spectrometric technique.

All methods are sensitive to the elements lithium to uranium, and secondary ion mass spectrometry also detects hydrogen. The capabilities of the methods are compared in Table 22.5 and in the individual appraisals of strength and limitations that follow.

X-Ray Photoelectron Spectrometry (XPS). This method features several strengths. First, its X-ray exciting beam is needed only for analysis. As a result, its ion gun (usually available) can be used exclusively for cleaning surfaces and depth profiling. At any time the sputtering action can be interrupted and XPS used to study details of sample composition. Another important advantage of XPS is that its photoelectron peaks are relatively free from interferences, that is, matrix effects. For that reason the technique has especially good capability for determining general chemical information about surfaces.

The limitations of XPS are several. The method has poor lateral resolution. Its exciting X-ray beam can be focused somewhat only with considerable loss of energy from its already relatively low intensity. Reduction of beam size by stopping it down to a small area improves resolution but also lowers S/N. The larger beam

Table 22.5 Intercomparison of Analytical Capabilities of Several Surface Spectrometric Methods[a,b]

	Characteristic	*AES*	*XPS*	*ISS*	*SIMS*	*RBS*
EL E MEN TAL	Elemental range;	Li and higher Z;	Li and higher Z;	Li and higher Z;	H and higher Z;	C (high Z on low Z)
	Specificity	Good	Good	Variable[c]	Good	
	Quantification	Semiquantitative from peak height; quantitative with standard or correction model			Standard or correction model necessary	Quantitative (from peak height)
A NAL Y SIS	Detection limits (atomic fraction)	10^{-2}–10^{-3}	10^{-2}–10^{-3}	10^{-2}–10^{-3}	10^{-3}–10^{-8} (quite variable)	10^{-2} (surface) 10^{-4} (bulk)
	Depth resolution/nm	0.3–2.5	1–3	0.3–1	0.3–2	10–2000
	Lateral resolution, μm	0.05	~1000	100	1	100
COM POUND	Identification	(Restricted)	Yes	No	Yes	No
	Structural information	No	No	Yes	No	Yes
A NAL Y SIS	Organic samples	No	Yes	No	Yes	No
	Insulator samples	Charge compensation	Easy	Possible with charge compensation		Easy
	Destructiveness of method	Small	Minimal	Small	Static: small Dynamic: large	Small

[a]After H. W. Werner in *Electron and Ion Spectroscopy of Solids*, L. Fiermans, J. Vennick, and W. Dekeyser (Eds.). New York: Plenum, 1978; C. J. Powell in *Applied Electron Spectroscopy for Chemical Analysis*, H. Windawi and F. Ho (Eds.). New York: Wiley-Interscience, 1982.

[b]Identifications: AES, Auger electron spectrometry; XPS, X-ray photoelectron spectrometry or ESCA; ISS, ion scattering spectrometry; SIMS, secondary ion mass spectrometry; RBS, Rutherford backscattering spectrometry or HEIS.

[c]Specificity depends on ratio R where R = mass target atom/mass probing ion; specificity is good if $R < 3$ and poor if $R > 3$.

diameter also makes depth profiling less reliable since a sizable crater must be formed. If adsorbed species are of interest, care should be taken to be sure they are not desorbed by X-ray bombardment. Any hydrocarbon vapors present (from an oil-diffusion pump) may polymerize on the surface of a sample and carbon buildup may occur under the action of X rays, leading to unrepresentative XPS spectra.

Auger electron spectrometry (AES). This technique also offers the advantage of an ion beam that can be limited to providing cleaning and depth profiling without affecting analytical measurements. A further strength is the capability of fine focusing of its electron beam and consequent high lateral resolution in elemental analysis. Also, AES is fast, in part because its electron beam is as many as six orders of magnitude more intense than the photon beam in XPS. This high intensity also ensures that an Auger spectrum can be displayed in real time. Finally, with scanning Auger capability images of a surface can be displayed for different analytes.

Some limitations are the following. In an Auger spectrometer it is possible that the ion beam used for sputtering and the electron beam may strike the surface at different angles, leading to possible spurious results. Also, if the electron beam strikes at too low an angle, surface roughness may lower the S/N ratio.

Electron bombardment may also cause surface compositional or bonding changes. Adsorbed species can seldom be studied by AES since they tend to desorb under electron bombardment. Samples are avoided that are known to be destabilized or degraded by intense electron beams, such as certain ionic substances like NaCl and KCl and various oxide surfaces. The high flux of the beam is also likely to lead to local heating and to changes in diffusion profiles. Finally, charging effects are more serious. There is no simple experimental strategy that offers help. If the whole spectrum is shifted by a determinable amount along the abscissa, a correction can, of course, by made. Referencing peaks have also been suggested.

Ion scattering spectrometry (ISS). This method offers the strong advantage of sensitivity to *the* top layer of atoms in a specimen. Relative to XPS and AES it can provide detailed information about the location of atoms. Thus, ISS is an effective technique for the study of adsorption of the first monolayer on a surface. Its beam intensity can be kept low enough to minimize desorption. It is, nevertheless, one of the slower techniques, offers poor lateral resolution, and has generally low sensitivity.

Secondary ion mass spectrometry (SIMS). This technique has strength in many areas. It offers parts per million detection limits for many elements and per billion for a few. Its isotopic and mass resolution is limited only by the performance of its mass analyzer. The technique is fast since response is obtained as soon as sputtered ions are collected. Microanalysis is possible since the ion beam may be focused to a narrow point.

Some weaknesses of SIMS are that the wide variation in ionization efficiency of elements causes sensitivity to vary over 4–6 orders of magnitude across the periodic table. One way to reduce this variation is by use of oxygen ion or cesium bombardment since these ion beams increase sensitivity for many elements. When oxygen is used, sputtering yields generally are smaller. To some degree the large variation in ionization efficiency has been overcome by use of either a glow discharge or a rf excited plasma discharge as the means of ionizing some of the neutral species ejected. Both methods do reduce the range of sensitivity as well as matrix effects. Matrix effects are a serious problem in quantifying data.

REFERENCES

General volumes on an intermediate level are:

1. A. W. Czanderna (Ed.), *Methods of Surface Analysis.* New York: Elsevier, 1977.
2. D. P. Woodruff and T. A. Delchar, *Modern Techniques of Surface Science.* Cambridge: University Press, 1986.

Volumes dealing with one or more methods are:

3. A. Benninghoven, F. G. Rudenauer, and H. W. Werner, *Secondary Ion Mass Spectrometry: Basic Concepts, Instrumental Aspects, Applications and Trends.* New York: Wiley, 1987.
4. C. R. Brundle and A. D. Baker (Eds.), *Electron Spectroscopy: Theory, Techniques and Applications*, Vol. 1, New York: Academic, 1977.
5. J. H. D. Eland, *Photoelectron Spectroscopy.* New York: Wiley, 1974.
6. L. Fiermans, J. Vennik, W. Dekeyser (Eds.), *Electron and Ion Spectroscopy of Solids.* New York: Plenum, 1978.
7. J. R. Fryer, *The Chemical Applications of Transmission Electron Microscopy.* New York: Academic, 1979.
8. P. K. Ghosh, *Introduction to Photoelectron Spectroscopy.* New York: Wiley, 1983.
9. J. I. Goldstein, D. E. Newbury, P. Echlin, D. C. Joy, C. Fiori, and E. Lifshin, *Scanning Electron Microscopy and X-Ray Microanalysis: A Text for Biologists, Materials Scientists, and Geologists.* New York: Plenum, 1981.
10. D. M. Hercules and S. H. Hercules, Analytical chemistry of surfaces, Parts I, II, III, *J. Chem. Educ.*, **61,** 402, 483, and 592 (1984).
11. D. C. Joy et al., *Principles of Analytical Electron Microscopy.* New York: Plenum, 1986.
12. D. E. Newbury et al., *Advanced Scanning Electron Microscopy and Microanalysis.* New York: Plenum, 1986.
13. M. Thompson et al., *Auger Electron Spectroscopy.* New York: Wiley-Interscience, 1985.
14. H. Windawi and F. F. L. Ho (Eds.), *Applied Electron Spectroscopy for Chemical Analysis.* New York: Wiley-Interscience, 1982.

A lucid volume covering most methods on an intermediate to advanced level is:

15. G. Ertl and J. Kueppers, *Low Energy Electrons and Surface Chemistry*, 2nd ed. (1st ed. under another publisher) Weinheim, West Germany: VCH, 1985.

Also of interest are:

16. C. J. Powell, Recent progress in quantification of surface analysis techniques, *Appl. Surf. Sci.*, **4,** 492 (1980).
17. J. T. Yates, Jr. and T. E. Madey (Eds.), *Vibrational Spectroscopy of Molecules and Surfaces. New York: Plenum, 1987.*

EXERCISES

22.1 Demonstrate that resolution is improved for the hemispherical particle energy analyzer in Fig. 22.13 by having a retarding grid at its entrance. Assume electrons arrive with energies of 2 keV but are slowed to energies of 20 eV by such a grid. What are the relative resolutions achieved with and without the grid?

22.2 In Section 22.5 use of a thin aluminum window is reported to ensure substantially monochromatic radiation for an X-ray tube with aluminum or magnesium anode. Refer to Sections 21.2 and 21.4 and sketch (a) the physical arrangement of tube, window, and so on, (b) the X-ray emission spectrum produced by an Al anode, (c) that emission spectrum after passing through an Al window (Al–K absorption edge 1562 eV and 0.795 nm), and (d) the emission spectrum of a magnesium anode X-ray tube after filtering by a thin aluminum window. (e) Discuss the results.

22.3 What limiting resolution should be attainble in XPS if aluminum $K\alpha_{1,2}$ X-ray exciting radiation is used?

22.4 Both electron and photon beams produce Auger electrons, yet Auger electron spectrometers always use electron gun sources. What advantages does electron excitation offer?

22.5 Compare the energy distribution $N(E)$ versus E curves in Figs. 22.3 and 22.23. Interpret the different regions of the $N(E)$ curve in the latter figure as completely as possible in terms of processes experienced by electrons. What instrumental factors contribute to the width of peak C in Fig. 22.3?

22.6 Suggest an electrostatic argument for the greater abundance of singly charged secondary ions as compared with doubly charged ions in SIMS.

22.7 In Examples 22.11 and 22.12 relative abundances were given for ionic species generated in secondary ion mass spectrometry (SIMS). No account was taken of natural isotope ratios and abundances. (a) On the basis of Example 22.11 list of all the different positive monatomic and diatomic ions that should appear. Give the total mass number and a rough relative abundance for each. (b) How much more complex would a SIMS spectrum of a chromium sample be likely to be, assuming similar behavior?

Chapter 23

METHODS USING RADIOISOTOPES

23.1 INTRODUCTION

More than 1000 unstable, that is, radioactive, isotopes are now known. Their nuclei have excess energy and degrade through the process of radioactive decay to a lower energy level at a characteristic rate. Usually the excess energy is discharged by the ejection of a highly energetic particle, such as an α or β particle. In many cases electromagnetic radiation of the γ-ray type also appears, either accompanying the particle or separately. The instrumental interest in radioactivity centers in the contrast between chemical behavior, which relates to the orbital electrons, and nuclear instability. In so far as the chemical properties are concerned, nuclear instability generally makes very little difference; at the moment of the disintegration of each radioactive atom, however, both its presence and location (within limits) can be instrumentally determined.

As a result of this combination of properties, unstable isotopes play an extremely important role as tracers. They are ordinarily added to a system in the chemical form most suitable to the study. For example, ^{32}P is often added as $(NH_4)_3\ ^{32}PO_4$. In what may be called a physical application such as the testing of engine wear, the tagged or labeled material moves with a given phase or substance and serves to establish its location and concentration. In a chemical application such as a study of the mechanism of a reaction, the tagged molecules undergo reactions. In this instance both the level of radioactivity in each product and the molecular location of the radioactive atoms after the change are of interest.

Under favorable conditions, radioactivity may be determined with extreme sensitivity; as few as several thousand radioactive atoms are detectable in many cases. Even under ordinary conditions, radiochemical techniques provide the possibility of analyzing concentrations as low as 10^{-10} to 10^{-14} mol mL^{-1}. Another consequence of the ease of detection is that no more than a small quantity of radioactive material, which is usually expensive, is required for studies. Since the widely used radioactive isotopes emit β and γ radiation, the discussions that follow concentrate on β- and γ-ray techniques.

23.2 ACTIVITY LEVEL AND DECAY RATE

We define the activity or decay rate of a substance in terms of the number of atomic disintegrations occurring in the bulk per second. A *curie* (Ci) is the quantity of a radioactive substance that gives rise to 3.700×10^{10} disintegrations per second. (Earlier the curie was considered the quantity that gave the same number of disintegrations per second as 1 g of radium.) In practice this unit is inconveniently large and is displaced by the metrically related *milli*curie and *micro*curie. For example, tracer studies are carried out at the μcurie level. The SI unit based on one disintegration per second is called the becquerel (Bq).

The rate at which radioactive atoms decay is directly proportional to the number N of such atoms present. The rate of disintegration $-dN/dt$ is then defined by the expression

$$-\frac{dN}{dt} = \lambda N, \tag{23.1}$$

where λ is the *decay constant* for the particular species. For Eq. (23.1) to be valid, we must assume that the number of atoms disintegrating in a unit of time is small compared with the total number of atoms present. According to the equation, λ is just the fraction of the total number of atoms that decay per unit time. Interestingly, the decay constant has been found invariant to all ordinary changes in conditions such as temperature, pressure, electric and magnetic fields, and, of course, even to chemical change.* When we integrate Eq. (23.1) between N_0, the number of active atoms initially, and N, the number present at time t, we obtain

$$\ln(N/N_0) = -\lambda t \tag{23.2}$$

or $N/N_0 = e^{-\lambda t}$. Since radioactive decay is a unimolecular process, Eq. (23.2) has the form of the rate law for a first-order reaction. Conventionally, the rate of disintegration is given in terms of the *half-life* $t_{1/2}$, the time required for a large number of atoms to decay to half the original number. Substituting this condition, $N/N_0 = 1/2$, in Eq. (23.2), one obtains $\ln 1/2 = -\lambda t_{1/2}$. Thus,

$$t_{1/2} = \ln 2/\lambda \approx 0.693/\lambda. \tag{23.3}$$

In work with radioisotopes it is actually not the number N_i of radioactive atoms of species i present at a particular time that is measured but a proportional quantity A_i, the *activity* of the species. It is related to N_i by the equation $A_i = c\lambda_i N$, where the coefficient c depends on the efficiency of measurement. It is usual to treat activity data by making observations over a period of time and then plotting the logarithm of A versus the time. If we have maintained conditions constant so that the detection efficiency has not varied, we may easily obtain the value of the half-life from this graph by inspection. Then λ can be evaluated as suggested above.

*Only the type of nuclear decay that results in K-shell electron capture can be influenced. Since an electron outside the nucleus is captured, any chemical alteration that influences the electron density near the nucleus does affect the disintegration rate. This change is an infrequent type of decay.

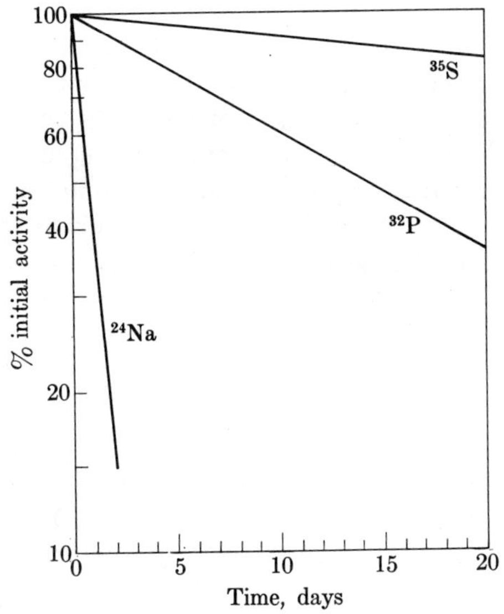

Fig. 23.1 Decay curves for three radioisotopes.

Figure 23.1 shows plots of this type, called *decay curves*, for three representative isotopes used in tracer work.

When several independent activies are present, the total activity is the sum of the separate activities $A = A_1 + A_2 + \cdots$. The activity for the mixture may be plotted as in Fig. 23.2. In this case, we always obtain a curve that is concave upward. The shorter-lived components contribute less to the total disintegration as

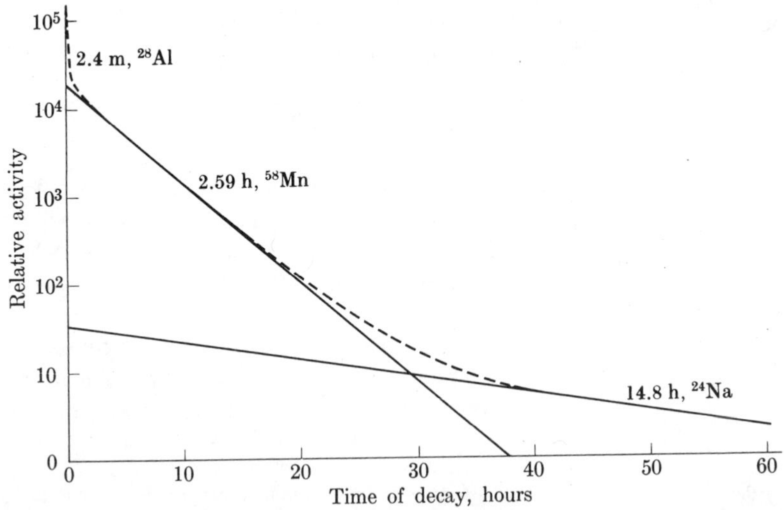

Fig. 23.2 Decay curve (broken line) for mixture of three independently disintegrating activities. The sample is a neutron-activated Mn–Al–Na alloy. The nuclides and half-lives are indicated on the curve.

time passes. After a time equal to several half-lives of the more active components, the longest-lived species predominates, and the curve is a straight line. From the linear region, the half-life of the last substance to decay can be found by inspection. Then if we extrapolate this portion to $t = 0$ and subtract it from the original curve, the residual curve should represent the decay of the other components. The new curve should also display a linear region, allowing the determination of the half-life of the next shorter-lived isotope. In principle, it should then be possible to repeat the process indefinitely. In practice, the resolution is seldom adequate to find $t_{1/2}$ by this procedure for more than three species. Figure 23.2 gives a good illustration of the method.

Example 23.1. When one measures the radioactivity level of a sample containing more than one radioisotope, one obtains only a total measurement. A graphical way to obtain the activities of the component radionuclides was suggested in Fig. 23.2. Is there a mathematical method for doing so? Take the case of a sample containing nuclides 1 and 2 with decay constants λ_1 and λ_2.

Clearly, at time t, the total sample activity A can be represented as $A = A_1^0 e^{-\lambda_1 t} + A_2^0 e^{-\lambda_2 t}$. The equation can be simplified for nuclide 1 by multiplying through the expression by $e^{\lambda_1 t}$ to obtain

$$Ae^{\lambda_1 t} = A_1^0 + A_2^0 e^{(\lambda_1 - \lambda_2)t}.$$

Since λ_1, λ_2, and the value of A at time t are known, the unknowns remaining are $Ae^{\lambda_1 t}$ and $e^{(\lambda_1 - \lambda_2)t}$. The equation defines a straight line whose intercept is the activity A_1^0 and whose slope is A_2^0.

Decay Schemes. Each type of radioactive nucleus, hereafter called a *radioactive nuclide* or simply *radionuclide*, exhibits a distinctive decay pattern. Beta and γ decay are the most common types of decay. Alpha emission is really probable only for the nuclides of elements above lead ($Z > 82$). For most nuclei, there is still excess energy after a particle is ejected and one or more γ photons are also emitted before radioactive decay ceases. For example, a ^{24}Na nuclide (see Table 23.2) decays by first emitting a β particle and then releasing in sequence a 2.76-MeV and a 1.38-MeV γ photon. Indeed, except for very light nuclides, equally or more complex decay patterns are the rule: there appear to be few simple disintegration patterns such as a single β emission. Their complexity has unfolded gradually as instrumentation has become more refined and better measurements have become possible. It should be noted also that the product of many disintegrations is another radionuclide. Notable examples are the isotopes of naturally occurring radioactive elements, which fall into chains of successive decays called families or series.* Whenever a product of a disintegration is also unstable, the activity of the daughter nuclide will be measured along with that of the parent.

*The parent radionuclides of the families are ^{238}U, ^{232}Th, and ^{235}Ac; each undergoes α emission to start its family line. Members in the first chain have mass numbers given by $4n + 2$, where n is an integer, the second by $4n$, and the last by $4n + 3$.

For alpha and gamma decay, particles are emitted with definite energies characteristic of the radioisotope. By contrast, in beta decay (the emission of either electrons or positions or the capture of an electron) energies for a given radioisotope fall in a characteristic, continuous distribution from zero to a characteristic *maximum* value. The energy ΔE expected from the initial and final nuclear states is shared in variable fashion between a beta particle and a neutrino.

23.3 INTERACTION OF RADIATION WITH MATTER

Nuclear radiations interact strongly with matter by virtue of their high energies. The nature of the process and the distance of travel in a medium before the energy falls to the thermal level vary with the mass, charge, and energy of the radiation as well as with the same properties of the medium.

Ionization and scattering are the most common kinds of interaction with matter. For the former, on collision with a molecule an electron is ejected and a positive fragment left behind: the result is an ion pair. The energy of formation of an ion pair in air averages about 35 eV. In almost all other gases the figure is smaller, ranging down to about 20 eV. Usually, less than half this energy goes into the actual ionization; the remainder becomes the kinetic energy given the ejected electron. A small percentage of these secondary electrons have energies greater than the ionization potential of the medium and act to produce further ionization. Indeed, some 60–80% of the total ionization created by α or β particles results from the collisions of energetic *secondary* electrons.

Ionization. *Alpha* particles [${}_2^4He^{2+}$] have high kinetic energies (3–8 MeV). Their paths in matter are uniform and short (about 6 cm in air and 2 μm in lead). In air in alpha particle produces from 30,000 to 50,000 ion pairs per centimeter. *Beta* particles [electrons or positrons] have a characteristic *maximum* energy that ranges widely (from 18 keV for ${}^3H$ to 4.81 MeV for ${}^{38}Cl$), have longer and irregular paths in matter, and produce 100–700 ion pairs per centimeter in air. *Gamma* rays [photons] produce from 1 to 70 ion pairs per centimeter in air and they travel very long straight paths. Ionization results from the photoelectric effect, which is most commonly shown by gammas that have lost a large part of their initial energy. *Neutrons* range widely in energy, produce negligible ionization, and travel paths of variable lengths.

Scattering. For *alpha* particles scattering by molecules and atoms is small but increases with target atomic number Z. *Beta* particles are effectively scattered by electron clouds of atoms, an effect that also increases with Z. They also produce X rays with some probability at energies <1 MeV and for atoms of $Z > 40$. For γ *rays* Compton scattering in which part of the gamma photon energy is transferred to an orbital electron in the path is common. Pair production (formation of a positron and electron) by an γ-ray photon occurs occasionally when the energy is greater than 1.02 MeV. *Neutrons* are scattered both elastically and inelastically. Capture by a nucleus is their common fate.

23.4 DETECTORS

The qualitative and quantitative measurement of nuclear radiation depends on its interaction with the matter in a detector. Nearly all detectors for energetic particles and photons were introduced and described earlier in Sections 21.5 and 22.5. The reader will want to examine that material carefully. Here it will be extended appropriately to reflect the fact that in studies with radioisotopes precision and sensitivity in counting will be the main focus. For example, for sensitive measurements it is appropriate to use detectors with large solid angles of viewing (4π maximum). Further, for accurate measurement of weakly radioactive samples background must be reduced drastically. Some properties of different radioactivity detectors are compared in Table 23.1. Detectors such as neutron detectors and others needed in the study of nuclear energy levels and nuclear reactions are omitted as beyond the scope of the text.

Gas Ionization Detectors. The proportional counter and Geiger–Mueller counter are commonly used nuclear radiation detectors whose response is based on the ionization of a gas. The *proportional counter* or *detector* offers versatility and precision and is employed both as a closed-tube type and as a flow counter. To achieve good sensitivity, the detector is usually designed with at least 2π-geometry, that is, planned so that its active volume will intercept 50% of the radiation emanating from a thin sample. A commercial flow counter is pictured in Fig. 23.3.

In this windowless or thin-window design a continuous stream of counter gas is allowed to flow, and the device is operated at atmospheric pressure or slightly above. Volatile samples such as those containing $^{14}CO_2$ may be carried into the counter on the gas stream. On the other hand, solid or liquid samples are mounted

Table 23.1 Characteristics of Some Nuclear Radiation Detectors

Type	Recovery Time (μs)	Required Electronic Amplification	Efficiency[a]		
			α	β	γ
Proportional, flow type[b]	1	Medium	High	High	Low
Geiger–Muller tube[c]	100–300	Low	High	High	High
Semiconductor Ge(Li) crystal	5	High	0	0	Medium
Scintillation					
ZnS–Ag phosphor	NA	NA	Medium	0	0
Liquid	10^{-3} to 1	Medium	0	High	Low
NaI (T1) crystal	10^{-3} to 1		0	0	Medium

[a]Given for the types of emanation listed.
[b]Methane fill gas, metalized Mylar window of 0.15–0.9 mg cm^{-2} thickness.
[c]Halogen-quenched, mica window of 1.4 mg cm^{-2} thickness.

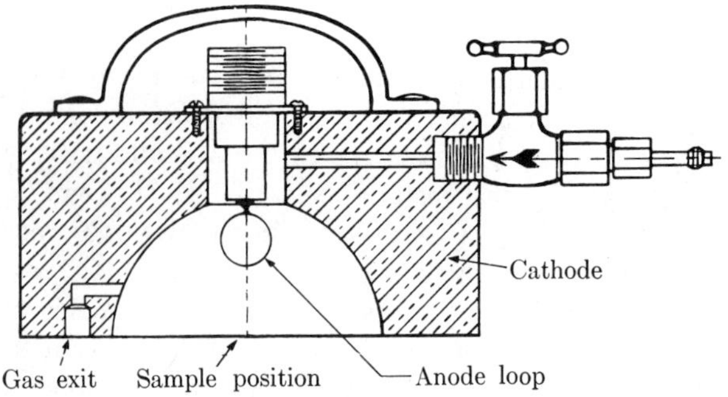

Fig. 23.3 Cross section of the upper half of a windowless flow counter. Note the gas entrance valve, the gas exit at the lower left, the hollow hemispherical cathode, and the anode in the form of a loop of thin wire. In use, a nearly identical lower half is seated firmly against the section shown to make a counter with 4π geometry. Samples are placed in the center so that the counter responds to virtually all radiation emitted.

on a slide and introduced either through an airlock (if the counter is windowless) or directly under a thin aluminum or aluminum-coated plastic (e.g., Mylar) film. The flow of gas minimizes the effect of leaks and gives the tube an indefinite lifetime. The filling gas may be a pure gas such as methane, but is more commonly a mixture such as 90% argon and 10% methane.

Since a constant amplification factor is essential in precise work, care is taken in the construction of proportional counters to ensure that the field is uniform along the length of the anode. Usually, the terminals of this electrode are shielded by some sort of guard rings (separate electrodes at the same potential) or, as in this case, a loop of wire is used to avoid field inhomogeneities that would otherwise occur at the ends of the wire. A more serious need is that the power supply be well regulated; as Fig. 21.11 indicates, a small change in detector voltage in the proprotional counting region will alter the amplification factor significantly.

Under the conditions of voltage and gas pressure in the counter the avalanche of ionization is largely confined to a small cylindrical volume around the radiation track. As a result, only a small fraction of the tube volume is inactivated by the ionization. The recovery time need be no longer than the time *necessary to collect the electrons*, about 1 μs. Even though the positive ions formed must be collected at the cathode, it is unnecessary to wait 100–200 μs for this to happen, for all portions of the gas around the anode except the sheath around the radiation tracks are still sensitive to radiation. The output pulses of the counter are small and pulse-shaping amplifiers are necessary to develop pulses of magnitude sufficient for stable counting. There are both flow and fixed-volume varieties of proportional counter.

At a given voltage setting the heights of α-particle pulses (about 5 MeV) will be large compared with those of β- and γ-particle pulses ($\leq$ 1.5 MeV). Either α- or β-γ radiation may be counted separately by discriminating against the other type electronically, as discussed in the next section. Usually, the counter voltage is adjusted so that the type of radiation of interest gives pulses of acceptable height.

For example, a representative proportional flow counter may be stably operated at 900 ± 50 V for counting α's and at 1300 ± 50 V for β-γ's.

Since the detector offers high internal amplification and the sample may be introduced directly into the active volume of the counter, low-energy radiations of all types or very low sample activities may be determined. For example, tritium ^3H (with its 0.018 MeV β-ray maximum energy) is often counted with a proportional device. Proportional counters also have quite low background. Only those background radiations in a small energy range are registered, with the result that the background count may be about half that for a Geiger detector.

In addition, high activity levels may be handled with a minimum correction of the data. The resolving time of the counter is essentially its recovery time, 1 μs. Although electronic amplifiers capable of equally rapid operation are available for use where high precision is required, amplifiers with resolving times of 5 μs are ordinarily used. With either type, several hundred thousand counts per minute can be registered with less than 2% coincidence loss (see Section 21.8).

An especial advantage is the linear correlation between pulse height and the energy of the radiation. By virtue of this property, proportional counters lend themselves readily to analysis of the energy spectrum of a radiation source.

The *Geiger–Mueller counter* enjoys general use but is a more limited ionization detector. A simple design of Geiger–Mueller tube is shown in Fig. 23.4a. With a thin mica window of 1-mg cm^{-2} thickness the tube will admit all but the lowest-energy β rays, such as those produced by ^3H. Many modifications of this simple design are available for different applications, but all retain the coaxial electrode arrangement shown in the figure. As shown in Fig. 23.4b, Geiger operation occurs on a voltage plateau at voltages higher than 1 kV and the potential difference need

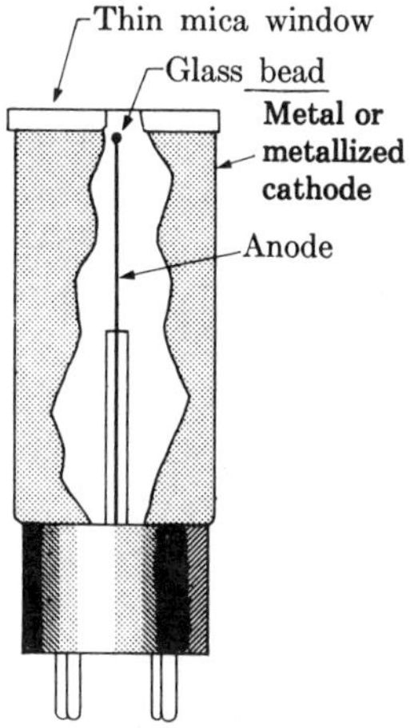

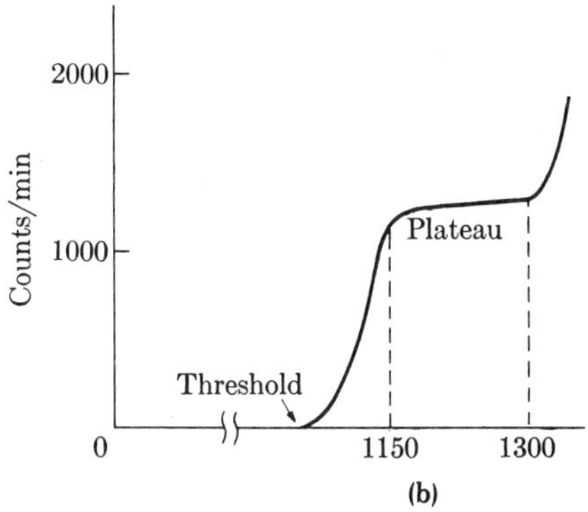

Fig. 23.4 A Geiger–Mueller counter and its response curve as a function of potential difference. (a) Construction of an end-window Geiger counter. The thin mica window allows ready entrance of nuclear radiation. (b) Characteristic response curve, obtained by observing a given radioactive sample with a fixed geometry.

be only moderately well regulated to obtain reproducible readings. Voltages still higher than the operating range are to be avoided, however, since tube damage is quite likely. As before, nuclear particles create ion pairs in the detector. At these voltages photons resulting from vigorous electron inpacts or recombinations of ion pairs spread ionization throughout the entire active volume. The resulting degree of amplification causes the height of output pulses to be determined solely by *conditions in the tube*. The filling gas is commonly a mixture that consists of about 90% inert gas such as argon and up to 10% polyatomic gas such as ethanol or a halogen, which acts as a discharge quenching agent. In return for a more robust character, what is lost in a Geiger tube, since all output pulses are essentially the same height, is proportionality between pulse height and particle energy.

Geiger tubes are mainly used today for laboratory monitoring and for many measurements involving radiotracers in which chromatographic or electrophoretic methods are used to separate components. Subsequent scanning produces plots of the distribution of radioactivity. In spite of their sensitivity, lower cost, and the simplicity of associated processing equipment, their use is limited because they are inherently less reproducible, have too long a recovery time for efficient counting at rates above 10,000 counts min^{-1}, and have little versatility.

Scintillation Detector. The scintillation counter is a widely used detector for α and β emitters. It consists of a fluorescing agent (scintillator) and photomultiplier tube together with associated electronic circuitry. The fluorescer is either a crystalline or a liquid material. Since the number of photons released per nuclear particle is moderate, the output must be amplified. Accordingly, the scintillator is coupled to the photocathode face of a multiplier phototube either directly or by an optical coupler. The scintillator and phototube face must, of course, be transparent to the fluorescence, which is usually in the blue or near UV. It is also essential that the scintillator volume be large enough to absorb all the energy of the ionizing radiation so that its output is proportional to energy. This detector offers a fast response to radiation (rise times from 10^{-9} to 10^{-6} s) and an output proportional to particle energy. Disadvantages are its poor resolution (about 30 eV is consumed for each photopeak as a result of the two-stage process) and high background (50 cpm cm^{-3} of scintillator). What adaptations are desirable for different kinds of radiation?

Alpha-particle detection may best be accomplished by using a very thin (1 mm maximum) layer of transparent *scintillation phosphor*. It has been found satisfactory to use a transparent ZnS screen with silver as an activator.

For β-particle counting liquid scintillation systems are commonly used, though single crystals of anthracene, transstilbene, or NaI (Tl-activated) are also employed. Because the sample is dissolved or suspended in the scintillator medium, the system offers a geometry that is especially advantageous for detection of low-energy beta particles from isotopes such as carbon-14 and tritium. The liquid scintillator has two main components, a solvent (usually toluene or *p*-xylene, though dioxane is valuable with aqueous systems) and a dissolved phosphor, termed a primary solute or fluor, for example, 2,5-diphenyloxazole (PPO). If fluorescence is at too short a wavelength for efficient absorption by the photocathode, 1,4-di-[2-(5-phenyloxazoyl)]-benzene (POPOP) may be added as a wavelength shifter that will absorb the fluorescence and itself fluoresce at a longer wavelength. New compounds such as BUTYL-PBD that serve as primary fluors and emit at wavelengths suitable for detection are also being used. The particular combination of components is often called a cocktail. Since we do not know the exact amount of *fluorescence quenching*, we must always determine the counting efficiency of a liquid scintillation counter. The more frequent the standardization, the more reliable the results. On the other side are several advantages. As compared with the Geiger tube, much higher counting rates are possible. In addition, with scintillation counting, one generally has at least 50% efficiency, straightforward sample preparation, and the possibility of use of multilabeled systems.

For γ-ray detection, NaI crystals activated with 0.1–0.2% Tl (designated NaI (Tl)) are common scintillators. This phosphor is efficient in absorbing γ rays as a result of the large atomic number of I and the high density of NaI. To cope with high-energy γ photons large cylindrical crystals 7.5 cm in diameter and 7.5 cm high are often used. In addition to photopeaks, the NaI(Tl) detector commonly yields escape peaks and Compton scattering peaks that must be sorted out (see Section 21.5).

Considerable precautions must be taken to reduce the background count, which is as much as 10 times that of a proportional counter. Regular shielding is supplemented by special discriminating circuits and by anticoincidence counting arrangements, such as those shown in the representative arrangement of modules in a liquid scintillation system in Fig. 23.5. The scintillation system and detectors are mounted in a light tight, very-low-background-radioactivity housing. By requiring coincidence of pulses from the photomultiplier tubes, much dark current and noise are blocked from the instrument channel. Discriminators are described in Section 21.6.

Example 23.2. How effective is the use of dual-labeled samples for tracer studies? For instance, can ^{3}H and ^{14}C tracers be used in the same experiment?

The average energies of these radioisotopes are sufficiently different that a dual pulse-height analyzer can be used. Alternatively, two channels of a multichannel analyzer can

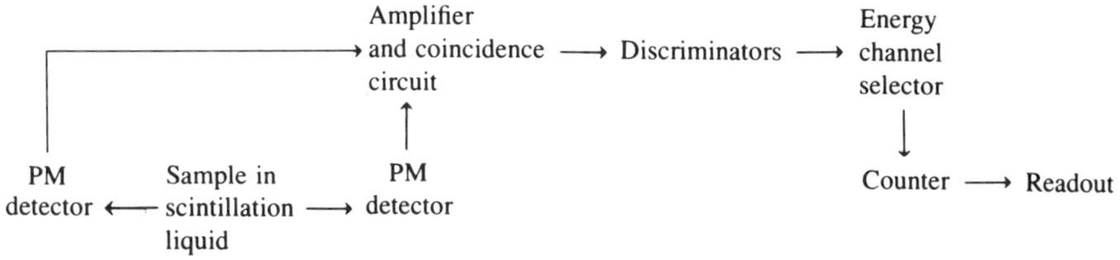

Fig. 23.5 Representative modular diagram for a liquid scintillation counter. PM = photomultiplier tube. The use of two detectors operated in coincidence ensures that dark current pulses or noise pulses arising in the photomultiplier tubes are generally ignored since they tend not to be in coincidence.

serve. One channel is set for the lower-energy beta particles, the other for the higher. Other pairs that can be used are $^{3}H + {}^{35}S$ and $^{32}P + {}^{35}S$.

Semiconductor Detectors. Solid-state detectors of specially prepared germanium are now widely used in γ-ray spectrometry. Most modern γ-ray spectrometry utilizes the Ge(Li) detector or sometimes the Si(Li) detector. They make possible a resolution of 1.7 keV for 1 MeV γ rays [as much as a 10-fold improvement over the NaI(Tl) scintillation counter].

A schematic cross section of a lithium-drifted semiconductor detector was shown in Fig. 21.13. It behaves essentially as a solid-state ionization chamber by collecting holes and electrons formed by ionizing events within its active volume. When reverse-biased as shown, the crystal itself shows minimal conduction. Ionization can be caused within the intrinsic region by γ-ray photons and the holes and electrons formed are quickly collected as a pulse of current. Since only 2.9 eV of energy is required in germanium to produce an electron-hole pair, a γ ray yields a great many pairs and therefore a large current pulse. A favorable S/N is realized at the 77 K temperature required to preserve the detector since thermal noise is low.

A range of sizes and shapes of Ge(Li) detectors is available but the larger ones are quite responsive. Use of a Ge(Li) detector to determine low-energy γ peaks is not recommended since Compton scattering may obscure them. With NaI(Tl) scintillators a much greater detector volume is possible, leading to stronger γ peaks and therefore to relatively less pronounced Compton peaks.

Photographic Emulsion Detector. Very thin photographic emulsions find use as two-dimensional or multichannel detectors in *radioautography*. Thin sections of tissue or metal in which portions have taken up radioisotopes selectively are placed in contact with the emulsion in the dark. After exposure and chemical development, the silver deposit provides a "map" of the areas that have become tagged. High resolution is possible (about 2 μm), but the method is not so easily adapted to quantitative determinations.

23.5 SCALERS, RATE METERS, AND MULTICHANNEL ANALYZERS

When pulses emerge from an amplifier associated with a nuclear particle detector, they are randomly spaced and may occur at rates up to 1000 s^{-1} and more. In the common mode of photon counting in which the information of interest will be the number of pulses and the time interval, an electronic scaling device will be interposed between the counter and display module. The scaler receives the input pulses, and after each p of them drives the display forward one unit. As a result of the mode of operation of a scaler, p is either

$$\text{(a) } p = 2^m \quad \text{or} \quad \text{(b) } p = 10^m, \tag{23.4}$$

where m is an integer denoting the number of stages in the scaler. If binary scaling (a) is used, m may range from 6 to 10; if decimal scaling (b) is used, m is usually 2 or 3.

In a binary scaler, the basic electronic unit is a flip-flop. A representative unit is discussed in Section 5.6. For scaling circuits, the resolving time is about 5 μs; still faster action can be realized with other arrangements. Precision and accuracy will depend on minimizing statistical error, allowing for background and correcting for or minimizing coincidence error (see Section 21.8 for a full discussion).

Rate Meter. Where the approximate rate of disintegration of a radioisotope is sought, it is useful to employ some sort of electronic integrating device and develop an analog signal. In this situation, it is essential that the incoming pulses be similar in shape and size and of equal duration. Pulses can be tailored for counting by having the entering pulses drive a circuit such as a monostable multivibrator that produces a single square pulse of characteristic duration when it is triggered by an incoming pulse (see Section 5.6).

Pulse-Height Analyzers. Analysis of the emission spectrum of a radioisotope for identification or characterization purposes requires a detector to furnish output pulses whose height is proportional to the energy of the radiation. *Discriminator* circuits may be used. A single circuit is frequently used to reject pulses below a certain minimum, such as noise pulses. A pair of discriminators set to slightly different voltages V_1 and V_2 and used in an *anticoincidence* configuration becomes a *single-channel analyzer*. The pair if wired to give no output if both discriminators or neither discriminator is activated; there is only an output if the lower discriminator is activated and not the upper. The difference $V_2 - V_1$ defines a *voltage or energy window* in which pulses are passed to a scaler and display unit. If the discriminator settings are moved up and down the range of pulse heights together (a 0- to 10-V difference is common), a spectrum is recorded. As noted above, the discussion of counting error in Section 21.8 is a necessary background for such measurements.

Multichannel Analyzers. To record a spectrum rapidly analyzers featuring as many as 4000 channels are available. Detector pulses are amplified and shaped, go to an AD converter and thence to the appropriate channel in the memory of a computer. Through selection of available programs the user asks appropriate manipulation and display of the data.

23.6 SOURCES OF ERROR

Several types of error occur in measurements of the radioactivity of samples: statistical fluctuations in the counting rate, coincidence errors because of the randomness of arrival of radioactive particles in the detector, error arising from unknown detector geometry and efficiency, and errors associated with the physical condition of the sample. A detailed coverage of the first two types was given earlier in

Section 21.8 and only the results will be summarized below. Two other possible sources of uncertainty, isotopic exchange and interference from other radioisotopes, are not general problems. Where they do occur, we may often devise special procedures to minimize them.

Statistical Uncertainty. Since the disintegration of radionuclides is a random process, counting rates are subject to statistical fluctuation. It can be shown that if the number of counts is reasonably large ($N > 100$) and the sample is observed for a short time compared with its half-life, the standard deviation σ_N in the total count is $\sigma_N = \sqrt{N}$ and the standard deviation in the counting rate is $\sigma_r = \sqrt{N}/t$. Clearly the precision of both the total count and the rate may be improved indefinitely by increasing the time of observation. For example, there will be approximately a 10% uncertainty if 100 counts are recorded, but only a 1% uncertainty if 10,000 counts are noted.

In general, the net count R for a sample, $R = S - B$, where S is the total count and B is the background count, will be the chief quantity of interest. The standard deviation in R is given by

$$\sigma_R = \sqrt{\sigma_S^2 + \sigma_B^2}, \tag{23.5}$$

where standard deviations of all three quantities appear.

Example 23.3. How should an experiment in radiochemistry be designed where an accuracy of $\pm 1\%$ is sought?

It is good practice to plan the experiment so that the standard deviation of the total counting rate is several times the standard deviation of the background rate B. According to Eq. (23.5) if this ratio is maintained, the background level need not be known with high precision. In the interest of time, some such arrangement is almost essential, for an unduly long period would be required to obtain 10,000 counts for background, using a well-shielded detector.

Example 23.4. How should an experiment be designed for cases in which sample activities are very small?

It appears now that the length of time required to obtain a precise count will be great. The principal difficulty is that the sample and background activity are nearly the same. The error in the difference between the two, the net rate, will be large unless a large number of disintegrations are counted. In this situation, it is imperative that the shielding be excellent and, if feasible, other circuitry such as a coincidence arrangement should be devised to minimize the background rate still further.

Geometry. The sensitive volume of a counter will subtend a certain solid angle with respect to the sample. For example, the end window of a Geiger tube may subtend a solid angle of several degrees with respect to material placed 5 cm from the window. Since the radiation from a sample emanates equally in all directions, we may calculate the percentage of the total that passes through the window into the tube from the angle subtended. Rather descriptively, the calculation is said to be a determination of the geometry of the system.

If we are to compare measurements, the geometry must be reproducible, which means that an arrangement for placing a sample precisely with respect to the detector must be provided. The sample–counter distance may be varied, but cannot be less than 1 cm. When samples are very close to a counter window, many rays enter the detector obliquely but pass through so little of the active volume they are not detected. Naturally the sample and detector must be properly shielded from background radiation.

Another factor conveniently considered under geometry is the *back scattering* of radiation from (a) the gas between the sample and the detector and (b) all solid parts. Most of the rays are diverted away from the detector on scattering, but some, such as part of those scattered from the walls of the holder, may be directed toward it. Since the back scattering increases with increasing atomic number Z, counter shields, which are invariably of a heavy metal such as iron ($Z = 26$) or lead ($Z = 82$), are placed several centimeters from the sample to reduce the percentage of scattering into the detector. The dependence on Z suggests that with weak emitters the sensitivity of detection may be much improved if helium is used in the counting chamber instead of air.

Absorption in the gas and in the detector window further reduces the number of ionizing particles or rays. Both absorption and back scattering are often said to reduce the geometry of the system.

Sample Preparation Errors. In considering radiation losses, we cannot neglect the absorption (and scattering) by the sample itself. Ordinarily, we prepare a sample for counting by spreading a given weight of substance in a planchet (a round, shallow dish of 2 or 3 cm diameter) to give a layer of uniform thickness. Both solids and dissolved solids (after evaporation of solvent) are examined in this fashion. Although relatively thick layers of a substance display a characteristic absorption per unit of weight, with thinner layers the actual absorption varies with thickness and density of packing. This behavior may be understood by assuming that a uniform sample comprises a series of very thin layers of equal thickness. By itself the layer closest to the detector gives a total count independent of thickness and representative of the true activity of the particular material. There is negligible absorption. When a sufficient number of layers have been taken so that all particles emitted from the most distant layer are absorbed within the nearer layers, we again obtain a count independent of thickness. In this case, however, the activity observed is per unit of weight. Samples of intermediate thickness give counting rates that are dependent on thickness and that are difficult to reproduce. We may best avoid errors of this type by preparing samples of definite weight in reproducible fashion.

Coincidence Loss. We may trace uncertainty in detection either to coincidence loss or to poor counter efficiency once a particle has entered a properly operating detector. Associated with any counter and its electronic circuitry is a certain pulse-resolving time. Since this time represents an interval during which a counter and associated circuit will fail to respond to new particles that enter the detector, there will be some *coincidence error* in general. The observed counting rate will be low and a correction of the form described in Section 21.8 will be needed.

Counter Efficiency Error. Finally, the possibility arises that some particles will exit from a detector before giving up all their energy even when the detector is capable of responding. In pulse-height determinations, this type of behavior produces spurious results. Clearly, detector efficiency will depend on the kind of radiation and its energy. Note that in simple activity determinations we must know the detector efficiency to estimate the absolute rate, but we need only know it to be constant to obtain reliable relative measurements. Table 23.1 listed the relative efficiencies of some representative detectors.

23.7 TRACER TECHNIQUES

When a stable atom in any chemical species (atom, ion, or molecule) is replaced with a radioactive atom, the entity becomes a tracer. In any sort of process (biological, chemical, or physical) its subsequent behavior may then be followed by measurements of its radioactivity. For example, we can find the effectiveness of separation of sodium and potassium ions by ion exchange by ''labeling'' a fraction of the sodium ions and then continuously observing the level of radioactivity in the effluent from the resin column. Both the presence of sodium and its concentration in the effluent can be determined at any stage of the process. Under optimum conditions, first the sodium and then the potassium is eluted. In more complex systems, separation procedures are usually called for since the activity in each phase or product must be ascertained. Note that tracer techniques offer advantages over other procedures only in certain types of study; we may best understand these by considering representative applications and some inherent limitations on the use of tracers. An important application, activation analysis, is discussed later (see Section 23.8). For coverage of radioimmunoassay the reader is referred to the bibliography.

Tracers. Radioisotopes are available from the Oak Ridge National Laboratory and various commercial concerns either in elemental form or in tagged compounds. Table 23.2 lists a few of the isotopes that see considerable use as tracers. Oxygen and nitrogen are conspicuous by their absence. Both have radioisotopes of such short half-life (at most a few minutes) that they decay too quickly to allow general use. The situation is considerably relieved, however, by the fact that both elements have stable isotopes (^{18}O, ^{17}O, ^{15}N) that can be used in tracer work and detected by a mass spectrometer.

If compounds rather than elements are being studied, radioisotopes must somehow be introduced into the desired position in the compound. Commonly, chemical synthesis is used.

Table 23.2 Characteristics of Some Widely Used Radioisotopes

Nuclide	*Half-Life*	*Maximum β-Energy (MeV)*	*Other Radiation*
^{3}H	12.5 years	0.0189	None
^{14}C	5720 years	0.155	None
^{24}Na	14.9 hours	1.39	1.38 and 2.758 MeV γ's
^{32}P	14.3 days	1.712	None
^{35}S	87.1 days	0.166	None
^{45}Ca	152 days	0.254	None
^{55}Fe	2.91 years	None	K X rays, Auger electrons[a]
^{60}Co	5.3 years	0.31	1.17 and 1.33 MeV γ's
^{131}I	8 days	0.60 (89%)	0.08 and 0.72 MeV γ's

[a] A K-electron capture may cause the release of an Auger electron.

Example 23.5. A simple, classic synthesis is used to prepare carboxyl-labeled acetic acid, $CH_3{}^{14}COOH$. $Ba^{14}CO_3$, a commercially available starting material is decomposed to $^{14}CO_2$. An excess of the CO_2 is then introduced into the Grignard reaction:

$$CH_3MgBr + {}^{14}CO_2 \rightarrow CH_3{}^{14}COOMgBr \xrightarrow{H_2O} CH_3{}^{14}COOH + Mg(OH)Br.$$

The labeled acetic acid is easily isolated.

Other procedures that are sometimes feasible are biological synthesis, chemical exchange of atoms, and irradiation of the compound to transform one or more stable atoms into radioactive atoms. Where we can use more than one method, the choice is governed by the ease of control of the reaction, its efficiency, the specificity of the position of the radioactive atom, the level of activity obtained in the labeled compound, and the difficulty of the separations. A wide variety of labeled compounds is available commercially.

In a representative tracer study, we use as low a level of activity as possible. The substance to be followed is tagged by adding a small amount of its labeled counterpart. The level of activity must be high enough so that detection difficulties do not arise because of necessary dilutions at any point in the procedure. Then an initial precise assay of activity is made. As the experiment is carried out, the radioactivity of the various phases or substances (after separation) is observed. (The details of such an *analysis* are considered in the next section.) If there is no isotope effect, chemical exchange, or other complication (see below), we need only correct the counting rates for dilution and decay before we interpret in terms of concentrations. The decay correction will be significant only if the duration of the study is greater than 0.01 $t_{1/2}$. Note that we may apply tracer methods to systems where no specific knowledge of the changes occurring is available.

Isotope Dilution. Certainly the method of isotope dilution analysis is one of the most valuable tracer procedures. It is used particularly in the analysis of complex inorganic and organic mixtures that are resolvable only with difficulty. For example, vitamin B_{12} can be assayed by isotope dilution in mixture with other intricate organic molecules. The method may be found in the *U.S. Pharmacopoeia*, Supplement XV.

The steps in isotope dilution are as follows: we mix a known weight S of pure labeled compound (chemically identical with the substance of interest) of specific activity A_1 (specific activity is defined as counts per unit of time per gram) thoroughly with the sample, which contains unknown weight x of the sought constituent. We then abstract a small amount of the desired constituent in pure form from the mixture. Whether there is appreciable loss in the separation is unimportant *as long as enough pure substance is isolated for an activity determination*. Let the specific activity of the isolated material be A_2. Then the following relation must hold:

$$\frac{A_1 S}{S + x} = A_2 \quad \text{or} \quad x = S\left(\frac{A_1}{A_2} - 1\right). \tag{23.6}$$

Note that we must know the weight isolated to calculate A_2. Depending on the specific activity of the tracer molecules, isotope dilutions may run as high as 1 to 1000 or indeed 1 to 10,000 when a precision of 1% is sought.

Limitations. We must always design the procedures for using and handling tracers with possible restrictions in mind. The following effects represent potential sources of interference in tracer work: isotope fractionation, chemical exchange (of atoms), radiation from radioactive daughters, and formation of radiocolloids. We should always investigate the chance that one or more of these effects will occur in a study. If we cannot allow for them by calibration or the use of a blank, we may have to abandon the method.

By definition, *isotope fractionation* is the change in isotopic abundances in different parts of a system that results from the carrying out of a process. The effect is negligible except in instances in which there are differences of 10% or more in the weights of isotopes (in this case, between the mass number of the radioisotope and that of the common stable isotopes). Thus, isotope fractionation is noted for radioisotopes such as ^{14}C and ^{45}Ca (^{12}C and ^{40}Ca are the common stable isotopes), but is not marked for radioisotope ^{131}I (cf. ^{127}I, which is stable). In general, fractionation is significant only in chemical kinetic studies and in measurements involving equilibria. In kinetics, the effect enters because of the differences in the ease of rupturing bonds. It has been shown by experiment and theoretical treatments that a bond that involves a lighter isotope, for example, a ^{12}C–^{12}C bond, will break more easily than the same bond with a heavier isotope, for example ^{12}C–^{14}C or ^{14}C–^{14}C. In equilibria, fractionation may generally be traced either to the same source or to differences in the ease of transporting molecules. For example, the vapor over ordinary water will be richer in ^1H than is the liquid, and somewhat poorer in ^2H. In most analytical measurements with tracers we may neglect the fractionation effect, as it is presumed that quantitative reactions and separation procedures are used. All bonds of one kind are broken or all of a substance is evaporated. If necessary, a calibration may be made to correct for fractionation.

Chemical exchange reactions in which atoms in molecules or ions interchange with chemically identical atoms in other species are often likely and must be allowed for in some tracer studies. For example, bromine exchangers readily between Br_2 and $AsBr_3$ in carbon tetrachloride. It is not necessary that ionization occur for atoms to exchange; the process often proceeds through intermediate states following a collision. Although prediction of the likelihood of exchange in a new system is uncertain, exchange with a sufficient number of compounds and ions has been studied so that the possibilities of exchange in the system under investigation may usually be checked in advance by reference to the literature.

Some interesting problems arise when the daughter atoms formed during the decay of the tracer are also radioactive. Note that a similar situation arises whenever two or more radioisotopes are present in a sample. It is, of course, impossible to admit to the detector only the radiation from one of the radionuclides unless the radiation from the other is much more easily stopped by an absorber. A significant degree of discrimination can often be achieved by the use of a proportional or scintillation counter with a pulse-height analyzer. The problem of daughter activity does not arise with most of the commonly used tracers. Where it appears, a delay in counting sufficient to allow the shorter-lived activity to decay is often desirable. When pure radioactive materials are present in solution at very low concentrations (10^{-4} to 10^{-6} M), they tend to form colloids. As a result, this material may be difficult to transfer or collect quantitatively. We may minimize the effect by adding a carrier, as long as the presence of the carrier is unobjectionable. For example, radiostrontium ion may be carried successfully by calcium ion, since the ions behave similarly. At these low concentrations, carriers are also highly desirable to avoid losses from adsorption on glass, filter paper, and so on, or by coprecipitation during chemical separation.

23.8 ACTIVATION ANALYSIS

About 60 elements have one or more stable isotopes that may be activated, that is, made radioactive, by neutron bombardment. Some activations have been accomplished by charged-particle bombardment, although relatively little application of this approach has been made. The overall nuclear reaction for neutron activation is usually of the (n, γ) type. Each nucleus that absorbs one of the bombarding neutrons becomes unstable, and releases γ rays to regain stability. Thermal neutrons are used ordinarily, since they are available at high flux density in nuclear reactors. Moderate flux densities are, however, available from radioisotopic sources. By making use of (a) known differences in neutron activation sensitivity, (b) different levels of neutron flux, and (c) different radiation times, we can restrict the activation to one or at most to a few of the elements in a complex sample. There is thus a basis for analysis, since the resulting radioactivity may be observed to determine quantitatively the concentrations of elements.

Activation procedures have wide applications, particularly in the determination of trace quantities. The sensitivity for certain elements is so great that 10^{-14} g of the element can be determined in a milligram of sample. It is convenient to express the sensitivity in terms of the number of grams x of the sought element that must be present in a sample to attain a particular specific activity. This leads to the defining expression

$$x = \frac{AW}{N_0 f \sigma_{at}(1 - e^{-\lambda t})}, \tag{23.7}$$

where A is the activity in disintegrations per minute, W is the atomic weight, N_0 is Avogadro's number, f is the neutron flux in neutrons per square centimeter per second, σ_{at} is the cross section of the absorbing species (nuclear cross section for activation multiplied by the isotopic abundance of the species), λ is the decay constant, and t is the time of exposure to the neutron flux. We may better understand the significance of the saturation factor, the parenthetical expression, by substituting $0.693t/t_{1/2}$ for $e^{-\lambda t}$: The degree of saturation with activated nuclei depends on the relative length of the bombardment time t and decay half-life $t_{1/2}$ of the isotope formed. If a given specific activity (proportional to AW/N_0) of a particular element is developed each time, this expression indicates that the number of grams we may detect is inversely proportional to the neutron flux. Large fluxes must be used for greatest sensitivity. Likewise, the number of grams detectable is smaller, the greater the nuclear cross section of the isotope and the longer the half-life of the nuclide produced.

Figure 23.6 is a table of the elements that we may detect with a sensitivity of 10^{-7} g or less for a moderate nuclear reactor flux.

Activation Procedure. The thermal neutron flux necessary to activate in a period as short as a week (a flux of 10^{11} neutrons cm^{-2} s^{-1} or higher) is ordinarily available only in a nuclear reactor. Neutron levels of 10^{14} cm^{-2} s^{-1} or higher are available at many nuclear reactors. In studies that involve higher concentrations or isotopes that are more easily activated, it is also possible to use sources with fluxes in the range of from 10^6 to 10^{10}, such as Van de Graaff generators and Sb–Be mixtures.

Analytical Procedures. Because of the inhomogeneity of the neutron flux in most reactors, the uncertainty in the neutron activation cross sections, and the difficulty of controlling other variables, it is generally desirable to perform activation analyses on a relative basis. A typical procedure consists of irradiating a standard sample and one or more samples that contain unknown amounts of an element and then comparing the activity produced in each. Since several radioisotopes are usually formed, a chemical separation is ordinarily essential. Its modest goal is that of isolating in pure radioactive form the single nuclide of interest. It is not necessary that the separation be chemically complete. If small amounts of a radioisotope are involved, the addition of a nonradioactive carrier facilitates the oper-

Li																**F**	
Na												Al		P		Cl	**Ar**
K		Sc	Ti	V	Cr	Mn		Co	Ni	Cu	Zn	Ga	Ge	As	Se	Br	**Kr**
Rb	Sr	Y	Zr		Mo		Ru	**Rh**	Pd	Ag	Cd	In	Sn	Sb	Te	I	**Xe**
Cs	Ba	La*	Hf	Ta	W	Re	Os	Ir	Pt	Au	Hg	Tl	**Pb**	Bi			
		†															

*Lathanides	Ce	Pr	Nd		Sm	Eu	Gd	Tb	Dy	Ho	Er	Tm	Yb	Lu

†Actinides	**Th**		U						

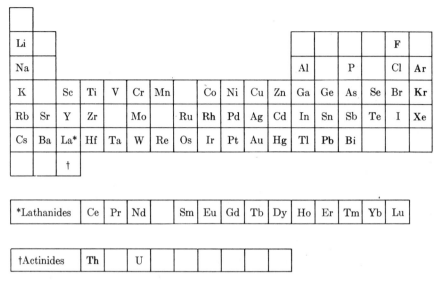

Fig. 23.6 Elements with thermal neutron activation sensitivities of 10^{-7} g or less in a neutron flux of 10^{12} neutrons cm^{-2} s^{-1}. Those elements with an activation sensitivity of 10^{-10} g or less under the same conditions are Dy, Eu, Ho, In, Ir, Lu, Mn, Sm, and Re.

ation. Then to be certain that unknown contaminants are not affecting the results, absorption curves (activity versus concentration) for the material isolated from the unknown and for the standard sample ought to be compared. Actually, a simultaneous decay determination may also be desirable. Any differences in the slope of the activity–time curves are indicative of contaminants.

Example 23.6. The concentration of sodium (all atoms present are ^{23}Na) in aluminum alloys may be determined by activation analysis. Samples are irradiated in standard aluminum sample cans for one week at a flux of 10^{11} neutrons cm^{-2} s^{-1}. The standard in this analysis may conveniently be either sodium bicarbonate or sodium carbonate. After irradiation, a suitable carrier for the radionuclide ^{24}Na is added, and a chemical separation of carrier and ^{24}Na from the rest of the alloy is effected. The activity of the separated material is then determined and compared with that of the standard. A precision of about 1% is obtained in the range of from 0.01 to 0.04% sodium.

We may also induce activation by bombardment with other particles. For example, surfaces of materials have been studied with a Van de Graaff generator as a particle source. In the investigations, energetic protons or deuterons beamed at surfaces were found to give reactions of the (n, p) or (d, p) types in general. Sensitivities for individual elements with this technique are in the range of from 10^{-6} to 10^{-8} g cm^{-2}. The method appears particularly useful in studying surface films and corrosion effects. A depth resolution of the order of 0.01 μm can be obtained.

23.9. MISCELLANEOUS METHODS

We can also use the *absorption* or *scattering* of nuclear radiation as the basis of measurements. Most techniques involve β particles, which have moderate penetration yet do not present complex problems. A *collimated source*, one that produces a nearly parallel beam of radiation, is essential for meaningful results, and is placed on one side of the sample, and a detector whose response has been calibrated by the use of standards is placed on the other side.

In addition, many absorption devices that allow determinations of the thickness of a product or its density of packing are in use commercially. Much of the metal and plastic sheeting currently produced is monitored by β-ray gauge to ensure constant thickness.

Procedures have also been developed to use the back scattering of β rays from compounds for the purpose of establishing the identity of materials. Although the methods are nonspecific, if the composition of a substance is known or suspected, we can verify its identity in a few minutes by means of the measurements. For the observation, we locate the source between the sample and the detector but shield it from the detector so that there is no direct penetration of the detector by primary radiation. The shield is so designed, however, as to pass a large percentage of the scattered radiation traveling toward the counter. We can apply the technique to organic or inorganic solids, liquids, and solutions. As long as a volume of sample greater than a determinable minimum is used, the results are independent of density. Since back scattering increases with atomic weight, the precision improves with material of higher Z. For example, chlorinated organic compounds may be identified more easily than fluorinated ones.

By using *radioactive tracers* we have gained information about reaction kinetics and mechanisms that we can presently obtain in no other way. The results have been particularly noteworthy in physiology and biochemistry. A good example is the use of ^{14}C in unraveling the complex reaction path by which atmospheric CO_2 is utilized. Although in many reaction rate studies other analytical methods might serve equally well, we can determine the rates under equilibrium conditions only by employing radioisotopes.

Whenever we can measure the exchange between a radioisotope and a stable combined isotope, we can obtain information about stability, bond strength, and to a degree, about bond type. For example, the absence of exchange between the sulfur atoms in sodium thiosulfate certainly indicates the separate chemical character of the two sulfur atoms as well as the stability of the complex. In a somewhat analogous fashion, the path of electron transfer in oxidation–reduction reactions based on complex ions has been investigated. For example, in some oxidations such as $S \rightarrow SO_4^{2-}$ and $NO_2 \rightarrow NO_3^-$, the solvent has been found to supply all or part of the oxygen; in other cases an oxygenated oxidant loses oxygen directly to the nonmetal.

Many physical processes scarcely susceptible to any other approach have been studied with the use of tracer techniques. In particular, we may cite the observation of rates of diffusion in solids and of the self-diffusion of ions in solutions, although the results on the latter have been less conclusive.

23.10 RADIATION HAZARDS

In carrying out radiochemical processes, there are three types of handling problems:

1. shielding (to safeguard personnel),
2. avoiding contamination of equipment, and
3. disposing of radioactive wastes.

This list should suggest that good procedure is as essential to the attainment of accurate experimental results as it is to the maintenance of healthful conditions for work.

Shielding. In order to ensure the safety of personnel, routine protective measures are highly recommended. Where there is likelihood of regular exposure over a long period, some sort of a health-physics device such as a film badge or pocket (pencil) ionization chamber is recommended. A few commercial sources lease and "read" these devices. Where there is a possibility of sudden exposure to an overdose of radiation, devices that immediately report the level of intensity must be used as well.

The type of routine precautions observed should be formulated in terms of the kind of radiation and the level of activity. To simplify shielding and handling measures, it is desirable to perform all radiochemical operations in a separate area and indeed to use different working areas for different levels of activity. A hood with a good exhaust fan is the preferred site for all manipulations to avoid danger to personnel that might result from any volatilization of radioactive materials.

Arrangements that may be sensible at the millicurie level will appear ridiculous at the microcurie level. If only weak β emitters such as ^{14}C and ^{35}S are used, the walls of laboratory glassware are thick enough to protect the experimenter even at the millicurie level. More vigorous β emitters, for example ^{32}P, should be used behind a layer of plastic at the millicurie level. Note that γ emitters require shielding even at the microcurie level. It is good practice never to handle glassware or other equipment directly, but to wear surgeon's gloves when handling weak emitters at the lowest level and to turn to tongs, beaker holders, and remote pipets at high levels or when the rays are more penetrating.

Contamination. The use of equipment contaminated with radioactivity may easily invalidate the results of a radiochemical experiment. It is necessary to use a separate stock of glassware for different levels of activity and to check all equipment with a portable radiation monitor before use. Any spills of radioactive material should be cleaned up immediately to avoid contaminating other areas as well as endangering personnel. In some cases, special decontamination methods may be advisable. Absorption of activity on glassware is always a problem and should be guarded against.

Disposal. The liquid radioactive waste from most experimentation may be disposed of by diluting the material to a safe level of activity and, if possible, adding carrier before pouring into a drain. The presence of the carrier is desirable where there may be some concentration of the active material either by a physical process or by aquatic organisms. Certain types of solid wastes may be incinerated, but usually solid radioactive material must be stored or buried.

REFERENCES

A good introductory volume is:

1. D. Brune, B. Forkman, and B. Person, *Nuclear Analytical Chemistry*. Deerfield Beach, FL: Verlag Chemie, 1984.

Well-written intermediate-level general volumes are:

2. G. R. Choppin and J. Rydberg, *Nuclear Chemistry Theory and Applications*. Oxford: Pergamon, 1980.
3. P. J. Elving, V. Krivan, and I. M. Kolthoff (Eds.), *Nuclear Activation and Radioisotopic Methods of Analysis*, Vol. 14, Part I, *Treatise on Analytical Chemistry*, 2nd. ed., New York: Wiley-Interscience, 1986.
4. G. Friedlander, J. W. Kennedy, E. S. Macies, and J. M. Miller. *Nuclear and Radiochemistry*, 3rd ed. New York: Wiley, 1981.

More specialized treatments of instrumentation are found in:

5. P. N. Cooper, *Introduction to Nuclear Radiation Detectors*. Cambridge: University Press, 1986.
6. A. Dyer, *An Introduction to Liquid Scintillation Counting*. Philadelphia: Sadtler, 1974.
7. F. S. Goulding and A. H. Pehl, Semiconductor radiation detectors, in J. Cerny (Ed.), *Nuclear Spectroscopy and Reactions*, Vol. A. New York: Academic, 1974.
8. J. Krugers (Ed.), *Instrumentation in Applied Nuclear Chemistry*. New York: Plenum, 1973.

Also of interest are:

9. D. Breitag and K. H. Voigt, Radioimmunoassay, in Elving, Krivan, and Kolthoff (Eds.), *ibid.*
10. G. Erdtmann and H. Petri, Nuclear activation analysis: fundamentals and techniques, in Elving, Krivan, and Kolthoff (Eds.), *ibid.*
11. K. H. Lieser, Fundamentals of nuclear activation and radioisotopic methods of analysis, in Elving, Krivan, and Kolthoff (Eds.), *ibid.*
12. J. L. Travis, *Fundamentals of Radioimmunoassay and Other Ligand Assays*. Anaheim, CA: Scientific Newsletter, 1977.
13. C. H. Wang, D. L. Willis, and W. D. Loveland, *Radiotracer Methodology in the Biological, Environmental, and Physical Sciences*. Englewood Cliffs, NJ: Prentice-Hall, 1975.

EXERCISES

23.1 A petroleum crude is to be analyzed for its naphthalene content by isotope dilution. To a 110.0-g sample of the crude is added 0.15 g naphthalene of specific activity 0.10

microcurie g^{-1} (as a result of labeling with ^{14}C). After thorough mixing, 0.250 g of naphthalene is separated in pure form and is found to give 3060 counts min^{-1}. Calculate the percentage of naphthalene in the crude.

23.2 What is the absolute activity of a 1.00 g sample of ^{35}S, whose half-life is 87 days?

23.3 How may an end-window Geiger tube be adapted to count only γ rays?

23.4 Discuss the use of a wire loop as the anode instead of a straight length of wire in the hemispherical proportional counter.

23.5 It is desired to count only the ^{35}S in a sample that contains both that isotope and an equal activity of ^{38}Cl. If the analysis may be deferred without inconvenience, discuss the factors that would determine how long the material might be held to allow for the decay of the ^{38}Cl. How many days must elapse before the ^{38}Cl activity is only 0.1% of the ^{35}S activity?

23.6 Discuss the relative advantages of the common γ-ray detectors: NaI(Tl) and Ge(Li). What factors restrict the relative efficiency of Ge(Li) detectors? Which detector would probably be chosen for an (n, γ) activation analysis of small, rare samples?

23.7 The degree of exposure of personnel to radioactivity is commonly monitored at present by thermoluminescent dosimeters. The devices contain a phosphor like LiF. When radiation impinges, electrons are excited in the phosphor and a certain fraction is trapped. Subsequent heating of the dosimeter causes the return of these electrons to the ground state and the emission of light. What properties should a phosphor have to be widely useful in this application?

ANSWERS

23.1 2.3%. *23.2* 1.59×10^{15} counts s^{-1}.

CHROMATOGRAPHIC METHODS

Chapter 24

GENERAL PRINCIPLES OF CHROMATOGRAPHY

Raymond P. W. Scott

24.1 INTRODUCTION

The first scientist to produce an effective chromatographic separation was the Russian botanist Tswett. In his original paper published in 1903, he described the results he had obtained by using a rather simple form of liquid–solid chromatography to separate a number of different plant pigments. Probably the most significant early publications in the field of chromatography were those by Martin and Synge* in which they first suggested the possibilities of gas chromatography and the epic paper by James and Martin† describing the first gas chromatograph. The rate of development of gas chromatography in the period that followed was unique in the history of analytical instrumentation because the basic chromatograph was relatively simple to construct and detecting systems were not nearly as difficult to develop as their counterparts in liquid chromatography. The reason is to be found in the fact that any analyte significantly modifies the physical characteristics of the carrier gas and thus can be detected by simple measurement of basic gas properties such as thermal conductivity, density, or specific heat.

Development of Liquid Chromatographic Instrumentation. In the early 1960s, however, a renaissance took place in the field of liquid chromatography. The difficulties that had to be overcome were twofold. First, detectors with limits of detection as low as those used in gas chromatography were needed. Second, high-efficiency columns were required to effect the desired separations.

The problems of detection were much more difficult. In liquid chromatography the presence of small amounts of solute does not substantially modify the physical properties of the liquid carrier and detection methods based on the measurement of an overall physical property of the eluent are insensitive. Nevertheless, the problem of detection was solved fairly adequately in the early 1960s by the introduction and development of the refractive index detector and the UV detector. Both these devices provided sufficiently low limits of detection to permit useful analyses to be accomplished and also to examine the properties of columns and allow the development of column technology.

In the early 1970s Majors and Kirkland separately developed techniques for packing

*A. J. P. Martin and R. L. M. Synge, *Biochem. J.*, **35,** 1358 (1941).
†A. T. James and A. J. P. Martin, *Biochem. J.*, **50,** 679 (1952).

very small particles (the so called microparticulate packings) to provide efficiencies of many thousands of theoretical plates. Since that time, in fact during the last decade, the development of liquid chromatography has assumed the same inpetus as that of gas chromatography during the 1960s. Today many thousands of theoretical plates are available in liquid chromatography and modern detectors are providing limits of detection between 10^{-8} and 10^{-10} g L^{-1}.

Column technology is now in an advanced state of development and in fact columns can be designed to have a performance far in excess of that which the instrumentation can support. Since the majority of the compounds of interest to chemists generally are involatile, and furthermore, those of interest to the biochemist and biologist are largely polar and of high molecular weight, much remains to be developed in liquid chromatography. The next decade may provide chromatographic systems based on liquid chromatography that will provide impressive separations of high-molecular-weight polar compounds such as proteins that are completed in a few minutes or maybe in an hour, while at the same time achieving hundreds of thousands of theoretical plates.

In this chapter the discussion of chromatography will be confined to column chromatography whether it be gas or liquid chromatography. The discussion of alternative forms of liquid–solid and liquid–liquid chromatography, where the separation medium is laminar in shape, such as thin-layer and paper chromatography, lies outside the scope of this book.

24.2 THE BASIC CHROMATOGRAPHIC PROCESS AND CLASSES OF CHROMATOGRAPHY

Chromatography has been classically defined as a process in which separation of substances is achieved by distributing them between two phases, a *mobile phase* and a *stationary phase*. The solute molecules will only migrate through the column while they are in the mobile phase; whereas when they are distributed in the stationary phase they will remain at rest. Thus, the more they are distributed in the mobile phase (the smaller the distribution coefficient with respect to the stationary phase), the faster they will migrate through the chromatographic system. Conversely, solutes that are mostly distributed in the stationary phase (the distribution coefficient of the solute with respect to the stationary phase is large) will be retained longer in the chromatographic system and thus move more slowly through it.

It should be clearly understood that chromatography is solely a separation process. The chromatographic system accepts a mixture of substances and retains them to different extents so that, ideally, they are eluted as individual components from the system. However, if the eluent from the chromatographic system is monitored by an appropriate detection device which responds to solute mass or concentration, the quantity of each component present in the original mixture can be determined, thus producing an analysis. The determination, however, can *only* be obtained with the aid of the detector; the chromatographic system can *only achieve a separation*.

The phases that are employed in the chromatographic system may take any of the three physical forms that are feasible. The mobile phase can be a gas or a liquid, the stationary phase can be a supported liquid or solid. Indeed, the physical form of the phases employed in the chromatographic system is used as a basis for the classification of the types of chromatography. As shown in Fig. 24.1, chromatography can be divided into two primary classes based on the physical nature of the mobile phase. In gas chromatography (GC) and liquid chromatography (LC) the mobile phases are a gas and liquid, respectively. Two further classes of chromatography arise from varying the physical nature of the stationary phase. Employing a solid as the stationary phase gives rise to gas–solid chromatography (GSC) and liquid–solid chromatography (LSC). Conversely, employing a liquid as the stationary phase (usually contained on a suitable inactive support) gives rise to gas–liquid chromatography (GLC) and liquid–liquid chromatography (LLC).

It has to be said, however, that although the difference between GSC and GLC is clearly defined and the terminology is a realistic description of the actual physical system, the terms LSC and LLC do not truly represent the physical systems involved. This ambiguity results because in *liquid–solid chromatography*, where the stationary phase will be either silica, alumina, or a bonded phase, the surface of the stationary phase is covered by at least one layer and often two layers of solvent molecules. This adsorption of solvent will be discussed later, but at present it is sufficient to say that the solute molecules are often not interacting directly with the solute surface but with a layer of adsorbed solvent molecules, that is, the distribution could be considered more liquid–liquid than liquid–solid. Nevertheless, for convenience the chromatographic systems involving silica and bonded phases are normally referred to as liquid–solid chromatography systems. Finally, one other system, *supercritical fluid chromatography* (SCFC), has recently been developed into an effective separation process. This system utilizes supercritical fluids, that is, substances at temperatures above their critical temperature and pressure, as the mobile phase. In effect the system is a hybrid between LC and GC. Since the mobile phase is really a very dense gas, there are significant interactions between the solute and the mobile phase as in liquid chromatography, and SCFC has been placed in a classification of its own.

In general, each class of chromatography has a specific area of application. There are some regions of overlap, but only for SCFC, for which there is presently insufficient information, is the field of application somewhat uncertain. The various areas of application for different classes of chromatography are depicted in Fig. 24.2 and described somewhat further below.

Gas–Solid Chromatography. With a stationary phase such as silica gel, alumina, or molecular sieves, GSC is employed almost exclusively for the analysis of permanent gases such as O_2 and N_2 and low-boiling hydrocarbons. The use of GSC for separation of high-molecular-weight substances, or substances that even have slight polarity, is unattractive since they will be attached so strongly to a solid surface that they will not migrate through a column in a reasonable time. Attempts have been made to reduce the activity of absorbants such as alumina by suitable treatment, and this has permitted the separation of the nonpolar normal hydrocarbons with carbon numbers up to 40 or 50 at operating temperatures of 300–

Mobile phase: gas
GAS CHROMATOGRAPHY (GC)
- Stationary phase liquid—GAS–LIQUID CHROMATOGRAPHY (GLC)
- Stationary phase solid—GAS–SOLID CHROMATOGRAPHY (SGS)

Mobile phase: supercritical fluid
- Stationary phase solid—SUPERCRITICAL-FLUID CHROMATOGRAPHY (SFC)

Mobile phase: liquid
LIQUID CHROMATOGRAPHY (LC)
- Stationary phase solid—LIQUID–SOLID CHROMATOGRAPHY (LSC)
- Stationary phase liquid—LIQUID–LIQUID CHROMATOGRAPHY (LLC)

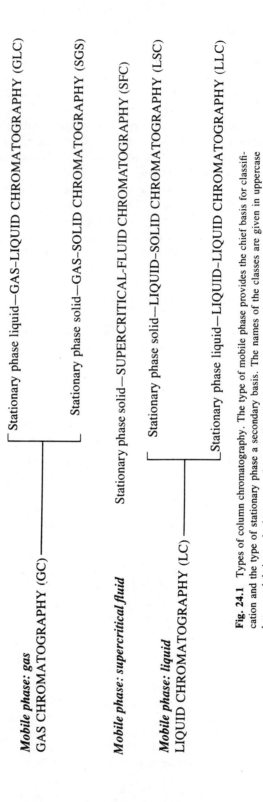

Fig. 24.1 Types of column chromatography. The type of mobile phase provides the chief basis for classification and the type of stationary phase a secondary basis. The names of the classes are given in uppercase letters and their standard acronyms in parentheses.

	Stationary Phases	*Applications*
Gas–solid chromatography (GSC)	Silica gel Alumina Molecular sieves	Analysis of permanent gases and low-boiling hydrocarbons
Gas–liquid chromatography (GLC)	High boiling hydrocarbons Polymeric alcohols and esters (PEG) Polymeric silicones	Analysis of a wide range of volatile, medium polarity, low boiling substances (e.g., essential oils, chlorinated solvents, fatty acid esters, etc.)
Liquid–solid chromatography (LSC)	Silica gel, bonded phases	Analysis of solute mixtures covering a wide range of molecular weights and polarities. Some limitations with very high-molecular weight materials such as proteins
Liquid–Liquid chromatography (LLC)	Appropriate solvents not miscible with the mobile phase	Preparative separations of labile materials of biological origin

Fig. 24.2 Some stationary phases and representative applications of different forms of column chromatography.

400°C. However, such deactivation is not sufficient to reduce the retention of polar substances.

Gas–Liquid Chromatography. This type of chromatography is used to separate all kinds of volatile materials provided they are thermally stable and are not excessively polar. Its inability to separate volatile, highly polar materials is largely due to the residual adsorptive capacity of the support material. In GLC, the stationary liquid phase has to be held on a support, for example, a diatomaceous earth such as Celite or the walls of a capillary tube. Although relatively inert, even coated Celite still has some residual adsorption sites that can retain strongly polar materials in the system and can render separation impossible or at least of very poor quality. GLC is also limited to relatively small-molecular-weight materials since substances must have a significant vapor pressure for gas chromatographic separations to be effective.

Liquid–Solid Chromatography. LSC has a very wide field of application, particularly if one includes the reversed-phase materials as proper liquid–solid systems. LSC systems have been applied to substances that have a wide range of molecular weights and a wide range of polarities and are in fact, effective even for the separation of substances that are normally separated by gas–liquid chromatography. At this time, however (and this is a transient situation), because insufficient development work has been carried out, liquid–solid chromatography has some limitations for very high-molecular-weight material such as proteins. Nevertheless, work is already in hand to produce effective separations even of these materials and it is likely that liquid–solid chromatography will be capable of handling the whole gamut of separation problems in due course.

Liquid–Liquid Chromatography. This method was originally used to separate biological materials which were somewhat labile, because the early absorbants such as silica and alumina had catalytic activity and unfortunately tended to decompose many of these materials. Since the advent of bonded phases, both reversed and polar, substances of biological origin, including those that are labile, are now being separated by liquid–solid chromatography. In the future the main application area of liquid–liquid chromatography appears likely to be in the field of preparative chromatography. Preparative applications will probably be in those areas where large quantities of biological materials are required, which normally precludes the economic use of the very expensive bonded phases. LLC is sufficiently gentle to handle such materials in an economic manner.

24.3 MECHANISMS OF RETENTION

Retention in a chromatographic system occurs when the distribution of the solutes favor the stationary phase. If the distribution is largely into the mobile phase, the solutes elute rapidly. If they are largely contained in the stationary phase, they are held in the column for a longer period of time.

The mechanisms by which the solutes are held in either phase in chromatography are the result of interactions between the solute molecules themselves and with the molecules of each phase. In GC the only molecular interactions are with the stationary phase since interactions with the gas phase are virtually negligible. Therefore, the more strongly the solutes interact with the stationary phase in GC, the more they will be retained. In GC, only by changes in the nature of the stationary phase it is possible to change solute selectivity.

In LC, further selectivity can be achieved by changing the nature of either the stationary phase or that of the mobile phase. Thus, solute interactions with the mobile phase as well as those with the stationary phase can be exploited in LC. One of the advantages of LC over GC is that it offers a greater degree of freedom in the way in which selectivity is established.

The interactions between solutes and different phases are, in general terms, based on three main types of interaction.

Ionic Interactions. Ionic interactions take place between a solute molecule and a molecule of the appropriate phase when each carries a permanent charge, that is, exists as an ion. In fact, ion-exchange chromatography results from the interactions of the solute ions with the ions of the phase system. To retain anions, cations would be obviously located in the stationary phase, and vice versa. Ionic interactions, so far, are used exclusively in LC and not in GC.

Polar Interactions. Polar interactions result not from net charges on molecules but from unsymmetrical distributions of charge within neutral molecules, i.e., dipoles. These interactions are between polar molecules with permanent dipoles or between a polar molecule and a polarizable molecule in which the polar molecule has induced a dipole as it has approached. For instance, the interaction of polar molecules of acetone and dimethyl sulfoxide would be an example of interaction between molecules having permanent dipoles and that of molecules of acetone with benzene, an example of a polar molecule interacting with a polarizable molecule in which it had induced a dipole. Polar interactions are exploited in both GC and LC. In GC they would be typified by the interaction of an ester with a polyethylene glycol stationary phase and in the case of LC they can be represented by the interaction of an alcohol with silica gel.

Dispersive Interactions. Interactions based on London dispersion forces are more difficult to define. They are typified by the interaction of one hydrocarbon with another hydrocarbon. To some extent they are a function of molecular volume and therefore are useful in exploiting molecular weight differences between different solute molecules as a basis for chromatographic selectivity. Dispersive interactions are employed in both GC and LC; in GC they are typified by the interactions of a solute with substances such as squalane or the silicone polymer OV101 and in LC by the interaction of a homologous series of alcohols with a reversed phase.

Molecular interactions form the basis of selectivity in all chromatographic systems. Without them almost every kind of chromatography except size-exclusion chromatography (see discussion below) would be severely limited as a separation technique. A chromatographic separation may therefore also be defined as one that is achieved by exploiting the different intermolecular forces that are exerted on solutes when they are distributed between a mobile and stationary phase. Those substances that interact more strongly with the molecules of the mobile phase pass through the system more rapidly than those that interact more strongly with the molecules of the stationary phase. Thus, the individual substances will move through or from the system in order of the increasing forces that exist between them and stationary phase.

Size-Exclusion Interactions. One further method of retention that results from the microstructure of the stationary phase as opposed to its chemical character should be mentioned. Solute retention does not solely depend on the intermolecular forces between the solute and the stationary phase, but also on the porosity of the stationary phase. In the case of silica gel, the material can have pores ranging from 0.3 to 100 nm (3 to 1000 Å) in diameter, depending on how the silica gel was made. Obviously a molecule of a given size can only enter a certain proportion of the pores in the material and will consequently be excluded from others. The solute molecules will only be retained when they enter the stationary phase, that is, pass into the apertures in the porous silica. Thus, smaller molecules will enter many more pores than larger molecules and will be retained to a greater extent. This retention mechanism based on molecular size is called *exclusion chromatography*.

In general, although silica gel is a good example of how exclusion processes work, it has somewhat limited use in analytical exclusion chromatography. Special media are normally employed, prepared from appropriate gels that are specifically designed to separate molecules of a particular size. It should be remembered, however, that in normal LSC using silica gel, although the major factor affecting retention will be molecular interaction between the solute and stationary phase, because silica has exclusion properties, exclusion will also play a significant part. For example, the larger the molecule, the less the number of pores it will penetrate and although the interaction between it and the silica surface may be the same as that for a smaller molecule, it will be eluted relatively early because it has less stationary phase available to interact with, because it will be excluded from many of the pores.

Exclusion processes are used in both GC and LC. In GC the molecular sieves are a typical exclusion type material that retain solutes on the basis of their pore structure in the separation of the permanent gases. In LC Sephadex gels and to a limited extent silica gels are commonly used for exclusion separations.

24.4 METHODS OF DEVELOPMENT

A given chromatographic system can provide a range of different types of separations depending upon how the sample is placed on the column and how the mobile phase is passed through the column. The combined process of placing the sample on the column and eluting the material from the column is called the *development of the chromatogram*. There are three basic systems of chromatographic development, frontal analysis, displacement development, and elution development. Of the three methods elution development in which a discrete sample is placed on a column and mobile phase is passed continuously through it is by far the most common method because the separation can be complete and discrete bands of each individual solute obtained from the chromatographic system.

Frontal Analysis. There are two important ways in which frontal analysis differs from elution development. First, in frontal analysis the sample is continuously fed onto the column during development (usually as a solution in the mobile phase), whereas in elution development a discrete sample is placed on the column and the chromatogram is subsequently developed. Second, elution development can, under the right circumstances, be made to separate completely all the components in the mixture, whereas in frontal analysis only part of the first component is eluted in a relatively pure state, each subsequent component being mixed with those previously eluted.

Consider a three-component mixture containing equal quantities of each component being fed continuously onto a column; because of the forces between solute and stationary phase, each solute will be retained to a different extent as it comes into equilibrium with the stationary phase in the column. The first component to elute will be that which is held least strongly in the stationary phase, then the second component will elute, but in conjunction with the first component, and finally the most strongly held of the three will elute in conjunction with the first and second components. Subsequently, there will be no change in concentration of solute in the mobile phase and the concentration of each respective solute will be the same as the feed mixture. The concentration profile resulting from such frontal analysis when operated under ideal conditions is shown in Fig. 24.3. The continuous curve shows the total concentration of the solute in the eluent plotted against the volume of mobile phase passed through the column and the dotted curve represents a similar concentration profile but for the individual components of the mixture.

Frontal analysis was employed as a development procedure in the early days of LC before detection procedures were fully effective. It is not often used in LC today and virtually never in GC because no individual component is completely separated from the others in the mixture, except possibly the first component, and even this may contain significant quantities of other materials unless the efficiency of the column is high.

Displacement Development. Displacement development depends upon the competition among solutes for the active sites of the absorbant and is only really effective in separating very strongly absorbed materials. Displacement development can be used in either GSC or LSC. In this method of development all the substances in a sample will be held on the stationary phase so strongly that they can not be eluted by the mobile phase but can be displaced by substances that will be held on the surface by even stronger forces. Consequently, in displacement development there will be a competition between individual solutes. When a sample is placed on a column the immediately available active sites of the absorbant will be occupied by the most strongly held solute. As the band of sample is moved down the column the next available sites will be occupied by the next most strongly retained component. Thus, all the components arrange themselves along the column in order of their absorbance strength or, in fact, in order of the magnitude of the forces between them and the absorbant.

To develop the chromatogram another substance called the displacer is introduced into the mobile phase stream: it must be a substance with an even greater affinity for the absorbant than any of the components to be separated. On coming in contact with the first sites, it will displace the most strongly absorbed component into the mobile phase. It will move to the group of sites occupied by the next component and will, in turn, displace that component. Thus a displacer forces the absorbed components progressively along the col-

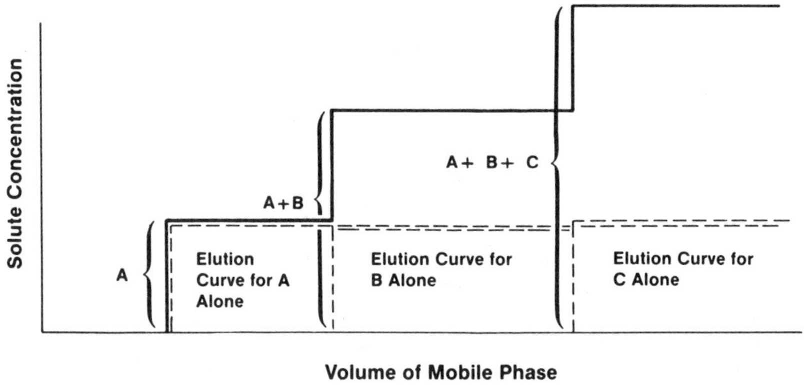

Fig. 24.3 A graphical representation of a chromatogram obtained by frontal development. In this procedure, a mixture containing components A, B, and C is fed continuously into a column. Only component A, which is the one least strongly held, can be obtained pure.

umn. Components will be eluted in the same spatial order as they were absorbed on the column, the least polar being eluted first. The concentration profile at the end of the column will take the form shown in Fig. 24.4. The order in which the solutes emerge will characterize the individual components and the *length* of band, not the height, will be proportional to the concentration of each individual component.

Displacement chromatography has very limited application as a separation technique. It depends on the substance being separated having significantly different response characteristics to the monitoring device and furthermore does not produce discrete local concentrations of individual components. It has however, been very successfully used in preparative GC and LC.

Elution Development. Elution development can be best described as a series of extraction and absorption processes that operate continuously during the passage of a solute down a chromatographic column. After injection the solute is absorbed into the stationary phase, and when fresh, solute-free, mobile phase passes over the point of absorption, the solute is extracted back into the mobile phase. At a point further along the column it will be reabsorbed into the stationary phase and reextracted. This process will be indefinitely repeated until elution from the end of the column. The concentration profile for ideal elution development, where no spreading of the solute band occurs in the column, is depicted in Fig. 24.5a. Each solute is separated completely from its neighbors, all the bandwidths are identical, and the bandwidth of any solute is equal to the bandwidth of the injected sample.

In practice, however, the exchange between the mobile phase and the stationary phase is not instantaneous. This nonideality of solute transfer, together with longitudinal diffusion and other factors, causes each band to spread or suffer dispersion. The combination of these factors results in a chromatogram with the shape shown in Fig. 24.5b: a typical Gaussian curve is produced for each solute. Elution development is by far the most effective method of development and is used extensively in both GC and LC. In fact, virtually all analytical chromatography of any kind is carried out using elution development.

Other Forms of Elution Development. The process of elution development can also be modified both in GC and LC to extend the range of solutes that can be

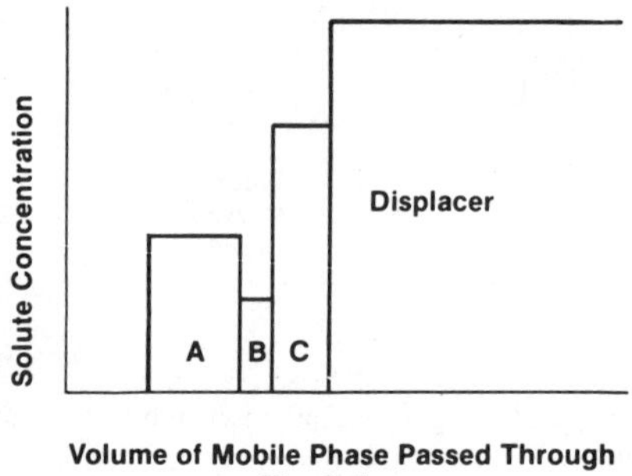

Fig. 24.4 A graphical representation of a chromatogram obtained by displacement development. After a sample containing components A, B, and C is placed on the column, a substance called the displacer, which will be held more strongly than any component, is fed continuously into the column until it appears in the eluent. At that time all components have been eluted.

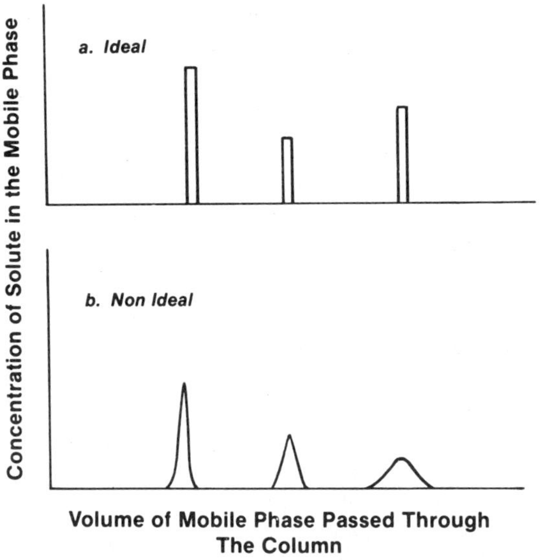

Fig. 24.5 A graphical representation of a chromatogram obtained by elution development. A sample containing a three-component mixture that is to be separated is initially introduced into the mobile phase. Additional pure mobile phase is admitted continuously until all components have been eluted.

chromatographed. Both the polarity and molecular-weight range of solutes can be considerably increased. In GC *temperature programming* is the most effective method for handling a wide range of solute molecular weights. While elution development is proceeding the temperature of the column is continously raised so that the classes of solutes that are ordinarily more retained, the more polar and/or the higher boiling materials, are eluted in a reasonable time. Raising the column temperature increases the distribution of these solutes into the mobile phase and causes them to be eluted more rapidly.

In LC *gradient elution*, whereby the composition of the solvent is continuously changed during development, is the best method for increasing the polarity or molecular-weight range of the solutes that can be separated. Normal-phase systems employing silica gel as the stationary phase will require a mobile phase that continuously increases in polarity during gradient elution. Conversely, gradient elution with reversed-phase systems will call for the dispersive characteristics of the mobile phase to be continuously increased.

Both temperature programming and gradient elution are very common ways of modifying elution development in GC and LC, respectively, to allow solutes having a wide range of polarity or molecular weight to be eluted in a reasonable time. Gradient elution can never be employed in GC, though on occasion temperature programming has been employed in LC.

Flow Programming. A further way to extend the solute range in elution development that sees very occasional use is flow programming. It is achieved by progressively increasing the mobile phase flow rate during a separation. This procedure accelerates the movement of the more retained solutes, allowing them to elute in a reasonable time. Flow programming is attractive in the separation of labile materials which could decompose at elevated temperatures or when in contact with strongly polar solvents. It is relatively rarely used because temperature programming and gradient elution tend to increase the solute migration rate in an almost exponential fashion in contrast to an essentially linear growth of the rate of solute migration produced with flow programming.

24.5 THE COLUMN

The chromatographic column is the heart of the chromatograph and constitutes the module in which the separations are carried out. It usually takes the form of a tube which can vary from a short liquid chromatography column 3 or 4 cm long to a capillary column 50–100 m long for use in GC. In LC the vast majority of columns to date are packed with microparticulate material ranging from 3 to 20 μm in diameter, depending on the length and the pressure available to operate the chromatographic system. Most of the packed columns have lengths between 3 and 25 cm, although some as long as 10 m or more have been constructed.

In GC the particles are usually significantly larger, 80 or 100 μm in diameter or perhaps more, and serve as supports for a layer of the stationary phase. In GSC the particles consist of native alumina or silica or of molecular sieves of commensurate size, 100 μm or more in diameter. Many GC separations are still carried out with packed columns, most under 3 m in length.

Capillary or open tubular columns are now becoming very popular in GC. Modern capillary columns are usually made of glass or fused silica tubing 0.25–0.5 mm i.d. and from 20 to 100 m long. Normally, the internal surface is deactivated (sometimes by a procedure that appears more akin to magic than science) and then coated with an appropriate stationary phase, such as a high boiling liquid or an appropriate polymer. Capillary columns are employed in GC mostly where complex mixtures are to be separated and high efficiency is needed. They are rarely used in LC today, but are being extensively examined by a number of workers in the field and may be developed into a more useful LC column system in the future.

The column performs two basic functions in any chromatographic process. First, as a result of the selectivity of the phase system individual solutes are moved apart in the column. Such movement depends of course on the choice of phases and, as already described, results from the differing molecular interactions of solutes with the two phases. The exclusion characteristics of the stationary phase may also play some part, as previously discussed. To maximize the possibility that the individual solutes can be eluted discretely, columns are designed and packed in such a way that band dispersion is minimized and solute bands are kept narrow. The extent to which this goal is met is measured in terms of theoretical plates and called *column efficiency*. The more narrow the bands, the better the resolution and the higher the column efficiency. The significance of these measurements will be discussed in due course. Thus, the column has to achieve two goals to separate mixtures successfully: it must provide conditions that allow peaks to *move apart* and *at the same time remain narrow.* In this way it will allow the individual solutes to be eluted discretely.

24.6 THE CHROMATOGRAM: TERMS AND NOMENCLATURE

A normal chromatogram, whether it be obtained from a gas or liquid chromatograph, is an analog curve relating concentration of solute in the column eluent to time. If a constant column flow is maintained during the development of a chromatogram, the time axis will be directly proportional to the volume of eluent passed through the column. A typical chromatogram for a single component eluted from a column is shown in Fig. 24.6. The elution curve has the form of an error function curve (discussed earlier in Section 10.2), and because of this it is worthwhile here to consider further the characteristics of such curves and their derivatives. The basic error function is described by the equation

$$X = \frac{X_0 e^{-W^2/2n}}{\sqrt{2\pi n}},$$

where X is the dependent variable and corresponds to the solute concentration in the column eluent, W is the independent variable and corresponds to the column flow, and X_0 and n are constants. This form of the error function is chosen because

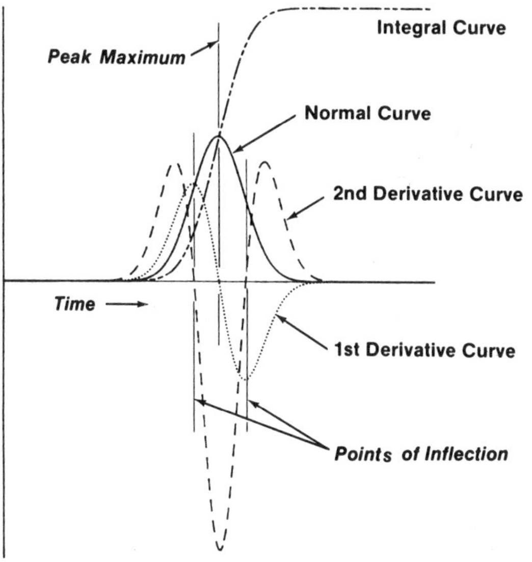

Fig. 24.6 Forms of a chromatographic peak. They are identified as follows: normal peak ——, first derivative · · ·, second derivative – – –, and integral curve – · · – · · –. The latter three curves aid in identifying key aspects of the normal curve. The height of the integral curve is proportional to the area under the normal error function curve. The first derivative becomes zero at the maximum of a peak. The points of inflection of the normal curve are identified by zero values of the second derivative.

it can be directly derived as the equation for the elution curve from the plate theory of chromatography as will be seen later.

The first and second derivatives of the equation are

$$\frac{dX}{dW} = -\frac{X_0}{\sqrt{2\pi n}} \cdot \frac{W}{n} e^{-W^2/2n} \quad \text{and} \quad \frac{d^2X}{dW^2} = \frac{X_0}{\sqrt{2\pi n}} \cdot \frac{(W^2 - n)}{n^2} e^{-W^2/2n}$$

Graphs of these three equations are included together with the integral of the error curve in Fig. 24.6. Each is named in the figure. It is important to note some uses of these curves. For example, the height of the integral curve is proportional to the area under the normal curve; in quantitative analyses the integral curve will yield the total area of each peak in the chromatogram. In addition, the first derivative curve has useful characteristics for the measurement of peak retention times. The position of the maximum of a peak can be obtained from the first derivative where the signal goes through zero to a negative value. It should be pointed out, however, that in differentiating a signal, if any noise is present, the S/N, and thus the sensitivity of the measurement, will decrease.

Since most GC and LC is carried out employing elution development, normal chromatograms are commonly obtained. Figure 24.7 depicts a chromatogram for two substances and also presents and identifies graphically the major terms used in describing chromatographic peaks. While the figure is information-dense, it can be interpreted straightforwardly by examining it step by step. Along the vertical axis will of course be the response of the detector to the eluent emerging from the column. Note that there are alternative choices for the variable to be plotted along the abscissa. Since the mobile phase passes through the system at a steady rate either time (the lower horizontal coordinate in Fig. 24.7) or the volume of the mobile phase passed (the upper horizontal coordinate) may be plotted along the abscissa. Each chromatographic separation begins with the injection of a sample of the mixture to be resolved. When it is injected, a mark is usually made on the chromatogram to define the beginning of the analytically significant portion; this point is usually called the injection point.

Fortunately, a substance that is not retained in the column is ordinarily present in a sample and if not, one is added to the mixture to aid in later peak identification. As in the figure, the time at which this unretained substance elutes is usually called the *dead time* or the dead point and is designated t_m. Its significance is that it also represents the volume of mobile phase that has passed through the column between the injection point and the dead time. This quantity is called the *dead volume* of the column, V_m. This volume in fact includes all the volume of the system between the injection point and the detector that is not occupied by stationary phase or support as long as two assumptions hold. The first is that the two phases are in equilibrium, and the second, that there is no mobile phase isolated from the bulk of the mobile phase by the stationary phase. The dead volume is obtained by multiplying the dead time by the flow rate.

The *baseline* is the portion of a chromatogram recorded when only mobile phase is emerging from the column. The point at the maximum concentration of any

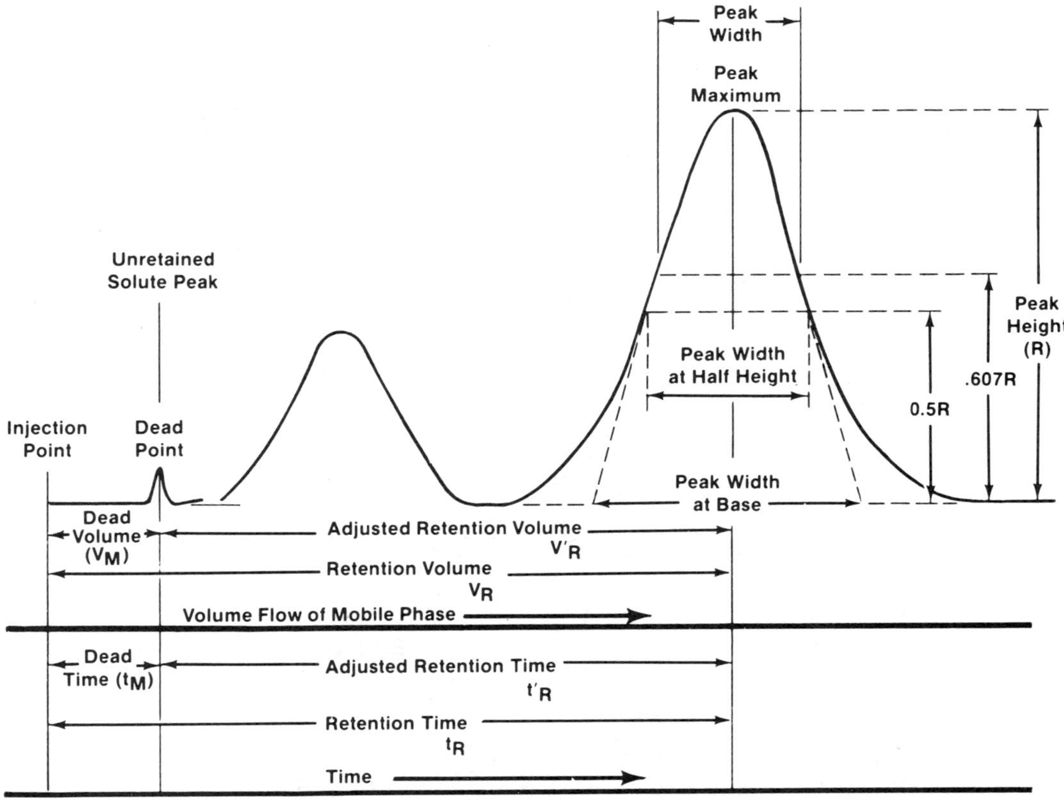

Fig. 24.7 A chromatogram showing complete separation (resolution) of two substances as a function of time and the volume of mobile phase that has passed through the column.

eluted peak is the peak maximum and the distance between the peak maximum and a line joining the base of the peak by extrapolation of the baseline is the peak height. There are various measurements used for the peak width. The width that has a theoretical significance and which should be used wherever possible is twice the standard deviation of the elution curve; it is measured at 0.607 of the peak height. The product of that width and the peak height will always give 79.8% of the total peak area for a true error function curve.

For convenience, in quantitative analyses the peak area is often taken as the product of the peak height and the peak width at half the peak height. This parameter is given the term peak width at half height. Another measure of the peak width, which is sometimes used, involves constructing the tangents to the points of inflection of the normal curve and measuring the distance between their intersection with the baseline produced. This is called the peak width at the base and is, in fact, equivalent to twice the peak width at 0.607 of the peak height.

Those characteristics of a chromatogram that have a theoretical significance will be discussed later, but the common measurements that are made are also illustrated by the right-hand peak in Fig. 24.7. The time between the injection point and the peak maximum of any peak is called the *retention time*, t_R, and the time between the dead time and the peak maximum is called the *adjusted retention time*, t'_R. If the retention time is multiplied by the flow rate, one obtains the retention volume V_R and similarly, multiplying the adjusted retention time by the flow rate gives the *adjusted retention volume* V'_R.

In order to obtain some rationalization between GC and LC the symbols used for chromatographic parameters given above are taken from those recommended for gas chromatography by the Gas Chromatography Discussion Group. Since the physical processes involved in GC are identical with those in liquid chromatography, the use of the same symbols for both techniques is permissible and will produce no confusion.

One further parameter ought to be mentioned, which to some extent is arbitrary but which is a useful measure of the column performance, and that is *resolution* R. The significance of resolution will also be discussed later. The value determined from the chromatogram, which is normally taken as a measure of resolution, is

$$R = (t_{R_2} - t_{R_1})/(W_{0_2} + W_{0_1}),$$

the ratio of the distance between the peak maxima of the two respective peaks and the sum of their peak widths at 0.607 of the peak height.

Unfortunately, pure Gaussian or error-function curves are not always produced for solutes in chromatographic separations. Some other common types of peaks are identified and described in Fig. 24.8.

24.7 PLATE THEORY

Since chromatography has such a large number of operating variables and such a wide range of application, a clear understanding of the theory of the process is

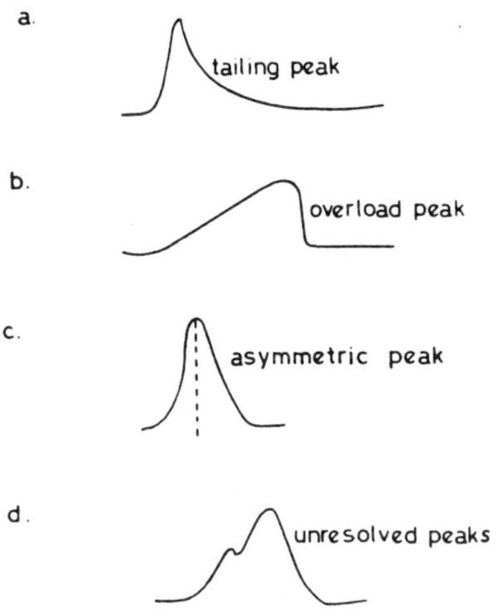

Fig. 24.8 Some non-Gaussian types of peaks produced in column chromatography. A *tailing peak* is characterized by a sharp front and a long tail while an *overload peak* has a sloping front and a sharp tail. Asymmetry in a peak cannot be related directly to a specific process. Unresolved peaks result from incomplete separation.

more essential for its efficient operation than with many other techniques. The theory of chromatography is now fairly well understood and can indeed show how solute characteristics and operating conditions affect the speed, resolution, and loading capacity that can be obtained from a given column system. It can also indicate how the system may be optimized to give any one of these attributes in excess. To do this the theory provides the basis for choosing a suitable distribution system, gives the necessary details for column design, and predicts the operating conditions necessary to effect a given separation.

There are two approaches to the theory of chromatography, the plate theory and the rate theory. The former was introduced by Martin and Synge and is applicable to all forms of partition chromatography. The rate theory, which was developed for GC by Van Deemter, Zuiderweg, Klinkenberg, and others will be discussed in the following section.

The approach to the plate theory is more straightforward and involves relatively simple mathematics, and for this reason will be considered first. The equations derived are quite general and can apply to either GC or LC.

Primarily, the plate theory provides an *equation for the elution curve* of a solute relating the concentration of solute at any point in the column to the volume of mobile phase that has passed that point. It is from the elution curve equation that the various characteristics of a chromatographic system can be determined using the data provided by the chromatogram.

Let the chromatographic column be considered to be divided into a number of cells or plates such that the solute can be assumed to be in equilibrium with the two phases in each plate. Thus, in each plate,

$$X_s = KX_m, \tag{24.1}$$

where X_m and X_s are the concentrations of the solute in the mobile and stationary phases, respectively, and K is the distribution coefficient of the solute between the two phases and is a dimensionless constant. In gas–liquid and liquid–liquid systems X_s and X_m are conveniently measured as mass of solute per unit volume of phase and in gas–solid and liquid–solid systems as the mass of solute per unit mass of phase.*

Equation (24.1) merely states that the general distribution law applies to the system and that the absorption isotherm is linear. At the concentrations normally used in chromatography this will be true for all gas–liquid and liquid–liquid systems and for most gas–solid systems. Only if the absorption isotherm is very close to linear will a system be practically useful, since systems with nonlinear isotherms produce asymmetric peaks.

Consider three consecutive plates in the column, P − 1, P, and P + 1. These plates are depicted in Fig. 24.9. Let there be a total of n plates in the column. Further, let the volumes of the mobile phase and stationary phase in each plate be v_m and v_s, respectively, and the concentration of solute in the mobile and stationary phases in each of the three plates be $X_{M(P-1)}$, $X_{s(P-1)}$, $X_{m(P)}$, $X_{s(P)}$, $X_{m(P+1)}$, and $X_{s(P+1)}$, respectively. Consider that a volume of mobile phase dV passes from plate P-1 into plate P and at the same time displaces a like volume of mobile phase from plate P to plate P + 1. As this happens there will be a change of mass of solute in plate P that will equal the difference between the mass of solute entering plate P from plate P − 1 and the mass of solute leaving plate P and entering plate P + 1. Accordingly, the change of mass of solute in plate P, dm is

$$dm = (X_{m(P-1)} - X_{m(P)})\, dV. \tag{24.2}$$

If equilibrium is maintained in plate P, the change of mass dm will distribute itself between mobile and stationary phases in the plate. As a result there will be a change of solute concentration in the mobile phase of $dX_{m(P)}$ and in the stationary phase of $dX_{s(P)}$. Thus,

$$dm = v_s dX_{s(P)} + v_m dX_{m(P)}. \tag{24.3}$$

Plate P-1	Plate P	Plate P + 1
$X_{m(P-1)}$ v_m	$X_{m(P)}$ v_m	$X_{m(P+1)}$ v_m
$X_{s(P-1)}$ v_s	$X_{s(P)}$ v_s	$X_{s(P+1)}$ v_s

Fig. 24.9 Compositions of successive plates in a chromatographic column. For each plate equilibrium exists between its mobile (upper half) and stationary (lower half) phase portions.

*An excellent discussion on the meaning of K by Karger et al. is recommended to those wishing to pursue the meaning of the distribution coefficient further: S. Karger et al., *An Introduction to Separation Sciences*. New York: Wiley, 1973, p. 11.

The last equation can be simplified. First, a value can be obtained for $dX_{s(P)}$ from Eq. (24.1). On labeling terms for plate P in that equation and differentiating one obtains $dX_{s(P)} = KdX_m$. With this substitution, Eq. (24.3) becomes

$$dm = (v_m + Kv_s)\, dX_{m(P)}. \tag{24.4}$$

Equating Eqs. (24.2) and (24.4) and rearranging,

$$\frac{dX_{m(P)}}{dV} = \frac{X_{m(P-1)} - X_{m(P)}}{v_m + Kv_s} \tag{24.5}$$

To aid in algebraic manipulation it is convenient to effect a change of variable. Consider that the volume of mobile phase will be measured in units of $v_m + Kv_s$ instead of milliliters by defining a new variable v as

$$v = \frac{V}{v_m + Kv_s}. \tag{24.6}$$

The function $v_m + Kv_s$ is called the *plate volume* and is just the volume of mobile phase necessary to contain all the solute at the equilibrium concentration in the plate in the mobile phase. For the present the flow of mobile phase will be measured in plate volumes. Differentiating Eq. (24.6) gives $dv = dV/(v_m + Kv_s)$. Substituting for dV from this equation into Eq. (24.5) gives

$$\frac{dX_{m(P)}}{dv} = X_{m(P-1)} - X_{m(P)}. \tag{24.7}$$

The solution to this equation can be shown by differentiation to yield for $X_{m(N)}$, the concentration at the Nth plate, which is taken to be the one at the end of the column, and that monitored by the detector,

$$X_{m(N)} = [X_m^0 e^{-v} v^N]/N!, \tag{24.8}$$

where X_m^0 is the initial mobile phase concentration at the point of sample introduction. Equation (24.8) is the equation of the elution curve since the detector at the end of the column measures the concentration leaving the Nth plate. The chromatogram is a graph of this equation. Though Eq. (24.8) is a Poisson function, it can be shown that if N is large, the function closely approximates the normal error or Gaussian function. In chromatography, since N is always greater than 100, and for the conditions assumed for this equation, it would be expected that all chromatographic peaks would be Gaussian or nearly Gaussian in shape. It should now be possible to obtain from plate theory equations defining important chromatographic parameters such as retention volume, column efficiency, and resolution.

Retention Volume. A chromatogram containing two peaks to illustrate the definition of parameters is shown in Fig. 24.10. The abscissa is labeled according to the elution curve equation. The dead time (point) on the chromatogram is the position of the maximum for a completely unretained solute. The volume of mobile phase passed through the column between the injection point and the dead time will be equivalent to the dead volume of the column. The retention volume of any substance is the volume of mobile phase that has passed through the column up to the elution of the peak maximum. The equation for the retention volume can, therefore, be obtained by differentiating the elution curve equation with respect to volume and setting the derivative equal to zero to obtain the position of the maximum:

$$\frac{dX_{\mathrm{m}(N)}}{dv} = \frac{X^0_{\mathrm{m}}}{N!} e^{-v} v^{(N-1)}(N - v) = 0.$$

This equation will equal zero when $v = N$. Since v is measured in plate volumes, at the peak maximum, N plate volumes of mobile phase will have passed through the column. Thus, in Eq. (24.6) let $V = V_{\mathrm{R}}$ and $v = N$. The retention volume V_{R} in milliliters of mobile phase is given by

$$V_{\mathrm{R}} = N(v_{\mathrm{m}} + Kv_{\mathrm{s}}). \qquad (24.9)$$

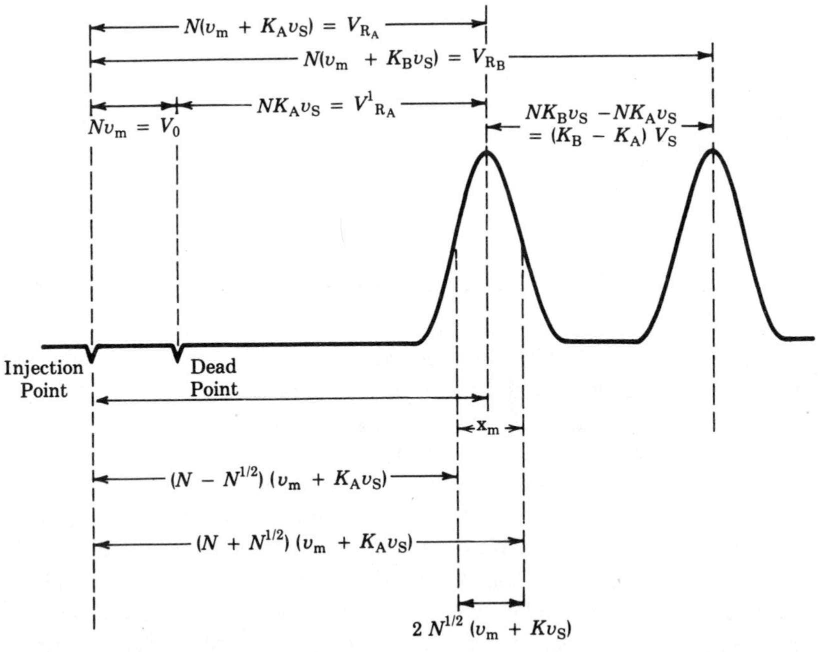

Fig. 24.10 A chromatogram containing two peaks. Functional relationships are overlaid.

Now if there are N plates in the column and v_m and v_s are the volumes of mobile phase and stationary phase, respectively, in each plate, then $Nv_m = V_m$ and $Nv_s = V_s$ are the total volumes of mobile phase and stationary phase in the column, respectively. Equation (24.9) can be rewritten $V_R = V_m + KV_s$. It should be remembered in practice that measured values of V_m include the total free volume of the column system and will include all the mobile phase between the point of injection and the detector. Consequently, volumes associated with detector connections will be included in V_m. If the solute is completely unretained, $K = 0$, and thus the dead volume of the column is given by $V_0 = V_m$. Since the adjusted retention volume $V'_R = V_R - V_0 = V_R - V_m$, it follows that

$$V'_R = V_m + KV_s - V_m = KV_s. \tag{24.10}$$

Since the value of K will be characteristic of the substance eluted, for a column carrying a given quantity of stationary phase V_s, the adjusted retention volume ($V'_R = KV_s$) will provide a parameter by which a solute could be identified.

Furthermore, since V'_R is a function of K, and K is a function of temperature, a change in temperature will modify V'_R in the same way that it modifies K, and thus measurement of the corrected retention volume over a range of temperatures can provide useful thermodynamic data for a given solute–solvent system.

Column Efficiency. In order to separate two substances, their bands must be moved apart as they pass through a chromatographic column. Further, as the two bands move apart, their peak widths must remain sufficiently narrow to permit the discrete elution of each peak. Thus, the peak width of a substance eluted from a column could give a measure of the separating power of the column. It will be useful to obtain the peak width at the points of inflection of the elution curve, which will correspond to twice the standard deviation of the curve, as observed earlier.

At the points of inflection, the second derivative with respect to plate volume must be zero. When the second derivative of Eq. (24.8) is equal to zero one obtains $v^2 - 2Nv + N(N - 1) = 0$. This equation can be solved to give $v = N \pm N^{1/2}$. According to this result, the points of inflection occur after $N - N^{1/2}$ and $N + N^{1/2}$ plate volumes of mobile phase have passed through the column. Further, the volume of mobile phase that passes through the column between the points of inflection will be $(N + N^{1/2}) - (N - N^{1/2})$ plate volumes. This volume is precisely the peak width in milliliters of mobile phase:

$$\text{peak width} = 2N^{1/2}(v_m + Kv_s). \tag{24.11}$$

Because the peak width at the points of inflection of the curve is twice the standard deviation measured in plate volumes $(v_m + Kv_s)$, the variance of the curve (the square of the standard deviation) will be N. Recall that each theoretical plate represents one equilibration of stationary and mobile phase, and the separating ability of a column increases with the number of equilibrations or theoretical plates N.

Thus the variance is a measure of the column separating power and is properly called the *column efficiency*, as noted earlier.

The variance of the solute curve has been chosen as a measure of the separating power of a column as opposed to the apparently more logical, standard deviation, or peak width, for a very good reason. The overall spread of a solute peak is attributable to a number of individual, unrelated band-spreading processes that are random in nature and together contribute to the final peak width. If the individual spreading processes are determined, they must be added together to account for the final bandwidth. As was discussed in Chapter 10, it is possible to add the variance of a number of random processes that contribute to overall variance to provide the final variance—but their standard deviations cannot be used in this way.

To determine the separating power of any column for a particular solute it is clearly important to be able to estimate the number of theoretical plates. It proves straightforward to calculate this quantity. Let the distance between the injection point and the peak maximum of a given solute band, measured on a chromatogram, be y cm and the peak width at the points of inflection be x cm (see Fig. 24.10). From Eqs. (24.6) and (24.9) we calculate the ratio y/x as

$$\frac{\text{ret. distance}}{\text{peak width}} = \frac{y}{x} = \frac{N}{2N^{1/2}}\frac{(v_{\text{m}} + Kv_{\text{s}})}{(v_{\text{m}} + Kv_{\text{s}})} = \frac{N^{1/2}}{2}.$$

Hence the number of theoretical plates for any solute peak calculated from measurements made directly on the chromatogram is

$$N = 4(y/x)^2. \tag{24.12}$$

The various characteristics of a chromatogram that have so far been derived from plate theory are all included in Fig. 24.10.

Column Resolution. It has already been stated that for two peaks to be resolved they must be moved apart in the column and at the same time be maintained sufficiently narrow to permit them to be eluted as discrete peaks. The criterion for two peaks to be resolved will be arbitrary but, if the areas are to be measured to give reasonable quantitative accuracy, the peak maxima of the two solutes must be at least two peak widths apart.*

Equating the separation between two peaks A and B to the peak width at the base of peak A gives $4N^{1/2}(v_{\text{m}} + K_{\text{A}}v_{\text{s}}) = N(v_{\text{m}} + K_{\text{B}}v_{\text{s}}) - N(v_{\text{m}} + K_{\text{A}}v_{\text{s}}) = Nv_{\text{s}}(K_{\text{B}} - K_{\text{A}}) = Nv_{\text{s}}K_{\text{A}}(\alpha - 1)$, where the quantity α, called the *separation ratio* of peaks A and B, has been introduced. Here

$$\alpha = V_{\text{B}}^{1}/V_{\text{A}}^{1} = Nv_{\text{s}}K_{\text{B}}/Nv_{\text{s}}K_{\text{A}} = K_{\text{B}}/K_{\text{A}}$$

*Two adjacent peaks from solutes of different chemical type will not necessarily have precisely the same peak widths. The difference is often not significant, however, and in the following discussion will be assumed to be negligible.

Defining the *capacity ratio* k' of a solute as V^1/V^0 or $k' = Nv_sK/Nv_m$ and substituting above gives $4N^{1/2}(1 + k'_A) = Nk'_A(\alpha - 1)$ or

$$N = 16(1 + k'_A)^2/k'^2_A\,(\alpha - 1)^2 \tag{24.13}$$

Thus, the efficiency required to effect a separation can be calculated from published retention ratios and data obtained by chromatographing one of the solutes under chosen conditions.

It is evident from Eq. (24.13) that the higher the value of the capacity ratio k'_A for a given value of α, the fewer the number of column plates N required to achieve a separation. Further, k' must be greater than unity to keep N small.

The Maximum Sample Volume. In order to produce an effective separation it has been shown that the solutes must be significantly retained in the column, that is, $k' \gg 1$. However, this will mean that the distribution of the solutes is strongly in favor of the stationary phase, which can result in a relatively small absolute solubility of the solutes in the mobile phase. In LC a problem may arise as a consequence. To meet the sensitivity requirements of the detector in LC, it may be necessary to inject a substantial amount of sample and to have it contained in the largest volume of solution that can be placed on the column without seriously impairing the column efficiency. Such a procedure will increase elution bandwidth and degrade resolution. How great an increase in elution bandwidth can be tolerated is obviously a matter of choice. However, the generally accepted maximum increase in bandwidth due to sample volume or, for that matter, any extra column band broadening or dispersion process is 5%, a contribution that will reduce column efficiency by approximately 10%.

Consider a volume of sample V_i injected on to the column forming a rectangular distribution at the front of the column. The variance of the final peak will equal the sum of the variance of this sample volume plus the normal variance of the peak for a small sample. Now the variance of the rectangular distribution of a sample of volume V_i at the beginning of the column is $V_i^2/12$ and, assuming the eluting peak width is increased by 5% by the sample volume,

$$V_i^2/12 + N(v_m + Kv_s)^2 = \left[1.05\sqrt{N}(v_m + Kv_s)\right]^2.$$

Thus $V_i^2/12 = N(v_m + Kv_s)^2\,(1.05^2 - 1) = N(v_m + Kv_s)^2\,0.102$. Further, $V_i^2 = N(v_m + Kv_s)^2\,1.23$ and

$$V_i = \sqrt{N}(v_m + Kv_s)\,1.1. \tag{24.14}$$

Equation (24.14) gives an expression for V_i, the maximum permitted sample volume, in terms of the efficiency N, the volume of mobile phase and stationary phase per plate (v_m and v_s), respectively, and the distribution coefficient of the solute K. However, most of these variables are not usually known and a more useful expression for V_i can be derived from Eq. (24.14) as follows: the retention volume V_R given by Eq. (24.9) is $V_R = N(v_m + Kv_s)$ or $V_R/\sqrt{N} = \sqrt{N}(v_m + Kv_s)$. Substituting $V_R/\sqrt{N}$ for $\sqrt{N}(v_m + Kv_s)$ in Eq. (24.14)

$$V_i = 1.1\,V_R/\sqrt{N}. \tag{24.15}$$

24.8 RATE THEORY

An elution curve or chromatogram can be expressed using parameters other than the volume flow of mobile phase as the independent variable. Instead, the distance travelled by the solute band along the column or the time of elution can serve as the independent variable and proportionally the same chromatogram will be obtained, as shown in Fig. 24.11.

The curves are describing the same chromatogram, because in the units in which all three independent variables are defined, the ratios of the variance to the square of the retention variable will be equal. Thus

$$\sigma_v^2/V_R^2 = \sigma_x^2/L^2 = \sigma_t^2/t_R^2, \tag{24.16}$$

where σ_v, σ_x and σ_t are the standard deviations of the elution peaks when related to the volume flow of mobile phase, distance traveled by the solute along the column, and time, respectively.

From statistics it is well known that variances from different sources add as long as the contributing processes are independent of one another. Using this idea, an important connection between the variances expressed in Eq. (24.16) and factors that contribute to performance in chromatographs can be established. In every calculation or experimental test the measure of performance will, of course, be the breadth of chromatographic peaks. First, a general relationship between the quantities in Eq. (24.16) and the efficiency of a column must be formulated. In Eqs. (24.9) and (24.11) it was shown that $\sigma_v = \sqrt{N}(v_m + Kv_s)$ and $V_R = N(v_m + Kv_s)$.

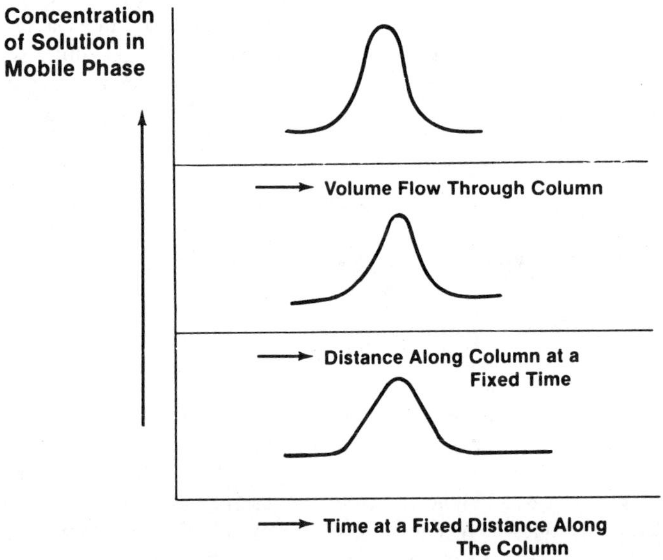

Fig. 24.11 Expression of a chromatogram in terms of different, essentially equivalent parameters.

Substituting these quantities in Eq. (24.16) gives

$$\sigma_x^2/L^2 = N(v_m + Kv_s)^2/N^2(v_m + Kv_s)^2.$$

On simplifying one obtains $\sigma_x^2/L^2 = 1/N$ or $\sigma_x^2/L = L/N$. The ratio L/N, the column length divided by the number of theoretical plates, has for obvious reasons been called the *height equivalent to a theoretical plate* (HETP) and given the symbol H:

$$H = L/N = \sigma_x^2/L. \tag{24.17}$$

Equation (24.17) emphasizes that when we are concerned with the band-spreading processes that occur in a column it is more useful to think of H as the variance per unit length of column. It is, therefore, independent of column length and thus allows the "goodness" of the construction of columns to be compared for columns of any length.

Whether the assumption of the independence of the individual spreading processes is valid is still to some extent a matter of opinion. Nevertheless, whatever interaction may be present is sufficiently small to allow the rate theory to predict fairly accurately the overall band spreading that will occur in a given column.

With this background the *rate theory* may now be examined. Assuming there are n noninteracting random peak-broadening processes in a column, any process (p) acting alone will produce a Gaussian elution curve having a variance σ_p^2. Together, by the principle of the summation of variances, they give for the variance of a peak eluting from a chromatographic column,

$$\sigma_c^2 = HL = \sigma_1^2 + \sigma_2^2 + \sigma_3^2 + \cdots + \sigma_n^2. \tag{24.18}$$

The rate theory considers each individual process that is likely to occur in the column, determines its contribution to the variance per unit length of the column, and then sums the contributions to provide the final value for H.

The first HETP equation was derived by Van Deemter et al.* in 1956 and experimental support for the equation was published by Keuleman and Kwantes† at the first Gas Chromatography Symposium held in London the same year. However, when the equation was applied to LC, it was found that the relationship predicted by van Deemter et al. did not appear to hold. This failure was probably due to artifacts such as those caused by extra column dispersion, large amplifier time constants, and so on. Nevertheless, this poor agreement between theory and experiment provoked a number of workers in the field to develop alternative HETP equations. In 1961 Giddings‡ produced an HETP equation, of which the van

*J. J. Van Deemter, F. J. Zinderweg, and A. Klinkenberg, *Chem. Eng. Sci.*, **5**, 271 (1956).

†A. I. M. Keuleman and A. Kwantes in D. H. Desty and C. L. A. Harbourn (Eds.), *Vapor Phase Chromatography*. London: Butterworth, 1956, p. A10.

‡J. C. Giddings, *J. Chromatogr.*, **5**, 46 (1961).

Deemter equation was a special case, valid when the mobile phase velocity was sufficiently high. At practical mobile phase velocities, the other functions in the equation were similar to those of Van Deemter.

Additional HETP equations were developed by Huber and Hulsman,* Knox and co-workers,* and Horvath and Lin.* As a result of the plethora of HETP equations in the literature, Katz, Ogan, and Scott† carried out an extensive series of carefully controlled experiments to determine H over a wide range of mobile phase velocities for solutes of different k' value and different diffusivities and for columns packed with particles of different diameter. Their results clearly demonstrated that the van Deemter equation not only accurately describes dispersion in a packed bed in GC, but also for LC. Consequently, only the van Deemter equation will be considered here.

Van Deemter and his colleagues considered that four basic processes contributed to the band variance in a chromatograph column, the multipath effect, longitudinal diffusion, and resistance to mass transfer in the mobile and stationary phases.

The Multipath Process. In a packed column the solute molecules describe a tortuous path between the interstices of the support. It is fairly obvious that some molecules will randomly travel shorter paths than others. Those that on an average pass along the shorter path will move ahead of the maximum of the concentration profile, while those molecules that pass along paths of greater length will lag behind the maximum. The variance contributed by this source of spreading the total band variance to H was deduced by van Deemter to be proportional to the particle diameter of the support d_p and a constant depending on the physical nature of the packing γ. Consequently, the multipath contribution to H is given as $\gamma\, d_p$.

Longitudinal Diffusion. If a local concentration of solute is placed at the midpoint of a long tube filled with either a liquid or a gas, the solute will slowly diffuse to either end of the tube. It will first produce a Gaussian distribution with a maximum concentration at the center of the tube and finally, when the solute vapor reaches the end of the tube, end effects will occur and the solute will diffuse until there is a constant concentration of solute throughout the whole tube. The latter effect is never realized in chromatography, but the initial spreading process does occur in the mobile phase of a column. The degree of spreading will obviously be proportional to some function of the time that the solute exists in the mobile phase and thus, if the mobile phase is flowing through the column at a linear velocity of u, the extent of the spreading process will be some function of $1/u$. The variance due to this spreading process will also be directly dependent upon the diffusivity of the solute in the mobile phase and the geometry of the spaces occupied

*J. F. K. Huber, J. A. R. J. Hulsman, *Anal. Chim. Acta*, **38,** 305 (1967); G. J. Kennedy and J. H. Knox, *J. Chromatogr. Sci.*, **10,** 606 (1972); Cs. Horvath and H. J. Lin, *J. Chromatogr.*, **126,** 401 (1976) and **149,** 43 (1978).

†E. Katz, K. L. Ogan, and R. P. W. Scott, *J. Chromatogr.*, **270,** 51 (1983).

by the mobile phase. Van Deemter et al. showed that the variance caused by this process in the packed columns used for GC was $\lambda D_m/u$ where λ is a constant depending upon the column packing and D_m is the diffusivity of the solute in the mobile phase.

Resistance to Mass Transfer in the Mobile Phase. During the movement of a solute band along a column, the solute molecules are continually transferring from the mobile phase into the stationary phase and back from the stationary phase into the mobile phase. This transfer process is not instantaneous, because a finite time is required for the molecules to traverse the mobile phase and enter the stationary phase. Thus, molecules close to the stationary phase will enter it almost immediately, whereas those molecules some distance away from the stationary phase will find their way to it a significant time interval later. However, as the mobile phase is traveling at a given velocity along the column, the solute molecules will move a finite distance along the column during this time interval. Thus, they will be absorbed into the stationary phase further along the column than those that were originally in close proximity to the stationary phase. The result of this delayed transfer results in band spreading and is termed the band dispersion due to the resistance to mass transfer in the mobile phase. Van Deemter showed that the variance of this dispersion was a function of the number of molecules that were present in the mobile phase at any time, that is, a function f_m of capacity ratio k', the particle diameter d_p, the diffusivity of the solute in the mobile phase D_m, and the mobile phase linear velocity u. The contribution that this dispersion effect made to h, the variance per unit length, was deduced to be

$$f_m(k')d_p^2u/D_m.$$

Resistance to Mass Transfer in the Stationary Phase. The dispersion resulting from the resistance to mass transfer in the stationary phase is analogous to that in the mobile phase. Those solute molecules close to the surface of the stationary phase will leave the surface and enter the mobile phase before others that have diffused somewhat into the stationary phase. This again will produce band spreading and contribute to the total variance per unit length. The contribution to H from the dispersion due to the resistance to mass transfer in the stationary phase was deduced by van Deemter to be a function f_s of the capacity ratio k'

$$f_s(k')d_f^2u/D_s,$$

where d_f is the film thickness of stationary phase and D_s is the diffusivity of the solute in the stationary phase. Combining the variances due to the four-band spreading processes to give the final variance per unit length will result in the overall van Deemter equation for plate height H

$$H = \gamma d_p + \lambda D_m/u + f_m(k')\, d_p^2u/D_m + f_s(k')\, d_f^2u/D_s. \quad (24.19)$$

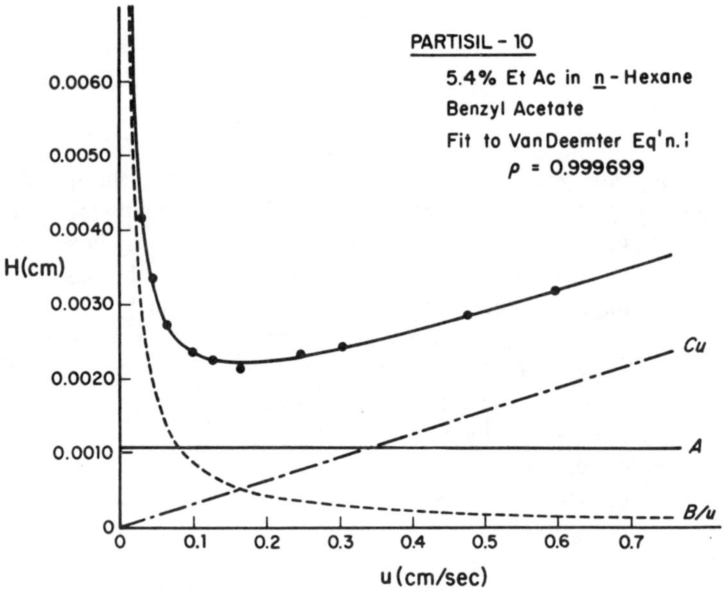

Fig. 24.12 A van Deemter plot (heavy line) for a liquid chromatographic column. The height equivalent to a theoretical plate H is plotted as a function of the average mobile phase velocity. The contributions of different rate processes to H are also shown: A, multipath processes; B/u, longitudinal diffusion; and Cu, mass transfer involving both phases.

A representative plot of H for a packed LC column is shown in Fig. 24.12. The points are experimental and the line through the points is from a statistical fit of the van Deemter equation to the data. In addition, the contributions arising from the multipath effect, longitudinal diffusion, and the combined resistance to mass transfer in the mobile and stationary phases are all illustrated in the figure. They are plotted as a function of mobile phase velocity using values obtained from the constants derived from the curve-fitting procedure.

Band Dispersion in an Open Tubular or Capillary Column. The capillary column does not contain packing material and the mobile phase is coated on the walls of the tube as a thin film. Often these columns are called wall-coated open tubes (WCOT). Dispersion effects in such a column occur from longitudinal diffusion and resistance-to-mass transfer in the mobile and stationary phases that are complementary to the same effects in a packed bed. In the absence of packing material there can, of course, be no multipath contribution to band dispersion. The HETP equation for a capillary column was derived by Golay in 1958 and took the following form:

$$H = \frac{2D_m}{u} + \frac{(1 + 6k' + 11k'^2)\, r^2 u}{24(1 + k')^2 D_m} + \frac{2k}{3(1 + k')^2} \frac{d_f^2\, u}{D_s}$$

where r is the radius of the capillary bore. The first term describes the variance due to longitudinal diffusion and the second and third terms the variance due to the resistance-to-mass transfer in the mobile and stationary phases, respectively.

The Golay equation provides a curve relating H and u having very similar form to that given by the van Deemter equation for a packed bed. Some differences are also evident. As a result of the simple geometric form of a capillary column it contains no arbitrary constants such as γ and λ, as were necessary in the van Deemter equation. It is also seen that when the mobile phase is a gas, D_m is significantly dependent on pressure and thus changes as u changes. Consequently, with a long column operating a high-inlet pressure (200 psi) the curves relating H and u are not accurately described by the Golay equation unless a particular value of u is employed.

Extra-Column Dispersion. Unfortunately, band dispersion does not take place solely within the chromatographic column. Band dispersion can occur in the injection device, in the connecting tube between injection device and column, in the connecting tube between column and detector, and in the detector itself. Further sources of dispersion that can be very significant in high-speed chromatography are the rate of response of the detector sensing system, the time constant of the detector electronics, and even the speed of response of the recording device. In dealing with extra column dispersion, it is important to note that these sources of dispersion have been far more significant in LC than in GC.

The major contribution to extra-column dispersion in any chromatographic system results from the parabolic velocity profile that accompanies the flow of any fluid through a tube. The profile in turn depends on the fluid viscosity and the diffusivity of the solute in the carrier fluid. If the fluid is a gas, the viscosity of the fluid is low and the solute diffusivity is relatively high. Consequently, extra-column dispersion in GC is usually relatively small in the fluid system of the chromatograph. Furthermore, on-column injection is frequently employed in GC and there is close column-detector association, both of which reduce the possibility of significant extra-column dispersion. Nevertheless, in GC, early peaks are often eluted rather fast and consequently the speed of response of the overall detecting–recording system can be important. An excellent discussion of extra-column dispersion that can take place in a gas chromatographic system has been given by Sternberg.*

Extra-column dispersion in the fluid system of a liquid chromatograph, however, can affect column performance significantly. In LC the viscosity of the mobile phase is much greater than that of a gas and the diffusivity of the solute in the mobile phase is four to five orders smaller.

To minimize extra-column dispersion in a liquid chromatographic system, several precautions have to be taken. First, the sample volume should be made as small as possible commensurate with the column dimensions, and sample valves with internal sample loops (*not* external loops) should be employed. Second, all connecting tubes should be made as short as possible and have the smallest internal diameter that is practical. In the extreme, if a small-bore column is used, the column should be connected directly into the sample valve and if possible the other end of the column connected directly into the detector. Alternatively, low-dispersion, serpentine connecting tubing can be used to connect sample

*J. C. Sternberg, *Adv. Chromatogr.*, **2,** 205, (1966).

valve to column and column detector.* Third, the detecting cell volume should be made as small as possible (preferably less than 1.5 μL), and if a cylindrical cell is employed the cell aspect ratio (cell length/cell radius) should be made as large as possible, commensurate with the need to provide a maximum S/N output from the detector. It should be remembered that the extra-column fluid dispersion controls both the mass sensitivity of the system and the solvent economy.

Modern liquid chromatographs are designed to have very small fluid dispersion so that columns of very small diameter can be used. Small-diameter columns will provide high mass sensitivity and very low solvent consumption. The overall fluid dispersion of a chromatograph is therefore an important specification that should be provided by the manufacturer and can permit the minimum column diameter for satisfactory operation to be calculated, as will be discussed in Chapter 26. A well-designed liquid chromatograph should provide an extra column-dispersion peak standard deviation at a flow rate of 1 mL per min of less than 1 μL (a dispersion variance of 1 μL^2).

24.9 QUANTITATIVE AND QUALITATIVE ANALYSIS

No introductory chapter to the techniques of GC and LC would be complete without a section dealing with quantitative analysis. Some examples of quantitative analysis both for GC and LC will be included in the subsequent two chapters, as well as detailed procedures that are unique to each technique, but the general principles and procedures will be given here.

Actually, identification of each separated analyte must precede its quantitative measurement by chromatography. Once analytes are identified, the appropriate standards for calibration can be chosen. In most cases the initial qualitative analysis is a matter of matching the retention behavior of analytes directly with retention times of known compounds, or especially in the case of LC, with their capacity factors. It is sometimes necessary, however, to confirm such an initial analysis by an independent method.

The accurate identification of chromatographic peaks is a critical issue in trace analysis, for example, for forensic or environmental studies. Corroborating each identification by independent evidence such as IR spectra or even by data secured by a second type of chromatographic detector proves essential. A more detailed treatment of qualitative analysis for both GC and LC will be given in Chapters 25 and 26.

To carry out quantitative chromatographic analysis the detector must have a linear response, as will be discussed in the sections dealing with LC and GC detectors. No LC detector presently used provides the same or a similar response to all compounds. It follows that a normalization procedure, where the percentage composition of a component is obtained by expressing the area of a peak as the percentage of the total area of all the peaks, cannot be employed. Nevertheless, in GC, when employing the flame ionization detector, approximate analyses can

*E. Katz and R. P. W. Scott, *J. Chromatogr.*, **268**, 169 (1983).

be obtained, though this procedure is not recommended. In general, a detector should be calibrated for all components to be quantitatively determined.

There are two general methods for obtaining quantitative measurements from a chromatogram, one using peak heights and the other peak areas. The relative merits of these methods have been, over the years, the subject of some controversy.

Quantitative Measurements Using Peak Heights. Many detectors are concentration-sensitive devices and, therefore, for quantitative accuracy the concentration at the peak maximum must reproducibly relate to the total mass of solute if peak heights are to be used as the basis of measurement. Now the peak height is inversely related to the peak width, and thus the peak widths must also be reproducibly constant relative to other peaks in the chromatogram. It follows that the k' of the solute must remain constant, which entails careful control of the stationary phase activity, and in the case of LC, maintaining a constant mobile phase polarity together with good temperature control. The method of sample injection must also be highly reproducible since poor injection can broaden early peaks in the chromatogram relative to peaks that are eluted later, thus falsely reducing their height. Because of the general shape of the HETP curve, control of the mobile phase need not be excessively precise since the *relative* peak heights will not change significantly with small changes in flow rate. Thus, peak heights will give accurate quantitative measurements providing the phase system is well controlled with respect to stationary phase activity and column temperature, and a reproducible injection technique is employed by adequately skilled personnel.

Quantitative Measurements Using Peak Areas. If the peak concentration (peak height) is integrated with respect to the volume flow of mobile phase across the peak, the integral will provide a value that is relative to the absolute mass of solute represented in the peak. Furthermore, because the peak height is always inversely related to the peak width in volume flow of mobile phase, the area integral will tend to be independent of changes in bandwidth resulting from differing phase selectivities, temperature variation, or even changes resulting from poor reproducibility of injection. Thus, it is no longer necessary to control the stationary phase activity, mobile phase polarity, or column temperature carefully or even to employ a reproducible injection system except to ensure that the required resolution is maintained.

A possible source of error in obtaining the concentration/volume integral arises in practice. In all cases in which the peak area is measured against a time axis, such as instances in which it is obtained from a chart, the retention axis value is not directly related to volume flow of mobile phase but to time. It follows that, for accurate quantitative analysis, volume flow of mobile phase must be precisely related to elapsed time and thus a precisely controlled mobile phase flow rate must be ensured. Therefore, if peak areas are to be used for quantitative analysis, a relatively costly gas-flow controller or an expensive pump must be used to provide precisely controlled and constant flow rates. Under these circumstances, carefully controlled chromatographic conditions and highly reproducible injection procedures are no longer essential. Thus, both peak height and peak area measurements can provide accurate quantitative analysis and the choice will depend on the apparatus available and the skill of the personnel involved. It should be pointed out, however, that if deconvolution of partially resolved peaks is carried out, then areas integrated over specific limits through the envelope of the unresolved peaks as determined by a computer will give far greater accuracy than the measurement of peak heights.

Procedures for Quantitative Analysis. The procedures outlined below are given for peak area measurements, but apply equally to peak height determinations. In the equations given, appropriately designated terms for Y (peak area) can be replaced by corresponding terms of H (peak height) and the resulting equations will be applicable for quantitative analysis using peak height measurements.

In the method of *internal standards*, to determine the percentage of a particular solute, its area must be compared with that of a standard that has been added to the sample in a known concentration. The standard must be quite pure and must be so chosen that it is eluted at a point in the chromatogram where it does not overlap the peaks of other substances present.

Consider the determination of a substance A in a particular sample using a standard B. Let a two-component mixture be made up have X_1^A% of compound A and X_1^B% of compound B and let the sample be injected on the column providing two peaks of area Y_1^A and Y_1^B, respectively. If the detector response factors for substances A and B are a and b, respectively, then

$$X_1^A/X_1^B = aY_1^A/bY_1^B = \psi Y_1^A/Y_1^B$$

where $\psi = a/b$ and is the *relative response factor* of substance A to that of substance B. Hence

$$\psi = X_1^A Y_1^B / X_1^B Y_1^A$$

and the proportionality factor ψ for the solutes A and B can be calculated. To determine substance A in an unknown sample add to the sample solute B at a percentage concentration X_2^B. Let the chromatogram for this mixture provide peak areas for A and B of Y_2^A and Y_2^B, respectively. Then the concentration of A in the sample, X_2^A, will be given by

$$X_2^A = \psi Y_2^A X_2^B / Y_2^B.$$

The same principle can be used for estimating any number of solutes in a mixture providing calibration samples are prepared and chromatographed to determine the relative response factors ψ for the standard and each of the solutes to be measured.

For simple mixtures having a limited number of solutes that are completely resolved on the chromatogram, the absolute response factor for each solute can be determined, each peak area corrected for its response factor, and a normalization procedure carried out. For example, if a mixture contains solutes A, B, C, D, . . . that are completely resolved by the column and their respective absolute response factors relative to a given standard are $a, b, c, d, \ldots$, then the percentage of any solute (e.g., C) in the mixture will be given by

$$(\mathrm{C}c/\mathrm{A}a + \mathrm{B}b + \mathrm{C}c + \mathrm{D}d + \cdots) \times 100$$

Unfortunately, the condition for complete resolution of all components in the mixture does not often arise in practice so that the latter method for quantitative analysis frequently cannot be used.

REFERENCES

Books dealing with the general subject of chromatography are a little scarce as authors tend to limit their treatment to gas chromatography, liquid chromatography, or some specialized aspect of them. The following texts are recent and well written and tend to cover the subject of chromatography in a general way.

1. E. Heftmann (Ed.), *Chromatography: Fundamentals and Applications of Chromatographic and Electrophoretic Methods, Part A Fundamentals, Part B Applications*, New York: Elsevier, 1983.
2. P. Sewell and B. Clarke, *Chromatographic Separations.* New York: Wiley, 1988.
3. E. Katz, *Quantitative Analysis Using Chromatographic Techniques.* New York: Wiley, 1987.
4. J. M. Miller, *Chromatography: Contrasts and Concepts.* New York: Wiley, 1988.
5. F. Bruner (Ed.), *The Science of Chromatography.* New York: Elsevier, 1985.
6. J. A. Jonsson (Ed.), *Chromatographic Theory and Basic Principles.* New York: Dekker, 1987.
7. A. S. Said, *Theory and Mathematics of Chromatography.* New York: Huthig, 1981.

Chapter 25

GAS CHROMATOGRAPHY

John D. Walters

25.1 INTRODUCTION

Gas chromatography is a columnar separation technique based on the differential interaction of volatized sample components with a liquid, or sometimes a solid, stationary phase. As sample components are transported through a column containing the stationary phase by an inert carrier gas mobile phase, they distribute themselves between the two phases. Along each segment of a column each component achieves equilibrium between the mobile and stationary phases a very large number of times. All compounds spend the same time in the mobile phase. Any analyte that is weakly retained by the stationary phase will spend less time in this phase and therefore be propelled to the end of the column at nearly the velocity of the carrier gas. However, a compound that interacts strongly with the stationary phase will spend relatively more time in it and emerge considerably later from the column.

As described in the last chapter, components of mixtures ideally appear in separated bands at the end of the column. The reader will wish to draw heavily on that chapter for description and the theory of the chromatographic process which included (a) plate theory to describe the equilibration of components between phases that leads to band formation, (b) rate theory to interpret the several band-broadening mechanisms, and (c) the basis for quantitative chromatographic measurements.

Basically there are two classes of gas chromatography (GC). In *gas–solid chromatography* (GSC) the stationary phase consists of silica or some other adsorptive substrate such as alumina or molecular sieves. Compounds are retained by adsorption on the substrate surface or by diffusion into its pores. While much of the early work in GC was performed on such columns, today GSC is primarily employed to separate mixtures of low molecular weight gases such as nitrogen, hydrogen, and oxygen and some low boiling hydrocarbons. Occasionally, the combined application of gas–solid and gas–liquid chromatography can be useful.

Gas–liquid chromatography (GLC) is by far the more common form. In this technique relatively high boiling liquids are coated on a substrate such as diatomaceous earth particles and these are packed into a column. If the diatomaceous earth support is omitted and the wall of a small diameter tube is coated with the liquid phase the system is termed an open-tubular or "capillary" column. In either case the analytes interact with the stationary phase by forces that range from hydrogen bonding to London dispersion forces. It is the combined interactions of each component that establish its relative retention time in the column.

The development of thermally stable liquid phases of known chemical nature and minimum volatility, whether for packed or open tubular columns, has led to the wide application of gas chromatographic separations of mixtures of volatile substances. All components of a mixture must have a boiling point no more than 100° above the highest useful column temperature (about 350°C) and although this restricts the separations possible, it has been estimated that the natural domain of GLC includes the analysis of roughly one-fifth of all types of substances. Successful separations in which volatility is not a problem are likely to continue to be performed by GC. In many other cases, however, high performance liquid chromatographic procedures will probably continue to replace gas chromatographic procedures, as will be explored in the following chapter.

25.2 MODULES IN A GAS CHROMATOGRAPH

A view of a gas chromatograph from the modular perspective is provided in Fig. 25.1. The first block represents a source of pressurized carrier gas such as helium. Next is the sample injection device, which is located close to the top of the chromatographic column so that the sample is carried easily into and through the column. Next there is a detector to sense separated components of a sample as they emerge. The injector, column, and detector are placed in an oven, whose temperature can be maintained constant or programmed to vary with time as needed. The detector output signal is amplified, and in all but the simplest systems, a microcomputer receives and stores that output. Either this microcomputer or an external integrator–computer will calculate component peak areas and retention times from the data. Finally a readout module will display the chromatogram of the mixture.

Manufacturers of gas chromatographs generally attempt to integrate all of the elements described above to provide a fairly optimum device capable of the widest range of instrumental conditions. For example, nearly always a microprocessor will also provide versatile and reliable control of operating variables such as gas flow and oven temperature. Other features such as special sampling devices and column-switching valves may also be incorporated into the instrument to solve specific analytical problems.

Sample
↓
Carrier gas source → Injection device → Column →
Detector → Processing modules → Display

Fig. 25.1 The modules of a gas chromatograph. On the left is a source of pressurized gas and pneumatic controls. Each sample is introduced through an injection port, resolved into components in the column (in an oven). Its components are sensed by the detector. In most cases the detector measures the response of column eluate relative to that of carrier gas under the same conditions.

Carrier Gas. Usually the mobile-phase gas supply is a cylinder of compressed gas provided with a two-stage regulator. The choice of carrier gas has important consequences for column efficiency, flow rate, analysis time, and other operational parameters. Naturally, a gas should be selected that is also compatible with the detector and types of samples expected, as will be discussed in Section 25.5. Nitrogen, argon, helium, hydrogen, and a mixture of argon and methane are common choices. In the United States, where helium is relatively abundant, it is the most widely used carrier gas, and in Europe and other areas, hydrogen and nitrogen are more common.

The curves in Fig. 25.2 illustrate the influence of gas density on column efficiency and flow rate. The minima in the van Deemter plots show that the optimum flow rate is a little higher for the low-density gases He and H_2. Note that the linear part of the N_2 curve has a steeper slope than do the linear portions of He and H_2 curves. As a result, a useful tradeoff is possible. By using a light gas, the flow rate can be increased significantly with little loss of separation efficiency and the analysis time can be significantly shortened. In general, practical flow rates are greater than the optimum value. Measurements that have shown a 65% reduction in analysis time can be achieved with *only a 20%* loss of separation efficiency.

Gas Purity. The carrier gas should also be the purest available, usually, 99.995% pure. Some common impurities, for example, hydrocarbons, fluorocarbons, or oxygen, can in-

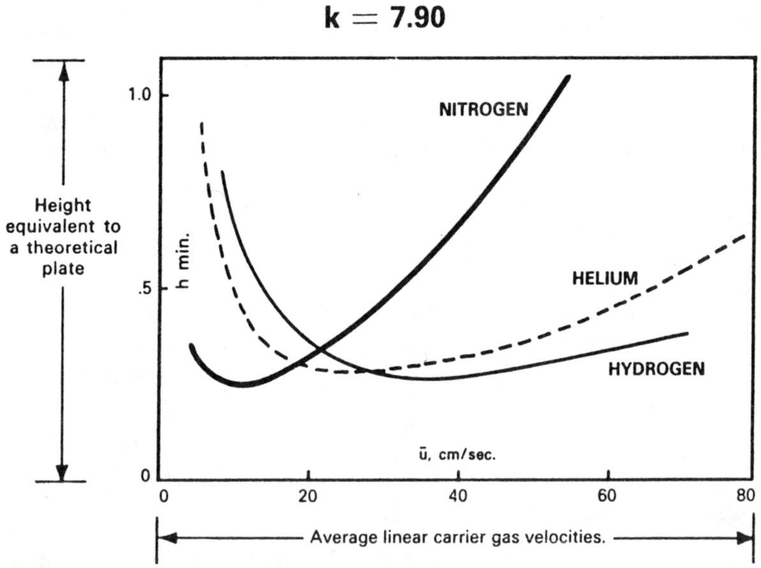

Fig. 25.2 The effect of mobile phase density on column efficiency. Van Deemter plots for three common carrier gases for a column of capacity factor k′ = 7.90. The low density gases (H_2 and He) have optimum efficiencies (HETPs) at slightly higher flow rates than N_2. The much lower slopes of the hydrogen and helium curves allows them to be used at substantially higher flow rates (e.g., compared with nitrogen) with very little loss of separation efficiency. (Courtesy of J & W Scientific, Inc.)

terfere with analytes. Others such as water, oil, or particulates can contaminate and build up on valves, flow controllers, and especially columns and detectors and can lead to failure. For this reason absorbing traps containing silica gel, Drierite, or activated charcoal are usually installed between the tank and the instrument as a final trap to scrub the carrier gas.

The Pneumatic System. Since the flow of the mobile phase links all the modules of a gas chromatograph through the detector, it is valuable to examine the behavior of this part of the instrument as a pneumatic system. For example, the carrier gas must pass through pressure and flow controls.

In simple GCs, and usually in those with open tubular columns, a two-stage pressure regulator provides gas at a constant inlet pressure and a gauge such as a soap-bubble flowmeter (see Fig. 25.3) can be employed to monitor the flow rate. There are really no subsequent flow controls since *pressure control* of the carrier gas suffices to maintain a constant flow.

Actually the bubble meter and many other meters measure the *volume flow* rate of gas in cubic centimeters per minute and not the mean linear gas velocity ($\bar{u}$) in centimeters per second often called for in chromatographic theory. These volume flow values are sufficient to establish reproducible measurements; if needed, absolute linear flow rates can be calculated by making corrections for water vapor pressure, temperature, and gas pressure.

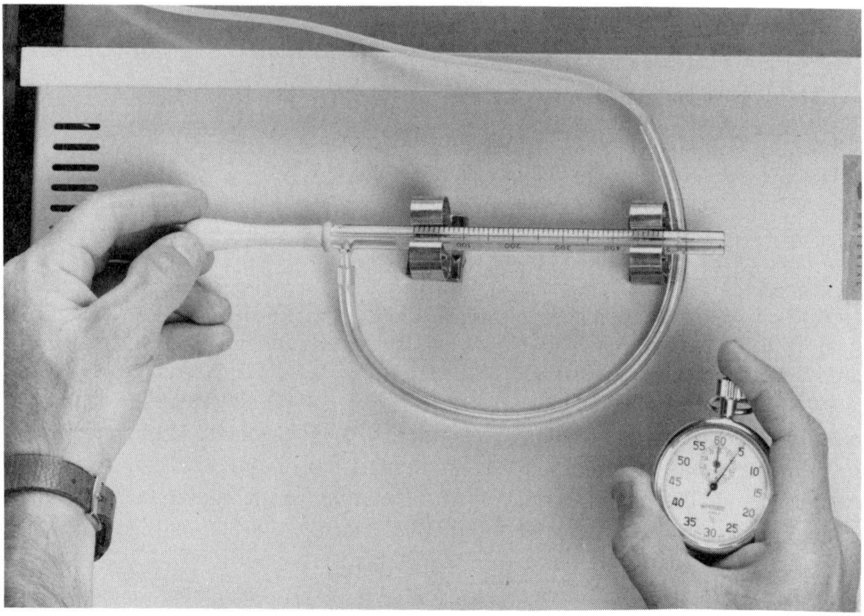

Fig. 25.3 A bubble meter for the measurement of volumetric column flow. The time it takes a gas bubble eluting from the column to move through a known volume of soap solution is timed. For instance, if a bubble passes through 10 cm^3 in 10 s, the flow rate is 60 cm^3 min^{-1}. (Courtesy of Supelco, Inc.)

Simple pressure control is inconvenient in many applications. Changes in the temperature of the oven or column and sometimes in room temperature will require readjustment of the regulator, a process that is time consuming and not easily reproducible.

As a result, some, GCs are provided with a *flow controller* to ensure constancy of flow over a range of values. Their control is precise. For example, when one is used, the flow rate measured by a bubble meter located just after the sample injector (see below), when there is no column in place, will equal the rate observed when a column is installed. Since flow controllers are not commercially available to handle flows as low as 0.5–1 $cm^3\ min^{-1}$, they are seldom employed with open tubular columns.

Example 25.1. How may gas flow be insulated from room temperature changes, especially when autosamplers are used to make injections, and assays are performed over several hours or days? Any significant variation in room temperature will result in changes in the flow of carrier gas and changed retention times.

In general, the best solution is to thermostat the pneumatic system itself. This step will actually be desirable whenever retention time reproducibility is a primary consideration, as in qualitative analysis. When maximum retention time reproducibility is not required, the extra thermostat can be eliminated, but the effect of ambient changes should not be forgotten since it can be the cause of unexpected variations.

Example 25.2. How will temperature programming of a GC affect its flow rate? In this type of elution development the temperature of the oven is increased in a systematic manner, so that at later times the column is at higher temperatures and sample components that would otherwise elute slowly will emerge from the column more quickly. As will be discussed below, this change shortens the time for a chromatogram considerably. How will a temperature change affect carrier gas flow and with what consequences? Will flow control eliminate whatever difficulties might otherwise arise?

In general, gas viscosity increases with temperature; such an increase will reduce gas flow rate. Indeed, it is possible to reduce the flow very signficantly and to have late-eluting compounds remain for a long time in the column. Flow control will provide a constant carrier flow and ensure reproducible chromatograms. If a packed-column chromatograph is used, it should be equipped with a flow controller for temperature-programmed studies.

Sample Injection Systems. The inlet system of a gas chromatograph must provide an efficient means of introducing a sample as a discrete narrow plug to the head of the column. The inlet must be heated sufficiently (to a temperature up to 500°C, but in any case 25–35°C above column temperature) to flash volatilize liquid samples. It must also provide space for the gas that is formed.

A variety of injection devices must be available to take care of sample types that range from gases to low-volatility liquids. The requirements are stringent: the device must introduce a sample (a) in a minute amount since its bulk can be no more than a fraction of the column capacity to avoid overloading, (b) quickly to give a narrow plug since its length in time establishes the intrinsic width of a chromatogram peak, and (c) reproducibly to permit quantitative results to be secured.

Precision syringes in conjunction with a rubber septum inlet are generally used for injection of liquid samples. A septum is a thin layer of self-sealing, silicone rubber that caps the top of a column and allows introduction of a sample by syringe. The most common syringe sizes are 1–10 μL and most sample volumes range from 0.1 to about 5 μL of liquid. Representative syringes are shown in Fig. 25.4. A septum can be used a great many times without leaking. Gas-tight syringes are also available as an alternative to injection with a gas-sampling valve; their advantage is convenience.

The quality of the injection technique determines the narrowness of the sample plug and its reproducibility. Volatility and viscosity of the sample or solvent are the principle drawbacks to the syringe (see the example relating to headspace sampling as a way to eliminate this difficulty). If the sample is very volatile, it can be difficult to contain the sample prior to injection. For samples of low volatility, such as waxes and steroids, significant errors can be introduced into the assay by the injection technique. For such compounds a flash vaporizing injector of the design shown in Fig. 25.5 is necessary.

Sampling valves of the type shown in Fig. 25.6 will handle gaseous samples with good reproducibility (better than $\pm 1\%$). Any pressure or temperature variations in the sampling loop will introduce error.

If an injection system is connected in such a way that the tip of a syringe will reach the head of the column, as in Fig. 25.7 the system is called an *on-column injector*. This arrangement ordinarily ensures the narrowest possible sample slug and maximum efficiency. This injector is valuable both for mixtures containing

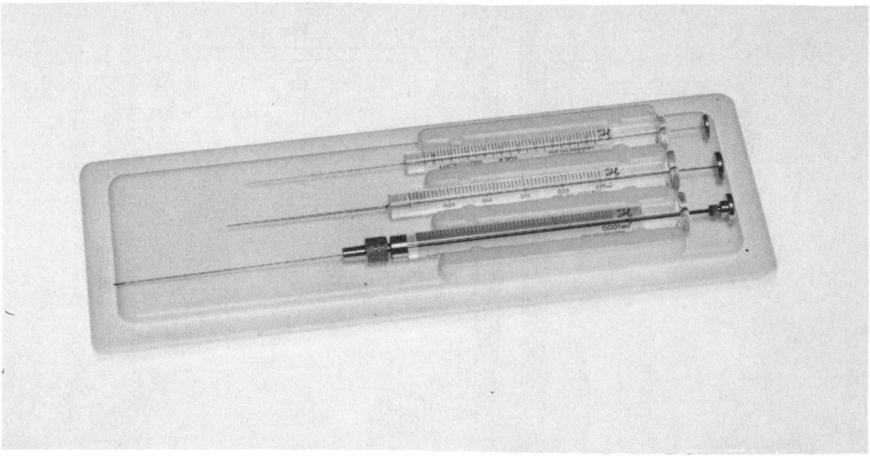

Fig. 25.4 Syringes (from top to bottom) of 10 μL, 25 μL gastight, and 1 μL capacity. Samples are delivered to a column by filling a syringe by pulling the plunger upward, expelling air bubbles, adjusting the volume, and plunging the syringe tip through the column septum, and injecting. In the 1 μL size a carefully matched wire slides in the needle, which holds the entire sample; samples as small as 0.1 μL are possible. The gas-tight syringe has a Teflon-tipped plunger which lightly seals the sample in the barrel. (Courtesy of Hamilton Company.)

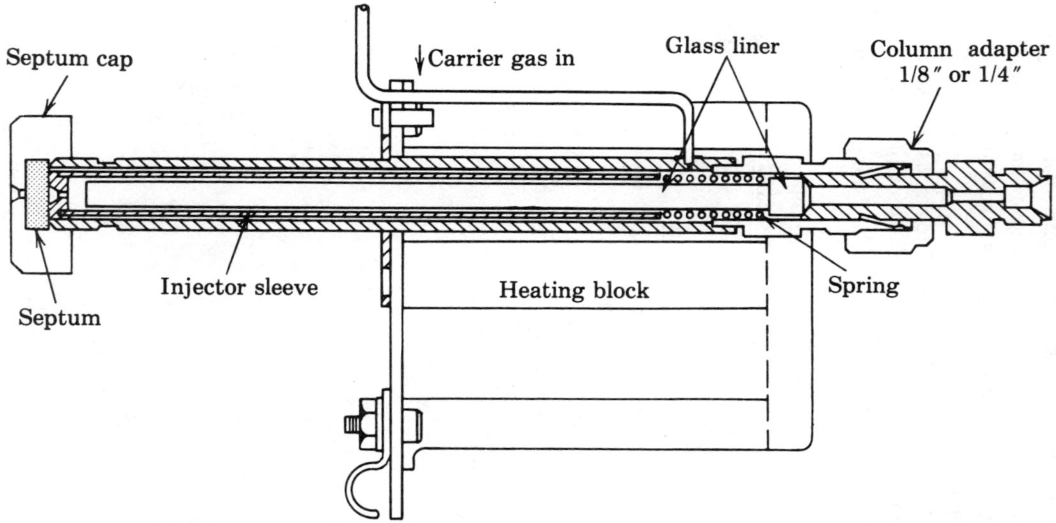

Fig. 25.5 A schematic diagram of a flash vaporizing injector. A sample is vaporized in the heated chamber through which carrier gas (usually preheated) flows. While the injector volume determines the size of sample that can be injected, an injector volume slightly larger than necessary is usually provided to avoid contamination of the pneumatic system by having sample back up into it. (Courtesy of Perkin-Elmer Co.)

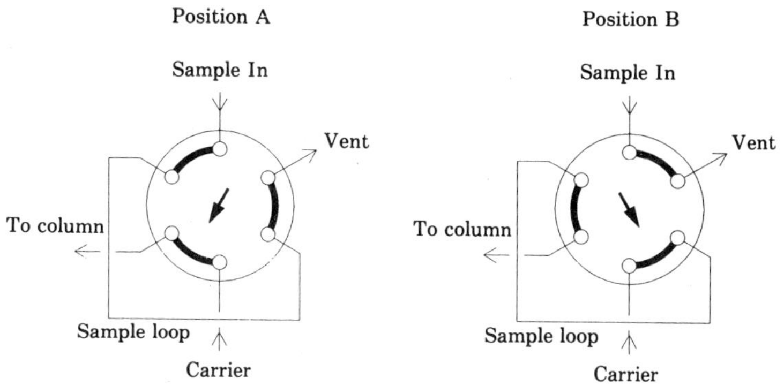

Fig. 25.6 A gas-sampling valve. In one position of the rotor the sample loop is filled while carrier gas flows to the column directly. In the other position the flow of carrier gas propels the sample into the column. While avoiding any interruption of carrier gas, this valve allows fast injection of a known, reproducible volume of sample. (Courtesy of Valco Instrument Company.)

thermally unstable compounds at low concentration since the hot injection port is bypassed and for high-boiling compounds when good precision is needed. Some disadvantages are that automation is difficult with this mode of injection and the column can be overloaded if samples contain components at concentrations greater than 50–100 ppm. Further, if a sample contains any particulate or nonvolatile components, they are also delivered to the column and after several injections, the column can become fouled with resulting degradation of performance.

Headspace sampling is often a good sampling procedure for a mixture that has a reasonably volatile analyte (one that boils below 200°C) in a mixture that is nonvolatile or difficult to manipulate. In this procedure a sample is first brought to equilibrium with its vapor in a container at a known temperature. According to Henry's law concentration of analyte in gas and condensed phases will be given by the expression $C_g = KC_l$, where C_g and C_l are concentrations in the two phases and K is the partition coefficient. A gas sample withdrawn with a gas-tight syringe is then chromatographed. Some determinations amenable to head-space analysis have been vinyl chloride monomer in polyvinyl chloride resin, halocarbons in drinking water, residual paint solvents on coated packaging foils, and residual ethylene oxide in suture materials.

A *split/splitless sample injector*, a quite different type of inlet module, can in its splitting mode deliver the minute samples (about 0.001–0.01 μL) required by most narrow-bore (0.2–0.32 mm i.d.) open tubular columns, as will be discussed in the next section. While these inlets take about 0.1 μL of sample, in the split mode they introduce only a fraction of it onto a column. A schematic diagram of a such an injector is shown in Fig. 25.8. The column is usually inserted directly into this inlet arrangement to reduce unswept dead volume. The sample is introduced into the upper portion of the injector and vaporizes. Once the vaporized

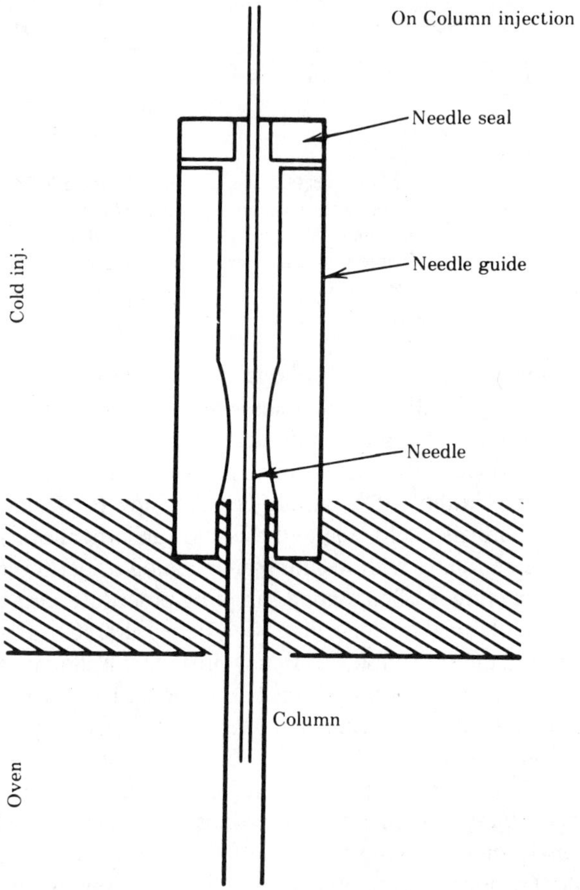

Fig. 25.7 An on-column injector that introduces a sample at the top of the column. The syringe should reach the top of the packing to minimize both band spread and thermal degredation of the sample. (Courtesy of Spectra Physics.)

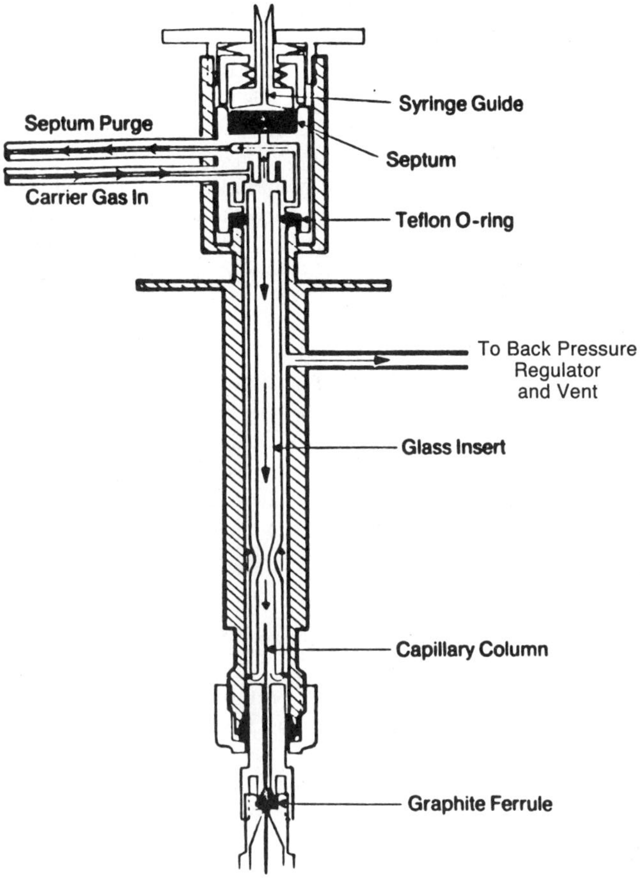

Fig. 25.8 Schematic of a modern split/splitless injector. A septum seals the top of the injector and a glass tube is inserted into its barrel (the insert shown is for split injection). After injection into the upper portion of the injector, a sample vaporizes, expands, and is mixed with carrier gas. (a) In the split mode most of the diluted sample passes through a vent; only a small portion passes into the capillary column that is attached. The vent line is heated to prevent condensation, and residual sample leaving through it is removed by an activated charcoal trap. At the end of the vent a solenoid-operated needle valve restrictor controls pressure in the line to ensure constant carrier velocity in the column. (b) In the splitless mode the glass liner is changed to one that allows no sample diversion. The septum purge is activated to ensure a clean injection. (Courtesy of Spectra Physics.)

sample has been homogeneously mixed with the carrier gas, most of it passes around the column and out through a vent, while a small fraction of it flows into the column. For a regular packed column this injector is fitted with a different glass liner and used in the splitless mode.

Example 25.3. In selecting an injector for an open tubular chromatograph, what bearing on operation will the characteristics of samples have? How important will properties such as volatility (boiling range), expected concentrations of analyte, thermal stability, and the presence of nonvolatile components be?

For capillary columns a split ratio of 100:1 is needed whenever analyte concentrations are above about 500 ppm. If a split ratio of 20:1 is used, components in the 50- to 100-ppm range can be determined. Larger samples or components at high concentration may otherwise overload the column.

The boiling-point range of a sample like gasoline or a wax must also be considered. Quantitative analysis of samples containing components which cover a wide boiling range can be affected by use of a splitting injector. Such an inlet may favor low-boiling compounds and discriminate against higher-boiling, higher-molecular-weight compounds, giving nonlinear splitting.

Split Ratio. For a split-sample injector, the fraction of total gas entering the column must be known. Its split ratio is defined as the ratio of the volume of gas entering the column to that being vented. The former is basically the flow in the column, which can be found from the unknown column volume. The column flow rate is in cubic centimeter per minute and is given as the $V/t_0 = \pi r^2 L/t_0$, where symbols identify column properties: V, volume; r internal radius; L, length; and t_0, elution time of an unretained compound such as methane. One measures the column exit flow with a bubble meter or similar device.

Example 25.4. What is the split ratio for a column of 0.25 mm i.d., 10 m length, and for which $t_0 = 29.4$ s if the vent flow is 99 cm^3 min^{-1}?

One first obtains for the column volume: $3.14(0.025\ \text{cm}/2)^2\ 1000\ \text{cm} = 0.49\ \text{cm}^3$. Further, $t_0 = 29.4\ \text{s} = 0.49$ min. The column flow is simply

$$V/t_0 = 0.49\ \text{cm}^3/0.49\ \text{min} = 1.0\ \text{cm}^3\ \text{min}^{-1}$$

Now the split ratio is calculated:

$$\text{split ratio} = 1.0\ \text{cm}^3\ \text{min}^{-1}\text{: } (99\ \text{cm}^3\ \text{min}^{-3} + 1.0\ \text{cm}^3\ \text{min}^{-1}) = 1\text{:}100.$$

One percent of the injected sample enters the column.

Splitless Injection. For determinations of trace-level components, a method called splitless injection has been described by Grob.* The procedure allows a larger sample to enter the column to ensure detection of one or more trace components. The sample is injected into a relatively small vaporization chamber under conditions of low carrier gas flow (25 cm s^{-1} or 1 mL min^{-1}) with the vent line closed so that all the sample is forced to enter the column. After a period of time when the split vent is opened, and gas flow rate increases, and any sample remaining in the injector is swept out. In this way, about 90% of the sample reaches the column as opposed to 1–5% in split injection.

Unfortunately, low carrier gas flows will increase band spreading. Grob describes a "solvent effect" that will reduce the undesirable band spreading. A solvent is used that has a boiling point about 20°C below the boiling point of any trace components. The oven

*R. L. Grob, *Modern Practice of Gas Chromatography.* New York: Wiley-Interscience, 1977, p. 312ff.

temperature is usually set about 20°C below the solvent boiling point and increased after injection.

The Oven. An oven whose temperature can be closely controlled is essential to house the column. The reasons are evident: both vapor pressures of analytes and the distribution of solutes between mobile and stationary phases are strongly temperature dependent. As column temperature rises the retention of components in the stationary phase drops rapidly.

What sort of *temperature control* should an oven have? Several types of specification must be considered: temperature range, precision and accuracy of setting, constancy of temperature, dependence of oven temperature on ambient conditions, and the type of temperature programming available. The *range* of operating temperatures must be broad enough to cover desired applications. For example, a low-boiling mixture such as natural gas or a butane mixture may require an oven to be thermostated about at room temperature while a crude oil sample may need to be chromatographed up to 350–400°C to ensure that all components will elute. Most commercial ovens operate at temperatures from 10°C above ambient to 400°C. If needed, the operative range can be extended downward with liquid coolants, for example, to −60°C with liquid CO_2 and to −99°C with liquid N_2.

Microprocessor control makes *setting* the temperature to 1°C possible. In some applications it is a disadvantage if an instrument can be set only in 10 or even 2 or 5° increments. Representative specifications for constancy of oven temperature are ±0.1% of the set point on older models. With modern high-performance instruments a somewhat more stringent specification of ±0.2°C up to 400°C holds. In addition, the accuracy of setting for a modern instrument is ±1% above 100°C.

Equally important is the stability of the oven temperature as the room temperature changes. Sometimes this is called "ambient rejection." Modern high-performance instruments have a rejection of 50:1 (< 0.02°C change in temperature per degree Celsius change in external temperature). Retention time repeatability is directly affected by the ambient rejection of the chromatograph since an instrument that fails to discriminate against ambient conditions will tend to drift with changes in external conditions.

Temperature programming allows a chromatogram to be developed along a temperature as well as the time axis. This flexibility is important for the efficient and accurate separation of mixtures with a wide boiling-point range of components (greater than 30°C). By increasing column temperature during elution even the least volatile components may be separated efficiently. At the beginning of programmed chromatography, usually a program provides an isothermal period of length sufficient to let the least strongly held components elute with good resolution. There follows a program-timed linear increase of the temperature of the oven to a set high temperature and perhaps a second isothermal interval at the final temperature. Intermediate isothermal plateaus can sometimes be provided.

A simple example of the effectiveness of temperature programming is shown in Fig. 25.9 in which chromatograms are obtained on two identical samples of a C_6–

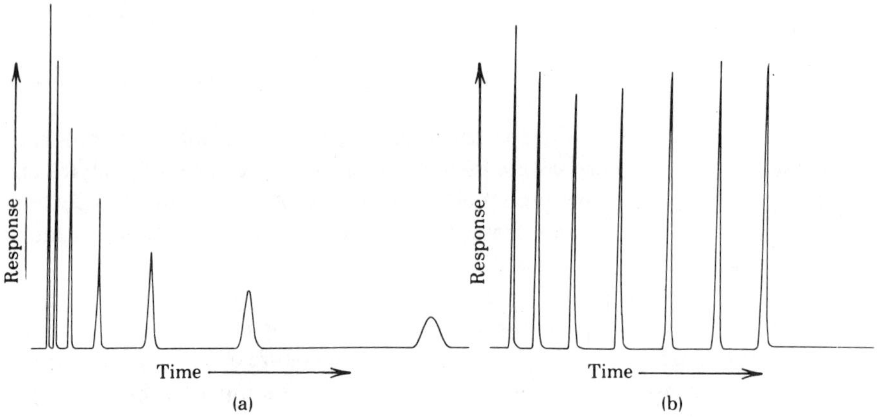

Fig. 25.9 A comparison of gas chromatography under isothermal and temperature-programmed elution for the normal hydrocarbons C_6–C_{12}. A packed column (t_0 = 0.5 min) was used with He flowing at 20 cm^3 min^{-1}, a flame ionization detector (FID), and a chart speed of 10 mm min^{-1}. (a) Isothermal chromatogram [t_{oven} = 130°C; FID attenuation 256]. In this chromatogram, while peaks eluting earlier are sharp and closely spaced, later peaks are broader and spaced successively farther apart: t_R's: C_6 = 0.89 min; C_{12} = 22.44 min. (b) Temperature-programmed chromatogram [t_{oven} = 100°C for 1 min, temperature increased 10°C min^{-1} to 200°C and held at 200°C for 10 min; FID attenuation 512]. Now the peaks obtained are of nearly equal width and are about equally spaced: t_R's: C_6 = 1.28 min; C_{12} = 10.97 min.

C_{12} normal hydrocarbon mixture (boiling range is 69–216.3°C). The chromatogram in Fig. 25.9a was obtained under isothermal conditions and that in Fig. 25.9b under appropriate temperature programming. As is evident, temperature programming provides a much faster assay when analytes boil over a wide range. It also offers improved quantitative reliability because late-eluting peaks are sharper and more easily measured. Column resolution (in terms of theoretical plates) is discussed in Example 25.5.

A trade-off between resolution and speed of running samples is involved in the use of temperature programming. It arises because time must be allowed for recycling the oven temperature after a run. In general, isothermal operation provides a faster rate of handling for mixtures of narrow boiling range. With a sample of intermediate range of boiling points, an analyst must determine if the resolution and better precision gained in temperature programming is sufficient to justify a probably longer recycle time.

25.3 COLUMNS

Both packed columns filled with sized solid particles and open tubular or capillary columns are commonly used in GC. In the former case the solid particles, and in the latter the interior wall of the column, provide a surface that can be coated with the high-boiling liquid that will serve as the stationary phase.

Packed Columns. For packed columns the tubing may be of aluminum, copper, stainless steel, or glass. Active oxides on the metal surface may rule out the first two for use with labile analytes such as steroids. Columns are almost always packed with pretreated diatomaceous earth particles (see discussion below) since their large surface area and porosity allow high stationary phase ''loading'' and good gas permeability. The size of particles is also an important characteristic. High column efficiency, that is, a large number of plates, is associated with small particle size [see Eq. (24.19)], and vice versa. Standard sizes of packing in GC columns are 30 to 200 mesh with the most common sizes being 60 to 120 mesh.* Unfortunately, the smaller the particle size, the greater the pressure drop of a column. To reduce this pressure drop, long wide-bore columns (10 m, 6 mm o.d.) are usually packed with larger particles (40/50 mesh). The *packing efficiency* (lack of voids) also falls as the particle size decreases; it is difficult to pack a long column evenly with fine particles. In practice, the selection of an appropriate size of packing involves a trade-off among these factors.

For the infrequently used method of gas–solid chromatography, sometimes called adsorption chromatography, a relatively active packing material is employed. Common choices are molecular sieves (a sodium–aluminum silicate), silica gel, activated carbon (Carbosieve), and alumina. Conditioning of these materials affects their performance and the pretreatment that should be used is specified in many methods.

Diatomaceous Earth. This natural material consists of the skeletal remains of microscopic single-cell algae called diatoms. Their complex microscopic structure accounts for their high surface area and fragility. Chemically they are mainly amphorous silica with trace-level impurities of metal oxides. In general, the particles are pretreated to produce a stable and inert support by any or all of the following processes: flux calcining, acid and/or base washing, and silane treatment. Their substantial adsorptive character and fragility (leading to production of fragments called fines) contribute to peak tailing and qualitative and quantitative chromatographic errors.

Open Tubular or Capillary Columns. Wall-coated open tubular columns,† which are given the acronym WCOT, are commonly of fused silica.‡ Two features make silica tubes almost a perfect support. High-purity fused silica tubing contains minimal alkali, alkaline, or transition metal impurities. In addition, when the outside of the tube is coated with a polyimide, the tube becomes reasonably resistant to

*U.S. standard sieves have meshes defined by the number of wires per inch. In sifting particles two screens are customarily used. If a packing is specified as 60/80, for example, all of it will pass through a 60-mesh screen and be retained by an 80-mesh one.

†The name open tubular columns better accommodates the range of tubing diameters in use today and emphasizes that the stationary phase is applied to their walls.

‡Support-coated open tubular columns (SCOT) employ stainless-steel tubing and are occasionally used. Particles are spread over the inside walls to increase the surface area available for coating. They have the disadvantage of high activity of both the particles and of the stainless-steel column. This adsorptive activity leads to tailing of more polar components.

abrasion and quite flexible because moisture is excluded and the silica does not hydrate. Fortunately the advance of stationary-phase coating technology has provided a way to increase the film thickness somewhat. Today most coatings are polymerized *in situ* and are caused to adhere to the silica by reactions widely used in preparing bonded phases for LC columns. These columns have thin films of stationary phase, and samples must have analyte concentrations less than 50–100 ng μL^{-1} or the sample capacity of the column will be exceeded.

The development of wider diameter fused silica tubing (0.53 and 0.75 mm i.d.) has also made possible thicker films and higher column capacities. In columns of this type thicker films of stationary phase are employed and samples with a component mass of up to 15 μg may be injected directly without loss of efficiency. Wide-bore open tubular columns can also be used with sample injectors for packed columns and flow rates from 2 to 6 $cm^3\ min^{-1}$. These columns approach packed columns in capacity and are therefore suitable for use with thermal conductivity or other detectors with relatively high limits of detection. Sample capacity is also much higher with these columns than with the more traditional thin-film (0.25 μm) columns. Representative column dimensions, efficiencies, and pressure drops for open tubular and packed columns are given in Table 25.1.

Useful capillary columns range in length from 5 to 100 m and internal diameters range from 0.1 to 0.75 mm. Longer columns ($L > 50$ m) result in high numbers of theoretical plates and are used for the most complex separations. While long columns imply long analysis times, the tradeoff of adequate resolution for shorter analysis time is often accepted. Typical internal diameters from 0.25 to 0.5 mm are used for most work. Narrow columns have high efficiency, but intrinsically permit only thin films. This results in lower capacity factors and smaller sample sizes.

Stationary Phases. In modern GC the liquid stationary phases have been reduced to a relatively few materials. Some of the most common are listed in Table 25.2. Naturally, the same materials are used in packed and open tubular columns. Two factors affect the choice of stationary phase, efficiency N and solute selectivity. Usually the analyst identifies a sample as polar or nonpolar and selects a column with a stationary phase that is similar. On the basis of the rule of thumb, "like dissolves like," it would be expected that nonpolar molecules such as hydrocarbons and halogenated compounds are more soluble in a nonpolar liquid phase such as methyl silicone than are polar molecules such as organic acids and esters. As

Table 25.1 Some Column Characteristics

	Theoretical Plates				
Column Type	*Per Meter*	*Total*	*Length (m)*	*Internal Diameter (mm)*	*Pressure Drop (psig)*
Packed	600–3000	20,000	1–6	6	10–60
Open tubular or capillary	2000–4500	75,000–150,000	10–100	0.1–0.75	5–50

Table 25.2 Some Properties of Representative Stationary Phases[a]

Common Designation	Maximum Temperature (°C)	Chemical Formula and Structure	McReynolds Constants[b]					$\Sigma \Delta I$[c]	Common Uses
			X'	Y'	Z'	U'	S'		
Squalane	100	2, 6, 10, 15, 19, 23-Hexamethyltetracosane	0	0	0	0	0	0	Reference; hydrocarbons
OV-1, SE-30	350	Poly(methyl dimethyl siloxane)	16	55	44	65	42	222	General purpose: BP separation
OV-7	350	Poly(phenyl methyldimethyl siloxane)	69	113	111	171	128	592	General purpose: aromatic compounds and olefins
OV-17, DC-710	300	Poly(phenyl methyl siloxane)	119	158	162	243	202	884	Fatty acids, esters, aromatics
OV-25	350	Poly(phenyl methyldiphenyl siloxane)	178	204	208	305	280	1175	Fatty acids, esters, alcohols
Silan 5CP, SP 2300	250	Poly(cyanopropyl-phenyl siloxane)	317	495	446	637	530	2425	Fatty acids, esters
Carbowax 20M	225	Poly(ethylene glycol)	322	536	368	572	510	2308	Alcohols, ethers, ketones
DEGS	200	Diethylene glycol succinate	496	746	590	837	835	3504	Fatty acids, esters, steroids

[a]After T. L. Isenhour et al., *J. Chromatog. Sci.*, **11,** 201 (1973). Most of these compounds are in Isenhour's preferred list.

[b]A McReynold constant I is defined as the increment in the Kovats retention index for a test compound in a column with the listed stationary phase compared with the index it has in a column with a squalane stationary phase. Each test compound (X' = benzene, Y' = butanol, Z' = 2-pentanone, U' = nitropropane, and S' = pyridine) is intended to be distinctive in the main intermolecular force by which it interacts with a stationary phase. [W. O. McReynolds, *J. Chromatog. Sci.*, **8,** 685 (1970).]

[c]This column gives the sum of the McReynolds constants and provides an overall polarity index for each stationary phase. These constants provide a guide in phase selection.

Table 25.2 implies, a scale of liquid phase polarity developed by McReynolds is an aid in discriminating among stationary phases. This scale is based on the use of test compounds for each type of column at a single oven temperature. The sum of the values $\Sigma\Delta I$ for a given stationary phase becomes a polarity index for columns with that stationary phase. Low values of *McReynold's constants* imply a nonpolar stationary phase and high values, polar phases. Values of the McReynolds constants are found in most catalogs of columns and in the literature.

As the constants increase in value the tendency of the liquid phase to retain compounds exhibiting the relevant characteristic increases. For example, aromatic and olefinic compounds are more strongly retained on a phenyl methyl column such as OV-17 than on OV-1, as can be predicted by examining the value under X' for each phase.

Liquid stationary phases are coated on the packing in weight percentages from 1% or less to 30%. Even coating is essential to minimize exposure of active areas. Exposed silica surfaces, such as those produced by fracturing of support with light loading, can lead to severe peak tailing. Over the lifetime of a column there is a gradual loss of coatings as a result of evaporation, a process called bleeding; it is worst at higher temperatures and column capacity is gradually lost.

In modern open-tubular columns stationary phases are polymerized *in situ* and bonded directly to the silica tube as described earlier. Film thicknesses range from 0.1 to 5 μm. Anchoring the stationary phase dramatically reduces column bleed and peak tailing and further minimizes substrate interaction with sensitive compounds such as amines, acids, and alcohols.

A Comparison of Columns. A direct comparison of some aspects of the two column types can be obtained from Table 25.1. The efficiency of a packed column appears about an order of magnitude smaller than for an open tubular column of the same length. The lower pressure drop in open tubular columns is evidence of their smaller flow resistance, but this alone does not explain differences in performance or practical applications between the two types of columns.

If high resolution is sought, a good packed column may not be long enough to provide it, and even if long enough to do so, may fail to provide it in reasonable time. As implied earlier in Eqs. (24.11) and (24.18) since resolution R is a function of peak width, $R \propto \sqrt{L}$, where L is the length of the column. To double the resolution of a pair of peaks, column length must be increased four times. There must be a corresponding geometric increase in the pressure to force the carrier gas through the column! From a purely practical point of view, the instrumental requirements for very long *packed* columns are difficult to achieve in laboratory equipment.

Yet as the table also shows, very long *open tubular* columns can be operated under conditions comparable to those required for average length packed columns at considerable savings in cost and complexity of instrumentation.

Additional practical and theoretical considerations contribute to the improved performance of open tubular columns over packed columns. One is that by eliminating the packing, the part of band broadening attributable to the multipath dispersion effect in the van Deemter equation [Eq. (24.19)], is removed and resolution is improved. Figure 25.10 shows the relative difference in performance of packed columns and of open tubular columns as expressed by the restated equation of Golay:

$$H = B/u + Cu.$$

It should be noted that the minimum of the plot is not the only distinction between packed columns and open tubular columns. The portion of the curve rising to the right with higher values of linear gas velocity is flatter for open tubular columns. The Golay equation shows that efficiency is lost more slowly at higher linear gas velocities in open tubular columns than in packed columns. This difference has practical consequences for improving the speed of analysis.

Since capillary columns can be very long and generate very high plate counts, it is possible to separate very complex mixtures readily with these columns. However, if maximum resolution is not required, a shorter column can be used and thus reduce the analysis time in proportion.

For example, if an assay is performed in 30 min on a 50-m column, it will take only 15 min on a 25-m column. Resolution on the shorter column is reduced to $\sqrt{1/2} = 0.7$ of the previous resolution. If resolution for a critical pair of peaks was 1.5 on the long column, it will now be 1.05 and peak separation will be about 94% complete.

Further, gas flow rates can be increased with open tubular columns with less loss of efficiency, as noted above. As was discussed in Chapter 24, retention times are inversely proportional to flow velocity. Therefore, flow-rate increases will shorten analysis times. It is often possible to use an average open tubular column to provide the same separation in less time than was achieved on a moderate-quality packed column.

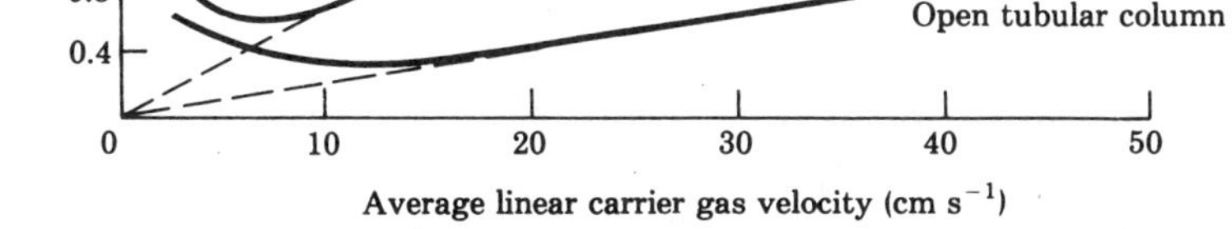

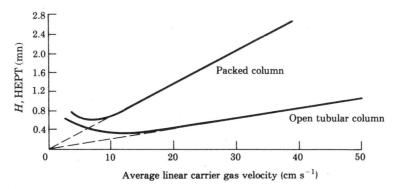

Fig. 25.10 Van Deemter plots comparing the efficiencies for packed and open tubular columns. The minimum in the plot for an open tubular column is about half the value for a packed column and for an open tubular column the linear (right-hand) portion of the curve has a smaller slope. As the curves imply, at the higher carrier gas velocities used in practice, open tubular column performance proves about three or four times better than packed column performance.

Nevertheless, limitations of open tubular columns must also be taken into account. They offer much lower capacities since there is less total stationary phase and for that reason can only be used with minute samples. In addition, there are drawbacks in their use in trace analysis. In this application a minute sample may fail to produce analyte concentrations at the detector that are above its limits of detection. Preconcentration will then be required.

Example 25.5. Estimate the number of effective plates generated for the C_{10} and C_{11} peaks by the column in the isothermal chromatogram of Fig. 25.9a. Use peak height and width at half height. How well can columns be compared by such estimates?

The number of plates can be found from the equation $N = 5.545\ (t_r/W_h)^2$. It can be estimated that the corrected retention times for C_{10} and C_{11} are 6.69 min and 12.19 min, respectively. Similarly, the half widths of their peaks are 2.75 mm (0.275 min) and 4.50 mm (0.450 min), respectively. Substituting in the above equation, one obtains for C_{10}, $N = 5.545\ [6.69/0.275]^2 = 5.545\ (24.3)^2 = 3280$ and for C_{11}, $N = 5.545\ [12.19/0.450]^2 = 5.545(27.1)^2 = 4070$.

Several points should be made about these calculations. If peak measurements are made manually, care must be taken. Chart paper should be of high quality and measurements should be made on peaks of reasonable height. Further, the width measured at the half height is the total width minus two pen-line widths. In the measurements it is helpful to use a comparator (an ocular reticule with an engraved scale).

Clearly, a chromatograph with a microcomputer can be programmed to display such values and even to calculate N. Measurements aside, the most important thing to understand is that column efficiency is dependent on the column length, the instrumental conditions, and the compound chosen.

For plate counts to be a useful basis of comparison of columns, an analyst must be sure that performance of different columns is compared using compounds with the capacity factor k' ($k' = t_R'/t_m$). For instance, as just demonstrated, a C_{10} hydrocarbon cannot be used in one column and a C_{11} hydrocarbon in another.

25.4 DETECTORS

The elution of analytes from a GC column can be monitored by a variety of detectors. Two essentially universal detectors, one based on gas thermal conductivity and the other on flame ionization, are the most widely used. Many selective types of detectors have also been developed and a few that have become important will be described.

Ideally the *response* of a detector, the increase in its output signal as analyte concentration is increased, should be linear. The slope of the response curve will be called the *sensitivity* of the detector and the region of linear response its *dynamic range*. At trace-level concentration detector noise will of course interfere and the *limit of detection* of a detector will generally be taken as two or three times the average detector noise level when it is expressed in the same units as the detector signal. Most GC detectors have a dynamic range of four to five orders of magnitude; their limits of detection extend downward to the parts per billion range and below.

Thermal Conductivity Detector. This classical detector is perhaps still the most commonly used in gas chromatography. It measures the small change in thermal conductivity that occurs in the eluting mobile phase or carrier gas when it contains an analyte. This device is a *universal detector* in that any compound eluting from a column will be detected as long as its heat-conducting properties differ from those of the mobile phase. It is also a *differential detector* in that the thermal conductivity of the column eluent is nearly always compared with that of eluting pure carrier gas.

The thermal conductivity detector (TCD) incorporates two or four carefully matched lengths of tungsten–rhenium wire called filaments wired as the arms of a Wheatstone bridge circuit and mounted in a thermostated block. A representative arrangement showing both the cells and the bridge is sketched in Fig. 25.11. The reference filament(s) is continuously surrounded by pure carrier gas and the sample filament(s) by the column eluent. Both flow rates and gas temperatures for the two sets must also be matched. The detector operates on either a constant-current, constant-temperature, or constant-voltage basis. In the circuit shown in the figure, the Wheatstone bridge is energized by application of a constant voltage across the terminals and the I^2R dissipation in both filaments heats them well above the gas temperature. Depending on the thermal conductivity of the gas, heat will be transported quickly or slowly from the hot filaments to the cell walls. Since the filament resistance is temperature dependent, the bridge will respond electrically to the cooling of filaments. When only carrier gas surrounds all the filaments, they are cooled to the same extent by bombardment of molecules of their surrounding gas, and the bridge is balanced, and the recorder will trace a straight baseline.

When carrier gas diluted by an analyte surrounds the *sample* filament, it will be cooled to a different extent than will the reference filament. Thus, the resistance and current in this arm of the Wheatstone bridge will be different from that in the reference arm and the tracing of the recorder will depart from baseline. An eluting component will usually lower the thermal conductivity of the carrier gas. With a TCD it is advantageous to choose either helium or hydrogen as the carrier gas since their thermal conductivities are substantially higher than those of most other substances and they afford the largest differential measurement.* Ordinarily helium is the choice for routine GC in the United States, where it is more abundant, and hydrogen and nitrogen are often used in other countries. Use of H_2 and He has the advantage that all solute peaks will be on the same side of the baseline since all analytes will lower the thermal conductivity of the carrier gas.† While nitrogen or argon can also be used with thermal conductivity detectors, both give higher limits of detection and inverted peak shapes with a few analytes. Either may still be attractive where the expense of helium is substantial.

Thermal conductivity detection is a simple, reliable method that is applicable

*Gaseous thermal conductivity is a function mainly of the rms molecular velocity, which at constant temperature varies inversely with the square root of molecular weight. Thermal conductivities for hydrogen and helium at 273 K are 42×10^{-5} and 35×10^{-5}, respectively.

†If hydrogen is to be determined, for best accuracy argon is used as the carrier gas.

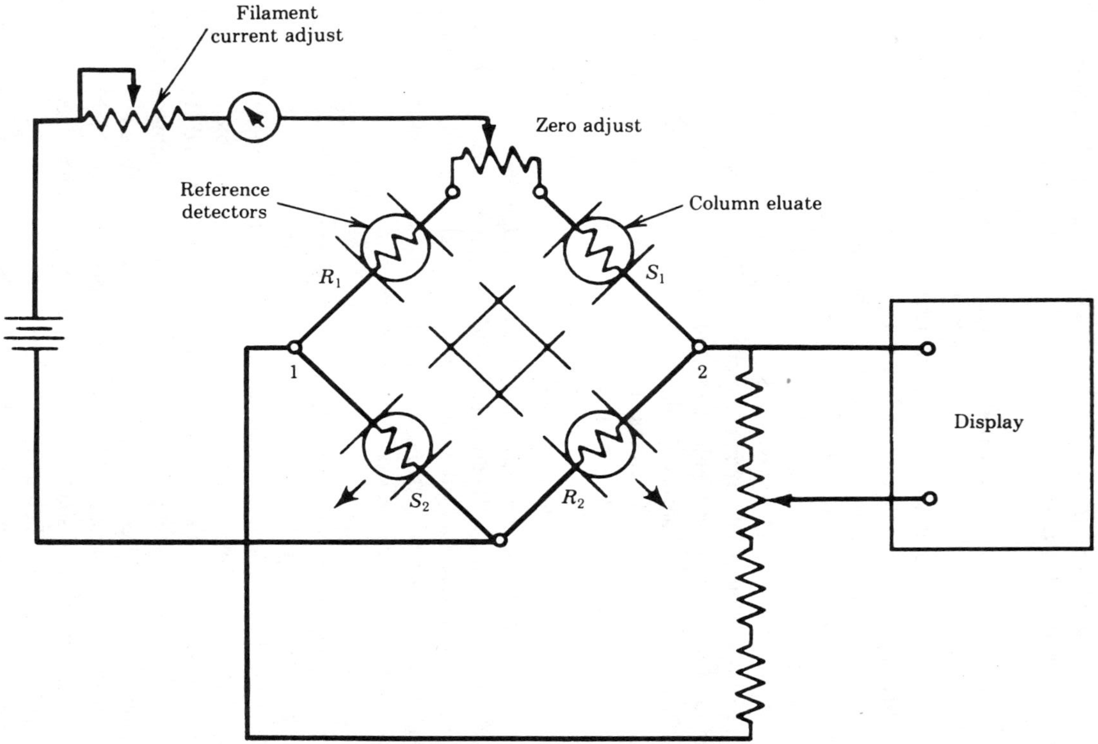

Fig. 25.11 A schematic of a representative four-wire thermal conductivity cell. In this Wheatstone bridge configuration, the four matched detectors form the bridge arms. When matched, the set compensates for changes in current, temperature, and flow. The eluent from the chromatographic column is directed equally to detectors S_1 and S_2 and carrier gas alone to reference detectors R_1 and R_2. (Courtesy of Gow–Mac Instrument Co.)

for analytes at concentrations from many percent down to 10–100 ppm. A TCD has a dynamic or working range of about 10^5.

Any leak or change in flow in the instrument channel will reduce the analyte concentration and lessen the detector response. For that reason, flow controllers (see below) are usually recommended with thermal conductivity detectors, but under some conditions pressure control alone is satisfactory. Slow changes in detector temperature will lead to baseline drift. For accurate results the detector temperature must be constant.

Flame Ionization Detector. The flame ionization detector (FID) offers lower limits of detection and a greater dynamic range than the thermal conductivity detector, retaining good accuracy, wide applicability, and reasonable speed. It can be used for the majority of *organic* compounds. Specifically, the detector measures only C–H containing molecules and gives no response for water, carbon dioxide, or other inorganic compounds. With a limit of detection 100–1000 times smaller than that of a TCD, it is also ideally suited for use with open tubular columns with their relatively low sample capacity. It has the further advantage of being a mass-sensitive detector. As a result, changes in carrier flow rate will have little or no effect on response and use of this detector will permit operation of a chromatograph as a single-channel device, that is, without a reference channel. While operation with temperature programming often introduces drift in the baseline, ways to minimize this effect are discussed below in a note.

As shown in Fig. 25.12, the FID consists of a small gas burner producing a hydrogen flame, a polarizing voltage source, and an ion collector. The ion current is directed through a shielded lead to a high input impedance op-amp for measuring the ion current. Because the device measures an ion current that depends on the mass of ions entering the detector per unit time and not on analyte concentration it operates as a *mass* sensitive detector. Note two important design features: the mixing of column eluent and hydrogen just before the flame jet and the introduction of air at the jet; these will be taken up in Example 25.6. As a mass detector it can accommodate the addition of the extra gas. An organic compound in this flame under steady conditions yields ions in proportion to the mass of carbon it contains. By contrast, the combustion of hydrogen produces few ions. Background currents are minute (10^{-13} A) and a dynamic range of 10^8 is not uncommon. The minimum detectability of the FID is usually specified as about 5×10^{-12} g C s^{-1} for simple hydrocarbons such as butane. Organic compounds containing oxygen, chlorine, and other types of atoms give similar linear responses but have smaller slopes and the detector must be calibrated for such species. The choice of carrier gas will be unimportant as long as it is hydrocarbon free.

The FID is reliable and convenient and has found use in almost every application area from routine control of product purity to environmental studies to biochemistry and petroleum chemistry. It overlaps the thermal conductivity detector in analytical range, but is far superior to it at trace levels and slightly inferior at high concentrations. The FID is usually the detector of choice for use with an open tubular column.

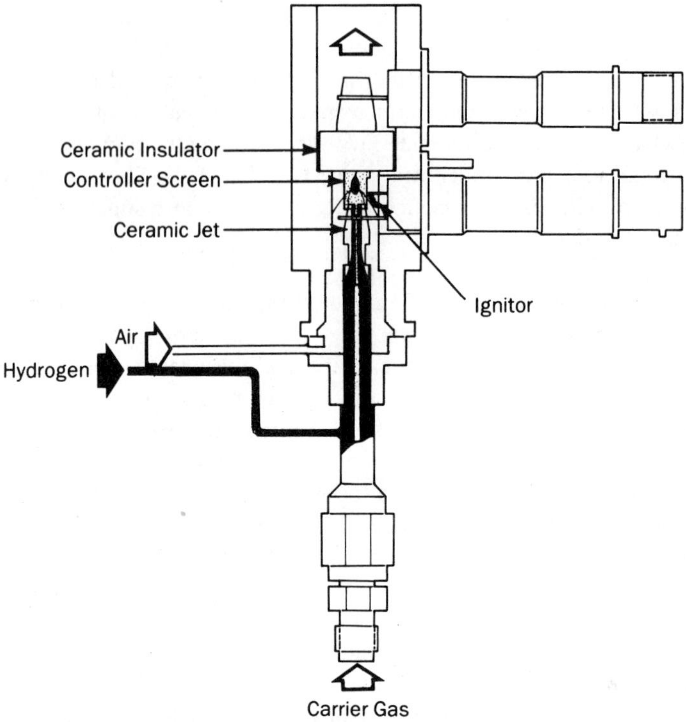

Fig. 25.12 A schematic of a flame ionization detector. Eluate from a column enters from below and is mixed with air and hydrogen where shown. As organic analytes burn in the hydrogen flame they form ions and electrons. Usually the collar-shaped, cylindrical collector is charged positively to collect electrons (the wire-mesh controller screen enhances their collection) while the burner collects cations. Since ion currents are minute, collector leads are shielded to minimize stray currents. (Courtesy of Spectra Physics.)

Example 25.6. Why is an FID commonly chosen for an open tubular or capillary column? What advantages does it offer in this application?

Certainly one advantage is the low limit of detection. With the limited capacity of open tubular columns, this property is important. Another advantage is the independence of detection on gas volume, that is, concentration of analyte. With most open tubular column GCs detector cells must have minute volumes to keep concentrations for detection reasonable, while this consideration is ignored for the FID.

Also, the addition of make-up carrier gas helps move the eluent from the column into the flame for detection. What might otherwise be a loss of sensitivity because of volume dilution by expansion of column eluant into the flame is largely overcome.

Temperature Programming with the FID. While a flame ionization detector will produce a steady signal without a reference channel, temperature programming will often produce a shifting baseline that can be traced to increased bleeding of the stationary liquid phase as the temperature increases. This drift can be compensated for by referencing the output to that of a second flame detector employed with a column similar to the analytical column. Alternatively, the operator can compensate for changes in the baseline by the use of a computer program that will subtract representative current increases. In addition, newer, bonded stationary phases for open tubular columns have reduced the amount of bleeding, making the baseline shift less of a problem.

Electron Capture Detector. The electron capture detector (ECD) is one of the most important selective detectors and has a high sensitivity for compounds containing highly electronegative atoms such as chlorine. A simplified design is shown schematically in Fig. 25.13. The detector contains a radioactive beta emitter such as tritium or nickel-63. The electrons it emits are collected at an anode, giving the detector a current output. When a chlorine-substituted molecule enters the detector and captures emitted electrons, fewer are collected at the anode and a component peak can be recorded. An advantage of the detector is that it is nondestructive and can be operated in series with other detectors.

In addition to selectivity these detectors have very low limits of detection (usu-

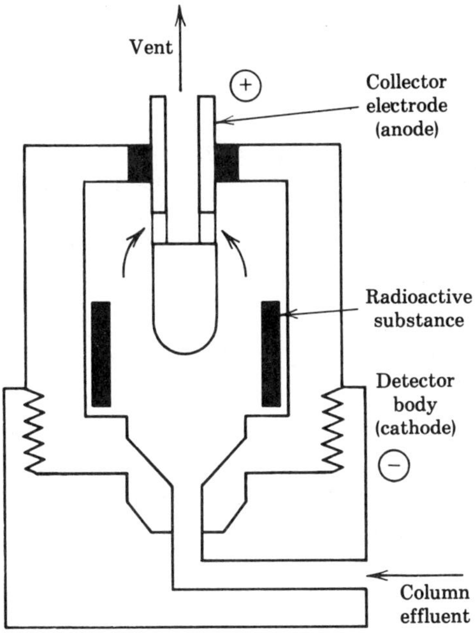

Fig. 25.13 A simplified schematic of an electron capture detector. Most modern instruments use ^{63}Ni as the radioactive source furnishing β particles. The anode (labelled +) attracts these high speed electrons. As eluting analyte mixed with carrier gas enters the chamber, it absorbs a small portion of the β-radiation. The carrier gas, usually an argon–methane mixture, also performs the role of moderator, slowing emitted electrons to improve their absorption by a sample. The anode is pulsed and the change in frequency required to maintain a constant electron flux as an analyte elutes is measured. While the absolute absorption of electrons could be measured, the frequency change versus analyte concentration gives a more linear plot.

ally lower by three orders of magnitude than the FID). These limits are nearly always given for a particular compound, implying that calibration will be necessary for each compound to be measured. For example, commercially available ^{63}Ni ECDs have a minimum detectable amount of about 1 pg of lindane under specific chromatographic conditions. Most compounds experience a loss of linearity in the range of 10–150 ng.

Until recently it was considered difficult to temperature program the oven when an ECD was used because the increase in bleed of the column would cause an unacceptable shift in the baseline. Again the advent of fused-silica capillary columns with highly stable thin-liquid phase films has reduced the bleed to a minimum and made temperature programming much easier, thus extending the utility of the ECD.

ECD Operation. Early ECDs operated with the anode at constant voltage, but the detector had only two or three decades of linearity and required a special carrier gas that ruled out the possibility of operation with other detectors. More recent versions operate in a constant-current mode. The cell voltage is pulsed in such a way that a constant current is maintained at the anode. To achieve constant current, the pulse frequency is modulated. A measurement consists of measuring the change in frequency that occurs when an analyte enters the cell. The analytical curve is linear over four to five decades.

Example 25.7. An electron capture detector, which is selective for halogen-containing compounds will respond to a few picograms of the chlorinated compound lindane and ignore any unchlorinated hydrocarbon of the same structure that is present. Note that the same weight of lindane will produce no discernable signal in a flame detector.

In testing for insecticides on crops, it is unlikely that extraction of a sample of a farm crop will yield chlorinated compounds other than the pesticides applied to the crop. Therefore, these compounds can be determined using the appropriate detector with a reduced requirement of prior isolation of the compound from the matrix. An FID, however, would respond strongly to many of the other substances extracted and make the determination difficult or impossible.

Other ECD Applications. The ECD has found use mostly for ultratrace analysis. Since it is selective, it has limited linear range and is sensitive to compounds in picogram quantities. The ECD has been the detector most often selected for work on environmental pollution problems as mentioned previously because of the large number of halogenated compounds which have been identified as hazardous.

It has been used to determine chlorinated pesticides in foods, body fat, and other tissues. Polychlorinated biphenyls and polybrominated biphenyls have been determined in many different matrices with the ECD. No other simple detector is in general use that has the sensitivity and selectivity of the ECD.

Range of Response. One limitation of the ECD is its lack of sharp selectivity. Several classes of compounds including halogenated, nitrosubstituted, oxygenated, and aromatic compounds will give a response.

In general, the more highly halogenated a compound, the greater its response will be. On one extreme we see vinyl chloride, which has a very poor response with the ECD (it is usually determined with an FID), and on the other extreme are the chlorinated compounds like lindane and polychlorinated biphenyls, which have limits of detection in the picogram range. This variation has the consequence that the response characteristics for any substance to be determined must be investigated during method development. The detector must be carefully calibrated and the analyst must be alert to possible interferences which could be misconstrued as the compounds of interest. Early work on chlorinated pesticides suffered from some of this confusion until prior isolation techniques were developed to make identification more reliable.

Cleanliness. Another drawback to the ECD is that the analyst must be scrupulous in maintaining a clean system: the detector, and especially its anode, are easily contaminated. First, the analyst should be sure that the carrier gas is especially pure. Its additional treatment with molecular sieves, silica gel or other drying agent, and an oxygen trap are all highly recommended. Second, samples should be cleaned up and columns should be carefully conditioned prior to use to avoid excessive bleeding of stationary phase into the detector. Since the ECD is a concentration-dependent detector, a constant flow is important for good reproducibility and of course flow-controlled operation is recommended. Third, leaks that result in oxygen entering the system produce a high background signal and drift.

Other Specific Detectors. Gas chromatographs have been coupled to both mass spectrometers and Fourier transform IR spectrometers. Whether the combination is the choice of the chromatographer who sees these devices as elegant, specific detectors or of the spectroscopist who is delighted by being able to separate components of mixtures and then examine their spectra without interferences, is uncertain; in any event, the results have been gratifying. These sophisticated instruments (with acronyms GC/MS or GC/FTIR) utilize the GC as a *sample channel* to isolate components quantitatively and present them one at a time to the associated spectrometer for extraction of both qualitative and structural information. These instruments were taken up in Sections 17.9 and 20.8 in the context of discussion of FTIR and mass spectrometers, respectively. The instrument to which the chromatographic sample channel is coupled must always be one with high-powered qualitative capability.

In addition, simpler dedicated spectrometric detectors have been developed for GCs. One example is the ion trap, a mass spectrometric detector (see Section 20.8). These devices serve as chromatographic detectors in the classical sense.

25.5 QUALITATIVE METHODS

Analyte identification by GC is usually based upon the fact that, for a given compound, its corrected retention time t'_R or retention volume V'_R will be unique on a given column under specific conditions of temperature and flow. Since column conditions are difficult to reproduce, ordinarily chromatographic measurements should be made relative to standards, as will be discussed below. One evident problem with reproducibility of column conditions is that the volume of liquid or stationary phase will be constantly changing with column use.

Example 25.8. Can one always take the presence of a peak at a given retention time under controlled conditions as confirmation of the presence of a particular analyte? Assume that measurements with standards have confirmed the reproducibility of the GC and column for the analyte.

Actually, it can be shown that two compounds can have the same retention time under particular conditions: they would fail to be resolved by a given column. For that reason, while the presence of a peak of given retention time is strong evidence for the presence of a particular compound, confirmation by an independent means is logically necessary. Nevertheless, the absence of a peak at that retention time can be accepted as proof that the suspect compound is not present. Therefore, identification of a compound by GC is more often a *negative* test.

Two forms of relative retention time are also used in analyte identification in GC–relative retention time and retention index. The *relative retention* for a given column is calculated from the equation

$$r = t'_{Ri}/t'_{Rs},$$

where t'_{Ri} is the adjusted retention time of the component of interest in the mixture and t'_{Rs} is the adjusted retention time of the standard or known component in the mixture. Relative retention is useful for a quick survey of a sample and for limited sample types. It suffers from the difficulty of finding a universal standard and from the need to find a standard close to the analyte for greatest accuracy. Clearly, the possibility of parallel columns with different stationary phases introduces a useful variation, it allows relative retention to be calculated for an analyte under two different column conditions. The same limitations will apply.

To overcome these drawbacks the concept of a *retention index* was proposed by Kovats in 1958. In his system retention time (or retention volume V_R) is related to that of the normal hydrocarbon eluting just before and after it. Each *n*-paraffin is assigned an index I, where $I = 100z$, where z is the carbon number. The retention index of the analyte (I) is expressed by the equation

$$I = 100z + \frac{100\,(\log t'_{Ri} - \log t'_{Rz})}{(\log t'_{Rz+1} - \log t'_{Rz})}$$

where z and $z + 1$ are the carbon numbers of the normal paraffins eluted before and after analyte i. The difference of adjusted retention times for the analyte and hydrocarbon z appears in the numerator and difference between the retention times of the normal paraffin of carbon number $z + 1$ and z in the denominator. A basic assumption is that hydrocarbon variations in retentions under constant temperature will be related to such variations for other compounds. The dependence of $\log t'_R$ on carbon number is found experimentally.

The retention index is independent of all instrumental parameters except temperature and the nature of the stationary phase. In addition, any homologous series of compounds can be used to produce the index for a given laboratory. Until recently this task was beyond the average laboratory, but the use of computers in chromatography and the development of powerful chromatographic data systems has made a library of indexes useful and possible for even small labs. Sadtler reference libraries of retention indexes are now commercially available. Comparison of unknowns to the reference index is a powerful qualitative tool. For an analyte, a corroboration of identity can be provided by measuring the analyte and appropriate indexing compounds on several columns. Index measurements with test compounds can also be used to validate the condition of a column, since only a change in the column or a change in the oven temperature will alter the index.

25.6 QUANTITATIVE METHODS

Gas chromatography has achieved the high level of importance in analytical chemistry that it enjoys today as a result of its power to gather, in one relatively short determination, quantitative information about components in the absence of almost all matrix effects. Quantitative data are obtained from a chromatogram when a linear detector is used to identify the eluting compounds and produce a chromatogram. While integrating detectors have been used in gas chromatographs, today detectors are arranged to provide the typical peak which represents the so-called normal chromatogram (see Chapter 24) in which the peak area is proportional to the concentration of the component.

In using GC for analysis of samples, it must be recognized that only the compounds that pass through the detector are determined. Nonvolatile or poorly volatilized substances will not produce useful responses. Further, the detector must respond to components if a selective detector is used.

The methods of quantifying most widely used in GC are area percent, external standard, and internal standard. Table 25.3 shows the equations used in each for reference. These methods are described in detail in Chapter 24; however, the following discussion will assess the merits of each technique. The method of standard addition is also useful, but is less often employed. It is a method of choice when an alternate standard is not readily available.

Area Percentage. This method offers a quick and easy way to determine the relative distribution of compounds in a mixture. It calls for taking the ratio of the area of each component's peak to the total area and translating the ratio into an estimated relative concentration. It ignores the variation in the detector response of different compounds, the presence of nonvolatile components in the mixture, and even sample size, which is unimportant because results are normalized. This approach is used chiefly with thermal conductivity detectors, which have similar responses for most compounds.

The chief applications for the area percentage method of calculation are to survey an unknown sample for approximate composition in situations where exact concentrations are not needed or are difficult to achieve in other ways. Samples that are often determined this

Table 25.3 Equations for Calculating Analyte Concentrations C_i by Different Methods

Method	*Equation*	*Identification of New Terms*
Area percentage (A%)	$C_i = A_i / \sum_{i=1}^{n} A_i$	A_i = area of analyte peak
External standard	$C_i = \dfrac{R_{fi} \times A_i \times F}{wt_{Sx}} \times 100$	R_{fi} = absolute response factor F = dilution factor wt_{Sx} = weight of sample
	$R_{fi} = C_{Es}/A_{Es}$	C_{Es} = concentration of external standard A_{Es} = area of external standard peak
Internal standard	$C_i = \dfrac{A_i}{A_{Is}} \times RR_f \times C_{Is}$	C_{Is} = concentration of internal standard
	$RR_f = \dfrac{A_{Is} \times wt_i}{A_i \times wt_{Is}}$	A_{Is} = area of internal standard peak RR_f = relative response factor wt_i = weight of analyte wt_{Is} = weight of internal standard

way are mixtures of essential oils where the sample complexity and similar chemical nature of many of the compounds make use of other calculations unwarranted.

External Standard Procedure. This procedure has the following advantages: (a) it determines the species of interest directly; (b) it corrects for the system response characteristics; and (c) it is easy to perform. A known amount of pure analyte is injected into the instrument and its detector response factor is calculated. A response curve can be generated and this can be used where slight nonlinear response behavior is found. Correction of the response factor for the purity of the standard can also be incorporated.

There are two drawbacks to the external standard procedure. One is that the weight (or volume) of sample must be known as well as the weight of the standard, which means that this method is highly technique-dependent. Skilled operators can vary as much as 3–5% in sample delivery technique, which may constitute a serious source of error in a determination. The second disadvantage of the technique is that frequent calibration is required. Otherwise, changes in instrument performance resulting from loss of stationary phase from the column, detector malfunction, or leaks in the pneumatic system will alter the response and may introduce serious error.

An external standard calculation is used often when a one-time assay is needed or when errors other than those described are more likely. When demands of this kind are made, some uncertainty resulting from instrumental considerations may be acceptable. A good example of the latter case is the determination of pollutants in the environment where a small error from the instrument system is much less than the sampling and clean-up errors, which may be 20% or more. The errors resulting from injection technique can be substantially reduced with the use of automatic sampling systems or gas-sampling valves with a constant volume loop. In fact gas analyses are performed using the external standard method because there is no convenient way to prepare an internal standard routinely for gas samples.

Internal Standard Method. In this method a compound which is not normally a component of the mixture of interest is added to the sample and the compounds are determined together. This method relies on the use of the ratio of the response of the added internal standard to that of the analyte to eliminate both the dependence on technique found in the external standard method and to minimize the effect of changes in instrument performance from run to run. Since a standard compound is added to both the calibration mixture and the samples, the system always sees both compounds and presumably changes in the system will affect the response of the internal standard in the same way as the components of interest. The chief disadvantages of this technique are that the search for an internal standard can be difficult and that it is often hard to find an uncluttered portion of the chromatogram to insert a compound. This sort of problem is a major reason for the lack of applicability of the technique to samples from environmental sources that may contain unpredictable interferences.

REFERENCES

The following books deal with gas chromatography in a general manner. Some are older books, but nevertheless cover the subject well.

1. J. Willet, *Gas Chromatography*. New York: Wiley, 1987.
2. J. C. Giddings, *The Dynamics of Chromatography*. New York: Dekker, 1988.
3. J. A. Perry, *Introduction to Analytical Gas Chromatography: History, Principles, and Practice*. New York: Dekker, 1980.
4. M. L. Lee, F. Y. Yang and K. D. Bartle, *Open Tubular Gas Chromatography: Theory and Practice*. New York: Wiley, 1984.

The following books deal with special aspects of gas chromatography.

5. M. Dressler, *Selective Gas Chromatography Detectors*. New York: Elsevier, 1986.
6. B. Kolo (Ed.), *Applied Head Space Analysis*. New York: Wiley, 1980.
7. G. M. Message, *Practical Aspects of GC/MS*. New York: Wiley, 1984.
8. S. A. Liebman and E. J. Levy (Eds.). *Pyrolysis and Gas Chromatography*. New York: Dekker, 1983.

EXERCISES

25.1 Two GC columns with the same stationary phase are available and are used under identical, isothermal conditions. A C_{13} normal hydrocarbon generates 3800 effective plates on one column and 4200 on the second column. (a) Which column should an analyst use? Explain. (b) If the second column is also twice as long as the first, should the same decision be made?

25.2 Describe the effect on resolution of introducing (a) a small sample into a GC with a sample injection inlet of large volume and (b) a large sample into another GC that has a small volume inlet.

25.3 Corrected retention times for isothermal elution of ethanol, 1-propanol, and 1-butanol on a packed column with a nonpolar silicone oil stationary phase, are 0.89, 1.71, and 3.77 min. (a) Predict retention times for the isothermal election of 1-pentanol and hexanol. (b) Would retention times be longer or shorter if a stationary phase of moderate polarity such as poly(phenylmethyldiphenylsiloxane) were used instead? Explain.

25.4 A compound has a Kovats index of 623 on a new column run under particular conditions. If the Kovats index of this compound changes to 541 after 3 w use of the column, what conclusion can be made?

25.5 The areas and retention times for the peaks in Fig. 25.9b are given below. (a) Calculate the concentration of the C_7 component by area percent. (b) Consider that the data were obtained using a 0.25 g standard sample of C_7. Calculate the percent of the C_7 component in a 1.0 g sample for which the area of the C_7 peak is 306.7. Equal volumes of sample and standard are injected. (c) Assume that the data below are for a standard sample and C_6 is the internal standard and that its amount in the sample and the standard is the same. If the sample weight is 1.0 g, the peak area for C_6 is 290.0, the amount of C_7 in the standard is 0.25 g, the peak area for C_7 is 280.15, how much C_7 is present?

Hydrocarbon	Retention Time	Peak area
C_6	1.28	290.0
C_7	2.18	340.8
C_8	3.62	376.4
C_9	5.38	409.6
C_{10}	7.27	434.2
C_{11}	9.16	456.0
C_{12}	10.97	468.4

ANSWERS

25.2a: More band spreading will occur than necessary. *b* The sample will be likely to back up into the carrier gas supply. *25.3a*: t'_R for pentanol, hexanol would be about 8.2 and 18.5 min. *b* The retention time would be longer since the alcohols would be attracted more strongly. *25.4* The column chemistry has changed. It may have been damaged by contamination, oxygen, or other source of degradation. *25.5a*: Concentration of $C_7 = [A_{C_7}/A_{C_i}] \times 100 = [340.8/2775.4] \times 100 = 12.3\%$; *b*. With this external standard, concentration of $C_7 = 22.5\%$. *c* $C_i = A_i/A_{Is} \times RR_f \times C_{Is}$; $C_{Is} = 1.0$; $RR_F = (A_{Is}w_{ti})/A_iw_{tIs} = 290.02 \times 0.25/340.8 \times 1.0$; $C_i = 280.15/290.19 \times (290.02 \times 0.25)/340.8 \times 1.0 = 0.206$ g C_7 or 20.6% C_7.

Chapter 26

LIQUID CHROMATOGRAPHY

Kenneth Ogan

26.1 INTRODUCTION

High-performance liquid chromatography, or liquid chromatography (LC) as it is more simply called, has developed into one of the premier analytical techniques. Its development came after gas chromatography, which was discussed in the last chapter. Liquid chromatography is similar to GC both in common underlying principles and basic instrumentation. Its unique characteristics arise from the distinctive properties of liquids—values of diffusion, viscosity, polarizability, and acidity are orders of magnitude different for liquids than for gases. As a consequence, in LC the mobile as well as the stationary phase can be used for differential interaction with the components of mixtures. The differences also greatly affect the design and construction of modules for LC. The situation can be summarized by stating that while the underlying principles of chromatography (see Chapter 24) are similar for the two techniques, the manner in which they are reduced to practice follows very different lines. Or to put it more generally, LC is truly complementary to GC.

The modular layout of a liquid chromatograph is shown in Fig. 26.1. As the liquid phase is pumped to the column, a sample is introduced by an injection system just before the column. The mobile phase containing a thin "plug" of sample enters a narrow, cylindrical packed chromatographic column. As the plug moves through the column, sample components are separated, and on emerging, their presence is sensed by a detector. Processing modules establish the chromatogram of the sample, allowing the solutes to be tentatively identified on the basis of elution times (or elution volumes). The amount of each is determined from the height or area of its peak in the chromatogram.

To plan a separation by LC, the user must select both a type of column and a mobile phase appropriate to the analytes in the samples at hand. In addition, the user must identify a chromatographic system that will maintain the sharpness of analyte bands as a sample moves through the column to the detector. Both of these aspects will be taken up in the following sections on the types of LC and the basic modules and systems aspects of a liquid chromatograph. The chapter will conclude with a brief discussion of qualitative and quantitative anaysis by LC.

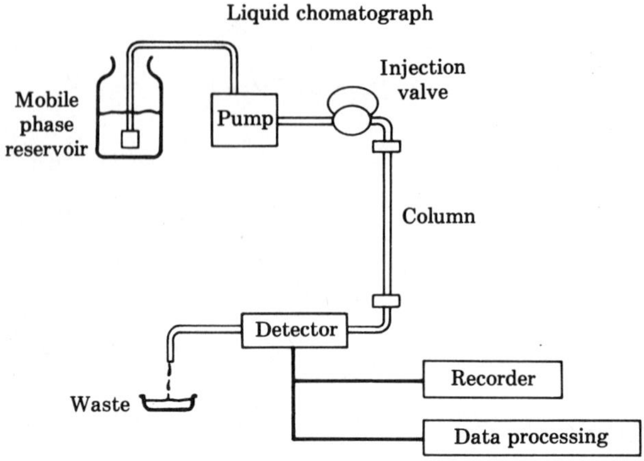

Fig. 26.1 Schematic diagram of the liquid chromatographic system.

26.2 LIQUID CHROMATOGRAPHIC SEPARATION OF COMPOUNDS

Separation in LC is achieved by means of differences in the interactions of the analytes with both the mobile and stationary phases. The mobile phase must be chosen to ensure solubility of the sample solutes. For the stationary phase, microparticulate silica (bare or chemically modified) is used almost universally because its high surface area accentuates the differences in solute stationary-phase interactions. The use of a stationary phase that interacts strongly with solutes relative to solute mobile-phase interactions will result in very long retention times, a situation which is not analytically useful. Hence, the stationary phase must be selected so as to provide weak to moderate solute interactions relative to those in the mobile phase. As a consequence, the nature of the sample governs the type of LC selected; the stronger interactions should occur in the mobile phase to ensure sample solubility and ready elution, while the stationary phase should be responsive to more subtle differences among the solutes. For example, polar neutral compounds are usually better analyzed using a polar mobile phase together with a nonpolar stationary phase that distinguishes subtle differences in the dispersive character of the solutes.

The conventional classification of separation modes in LC is that of liquid–solid, liquid–liquid, ion exchange, and size-exclusion chromatography. In reality, the precise classification of a system into one of the first two categories can be difficult because the experimental evidence often suggests a mixture of retention mechanisms. Consequently, the classifications used in this chapter will be based instead on the type of solute interaction. Four general classes of LC can be distinguished from this perspective:

1. *Normal-phase chromatography* calls for the use of a polar stationary phase in conjunction with a nonpolar, or as it is often called, dispersive mobile phase.

2. *Reversed-phase chromatography*, the opposite possibility, calls for the use of a nonpolar stationary phase and a polar mobile phase. The vast majority of chromatographic analyses fall into one of these two classes.

3. *Ion-exchange chromatography* involves ionic interactions. In this case the mobile phase must support ionization to ensure solubility of ionic solutes. The stationary phase must also be partially ionic to promote some retention. Hence, the interactions with the stationary phase are strong, and this is usually reflected in longer analysis times and broad peaks.

4. *Size-exclusion chromatography* involves separations based on molecular size alone and ideally requires that there be no energetic interaction of the solutes with the stationary phases.

Historically, high-performance liquid chromatography used silica or alumina as the stationary phase in conjunction with hydrocarbonaceous mobile phases. Separations were achieved on the basis of polar interactions with the stationary phase and dispersive interactions in the mobile phase, as was described in Chapter 24. There were also cases in which separations were carried out on silica or alumina coated by a strong organic modifier that had been added to the mobile phase. The latter approach provided a liquid–liquid chromatographic system with a reversal of the types of interactions: solutes experienced dispersive interactions with the stationary phase (the adsorbed organic modifier) and strong polar interactions with the polar mobile phase. Because of this reversal of interactions and type of mobile phase relative to the original form of LC, this approach was called "reversed-phase" chromatography. For that reason LC using silica as stationary phase and nonpolar mobile phases came to be called normal-phase chromatography. As will be seen below, both types have continued to evolve.

Normal-Phase Liquid Chromatography. Normal-phase systems are characterized by use of a polar stationary phase and a nonpolar mobile phase that promotes solute–solvent interactions of the London dispersion type. Actually, normal-phase systems now extend well beyond silica or alumina stationary phases to include polar *bonded-phase silicas*. They may, for example, have diol functionality bonded to the silica. They will also be used with nonpolar mobile phases. Normal-phase systems are particularly useful in the separation of weakly polar compounds, especially isomers. For example, isomeric substitution of benzene or other aromatic compounds with various groups can lead to significantly different dipole moments and polarizabilities, differences that are readily distinguished by normal-phase chromatography.

The use of bare silica as a stationary phase has significantly declined since the introduction of the bonded phases. Dry silica is highly polar and will strongly adsorb even traces of polar impurities such as water from mobile-phase solvents and thereby alter chromatographic activity. Very polar compounds in the sample are also strongly adsorbed, which again can change the activity of the silica. The availability of new, polar bonded-phase silicas has reduced this problem of alter-

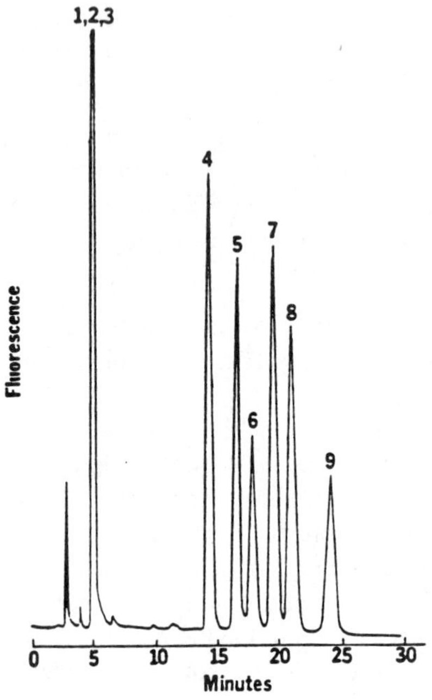

Fig. 26.2 Chromatogram of the dansyl derivatives of equine estrogens on a LiChrosorb Si-60 column with 50% chloroform in heptane as mobile phase. Peak identification (dansyl derivatives): 1, estrone; 2, equilin; 3, equilenin; 4, α-estradiol; 5, α-dihydroequilin; 6, α-dihydroequilenin; 7, β-estradiol; 8, β-dihydroequilin; 9, β-dihydroequilenin.

ation of stationary phase activity; at the very least, the regeneration and reequilibration of these phases is much easier and quicker.

An example of a normal phase separation is given in Fig. 26.2. The estrogens being separated are quite similar in structure, differing only in the degree of ring saturation and keto versus hydroxyl substitution.

There are two main theories of retention in normal-phase liquid chromatography, one due to Snyder and Soczewinski* and one due to Scott and Kucera.† Neither of these theories has been demonstrated to be universally applicable for normal-phase chromatography. The two differ primarily in the relative importance assigned to solute mobile-phase interactions. The Snyder–Soczewinski theory assumes strong specific interactions of the solute with sites on the stationary phase and ignores any solute mobile-phase interactions. The retention behavior of a series of standard solutes are used to generate numerical values for coefficients in key equations. The fundamental retention relationship in this theory is given by

$$\log k' = B + a'A_s e^0, \tag{26.1}$$

where e^0 is a constant characteristic of the mobile phase in use with the specific adsorbant, A_s is the area of the solute molecule, and k' is the chromatographic

*L. R. Snyder, *Principles of Adsorption Chromatography*. New York: Dekker, 1968, *Anal. Chem.*, **46,** 1384 (1974).

†R. P. W. Scott, *J. Chromatogr.*, **122,** 35 (1976); R. P. W. Scott and P. Kucera, *J. Chromatogr.*, **149,** 93 (1978).

capacity factor. If a wide variety of compounds is to be separated, additional terms and adjustments to e^0 and A_s are usually required to fit the retention data obtained.

The Scott–Kucera model is based upon the assumption that the strong organic component in the mobile phase is attracted to the surface of the adsorbent in a Langmuir adsorption isotherm process and that the solute molecules then interact with this adsorbed organic layer, which is called a modifier. According to this theory, the basic retention relationship is given by

$$1/V'_R = A + Bc_p, \qquad (26.2)$$

where V'_R is the adjusted retention volume, $(V_R - V_m)$, and c_p is the concentration of the strong organic modifier in the mobile phase. This model has not received as much attention as the Snyder–Soczewinski model, and its applicability is not as thoroughly tested. A statistical–mechanical treatment of retention by adsorption has also been recently published.*

Reversed-Phase Liquid Chromatography. As noted above, the original reversed-phase systems were generated by an organic modifier strongly adsorbed onto silica from the mobile phase. Such systems were useful for many separations but were difficult to reproduce and use. As a consequence, reversed-phase chromatography was not widely used until bonded-phase silica supports were developed in the late 1960s and early 1970s. Bonded phases are prepared by reacting selected silanes with the silica surface. Halogen- or methoxy-substituted organosilanes will react with silanols on that surface, yielding an Si–R functionality attached to the silica surface through an Si–O–Si bond, as outlined in Fig. 26.3. (Hydrochloric acid or methanol is the byproduct.) It is possible to produce silicas with a wide variety of organic groups bound to the silica surface, but octadecyl silica [$R = (CH_2)_{17}CH_3$ in the figure] is by far the most commonly utilized bonded phase.

The interactions of solutes with an alkyl-bonded phase such as the C_{18} bonded-phase silica are dispersive, and a polar mobile phase is conventionally used in

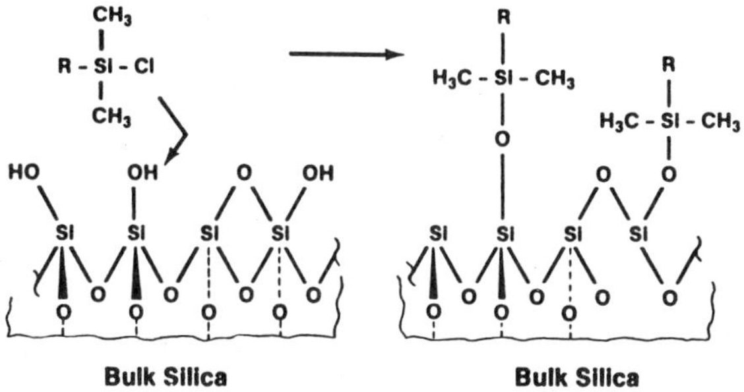

Fig. 26.3 Reaction between chlorodimethylalkysilane and a silica surface.

*R. E. Boehm and D. E. Matire, *J. Phys. Chem.*, **84**, 3620 (1980).

conjunction with these stationary phases. The nonpolar stationary phase makes these systems very useful for separating organic compounds with slight differences in the backbones or side chains, differences as little as a single methylene group in some cases. The alkyl bonded-phase supports have proven to be very useful chromatographic materials, with several desirable characteristics. Mobile phase changes can be accomplished easily and quickly, which is especially useful for reequilibration after a gradient run. The lack of excessively strong solvent–surface interactions facilitates the use of a wider range of mobile phases, which allows for greater flexibility in the control of selectivity as well as more options in the choice of a suitable solvent for a sample. Selectivity and retention can be adjusted by altering the type of group bonded to the silica, and the degree of retention for many compounds can be regulated by controlling the extent of bonding. Another useful aspect of bonded phases is the greater homogeneity of surface energy, relative to bare silica or alumina, which not only minimizes sample loss due to irreversible adsorption at highly energetic sites, but also reduces peak broadening arising from heterogeneous interactions.

The theory of retention for reversed-phase chromatography, or more specifically, for retention on alkyl-bonded phases, has received considerable study, but is as yet less than totally successful at predicting the retention behavior of a particular solute. There have been several papers dealing with the "solvophobic theory" of reversed-phase chromatography,* which treats the energetics of the state process of (a) transferring the solute molecule from the mobile phase to the gas phase, (b) opening a "hole" in the mobile phase that solvates the alkyl chains of the stationary phase, and (c) transferring the solute molecule into this hole. This theory predicts that retention of a solute depends on the area of contact between the solute and an alkyl chain of the stationary phase, and the energy required to remove the mobile phase from this area of the alkyl chain (the net energy expended in creating the hole along the alkyl chain). This theory has had some qualitative success in that it can interpret trends in retention of homologous series or of specific classes of solutes. However, several factors which are needed in the quantitative expression of the retention are not readily available or known a priori, factors such as the area of contact between the solute molecule and an alkyl chain of the stationary phase.

An example of a reversed-phase separation is shown in Fig. 26.4. This chromatogram demonstrates the separation of polycyclic aromatic hydrocarbons on a C_{18} bonded-phase column using gradient elution (from 40% acetonitrile in water to pure acetonitrile). These compounds elute in the order of increasing molecular weight, reflecting decreasing solubility in the mobile phase (and hence, progressively weaker solute mobile-phase interactions).

Ion Chromatography. Charged solutes can also be separated by chromatographic means. Aqueous mobile phases are generally needed in order to dissolve the ionic solutes. Most ionic solutes of current interest will not interact with hydrocarbon-

*C. Horvath, W. Melander, and I. Molnar, *J. Chromatogr.*, **125**, 129 (1976).

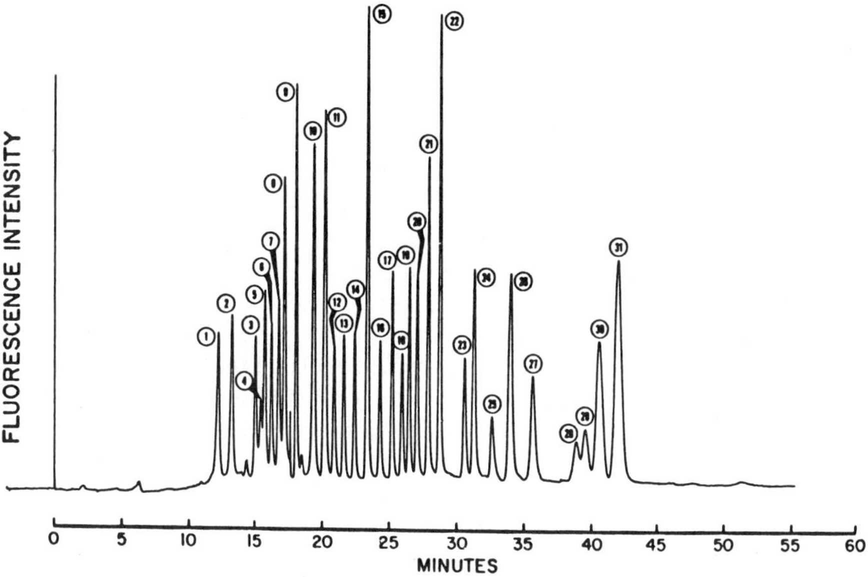

Fig. 26.4 The reversed-chromatographic separation of polycyclic aromatic hydrocarbons on a bonded-phase C_{18} column (Perkin-Elmer PAH/10, 0.26 cm i.d., 25 cm length). Mobile phase: 40% acetonitrile–water, linear gradient to 100% acetonitrile in 20 min. Flow rate: 0.5 mL/min. Fluorescence detection: excitation/emission 280/340 nm initially; excitation/emission = 305/430 nm after 17.6 min. Peak identification:

1 indene
2 naphthalene
3 1-methylnaphthalene
4 2-methylnaphthalene
5 acenaphthene
6 fluorene
7 1,4-dimethylnaphthalene
8 phenanthrene
9 anthracene
10 fluoranthene
11 pyrene
12 9,10-dimethylanthracene
13 2-methylanthracene
14 benzo[a]fluorene
15 Benz[a]anthracene
16 chrysene
17 7,12-dimethylbenz[a]-anthracene
18 benzo[e]pyrene
19 benzo[b]fluoranthene
20 dibenz[a, c]anthracene
21 benzo[k]fluoranthene
22 benzo[a]pyrene
23 dibenz[a, h]anthracene
24 benzo[ghi]perylene
25 indeno[1,2,3-cd]-pyrene
26 dibenzo[a, e]pyrene
27 benzo[b]chrysene
28 dibenzo[b, h]chrysene
29 picene
30 *p*-quarterphenyl
31 coronene

aceous stationary phases (e.g., alkyl bonded phases), and the interactions of ions with polar stationary phases such as silica would be comparable or greater in strength than the interactions with the polar mobile phase. For this reason, most analyses of ionic species have utilized an aqueous mobile phase containing electrolytes and a stationary phase with weak ionic sites to achieve retention and thereby separation, that is, to effect ion-exchange chromatography. A displacement type of development of the chromatogram is used. The solute ions first displace the eluent ion from the ionic sites on the stationary phase, and then in turn are displaced by the eluent and move down the column in bands in a fashion completely analogous to that for normal- or reversed-phase chromatography. The kinetics of this exchange process are often not as fast as for the other types of chromatography and there can be considerable band broadening. Furthermore, most ion-exchange columns display a range of exchange-site energies, which also contributes to band broadening. The strong interaction of solute ions with the stationary phase also increases retention times, thereby lengthening separations.

Three major approaches to ion chromatography exist, differing largely on the basis of the detection method. The use of a conductivity detector is a natural choice for the detection of the ions. Many ions of analytical interest do not adsorb appreciably in the UV, nor do they induce a sufficiently large refractive index change to be readily detected. Unfortunately, the salt and buffers required to carry out an ion-exchange separation ordinarily generate an overwhelming background signal in the conductivity detector. A clever solution to this problem was developed by Small,* who mounted a second column in series with the separation column which served to *neutralize* the added buffer salts, thereby reducing the background conductivity. This reduction enabled the sensitive detection of the conductivity changes caused by the eluting bands of ions. This approach has had wide success. It clearly has its limitations in that each buffer system must be compatible with the second "suppressor" column and still be capable of effecting the separation, but it has made possible a number of separations such as the separation of simple anions such as Cl^- and SO_4^{2-} not previously feasible.

The second approach to the problem came about with the extensive development of bonded-phase columns. In this method a single column is used with a mobile phase of lower ionic strength in which a conductivity detector can be used directly. The column is one of two types: either a static ionic exchange column with a low density of exchange sites or a reversed-phase column with ion-exchange sites dynamically generated by additives to the mobile phase.

Each of these two approaches has its advantages and disadvantages. A suppressor column enables the detection of ionic species at lower concentration, but itself contributes signficantly to band broadening and thereby lessens the peak signal. The single-column system is simple to use and gives much greater flexibility in choice of mobile phase, but its higher background conductivity does restrict the sensitivity.

The third detection method, indirect UV detection, has recently been developed. A UV adsorbing compound is added to the mobile phase and generates a continuous signal in a UV detector. Elution of the separated ionic species causes dilution of the adsorber and leads to a smaller signal from the UV detector. A chromatogram actually appears as a series of *negative* peaks. Fortunately, the degree of dilution and reduction in UV absorption signal is proportional to the concentration of the ionic species and detection for these ionic compounds is linear. With the proper choice of background absorber and operating conditions, this tech-

*H. Small, T. S. Stevens, and W. C. Baumann, *Anal. Chem.*, **47**, 1801 (1975).

nique has in some cases proven to be more sensitive than monitoring the conductivity.

The peak-broadening and lengthy retention times experienced with ion-exchange chromatography have led to exploration of alternative means of analysis and some clever and successful methods have been developed. In the analysis of samples of weak acids and bases one of the simplest solutions has been adjustment of the pH of the mobile phase to neutralize the ionic species and generate a neutral form that can be separated by reversed-phase chromatography. Obviously, this approach is only successful for the analysis of a limited number of acids or bases which have sufficiently close pK values in a single run. An approach with more general applicability is that of "ion pairing." In this method a counter ion that will interact sufficiently strongly with the ions of interest to generate a complex that can be chromatographed as a neutral species is added to the mobile phase.* (Adjustment of pH is clearly a subset of this approach in which the "complex" is the protonated acid anion.) The procedure has been successfully applied to the separation of many ionic species.

Size-Exclusion Chromatography (SEC). SEC is unique (or perhaps anomalous) in the sense that the separation does not involve direct energetic interactions. Instead, separation is achieved with specially designed stationary phases that have

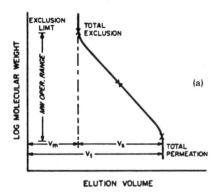

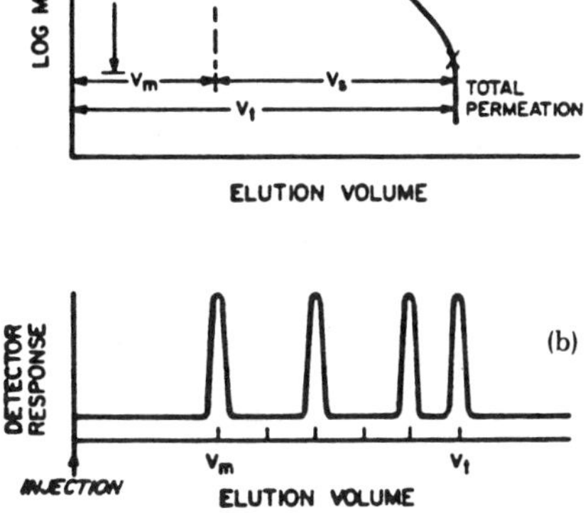

Fig. 26.5 A molecular-weight calibration chromatogram for size-exclusion chromatography with four distinct molecular-weight standards. (a) Calibration curve. (b) The chromatogram of the standards. This curve provides values of V_R for the abscissa in (a).

*Consideration of the equilibrium constants and kinetics of the complexation process, and further experiments with a variety of ion-pairing reagents has suggested that it is not always the complex of the counter ion and analyte ion that is actually being chromatographed, but rather that the counter ion is serving as a dynamically generated ion-exchange site with a true ion-exchange process taking place.

rigid channels or pores of relatively similar diameter available to molecules. This results in a smaller elution volume for large molecules that are excluded from the pores. A continuous distribution of pore sizes will provide a continuous range of elution volumes, with progressively *increasing* elution volume for progressively *decreasing* molecular size. Completely resolved peaks are rarely obtained for real samples; instead, a rather complex envelope is obtained, arising from the overlap of sequentially eluting molecular size fractions. Calibration of the elution volume with a set of standards of known size then enables the distribution of molecular sizes of an unknown sample to be determined.

Molecular weight, which is of considerable analytical interest and importance, can be related to molecular size. Excellent correlations between elution volume and molecular weight are empirically obtained for individual classes of compounds. Calibration of either a single SEC column or a set of serially coupled SEC columns with standards then generates a calibration graph, as shown in Fig. 26.5, and enables an analyst to determine the molecular weight from the elution volume. Because of its ability to provide molecular weight distributions, SEC is one of the principal analytical techniques in a number of industrial areas.

26.3 MODULES IN A LIQUID CHROMATOGRAPH

The column is the heart and raison d'etre for a liquid chromatographic system. Accordingly, a consideration of its main parameters and behavior will open the section followed by discussion of other modules such as pumps, sample injectors, and detectors.

The Column. The successful and efficient resolution of solutes into bands in a column is described by the chromatographic theory presented in Chapter 24. Both (a) thermodynamic factors that lead to resolution of analytes into bands in a flowing mobile phase and (b) kinetic factors that affect the sharpness and hence the distinctness of such bands are taken up. That discussion will be essential background for understanding the behavior of LC columns.

For each of the types of chromatography described in the last section, columns are available. Given quantitative recovery of components from a mixture as a minimum requirement, the user usually seeks an optimum resolution of components and speed. These aspects will be taken up in Section 26.5; here the focus will be the physical parameters of columns.

Chromatographic theory in particular defines column performance by what might be described as a two-step process. Recall that the efficiency of column performance is described by the number of theoretical plates N. Theory also defines a plate height H, which is a measure of band broadening that is linked to column length L and plate height by the equation

$$N = L/H. \tag{26.3}$$

The plate height is in turn related by the van Deemter equation [Eq. (24.19)] to particle size, linear velocity, solute diffusivities, and the degree of retention in a column. Since this equation includes terms both reciprocal and linear in u, the linear velocity (in addition to a constant term), H goes through a minimum as u is varied (see Fig. 24.12). Thus, there is a single, optimum value of u for which H is a minimum.

It is now important to deal with physical parameters of columns. From the very beginning of high-performance liquid chromatography, it has been evident that column resolution could be improved by reducing particle size. The connection becomes clear on examining the expression for the minimum plate height, derivable from the van Deemter equation:

$$H_{\text{min}} = A + 2\sqrt{BC} = \lambda d_{\text{p}} + 2d_{\text{p}}\sqrt{f_{\text{m}}(k')}. \qquad (26.4)$$

Note that the stationary phase contribution to the resistance-to-mass transfer was omitted in the derivation. The linear dependence of H_{min} on the particle size implies that smaller particles will yield columns of higher efficiency and improved resolution.

Unfortunately, the pressure drop per unit column length is inversely proportional to the *square* of the particle size. A reduction in size must therefore be accompanied by a sharp increase in the operating pressure. Chromatographic apparatus has had to be redesigned for the use of smaller beads. In fact, careful examination of the theory and experimental data demonstrates that for a given operating pressure and a desired separation, that is, a particular efficiency [see Eq. (26.3)], there is a unique particle size which will generate this separation in a minimum time. This size allows the optimum trade-off between the decrease in H and increase in pressure. Particles smaller than the optimum value cannot be used to generate the required efficiency because of pressure limitations. While the required efficiency could be generated with larger particles in longer columns, analysis times would also be longer.

The consequences of these findings are of major importance. Small diameter particles are suitable for very fast, *simple* separations, while more complex separations require *larger* particles, longer columns, and longer analysis time. This result is not widely understood nor appreciated. Finally, the appropriate value of the column *radius* is set by the contribution to band dispersion of the rest of the chromatograph, that is, the extra-column dispersion. It is important that the column contribution to resolution predominate over the detrimental effects of the extra-column system.

In reality, there are only a limited number of column lengths and diameters commercially available to the analyst more because of the pressure restrictions encountered as the particle size is reduced than because of theoretical considerations. Typical dimensions of commercial columns presently available are given in Table 26.1. The 0.1-cm i.d. columns in this table are sometimes referred to as microbore columns, because of their much smaller bore compared with the conventional 0.46-cm i.d. columns. These columns require a very low dispersion chromatograph for their proper use. Such columns do offer greatly reduced solvent consumption and greater mass sensitivity.

Table 26.1 Commercial LC Columns

Particle Size (μm)	*Column Length (cm)*	*Column Diameter, i.d. (cm)*
10	25, 30	0.10, 0.20, 0.46
5	5, 10, 15, 25	0.2, 0.46
3	3, 5, 10	0.46, 0.6

The achievement of reproducible analytical results requires that the column temperature be controlled. Changes in temperature alter the equilibrium distribution of solute between the mobile and stationary phase, and thereby the capacity factor. The last effect can in some cases be pronounced. (It has been recently demonstrated that viscous dissipation, the energy expended in forcing a liquid through small holes, can cause as much as a 24°C temperature increase in well-thermostated wide-bore columns packed with small particles when flow rates are high.* A temperature change of this magnitude can alter k' by 8% or more, a change sufficiently large to cause misidentification of a peak.) Changes in column temperature will also affect the mobile-phase viscosity and solute diffusivities, leading to changes in H, and therefore, in peak resolution as well.

Mobile Phase Pump Systems. Modern LC requires the use of a pump to force the mobile phase through the column. The pump must (a) develop pressures as high as several tens of megapascals (several thousand psi) and (b) provide a steady, precise, and accurate flow of fluid.

Syringe pumps of the type shown in Fig. 26.6 were widely used in early LC instruments. The pressure generated by the pump is governed by the ratio of the force delivered by the motor (through the gear train) to the cross-sectional area of piston. The steady, pulseless flow delivered by these pumps is one of their major attributes, one that is particularly useful with detectors that are highly flow-sensitive, such as electrochemical and refractive index detectors. A precise, accurate fluid flow requires that a precise and accurate drive motor, gear train, and screw drive be used. A pulseless flow is a sufficient attraction to account for the recent introduction of syringe pumps in the new small volume LC systems.

Syringe pumps suffer three disadvantages in comparison with reciprocating pumps, the other major pump type. First, in early syringes considerable time and difficulty was encountered in changing mobile phases and refilling the syringe. The piston had to be retracted and the reservoir dismantled and flushed. Second, in earlier syringes an accurate piston drive system was expensive and large mainly because of the large volumes required for elution. In the newer, small-volume systems designed for small-bore (microbore) columns these difficulties are much reduced. Third, a critical problem occurs with syringe pumps in gradient generation with solvents of considerably different compressibilities. In these cases, an increase in the viscosity of the solvent mixture as the gradient is developed leads

*M. J. E. Golay and J. G. Atwood, *J. Chromatogr.*, **186**, 353 (1979).

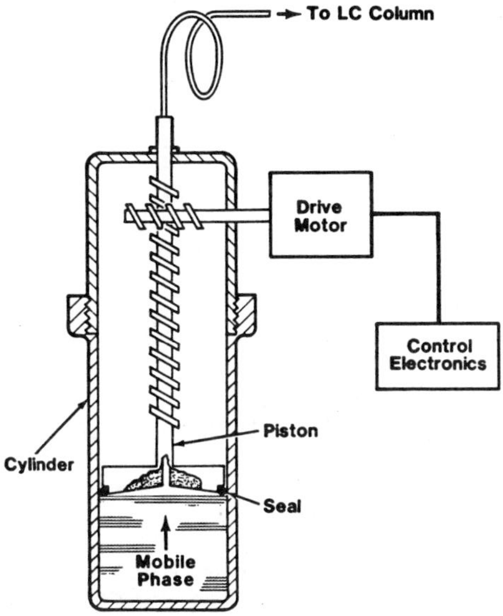

Fig. 26.6 Cross section of a syringe pump. The piston drive motor forces the piston down into the reservoir cylinder by means of a screw drive. The mobile phase is forced out through a channel up the center of the screw gear.

to an increase in back pressure from the column and a diminution of the flow of the more compressible solvent until its pump comes up to the new operating pressure. Conversely, a decrease in pressure will result in a temporarily increased flow of the more compressible solvent. In an extreme case, flow reversal can occur when only a small volume percentage is being delivered by one pump and the pressure change induces a volume change larger than the volume delivered by movement of the piston.

Reciprocating pumps are by far the most widely used type in LC today. The basic element of the pump is a small piston driven in a reciprocating motion, as is shown in Fig. 26.7. The check valves, in addition to their role in directing solvent flow, serve to prevent any back flow due to compressibility differences. Only the volume of solvent in the lines from the outlet check valve to the mixing point undergoes a volume change induced by a change in back pressure, as opposed to the entire reservoir of solvent in the case of the syringe pump. Hence, the actual volume change occurring in gradient formation in the reciprocating pump is orders of magnitude smaller.

Check Valve Design. The check valve is the Achilles heel of the reciprocating pump. Functionally, the check valve is usually a matched ball and seat. With proper design and maintenance, its performance is acceptable, but it is frequently the source of faulty operation. At the end of piston travel on the delivery stroke as the piston begins to reverse direction, the pressure in the pump head will drop below that in the outlet line. While the ball in the outlet check valve is being forced back against the seat, there will be a small

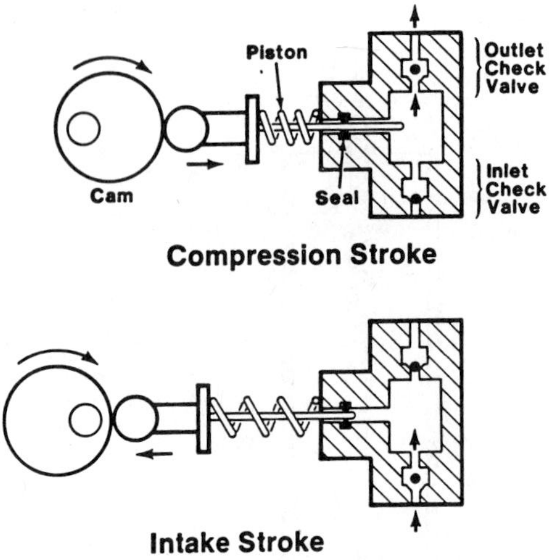

Fig. 26.7 Cross section of a reciprocating pump. In the intake stroke (bottom diagram) as the spring pushes the piston against the cam, the mobile phase flows into the pump head from a reservoir through the intake check valve (see arrows). (Pressure in the chromatographic column causes the output check valve to close.) The cam continues around, forcing the piston into the pump head, thereby forcing mobile phase in the pump head through the outlet check valve (top diagram). (Pressurization of the liquid by the incoming piston forces the inlet check valve closed.)

backflow of solvent into the pump. Valve design must minimize its closure time. Small particles that lodge in the seat will prevent it from closing completely, causing erratic and inaccurate column flow rates; hence, it is important that only clean, carefully filtered solvents be used. Wear products from the piston seal are another source of particles. As the piston is withdrawn from the cylinder, a new portion of solvent enters the pump head. The flow resistance offered by the inlet check valve, the inlet filter, and inlet plumbing must be minimal to avoid cavitation and incomplete filling of the pump head. (This defines the maximum permissible viscosity of the solvent.)

By its design, the reciprocating pump delivers solvent in a discontinuous fashion. Flow and pressure change sinusoidally, a pattern poorly suited to chromatography. The simplest solution is to use a pulse damper to smooth these fluctuations. Unfortunately, such dampers add additional volume and place pressure restrictions on the system. Another way to reduce pulses is to use a second, or even a third, pump head that operates out of phase with the first so that while one pump is refilling, solvent is being delivered from another. This approach results in significantly less fluctuation, but the additional parts add both mechanical complexity and cost. In any case, it is helpful to design the cam and motor controls to permit a fast refill cycle; then solvent delivery will account for a larger proportion of each cycle.

Gradient Development. Complex samples often include solutes with a wide range of retention times. To ensure resolution of the less retained compounds, a weak mobile phase is frequently required, with the result that strongly retained compounds elute only after a prohibitively long time. This problem can be solved by gradient development, a process in which a strong solvent is mixed in increasing proportions with the weak phase to produce a progressively stronger mobile phase. With gradient elution, weakly retained compounds are resolved under conditions of the weak mobile phase and the subsequent increase in the mobile phase strength shortens the retention time of compounds that interact more strongly. The use of gradient elution in LC is exactly analogous to temperature programming in GC separations.

The mobile phase gradient can be generated discontinuously by a series of discrete additions or continuously (and possibly in nonlinear fashion). The process must be carried out carefully and reproducibly for quantitative results. There are two general approaches. In the first, called *low-pressure gradient formation*, changes in mobile-phase composition are made prior to pump pressurization, as shown in Fig. 26.8. The actual changes in phase composition can be accomplished by (a) sequentially sampling from a series of reservoirs, each containing a steadily increasing strength mobile phase, (b) use of an exponential mixing volume, or (c) precision metering from different solvent reservoirs, as is done in modern LC instrumentation. The other general approach is *high-pressure gradient generation* in which the weak and strong mobile phases are individually pressurized in separate pumps and then mixed in a small volume immediately ahead of the sample injection valve and LC column. This arrangement is sketched in Fig. 26.9.

Each approach has distinct advantages and disadvantages. High-pressure gradient generation requires a separate pump for each solvent used, which becomes prohibitively expensive when more than two are required. It has the advantage, however, that the volume of the system between the mixing point and the column can be made quite small. Consequently, the volume delay between the initiation

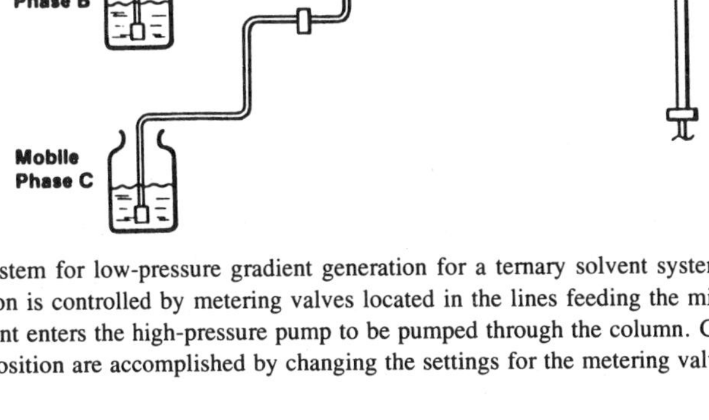

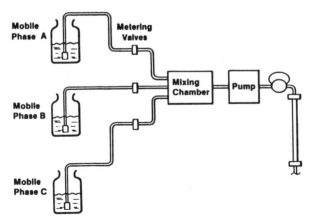

Fig. 26.8 A system for low-pressure gradient generation for a ternary solvent system. The mobile phase composition is controlled by metering valves located in the lines feeding the mixing chamber. The mixed solvent enters the high-pressure pump to be pumped through the column. Changes in mobile-phase composition are accomplished by changing the settings for the metering valves.

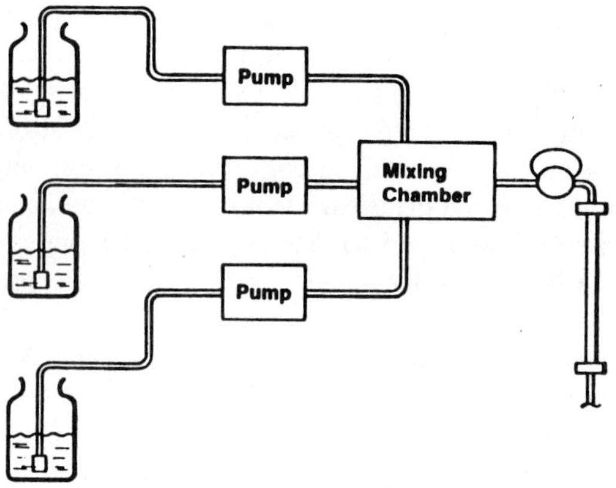

Fig. 26.9 A system for high-pressure gradient generation. (A ternary system is again shown for comparison.) The variable components of the mobile phase are introduced to the mixing chamber under high pressure by individual pumps. The output of the mixing chamber is connected directly to the injection valve.

of a composition change and its entrance into the column can be very short. This small delay volume does make possible an accurate and reproducible gradient generation in the midcomposition range when each pump is contributing an appreciable portion to the total flow. At either extreme of the gradient profile, when one pump is delivering a small portion of the total flow, the reproducibility and accuracy of gradient generation can suffer. Deviations in this region are in part caused by the compressibility effects noted earlier. A more significant factor is the minimum step volume delivered by the reciprocating pump. These discrete steps are the smallest volumes the pump can deliver. The smaller this step, the smaller the volume from the pump can be before the discontinuous nature of the delivery becomes apparent.

The use of precision metering valves in low-pressure gradient generation does permit a more nearly smooth and uniform delivery in the low percentage range of any solvent. The method is also much more amenable to use with multiple solvent systems since it is easier and cheaper to add a precision metering valve and reservoir than it is to add a complete pump for every additional solvent. It is essential, however, that solvents be thoroughly mixed prior to entering the high-pressure pump. A disadvantage of low-pressure gradient formation is that the delay volume between the point of mixing and the column now includes the pump head and is almost invariably larger than in the high-pressure gradient generation system. The larger delay volume is a particular problem with microbore or high-speed columns.

Sample Injectors. The ideal sample injection system places the sample at the top of the column in a sharp, well-defined plug of minimal thickness.* While real

*There is some evidence that sharper peaks can be achieved if the sample is placed just in the center, or even spotted into the beginning of the column.

injection devices never achieve an ideal plug because of the finite-volume tubing connecting the injector and the column, such deviations from ideality can be made insignificant with careful attention to system design.

Historically, three major sample injection systems have been used: stop-flow, septum injectors, and valve injectors. Only the last will be discussed in detail.

Stop-Flow Injection. This early technique involved stopping the flow of solvent through the column, opening the top of the column to add sample, closing the column, and restarting the pump. While no special equipment is needed for this system, starting and stopping the pump presents problems with the reproducibility of retention times but not volumes, and of resolution, because nonlinear effects occur in peak width. Except in cases when other injection methods are not feasible, such as in capillary LC research, this technique is rarely used.

Septum Injection. This system was modeled after the septum injector used in GC. A needle port at the top of the column is sealed with a circular septum. The sample is then introduced at the top of the packed bed into the flowing mobile phase by a syringe which penetrates the septum. This device comes closest to providing an ideal injection. It suffers from a limited operating pressure range and is ineffective at the high pressures used in LC. Further, since septum injectors in LC are manual devices, the reproducibility of the sample volumes is very much a function of the skill and care exercised by the operator. As a result, septum injection is rarey used in laboratories today.

In modern analytical laboratories an *injection valve* is almost universally used to introduce a sample. The valve is constructed with a rotor and stator designed to operate at high pressures. Rotation of the valve results in a small length of tubing (the sample loop) being either inserted or removed from the flow path of the mobile phase just before it enters the head of the column, as shown in Fig. 26.10. The volume of sample injected is fixed by the volume of the sample loop (plus that of connecting channels), which ensures a virtually constant injection volume every

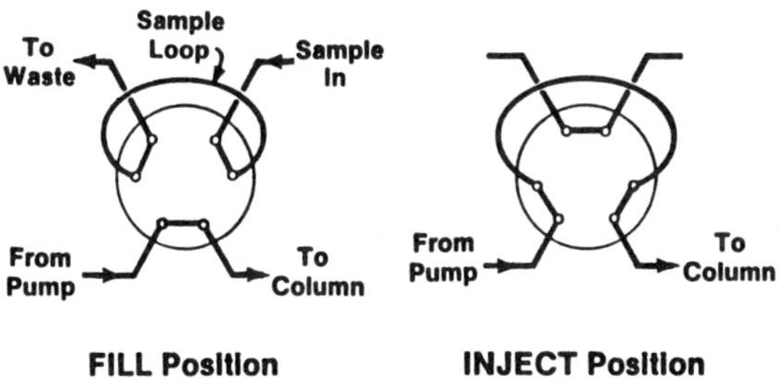

Fig. 26.10 A sample injection valve. The six-port valve shown allows three pairs of ports to be switched synchronously. In the FILL position the mobile phase flows directly to the column through one pair of ports, while the other ports allow new sample to flush out the sample loop. Rotation of the valve to the INJECT position directs the mobile-phase flow into the sample loop, leading to injection of sample onto the column.

time. Other factors that have contributed to its wide use are its simplicity and adaptability to automated forms.

The actual shape of the sample plug introduced into the column by the valve can deviate considerably from the ideal. Most of the deviation is caused by the development of a parabolic flow profile in the injection loop and connecting tubing. The viscous drag of the liquid on the walls of the tubing causes an initially sharp sample band to spread because the fluid in the center moves ahead of fluid in the more slowly moving layers near the wall.

How serious is this loss of ideality in the sample plug? If peak broadening caused by processes in the chromatographic column greatly exceeds that caused by the injector and other extra-column sources, or if the resolution achieved by the column is much greater than 1.5, nonideal injection profiles will not present serious problems. As implied by this comment, adjustment of aspects of the overall chromatographic system (to be examined in the next section) will ordinarily take care of the nonideality. For a fully optimized separation system, however, such as might be developed for quality control or industrial process monitoring, or for complex measurements such as trace analyses, the nonideality of the injection process can pose a serious limitation: the ultimate speed and sensitivity possible would be affected. In these cases care must be exercised to attach the injection valve as close as possible to the column, using small-bore tubing (typically 0.2–0.25 mm i.d.).

An interesting alternative means of reducing the dispersion contribution arising in connecting tubing has recently been described, the use of *serpentine tubing*.* The viscous drag of the liquid on the walls can be reduced by bending a tube continuously. In a sharply bent tube, inertial effects will cause the center fluid to move toward the wall, displacing liquid at the periphery and causing it to flow back toward the center, thereby establishing a circular secondary flow pattern. With a continuous series of bends of appropriate size, layers in the center and those near the wall are mixed continuously and intimately. The use of serpentine tubing permits longer lengths of connecting tubing when they are absolutely necessary.

Two problems encountered with injection valve systems are leakage and wear. The high-pressure seal is achieved by applying large mechanical forces to bring the stator and rotor into tight contact. Such forces lead to accelerated wear, but most modern designs have greatly reduced it. Further, misalignment of parts or the introduction of dirt particles during disassembly and reassembly result in leakage.

Detectors. A great range of detectors is available to monitor the elution of analytes from an LC column. The refractive index detector and UV absorption detector are the most widely used and will be described along with a few other useful types. Unfortunately, in describing detector performance the terms used by manufacturers and authors have not always been consistent. Here the terminology of

*E. Katz and R. P. W. Scott, *J. Chromatogr.*, **268,** 169 (1983).

Scott [5] will be mainly employed. The *response* of a detector will be its output signal as a function of analyte concentration, and the range over which the response is linear, the *linear dynamic range* of the detector. Its *limit of detection* will be the minimum response that can be taken as indicating the presence of analyte (see Section 10.8). It will be defined as twice, or preferably three times, the detector noise (expressed in the same units as the detector signal). The term sensitivity will be reserved for the slope of the response curve.

Refractive index (RI) detectors respond to the change in the refractive index of the (transparent) mobile phase solvent that occurs when a solute is eluting. The fact that most substances differ in refractive index will make this sensor a *universal detector*. A dilute solution of almost any analyte will have a refractive index slightly different from that of the solvent in use. Standard differential refractive index techniques (see Section 19.4) can be used to distinguish what are usually minute increments in refractive index. The universality is a major advantage of the RI detector. However, this general responsiveness makes the RI detector unattractive for gradient chromatography in which the mobile phase refractive index changes steadily as a gradient is developed.

There are two major types of RI detector, reflective and Fresnel. The reflective RI detector utilizes two prismoidal cells joined along their faces, as shown in Fig. 26.11a. In the Fresnel detector the cell is a small channel at the base of an optical prism, as shown in Fig. 26.11b. An advantage of the Fresnel type is its much smaller cell volume.

Since the refractive index of liquids is strongly temperature dependent, detector temperatures must be controlled to $\pm 0.001\,^\circ\mathrm{C}$. With themostating, commercial RI detectors can detect changes as small as 10^{-7} refractive index units, which corresponds to solute levels typically in the microgram range. Consequently, the RI detector is not particularly useful for trace-level components. This detector also responds to the fluctuations in eluent flow; unless a stable, pulse-free pump system is used, the sensitivity of detection will be degraded.

The *UV detector* has been the workhorse of LC detection. It is based on the UV absorption of solutes (see Section 16.3) as they elute from the column. Since many types of organic compounds absorb at wavelengths below 300 nm and several useful chromatographic solvents are transparent in this region, these detectors can be used for many types of samples. For mixtures of substances that absorb in the UV, this sensor is often the detector of choice.

The application of UV absorption to LC requires careful design. Because low limits of detection are sought, the small volumes involved in LC systems present a formidable challenge. The proportionality between absorbance and path length in Beer's law suggests that cells be long, thin cylinders in which the light beam passes down the axis. Only with careful design is high throughput ensured. Volumes of commercial UV detectors range from 12 μL to as little as 0.5 μL. Noise levels are typically of the order of 0.00002 absorbance unit.

Both fixed and variable-wavelength UV detectors are in use. They are distinguished by whether a discrete source, such as a low pressure mercury arc lamp that emits only a few wavelengths, or a nearly continuous source such as a high

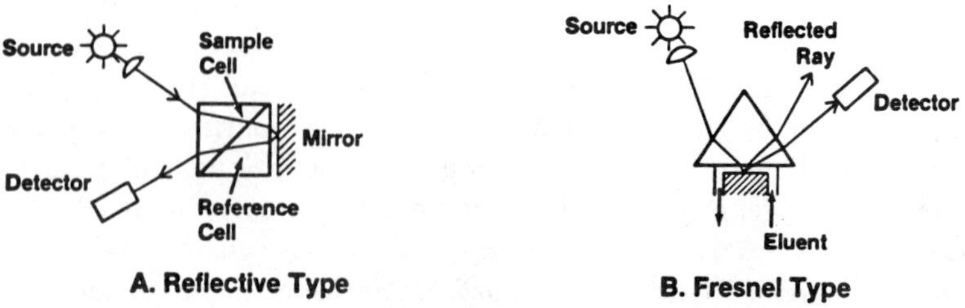

Fig. 26.11 Refractive index detectors. (a) Reflective type. A light beam directed into the first cell (typically the cell through which the column eluent is flowing) is refracted by an amount characteristic of the relative refractive index of the eluting solution. The beam then passes into the second cell (usually the cell containing only mobile phase) in which it is again refracted. The beam is reflected from the rear mirror to return through both cells where it is again twice refracted. The detector is tuned by rotating the cell until the appropriate baseline signal is obtained. The appearance of a solute in the first cell leads to a change of refractive index that changes the position at which the beam leaves the cell assembly. (b) Fresnel type. A beam is refracted twice at the prism–eluent interface, once as it enters the eluent channel, and a second time, on leaving. The bottom of the eluent channel is made highly scattering by covering it with roughened titanium oxide or another scatterer that causes the light beam to be diffusely reflected back into the eluent. It is the diffusely reflected light and not the specularly reflected beam that is monitored.

pressure xenon lamp is used. *Fixed-wavelength UV detectors* have the advantage that emitted UV energy is concentrated into a few lines. Furthermore, the wavelengths are fixed and reproducible, which makes possible simpler, more stable detectors. *Variable-wavelength detectors* permit, in combination with a monochromator or an interference filter, detection at any selected wavelength and can be tuned to the wavelength of maximum absorbance for trace substances, greatly reducing detection limits for those compounds.

With a continuous source and a *diode array UV polychromator* rapid repetitive scanning of eluate absorption over the UV–VIS range of interest is possible. In this way several scans can be made during the time of appearance of all but the narrowest peaks. Not only are such spectra helpful in identifying eluting species but with multiple spectra per peak, overlapping chromatographic peaks may be detected. Any change in spectra from the beginning to the end of an LC peak will imply unresolved analytes.

The *fluorescence detector*, like the UV detector, is based on an optical spectrometric measurement (see Section 15.2). Fluorescence detection is highly attractive because limits of detection are lower than for absorption. Recall that fluorescence intensity is directly proportional to source intensity. Indeed, detection limits as low as parts per trillion can be achieved with a broad-band excitation source. With laser excitation, the limit of detection can be an order of magnitude lower.

Another distinct advantage of fluorescence detection for LC is that the detector can often be a filter fluorometer as long as analytes are chromatographically resolved; the extra throughput secured will further lower the limit of detection. Further, fluorescence optics can be more readily adapted to small volume LC systems. The major drawback is the limited number of compounds that fluoresce. Fortunately, many substances of biological interest as well as many pollutants fluoresce and can be sensitively detected. Even this selectivity can be turned to advantage, however, since a fluorescent solute can be identified among nonfluorescent co-eluting solutes.

Electrochemical detectors are finding increasing use as LC detectors because of their low limits of detection and selectivity. Those detectors designed for amperometric measurements (see Section 29.11) at constant potential are the most widely used. Ordinarily their cells have the three-electrode arrangement shown in Fig. 26.12 and produce a current signal, which is converted to a voltage output by an operational amplifier.

The choice of the working electrode is important: it affects the range of cathodic or anodic behavior, the efficiency of the electrolysis reaction, and the detector noise level. The most commonly used electrode materials are carbon paste, glassy carbon, and mercury. Their response is also strongly influenced by the state of the electrode surface. If samples that tend to foul the electrode are chromatographed, its surface should be cleaned frequently and the detector recalibrated, a distinct disadvantage.

Selectivity arises because analytes must be electroactive to be detected, that is, be oxidizable or reducible at the working electrode. The degree of analyte response can be varied somewhat by changing the working electrode potential. Selectivity can also be extended by the use of multiple electrodes.

An electrochemical LC detector is also quite sensitive to mobile phase flow rate. If there is no flow, the diffusion of electroactive solute from the bulk solution to the electrode surface will limit the cell current. In this case the detector response (current) will be large when the electrode potential is first applied, but will fall exponentially to zero as the solute concentration is depleted in the thin diffusional layer near the electrode. If the bulk solution is flowing, however, the diffusional layer will be thinner since the neighboring layer will be continually replenished.

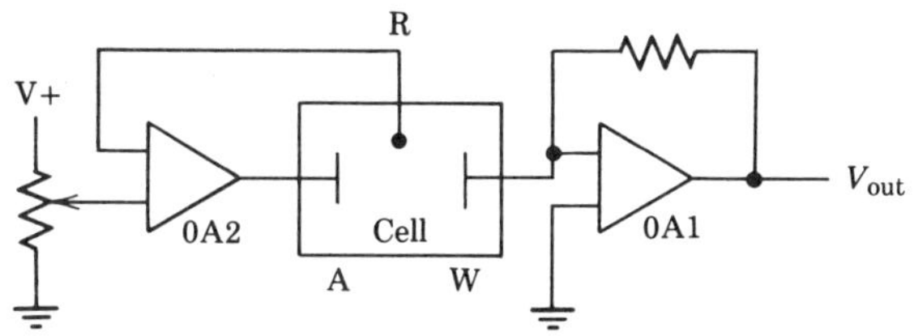

Fig. 26.12 Amperometric electrochemical detector using a three-electrode cell: R, reference electrode, W, working electrode, and A, auxiliary electrode. When this detector is used, an inert electrolyte is added to the mobile phase to carry current. The potential difference between the working electrode W and the reference electrode R is set to provide oxidation or reduction of selected components in the mixture being chromatographed. To accomplish this, W is held essentially at ground by operational amplifier OA1 and the voltage of R is held at the potential obtained from the voltage divider. Operational amplifier OA2 actually provides the current through the auxiliary electrode to maintain this desired voltage, ensuring that the cell IR drop will not be between R and W, where it would affect the operating potential. Current exists in the cell only on elution of a solute that can be electrolyzed at the potential of the working electrode (relative to that of the reference electrode). This current is changed to a voltage output signal by the circuitry at OA1.

With a flowing solution, the peak current is still proportional to the solute concentration in the bulk solution, but the proportionality constant itself depends on the rate of bulk solution flow. Constancy of flow rate will be important for reliable measurements.

Theoretical treatments and confirming experimental studies have demonstrated that improved S/N values are obtained by reducing the dimensions of the electrode, which makes this detector ideal for small-volume LC systems. Several electrochemical detectors with cells with volumes as small as 1 nL have been described in the literature.

Detection limits in the parts per trillion range have been reported with LC electrochemical detectors. In general selectivity works to advantage in that it does enable the detection of electrochemically active compounds in the presence of coeluting compounds that are not electrochemically active.

Processing Modules. There is an intimate connection between the processing modules of a chromatograph and the quality and amount of information that can be extracted from a chromatogram with facility. Today digital data handling techniques are generally used since they are cost effective. The simplest form of digital data handling is that of a simple integrator in which the areas under the chromatographic peaks are progressively accumulated. Slightly more sophisticated integrators report individual peak areas and retention times directly, but again only these two parameters are stored and reported.

At the other extreme, the entire chromatogram is stored point by point in a computer and then retrieved later for data processing. This latter approach permits the repeated processing of the raw chromatogram with different peak threshold parameters and baseline parameters.

26.4 THE CHROMATOGRAPHIC SYSTEM

Now that individual LC modules have been described and their functions discussed, attention must be given to how best to assemble them into a liquid chromatograph. The determination of the proper column length and particle size was discussed earlier on the basis of parameters such as stationary phase particle diameter and solvent. Yet to be treated are diameters and volumes that relate to the hydrodynamic or pneumatic system, the diameter and volume of the column, sample injection valve, detector, and connecting tubing. Since this review is a system analysis, response times for the detector and other modules will be included.

A useful criterion for assessing the impact of all these factors is its contribution to band broadening. Clearly the resolution achieved by the chromatographic column should be degraded as little as possible by the other modules in the signal channel.

In chromatography band dispersion is a good measure of the quality of analyte separation. Band dispersion in an LC system can also be defined mathematically as the variance of the chromatographic peak since peaks are ordinarily Gaussian in shape. We have seen that both column and extracolumn contributions are made to that dispersion, that is, to peak broadening. The contributions of different modules to the variance of the chromatographic system should be measurable using the principle of summation of variances expressed in Eq. (24.18). For a liquid chromatograph the contributions to the overall variance of the chromatograph are given by the expression

$$\sigma_{\text{tot}}^2 = \sigma_{\text{c}}^2 + \sigma_{\text{i}}^2 + \sigma_{\text{t}}^2 + \sigma_{\text{d}}^2 + \sigma_{\tau}^2, \tag{26.5}$$

where the column contribution to variance is σ_{c}^2, and the other contributions are σ_{i}^2 from the sample injection system, σ_{t}^2 from the connecting tubing, σ_{d}^2 from the detector, σ_{τ}^2 from the time constant of the detector and processing circuitry. Recall from Section 24.8 that the system variances can be expressed either in units of length^2, time^2, or volume^2. Here variances will be expressed in volume units of μL^2 since the elution volume is invariant with mobile phase flow.

Because of the central role of the column in effecting separations, one strategy is to design all other modules and the coupling to give minimum variances. Actually, the result may be a column with inadequate performance. It proves much better to let the choice of column diameter be governed by extracolumn dispersion (broadening) contributions; the column contribution to dispersion should dominate the dispersion caused by extracolumn sources. Since only a few limited column sizes are currently available from manufacturers, it has been recommended that design should be altered in any appropriate way so that at maximum a 5% decrease in resolution is contributed by extracolumn sources. It can be shown that a 5% loss in resolution corresponds to a 10% increase in peak variance.

Column Variance. The variance contribution of the column must thus be calculated or estimated. With data in hand the design of the rest of the instrument quickly follows. The column variance is directly related to the plate height H, as given in Eq. (24.18): $\sigma_{\text{c}}^2 = HL$. Expressed in volume units of μL^2, it is

$$\sigma_{\text{c}}^2 = HV_{\text{R}}^2/L = V_{\text{R}}^2/N, \tag{26.6}$$

where V_{R} is the retention volume.

In Table 26.2 the impact of different column parameters on column variance is illustrated. Calculations were made for an unretained peak ($k' = 0$) and a peak at $k' = 2.0$ for typical LC columns (operated at their optimum linear velocities). From the table it is clear that a reduction in bead diameter d_p and column length L greatly reduces the column peak variance, which in turn reduces the total extracolumn variance that can be tolerated without significant loss in peak resolution. Some notes about contributions from various extracolumn sources follow.

Variance from the Sample Injector. The dispersion contribution from the injector system can be estimated by assuming that the injection is truly a square-plug type. In this case the variance is given by $\sigma_{\text{i}}^2 = V_{\text{i}}^2/12$, where V_{i} is the volume of sample injected. For an

Table 26.2 Contributions of the Column to Variance[a]

Column Parameters			*Peak Variance (μL^2)*	
d_p (μm)	L (cm)	i.d. (cm)	$k' = 0$[b]	$k' = 2.0$
10	25	0.46	677	6090
5	10	0.46	135	1217
3	3	0.46	24	218

[a]Well-packed columns with $H = 2d_p$ and $\epsilon = 0.7$ (porosity) that are operated at optimum linear velocity. The symbols used are d_p, diameter of packing, and L, column length.
[b]From the table the maximum permissible extracolumn variance would range from 67 μL^2 to as little as 2.4 μL^2.

injection of 20 μL, the peak variance is calculated to be 33 μL^2, for a 10-μL injection the variance is 8.3 μL^2, for a 5-μL injection there is a 2.1-μL^2 variance, and finally, for 1 μL injection a 0.08 μL^2 variance. It is seen that the 20-μL injection volume is suitable for a column with 10-μm beads. The 10-μL injection volume can be used with 5-μm and larger beads, while the 5-μL injection volume can be used with all sizes of beads in columns. In practice, it is found that the calculated variances are lower than the variances found experimentally, reflecting the nonideality of the injection system. In actuality, the injection volume is not placed at the head of the column as a square plug, but rather as a smoothed plug, which results in the larger variance found experimentally for the injection step.

Variance from the Connector Tubing. Taylor's law for dispersion attributable to flow in a tube provides an order-of-magnitude estimate of the peak dispersion from connecting tubes. Comparison of the calculated values with those in Table 26.2 shows that 3-cm lengths of 0.38 mm i.d. connecting tubing, which is about the minimum length that is practical, at either end of the column will contribute sufficient extra peak variance to broaden even the unretained peak from the 25-cm column packed with 10 μm particles. Hence, the connecting tubing that is invariably used must be as short and as narrow as possible to avoid excessively broadening the chromatographic peaks.

Variance from the Detector. A very rough estimate of the dispersion contribution from the detector can be obtained by treating the detector cell as a short length of tubing. Several problems arise: the detector cells are too short to be accurately described by Taylor's law; the detector inlet and outlet geometry often causes considerable mixing of the sample stream, which can reduce the dispersion of the system; and the design and geometry of detectors will significantly affect the actual value. Assuming variances for cell volumes used in UV detectors as only 10% of the value predicted by Taylor's law gives a figure of 142 μL^2 at a flow rate of 1 mL min^{-1} for a detector of 8 μL volume and a variance of only 7 μL^2 at the same flow rate for a 1.4-μL detector. Comparison of these figures with those in Table 26.2 suggests that the 8-μL detector could be used for solutes eluting above $k' = 2$ only with the 10-μm column if good performance is sought. Even the smallest detector, that with 1.4 μL volume, would be suitable only for solutes eluting above $k' = 1$ with the short 3-μm column, although it would be suitable for the larger particle size columns in Table 26.2.

Variance from Processing Electronics. Peak distortion can also occur if the processing electronics are not sufficiently fast to follow accurately the signal developed as a rapidly

Table 26.3 Maximum Processing Electronics Time Constants when Standard LC Columns Are Used

Column Type[a]		*Time Constant* τ (s)	
Bead Diameter (μm)	Length (cm)	$k' = 0$	$k' = 2$
10	25	0.70	2.1
5	10	0.31	0.94
3	3	0.13	0.40

[a]All columns are 0.46 cm i.d.

eluting component passes through the detector. This effect is also evaluated in terms of the change in peak variance, but the analysis for this case is very complex. Modeling of the electronic response as a passive RC circuit results in the characterization of the system in terms of an exponential relaxation time. The requirement of no greater than 5% loss of resolution of two adjacent peaks then imposes the condition $\sigma_\tau^2 < 0.1\sigma_c^2$, where σ_c, the column variance, is now expressed in units of seconds^2 and equals $(1 + k')^2 v_0^2 h/F^2 L$.

Table 26.3 lists the time constants calculated for an unretained peak ($k' = 0$) and one for which $k' = 2$ for the columns used in Table 26.2. Not unexpectedly, the shortest column demands the shortest processing time constant. It is also seen from Table 26.3 that values of $\tau > 2$ s would not be widely needed, and a time constant 0.1 s or less is necessary when using the newer small columns packed with small-diameter particles, since flow rates in excess of 1 mL min^{-1} are typically used with these columns.

Selection of Chromatographic Modules. The total extracolumn dispersion could be divided among the several modular sources in any number of ways, but some arrangements suit the needs of an analyst better than others. The variance contribution of the detector is an important specification when selecting a new detector, and the smallest possible value should be sought while retaining adequate sensitivity. The sensitivity requirement of the analysis is a key question. If low limits of detection are not needed, the detector should be the major source of extracolumn dispersion and the injection volume reduced to a value such that its variance contribution is comparatively negligible. The appropriate value of the column radius can then be calculated; it will be governed primarily by the dispersion of the detector. If low limits of detection are needed, the injection volume must be increased appropriately. This change increases the magnitude of the extracolumn dispersion, and therefore results in a larger diameter column being required. In all cases, the variance contribution due to connecting tubing must be minimized, either through the use of short lengths of serpentine tubing, or by coupling the column directly to the injector and the detector.

26.5 ANALYTICAL MEASUREMENTS

The process of identifying and determining analytes by LC begins with the choice of a type of chromatography and continues with the selection of an appropriate mobile phase-stationary phase combination. These choices are largely governed by sample characteristics and requirements on the measurements, as was discussed in the introduction. Choices for some types of sample are straightforward, for example, ion chromatography is the first thing to try for ionic solutes. It is tacitly assumed that the sample is already dissolved or in liquid form. In most cases, sample collection and sample preparation steps precede the actual chromatographic separation step. The solvent used for the sample must be compatible with that chosen as mobile phase; this requirement may affect the type of chromatography selected or the sample preparation procedure.

The analyst must next determine the mobile phase composition required to achieve the chromatographic separation. As yet, this is still very much an empirical process, drawing heavily on the user's prior experience, although progress is being made in the elucidation of the physical and chemical factors controlling retention in the various types of chromatography. The resolution R_s of the peaks of a pair of analytes, expressed in terms of the capacity factor k'_B of the later eluting member of the pair, their selectivity ratio α (which is the ratio of the capacity factors, k'_B/k'_A), and the column efficiency N, is just

$$R_s = [(\alpha - 1)\, k'_B]\, \sqrt{N}/4\alpha\, (1 + k'_B).$$

Changes in mobile phase composition have a very pronounced effect on α, a moderate effect on k', and only a minor effect on N.

Resolution relates to the separation of a single pair of compounds, and hence a resolution value can be calculated for every pair of neighboring peaks in the chromatogram. The effects of changing the mobile phase composition are easily followed in the case of samples containing only a few compounds (<5), but as the number of compounds increases, the possibility of significant peak overlap increases rapidly, and the tracking and unambiguous identification of peaks becomes more difficult. Indeed, Davis and Giddings* have shown that considerable overlap can be expected in samples with large numbers of compounds. Changes in mobile-phase composition designed to improve resolution for one particular pair of peaks frequently results in a decrease in resolution for another peak pair in such systems. The time required to determine the optimum mobile-phase composition often increases geometrically with the number of compounds present in the sample. One useful empirical means of identifying the best compromise is the "window diagram" approach of Laub and Purnell.†

After resolution has been optimized by mobile-phase adjustment, examination of the chromatogram will generally reveal the pair of peaks having the *least* resolution. This pair of compounds is called the critical pair since the tuning of instrumental factors to achieve full resolution of this pair will result in a fully resolved chromatogram. The efficiency or number of theoretical plates required to achieve full resolution of the critical pair can be found by rearranging the equation above and solving for N:

$$N = 16R_2^2\alpha^2(1 + k'_B)^2/(\alpha - 1)^2k'^{\,2}_B.$$

*J. M. Davis and J. C. Giddings, *Anal. Chem.*, **57,** 2168, 2178 (1985).
†R J. Laub and J. H. Purnell, *J. Chromatogr.*, **112,** 71 (1975).

This equation provides the value of N, the number of theoretical plates or column efficiency needed to resolve the critical pair fully.

Complex samples frequently contain compounds that display a wide retention range on the selected chromatographic system. A weak mobile phase is required to achieve separation of the weakly retained compounds, but this then results in large retention volumes for the more strongly retained compounds. Gradient elution is often used in such cases, in which the concentration of the stronger mobile phase component is progressively increased in order to hasten the elution of the more strongly retained compounds once the weakly retained compounds have been separated. A simple linear gradient is not necessarily the best solution since elution volume versus solvent strength characteristics will differ for the various compounds. Many commercial instruments offer the capability of generating nonlinear or multisegment linear gradients for this purpose. The determination of the proper gradient profile is largely empirical at the present time.

Second, a detector must be chosen to monitor elution that responds to all the compounds of interest. Almost all LC detectors are selective and cannot detect all compounds. The detector output signal should also vary linearly with solute concentration. If the output is nonlinear, it must be transformed by analog or digital electronics to provide a signal that is linear with solute concentration. When the correlation is linear, quantitative data can be extracted from chromatographic peaks by means of peak height or peak area.

Identification of each compound must precede its quantitative measurement. For *qualitative analysis* the solute retention behavior of an unknown is matched either directly with the retention times or capacity factors for known compounds. Indeed, the accurate identification of solute peaks is a critical issue in trace analyses (e.g., for environmental or metabolite studies). Two factors are crucial in this process, establishing stable instrumental conditions and independently corroborating the identification.

Employing a series of detectors with one LC also provides the additional data needed to establish an accurate identification of solutes whether the detectors are based on different principles (e.g., UV and electrochemical detection) or the use of two or more facets of one detection principle (e.g., detection at two different wavelengths in the UV). Although this approach will increase the data processing demands on the processing modules of the chromatograph, it will usually lessen the number of different instruments that will be needed for good results. If selective detectors are available, they can be used to simplify an analysis by reducing the number of interfering peaks.

Attempts to establish a retention index procedure for LC similar to that used in GC (see Section 25.5) have been reported, but have not met with great success as yet. One major difficulty is that there is another dimension to the problem in LC in that solute–solvent interactions can play a major role in governing retention, a factor not present in GC.

Quantitative measurements require the determination of the response of the analytical system, that is, the slope of the plot of output signal (peak height or peak area) against the amount of solute injected. This procedure is typically carried out using standards made up in the mobile phase. A similar procedure, but with the standards made up in the sample (the method of standard additions) provides additional confirmation of peak retention time matching. The minimum detectable concentration or limit of detection is the concentration corresponding to a signal twice the baseline noise or better, three times the standard deviation of the noise, as has recently been recommended (see Section 10.8).

The use of the injection valve in LC systems simplifies the generation of the response plot (relative to GC) in that the injection volume is fixed and reproducible, provided that the loop is *generously overfilled* each time. An *internal standard* is not as widely used in LC as it is in GC, largely because the sample volume introduced with the injection valve is sufficiently reproducible not to require its use. When a partially filled loop is being used, or when a syringe *injector* is being used, an internal standard is recommended.

In LC, a minimum of three samples should be analyzed, preferably more. The relative precision of retention times should be 0.5% or better, while peak area relative precision will typically be of the order of 1–5%. This precision will vary according to the S/N, the degree of peak resolution, detector stability, and the skill of the operator. Retention time reproducibility will also be affected by the constancy of the state of the stationary phase. Very strongly retained compounds will accumulate on the packing in repeated injections, ultimately altering the nature of the stationary phase. Incomplete purging or equilibration of the column in gradient analyses will also lead to variable retention times.

Monitoring the Efficiency of a Sample Preparation Step. Liquid chromatography with *interal standards* is widely used to monitor the efficiency of an extraction step, or other sample preparation procedures, that is, for nonchromatographic purposes. A series of injections of progressively more dilute standard solutions then provide a series of decreasing peaks, and a calibration curve such as shown in Fig. 26.13 is generated. This curve should be linear across the solute concentration range to be monitored.

Example 26.1. How is it possible to determine the important vitamins by liquid chromatography? Can a single column be used or will several be required?

There is considerable interest in the determination of the major vitamins by LC as is evident from the questions. The vitamins are classified as either water-soluble: B-1, B-2, B-6, B-12, folic acid, pantothenic acid, C, niacin, and niacinamide; or fat-soluble: A, D, and E. Reversed- and normal-phase chromatography, respectively, are natural first-candidate systems for each of these vitamin classes. Trial separations of the water-soluble vitamins were begun with methanol–water mobile phases on a C_8 silica. Retention was low, even with low methanol content, so a variety of techniques, such as pH adjustment, use of surfactants, and use of ion-pairing agents, were explored as a means of increasing retention and resolution. Other bonded-phase silicas were also evaluated, including C_2- and amino-bonded phases. Vitamin B_6 was often the most troublesome, yielding tailing peaks in neutral or alkaline mobile phases. Peak shape was much improved with acidic mobile phases, but retention was low and resolution was poor. The addition of hexanesulfonic acid, together with gradient elution, on a C_8 column resulted in acceptable peak shape, as shown in Fig. 26.14a. Unfortunately, now the peak shape for niacinamide has deteriorated; however, it is typically present at high concentrations in tablets, which would help improve the quan-

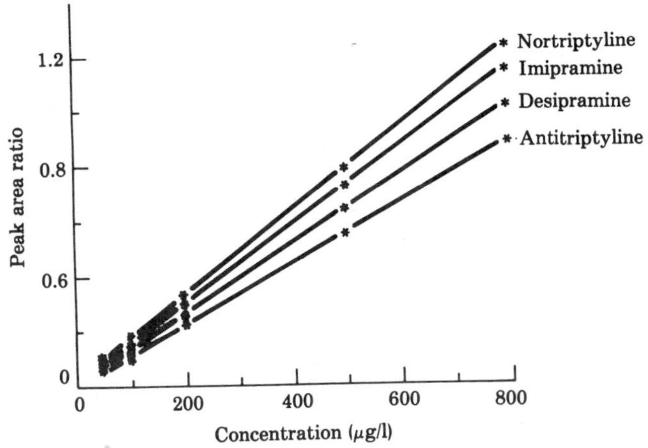

Fig. 26.13 Calibration curve for four common tricyclic antidepressants: nortriptyline, imipramine, desipramine, and antitriptyline. The mobile phase is 0.7% aqueous ammonia in acetonitrile with UV detection at 210 nm.

titative accuracy. The chromatogram from the assay of a tablet is shown in Fig. 26.14b. It is seen that good separation is achieved, with a small shoulder on the niacinamide peak. This small, side peak is present in all tablet assays, and it is presumed to be due to a matrix component, and reproducible niacinamide values are obtained in spite of it.

Two water-soluble vitamins, pantothenic acid and B_{12}, are missing from Fig. 26.14. Far-UV detection (<200 nm) is required for the sensitive determination of these two vitamins, and since the cutoff wavelength for methanol is 205 nm, they were not included in these assay procedures. The best compromise wavelength for the remaining set of vitamins is 272 nm.

An analytical procedure was desired to analyze the fat-soluble vitamins using the same hardware system employed for the water-soluble vitamins, such that these vitamins could be assayed by only making minor changes in mobile-phase composition. It was found that acceptable separation of these vitamins could be achieved with 93.5% methanol–water. (Injection of tablet extracts revealed other peaks which could interfere with the vitamin peaks; this mobile-phase composition provided the best resolution.) Spectral scans showed that the best compromise wavelength was 265 nm for the fat-soluble vitamins.

Sample preparation for the water-soluble vitamins consisted of crushing the tablet and adding 10 mL of dimethylsulfoxide. This mixture is sonicated briefly, then 90 mL of water is added, and the mixture sonicated again. The mixture is centrifuged and filtered, and 10 μL is injected. Sample preparation for fat-soluble vitamin assays began by crushing the tablet, mixing with 5 mL of 80% ethanol in water (containing 1% citric acid to minimize degradation), and then extracting this mixture with 1 mL hexane. The mixture was centrifuged, and then 20 μL injected.

The assay procedure consists of equilibration of the column at 10% methanol–water, injection of the water-soluble sample extract, and initiation of the gradient. After vitamin B_1 has eluted the mobile phase is changed to 93.5% methanol–water and the column is equilibrated. The fat-soluble sample extract is then injected. A sample chromatogram is shown in Fig. 26.15. Data for this example have been taken from F. L. Vandemark and G. J. Schmidt, *J. Liq. Chromatogr.* **4,** 1157 (1981).

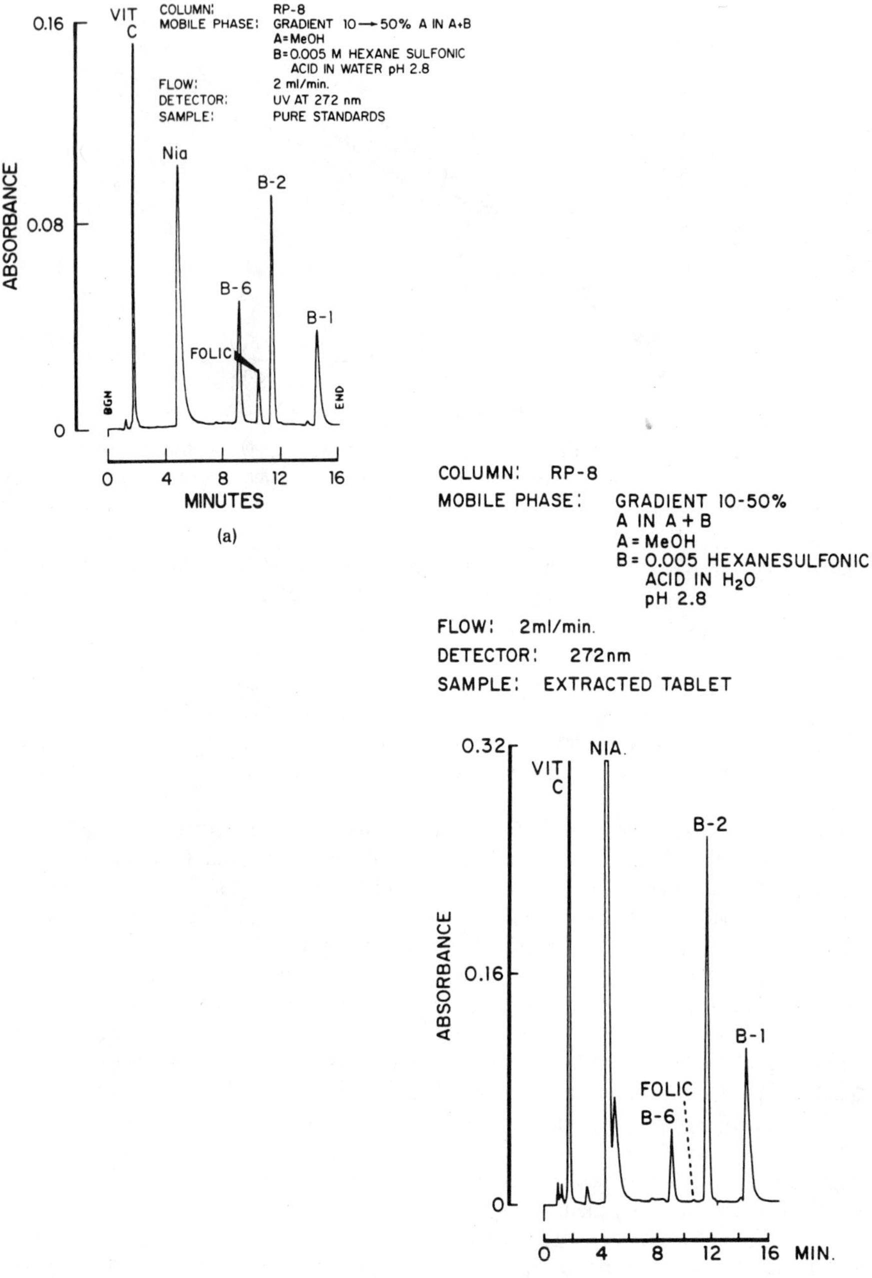

Fig. 26.14 Chromatograms of water-soluble vitamins on RP-8 column (C_8 bonded-phase silica). (a) Separation of vitamin standards. (b) Separation of tablet extract. Mobile phase in methanol/water, with 0.005 *M* hexanesulfonic acid in the water (pH = 2.8), with a linear gradient run from 10 to 50% (at 2 mL/min). Ultraviolet detection is at 272 nm.

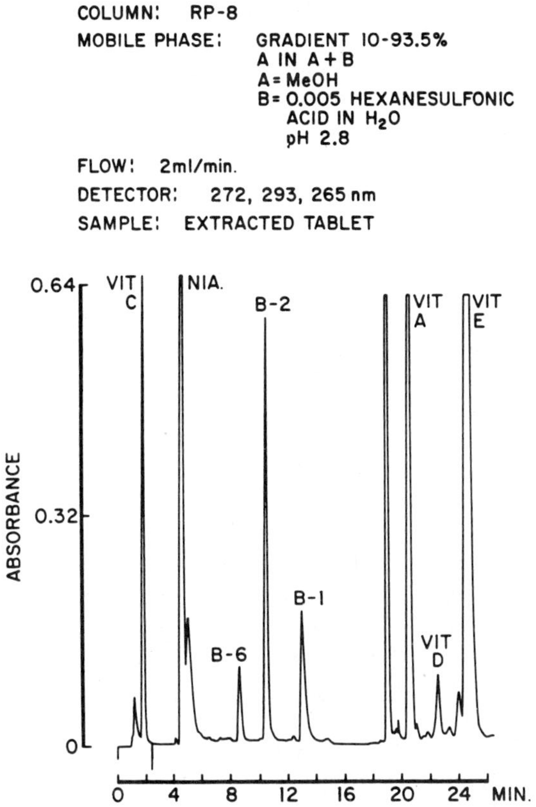

Fig. 26.15 Chromatogram from the total-vitamin assay of a tablet on an RP-8 column. The mobile phase is the same as in Fig. 26.14. A linear gradient is employed from 10 to 50% and a rapid gradient to 93.5% methanol. Composition is then held constant. Detection is at 272 nm for the water-soluble vitamins and 265 nm for the later-eluting fat-soluble vitamins.

REFERENCES

The following recently published books are well written and cover the general aspects of liquid chromatography well.

1. P. Brown and R. A. Hartwick, *High Performance Liquid Chromatography.* New York: Wiley, 1988.
2. S. Lindsey, *High Performance Liquid Chromatography.* New York: Wiley, 1987.
3. C. F. Simpson (Ed.). *Techniques in Liquid Chromatography.* New York: Wiley, 1982.
4. V. Meyer, *Practical High Performance Liquid Chromatography.* New York: Wiley, 1988.

The following books dealing with special aspects of liquid chromatography are recommended.

5. R. P. W. Scott, *Liquid Chromatography Detectors*, 2nd ed. New York: Elsevier, 1986.
6. T. M. Vickrey (Ed.). *Liquid Chromatography Detectors.* New York: Dekker, 1983.

7. R. P. W. Scott (Ed.). *Small Bore Liquid Chromatography Columns.* New York: Wiley, 1984.
8. P. R. Brown and A. M. Krstulovic, *Reverse Phase High Performance Liquid Chromatography.* New York: Wiley, 1982.
9. J. C. Berridge, *Techniques for the Automated Optimization of HPLC Separations.* New York: Wiley, 1985.
10. J. T. Tarter (Ed.). *Ion Chromatography.* New York: Dekker, 1987.
11. M. T. Hearn (Ed.). *Ion-Pair Chromatography.* New York: Dekker, 1985.
12. J. S. Krull (Ed.). *Reaction Detection in Liquid Chromatography.* New York: Dekker, 1986.

EXERCISES

26.1 What is the minimum number of theoretical plates needed to produce baseline resolution ($R = 1$) of two peaks whose retention times differ by 0.2 min? The first peak elutes at 8 min.

26.2 An LC system provides for the separation of two components which elute at 12.6 and 13.5 min with a resolution of 0.8. The elution time for an unretained component in this system is 1.4 min. (a) Calculate the values of k', α, and N. (b) What *independent* changes in separation parameters N and α are required to double the resolution?

26.3 Predict the elution order of the following solute pairs in: (a) a reversed-phase LC system employing an octadecyl bonded stationary phase and a methanol–water mobile phase and (b) a normal-phase LC system consisting of a silica gel stationary phase and isopropanol–heptane mobile phase. Solute pairs: (1) bromobenzene and iodobenzene, (2) phenol and benzyl alcohol, (3) nitrotoluene and 2,4-dinitrotoluene, (4) *ortho-* and *para*-ethylphenol, (5) *syn-* and *anti* thiophene oxime.

26.4 What is the magnitude of $\Delta(\Delta G°)$ for a chromatographic separation that results in the resolution of two sample components with a selectivity ratio α of 1.05? This separation is accomplished using an LC system operated at 35°C.

26.5 An LC solvent gradient can be simply generated by using the logarithmic dilution afforded by mixing one solvent at a constant rate into a cell containing another solvent. The differential equation describing this dilution cell is $X_A = e^{-ft/V}$ where X is the fractional concentration of solvent A, f is the flow rate, t is time, and V is the volume of the dilution reservoir. How long will it take to attain a concentration of 5% solvent A (the original solvent in the reservoir) using a reservoir of 30 mL at a flow rate of 1.5 mL min^{-1}?

26.6 What is the pressure drop (in psig) produced by a 50-cm, 1-mm i.d. LC microbore column packed with 10-μm diameter silica gel packing and used with a methanol mobile phase at a flow rate of 2.0 mL min^{-1}? What is the minimum particle diameter packing that may be used when employing an LC pump with a 6000-psig pressure limit? The equation for the pressure drop generated by the resistance of the column to liquid flow is $\Delta P = \eta L v / \theta d_p^2$, where η is the viscosity, L is the column length, v is the linear flow rate, θ is a dimensional structural constant (approximately 600 for packed LC beds), and d_p is the particle diameter.

26.7 Assuming approximate additivity in $\epsilon°$ values in the eluotropic series, find the composition of a cyclohexane–chloroform mixture that will be equivalent to a 20% isopropanol–80% isooctane mobile phase.

ANSWERS

26.1: Assuming equal peak widths for the two peaks, the resolution equation, $R = 2\Delta t/(w_2 + w_1)$, reduces to $\approx R = \Delta t/w_2$. Therefore the peak width required for a resolution of 1 is, $1 = 0.2/w_2$ and $w_2 = 0.2$ min. Applying this value to the form of the column efficiency equation that uses the base peak width $[N = 16(t_r/w)^2]$ we have $N = 16(8/0.2)^2 =$ 25,600 plates. *26.2a:* The equation that relates resolution to efficiency, selectivity, and the capacity factor is required, $R = (1/4)\sqrt{N}[(\alpha - 1)/\alpha][k_2'/(k_2' + 1)]$. First, k', α, and N for the given system must be calculated. For component 1, $k_1' = (12.6 - 1.4)/1.4 =$ 8.0 and for component 2, $k_2' = (13.5 - 1.4)/1.4 = 8.64$. Second, $\alpha = k_2'/k_1' = 1.08$. Third, with R, α, and k' known we may calculate N from the above equation for resolution. Solving for N we obtain $N = 16R^2[\alpha/(\alpha - 1)]^2[(k_2' + 1)/k_2']^2 = [4 \times 0.8(1.08/0.08)(9.64/8.64)]^2 = 2323$; *26.2b:* Doubling of R by changing N alone will require a fourfold increase: $N_2 = 4N_1 = 9292$. To double resolution by changing α alone, the ratio $[(\alpha - 1)/\alpha] = 0.0741$ must be doubled to 0.148. Thus the new value of α is 1.17; and it must be increased by 0.09. *26.3:* The compound eluting first in the trials is: (a1) bromobenzene, (b1) iodobenzene, (a2) phenol, (b2) benzyl alcohol, (a3) 2,4-dinitrotoluene, (b3) nitrotoluene, (a4) *p*-ethylphenol, (b4) *o*-ethylphenol, (a5) *anti*-thiophene oxime, (b5) *syn*-thiophene oxime. *26.4:* Since $\Delta G° = -\mathrm{RT} \ln \mathrm{K}$ and $\mathrm{K} = k'(V_s/V_m)$, we can construct an equation for $\Delta(\Delta G°)$ as follows: $\Delta G_1° = -\mathrm{RT} \ln k_1'(V_s/V_m)$ for component 1, $\Delta G_2° = -RT \ln k_2'(V_s/V_m)$ for component 2. Since $\alpha = k_2'/k_1'$, we have $\Delta G_2° - \Delta G_1° = -RT \ln k_2'/k_1' = \Delta(\Delta G°) = -RT \ln \alpha$ and $\Delta(\Delta G°) = -(8.31\ \mathrm{J\ mol^{-1}\ K^{-1}})$ $308\ \mathrm{K} \ln 1.05 = 124.5\ \mathrm{J\ mol^{-1}}$. *26.5:* Solving for t we obtain $t = -v/\mathrm{f}(\ln X_A)$. Substitution in this equation gives $t = [-30\ \mathrm{mL}/1.5\ \mathrm{mL\ min^{-1}}] \ln(0.05) = 60$ min.

ELECTROANAYTICAL METHODS

Chapter 27

INTRODUCTION TO ELECTROANALYTICAL CHEMISTRY

William R. Heineman

27.1 INTRODUCTION

Electrochemistry involves the measurement of electrical signals associated with chemical systems that are incorporated into an electrochemical cell. The cell consists of two or more electrodes that function as transducers between the chemical system and an electrical system in which the electrical parameters of voltage and current can be measured or controlled. Electrochemistry plays key roles in many areas of chemistry: analysis, thermodynamics, synthesis, kinetics, energy conversion, biological electron transport, and nerve impulse conduction to name a few.

Electroanalytical chemistry makes use of electrochemistry for the purpose of analysis. In this application the magnitude of a voltage or current signal originating from an electrochemical cell is related to the activity or concentration of a particular species in the cell. A host of electroanalytical techniques has been developed for this purpose. These techniques have certain features that in some situations make electroanalysis advantageous in comparison with other analytical methods. Excellent detection limits coupled with a wide dynamic range are exhibited by many techniques. An operating range of 10^{-8}–10^{-3} M is typical of some techniques. Methods are available for the rapid determination of the relatively high concentrations found for species such as blood electrolytes (Na^+, Cl^-, HCO_3^-), as well as trace levels of species such as heavy metals in crop samples and drug metabolites in blood and urine samples. Measurements can generally be made on very small volumes of sample, for example, in the microliter range. The combination of low detection limits with microliter volume samples enables amounts of analyte at the picomole level to be measured routinely in some instances. Measurements on nanoliter volume have been demonstrated in research laboratories! Electroanalysis is an inexpensive technique in comparison with many instrumental methods. Electrochemistry benefits from the feature that the signals originating from an electrochemical cell are electrical in nature. By comparison, many analytical techniques require conversion of the signal of interest (such as light intensity in spectroscopy) into an electrical signal for measurement. Electroanalysis lends itself to measurements made in vivo. For example, miniature electrochemical sensors are used to measure pH and pO_2 in the bloodstream of patients with in-dwelling catheters.

The utility of many electrochemical techniques extends beyond qualitative and quantitative analysis into the fundamental characterization of chemical processes. For example, electroanalytical methods are commonly used to study the redox properties of inorganic, organic, and biological compounds.

The purpose of this chapter is to provide a fundamental background for understanding the modern electroanalytical techniques that are discussed in Chapters 28–32.

27.2 ELECTROCHEMICAL CELL: CONCEPTS, TERMS, AND SYMBOLS

An *electrochemical cell* typically consists of two *electrodes* immersed in a solution of ions in solvent, that is, an *electrolyte* solution. A representative cell that serves as a suitable illustration is shown in Fig. 27.1. The cell consists of two *half-cells*: an electrode of Zn immersed in an aqueous solution of $ZnSO_4$ and an electrode

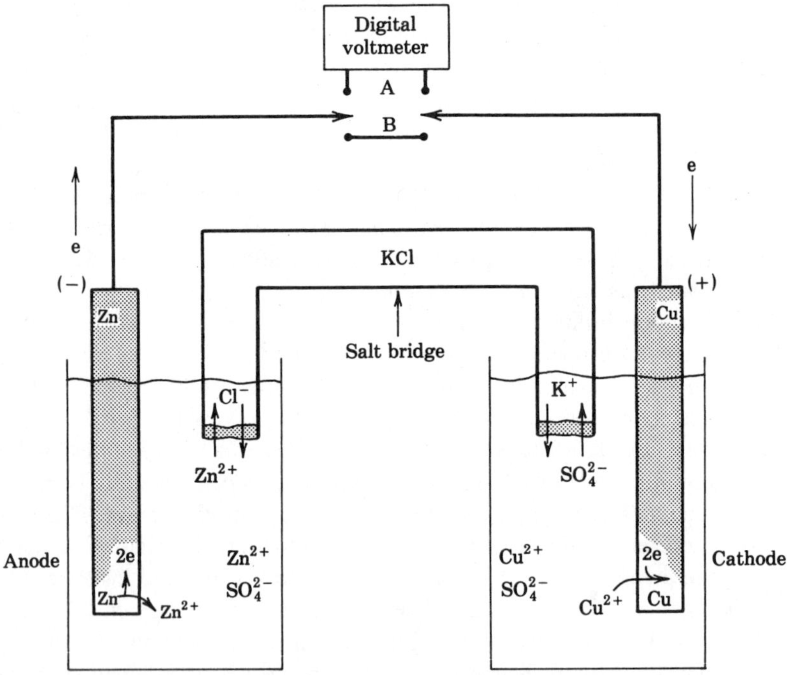

Anode reaction: oxidation	$Zn \rightleftharpoons Zn^{2+} + 2e$
Cathode reaction: reduction	$Cu^{2+} + 2e \rightleftharpoons Cu$
Overall cell reaction:	$Zn + Cu^{2+} \rightleftharpoons Zn^{2+} + Cu$

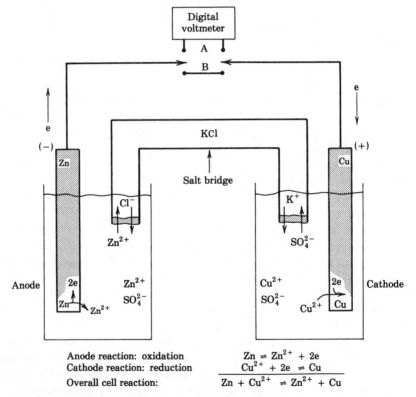

Fig. 27.1 Electrochemical cell consisting of a zinc electrode in 0.1 *M* $ZnSO_4$, a copper electrode in 0.1 *M* $CuSO_4$, and a salt bridge. (a) Galvanic cell.

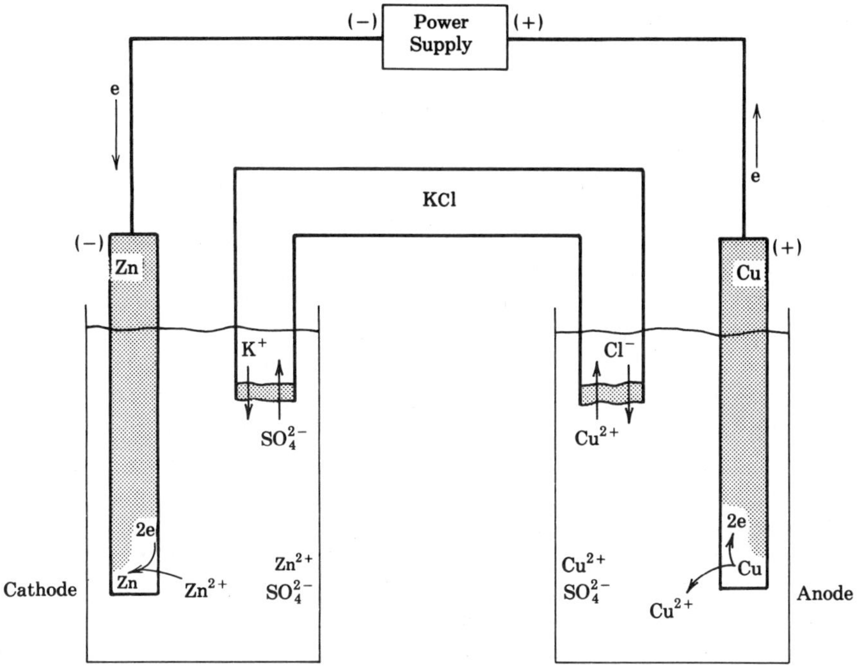

Anode reaction: oxidation	$Cu \rightleftharpoons Cu^{2+} + 2e$
Cathode reaction: reduction	$Zn^{2+} + 2e \rightleftharpoons Zn$
Overall cell reaction:	$Cu + Zn^{2+} \rightleftharpoons Cu^{2+} + Zn$

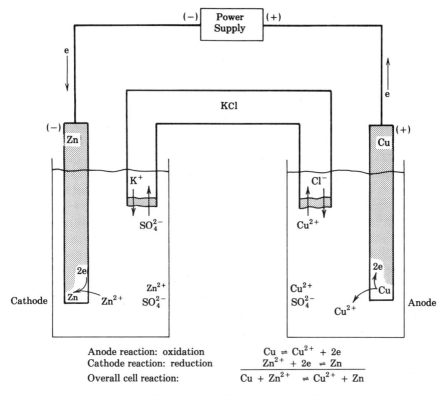

Fig. 27.1 (*Continued*) (b) Electrolytic cell.

of Cu in an aqueous solution of $CuSO_4$. The two sulfate solutions are connected by a *salt bridge*, which consists of a tube containing an aqueous solution of concentrated KCl. The ends of the tube are plugged with porous frits that allow the passage of ions, but prevent the contents of the tube from rapidly draining out. The purpose of the salt bridge is twofold: to allow electrical contact based on ionic conductance to be maintained between the electrolytes in the two containers and to prevent the two solutions from mixing. The wires attached to the electrodes in Fig. 27.1a can be connected to a digital voltmeter (position A) or to each other (B), whereas the electrodes in Fig. 27.1b are connected to an external power supply. The behavior of the electrochemical cell can be examined by performing the three experiments outlined below.

Cell Potential. In the *first experiment* the *potential of the electrochemical cell* (E_{cell}) is measured by connecting the wire leads to the digital voltmeter (position A). Since this voltmeter requires negligible current, the potential measurement can be made without perturbing the electrochemical cell (i.e., without allowing any electrochemical reactions to occur).

The cell potential is a measure of the difference in electron energy between the two electrodes. The *electron energy* of each electrode is related to the driving force for specific redox reactions that occur at the *electrode–solution interface*. For the example in Fig. 27.1a, one electrode reaction (as shown at the bottom of the Zn electrode) is the oxidation of Zn to form Zn^{2+}. The electrons released by this process contribute to an excess negative charge on the Zn electrode as indicated by the (−) sign at the top cf the electrode. The other electrode reaction (as shown at the bottom of the Cu electrode) is the reduction of Cu^{2+} to form Cu. The electrons consumed by this process contribute to a positive charge on the Cu electrode, as indicated at the top of the electrode. Thus, the potential of an electrode depends on an excess or deficit of charge on the electrode. The energy required to add or subtract an electron from the electrode can be expressed in terms of this potential. With an excess of negative charge, the energy of electrons is high, and the electrode potential is negative. An excess in positive charge corresponds to low electron energy and a positive potential. The digital voltmeter responds to the magnitude of the electron energy *difference* between the two electrodes, which is E_{cell}.*

The difference of potential between two points is measured by the work necessary to carry a unit positive charge from one point to the other. The unit in which potential is generally expressed is the *volt* (V) or *millivolt* (mV). Typical values for electrochemical cells range from a few tenths of a millivolt to a few volts. A volt is that potential difference against which one joule of work is done in the transfer of one coulomb of charge. *Ohm's law* states that the volt is the electrical potential which when applied to a resistor of one ohm causes a current of one ampere. These and other terms, units, constants, symbols, and conversions in electrochemistry are listed in Table 27.1.

Electrolysis in a Galvanic Cell: Current, Charge, and Faraday's Law. In the *second experiment* the two wire leads are connected (position B). This completion of the electrical circuit for the electrochemical cell is accompanied by the onset of several phenomena: *electrolysis* at the two electrodes, electrons flowing through the wire leads, and ions flowing through the salt bridge. The cell is now functioning as a *galvanic cell*, which means that it is producing energy spontaneously. The electrolysis reactions at the two electrodes consist of reduction–oxidation (*redox*) reactions that can be described as shown at the bottom of Fig. 27.1a. The electrode at which *oxidation* occurs is termed the *anode*, which in this cell is the Zn electrode. The electrode at which *reduction* occurs is termed the *cathode*, which is the Cu electrode. The electrochemical cell can be thought of in terms of the two *half-cells*, namely the anode compartment and the cathode compartment. The overall chemical reaction is simply the sum of the two *half-cell* reactions. In this cell the overall chemical reaction is the reduction of copper ion to copper metal by zinc metal. Since Zn reduces Cu^{2+} to Cu, Zn is termed the *reductant* in the redox reaction. Conversely, Cu^{2+} is termed the *oxidant*, since it oxidizes Zn to Zn^{2+}.

*The reader is referred to ''Understanding electrochemistry: some distinctive concepts,'' L. R. Faulkner, *J. Chem. Ed.* **60,** 263 (1983), from which some of this discussion was taken.

Table 27.1 Electrochemical Terms, Units, Constants, Symbols, and Conversions

Term	*Symbol*	*Unit or Constant*	*Symbol*	*Conversion or Value*
Potential (standard potential) (formal potential)	E (E^0) $(E^{0'})$	volt	V	$V = JC^{-1}$ $V = iR$
Current	i	ampere	A	$A = Cs^{-1}$ $1A = 1.05 \times 10^{-5}$ mol charge s^{-1}
Charge	Q	coulomb	C	$C = As$ $1C = 1.05 \times 10^{-5}$ mol charge
Resistance	R	ohm	Ω	
Time	t	second	s	
Temperature	T	kelvin	K	
Activity	a	moles per liter	mol L^{-1}	
Concentration	C	moles per liter (or moles per cubic centimeter)	mol L^{-1} (mol cm^{-3})	
Area	A	square centimeters	cm^2	
Diffusion coefficient	D	square centimeters per second	$cm^2\ s^{-1}$	
		gas constant	R	8.31441 J mol^{-1} K^{-1}
		Faraday constant	F	9.64846×10^4 C mol^{-1}
		number of electrons in electrode or redox reaction	n	

Examination of the wire leads shows electrons to be flowing from the Zn anode to the Cu cathode as a result of the two half-cell reactions and in response to the difference in electron energies of the two electrodes. The flow of electrons is termed *current*, which is the rate of transfer of electricity. Current is an expression of the rate at which the two electrode processes are occurring. Since the conversion of each Zn atom to a Zn^{2+} ion releases 2e, the current is a direct measure of the rate at which the oxidation process is occurring. The same argument applies to the rate of reduction of Cu^{2+} at the Cu electrode.

The practical unit for current is the *ampere* (A), which is the transfer of one coulomb per second. This corresponds to the passage of 1.05×10^{-5} moles of electrons per second. Since the current involved in most electroanalytical techniques is very small, milliamperes (mA), microamperes (μA), and nanoamperes (nA) are commonly used units.

Charge is a quantity of electricity. The charge passed during a period of time can be obtained by integration of the current during that period:

$$Q = \int_0^t i\,dt. \tag{27.1}$$

The unit for charge is the *coulomb* (C), which corresponds to 1.05×10^{-5} moles of electrons.

In electroanalytical chemistry, it is often important to relate the amount of electrical charge passed through an electrochemical cell to the quantity of material that has undergone electrolysis. *Faraday's law* gives the relationship

$$Q = nFN, \tag{27.2}$$

where F is the Faraday constant (96,485 C/mol), N is the number of moles electrolyzed, and n is the number of electrons involved in the electrode or redox reaction (n can be taken as a dimensionless parameter that defines the electron stoichiometry of the reaction). Differentiation of Eqs. (27.1) and (27.2) gives the expression

$$i = \frac{dQ}{dt} = nF\frac{dN}{dt}, \tag{27.3}$$

which clearly delineates the direct relationship between the rate at which electricity is moved across the electrode–solution interface (dQ/dt) and how fast the chemistry is accomplished at that interface (dN/dt).

Example 27.1. Assume that a steady current of 1.00 mA is measured for 5 min in the cell shown in Fig. 27.1. Calculate the charge passed, the total weight of Cu deposited at the cathode, and the rate per second of Cu deposition.

Since the current is constant, Eq. (27.1) reduces to $Q = it = (1.00 \times 10^{-3}\ \text{A})\ (5\ \text{min})(60\ \text{s min}^{-1}) = 0.300$ C of charge. Substitution into Eq. (27.2) gives $0.300\ \text{C} = 2(96{,}485\ \text{C mol}^{-1})(N)$ which gives $N = 1.55 \times 10^{-6}$ mol. The weight of Cu deposited is $(1.55 \times 10^{-6}\ \text{mol})(63.55\ \text{g mol}^{-1}) = 9.88 \times 10^{-5}$ g. The rate of Cu deposition $= 9.88 \times 10^{-5}\ \text{g}/300\ \text{s} = 3.29 \times 10^{-7}\ \text{g s}^{-1} = 329\ \text{ng s}^{-1}$.

In an electrochemical cell, the current at the anode must equal the current at the cathode or huge potential differences would result from the loss of electroneutrality. Thus, the electrons removed from the anode compartment are replaced in the cathode compartment. In order to maintain electroneutrality in each half-cell, the flow of charge by the movement of electrons in the electrodes and wire leads of the cell must be accompanied by the flow of ions in the solution components of the cell, that is, the *electron current* must equal the *ionic current*. The salt bridge provides the path for ion flow to maintain electrical neutrality in each half-cell compartment. The movement of ions in this particular cell is shown diagrammatically at the two ends of the salt-bridge tube (Fig. 21.1a). In the Zn^{2+} compartment, the development of a positive electrostatic charge from the electrolytic production of Zn^{2+} is prevented by the movement of Zn^{2+} into the salt bridge and of Cl^- out of the bridge. At the other end of the bridge, K^+ moves out and SO_4^{2-} moves into the bridge to prevent the buildup of a negative electrostatic charge in the Cu^{2+} half-cell compartment due to the loss of Cu^{2+} by reduction to Cu. An overall movement of positive ionic charge from the anode half-cell to the cathode half-cell and of negative charge in the reverse direction occurs to maintain electrical neutrality in both half-cell compartments and the salt bridge. The proportion of ionic charge carried by each type of ion (K^+, Cl^-, Zn^{2+}, Cu^{2+}, and SO_4^{2-}) depends on the charge, mobility, and concentration of each ion. From a practical point of view, removing the salt bridge halts electrolysis in the electrochemical cell just as effectively as disconnecting the two wire leads. In one case the path for ions is disrupted, in the other the path for electrons is disrupted. Both are essential for completion of the circuit for the cell.

As the electrolysis in the electrochemical cell proceeds, the overall chemical reaction is approaching a position of equilibrium. The attainment of equilibrium is signaled by the measurement of zero voltage and current in what is now a ''dead'' battery.

Electrolysis in an Electrolytic Cell. The *last experiment* involves the application of a potential to the electrochemical cell from an external power supply, as shown in Fig. 27.1b. If an external voltage of the appropriate polarity is impressed on the electrochemical cell the overall chemical reaction can be forced in the opposite direction of that observed for the galvanic cell, as shown at the bottom of Fig. 27.1b. In so doing, we find that the Cu electrode is now the anode and the Zn electrode the cathode. The electrochemical cell is now functioning as an *electrolytic cell* in that it is consuming electrical power, which is driving the chemical reaction. The external power supply is forcing the electron energy of the Zn electrode to a sufficiently high level to cause reduction of Zn^{2+} at the electrode; correspondingly, the electron energy of the Cu electrode is diminished to the point that electrons can be removed to form Cu^{2+}.

Schematic Representation of Cells. The portrayal of an electrochemical cell can be simplified by a schematic representation based on symbols. The commonly used symbols are as follows: (,) two species in the same phase or of the same phase type but where no potential is developed; (/) phase boundary at which a potential may develop; and (//) salt bridge, which has two phase boundaries at which potentials may develop. The chemical components of the cell are identified by the usual symbols with appropriate activities or concentrations indicated in parentheses. By convention, the anode half-cell is written left of the salt-bridge symbol. Also by convention the anode is considered the negative (−) electrode in a galvanic cell, but it is the positive (+) electrode in an electrolytic cell.

Example 27.2. The cell in Fig. 27.1a is written as

$$(-)\ Zn/Zn^{2+}\ (0.1M),\ SO_4^{2-}(0.1M)//SO_4^{2-}\ (0.1M),\ Cu^{2+}(0.1M)/Cu\ (+)$$

anode cathode

27.3 CELL POTENTIALS AND THE NERNST EQUATION

Free Energy and Cell Potential. The *free-energy change* (ΔG) for a chemical reaction is a measure of the thermodynamic driving force for the reaction to occur. The free-energy change for the chemical reaction of an electrochemical cell is related to the equilibrium cell potential by the equation

$$\Delta G_{\text{cell}} = -nFE_{\text{cell}}. \tag{27.4}$$

By thermodynamic convention, the cell reaction is spontaneous when E_{cell} is positive and ΔG_{cell} is negative. If E_{cell} is negative and ΔG_{cell} is positive, the reaction is not spontaneous as written and will proceed in the reverse direction. The cell is at equilibrium when E_{cell} and ΔG_{cell} are zero.

The free-energy change of a reaction and the potential of an electrochemical cell are said to be the *standard free-energy change* ΔG^0_{cell} and the *standard cell potential* E^0_{cell}, respectively, when all of the reactants and products are in the standard state, which is defined as unit activity. The standard free-energy change is related to the standard cell potential by the equation

$$\Delta G^0_{\text{cell}} = -nFE^0_{\text{cell}}. \tag{27.5}$$

Since analytical measurements are rarely, if ever, made under the conditions of standard state, an expression that relates cell potential to activity is of utmost importance.

Nernst Equation. Consider a general chemical reaction that is comprised of two half-cell redox reactions as shown below:

$$a\text{A} + ne \rightleftharpoons c\text{C} \tag{27.6}$$

$$b\text{B} \rightleftharpoons ne + d\text{D} \tag{27.7}$$

$$a\text{A} + b\text{B} \rightleftharpoons c\text{C} + d\text{D} \tag{27.8}$$

The free-energy change for the overall reaction, Eq. (27.8), can be related to the standard free energy change by

$$\Delta G_{\text{cell}} = \Delta G^0_{\text{cell}} + RT \ln \frac{a_{\text{C}}^{\text{c}} a_{\text{D}}^{\text{d}}}{a_{\text{A}}^{\text{a}} a_{\text{B}}^{\text{b}}} \tag{27.9}$$

where a is the activity of the constituent of the cell reaction, R is the gas constant, and T is temperature. Substitution for ΔG_{cell} and ΔG^0_{cell} from Eqs. (27.4) and (27.5) and division by the term $-nF$ gives the *Nernst equation*,

$$E_{\text{cell}} = E^0_{\text{cell}} - \frac{RT}{nF} \ln \frac{a^c_C a^d_D}{a^a_A a^b_B}, \tag{27.10}$$

which is one of the most important equations in electroanalytical chemistry! This general expression can be converted into a commonly used form of the Nernst equation at room temperature (25°C) by conversion to a base ten logarithm and substitution for the constant terms (R = 8.314 J mol^{-1} K^{-1}, T = 298 K, F = 96,485 C mol^{-1}) to give a constant in front of the log of $0.0591/n$ with units of J C^{-1} or V

$$E_{\text{cell}} = E^0_{\text{cell}} - \frac{0.0591}{n} \log \frac{a^c_C a^d_D}{a^a_A a^b_B}. \tag{27.11}$$

The Nernst equation enables the standard potential of a cell to be used to calculate the cell potential for reactant and product activities that are different from those of the standard state. The activities of soluble ionic or molecular species should be expressed in terms of moles per liter. The activity for solvent and a pure solid are taken as unity; the activity of a gas is taken as the partial pressure in atmospheres.

Figure 27.2 (left ordinate) shows how the potential of an electrochemical cell at 25°C is influenced by the ratio of activities in the logarithmic term. When the activities are such that the logarithmic term is zero, the cell potential is the standard potential E^0_{cell}. As the logarithmic term increases, the cell potential becomes more negative, and, conversely, as the logarithmic term decreases, the cell potential increases. The slope of the plot, $-0.0591/n$, is determined by the electron stoichiometry of the redox reaction.

Example 27.3. Fig. 27.2 (right ordinate) shows the potential-activity behavior for a cell the overall reaction of which is described by Eq. (27.8). What information can be obtained about the cell from this diagram? E^0_{cell} is 0.0780 V, which corresponds to the condition of standard state in which $\log (a^c_C a^d_D / a^a_A a^b_B) = 0$, and $n = 1$ from the slope. If a positive cell potential were measured, the cell reaction would be as written in Eq. (27.8), whereas a negative potential would indicate that the cell reaction is in the opposite direction. If the cell is connected and allowed to discharge to equilibrium (i.e., E = 0 V), $\log (a^c_C a^d_D / a^a_A a^b_B) = 1.33$, as shown by the dotted line.

Example 27.4. Calculate the potential of the following electrochemical cell

$$(+)\ \text{Fe}/\text{Fe}^{2+}(0.1\ M)//\text{Cd}^{2+}(0.001\ M)/\text{Cd}(-)$$

for which a potential of +0.037 V was measured at 25°C under conditions of unit activity for reactants and products. Activities are given in parenthesis.

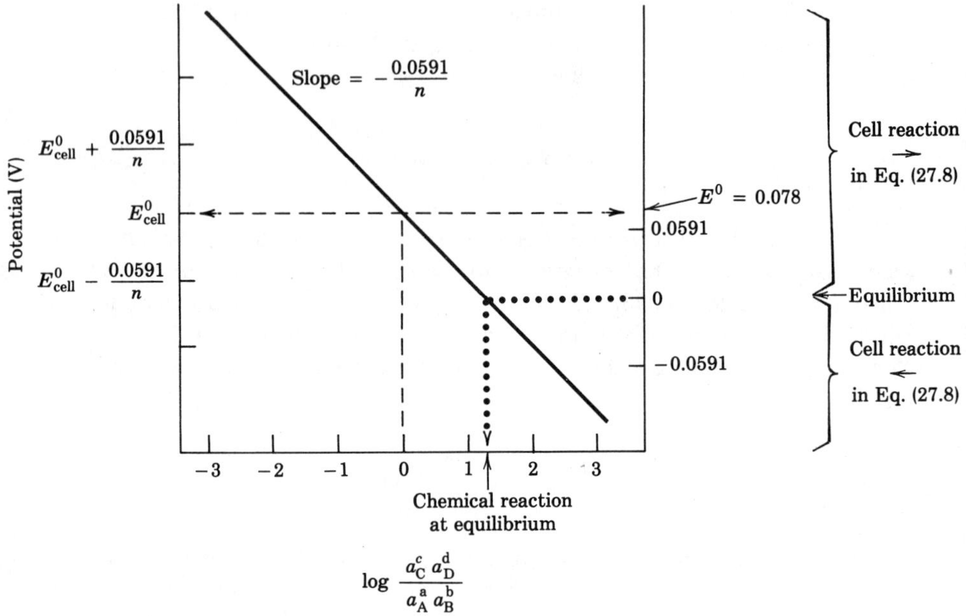

Fig. 27.2 Graphical representation of Nernst equation as written in Eq. 27.11. Left ordinate: general case. Right ordinate: cell with $E^0 = 0.078$ V and $n = 1$ (See Example 27.3).

A cell potential measured under conditions of unit activity is E^0_{cell}, which is +0.037 V. According to the schematic representation, the iron electrode is to be considered the anode and, therefore, is being oxidized. This means that the overall chemical reaction should be written as $Fe + Cd^{2+} \rightleftharpoons Cd + Fe^{2+}$, for which the Nernst equation is

$$E_{cell} = E^0_{cell} - \frac{0.0591}{2} \log \frac{a_{Cd}a_{Fe^{2+}}}{a_{Fe}a_{Cd^{2+}}}$$

Substitution for E°_{cell} and the appropriate activities for Fe^{2+} and Cd^{2+} and unity for the solids gives

$$E_{cell} = 0.037 - \frac{0.0591}{2} \log \frac{0.1}{0.001} = -0.0221 \text{ V}$$

The negative potential means that the chemical reaction for this cell under these conditions actually proceeds in the reverse of the direction as written above, that is, Cd is oxidized rather than Fe. This example illustrates the point that one does not know the correct direction in which an electrochemical cell reaction will proceed until the potential of the cell is either calculated or measured.

Formal Cell Potential. Since analytical chemists are generally more interested in measuring concentration than activity, it is useful to express activity in terms of concentration C and activity coefficient γ where

$$a = \gamma C, \tag{27.12}$$

which then substituted into Eq. (27.10) gives

$$E_{\text{cell}} = E^{0}_{\text{cell}} - \frac{RT}{nF} \ln \frac{\gamma^{\text{c}}_{\text{C}}\gamma^{\text{d}}_{\text{D}}}{\gamma^{\text{a}}_{\text{A}}\gamma^{\text{b}}_{\text{B}}} - \frac{RT}{nF} \ln \frac{C^{\text{c}}_{\text{C}}C^{\text{d}}_{\text{D}}}{C^{\text{a}}_{\text{A}}C^{\text{b}}_{\text{B}}}. \tag{27.13}$$

This general expression can be converted into a commonly used form of the Nernst equation by combining the activity coefficient terms and E^{0}_{cell} to give the *formal cell potential* $E^{0'}_{\text{cell}}$

$$E_{\text{cell}} = E^{0'}_{\text{cell}} - \frac{RT}{nF} \ln \frac{C^{\text{c}}_{\text{C}}C^{\text{d}}_{\text{D}}}{C^{\text{a}}_{\text{A}}C^{\text{b}}_{\text{B}}}. \tag{27.14}$$

The appropriate substitutions for C are concentration (mol L^{-1}) for soluble ionic and molecular species, partial pressure (atm) for gases, and unity for solvent and pure solids.

Equilibrium Constants. As mentioned above, electrolysis in a cell with the two electrodes connected will proceed until the position of equilibrium for the overall chemical reaction has been reached. At equilibrium, the free energy of the cell has been expended so that $\Delta G_{\text{cell}} = 0$ (and $E_{\text{cell}} = 0$), and the logarithmic term of activities becomes the *equilibrium constant* K for the reaction. Equation (27.9) can then be written

$$0 = \Delta G^{0}_{\text{cell}} + RT \ln K. \tag{27.15}$$

Substitution of Eq. (27.5) gives

$$\ln K = \frac{nFE^{0}_{\text{cell}}}{RT} \tag{27.16}$$

The equilibrium constant for a chemical reaction can be calculated from E^{0}_{cell} of the appropriate electrochemical cell.

Example 27.5. Calculate the equilibrium constant for the electrochemical cell described in Fig. 27.2 (right ordinate).

The cell is at equilibrium when $E_{\text{cell}} = 0$ V, which corresponds to a value of 1.33 on the abscissa, as indicated by the dotted line.

$$\log \frac{a^{\text{c}}_{\text{C}}a^{\text{d}}_{\text{D}}}{a^{\text{a}}_{\text{A}}a^{\text{b}}_{\text{B}}} = \log K = 1.33; \qquad K = 2.1 \times 10^{1}.$$

Alternatively, Eq. (27.16) in which $E^0_{cell} = 0.078$ V, $T = 298$ K, and $n = 1$ can be used:

$$\ln K = \frac{(1)(96{,}485 \text{ C mol}^{-1})(0.078 \text{ V})}{(8.314 \text{ J mol}^{-1}\text{K}^{-1})(298 \text{ K})}; \qquad K = 2.1 \times 10^1.$$

Calculation of Cell Potential from Half-Cell Potentials. An electrochemical cell based on the redox reaction in Eq. (27.8) consists of two half-cells with the half-cell reactions described by Eqs. (27.6) and (27.7). The free-energy change for the overall reaction in Eq. (27.8) is equal to the sum of the free-energy changes for the two half-cell reactions

$$\Delta G_{cell} = \Delta G_{A,C} + \Delta G_{B,D}. \tag{27.17}$$

Since tables of standard redox potentials tabulate E^0 in terms of a reduction reaction, it is convenient to express the free-energy for the oxidation reaction $\Delta G_{B,D}$ in terms of a reduction reaction ($-\Delta G_{D,B}$), which has the opposite sign. Using this convention, one would express the half-cell reactions for Eq. (27.8) as follows

cathode:	$a\text{A} + ne \rightleftharpoons c\text{C}$	$\Delta G_{A,C}$	(27.18)
anode:	$-(d\text{D} + ne \rightleftharpoons b\text{B})$	$-\Delta G_{D,B}$	(27.19)
overall	$a\text{A} + b\text{B} \rightleftharpoons c\text{C} + d\text{D}$	ΔG_{cell}	(27.20)

so that

$$\Delta G_{cell} = \Delta G_{A,C} - \Delta G_{D,B}. \tag{27.21}$$

Substitution of Eq. (27.4) and division by $-nF$ gives

$$E_{cell} = E_{A,C} - E_{D,B} \tag{27.22}$$

or

$$E_{cell} = E_{cathode} - E_{anode}. \tag{27.23}$$

Because the electrode to be considered as the anode is written as the left half-cell in the schematic diagram of an electrochemical cell, Eq. (27.23) can be expressed in more general terms as

$$E_{cell} = E_{right} - E_{left}. \tag{27.24}$$

Equation (27.24) is the convention established by IUPAC for the calculation of cell potentials. In this convention the half-cells are not specifically identified as being the anode or cathode. This is advantageous in that the true identity of the anode is generally not known until the result of the calculation is known or the cell is actually constructed and its potential measured. Thus, couching the calculation of E_{cell} in the framework of E_{right} and E_{left} for a schematically represented cell is generally less confusing than dealing with $E_{cathode}$ and E_{anode}, which may have to be reassigned after the results of the calculation are known.

The potential of a cell is easily calculated by first calculating the potential of each half-cell, *both written in terms of reduction reactions*. For example, the appropriate expressions for E_{right} and E_{left} for the general case considered in Eqs. (27.18–27.20) would be

$$E_{right} = E^0_{A,C} - \frac{RT}{nF} \ln \frac{a_C^c}{a_A^a} \tag{27.25}$$

$$E_{left} = E^0_{D,B} - \frac{RT}{nF} \ln \frac{a_B^b}{a_D^d} \tag{27.26}$$

where E^0 is the standard electrode potential for each half-cell. Once E_{right} and E_{left} are calculated, Eq. (27.24) is applied to calculate the cell potential.

Tables of Standard Electrode Potentials. As described in the previous section, the potential of an electrochemical cell can be calculated by taking the difference in the potentials of the two half-cells. For the purpose of calculating half-cell potentials, it is useful to have a table of standard electrode potentials, E^0, for various redox systems. One difficulty of a fundamental nature in obtaining data for such a table is the inability to measure the potential of a single half-cell. One can only measure the *potential difference between two half-cells!* The question "What is the potential of the half-cell Cu^{2+} (0.1 *M*)/Cu?" is somewhat analogous to the question "How far away is Cincinnati?" In answering the latter question one must ask "How far away is Cincinnati from where: Lubbock, Durham, the planet Jupiter?" In other words one must establish a point of reference from which the distance is to be measured. The same is true of the first question, which should also be rephrased: 'What is the potential of the half-cell Cu^{2+} (0.1 *M*)/Cu with respect to some other half-cell?' In the case of electrochemical cells, it is useful to select a half-cell against which other half-cells can be measured or referenced. Such a half-cell is called a reference electrode.

A short tabulation of E^0 values for representative half-cell reactions is shown in Table 27.2. The half-cell reaction for the reduction of H^+ is assigned a value of 0 V. As such, this couple is the "reference electrode" for the table and is called the *standard hydrogen electrode* (SHE). The measurement of half-cell potentials for other redox couples consists of preparing the half-cell for each redox couple under the conditions of standard state. The potential of the cell is then measured with respect to the SHE. By convention, the potential is assigned a positive sign if the half-cell functions as a cathode versus the SHE. A negative sign is given if the half-cell functions as an anode. Also by convention, all of the half-cell reactions in the table are written as reductions. The sign tells how the redox couple reacts with respect to the SHE. A positive sign indicates that the reaction for the

Table 27.2. Some Standard and Formal Electrode Potentials[a]

Half-Cell Reaction	E^0(V)	$E^{0'}$(V)
$Ag^+ + e \rightleftharpoons Ag(s)$	+0.799	0.228, 1 M HCl; 0.792, 1 M $HClO_4$; 0.77, 1 M H_2SO_4
$AgBr(s) + e \rightleftharpoons Ag(s) + Br^-$	+0.073	
$AgCl(s) + e \rightleftharpoons Ag(s) + Cl^-$	+0.222	0.228, 1 M KCl
$Ag(CN)_2^- + e \rightleftharpoons Ag(s) + 2CN^-$	−0.31	
$C_6H_4O_2$(quinone) $+ 2H^+ + 2e \rightleftharpoons C_6H_4(OH)_2$	+0.699	0.696, 1 M HCl, $HClO_4$
$Cd^2 + 2e \rightleftharpoons Cd(s)$	−0.403	
$Ce^{4+} + e \rightleftharpoons Ce^{3+}$		1.70, 1 M $HClO_4$; 1.61, 1 M HNO_3; 1.44, 1 M H_2SO_4; 1.28, 1 M HCl
$Cl_2(g) + 2e \rightleftharpoons 2Cl^-$	+1.359	
$Cr^{3+} + e \rightleftharpoons Cr^{2+}$	−0.408	
$Cr_2O_7^{2-} + 14H^+ + 6e \rightleftharpoons 2Cr^{3+} + 7H_2O$	+1.33	
$Cu^{2+} + 2e \rightleftharpoons Cu(s)$	+0.337	
$Fe^{2+} + 2e \rightleftharpoons Fe(s)$	−0.440	
$Fe^{3+} + e \rightleftharpoons Fe^{2+}$	+0.771	0.700, 1 M HCl; 0.732, 1 M $HClO_4$; 0.68, 1 M H_2SO_4
$Fe(CN)_6^{3-} + e \rightleftharpoons Fe(CN)_6^{4-}$	+0.36	0.71, 1 M HCl; 0.72, 1 M $HClO_4$, H_2SO_4
$2H^+ + 2e \rightleftharpoons H_2(g)$	0.000	−0.005, 1 M HCl, $HClO_4$
$Hg^{2+} + 2e \rightleftharpoons Hg(l)$	+0.854	
$Hg_2Cl_2(s) + 2e \rightleftharpoons 2Hg(l) + 2Cl^-$	+0.268	0.244, sat'd KCl; 0.282, 1 M KCl; 0.334, 0.1 M KCl
$K^+ + e \rightleftharpoons K(s)$	−2.925	
$Na^+ + e \rightleftharpoons Na(s)$	−2.714	
$O_2(g) + 4H^+ + 4e \rightleftharpoons 2H_2O$	+1.229	
$O_2(g) + 2H^+ + 2e \rightleftharpoons H_2O_2$	+0.682	
$Pb^{2+} + 2e \rightleftharpoons Pb(s)$	−0.126	−0.14, 1 M $HClO_4$; −0.29, 1 M H_2SO_4
$Sn^{2+} + 2e \rightleftharpoons Sn(s)$	−0.136	−0.16, 1 M $HClO_4$
$Sn^{4+} + 2e \rightleftharpoons Sn^{2+}$	+0.154	0.14, 1 M HCl
$Tl^+ + e \rightleftharpoons Tl(s)$	−0.336	
$TlCl(s) + e \rightleftharpoons Tl(s) + Cl^-$	−0.577	
$UO_2^{2+} + 4H^+ + 2e \rightleftharpoons U^{4+} + 2H_2O$	+0.334	
$Zn^{2+} + 2e \rightleftharpoons Zn(s)$	−0.763	

[a]Sources for E^0 and $E^{0'}$ values: G. Milazzo, S. Caroli, and V. K. Sharma, *Tables of Standard Electrode Potentials*. Wiley: New York, 1978. Source of formal potentials: E. H. Swift and E. A. Butler, *Quantitative Measurements and Chemical Equilibria*. W. H. Freeman and Company: San Francisco. Copyright 1972.

redox couple is a reduction when connected to the SHE, whereas a negative sign denotes that an oxidation would take place.

Some formal electrode potentials, $E^{0'}$, are also given in the table. Note that the electrolyte concentration must be stipulated for each value of $E^{0'}$.

Example 27.6. Calculate the potential of the following cell at 25°C:

$$Fe/Fe^{2+}\ (a = 1.00 \times 10^{-2}\ M)//Cd^{2+}\ (a = 1.00 \times 10^{-3}\ M)/Cd.$$

First, calculate the potential of each half-cell with the half-cell reactions written as reductions. Obtain values for E^0 from Table 27.2.

right half-cell reaction: $Cd^{2+} + 2e \rightleftharpoons Cd$

$$E_{right} = E^0_{Cd^{2+},Cd} - \frac{0.0591}{2} \log \frac{a_{Cd}}{a_{Cd^{2+}}}$$

$$= -0.403 - \frac{0.0591}{2} \log \frac{1}{1.00 \times 10^{-3}} = -0.492\ V$$

left half-cell reaction: $Fe^{2+} + 2e \rightleftharpoons Fe$

$$E_{left} = E^0_{Fe^{2+},Fe} - \frac{0.0591}{2} \log \frac{a_{Fe}}{a_{Fe^{2+}}}$$

$$= -0.440 - \frac{0.0591}{2} \log \frac{1}{1.00 \times 10^{-2}} = -0.499\ V$$

Then, calculate the potential of the cell

$$E_{cell} = E_{right} - E_{left} = -0.492 - (-0.499) = 0.007\ V.$$

Note that the calculated value of E_{cell} is positive, which indicates that the iron electrode is in fact the anode for this cell under these conditions and that the overall cell reaction is Fe + $Cd^{2+} \rightleftharpoons Fe^{2+}$ + Cd.

Example 27.7. The potential of the following electrochemical cell is found to be 1.115 V at 25°C:

$$Cd/Cd(CN)_4^{2-}(a = 8 \times 10^{-2}\ M),\ CN^-(a = 10^{-1}\ M)//SCE$$

Calculate the formation constant of $Cd(CN)_4^{2-}$. ($E_{SCE} = 0.244$ V vs. SHE.)

This problem is a good example of the use of electrochemistry to give information about a chemical reaction that is not a redox reaction. The complexation reaction is

$$Cd^{2+} + 4CN^- \rightleftharpoons Cd(CN)_4^{2-}$$

and the formation constant is

$$K_f = \frac{a_{Cd(CN)_4^{2-}}}{a_{Cd^{2+}}a_{CN^-}^4}$$

The electrochemical cell is made so that the values of a $a_{Cd(CN)_4^{2-}}$ and a_{CN^-} are known and $a_{Cd^{2+}}$ can be calculated from E_{cell}.

Using the convention in Eq. (27.23), we can substitute the values for E_{cell} and E_{SCE} to give

$$1.115 = 0.244 - E_{left}; \qquad E_{left} = -0.871 \text{ V}.$$

The left half-cell reaction written as a reduction can be considered as $Cd^{2+} + 2e \rightleftharpoons Cd$ for which the Nernst equation is

$$E_{left} = E^0_{Cd^{2+},Cd} - \frac{0.0591}{2} \log \frac{1}{a_{Cd^{2+}}}.$$

Substitution for E_{left} and $E^0_{Cd^{2+},Cd}$ gives

$$-0.871 = -0.403 - \frac{0.0591}{2} \log \frac{1}{a_{Cd^{2+}}}; \qquad a_{Cd^{2+}} = 1.45 \times 10^{-16} \text{ M}$$

Substitution into the equation for K_f gives

$$K_f = \frac{(8 \times 10^{-2})}{(1.45 \times 10^{-16})(10^{-1})^4} = 5.5 \times 10^{18}$$

Example 27.8. Calculate the solubility product constant of AgCl at 25°C.

The solubilization of AgCl can be represented as the summation of two half-cell reactions

$$\begin{array}{rl} \text{right:} & AgCl + e \rightleftharpoons Ag + Cl^- \\ \text{left:} & -(Ag^+ + e \rightleftharpoons Ag) \\ \hline \text{overall:} & AgCl \rightleftharpoons Ag^+ + Cl^- \end{array}$$

for which

$$K_{sp} = a_{Ag^+}a_{Cl^-}.$$

Using Eq. (27.24), one can calculate the standard cell potential

$$E^0_{cell} = E^0_{AgCl,Ag} - E^0_{Ag^+,Ag} = +0.222 - (+0.799) = -0.577 \text{ V}.$$

Substitution into Eq. (27.16) gives

$$\ln K = \frac{(1)(96{,}485 \text{ C mol}^{-1})(-0.577 \text{ V})}{(8.314 \text{ J mol}^{-1} \text{ K}^{-1})(298 \text{ K})}; \qquad K_{sp} = 2 \times 10^{-10}$$

27.4 REFERENCE ELECTRODES

A *reference electrode* is an electrochemical half-cell that is used as a fixed reference for the measurement of cell potentials. Ideally, a reference electrode should possess the following characteristics: a stable, easily reproducible half-cell potential; a reversible half-cell reaction; chemical stability of its components; and ease of fabrication and use. Three reference electrodes are discussed below, one is of fundamental significance and two are of practical importance.

Standard Hydrogen Electrode. The *standard hydrogen electrode* (SHE) has been chosen as the reference half-cell electrode upon which tables of standard electrode potentials are based. A schematic diagram of the SHE, its half-cell reaction, and short notation are shown in Fig. 27.3a. Hydrogen gas at a pressure of 1 atm is bubbled over a platinum electrode immersed in acid solution for which the activity of H^+ is unity. The platinum serves as a conducting electrode for the half-cell reaction. The platinum electrode consists of platinum foil covered with finely divided platinum (platinum black), which serves to increase the active surface area at which the half-cell reaction can occur. The hydrogen gas that is involved in the half-cell reaction is adsorbed on the platinum electrode where the redox reaction occurs. Note that the platinum electrode does not participate as a chemical species in the half-cell reaction, but merely serves as a conduit of electrons for the redox couple.

The potential of the SHE is defined as 0 V at all temperatures, and the potentials of other half-cell couples are referenced to this value. The potentials of other half cells are either negative or positive of 0 V depending on the thermodynamic driving force of the chemical reaction relative to that for the H^+, H_2 couple. Consequently, the SHE functions as either an anode or a cathode, depending on whether the other half-cell couple is capable of oxidizing H_2 or reducing H^+.

Although it is of fundamental importance in the organization of tables of standard potentials, the SHE is rarely employed in the everyday use of electrochemistry. Many electrochemists have, in fact, never even seen a SHE, much less actually used one! The necessity of having accurately prepared H^+ of unit activity and hydrogen gas (*explosive*!) at exactly 1 atm pressure has encouraged the development of other reference electrodes for practical, everyday use.

Saturated Calomel Electrode. A commonly used reference electrode is the *saturated calomel electrode* (SCE). Schematic diagrams of two common types of SCE's and the half-cell reaction are shown in Fig. 27.3b,c. The electrode consists of elemental mercury covered with a thin coating of calomel (Hg_2Cl_2) that is in contact with an aqueous solution saturated with KCl. It is apparent from the Nernst equation that in Fig. 27.3b,c the potential of the half-cell will be constant so long as the activity of Cl^- is invariant. The easiest way to set a_{Cl^-} to a fixed value that is easily verifiable is to saturate the solution with a chloride salt such as KCl. As long as crystals of KCl are present, the experimenter knows that the solution is saturated and that the activity of Cl^- is constant. Note that the other participants in the redox couple (Hg_2Cl_2 and Hg) are solids and, consequently, exhibit unit

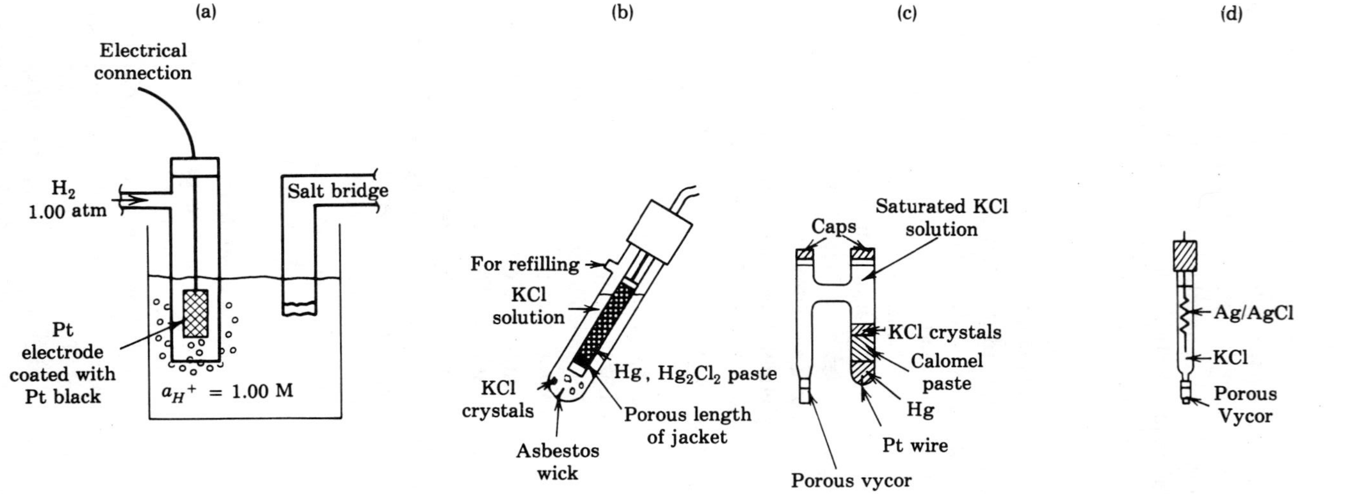

Pt, H_2(1.00 atm)/H^+(a = 1.00 M)

$2H^+ + 2e \rightleftharpoons H_2(g)$

$E_{SHE} = 0$ V

Hg, Hg_2Cl_2(KCl sat.)

$Hg_2Cl_2 + 2e \rightleftharpoons Hg + 2\,Cl^-$

$$E_{SCE} = E^0_{Hg_2Cl_2,\,Hg} - \frac{RT}{2F} \ln a^2_{Cl^-}$$

Ag, AgCl/NaCl

$AgCl + e \rightleftharpoons Ag + Cl^-$

$$E_{Ag/AgCl} = E^0_{AgCl,\,Ag} - \frac{RT}{F} \ln a_{Cl^-}$$

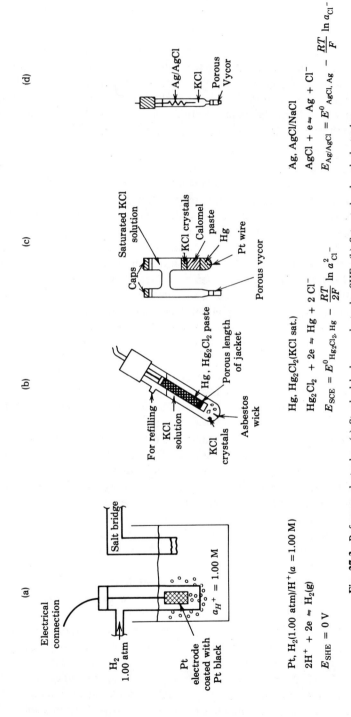

Fig. 27.3 Reference electrodes. (a) Standard hydrogen electrode, SHE. (b) Saturated calomel electrode, SCE, with asbestos wick for salt-bridge junction. (c) SCE of H-cell design with porous vycor glass for salt bridge junction. (d) Silver–silver chloride electrode, Ag/AgCl.

activity regardless of the amounts present in the cell. Thus, the SCE offers the extraordinary convenience of being fabricatable without the need for accurately preparing the activities of any of the components.

The potential of the SCE at several temperatures is shown in Table 27.3. The temperature dependence of the SCE is attributable to the temperature term T in the Nernst equation and the change in solubility of KCl with temperature. The SCE is commonly used in both potentiometric and voltammetric techniques, as outlined in Chapters 28–30. Sometimes NaCl is used rather than KCl in order to establish the activity of chloride in the SCE. This electrode is abbreviated *SSCE* for *sodium* SCE. The SSCE is commonly used for measurements on solutions containing perchlorate. Precipitation of $KClO_4$ at the junction is troublesome with the SCE. The SSCE does not have this problem since $NaClO_4$ is substantially more soluble than $KClO_4$.

Silver/Silver Chloride Electrode. Another commonly used reference electrode is the *silver/silver chloride electrode* (Ag/AgCl). A representative Ag/AgCl reference electrode and ancillary information are shown in Fig. 27.3d. The electrode is prepared by coating a silver wire with a thin film of AgCl. This is easily accomplished by dipping the wire into a concentrated solution of hydrochloric acid and making the wire an anode in an electrolytic cell in which the cathode is simply a platinum wire. In the presence of chloride, silver oxidizes to form a precipitate of AgCl on the silver wire. After coating with AgCl, the wire is rinsed and dipped in a solution of known chloride concentration to form the reference electrode. The potential of an Ag/AgCl electrode with unit activity Cl^- at various temperatures is shown in Table 27.3.

The Ag/AgCl reference electrode is commonly used, especially as the inner reference electrode in potentiometric membrane electrodes (Chapter 28). Compared to the SCE, the Ag/AgCl reference electrode offers the advantage of the absence of mercury in case of breakage. Additionally, the silver wire is more easily incorporated in some electrode designs than is liquid mercury.

Table 27.3 Potential of the SCE and the Ag/AgCl Electrode

Temperature (°C)	E_{SCE} *(V)*	$E_{Ag/AgCl}$ *(V)* *(Unit Activity* Cl^-*)*
0		0.23655
5		0.23413
10		0.23142
15	0.2508	0.22857
20	0.2476	0.22557
25	0.2444	0.22234
30	0.2417	0.21904
35	0.2391	0.21565

27.5 LIQUID JUNCTION POTENTIAL

As described in Section 27.3, the overall cell potential is determined by the values of the two half-cell potentials. Another contribution to the cell potential arises at the junction between two dissimilar liquids. This *liquid junction potential* (E_{lj}) develops at the interface between two liquids as a result of differences in the rates with which ions move from one liquid to the other. Thus, E_{lj} is a potential resulting from charge separation, rather than from a redox couple at an electrode. This source of potential should be added to the expression for E_{cell} from Eq. (27.24):

$$E_{cell} = E_{right} - E_{left} + E_{lj}. \tag{27.27}$$

Three types of liquid junction situations are shown schematically in Fig. 27.4. Fig. 27.4a shows a Type 1 junction between two liquids that contain the same ions (H^+ and Cl^-), but at different concentrations. The rate of movement of H^+ and Cl^- in solution is measured by their mobilities (μ), which in aqueous solution are $\mu_{H^+} = 3.625 \times 10^{-3}\ cm^2\ s^{-1}\ V^{-1}$ and $\mu_{Cl^-} = 7.912 \times 10^{-4}\ cm^2\ s^{-1}\ V^{-1}$. The movement of protons in solution is more than four times faster than the movement of chloride. This difference is shown schematically in Fig. 27.4 by the relative lengths of the arrows associated with the ions. Because of the difference in concentration of HCl in the two solutions, a net movement of HCl across the boundary from the more concentrated solution to the less concentrated solution occurs by diffusion. Since H^+ moves more rapidly than Cl^-, a charge separation develops at the junction with the less concentrated solution being positively charged by excess H^+ and the more concentrated solution being negatively charged by excess Cl^-. This is shown schematically in Fig. 27.4a. Fig. 27.4b illustrates a Type 2 junction between two solutions of identical concentration and containing the same anion but with different cations. In this case a junction potential develops because of the more rapid movement of H^+ than K^+ (μ_{K^+} $2.619 \times 10^{-7}\ cm^2\ s^{-1}\ V^{-1}$). A Type 3 junction with dissimilar cations, anions, and concentrations is illustrated in Fig. 27.4c.

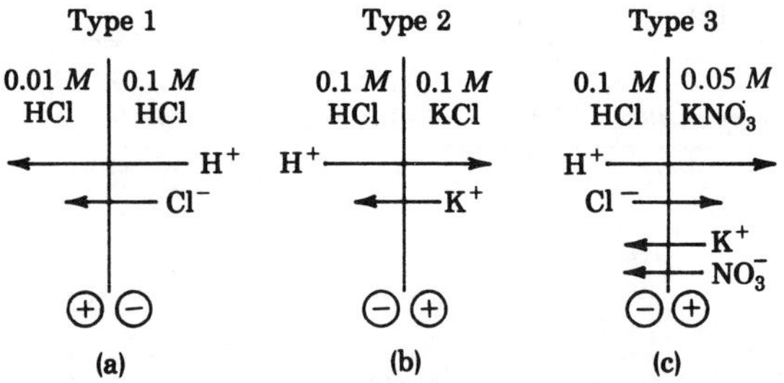

Fig. 27.4 Types of liquid junction. Arrows show the direction of net transfer for each ion, and their lengths indicate relative mobilities. The polarity of the junction potential is indicated in each case by the circled signs. (From J. J. Lingane, *Electroanalytical Chemistry*, 2nd ed., Wiley-Interscience, New York, 1958, p. 60, and A. J. Bard and L. R. Faulkner, *Electrochemical Methods*, Wiley, New York, 1980, p. 63, with permission).

Table 27.4 Liquid Junction Potentials of 0.1 *M* Concentrations of Electrolytes[a]

Junction	E_{lj} *Observed* (*mV*)	*Junction*	E_{lj} *Observed* (*mV*)
HCl : KCl	26.78	KCl : NaCl	6.42
HCl : NaCl	33.09	KCl : NH_4Cl	2.16
KCl : LiCl	34.86	NaCl : LiCl	2.62
HCl : NH_4Cl	28.40	NaCl : NH_4Cl	−4.21
KCl : LiCl	8.79	LiCl : NH_4Cl	−6.93

[a]From D. T. Sawyer and J. L. Roberts, Jr., *Experimental Electrochemistry for Chemists*. New York: Wiley-Interscience, 1974, p. 22.

As one can infer from the discussion above, the magnitude of E_{lj} is determined by the ionic composition of the two solutions in contact. Some representative values of E_{lj} are shown in Table 27.4. Note the variation in polarity and in the magnitude of E_{lj} depending on solution composition.

A typical point at which a liquid junction potential exists is at the interface between the salt bridge and the solution of a half-cell. Potassium chloride is a good choice of electrolyte for a salt bridge. The comparable mobilities of K^+ and Cl^- minimize the contribution of the salt bridge electrolyte to E_{lj}. Also, the contribution of KCl to both ends of a salt bridge of the type shown in Fig. 27.1 compensates since the two values of E_{lj} are of opposite polarity. The junction potential at the salt bridge interface can be minimized by increasing the concentration of KCl in the bridge so that transport at the junction is dominated by KCl. The effect of KCl concentration is shown in Fig. 27.5 for a junction with 0.1 *M* HCl.

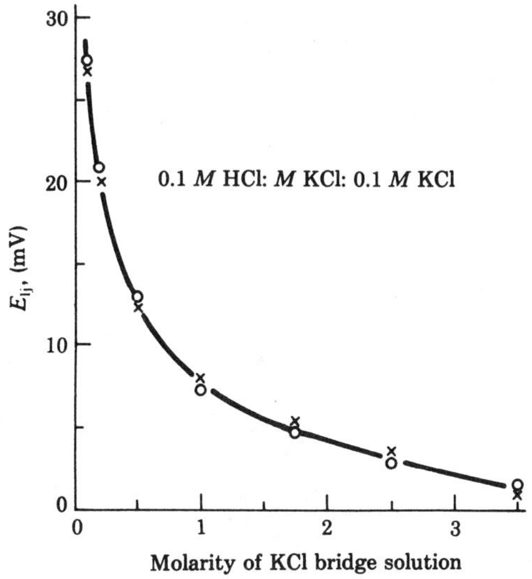

Fig. 27.5 Junction potential between a 0.1-*M* HCl solution and a 0.1-*M* KCl solution as a function of the concentration of KCl in the connecting salt bridge. (From D. T. Sawyer and J. L. Roberts, Jr., *Experimental Electrochemistry for Chemists*, Wiley-Interscience, New York, 1974, p. 25).

27.6 OHMIC LOSSES

The potential of an electrochemical cell when electrolysis is occurring is diminished by resistance of the cell to current. The electrolyte in the cell offers resistance to the movement of ions that is necessary for electrolysis to occur in the cell. The potential loss is related to the resistance of the solution in the cell (see Chapter 32) and the magnitude of the current as described by Ohm's law

$$E_{iR} = iR \tag{27.28}$$

where E_{iR} is the potential drop due to the ohmic losses or the *iR drop* in the cell and R is cell resistance. The equation for E_{cell} can now be further refined to give

$$E_{\text{cell}} = E_{\text{right}} - E_{\text{left}} + E_{\text{lj}} - E_{iR} \tag{27.29}$$

The magnitude of E_{iR} is minimized by increasing the conductivity of the solutions in the cell by the addition of electrolyte.

Example 27.9. Calculate the potential of the following cell, which has a resistance of 1.00 Ω, when it is producing a current of 0.300 A at 25°C:

$$\text{Zn}/\text{Zn}^{2+}\ (a = 0.010\ \text{M})//\text{Cu}^{2+}(a = 0.010\ \text{M})/\text{Cu}.$$

Substitution of $E^0_{\text{Zn}^{2+},\text{Zn}} = -0.763$ V and $E^0_{\text{Cu}^{2+},\text{Cu}} = +0.337$ V and the appropriate activities into Eqs. (27.25) and (27.26) gives $E_{\text{right}} = 0.278$ V and $E_{\text{left}} = -0.822$ V. Substitution of these values into Eq. (27.24) gives

$$E_{\text{cell}} = 0.278 - (-0.822) = 1.100\ \text{V},$$

which is the thermodynamic potential measured with no electrolysis occurring. Connecting the two electrodes together allows electrolysis to proceed, and is accompanied by a significant drop in cell potential.

$$E_{\text{cell}} = 1.100 - iR = 1.100 - (0.300\ \text{A})\,(1.00\ \Omega) = 0.800\ \text{V}$$

27.7 KINETICS OF ELECTRODE REACTIONS

Reduction and oxidation at electrode surfaces are accomplished by *heterogeneous electron transfer reactions*. They are termed heterogeneous since the electron transfer occurs at the *interface* between electrode and solution. By comparison a homogeneous electron transfer reaction involves solution reactants such as in the redox reaction between Ce^{4+} and Fe^{2+}. An important characteristic of heterogeneous electron transfer rections is the relationship between potential and rate of electrode reaction, which determines current.

Relationship Between Current and Heterogeneous Rate Constants. Consider the following reaction at an electrode immersed in a solution containing equal concentrations of O and R, namely C_O and C_R

$$\mathrm{O} + ne \underset{k_{b,h}}{\overset{k_{f,h}}{\rightleftharpoons}} \mathrm{R} \tag{27.30}$$

As shown schematically in Fig. 27.6a, this redox reaction occurs at the *interface* between the solid electrode, which supplies or removes the electrons, and the solution in which the chemical species that constitute the redox couple are dissolved. It is important to realize that O and R must be at the electrode surface (within a few angstroms) in order to participate in the redox reaction with the electrode. Any O and R that are distant from the electrode must be *transported* to the electrode surface before they can undergo the reaction. (These concepts of reaction at the electrode surface and transport to it will prove important in our discussion of voltammetric techniques in Chapter 29.) The discussion in this section is based on the assumption that the concentrations of O and R are homogeneous throughout the solution and, consequently, are not altered by the redox reaction at the electrode. However, to emphasize the surface nature of the reaction and to prepare for later discussion of techniques in which concentrations are inhomogeneous, the abbreviations C_O^s and C_R^s will be used to designate surface (s) concentrations of O and R.

Equation (27.30) has been expressed so that the rate of the forward reaction,

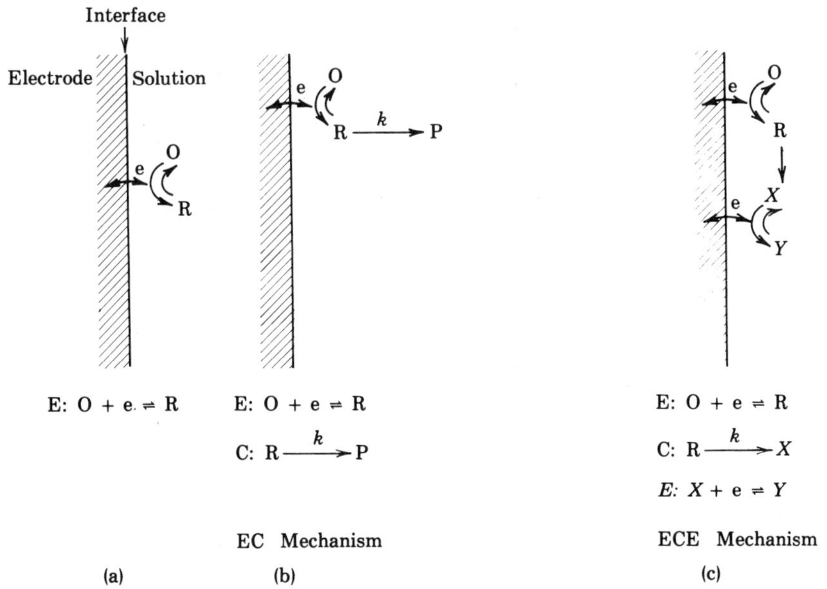

Fig. 27.6 Schematic diagram of electrode reaction mechanisms. (a) Simple heterogeneous electron exchange reaction at electrode–solution interface. (b) EC mechanism. (c) ECE mechanism.

that is, the reduction of O to R, is dependent on the heterogeneous forward rate constant, $k_{\mathrm{f,h}}$, and the surface concentration of O, $C_{\mathrm{O}}^{\mathrm{s}}$,

$$\text{rate of forward reaction} = v_{\mathrm{f}} = k_{\mathrm{f,h}} C_{\mathrm{O}}^{\mathrm{s}}, \tag{27.31}$$

whereas, the rate of the backward reaction, that is, the oxidation of R to O, is dependent on the heterogeneous back rate constant, $k_{\mathrm{b,h}}$, and the surface concentration of R, $C_{\mathrm{R}}^{\mathrm{s}}$;

$$\text{rate of backward reaction} = v_{\mathrm{b}} = k_{\mathrm{b,h}} C_{\mathrm{R}}^{\mathrm{s}}. \tag{27.32}$$

If the forward reaction prevails (i.e., $v_{\mathrm{f}} > v_{\mathrm{b}}$), the rate of formation of R from O is the difference in the rates of the forward and backward reactions,

$$\text{rate of formation of R} = \text{rate of loss of O} = v_{\mathrm{f}} - v_{\mathrm{b}}, \tag{27.33}$$

which can be expressed as

$$\frac{dN_{\mathrm{R}}}{dt} = -\frac{dN_{\mathrm{O}}}{dt} = k_{\mathrm{f,h}} C_{\mathrm{O}}^{\mathrm{s}} - k_{\mathrm{b,h}} C_{\mathrm{R}}^{\mathrm{s}}, \tag{27.34}$$

where dN/dt = rate of conversion per unit area (mol $\mathrm{cm}^{-2}\,\mathrm{s}^{-1}$);
C^{s} = solution concentration at the electrode surface (mol cm^{-3});
k = heterogeneous rate constant (cm s^{-1}).

Note the units used. In electrochemistry the rate of conversion of O to R depends on the surface area of the electrode. Consequently, it is useful to express conversion rates in terms of moles of material converted per square centimeter of electrode area per unit time (i.e., mol $\mathrm{cm}^{-2}\,\mathrm{s}^{-1}$). Consequently, C^{s} is expressed as mol cm^{-3} (not mol L^{-1}) and k has units of cm s^{-1}.

The *net* conversion rate of O to R determines the current that is measured at the electrode, as expressed earlier in Eq. (27.3). Current at an electrode of area A is related to the difference in the rates of conversion for the foward and back reactions by the following expression:

$$i = nFA\frac{dN_{\mathrm{R}}}{dt} = -nFA\frac{dN_{\mathrm{O}}}{dt}, \tag{27.35}$$

where A is the surface area of the electrode (cm^2) and the other terms have been defined (Table 27.1 and Eq. 27.34).

Combining Eqs. (27.34) and (27.35) gives an expression that relates the current at an electrode to the forward and backward heterogeneous rate constants:

$$i = nFA(k_{\mathrm{f,h}} C_{\mathrm{O}}^{\mathrm{s}} - k_{\mathrm{b,h}} C_{\mathrm{R}}^{\mathrm{s}}). \tag{27.36}$$

Note that when $k_{f,h}C_O^s = k_{b,h}C_R^s$, the current is zero, and the system is at equilibrium. When $k_{f,h}C_O^s > k_{b,h}C_R^s$, a positive current (*cathodic current*, i_c) that reflects the rate of reduction of O to R is obtained. When $k_{f,h}C_O^s < k_{b,h}C_R^s$, a negative current (*anodic current*, i_a) that is equivalent to the rate of oxidation of R to O is obtained.

Potential Dependence of Heterogeneous Rate Constants. A key feature of electrode kinetics is the potential dependence of the heterogeneous rate constants $k_{f,h}$ and $k_{b,h}$. This phenomenon is understood by considering the free-energy changes that accompany a heterogeneous electron transfer reaction. Figure 27.7 shows representation reaction coordinates for a redox reaction under two conditions. First consider Fig. 27.7a, which is representative of a state of electrochemical equilibrium between O and R. Curves O and R represent the potential-energy curves for species O and R, respectively. In order for O to react with an electron to form R, it is necessary to pass over the *activation free-energy barrier*, $\Delta G_f^\ddagger$, which is the free energy required to convert a mole of reactant to the activated state in the reaction. For the backward reaction, that is, conversion of R to O, the activation free-energy barrier $\Delta G_b^\ddagger$ must be crossed. At equilibrium $\Delta G_b^\ddagger = \Delta G_f^\ddagger$, as shown in Fig. 27.7a, and the probability of electron transfer is the same in each direction (assuming $C_O^s = C_R^s$). Consequently, $i_c + i_a = 0$ and no *net* current is observed. Although no net current is observed, it is important to recognize that the forward and backward reactions are occurring, albeit at identical rates. The *exchange current* i_0 is a measure of the electron exchange that is occurring at equilibrium, where $i_0 = i_c = -i_a$. The rates of the forward and backward reactions are determined by the magnitude of the energy barrier in proportion to $\exp(-\Delta G^\ddagger/RT)$. Thus, the exchange current will increase as $\Delta G^\ddagger$ decreases.

The application of a potential to the electrode changes the heterogeneous rate constants by changing the relative electron energy of the electrode. Negative po-

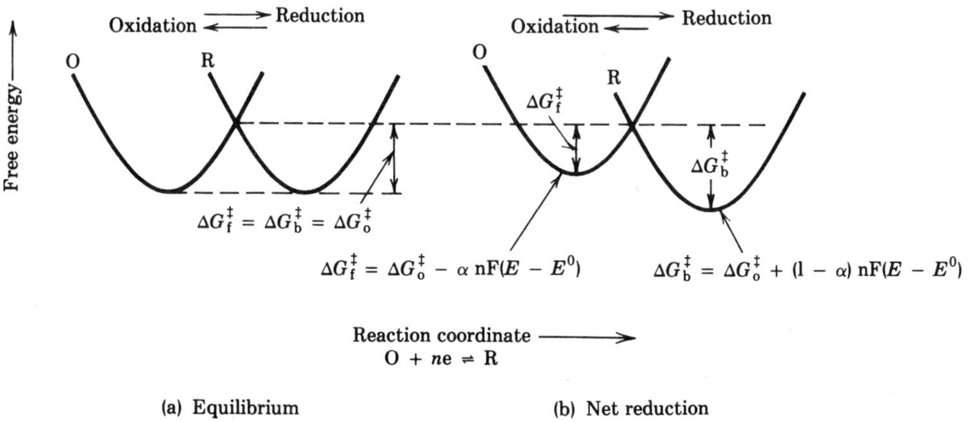

Fig. 27.7 Reaction coordinate diagram for an electron transfer reaction at an electrode. (a) Equilibrium. (b) Net reduction.

tential changes increase the electron energy and enhance k_f and lower k_b. A positive potential change has the opposite effect. In effect, the application of an external potential to an electrode changes $\Delta G_f^\ddagger$ and $\Delta G_b^\ddagger$ by shifting the potential energy curves. Suppose that the potential is moved to a value E that is negative of the equilibrium potential E^0. This application of a negative external potential shifts the reaction coordinate (Fig. 27.7b) to lower the energy barrier $\Delta G_f^\ddagger$ for the conversion of O to R and to raise the barrier $\Delta G_b^\ddagger$, for the backward reaction.

If the magnitude of the potential change is $E - E^0$, the free energy of the system is changed by an amount $-nF(E - E^0)$. Part of this energy change results from a decrease in the activation barrier for reduction (forward reaction),

$$\Delta G_f^\ddagger = \Delta G_0^\ddagger - \alpha nF(E - E^0), \quad (27.37)$$

whereas part results from an increase in the activation barrier for oxidation

$$\Delta G_b^\ddagger = \Delta G_0^\ddagger + (1 - \alpha)\, nF(E - E^0). \quad (27.38)$$

Thus, the activation free-energy barriers for forward and reverse reactions contain an ''electrochemical'' term which can be varied by simply applying a potential to the electrode. The *transfer coefficient* α is the fraction of the electrical energy that acts to decrease the height of the energy barrier for the forward reaction, and $1 - \alpha$ is the fraction that acts to increase the barrier for the backward reaction. The transfer coefficient is a measure of the symmetry of the energy barrier. A symmetrical energy barrier would have equal but opposite slopes of the energy curves at the point of intersection, and α would be 0.5. Values of α for real systems typically lie between 0.3 and 0.7.

The effect of potential change on the rate constants can be expressed in terms of equations in the Arrhenius form

$$k_{f,h} = k^0 \exp[-\alpha nF(E - E^0)/RT] \quad (27.39)$$

and

$$k_{b,h} = k^0 \exp[(1 - \alpha)\, nF(E - E^0)/RT], \quad (27.40)$$

where k^0 is the standard rate constant of the system, that is, the value of $k_{f,h}$ and $k_{b,h}$ when they are equal at E^0. These two equations clearly show the relationship between electrode potential and the rates of the heterogeneous electron transfer processes.

Relationship Between Current and Potential. The background has now been established to enable the relationship between current and potential to be drawn. Consideration of Eq. (27.36) shows that the net current i is the arithmetic sum of the forward, cathodic current i_c and the back, anodic current i_a where

$$i_c = nFAC_O^s k_{f,h} \quad \text{and} \quad i_a = -nFAC_R^s k_{b,h}. \quad (27.41)$$

These individual currents can be related to electrode potential by substitution of the potential dependence $k_{f,h}$ and $k_{b,h}$ from Eqs. (27.39) and (27.40) to give

$$i_c = nFAC_O^s k^0 \exp[-\alpha nF(E - E^0)/RT] \tag{27.42}$$

and

$$i_a = -nFAC_R^s k^0 \exp[(1 - \alpha)\, nF(E - E^0)/RT]. \tag{27.43}$$

The behavior of i_c and i_a as a function of potential E, as described by the two equations given above is shown by the dashed lines in Fig. 27.8. Note that i_c increases with increasingly negative potential, whereas i_a increases with increasingly positive potential.

At any potential, the net current i is equal to the arithmetic sum of the two opposing currents. Combining Eqs. (27.42) and (27.43) gives

$$\begin{aligned} i = i_c + i_a = nFAk^0\{C_O^s\, exp[-\alpha nF(E - E^0)/RT] \\ - C_R^s \exp[(1 - \alpha)\, nF(E - E^0)/RT]\}, \end{aligned} \tag{27.44}$$

which is a fundamental equation that relates current and potential by means of the heterogeneous rate constant k^0. The net current i as a function of potential is shown by the solid line in Fig. 27.8. Note that the overall redox reaction can be made to

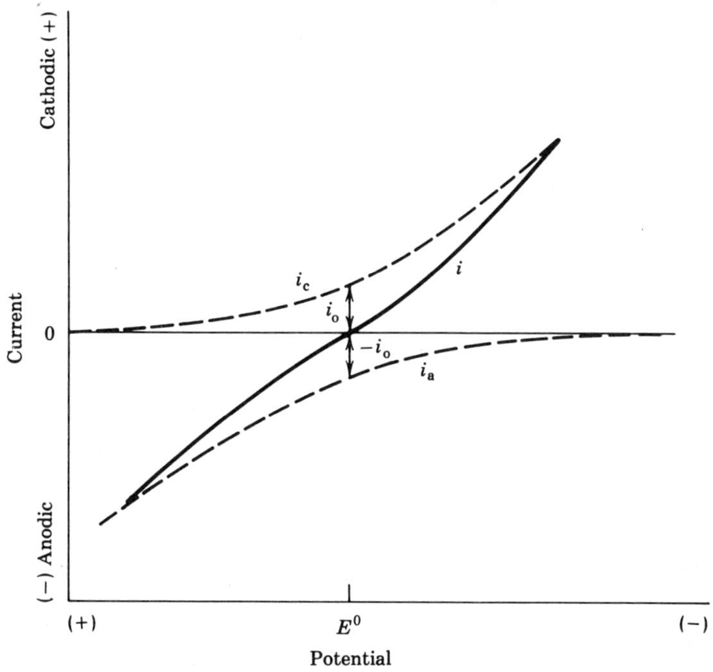

Fig. 27.8 Current–potential relationship for electrode process.

proceed forward by moving the potential to a sufficiently negative value, or backward by moving the potential to a sufficiently positive value. As the potential is moved further in either direction, the current continues to increase as the free energy for the dominant reaction is made more and more favorable.

The values of the standard rate constant, k^0, and the exchange current i_0 are a measure of the electrochemical redox system's ability to pass charge at the interface. Values of k^0 range from 1 to 50 cm s^{-1} for very fast reactions to 10^{-9} cm s^{-1} for very slow reactions. The corresponding range of i_0 values is from tens of amperes to picoamperes per square centimeter. Fast kinetics is generally associated with very simple redox reactions that might involve only solvation changes. Slow kinetics is generally associated with reactions involving major structural changes or multistep reactions. Redox systems that exhibit fast heterogeneous reactions are termed *electrochemically reversible* systems. Those that exhibit slow reactions are termed *electrochemically quasireversible* or *irreversible* systems.

Equilibrium: The Nernst Equation. At equilibrium, the net current is zero since $i_c = -i_a$. Since net current $i = 0$, Eq. (27.44) reduces to

$$C_O^s \exp[-\alpha nF(E - E^0)/RT] = C_R^s \exp[(1 - \alpha)\, nF(E - E^0)/RT] \quad (27.45)$$

or

$$\frac{C_R^s}{C_O^s} = -\exp[nF(E - E^0)/RT], \quad (27.46)$$

which becomes the familiar Nernst equation

$$E = E^0 - \left(\frac{RT}{nF}\right) \ln \frac{C_R^s}{C_O^s}. \quad (27.47)$$

Thus, the kinetic equations under the condition of equilibrium lead to the Nernst equation that was derived earlier by thermodynamics (Section 27.3).

The kinetic-based derivation of the Nernst equation gives some insight into the electrode process itself. It also emphasizes the fact that the Nernst equation is strictly valid only at equilibrium (i.e., $i = 0$). As will be shown in Chapters 29 and 30, many modern electroanalytical techniques operate at potentials other than the equilibrium potential. In such cases the Nernst equation can be used only if the heterogeneous rate constants are sufficiently large (i.e., the redox system is reversible) that equilibrium is quickly established at the electrode surface.

27.8 COUPLED CHEMICAL REACTIONS

Heterogeneous electron transfer reactions are often accompanied by chemical reactions that do not involve electron transfer with the electrode. These accompanying reactions are termed *coupled chemical reactions* since they are coupled to the primary electrode reaction. The overall sequence of reactions at an electrode is termed the *electrode reaction mechanism*. Some representative electrode mechanisms are shown in Fig. 27.6. A mechanism is named by the acronym that describes the sequence of reactions where by convention E is a heterogeneous electron transfer reaction and C is a coupled chemical reaction. An example of an ECE mechanism is the oxidation of catechol in the presence of thiol

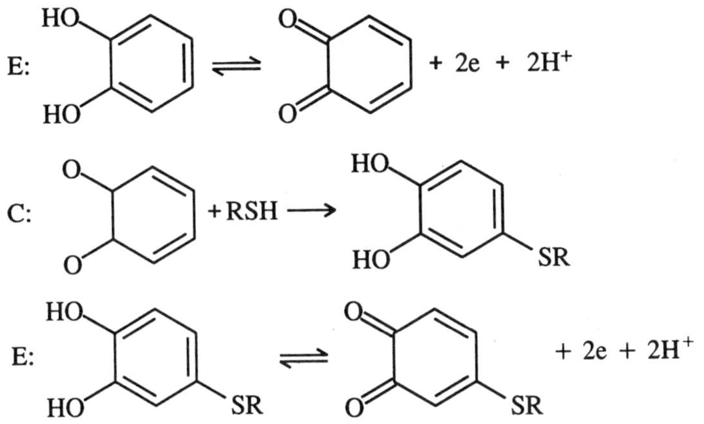

27.9 MASS TRANSPORT

By its nature, electrochemistry involving electrode reactions is an interfacial process. The electrochemical reduction or oxidation of a molecule occurs at the electrode–solution interface. A molecule dissolved in solution in an electrochemical cell must be *transported to the electrode surface* for the electrochemical event to occur. Consequently, the transport of molecules from regions in the solution phase to the electrode surface is an important aspect of many electrochemical techniques. The movement of material from one place to another is generally termed *mass transport* or *mass transfer*. Three mechanisms for mass transport play a role in electroanalytical techniques: *hydrodynamic movement*, *diffusion*, and *migration*.

Hydrodynamic Movement. *Hydrodynamic* mass transport involves the transport of dissolved chemical species by physical movement of the solution or the electrode relative to each other. Common means of implementing hydrodynamic mass transport include stirring the solution, rotating the electrode or flowing solution across the electrode by placing it in a tube through which solution is pumped. The movement of solution will act to transport new reactant to the electrode surface and product away from the surface. Electroanalytical techniques that rely on this means of mass transport are sometimes categorized as hydrodynamic techniques. A number of these are discussed in Chapter 29.

Diffusion. *Diffusion* is a process of spatial movement of ions or molecules due to their kinetic motion. Consider the hypothetical situation in which an impulse of solute molecules is created at $t = 0$ in a narrow region of solvent, as shown in Fig. 27.9a. The kinetic motion of both solute and solvent molecules results in a random drift of molecules that causes the impulse of solute molecules to spread as shown in Fig. 27.9b–e. Eventually, the concentration gradient created at $t = 0$ will be removed by the diffusional process and solute will be distributed homogeneously throughout the solvent. Two important points should be recognized. First, diffusion acts to remove a concentration gradient by the mass transport of molecules from a region of high concentration to a region of low concentration. Thus, diffusion increases the entropy of a system by maximizing the randomness with which solute molecules are distributed in a container. Second, diffusion is a random process.

A plot of concentration versus distance is termed a *concentration profile*. Concentration profiles for the impulse shown in Fig. 27.9a–e are shown in Fig. 27.9f–

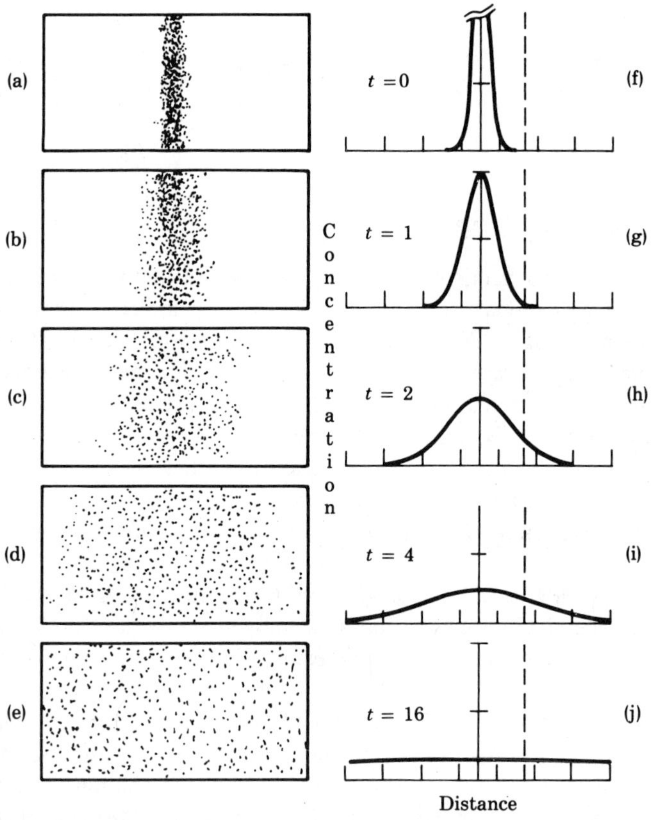

Fig. 27.9 Spreading of an impulse of molecules at various times after creation. (From *Laboratory Techniques in Electroanalytical Chemistry*, P. T. Kissinger and W. R. Heineman (Eds.), Dekker, New York, 1985, p. 14).

j. The concentration profile can be expressed in terms of the Gaussian function

$$C_{x,t} = \frac{C_0}{(4\pi Dt)^{1/2}} \exp\left(\frac{-x^2}{4Dt}\right), \tag{27.48}$$

where $C_{x,t}$ = concentration of solute molecules at x and t (mol cm^{-3});
C_0 = concentration of solute molecules in initial impulse (mol cm^{-3});
x = distance from center of impulse (cm);
t = time (s);
D = diffusion coefficient (cm^2 s^{-1}).

This equation describes the random *net* movement of solute molecules along the x axis as a function of time. The speed with which the solute molecules move is defined by the *diffusion coefficient* D. The population standard deviation σ is a measure of the distribution width and is defined as

$$\sigma = \sqrt{2Dt}. \tag{27.49}$$

The concept of an impulse of molecules generated in a specific region of solution with subsequent spreading by diffusion will be very useful in our consideration in Chapter 30 of some electroanalytical techniques in which the impulse is generated by electrolysis at the electrode surface and diffusion into solution occurs. Equation 27.49 is useful for estimating the distance that molecules will have diffused. Since σ represents one standard deviation of the concentration–distance profile, 68.3% of the population will have diffused less than σ, and 95.5% will have diffused less than 2σ.

The rate at which solute species is being transported to a particular place in solution, such as the electrode surface, is a useful parameter in electroanalytical chemistry. Consider the plane that is defined by the vertical dashed line in Fig. 27.9. The net number of moles of solute species crossing this imaginary plane per unit time is termed the *flux*, $J_{x_i t}$. The flux then is the rate of the transport of solute from one side of the plane to the other. The flux at a particular instant in time is determined by the slope of the concentration–distance profile at the position x of the plane. A steeper concentration–distance profile means a greater flux. A horizontal profile, that is, slope of zero, means no net transport of solute species and, hence, a flux of zero. This behavior is intuitively sensible since one would expect the transport of material from one point to another to occur at the greatest rate when the concentration difference between the two points is greatest. Consideration of Fig. 27.9f–j shows that the flux of solute species across the plane defined by the dashed line is initially zero (f), since no material has yet arrived at the plane to cross it. Note that the concentration–distance profile is horizontal and therefore has a slope of zero. As time progresses during the sequence g to j, the slope of the profile at the plane increases and then decreases until it is again zero for j. In the latter case, the solution has become homogeneous and there is now no net transport of material across the plane.

The relationship between flux and the slope of the concentration–distance profile of a solute is given by *Fick's first law*

$$J_{x_it} = -D\left(\frac{\partial C}{\partial x}\right)_{x_it} \qquad (27.50)$$

where J_{x_it} = flux (mol cm^{-2} s^{-1}), and
$(\partial C/\partial x)_{x_it}$ = concentration gradient of i at position x and time t.

Note that the flux is directly proportional to D, which is a measure of the speed with which a molecule diffuses in solution. Since a real electrochemical cell is three-dimensional, flux is defined in terms of an imaginary plane with a cross-sectional area of 1 cm^2. Thus, the units of flux are moles per square centimeter per second.

Migration. *Migration* is the third type of mass transport that can occur in an electrochemical cell. Migration is the movement of charged particles under the influence of an electric field. For example, a positively charged electrode attracts a negatively charged solute species, but repels a positively charged solute species. In some electroanalytical techniques, the effect of migration on the solute analyte is eliminated by adding a large excess of inert electrolyte such as potassium chloride. This supporting electrolyte increases the dielectric constant of the medium, which decreases the field strength near the electrode surface. In effect the electrostatic interaction is spread over a larger number of ions, thereby diluting its effect on any individual analyte ion to an insignificant level. The inert electrolyte is termed the *supporting electrolyte.*

27.10 ELECTRODE–SOLUTION INTERPHASE

An important aspect of electroanalytical chemistry that distinguishes it from many other chemical techniques is the existence of a phase boundary between electrode and solution at which the critical phenomenon of electron transfer occurs. Electrode processes occur at a surface. Considerable effort has been expended over the past 100 years to understand the nature of the electrode–solution interphase. One model of the electrode–solution interphase that has evolved is shown in Fig. 27.10a. In this example a potential has been applied to the electrode so that it is positively charged. The electrode surface is covered with a sheath of water-solvent molecules oriented so that their negative dipole contacts the positively charged electrode. Some large, poorly solvated anions in solution are also in direct contact with the electrode surface. Such anions are said to be *specifically adsorbed* on the electrode. The plane that passes through the center of these anions is termed the *inner Helmholtz plane* (IHP). The next layer of molecules consists of cations that are strongly solvated by water molecules. The plane passing through the center of this layer of solvated cations is termed the *outer Helmholtz plane* (OHP). The region beyond

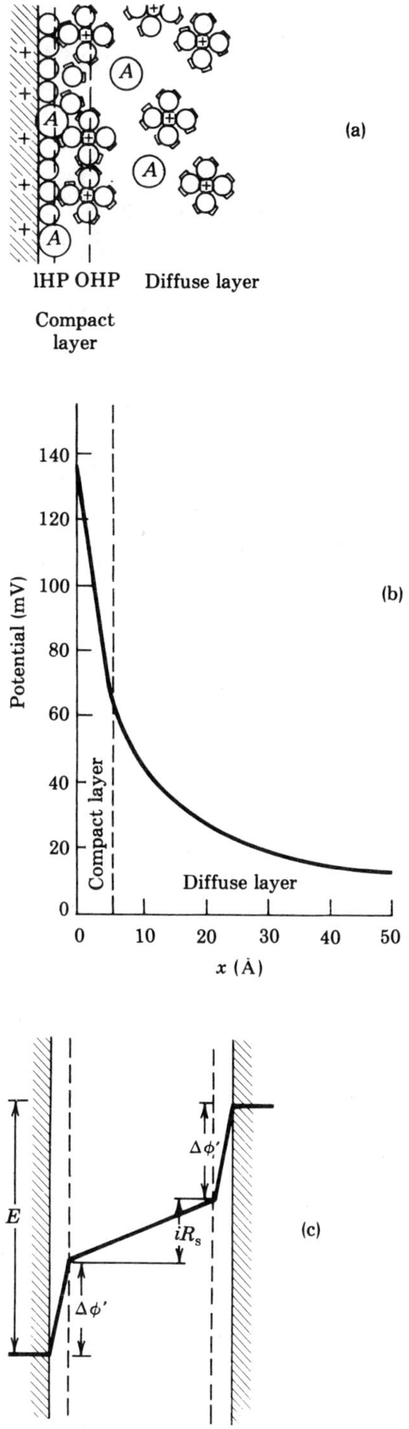

Fig. 27.10 (a) Model of the electrode-solution interphase showing anions (A), cations (+) and water molecules as described by Bockris, Devanathan, and Muller (see J. O'M. Bockris and A. K. N. Reddy, *Modern Electrochemistry*, Vol. 2, Chap. 7, Plenum, New York, 1970). (b) Potential gradient at interphase. (c) Potential gradient across electrochemical cell. E = potential applied between two electrodes, $\Delta\phi'$ = potential gradient across compact layer, iR_s = potential gradient across bulk solution between electrodes. Dimensions not to scale.

the OHP, termed the *diffuse layer*, contains a Boltzmann distribution of anions and cations in which the potential at the OHP determines whether an excess of cations or anions is present. In this region thermal agitation tends to counteract the attractive–repulsive forces between the OHP and anions and cations in solution. The IHP is populated mainly by anions that are weakly solvated and therefore can loose solvent molecules to become absorbed on the electrode. By comparison, most inorganic cations tend to be too strongly solvated to adsorb on the electrode and therefore the OHP defines their closest approach to the electrode. Organic molecules with low solubility can adsorb directly on the electrode.

The potential of an electrochemical half-cell causes a potential field to exist at the interphase. As shown in Fig. 27.10b the field drops rapidly in the region very close to the electrode because of the ionic composition of the electrolyte. Electroactive species that approach the compact layer are subject to a large potential field, which causes electron transfer to occur.

A schematic representation of the potential gradients in a two-electrode cell that is undergoing electrolysis is shown in Fig. 27.10c. As can be seen, significant potential drops occur across the double layer regions of the two electrodes. The balance of the potential gradient appears linearly across the cell in the form of *iR* drop (see Section 27.7).

27.11 ELECTROANALYTICAL TECHNIQUES

A large variety of electroanalytical techniques has been developed. Commonly used techniques that are described in this book are categorized in Fig. 27.11.

Electroanalytical techniques are divided into two main categories: *electrodics* and *ionics*. Electrodics consists of those techniques that involve the measurement of interfacial properties. Electrodics is divided into *static* and *dynamic* techniques. *Static techniques* are those in which the potential of an electrochemical cell is measured at equilibrium. In this case no electrolysis is occurring and hence the net current is zero. This field of electroanalytical chemistry is referred to as *potentiometry* and is described in Chapter 28. In potentiometry, the potential of the cell is used to determine the activity or concentration of the analyte.

Techniques in which electrolysis is occurring in the electrochemical cell are termed *dynamic techniques*. The most commonly used dynamic techniques are those in which a potential is applied to the electrochemical cell and the resulting current or charge is used to quantify the analyte. These controlled potential techniques can be conveniently categorized as forced convection or hydrodynamic techniques, which are described in Chapter 29, or quiescent solution techniques, which are described in Chapter 30.

Coulometry involves the quantification of an analyte by the measurement of the charge required for its complete electrolysis. This is accomplished by either controlled potential coulometry or a coulmetric titration in which a constant current is applied to the electrochemical cell for a precisely measured time interval. Both techniques are covered in Chapter 31.

Ionics involves the measurement of certain physical properties of solutions such

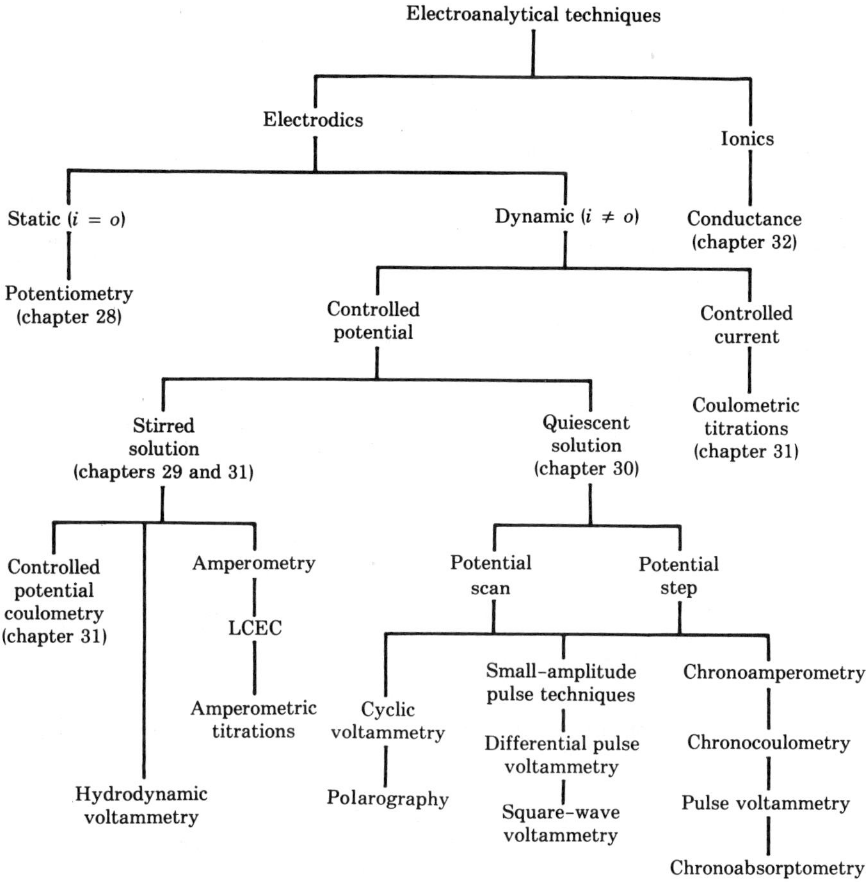

Fig. 27.11 Family tree of commonly used electroanalytical techniques.

as *conductivity*. The electroanalytical application of conductivity is described in Chapter 32.

REFERENCES

References providing a general coverage of electrode processes are:

1. R. N. Adams, *Electrochemistry at Solid Electrodes*. New York: Dekker, 1969.
2. W. J. Albery, *Electrode Kinetics*. Oxford: Clarendon Press, 1975.
3. A. J. Bard and L. R. Faulkner, *Electrochemical Methods*. New York: Wiley, 1980.
4. H. H. Bauer, *Electrodics*. New York: Wiley, 1972.
5. J. O'M. Bockris and A. K. N. Reddy, *Modern Electrochemistry*, Vols. I and II. New York: Plenum, 1970.
6. E. A. M. F. Dahmen, *Electroanalysis*. New York: Elsevier, 1986.
7. L. R. Faulkner, in *Physical Methods in Modern Chemical Analysis*, Vol. 3, T. Kuwana (Ed.). New York: Academic Press, 1983, pp. 137–248.

8. Z. Galus, *Fundamentals of Electrochemical Analysis*. London: Horwood, Halsted Press, 1976.
9. P. T. Kissinger and W. R. Heineman (Eds.), *Laboratory Techniques in Electroanalytical Chemistry*. New York: Dekker, 1984.
10. I. M. Kolthoff and P. J. Elving (Eds.), *Treatise on Analytical Chemistry*, Part I, Vol. 4, Section D-2. New York: Interscience, 1963.
11. J. J. Lingane, *Electroanalytical Chemistry*, 2nd ed. New York: Interscience, 1978.
12. D. D. Macdonald, *Transient Techniques in Electrochemistry*. New York: Plenum, 1977.
13. J. A. Plambeck, *Electroanalytical Chemistry*. New York: Wiley, 1982.
14. P. H. Rieger, *Electrochemistry*. Englewood Cliffs, NJ: Prentice-Hall, 1987.
15. D. T. Sawyer and J. L. Roberts, Jr., *Experimental Electrochemistry for Chemists*. New York: Wiley-Interscience, 1974.
16. Southampton Electrochemistry Group, *Instrumental Methods is Electrochemistry*. New York: Halsted Press (Wiley), 1985.
17. H. R. Thirsk and J. A. Harrison, *A Guide to the Study of Electrode Kinetics*. New York: Academic Press, 1972.
18. B. H. Vassos and G. W. Ewing, *Electroanalytical Chemistry*. New York: Wiley-Interscience, 1983.
19. A. Weissberger and B. W. Rossiter (Eds.), *Techniques of Chemistry*. Vol. I, Part II. New York: Wiley-Interscience, 1971.

Treatments of the practical aspects of performing electroanalytical experiments are found in references 9 and 15 above. A more advanced treatment of double layer and electrode kinetics may be found in:

20. P. Delahay, *Double Layer and Electrode Kinetics*. New York: Interscience, 1965.

The physical chemistry of electrolyte solutions is discussed on an advanced level in:

21. H. S. Harned and B. B. Owen, *The Physcial Chemistry of Electrolytic Solutions*, 3rd ed. New York: Reinhold, 1958.
22. R. A. Robinson and R. H. Stokes, *Electrolyte Solutions*, 2nd ed. New York: Academic, 1959.

Comprehensive discussions of advances in selected areas of electroanalytical chemistry appear approximately annually in the following continuing series:

23. *Advances in Analytical Chemistry and Instrumentation*, Vol. I. New York: Interscience, 1960.
24. *Advances in Electrochemistry and Electrochemical Engineering*, Vol. 1. New York: Interscience, 1961.
25. *Electroanalytical Chemistry. A Series of Advances*, Vol. 1. New York: Dekker, 1966.

Redox couples and standard potentials for the elements may be found in:

26. A. J. Bard, R. Parsons, and J. Jordan (Eds.), *Standard Potentials in Aqueous Solution*. New York: Dekker, 1985.

EXERCISES

27.1 Calculate the reduction potentials of the following half-cells and indicate whether they will behave as anodes or cathodes when coupled with an SCE as a galvanic cell. (a) Ag/Ag^+ ($a = 0.0150\ M$). (b) $Pt/Cr_2O_7^{2-}$ ($a = 10^{-4}\ M$), Cr^{3+} ($a = 10^{-3}\ M$), H^+ ($a = 10^{-2}\ M$). (c) Ag/AgSCN (sat'd), SCN^- ($a = 10^{-2}\ M$); (AgSCN, $K_{sp} = 1.0 \times 10^{-12}$).

27.2 Calculate the potentials of the following cells and indicate which electrode would be the anode in a galvanic cell. (a) Cd/Cd^{2+} ($a = 10^{-3}\ M$)//SCE; (b) Pb/Pb^{2+} ($a = 0.01\ M$)//Cd^{2+} ($a = 0.001\ M$)/Cd.

27.3 An excess in silver metal is added to a solution of $Cr_2O_7^{2-}$. (a) Write a balanced equation for the reaction. (b) Calculate the equilibrium constant for the reaction.

27.4 Compute E^0 for the half-cell reaction

$$Cd\,(NH_3)_4^{2+} + 2e \rightleftharpoons Cd + 4NH_3$$

where the formation constant of the complex is 1.3×10^7.

27.5 Calculate the following equilibrium constants: (a) K_{sp} for TlCl. (b) The formation constant of the complex $Ag(CN)_2^-$. (c) K_{eq} for the reaction $U^{4+} + 2\ Ce^{4+} + 2H_2O \rightleftharpoons UO_2^{2+} + 2\ Ce^{3+} + 4H^+$.

27.6 What is the dissociation constant of the acid HX if the following cell develops a potential of +0.832 V?

$$Pt/H_2\ (1\ atm),\ HX\ (0.02\ M),\ NaX\ (0.06\ M)//SCE$$

27.7 The potential of the following cell is 0.277 V: Cu/CuY^{2-} ($1.00 \times 10^{-4}\ M$), Y^{4-} ($1.00 \times 10^{-2}\ M$)//SHE. Calculate K_f of the EDTA (Y^{4-}) complex of Cu(II).

27.8 Calculate the potential of the cell

$$Hg/HgY^{2-}\ (2.00 \times 10^{-4}\ M),\ CaY^{2-}\ (5.00 \times 10^{-4}\ M),\ Ca^{2+}\ (10^{-1}\ M)//SCE$$

The solution is sufficiently alkaline that essentially all EDTA is complexed. The formation constants of the complexes are $K_{HgY^{2-}} = 6.3 \times 10^{21}$ and $K_{CaY^{2-}} = 5.0 \times 10^{10}$.

27.9 The following cell was found to have a potential of 1.034 V:

$$Cd/CdX_2\ (sat'd),\ X^-\ (0.0100\ M)//SCE$$

Calculate the solubility product of CdX_2.

27.10 Zinc ion is known to form a relatively stable complex with the ligand Y^{3-}: $Zn^2 + Y^{3-} \rightleftharpoons ZnY^-$. In order to measure the formation constant of the complex, the following cell was constructed: Zn/ZnY^- ($1 \times 10^{-4}\ M$), Y^{-3} ($0.05\ M$)//SCE. The potential of the cell was found to be 1.597 V with zinc being the anode. Calculate the formation constant K_f of the complex.

27.11 Explain why HCl is a poor choice of electrolyte for a salt bridge.

ANSWERS

27.1(c) $E = 0.208$ V, anode *27.3(b)* $K_{eq} = 8 \times 10^{53}$ at 25°C *27.6* $K_a = 3.36 \times 10^{-10}$ at 25°C *27.9* $K_{sp} = 8.0 \times 10^{-18}$ at 25°C

Chapter 28

POTENTIOMETRIC METHODS

William R. Heineman and Howard A. Strobel

28.1 INTRODUCTION

Measurements of the difference in potential between the two electrodes of a galvanic cell under the condition of zero current are described by the term *potentiometry*. Since no current passes through the cell while the potential is measured, potentiometry is an equilibrium method. Potentiometric techniques are important because they can provide accurate measurements of (a) either activities, concentrations, and/or activity coefficients of many solution species and (b) free-energy changes and equilibrium constants of many solution reactions.

Typical apparatus for potentiometry is shown in Fig. 28.1. The potential difference between the two electrodes immersed in solution is usually measured with pH/mV meter. One electrode, the *indicator electrode*, is chosen so that its half-cell potential responds to the activity of a particular species in solution whose activity or concentration is to be measured. The other electrode is a *reference electrode* whose half-cell potential is invariant (Section 27.4). The most commonly used reference electrodes for potentiometry are the saturated calomel electrode (SCE) and silver–silver chloride electrode (Ag/AgCl). The potential of the electrochemical cell, E_{cell}, is given by

$$E_{cell} = E_{ind} - E_{ref} + E_{lj}, \tag{28.1}$$

where E_{ind} is the half-cell potential of the indicator electrode, which is considered to be the right-cell or cathode in this example (See Eq. (27.27)), E_{ref} is the half-cell potential of the reference electrode, and E_{lj} is the liquid-junction potential. Liquid-junction potentials develop at the interface between two electrolytes (see Section 27.5). In potentiometric cells, E_{lj} is typically found at the junction of the solution in the cell with the reference electrode or a salt bridge.

The indicator electrode is of paramount importance in analytical potentiometry. This electrode should interact with the species of interest so that E_{ind} reflects the activity of this species in solution and not other compounds in the sample that might constitute an interference. The importance of having indicator electrodes that selectively respond to numerous species of analytical significance has stimu-

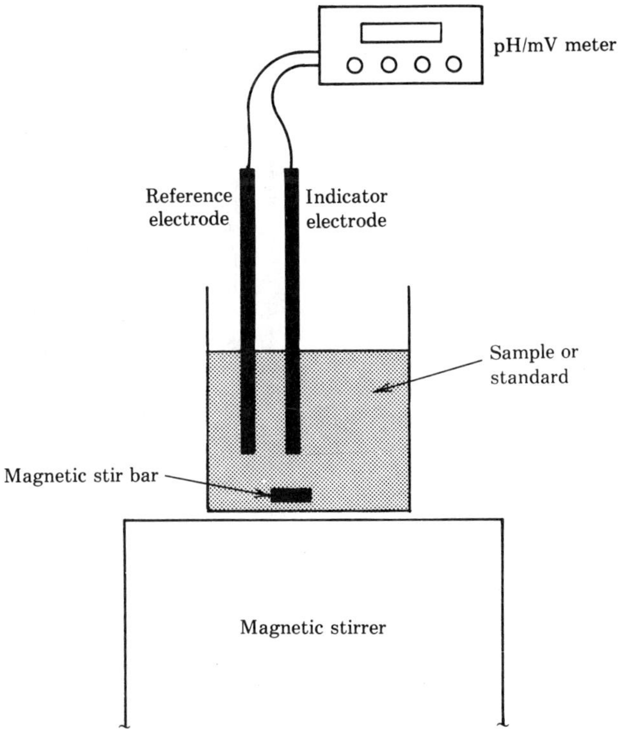

Fig. 28.1 Schematic diagram of apparatus for potentiometry.

lated the development of many types of indicator electrodes, as described in the following sections.

28.2 METAL INDICATOR ELECTRODES

Metal indicator electrodes fall into four categories, as shown in Table 28.1. A representative electrode reaction mechanism and Nernst equation is shown for each type.

Redox Electrodes. A *redox electrode* responds to the redox potential of a solution as established by one or more redox couples in the solution. An inert metal, such as platinum, responds to many redox couples. For example, a platinum electrode immersed in a solution containing Fe^{2+} and Fe^{3+} gives a potential that is, dependent on the ratio of activities of these two species, as shown by the electrode reaction mechanism and Nernst equation in Table 28.1. Electrodes of this type are commonly used for redox titrations (Section 28.15).

Electrodes of the First Kind. An *electrode of the first kind* consists of a metal in contact with a solution containing its ion. An example is a silver wire, which

Table 28.1 Metal Indicator Electrodes

Type	*Representative Electrode Reaction Mechanism*	*Nernst Equation*
Redox	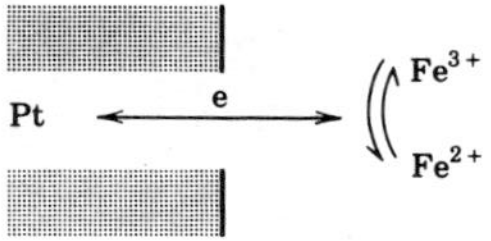	$E_{ind} = E^0_{Fe^{3+},Fe^{2+}} - \frac{RT}{F} \ln \frac{a_{Fe^{2+}}}{a_{Fe^{3+}}}$
First kind	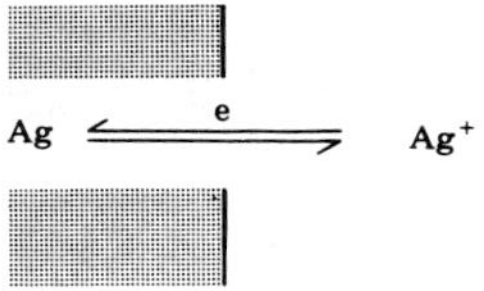	$E_{ind} = E^0_{Ag^+Ag} - \frac{RT}{F} \ln \frac{1}{a_{Ag^+}}$
Second kind	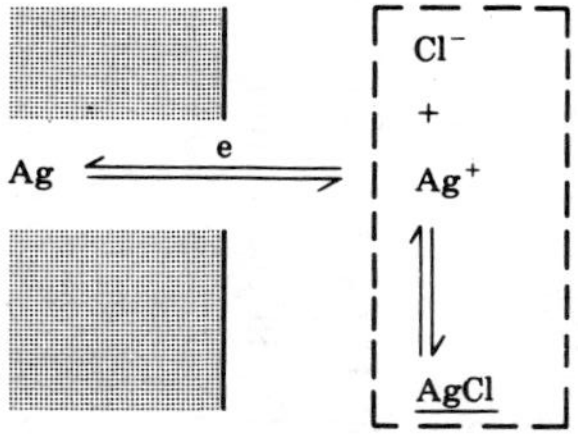	Substitute $K_{sp} = a_{Ag^+}a_{Cl^-}$ into above equation: $E_{ind} = E^0_{Ag^+,Ag} - \frac{RT}{F} \ln \frac{a_{Cl^-}}{K_{sp}}$ $= E^0_{Ag,AgCl} - \frac{RT}{F} \ln a_{Cl^-}$
Third kind	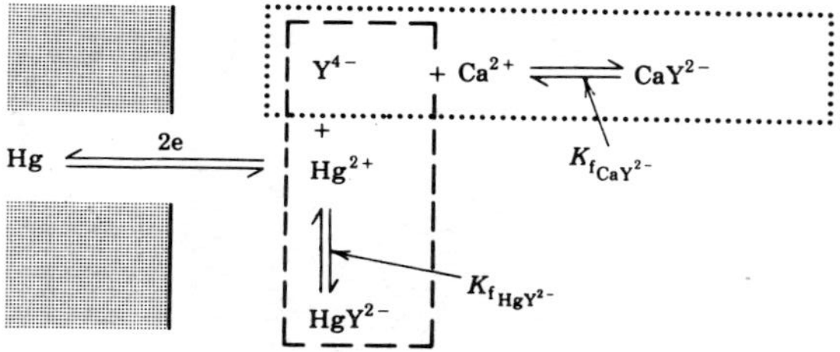	$E_{ind} = E^0_{HgY^{2-},Hg} + \frac{RT}{2F} \ln \frac{a_{HgY^{2-}}K_{f_{CaY^{2-}}}}{a_{CaY^{2-}}} + \frac{RT}{2F} \ln a_{Ca^{2+}}$

responds to a_{Ag^+}, as shown by the electrode reaction mechanism and Nernst equation shown in Table 28.1.

Example 28.1. The activity of Ag^+ in a sample was determined by measuring the potential of an Ag electrode (cathode) versus SCE at 25°C. E_{cell} was 0.324 V. Calculate a_{Ag^+}, assuming $E_{lj} = 5$ mV.

Subsitution of the Nernst equation in Table 28.1 for the silver electrode of the first kind into Eq. 28.1 gives

$$E_{cell} = E^0_{Ag^+,Ag} - \frac{RT}{F} \ln \frac{1}{a_{Ag^+}} - E_{SCE} + E_{lj};$$

$$0.324 = 0.796 - 0.0591 \log \frac{1}{a_{Ag^+}} - 0.244 + 0.005; \text{ and}$$

$$a_{Ag^+} = 1.14 \times 10^{-4} \text{ mol L}^{-1}.$$

Electrodes of this type are often not entirely selective for the oxidized form of the metal. A particular electrode is generally also sensitive to cations that are more easily reduced. Many metal–metal ion electrodes that might be suitable as selective electrodes cannot be used in acidic air-saturated aqueous solutions because the metal must be thermodynamically stable with respect to air oxidation, especially at low ion activities. Suitable electrodes are restricted to Ag/Ag^+ and Hg/Hg_2^{2+} in neutral solution. Removal of dissolved oxygen by aeration with nitrogen or helium makes the following electrodes feasible: Cu/Cu^{2+}, Bi/Bi^{3+}, Pb/Pb^{2+}, Cd/Cd^{2+}, Sn/Sn^{2+}, Tl/Tl^+, and Zn/Zn^{2+}.

Electrodes of the Second Kind. An *electrode of the second kind* consists of a metal in contact with a solution saturated with one of its sparingly soluble salts. Saturation can sometimes be accomplished by simply coating the salt on the metal itself. As shown for the Ag/AgCl system in Table 28.1, the Ag electrode still responds to a_{Ag^+} by means of electron transfer between the Ag^+, Ag redox couple. However, a_{Ag^+} is now controlled by a_{Cl^-} via the coupled precipitation reaction with Cl^-. Increasing a_{Cl^-} in solution consumes Ag^+ by precipitation as AgCl causing a decrease in a_{Ag^+} by the reaction enclosed in the dashed-line box. Consequently, the electrode potential responds to the activity of Cl^- even though Cl^- undergoes no electron transfer with the Ag wire. The Nernst equation for this half-cell can be derived by solving the solubility product expression for a_{Ag^+} and substituting it into the Nernst equation for the electrode of the first kind, as shown in Table 28.1. Combining the constant terms $E^0_{Ag^+,Ag}$ and K_{sp} gives the relationship between E_{ind} and a_{Cl^-} to be the last equation shown for this electrode of the second kind, which is the Nernst equation for the electrode reaction

$$AgCl + e \rightleftharpoons Ag + Cl^-. \qquad (28.2)$$

Electrodes of the second kind work for the determination of halides and other anions of slightly soluble salts of silver and other metal cations. The electrodes can be used only over a range of anion activities such that the solution remains saturated with the salt. Other anions that precipitate with the cation are interferents. The commonly used SCE and Ag/AgCl reference electrodes are electrodes of the second kind in which the halide activity is maintained constant so that the half-cell potential serves as an invariant reference potential. Electrodes of the second kind can also be established with ligands that form reversible complexes with the metal ion of an M^{n+}/M redox couple.

Example 28.2. According to the equations in Table 28.1 for the electrode of the second kind, $E^0_{Ag,AgCl}$ must be equal to $E^0_{Ag^+,Ag} + (RT/F) \ln K_{sp}$. From Table 27.2 we find that $E^0_{Ag^+,Ag} = 0.799$ and $K_{sp,AgCl} = 1.82 \times 10^{-10}$ at 25°C. Thus, $E^0_{Ag^+,Ag} + 0.0591 \log K_{sp} = 0.799 + (-0.576) = 0.223$ V. Compare this with $E^0_{Ag,AgCl} = 0.222$ V from Table 27.2.

Electrodes of the Third Kind. An *electrode of the third kind* involves two solution equilibria that are coupled to a M^{n+}/M redox couple. For example, a third-order electrode that responds to Ca^{2+} can be established with EDTA as a common ligand between Hg^{2+} and Ca^{2+}. The overall electrode reaction mechanism is shown in Table 28.1 where Y^{4-} represents EDTA. (It is assumed in this discussion that ligand and complexes are fully deprotonated and that hydroxy-metal complexes are not formed.) The heterogeneous electron exchange reaction involves the Hg^{2+}/Hg couple

$$Hg^{2+} + 2e \rightleftharpoons Hg, \tag{28.3}$$

for which the Nernst equation is

$$E_{ind} = E_{Hg^{2+},Hg} - \frac{RT}{2F} \ln \frac{1}{a_{Hg^{2+}}}. \tag{28.4}$$

The behavior of this electrode can be understood by first realizing that the dashed box encloses the coupled chemical reaction that makes the Hg^{2+}/Hg redox couple an electrode of the second kind for EDTA. Solving the formation constant equilibrium expression

$$K_{f_{HgY^{2-}}} = \frac{a_{HgY^{2-}}}{a_{Hg^{2+}} a_{Y^{4-}}} \tag{28.5}$$

for $a_{Hg^{2+}}$ and substituting into Eq. (28.4) gives

$$E_{ind} = E^0_{HgY^{2-},Hg} + \frac{RT}{2F} \ln a_{HgY^{2-}} - \frac{RT}{2F} \ln a_{Y^{4-}}, \tag{28.6}$$

where $E^0_{HgY^{2+},Hg} = E^0_{Hg^{2+},Hg} - (RT/2F) \ln K_{f_{HgY^{2-}}}$. The chemical reaction enclosed in the dotted box couples the activity of Y^{4+} to that of Ca^{2+}, thereby making the electrode an electrode of the third kind for Ca^{2+}. Solving the formation constant equilibrium expression

$$K_{f_{CaY^{2-}}} = \frac{a_{CaY^{2-}}}{a_{Ca^{2+}} a_{Y^{4-}}} \tag{28.7}$$

for $a_{Y^{4-}}$ and substituting into Eq. (28.6) gives

$$E_{ind} = E^0_{HgY^{2-},Hg} + \frac{RT}{2F} \ln \frac{a_{HgY^{2-}} K_{f_{CaY^{2-}}}}{a_{CaY^{2-}}} + \frac{RT}{2F} \ln a_{Ca^{2+}}. \tag{28.8}$$

When solution conditions are such that the ratio $a_{HgY^{2-}}/a_{CaY^{2-}}$ is essentially constant, Eq. (28.8) can be approximated by

$$E_{ind} \cong K + \frac{RT}{2F} \ln a_{Ca^{2+}} \tag{28.9}$$

where K represents the constant terms in Eq. (28.8). The species HgY^{2-} must be considerably more stable than CaY^{2-} and an excess of Ca^{2+} must exist for this electrode to function in this manner. (What would happen if CaY^{2-} were more stable than HgY^{2-}?) The selectivity for Ca^{2+} is governed by the selectivity of EDTA complexation for Ca^{2+} relative to other metal ions in solution. Since EDTA forms stable complexes with a wide spectrum of metal ions, its selectivity for Ca^{2+} is poor. Consequently, the applicability of this electrode is limited to the measurement of relatively pure solutions of Ca^{2+}.

Historically, indicator electrodes have been metals that respond to activity by means of half-cell redox reactions as described above. Whereas simple indicator electrodes of this type perform well for the analysis of relatively pure samples, such as in a potentiometric titration, they are often subject to interferences when applied to complex samples such as those of biological, environmental, and industrial origin.

28.3 ION-SELECTIVE ELECTRODES

A significant development in the methodology of potentiometry that paved the way for its use in complex samples is the *ion selective electrode* (ISE). Conceptually, the ISE involves the measurement of a membrane potential. The response of the electrochemical cell is therefore based on an interaction between the membrane and the analyte that alters the potential across the membrane. The selectivity of the potential response to the analyte depends on the specificity of the membrane interaction for the analyte.

A representative ISE is shown schematically in Fig. 28.2. The electrode consists of a membrane, an internal reference electrolyte (or internal solution) of fixed activity, $(a_i)_{internal}$, and an internal reference electrode. The ISE is immersed in sample solution that contains analyte of some activity, $(a_i)_{sample}$. An external reference electrode is also immersed in this solution. The internal and external reference electrodes constitute the two half-cells of the electrochemical cell. The potential measured by the pH/mV meter (E_{cell}) is equal to the difference in potential between the external ($E_{ref,ext}$) and internal ($E_{ref,int}$) reference electrodes, plus the membrane potential (E_{memb}), plus the liquid junction potential (E_{lj}) that exists at the junction between the external reference electrode and the sample solution

$$E_{cell} = E_{ref,ext} - E_{ref,int} + E_{memb} + E_{lj}. \tag{28.10}$$

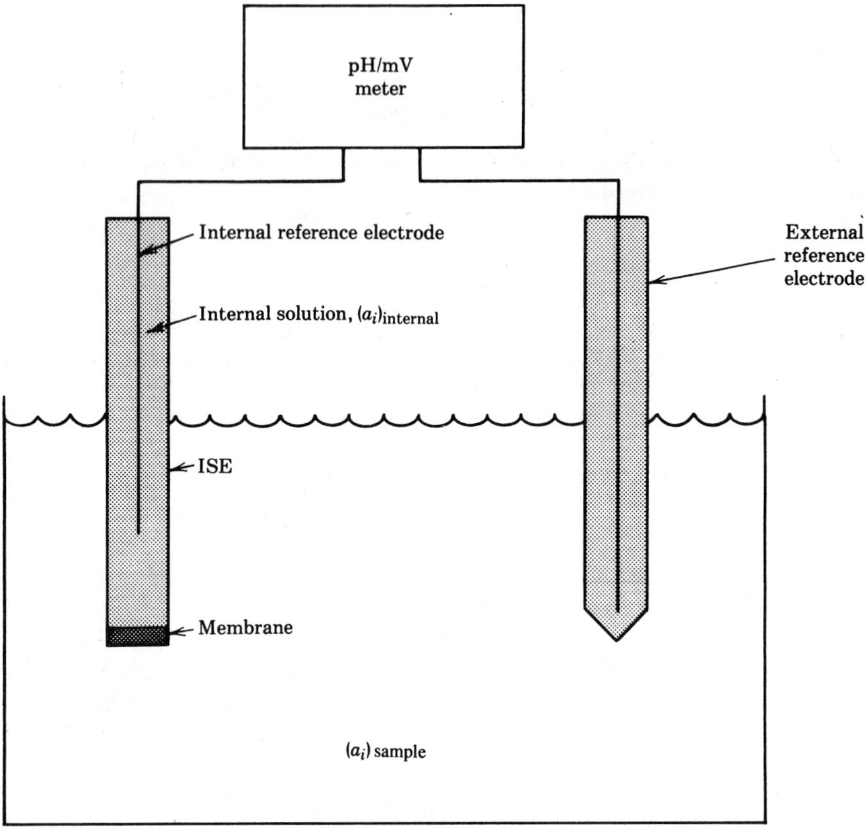

Fig. 28.2 Schematic diagram of ion-selective electrode (ISE).

If the membrane is permselective for a particular ion (i) a potential develops across the membrane which depends on the ratio of activities of the ion on either side of the membrane, as described by the Nernst equation

$$E_{\text{memb}} = \frac{RT}{zF} \ln \frac{(a_{\text{i}})_{\text{sample}}}{(a_{\text{i}})_{\text{internal}}}, \tag{28.11}$$

where z is the charge on the ion (cations: +1, +2, +3, etc.; anions: −1, −2, −3, etc.). Substitution of Eq. (28.11) into Eq. (28.10) gives an overall expression for the electrochemical cell that can be written as follows:

$$E_{\text{cell}} = E_{\text{ref, ext}} - E_{\text{ref, int}} + \frac{RT}{zF} \ln \frac{1}{(a_{\text{i}})_{\text{internal}}} + E_{\text{lj}} + \frac{RT}{zF} \ln (a_{\text{i}})_{\text{sample}}. \tag{28.12}$$

The half-cell potentials of the two reference electrodes are constant, sample solution conditions can often be controlled so that E_{lj} is effectively constant, and the composition of the internal solution can be maintained so that $(a_i)_{internal}$ is fixed. Consequently Eq. (28.12) can be simplified to give

$$E_{cell} = K + \frac{RT}{zF} \ln (a_i)_{sample}, \tag{28.13}$$

where K represents the constant terms in Eq. (28.12). This logarithmic relationship between cell potential and analyte activity is the basis of the ISE as an analytical device. A plot of E_{cell} versus log a_i for a series of standard solutions should be linear over the working range of the electrode and have a slope of 2.303 RT/zF or $0.0591/z$ for measurements made at 25°C.

Since membranes respond to a certain degree to ions other than the analyte (i.e., interferents), a more general expression than Eq. (28.13) is

$$E_{cell} = K + \frac{RT}{zF} \ln (a_i + k_{ij} a_j^{z/x}), \tag{28.14}$$

where a_j = activity of interferent ion, j,
x = charge of interferent ion,
k_{ij} = selectivity constant.

Small values of k_{ij} are characteristic of electrodes with good selectivity for the analyte i. A value of unit means the electrode responds equally well to the interferent j; values greater than unity mean the electrode responds better to the interferent.

The development of successful ISEs has hinged on the search for membranes that exhibit both sensitivity and selectivity for the species of interest. While selectivity is by far the more difficult of the properties to achieve, ISEs with selectivity for numerous cations and anions have been developed using three basic types of membranes: liquid and polymer, solid state, and glass. These membranes function by selectively incorporating the analyte ion into the membrane and thereby establishing a membrane potential. The reaction mechanisms are based on complexation, crystallization, or ion exchange. An ISE membrane must exhibit low solubility in the analyte medium to provide a durable electrode with a stable response. This requirement imposes a severe restriction on the materials that can be used for membranes. Also, the membrane must exhibit some electrical conductivity in order to function in an electrochemical cell. Since measurements are made under conditions of essentially zero current, the membrane conductance can be very low; resistances of over 100 megohms are possible. The scope of ISEs has been expanded to include the measurement of gases and neutral organic compounds by combining ISEs with gas-permeable membranes and layers of enzymes, bacteria, and tissues. These general categories of electrodes are considered in the following sections.

Liquid and Polymer Membrane Electrodes. A selective liquid membrane is the basis for a number of excellent ISEs. The liquid consists of a water-insoluble, viscous solvent in which is dissolved a hydrophobic organic ion exchanger or a neutral carrier ionophore that reacts selectively with the ion of interest. The liquid is typically soaked into a thin porous solid membrane such as cellulose acetate, which is then incorporated into the ISE.

The schematic diagram of the membrane portion of a liquid membrane ISE that is in contact with internal and sample aqueous solutions of analyte M^+ and its counter anion X^- in Fig. 28.3 shows the mechanism whereby an electrode responds to M^+ activity. The neutral carrier ionophore (R) reacts with M^+ at each membrane–solution interface and extracts M^+ into the membrane as MR^+. The hydrophobicity of R constrains R and MR^+ to the membrane phase. The extraction of M^+ into the membrane generates a positive membrane potential at each interface due to the charge separation that occurs when (a) M^+ is extracted into the membrane in the form of MR^+ and (b) the counter anion X^- remains in the aqueous solution. When the solution activity of M^+ increases, the activity of MR^+ in the membrane increases, and the membrane potential increases. Thus, the potential response of the electrode is controlled by the interfacial reaction shown in Fig. 28.3. This reaction occurs at both the outer membrane surface, which is exposed to the sample, and the inner membrane surface, which contacts the inner solution of the ISE. The potential of the inner membrane surface, $E_{\text{memb(internal)}}$, is kept constant by maintaining a constant activity of M^+ in the internal solution so that the internal surface reaction is fixed at some equilibrium position. An internal solution of aqueous MCl serves this purpose as well as providing a fixed chloride activity to poise the potential of an Ag/AgCl internal reference electrode. Thus, the only potential change measured in the circuit is the potential of the membrane surface contacting the sample, $E_{\text{memb(sample)}}$.

The availability of liquid and polymer membrane electrodes for a variety of ions is due to the development of different neutral carrier ionophores and liquid ion exchangers that react selectively with particular ions. In the surface equilibria shown in Fig. 28.3, any ionic species other than M^+ that react to an appreciable

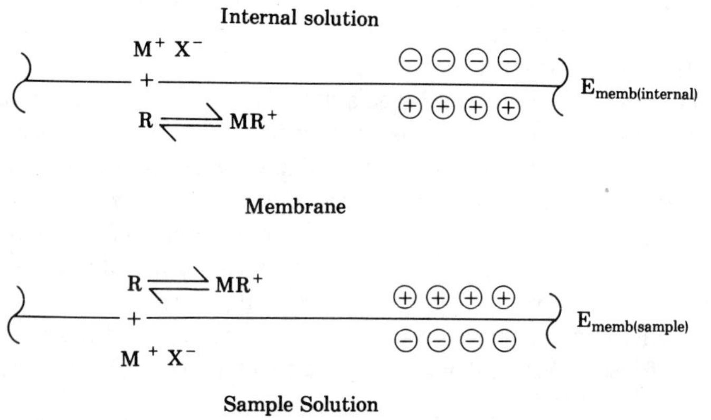

Fig. 28.3 Schematic diagram of response mechanism for liquid membrane ISE for M^+. Interfacial reaction between M^+ and R and resulting membrane potential.

degree with R will also generate a membrane potential and thereby cause an interference. The selectivity of the electrode for M^+ is therefore determined by the relative formation constants between R and M^+ and R and the various interferent ions in the sample.

Some of the ionophores and organic ion exchangers that have been used for ISEs are shown in Fig. 28.4. Neutral lipophilic *ionophores*, or ion-carrying agents, have been found to selectively transport cations across nonaqueous membranes. The ionophores provide coordinating atoms positioned such that the ionophore surrounds the cation. Long alkyl chains, benzene molecules, or cyclohexyl rings impart hydrophobicity to the ionophores. When it reacts with an ionophore, a cation is essentially inserted in a hydrophobic capsule that renders the cation compatible with a nonaqueous membrane medium. Selectivity for a particular cationic species is controlled by providing an optimum coordination environment in terms of number and position of coordinating atoms. The antibiotic valinomycin, which was used for the first electrode of this type, exhibits excellent selectivity for K^+. A model of the K^+ complex of valinomycin* shows the K^+ cation to be surrounded by coordinating oxygen atoms in a snug cavity. The alkyl groups in the complex are directed outward, which makes the complex hydrophobic. The electrode exhibits excellent selectivity for K^+ against Na^+ because the smaller Na^+ ion is bound less tightly in a cavity that is too large for it. This feature is of considerable practical importance in the clinical determination of K^+ in blood serum, which contains higher concentrations of Na^+ than K^+. Ionophores for NH_4^+, Ca^{2+}, Na^+, Li^+, and Mg^{2+} electrodes are shown in Fig. 28.4.

Hydrophobic ion exchangers that react with the analyte ion to form neutral complexes or ion pairs have been used successfully in ISE membranes. A selective electrode for Ca^{2+} is based on complexation between the phosphate group of di-(n-decyl)phosphate, R^-, and Ca^{2+} to form neutral CaR_2. The decyl groups provide hydrophobicity to maintain the negatively charged R^- in the membrane solvent, dioctylphenylphosphonate. Electrodes for Cu^{2+} and Pb^{2+} are based on reaction with ion exchangers of the form $^-OOC–CH_2–S–R$ in which the sulfur and carboxylate groups form five-membered chelate rings with the metal ion, $M(OOC–CH_2–S–R)_2$. Anions such as NO_3^-, ClO_4^-, and BF_4^- are detected on the basis of ion-pairing reactions with cationic nickel o-phenanthroline complexes. An electrode for Cl^- is based on ion pairing with a quaternary ammonium ion. A pH electrode is based on the protonation of tri-n-dodecylamine. Electrodes based on ion-pairing reactions have been developed for numerous organic compounds. New ion exchangers and neutral carriers are continually being evaluated in an effort to improve selectivity of existing electrodes and to develop electrodes for other ions [38, 39].

Example 28.3. How can a liquid membrane ISE be made to respond to an organic compound?

*D. Ammann, *Ion-Selective Microelectrodes*, Springer Verlag, New York, 1986, p. 15.

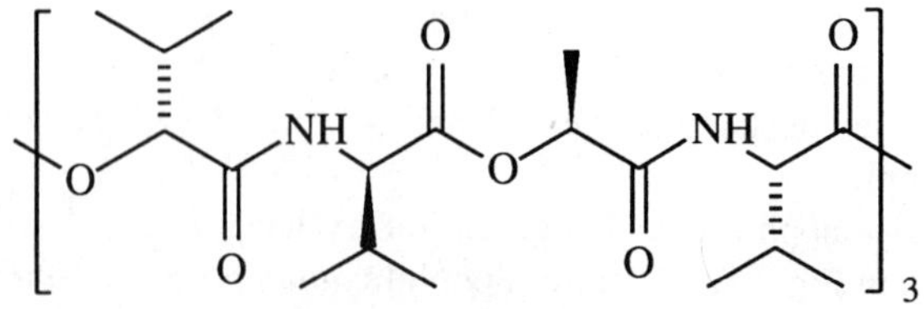

Valinomycin

K^+

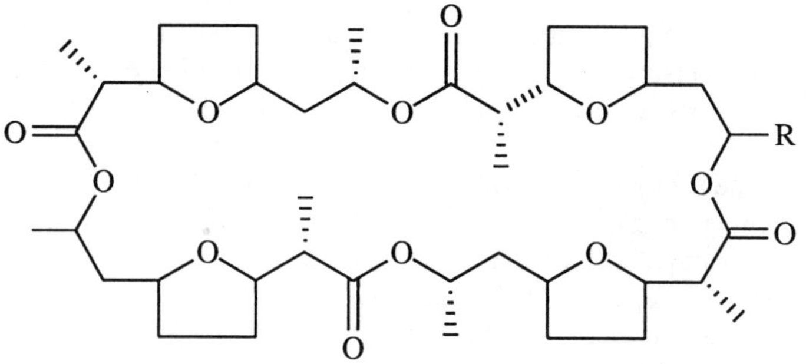

R = CH_3: nonactin

R = C_2H_5: monactin

NH_4^+

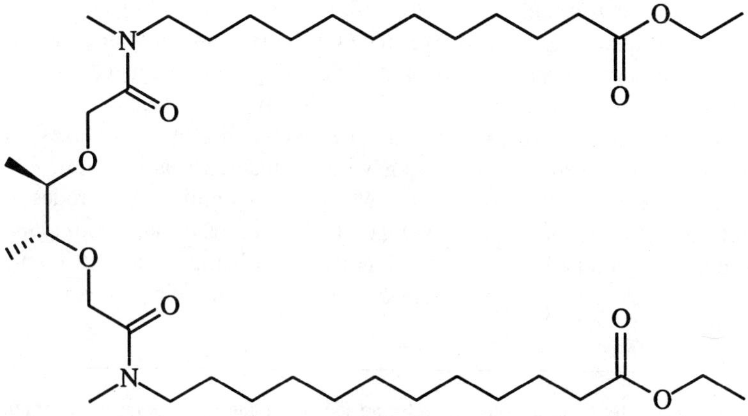

ETH 1001

Ca^{2+}

Fig. 28.4 Ionophores (20) and ion exchangers for liquid and polymer membrane ISEs.

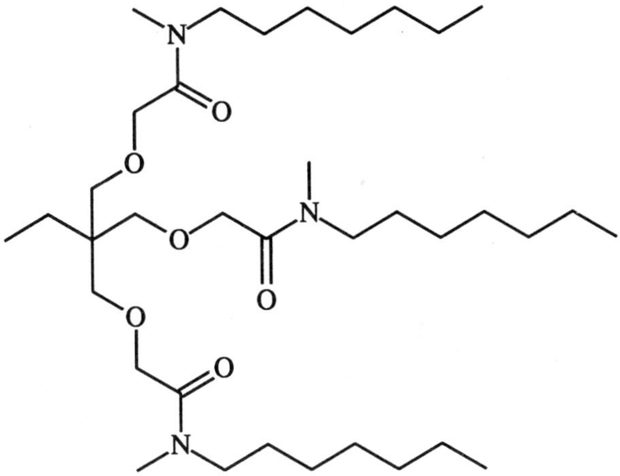

ETH 227
Na^+ ($Na^+ > K^+$)

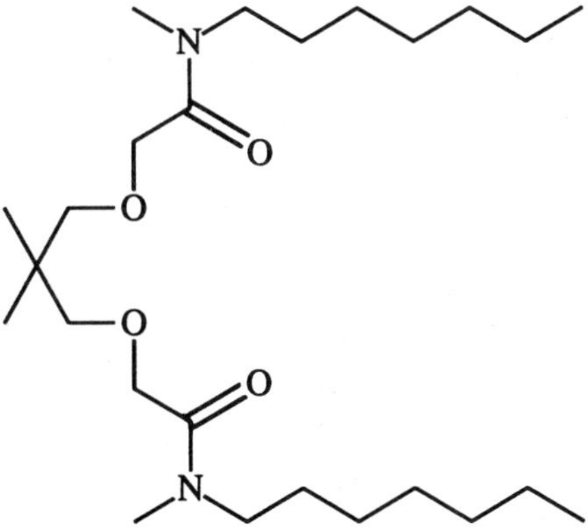

ETH 149
Li^+

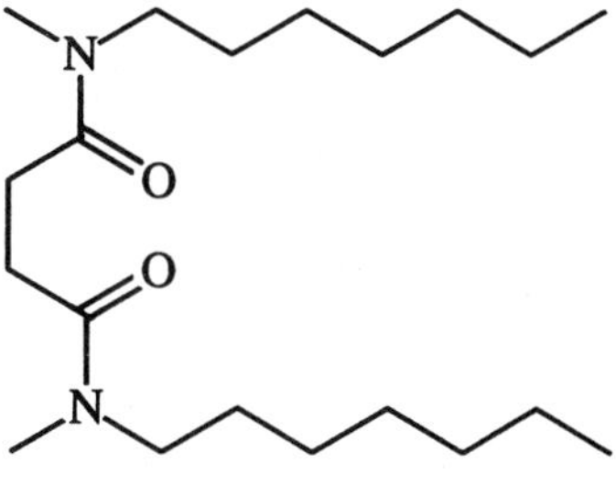

ETH 1117
Mg^{2+}

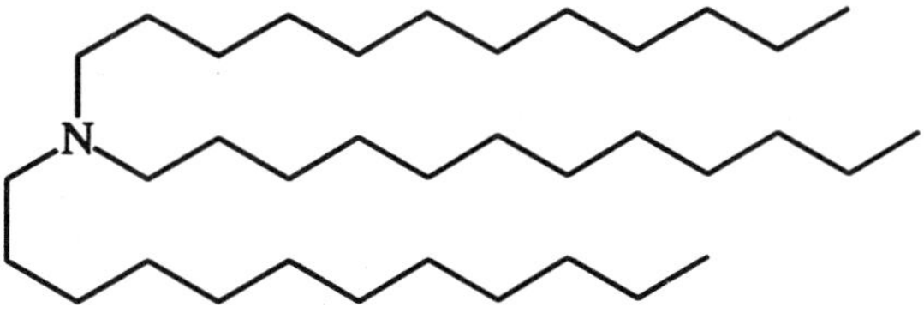

Tri-*n*-dodecylamine
H^+

Fig. 28.4 (*Continued*)

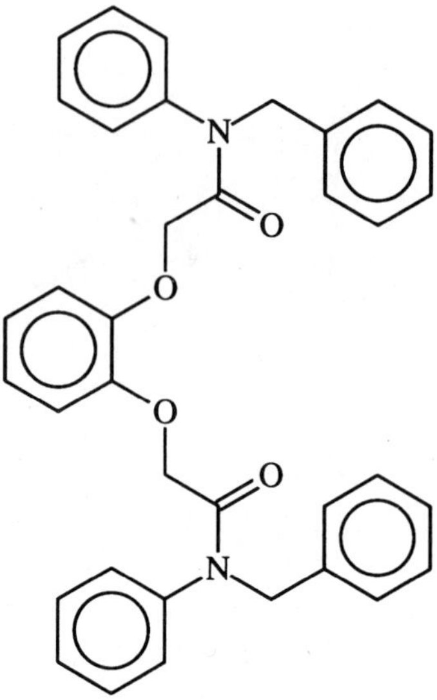

ETH 157

Na^+

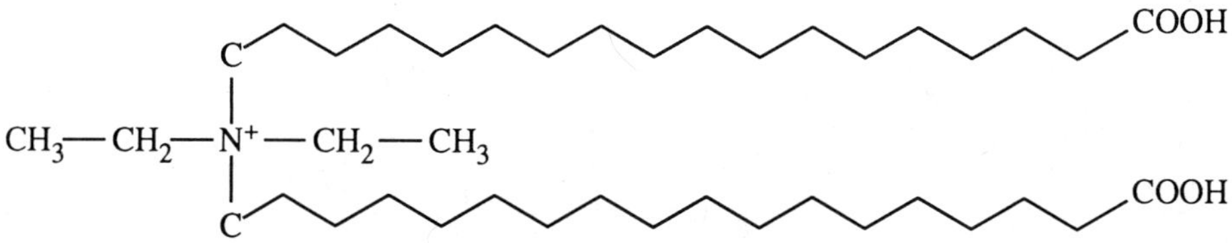

Diethyldistearyl ammonium ion

Cl^-

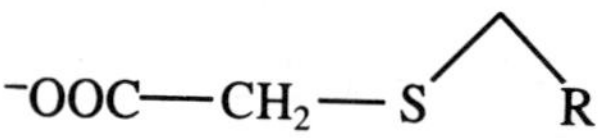

Cu^{2+}, Pb^{2+}

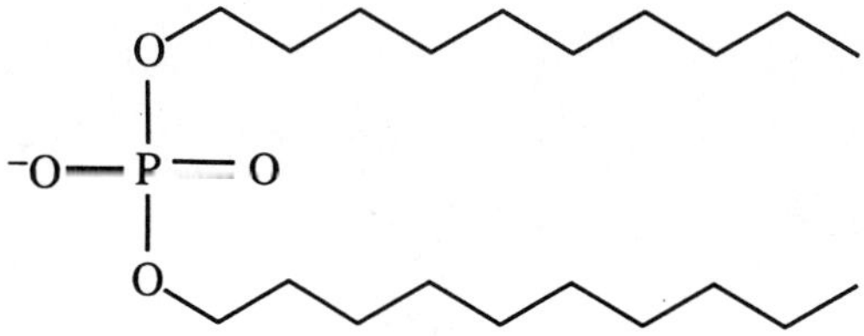

Di-(n-decyl)phosphate

Ca^{2+}

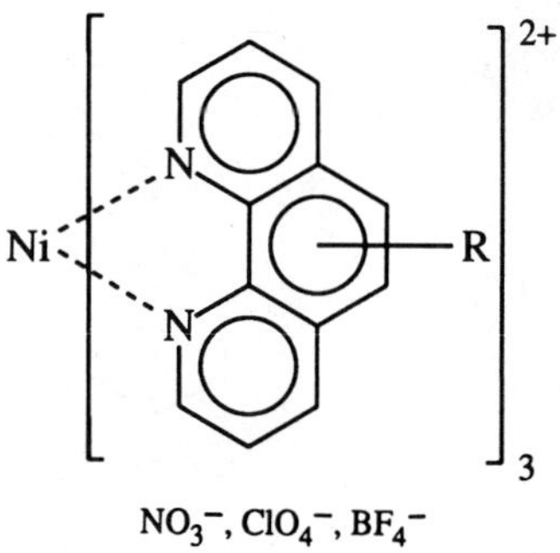

NO_3^-, ClO_4^-, BF_4^-

Fig. 28.4 (*Continued*)

Electrodes for organic compounds are based on the formation of ion pairs between an ionic form of the organic compound and an ion-pairing reagent. An example is an electrode for the antiepileptic drug phenytoin (5,5-diphenylhydantoin) based on the ion-pair complex between the 5,5-diphenylhydantoinate anion and the quarternary ammonium cation, tricaprylmethyl ammonium, which is immobilized in poly(vinyl chloride).

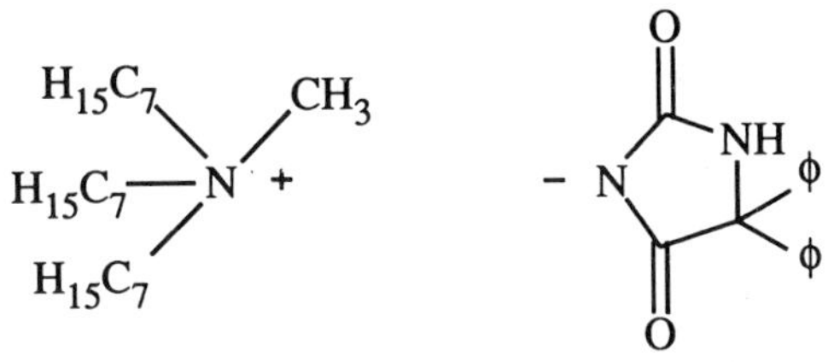

The electrode shows a near-Nernstian response over the range of 10^{-1} to 10^{-4} mol L^{-1} and has a detection limit of 1.5×10^{-5} mol L^{-1}. The electrode can be used to determine phenytoin in tablets and capsules. [See V. V. Cosofret and R. P. Buck, *J. Pharm. Biomed. Anal.*, **4**, 45 (1986).]

The membranes for ISEs of the liquid membrane type are now prepared predominantly by casting the carrier or ion exchanger in a plasticizer such as poly(vinyl chloride). Small, thin disks of this material are used as the membrane. These polymer-membrane ISEs exhibit response characteristics that are similar to the "wet" liquid membranes described above and are generally more durable.

The preparation of a polymer–membrane electrode can be simplified by coating polymer-containing ion exchanger or ionophore on a wire such as copper, platinum, or gold. This coated-wire design eliminates the internal reference solution in the conventional ISE and thereby enables smaller, more rugged electrodes to be more easily fabricated. Detection limits, selectivities, and operating ranges of coated-wire electrodes are generally comparable to their conventional ISE analogs. However, the lack of a defined reference potential at the membrane–metal wire interface results in drifting of the electrode potential, which necessitates frequent calibration of the electrode.

Liquid–polymer membrane ISEs have seen extensive use for the measurement of ions in samples of diverse origin. The electrodes have been especially successful in clinical laboratories and are now used routinely for Ca^{2+}, K^{+}, Na^{+}, and Cl^{-} in serum. The response characteristics for liquid–polymer membrane ISE systems commonly used in biomedical investigations are shown in Table 28.2.

Solid-State Membrane Electrodes. Solid-state membranes consist of single crystals or pressed pellets of salts of the ions of interest. The crystal or pellet must have some degree of electrical conductivity and exhibit very low solubility in the solvent in which the electrode is to be used—usually water. An excellent ISE for F^{-} uses LaF_3 that is doped with Eu^{2+} to provide electrical conductivity. The membrane potential is generated by a selective surface reaction between LaF_3 and F^{-}

Table 28.2 Common ISE Systems Used in Biomedical Investigations[a]

	Analyte	*Membrane composition*	*Linear response range*	*Selectivity constants*
Glass	H^+ (Corning 015)	72.17% SiO_2, 6.44% CaO 21.39% Na_22 (mol %)	$10^{-12} - 10^{-2}$ *M*	$K_{HNa} \approx 10^{-11}$
	Na^+	11% Na_2O, 18% Al_2O_3, 71% SiO_2 (NAS 11-18)	$10^{-6} - 10^{-1}$ *M*	$K_{NaK} \approx 10^{-2}$ $K_{NaAg} = 4 \times 10^2$ $K_{NaNH_4} = 6 \times 10^{-5}$
Solid state	F^-	LaF_3 crystal	10^{-6}-saturated solution	$K_{FOH} = 10^{-1}$ $K_{FBr} = 10^{-4}$ $K_{FCl} = 10^{-4}$ $K_{FHCO_3} < 10^{-3}$
Liquid or polymer membrane	Na^-	Na^+ ionophore (ETH 227), 2-nitrophenyloctyl ether, sodium tetraphenylborate	$10^{-3} - 10^{-1}$ *M*	$K_{NaH} = 2.5$ $K_{NaK} = 2 \times 10^{-2}$ $K_{NaCa} = 0.36$ $K_{NaMg} = 4 \times 10^{-3}$
	Cl^- (Corning 477315)	Tri-*n*-octylpropylammonium chloride, decanol	$10^{-3} - 10^{-1}$ *M*	$K_{ClOH} = 2.5$ $K_{ClBr} = 0.40$ $K_{ClF} = 0.67$ $K_{ClAc} = 4.55$
	K^+	Valimonycin, dioctyladipate, PVC	$3 \times 10^{-5} - 1$ *M*	$K_{KNa} = 6 \times 10^{-5}$ $K_{KNH_4} = 1.3 \times 10^{-2}$ $K_{KCu} = 5 \times 10^{-4}$ $K_{KMg} = 4 \times 10^{-5}$

	Ca^{2+}	Ca^{2+} ionophore (ETH 1001), 2-nitrophenyloctyl ether, sodium tetraphenylborate	$10^{-7} - 10^{-2}$ *M*	$K_{CaNa} = 3.2 \times 10^{-6}$ $K_{CaK} = 4 \times 10^{-6}$ $K_{CaMg} = 1.3 \times 10^{-5}$
	Ca^{2+}	Calcium di-(*n*-decyl)phosphate, di-(*n*-octylphenyl)phosphonate, PVC	$3 \times 10^{-5} - 1$ *M*	$K_{CaNa} = 5.8 \times 10^{-5}$ $K_{CaK} = 3 \times 10^{-5}$ $K_{CaMg} = 0.14 - 0.025$
	acetylcholine	Tetra(*p*-chlorophenyl)borate, 3-nitro-o-xylene	$10^{-5} - 10^{-1}$ *M*	$K_{AcNa} = 10^{-4}$ $K_{AcNH_4} = 10^{-3}$ $K_{AcK} = 10^{-3}$ $K_{Ac/choline} = 6.6 \times 10^{-2}$
Gas sensors	CO_2	Combination glass pH electrode, 0.01–0.1 *M* $NaHCO_3$-NaCl filling solution; behind silicone rubber membrane	$10^{-4} - 10^{-1}$ *M*	Interferents are organic acids with similar solubilities to CO_2 in gas-permeable membrane
	NH_3	Combination glass pH electrode, 0.1*M* NH_4Cl filling solution; behind porous Teflon gas-permeable membrane	$10^{-5} - 5 \times 10^{-2}$ *M*	Interferents are low-molecular-weight volatile amines

[a]From M. E. Meyerhoff and W. M. Opdycke, Ion-selective electrodes, in H. E. Spiegel (Ed.), *Advances in Clinical Chemistry*, Vol. 25. New York: Academic Press, 1986, p. 1, with permission.

in which solution F^- is incorporated into vacancies in the crystal lattice. The selectivity is very good since other anions do not fit well into the crystal structure. The properties of the F^- ISE are shown in Table 28.2. Electrodes that are based on cast-disk or pressed-pellet membranes of the ionic conductor Ag_2S and another silver or sulfide salt have been formed for the following anions and cations: Cl^- ($AgCl/Ag_2S$), Br^- ($AgBr/Ag_2S$), CN^- and I^- (AgI/Ag_2S), SCN^- ($AgSCN/Ag_2S$), S^{2-} (Ag_2S), Ag^+ (Ag_2S), Cu^{2+} (CuS/Ag_2S), Pb^{2+} (PbS/Ag_2S), and Cd^{2+} (CdS/Ag_2S).

Glass-Membrane Electrodes. The first ISE was the glass-membrane electrode for pH. Glasses of certain compositions respond to pH due to a membrane potential generated by an ion-exchange mechanism with H^+ that occurs in the thin, hydrated outer layer of a glass membrane that has soaked in solution. The outstanding properties of the glass pH electrode are due to the remarkable selectivity of this surface reaction for H^+.

The basic design of the glass electrode for pH is shown in Fig. 28.5. The electrode consists of a glass or plastic tube with a thin, pH-sensitive glass membrane sealed in the tip. Ordinarily the membrane is only about 50 μm thick (and hence is very fragile) and has a resistance of 10^6–10^9 Ω. The bulb at the end contains an internal solution consisting of 0.1 *M* HCl into which is dipped a silver wire coated with AgCl, which keeps the solution saturated with AgCl. This solution of fixed composition establishes a stable potential for the internal Ag/AgCl reference electrode and maintains a fixed activity of H^+ to which the internal surface of the membrane is exposed. A shielded cable in electrical contact with the internal Ag wire is plugged into the pH meter.

The pH response of the glass membrane is determined by the composition of the glass. Corning 015 glass has been used for many years in pH electrodes. The glass consists of approximately 22% Na_2O, 6% CaO, and 72% SiO_2. Pure SiO_2 is

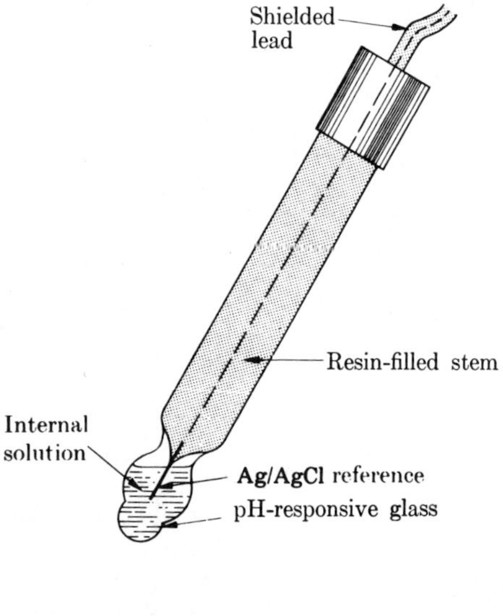

Fig. 28.5 Representative pH electrode. Usually only the membrane at the tip of the electrode bulb is of H^+ ion-selective glass.

essentially an insulator that is unresponsive to pH. The addition of Na_2O to the glass formulation disrupts the silica structure so that negatively charged oxide sites are paired with Na^+. The mobility of Na^+ in the lattice renders the glass membrane slightly conductive to electrical charge. The negative oxide ions serve as ion-exchange sites when the glass is hydrated, which provides the basis for pH response. The pH response of an electrode with a Corning 015 membrane is shown in Fig. 28.6. The potential response to pH is extraordinarily accurate over a pH range of 0 to 14. At pH's above about 9–10 the electrode exhibits a significant response to other monovalent cations such as Na^+ and K^+, as illustrated. This response to alkali cations at high pH is termed the "alkaline error." It can be minimized by replacing Na_2O and CaO to a certain extent with Li_2O and BaO. Consequently, newer formulations of this kind have largely replaced Corning 015 to provide electrodes with improved selectivity.

The membrane response to H^+ is attributed to an ion-exchange process that occurs in the vicinity of the membrane solution interface. Upon immersion of a dry glass membrane in an aqueous solution, the membrane surface becomes hydrated during the course of a few hours. This imbibing of water leads to a gradual dissolving of the glass; this process generally determines the useful lifetime of an electrode. Surface hydration is essential for electrode function; new electrodes immersed in solution respond poorly until adequately soaked. Soaking establishes a hydrated, or gel, layer that is only 10^{-4} mm or less in thickness. This hydrated layer then functions as a cation-exchange layer in which negatively charged oxygen sites are covalently linked to the glass matrix ($\sim O^-$) but in which Na^+ is mobile. Soaking the electrode in acid, for example, results in the replacement of

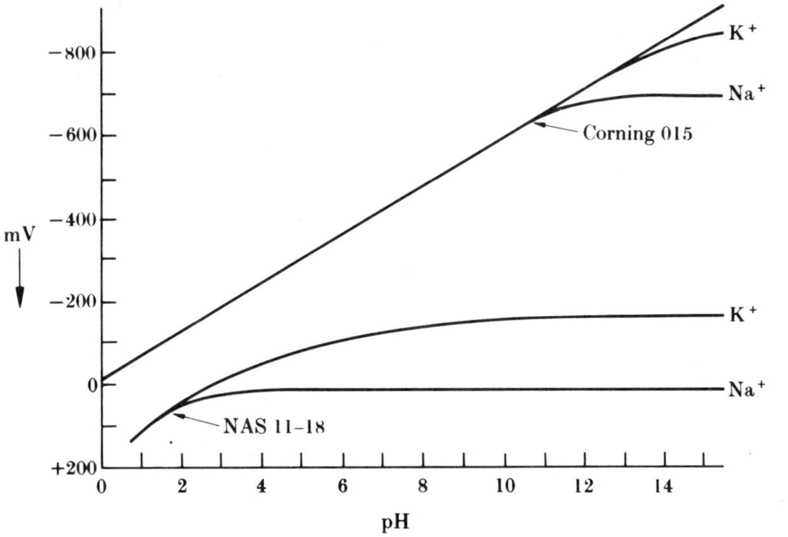

Fig. 28.6 Voltage response of a H^+-selective glass (Corning 015) and a Na^+-selective glass (NAS 11-18) in 0.1 *M* solutions of the indicated cation.

Na^+ with H^+. The membrane response to H^+ can be understood in terms of a surface potential resulting from the ion-exchange reaction of $\sim O^-$ with H^+ in the hydrated gel, as shown pictorially in Fig. 28.7. Immersion of the electrode membrane in basic solution shifts the surface equilibrium to the right as membrane H^+ moves into solution, and a negative membrane potential is established. The inner membrane potential is held constant by exposure to a fixed activity of H^+ in the internal solution. This response mechanism is somewhat analogous to that shown in Fig. 28.3 for the ionophore-based electrode.

Although H^+ has a strong affinity for $\sim O^-$ in the membrane, immersion in basic solution in which H^+ activity is very low enables other monovalent cations to react with $\sim O^-$. The resulting potential generated is termed the alkaine error. Univalent cations that enter can also diffuse slowly through the hydrated layer; divalent cations that penetrate diffuse scarcely at all. The poor response of a glass membrane to cations of 2+ or 3+ charge is not the result of their exclusion in the exchange step, but appears caused by their very slow diffusion once they have penetrated. Through the dry region of the membrane current is transported mainly by the cation of the lowest charge (other than H^+), in a kind of chain displacement process. Glass electrodes also function in other polar solvents such as liquid ammonia, methanol, and molten salts.

Electrodes for Na^+, Ag^+, and NH_4^+ have been developed by varying the composition of the glass. In general, glass membranes are made sensitive to ions other than H^+ by addition of a metal whose coordination number is greater than its oxidation number. For example, aluminum (coordination number 4, oxidation state 3) is used in the preparation of a Na^+-selective glass (NAS 11-18: 11.0% Na_2O, 18.0% Al_2O_3, 71% SiO_2) that is essentially unresponsive to pH over the range of 3 to 14, as shown in Fig. 28.6. This electrode is much less responsive to K^+ and other common cations.

Characteristics of glass H^+ and Na^+ electrodes are shown in Table 28.2.

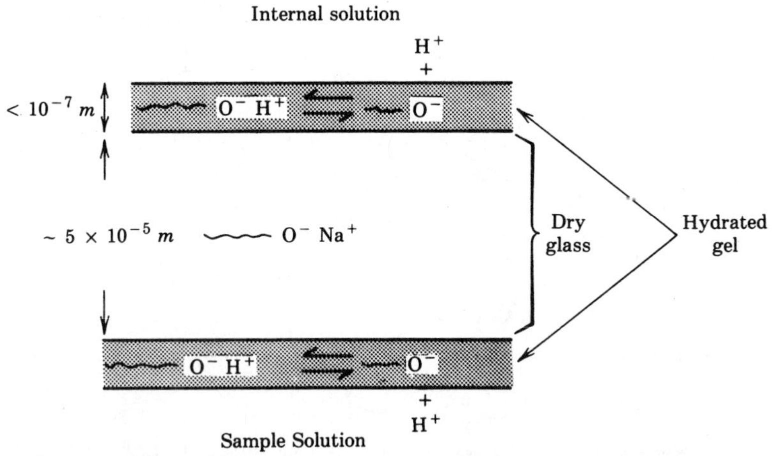

Fig. 28.7 Response mechanism of glass membrane electrode for pH.

Glass membranes have such a high resistance that special measurement problems are introduced. If they are to function as a voltage source, the current drawn must be kept smaller than 10^{-11} or 10^{-12} A or response will be in error by more than 1 mV. A further problem is that resistance increases rapidly as temperature decreases. For that reason, the use of a glass electrode at temperatures below those for which it is intended is risky.

Where resistance is very high, there are always stray currents, which must be minimized in precise work. Adequate shielding of the glass electrode and its lead can virtually eliminate stray capacitive currents to ground. It is possible to minimize surface conduction along the glass by coating it externally with a water-repellent silicone fluid (it appears not to interfere with the membrane response), and filling most of the interior with a resin or nonconducting fluid. Note that these steps have been taken in the electrode shown in Fig. 28.5.

28.4 GAS-SENSING ELECTRODES

Gas-sensing electrodes consist of an ion-selective electrode in contact with a thin layer of aqueous electrolyte that is confined to the ISE surface by an outer membrane, as shown schematically for a CO_2 electrode in Fig. 28.8a. The outer membrane is chosen to be permeable to gases and is made very thin (0.01–0.1 mm) to enable rapid permeation of gas through it. Gas that passes through the membrane dissolves in the thin layer of electrolyte where it shifts the equilibrium position of a chemical reaction. The shift is detected by the ISE and related to the partial pressure of the gas in the sample. The "external" reference electrode for the ISE is positioned behind the membrane.

The mechanism of the electrode is illustrated by the detection of CO_2 in Fig. 28.8b, which magnifies the membrane and thin layer of electrolyte. Since the detection of CO_2 is based on a change in pH, a glass pH electrode is the ISE. Exposure of the electrode to a sample containing CO_2 causes CO_2 gas to pass through the membrane, which is silicone rubber. The thin layer of solution rapidly equilibrates with the CO_2 in the sample. Dissolution of the CO_2 in the thin layer of electrolyte causes a change in pH due to a shift in the equilibrium position of the chemical reaction shown in Fig. 28.8b. The equilibrium constant for the reaction is given by

$$K = \frac{a_{H^+} a_{HCO_3^-}}{pCO_2}. \qquad (28.15)$$

The activity of HCO_3^- in the thin layer is not altered significantly by the CO_2 from the sample if the concentration of HCO_3^- in the electrolyte is made high. Equation (28.15) can then be rearranged to give

$$\frac{a_{H^+}}{pCO_2} = \frac{K}{a_{HCO_3^-}} = \text{constant}$$

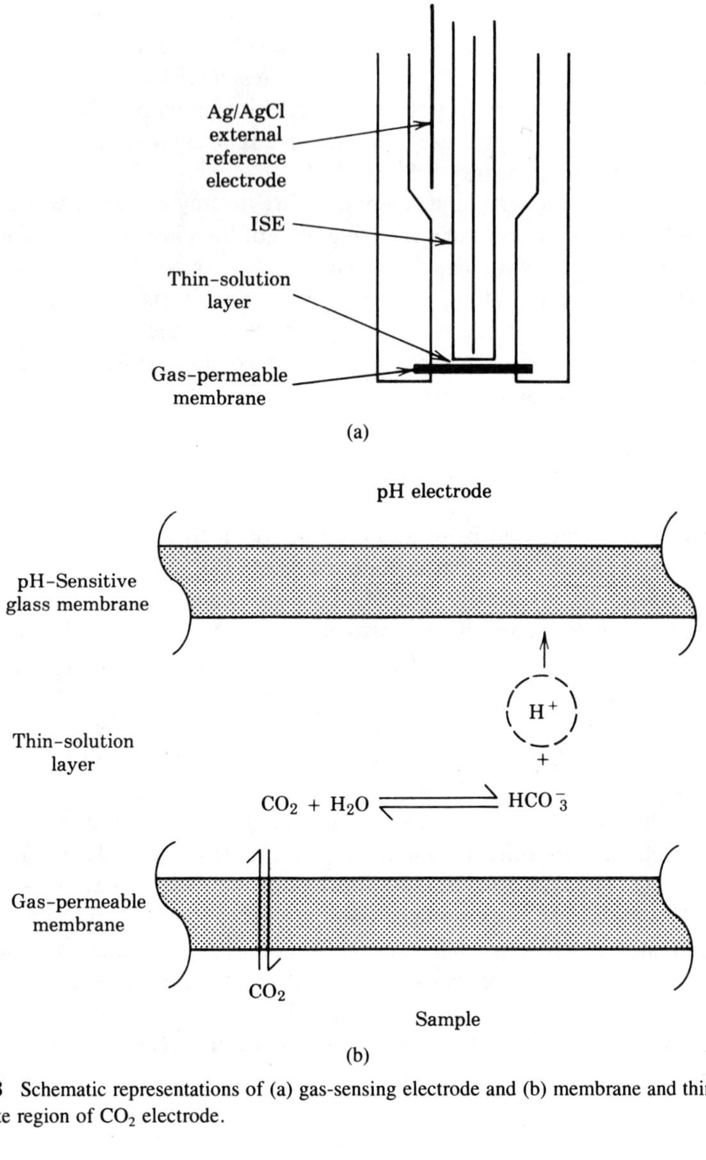

Fig. 28.8 Schematic representations of (a) gas-sensing electrode and (b) membrane and thin layer of electrolyte region of CO_2 electrode.

so that

$$a_{H^+} = (\text{constant})\,(pCO_2) \tag{28.16}$$

The change in pH sensed by the pH electrode is in proportion to pCO_2 of the sample.

Electrodes for a variety of gases have been developed based on the internal solution equilibria and sensing electrodes shown in Table 28.3. Characteristics of

Table 28.3 Gas-Sensing Electrodes

Gas Electrode	Internal Solution Equilibrium	Sensing Element
CO_2	$CO_2 + H_2O \rightleftharpoons HCO_3^- + H^+$	Glass, pH
NH_3	$NH_3 + H_2O \rightleftharpoons NH_4^+ + OH^-$	Glass, pH
HCN	$HCN \rightleftharpoons H^+ + CN^-$	Ag_2S, pCN
HF	$HF \rightleftharpoons H^+ + F^-$	LaF_3, pF
H_2S	$H_2S \rightleftharpoons 2H^+ + S^{2-}$	Ag_2S, pS
SO_2	$SO_2 + H_2O \rightleftharpoons HSO_3^- + H^+$	Glass, pH

the CO_2 and NH_3 electrodes, which have been used extensively in bioanalytical chemistry, are shown in Table 28.2.

28.5 BIOCATALYTIC MEMBRANE ELECTRODES: BIOSENSORS

Biocatalytic membrane electrodes have an ISE or a gas-sensing electrode in contact with a thin layer of biocatalytic material, as shown in Fig. 28.9. The biocatalyst converts substrate (the analyte) into a product that is detectable by the electrode. Electrodes of this type are often referred to as *biosensors*. The biocatalytic layer consists of an immobilized enzyme that is chosen for its catalytic selectivity toward the analyte. Immobilized bacterial particles, a tissue slice that contains an enzyme

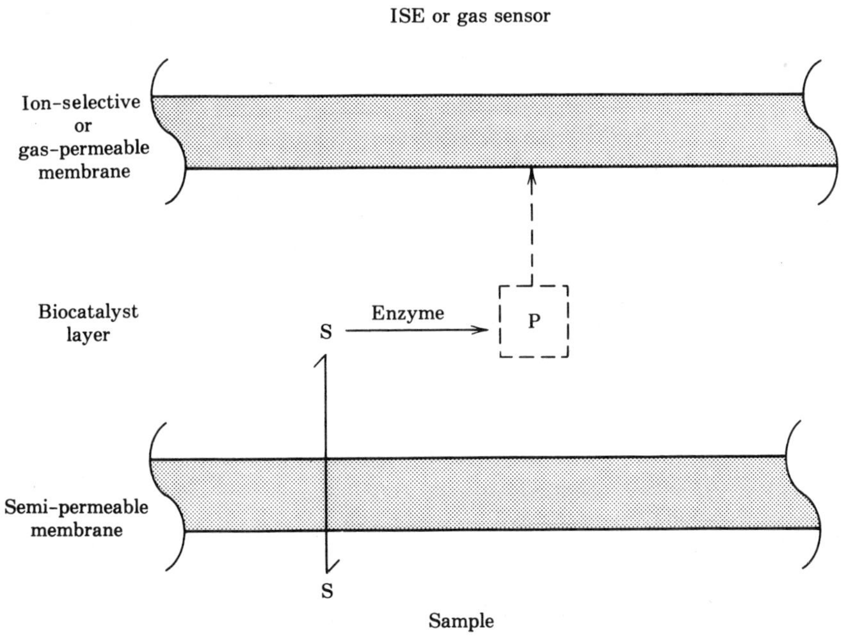

Fig. 28.9 Schematic diagram of biocatalytic electrode.

or, perhaps several enzymes that catalyze a sequence of reactions leading to a detectable product can also be used.

The first type of biocatalytic membrane electrode was based on the immobilization of an enzyme on the surface of an ISE. For example, urease immobilized in a polyacrylamide gel on the end of a glass ISE for NH_4^+ gives a selective electrode for the determination of urea. Urease catalyzes the production of NH_4^+ from urea according to the following reaction

$$\underset{\text{urea}}{\begin{array}{c} NH_2 \\ | \\ C = O \\ | \\ NH_2 \end{array}} + 2H_2O \xrightarrow{\text{urease}} 2NH_4^+ + CO_3^{2-}$$

When the electrode is immersed in a solution containing urea, urea diffuses into the enzyme layer where urease catalyzes the production of NH_4^+, which diffuses to the glass NH_4^+ electrode. The concentration of NH_4^+ detected is proportional to the concentration of urea in the sample. Biosensors based on numerous enzymes have been reported, some of which are listed in Table 28.4.

Intact bacterial particles and thin slices of tissue have been immobilized on electrode surfaces. A representative example is an electrode for L-arginine for which the bacterium *streptococcus faecium* is immobilized on the gas-permeable membrane of an ammonia electrode.* Upon immersion of the electrode in a solution of L-arginine, L-arginine diffuses into the biocatalytic layer where arginine deaminase in the bacteria catalyzes the following reaction

$$\text{L-arginine} + H_2O \xrightarrow[\text{deaminase}]{\text{arginine}} \text{citrulline} + NH_3.$$

The rate of NH_3 formation is proportional to the concentration of L-arginine in the sample. Enzyme-generated NH_3 diffuses to the surface of the NH_3-permeable membrane of the ammonia electrode where it is detected. The electrode response to NH_3 is proportional to the concentration of L-arginine in the sample. Nernstian plots give slopes of 40–45 mV with a linear range of 1×10^{-4} to 6.5×10^{-3} M. Other examples of bacterial-particle- and tissue-slice-based electrodes are listed in Table 28.4.

Biocatalytic membrane electrodes significantly expand the scope of direct potentiometry by enabling biosensors that respond to a whole host of organic substrates to be made. The selectivity of these sensors is a combination of the selectivity of the biocatalyst for the substrate and the ISE for other constituents in the sample that might reach the ISE surface membrane. Electrodes such as that for L-arginine exhibit excellent selectivity due to the high specificity of the enzyme ar-

*G. A. Rechnitz, R. K. Kobos, S. J. Riechel, and C. R. Gebauer, *Anal. Chim. Acta*, **94,** 357 (1977).

ginine deaminase for L-arginine and the excellent selectivity of the ammonia electrode for NH_3. The working range of biosensors of this type is typically only 2–3 orders of magnitude with a detection limit of 10^{-5}–10^{-4} M.

Cell-Based Biosensors. Electrodes based on living cells have comparable response characteristics to enzyme electrodes, but offer several advantageous features.* (a) Strains of bacteria are generally less expensive than isolated enzymes. A specific sterile bacterial culture is not needed to prepare an electrode. For example, human dental plaque suffices to prepare electrodes for D(+) glucose, D(+) mannose, and D(−) fructose. (b) Enzyme activity is often enhanced in bacterial cells and remains longer owing to the optimal environment. Consequently, lifetimes of bacterial electrodes average around 20 days compared with 14 days for enzyme electrodes. Cells can be regrown if the catalytic activity is lost. (c) Bacteria can readily effect complex reaction sequences requiring cofactors. For example, an electrode for nitrilotriacetic acid (NTA) is practical only because *pseudomonas* bacterial cells contain all of the enzymes and cofactors necessary to execute the following sequence of reactions on which the electrode response is based [R. K. Kobos and H. Y. Pyon, *Biotechnol. Bioeng.*, **23**, 627 (1981)]:

$$\text{NTA} + \text{NADH} + O_2 \xrightarrow[\text{monooxygenase}]{\text{NTA}} (\alpha\text{-hydroxyl NTA})$$

$$\text{iminodiacetate} + \text{glyoxylate}$$

$$\text{iminodiacetate} \longrightarrow \text{glycine} + \text{glyoxylate}$$

$$\text{glycine} \xrightarrow[\text{decarboxylase}]{\text{glycine}} \text{serine} + CO_2$$

$$\text{serine} \xrightarrow[\text{deaminase}]{\text{serine}} \text{hydroxy pyruvate} + NH_3$$

The underlying gas electrode for ammmonia responds to NH_3 formed in the last reaction, which is proportional to the concentration of NTA in the sample.

Cell- and tissue-based electrodes are generally less selective than enzyme-based electrodes since the many constituents of cells result in response to many possible substrates. This problem may be solved by the use of specific inhibitors to block interfering reactions. These electrodes also suffer from long recovery times (3–4 hr) after a measurement has been made. However, such electrodes may be sufficiently cheap to be disposable. Most biocatalytic membrane electrodes are susceptible to irreversible loss of activity when exposed to samples containing denaturing or toxic agents. The difficulties associated with electrode longevity, both during use and storage, have stymied the commercial development of biocatalytic membrane electrodes.

28.6 DIRECT POTENTIOMETRIC MEASUREMENTS

In *direct potentiometry* the appropriate indicator electrode and a reference electrode are immersed in the solution to be analyzed, and the activity or concentration

*C. A. Corcoran and G. A. Rechnitz, *Tr. Biotech.*, **3**, 92 (1985).

Table 28.4 Representative Biocatalytic Membrane Electrodes[a]

Biocatalyst Category	*Substrate*	*Biocatalyst*	*Detected Substance*
Enzyme	Urea	Urease	NH_4^+, NH_3,CO_2, or H^+
	Glucose	Glucose oxidase and peroxidase	I^-
		Glucose oxidase	H^+
		Glucose oxidase	F^-
	Amygdalin	β-glucosidase	CN^-
	Pencillin	Penicillinase	H^+
	L-phenylalanine	L-amino acid oxidase	NH_4^+
		L-amino acid oxidase and horseradish peroxidase	I^-
		Phenylalanine ammonia lyase	NH_3
		Phenylalanine decarboxylase	CO_2
	Uric acid	Uricase	CO_2
	Acetylcholine	Acetylcholinesterase	H^+
	D-gluconate	Gluconate kinase and 6-phospho-D-gluconate dehydrogenase	CO_2
	Acetaldehyde	Aldehyde dehydrogenase	H^+
	Oxalate	Oxalate decarboxylase	CO_2
	Flavin adenine dinucleotide	Alkaline phosphatase and adenosine deaminase	NH_3
	Methotrexate	Dihydrofolate reductase and 6-phosphogluconic dehydrogenase	CO_2
	Salicylate	Salicylate hyroxylase	CO_2
Bacterial particle	L-arginine	*Streptococcus faecium*	NH_3
		Streptococcus lactis	NH_3
	L-aspartate	*Bacterium cadaveris*	NH_3
	L-glutamine	*Sarcina flava*	NH_3
	NAD^+	*Escherichia coli*/NAD*ase*	NH_3
	Nitrilotriacetate acid	*Pseudomonas* sp.	NH_3
	L-tyrosine	*Aeromonas phenologenes*	NH_3
	L-histidine	*Pseudomonas* sp.	NH_3
	L-serine	*Clostridium acidiurici*	NH_3
	Nitrate	*Azotobacter vinelandii*	NH_3
	Uric acid	*Pichia membranaefaciens*	CO_2
	L-glutamic acid	*Escherichia coli*	CO_2

Table 28.4 (*Continued*)

Biocatalyst Category	*Substrate*	*Biocatalyst*	*Detected Substance*
Bacterial particle	Pyruvate	*Streptococcus faecium*	CO_2
	L-cysteine	*Proteus morganii*	H_2S
	Sugars	Bacteria from human dental plaque	H^+
Tissue	Glucosamine 6-phosphate	Porcine kidney tissue	NH_3
	L-cysteine	Treated plant leaf	
	Adenosine	Mouse small intestine mucosal cells	NH_3
	Adenosine 5′-monophosphate	Rabbit muscle	NH_3
	L-glutamate	Yellow squash	CO_2
	Glutamine	Porcine kidney cortex	NH_3
	Guanine	Rabbit liver	NH_3

[a]Compiled from C. E. Lunte and W. R. Heineman, Electrochemical Techniques in Bioanalysis in *Topics in Current Chemistry*, Vol. 143, E. Steckhan (Ed.), Springer-Verlag, Berlin Heidelberg, 1988, p. 8.

of the analyte in the sample is calculated from the measured potential by means of the Nernst equation. If activity is to be measured, an equation of the general form

$$E_{\text{cell}} = K + \frac{RT}{zF} \ln a_{\text{M}^{z+}} \tag{28.17}$$

is used where K is a constant that contains E_{lj}. K may be determined from measurements made on solution(s) of known activity. When concentration is to be measured, $C_{\text{M}^{z+}}$ and $\gamma_{\text{M}^{z+}}$ [see Eq. (27.12)] can be substituted into Eq. (28.17) to give

$$E_{\text{cell}} = K + \frac{RT}{zF} \ln (\gamma_{\text{M}^{z+}} C_{\text{M}^{z+}}) \tag{28.18}$$

If solution conditions are such that $\gamma_{\text{M}^{z+}}$ is constant, its contribution to potential can be added to the constant term to give

$$E_{\text{cell}} = K' + \frac{RT}{zF} \ln C_{\text{M}^{z+}}, \tag{28.19}$$

where the constant K' now also contains the term $(RT/zF) \ln \gamma_{M^{z+}}$ and must be determined from measurement(s) on solution(s) of known concentration.

Calibration Techniques. Calibration of the electrode (i.e., of the electrochemical cell) is achieved by measuring K or K' with one or more standard solutions. This measurement can be made by any of several methods. One approach is to simply measure E_{cell} for a single standard solution of known $a_{M^{z+}}$ or $C_{M^{z+}}$ and calculate K or K', respectively. Once the constant is known, one can then measure E_{cell} for the samples and calculate $a_{M^{z+}}$ or $C_{M^{z+}}$ from Eq. (28.17) or (28.19). This approach is based on the assumption that K or K' is in fact constant over the activity or concentration range spanned by the samples.

Example 28.4. The potential of an electrochemical cell with a F^- ISE is found to be 0.214 V with 1.00×10^{-3} M standard F^- solution at 25°C. A sample gave a potential of 0.235 V. Calculate the concentration of F^- in the sample.

The appropriate Nernst expression is Eq. (28.19) with $z = -1$, which gives at 25°C

$$E_{cell} = K' - 0.0591 \log C_{F-}.$$

First, K' is calculated:

$$0.214 = K' - 0.0591 \log (1.00 \times 10^{-3}); \qquad K' = 0.037 \text{ V}.$$

Then C_{F-} is calculated:

$$0.235 = 0.037 - 0.0591 \log C_{F-}; \qquad C_{F-} = 4.46 \times 10^{-4} \text{ mol L}^{-1}.$$

If the range of sample activities is greater than an order of magnitude, it is prudent to calibrate the cell with a series of standards whose range encompasses the range of the samples. This is usually accomplished by constructing a *calibration curve* (see Section 11.3) that consists of a Nernstian plot of Eq. (28.17) or (28.19), as shown in Fig. 28.10. The calibration curve is then used to convert E_{cell} from the samples to activities or concentrations. A calibration curve for activity (as shown in Fig. 28.10) should be linear with a slope of RT/zF mV over the Nernstian range of the electrode. The equation for a linear least-squares plot of the Nernstian region can be used by a computer to calculate results. Such a plot for concentration is nonlinear because of the changing value for $\gamma_{M^{z+}}$ as concentration of M^{z+} increases in the standard solutions. $\gamma_{M^{n+}}$ is affected by the ionic strength of the calibration solutions, and K', which contains $\gamma_{M^{z+}}$, is therefore not a constant over the concentration range of the standards. This situation can be improved by adding an innocuous salt to all of the standards (and samples) to adjust their ionic strength to a common value.

Perhaps the most commonly used calibration technique is the method of *standard additions* (see Section 11.4). The significant advantage of this method is that the cell is calibrated while the electrodes are immersed in the sample. After measuring E_{cell} of the sample, one then adds a small aliquot of standard solution and then measures E_{cell} again. The concentration of the sample can be calculated from

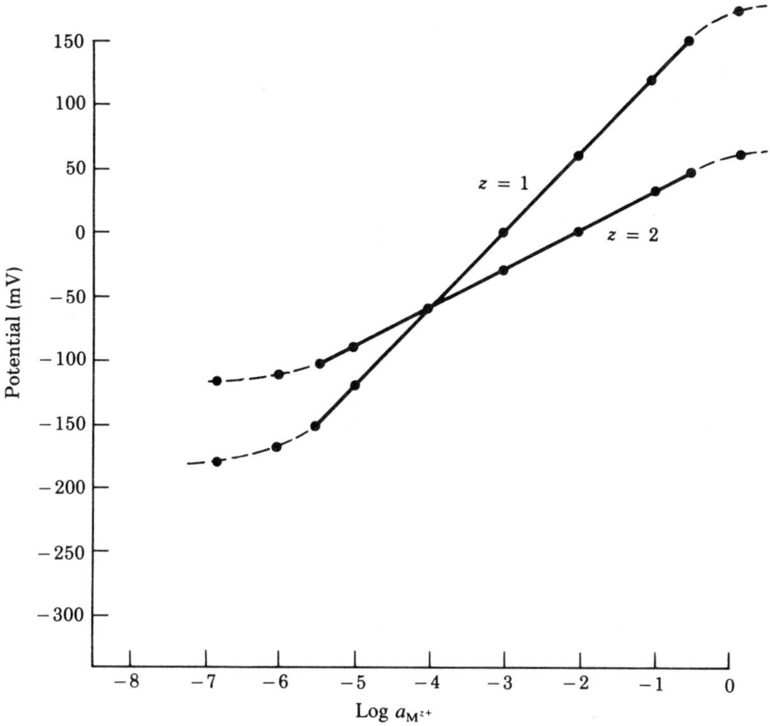

Fig. 28.10 Calibration curves for direct potentiometry of z (or n) = 1 and 2 species.

the difference in potential between sample and sample + standard according to the following equation:

$$C_x = \frac{C_s V_s}{(V_x + V_s)(10^{-n\Delta E/0.0591} - V_x)}, \tag{28.20}$$

where C_x = concentration of unknown, C_s = concentration of standard, V_x = volume of unknown, V_s = volume of standard, and ΔE = (E after addition of standard) − (E sample). The most accurate determinations are made when the concentration change is such that the total concentration is approximately doubled. The addition of a small volume of concentrated standard minimizes dilution of the sample so that the standardization step is accomplished in essentially the same solution *matrix* as was the sample measurement.

In the method of *sequential standard additions*, several aliquots of concentrated standard are added sequentially to the sample with the potential being measured after each addition. A plot of the type shown in Fig. 11.4 is then constructed and the concentration of the unknown calculated.

Example 28.5. A 50-mL sample is analyzed for K^+ with a K^+ ISE by the method of sequential standard additions. Four 50-μL aliquots of 1.00 *M* KCl standard are added sequentially to the sample at 25°C and the following potentials are measured after addition of each aliquot: 0 μL, −0.145 V; 50 μL, −0.120 V; 100 μL, −0.108 V; 150 μL; −0.0994 V, 200 μL, −0.0931 V. The ISE-reference electrode system is known to obey Eq. (28.19) for solutions of ionic strength comparable to the sample.

In order to obtain a plot as shown in Fig. 11.4, Eq. (28.19) is linearized by first dividing by the slope ($S = RT/zF$) and then taking the antilog$_{10}$ of the equation to give

$$10^{E/S} = 10^{K'/S}\,C$$

A plot of $10^{E/S}$ versus C is linear if K' is constant. A plot of $10^{E/S}$ versus ΔC *where* $S =$ 0.0591 V and ΔC is the cumulative incremental increase in concentration resulting from the sequential addition of standards is shown in Fig. 28.11. The intercept on the x axis is $\Delta C'$; $-\Delta C'$ is equal to the original sample concentration C, which in this example is 6.00×10^{-4} mol L^{-1}. (Note: Correction for dilution of the sample by 200 μL of standard is unnecessary unless accuracy of better than 0.5% is required.)

Response to Free Ion. It is important to realize that most direct potentiometric measurements involve indicator electrodes that respond to the free, solvated form of the ion and not to other forms of the ion. Thus, equilibrium activities of a free ion can be measured in the presence of other forms of the ion. For example, the fluoride ISE responds to free fluoride F^- in the following equation and not to the protonated form, hydrofluoric acid HF:

$$HF + H_2O \rightleftharpoons H_3O^+ + F^-.$$

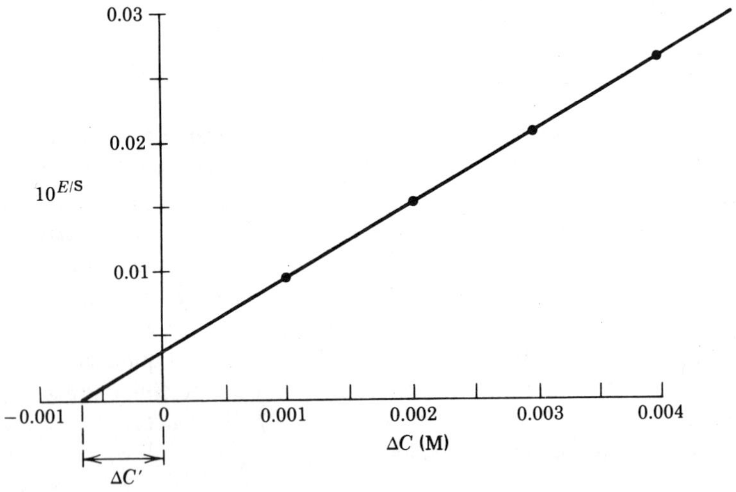

Fig. 28.11 Determination by method of sequential standard additions.

If one wants to measure total fluoride in a sample, the equilibrium given above must be shifted to the right so that all fluoride is in the chemical form of F^-. This can be accomplished by adding a sufficiently basic buffer to cause complete dissociation of HF to F^-. ISEs generally do not respond to complexed forms of an ion. For example, a calcium electrode is unresponsive to the EDTA complex of Ca^{2+}.

Sources of Error. Direct potentiometry is susceptible to several sources of error that are generally not readily apparent while analyzing samples. Perhaps the most important source of error is lack of *selectivity* against a nonanalyte ion in the sample. As the selectivity coefficients in Table 28.2 show, even the best ISEs are not 100% selective. It is therefore important to know the selectivity properties of the electrode being used and to know if nonanalyte ions to which the electrode responds are in sufficiently low concentration as to not constitute an interference. That is to say, the $k_{ij}a_j^{z/x}$ term in Eq. (28.14) should be negligibly small relative to a_i. Components in certain samples can change the sensitivity of an electrode by adsorbing on its surface and thereby blocking the analyte's access to the surface. Such electrode passivation can be a problem in samples containing surface-active species such as proteins.

As shown in Eq. (28.17), the *liquid-junction potential* is assumed to be constant during the course of a determination. This is correct only if the standards and samples are of similar ionic composition. Variations between standards and samples and among the samples can cause changes in E_{lj} of several to tens of millivolts (See Table 27.4). Variations of this magnitude corresponds to large errors in concentration or activity because of the logarithmic relationship with E_{cell} as shown in Table 28.5 for $z = 1$ and $z = 2$ species. Note that a twofold error in potential translates through the logarithmic function into a 10-fold error in activity. Thus, the logarithm function amplifies errors in potential. In order to achieve only a 2% relative accuracy in activity, all errors in the potential measurement must be kept to 0.5 mV ($z = 1$). This includes both the instrumental accuracy of measuring E_{cell} and the degree to which K is a constant. This underscores the importance of matching the ionic strength of samples and standards to minimize variations in E_{lj}. This same argument applies to the importance of holding *activity coefficients* con-

Table 28.5 Relationship between Error in E_{cell} and Error in Activity

	Relative Error in Activity (%)	
Error in Potential (mV)	$z = 1$	$z = 2$
0.1	0.4	0.8
0.5	1.9	3.8
1.0	3.8	7.5
2.0	7.5	15.0

stant when measurements of concentration are being made. As shown in Eqs. (28.18) and (28.19), the activity coefficient is assumed to be constant. Since pH/mV instruments can easily measure E_{cell} with an accuracy of 0.1 mV or better, the main source of error in direct potentiometry is due to variations in E_{lj}, γ, and an insufficiently small k_{ij}. *Temperature* variations can also be a source of error since the Nernst equation is temperature-dependent.

28.7 PROCEDURE FOR pH MEASUREMENTS

In aqueous solution direct potentiometry with a glass electrode-saturated calomel electrode pair and a pH meter is used to routinely determine the hydrogen ion. The technique is nondestructive and has the further advantage of very wide concentration range (e.g., 14 pH units in aqueous media at 25°C), simplicity, and reasonable speed (2- to 30-s response time). The sensitivity of a glass electrode to alkali cations, especially Na^+, may reduce its response below the theoretical at pH values above 10. In solutions of pH less than 1, low water activities give rise to error. It is also important to realize that there is at least one liquid junction in the cell when the usual reference electrode (SCE) is used. This reference incorporates a saturated KCl salt bridge, which establishes the junction with the sample solution. This junction potential is usually unknown, though it may be estimated. Though the precision of most measurements is only moderate, it is adequate for a trace method.

Example 28.6. Precision may be illustrated by translating pH specifications into activity or concentration units. Measurements reliable to ±0.1 pH such as might be made with an inexpensive meter yield a H^+ ion concentration accurate to about ±20%; those reliable to ±0.01 pH, concentration to ±2.5%. Under circumstances where conditions are precisely controlled and a very sensitive pH meter is used, accuracies of the order of ±0.001 pH unit (concentration to ±0.2%) become possible.

Operational Definition of pH. In the case of pH the formal definition $\text{pH} = -\log a_{\text{H}+}$ does not lead to a well-prescribed experimental measurement. Two factors are mainly responsible: It is especially difficult to interpret data from a cell with a liquid junction, and, further, it is not possible to obtain from thermodynamics a rigorous definition of the activity coefficient of a single species. These difficulties are surmounted experimentally by devising an appropriate convention. This permits the assignment of internally consistent pH values to a number of easily reproducible buffers; pH values of other solutions can then be related operationally to these standards.

Accordingly, after standardizing the scale of a pH meter with a buffer solution, the pH of any other solution is given by the operational definition

$$\text{pH} = \text{pH(B)} + \frac{(E - E_{\text{B}})F}{2.303\,RT}. \tag{28.21}$$

Here pH(B) and E_B are the pH and cell potential, respectively, with buffer solution in the cell, and pH and E are the values with the sample present.

An important additional advantage of the operational approach is that it provides a basis for circumventing day-to-day instrument and electrode drift, most junction potential error, and changes in E^0 from glass electrode to electrode. Standardizing cell and pH meter compensates these variations.

To what extent does Eq. (28.21) allow one to ignore differences in concentrations, constituents, and temperature, which might exist between buffer and sample solutions? Also, to what degree does this definition solve the problem of the unknown liquid junction potential? The answers lie in two important assumptions made whenever Eq. (28.21) is used.

First, the liquid junction potential in sample solutions is assumed to be identical with that in the standardizing buffer. Should this not be true, the pH reading for the sample solution will include the difference between its liquid junction potential and that of the buffer. Fortunately, the use of a saturated KCl bridge minimizes junction potentials (see Section 27.5); but the greater the difference in ionic strength or pH between sample and buffer, the larger the error, nevertheless. For that reason, buffers of high ionic strength are preferable whenever pH values are to be determined in concentrated solutions, and buffers should if possible be within 2 pH units of a sample.

Second, the buffer and unknown solutions are assumed to be at the same temperature.* For work at room temperature, this requirement seldom presents any problem; at other temperatures thermostating or measuring the temperature (for compensation) of all solutions is necessary.

Buffers. Many standard buffers of known composition and defined pH are available commercially or may be prepared easily in the laboratory; Table 28.6 includes five supplied by the National Bureau of Standards.

Procedure. Measurements are rapid if the electrodes have been conditioned in distilled water. Following a calibration of the meter and cell against one or two buffer solutions, the electrodes are dipped into the sample, and then the meter is balanced and read. About 10 mL of sample are adequate in most cases, and with microelectrodes amounts as small as one drop can be used.

Standard pH Scale in Other Media. The operational definition of pH can be extended to media other than water. In a nonaqueous solvent, heavy water, or even a mixed solvent, Eq. (28.21) becomes

$$\mathrm{pH}^*(x) = \mathrm{pH}^*(\mathrm{B}) + \frac{(E_x - E_\mathrm{B})F}{2.303\,RT}, \tag{28.22}$$

*Strictly speaking, there is little connection between pHs at different temperatures. By definition, the standard potential of the hydrogen electrode is zero at all temperatures. The consequence is that a new pH scale is established at each temperature.

Table 28.6 Standard Buffers

Buffer Solution	*pH at 25 °C*
(a) NBS Primary Standards of pH[a]	
Saturated KH tartrate	3.557
0.05 *M* KH phthalate	4.008
0.025 *M* KH_2PO_4; 0.025 *M* Na_2HPO_4 (mixture)	6.865
0.008695 *M* KH_2PO_4; 0.03043 *M* Na_2HPO_4	7.413
0.01 *M* borax	9.180
(b) Secondary and Supplementary Standards[b]	
0.05 *M* $KH_3(C_2O_4)_2 \cdot H_2O$	1.679
Saturated $Ca(OH)_2$	12.454

[a]Known pH values are given rather than potentials, since most pH meters are calibrated to read in pH units. In addition, for the user the quotation of a pH value avoids the necessity of knowing the calomel electrode potential exactly. The advantage results from the fact that the saturated calomel electrode is not readily reproducible in its half-cell potential to the precision required for measurement to 0.001 pH unit. For values at other temperatures see Bates [5].

[b]These standards may be used with care outside the intermediate pH range (see reference 5).

where the asterisk identifies the pH scale as applying only in the particular medium. Thus, pH*(x) and E_x are values of pH and potential for a sample solution in the medium, and pH*(B) and E_B are values of pH and potential for a pH buffer in the medium. No attempt is made to refer the pH* scale back to water [5]. Scales of this kind that apply uniquely to a given medium have already been devised for alcoholic media and heavy water [32] and seem certain to be extended to other solvents.

28.8 pH METERS

By usage, the term *pH meter* defines a class of voltmeters that is used to measure the potential of very-high-resistance galvanic cells. In most cases the unusual resistance is attributable to the pH indicator electrode alone. As a variety of additional ion-selective electrodes of high resistance have appeared the designation pH has persisted, though the name pION or ion meter would be more appropriate. The terms pH meter and ion meter will be used interchangeably in this discussion. The pH meters are also used to measure the potential of electrochemical cells in millivolts. Thus, the pH meter is a general purpose measuring instrument for potentiometry.

It is useful to consider the operation definition of pH (Eq. 28.21) before dealing with the workings of a pH meter. Since $F/2.303\ R$ are constants, they can be replaced by constant K and the equation recast in terms of two buffers, B_1 and B_2, to give

$$\mathrm{pH}(B_2) = \mathrm{pH}(B_1) + \frac{(E_{B_2} - E_{B_1})\,K}{T}. \qquad (28.23)$$

Figure 28.12 shows schematically how a pH meter is calibrated with two buffer solutions to conform to Eq. (28.23). A measurement is made on the first buffer B_1, usually pH 7.00, and the meter is adjusted with the standardize control to read 7.00 after the temperature control has been set to the temperature of the buffer. Standardize is simply an offset control that compensates for the differences in zero potential of different electrode systems. It moves the calibration curve in the vertical direction. A second buffer B_2, acidic or basic depending on the pH range of the samples to be measured, is then measured. The *slope* control is used to adjust the slope of the line by rotating it about the 0 mV position so that the meter reading coincides with the pH of the buffer B_2, which is 5.00 in this example. At this point, the constant K/T in Eq. (28.23) has been adjusted for the particular temperature of the two buffers, which is 25°C. Ideally, K should equal $F/2.303R$, but most indicator electrodes do not exactly follow Nernstian behavior. The temperature control enables the instrument calibration to be converted to other temperatures. As shown in the figure, the temperature control changes the slope to conform to the appropriate value of T. Most instruments allow temperature compensation from 0 to 100°C. This can be done automatically by immersing a thermocouple temperature probe in each sample prior to the pH measurement.

Modern pH meters measure pH from 0 to 14 and potential from 0 to ±1999 mV. They have the necessary controls and/or connectors to compensate manually or automatically for the effects of solution temperature; to correct electrode response to that predicted by Nernstian theory; to standardize the meter with a buffer; and to attach a laboratory recorder.

The circuit block diagram of a representative pH meter is shown in Fig. 28.13. This instrument is a high-input impedance dc millivolt meter designed to measure millivolts and pH. Measurements are made with an electronic nullbalance system. The circuit includes an

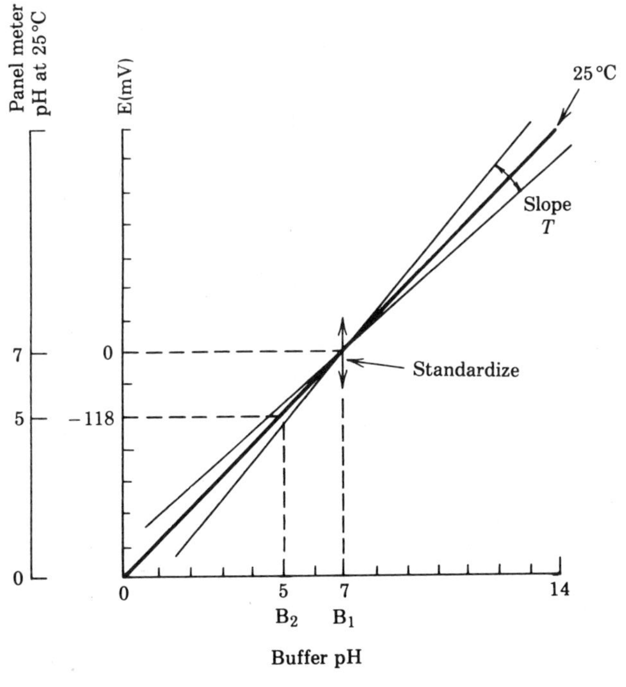

Fig. 28.12 Calibration of a pH meter with two buffers, B_1 and B_2.

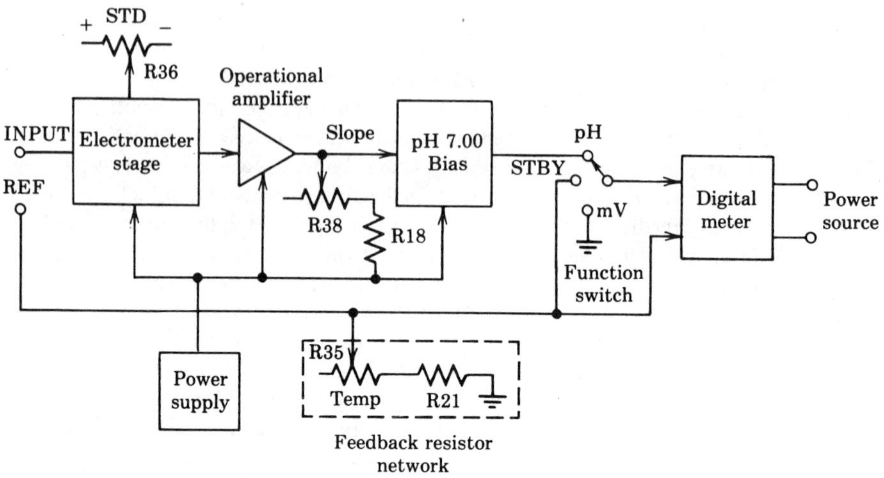

Fig. 28.13 Circuit block diagram of representative pH meter. (From Instruction Manual: Fisher Model 610A Accumet® pH Meter, reprinted with permission Fisher Scientific Company.)

electrometer stage, operational amplifier with feedback resistor network, millivolt-to-pH conversion circuit, digital readout meter, and power supply.

During operation, the cell potential is applied across the INPUT and REF jacks and directly into the high-impedance electrometer stage. The electrometer consists of two matched-pair field-effect transistors and is compatible with electrode systems having resistances of up to 10,000 MΩ. The output of the electrometer is applied to the operational amplifier. The amplified signal flows through SLOPE control R38, resistor R18, potentiometer R35 (temperature control), and resistor R21. Potentiometer R35 and resistor R21 form a feedback resistance network. The voltage drop across them is equal and opposite in polarity to the input voltage plus or minus the offset established by the standardize potentiometer R36. This voltage is fed back to the input to maintain this condition.

In the millivolt mode, the voltage developed across the feedback resistor network is applied to the digital meter.

In the pH mode, the millivolt output of the operational amplifier is converted to millivolts that are the pH equivalent of the measured voltage. In addition, an offset voltage must be added to compensate for the fact that the glass electrode measuring system has an output of zero millivolts at 7.00 pH and at 25°C. The resistance values of resistors R18, R35, and R21 are selected so that the millivolts appearing across resistor R18 and R38 are the pH equivalent of the millivolts appearing across potentiometer R35 and resistor R21. This voltage plus the offset voltage provided by the pH 7.00 bias circuit is applied to the digital meter which indicates directly in pH.

Since the output of the amplifier is proportional to the relationship between the total feedback resistance and the total load resistance, the settings of both the TEMPERATURE and the SLOPE controls affect the indicated pH reading. These controls allow for corrections to electrode temperature and non-Nernstian response, respectively. The STANDARDIZE control functions electrically as a dc offset control. In making potential measurements in millivolts, the control is set to provide a zero reference. When the instrument is used to measure pH units, the STANDARDIZE control compensates for electrode zero potential since it is used to set the digital display indication to agree with the value of a standardizing pH buffer solution.

28.9 MINIATURE ELECTRODES AND IN VIVO MEASUREMENTS

A significant characteristic of the ISE is the feasibility of fabricating miniature sensors for measurements in samples of very small volume or in vivo. Microelectrodes with tip diameters ranging from 0.1 to 200 μm have been devised using glass electrodes for H^+, K^+, Na^+; solid-state electrodes for Ag^+, Cu^{2+}, I^-, Cl^-, Br^-, S^{2-}; and liquid membrane electrodes for K^+, Cl^-, Na^+, Mg^{2+}, H^+, acetylcholine. Perhaps the most widely used of these are the liquid membrane electrodes, which exhibit response characteristics comparable to their conventional-sized analogs.

Microelectrodes can be made sufficiently small to measure intracellular ion activities, as shown schematically in Fig. 28.14. The micro ion-selective electrode is inserted into a single cell by means of a micromanipulator. The intracellular activity of the analyte ion is then monitored by measuring the potential difference between the ISE and a reference electrode. Another reference microelectrode can be placed in the bathing solution outside the cell, and the potential between the two reference electrodes measured in order to simultaneously monitor the membrane potential E_m of the cell. Some representative applications of ion-selective electrodes for intracellular measurements are Ca^{2+} and K^+ in frog heart ventricle cells, Ca^{2+} in neurons of giant snails and molluscs, H^+ in squid axons and vacuoles of plant cells, and Cl^- in renal tubules of bull frogs.*

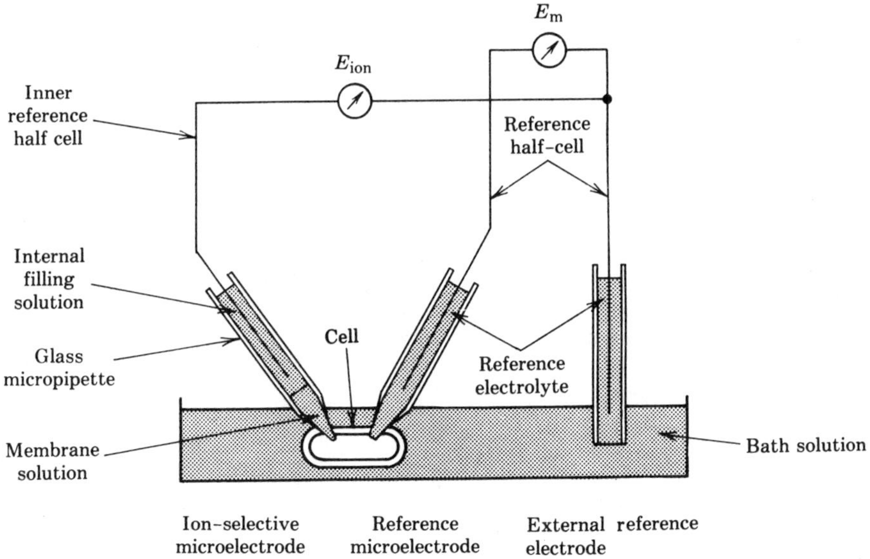

Fig. 28.14 Schematic diagram of microelectrode arrangement for intracellular measurements of ion activity. (From D. Ammann, *Ion-Selective Microelectrodes*, New York: Springer Verlag, 1986, with permission).

*M. E. Meyerhoff and W N. Opdycke, Ion-selective electrodes, in H. E. Spiegel (Ed.), *Advances in Clinical Chemistry*, Vol. 25. New York: Academic Press, 1986, p. 1.

28.10 CLINICAL APPLICATIONS

ISEs have been developed to the point that accurate measurements of ions in complex matrices such as serum and urine are routine. Blood electrolytes [Na^+, K^+, Ca^{2+} (ionized and total), pH, pCO_2, total CO_2 and Cl^-] are now routinely measured in clinical laboratories by automated procedures based on detection of each species by an ISE. Numerous instruments are now commercially available. The membranes used for each electrode are primarily those listed in Table 28.2.

A common design for a clinical analyzer incorporates the ISEs in a flow-through configuration. Calibrants, samples, and flush solutions are pumped across the electrode surfaces, which are placed in series. A single reference electrode is used for all ISEs in a system with the exception of the CO_2 sensor, which has its own reference electrode behind the gas-permeable membrane. Calibrants are constituted with constant total ionic strength, which closely matches that of the physiological samples to minimize errors involving liquid-junction potential differences between samples and standards.

Kodak has developed a unique approach for its Ektachem analyzer, which is based on disposable ISE slides. Figure 28.15 shows a diagram of the electrodes found in an Ektachem slide for K^+. The electrode consists of a support base, a Ag/AgCl layer that serves as the reference electrode, and a salt bridge made of KCl deposited in a film. On top resides the K^+ selective membrane containing valinomycin in a hydrophobic medium. The slide has two such electrodes joined by a conducting bridge. Sample is deposited in one opening and reference KCl solution in the other; the K^+ concentration of the sample is determined from the difference in potential between the two half-cells. The dimensions of the slide are 2.8 $\times$ 2.4 cm by 150 μm thick. Only 10 μL of sample and reference solution are needed. Similar Ektachem slides have been developed for Na^+, Cl^-, and CO_2.

ISEs have been used to monitor patients during open-heart surgery and in intensive care

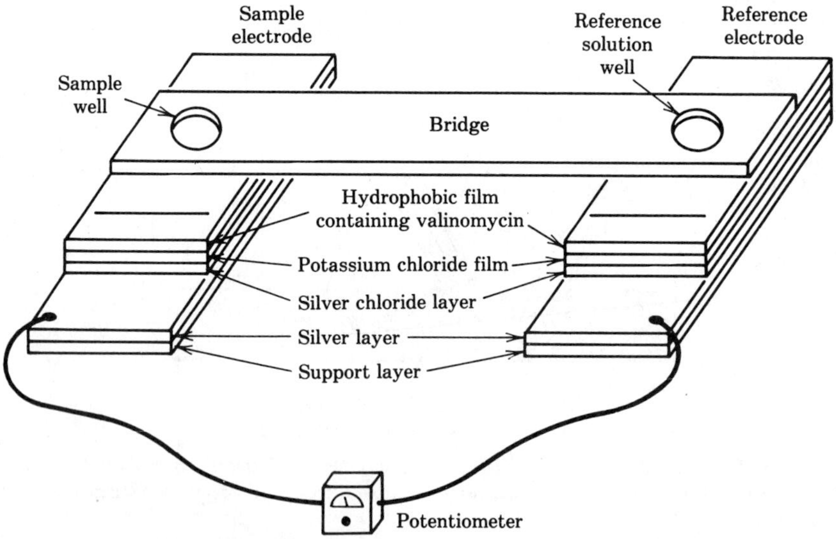

Fig. 28.15 Diagram of Eastman Kodak Ektachem slide for K^+ determination. [From B. Walter, *Anal. Chem.*, **55,** 88A (1983), with permission.]

units. Potassium concentrations in blood can vary during open-heart surgery. The normal physiological concentration should be present before the heart is reactivated. Continuous near-instantaneous (within 20 s) determination of K^+ have been made with ionophore-based ISEs in whole blood during open-heart surgery. On-line continuous monitoring of K^+ in the urine of intensive care patients has also been demonstrated. The electrode systems require a minimal amount of space in the usually crowded intensive care unit.

28.11 POTENTIOMETRIC TITRATIONS

Potentiometry is used widely as an end-point method in titrations. In the concentration range down to 10^{-2} *M* a potentiometric titration is one of the common precise instrumental techniques for analysis of inorganic samples. Some organic samples are also so analyzable. The reasons are not hard to find. Virtually any "complete" reaction can be followed potentiometrically whether it is a redox, precipitation, complexation, or acid–base type of reaction. One need only to find an indicator electrode or ISE that responds to the analyte, the titrant, or the product.

Determination of End Point. Representative potentiometric titration curves are the titration of monoprotic acids with NaOH shown in Fig. 28.16. Solution pH is monitored with a pH electrode during the course of the titration. The pH changes abruptly at the equivalence point, which is a characteristic feature of the potentiometric titration.

The magnitude of the potential change at the equivalence point varies depending on the titration. For example, the strong acid titration in Fig. 28.16 gives a much sharper break for the end point than the weakest acid (pK_a = 9.0) does. The

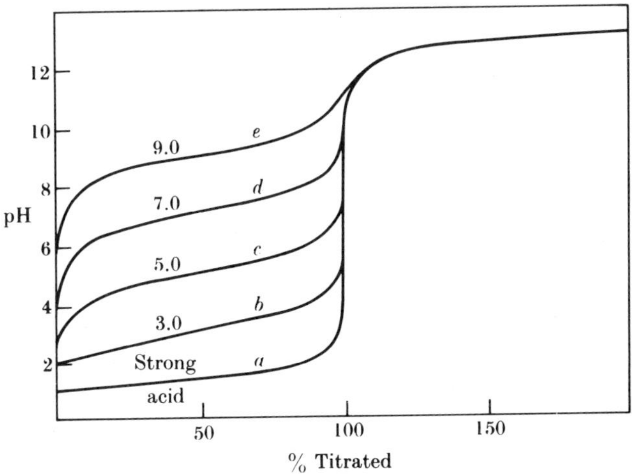

Fig. 28.16 Titration curves of 0.1 *M* monoprotic acids with 0.1 *M* NaOH. The pK_a of the acid is indicated on each curve.

precision of a potentiometric titration is closely related to the rate of change of the indicator electrode at the end point. The titration curve should have a steep slope in the end-point region to ensure high precision. The end point of the titration is generally taken as the point of inflection in the potential versus volume curve (i.e., the point of steepest slope). Figure 28.17 shows graphical procedures for identifying this value. If there is a distinct break in a titration curve, it is easy to establish the end point by inspection. For example, in the titration of a strong acid by a strong base (see curve *a* in Fig. 28.16) one can locate the end point visually with a precision of 1% or better. The majority of titrations yield curves more like that in Fig. 28.17a, however, where the "break" is a long region of moderate slope.

Even in these cases, it is easy to locate the inflection point by graphing the first derivative dE/dv versus v as in Fig. 28.17b. Since the derivative is just the slope of the curve, the maximum corresponds to the point of inflection. The graph can be prepared manually, in which case $\Delta E/\Delta v$ is calculated for several points on either side of the estimated end point and plotted, or automatically by electronic differentiation of cell potential during the titration or by computer calculation.

It is possible to locate the end point still more precisely by obtaining the second derivative d^2E/dv^2 and graphing it versus v as in Fig. 28.17c. Since the rate of change of slope at a point of inflection is zero, its value falls to zero at the end point. Usually, since it is tedious to take points sufficiently close for a good calculation, manual procedure fails to give a good second derivative curve. Electronic differentiation or computer calculation is eminently successful, however.

Does the end point, which was mathematically defined, ever fail to coincide with the *chemical equivalence* point? Actually, only if the titration curve is symmetrical about the end point will the equivalence point of the chemical reaction be identical with the inflection point. For symmetry in the end-point region, the indicator electrode must function reversibly and equal numbers of molecules of reactant and titrant must be consumed in the reaction. For example, the neutralization reaction $HCl + NH_3 \rightarrow NH_4Cl$ gives a curve that is symmetrical and the precipitation reaction $2Ag^+ + CrO_4^{2-} \rightarrow Ag_2CrO_4$ does not. The lack of symmetry is seldom serious; agreement between equivalence and inflection points is about

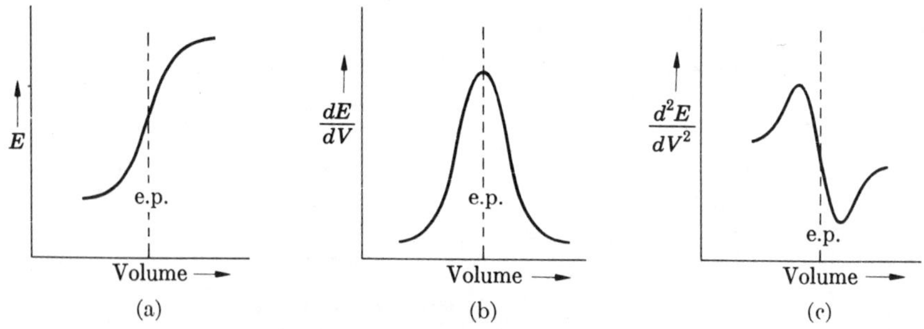

Fig. 28.17 E_{cell} versus volume of titrant added for the end-point portion of a representative potentiometric titration. (a) Direct plot. (b) First derivative plot. (c) Second derivative plot.

0.01% for any quite complete reaction. Even if a reaction is only moderately complete, agreement will probably be 0.1%. Usually this source of error is unimportant, but it may be allowed for if necessary (Section 28.15).

28.12 POTENTIOMETRIC TITRATORS

Manual Titrations. In its simplest form, the instrumentation for a potentiometric titration is the apparatus shown in Fig. 28.1 with the addition of a buret for the accurate and precise delivery of titrant to the sample solution. Titrant can also be added from a syringe, the plunger of which is motor driven or from a reservoir by pumping. In the first case the volume of titrant is known from the position of the lead screw at beginning and end of the titration. In the second method, both rate and duration of pumping must be known. Excellent results can be obtained with a *manual* titration if good experimental technique is employed.

Automatic Titrators. The addition of titrant with the detection of a potentiometric end point may be carried out automatically. A schematic representation of a modern, computer-controlled automatic titrator is shown in Fig. 28.18. The automatic piston buret can add titrant rapidly in continuous addition or in increments as small as 1 μL. The buret volume is divided into 10,000 pulses for precise addition of titrant from 0.001 to 99.99 mL. The titrant is delivered into the sample from a capillary whose tip is beneath the surface of the solution in the cell. The cell consists of the usual indicator and reference electrodes immersed in the sample,

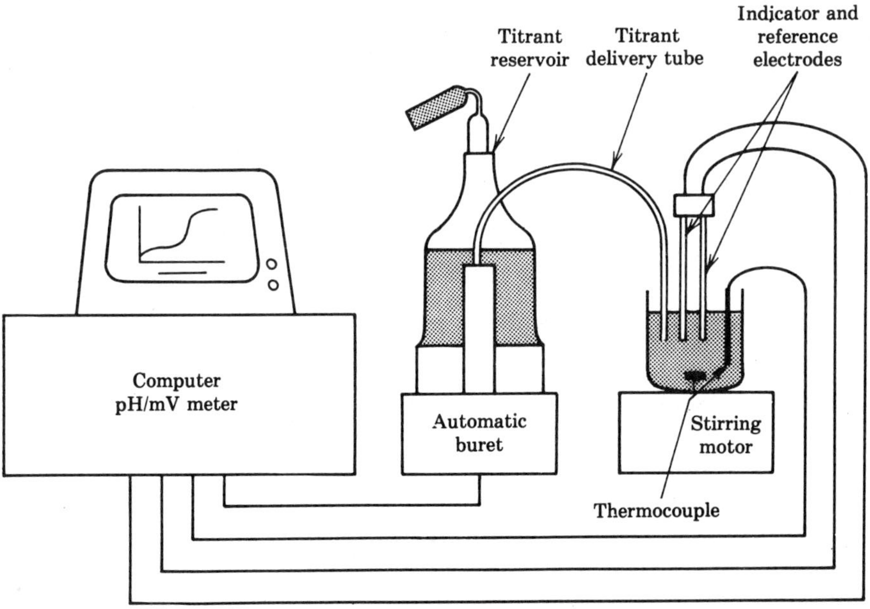

Fig. 28.18 Automatic titrator.

which is stirred. A thermocouple provides temperature monitoring. The cell potential is measured with a pH/mV meter. The computer controls the components of the apparatus and collects the data. The titration curve is displayed on a video terminal as the titration proceeds. The instruments also records the first derivative mode and automatically adapts titrating speed to the slope of the curve by control of the pump. Thus, titrant is added more and more slowly as the end point is approached. A plotter provides a copy of the titration curve, titration parameters, end-point volumes, and calculated results. The end point can be determined from the inflection point of the titration curve or by titration to an end-point potential set by the operator.

28.13 ACID–BASE TITRATIONS

This section is the first of four devoted to different types of titrations. Most potentiometric titrations of acids or bases are carried out using a glass–calomel electrode pair. If the titration is being performed for the first time, one should obtain the whole curve to be sure that no useful detail is lost and that all important variables are under control. Otherwise, titrant need be added only to the end point.

What limits, if any, does the requirement of a precisely detectable end point impose on the concentration and strength of a substance that is to be determined? In an aqueous acid–base titration a precise analysis is possible if (a) an acid or base is appreciably stronger than water (as acid or base, respectively) and (b) the acid or base is substantially more concentrated than the hydrogen ion or (essentially) hydroxyl ions from the water. With these criteria, approximate lower limits on acid and base strength for a reliable determination can be given. To be titrable with adequate precision when 0.1 *M* titrant and 0.1 *M* sample are used, it is necessary that

1. for an acid, $pK_a < 8$ or for a base, $pK_b < 8$;
2. for a salt of weak acid, $pK_a > 4$; and
3. to distinguish a strong and a weak acid or a strong and a weak base $pK_{strong}/pK_{weak} > 10^3$.

In Fig. 28.18 the basis for these limits* is made quite clear.

Example 28.7 As may be seen from a comparison of curves *c* and *d* in Fig. 28.16 the titration of acetic acid ($pK_a = 4.76$) should yield a much sharper end point than the titration of H_2S ($pK_a = 7.00$) to its first end point. A weak acid reacts incompletely with a strong base, and some unneutralized base is present at the equivalence point. As a result, the pH at this point is greater than 7, and the characteristic end-point region is short.

If one titrates a very weak acid, for example, boric acid ($pK_a = 9.23$), a curve like *e*

*If sample concentrations fall below 0.1 *M*, the first limit is subject to revision in the direction of smaller p*K* values, and the second to larger pK_a values.

in Fig. 28.16 is obtained. There is enough unreacted base at the end point to make the pH nearly that of the strong base titrant. The end point cannot be located with any reliability.

Polyprotic acids (e.g., H_3PO_4 or H_2CO_3) or mixtures of a strong and a weak acid or a strong and a weak base may be titrated if the end-point regions are clearly distinguishable. A quantitative criterion was suggested above. The existence of multiple end points for a polyprotic acid make its determination more flexible, a fact that may often be used to advantage in analyzing mixtures of acids. Thus, malonic acid ($pK_1 = 2.85$ and $pK_2 = 5.66$) may just be potentiometrically titrated in the presence of benzoic acid ($pK_a = 4.20$) by titration to the first end point.

Clearly, atmospheric CO_2 should be excluded from a basic sample before and during titration. If a strong acid is being titrated, CO_2 does not interfere, since the end point for the strong acid can be distinguished from that of the carbonic acid. On the other hand, CO_2 should be excluded from a weak acid or polyprotic acid sample.

Procedure. The rate of addition of titrant must allow adequate time for mixing and response. Thus, an appropriate rate depends strongly on such factors as the speed of stirring, solution volume, location of the indicator electrode, and nature of the liquid junction. For example, if one uses a power stirrer, as in an automatic titrator, the cell potential is erratic unless a reproducible liquid junction is established during the stirring.

Perhaps a 30-s delay before reading the cell potential is adequate for most manual titrations. A reasonable rule is to wait until the pH reading does not change more than 0.02 pH units per minute.

It should also be noted that some systems react slowly with the acid or base. Slurries of any kind and some very weak acids fall in this category. In a new system, the rate of response should always be checked.

Qualitative Analysis. It is possible to deduce visually from a titration curve the approximate strength of an acid or base. Ordinarily, the pH value at half-neutralization is used, since in aqueous solution it gives the apparent pK_a of a monoprotic acid directly.* Thus, in curve *b* in Fig. 28.16 the half-neutralization pH observed is about 3. For a "monoacidic" base the pOH value at half-neutralization equals pK_b.

Such information is particularly helpful in characterizing organic compounds with acidic or basic properties, for it furnishes an index as to which of several kinds of groups is present. For example, information about acid strength would allow sulfonic acid ($pK_a \sim 3$), carboxyl ($pK_a \sim 5$), and phenol ($pK_a \sim 10$) groups to be distinguished. If a substance contains more than one type of acidic or basic group, each half-neutralization pH may be used to establish a rough value of dissociation constant.

*To obtain the thermodynamic value of pK_a, a correction for activity effects would be needed. Most acid and base dissociation constants are determined potentiometrically.

28.14 PRECIPITATION AND COMPLEXATION TITRATIONS

Conditions that must be satisfied if an ion is to be determined quantitatively by a potentiometric precipitation or complexation titration are:

1. The ion must readily form a salt of low-solubility product ($K_{sp} < 10^{-8}$ for a binary compound) or a stable complex of high formation constant.
2. A reliable electrode must be found for one of the ions in the reaction.
3. The reaction must be relatively fast.

If these conditions hold, at the equivalence point the concentration of some reactant or soluble product changes very rapidly with volume of reagent added, and there is a distinct potential break.

Precipitation Titrations. Insoluble salts that meet the requirements given are unfortunately not numerous. They are mainly those containing silver and mercury and less often, copper, zinc, and a few other metals. Anions determined in this way include Cl^-, Br^-, I^-, CN^-, SCN^-, SO_4^{2-}, S^{2-}, and those of some organic acids. Fig. 28.19 graphs three curves representing the formation of salts of varying solubility.

The indicator electrode may be the metal participating in the reaction, a noble metal, an ISE, or an electrode of the second kind. For example, in the titration of iodide ion by $AgNO_3$, a silver wire or even a piece of silver-plated platinum may be used. Alternatively, a noble metal like platinum can often be used. (The half-cell potential of the platinum electrode would be determined by the iodine–iodide ion couple.) The reference electrode must, of course, be appropriate to the titration but is usually glass, calomel, or a metal electrode. The reference electrode should be isolated from the titration cell by means of a salt bridge if constituents leaking

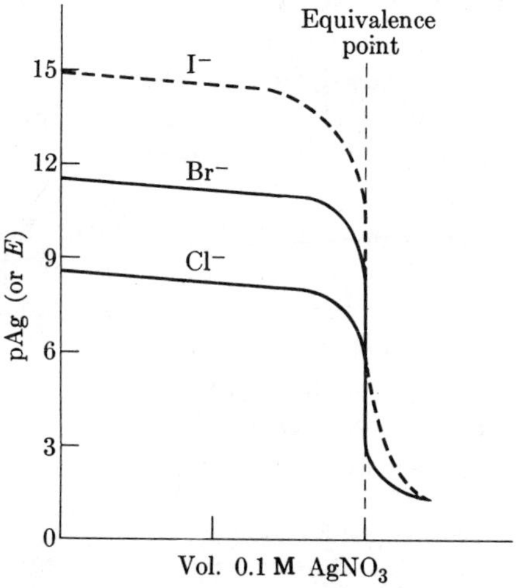

Fig. 28.19 Titration curves for the determination of I^-, Br^-, and Cl^- by titration with $AgNO_3$ solution. A silver indicator electrode is used. Absorption effects cause the "rounding off" of the I^- curve, which is represented by a broken line.

from the reference electrode junction constitute an interference in the titration reaction.

Example 28.10. Fluoride ion may be determined by titration with standard La^{3+} solution by the reaction $3F^- + La^{3+} \rightleftharpoons LaF_3$. A F^- ISE may be used as the indicator electrode.

The occurrence of *coprecipitation*, *adsorption*, or even *simultaneous precipitation* may lead to uncertainty in the detection of the end point. The first process can result in a very slow attainment of equilibrium. In this case, one might easily appear to have reached the end point too soon. One measure often taken to reduce coprecipitation is to maintain the solution above room temperature. Naturally, this means is feasible only if any accompanying increase in solubility can be tolerated. The second factor, adsorption on the precipitate, must be considered separately for each precipitate, and cannot be eliminated readily. Figure 28.19 shows its effect on the titration of I^-. Simultaneous precipitation, the third undesirable behavior, occurs in the titration of mixtures such as in the titration of Br^- and I^-. Since some AgBr precipitates before all the AgI has come out of solution, the effect of simultaneous precipitation is to give a high end point for the I^- ion. A similar situation exists for the Br^- ion. Where these three phenomena cannot be eliminated it is often possible to estimate corrections to apply to the actual data by titration of knowns under identical conditions.

Mechanical stirring is highly desirable, in order to hasten equilibration. It is advantageous to form the least soluble precipitate of an ion possible to secure a sharp end-point break. For example, titration of silver with iodide ion ($K_{AgI} \sim 10^{-16}$) is preferable to titration with chloride ($K_{AgCl} \sim 10^{-10}$). If only a moderately insoluble precipitate forms, close temperature control is necessary to ensure reproducibility and accuracy. Sometimes ethanol or methanol may be added to lower the solubility sufficiently to make regulation unnecessary. Indeed, it may be necessary to add nonaqueous solvents if the salt is to be insoluble enough for the titration to be run.

If the titration involves the anion of a weak acid, such as oxalate, pH control is also necessary; otherwise the concentration of the anion varies with acidity, and the solubility is affected. Commonly a noninterfering buffer is added.

Complexation Titrations. The avenue to the potentiometric determination of a number of metals has been opened by the possibility of titration using their reaction with polydentate complexing reagents (i.e., chelating agent). The best known of these is ethylenediaminetetraacetic acid (EDTA), which is used in a partially neutralized form. A typical complexation reaction is of the form

$$M^{2+} + H_2Y^{2-} = MY^{2-} + 2H^+,$$

where H_2Y^{2-} represents the anion of the disodium salt of EDTA. Note that a single, well-defined complex is shown as being formed. Stability constants are high except for the alkali metals and a few other metals.

In this type of titration, the end point can be detected with an ISE that is sensitive to the metal ion being complexed or with an electrode of the third kind (Table 28.1).

28.15 REDOX TITRATIONS

The potentiometric method is ideal for monitoring a redox reaction. In such a system, the best indicator electrode is a piece of metal sufficiently noble to be unaffected by the oxidizing agent. Platinum is widely used. Either a short length of wire or a small square is adequate. Noble metals give a somewhat faulty response when both a strong oxidizing agent and F^- ion or Cl^- ion are present in moderate to high concentration. Noble metals are no longer inert when their complex halide ions, for example, $PtCl_6^{2-}$, may be formed. In redox titrations, a calomel or glass electrode may be used as a reference.

Titration Curves. The calculation of cell potential during a redox titration is simple because only the potential of the indicator electrode will vary. Accordingly, one need calculate only E_{ind}. Then E_{cell} can be found from Eq. (28.1). Consider the system in which reductant red_1 is titrated with oxidant ox_2. The reaction can be represented as

$$red_1 + ox_2 = ox_1 + red_2. \qquad (28.24)$$

It is convenient to write the two half-reactions of the system as reductions

$$ox_1 + e = red_1,$$

$$ox_2 + e = red_2.$$

When the indicator electrode is in equilibrium with the solution, its potential will be given at 25°C by the equation

$$E_{ind} = E_1^0 - 0.0591 \log \frac{[red_1]}{[ox_1]} = E_2^0 - 0.0591 \log \frac{[red_2]}{[ox_2]}, \qquad (28.25)$$

where E_1^0 and E_2^0 are the two standard reduction potentials and brackets denote activities. Note that Eq. (28.25) must follow since the indicator electrode is in equilibrium with both redox couples yet it can have only one potential. How does E_{ind} vary during a titration? Consider again the general reaction given in Eq. (28.24).

Prior to the equivalence point, $[ox_2] \cong 0$ and only the redox system associated with the substance titrated, $ox_1 + e = red_1$, contributes. If f is the fraction of red_1 titrated,

$$E_{ind} = E_1^0 - 0.0591 \log (1 - f)/f \quad (\text{when } f < 1). \qquad (28.26)$$

At the equivalence point, activities of products of the titration will be equal: $[ox_1] = [red_2]$. It also follows that $[red_1] = [ox_2]$. The solution of Eq. (28.25) for this case proves especially interesting. From Eq. (28.25)

$$E_{ind} = E_1^0 - 0.0591 \log [red_1]/[ox_1]$$

and

$$E_{ind} = E_2^0 - 0.0591 \log [red_2]/[ox_2].$$

Adding the equations and substituting the equalities noted leads to the desired solution. One obtains

$$2E_{ind} = E_1^0 + E_2^0 - 0.0591 \log \frac{[red_1][red_2]}{[ox_1][ox_2]}$$

$$= E_1^0 + E_2^0 - 0.0591 \log [1] = E_1^0 + E_2^0$$

and

$$E_{ind} = (E_1^0 + E_2^0)/2 \qquad (\text{when } f = 1). \tag{28.27}$$

In an unsymmetrical titration, Eq. (28.27) becomes

$$E_{ind} = (n_1 E_1^0 + n_2 E_2^0)/(n_1 + n_2),$$

where n_1 and n_2 are the number of moles of electrons involved in reactions for which the standard potentials are E_1^0 and E_2^0, respectively. It is left to the reader in Exercise 28.13 to establish that this equation is valid.

Beyond the equivalence point, $[red_1] \cong 0$ and E_{ind} is determined mainly by the second redox system, $ox_2 + e = red_2$. Now Eq. (28.25) becomes

$$E_{ind} = E_2^0 - 0.0591 \log 1/(f - 1) \qquad (\text{when } f > 1), \tag{28.28}$$

where f is again the fraction of red_1 titrated. In this range $f > 1$.

Formal Potentials. Since the ionic strength of solutions in redox titrations is usually high (e.g., 1 *M* H_2SO_4 may be present), it is impossible to predict activity coefficients of species. A useful empirical approach that still permits calculations of E_{ind} to be made is the use of formal potentials (see Section 27.3).

The break in a titration curve is determined by the difference in redox potentials of reductant and oxidant. Consider the typical potential shift during a redox change. In Fig. 28.20 the variation of electrode potential with fraction reduced is shown for several species that are commonly employed as oxidizing agents. To determine the amount of such substances accurately (to $\pm 0.1\%$) at the conclusion of a reduction the potential must be as small as $E^0 - (2.303\, RT/nF) \log 1000 = E^0 - 0.18/n$. For the oxidation of the titrant a similar result follows: the final potential

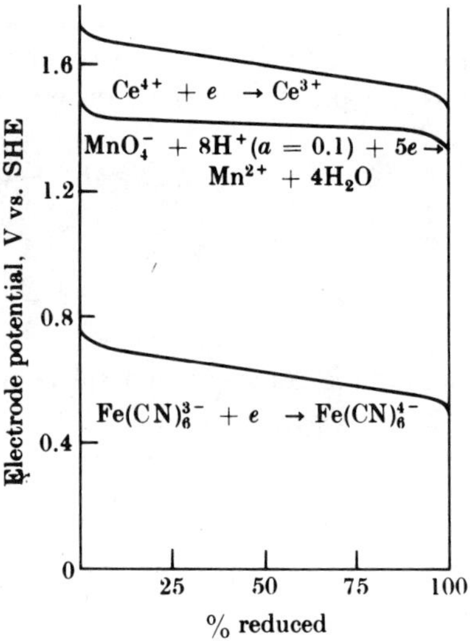

Fig. 28.20 Electrode potential changes during reduction of species shown. The slope of the flat central portion of each curve is determined by n, the number of electrons per mole of reactant involved in the reduction.

must be as large as $E^0 + 0.18/n$. If this be so, how strong a reducing agent must one use as the titrant? The answer is straightforward. If $n = 1$ for both, the standard potentials of oxidant and reductant must differ by 0.36 V; if $n = 2$, by 0.18 V, and so on.

Conditions. Most redox reactions are inherently slow because of the complexity of this type of chemical change. For example, the reduction of permanganate ions involves several consecutive reactions. As would be expected, the use of a catalyst and carrying out the titration at a higher temperature often speed up these reactions sufficiently so that they can be followed potentiometrically. Nearly always, the acidity of the solution must be maintained within limits to avoid incomplete reactions. A particular redox change often occurs only within a restricted pH range. A buffer may be unnecessary; it may suffice to add excess acid or base. A buffer is essential, however, if one or more anions of weak acids are involved in the reaction.

In many instances air (or oxygen) must be excluded from a titration. If good precision is required (better than 1%) or if very active reductants such as the chromous ion are used, the titrant and the titration vessel should be protected by an atmosphere of an inert gas. In addition dissolved oxygen must be removed from the sample solution prior to titration by bubbling the inert gas through.

Example 28.8. Ferrous ion may be determined quantitatively by titration with permanganate ion in a solution strongly acidified with sulfuric acid (concentration of $H_3O^- \sim 1\ M$). During the titration, the iron is oxidized to the 3^+ state, and the manganese is reduced to the $2+$ state:

$$5Fe^{2+} + 8H^+ + MnO_4^- = 5Fe^{3+} + Mn^{2+} + 4H_2O.$$

Although the reaction involves one permanganate ion for every five ferrous ions and is highly asymmetrical (Section 28.11), the rate of change of potential as titrant added is so great that no appreciable error results if we take as the end point the point at which the slope has its maximum value.

28.16 ACID–BASE TITRATIONS IN NONAQUEOUS SYSTEMS

Although a great many organic substances possess acidic or basic character, they do so to so slight a degree that they cannot be titrated potentiometrically in water. Water fails as a medium for titrating a very weak acid because of its own weakly acidic character. It will compete with the weak acid of interest on titration with a base and thus diminish the sharpness of the end point. Clearly, substitution of a less acidic solvent like dimethylformamide or an inert one like acetone will correct this situation. Note that the degree of basic character of the nonaqueous solvent is of no consequence.

Just the opposite criterion applies in selection of a nonaqueous solvent suitable for potentiometric titration of a very weak base B: the solvent must be less basic than water. For example, acetic acid is commonly used. It is desired that the reaction

$$B + H^+_{\text{titrant}} \rightarrow BH^+$$

be essentially complete. In acetic acid completeness is more nearly assured than in water since the competing protonation of solvent SH

$$SH + H^+ \rightarrow SH_2^+$$

is comparatively slight.

Titration curves obtained in nonaqueous media are similar in shape to those obtained in water. The only important difference is that only potential data are available in nonaqueous media.*

Titrants are somewhat standardized. The acid or base used should be as strong as possible so that it will drive the neutralization essentially to completion. Thus, in the titration of weak bases, the very strong acid perchloric acid is preferable to HCl, HNO_3, or HBr. Similarly, the strongest bases are used for titrations of weak acids. These substances include materials such as sodium or potassium methoxide, tetrabutylammonium hydroxide, and sodium aminoethoxide, $NaOCH_2CH_2NH_2$. The titrant is ordinarily dissolved in the solvent to be used. Such reagents are more sensitive to contamination and may be less constant in titer because of high solvent vapor pressure.

*A unique pH scale can be established for each solvent (Section 28.7), but there is no theoretical basis on which to relate such pHs to aqueous values [32]. One problem is that it is not known how to determine the liquid-junction potential between solutions in different solvents.

Example 28.9. Vanillin, a substituted phenol of formula

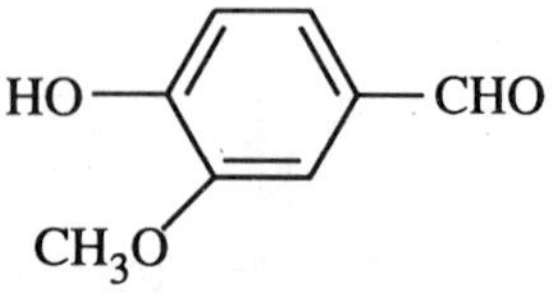

is a somewhat stronger acid than phenol and may be titrated in ethylenediamine or dimethylformamide. A 0.1-*M* sodium aminoethoxide solution can be used as the titrant and is prepared by dissolving sodium in ethanolamine. Anhydrous solvent must be used. A double set of antimony electrodes can be used, one in the tip of the buret, the other in the solution; these are formed by casting high-grade metal into rods, using glass tubing as a mold. There is a drop of more than 150 mV in the end-point region. (A glass indicator electrode cannot be used with this titrant, apparently because it functions as an ion-selective electrode for sodium.)

Conditions. In most titrations of weak bases, water can be tolerated in concentrations of up to 3% of the weight of the original solvent. Greater concentrations lead to poor end points and inaccuracy. Alcohols also interfere because of their basic properties. However, only if the system is free of water is it possible to carry out most titrations of weak acids. The presence of water is the cause of most interferences with very weak acids and bases. Other substances should be checked for possible interference. As examples, (a) CO_2 is readily absorbed by basic solvents unless air is excluded, and (b) most inorganic cations interfere in titrations in acetic acid, since they release hydrogen ions as they form slightly dissociated acetates.

Electrodes. In nonaqueous titrations in acidic media, the indicator electrode is commonly the glass pH electrode. To function with any speed, however, its surface layers must be hydrated; presoaking it in water does this.* Minimal dehydration occurs during a regular nonaqueous titration. Electrode response may, however, be sluggish.

Either a calomel reference electrode with an integral sleeve type of salt bridge or a silver–silver chloride reference electrode may be employed. If a regular commercial calomel electrode is used, it may require periodic soaking in water. Since KCl is not as soluble in most nonaqueous solvents as in water, it tends to precipitate from solution at the junction of the salt bridge. These crystals will block ionic migration at the boundary unless periodically redissolved. A good practice is to isolate the calomel electrode with a salt bridge that contains a suitable nonaqueous solvent and supporting electrolyte.

*The presence of water precludes attaining an absolute value of potential in the medium. Absolute potentials, if needed for activity determinations, may be obtained with a dry-glass electrode providing sufficient time is allowed for equilibration.

Unfortunately, the glass electrode does not operate with success in some basic titration media. Instead, the hydrogen or antimony–antimonious oxide electrode has been used together with a calomel reference.

Bimetallic electrode systems, that is, systems employing two identical metal electrodes, are also common. In this case one electrode is inserted in the titrant stream, and the tip of the other is placed in the bulk of the solution. Examples are "double" antimony and "double" platinum electrodes. The analogous "sheltered" reference, in which one electrode of an identical pair is made a reference electrode by inserting it in a small, isolated part of the system being titrated, is used as well.

The measurement of cell potentials in these systems is more difficult because solution resistances tend to be large. As a result, it is essential to employ some type of pH meter (Section 28.8) or titrimeter that can cope with the high resistance. The reproducibility of electrode response tends to be poorer than in aqueous media, and it is generally desirable to obtain the end point from the titration curve rather than by titrating to a preset voltage.

Qualitative Analysis. With some experience, it is possible to estimate approximate acid and base strengths from nonaqueous titration curves. The results are relative at best.

REFERENCES

General introductions to potentiometry are available in:

1. N. H. Furman, Potentiometry, in *Treatise on Analytical Chemistry*, Vol. 4, Part I, I. M. Kolthoff and P. J. Elving (Eds.). New York: Interscience, 1963.
2. I. M. Kolthoff and H. A. Laitinen, *pH and Electro Titrations*, 2d ed. New York: Wiley, 1941.
3. J. J. Lingane, *Electroanalytical Chemistry*, 2d ed. New York: Interscience, 1958.
4. C. Tanford and S. Wawzanek, Potentiometry, in *Physical Methods of Organic Chemistry*, 3d ed., Part 4, A. Weissberger (Ed.). New York: Interscience, 1960.

The theory, instrumentation and experimental procedure in determination of pH are set forth in detail in:

5. R. G. Bates, *Determination of pH; Theory and Practice*. New York: Wiley, 1964.
6. G. Matlock, *pH Measurement and Titration*. New York: Macmillan, 1961.

Glass and other ion-selective electrodes or reference electrodes are treated in:

7. P. L. Bailey, *Analysis with Ion-Selective Electrodes*, 2d ed. London: Heyden, 1980.
8. R. P. Buck, Electrochemical methods: Ion-selective electrodes, in *Water Analysis*, Vol. II. New York: Academic Press, 1984, Chapter 6.
9. R. P. Buck, Electrochemistry of ion-selective electrodes, in *Comprehensive Treatise of Electrochemistry*, Vol. 8, R. E. White, J. O'M. Bockris, B. E. Conway, and E. Yeager (Eds.). New York: Plenum Press, 1984, Chapt 3.
10. A. K. Covington (Ed.), *Ion-Selective Electrode Methodology*, Vols. I, II. Boca Raton, Florida: CRC Press, 1979.

11. R. A. Durst (Ed.), *Ion-Selective Electrodes*. Washington: U.S. Government Printing Office, 1969.
12. G. Eisenman (Ed.), *Glass Electrodes for Hydrogen and Other Cations*. New York: Dekker, 1967.
13. H. Freiser (Ed.), *Ion-Selective Electrodes in Analytical Chemistry*, Vols. I, II, New York: Plenum, 1978, 1980.
14. J. Koryta, *Ion-Selective Electrodes*. Cambridge: Cambridge University Press, 1975.
15. G. J. Moody and J. D. R. Thomas, *Selective Ion Sensitive Electrodes*, Watford England: Merrow Publishing Co. Ltd., 1971.
16. W. R. Morf, *The Principles of Ion-Selective Electrodes and of Membrane Transport*, New York: Elsevier, 1981.

The use of ion-selective electrodes in biology and medicine is described in:

17. M. Kessler, L. C. Clark, Jr., D. W. Lubbers, I. A. Silver, and W. Simon (Eds.), *Ion and Enzyme Electrodes in Biology and Medicine*. Baltimore: University Park Press, 1976.
18. J. Koryta (Ed.), *Medical and Biological Applications of Electrochemical Devices*. New York: Wiley, 1980.
19. M. E. Meyerhoff, W. N. Opdycke, Ion-selective electrodes, in *Advances in Clinical Chemistry*, Vol. 25, H. E. Spiegel (Ed.). New York: Academic Press, 1986, p. 1.
20. W. Simon, D. Ammann, W. Bussmann, and P. C. Meier, Ion selective electrodes in biology and medicine, in *Frontiers of Chemistry*, K. J. Laidler (Ed.). New York: Pergamon Press, 1982, p. 217.
21. R. L. Solsky, Ion-selective electrodes in biomedical analysis, in *CRC Critical Rev. Anal. Chem.*, **14,** 1 (1982).

Ion-selective microelectrodes are described in:

22. D. Ammann, *Ion-Selective Microelectrodes*. New York: Springer-Verlag, 1986.
23. H. J. Berman and N. C. Hebert (Eds), *Ion-Selective Microelectrodes*. New York: Plenum Press, 1974.
24. E. Sykova, P. Hnik, and L. Vyklicky (Eds.), *Ion-Selective Microelectrodes and Their Use in Excitable Tissues*. New York: Plenum Press, 1981.
25. J. L. Walker, *Anal. Chem.*, **43,** 89A, 1971.
26. T. Zeuthen (Ed.), *The Application of Ion-Selective Microelectrodes*. New York: Elsevier/North-Holland Biomedical Press, 1981.

The theoretical and phenomenonological aspects of pH and acid–base titrations are treated in:

27. E. J. King, *Acid–Base Equilibria*. New York: Macmillan, 1965.
28. I. M. Kolthoff and S. Bruckenstein, Acid–bases in analytical chemistry, a reprint from *Treatise on Analytical Chemistry*, Vol. I, Part 1, I. M. Kolthoff and P. J. Elving (Eds.). New York: Interscience, 1959.
29. J. E. Ricci, *Hydrogen Ion Concentration*. Princeton: Princeton University Press, 1952.

General introductions to titration in nonaqueous solvents are given in:

30. W. Huber, *Titrations in Nonaqueous Solvents*. New York: Academic, 1967.
31. J. S. Fritz, *Acid–Base Titrations in Non-aqueous Solvents*. Boston: Allyn and Bacon, 1973.

Other papers or chapters of interest that concern specialized aspects of potentiometry are:

32. R. G. Bates, Standardization of acidity measurements, *Anal. Chem.*, **40**(6), 28A (1968).
33. D. G. Davis, Potentiometric titrations, in *Comprehensive Analytical Chemistry*. Vol. IIA. C. L. Wilson and D. W. Wilson (Eds.). New York: Elsevier, 1964.
34. J. T. Stock and W. C. Purdy, Potentiometric electrode systems in nonaqueous titrimetry, *Chem. Rev.*, **57**, 1159 (1957).
35. E. P. Sergeant, *Potentiometry and Potentiometric Titrations*. New York: Wiley, 1984.

The basic theory and practice of complexation titrations are treated in:

36. T. S. West, *Complexometry with EDTA and Related Reagents*, 3d ed. Dorset, England: BDH Chemicals, 1969.
37. G. Schwarzenbach and H. Flaschka, *Complexometric Titrations*, 5th ed. New York: Barnes and Noble, 1969.

The field of ion-selective electrodes is reviewed biennially in the April review issue of *Analytical Chemistry*. The two most recent reviews are:

38. M. A. Arnold and R. L. Solsky, *Anal. Chem.*, **58**, 84R (1986).
39. R. L. Solsky, *Anal. Chem.*, **60**, 106R (1988).

Applications, theory and development of electrochemical sensors are reviewed regularly in the journal:

40. *Ion-Selective Electrode Reviews*.

EXERCISES

28.1 The following electrochemical cell is assembled for the direct potentiometric determination of Br^-:

$$-\text{SCE}//\text{Br}^- \text{ (unknown), AgBr/Ag } +$$

When a solution of Br^- was transferred into the cell, a potential of 0.128 V was measured. Calculate the activity of Br^- in the solution.

28.2 Explain why an electrode of the second kind for Cl^- is more practical than an electrode of the first kind would be.

28.3 An ion-selective electrode for NO_3^- is dipped into a standard 1.00×10^{-4} *M* solution of NO_3^- and a potential of -0.200 V versus SCE is measured. The electrode is then dipped into a sample solution and a potential of -0.150 V versus SCE is measured. Calculate the concentration of NO_3^- in the sample solution.

28.4 Strychnine is known to react with picrolonic acid to form a stable 1 : 1 water insoluble complex as shown below:

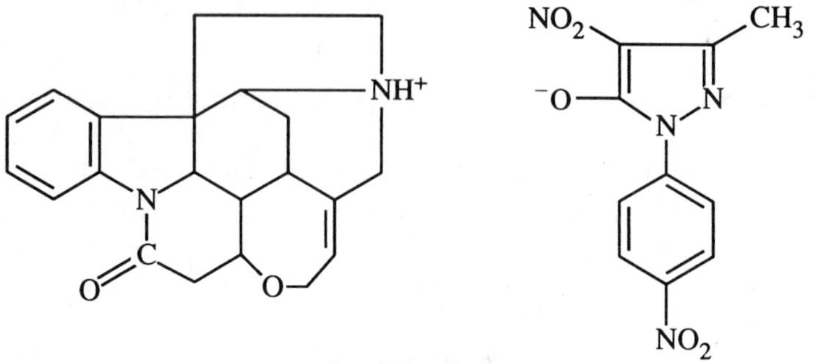

Strychnine picrolonate

Describe how an ion-selective electrode for determining strychnine in water could be made. Include in your discussion an explanation of how the electrode responds to strychnine (interfacial equilibrium involved), the appropriate Nernst equation, and the expected calibration curve.

28.5 An electrode has been developed which is entirely selective for M^{2+}. When immersed in a solution of 10^{-5} *M* M^{2+}, the potential of the electrode was 0.100 V versus SCE. Calculate the potential of the electrode after addition of ligand L^- to a total concentration of 0.2 *M*. Assume that there are no dilution effects and that the solution is buffered at pH = 5.0.
L^- complexes M^{2+} according to the following:

$$M^{2+} + L^- \rightleftharpoons ML^+, \qquad K_1 = 10^3$$

$$ML^+ + L^- \rightleftharpoons ML_2, \qquad K_2 = 10^4$$

HL is a weak acid with $K_a = 10^{-5}$.

28.6 Show that the potential of the antimony–antimonious oxide electrode is determined simply by the pH.

28.7 What is the relative error in the determination of the activity of hydrogen ion in a system when the pH is known with an accuracy of 0.05 pH unit?

28.8 A pH meter utilizing the saturated calomel–glass electrode pair is standardized with 0.05-*M* potassium tetroxalate buffer (pH = 3.56 at 25°C). (a) What E_{cell} does this pH correspond to? (b) An unknown solution gives an E_{cell} of 0.564 V. What is its pH?

28.9 It is desired to have the uncertainty in reading of the potential of an ion-selective electrode be no greater than 0.5 mV. If the total resistance of electrodes and cell is 100 MΩ, what is the maximum tolerable input current to a pH or ION meter used with the electrode?

28.10 A pH-sensitive glass electrode reads low in sodium hydroxide solutions. There is no error in reading in a tetramethyl ammonium hydroxide solution of the same composition. Discuss in terms of the model used to interpret glass electrode behavior.

28.11 An electrode-solution system is said to be poorly *poised* when electrode potential is strongly influenced by the magnitude of cell current. (a) Will the Pt; Ce^{4+}, Ce^{3+} electrode be poorly poised when only Ce^{4+} is present? When only Ce^{3+} is present? When 10^{-3} M Ce^{4+} and 10^{-2} *M* Ce^{3+} are present? (b) In a given aqueous solution, pH readings that differ by 0.04 pH unit are obtained with a particular glass electrode–SCE pair when the potential difference is read with different pH meters, yet the meters read the pH of standard buffers to within the manufacturer's specification of ± 0.01 pH unit. Buffers are well poised. Explain. [See S. Z. Lewin, *Anal. Chem.*, **33**(3), 41A (1961).]

28.12 Show that the half-neutralization pH for a weak acid equals pK_a.

28.13 Show that in an unsymmetrical redox titration the indicator potential at the equivalence point has the value $E_{ind} = (n_1E_1^0 + n_2E_2^0)/(n_1 + n_2)$. (See page 1045.)

28.14 Calculate the equivalence-point cell potential for the potentiometric titration of X^{2-} with a standard solution of Ag^+ using a silver indicator electrode. Treat the indicator electrode as the cathode; the reference electrode is a SCE. (Ag_2X, $K_{sp} = 32.0 \times 10^{-15}$.)

28.15 A sample of M^+ is titrated by precipitation with A^- to form the precipitate MA. The titration is monitored potentiometrically with an electrode of the metal M. The potential of the equivalence point is +0.500 V versus SCE (cathode). The sample also contains X^-, which forms the soluble complexes MX and MX_2^- with M^+. Calculate the K_{sp} of MA if the concentration of X^- is 10^{-1} *M* at the equivalence point.

$$M^+ + X^- \rightleftharpoons MX, \qquad K_1 = 10^7$$

$$MX + X^- \rightleftharpoons MX_2^-, \qquad K_2 = 10^5$$

$$M^+ + e \rightleftharpoons M, \qquad E^0 = +1.00V$$

28.16 A 100-mL sample of 0.1 *M* Cr^{2+} was titrated with 0.1 *M* M^{4+}. The titration reaction is

$$2\,Cr^{2+} + M^{4+} \rightleftharpoons 2\,Cr^{3+} + M^{2+}.$$

The potential measured after addition of 25 mL of M^{4+} was -0.41 V and after 100 mL was 0.100 V versus SHE. (a) What is $E^{0'}_{Cr^{3+},Cr^{2+}}$? (b) What is $E^{0'}_{M^{4+},M^{2+}}$? (c) Calculate the potential of the equivalence point of the titration.

28.17 A solution of 1.00×10^{-4} *M* F^- was titrated with standard La^{3+} solution according to the following precipitation reaction:

$$La^{3+} + 3F^- \rightleftharpoons La\,F_3.$$

The titration was monitored with a F^- ion-selective electrode. The initial potential (before addition of La^{3+}) was -0.200 V versus SCE; the potential at the titration equivalence point was -0.123 V versus SCE. Calculate the solubility product constant (K_{sp}) for LaF_3.

28.18 A solution of 10^{-3} *M* Pb^{2+} is titrated with a standard solution of CrO_4^{2-} by a precipitation reaction

$$Pb^{2+}\ (\text{sample}) + CrO_4^{2-}\ (\text{titrant}) \rightarrow PbCrO_4^{2-}.$$

The precipitation titration is monitored with a Pb^{2+} ion-selective electrode. The initial potential (before adding any CrO_4^{2-}) is found to be +0.200 V versus SCE. The potential at the equivalence point is 0.086 V. (a) Calculate the concentration of Pb^{2+} at the equivalence point of the titration. (b) Calculate the solubility product constant K_{sp} for $PbCrO_4$.

28.19 Compare direct potentiometry and potentiometric titrations with regard to optimum concentration ranges, precision, and ease of use. Explain the basis for differences.

ANSWERS

28.1 8.7×10^{-6} M *28.3* 1.4×10^{-5} M *28.8* (a) 0.210 V; (b) 9 *28.14* 0.295 V *28.18* (a) 1.35×10^{-7} M; (b) $K_{sp} = 1.8 \times 10^{-14}$.

Chapter 29

VOLTAMMETRY: BASIC CONCEPTS AND HYDRODYNAMIC TECHNIQUES

William R. Heineman

29.1 INTRODUCTION

Voltammetry is the measurement of current–potential relationships in an electrochemical cell. A variety of voltammetric techniques in which a *potential is applied* to an electrochemical cell and the resulting *current is measured* has been developed.* Since the potential applied to the cell can force changes to occur in the concentrations of species at the electrode surface by electrochemical reduction or oxidation (i.e., by *electrolysis*, see Section 27.2), voltammetry is an *active* technique. By contrast, the potentiometric techniques discussed in Chapter 28 are termed *passive* techniques since they are based on the measurement of a cell potential that is simply a response to the activity of an analyte in solution.

The basic concept of applying a potential to an electrochemical cell and measuring current resulting from electrolysis can be implemented in numerous ways. Some of the parameters of the measurement that are easily varied are:

1. The solution may be moving or stationary with respect to the electrode. *Hydrodynamic* voltammetric techniques, which involve forced convection of the solution, are dealt with in this chapter. *Stationary* (or *quiescent*) solution techniques and some "hybrid" techniques are covered in Chapter 30.

2. The waveform of the applied potential (i.e., how the applied potential varies with time) may be varied.

3. The timing sequence of the current measurement process with respect to the waveform of the excitation signal may be varied.

4. Numerous electrode and cell geometries, which affect the current response, have been developed.

All of these factors can dramatically influence the shape of the current–time response. Consequently, numerous voltammetric techniques (see Fig. 27.11) in

*Techniques in which current is applied to the cell and potential is measured are also voltammetric methods. However, these are rarely used in electroanalysis.

which the specific parameters of the measurement differ have been developed, both theoretically and experimentally. Although the output signals and the practical applications of these techniques are quite varied, they all share the common basis of applying E and measuring i or a parameter derived from i such as charge Q. The most useful of these techniques for analysis are explained and illustrated in this chapter and in Chapter 30. One objective of these chapters is to illustrate the relative advantages of the different types of voltammetry.

Research in the field of voltammetry has been literally a hotbed of activity since 1960. The theoretical underpinning and practical demonstration of some techniques have been developed by electroanalytical chemists. Voltammetry has become increasingly popular in all fields of chemistry as a means of studying redox chemistry. The electrode potential at which a drug, a metal ion or complex, or some other type of compound undergoes reduction or oxidation can be rapidly located by an electrochemical technique such as cyclic voltammetry. Furthermore, information regarding the stability of the electrogenerated product(s) can be obtained by electrochemical and spectroelectrochemical techniques. Such information is often useful in subsequent experiments involving electrosynthesis, corrosion, coulometry, amperometric titration, and trace analysis. In the last case, electrochemical techniques such as liquid chromatography with electrochemical detection, stripping voltammetry, pulse voltammetry, and differential pulse voltammetry have proved to be powerful, widely applicable techniques for analysis at the trace level. Voltammetric techniques benefit from the ease with which different potential waveforms can be generated and accurate measurements of very small current can be made with modern instrumentation.

29.2 POTENTIAL EXCITATION SIGNAL, MASS TRANSPORT, AND CURRENT RESPONSE SIGNAL

Voltammetry involves the application of a potential to an electrochemical cell, the transport of material to an electrode for electrolysis, and the measurement of current from the electrolytic process. Each of these aspects of voltammetry is considered individually in this section.

Potential Excitation Signal. The potential that is applied to the electrochemical cell and causes electrolysis to occur in a voltammetric experiment is referred to as the *excitation signal*. The potential can be changed during the experiment. The *waveform* is then the potential–time behavior of the excitation signal. Of the countless waveforms that are possible, a few have been used extensively in voltammetry. In Fig. 29.1 the waveforms are shown for the voltammetric techniques that will be discussed in Chapters 29 and 30. The forms commonly used in hydrodynamic voltammetry are a linear potential scan, a staircase step-wise "potential scan," and a constant potential. The waveform determines (in part) which voltammetric technique is being implemented. More will be said about each of these excitation signals.

The potential excitation signal is applied between the *reference electrode* and

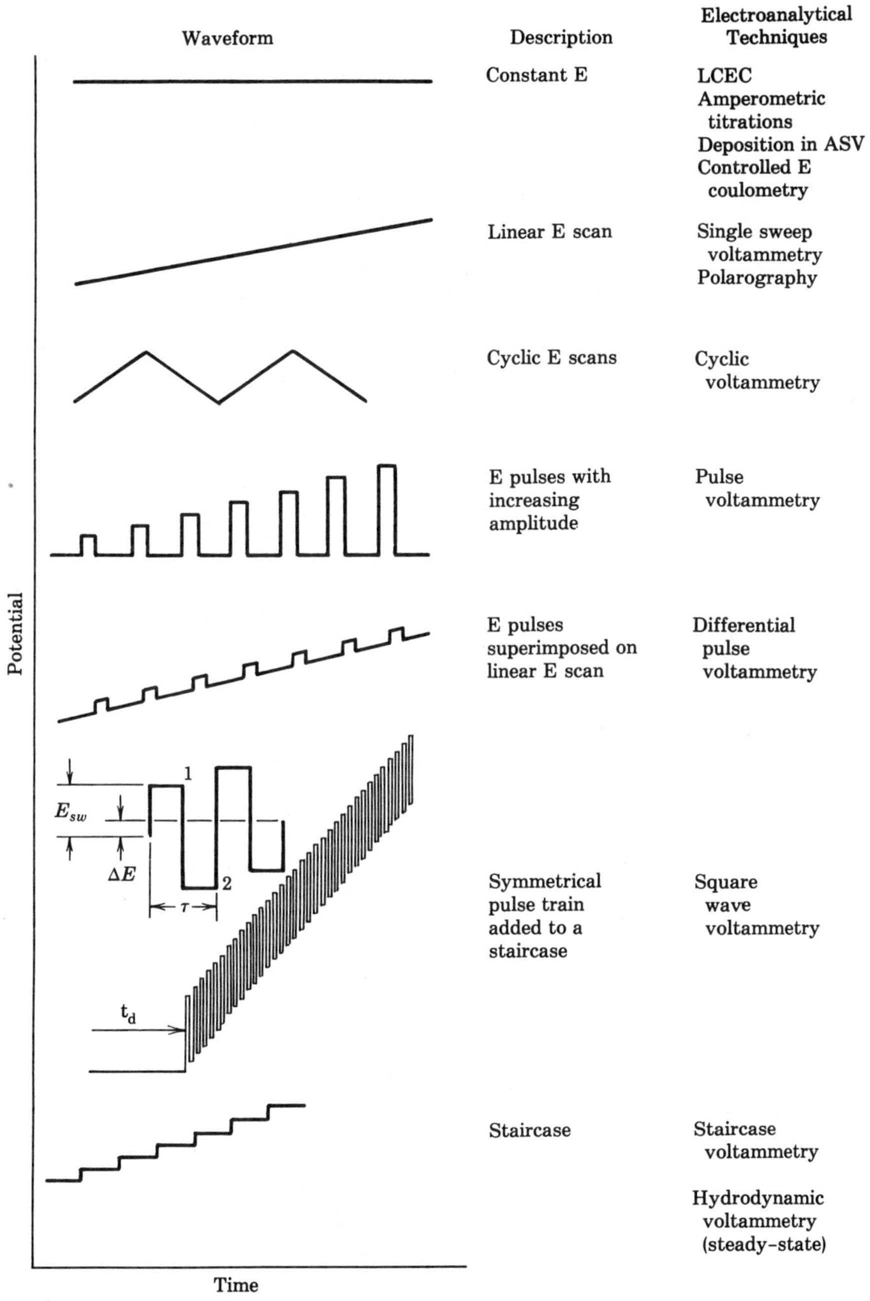

Fig. 29.1 Potential excitation signals for some common voltammetric techniques described in Chapters 29 and 30.

the *working electrode* of the electrochemical cell. Since the half-cell potential of the reference electrode is constant, the half-cell potential of the working electrode is forced to assume the value of the applied potential with respect to the reference electrode half-cell.

Potential Control and the Nernst Equation. A good understanding of voltammetric techniques is based on knowing the relationship between the potential applied to an electrode and the concentration of redox species at the *electrode surface*. Let us consider a metal electrode, such as a square of platinum foil, that is immersed in a solution containing equal concentrations of the electroactive species O and R. O can be reduced to R and R can be oxidized to O by means of the following reversible* electrochemical reaction:

$$\mathrm{O} + \mathrm{e} \rightleftharpoons \mathrm{R}. \tag{29.1}$$

Example 29.1. The symbols O and R are a convenient way to represent an electrochemical reaction for the purpose of deriving mathematical expressions. However, it can be good to have some specific chemical systems in mind while working through a "general" discussion. Two examples of Eq. (29.1) are

$$\mathrm{Fe(CN)_6^{3-}} + \mathrm{e} \rightleftharpoons \mathrm{Fe(CN)_6^{4-}}$$

and

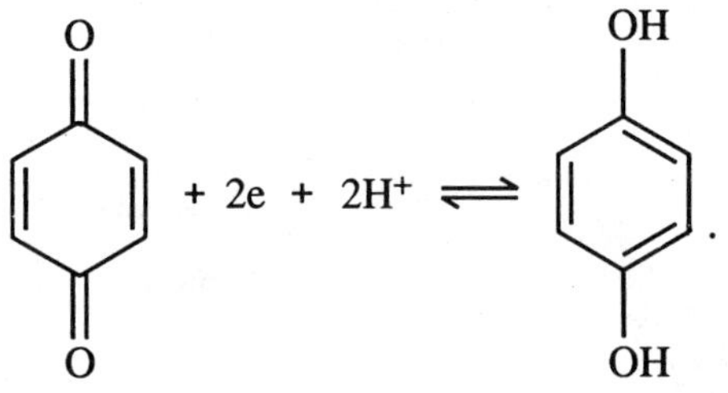

The fundamental equation that relates the potential E of the electrode and the concentration of electroactive species O and R at the *electrode surface* is the *Nernst equation:*

$$E = E^{0'}_{\mathrm{O,R}} - \frac{RT}{nF} \ln \frac{C^{\mathrm{s}}_{\mathrm{R}}}{C^{\mathrm{s}}_{\mathrm{O}}} \tag{29.2}$$

*A reversible electrochemical reaction is one that proceeds rapidly in both directions (see Section 27.7). This means that the surface concentrations of O and R respond essentially instantly to any change in potential.

where C_O^s is the *surface* concentration of O, C_R^s is the *surface* concentration of R, and other terms have been identified in Table 27.1. As described in Section 27.7, the electrochemical cell exhibits a potential that reflects the ratio of surface concentrations C_R^s/C_O^s. If by some external means the electrode potential is forced to a different value, the ratio C_R^s/C_O^s must change to the new value dictated by the Nernst equation. For example, if the potential is moved to a more negative value, the ratio C_R^s/C_O^s must become larger. The electrochemical system accomplishes this by converting some O to R by electrolytic reduction, as described by Eq. (29.1). On the other hand, if the potential is moved to a more positive value, the ratio C_R^s/C_O^s must become smaller. This is accomplished by oxidation of the appropriate amount of O to R at the electrode surface, that is, the reverse of Eq. (29.1).

Figure 29.2 illustrates how the ratio C_R^s/C_O^s changes as E is changed at 25°C. Note that the surface concentration can be altered to essentially all O (C_R^s/C_O^s < $1/1000$) by making E sufficiently positive of $E^{0'}$ or essentially all R (C_R^s/C_O^s > $1000/1$) by making E sufficiently negative of $E^{0'}$. This *million-fold* change in

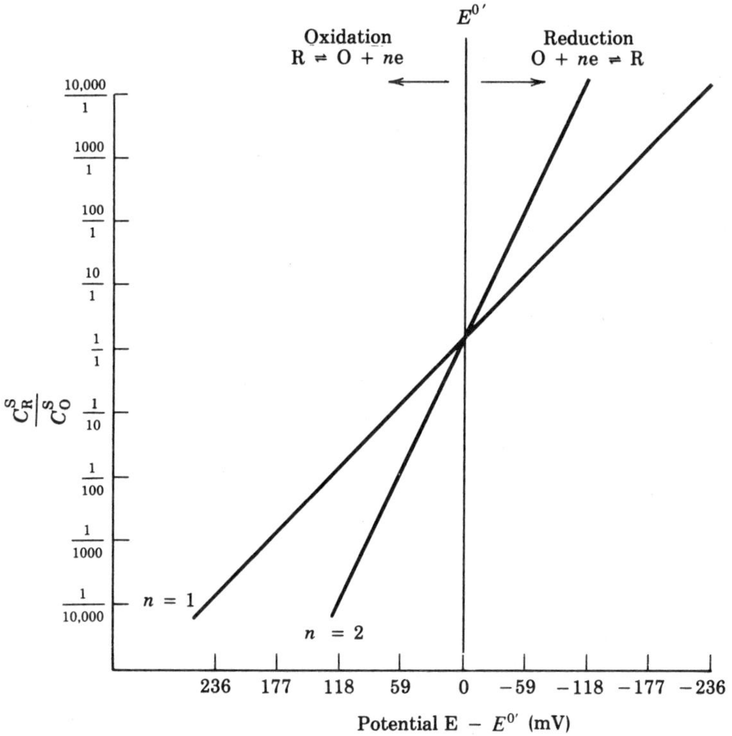

Fig. 29.2 Relationship between E and C_R^s/C_O^s, as described by the Nernst equation for $n = 1$ and $n = 2$ at 25°C: $E = E^{0'}_{O,R} - (0.0591/n) \log (C_R^s/C_O^s)$.

surface concentration ratio is accomplished by a mere 0.354-V change in potential for an $n = 1$ redox couple! Making E equal to $E^{0'}$ forces the surface concentrations of O and R to be equal ($C_R^s/C_O^s = 1/1$).

This variation of C_R^s/C_O^s as a function of E is the basis of all voltammetric methods. A good understanding of this concept is essential!

Example 29.2. It is useful to think of the electrode in terms of a device that can be made a reductant or an oxidant, depending on the value of E that is applied to it. Thereby, Fe(III) can be reduced to Fe(II) if the electrode is made a sufficiently strong reductant, and Fe(II) can be oxidized to Fe(III) if the electrode is made a sufficiently strong oxidant. The ease and precision with which the oxidizing or reducing power of an electrode can be varied over the range of strong oxidant to strong reductant by simply changing E is the *forte* of voltammetry!

Concentration–Distance Profiles and Mass Transport in a Stirred Solution. It is apparent from the preceding section that the concentrations of redox species at the electrode surface can be easily (and rapidly) changed by simply varying the potential applied to the cell. However, solution that is sufficiently far from the electrode surface ($x >$ *ca.* 0.1 cm) is unaffected by these changes, unless the electrolysis proceeds for a sufficiently long time to effect the change throughout the cell. Consequently, electrolysis occurring at the electrode surface causes a concentration imbalance to exist between the solution at the electrode surface and the solution at a distance from the electrode.

The concentration gradients of species in the solution adjacent to an electrode are illustrated by means of concentration (C)–distance (x) profiles (i.e., C–x profiles). Profiles for a planar electrode immersed in a stirred solution of O and R (1.00 mM each) are illustrated in Fig. 29.3a. The vertical axis represents concentration and the horizontal axis represents distance from the electrode into solution. The vertical line at $x = 0$ represents the interface, or boundary, between electrode and solution. Thus, a C–x profile is a graphical representation of the concentration of a certain species starting at the electrode and extending into solution. *A good grasp of concentration–distance profiles is critical to understanding voltammetric techniques!*

The solid and dashed horizontal lines in Fig. 29.3a show uniform concentrations of 1.00 mM for O and R from the electrode outward into the solution and describe the *initial condition* of the solution (prior to application of a potential). Consideration of the Nernst equation (29.2) shows that the application of a potential equal to $E^{0'}_{O,R}$ would not change the ratio of O/R from unity. Consequently, the C–x profiles in Fig. 29.3a also apply whenever the applied potential exactly matches the inherent potential of the electrochemical cell, E_{cell}.

Some understanding of liquid flow along a solid-solution interface is useful in understanding C–x profiles for an electrode immersed in a stirred solution. Three regions of solution flow can be identified.

1. Turbulent flow exists in the solution bulk.

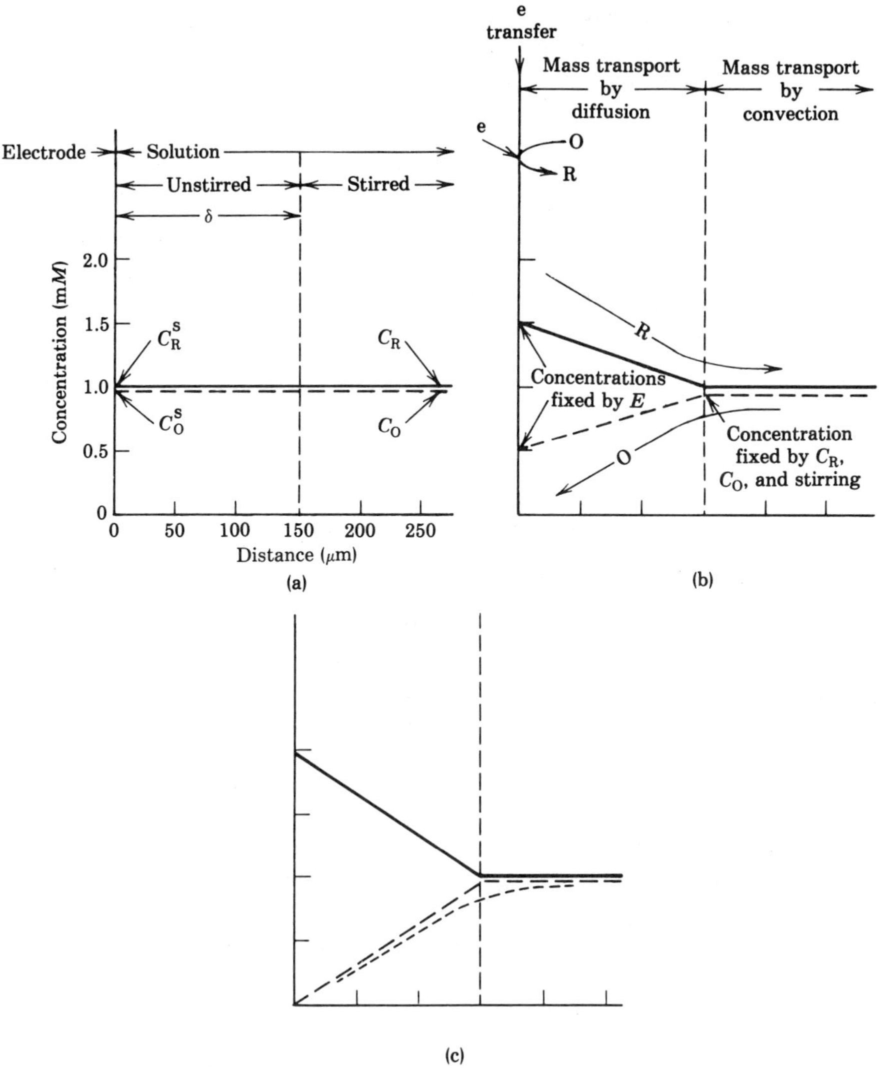

Fig. 29.3 Concentration–distance profiles for O and R at a planar electrode immersed in a stirred solution where $n = 1$. C_O and C_R are bulk solution concentrations of O and R, which are both 1.00 mM. (The profiles are offset slightly for clarity.) C_O^s and C_R^s are surface concentrations. (a) $E = E^{0'}_{O,R}$. (b) $E = E^{0'}_{O,R} - 28$ mV. (c) $E = E^{0'}_{O,R} - 118$ mV.

2. As the electrode surface is approached, a transition to laminar flow occurs. This is a nonturbulent flow in which adjacent layers of solution slide by each other parallel to the electrode surface.

3. The rate of this laminar flow decreases near the electrode owing to frictional forces, until a thin layer of stagnant solution is present immediately adjacent to the electrode surface. Within this layer diffusion is the predominant mode of mass transport. It is convenient, although not entirely correct, to consider this thin layer of stagnant solution as having a discrete thickness δ, as shown in Fig. 29.3. This layer is called the Nernst diffusion layer and is typically 10^{-2}–10^{-3} cm thick depending on the efficiency of stirring and solution viscosity.

Let us now consider the case in which E is moved 28 mV negative of $E^{0'}_{O,R}$. According to Fig. 29.2, the ratio of C^s_R/C^s_O must now be 3/1 for the Nernst equation to be satisfied. Consequently, some O must be reduced to R to change the ratio. This occurs instantly and is accomplished by a flow of electrons through the working electrode as a result of the electroreduction process. The new surface concentrations are $C^s_O = 0.50$ mM and $C^s_R = 1.50$ mM. These values satisfy the ratio set by the Nernst equation and the requirement of mass balance $C^s_O + C^s_R =$ 2.00 mM.

A concentration gradient now exists in the solution adjacent to the *interface* as shown by the C–x profiles in Fig. 29.3b. Two regions of this profile can be considered:

1. At distances greater than δ, concentrations are maintained homogeneous by the stirring action. So long as the electrode area is small relative to the solution volume and the experiment is not prolonged, the bulk solution concentrations will not be altered appreciably by the electrolytic conversion of O to R that is now occurring at the electrode surface.

2. The removal of O at the electrode surface by its reduction to R sets up a concentration gradient across the stagnant solution layer of thickness δ. Species O diffuses from the stirred region across this layer to the electrode surface where it is electrolyzed to R, which then diffuses back across the layer to the bulk solution (as shown in Fig. 29.3b, the species diffuse *down* the C–x profiles.)

Thus, the electrolysis process is controlled by a combination of (a) *mass transport* of O to the edge of the stagnant layer by *convection* and (b) subsequent *diffusion* of O across the stagnant layer under the influence of the concentration gradient caused by electrolysis of O to R at the electrode surface. Because R diffuses away from the interface and is then swept away by convection, a continuous electrolysis of O to R is necessary to maintain these surface concentrations. Convection also maintains a constant supply of O at $x = \delta$. Thus, a *steady-state* current is observed. (This is a good point to review the discussion of mass transport in Section 27.9.)

Example 29.3. What will happen to the steady-state current referred to above if the potential is maintained for a long time?

Gradually O will be reduced to R until the entire solution is adjusted to the ratio $C_R/C_O = 3/1$ (i.e., $C_R = 1.50$ m*M* and $C_O = 0.50$ m*M*). During this time the steady-state current will decay to zero, at which time the cell will have equilibrated with the applied potential.

However, this process will take a considerable amount of time if the electrode surface area is sufficiently small that the electrolysis proceeds slowly. For example, suppose that the steady-state current is 10.0 μA, $C_O = 1.00$ m*M*, and the cell volume is 20 mL. What will be the new concentration C_O after the current is maintained for 10 min? Use of Eqs.

(27.1) and (27.2) shows that $(10.0 \times 10^{-6}\ \text{A})(10\ \text{min})(60\ \text{s min}^{-1})/(96{,}485\ \text{C mol}^{-1}) = 6.22 \times 10^{-8}$ mol of O will be electrolyzed. The new amount of O in solution is $2.00 \times 10^{-5} - 6.22 \times 10^{-8} = 1.99 \times 10^{-5}$ mole, which corresponds to a concentration of 0.995 m*M*. Thus, it is reasonable to assume that the recording of a voltammogram does not significantly change the solution concentration of the sample if the current is sufficiently small.

Current. Current is a quantitative expression of how fast material is being electrolyzed at the electrode surface as expressed by Eq. (27.3). Current is the *response signal* to the potential excitation signal. The current at a planar electrode as a result of the reduction of O to R is related to the flux [rate of mass transport, see Eq. (27.50)] of material to the electrode by the following equation:

$$i_t = nFAD_O\left(\frac{\partial C_O}{\partial x}\right)_{x=0,t} \tag{29.3}$$

where the terms are given in Table 27.1 and C is in mol cm^{-3}. Thus, the current is directly proportional to the slope of the concentration–distance profile at the electrode surface, that is, to $(\partial C_O/\partial x)\ x = 0, t$. This equation can be expressed in terms of the Nernst diffusion layer concept by simply approximating ∂x by δ and ∂C by $C_O - C_O^s$ to give

$$i = nFAD_O\frac{C_O - C_O^s}{\delta}. \tag{29.4}$$

Examination of the profiles in Fig. 29.3b shows that this is a valid substitution for the slope of the profile at the electrode surface for the ideal case of an abrupt transition at $x = \delta$.

Figure 29.3c shows profiles for O and R when E is moved 118 mV negative of $E^{0'}_{O,R}$. According to Fig. 29.2, the ratio C_R^s/C_O^s for an $n = 1$ system must now be 100/1 for adherence to the Nernst equation. The surface concentrations are now $C_O^s = 0.0198$ m*M* and $C_R^s = 1.9802$ mM. Note that the surface concentration of O as shown in Fig. 29.3c is essentially zero. That is to say, essentially all of O that reaches the electrode is being reduced to R. Once this ratio has been reached, for all practical purposes C_O^s is zero, although in reality it approaches zero as the potential moves more negatively. Thus, the profile slope reaches a limiting value for which $C_O^s \sim 0$, and the appropriate current expression from Eq. (29.4) becomes

$$i_l = \frac{nFAD_O C_O}{\delta}, \tag{29.5}$$

where i_l is referred to as a *limiting current*.

Although the transition between stagnant and flowing solution is considered to be abrupt in this example, the transition is in reality gradual. Consequently, the profiles are rounded, as shown by the dotted line in Fig. 29.3c. However, the hypothetical situation of an abrupt transition is a useful approximation in mathematical treatments.

29.3 HYDRODYNAMIC VOLTAMMOGRAM

A *voltammogram* shows the current response to a range of potentials applied to an electrochemical cell. Voltammograms are sometimes called *current–potential curves*. If recorded in a stirred solution, they are referred to as *hydrodynamic voltammograms*. They are usually obtained by means of an instrument that scans the potential over the desired range, measures the current, and records the voltammogram on an x–y recorder. In this case the potential excitation signal is the linear potential scan shown in Fig. 29.1. Alternatively, one can apply a sequence of discrete potentials that cover the desired potential range, measure the steady-state current at each potential, and plot current versus potential to obtain a voltammogram. In this case the potential excitation signal is analogous to the staircase shown in Fig. 29.1.

Voltammograms are presented such that potential is always the abscissa and current the ordinate, as shown in Fig. 29.4. The convention to be used in this book has cathodic current up and negative potential right. (Note: Other conventions have negative potential left and cathodic current down. It is important to note the convention used when reading any paper on voltammetry!) As shown below the abscissa, it is convenient to think in terms of the electrode becoming a stronger reductant as the potential becomes more negative and a stronger oxidant as the potential becomes increasingly positive. However, whether the electrode functions as a reductant or an oxidant depends on the applied potential *relative* to the $E^{0'}$ of the redox couple and the concentrations of O and R!

A hydrodynamic voltammogram for a solution of 1.00 mM O and R is shown in Fig. 29.4. The voltammogram has a point at which current is zero ($E^{0'}$ in this case), a region of cathodic current in which O is being reduced to R, and a region of anodic current in which R is being oxidized to O. Beginning at $E^{0'}$ ($E - E^{0'} = 0$ V), we see that cathodic current increases as the potential is scanned in the negative direction (that is, the electrode is being made a stronger reductant) and the electrolysis of O proceeds faster and faster. The cathodic current levels off when O is being converted to R as rapidly as it is being transported to the electrode surface. A similar behavior is observed for the electrolysis of R when the potential is scanned in the positive direction (that is, the electrode is being made a stronger oxidant). Anodic current increases as the rate of oxidation of R increases until the current becomes limited by the rate of transport of R to the electrode surface, at which point the current levels off.

The shape of the voltammogram can be further understood by applying concepts described in the previous section:

1. Control of surface concentration of O and R by the applied potential E, as determined by the Nernst equation.

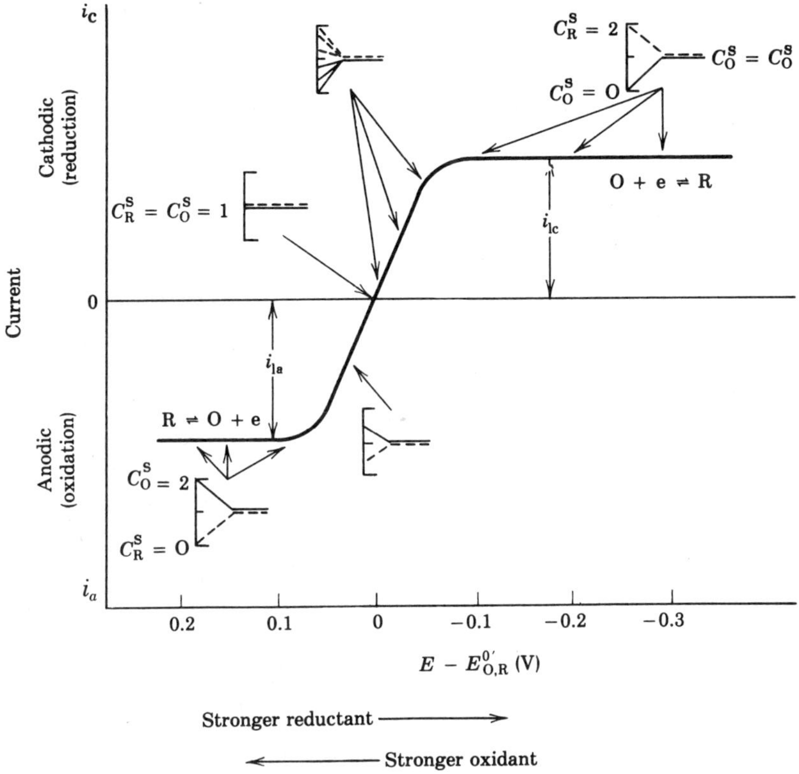

Fig. 29.4 Hydrodynamic voltammogram for the electrode process O + e ⇌ R including representative concentration–distance profiles for solution containing 1 m*M* each of O and R. $D_O = D_R$. $T =$ 25°C. (The profiles are offset slightly for clarity.)

2. The *C–x* profiles for O and R and attendant mass transport to and from the electrode surface.
3. The relationship between the current response and the slope of the concentration–distance profile at the electrode surface, that is, $(\partial C/\partial x)\ x = 0$.

Let us initiate the voltammogram at a potential equal to the cell potential. As discussed in Section 29.3, the application of a potential equal to $E^{0'}_{O,R}$ does not cause a change in C_R^s/C_O^s. This ratio is unity, and the concentration–distance profiles for O and R are horizontal, as shown in the inset of Fig. 29.4. Since the profile $(\partial C_O/\partial x)\ x = 0$ is zero, no current is measured and $i = 0$ is observed on the voltammogram.

Now let us assume that the potential is being moved in the *negative* direction (right from $E^{0'}$). This forces the surface concentrations to assume ratios of C_R^s/C_O^s that are increasingly larger. Observe how the slopes of the *C–x* profiles for O increase as the potential is made more negative. Cathodic current increases as the slope increases, as defined by Eq. (29.4).

Note how the C–x profiles for O cease to show any noticeable change once the potential is sufficiently negative that C_O^s approaches zero. This situation is achieved once C_R^s / C_O^s exceeds about 100/1. Thus, the profile slope reaches a limiting value for which $C_O^s \sim 0$, and the appropriate current expression from Eq. (29.5) becomes

$$i_{lc} = \frac{nFAD_O C_O}{\delta}, \tag{29.6}$$

where i_{lc} is referred to as the *limiting cathodic current* and D_O and C_O are for the species undergoing mass transport *to* the electrode, namely O. As shown by the voltammogram, the current levels off (or plateaus) once C_O^s approaches zero. At this point, O is being converted to R as rapidly as it reaches the electrode surface.

Figure 29.4 also shows what happens when the potential is moved *positive* (left from $E_{O,R}^{0'}$). In this case anodic current results from the oxidation of R to O. When the potential becomes sufficiently positive, $C_R^s \sim 0$ and a *limiting anodic current*, i_{la}, is reached

$$i_{la} = -\frac{nFAD_R C_R}{\delta}, \tag{29.7}$$

where D_R and C_R now represent species R, which is being transported to the electrode surface.

It is apparent from Eqs. (29.6 and 29.7) that i_{lc} and i_{la} are directly proportional to the concentrations of O and R, respectively, and can therefore be used for the analytical determination of these two species. Quantitative analysis by voltammetry will be discussed later in Section 29.7.

Example 29.4. A hydrodynamic voltammogram for a solution of 1.00 m*M* Fe^{3+} in 1 *M* $HClO_4$ at a platinum electrode versus SCE is shown in Fig. 29.5. The voltammogram is calculated from the following parameters: $A = 1.00$ cm^2, $D_{Fe^{3+}} = 5.0 \times 10^{-6}$ cm^2 s^{-1}, and $\delta = 0.01$ cm. $E^{0'}_{Fe^{3+},Fe^{2+}}$ in 1.0 *M* ClO_4 versus SCE is 0.488 V [0.732 V (from Table 27.2) − 0.244 V (SCE)], which gives the Nernst equation shown below the figure. The table shows the surface concentrations for Fe^{3+} and Fe^{2+} that exist at a series of applied potentials as calculated from the Nernst equation. The C–x profiles that exist for each potential are shown at the bottom of each column. The current is calculated with Eq. (29.4) from the slope of the concentration distance profile and the appropriate value of D and C. Parameters for Fe^{3+} (e.g., $D_{Fe^{3+}}$, $C_{Fe^{3+}}$, C–x profile for Fe^{3+}) are used to calculate cathodic current, which is determined by the rate of reduction of Fe^{3+} (i.e., rate of mass transport of Fe^{3+} to the electrode.)

29.4 CURRENT–POTENTIAL RELATIONS

Reversible Reactions. The voltammogram in Fig. 29.4 exhibits a sigmoid-shaped region in which the current changes markedly with change in the applied potential.

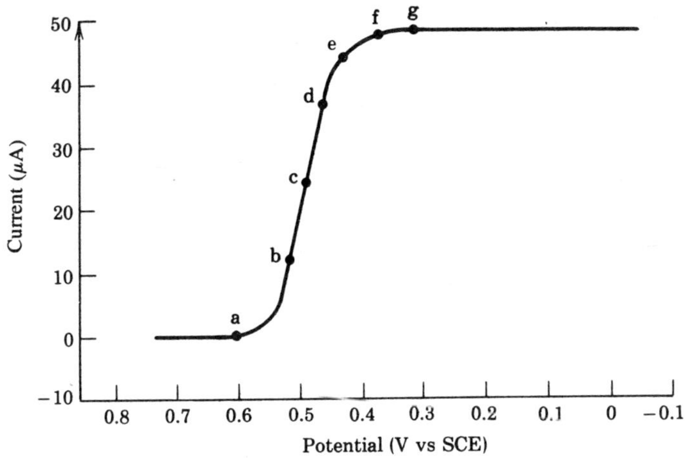

$$E = 0.488 - 0.0591 \log (C^s_{Fe^{2+}}/C^s_{Fe^{3+}})$$

	a	*b*	*c*	*d*	*e*	*f*	*g*
E(V)	0.606	0.516	0.488	0.460	0.429	0.370	0.311
$C^s_{Fe^{2+}}/C^s_{Fe^{3+}}$	1/100	1/3	1/1	3/1	10/1	100/1	1000/1
$C^s_{Fe^{2+}}$(mM)	0.01	0.25	0.50	0.75	0.91	0.99	0.10
$C^s_{Fe^{3+}}$(mM)	0.99	0.75	0.50	0.25	0.09	0.01	0.00
i(μA)	0.48	12.1	24.1	36.2	43.9	47.8	48.2

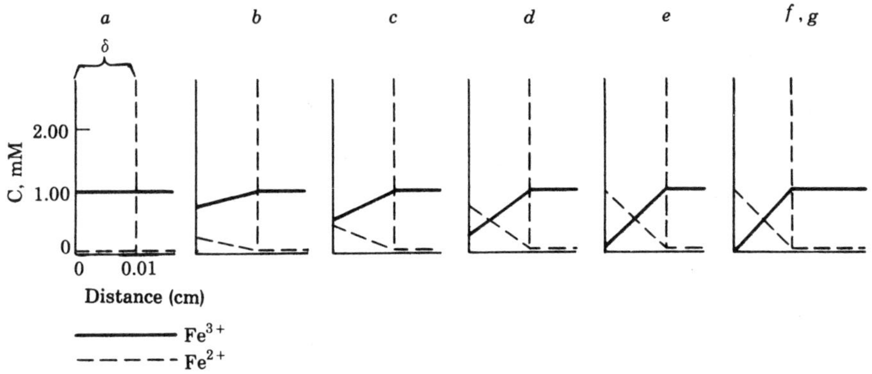

Fig. 29.5 Hydrodynamic voltammogram of 1.00 m*M* Fe^{3+} and Fe^{2+} in 1.0 *M* $HClO_4$ at 25°C. Recorded at Pt electrode (A = 1.00 cm^2) versus SCE.

This is a transition region between i_{lc} and i_{la} in which C^s_R and C^s_O change dramatically over a potential range of only a few tenths of a volt. An equation that describes this sigmoid region can be derived by combining the relationship between surface concentration and applied potential as described by the Nernst equation (29.2) and the relationship between surface concentration and current as described by Eq. 29.4 (for the reduction process). This approach is valid so long as the rate of heterogeneous electron exchanges is fast (i.e., a *reversible* system). The terms

n, F, A, D_O, and δ in Eq. (29.4) can be taken as constant for a given experiment, where the constant K_O is defined as

$$K_O = \frac{nFAD_O}{\delta}. \tag{29.8}$$

Substitution of K_O into Eq. (29.4) gives $i = K_O C_O - K_O C_O^s$. Noting that the term $K_O C_O$ is equal to i_{lc} (from Eq. 29.6), one obtains $i = i_{lc} - K_O C_O^s$, which can be rearranged to give an expession that relates the surface concentration of O to current

$$C_O^s = \frac{i_{lc} - i}{K_O}. \tag{29.9}$$

The relationship between surface concentration and current can also be expressed in terms of the C–x profile for R

$$i = -\frac{nFAD_R}{\delta}(C_R - C_R^s), \tag{29.10}$$

where the minus sign accounts for the difference in sign between the slopes of $(\partial C_O/\partial x)_{x=0}$ and $(\partial C_R/\partial x)_{x=0}$. By analogy to Eqs. (29.8–29.10), one can write for R,

$$K_R = \frac{nFAD_R}{\delta} \tag{29.11}$$

and the following equations: $i = -K_R C_R + K_R C_R^s$; $i = i_{la} + K_R C_R^s$; and

$$C_R^s = \frac{i - i_{la}}{K_R} \tag{29.12}$$

Equation (29.9) and (29.12) express surface concentrations C_O^s and C_R^s in terms of the two limiting currents i_{lc} and i_{la} and the current in the sigmoid region of the voltammogram, i. These equations can be substituted into the Nernst equation (29.2) to give an expression that relates the potential applied to the cell, E, and the current in the sigmoid region, i, of the voltammogram.

$$E = E_{O,R}^{0'} - \frac{RT}{nF}\ln\frac{K_O}{K_R} - \frac{RT}{nF}\ln\frac{i - i_{la}}{i_{lc} - i}. \tag{29.13}$$

Half-Wave Potential. The *half-wave potential* $E_{1/2}$, is the potential that corresponds to a current that is exactly between i_{lc} and i_{la}, that is,

$$i = \frac{i_{lc} + i_{la}}{2}. \tag{29.14}$$

For this value of current, Eq. (29.13) becomes

$$E_{1/2} = E^{0'}_{\mathrm{O,R}} - \frac{RT}{nF} \ln \frac{K_{\mathrm{O}}}{K_{\mathrm{R}}}. \tag{29.15}$$

Inspection of Eqs. (29.8) and (29.11) shows that n, F, A, and δ would be constant in a given experiment. Consequently, the ratio $K_{\mathrm{O}}/K_{\mathrm{R}}$ is equal to the ratio of the diffusion coefficients $D_{\mathrm{O}}/D_{\mathrm{R}}$. Thus, Eq. (29.13) can be expressed as

$$E = E_{1/2} - \frac{RT}{nF} \ln \frac{i - i_{\mathrm{la}}}{i_{\mathrm{lc}} - i}, \tag{29.16}$$

where

$$E_{1/2} = E^{0'}_{\mathrm{O,R}} - \frac{RT}{nF} \ln \frac{D_{\mathrm{O}}}{D_{\mathrm{R}}}. \tag{29.17}$$

The term $E_{1/2}$ is a quantity characteristic of a given electroactive species in a given medium and for a given electrode. If the diffusion coefficients of the reduced and oxidized form of the couple are similar, $E_{1/2}$ will then be very close to the formal reduction potential $E^{0'}$ of the couple, as shown by Eq. (29.17) for which the ln term approaches zero as the diffusion coefficients approach equality. Therefore, $E_{1/2}$ is a measure of the "energetics" of the redox system, just as $E^{0'}$ is. As implied in Fig. 29.4, a more negative $E_{1/2}$ corresponds to greater energy required to reduce O, whereas a more positive $E_{1/2}$ corresponds to greater energy to oxidize R.

The $E_{1/2}$ values for a variety of electroactive species in several media are shown in Table 29.1. The medium can have a dramatic effect on $E_{1/2}$, especially if it

Table 29.1 Some Half-Wave Potentials for Different Supporting Electrolytes at the Mercury Electrode[a]

Metal Ion	*Half-Wave Potential (V vs. SCE)* 1 M KCl	1 M KCN	1 M NH_3 1 M NH_4Cl	1 M NaOH	7.3 M H_3PO_4	1 M Na_3 Cit. 0.1 M NaOH
Cd^{2+}	−0.64	−1.18	−0.81	−0.76	−0.71	−1.46
Co^{2+}	−1.20	−1.45	−1.29	−1.43	−1.20	−1.45
Cu^{2+}	−0.22	NR[c]	−0.51	−0.41	−0.09	−0.50
Fe^{2+}			−1.49	−1.46		
Pb^{2+}	−0.44	−0.72		−0.76	−0.53	−0.78
Ni^{2+}	−1.1 [b]	−1.36	−1.10		−1.18	NR
Zn^{2+}	−1.00	NR	−1.35	−1.53	−1.13	−1.43
Sn^{2+}				−1.22	−0.58	−1.12
Tl^{+}	−0.48		−0.48	−0.46	−0.63	−0.56

[a]Based on tabulation of L. Meites *Polarographic Techniques*, 2nd ed. New York: Wiley, 1965.
[b]0.1 *M* KCl.
[c]Not reducible.

contains a ligand that can alter the redox energetics of an electroactive species substantially by the formation of a complex. For example, the $E_{1/2}$ values for Cd^{2+} are significantly more negative in the KCN medium than in KCl as a result of stronger complexation with CN^- than with Cl^-. (This topic is discussed in more detail in Chapter 30.)

***Nernst Plots—Measurement of* $E_{1/2}$ *and* n.** Voltammograms of reversible systems of the type shown in Fig. 29.4 serve a number of useful functions, two of which are the measurement of $E_{1/2}$ and n. The electron stoichiometry of the electrode reaction affects the slope of the sigmoid region of a voltammogram, as shown in Fig. 29.6a for reversible $n = 1$ and $n = 2$ systems in which O is being reduced to R. The value of n and $E_{1/2}$ can be obtained from a voltammogram by means of a plot of Eq. (29.16), that is, a Nernst plot. Plots for $n = 1$ and $n = 2$ systems are shown in Fig. 29.6b. The potential at which $\log(i - i_{la})/(i_{lc} - i)$ is zero equals $E_{1/2}$ and the slope of the plot equals $-0.0591/n$ for 25°C.

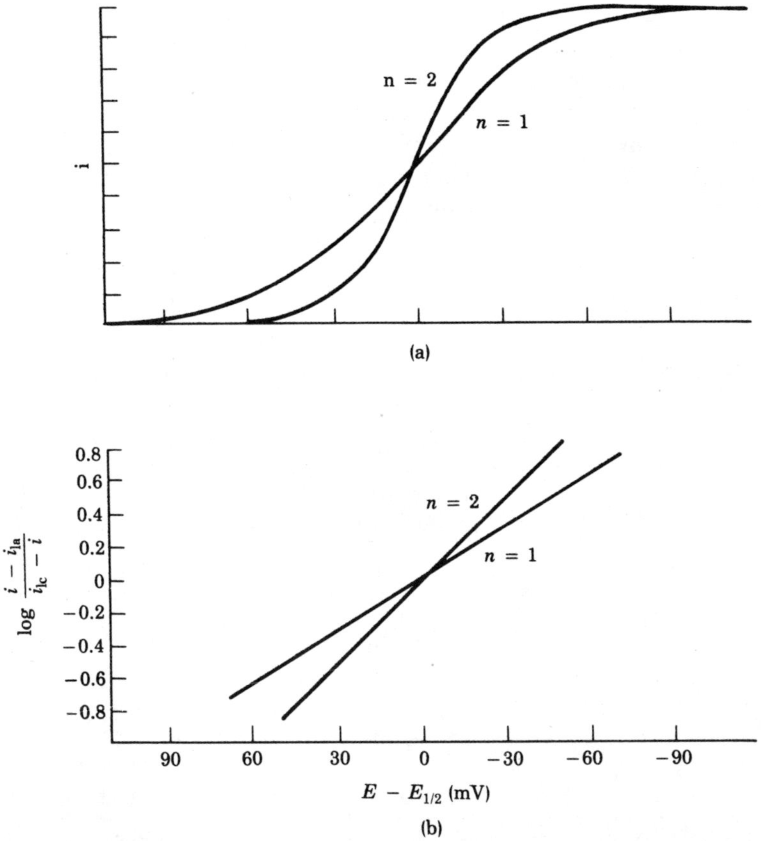

Fig. 29.6 Effect of n on (a) voltammetric waves for reversible system and (b) Nernst plots.

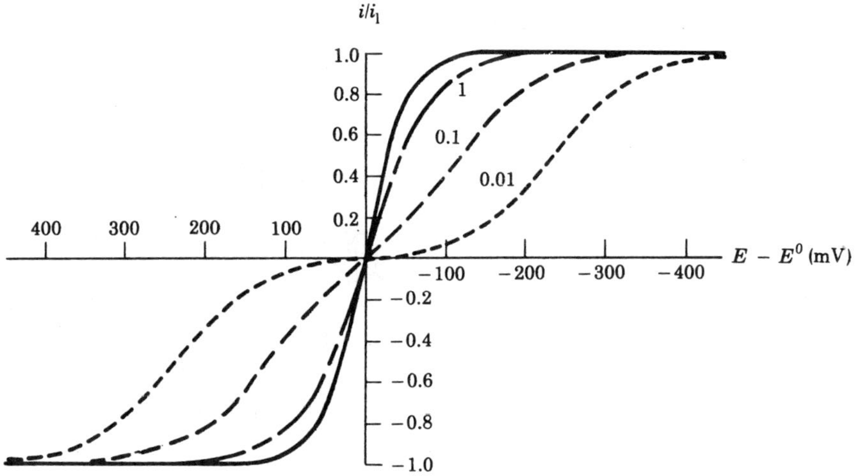

Fig. 29.7 Effect of slow heterogeneous electron transfer on voltammogram shapes for $n = 1$, $\alpha = 0.5$, and $T = 25°C$. Numbers by curves show i_o/i_l where $i_l = i_{lc} = -i_{la}$. (Adopted from A. J. Bard and L. R. Faulkner, *Electrochemical Methods.* New York: Wiley, 1980, p. 110).

Irreversible Reactions. The derivation given above is based on the assumption that the rate of the electrode reaction is sufficiently great that the relationship between the applied potential E and the surface concentrations of the redox couple, C_R^s/C_O^s, is Nernstian. That is to say, any change in E causes a rapid change in C_R^s/C_O^s as described by Eq. (29.2). However, many redox couples do not obey the Nernst equation because the rate constant for the heterogeneous redox reaction is slow.

Figure 29.7 shows the effect of increasingly slow heterogeneous electron transfer on the shape of a hydrodynamic voltammogram. As the exchange current, i_0 (see Section 27.7) diminishes in magnitude, the current in the sigmoid region of the voltammogram decreases. Voltammograms of this type are termed *electrochemically irreversible*, since the facile electron transfer required for adherence to the Nernst equation is not observed. Note that application of a sufficiently negative or positive potential still results in limiting cathodic and anodic currents of the same magnitudes as would be obtained if the system were reversible.

29.5 CELL AND INSTRUMENTATION FOR VOLTAMMETRY

Cell. Modern electrochemical cells for voltammetry utilize a three-electrode configuration. The desired potential is applied between a working electrode and a reference electrode. The *working electrode* is the electrode at which the electrolysis of interest takes place. The current required to sustain the electrolysis at the working electrode is provided by the *auxiliary electrode*. This arrangement prevents the reference electrode from being subjected to large currents that could change its potential. Some instrumentation is based on a two-electrode system.

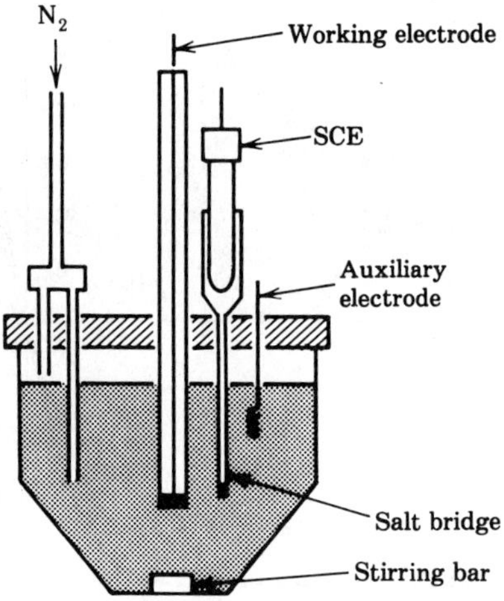

Fig. 29.8 Electrochemical cell for voltammetry.

Here the auxiliary electrode is absent and the reference electrode is subjected to the entire cell current.

A representative electrochemical cell for voltammetry is illustrated in Fig. 29.8. Such a cell usually consists of a glass container with a cap having holes for the positioning of electrodes and a tube for nitrogen. Provision is made for oxygen removal from solution by bubbling with nitrogen (or argon) gas. The cell is then maintained oxygen-free by passing nitrogen over the solution. The nitrogen gas itself may be deoxygenated by procedures such as passing it through a hot copper furnace or through solutions of strong reductants such as vanadous or chromous ion. The reference electrode is typically a SCE or a Ag/AgCl electrode, which is sometimes isolated from the solution by a salt bridge to prevent contamination of the sample by Cl^- and impurities in the reference electrode that slowly leak out at the junction. Reference electrodes are discussed in Section 27.4. In three-electrode systems, the auxiliary electrode is typically a coiled platinum wire that is generally placed directly in the solution. Since the limiting (or peak) current in any type of voltammetry is temperature-dependent, the cell should be thermostated for the most exacting work. A stir-bar can be added to the cell if stirring is necessary. Cells that require as little as 1–2 mL of solution for analysis are commercially available. A wide variety of cell designs has been developed.*

Working Electrodes. A variety of materials has been used successfully as working electrodes for voltammetry. The choice of which electrode to use depends on a number of factors such as potential range, residual current, geometry considerations (solid versus liquid), reproducibility, solvent compatibility, and, in some instances, specific catalytic properties or chemical reactivity for a species of interest.

*F. M. Hawkridge, Electrochemical Cells, in *Laboratory Techniques in Electroanalytical Chemistry*, W. R. Heineman and P. T. Kissinger (Eds.), Dekker, New York, 1985, pp. 337–366.

Working electrodes have certain properties in common. Good electrical conductance is of foremost importance. Consequently, working electrodes are metals and semiconductors. Chemical and electrochemical inertness is important in applications for which the electrode should function simply as an innocuous conveyor of electrons to and from species dissolved in solution. Inertness gives a wide potential region with minimal background contributions due to electrode and solvent redox properties in which the electrochemistry of the analyte(s) can be easily monitored.

Platinum, gold, glassy carbon, and pyrolytic graphite are commonly used materials for voltammetric electrodes. They exhibit good positive and negative potential ranges in most solvents and supporting electrolytes. These electrodes are commonly fabricated in the form of a disk that is press-fitted into a rod of inert material such as Teflon or Kel-F into which a contact wire is imbedded, as shown in Fig. 29.9a. Electrodes of this type have a well-defined surface area that can be easily measured. The carbon paste electrode consists of carbon particles mixed with a viscous oil such as Nujol. The paste is packed into a cavity in an inert container into which a wire for electrical contact is inserted. This unusual material has proved useful for oxidations of organic compounds. Semiconductors such as SnO_2 and In_2O_3 also exhibit good positive and negative potential ranges. These materials are generally used as thin films deposited on glass or quartz substrates.

Recently, voltammetric measurements have been made at extremely small electrodes termed ultramicroelectrodes.* One type of this electrode consists of a fiber of graphite, with a diameter of 10 μm or less. These electrodes enable electroanalytical measurements to be made on extremely small volumes of sample. By virtue of their small surface areas, ultramicroelectrodes exhibit currents that are sufficiently small to minimize problems with *iR* drop in highly resistive solvent supporting electrolyte systems.

Mercury has been used extensively in voltammetry for several reasons. It exhibits an excellent negative potential range due to large overpotential† for reduction of H^+. The surface of a mercury drop electrode is easily renewed by the extrusion of a new drop. Many metals dissolve in mercury by the formation of amalgams. This property can be advantageous for the electrochemistry of these metals. When employed for voltammetry, mercury can be used in the form of a *hanging mercury drop electrode* (HMDE). A suitable mercury drop can be extruded through a narrow glass capillary by means of a commercially available micrometer syringe as shown in Fig. 29.9b. A new drop is formed by simply dislodging the old one and extruding more mercury, the size of the drop being determined by the number of micrometer turns. Such devices are capable of producing drops of uniform area to within ~5%. The *static mercury drop electrode* (SMDE) utilizes a solenoid to control drop formation by mercury that is rapidly dispensed through a capillary. The mechanical details of the SMDE are shown in

*R. M. Wightman, *Science*, **240,** 415 (1988).

†Overpotential is the difference between the thermodynamic reduction potential of the redox couple and the potential at which electrolysis occurs at an electrode.

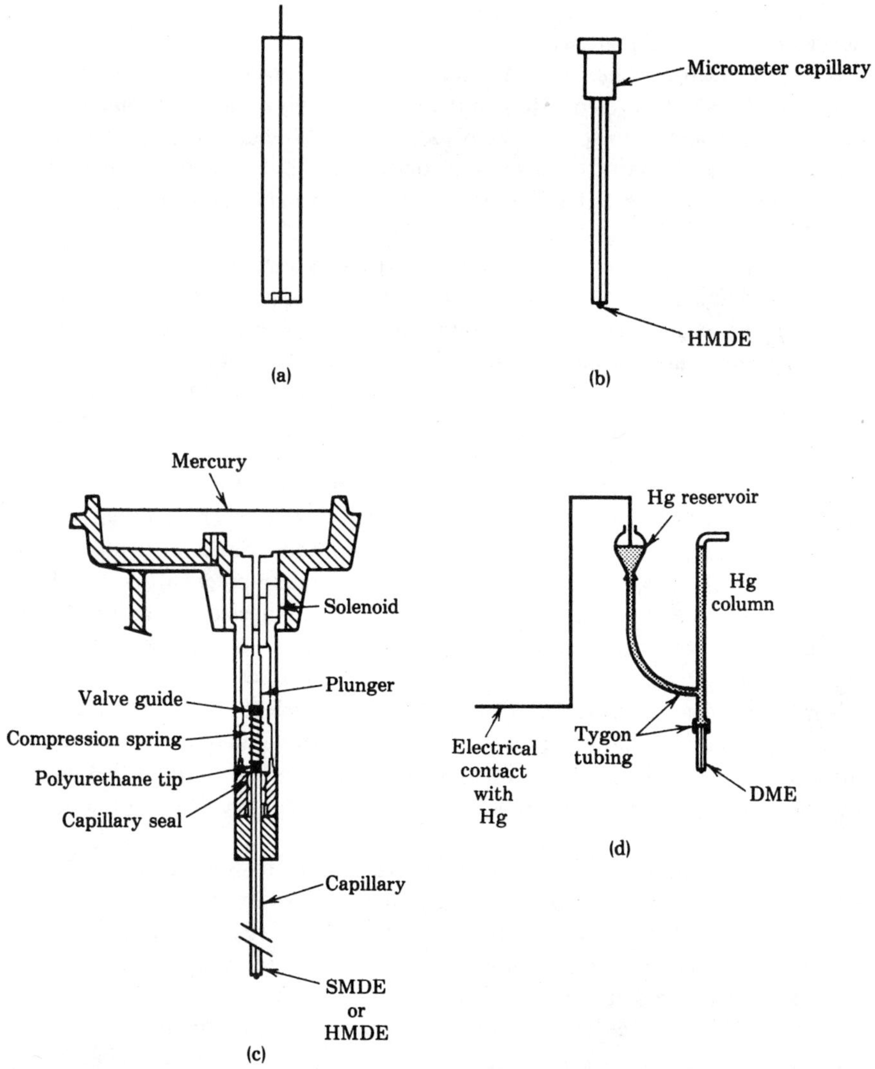

Fig. 29.9 Working electrodes. (a) Disk electrodes (Pt, Au, Ag, carbon paste, or glassy carbon), (b) HMDE. (c) SMDE. (e) DME.

Fig. 29.9c. Activation of the solenoid lifts the plunger, which allows mercury to flow from the large reservoir on top through the capillary to form a drop. The flow of mercury through the capillary is very fast owing to its relatively large bore (0.006 in.). Drop size is controlled by the length of time that the solenoid is activated. Solenoid deactivation stops the flow of mercury, resulting in a static drop. The SMDE can be controlled by the instrumentation for the electrochemical experiment. Mercury can also be used in the form of a *mercury film electrode* (MFE), which is prepared by the electrochemical reduction of Hg^{2+} to Hg^0 on a conductive substrate such as glassy carbon. The *dropping mercury electrode* (DME) is the working electrode for *polarography* (see Section 30.5). Mercury is forced by gravity through a very fine capillary to prove a continuous stream of identical droplets, each having a maximum diameter of between 0.5 and 1 mm. Each droplet expands, becomes too heavy to be suspended, and breaks loose from the capillary. The lifetime of a droplet is typically from 4 to 6 s. The actual drop time is determined in part by the rate of flow of mercury through the capillary. This can be varied by altering the height of the mercury head above the capillary by moving the mercury reservoir up or down. If the reservoir is moved up, mercury flows from the reservoir through the connecting flexible tube to the glass column until the mercury level is equal in the column and the reservoir. In order to increase the lifetime of a droplet, the mercury head should be moved down. The drop time can also be controlled by mechanical dislodging devices (drop knockers) that tap or vibrate the capillary at a fixed time. The SMDE can also function as a dropping electrode by reforming a mercury drop every few seconds. The SMDE differs from the DME in that a drop is rapidly formed to maximum size and then retained at that size until dislodgement, whereas with the DME the mercury drop grows until dislodgement.

In specific instances, chemical reaction between electrode and analyte is desirable. An example is the determination of Cl^- at a silver electrode in which electrogenerated Ag^+ reacts with Cl^- to form AgCl (see Section 29.6).

The chemical modification of electrode surfaces to produce specific chemical properties is an area of active research.* Chemical modification is the attachment of specific chemical compounds to the electrode surface in order to produce electrodes with different properties. *Chemically modified electrodes* can be prepared by several different attachment techniques: strong chemisorption onto the surface, covalent attachment, and polymer layers. For example, in the case of covalent attachment chemical reactions are carried out to form bonds between the substrate electrode and a molecule of interest. A metal or carbon can be oxidized so that the surface is covered with hydroxyl groups, as shown in Fig. 29.10. Such a surface can be silanized by reaction with an organosilane (reaction I) to give a pendant amine group. This surface can then be reacted with another molecule of interest (reaction II) to form the modified electrode.

Solvent-Supporting Electrolyte. The electrodes in an electrochemical cell are immersed in a medium that generally consists of a solvent in which supporting electrolyte is dissolved. The most important properties of an ideal solvent-supporting electrolyte system include *electrochemical inertness*, *electrical conductivity*, *good solvent power*, and *chemical inertness*.†

*R. W. Murray, Chemically modified electrodes, in *Electroanalytical Chemistry*, A. J. Bard (Ed.), New York: Dekker; 1984, pp. 191–368.

†A. J. Fry and W. E. Britton, Solvents and supporting electrolytes, in *Laboratory Techniques in Electroanalytical Chemistry*, P. T. Kissinger and W. R. Heineman (Eds.), New York: Dekker, 1984, pp. 367–382.

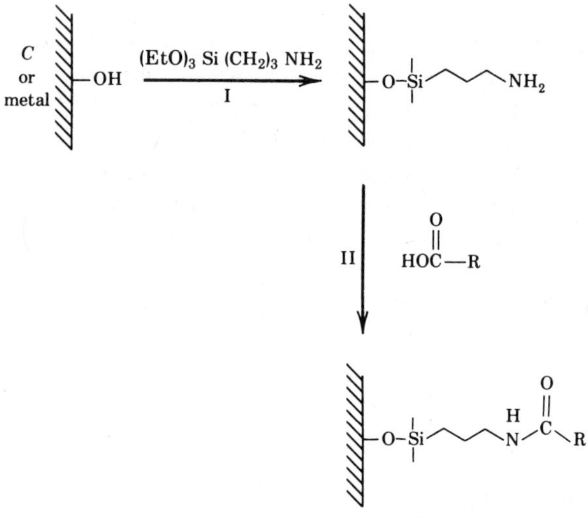

Fig. 29.10 General reaction scheme for covalent attachment to an electrode surface to produce a chemically modified electrode. Treatment I: oxidized carbon or metal electrode is treated with γ-aminopropyltriethoxysilane. Treatment II: the silanized electrode is treated with compound containing desired group to be attached: R. (From A. J. Bard, *J. Chem. Ed.*, **60**, 302 (1983)).

To enable the use of a wide range of potentials, the system should not undergo any electrochemical reaction from a very positive to a very negative potential. The potential at which the solvent or supporting electrolyte commences electrolysis is referred to as the *background limit*. The accessible potential ranges for several solvent-supporting electrolyte combinations are shown in Fig. 29.11. In practice, both extended positive and negative potential ranges are often unnecessary in a single set of experiments, and systems that exhibit either a very negative *or* a very positive background limit are often used.

In order to support the passage of current in a cell for voltammetry, the solvent should dissolve electrolytes well. Hence, solvents with a moderately high dielectric constant (≥ 10) are preferable. The most commonly used solvents and their dielectric constants are: water, 80; dimethylformamide (DMF), 36.7; acetonitrile (AN), 37.5; and dimethyl sulfoxide (DMSO), 46.7. Solvents of lower polarity have been used in voltammetry, although the prevalence of ion pairing of the supporting electrolyte results in low conductance. Examples are tetrahydrofuran (THF), 7.3, and methylene chloride (MC), 8.9. Uncompensated iR errors are especially troublesome in such high-resistance solvents.

A good solvent should dissolve a wide range of substances at acceptable concentrations. Supporting electrolytes should be soluble to the extent of at least 0.1 *M*, whereas electroactive material should be soluble at millimolar levels. Tetraalkylammonium perchlorates, hexafluorophosphates, and tetrafluoroborates exhibit good solubility in nonaqueous solvents and are often used as supporting electro-

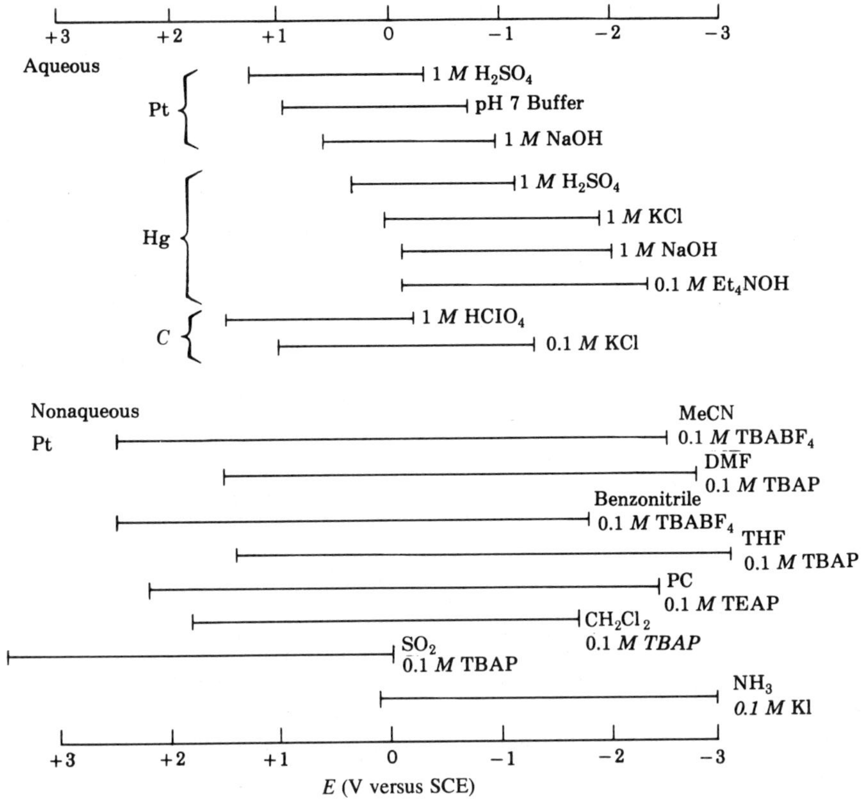

Fig. 29.11 Estimated potential ranges of various combinations of working electrode and solvent-supporting electrolyte systems. (From A. J. Bard and L. R. Faulkner, *Electrochemical Methods.* New York: Wiley, 1980, back cover.)

lytes. Examples are tetrabutylammonium hexafluorophosphate (TBAHF) or tetrafluoroborate (TBATFB). A variety of salts, acids, and bases can be used as supporting electrolytes in aqueous solvent. The chemical should be obtainable in a highly purified form, since contaminants may be electroactive and therefore constitute an interference. This is especially important when high concentrations of supporting electrolyte, such as 0.5–1.0 *M*, are used.

The solvent should be unreactive with the electroactive material to be investigated electrochemically. In studies of electrode mechanism, it is often preferable that the solvent and supporting electrolyte not react with electrogenerated products and intermediates.

A supporting electrolyte typically serves three purposes. First, it provides the ions necessary to enable electrical charge to pass through the electrochemical cell. The supporting electrolyte *lowers the resistance of the cell.* High resistance causes distortion of voltammograms due to loss of potential from iR drop in the cell. The potential actually applied to the electrochemical cell (E_{cell}) is the applied potential

($E_{applied}$) less potential loss due to iR drop (see Section 27.8), as described by the following equation:

$$E_{cell} = E_{applied} - iR. \tag{29.18}$$

The distortion is similar to that shown in Fig. 29.7 for slow heterogeneous electron transfer.

Second, supporting electrolyte minimizes the effect of migration (see Section 27.9) on mass transport of analyte to the electrode for electrolysis. Equations (29.6) and (29.7), which relate i_l to C, are based on convection–diffusion as the sole means of mass transport. The coulombic effect of migration can increase or decrease the limiting current as a result of electrostatic attraction or repulsion between the charged electrode and an ionic analyte. The magnitude of this effect on the analyte depends on the ionic strength of the solution relative to the analyte concentration. For example, the effect of supporting electrolyte concentration on the limiting current for reduction of 1.00 mM Cd^{2+} at a negatively charged electrode is shown in Fig. 29.12. As supporting electrolyte is added, the limiting current decreases until a concentration of about 0.1 M is reached, after which the change is negligible. At low ionic strength, the electrostatic attraction of the negatively charged electrode for Cd^{2+} significantly enhances its value of i_l by enhancing the mass transport of Cd^{2+} to the electrode. As the ionic strength of the solution is increased by adding supporting electrolyte, the coulombic attraction of the electrode is *dispersed* over an increasing population of ions. At a sufficiently high ionic strength, the electrostatic attraction for an individual ion is essentially

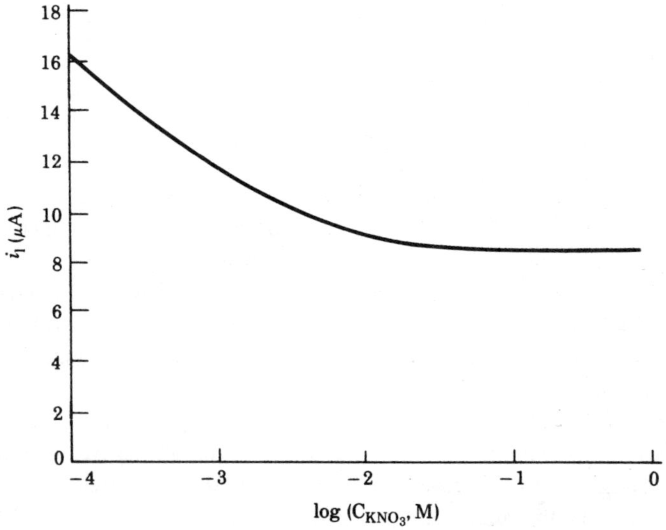

Fig. 29.12 Effect of supporting electrolyte concentration on limiting current for reduction of Cd^{2+} at a mercury electrode.

negligible in terms of an effect on i_l. This limiting value of i_l then corresponds to a mass transport situation *controlled* by diffusion–convection with an essentially negligible contribution from migration. In analysis it is important to operate under ionic strength conditions such that i_l is on the limiting plateau section of Fig. 29.12, so that migration is eliminated as a source of mass transport. Adjusting the ionic strength of all standards and samples to a value that is about 100× the analyte concentration eliminates ionic strength as a variable.

The third role of a supporting electrolyte is to influence the half-wave potentials of the analyte(s). As shown in Table 29.1, the half-wave potentials for the reduction of metal ions at a mercury electrode can be shifted substantially by changing the supporting electrolyte. Metal ions can be made more difficult to reduce by adding a complexing ligand. The half-wave potentials of many organic compounds are affected by the pH of the medium (see Section 29.6). The careful adjustment of supporting electrolyte is particularly important in the simultaneous determination of multiple analytes in which the $E_{1/2}$ values of the analytes must be sufficiently different that the individual voltammetric waves are resolved.

Instrumentation. Instrumentation is needed to apply the potential excitation signal and measure the resulting current response signal for the particular voltammetric technique that is to be implemented (see Fig. 29.1). Voltammetry requires a waveform generator to produce the excitation signal, a potentiostat to apply this signal to the electrochemical cell, a current-to-voltage converter to measure the resulting current, and an *XY* recorder or oscilloscope to display the voltammogram. A representative block diagram of this instrumentation is shown in Fig. 29.13a. The waveform generator, potentiostat, and current-to-voltage converter are normally incorporated into a single electronic device, as shown by the dashed line.

The potentiostat applies the potential excitation signal between the working electrode and the reference electrode without allowing charge to pass through the reference electrode (see Fig. 29.13b). The potential of the auxiliary electrode adjusts to an electrolytic process so that its current is equal but of opposite polarity

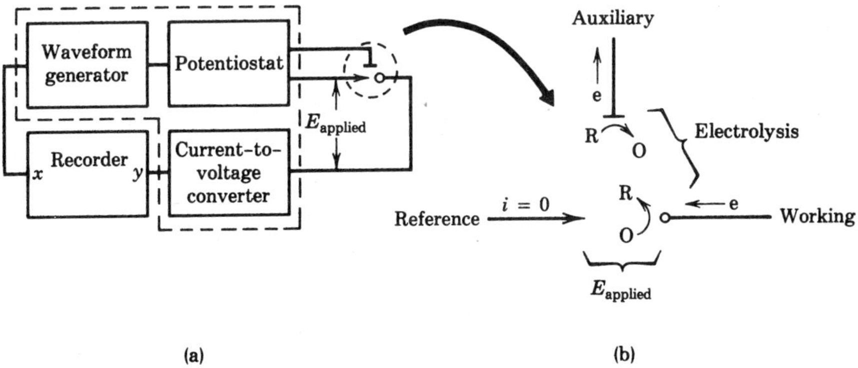

Fig. 29.13 (a) Instrumentation for voltammetry. Electrode designation: ○—, working; ⊢, auxiliary; ←, reference. (b) Operation of three-electrode cell.

to the current at the working electrode. Thus, the potential of the working electrode is controlled with respect to a reference electrode whose half-cell potential is invariant, but the actual electrolytic cell consists of the working and auxiliary electrodes. When the working electrode functions as a cathode, as shown in Fig. 29.13b, the auxiliary electrode is an anode, and vice versa.

Many voltammetric techniques can be implemented with circuitry based on *operational amplifiers* (OAs). Chapter 4 provides insight into the operation of these circuits. A potentiostat can be assembled from a single operational amplifier based on the circuit shown in Fig. 29.14a. The voltage applied at E_i will appear at E_o because of the operating characteristics of a differential amplifier (Chapter 4). This enables a voltage difference to be applied across resistor R_u. The current required by Ohm's law is provided by the output of the amplifier, the voltage of which (E_a) moves to the value required to provide $i = E_o/R_u$ (i.e., to $E_a = i(R_c + R_u)$).

Example 29.5. Assume that R_u in Fig. 29.14a is 2 Ω and 0.8 V is applied at E_i. Calculate i and E_a if R_c is 20 Ω.

$$i = 0.8\ \text{V}/2\ \Omega = 0.4\ \text{A}$$

$$E_a = 0.4(20 + 2) = 8.8\ \text{V}$$

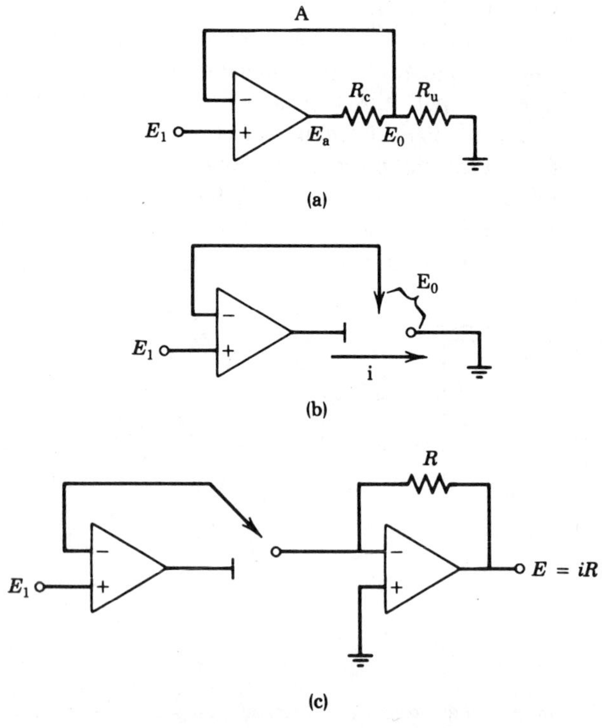

Fig. 29.14 Operational amplifier circuitry for potentiostat and current-to-voltage converter. (a) Basic circuit. (b) Potentiostat circuit. (c) Potentiostat and current-to-voltage converter.

Since essentially no current is observed at the input of the amplifier, no current exists in wire A. Thus, this circuit provides a means of applying a potential across a resistor without drawing current through one wire, but rather providing the required current from the output of the amplifier. Connecting the three wires to three electrodes as shown in Fig. 29.14b causes the potential E_i to be applied across the electrochemical cell consisting of reference and working electrodes. Charge cannot pass through the reference electrode, which is connected to the amplifier input. The current required for electrolysis at the working electrode is provided at the auxiliary electrode half-cell, which operates at potential E_a. R_u becomes solution resistance between the working and reference electrode. This "uncompensated resistance" is the source of iR_u drop in the cell. $R_u + R_c$ becomes the cell resistance between working and auxiliary electrodes. A current-to-voltage converter for measuring the current response signal is added by connecting another amplifier to the working electrode as shown in Fig. 29.14c.

A complete instrument for cyclic voltammetry, a technique in which the potential is cycled between two values (see Section 30.2), is shown in Fig. 29.15. The signal generator produces the triangular waveform excitation signal at the output of OA2. Ancillary circuitry enables scan rate and potential limits to be easily adjusted. This excitation waveform is then connected to the input of the potentiostat (OA3), which applies the potential excitation signal across the reference and working electrodes. The current-to-voltage converter with a selection of resistors in the feedback loop enables the current to be measured in terms of an output voltage that can be scaled by choice of resistor.

Electrochemical instrumentation has benefited substantially from computerization. The microprocessor has been incorporated in signal generation and data processing, while the potentiostat and current-to-voltage converter remain as shown in Fig. 29.14. A block diagram for one commercially available instrument is shown in Fig. 29.16. Microprocessor instruments provide flexibility in that different experiments can be implemented with software. For example, the instrument in Fig. 29.16 can be used for essentially all of the techniques described in Chapters 29 and 30.

29.6 VOLTAMMOGRAMS OF REPRESENTATIVE REDOX SYSTEMS

The voltammograms in Figures 29.4 and 29.5 are representative of those obtained for a single redox couple with one or both oxidation states of the couple present in solution. In this section voltammograms for several other electrode mechanisms that are encountered in electroanalytical chemistry are illustrated.

Solvent-Supporting Electrolyte. Before evaluating voltammograms of representative electroactive species, it is useful to consider voltammograms recorded on electrodes immersed in solvent-supporting electrolyte only. A voltammogram at a platinum electrode in 0.1 *M* KCl is shown in Fig. 29.17a. The dashed line shows a cathodic wave due to reduction of *dissolved oxygen* in the solution by the electrode reaction shown in the figure. The oxygen reduction wave interferes with the

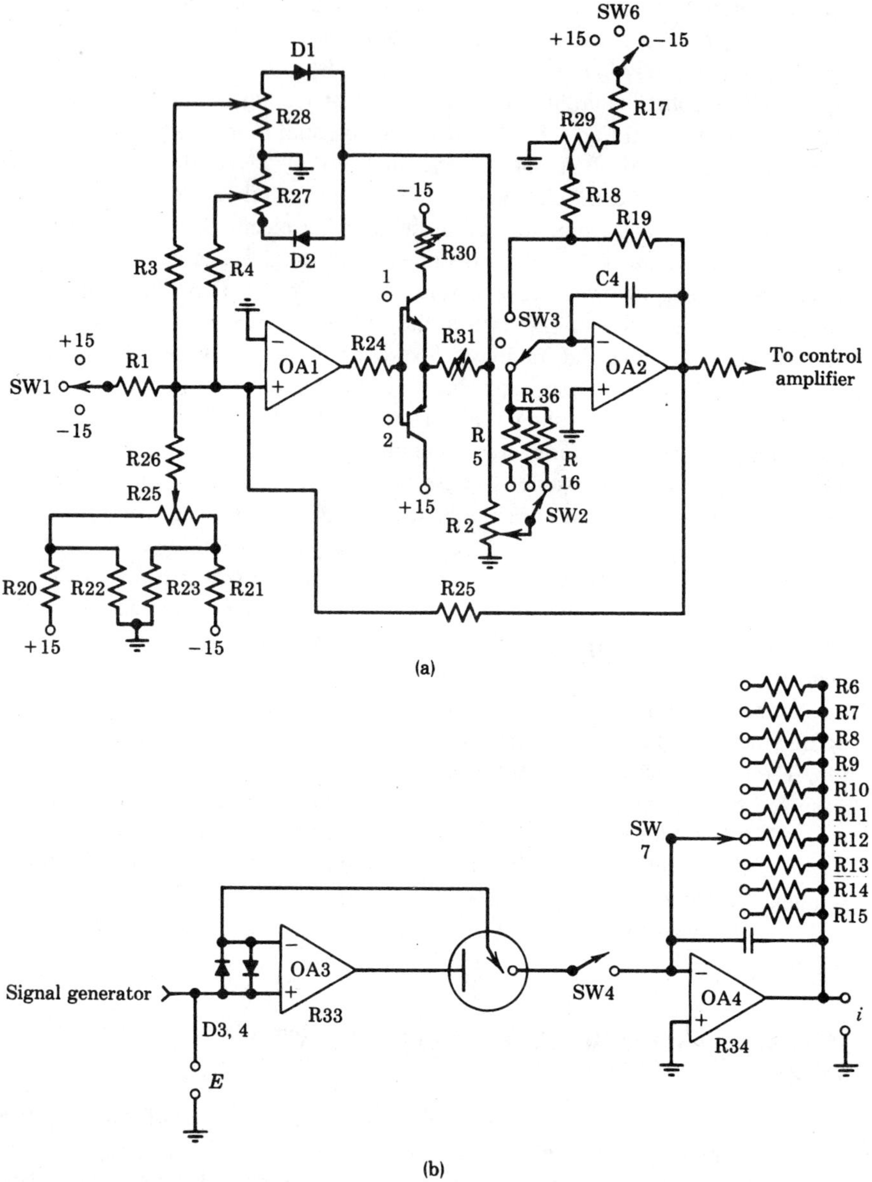

Fig. 29.15 Circuitry for operational amplifier instrument for cyclic voltammetry. (a) Signal generator. (b) Potentiostat and current-to-voltage converter. (From P. T. Kissinger, Introduction to analog instrumentation, in *Laboratory Techniques in Electroanalytical Chemistry*, P. T. Kissinger and W. R. Heineman (Eds.), Dekker, New York, 1985, pp. 180–181.)

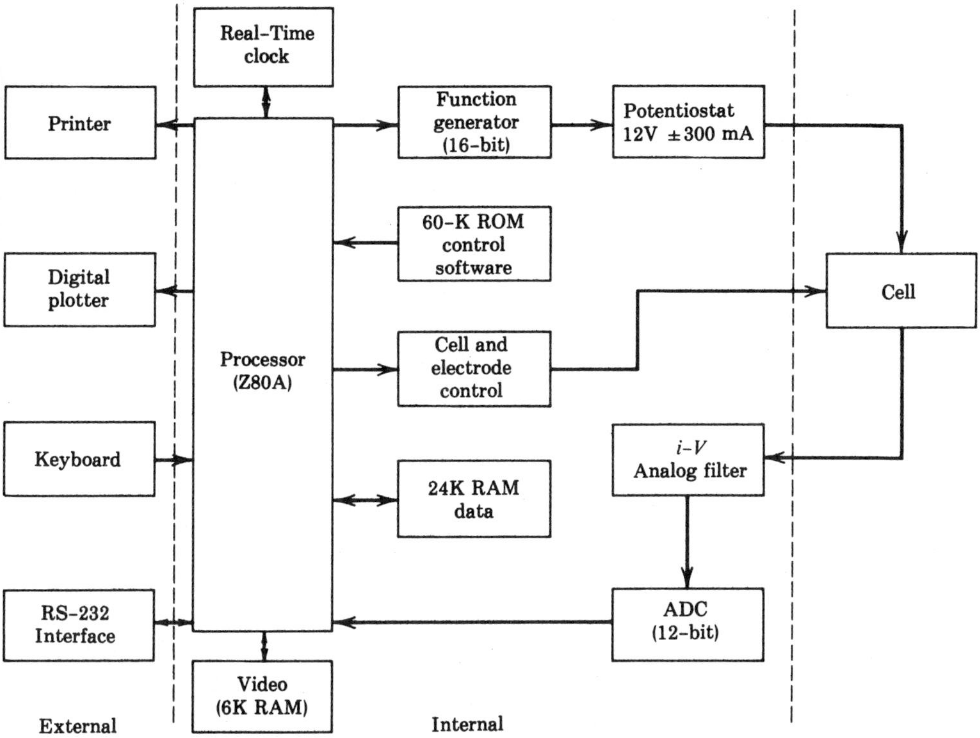

Fig. 29.16 Block diagram for BAS Electrochemical Analyzer. (From P. T. Kissinger, Introduction to analog instrumentation, in *Laboratory Techniques in Electroanalytical Chemistry*, P. T. Kissinger and W. R. Heineman (Eds.), Dekker, New York, 1985, p. 192.)

observation of other species whose reduction waves fall in this same potential range. Consequently, dissolved oxygen is usually removed from solution prior to recording a voltammogram of a species whose electrochemistry occurs in this negative potential region. This procedure is generally unnecessary if oxidations in the positive potential region are to be observed. Dissolved oxygen is usually removed by bubbling nitrogen or argon gas through the solution for 5–20 min. The gas flow is then diverted over the solution to blanket it while the voltammogram is recorded. The solid-lined voltammogram in Fig. 29.17a was recorded after removal of oxygen. A smooth baseline is now obtained in the negative potential range.

The accessible potential range for a voltammogram (the *potential window*) is limited in the negative and positive regions by large cathodic and anodic *background currents* due to reduction and oxidation, respectively, of components in the solvent-supporting electrolyte, as discussed in Section 29.5. The reduction of H^+ to evolve hydrogen gas and the oxidation of H_2O to evolve oxygen gas are two common electrode reactions that give rise to background currents in aqueous solutions, as shown in Fig. 29.17a. Since solvent-supporting electrolyte are present in very high concentration, scanning the potential into a potential region in

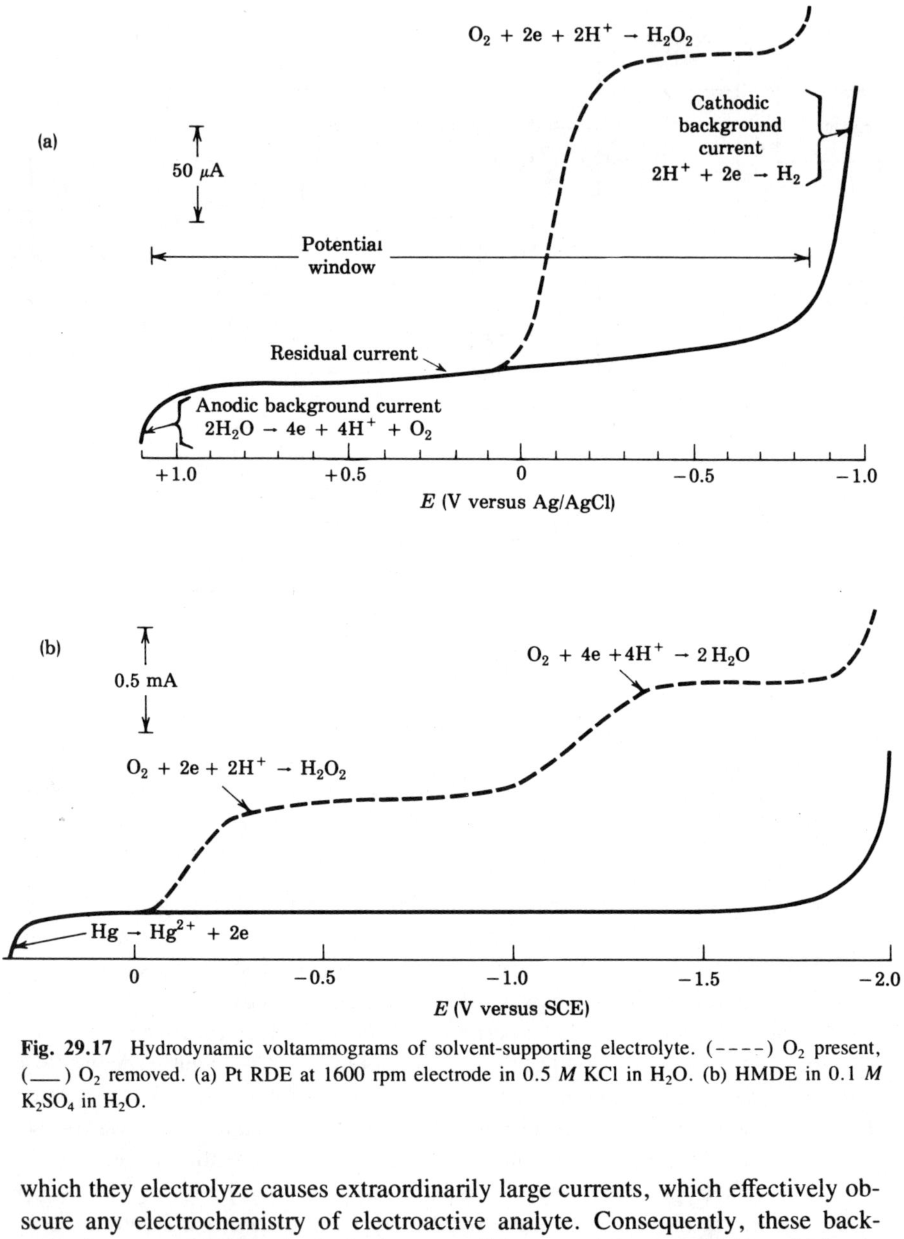

Fig. 29.17 Hydrodynamic voltammograms of solvent-supporting electrolyte. (----) O_2 present, (——) O_2 removed. (a) Pt RDE at 1600 rpm electrode in 0.5 *M* KCl in H_2O. (b) HMDE in 0.1 *M* K_2SO_4 in H_2O.

which they electrolyze causes extraordinarily large currents, which effectively obscure any electrochemistry of electroactive analyte. Consequently, these background currents effectively limit the accessible potential range within which experiments can be performed.

Some electrode materials such as mercury and silver are more easily oxidizable than components in the solvent-supporting electrolyte. The voltammogram for a mercury electrode is shown in Fig. 29.17b. The anodic background current is due to oxidation of the mercury electrode itself to form Hg^{2+}. The extended negative potential range on Hg due to the large overpotential for H^+ reduction enables two reduction waves for dissolved O_2 to be observed at Hg. In the first wave O_2 is reduced to hydrogen peroxide; reduction to H_2O occurs in the second wave.

The potentials of the cathodic and anodic background currents vary dramatically with solvent, supporting electrolyte, and electrode material. Examples of ''potential windows'' are shown in Fig. 29.11. Note the effect of pH on the background limits for electrodes in water; the limits shift to more negative potentials as pH increases.

The baseline between the two large background currents of the voltammograms on deoxygenated solutions in Fig. 29.17 consists of *residual current* due to the reduction–oxidation of electroactive impurities, the electrode itself and solvent-supporting electrolyte components which electrolyze only very slowly at these intermediate potentials. The residual current is potential-dependent, becoming larger as the background current limits are approached. When the potential is being scanned to obtain the voltammogram, the residual current also contains a capacitive component due to charging of the double layer (see Section 27.10). This component is absent if the voltammogram is recorded point by point by moving the potential stepwise along the potential axis and recording the steady-state current response after the capacitive component has decayed to zero.

It is always prudent to record a voltammogram on the electrode–solvent-supporting electrode system before adding the electroactive species of interest. This minimizes the inadvertent assigning of impurity waves to the analyte.

Multiple Redox Reactions. The voltammogram in Fig. 29.18 illustrates the behavior of a redox species that undergoes *two redox reactions*, $Cu(NH_3)_6^{2+}$. The first reduction wave corresponds to the reduction of $Cu(NH_3)_6^{2+}$ to $Cu(NH_3)_6^{+}$ and the second reduction wave corresponds to the reduction of $Cu(NH_3)_6^{2+}$ to Cu

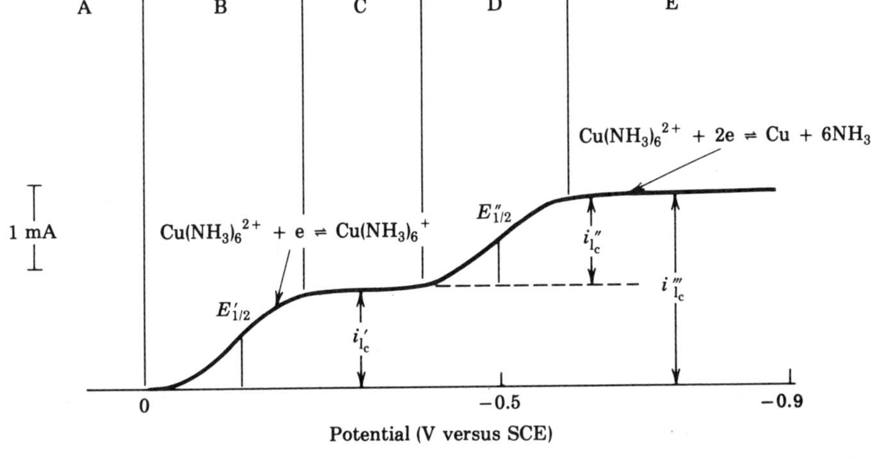

Fig. 29.18 Hydrodynamic voltammogram of 1.00 m*M* $Cu(NH_3)_6^{2+}$ in aqueous 1 *M* NH_4OH at HMDE, scan rate 10 mV s^{-1}.

metal. In order to interpret this voltammogram, let's start at O V and scan in the negative potential direction (to the right). The voltammogram is divided into regions to facilitate the discussion. In region A, the electrode is an insufficient reductant to reduce $Cu(NH_3)_6^{2+}$, and the current is negligible. In region B, the electrode potential becomes sufficiently negative to reduce $Cu(NH_3)_6^{2+}$ to $Cu(NH_3)_6^+$ and the current increases as $C^s_{Cu(NH_3)_6^{2+}}$ decreases. In region C, $C^s_{Cu(NH_3)_6^{2+}}$ has become essentially zero so that a limiting current plateau, i'_{lc}, is attained. Note that $E'_{1/2}$ for $Cu(NH_3)_6^{2+}$ reduction to $Cu(NH_3)_6^+$ is the potential at which $i = i'_{lc}/2$. In region D, the electrode is now a sufficiently strong reductant to reduce $Cu(NH_3)_6^{2+}$ all the way to Cu^0. The current increases as a greater fraction of $Cu(NH_3)_6^{2+}$ is reduced to Cu, rather than $Cu(NH_3)_6^+$. Note that $E''_{1/2}$ for $Cu(NH_3)_6^{2+}$ reduction to Cu^0 is the potential at which $i = i''_{lc}/2$, where the baseline for i''_{lc} is obtained by extrapolation of the first plateau region, as shown by the dashed line. In region E, essentially all $Cu(NH_3)_6^{2+}$ arriving at the electrode surface is being converted to Cu^0, which gives rise to the second plateau region, i''_{lc}. Note that $i'_{lc} = i''_{lc}$, both of which are characteristic of a one-electron (1-e) process, whereas i'''_{lc} if for a 2-e process and is consequently double i'_l and i''_l. It is apparent from the voltammogram that it is energetically easier to reduce $Cu(NH_3)_6^{2+}$ to $Cu(NH_3)_6^+$ than to Cu^0, as evidenced by the more negative value of $E''_{1/2}$ than $E'_{1/2}$.

Mixtures. Hydrodynamic voltammograms of *mixtures* are characteristically step-shaped in appearance, as shown in Fig. 29.19. The solution contains a mixture of two oxidizable species, caffeic and vanillic acids. As the potential is scanned in the positive direction, caffeic acid oxidizes first, as defined by $E'_{1/2}$. The first plateau, i'_{la}, is proportional in height to the concentration of caffeic acid. As the potential scans more positively, the electrode is able to oxidize vanillic acid at a potential characterized by $E''_{1/2}$. The component of the limiting current of this second plateau that is due to oxidation of vanillic acid is defined by i''_{la}, which uses a baseline defined by extrapolation of the first plateau. The magnitude of i''_{la} is proportional to the concentration of vanillic acid. It is important to realize that the oxidation of caffeic acid continues uninterrupted at the onset of oxidation of vanillic acid.

Electrode Oxidation. Anions or ligands that react chemically with a particular oxidation state of the metal of which the electrode is composed can influence the shape of a voltammogram. The shape is perturbed in proportion to the concentration of the anion or ligand such that these species can be determined electrochemically, even though they are themselves electroinactive. An example is the voltammogram of Cl^- at a silver electrode shown in Fig. 29.20. A voltammogram in the absence of Cl^- shows an anodic background current from oxidation of Ag to form Ag^+. Increasing the Cl^- concentration by addition of Cl^- gives rise to an anodic wave whose limiting anodic current is proportional to Cl^- concentration and whose $E_{1/2}$ is less positive than background current for silver oxidation. The ease with

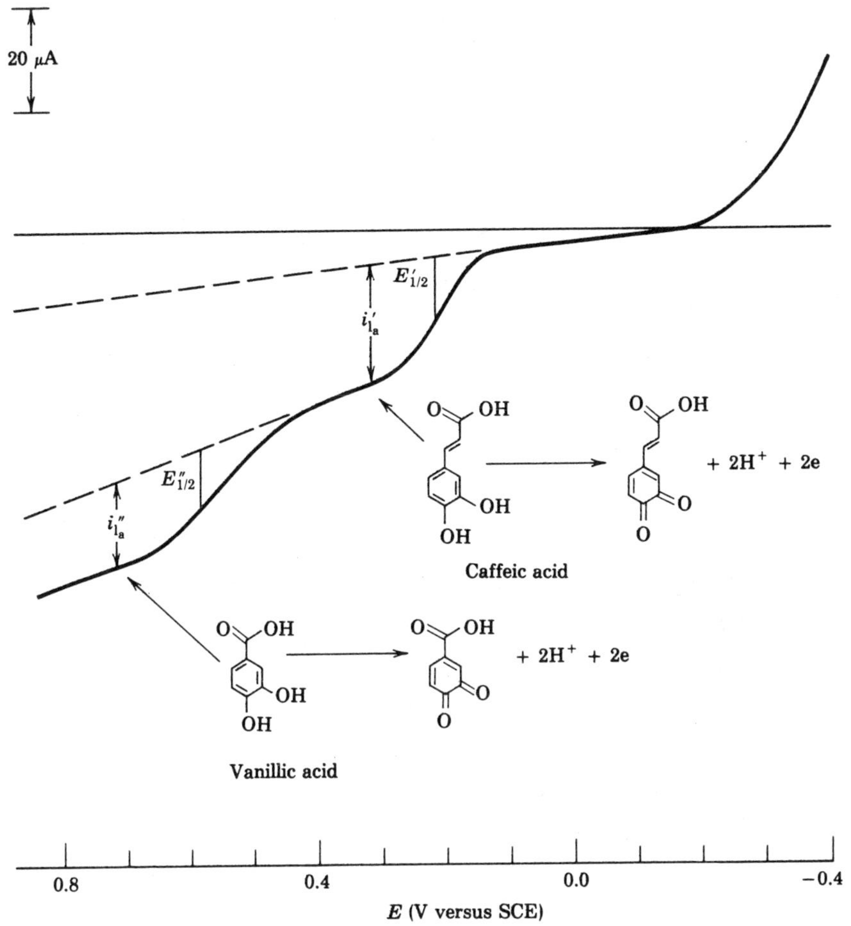

Fig. 29.19 Hydrodynamic voltammogram of 1.00 m*M* caffeic and vanillic acids in aqueous 0.5 *M* KCl, pH 7.0 phosphate buffer at Pt rotating disk electrode.

which silver can oxidize in the presence of chloride can be traced to the formation of AgCl(s), a precipitate that is more stable than Ag^+ (see electrode reaction in Fig. 29.20). In other words, Ag oxidizes more easily if Cl^- is present to precipitate AgCl(s). Since Cl^- must be present for this electrode reaction to occur, i_{la} for the anodic wave is proportional to Cl^- concentration and, hence, can be used for the quantitative analysis of Cl^-. As might be expected, the $E_{1/2}$ of the Cl^- wave is related to the K_{sp} of AgCl(s), which is a measure of its stability.* The less soluble the precipitate the less positive its $E_{1/2}$. A mixture of anions that precipitate with

*This point is related to the Nernst equation that is derived for an electrode of the second kind as shown in Table 28.1.

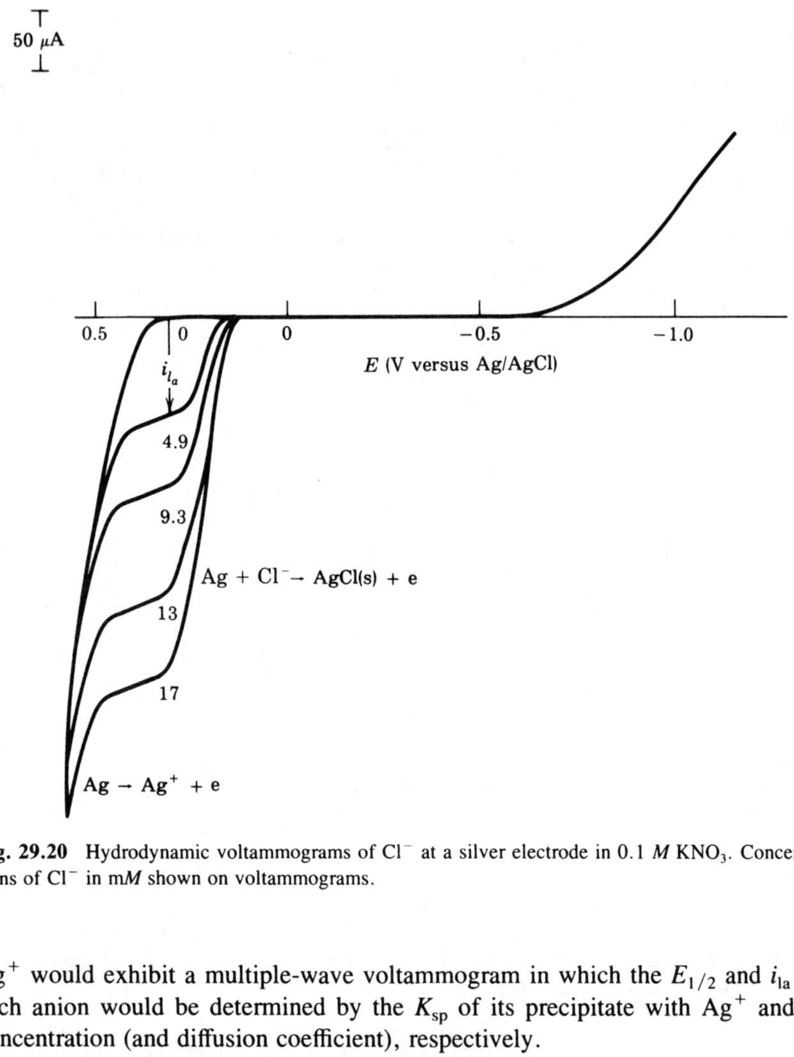

Fig. 29.20 Hydrodynamic voltammograms of Cl^- at a silver electrode in 0.1 *M* KNO_3. Concentrations of Cl^- in m*M* shown on voltammograms.

Ag^+ would exhibit a multiple-wave voltammogram in which the $E_{1/2}$ and i_{la} for each anion would be determined by the K_{sp} of its precipitate with Ag^+ and its concentration (and diffusion coefficient), respectively.

Anodic waves are also observed for ligands that form conplexes with the oxidized form of a metal electrode. Examples are CN^- at Ag and Hg electrodes and EDTA at a Hg electrode. In these cases, $E_{1/2}$ is related to the formation constant of the complex.

Electrode Reactions of Organic Compounds. Many organic compounds undergo reduction or oxidation at an electrode. Electroanalytical techniques have been used extensively for analytical determinations of organic compounds and for studying the mechanisms of their electrode reactions. In aqueous solution, the reduction of organic compounds is frequently a 2-e process accompanied by protonation, as shown below for the reduction of quinone (Q) to hydroquinone (H_2Q).

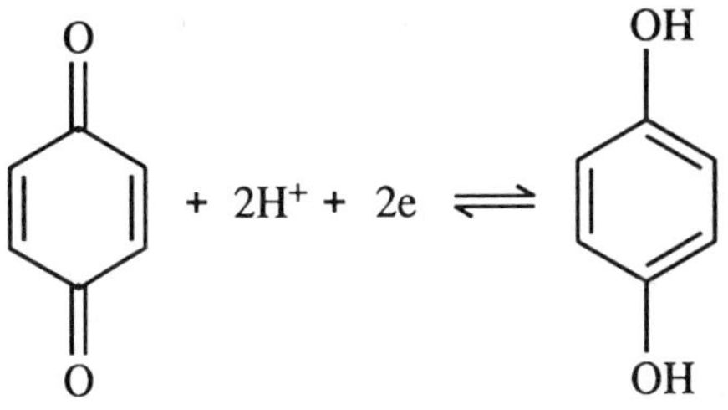

(29.19)

Q $\qquad$ H_2Q

Hydrodynamic voltammograms of quinone–hydroquinone are shown in Fig. 29.21. The voltammograms were obtained on solutions of Q, H_2Q, and an equimolar mixture of Q, H_2Q. In the latter case, both reduction and oxidation waves are obtained. Note that $E_{1/2}$ is independent of the concentrations of Q and H_2Q.

The $E_{1/2}$ for an electrode reaction that involves protonation is pH dependent. As the pH of the solution increases, $E_{1/2}$ shifts negatively since the decreasing availability of protons makes the reaction energetically more difficult. Consider the general case of a 2-e, 2-H^+ reduction

$$O + 2e + 2H^+ \rightleftharpoons H_2R \tag{29.20}$$

for which the Nernst equation is

$$E = E^{0'}_{O,H_2R} - \frac{RT}{2F} \ln \frac{C_{H_2R}}{C_O C^2_{H^+}}. \tag{29.21}$$

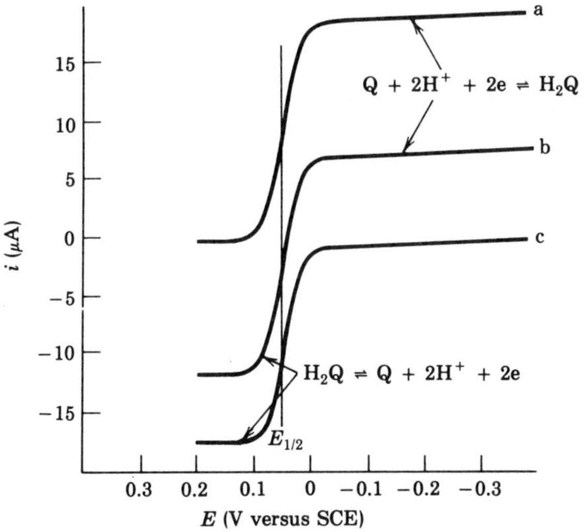

Fig. 29.21 Voltammograms obtained for a quinone–hydroquinone couple in a solution buffered at pH 7. (a) Quinone (Q) (b) Quinone and hydroquinone (H_2Q) mixture. (c) Hydroquinone.

Rearrangement of this equation yields

$$E = E^{0'}_{O,H_2R} - \frac{RT}{2F} \ln \frac{1}{C^2_{H^+}} - \frac{RT}{2F} \ln \frac{C_{H_2R}}{C_O} \qquad (29.22)$$

If $E_{1/2}$ is taken as the value of E for which $C^s_O = C^s_{H_2R}$ at the electrode surface, then

$$E_{1/2} = E^{0'}_{O,H_2R} - \frac{RT}{2F} \ln \frac{1}{C^2_{H^+}}, \qquad (29.23)$$

which for 25°C becomes

$$E_{1/2} = E^{0'}_{O,H_2R} - 0.591 \text{ pH}. \qquad (29.24)$$

It is apparent from this equation that a plot of $E_{1/2}$ versus pH has a slope of −0.0591 (i.e., the potential shifts −0.0591 V for each increase in pH unit).

The organic acid H_2R will undergo deprotonations at sufficiently high pHs, as determined by K_{a1} and K_{a2}. Consequently, at a sufficiently high pH the electrode reaction becomes

$$O + 2e + H^+ \rightleftharpoons HR^-, \qquad (29.25)$$

since HR^- is the stable form at this pH. In this case, a derivation of the expression for $E_{1/2}$ will show that the shift in $E_{1/2}$ with increasing pH unit is only −0.0295

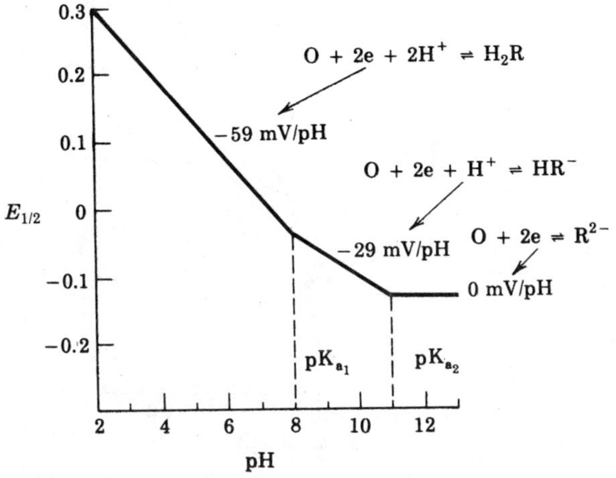

Fig. 29.22 Dependence of $E_{1/2}$ on pH for representative reduction of an organic compound. $O + 2e + 2H^+ \rightleftharpoons RH_2$.

V. At a pH above pK_{a2}, the electrode reaction becomes

$$O + 2e \rightleftharpoons R^{2-} \tag{29.26}$$

and $E_{1/2}$ loses its pH dependence. This behavior is summarized in Fig. 29.22. The inflection points in this $E_{1/2}$ versus pH plot correspond to the values of pKa_1 and pKa_2, which are 8.0 and 11.0 for the hypothetical example RH_2.

Reduction–oxidation reactions of organic compounds without the attendant protonation reactions discussed above can be carried out in nonaqueous solvents such as acetonitrile, dimethylformamide, tetrahydrofuran, dimethylsulfoxide, methylene chloride, propylene carbonate, benzonitrile, liquid NH_3, liquid salts (melts), liquid SO_2, nitrobenzene, hexamethylphosphoramide, and (with ultramicro electrodes) *n*-hexanes and toluene. Sometimes these solvents are used because of analyte solubility considerations.

29.7 QUANTITATIVE ANALYSIS

The limiting current of a hydrodynamic voltammogram is proportional to the concentration of species responsible for the current, either by its direct electrolysis or its shifting of the electrolysis of the electrode material as in the case of Cl^- at a Ag electrode. Equations (29.6) and (29.7) express the general relationship between limiting current and concentration, which may be expressed as

$$i_l = KC \tag{29.27}$$

Thus, a plot of i_l versus C gives a linear calibration plot.

Generally, one records entire voltammograms solely for the purpose of finding the limiting current plateau. A potential on the plateau is then selected for the analyses and the current is measured at that potential for all standards, from which a calibration plot is constructed, and samples. The concentration of analyte in a sample is calculated from the calibration plot. This technique is termed *amperometry*. Recording the current at a single, appropriate potential minimizes analysis time by saving the time required to obtain a complete voltammogram. A blank containing supporting electrolyte only gives the residual current, which is subtracted from each measurement.

In the case of samples with complex matrices that are quite different from the supporting electrolyte in standard solutions, the *method of standard additions* is advisable (see Section 11.4). In this procedure a current measurement is first obtained on the sample. A small aliquot of concentrated standard analyte solution is then mixed into the sample, and a second current measurement is made. The concentration of the unknown is calculated by the following equation:

$$C_u = \frac{C_s V_s i_{l_u}}{(V_u + V_s)\, i_{l_{u+s}} - V_u i_{l_u}}, \tag{29.28}$$

where s = standard, u = unknown, C = concentration, V = volume, and i_l = limiting current. Alternatively, the method of sequential standard additions can be used to generate a plot as shown in Fig. 11.5.

All of these analytical procedures share a common requirement—K in Eq. (29.27) should be constant throughout the measurements on standards and samples. Thus, measurements should be made on electrodes of the same surface area (preferably at the same electrode) and under identical hydrodynamic conditions so that δ in Eq. (29.6) is constant.

The optimum potential for quantitative measurements is usually on the plateau of the limiting current. This region gives maximum sensitivity in the current response. Precision is usually better on the plateau than on the rising portion of the voltammogram, where small variations in potential due to any drift in the measurement system will result in differences in current. Good results can be obtained for irreversible systems by selecting a potential in the limiting current region, even though the voltammograms are drawn out by slow heterogeneous electron transfer, as shown in Fig. 29.7.

Selectivity can be achieved by appropriate control of potential in some cases. For example, in Fig. 29.19 caffeic acid can be determined without interference from vanillic acid by holding the potential sufficiently positive to oxidize caffeic acid, but less positive than required to oxidize vanillic acid.

The instrumentation for hydrodynamic voltammetry can take a variety of formats. Its simplest form is a beaker to hold the solution in which the electrodes are immersed; the solution is stirred by a magnetic bar. With this apparatus the accessible concentration range is typically 10^{-6}–10^{-2} M with a precision of measurement on the order of 5% or less. Detection limits can be improved to about 10^{-8} M with methodology such as the rotating disk electrode, flow-injection analysis, and liquid chromatography with electrochemical detection. These three techniques are discussed later in this chapter.

29.8 ROTATING DISK ELECTRODE

Although hydrodynamic voltammograms are easily obtainable at an electrode immersed in a solution that is stirred with a stir bar, the hydrodynamics are better defined and more reproducible if convection is produced by rotating the electrode. The *rotating disk electrode* (RDE) is generally a disk of conductive material such as platinum imbedded in a rod of insulator such as Teflon, as shown in Fig. 29.23. This electrode assembly is carefully machined so that it can be rotated rapidly with minimal wobble.

The hydrodynamic flow pattern resulting from rapid rotation of the disk moves liquid horizontally out and away from the center of the disk with a consequent upward axial flow to replenish liquid at the surface, as shown by the arrows in Fig. 29.23a. A rigorous hydrodynamic treatment for the limiting current is

$$i_l = 0.620nFACD^{2/3}\nu^{-1/6}\omega^{1/2} \tag{29.29}$$

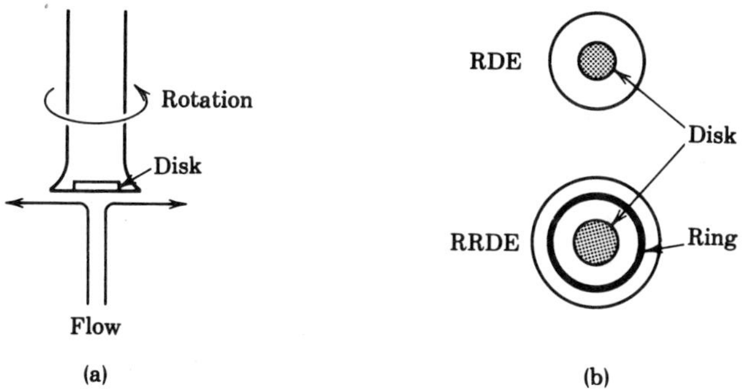

Fig. 29.23 (a) Rotating disk electrode (RDE) with hydrodynamic flow pattern. (b) Bottom view of RDE and rotating ring-disk electrode (RRDE).

where ω = angular velocity of the disk ($\omega = 2\pi N$, $N = rps$), ν = kinematic viscosity of the fluid (cm^2s^{-1}), i_l = limiting current (A), C = solution concentration (mol cm^{-3}), and other terms are identified in Table 27.1. Good electrode design is necessary for strict adherence to this equation.*

The shapes of voltammograms recorded at a rotating disk are identical in form to the stirred-solution voltammograms described earlier in this section. The improved definition of hydrodynamic flow across the electrode causes the voltammograms to be less noisy than those recorded in a solution stirred with a stir bar. Also, the Nernst diffusion layer decreases the faster solution flows across the electrode. Thinner Nernst diffusion layers give larger limiting currents since the slope of the concentration distance profile is greater. [See Eq. (29.4)].

Rotating Ring-Disk Electrode. The rotating ring-disk electrode (RRDE) is an extension of the rotating disk electrode in which the disk is encircled by a ring of conductive material that functions as a second working electrode. The configuration is shown in Fig. 29.23b. The disk is used to electrochemically generate a reactive species which is electrochemically monitored as it is swept by laminar flow past the ring. This technique has prove useful in the study of reaction mechanisms, since electroactive intermediates of coupled homogeneous chemical reactions can be monitored at the ring electrode.

The reduction of oxygen illustrates the use of rotating ring-disk voltammetry. While Fig. 29.24a shows the *disk* voltammogram obtained for the reduction of oxygen, Fig. 29.24b shows the *ring* voltammogram when the potential of the ring is maintained at a value for oxidation of H_2O_2. In each curve current is plotted as a function of the potential applied to the disk. In the potential region in which the disk is generating H_2O_2, the *anodic* current recorded at the ring results from oxidation of the H_2O_2 that is being swept past it by laminar flow. When the potential applied to the disk becomes sufficiently negative that O_2 is being reduced to OH^- at the disk, the ring current diminishes, since OH^- rather than H_2O_2 is being swept past. Such experiments can confirm the existence of a particular intermediate, such as H_2O_2 in this case, in an electrode reaction.

*Unless special precautions are taken (e.g., special cell designs or better rotors) 5000 rpm is approximately the upper limit for rotation rates in H_2O which can be implemented. Above this rate, turbulent flow begins to occur and observed currents become much higher.

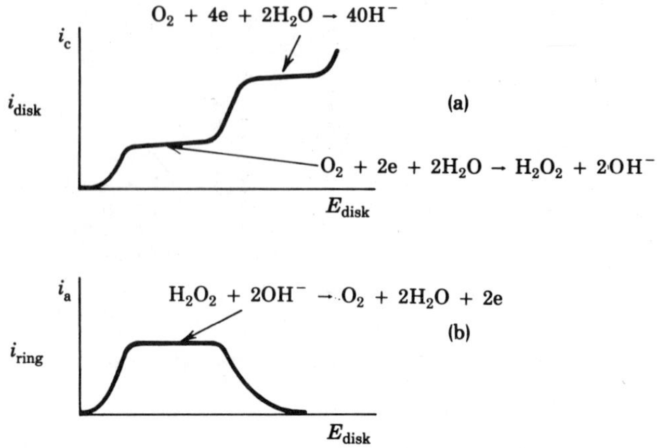

Fig. 29.24 Rotating ring disk electrode voltammograms. (a) Disk current for reduction of oxygen. (b) Ring current for oxidation of H_2O_2.

The rotating ring-disk electrode has been applied to a variety of electrode mechanisms. It is one of the most useful techniques for studying electrode reactions.

29.9 OXYGEN ELECTRODE

One of the most common applications of hydrodynamic voltammetry is the determination of dissolved oxygen levels; samples may be of the blood of a patient, water in a nuclear reactor system, or outdoor streams, ponds, and lakes. As shown by the voltammogram for dissolved oxygen in Fig. 29.17, a limiting current plateau is obtained for oxygen reduction at a platinum electrode. Although the oxygen waves are a nuisance in the general application of voltammetry to the study of redox systems, they are of significant analytical utility when O_2 itself is to be measured.

Oxygen can be determined in relatively pure samples by simply dipping an electrode such as platinum in the solution and measuring the limiting current which is proportional in height to $p\hat{O}_2$. However, this simple procedure is not suitable for many samples that contain other electroactive species that are reducible at this same potential and/or surface-active species that absorb on the electrode surface and inhibit the reduction of O_2. These interferences are effectively circumvented by protecting the electrode with an oxygen-permeable membrane. One type of oxygen electrode is shown in Fig. 29.25. The oxygen-sensing electrode consists of a platinum disk that is encircled by a silver ring-counter electrode. These electrodes are in contact with a thin layer of aqueous KCl solution that is bounded on the other side by a thin oxygen-permeable membrane such as Teflon or polyethylene. When the electrode is immersed in sample solution, O_2 in the sample permeates the membrane and diffuses to the Pt disk cathode where it is detected by reduction. The silver anode oxidizes in the presence of Cl^- to form AgCl, which

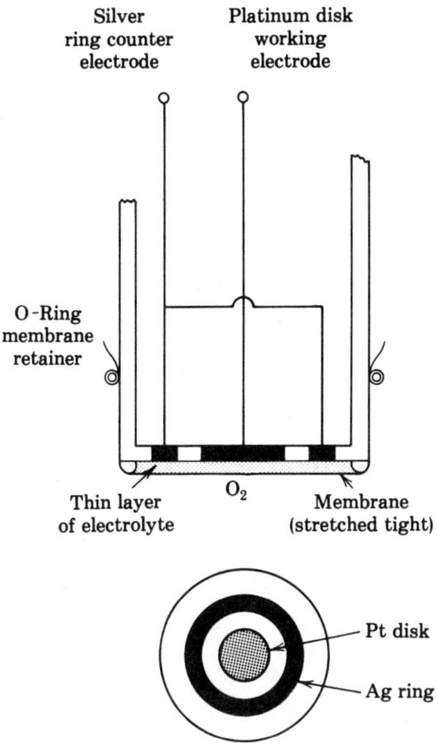

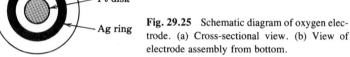

Fig. 29.25 Schematic diagram of oxygen electrode. (a) Cross-sectional view. (b) View of electrode assembly from bottom.

precipitates on the Ag electrode surface and therefore does not interfere with the detection of O_2 at the cathode.

The oxygen-permeable film imparts excellent selectivity for oxygen. Since it is only permeable to gaseous species, most of the components in a sample are excluded from the cell. Thus, electroactive species that would coreduce with O_2 and surfactants that would coat the electrode surface are effectively eliminated as interferences. Only gases such as O_2, N_2, and CO_2 can pass through the membrane, and of these only O_2 is reducible.

A series of voltammograms for solutions containing different partial pressures of O_2 is shown in Fig. 29.26a and a plot of i_l versus $p\hat{O}_2$ is in Fig. 29.26b. Standard plots of this type can be obtained with standard solutions of O_2 prepared by bubbling accurately known ratios of O_2/N_2 through solution. A two-point calibration is easily made from measurements on a solution that is equilibrated with air and then bubbled with N_2 to remove O_2. This gives one point that is based on the known concentration of O_2 in solution at a given altitude and temperature and a "zero" oxygen point.

Oxygen electrodes have been constructed for a variety of purposes. Rugged, battery-powered oxygen monitors have been used for the continuous recording of dissolved oxygen levels in, for example, streams of environmental interest. Microelectrodes enable oxygen levels to be measured in tissues and arterial or venous

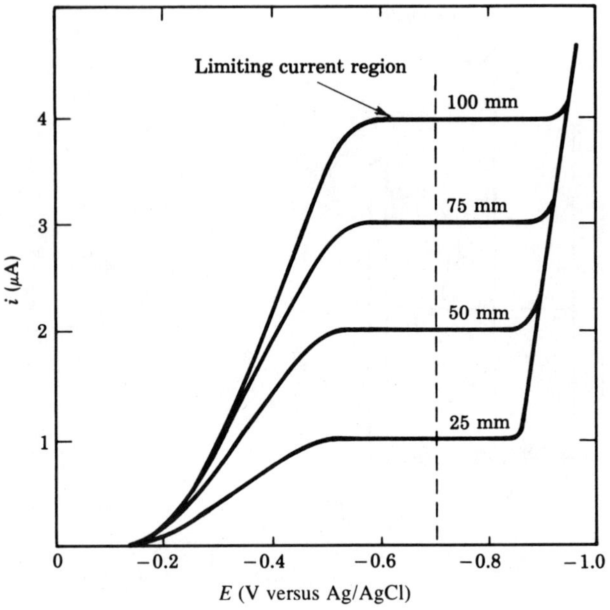

(a)

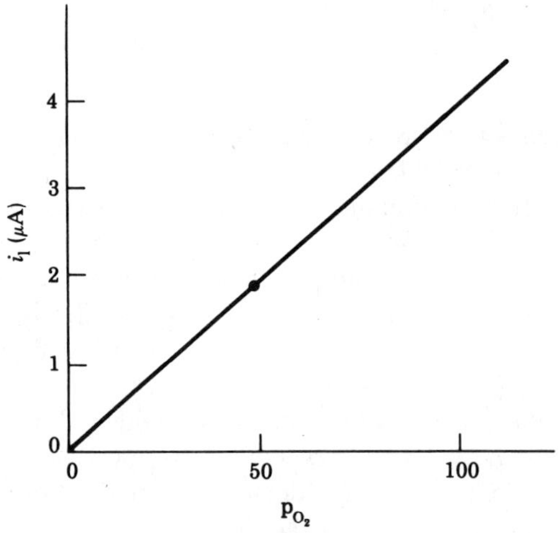

(b)

Fig. 29.26 Oxygen electrode. (a) Voltammograms on solutions of known p_{O_2}. (b) Standard plot of i_l versus p_{O_2}.

blood. Electrodes can be "threaded" through veins into the human heart to measure oxygen inside a particular cavity in the heart.

29.10 AMPEROMETRIC ENZYME ELECTRODES: GLUCOSE ELECTRODE

In the use of hydrodynamic voltammetry enzymes play an important role in measuring certain species. A good example is the glucose electrode. It is commonly used in clinical laboratories for the measurement of serum glucose levels. In this electrode, a thin layer of enzyme is immobilized in the electrochemical cell, as shown in Fig. 29.27. A magnified view of a section of the cell shows the immobilized enzyme, glucose oxidase, to be sandwiched between two different membranes. When the electrode is immersed in sample solution, glucose diffuses through the outer membrane into the enzyme layer where the enzyme-catalyzed reaction of glucose occurs. The H_2O_2 reaction product diffuses through the inner membrane to the electrode where it is detected by oxidation to water at a potential of about +0.7 V versus the Ag/AgCl counter electrode. Since the amount of H_2O_2 generated is proportional to the concentration of glucose in the sample, a calibration plot can be constructed by measuring the limiting anodic current for oxidation of H_2O_2 as a function of glucose concentration in standard solutions.

The selectivity of the electrode is determined by the specificity of the enzyme

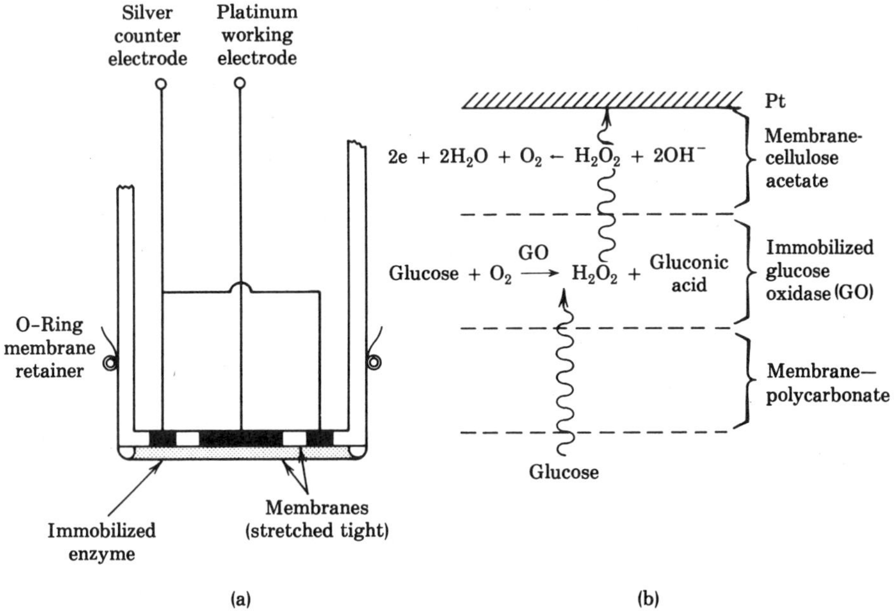

Fig. 29.27 Amperometric glucose sensor. (a) Cross section of electrode. (b) Membrane assembly and chemical reactions.

for glucose and the permeability of the membranes to species that would interfere. The outer membrane (polycarbonate) traps the thin layer of glucose oxidase and serves as an outer barrier to exclude large molecules such as proteins. The glucose oxidase is immobilized in a layer of glutaraldehyde. The inner membrane (cellulose acetate) is permeable only to very small molecules such as H_2O_2. It serves to keep glucose oxidase off the electrode itself and to exclude electroactive components in the sample such as uric acid, ascorbic acid, and acetaminophen (Tylenol) by preventing them access to the electrode.

29.11 ELECTROCHEMICAL DETECTION IN LIQUID CHROMATOGRAPHY AND FLOW-INJECTION ANALYSIS

One of the fastest growing uses of hydrodynamic voltammetry is electrochemical detection for liquid chromatography (LCEC) and flow-injection analysis. In these applications a small volume of sample is injected into a flowing stream of mobile phase. In liquid chromatography a sample passes through a chromatographic column where it is separated into its components (see Chapter 26). In flow-injection analysis a sample is also injected into a flowing stream of mobile phase, but no column is used to separate components of the sample. Flow-injection analysis serves as a convenient means for the analysis of small volumes (microliters) of sample.

The most commonly used electrochemical detector is a thin-layer cell in which the working electrode is positioned in a thin channel through which the mobile phase flows. A detector of this type is shown schematically in Fig. 29.28, which illustrates the detection of a "plug" of analyte R that has eluted from the column being detected by oxidation to O as it sweeps over the electrode in the thin channel. For maximum sensitivity the potential of the electrode is held on the limiting current region of a hydrodynamic voltammogram for oxidation of R. The resulting chromatogram shows a peak current response for the detection of R. The concentration of R can be quantified by measuring either the peak height i_p or integrating the peak and measuring the charge Q, both of which are proportional to C_R.

Thin-layer cell electrochemical detectors are available commercially in several designs. A commonly used, commercially available detector is shown in Fig. 29.29. The detector consists of two blocks separated by a thin gasket that defines the thickness and flow geometry of the channel through which solution flows. The upper block of stainless steel serves as the auxiliary electrode. The working electrode is imbedded in the bottom block of plastic. Dual working electrodes in either a parallel or series configuration are also available. The working electrodes are typically glassy carbon, carbon paste, or mercury film. An Ag/AgCl reference electrode is positioned downstream of the thin-layer detector. The active volume of the cell is typically 0.1–1 μL. Thin-layer detectors of this type are generally operated as *amperometric* devices, that is, the potential of the working electrode is held at a constant value and the current is measured as a function of time. Typically about 5% of the analyte is electrolyzed as it is swept past the electrode. Amperometric detectors can achieve detection limits as low as 10^{-9}–10^{-8} M of

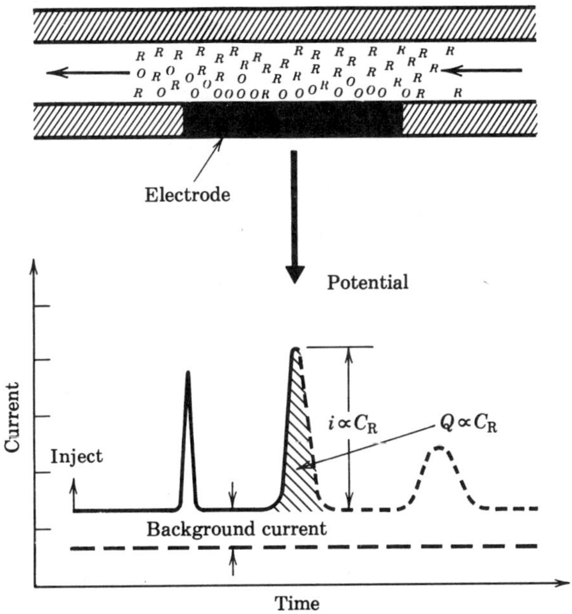

Fig. 29.28 Schematic view of thin-layer amperometric detection. A mass transport controlled oxidation is taking place (R $\rightleftharpoons$ O + ne). The current is recorded as a function of time at a single applied potential. Both the current (i) and charge (Q) are proportional to the amount of material injected. (Reproduced by permission of Bioanalytical Systems, Inc.)

injected anlayte. Coupled with an injected sample of, say, 10 μL, amperometry can detect as little as 10^{-16} mol (or 0.1 fmol!). Compared with voltammetric techniques in which the potential is scanned, amperometry is advantageous in that charging current is minimized by operating at a fixed potential.

Hydrodynamic voltammograms of eluting components are useful for selecting the appropriate detection potential when operating the detector at a fixed potential (i.e., amperometrically). Voltammograms also give qualitative information in the form of $E_{1/2}$ values that can be used in conjunction with chromatographic retention time for purposes of analyte identification. Hydrodynamic voltammograms can be obtained on eluting components by repetitive injection of the sample with the electrode potential held at a different value for each injection. Examples of hydrodynamic voltammograms obtained in this way are shown in Fig. 29.30a. An alternative approach for obtaining voltammograms is to rapidly scan the potential of the electrode as the peak(s) elutes across it. A technique with low detection limits as well as rapid scanning capability is required. Square wave voltammograms (See Section 30.4) taken rapidly enable a three-dimensional chromatovoltammogram such as the one shown in Fig. 29.31 to be constructed.

The choice of potential for amperometric detection influences both *sensitivity* and *selectivity* of analysis. Optimum sensitivity for a particular analyte is achieved by holding the potential in the limiting current region of the hydrodynamic voltam-

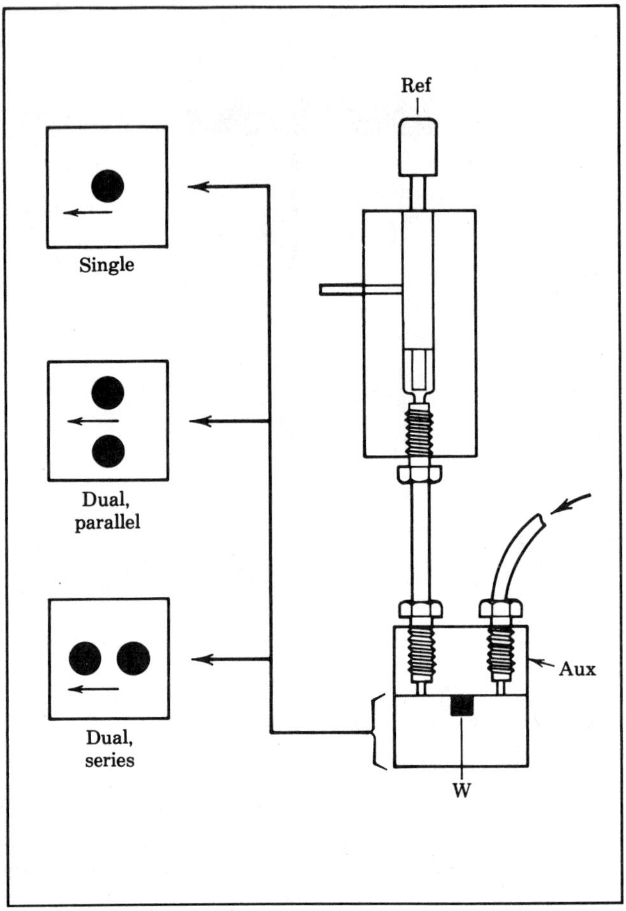

Fig. 29.29 Commercially available thin-layer amperometric detector cells. A reference electrode, auxiliary electrode, and one or more working electrodes are used. (From Bioanalytical Systems, Inc., West Lafayette, IN.)

mogram for the analyte in question. The chromatograms in Fig. 29.30b and c show the effect of detection potential on a mixture of biogenic amines and metabolites the hydrodynamic voltammograms of which are shown in Fig. 29.30a. Correlate the change in peak height (sensitivity) at the two potentials for each component with its respective voltammogram. Why is the peak for 5 HT essentially unchanged, but sensitivity for HVA dramatically different at the two potentials? If $E_{1/2}$ values for two electroactive components in a mixture are sufficiently different, it can be possible to improve selectivity by operating at a potential at which the "interfering" component is not detected by the electrode. Selectivity of this type in a detector can be advantageous for the analysis of a complicated mixture for which all of the components cannot be resolved chromatographically.

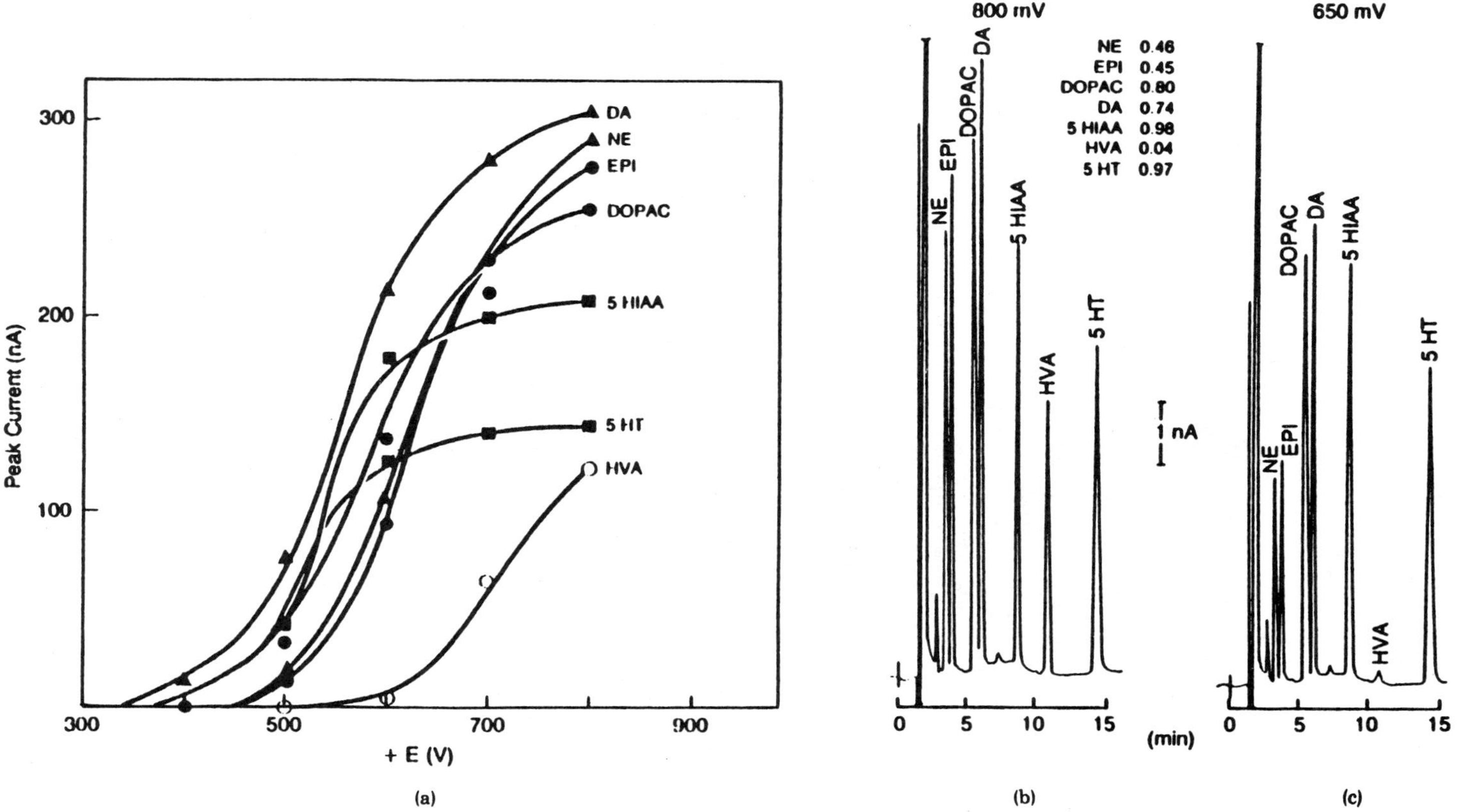

Fig. 29.30 (a) Hydrodynamic voltammograms for a series of biogenic amines and metabolites. Chromatograms at (b) 800 mV and (c) 650 mV. Conditions: glassy carbon electrode versus Ag/AgCl; Biophase ODS 5-μm column (250 × 4.6 mm); mobile phase, 1.9 parts THF, 3.5 parts CH_3CN, and 96.5 parts 0.15 monochloroacetate buffer, pH 3.0, containing 200 mg L^{-1} sodium octyl sulfate and 0.25 g L^{-1} Na_2EDTA; flow rate, 1.5 mL/min. Abbreviations: NE, norepinephrine; EPI, epinephrine; DOPAC, 3,4-dihydroxyphenylacetic acid; DA, dopamine; 5-HIAA, 5-hydroxyindole-3-acetic acid; HVA, homovanillic acid; and 5-HT, serotonin. [Adapted from D. A. Roston, R. E. Shoup, and P. T. Kissinger, *Anal. Chem.*, **54,** 1417A (1982).]

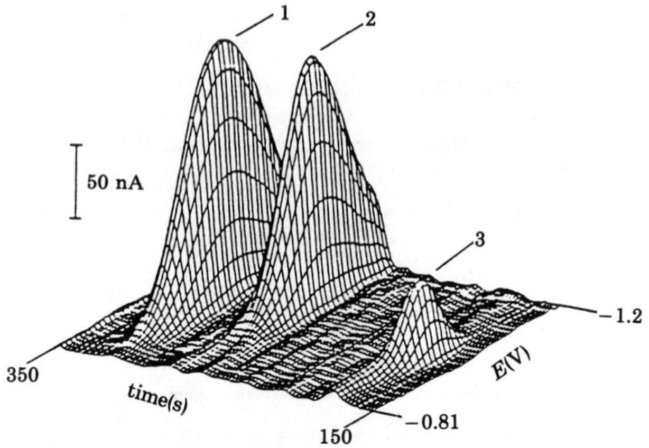

Fig. 29.31 Three-dimensional "chromatovoltammogram" for two nitrosamines (peaks 1 and 2) and an unknown impurity (peak 3). [From R. Samuelsson, J. O'Dea, and J. Osteryoung, *Anal. Chem.*, **52**, 2215 (1980).]

Example 29.6. Dual series electrodes can be useful in cases where the upstream electrode can be used as a "generator" electrode to convert the analyte into a more easily detected form. For example, disulfides can be detected by reduction at −1.1 V, although the detection limit is poor owing to high residual current at this potential. The reduction of a disulfide generates a thiol,

$$R - S - S - R + 2e + 2H^+ \rightarrow 2\ RSH,$$

which can be detected by oxidation at a downstream mercury film electrode at +0.15 V according to the following mechanism:

$$2\ RSH + Hg \rightarrow Hg(SR)_2 + 2e + 2H^+.$$

The low residual current on a mercury electrode at 0.15 V enables a significant improvement in detection limit.

LCEC has been used extensively for the analysis of samples containing small amounts of analyte(s) that are electroactive. Representative substances determined by LCEC are listed in Table 29.2.

29.12 AMPEROMETRIC TITRATIONS

A titration in which the end point is determined by the current resulting from a potential applied across two electrodes is an amperometric titration. The two types of amperometric titrations differ in the types of electrodes that are used.

Table 29.2 Representative Substances Determined by LCEC

Aminophenols	Indoles	Phenols
Anilines	NADH, NADPH	Phenothiazines
Anisoles	Narcotic alkaloids	Quinones
Ascorbic acid	Nitro compounds	Sulfhydryls
Azo compounds	Nitrosamines	Tryptophan and metabolites
Benzidines	Organometallics	Tyrosine and metabolites
Benzodiazepines		Uric acid
Carbamates		Vanillin
Estrogens		
Hydroquinones		

Working Electrode–Reference Electrode. In the case in which the potential of a working electrode is controlled relative to a reference electrode, the potential is applied so that a limiting current that is proportional to the concentration of one or more of the reactants or products of the titration is measured. A titration curve is obtained by plotting the limiting current as a function of volume of titrant added. The shape of the titration curve can be predicted from hydrodynamic voltammograms of the solution obtained at various stages of the titration. Figure 29.32a shows voltammograms recorded during the titration of Cl^- with Ag^+. The working electrode is silver metal. At 0% titration, the anodic limiting current is controlled by the mass transport of Cl^- to the electrode for the oxidation of Ag to form AgCl. As the titration proceeds, the decrease in this limiting current reflects the decrease in Cl^- concentration due to addition of Ag^+ titrant until the current becomes zero at the equivalence point (100% titration). The appearance of a cathodic limiting current after the equivalence point reflects the reduction of excess Ag^+ titrant. Fig. 29.32b shows amperometric titration curves for two values of applied potential, E, and E_2. Their shapes are determined by the behavior of the limiting current of the voltammograms at the particular potential during the titration.

Two Working Electrodes. A useful variation of the amperometric titration involves measuring the current resulting from a small fixed potential applied across two working electrodes. One electrode functions as an anode and the other as a cathode. Once again, the expected current behavior during a titration can be explained by means of hydrodynamic voltammograms. Position A in Fig. 29.32a shows the small potential, ΔE, applied across two Ag electrodes for the example of Cl^- titrated with Ag^+. Before the equivalence point is reached, the measured current is only the small residual current for reduction at the more negative electrode and oxidation at the more positive electrode. (Note that for a significantly larger current, the value for ΔE would have to be increased to a value where H_2 would evolve at the cathode and AgCl would deposit on the anode.) However, after the equivalence point is reached, the potential of the two electrodes shifts positively to position B, where the value of ΔE is sufficiently great to cause reduction of Ag^+ at the cathode and oxidation of Ag at the anode. The current is

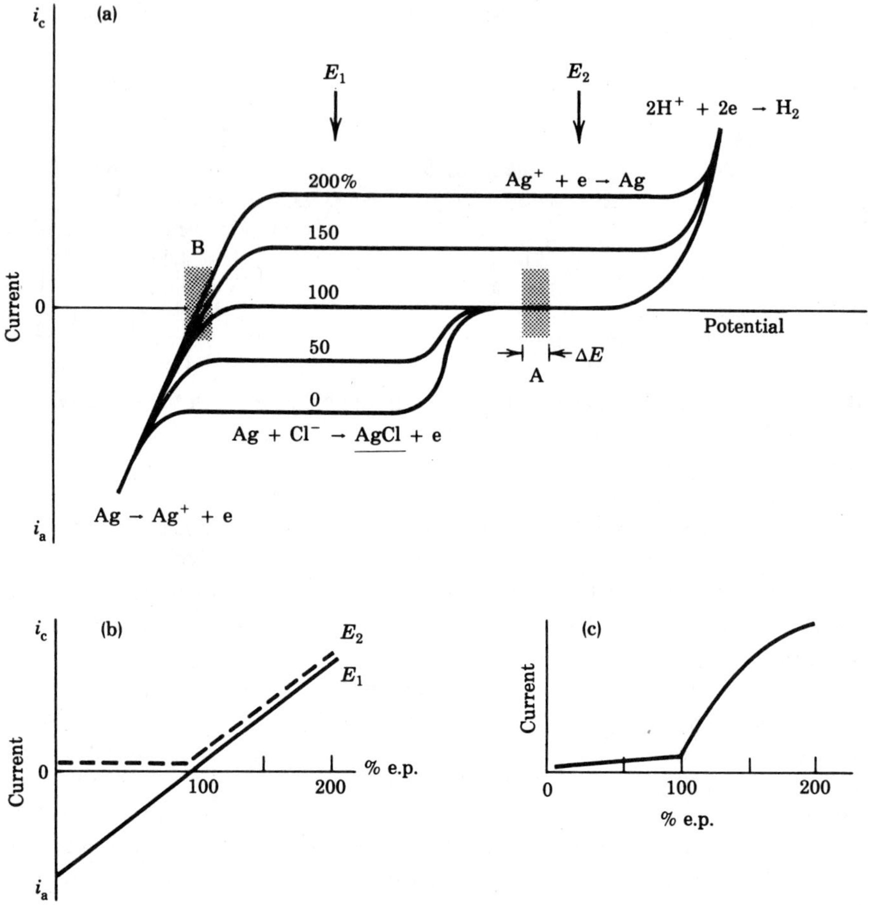

Fig. 29.32 Titration of Cl^- with Ag^+ based on the reaction $Cl^- + Ag^+ \rightleftharpoons AgCl$. (a) voltammograms at Ag electrode for 0, 50, 100, 150, and 200% of the equivalence point during the titration. (b) Amperometric titration curve for applied potential at E_1 and E_2. (c) Amperometric titration curve for two Ag electrodes. ΔE indicated by the shaded areas in (a). (From P. T. Kissinger and W. R. Heineman (Eds.), *Laboratory Techniques in Electroanalytical Chemistry*, Dekker, New York, 1985, p. 123. With permission.)

limited by the mass transport of Ag^+ to the cathode surface, which increases as additional excess Ag^+ is added. The resulting amperometric titration curve is shown in Fig. 29.32c.

It is important to realize that the potentials of the two electrodes are not fixed with respect to any reference electrode. Consequently, in Fig. 29.32a the ΔE increment is free to move on the potential axis. It locates itself at whatever potential is necessary for the current at the anode to balance the current at the cathode. This means that the ΔE increment will always straddle the potential at which the current voltage curve shows zero current (i.e., intercepts the potential axis).

One advantage of the amperometric titration is its ease of automation. A titrator can be signaled to shut off when a specified current level is reached. The advantage of the two working electrode variation is the elimination of a reference electrode, which can be troublesome in nonaqueous solvents.

29.13 POTENTIOMETRIC TITRATIONS

A titration in which the end point is determined from the potential of an electrochemical cell is termed a potentiometric titration. Although this type of titration was described in Sections 28.11–28.16, our knowledge of hydrodynamic voltammograms can now be used to explain the shape of the titration curves. As is the case with amperometric titrations, two types of potentiometric titrations exist.

Indicator Electrode–Reference Electrode. The classic potentiometric titration involves the measurement of the potential difference between an indicator and a reference electrode under the condition of zero current. Another set of voltammograms for the example used in Section 29.12 is shown in Fig. 29.33. The potential measured at a particular point during the titration is given by the potential at which current is zero on the appropriate voltammogram, that is, the potential at

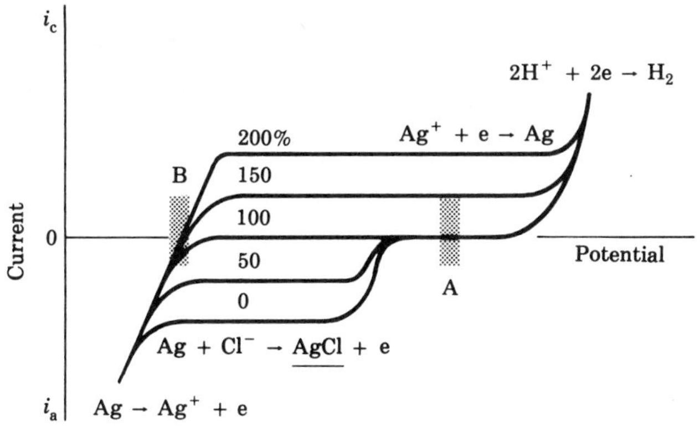

(a)

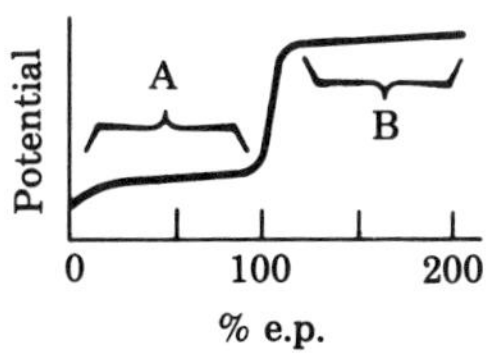

(b)

Fig. 29.33 Potentiometric titration for example in Fig. 29.32. (a) Voltammograms recorded during titration. (b) Potentiometric titration curve.

which the voltammogram intersects the potential axis. The voltammograms show that a shift in potential from region A to region B occurs at the equivalence point, as shown by the titration curve.

Two Working Electrodes. A variation on the potentiometric titration described above involves the measurement of potential across two working electrodes across which a small

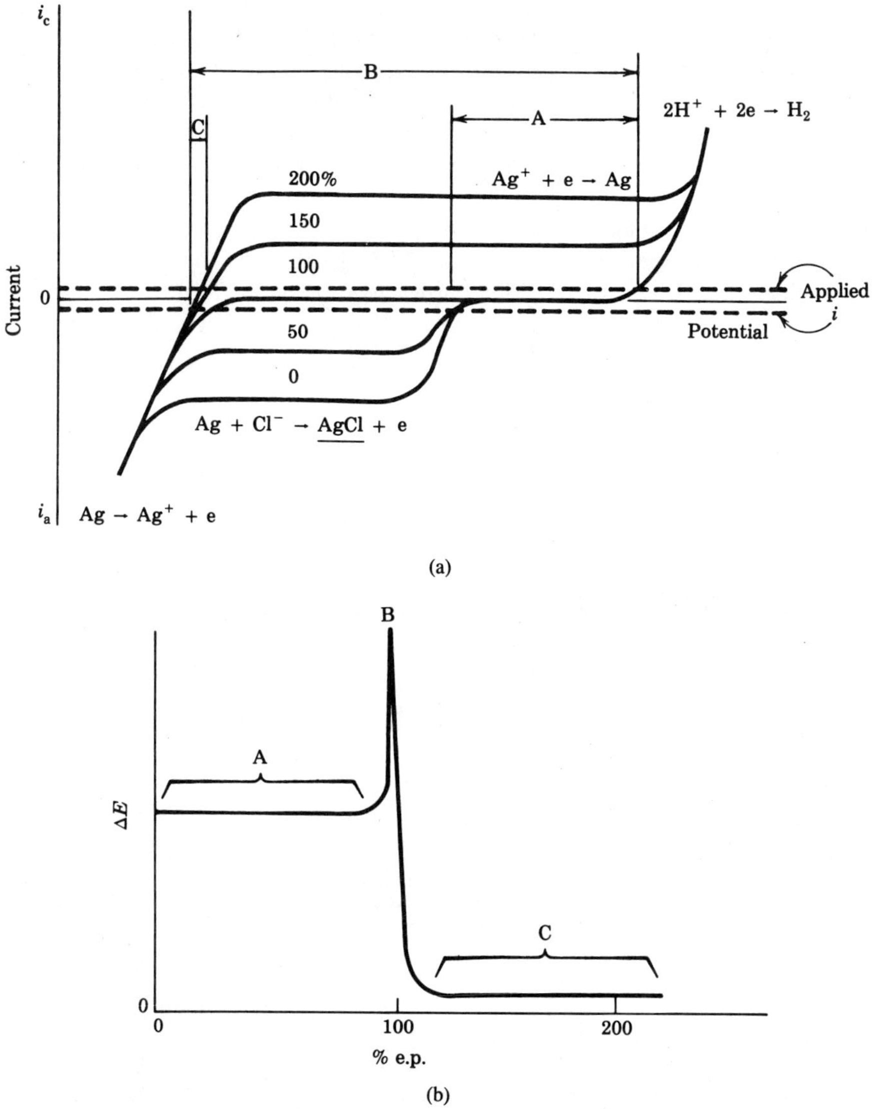

Fig. 29.34 Potentiometric titration with two working electrodes for example in Fig. 29.32. (a) Voltammograms recorded during titration. (b) Potentiometric titration curve.

current is impressed. The potential difference between the two electrodes is monitored as a function of volume titrant added to produce a titration curve.

As in the previous section, hydrodynamic voltammograms recorded during the titration can be used to predict the shape of the titration curve. Another set of voltammograms for the example used in Section 29.12 is shown in Fig. 29.34. The two dashed horizontal lines above and below the potential axis represent a small current impressed across two silver electrodes one of which is the cathode and the other the anode. The potential difference between the two electrodes is the cell potential required to provide a sufficient rate of electrolysis at both electrodes to support the applied constant current. This potential is simply obtained by measuring the difference between intersection points of the appropriate voltammogram and the cathodic and anodic impressed current lines. For example, at 0% titration, the cathode is at a potential to reduce H^+, whereas the anode is at a potential to oxidize Ag to AgCl, and the potential difference is defined as shown by A. This situation exists until 100% is reached at which point (B) the anode moves to a more positive potential to oxidize Ag to Ag^+, Cl^- having been removed from solution by the titration reaction. The addition of excess Ag^+ beyond the equivalence point enables the cathode's potential to shift dramatically in the positive direction due to the availability of easily reduced Ag^+. Since both halves of the Ag^+/Ag^0 couple are present at this point, only a small potential difference (C) is required to support the applied current. A plot of the potential difference as a function of added Ag^+ gives the titration curve shown in Fig. 29.34b. The endpoint is signaled by the abrupt rise and drop in potential.

As explained for the amperometric titration with two working electrodes, the potentials of the two electrodes are free to move along the potential axis as required to provide the necessary current. The two techniques share the same advantages.

REFERENCES

References providing a general coverage of voltammetry are:

1. R. N. Adams, *Electrochemistry at Solid Electrodes.* New York: Dekker, 1969.
2. A. J. Bard and L. R. Faulkner, *Electrochemical Methods.* New York: Wiley, 1980.
3. P. T. Kissinger and W. R. Heineman (Eds.), *Laboratory Techniques in Electroanalytical Chemistry.* New York: Dekker, 1984.
4. H. H. Bauer, *Electrodics.* New York: Wiley, 1972.
5. J. O.'M. Bockris and A. K. N. Reddy, *Modern Electrochemistry*, Vols. I and II, New York: Plenum, 1970.
6. E. A. M. F. Dahmen, *Electroanalysis.* New York: Elsevier, 1986.
7. P. Delahay, *New Instrumental Methods in Electrochemistry.* New York: Interscience, 1954.
8. Z. Galus, *Fundamentals of Electrochemical Analysis.* London: Horwood, Halsted Press, 1976.
9. I. M. Kolthoff and P. J. Elving (Eds.), *Treatise on Analytical Chemistry*, Part I, Vol. 4, Sec. D-2. New York: Interscience, 1963.
10. J. J. Lingane, *Electroanalytical Chemistry*, 2nd ed. New York: Interscience, 1978.
11. D. D. Macdonald, *Transient Techniques in Electrochemistry.* New York: Plenum, 1977.

12. J. A. Plambeck, *Electroanalytical Chemistry.* New York: Wiley, 1982.
13. D. T. Sawyer and J. L. Roberts, Jr., *Experimental Electrochemistry for Chemists.* New York: Wiley-Interscience, 1974.
14. Southhampton Electrochemistry Group, *Instrumental Methods in Electrochemistry.* New York: Halsted Press (Wiley), 1985.
15. H. R. Thirsk and J. A. Harrison, *A Guide to the Study of Electrode Kinetics.* New York: Academic Press, 1972.
16. B. H. Vassos and G. W. Ewing, *Electroanalytical Chemistry.* New York: Wiley-Interscience, 1983.
17. A. Weissberger and B. W. Rossiter (Eds.) *Techniques of Chemistry*, Vol. I: ''Physical Methods of Chemistry,'' Part II. New York: Wiley-Interscience, 1971.
18. State of the art symposium: Electrochemistry. *J. Chem. Ed.*, **60,** 258–340 (1983).
19. W. H. Smyrl and F. McLarnon (Eds.), *Proceedings of the Symposium on Electrochemistry and Solid State Science Education at the Graduate and Undergraduate Level.* Pennington: The Electrochemical Society, 1987.

Treatments of the practical aspects of performing voltammetric experiments are found in references 3 and 13.

Comprehensive discussions of advances in voltammetry appear regularly in the following continuing series:

20. *Advances in Analytical Chemistry and Instrumentation*, Vol. I, Interscience, 1960.
21. *Advances in Electrochemistry and Electrochemical Engineering*, Vol. 1, Interscience, 1961.
22. *Electroanalytical Chemistry. A Series of Advances.* Vol. 1, Dekker, 1966.

Two monographs on rotating disk and ring-disk electrodes are:

23. W. J. Albery and M. L. Hitchman, *Ring-Disc Electrodes.* London: Clarendon Press Oxford, 1971.
24. F. Opekar and P. Beran, Rotating disk electrodes, *J. Electroanal. Chem.*, **69,** 1-105 (1976).

EXERCISES

29.1 A hydrodynamic voltammogram is obtained at a Pt electrode versus SCE for 1.00 mM Fe^{2+} in 1 M $HClO_4$. ($E^{0'}_{Fe^{3+},Fe^{2+}} = 0.732$ V versus SHE, 1 M $HClO_4$). (a) Draw the voltammogram. (b) Calculate the values of $C^{s}_{Fe^{3+}}$ and $C^{s}_{Fe^{2+}}$ at $E = +0.549$ V, $+0.490$ V, and $+0.431$ V. (c) Draw the C–x profiles for Fe^{3+} and Fe^{2+} at these three potentials.

29.2 A potential of -0.400 V versus SCE is applied to a HMDE immersed in a stirred solution containing Tl^{+} ($E_{1/2} = -0.48$ V). (a) Calculate the ratio of surface concentrations $C^{s}_{Tl}/C^{s}_{Tl^{+}}$. (b) Draw the C–x profile for Tl^{+}. (c) Draw the C–x profiles for Tl as a function of time.

29.3 Calculate the effective thickness of the Nernst diffusion layer at a Pt electrode ($A = 0.150\ cm^2$), which in a 2.10-mM stirred solution of Fe^{3+} gave a 175 μA limiting current for reduction to Fe^{2+}. ($D_{Fe^{3+}} = 0.9 \times 10^{-5}\ cm^2\ s^{-1}$).

29.4 A hydrodynamic voltammogram recorded at a HMDE immersed in a stirred solution of 1 m*M* Pb^{2+} in 0.5*M* KCl gave a limiting cathodic current of 20 μA for the reduction of Pb^{2+}. Calculate the expected limiting current if the concentration of Pb^{2+} is decreased to 0.0065 m*M*.

29.5 Give an estimate of the percentage depletion of a 10-mL sample of a 1.00 μM Zn^{2+} solution during the 5-min application of a potential that gives a limiting current of 10 μA.

29.6 The following data were taken from a hydrodynamic voltammogram for a reversible reduction wave at 25°C:

E, *versus SCE (V)*	*i* (μA)	E, *V versus SCE*	i (μ*A*)
−0.560	0.60	−0.601	3.84
−0.575	1.53	−0.615	4.52
−0.583	2.32	−0.629	4.80
−0.591	3.02		

The limiting current was 4.95 μA. (a) What number of electrons was involved in the electrode reaction? (b) What is $E_{1/2}$? (c) What is the approximate $E^{0'}$ vs. SHE for the system? State the assumptions made in estimating the last quantity.

29.7 Explain why the limiting current in voltammetry is temperature-dependent.

29.8 Explain why the limiting current in Fig. 29.7 is unaffected by the rate of electron transfer.

29.9 (a) Contrast the role of the cell in an electroanalytical technique with that in a spectrophotometric method. (b) To what extent is there a useful analogy between varying the potential in an electroanalytical technique and the wavelength in a spectrophotometric method?

29.10 One cannot control both current and potential simultaneously. Is this statement correct or incorrect? Explain.

29.11 (a) Sketch the hydrodynamic voltammogram that would be obtained at a Pt electrode (versus SCE) for a solution of Tl^{3+}. ($E^0_{Tl^{3+},Tl^+} = 1.25$ V, $E^0_{Tl^+,Tl} = -0.336$ V versus SHE). (b) Sketch the concentration–distance profiles for the following applied potentials (versus SCE): 1.01 V, 0.50 V, −0.56 V.

29.12 Draw hydrodynamic voltammograms obtained at a HMDE for each of the following deoxygenated solutions. Where multiple waves are obtained, show the relative heights of the limiting currents. Write the electrode mechanism for each wave. (a) A mixture of equal concentrations of Tl(I), $E_{1/2} = -0.48$ V; Zn(II), $E_{1/2} = -0.81$ V; and Al(III), $E_{1/2} = -1.74$ V versus SCE in 1.0 *F* KCl. The metal ions are reduced to the metallic state at the HMDE. (b) A solution of Cr(III) in a buffer of 1 *F* NH_3, 1 *F* NH_4Cl. $E_{1/2} = -1.43$ V for reduction to Cr(II) and −1.71 V for reduction to Cr(0).

29.13 (a) Draw the voltammogram that is obtained for a stirred solution of Cl^- at a Ag electrode versus SCE. (b) Draw *C–x* profiles for Cl^- at each of the following potentials: (1) one that gives no current; (2) $E_{1/2}$, and (3) one that gives limiting current for Cl^-.

29.14 Derive the equation for the voltammogram obtained for a stirred solution of Cl^- at a Ag working electrode (see Exercise 29.13.)

29.15 Derive the equations for standard addition in voltammetry for the following two cases. (a) The standard aliquot is sufficiently large that dilution of the sample solution must be considered (Eq. 29.28). (b) The standard aliquot is sufficiently small that dilution of the sample solution is negligible.

29.16 The limiting current for reduction of Zn^{2+} in an unknown solution is 4.15 μA. On addition of 1.00 mL of 5.05×10^{-3} *M* Zn^{2+} solution to 25.0 mL of the unknown solution, the Zn^{2+} wave height increased to 5.32 μA. Calculate the concentration of Zn^{2+} in the unknown.

29.17 (a) Explain how an amino acid ($RCHNH_3{}^+COO^-$) could be determined *amperometrically* by an enzyme electrode based on the following enzyme-catalyzed reaction:

$$2RCHNH_3{}^+COO^- + O_2 \xrightarrow{\text{amino acid oxidase}} 2RCOCOO^- + 2NH_4{}^+.$$

(b) How could the measurement be made *potentiometrically*?

29.18 A sample solution containing an organic thiol (RSH) is titrated with iodine titrant by the following reaction:

$$2\ RSH + I_2 \rightarrow RSSR + 2\ H^+ + 2\ I^-.$$

A hydrodynamic voltammogram at a Pt electrode for a solution containing both I_2 and I^- is shown below. The sulfur-containing species are electroinactive.

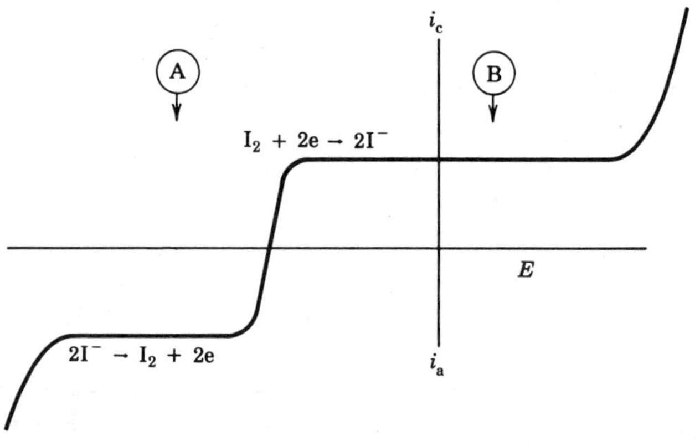

(a) Draw the voltammograms that would be obtained at 0, 50, 100, and 150% of the titration. (b) Draw the amperometric titration curves obtained at the potentials labeled A and B on the voltammogram shown above. Label the axes. (c) Draw the titration curve for an amperometric titration with two polarized Pt electrodes. Label the axes. (d) Draw the titration curve for a potentiometric titration with two polarized Pt electrodes. Label the axes.

29.19 A sample solution containing Pb^{2+} is titrated with a standard solution of X^{2-} by a precipitation reaction

$$Pb^{2+}(\text{sample}) + X^{2-}(\text{titrant}) \rightarrow PbX(s).$$

Stirred-solution voltammograms obtained on pure solutions of Pb^{2+} and X^{2-} at a HMDE are shown below. The precipitate PbX(s) is not electroactive.

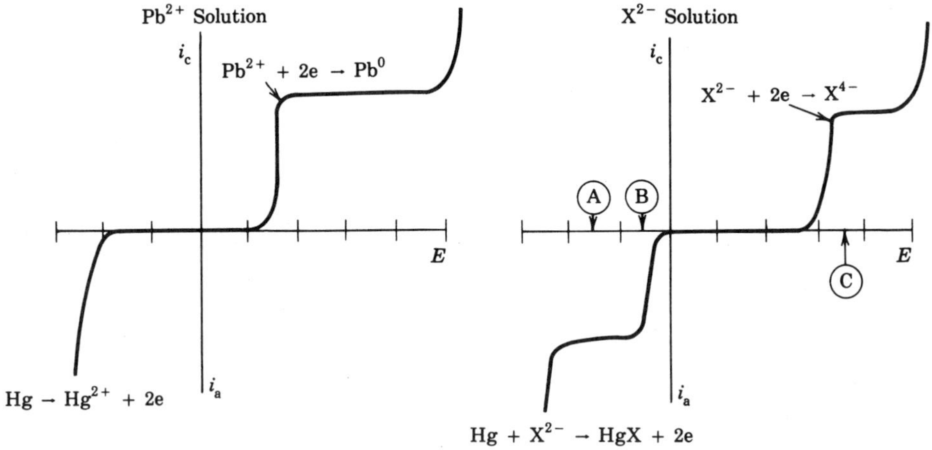

(a) Draw the voltammograms that would be obtained at 0, 50, 100, and 150% of the titration. (b) Draw the amperometric titration curves obtained at the potentials labeled A, B, and C on the voltammogram shown above. Label the axes. (c) Draw the titration curve for an amperometric titration with two polarized Hg electrodes. Label the axes. (d) Draw the titration curve for a potentiometric titration with two polarized Hg electrodes. Label the axes.

ANSWERS

29.3 0.0016 cm *29.4* 0.13 μA *29.6* (a) n = 2.02; (b) $E_{1/2}$ = −0.585 V vs. SCE; (c) $E^{0'}$ = −0.341 V vs. SHE for equal diffusion coefficients *29.16* 6.06 × 10^{-4} M

Chapter 30

VOLTAMMETRY: STATIONARY SOLUTION TECHNIQUES

William R. Heineman

30.1 INTRODUCTION

Stationary solution voltammetric techniques are implemented in quiescent solution so that diffusion is the only means of mass transport of electroactive species to the electrode surface. In comparison with hydrodynamic techniques, the electrogenerated product(s) is not swept away by forced convection. This aspect of stationary solution techniques enables electrogenerated species to be subsequently probed electrochemically at the same electrode at which they were formed. Thus, these techniques are especially useful for the investigation of mechanisms of electrode reactions.

Many of the basic principles described in Chapter 29 for hydrodynamic techniques apply to stationary solution techniques. The concepts of a potential excitation signal (Fig. 29.1) that controls the surface concentrations of a redox couple and a current response signal the magnitude of which is determined by the slope of a concentration–distance profile are applicable to stationary solution techniques. The cell and instrumentation are also essentially the same for both types of voltammetry.

30.2 CYCLIC VOLTAMMETRY*

Cyclic voltammetry (CV) is perhaps the most versatile electroanalytical technique for the study of electroactive species. Its versatility combined with ease of measurement has resulted in extensive use of CV in the fields of electrochemistry, inorganic chemistry, organic chemistry, and biochemistry. Cyclic voltammetry is often the first experiment performed in an electrochemical study of a compound, a biological material, or an electrode surface. The effectiveness of CV results from its capability for rapidly observing redox behavior over a wide potential range.

Fundamentals of Cyclic Voltammetry. Cyclic voltammetry consists of cycling the potential of an electrode, which is immersed in an unstirred solution, and mea-

*From P. T. Kissinger and W. R. Heineman, *J. Chem. Ed.*, **60**, 702 (1983).

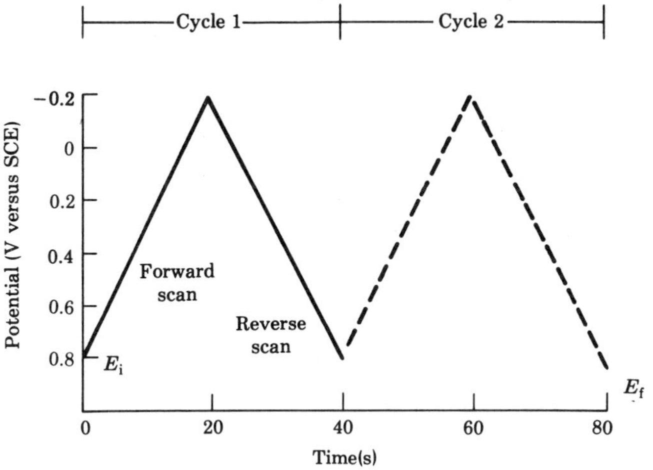

Fig. 30.1 Representative excitation signal for cyclic voltammetry—a triangular potential waveform with switching potentials at 0.8 and −0.2 V versus SCE.

suring the resulting current. The excitation signal for CV is a linear potential scan with a triangular waveform, as shown in Fig. 30.1. This triangular potential excitation signal sweeps the potential of the electrode between two values, sometimes called the *switching potentials*. The excitation signal in Fig. 30.1 causes the potential first to scan negatively from an *initial potential* (E_i) of +0.80 to −0.20 V versus SCE at which point the scan direction is reversed, causing a positive scan back to the original potential of +0.80 V. The scan rate, as reflected by the slope, is 50 mV s^{-1}. A second cycle is indicated by the dashed line. Single or multiple cycles can be used. Modern instrumentation (Figs. 29.15 and 29.16) enables switching potentials and scan rates to be easily varied.

A typical cyclic voltammogram is shown in Fig. 30.2 for a platinum working electrode in a solution containing 6.0 m*M* $K_3Fe(CN)_6$ as the electroactive species in aqueous 1.0 *M* KNO_3. The potential excitation signal used to obtain this voltammogram is that shown in Fig. 30.1. Thus, the vertical axis in Fig. 30.1 is now the horizontal axis for Fig. 30.2. The initial potential of 0.80 V applied at *a* is chosen to avoid any electrolysis of $Fe^{III}(CN)_6^{3-}$ when the electrode is switched on. The potential is then scanned negatively, *forward scan*, as indicated by the arrow. When the potential is sufficiently negative to reduce $Fe^{III}(CN)_6^{3-}$ to $Fe^{II}(CN)_6^{4-}$, cathodic current is observed as indicated at b. The cathodic current increases rapidly (*b–d*) until the concentration of $Fe^{III}(CN)_6^{3-}$ at the electrode surface is substantially diminished, causing the current to peak (*d*). The current then decays (*d–g*) as the solution surrounding the electrode is depleted of $Fe^{III}(CN)_6^{3-}$ owing to its electrolytic conversion to $Fe^{II}(CN)_6^{4-}$. The scan direction is switched to positive at −0.20 V (*f*) for the *reverse scan*. The potential is still sufficiently negative to reduce $Fe^{III}(CN)_6^{3-}$, so cathodic current continues even though the potential is now scanning in the positive direction. When the electrode becomes a sufficiently strong oxidant, $Fe^{II}(CN)_6^{4-}$, which has been accumulating

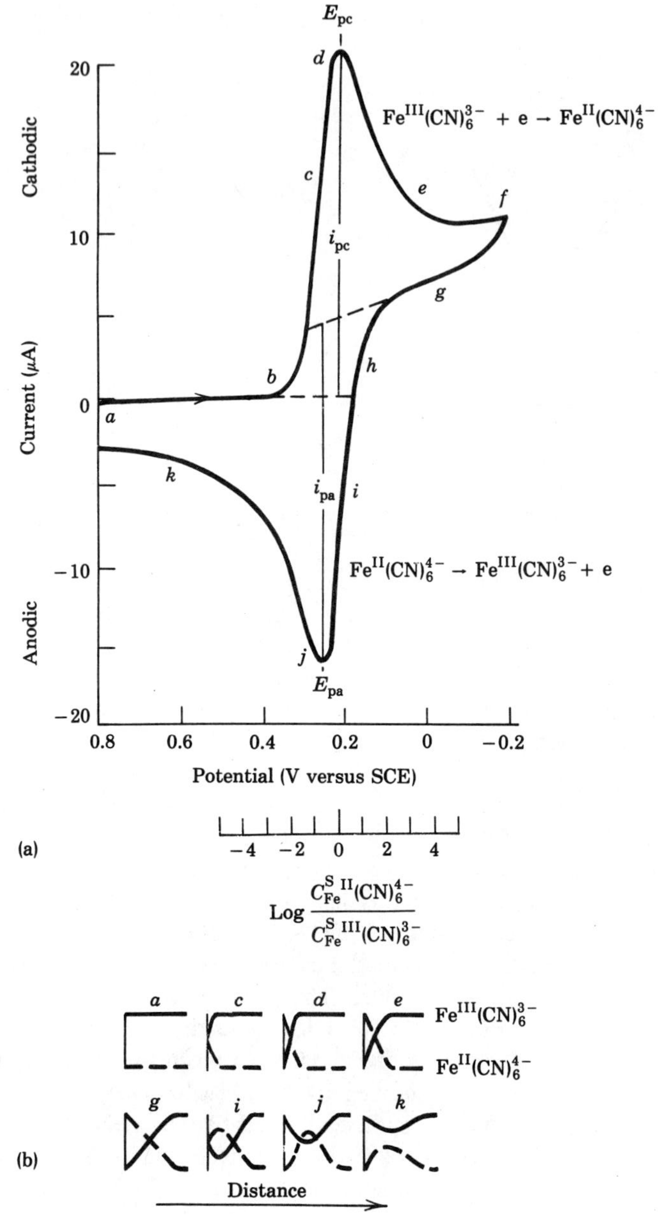

Fig. 30.2 (a) Cyclic voltammogram of 6 m*M* $K_3Fe(CN)_6$ in 1 *M* KNO_3. Scan initiated at 0.8 V versus SCE in negative direction at 50 mV s^{-1}. Platinum electrode, $A = 2.54\ mm^2$. (b) Concentration–distance (C-x) profiles a–k keyed to voltammogram.

adjacent to the electrode, can now be oxidized to $Fe^{III}(CN)_6^{3-}$. This causes anodic current (i–k). The anodic current rapidly increases until the surface concentration of $Fe^{II}(CN)_6^{4-}$ is diminished, causing the current to peak (j). The current then decays (j–k) as the solution surrounding the electrode is depleted of $Fe^{II}(CN)_6^{4-}$. The first cycle is completed when the potential reaches +0.80 V. (Now that the cyclic voltammogram is obtained, it is apparent that any potential positive of approximately +0.4 V would be suitable as an initial potential in that reduction of $Fe^{III}(CN)_6^{3-}$ would not occur when the potential is applied. This procedure avoids inadvertent electrolysis as a result of applying the initial potential.)

Simply stated, in the forward scan $Fe^{II}(CN)_6^{4-}$ is electrochemically generated from $Fe^{III}(CN)_6^{3-}$ as indicated by the cathodic current. In the reverse scan this $Fe^{II}(CN)_6^{4-}$ is oxidized to $Fe^{III}(CN)_6^{3-}$ as indicated by the anodic current. Thus, CV is capable of rapidly generating a new oxidation state during the forward scan and then probing its fate on the reverse scan. This important feature is characteristic of stationary solution techniques in which the electrogenerated species is not swept away as in hydrodynamic methods.

A more detailed understanding of the shape of the voltammogram can be gained by considering the Nernst equation and the changes in concentration that occur in solution adjacent to the electrode during electrolysis. The potential excitation signal exerts control of the surface concentrations of $Fe^{II}(CN)_6^{4-}$ and $Fe^{III}(CN)_6^{3-}$, as described by the Nernst equation for a reversible system at 25°C:

$$E = E^{0'}_{Fe^{III}(CN)_6^{3-},\,Fe^{II}(CN)_6^{4-}} - \frac{0.0591}{n} \log \frac{C^{s}_{Fe^{II}(CN)_6^{4-}}}{C^{s}_{Fe^{III}(CN)_6^{3-}}}. \tag{30.1}$$

An initial value of E that is sufficiently positive of $E^{0'}$ maintains a ratio in which $Fe^{III}(CN)_6^{3-}$ greatly predominates. Thus, application of +0.80 V as the initial potential causes negligible current. However, as E is scanned negatively, conversion of $Fe^{III}(CN)_6^{3-}$ to $Fe^{II}(CN)_6^{4-}$ by reduction is mandatory for satisfaction of the Nernst equation. The ratios of iron redox states that must exist at the electrode surface at several potentials during the scan are shown on the lower horizontal axis in Fig. 30.2. The logarithmic (log base 10, $T = 25°C$, $n = 1$) relationship between E and $C^{s}_{Fe^{II}(CN)_6^{4-}}/C^{s}_{Fe^{III}(CN)_6^{3-}}$ is reflected by a rapid rate of change in the region where $E = E^{0'}$, that is, $C^{s}_{Fe^{II}(CN)_6^{4-}}/C^{s}_{Fe^{III}(CN)_6^{3-}} = 1$. This causes the dramatic rise in cathodic current (b–d) during the forward scan.

The physical situation in the solution adjacent to the electrode during the potential scan is illustrated by the C–x profiles at the bottom of Fig. 30.2. Note that the application of E_i does not measurably alter the concentration of $Fe^{III}(CN)_6^{3-}$ at the electrode surface as compared to the solution bulk. As the potential is scanned negatively, $C^{s}_{Fe^{III}(CN)_6^{3-}}$ decreases. The profile for which the concentration of $Fe^{III}(CN)_6^{3-}$ at the electrode surface exactly equals the concentration of $Fe^{II}(CN)_6^{4-}$ corresponds to an E that equals $E^{0'}$ (versus SCE) of the couple. Once the potential has reached a value sufficiently negative that $Fe^{III}(CN)_6^{3-}$ is effectively zero at the electrode surface, the C–x profiles reflect the increasing depletion of the solution adjacent to the electrode surface. In other words, the diffusion layer extends further into solution (e, g).

The behavior of the current during the potential scan can be understood by carefully examining the *C–x* profiles in Fig. 30.2. The current is proportional to the slope of the *C–x* profile at the electrode surface as described by Eq. (29.3), just as was the case in hydrodynamic voltammetry (Section 29.3). The slope of profile *a* is zero and current is negligible at that potential. As the potential is then scanned negatively, $(\partial C/\partial x)_{x=0}$ increases for profiles *c–d*, and the cathodic current increases correspondingly. However, when profile d is reached, $(\partial C/\partial x)_{x=0}$ decreases, as shown by profiles *e* and *g*, because of depletion of $Fe^{III}(CN)_6^{3-}$ near the electrode. Correspondingly, the current now drops.

During the negative scan $Fe^{II}(CN)_6^{4-}$ accumulates in the vicinity of the electrode as can be seen by the *C–x* profiles for $Fe^{II}(CN)_6^{4-}$. After the direction of potential scan is switched at −0.20 V to a positive scan, reduction continues (as is evident by the cathodic current and the *C–x* profile for *g* until the applied potential becomes sufficiently positive to cause oxidation of the accumulated $Fe^{II}(CN)_6^{4-}$, as signaled by the appearance of anodic current. Once again, the current increases as the potential moves increasingly positive until the concentration of $Fe^{II}(CN)_6^{4-}$ becomes depleted at the electrode. At this point the current peaks and then begins to decrease. Thus, the physical phenomena that caused a current peak during the reduction cycle also cause a current peak during the oxidation cycle. This can be seen by comparing the *C–x* profiles for the two scans.

The important parameters of a cyclic voltammogram are the magnitudes of the *anodic peak current* (i_{pa}) and *cathodic peak current* (i_{pc}), and the *anodic peak potential* (E_{pa}) and *cathodic peak potential* (E_{pc}). These parameters are labeled in Fig. 30.2. One method for measuring i_p involves extrapolation of a baseline current as shown by the dashed lines in the figure. The establishment of a correct baseline is essential for the accurate measurement of peak currents. This is not always easy, particularly for more complicated systems.

The formal reduction potential ($E^{0'}$) for a reversible couple is centered between E_{pa} and E_{pc}

$$E^{0'} = \frac{E_{pa} + E_{pc}}{2}. \tag{30.2}$$

The number of electrons transferred in the electrode reaction (n) for a reversible couple can be determined from the separation between the peak potentials, which at 25°C is

$$\Delta E_p = E_{pa} - E_{pc} = \frac{0.059}{n}. \tag{30.3}$$

Thus, a one-electron process such as the reduction of $Fe^{III}(CN)_6^{3-}$ to $Fe^{II}(CN)_6^{4-}$ exhibits a ΔE_p of 0.059 V.

The peak current for a reversible system is described by the Randles–Sevcik equation for the forward sweep of the first cycle

$$i_p = (2.69 \times 10^5) n^{3/2} A D^{1/2} C v^{1/2} \tag{30.4}$$

where C is concentration of solution species (mol cm^{-3}), v is scan rate (V s^{-1}), and other symbols are identified in Table 27.1. Accordingly, i_p increases with $v^{1/2}$ and is directly proportional to concentration. The effect of the scan rate on the appearance of the CV can be seen in Fig. 30.3a. A plot of Eq. (30.4), i_p versus $v^{1/2}$, yields straight lines for both anodic and cathodic peak currents (Fig. 30.3b), the slopes of which can be used to determine diffusion coefficients of the electroactive species at an electrode of known area. The values of i_{pa} and i_{pc} should be of comparable magnitude for a reversible redox couple. That is

$$\frac{i_{pa}}{i_{pc}} = 1. \tag{30.5}$$

However, the ratio of peak currents can be significantly influenced by chemical reactions coupled to the electrode process, as discussed below.

Electrochemical irreversibility is caused by slow electron exchange of the redox species with the working electrode (Section 27.7). In this case Eqs. (30.2–30.4) are not applicable. Electrochemical irreversibility is characterized by a separation of peak potentials greater than indicated by Eq. (30.3), just as it causes skewing of hydrodynamic voltammograms.

Effect of Coupled Chemical Reactions. Many electrochemical reactions involve an electron transfer step which leads to a species that rapidly reacts with components of the medium via so-called coupled chemical reactions (Section 27.8). One of the most useful aspects of CV is its application to the qualitative diagnosis of these homogeneous chemical reactions that are coupled to the electrode reaction. Cyclic voltammetry provides the capability for generating a species during the forward scan and then probing its fate with the reverse scan and subsequent cycles,

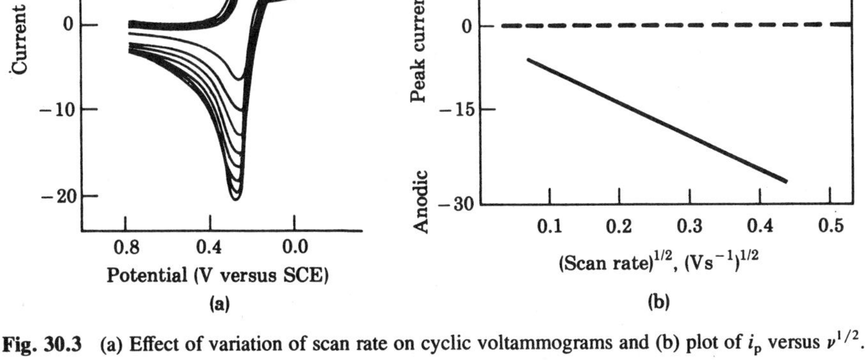

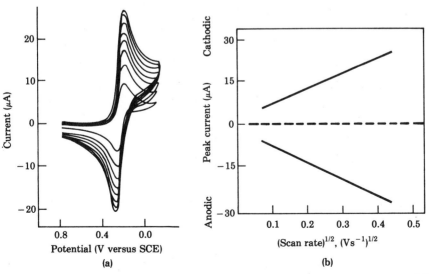

Fig. 30.3 (a) Effect of variation of scan rate on cyclic voltammograms and (b) plot of i_p versus $v^{1/2}$.

all in a matter of seconds or less. In addition, the time scale of the experiment is adjustable over several orders of magnitude by changing the potential scan rate, enabling some assessment of the rates of various reactions.

A representative example of a mechanism for an organic compound is the oxidation of *p*-aminophenol (PAP), which undergoes an EC mechanism (Section 27.8). A cyclic voltammogram of PAP is shown in Fig. 30.4. The positive scan (initiated at $E_i = 0.00$ V) causes an anodic peak as PAP is oxidized to benzoquinoneimine (BQI), which reacts chemically with water at a rate characterized by the homogeneous rate constant k to form quinone (Q). This homogeneous chemical reaction reduces the concentration of BQI in the reaction layer and thereby diminishes the amount of BQI available for reduction back to PAP during the subsequent negative scan. Thus, the current that corresponds to reduction of BQI during the subsequent negative scan is less than its value would be if some BQI had not been lost via reaction. Consequently, $i_{pc} < i_{pa}$ and $i_{pc}/i_{pa} < 1$.

The extent to which i_{pc} is less than i_{pa} depends upon two factors: the scan rate v and the magnitude of the rate constant k. The behavior of the ratio i_{pc}/i_{pa} as a function of k and τ, where $\tau \propto 1/v$, is shown in Fig. 30.5 for the EC mechanism. Consideration of this curve reveals two important correlations. (a) For a given scan rate, i_{pc}/i_{pa} decreases as the rate constant k of the chemical reactions increases. This behavior is reasonable since, the greater k is, the less BQI is left in the reaction layer for oxidation back to PAP. (b) For a given rate constant, i_{pc}/i_{pa} increases as the scan rate increases (τ decreases). Increasingly rapid scan rates allow

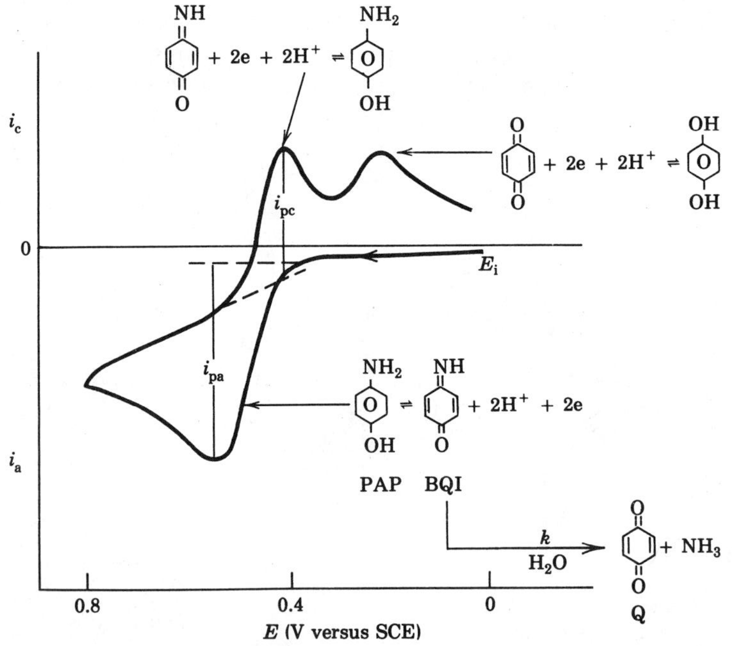

Fig. 30.4 Cyclic voltammogram of *p*-aminophenol in 0.01 *M* H_2SO_4 at a carbon paste electrode.

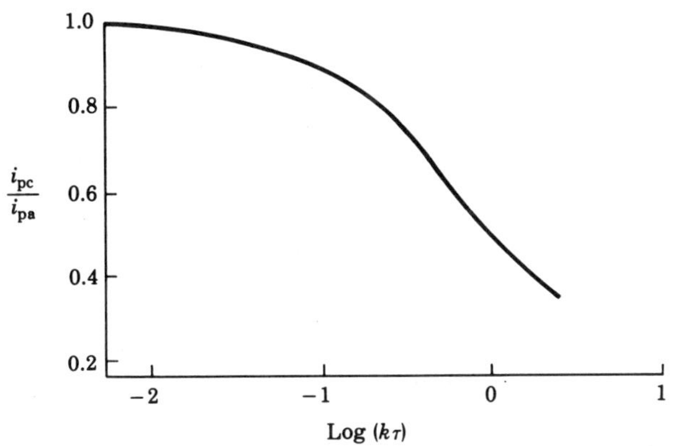

Fig. 30.5 i_{pc}/i_{pa} as a function of log ($k\tau$) for the EC mechanism. $\tau \propto 1/v$.

less time for the chemical reaction to occur between the positive and negative scans. In fact, for sufficiently fast scan rates the effect of the chemical reaction becomes negligible and i_{pc}/i_{pa} approaches unity. The value of the rate constant for a homogeneous chemical reaction can be determined from i_{pc}/i_{pa} for an experimental scan rate from a curve such as that in Fig. 30.5.

30.3 POTENTIAL STEP TECHNIQUES

Several electroanalytical techniques are based on stepping the potential from one value to another and measuring the associated current or charge. These experiments are carried out with an electrode immersed in an unstirred solution. A potential step excitation signal is the basis for two electrochemical techniques, chronoamperometry and chronocoulometry, in which the associated current or charge is measured, respectively. These techniques are commonly used for the measurement of n-values, electrode surface areas, diffusion coefficients, rate constants of coupled chemical reactions and the concentration of material adsorbed on an electrode surface. The concepts involved in these two techniques are useful for understanding the pulse voltammetric techniques in the next section.

Potential-Step Excitation. The excitation signal for potential step techniques is shown in Fig. 30.6a. The potential is stepped from an *initial potential*, E_i, to a *second* (or *step*) *potential*, E_s. This step is termed the *forward potential step*. If the experiment is terminated at this potential, it is called a *single potential step* experiment. If, on the other hand, the potential is then returned to a *final potential step*, E_f, after time τ, this step is called the *reverse potential step* of a *double potential step* experiment.

As in the case of the previously discussed voltammetric techniques, the potential applied to the electrochemical cell controls the surface concentrations of the redox species in solution. In many potential step techniques, the values of E_i and E_s are chosen so that essentially 100% conversion occurs at the electrode surface. That is to say, E_i and E_s are such

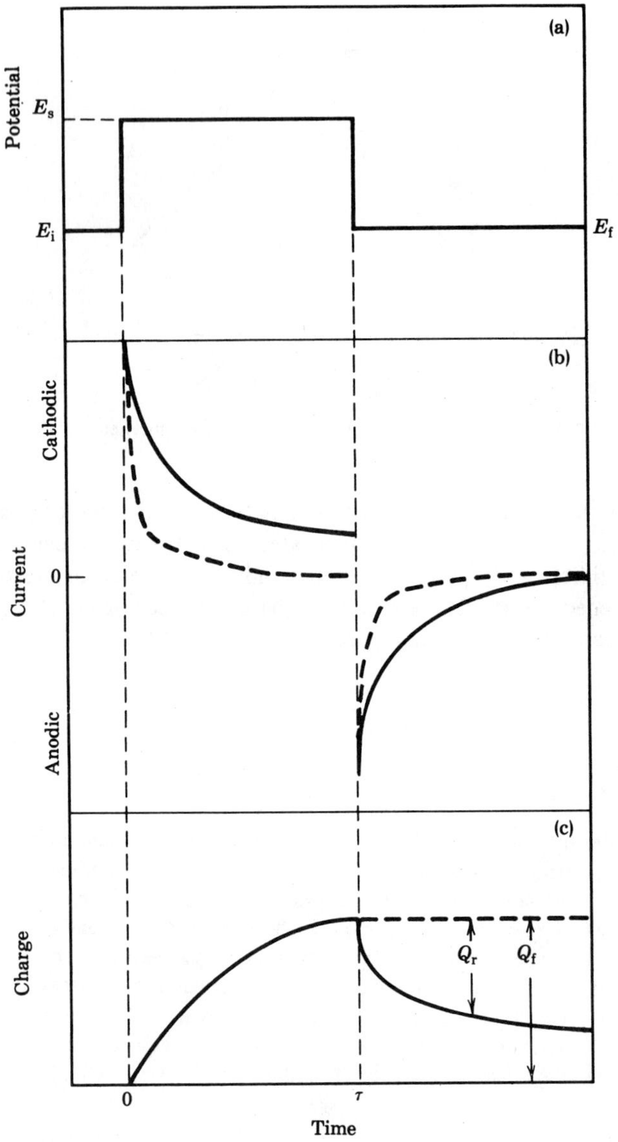

Fig. 30.6 Potential step techniques. (a) Excitation signal. (b) Chronoamperometry—current response signal. Dashed line shows charging current. (c) Chronocoulometry—charge response signal.

that C_R^s/C_O^s changes from $<1/100$ to $>100/1$ for a reduction and $>100/1$ to $<1/100$ for an oxidation.

The *C–x* profiles for a typical double-potential step experiment are shown in Fig. 30.7. Fig. 30.7a represents a solution containing O at the concentration of 1 m*M* and none of the other chemical form of the couple, R. The initial potential should be sufficiently positive of $E^{0'}_{O,R}$ that this profile is not distributed, that is, no measurable amount of O is reduced to

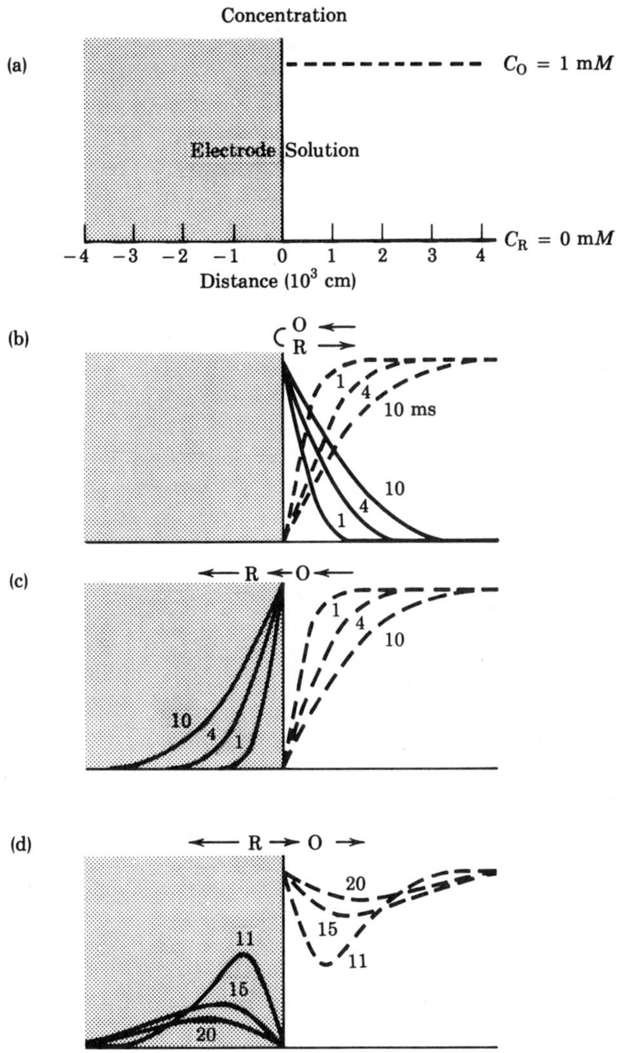

Fig. 30.7 Concentration–distance profiles during diffusion-controlled reduction of O to R at a planar electrode. $D_O = D_R$. (a) Initial conditions prior to potential step: C_O = 1 mm, C_R = 0. (b) Profiles for O (dashed line) and R (solid line) at 1, 4, and 10 ms after potential step to E_s. R is soluble in solution. (c) Profiles for O and R at 1, 4, and 10 ms after potential step to E_s. R is soluble in the electrode. (d) Potential stepped from E_s to E_f at τ = 10 ms so that oxidation of R to O is now diffusion controlled. Profiles are for 11, 15, and 20 ms after step to E_s. (From *Laboratory Techniques in Electroanalytical Chemistry*, P. T. Kissinger and W. R. Heineman (Eds.), Dekker, New York 1985, p. 53).

R, when E_i is applied to the cell. This condition is met for a reversible system with $n = 1$ if $E_i > 118$ mV positive of $E^{0'}_{O,R}$ (See Fig. 29.2). When the potential is stepped from E_i to a value for E_s, such that C^s_O is essentially zero ($E_s < -118$ mV negative of $E^{0'}_{O,R}$), the C–x profiles respond as shown in Fig. 30.7b. A concentration gradient is established near the electrode such that O diffuses to the electrode where it is reduced to R, which then diffuses into the bulk solution. As the electrolysis continues, the profile for O extends further into solution, since the vicinity of the electrode becomes further depleted of O; the profiles for R also extend further into solution as this electrogenerated material diffuses away from the electrode. Figure 30.7c represents the situation in which R diffuses *into* the electrode. This would be the case for reduction of metal ions such as Pb^{2+} and Tl^+ to their metallic state at a mercury electrode in which these metals are soluble.

The C–x profile response to the reverse potential step to E_f in a double potential step experiment is shown in Fig. 30.7d. The R in the electrode is now oxidized to O as rapidly as it diffuses to the interface. The C–x profiles relax to the original condition in Fig. 30.7a as R is converted back to O.

It is useful at this point to compare the sequence of C–x profiles shown in Fig. 30.7b for a potential step experiment in unstirred solution with one of the profiles in Fig. 29.3 for hydrodynamic voltammetry.

Chronoamperometry. Chronoamperometry is the technique in which the current response signal is recorded after a potential step excitation signal of the type shown in Fig. 30.6a. The term *chronoamperometry* is descriptive of the technique: *chrono* means time and *amperometry* refers to the measurement of current.

Since current is related to the slope of the C–x profile at the electrode surface, Eq. (29.3) can be used to predict from the C–x profiles the current–time response, the *chronoamperogram*, that accompanies the potential step. As shown in Fig. 30.7b, the slope of the profile for O is immediately very steep when the potential is stepped and then decreases as electrolysis continues. The predicted curent–time response is, thus, as shown in Fig. 30.6b (solid line): a large initial cathodic current with a subsequent decay as O is depleted in the region near the electrode by conversion to R. The current is determined by the flux of O at the electrode surface: $D_O(\partial C_O/\partial x)_{x=0,t}$

The reverse potential step to E_f causes a large anodic current as shown in Fig. 30.6b. This anodic current–time response can be understood by considering the slope of the C–x profile for R at the electrode surface since the current is determined by the flux of R at the electrode surface: $D_R(\partial C_R/\partial x)_{x=0,t}$.

The current–time response signal for the forward potential step is described for a planar electrode and linear diffusion by the Cottrell equation

$$i = \frac{nFAC_O D_O^{1/2}}{\pi^{1/2} t^{1/2}} = Kt^{-1/2}, \tag{30.6}$$

which shows the current to be inversely proportional to the square root of time. Thus, the product $it^{1/2}$ is equal to the constant K.

As shown by the dashed line in Fig. 30.6b, the current–time response also contains current due to charging the electrode to the new potentials, that is, *charging current*.* This

*Charging current is often referred to as *non-faradaic current*, since it does not involve the electrolysis of an electroactive species. Current attributed to an electrolysis process is termed *faradaic current*.

currently usually decays very quickly (milliseconds or less). It can be corrected for by performing a "blank" experiment in supporting electrolyte/solvent alone, and subtracting the blank current contribution from the chronoamperogram of the sample.

Although the current observed at any time is proportional to concentration of electroactive species, as shown by Eq. (30.6), chronoamperometry is rarely used as a technique for analysis. One example, however, is the determination of neurotransmitter concentration in a rat's brain into which a small electrode has been placed.* Chronoamperometry has been used mainly for the measurement of electrode surface area, diffusion coefficients, n-values, or the study of coupled chemical reactions. For example, the surface area A of an electrode can be calculated from $it^{1/2}$ if chronoamperometry is performed on a species for which D, n, and C are known. A well-characterized system is $K_3Fe(CN)_6$ for which $Fe(CN)_6^{3-}$ undergoes a 1-e reduction to $Fe(CN)_6^{4-}$, $D_{Fe(CN)_6^{3-}}$ is known in a number of supporting electrolytes, and solutions of known concentrations are easily prepared. Measurement of electrode area in this way is especially useful for electrode geometries that are difficult to measure by ruler or caliper, such as a minigrid electrode. Once the surface area of an electrode has been calibrated, chronoamperometry with this electrode can be used to measure the diffusion coefficient of a species for which n and C are known.

Chronocoulometry. *Chronocoulometry* is essentially chronamperometry that has been taken one step further. The i–t response is integrated to give a monitored response of *charge* (Q) versus time. Thus, chronocoulometry has the identical excitation signal of one or more potential steps and response signal of current as described above. The equation that describes the resulting Q–t curve for the forward step is the integral of Eq. (30.6):

$$Q = \frac{2nFAC_O D_O^{1/2} t^{1/2}}{\pi^{1/2}} = 2Kt^{1/2}. \qquad (30.7)$$

A *chronocoulogram* for double potential step chronocoulometry is shown in Fig. 30.6c. This monitored response is simply the mathematical integral of Fig. 30.6b. As predicted by Eq. (30.7), Q increases as a function of $t^{1/2}$ during the forward potential step. The value of Q at any time reflects the total amount of O that has been reduced up to that point (which is equal to the amount of R produced). Thus, whereas i is a measure of the rate of electrolysis at a given time, Q is a measure of the amount of material electrolyzed up to that time. The reverse potential step at time τ initiates an anodic electrode reaction. Consequently, the charge for the oxidation of R subtracts from the final value of Q in the forward step (Q_f in Fig. 30.6c), which becomes the baseline for the measurement of reverse-step charge (Q_r). The equation that applies to the reverse step is

$$Q_r = \frac{2nFACD^{1/2}}{\pi^{1/2}} \left[\tau^{1/2} + (t - \tau)^{1/2} - t^{1/2}\right]. \qquad (30.8)$$

Chronocoulometry is especially useful for studying electroactive material that is adsorbed on an electrode surface. The component of charge that electrolyzes adsorbed species is conveniently distinguished from the component that electrolyzes solution species. The distinction is based on the fact that the adsorbed species is on the electrode surface and is therefore electrolyzed immediately, whereas the solution species must diffuse to the elec-

*A. G. Ewing, J. C. Bigelow, R. M. Wightman, *Science*, **221**, 169 (1983).

trode in order to react. The total charge (Q_{total}) measured as a result of a potential step excitation signal comes from three sources: diffusing component (Q_{diff}), adsorbed component (Q_{ads}), and electrode charging (Q_{dl}):

$$Q_{total} = Q_{diff} + Q_{ads} + Q_{dl} \tag{30.9}$$

$$Q_{total} = \frac{2nFAC_O D_O^{1/2}}{\pi^{1/2}} t^{1/2} + nFA\Gamma_O + Q_{dl}, \tag{30.10}$$

where Γ_O is the surface concentration of adsorbed species, mol cm^{-2}, and Q_{dl} is double layer changing, C. The individual contributions of Q_{diff}, Q_{ads}, and Q_{dl} to Q_{total} in a typical chronocoulometry experiment are shown in Fig. 30.8a. The mathematical expression for Q_{diff} is simply Eq. (30.7) which shows the $t^{1/2}$ behavior of Q_{diff}. Consequently, a plot of Q_{diff} versus $t^{1/2}$ is a straight line with a slope of $2nFAC_O D_O^{1/2} \pi^{-1/2}$. The second term in Eqs. (30.9) and (30.10) is the charge associated with absorbed electroactive species. Since the adsorbed species is on the electrode surface, it electrolyzes essentially instantly when the potential is stepped. Consequently, this term is not time dependent, as shown by the constant contribution of Q_{ads} in Fig. 30.8a. Here Q_{dl} is the charge required to move the cell potential from E_i to E_s (i.e., the double layer charge). This is also an essentially instantaneous event, as shown by the curve label Q_{dl}. The behavior of Q_{total} is obtained by summing the contributions of Q_{diff}, Q_{ads}, and Q_{dl}, as shown by curve Q_{total}. As expressed in Eq. (30.10), a plot of Q_{total} versus $t^{1/2}$ should be linear with an intercept that equals $nFA\Gamma_O + Q_{dl}$, as shown in Fig. 30.8b. From the intercept, Γ_O can be calculated for an electrode of known A if Q_{dl} is known. Usually Q_{dl} for the potential step E_i to E_s is measured in a separate experiment on supporting electrolyte alone. Thus, chronocoulometry provides a convenient means for extracting Q_{ads} from the response signal to a potential step excitation signal. As shown in Eq. (30.10) the surface concentration of adsorbed species Γ_O is easily calculated once the electrode surface area A and n are known.

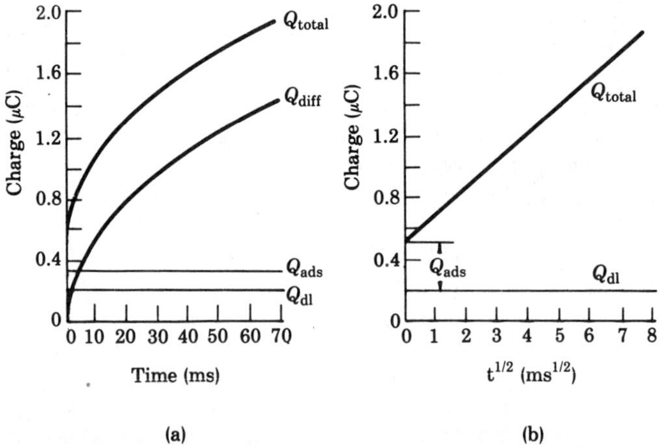

Fig. 30.8 Effect of adsorption of electroactive species on chronocoulometry. (a) Charge-time curves for individual components contributing to total charge (Q_{total}) as in Eq. (30.9). (b) Plots of charge versus $t^{1/2}$ for Q_{total} and Q_{dl}.

Chronocoulometry is an effective technique for studying homogeneous chemical reactions that are coupled to the heterogeneous reduction–oxidation reaction at the electrode. Such coupled chemical reactions often perturb the Q–t response in a single- or double-potential step experiment in a predictable and measurable way so that information about the mechanism and rate of the coupled reaction can be obtained.

A good example of how chronocoulometry is used to study a coupled chemical reaction is given by the EC mechanism shown below.

$$\text{E: O} + \text{e} \rightleftharpoons \text{R} \qquad (30.11)$$

$$\text{C: R} \xrightarrow{\text{k}} \text{P} \qquad (30.12)$$

R that is electrogenerated by reduction of O due to the potential step from E_i to E_s is a reactive species which forms P at a rate characterized by rate constant k. The homogeneous chemical reaction of R to form P can be studied by means of double-potential step chronocoulometry. A series of double-potential step chronocoulograms for the EC mechanism for different values of the rate constant k is shown in Fig. 30.9. Note that the forward step of the chronocoulogram is unaffected by the presence of the coupled chemical reaction. This is to be expected since the mass transport of O to the electrode surface is dependent on C_O and D_O and is not affected by the fate of the electrogenerated R (unless, of course, it reacts to form more O—but, this would be another mechanism). The reverse step of the chronocoulogram is sensitive to the value of k. The magnitude of k can be calculated from a dimensionless working curve that has been calculated from theory. Such a working curve for the EC mechanism is shown in Fig. 30.10. The curve shows how the ratio Q_r/Q_f varies as a function of $\sqrt{k\tau}$. One simply measures Q_r/Q_f for a potential step of duration τ, reads the corresponding value of $\sqrt{k\tau}$ from the working curve, and then calculates k. The shape of the working curve is dependent on the mechanism of the coupled chemical reaction(s). Note how the magnitude of Q_r decreases as the rate constant for the reaction increases. R that is converted to P by the reaction is unavailable for reoxidation to O during the reverse potential step—the faster the chemical reaction, the less R present and, consequently, the smaller Q_r is as shown in Fig. 30.9.

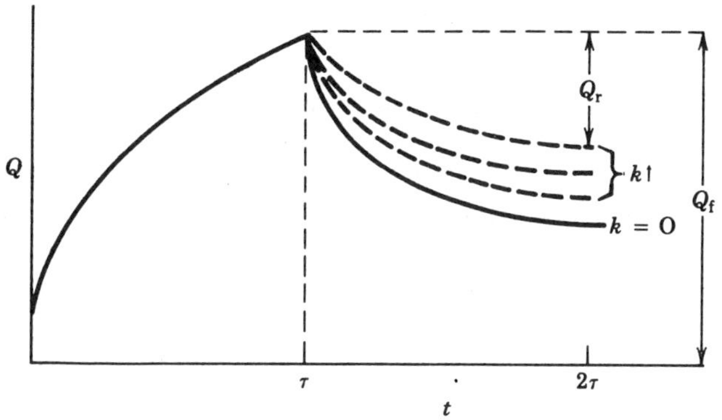

Fig. 30.9 Chronocoulograms for double-potential step chronocoulometry of EC mechanisms for different values of k.

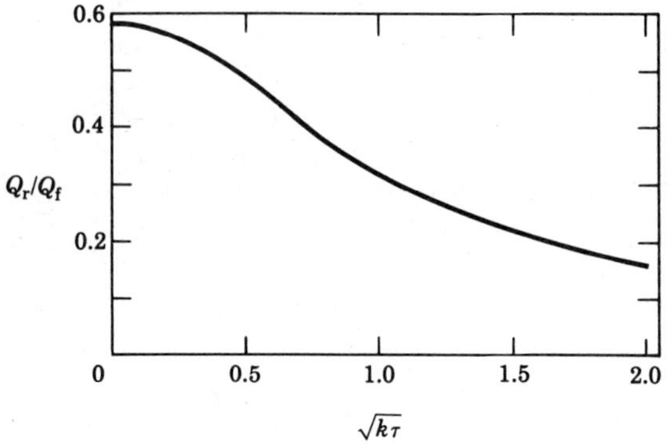

Fig. 30.10 Double-potential step chronocoulometry EC mechanism working curve.

30.4 PULSE VOLTAMMETRIC TECHNIQUES

Several electroanalytical techniques scan the cell potential by means of potential pulse excitation signals. These techniques have good detection limits resulting from discrimination against double-layer charging current.

Large-Amplitude, "Normal" Pulse Voltammetry. *Pulse voltammetry* consists of the application of a series of potential pulses of increasing amplitude for which the current response is measured near the end of each pulse. The experiment is carried out at an electrode in an unstirred solution.

The potential excitation signal is shown in Fig. 30.11a. The potential is pulsed from an initial potential E_i, with a sequence of pulses of increasing amplitude. The duration of each pulse, τ, is typically 1–100 ms, and the interval between pulses is usually 0.1–5 s. This excitation signal is essentially a repeating chronoamperometry experiment in which each successive potential step is greater than the last and E_i is usually chosen to be a potential at which no electrolysis occurs when it is first applied to the cell. The sequence of potential pulses then increases to a potential at which electrolysis occurs. One can think of the pulses as a stepwise potential scan from E_i to E_f, as in cyclic voltammetry. Fig. 30.11b shows the forward scan of a cyclic voltammogram for the reduction of O to R. Superimposed are arrows showing four representative pulses of the excitation signal.

The *C–x* profiles for the representative pulses are shown in Fig. 30.11c for species O. Pulse A is a small pulse that does not extend into a potential range in which electrolysis occurs. Consequently, the profile for species O does not change and only charging current is observed as shown in Fig. 30.11d. Pulse B extends into the base of the voltammetric wave. Reduction of O occurs as reflected by its *C–x* profile. The resulting faradaic current is given by

$$i = \frac{nFAC_O D_O^{1/2}}{\pi^{1/2} t^{1/2}} \left[1 + \frac{C_O^s}{C_R^s}\right]^{-1} \qquad (30.13)$$

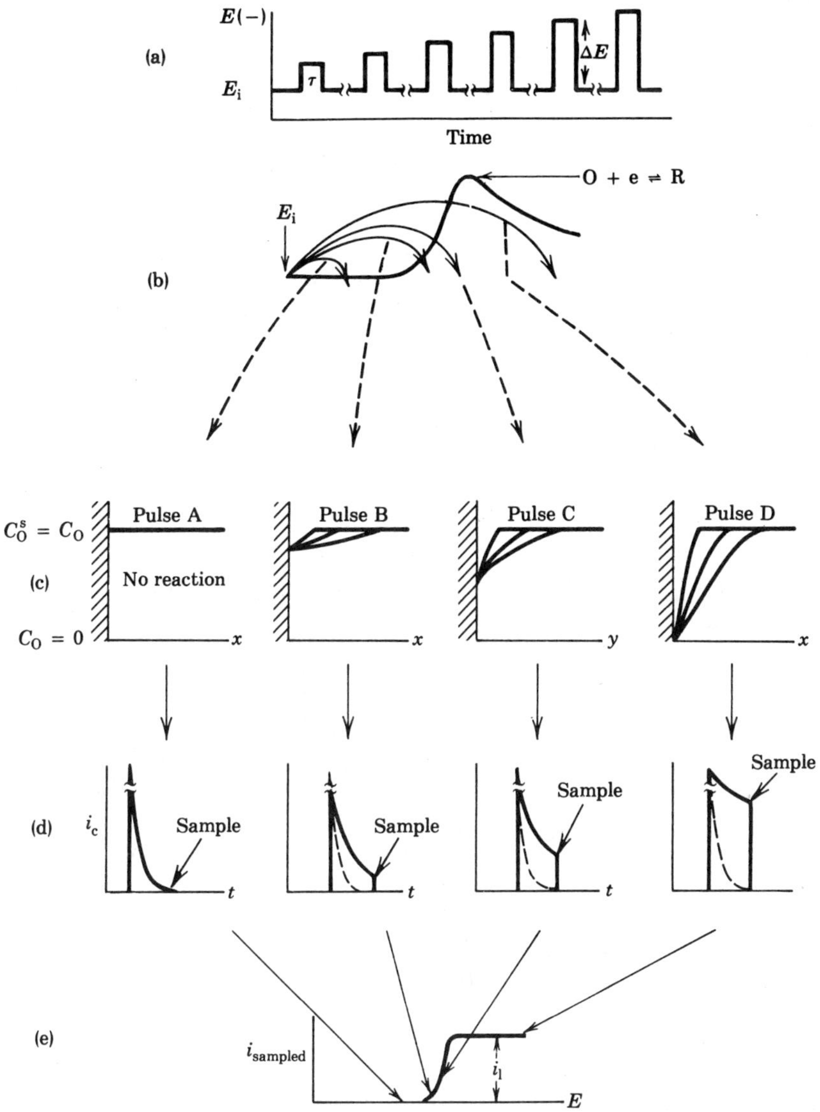

Fig. 30.11 Normal pulse voltammetry. (a) Excitation signal. (b) Representative pulses superimposed on cyclic voltammogram. (c) Concentration–distance profiles for representative pulses. (d) Current response. Dashed line shows charging current. (e) Voltammogram.

This is a more general form of the Cottrell equation in which the current at time t is a function of the surface concentration ratio at the electrode surface, C_O^s/C_R^s. Pulse C is a potential step to $E^{0'}$ of the redox system, so that $C_O^s/C_R^s = 0.5$. Pulse D brings the potential into the diffusion-limited region so that C_O at the electrode surface is essentially zero, and the reduction of O is occurring at its maximum rate.

The current responses to the four potential pulses show an increase in faradaic current as the sequence of potential pulses ''scans'' through the voltammetric wave. Pulsing to potentials beyond pulse D gives no further increase in faradaic current since the diffusion-limited region of the voltammetric wave has been reached. The individual pulses are now each chronoamperometry experiments, as described in Section 30.3. A primary objective of pulsing the potential is to enable the interfering non-faradaic current due to electrode charging to be separated from the faradaic current due to electrolysis of the redox species. As shown in Fig. 30.11d, the non-faradaic current (dashed lines) decreases more rapidly than does the faradaic current. Consequently, sampling current toward the end of each pulse (typically for 10–40 ms) improves the ratio $i_{\text{farad}}/i_{\text{non-farad}}$, which leads to improved detection limits for the analyte. For example, pulse voltammetry is applicable down to 10^{-7} M, compared with 10^{-5} M for cyclic voltammetry.

A voltammogram (Fig. 30.11e) is constructed by plotting the sampled current versus the potential to which the pulse is stepped. The appearance of the voltammogram is analogous to that of a hydrodynamic voltammogram and the $E_{1/2}$ and limiting current, i_l, have the same uses.

Differential Pulse Voltammetry. Differential pulse voltammetry is analogous to pulse voltammetry in that a voltammogram is obtained by scanning the potential with a sequence of pulses. However, in differential pulse voltammetry this is accomplished by a constant amplitude pulse, as shown by the excitation signal in Fig. 30.12a (left). For each pulse the electrode is stepped by the amount ΔE_p (pulse amplitude) to a new potential. After t_p (pulse time), the potential is stepped back, but by an amplitude less than ΔE_p. The result of this difference in the forward and reverse steps causes the potential pulse to ''scan'' by an amount equal to ΔE_s (step potential) for each pulse. The scan rate is determined by the magnitude of ΔE_s in conjunction with the pulse repetition period τ. A comparable excitation signal can also be obtained by superimposing the potential pulses on a ramp potential scan, as shown in Fig. 30.12a (right). The excitation signal on the left is used by many digitally based instruments, whereas the one on the right is used by most analog instruments.

The current response to this excitation signal is analogous to double-potential step chronoamperometry of small amplitude that is repeated along the potential axis of the voltammogram. Figure 30.12b (left) shows the current response obtained in the potential region before a voltammetric wave (the forward scan of a cyclic voltammogram). Since the potential steps cause no electrolysis, the current response is simply non-faradaic current caused by double layer charging. The current response in the vicinity of $E^{0'}$ of a reversible redox couple is shown in the center drawing. Now a substantial faradaic current is caused by reduction during the forward step and the opposite, oxidation during the reverse step. The magnitude of these faradaic currents increases as the potential scans from the baseline toward $E^{0'}$ of the couple. The current response when the potential is in the limiting current region of the voltammogram is now simply double-layer charging again as shown in the right-drawing. In this case, electrolysis is already proceeding at a diffusion-controlled rate, which is not influenced by the small potential step.

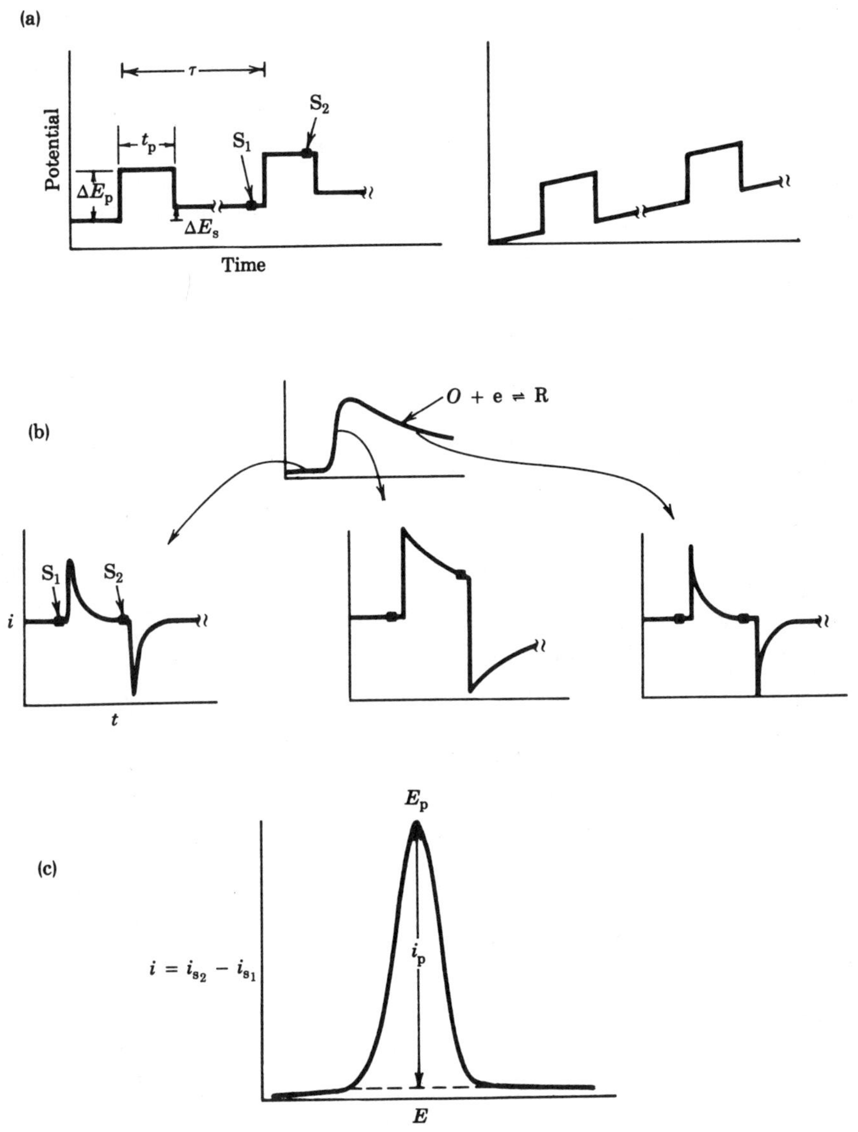

Fig. 30.12 Differential pulse voltammetry. (a) Excitation signals. (b) Current response signal obtained at different positions on a cyclic voltammogram. (c) Voltammogram.

As the name of the technique implies, the displayed current is obtained by measuring current at two points for each pulse and recording the difference. The current is first sampled at S_1, which is just prior to application of a pulse, and then at S_2, which is at the end of a pulse. These points are chosen to allow charging current to have decayed substantially, thereby improving the ratio $i_{\text{farad}}/i_{\text{non-farad}}$. A voltammogram is constructed by plotting the current difference ($\Delta i = i_{S_2} - i_{S_1}$) for each pulse as a function of the potential to which the pulse is stepped. A representative voltammogram is shown in Fig. 30.12c. Note that the measurement of current difference (Fig. 30.12b) is greatest in the vicinity of $E^{0'}$ for reversible couples and essentially zero in the baseline and limiting current regions of the voltammogram. Consequently, the voltammogram is peak-shaped. One can think of the shape as being the first derivative of the voltammogram in Fig. 30.11d with the peak corresponding to the maximum slope in the sigmoid region.

The peak is characterized by the peak potential E_p, which is related to $E_{1/2}$ by the following expression:

$$E_p = E_{1/2} = \frac{\Delta E_p}{2}, \qquad (30.14)$$

where E_p approaches $E_{1/2}$ as ΔE_p decreases. For reversible systems the peak current, i_p, is directly proportional to concentration. Because of its effective discrimination against non-faradaic current, differential pulse voltammetry is often used for trace analysis. Detection limits are typically 10^{-7}–10^{-8} M.

In most instruments some of the parameters in the excitation waveform can be varied, whereas others are fixed. The pulse width is usually 50 ms, and the current is sampled over two 16.7-ms windows—one immediately preceding the pulse and one at the end of the pulse. A 30-ms pulse is sufficiently long to enable the double-layer charging current interference to decay substantially, but not so long that faradaic current sensitivity diminishes substantially. The 16.7-ms sampling period minimizes the effect of 60-Hz line interference. The pulse amplitude is usually variable over a range of 5 to 50 mV. The effective scan rate of the technique is quite slow, typically 1–10 mV/s. Pulses are generally applied at an interval of 0.5–5 s.

Square Wave Voltammetry. An increasingly popular voltammetric technique is square wave voltammetry, which combines the advantages of a pulse waveform with judicious current sampling to minimize interference from non-faradaic current with a fast scan rate. As shown in Fig. 30.13a, the excitation signal consists of a symmetrical pulse train of amplitude $2E_{sw}$ and period τ, that is superimposed on a staircase with a step height of ΔE_s and a period of τ. The result is a square-wave excitation signal that scans at a rate determined by the values of ΔE_s and τ. The current is sampled at the end of each forward and reverse pulse.

The current signals in response to this potential signal are shown in Fig. 30.13b for three positions on a representative cyclic voltammogram (forward scan only). As in the case of differential pulse voltammetry, the potential steps cause only non-faradaic charging current in the potential region prior to the voltammetric wave. Also, as for the other pulses techniques, the current measurement window is delayed to the end of the pulse to allow substantial decay of the non-faradaic current. When the potential steps are occurring in the ascending portion of the voltammetric wave, two types of faradaic currents are observed—the *forward current* due to the forward step of the square wave and the *reverse current* caused by the reverse step of the square wave. This is analogous to double-potential step

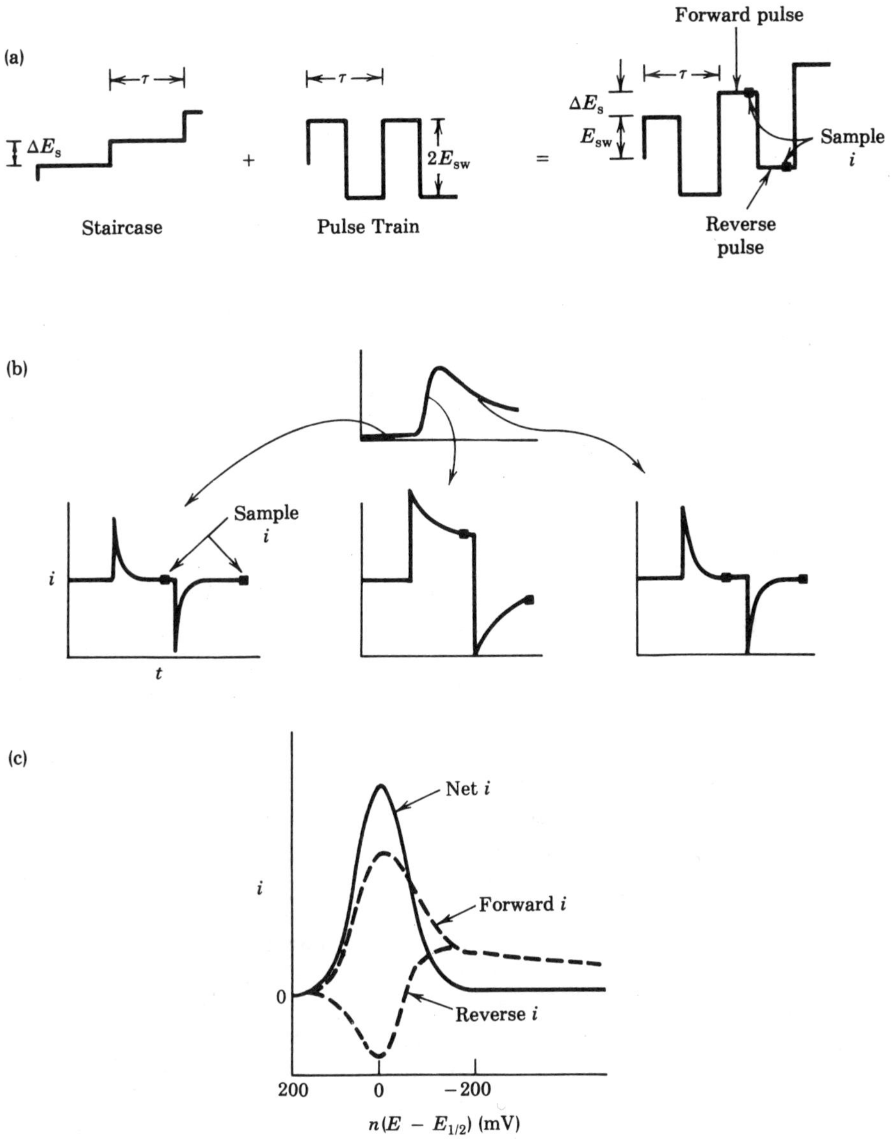

Fig. 30.13 Square wave voltammetry. (a) Excitation signal. (b) Current response signal. (c) Voltammograms.

chronoamperometry. In potential regions of the plateau of the voltammogram, the potential steps do not influence the redox process, which is diffusion controlled at this point.

Voltammograms are constructed in any of three ways, as shown in Fig. 30.13c, by plotting the forward current, the reverse current, or the net current ($\Delta i = i_{\text{forward}} - i_{\text{reverse}}$) as a function of the potential of the excitation signal. Note that the forward current is larger in absolute magnitude than the reverse current for the same reason that i_f is larger than i_b in double-potential step chronoamperometry (see Section 30.3), namely that some product generated during the forward step is not recovered during the reverse step since it diffuses away from the electrode. This difference causes the net current response to have a peak-shaped wave. Note that forward i and reverse i merge to a common value when the plateau of the hydrodynamic voltammogram is reached. This value is the limiting current for the overall electrode process, which is not influenced by the square wave at this point.

Square wave voltammetry is undergoing increased use as an analytical technique. The peak net current (i_p) is proportional to the concentration of analyte and E_p is equal to $E_{1/2}$ for reversible systems. Detection limits of 10^{-7}–10^{-8} M are usually achieved. Perhaps the most significant feature of square-wave voltammetry is the rapid scan rate. As described in Section 30.5, complete voltammograms can be recorded on a single drop of a dropping mercury electrode or on a peak of a liquid chromatogram (Fig. 29.31).

30.5 POLAROGRAPHY

Polarography is a voltammetric technique in which the working electrode is a continually renewable surface such as the dropping mercury electrode (DME) or the static mercury drop electrode (SMDE), which are described in Section 29.5. In modern nomenclature, the term polarography is reserved specifically for techniques used with a dropping mercury electrode.

The classical form of polarography—now called *DC polarography*—was discovered in 1922 by Jaroslav Heyrovsky in Prague, Czechoslovakia. Polarography revolutionized the determination of metal ions and, consequently, its discoverer was recognized with the Nobel Prize.

The excitation signal for dc polarography is a slow potential scan (see Fig. 29.1), and the resulting current is displayed versus the applied potential to give a *polarogram*. The solution is quiescent while a polarogram is recorded, except for the localized stirring caused by the drop falling off of the capillary.

A polarogram for the reduction of Pb^{2+} in KCl supporting electrolyte is shown in Fig. 30.14. The potential scan is in the negative direction beginning at 0.00 V versus SCE. As the potential is scanned, the onset of Pb^{2+} reduction to Pb^0 is signalled by the increase in cathodic current at the *decomposition potential*. The current increases until Pb^{2+} reduction is limited by diffusion of Pb^{2+} to the surface of the expanding mercury drop, at which point the current plateaus to the *diffusion current*, i_d. The voltammogram is sigmoid in shape, as are hydrodynamic voltammograms (see Section 29.3). As in the case of hydrodynamic voltammetry, the current is determined by the slope of the C–x profile of the electroactive species at the electrode surface, $(\partial C/\partial x)\, x = 0$. Profiles for representative potentials are shown in Fig. 30.14. Compare the profiles for Pb^{2+} in this figure with those for O in Fig. 29.5. Note the correlation between the slopes of the profiles and the

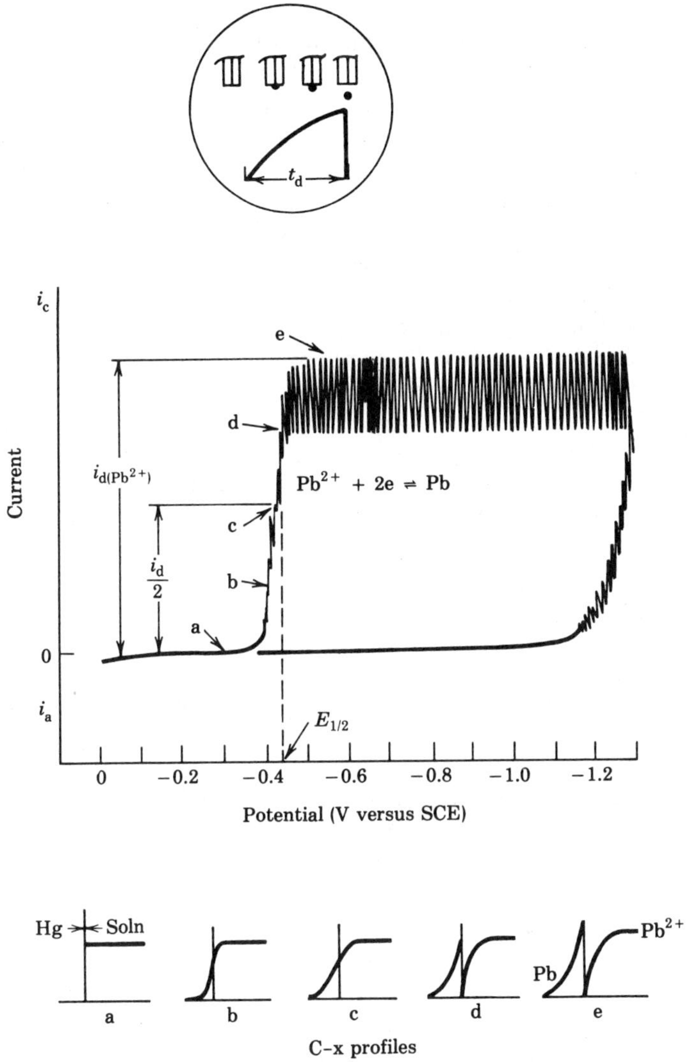

Fig. 30.14 Polarogram of 1.00 m*M* Pb^{2+}, 0.1 *M* KCl, deoxygenated. Representative *C–x* profiles, are shown below the polarogram. Correlation of current response to size of an individual drop is shown above the polarogram.

currents in the polarogram and the hydrodynamic voltammogram. In both cases, the current reaches a plateau when the concentration of electroactive species is essentially zero at the electrode surface. The reduction product Pb^0 dissolves in the mercury drop. Hence, the *C–x* profiles for Pb^0 extend into the electrode. Compare the profiles for Pb^0 with those for R in Fig. 29.5.

The repetitive current fluctuations correspond to the repetitive growth and dislodgment of mercury drops, which is characteristic of the DME. As shown in the insert in Fig 30.14, the current increases as the drop grows and declines abruptly the moment the drop falls off the mercury capillary. The current–time behavior of an individual drop is determined by the increasing area of the drop ($t^{2/3}$) combined with the decrease in current associated with depletion of Pb^{2+} in the adjacent solution ($t^{-1/2}$), which is analogous to chronoamperometry at an expanding area electrode. The combination of these two opposing phenomena gives a current that increases at a $t^{1/6}$ rate during the lifetime of each drop. The current drops abruptly when a drop falls due to the instantaneous decrease in electrode surface area. The falling drop stirs the solution at the capillary tip enough to remove almost all of the concentration depletion of Pb^{2+} caused by electrolysis at the drop. This "microstirring" causes the polarogram to level off at a plateau, rather than peak and then drop off as with cyclic voltammetry. The reader will have realized that polarography is analogous to the forward scan of cylic voltammetry, except that the DME influences the resulting waveform dramatically. In fact, cyclic voltammetry was initially termed "stationary electrode polarography."

The *Ilkovic* equation describes the diffusion-limited current i_d:

$$i_d = knm^{2/3}D^{1/2}t_d^{1/6}C \tag{30.15}$$

where i_d = diffusion current (μA), m = rate of flow of mercury from the electrode (mg s^{-1}), t_d = drop time (s), k = constant (708 for maximum current, 607 for average current), C = concentration (mmol L^{-1}), and other terms are defined in Table 27.1. The constant k is taken as 708 if i_d is measured to the maximum current fluctuations, as shown in Fig. 30.14, or as 607 if i_d is taken as the midpoint between maximum and minimum fluctuations. The residual current occurring in the absence of Pb^{2+} can be used as a baseline for measuring i_d, as shown in Fig. 30.14. Since i_d is directly proportional to the concentration of electroactive species, it is useful for determining the concentration of electroactive species in solution. The optimum concentration range for dc polarography is 10^{-2}–10^{-4} M. Reproducibility is in the order of $\pm 3\%$.

Polarographic maxima are characterized by a large, sometimes erratic, current that diminishes abruptly during the potential scan. One type of maximum occurs at the $E_{1/2}$. Maxima can result from hydrodynamic flow of solution around the expanding mercury drop. Maxima are typically eliminated by the addition of a small amount of surfactant, such as Triton X-100, to the solution.

The drop time of the DME varies during the potential scan. Figure 30.15 shows an electrocapillary curve, which is the variation of drop time t_d with potential. The changing drop time reflects the change in surface tension of the mercury drop depending on the interfacial charge. The potential at which the longest drop time occurs is the *electrocapillary maximum*. Note how the residual current increases as the potential deviates from the electrocapillary maximum. To avoid this vari-

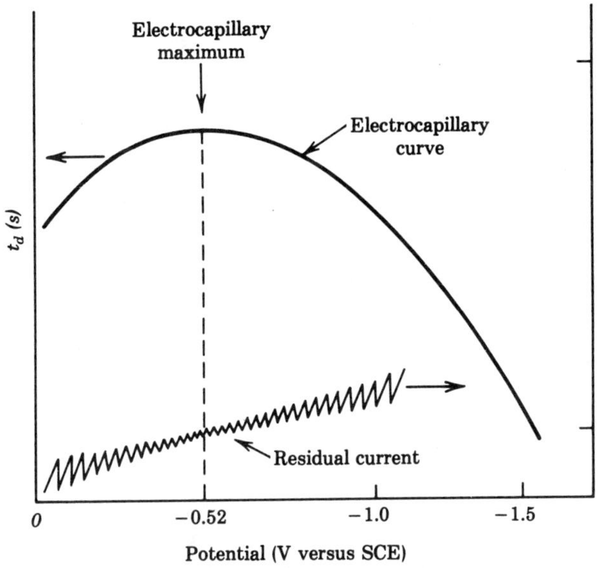

Fig. 30.15 Representative electrocapillary curve and residual current.

ability in drop size, the mercury drop can be mechanically dislodged for polarographic measurements at a constant time interval.

In addition to its extensive use for analytical purposes, dc polarography has been widely employed for the measurement of formation constants of metal-ion complexes, amperometric titrations, and adsorption of species on mercury surfaces. Many of these measurements can now be made more effectively by the more recently developed potential step, potential scan, and hydrodynamic methods. Polarographic techniques derive their main benefits from the clean, geometrically reproducible surface of pure mercury obtainable at a DME.

More modern polarographic techniques include pulse polarography, differential pulse polarography, and square wave polarography. These pulse techniques have been described in Section 30.4. When applied to the DME, the potential pulse is sequenced with the mechanically controlled droptime of the DME so that the pulse is applied at the end of droplife when surface area is greatest for maximum current and the *rate* of growth of surface area is less to minimize charging current.

Effects of Complexation of Reduction of Metal Ions. Polarography has been used to investigate labile metal-ion complexes since the characteristic half-wave potential ($E_{1/2}$) of a simple metal ion is shifted when the metal ion undergoes complex formation. The extent of this shift in $E_{1/2}$ depends on the concentration of the complexing agent and the formation constant of the complex. By measuring the shift in $E_{1/2}$ as a function of the concentration, it is possible to obtain information concerning both the formula and the stability of the metal complex. As mentioned in Section 29.5, the addition of an appropriate ligand to shift

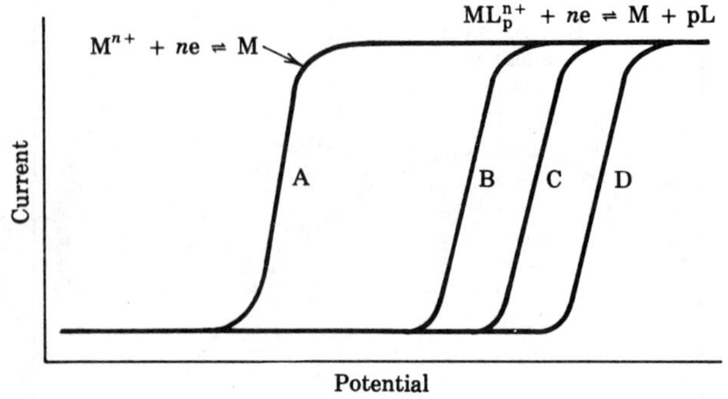

Fig. 30.16 Polarograms for reduction of M^{n+} in the presence and absence of ligand L. (a) M^{n+} in 0.1 *M* KNO_3. (B–D) Increasing concentration of L.

$E_{1/2}$ in order to better resolve voltammetric waves in a mixture is sometimes useful in multielement analysis.

Polarograms for the reduction of M^{n+} in the absence of ligand and in the presence of increasing concentrations of ligand L are shown in Fig. 30.16. A more negative potential is required to reduce the complex ML_p than free (aquated) M^{n+}, hence the negative shift in $E_{1/2}$ with added ligand.

An equation can be derived that relates the shift in $E_{1/2}$ with complex stoichiometry, p, and the overall formation constant, β. The following derivation is based on the assumption that the polarographic waves for both M^{n+} and ML_p^{n+} are reversible, that the diffusion coefficients of M^{n+} and ML_p^{n+} are equal, that the complex is labile, that the reduced form M^0 does not complex with L, and that the concentration of ligand, C_L, is substantially greater than the concentration of complex, $C_{ML^{n+}}$, so that C_L can be taken as equal to the total concentration of ligand added to solution.

The reduction of simple metal cation and the corresponding Nernst equation (base 10 log, $T = 25°C$) are

$$M^{n+} + ne \rightleftharpoons M^0 \tag{30.16}$$

$$E = E^{0'}_{M^{n+},M^0} - \frac{0.0591}{n} \log \frac{C_{M^0}}{C_{M^{n+}}}. \tag{30.17}$$

(In the case of reduction to metal, M^0, the experiment would be performed at a mercury electrode for which C_{M^0} refers to concentration of amalgamated M^0 in liquid mercury.) The reaction between ligand L and the metal ion and the formation constant of the resulting complex are

$$M^{n+} + pL \rightleftharpoons ML_p^{n+} \tag{30.18}$$

$$\beta = \frac{C_{ML_p^{n+}}}{C_{M^{n+}} C_L^p}. \tag{30.19}$$

Solving Eq. (30.19) for $C_{M^{n+}}$ and substituting this result into the Nernst equation yields

$$E = E^{0'}_{M^{n+},M^0} - \frac{0.0591}{n} \log \frac{\beta C_L^p C_{M^0}}{C_{ML_p^{n+}}} \tag{30.20}$$

Since at the half-wave potential the surface concentration of the oxidized form and reduced form are equal, from Eq. (30.17) for the metal ion one obtains, when $C_{M^{n+}} = C_{M^0}$ (assuming equal diffusion coefficients),

$$E_{1/2} = E^{0'}_{M^{n+},M^0}, \tag{30.21}$$

and from Eq. (30.20) for the reduction of the complex when $C_{ML_p^{n+}} = C_{M^0}$,

$$(E_{1/2})_{\text{complex}} = E^{0'}_{M^{n+},M^0} - \frac{0.0591}{n} \log \beta C_L^p. \tag{30.22}$$

The difference in half-wave potentials is then given by

$$(E_{1/2})_{\text{complex}} - E_{1/2} = -\frac{0.0591}{n} \log \beta - \frac{0.0591}{n} p \log C_L. \tag{30.23}$$

Experimentally one obtains $E_{1/2}$ for the simple metal ion and for the complex in a series of solutions containing a given concentration of metal ion and various concentrations of the ligand. (Twentyfold and greater excess is employed so that the concentration of L at the electrode reaction will remain essentially constant and of known value.) A plot of $(E_{1/2})_{\text{complex}} - E_{1/2}$ versus C_L gives a slope of $(-0.0591/n)p$ from which p can be calculated and an intercept of $(-0.0591/n) \log \beta$ from which β can be calculated.

In the case of reactions that involve protons, it is important that the solution be well buffered so that the pH in the Nernst diffusion layer is not altered to a different value by the electrode reaction, since β will be dependent on pH.

30.6 STRIPPING VOLTAMMETRY

Stripping voltammetry has the lowest detection limit of the commonly used electroanalytical techniques. Analyte concentrations as low as 10^{-10} *M* have been determined. The technique consists of two steps. In the first, analyte is deposited at the electrode by controlled-potential electrolysis. This step serves to preconcentrate the analyte either by electrochemically extracting it into a mercury electrode or depositing it as a film on the electrode surface. This preconcentration feature is responsible for the low detection limits of stripping voltammetry. In the second step the deposited analyte is removed or "stripped" from the electrode by an appropriate potential scan, and the resulting current signal is used to quantify the analyte. Any of a number of voltammetric techniques can be used for the stripping step. If the stripping step gives anodic current, the technique is termed *anodic stripping voltammetry* (ASV); if cathodic current is obtained, the term *cathodic stripping voltammetry* (CSV) is used.

Anodic stripping voltammetry is used primarily for the determination of heavy metals that are soluble in mercury. The technique is illustrated in Fig. 30.17 for the determination of Pb^{2+} at a HMDE. Figure 30.17 shows the potential excitation signal and an anodic stripping voltammogram. The potential of the electrode is first maintained at a negative value for several minutes to concentrate some of the Pb^{2+} from the analyte solution into the mercury drop as amalgamated lead. The

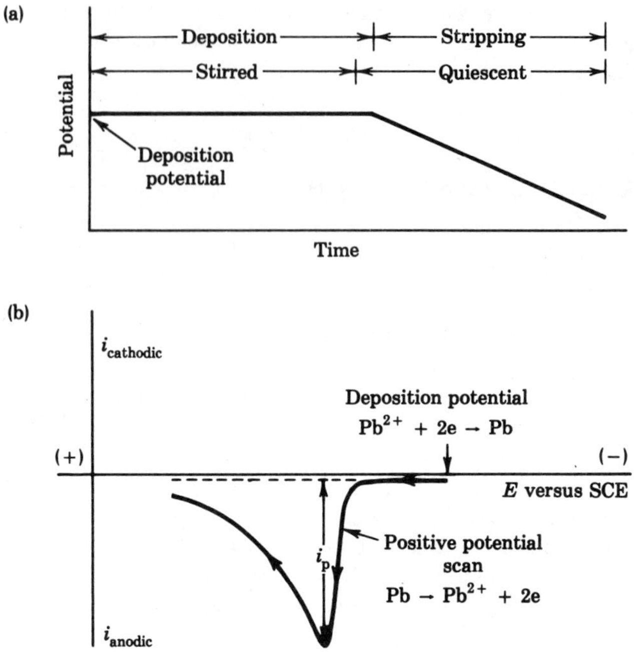

Fig. 30.17 Stripping voltammetry. (a) Excitation signal. (b) Stripping voltammogram for the determination of Pb^{2+}.

sample ions are electrochemically extracted as metal atoms into a mercury electrode, the volume of which is considerably less than the volume of sample solution in the electrochemical cell. The resulting "solution" of the metal atoms in the liquid mercury is substantially more concentrated than the solution of metal ions being analyzed, by a factor of up to 10^6. During this step the solution is stirred to maximize the rate of accumulation of lead in the electrode. After discontinuing the stirring, the potential is scanned positively, causing the amalgamated lead to be oxidized back into the solution (i.e., stripped out of the electrode). This oxidation of lead gives a current peak i_p, the magnitude of which is determined by the concentration of Pb(Hg) in the mercury electrode, which is in turn proportional to the concentration of Pb^{2+} in the sample. Examples of metal ions that have been determined by ASV at a mercury electrode are Bi^{3+}, Cd^{2+}, Cu^{2+}, Ga^{3+}, Ge^{4+}, In^{3+}, Ni^{2+}, Pb^{2+}, Sb^{3+}, Sn^{2+}, Tl^{+}, and Zn^{2+}. Solid electrodes such as graphite enable Hg^{2+}, Au^{3+}, Ag^{+}, and Pt^{6+} to be determined by ASV. In this case, the metal is preconcentrated on the surface of the electrode as a metallic film and then stripped off by a positive potential scan.

Cathodic stripping voltammetry can be used to determine a variety of anions that can be deposited on the surface of a mercury electrode in the form of an insoluble mercury salt. A potential is applied so that an electrode reaction of the following general type occurs where X^- is the anion:

$$2Hg + 2X^- \rightleftharpoons Hg_2X_2 + 2e \qquad (30.24)$$

After a sufficient amount of the anion is deposited (concentrated) on the electrode, it is stripped off via a negative potential scan that reduces the mercury salt to Hg and X^-. The height of the cathodic peak is proportional to the concentration of the anion in solution. This technique has been applied to the determination of Cl^-, Br^-, I^-, S^{2-}, CrO_4^{2-}, WO_4^{2-}, MoO_4^{2-}, VO_3^-, SO_4^{2-}, oxalate, succinate, dithizonate, and diethylthiophosphate. Metal ions such as Ce^{3+}, Tl^+, Mn^{2+}, and Fe^{2+} can be concentrated on an electrode as insoluble hydroxides by oxidation at a graphite electrode. The hydroxides are then stripped from the electrode by a negative potential scan, giving characteristic reduction peaks.

The deposition potential should be selected so that species are reduced as rapidly as they are transported to the electrode surface. Correct selection of deposition potentials can be made by considering voltammograms obtained on solutions containing metal ions of the type to be analyzed. Figure 30.18 shows a polarogram for a solution of Cu^{2+}, Pb^{2+}, Cd^{2+}, and Zn^{2+}. When the current levels off on the first plateau, the Cu^{2+} in solution is being reduced to Cu^0 as rapidly as it is transported to the electrode surface. For the determination of Cu^{2+} only, the maximum efficiency for deposition is achieved if the deposition potential is maintained in the region of limiting current i_l. The Pb^{2+}, Cd^{2+} and Zn^{2+} are not determined since they are not deposited at this potential. To determine all four ions, the deposition potential should be maintained on the fourth region of limiting current i_l''', where all four ions are being reduced as rapidly as they arrive at the electrode surface. The bottom curve in Fig. 30.18 shows the stripping voltam-

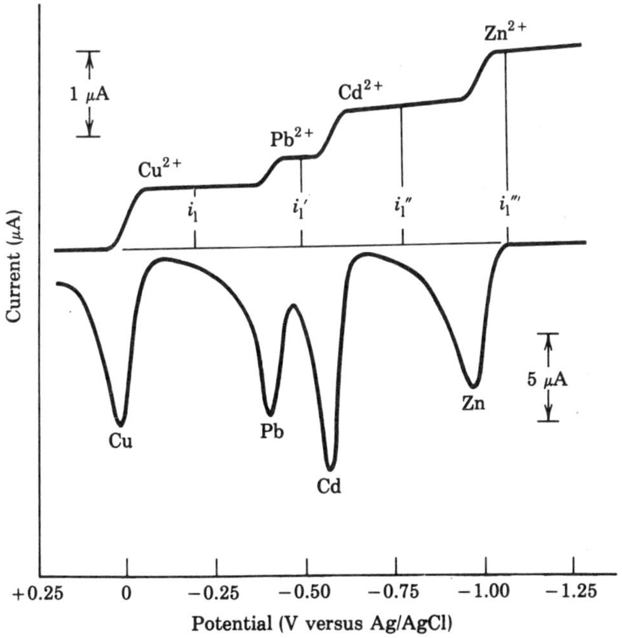

Fig. 30.18 Polarogram at a DME (top) and anodic stripping voltammogram at a HMDE (bottom) of 2.5 ppm Cu^{2+}, Zn^{2+} and 5 ppm Pb^{2+}, Cd^{2+} in 0.1 M sodium acetate.

mogram that would be obtained after deposition at a potential corresponding to the fourth limiting-current plateau. Note the correlation between the $E_{1/2}$ of each polarographic wave and the E_p of each stripping wave.

In stripping voltammetry some fraction of the total analyte is deposited into the mercury electrode by electrolysis during the preconcentration step. Complete deposition of all of the analyte into the electrode is time-consuming and generally unnecessary, since adequate concentrations can usually be deposited into the electrode to give a satisfactory stripping signal in much shorter times. Since the deposition is not exhaustive, it is important to deposit the same fraction of analyte for each stripping voltammogram. The parameters of electrode surface area, deposition time, and stirring must be carefully duplicated for all standards and samples. Deposition times vary from 60 sec to 30 min, depending on the analyte concentration, the type of electrode, and the stripping technique. The less concentrated solutions require longer deposition times to give adequate stripping peaks; the mercury film electrode requires less time than does the hanging mercury drop electrode; differential pulse voltammetry requires less time than linear sweep voltammetry as the stripping technique.

Differential pulse voltammetry has become very popular as the stripping technique because its discrimination against interference from double-layer charging current enables considerably improved analytical detection limits for a given deposition time. A second feature that enhances sensitivity is the redeposition that occurs when the potential is returned to the ramp at the end of each pulse. Some of the oxidized material is redeposited in the electrode, to be stripped out again during the next pulse. This partial cycling of analyte between electrode and solution also contributes to improved detection limits. A variety of other stripping techniques have been successfully utilized in ASV. Of these, square wave voltammetry, which shares the low detection capability of differential pulse voltammetry but at a faster scan rate, is being used most often.

Electrodes. The two most commonly used electrodes for ASV are the hanging mercury drop electrode (HMDE) and the mercury film electrode (MFE). (See Section 29.5 for a description of these electrodes). The main advantages of the HMDE are the very low residual current, the excellent negative potential range that pure mercury exhibits, and the ease and reliability with which reproducible drops can be extruded with a micrometer capillary or a static mercury drop electrode (although clogging of the capillary in some solvents is occasionally an annoying problem). Disadvantages of the HMDE include a limitation to rather slow stirring rates during the deposition step and restrictions on rotating the electrode without dislodging the drop; a low surface area/volume ratio, which necessitates longer deposition times; broader stripping peaks, which can lead to loss of resolution in multielement analysis; and possibility of some sample loss by diffusion into the mercury thread with the micrometer HMDE. In many analyses, these disadvantages are unimportant, and the HMDE serves as an excellent electrode.

The MFE exhibits certain characteristics that provide substantial impetus for its usage. An advantage of the MFE results from the greater electrode surface area/

volume ratio of a film as compared with the spherical HMDE. The larger surface area enables more sample to be concentrated into a given amount of mercury during a specified deposition time. This means greater analytical sensitivity for the MFE compared with the HMDE (or a shorter deposition time for the MFE for equal sensitivity with the HMDE). The MFE is also compatible with stirring rates that would dislodge a HMDE. The ability to use greater stirring rates or to rapidly rotate the electrode improves mass transport during the concentration step, which further reduces the time required to deposit a given quantity of sample.

Another advantage of the MFE is a result of its thinness compared with the HMDE. In a thin film, the electrodeposited sample is constrained near the interphase to which it must diffuse for oxidation during the stripping step. By comparison, metals deposited into a HMDE can diffuse to the center of the spherical electrode, which greatly lengthens the return trip to the interphase for oxidation. The practical consequence of this is much sharper stripping peaks at a MFE than a HMDE. This minimizes the overlap of stripping peaks in multielement analysis, reducing the uncertainties in baseline estimation for the measurement of peak current.

30.7 SPECTROELECTROCHEMISTRY

Electrochemistry can be combined with a quite different technique, spectroscopy, for studying the redox chemistry of inorganic, organic, and biological molecules. Oxidation states are changed electrochemically by addition or removal of electrons at an electrode while spectral measurements on the solution adjacent to the electrode are made simultaneously. Such *spectroelectrochemical* techniques are a convenient means for obtaining spectra of electrogenerated species and redox potentials and for observing subsequent chemical reactions of electrogenerated species. Several commonly used spectroelectrochemical methods are summarized in Fig. 30.19.

Optically Transparent Electrodes. The initial development of spectroelectrochemistry was stimulated by the availability of *optically transparent electrodes* (OTEs), which enable light to be passed directly through the electrode and adjacent solution. Electrode transparency is necessary for several spectroelectrochemical techniques. One type of OTE consists of a very thin film of conductive material such as Pt, Au, SnO_2, C, or Hg-coated Pt that is deposited on a transparent substrate such as glass or plastic (visible), quartz (UV-visible), or Ge (infrared), depending on the spectral region of interest. The transparency (20–85%) of these electrodes is due to the thinness (100–5000 Å) of the conducting film. The transparency of a second type of OTE is due to small holes in the electrode. Two examples are the minigrid electrode, which consists of a metal (Au, Ni, Ag, or Hg-coated Au) micromesh of 100–2,000 wires per inch, and porous reticulated vitreous carbon, which is a carbonized foam.

Thin-layer Spectroelectrochemistry. One of the most generally useful spectroelectrochemical techniques involves observation of a thin layer of solution that is confined next to a transparent electrode, as shown in Fig. 30.19a. The optical beam of the spectrophotometer

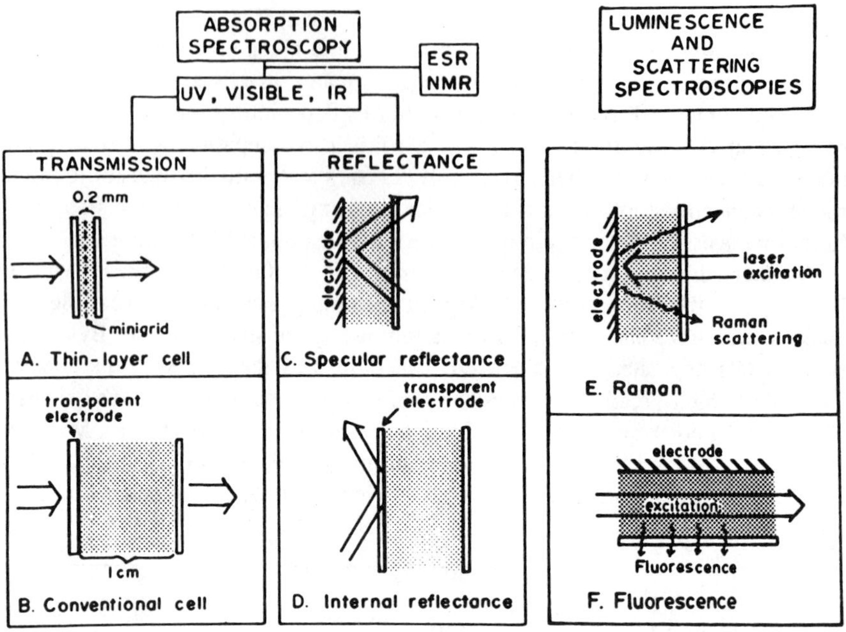

Fig. 30.19 Spectroelectrochemical techniques. [From W. R. Heineman, *J. Chem. Ed.*, 60, 305 (1983), with permission.]

is passed directly through the transparent electrode and the solution. An easily constructed *optically transparent thin-layer electrode* (OTTLE) is shown in Fig. 30.20. This OTTLE consists of a transparent gold minigrid electrode sandwiched between two ordinary microscope slides that are separated 0.01–0.03 cm by strips of Teflon tape spacers. To use the cell, the bottom edge is dipped into a small cup containing a few milliliters of the solution to be investigated. Reference and auxiliary electrodes are also immersed in this cup. Solution drawn into the OTTLE by application of suction at the top corner maintains its level above the minigrid by capillary action. Electrochemical experiments can then be performed on the thin layer of solution surrounding the minigrid.

An attractive feature of the thin-layer technique is the speed with which complete electrolysis can be achieved. The solution volume that undergoes electrolysis is the thin layer of solution between the microscope slides that is defined by the area of the minigrid. The volume of this "cell" is only 30–50 μL, and complete electrolysis occurs in 30–60 s.

Since the minigrid electrode is transparent to light, optical spectra of the solution in the thin-cell surrounding the minigrid can be recorded by passing light directly through the minigrid as shown in Fig. 30.20. Spectra of electroactive species in different oxidation states can be obtained by placing the OTTLE in the sample compartment of a spectrometer. Spectra are recorded after the electroactive species has been converted to the desired oxidation state by applying an appropriate potential to the minigrid.

The use of the OTTLE for determining formal redox potentials, *n* values, and spectra of redox couples is illustrated by measurements on a representative inorganic metal complex $[Re^{III}(dmpe)_2Br_2]^+$, where dmpe is 1,2-bis(dimethylphosphino)ethane, which is reduced in DMF by one electron to $[Re^{II}(dmpe)_2Br_2]^0$.

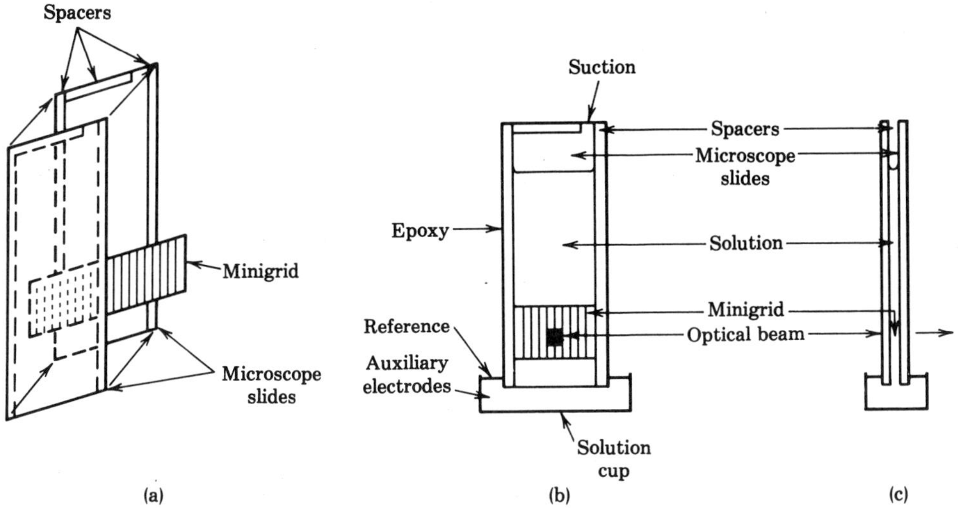

Fig. 30.20 Optically transparent thin-layer electrode. (a) Assembly of the cell. (b) Front view. (c) Side view.

Potential scan techniques such as cyclic voltammetry (see Section 30.2) are exceedingly useful for locating the redox potentials of electroactive species in a thin-layer cell, just as they are in conventional electrochemical cell configurations. Fig. 30.21a shows a cyclic voltammogram for $[Re^{III}(dmpe)_2Br_2]^+$ in the OTTLE. The rapid drop in current after the cathodic and anodic peaks coincides with the complete electrolysis of the metal complex in the thin solution layer and is a characteristic feature of thin-layer voltammetry. Just as in the case of conventional cyclic voltammetry, a formal redox potential for a reversible couple can be determined from the average of the cathodic (E_{pc}) and anodic (E_{pa}) peak potentials (Eq. 30.2), which gives $E^{0'} = -0.297$ V for the rhenium complex.

Thin-layer electrochemistry offers a simple way of controlling the oxidation state of redox species in a very small volume of solution for simultaneous spectral observation. The redox potential of the thin layer of solution is adjusted precisely by the applied potential as determined by the Nernst equation for a reversible system (Fig. 29.2). Although the applied potential E controls the ratio C_R^s/C_O^s at the electrode surface, the ratio in the thin solution layer quickly adjusts to the same ratio by electrolysis so that $C_R^s/C_O^s = C_R/C_O$.

A *spectropotentiostatic* technique has been developed for obtaining spectra, formal reduction potentials ($E^{0'}$), and electron staoichiometries (n values) of redox couples. The redox couple is converted incrementally from one oxidation state to another by a series of applied potentials for which each corresponding value of C_R/C_O in the thin solution layer is determined spectrally. Each potential is maintained until electrolysis ceases so that the equilibrium value of C_R/C_O is established as defined by the Nernst equation. A Nernstian plot can then be made from the values of E and the corresponding values of C_R/C_O.

Figure 30.21b shows spectra of the rhenium complex in an OTTLE for a series of applied potentials. The spectrum recorded after application of $+0.000$ V maintains essentially all of the complex in the oxidized form ($C_R/C_O < 1/1000$). The spectrum recorded at -0.600 V causes complete reduction ($C_R/C_O > 1000/1$). The intermediate spectra correspond to intermediate values of E. Since the absorbance at 467 nm reflects the amount of complex in the oxidized form (via Beer's law), the ratio C_R/C_O that corresponds to each

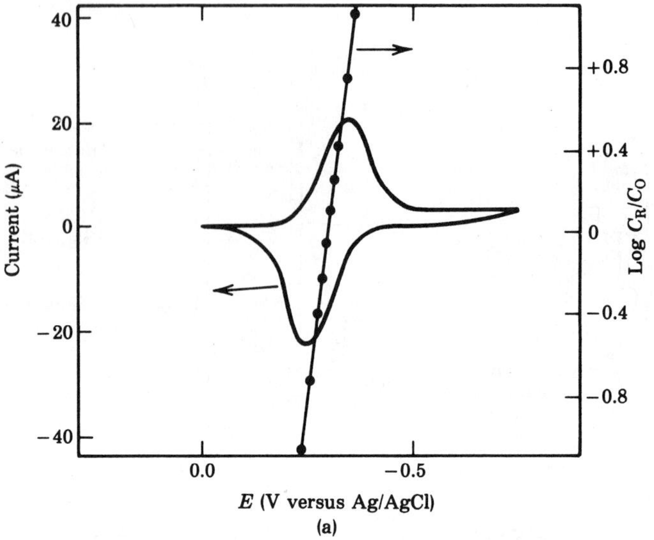

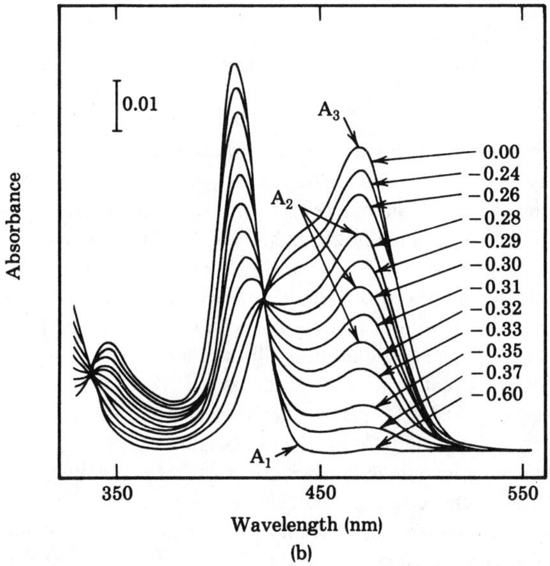

Fig. 30.21 Thin-layer spectroelectochemistry of 0.75 m*M* $[Re(dmpe)_2Br_2]^+$, 0.5 *M* TEAP in DMF. (a) Cyclic voltammogram, scan rate 2 mV s^{-1}, and Nernst plot of data in *b* at 467 nm. (b) Spectra recorded during OTTLE spectropotentiostatic experiment.

value of E can be calculated from the spectra by Eq. (30.25)

$$\frac{C_R}{C_O} = \frac{A_3 - A_2}{A_2 - A_1}. \tag{30.25}$$

As shown in Fig. 30.21b, A_3 is the absorbance when the complex is completely oxidized, A_1 is the absorbance when entirely reduced, and A_2 is the absorbance for a mixture of oxidized and reduced forms.

Figure 30.21a shows a plot of E versus log (C_R/C_O) for the data in Fig. 30.21b. The plot is linear, as predicted by the Nernst equation. The slope of the plot corresponds to an n value of 0.97, and the intercept gives $E^{0'} = -0.305$ V versus Ag/AgCl.

Thin-layer spectroelectrochemical measurements have been made on organic molecules and inorganic metal complexes in aqueous and nonaqueous solvents and on biological materials.

Chronoabsorptometry. *Chronoabsorptometry* is the spectral analog of chronoamperometry and chronocoulometry (Section 30.3). An optical beam of light is directed perpendicularly to an OTE as shown in Fig. 30.19. The cell for this technique is similar to a conventional electrochemical cell in that the electrode is in contact with an electrolyte solution that is much thicker than the diffusion layer adjacent to the electrode. Such a cell is analogous to a standard 1-cm cuvette for UV-visible spectroscopy, with one of the optical faces being a transparent electrode. The excitation signal is a potential step (Fig. 30.22a), and the response signal is an absorbance–time curve (Fig. 30.22b).

Consider the situation in which species O is present in an unstirred solution that is contacting the OTE. Species R is electrogenerated from O by application of the appropriate potential step. If the wavelength is such that R absorbs light, an absorbance-time response of the type shown by the top curve ($k = 0$) in Fig. 30.22b is obtained. The increase in absorbance reflects the generation of R at a rate determined by the diffusion of O to the electrode surface. Essentially, the optical beam monitors the area under the concentration–distance profile for R (Fig. 30.7b). The appropriate Beer's law expression that takes into account the inhomogeneity of concentration of R and the increasing thickness of the "optical cell" (diffusion layer) is

$$A_t = \epsilon_R \int_0^\infty C_{R,x,t}\, dx, \qquad (30.26)$$

where ϵ_R is the molar absorptivity (M^{-1} cm^{-1}) of R and $C_{R,x,t}$ is the concentration distance profile of R, which is changing during electrolysis. Substitution of the diffusion equations that describe $C_{R,x,t}$ gives the following absorbance–time behavior:

$$A = \frac{2}{\pi^{1/2}} \epsilon_R C_O D_O^{1/2} t^{1/2}, \qquad (30.27)$$

where C_O and D_O are the solution concentration and diffusion coefficient, respectively, of the species O from which R is being electrogenerated.

Chronoabsorptometry has been used primarily to measure fast homogeneous chemical reactions of an electrogenerated species. Consider the following EC mechanism:

$$\text{E: } O + e \rightleftharpoons R \qquad (30.28)$$

$$\text{C: } R + Z \xrightarrow{k} O + Z'. \qquad (30.29)$$

The mechanism consists of generation of R at the electrode with reaction of R as it diffuses away from the electrode and encounters Z (Eq. 30.29). When the potential-step experiment

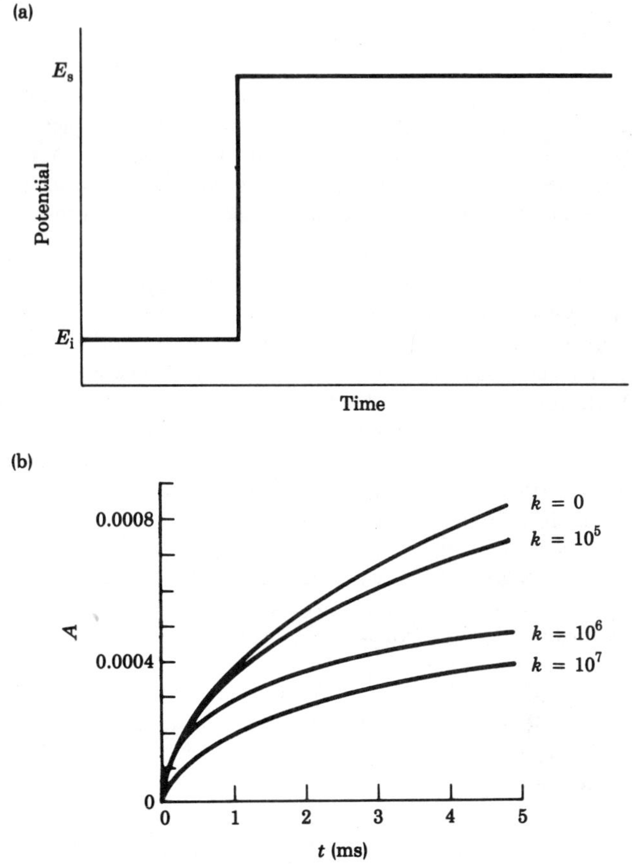

Fig. 30.22 Chronoabsorptometry. (a) Excitation signal. (b) Response signal of absorbance–time curves for the EC mechanism in Eqs. (30.28, 30.29) for various values of k, $M^{-1}\ s^{-1}$. R monitored optically, $D_O = 10^{-6}\ cm^2\ s^{-1}$, $\epsilon_R = 10^4\ L\ mol^{-1}\ cm^{-1}$, concentrations of A and $Z = 1.0\ mM$.

is performed in the presence of Z, the absorbance response caused by generation R is less, owing to its reaction with Z. The absorbance due to R diminishes for increasingly fast reactions, as shown in Fig. 30.22b. The value for k can be calculated from the magnitude by which the absorbance is decreased by the reaction in a manner analogous to that described for chronocoulometry (Section 30.3).

Although this discussion has focused on optical observation of the reaction by a light beam passing through an OTE and the solution, other optical modes can be used to monitor such reactions. In reflectance spectroscopy (Fig. 30.19C) the beam is passed through the solution and reflected from the electrode surface back through the solution. This technique has been used with FTIR to obtain IR spectra of species absorbed on the electrode surface. Internal reflectance spectroscopy (Fig. 30.19D) involves introducing the optical beam through the back side of a transparent electrode at an angle greater than the critical angle so that the beam is totally reflected. Spectral changes in solution near the electrode are observable owing to the small penetration (~500 nm for 500 nm light) of the electric field vector into the solution.

These techniques are capable of monitoring extremely fast reactions with rates up to the diffusion-controlled limit. A variety of chemical reactions have been studied by the various spectroelectrochemical techniques.

Other Spectroelectrochemical Techniques. A beam of excitation light can be passed through an electrochemical cell and the resulting fluorescence (Fig. 30.19F) of electrogenerated species observed. The greater sensitivity of fluorescence enables measurements to be made at lower concentrations of fluorescence species than is possible with absorption spectroscopy.

Spectroscopic techniques based on light scattering have also been coupled with electrochemistry. In Raman and resonance Raman spectroelectrochemistry (Fig. 30.19E) the excitation is by a laser beam directed through the solution at an electrode, and the Raman back-scattering is observed. A particularly important aspect of Raman spectroelectrochemistry is the structural information about electrogenerated species which is contained in a Raman spectrum. The surface-enhanced Raman signal is especially useful for studying species adsorbed on the electrode surface.

Thin-layer electrochemical cells are commonly placed in the sample cavity of an electron spin-resonance spectrometer to record the spectra of electrogenerated radicals.

REFERENCES

References providing a general coverage of voltammetry are:

1. R. N. Adams, *Electrochemistry at Solid Electrodes*. New York: Dekker, 1969.
2. A. J. Bard and L. R. Faulkner, *Electrochemical Methods*. New York: Wiley, 1980.
3. P. T. Kissinger and W. R. Heineman (Eds.), *Laboratory Techniques in Electroanalytical Chemistry*. New York: Dekker, 1984.
4. H. H. Bauer, *Electrodics*. New York: Wiley, 1972.
5. J. O'M. Bockris and A. K. N. Reddy, *Modern Electrochemistry*, Vols. I and II. New York: Plenum, 1970.
6. E. A. M. F. Dahmen, *Electroanalysis*. New York: Elsevier, 1986.
7. P. Delahay, *New Instrumental Methods in Electrochemistry*. New York: Interscience, 1954.
8. Z. Galus, *Fundamentals of Electrochemical Analysis*. London: Horwood, Halsted Press, 1976.
9. I. M. Kolthoff and P. J. Elving (Eds.), *Treatise on Analytical Chemistry*, Part I, Vol. 4, Sec. D-2. New York: Interscience, 1963.
10. J. J. Lingane, *Electroanalytical Chemistry*, 2nd ed. New York: Interscience, 1978.
11. D. D. Macdonald, *Transient Techniques in Electrochemistry*. New York: Plenum, 1977.
12. J. A. Plambeck, *Electroanalytical Chemistry*. New York: Wiley, 1982.
13. D. T. Sawyer and J. L. Roberts, Jr., *Experimental Electrochemistry for Chemists*. New York: Wiley-Interscience, 1974.
14. Southampton Electrochemistry Group, *Instrumental Methods in Electrochemistry*. New York: Halsted Press (Wiley), 1985.

15. H. R. Thirsk and J. A. Harrison, *A Guide to the Study of Electrode Kinetics*. New York: Academic Press, 1972.
16. B. H. Vassos and G. W. Ewing, *Electroanalytical Chemistry*. New York: Wiley-Interscience, 1983.
17. A. Weissberger and B. W. Rossiter (Eds.), *Techniques of Chemistry*, Vol. I: Physical Methods of Chemistry, Part II. New York: Wiley-Interscience, 1971.
18. State of the Art Symposium: Electrochemistry. *J. Chem. Ed.*, **60,** 258–340 (1983).
19. W. H. Smyrl and F. McLarnon (Eds.), *Proceedings of the Symposium on Electrochemistry and Solid State Science Education at the Graduate and Undergraduate Level*. Pennington: The Electrochemical Society, 1987.

Treatments of the practical aspects of performing voltammetric experiments are found in references 3 and 13 given above.

Comprehensive discussions of advances in voltammetry appear regularly in the following continuing series:

20. *Advances in Analytical Chemistry and Instrumentation*, Vol. I, Interscience, 1960.
21. *Advances in Electrochemistry and Electrochemical Engineering*, Vol. 1, Interscience, 1961.
22. *Electroanalytical Chemistry. A Series of Advances*, Vol. 1, Dekker, 1966.

Thorough discussions of the principles and techniques of polarography are available in:

23. A. M. Bond, *Modern Polarographic Methods in Analytical Chemistry*. New York, Dekker, 1980.
24. J. Heyrovsky and J. Kuta, *Principles of Polarography*. New York: Academic Press, 1966.
25. I. M. Kolthoff and J. J. Lingane, *Polarography*. 2d ed., Vol. 1. New York: Interscience, 1952.
26. L. Meites, Voltammetry at the dropping mercury electrode, in *Treatise on Analytical Chemistry*, Pt. I, Vol. 4, I. M. Kolthoff, P. J. Elving, and E. B. Sandell (Eds.). New York: Interscience, 1963.

Some references dealing with specialized aspects of polarography are:

27. P. Zuman, *Organic Polarographic Analysis*. New York: Macmillan, 1964.
28. W. F. Smyth, *Polarography of Molecules of Biological Significance*. New York: Academic Press, 1979.
29. S. G. Mairanovskii, *Catalytic and Kinetic Waves in Polarography*. New York: Plenum, 1968.

References on stripping voltammetry are:

30. E. Barendrecht, Stripping voltammetry, in *Electroanalytical Chemistry*, Vol. 2. A. J. Bard (Ed.). New York: Dekker, 1967.
31. F. Vydra, K. Stulik, and E. Julakova, *Electrochemical Stripping Analysis*. New York: Halsted Press, 1976.
32. J. Wang, *Stripping Analysis: Principles, Instrumentation, and Applications*. Deerfield Beach, Fl: VCH Publishers, 1984.

References on spectroelectrochemistry are:

33. T. Kuwana and N. Winograd, Spectroelectrochemistry at optically transparent electrodes. I. Electrodes under semi-infinite diffusion conditions, in *Electroanalytical Chemistry*, Vol. 7, A. J. Bard (Ed.). New York: Dekker, 1974.
34. W. R. Heineman, F. M. Hawkridge, and H. N. Blount, Spectroelectrochemistry at optically transparent electrodes. II. Electrodes under thin layer and semi-infinite diffusion conditions and indirect coulometric titrations, in *Electroanalytical Chemistry*, Vol. 13, A. J. Bard (Ed.). New York: Dekker, 1984.
35. R. L. McCreery, Spectroelectrochemistry, in *Physical Methods in Chemistry*, Vol. 2, B. Rossiter (Ed.). New York: Wiley, 1987.

EXERCISES

30.1 Iodate (IO_3^-) undergoes a 6-e reduction to iodide at a HMDE. Calculate i_{pc} for the forward scan of a cyclic voltammogram for $D_{IO_3^-} = 1.0 \times 10^{-5}$ cm^2 s^{-1}, $C_{IO_3^-} = 1.40$ mM, $A = 3.0$ mm^2, and $\nu = 6$ V min^{-1}.

30.2 Discuss the uncertainties in measuring an n value by cyclic voltammetry.

30.3 A cyclic voltammogram for *p*-aminophenol in 0.01 *M* H_2SO_4 at a carbon paste electrode at a scan rate of 3.33 V/min is shown in Fig. 30.4. Calculate the first-order rate constant for hydrolysis of the electrogenerated quinoneimine from the working curve shown in Fig. 30.5. Assume $\tau = 0.1$ V/ν.

30.4 4-hydroxy-propranolol is a metabolite of the β-adrenergic blocking agent propranolol. A cyclic voltammogram of 4-hydroxy-propranolol in 0.1 *M* $HClO_4$ at a carbon paste electrode at a scan rate of 200 mV s^{-1} is shown below.

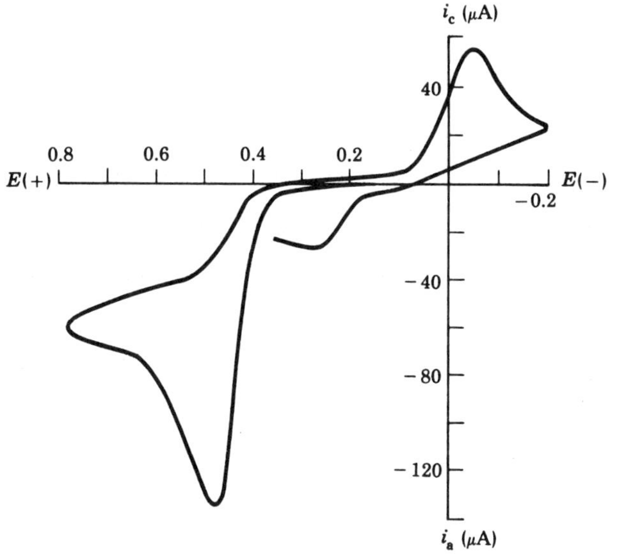

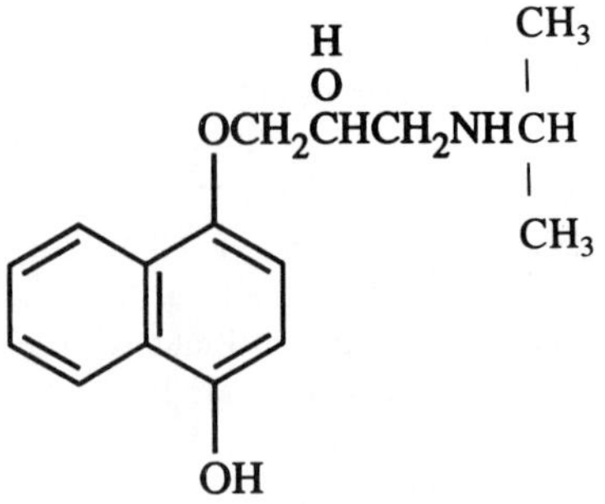

4-Hydroxy-propanolol

What electrode mechanism would be consistent with this voltammogram? Label the peaks on the voltammogram to indicate the electrode reactions and any coupled solution reactions.

30.5 Given the following mechanism for the oxidation of the catecholamine epinine in 0.1 *M* phosphate buffer, pH 7.36 at a carbon paste electrode,

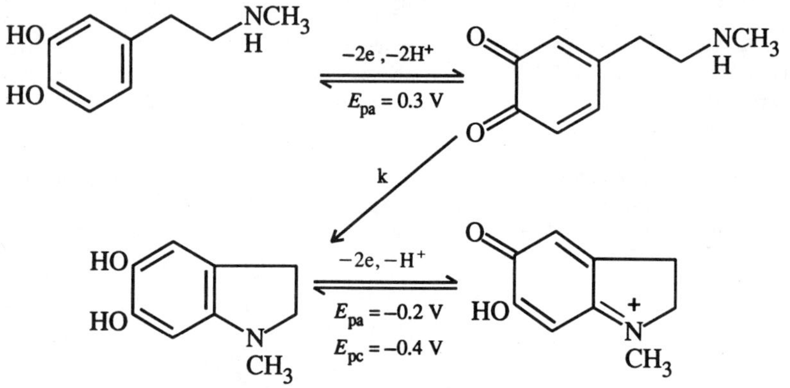

Draw the expected cyclic voltammogram.

30.6 Explain how cyclic voltammetry might be used to distinguish between an electroactive species that is adsorbed on the electrode surface and an electroactive species that is dissolved in solution.

30.7 Calculate the value of $it^{1/2}$ expected for a chronoamperometry experiment performed on an electroactive species assuming $D = 1.0 \times 10^{-6}$ cm^2 s^{-1}, $C = 1.0$ m*M*, $A = 1.00$ cm^2, and $n =$ 1e.

30.8 A chronoamperometry experiment was performed at a hanging mercury drop electrode. Do you expect the current observed at a spherical electrode to be larger, smaller, or the same in comparison to a planar electrode of the same surface area? Explain.

30.9 The cyclic voltammogram obtained on a compound is electrochemically irreversible as shown in Fig. 30E.9 (——). A comparison reversible voltammogram is also shown (----).

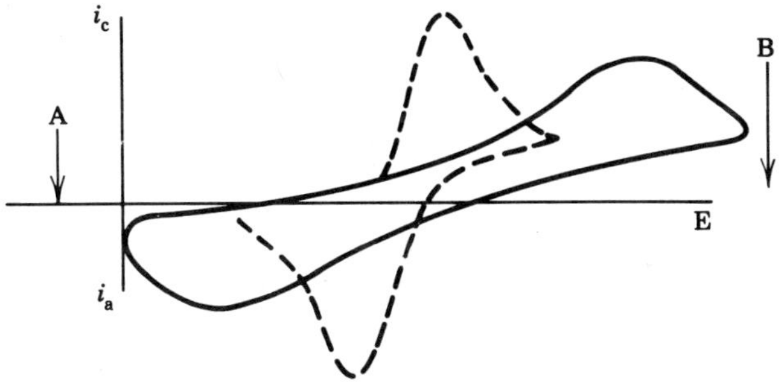

A chronoamperometry experiment is performed by stepping from potential A to B. Do you expect the resulting chronoamperogram to obey the Cottrell equation? Explain.

30.10 Calculate the value of Q_{ads} that would be expected for chronocoulometry of an electroactive species that absorbs on the electrode to give monolayer coverage assuming the area covered by a single absorbed molecule is defined by a rectangle with the dimensions 4.4×11.0 A, $A = 0.027\ cm^2$, and $n = 1$.

30.11 Devise a chronocoulometric experiment for measuring the diffusion coefficient of Tl in mercury.

30.12 The pulse polarogram in Fig. 30E.12 is obtained for a solution containing 1 mM each of L^- and Br^- in 0.5 M Na_2SO_4 supporting electrolyte.

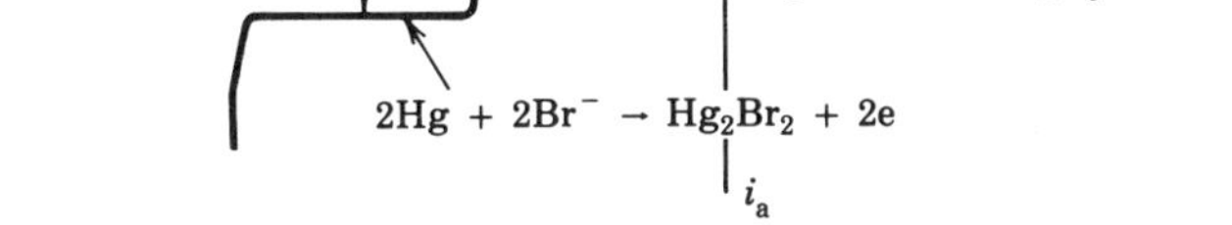

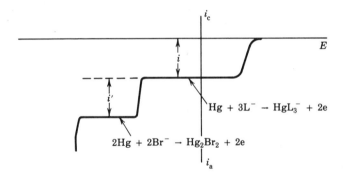

Show what the relative magnitudes of i and i' are.

30.13 A DME in the presence of hydroxide gives a reversible anodic wave in pulse polarography. The electrode process is $Hg + 2OH^- \rightarrow Hg(OH)_2 + 2e$ where $Hg(OH)_2$ is soluble for low OH^- concentrations. Derive the i-E equation for the polarographic wave. Show the equation for $E_{1/2}$. How does $E_{1/2}$ shift with decreasing OH^- concentration?

30.14 A 5×10^{-4} M solution of Pb^{2+} in 1.0 M KCl solution gives a polarogram with an average diffusion current of 2.81 μA. The dropping rate of the DME was found to be 24 drops a minute and when 20 drops were collected they weighed 0.0750 g. Calculate the diffusion coefficient D of Pb^{2+}.

30.15 Sketch the amperometric titration curve obtained at a dropping mercury electrode for the titration of a Ni(II) sample with dimethylglyoxime, DMG, according to the following precipitation reaction: $Ni^{2+} + 2\ DMG \rightarrow Ni(DMG)_2(s)$. The applied potential is −1.85 V. The $E_{1/2}$ for reduction of Ni(II) is −1.1 V; the $E_{1/2}$ for reduction of DMG is −0.8 V. The precipitate $Ni(DMG)_2(s)$ is electroinactive.

30.16 A 25-mL solution containing an unknown Pb^{2+} concentration gave a differential pulse polarographic diffusion current of 2.00 μA for Pb^{2+} reduction. Direct addition of 5 mL of 1.00 m*M* Pb^{2+} standard to the unknown sample in the polarographic cell gave a new diffusion current of 2.82 μA. Calculate the Pb^{2+} concentration in the sample.

30.17 Draw a square wave polarogram for a mixture of 1 m*M* Cd^{2+} ($E_{1/2} = -0.64$ V versus SCE) and 3 m*M* Pb^{2+} ($E_{1/2} = -0.44$ V versus SCE) in 1.0 *M* KCl. Label E_p and i_p for each wave. Clearly show the relative values of i_p. Write the electrode reaction for each wave.

30.18 The half-wave potential for the reduction of Pb^{2+} to lead amalgam by polarography is −0.405 V versus SCE. What would be the half-wave potential of the complex PbY^{2-} if a 1.00×10^{-4} solution of Pb^{2+} were made 1.00×10^{-2} *M* in Y^{4-} ion as well? (Y^{4-} is the ligand EDTA; the formation constant for PbY^{2-} is 1.1×10^{18}.)

30.19 The $E^{0'}$ and *n* of a metal complex are to be determined by the optically transparent thin-layer electrode technique. The spectra in Fig. 30E.19 were obtained at the potentials indicated on the figure.

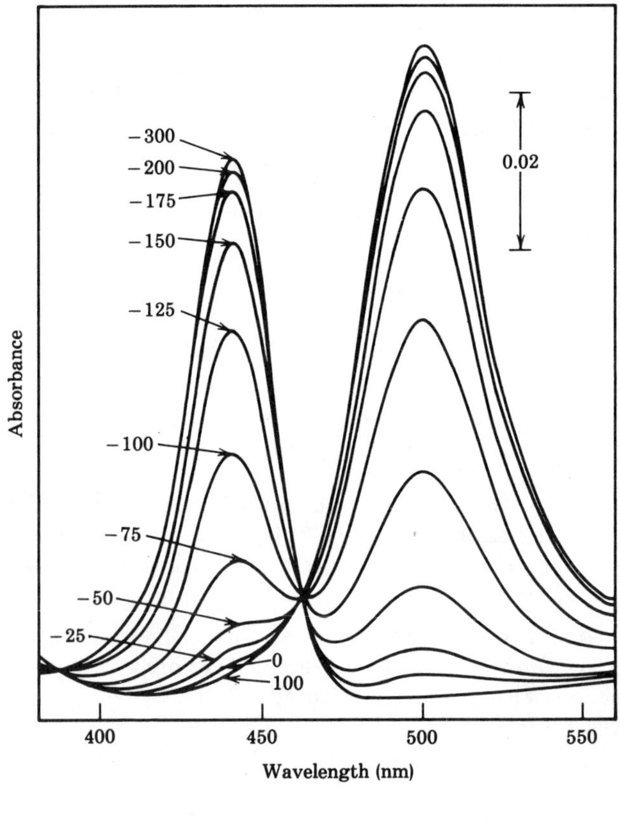

Draw the appropriate Nernstian plot and calculate $E^{0'}$ and n for the complex.

30.20 The results of two chronoabsorptometry experiments are shown below. Curve a represents the reduction O + $ne \rightleftharpoons$ R where R is optically monitored. Species Z was then added to solution and curve b obtained. In both cases the concentation of O is 1 m*M*. For curve b the concentation of Z is 1 m*M*. A chronoamperometry experiment performed on a 1-m*M* solution of O gave a diffusion coefficient of 1.00×10^{-5} cm^2 s^{-1}. The kinetic working curve for the EC mechanism for the reaction of Z with electrogenerated R is shown below. (a) Calculate ϵ_R. (b) Calculate the rate constant for the reaction of R with Z.

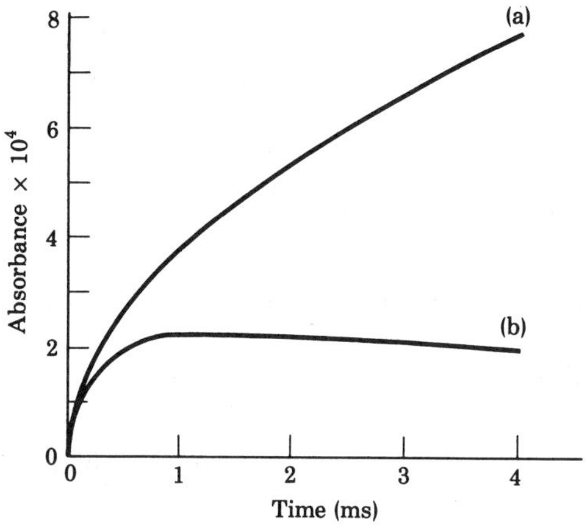

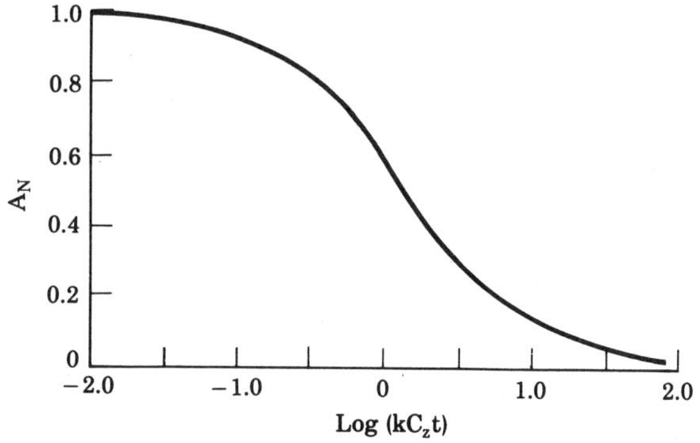

ANSWERS

30.12 i/i′ = 2/3 *30.14* 9.20×10^{-6} cm^2 s^{-1} *30.16* 0.289 mM *30.18* −0.879 V versus SCE

Chapter 31

COULOMETRIC METHODS

31.1 INTRODUCTION

This chapter deals with three procedures based on exhaustive electrolysis—constant-current coulometry, controlled potential coulometry, and electrogravimetry. Contrasting these techniques with voltammetry, we find they not only call for complete reaction of the substance(s) of interest but also require much larger currents. They are capable of much higher precision, $\pm 0.1\%$ or better, but are more limited in scope because of the restrictions imposed on the electrolysis process.

The effectiveness of coulometric techniques stems from their basic reliance on electrolysis. Recall that Faraday's laws of electrolysis relate the quantity Q of charge used to the weight W of substance that is electrolyzed by the expression

$$W = \frac{QM}{nF} \tag{31.1}$$

where M is the formula weight, n is the number of electrons required per molecule, and F is the Faraday (96,487 C mol^{-1}). Charge Q is in coulombs. In a legitimate sense, we can regard a "coulometric" reaction as a titration of an electroxidizable or reducible species in which electrons are the titrant. We measure the quantity of titrant as the number of coulombs. With the passage of each Faraday of electricity, one equivalent weight of electroactive species reacts.

To make a coulometric analysis feasible, the following basic conditions must be satisfied:

1. The substance of interest must be electrolyzed with 100% current (reaction) efficiency.
2. The number of coulombs (or the weight of product) must be precisely determined.

It is the first condition that essentially limits the range of substances determinable by coulometric procedures. It is discussed at some length in the following section. The second condition calls for an appropriate instrumental solution. Since the number of coulombs Q is given by $Q = \int i\, dt$, either the integral must be obtained directly or i or t separately determined and then combined. These two approaches have given rise to controlled-potential coulometry, which measures the integral, and constant-current coulometry, which determines the time of electrolysis with a steady current.

Particular advantages of coulometry over other techniques are (a) high accuracy, (b) the possibility of generating and working with reagents like iron (II) that are unstable under ordinary conditions, (c) simplicity of instrumentation, and (d) good sensitivity. High accuracy is possible because of the electrical character of the measurements. Such determinations are intrinsically accurate. It is always possible to determine the number of electrons exchanged by a measurement external to the cell, and no standardization is required. Indeed, precision in coulometry is high enough for the Faraday to have been seriously proposed as the primary chemical standard. Standardization of solution by coulometry is capable of accuracy at least as great as, and probably greater than, by classical wet methods.

The simplicity of much coulometric instrumentation is a great advantage. The end point need only be detected and the circuit broken. Finally, the sensitivity of coulometric measurements is good, for by use of small currents and microapparatus, the precision of results is good even at sample concentrations of 10^{-4} M and below.

31.2 COULOMETRIC RELATIONSHIPS

In considering the fundamentals on which coulometry rests it is convenient to limit discussion to reductions though the approach is equally valid for oxidations. To convert from reductions to oxidation sometimes requires a change in the working electrode (e.g., substitution of platinum for mercury in order to achieve sufficiently positive potentials). It also calls for current reversal, but this will occur naturally if the potential is altered in a positive direction.

The theory of electrode processes and mass transport developed in Sections 29.2–29.4 is essential background to this chapter. Two basic questions must be dealt with. In carrying out quantitative coulometry

1. What potential control is necessary to reduce a species of interest with minimal interference?
2. How is it possible to control the system to bring the time of electrolysis down to a half-hour or less?

The questions are relevant both to controlled-potential and constant-current coulometry. Other questions will mainly relate to instrumentation and will be treated in the following sections.

Electrode Potential Change During Electrolysis. During an exhaustive reduction, a working cathode becomes steadily more negative. How great a change in potential we need to electrolyze a substance quantitatively when a constant, small current flows we may compute from the Nernst equation (see Section 29.2). While this situation may appear idealized, it corresponds to the final condition that obtains in controlled-potential coulometry. Attention is given later to the effect on potential of the larger currents usually employed in constant-current coulometry.

Let the initial concentration of the oxidized form of the analytical species be C_0. From the Nernst equation, the initial potential E' of the working electrode at 25°C is given approximately by

$$E' = E^{0'} + \frac{0.0591}{n} \log \frac{C_0}{10^{-3}C_0} = E^{0'} + \frac{0.177}{n}. \tag{31.2}$$

Two approximations have been made. Activity coefficients have been neglected since a formal potential (see Section 27.3) $E^{0'}$ is substituted for the standard potential. Also, a meaningful equilibrium potential was assumed first established when the concentration of the reduced form reaches $10^{-3} \times C_0$. After completion of reduction the concentration ratio of the analytical species is reversed giving

$$E' = E^{0'} + \frac{0.0591}{n} \log \frac{10^{-3}C_0}{C_0} = E^0 - \frac{0.177}{n}. \tag{31.3}$$

The range of voltage ΔE for electrolysis is thus 0.35 V for a one-electron reduction, 0.18 V for a two-electron reduction, and 0.12 V for a three-electron reduction, assuming equal concentrations of each species.

Calculating the final electrode potential that must be reached for an electrolysis to be quantitatively complete has immediate value in determining whether another reducible species present will interfere. The interference takes the form of a reduction of a small quantity of a second constituent during the determination of the first. How great an error results depends on the difference between their formal potentials (or standard electrode potentials) and on their relative concentrations, as indicated in the following example. During coulometry the potential of the working electrode need not be held constant but rather must not be allowed to exceed a value at which there will be sufficient reduction of the next most easily reduced species to constitute a serious error.

Example 31.1. Metal B is to be determined coulometrically in an alloy of metals A and B containing about 95% A and 5% B. Pertinent data are that in 1 M H_2SO_4 $E^{0'}$ for $A^{2+} + 2e \rightarrow A\,(Hg)$ is −0.50 V and $E^{0'}$ for $B^{2+} + 2e \rightarrow B\,(Hg)$ is −0.362 V. What error will be incurred in the determination if the potential of the working electrode is allowed to become as negative as −0.450 versus SCE during electrolysis?

The potential selected ensures a quantitative determination of B since $E^{0'} - E_{\text{final}} =$ 0.088 V, half the range required for a two-electron reduction. Because the final potential falls close to $E^{0'}$ for A, some A will also be reduced. The fraction of A reacting can be deduced from the Nernst equation. Let [ox]/[red] now refer to A only. Then $-0.45 = -0.50 + (0.059/2) \log ([ox]/[red])$, and $\log ([ox]/[red]) = 0.05/0.0296 = 1.7$, or $[ox]/[red] = 50$. About 2% of A was reduced: since its initial concentration was about 5%, that of metal B, the error from this source in determining B is about $0.02 \times 0.05 =$ 0.001 or 0.1%.

Current Efficiency and Redox Buffering. When we have to maintain a relatively large current (10–500 mA), as in constant-current coulometry, the working electrode swings to a still more extreme voltage than the Nernst equation predicts before an electrolysis is completed. The reason is that we must increase the potential of the electrode above its equilibrium voltage (see Section 27.7). As a result, it is common to find toward the end of a determination that electrolysis of a second species has begun and that current efficiency has dropped seriously below 100%.

A good example of this situation is the direct electrolytic oxidation of iron (II) at a platinum anode. The E^0 for the reaction $Fe^{2+} \rightarrow Fe^{3+} + e$, is +0.53 V versus SCE. It proves to be the case that the direct electrolysis of ferrous ion cannot be completed without oxidizing some water.*

Example 31.2. Current-voltage curves for the coulometric oxidation of iron (II) are graphed in Fig. 31.1. Only curves *A* and *C* are applicable unless cerium (III) salt has been added. Further, these are an initial set of curves; as iron is oxidized, curve *A* sinks steadily nearer the abscissa. When all iron (II) has disappeared, the iron wave vanishes and only curve *C* remains.

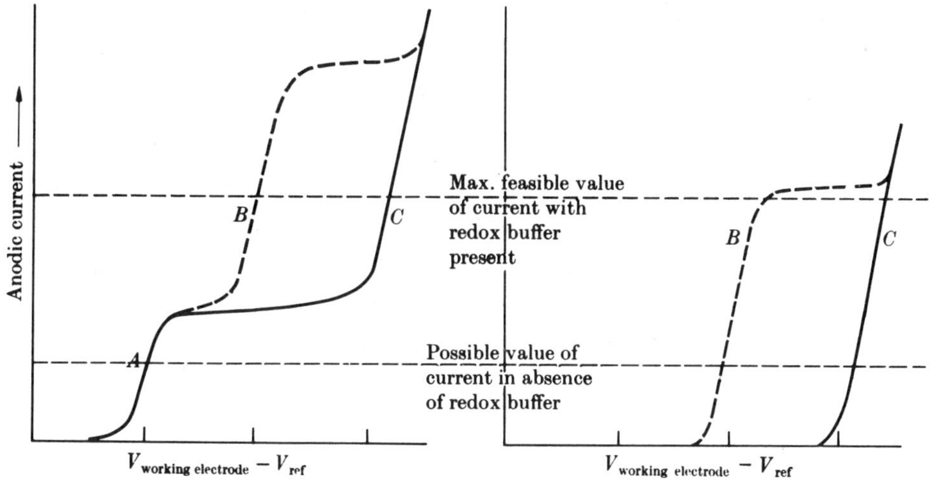

Fig. 31.1 Current–voltage curves applicable to the determination of Fe^{2+} by aqueous constant-current coulometry. Curve A: Fe^{2+}, Fe^{3+} couple; curve B: Ce^{3+}, Ce^{4+} couple; curve C: H_2O, O_2 couple. As electrolysis of Fe^{2+} proceeds the set of curves, in effect, moves down toward the abscissa. Two cases can be distinguished. (1) When no Ce^{3+} is added to the system, curve A–C is applicable. If current is fixed at the value given by the lower broken line, only Fe^{2+} is oxidized at first as in (a) but water is also oxidized before all Fe^{2+} is electrolyzed as in (b). (2) When Ce^{3+} is added, A–B is the appropriate composite curve. The electrolysis current can now be maintained at the value of the upper broken line without bringing about electrolysis of water.

*It is possible for the oxygen produced to react with remaining ferrous ion but improbable that it will do so rapidly enough to be helpful in this situation.

We can now show why it is impossible to complete this direct oxidation with 100% current efficiency. If the current is set at the lower dashed line in Fig. 31.1 initially, it is only a matter of time before curve *A* moves downward enough for the current line to intercept curve *C*. Then any iron (II) remaining is oxidized simultaneously with water. If this occurs, the value of iron determined is high.

As a consequence of the nearly certain loss of current efficiency in direct methods, nearly all constant-current methods call for the electrical generation of an intermediate species with 100% current efficiency. It must be capable of quantitative reaction with the analytical species. If an oxidation is sought, the intermediate must be added initially in its reduced form. It is of interest that now the potential of the working electrode remains almost constant. The added redox system is called a *redox buffer* and the overall method now becomes *secondary* constant-current coulometry.

The role of a redox buffer may be clarified by returning to the example of coulometric oxidation of ferrous ions represented in Fig. 31.1. Assume the redox buffer is Ce^{3+}, Ce^{4+}. Then a large concentration of cerous ion Ce^{3+} is added to the solution before beginning the titration. In Fig. 31.1 curve *A* is now followed by curve *B*. As current is passed, ceric ion, Ce^{4+}, is electrically generated and easily oxidizes the iron (II) according to the equation $Ce^{4+} + Fe^{2+} \rightarrow Fe^{3+} + Ce^{3+}$. The current can now be raised to the level of the upper dashed line in the figure. Clearly, the larger the concentration of Ce^{3+} added, the higher this current may be. Not only is the time of electrolysis shortened substantially, but current efficiency remains at 100%. As Fig. 31.1*b* shows, the working electrode is at all times at a potential below that required to produce oxygen from water. The number of coulombs used in secondary coulometry is, of course, identical to that required in the direct oxidation of iron.

The few exceptions to the use of an intermediate couple in constant-current coulometry are cases in which hydrogen or hydroxyl ions are the generated titrant and there is no concentration polarization. If the substance to be electrolyzed behaves irreversibly at the electrode, it may be necessary to titrate it entirely with generated reagent. Note that precipitation and other types of titrations also call for the generation of a titrant.

In constant-current coulometry, the current must be terminated when the desired electrolysis is complete. The necessary end-point determination is discussed in the next section.

Current Change at Constant Potential. Since it is possible to predict a potential suitable for coulometry of a species, it is also important to consider the current variation if this potential is held throughout. Where a single electrode reaction is occurring, not only is the form of the dependence of current electrode potential of interest but the factors that determine the magnitude of the current need to be identified. If we know these, we can shorten the electrolysis time. The feasibility of a particular controlled-potential coulometric determination often depends on the time required.

When the working electrode is adjusted to a potential at which a single-reaction occurs at a fast rate, electrolysis current is diffusion-controlled. With rapid stirring the diffusion layer is thin, and it can be shown that the current i_t at a given time t is given by the equation

$$i_t = i_0 \times 10^{-kt}, \tag{31.4}$$

where i_0 is the initial current and k is a constant defined [6] as

$$k = 0.43 \frac{DA}{v\delta}.$$

Here A is the area of electrode, D is the diffusion coefficient of the species being reduced, v is the solution volume, and δ is the thickness of the diffusion layer. Since the parameters D and δ depend on a great many variables, such as the temperature and rate of stirring, the constant k is best evaluated for each experimental situation. The time required for a quantitatively complete electrolysis can be estimated from Eq. (31.4).

Example 31.3. It is sought to estimate the length of time required to effect 99.9% completion of a controlled-potential electrolysis for which $k = 0.0025\ \text{s}^{-1}$. A criterion by which the percentage of reaction can be related to the electrolysis current must first be formulated. As discussed in Section 29.2, for a thin diffusion layer the electrolysis current i is proportional to the bulk concentration of the electroactive species if all the active ions are reduced as fast as they arrive. The current at a given time t is then a direct measure of the concentration of the reducible species left in the bulk of the solution at that time, and the current i_0 should be a measure of the original concentration. Then the percentage remaining unelectrolyzed should be given by the ratio i_t/i_0.

The time for 99.9% complete-electrolysis corresponds to a value of i_t/i_0 of 0.001. This value can be substituted in Eq. (31.4) and the magnitude of t determined. The substitution yields

$$0.001 = 10^{-0.0025t}, \qquad \text{or } 3 = 0.0025t,$$

and t = 1200 s or 20 min. To obtain 99.99% completion, about 27 min would be required, a time which is a third again as long.

From Eq. (31.4) it is apparent that the larger the value of k, the shorter the electrolysis time. Accordingly, for fast electrolysis it is advantageous to use large, closely spaced electrodes. Thus area A is large and volume v is small. Higher temperatures increase D and, up to a point, more rapid stirring reduces δ. Changes in the direction stated shorten an electrolysis. As the above example suggests, a practical criterion for the termination of this type of coulometric determination is that the final current be at least three orders of magnitude less than the initial value.

31.3 CONSTANT-CURRENT INSTRUMENTATION

The essential components for a coulometric titrator are shown in Fig. 31.2 as a constant-current supply with timer, an electrolysis cell, and an end-point detection device. During a determination the electrolytic generation of titrant continues unabated until the end point of the desired reaction is sensed by the end-point detector. Then current and timer are stopped.

A simple constant-current source can be made from a regulated dc power supply that furnishes a current at least 250 mA at a constant voltage of 200 V. It becomes a constant-current generator when used in series with a resistance much larger than that of the electrochemical cell. Virtually all of the iR drop appears across it and no more than 0.5–3 V across the cell. Usually there is a change of 0.2–0.4 V in cell voltage during a determination as the working electrode potential shifts. Since this represents no more than 0.1–2.0% of the output voltage of the power supply the cell current varies by no more than this figure.

Ordinarily the current used in smaller than 200 mA to avoid the generation of excessive heat. The voltage drop across the cell is self-adjusting as long as its internal resistance is small ($< 10\ \Omega$) and, within limits, is whatever value is necessary to initiate and maintain electrolysis. If feasible, we select a current that allows the electrolysis of the analyte to be completed in from 10 to 200 s. If it is possible to adjust the current to values of from 0.1 to 100 mA, it is possible to electrolyze from 1×10^{-9} to 1×10^{-6} equivalents of a substance per second. The size and concentration of sample that can be accommodated we can quickly estimate on the basis of the time and current restriction. It is clear that the method is particularly well adapted to quite dilute systems. The current may usually be ascertained to 0.1% or better and the time to 0.1 s.

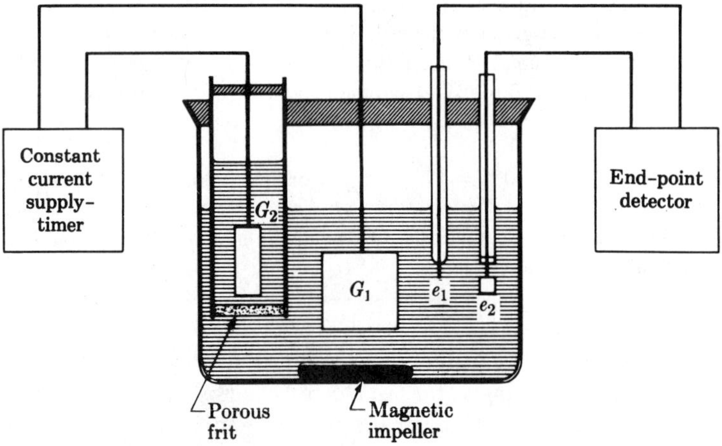

Fig. 31.2 Cell and electrodes for coulometry with potentiometric detection of end points. Potentiometric electrodes e_1 and e_2 monitor the concentration of reactant through the end-point detector circuitry. Coulometric electrodes G_1 and G_2 are the generating and auxiliary electrodes, respectively, operating from the constant current supply and timer.

End-Point Detector. The sensitivity and precision of the end-point device is often crucial to the successful operation of a coulometer. As we have already noted, it is not difficult to maintain and to determine the value of the current and time to within 0.1%. The end-point device that shuts off the current must operate with comparable precision or the reliability of the other components is wasted. While any standard volumetric end-point detector may be installed only potentiometric and amperometric detectors are used to any extent because they permit straightforward feedback control of electrolysis current. Potentiometric detectors are commonly employed in redox and acidimetric reactions (see Sections 28.13, 28.15, and 29.13), but adaptation for amperometric control is straightforward and follows expected lines. Amperometric detectors are more sensitive and find use especially in more dilute systems (see Section 29.12).

Cell and Generating Electrode. Figure 31.2 shows the cell for coulometric determination in more detail. One of the monitoring electrodes (e_1 or e_2) should be an indicator electrode sensitive to the oxidized or reduced form of the electroactive species, and the other a suitable reference electrode (see Section 27.4).

Usually the platinum generating electrode is used because of its inertness. If halides are being determined, the generating electrode may be silver, and the precipitation of halide ion by silver ion may be the basis for the reaction. In any case, the auxiliary or counter electrode is carefully shielded from the solution by the frit to avoid unneccessary side reactions and any loss of reactant through mechanical trapping. Ordinarily the chamber surrounding the auxiliary electrode is even filled with an electrolyte different from that in the bulk of the solution to isolate it more effectively. Since the generated titrant is often unstable, it is preferable to carry out most of these titrations in a closed vessel.

In some instances we cannot maintain 100% current efficiency with the arrangement shown in Fig. 31.2 because of undesirable side reactions occurring when the titrant is generated in the solution. The best alternative is external generation. In this procedure, the generating electrode is in a capillary tube that extends into the solution. The earliest design (Fig. 31.3) is representative of the type of arrangement commonly used. Most designs deliver only an oxidant or a reductant and are better capable of dissipating heat associated with large currents. The generated titrant, as in Fig. 31.3, is moved from the appropriate outlet of the capillary into the analysis flask by a flow of solution, usually by gravity, through the capillary at a constant rate of 0.1 mL min^{-1} or greater. One disadvantage of external generation is that a small amount of unreacted titrant is always in the capillary when the reaction is stopped. It is not difficult to determine the appropriate correction by running a blank, however, and precision can still be good.

Automatic Titrators. The operation of the device in Fig. 31.2 may be made automatic rather simply. If the electrode and other reactions are fast, all that is basically necessary is to add a relay that stops the current and the timer when the potential reaches the end-point value. Note the contrast between this operation and the termination of a titration performed by an automatic potentiometric or photo-

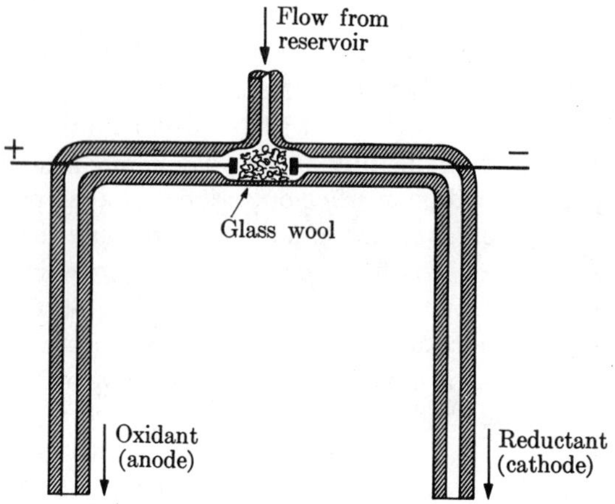

Fig. 31.3 An arrangement for external generation of reagent. The appropriate outlet is inserted in the titration flask.

metric titrator. Since the flow of a solution need not be cut off in coulometry, instrumentation can be simpler and more reliable. In effect, the coulometric apparatus has taken the place of the buret and standard solution required in other titrations.

Unfortunately, it is impossible to avoid some additional complexity if a coulometric titrator must also cope with reactions in which the reactant is slow to combine with the titrant. To achieve good precision ($\pm 0.5\%$) with these systems we have to add some sort of anticipation device (see Section 28.12). Its function, as in a potentiometric titrator, is to interrupt the generation of titrant briefly when the potential of the indicating electrode e_1 nears its end-point value. When the electrode voltage changes as the result of further reaction, the current is started again, and new titrant is generated. There may be several interruptions before the true end point is attained.

31.4 CONTROLLED-POTENTIAL INSTRUMENTATION

Straightforward designs are required for two modules: a precise current integrator with a wide dynamic range and a precise automatic device to control the potential of the working electrode without interference from other elements in the cell.

Current Integrator. The operational amplifier integrator discussed in Section 4.7 is the basis for a good coulometric integrator. With a good-quality integrating capacitor (e.g., C in the circuit of Fig. 4.18), adequate shielding of the input lead, and careful control of dc offset of the amplifier, we can obtained precise integration. In coulometry we must give particular attention to sources of error such as faradaic current from electrolysis of impurities and other errors that arise from the electrolysis cell.

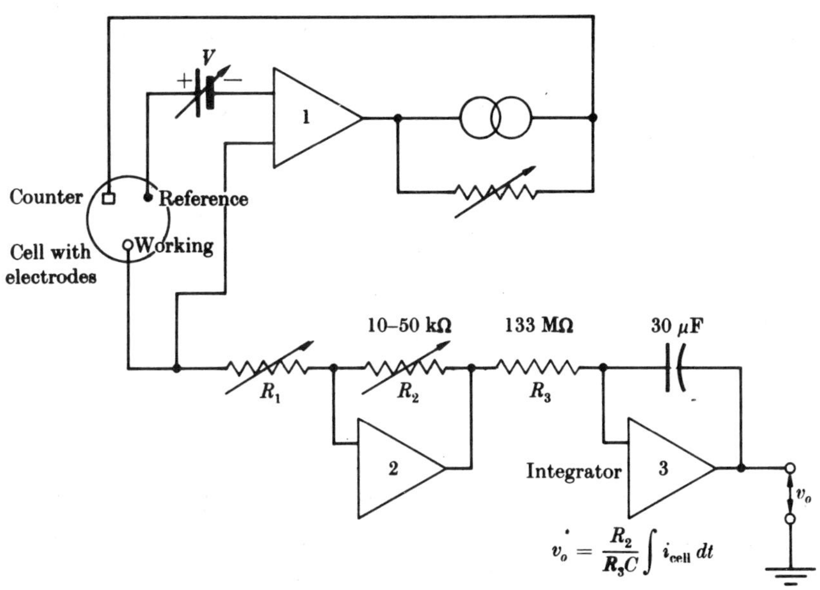

Fig. 31.4 Schematic diagram of a circuit for controlled potential coulometry. The potentiostat (upper circuit) maintains the working electrode at exactly—V volts relative to the reference electrode. The lower circuit amplifies the cell current, isolates the cell (OA2 plus R_1 and R_2 serve these functions), and integrates the varying current with respect to time (OA3 plus R_3 and the capacitor perform this operation). In the upper circuit the overlapping circles symbolize a current source, perhaps a booster amplifier. In the lower circuit the noninverting terminals (not shown) of OA2 and OA3 are grounded. (G. L. Bowman, *Anal. Chem.* 29, 213 (1957); courtesy of *Analytical Chemistry*.)

Potentiostat. Operational amplifier devices for maintaining a fixed potential were treated in Section 4.4. Figure 31.4 shows the circuit of Booman for a controlled-potential coulometric system. Note that the potentiostat feedback loop (upper branch of circuit) provides the control. In effect, amplifier 1 provides sufficient current through the counter electrode to maintain electrolysis and keep the potential of the working electrode at $-V$ volts relative to the reference electrode (about at ground potential).

31.5 COULOMETRIC PROCEDURES

Table 31.1 partially indicates the scope of constant-current coulometry; it lists some electrically generated reagents that have seen a significant amount of use. To perform a particular analysis, it is necessary to select generating electrodes, end-point procedure and electrodes, the value of the electrolysis current, and the sample size or concentration. Since the last two variables are interrelated, some latitude is possible in each. The end-point detection arrangement chosen must provide adequate sensitivity at the concentration level decided on. If a dilute solution

Table 31.1 Some Electrolytically Generated Titrants

Oxidizing Agents	*Reducing Agents*	*Precipitants*	*Acidimetric Agents*
Br_2	$CuCl_2^-$	Ag^+	OH^-
I_2			
Cl_2	Fe^{2+}	Hg_2^{2+}	H^+
BrO_3^-			
Ce^{4+}	Ti^{3+}	$Fe(CN)_6^{4-}$	
Fe^{3+}			
Ag^+			
Mn^{3+}			

or a solution with a large portion of nonaqueous solvent is being analyzed, some electrically indifferent electrolyte such as KCl must be added until the cell resistance is decreased to about 10 Ω. (A great many good coulometric procedures have been described by Milner and Phillips [6].) Two brief examples are given here as illustrations of typical procedures.

Example 31.4. The mercaptan content of petroleum stocks may be measured in straightforward fashion by coulometry. (If hydrogen sulfide is also present, additional steps are required.) The hydrocarbon sample is dissolved in a mixture of aqueous methanol and benzene to which ammonia and ammonium nitrate are added to buffer the solution and lower its resistance sufficiently to allow the use of moderate electrolysis currents.

A silver anode generates the titrant, Ag^+ ions, and a platinum electrode immersed in an ammonium nitrate solution serves as an auxiliary or counter electrode. As Ag^+ ions are generated, they precipitate the mercaptide ions quantitatively according to the equation

$$Ag^+ + RS^- \rightarrow AgSR.$$

The end point is located by a silver–silver sulfide, glass electrode pair that sensitively monitors the concentration of Ag^+ ion. The relay is set to terminate the titration. The end point corresponds to an emf of 0.025 V across the monitoring electrodes. When this value is reached, the electrolysis is terminated.

Example 31.5. The determination of arsenite ion (AsO_3^{3-}) by oxidation to arsenate (AsO_4^{3-}) can be performed coulometrically by generating bromine. In this representative inorganic titration, the supporting electrolyte is an aqueous solution about 0.1 *M* in sulfuric acid and 0.2 *M* in sodium bromide. The titrant Br_2 is produced from the Br^- ions at a platinum anode.

Amperometry with two polarized electrodes (see Section 29.12) is often employed to determine the end point in this determination. When this procedure is used, two platinum electrodes of small surface area are used as the amperometric detector, and a steady potential of about 0.1 or 0.2 V is applied across them. The end point is marked by a sudden drop in the current between the two platinum electrodes.

Procedures in controlled-potential coulometry are not essentially different in this chemical aspect. (An extensive tabulation of systems available to this type of analysis is given by Milner. [6]) In that no end-point detection arrangement is required, operating procedures are less elaborate. Further, no intermediate reagent need be added. Usually with each reference to a method we take care to specify the electrolytes, working and reference electrodes, and the potential at which the working electrode must be maintained.

31.6 ELECTROGRAVIMETRY: INSTRUMENTATION AND PROCEDURES

We can analyze a wide variety of metals whose activity is below that of manganese by *electrodeposition* at controlled potential. This type of analysis was one of the first high-precision methods and is still in common use. As its earlier and less descriptive name, *electroanalysis*, implies, the method measures the amount of substance present by the increase in weight of the working electrode when the analytical substance is quantitatively plated out. A successful electrogravimetric analysis of a given sample is possible if the substance sought can be

1. Plated as an adherent coating when a current density high enough to complete the electrolysis in a "reasonable" time is used (times shorter than one hour are preferred) and
2. Electrolyzed quantitatively at a potential lower than that at which any other species is *plated out*.

Note that current efficiency need *not* be 100%. If any other substance is reduced at the working electrode during the analysis, however, it must not be trapped in the plated layer.

A discussion of factors that control the degree of adherence of an electroplated substance is beyond the scope of this text, although some mention is made of conditions under which we obtain suitable deposits in particular analyses. Given the chemical characteristics of a sample, the possibility of attaining the remaining objectives of this technique depend largely on how closely we can regulate the potential of the working electrode.

Potential Control. During an exhaustive reduction, a working cathode becomes steadily more negative or cathodic in potential. How great a change in potential is required to electrolyze a substance quantitatively we may readily compute from the Nernst equation (see Section 27.3).

Such a calculation has value in determining whether other metals may interfere. Figure 31.5 shows the minimum difference in electrolysis curves that allows a clean separation of two metals. The potential of the working electrode need not be held constant, but must be allowed to exceed E_1, the value at which electrolysis of the metal that is more difficult to reduce begins. In order to meet the requirements that the time of electrolysis be minimal, the voltage of the electrode is or-

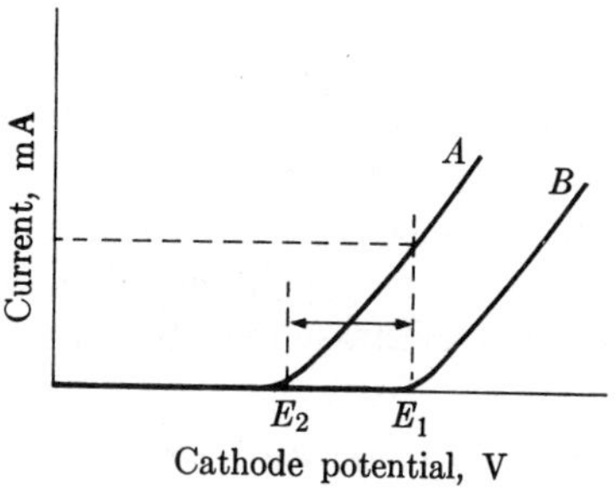

Fig. 31.5 Selective electrolysis by control of potential of working electrode. Substance A may be reduced without interference from substance B if the cathode potential (relative to a reference electrode) is kept between the values represented by the broken lines.

dinarily set at the most negative value possible without (a) producing interference (e.g. E_1 in the case discussed) or (b) depositing metal too rapidly to give an adherent coating.

Finally in this consideration of potential control, we look at the anode. In certain electrolyses the product of the regular oxidation process at the anode interferes with the desired reduction. The possibilities for such interference are inherent in the open design of the cells used in electrogravimetry and in the selection of platinum as the most suitable anode material. If chlorine or oxygen is produced at the anode, as often happens in the absence of countermeasures, some of the gas is usually swept to the cathode where it reoxidizes the metal being deposited. We can avoid the production of these oxidizing agents by adding a substance such as hydrazine dihydrochloride, which functions as an *anodic depolarizer*. By definition, such as substance is oxidized at potentials below those required for chlorine or oxygen evolution and yields a product that is chemically inoffensive (hydrazine oxidizes to N_2). The anodic depolarizer acts to stabilize the anode at a low potential.*

Current and Time. If the potential of the working electrode is maintained at a fixed value, as is generally done where close separations are necessary, the electrolysis current decreases continuously with time. Where a single reaction is occurring, the efficiency is, of course, 100%, and the current finally falls nearly to zero. The time needed for any percentage completion of deposition may be calculated by use of Eq. (31.4). In some cases the desired reduction at the working electrode may be accompanied by the evolution of hydrogen. Then the current decreases to some small value other than zero.

Separations. Metal ions that interfere with the electrolysis of an analyte can often be separated expeditiously by electrochemical means. A mercury cathode with a large, stirred surface has been widely adopted for this purpose. Since the cathode is mercury, the potential required to effect a particular separation can ordinarily be decided by reference to the extensive literature on polarographic half-wave po-

*In a real sense, an anodic depolarizer can also be termed a redox buffer (see Section 31.2) if its oxidation occurs reversibly at the electrode. Hydrazine does not quite qualify, since it oxidizes irreversibly.

tentials or by performing a preliminary polarographic determination. (See also Chapter 29.) The voltage selected should be about 0.15 V more negative than the $E_{1/2}$ of the offending element to give complete separation.

Electrogravimetric Procedures. Under optimum conditions, electrogravimetric procedures result in clean separations and deposition with better than 99.99% completeness. Their accuracy is generally of the order of 0.1% but depends on the weight of metal deposited and the sensitivity of weighing. Trace amounts of metals in alloys can be successfully analyzed with semimicro equipment. If a metal to be determined has a half-wave potential well separated from that of the next most easily reduced species, ($\Delta E^0 = 0.5$ V), we can carry out the analysis with little voltage control. For example, copper (II) ($E^0 = +0.34$ V) can easily be analyzed in the presence of nickel (II) ($E^0 = -0.25$ V). A constant-total applied potential in this case ensures precise results. Only when other metals that deposit at potentials close to the favorable potential are present is there a need for potentiostatic control.

Note that the presence of interfering metals need not necessarily preclude analysis by this technique. There are many possibilities for altering the potentials at which substances are electrolyzed by changing the pH or by adding complexing agents (see Table 29.1, p. 1069).

REFERENCES

A comprehensive general coverage of coulometric techniques is provided by:

1. K. Abresch and I. Claassen, *Coulometric Analysis*. London: Chapman and Hall, 1965.
2. A. J. Bard and L. B. Faulkner, *Electrochemical Methods*. New York: Wiley, 1980, Chapter 10.
3. E. Bishop, *Coulometric Analysis*, Vol. IID of *Comprehensive Analytical Chemistry*, C. L. Wilson and D. W. Wilson (Eds.). New York: Elsevier, 1975.
4. D. J. Curran, Constant-current coulometry, in *Laboratory Techniques in Electroanalytical Chemistry*. P. T. Kissinger and W. R. Heineman (Eds.). New York: Dekker, 1984, Chapter 20.
5. J. E. Harrar, Techniques, apparatus, and analytical applications of controlled potential coulometry, in *Electroanalytical Chemistry*, vol. 8, A. J. Bard (Ed.). New York: Marcel Dekker, 1975, pp. 1-167.
6. G. W. C. Milner and G. Phillips, *Coulometry in Analytical Chemistry*. New York: Pergamon, 1967.

EXERCISES

31.1 What length of time would be required to titrate 0.15 g methanethiol, CH_3SH, in alkaline solution, using bromine electrolytically generated by a constant current of 100 mA? One mole of bromine oxidizes two moles of thiol.

31.2 An electric timer reliable to ± 0.1 s is to be used in constant-current coulometric titrations. Samples of the order of 10^{-4} equivalents are to be electrolyzed. What is the maximum current that can be used if the uncertainty in the determination of the sample is to be no greater than $\pm 0.1\%$?

31.3 For routine titrations with a constant-current coulometer it is found that currents of either 1 or 10 mA are adequate. A timer graduated to the nearest 0.1 s is used, but it is desired to mark its scale so that it is direct-reading in milliequivalents or microequivalents electrolyzed. For the two currents used, what labels should be assigned to the time scale at the intervals corresponding to 20 s and to 100 s?

31.4 Samples of As_2O_3 are to be determined by reaction with coulometrically generated bromine as outlined on p. 1164. Sketch a suitable cell and label all elements clearly. Be sure to include a practicable type of end-point detector. Explain the basis for its choice.

31.5 Show that the output voltage of the integrator of Fig. 31.4 is given by the formula

$$v_0 = \frac{R_2}{R_3 C} \int i_{\text{cell}}\, dt.$$

31.6 In the constant-potential coulometric determination of iron [by primary oxidation of iron (II) in acid solution] it was found that the current-time curve followed Eq. (31.4). (a) Show that when the equation holds, a plot of log i_t versus t gives a straight line of slope $-k$ and intercept log i_0 *and* that the area under the current versus time curves is $i_0/2.303k$. (b) The approach implied in part (a) is used in the determination of an iron sample. Current is determined precisely at two known times and the electrolysis is stopped. The data recorded are, when $t = 35.5$ s, $i = 75.3$ mA and when $t = 74.0$ s, $i = 40.5$ mA. Find k and i_0. (c) What weight of iron was present in the sample?

31.7 In an electrogravimetric cell why cannot external control of anode potential be used instead of internal control (achieved by the addition of an anodic depolarizer)? Presumably external control would regulate the anode potential relative to a reference electrode.

ANSWERS

31.1 3010 s. *31.6* (b) $k = 7.00 \times 10^{-3}$ s^{-1}; $i_0 = 135.5$ mA; (c) 4.86 mg.

Chapter 32

CONDUCTOMETRIC METHODS

32.1 INTRODUCTION

Measurements of the electrical conductance of solutions fall in the electroanalytic category of ionics. This technique stands in contrast to the electrodics techniques described in Chapters 28–31 that involve the measurement of interfacial properties as redox changes occur at an indicator or working electrode. Conductance, by contrast, is a nonspecific property of electrolytes that has been useful in measurements of concentrations of ionic species, in elucidating the extent to which ionogenic substances dissociate in solvents, and in the development of electrolyte solution theory.

The applications of conductometric methods extend from systems of small conductance and very low concentration (e.g., a saturated aqueous solution of AgCl at 25°C) to those of high conductance and concentration (e.g., the fused salt mixture NaCl–KCl at 800°C). When a single strong electrolyte is present in dilute solution (in a pure solvent) its concentration can commonly be found directly from a conductance observation. At concentrations above 10^{-3} M, conductance may still be used to measure the concentration if a calibration curve is first determined.

In the usual sample, several electrolytes (some of which are impurities) are present. The contribution of an individual analyte to the conductance of such a solution cannot be deduced in any simple, exact way. Even in this situation, however, the concentration of an ionic analyte single species may be determinable conductometrically if it is the only species varying in concentration. For example, this situation ideally holds for the elution of species in ion chromatography. Alternatively, many analytes may be measured by conductometric titration. With ordinary care, a precision of about $\pm 1\%$ is possible in a conductometric determination; with precision equipment and control of variables it may be extended downward to $\pm 0.1\%$.

32.2 CONDUCTANCE RELATIONSHIPS

Conductance is the reciprocal of resistance, a more fundamental property. A measured resistance depends on the spacing and area of a pair of electrodes and the volume of solution between them. For example, if a sample of regular shape is placed between a pair of parallel electrodes, the resistance measured increases linearly with sample length l and decreases linearly with cross-sectional area A.

We remove the dependence on shape and size by defining the *specific resistance* ρ as the resistance of a cube of sample 1 cm on edge. In terms of ρ the measured resistance R of a sample is given by the expression $R = \rho l/A$. Since R is in ohms, ρ must have units Ω cm. The reciprocal of ρ, the *specific conductance* κ, is the quantity of interest here. It is defined by the equation

$$\kappa = l/\rho = l/AR \tag{32.1}$$

and has the units Ω^{-1} cm^{-1}. For measurements on solutions the ratio l/A is fixed by the spacing and size of electrodes in the conductance cell (see Section 32.4).

In dealing with dissolved electrolytes, it is convenient to define also an equivalent conductance as the conductance associated with one Faraday of charge. This is taken as the conductivity of a slab of solution 1 cm thick and of sufficient breadth and length to hold the volume of solution that contains one equivalent of the electrolyte.* The equivalent conductance Λ is related to specific conductance by the formal expression

$$\Lambda = 1000\ \kappa/C \tag{32.2}$$

where C is the normality of the solution. Since both positive and negtive ions will share in carrying the current, we can write Eq. (32.2) in terms of the equivalent ionic conductances λ^+ and λ^-: $\lambda^+ + \lambda^- = 1000\ \kappa/C$. Only at infinite dilution are the ionic conductances known precisely. Table 32.1 gives a compilation of limiting or infinite-dilution conductances λ_0.

How should we relate conductances to the discussion of mass-transport processes in Section 27.9? As noted there, conductance is the experimental measure of the transport process called migration. When we apply a potential difference across a pair of electrodes, ions first move to set up electrical double layers at the

Table 32.1 Limiting Equivalent Conductivities of Ions at 25°C[a]

Ion	λ_0	*Ion*	λ_0
H^+	349.8	OH^-	198.6
Li^+	38.7	F^-	55.4
Na^+	50.1	Cl^-	76.4
K^+	73.5	I^-	76.8
NH_4^+	73.5	NO_3^-	71.5
$CH_3NH_3^+$	58.7	ClO_3^-	64.6
Mg^{2+}	53.0	CH_3COO^-	40.9
Ca^{2+}	59.5	SO_4^{2-}	80.0
Ba^{2+}	63.6	CO_3^{2-}	69.3

[a]Data taken from the compilation of Robinson and Stokes [5, Chap. 23].

*Experimentally, it might be measured by using planar electrodes 1 cm apart and of sufficient surface area just to contain the required volume of solution.

electrode surface. If the potential is sufficiently large, the oxidation or reduction of electroactive species also begins. As ions are removed by reaction, additional ions move toward the electrodes. Figure 32.1 shows in simplistic terms the mechanism by which conductance occurs.

The current through a unit cube of solution may now be expressed in terms of mobilities. Assume for simplicity that a single electrolyte has completely dissociated. For the positive ions let N_+ be their number per cubic centimeter, u_+ be their mobility (see below), and z_+ be their charge. The related quantities for the negative ions will be denoted by N_-, u_-, and z_-. The total charge arriving at the negative electrode per unit area per second is then $N_+u_+z_+eE$. To convert to cell current, we must include charge arriving at the positive electrode and multiply by electrode area A. The equation obtained is

$$i = (N_+u_+z_+ + N_-u_-z_-)\,eEA. \tag{32.3}$$

By use of Ohm's law and Eq. (32.1), we can formulate the specific conductance from Eq. (32.3). We obtain

$$\kappa = (n_+u_+z_+ + N_-u_-z_-)\,e. \tag{32.4}$$

Concentration Dependence. At very low concentrations, ions behave essentially independently. Any given ion moves in a medium where other ions are so distant that they fail to influence its velocity or physical behavior. But from concentrations of the order of 10^{-6} upward ions approach each other sufficiently often that inter-

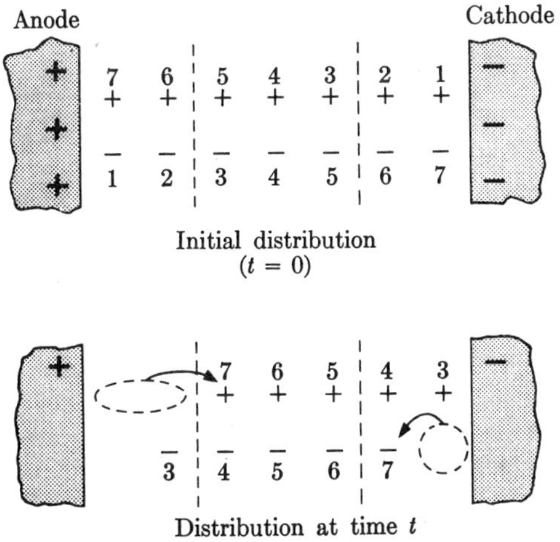

Fig. 32.1 Transport of charge by migration. Electrical double layers at electrode surfaces are omitted. While initially there are equal numbers of positive and negative ions through the solution, at time t two cations and two anions have been removed by electrolysis. If cation mobility is assumed twice that of anions, while two cations are replaced at the cathode only one anion is replaced at the anode. Ionic species not electrolyzed simply tend to accumulate at the electrodes.

ionic forces are important. In addition many kinds of ions begin to associate. Where such a process occurs, conductance decreases proportionally. While it is beyond the scope of this book to treat ion association processes, some interactions of ions and of ions and solvent are of sufficient importance to warrant a brief description.

Mobility. By definition, the mobility u of an ion is its velocity v under an electric field strength E of 1 V cm^{-1}. The defining equation is v = uE. A force z_ieE acts upon each ion, where z_i is the charge on the ion and e the electron charge. As a result of this force, an ion accelerates very rapidly until its motion is just offset by the frictional resistance of the solution. Mobility is a measure of its steady-state motion. So rapidly is a limiting velocity attained by ions that in audio-frequency conductance measurements ions may be assumed to travel at a constant velocity, even though the field is reversing a great many times per second.

The limiting velocity or mobility of an ionic species is determined by the viscosity of the solvent, the solvated size of the ion, the concentration of the solution, and the potential gradient. It has been found that for ions whose radius is about 0.5 nm (5 Å) or greater, Stokes's law describes approximately the relation between the force on the ion F, bulk viscosity η, and mobility u. The statement of the law is

$$u = F/6\pi\eta r \tag{32.5}$$

where r is the radius of the ion. Equation (32.5) is of limited validity, but does apply to spherical ions moving in a solvent whose molecules are considerably smaller than the ions.

Remember that most ions are of the same general size as solvent molecules. They therefore share in the general thermal agitation and have at best a randomly directed type of progress. The instantaneous velocity of any ion in a liquid is of the order of 10^4 cm s^{-1}, but its mean free path is so short that its average velocity toward an electrode is no more than 10^{-3} to 10^{-4} cm s^{-1} when the electrical field strength is of the order of 1 V cm^{-1}.

Electrophoretic Effect. In an electrolytic solution, any ion is surrounded by a sheath of solvent molecules, each held with reasonably strong ion-dipole forces. When an ion moves, its solvation sheath tends to accompany it. In this connection remember there is a continuing interchange between "bound" and "free" sovlent in the sheath. Since ions of opposite charge move toward different electrodes, a given ion experiences a drag as solvent "bound" to ions of opposite charge moves past. For example, any negative ion moves through solvent that is not stationary but is actually flowing in the opposite direction since it loosely accompanies positive ions.

Relaxation Effect. Another important property of an electrolyte solution that affects ion migration arises from the tendency toward electroneutrality. Any ion may be regarded as surrounded by an atmosphere of ions whose net charge is equal to its charge but opposite in sign. For an ion of charge +1, the essentially spherical atmosphere will include both positive and negative ions but will have an overall charge of −1. The dimensions of an *ion atmosphere* can be shown to be inversely proportional to the ionic strength of the solution. At high dilution the radius of the atmosphere is large; in concentrated solutions it may be only a few times the radius of the central ion.

When an ion moves, it tends to leave its atmosphere and a finite time will be required for the thermal and electrical forces to re-establish the randomly arranged atmosphere. Each ion is therefore subjected to a transient restoring force exerted by its old atmosphere as it decays. The opposing force tending to return the central ion to its original location is small at best and because of its time-dependent behavior is termed a relaxation effect. This force also tends to diminish conductance.

According to the general electrostatic theory of electrolytes developed by Fuoss and Onsager we can represent conductance as a function of concentration C by the following equation

$$\Lambda = \Lambda_0 - SC^{1/2} + EC \log C + JC, \quad (32.6)$$

where Λ_0 is the equivalent conductance at infinite dilution, S is the Onsager coefficient of the limiting conductance law, E is a constant,* and J is a factor dependent on ion size. Appropriate modifications to Eq. (32.6) can be made to cover association of electrolytes. The equation has been shown to hold generally up to concentrations of about 0.1 N.

32.3 ALTERNATING-CURRENT MEASUREMENTS

Because conductance involves the transfer of mass, both solution and electrodes are altered during a measurement. If we impose a dc voltage across a conductance cell there result two immediate, undesirable effects. The electrodes polarize slightly as (a) the solution layer near the electrodes tends to become depleted in any species being oxidized or reduced and (b) the electrode surfaces are altered by the products of electrolysis. The effects are not serious if the current is kept small ($<10^{-7}$ A), but attention must customarily be given to them. If a larger current flows, a dc conductance measurement may well be invalid.

When an *audio frequency* voltage is applied, the changes described are largely minimized. Because of the frequency reversal of electrolysis, the ionic movement and electrolysis that take place during one half of a cycle can be completely or nearly completely destroyed during the second half of each cycle. Concentrations are maintained essentially constant even though a current exists. The conductance of the solution and the current density at the electrode for a given applied voltage are key variables in arriving at an optimum frequency. If solutions of extremely low conductivity ($\kappa < 10^{-7}$) are being studied or the current density is very low, even dc measurements can be accurate. If the conductance is slightly larger, 60 Hz line current may allow precise measurements. Usually, however, a frequency of about 1000 Hz is preferable. Where great precision is required, we find the conductance at several frequencies in the audio range and extrapolate to infinite frequency [1].

*Both S and E are determined by the absolute temperature, dielectric constant and viscosity of the solvent, valence type of the solute, and universal constants. S also includes Λ_0.

Note that larger conductance values are usually found at radio frequencies (10^5–10^7 Hz). The change is a direct consequence of the increased importance of circuit capacitances and inductances. At radio frequencies, the interpretation must thus be broadened to include the bulk capacitance of the cell as well as the resistance of the solution. Separate consideration will be given to this type of measurement in Section 32.7.

A further aid in the elimination of surface polarization effects is the use of platinized platinum electrodes. These are electrodes on which finely divided platinum has been deposited in a thin, adherent layer by electrolysis. As a result of the greatly increased surface area, the reunion of liberated hydrogen and oxygen appears to be catalyzed. The polarization from this source is thus minimized. The large surface area also eliminates concentration polarization.

32.4 CONDUCTANCE CELLS

The usual envelopes for cells with electrodes are made of hard glass. Where ruggedness is required (e.g., in many field and plant applications), other inert, stable dielectrics such as hard rubber and some of the plastics are also in common use. The electrodes are generally square pieces of stiff platinum foil aligned parallel to each other. It is essential that the electrodes be rigidly supported at the desired spacing; by proper design they may be self-supporting like those shown in Fig. 32.2.

Leads. Special attention is always given to the arrangement of the leads to the electrodes, except where an accuracy of from 2 to 5% is adequate. If leads are bare and are brought out close together through the solution, stray electrolytic and capacitive current will pass between them. Accordingly, it is good practice to use insulated lead wires and bring them out of the electrode chamber in opposite directions. Two different designs of conductance cells are illustrated in Fig. 32.2. Note that all obstructed spaces where mixing will not occur readily have been eliminated in the cells of types (a) and (b).

Cell Constant. The resistance of a solution between the electrodes of a cell is a function not only of solution specific conductance κ but also of the volume of

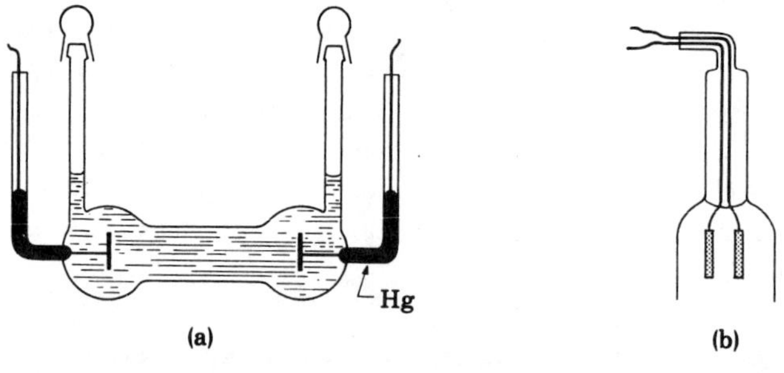

Fig. 32.2 Some types of conductance cells. (a) Jones and Bollinger precision cell. (b) Dip-type cell.

conducting solution between the electrodes. For a pair of parallel electrodes of area A and spacing l, κ may be obtained by rewriting Eq. (32.1) as $R = l/A\kappa$. In practice we determine the ratio l/A, termed the *cell constant*, for each cell by measuring its resistance when filled with a conductance standard. Solutions of potassium chloride of known concentration are primary standards, their conductances having been accurately determined in cells of known electrode geometry.

For accurate conductance work over a range of concentration it is desirable to use cells of different cell constant. In aqueous work cell constants from about 0.1 to 10 are needed. In nonaqueous media other ranges are called for. The reason is that a bridge of conventional design (see Section 32.5) is capable of greatest accuracy if the cell resistance falls in the range from 1 to 30 kΩ.

Thermostating. Control of temperature is indispensible if reliable conductance measurements are sought. The specific conductance of electrolytes increases on the average about 2% per degree Celsius. To reduce the error from this source to 1% therefore requires regulation to ±0.5°C; to reduce the error to 0.01% requires regulation to ±0.005°C. A constant-temperature bath filled with a light transformer oil is often used to achieve the desired regulation. Water is seldom used as the fluid because of accompanying undesirable capacitance effects between cell and ground.

32.5 THE AC WHEATSTONE BRIDGE

This bridge is the basic instrument for determining conductance. The dc version of the Wheatstone bridge was treated in Section 3.4; that background is now assumed familiar. In Fig. 32.3a a schematic circuit of the dc bridge has been repeated

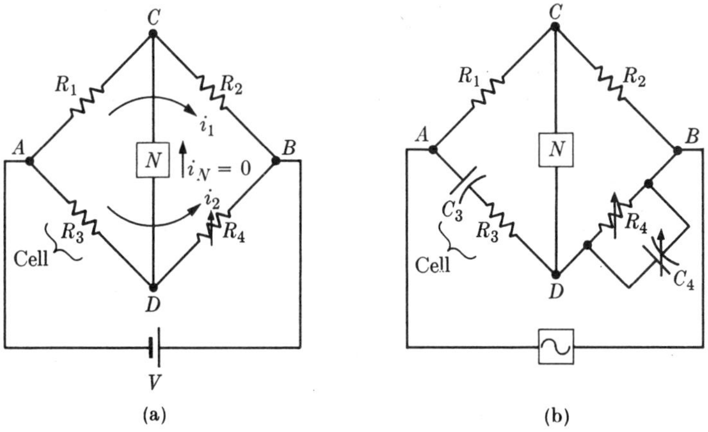

Fig. 32.3 Wheatstone bridge circuit in two versions. In each, R_1 and R_2 are ratio arms and N is a suitable null detector. (a) Simple bridge. (b) AC conductance bridge. The C_4–R_4 combination provides adequate compensation for both resistance and capacitance of the conductivity cell under most conditions.

to permit comparison with the common form of ac Wheatstone bridge shown in Fig. 32.3b. Recall that the condition of balance of the dc bridge is that the potential at points C and D must be equal, yielding the equation

$$\frac{R_1}{R_2} = \frac{R_3}{R_4}. \tag{32.7}$$

Conductance is then obtained by taking the reciprocal of R_3. It is of particular interest that this condition also holds for balance of the ac bridge to within $\pm 0.1\%$. Some variation is to be expected since the ac bridge is properly an *impedance bridge*. Sources of error in the ac bridge will be considered below.

Range of Measurement. The range of *resistance* measurable may be deduced from Eq. (32.7). If $R_1 = R_2$, unknown resistance R_3 can be measured by the bridges shown in Fig. 32.3 when its value falls within the range $0 < R_3 \leq R_4$. Since this span is short, ways to extend it are important. One method is to vary the ratio R_1/R_2 as well as R_4. Bridges offering several set ratios of R_1/R_2 from 0.01 to 100 are common. Range is traded off for accuracy in these bridges; they are accurate at best to about $\pm 1\%$.

Alternatively, range can be extended in conductance measurements by the strategem of use of cells of different cell constant. This approach permits use of equal values for R_1 and R_2, which is necessary to the construction of precision conductance bridges. If R_1 and R_2 have equal values and have been carefully constructed of stable, low-temperature-coefficient alloy, we can assume their resistances will change in like amount with time and temperature and keep the ratio invariant.

Sources of Error. Contact resistance in switches and in leads to the cells, a major potential source of error, can be minimized by keeping every contact possible in series with the power supply (V) or detector (Section 3.4).*

We must now consider certain sources of error peculiar to the use of ac. The resistors that comprise the bridge arms possess distributed inductance and capacitance, and we must regard the cell itself as a combination of capacitances and resistances. Second, there is a considerable number of possible stray current paths in an ac bridge. These are of two types. Any part of the bridge has some capacitance with respect to ground and offers a leakage path. Also, by virtue of the inductance of the resistance coils, there exists the possibility of inductive pickup of stray ac currents from power lines or from the oscillator that supplies the bridge power.

The contribution of error from these sources may be reduced considerably by proper resistance, shielding, and physical arrangement [8]. The resistors should be noninductively wound. The bifilar winding, in which the length of wire required to obtain the desired resistance is doubled back on itself and then wound on the

*In any event, there should be sufficient resistance in the power circuit to ensure that bridge resistors dissipate less than the maximum allowable power.

form, is widely used to minimize inductance. It is advantageous to have enough residual capacitance so that the capacitive reactance will nearly cancel the inductive reactance at the operating frequency. The cell capacitance C_3 can be compensated by placing a variable capacitor C_4 in the bridge parallel with resistance R_4, as shown in Fig. 32.3b. For examination of the more difficult problem of eliminating stray leakage paths the reader may consult references 6–8.

Power Sources. Some industrial and field conductivity instruments operate on 60 Hz ac stepped down from a power line. Much better accuracy is generally secured by operation at audio frequencies in the range of from 500 to 4000 Hz. In this case, a sine-wave ac signal generator is usually employed. Its output should ideally be of a single frequency (a pure sine wave) and should be variable in amplitude from zero to several volts. If the harmonic content is minimized, a more precise balance can be obtained, for the problem of phase shifts will be simplified (see below). A variable voltage output allows flexibility of operation.

Phase Relationships. For a true bridge balance, the ac waves must be in phase at points C and D (Fig. 32.3b). This condition requires either no phase change in either arm or the same phase change in each. Only the latter is a possible solution to the requirement. The capacitance and inductance of the resistors and cell can be minimized but not eliminated.

If accuracies of the order of 1% are satisfactory we may ignore the phase difference, providing the bridge resistors have been wound with reasonable care. For most nonresearch measurements and conductometric titrations, the phase difference can be neglected. On the other hand, work in which the precision must be 0.1% or better calls for a careful examination of the phase dependence of the arms. It is customary to simplify the problem by using matched resistors for the ratio arms R_1 and R_2 so that not only are the resistances equal, but the phase behavior is identical. There remains the question of whether the phase difference introduced by the cell in arm 3 will be equal to that caused by the parallel R_4-C_4 combination. A thorough discussion is beyond the scope of this book, but a limiting case can be considered. In general, if the R_4-C_4 combination introduces a phase shift of less than about 10 minutes of arc, R_4 can be taken as equal to the cell resistance R_3 within 0.1%.

Bridge Amplifiers. An amplifier designed to give a logarithmic response offers a definite advantage, for it will produce the greatest response where the signal is smallest, near the point of balance. Some protection against overload must also be provided and is not difficult to incorporate.

Amplifiers inevitably have considerable capacitance to ground. Since the possibility of picking up stray signals is therefore large, electrostatic shielding of the amplifier is essential, and transformer coupling between bridge and amplifier or between stages of the amplifier should be avoided. Where transformers must be used, they should be small and carefully shielded magnetically. It will also relieve the situation considerably if the bridge points C and D can be operated at zero potential by means of a Wagner ground or a similar device.

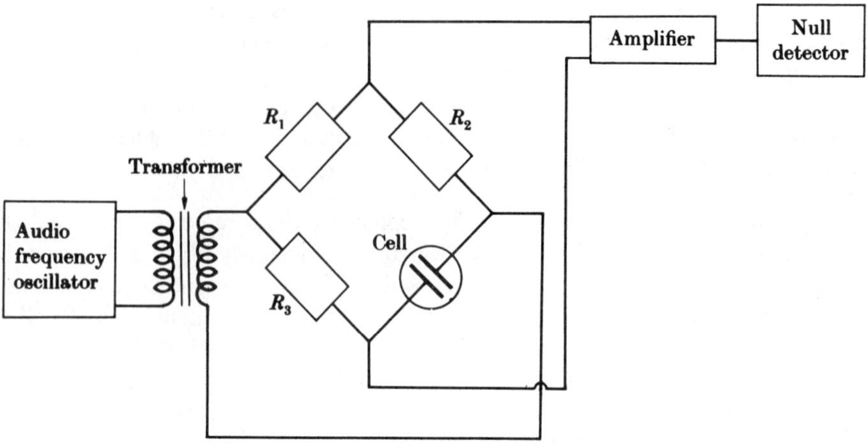

Fig. 32.4 Block diagram of an ac conductance bridge.

Figure 32.4 is a block diagram of an ac conductance bridge with amplifier. Many commerical bridges use 60 Hz line voltage in lieu of an audio frequency of 500–1000 Hz such as the oscillator would supply. The null detector may be an electron ray tube (magic eye), oscilloscope, or other suitable devi‹ › to register the level of ac voltage across the bridge.

Conductometric Titrator. Where only changes in conductance and not absolute values are of interest, a Wheatstone bridge is not mandatory. Ohm's law, $V = iR$, suggests that changes in the resistance of a cell might be measured by determining changes in i when a constant voltage V is imposed, $\Delta i = V/\Delta R$. It is easy to instrument this approach using operational amplifiers, as the block diagram in Fig.

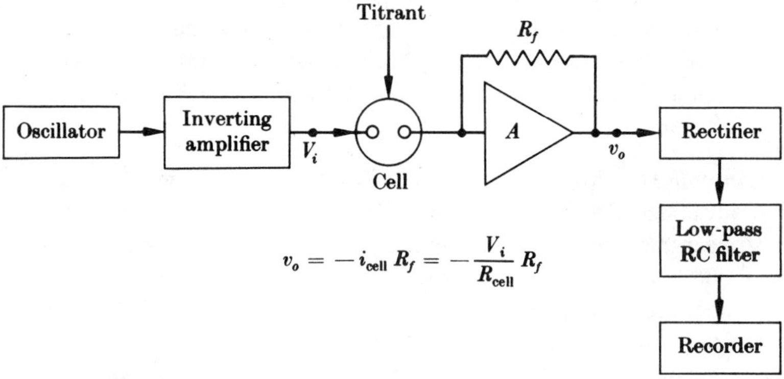

Fig. 32.5 Operational amplifier conductometric titrator. The inverting amplifier serves to isolate the oscillator and thus avoid loading by the cell, which might otherwise cause changes in voltage V_i as the cell conductance varies. Op-amp A is a current amplifier with an output directly proportional to the cell conductance as long as V_i is constant. The rectifier converts the ac signal to dc plus ripple, following which the RC filter smooths the signal to avoid any spurious dc offset at the recorder.

32.5 shows. In the circuit, operational amplifier A serves as a current amplifier (see Section 4.6) and provides an output voltage proportional to i.

The instrument in Fig. 32.5 is especially suitable for conductometric titrations, for which we need only relative values of conductance. It is also usual to provide for attenuation in the inverting amplifier by adjustments of the ratio of feedback to input resustance (R_f/R_{cell}) that determines the gain of this device. We should estimate in advance the maximum value of conductance expected during a titration and use it as a basis for setting the attenuation to the level to be employed throughout the titration.

32.6 CONDUCTOMETRIC TITRATIONS

In solution, chemical reactions that involve electrolytes are accompanied by a conductance change. If the change is sufficient, we may often determine the end point of the reaction simply by monitoring the conductance. We can best understand the origin of the variation by inspection of a representative ionic reaction:

$$\underset{\text{unknown}}{A^+B^-} + \underset{\text{titrant}}{C^+D^-} \rightarrow AD + C^+B^-.$$

In the equation, CD is taken as the titrant and we assume AD, one of the products, to be a weakly ionized species. Up to the end point, the equivalents of C^+ ion in the solution at any time are essentially equal to the equivalents of A^+ that have been used up to form species AD. The concentration of B^- ions does not change. The conductance attributable to C^+ ions increases gradually during titration while that of A^+ ions decreases. After the end point, further addition of titrant sends the conductance upward in proportion to the volume added, since the concentration of C^+ and D^- ions in the solution grows steadily.

Example 32.1 The titration of 0.01 M HCl by 0.1 M NaOH gives the V-shaped conductance curve in Fig. 32.6. As long as rapidly moving hydrogen ions are being replaced by much more slowly moving sodium ions, the conductance falls. After the reaction is complete, further addition of NaOH adds both sodium ions and fast-moving hydroxyl ions and conductance rises sharply and linearly.

In a manual titration we take conductance values periodically after addition of titrant and mixing. Amounts of titrant should be roughly calculated on the basis of estimated end point and the need to obtain four or more points on each branch of the conductometric titration curve well away from the end-point region. We eliminate any appreciable dilution error resulting from the increase in solution values at each point by multiplying the observed conductance by the ratio $(V + V_o)/V_o$, where V_o is the original solution volume and V is the volume of titrant added.* We then plot the data and draw the best straight line through each set of

*Use of a titrant 10–20 times more concentrated than the species being titrated ordinarily allows this correction to be omitted.

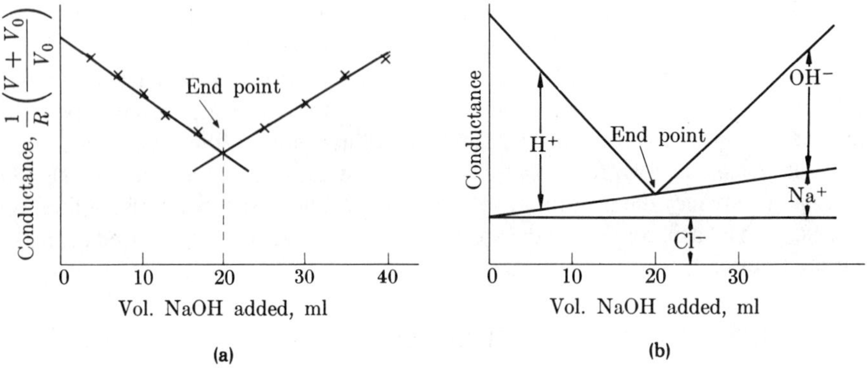

Fig. 32.6 Conductometric titration of 0.01 *M* HCl by 0.10 *M* NaOH solution. (a) Titration curve. Note that the ordinate has been corrected for volume change during titration. (b) Ionic interpretation of the titration curve. While the conductance attributable to each species is shown separately, at any point in the titration (V-shaped line), the solution conductance is the sum of these contributions.

points. As Fig. 32.6 indicates, we should take the point of intersection as the end point.

With an operational amplifier system like that in Fig. 32.5, we obtain a titration curve automatically, provided we add titrant slowly and secure complete reaction at all times. Note there is no provision to correct for dilution as the curve is recorded. Linear parts of the curve should be extrapolated to their intersection to locate the end point.

Conductance data near the end point are less valuable since in this region there is little excess of any common ion and reactions may not be complete. For example, near the end point a weak acid or a weak base is more fully dissociated than at other times. Similarly, if a slightly soluble precipitate forms in a titration, it will be more soluble near the end point. Any incomplete reaction leads to curvature in the plot.

Since we can determine an end point by reliance on data far from the end-point region, we can follow conductometrically many reactions that are too incomplete for their end point to be located potentiometrically. For example, phenol, boric acid, and other quite weak acids can be successfully titrated in aqueous solution conductometrically but not potentiometrically. We can also apply the conductometric method to very dilute solutions and to some nonaqueous solutions [13] with good precision if we use sufficiently sensitive bridges and good thermostating. Against these advantages we must set a number of limitations considered later.

In any analysis it is valuable to be able to interpret the form of the characteristic curve obtained. This way we can gain both insight into the behavior of a system and assurance that a reaction is progressing. Since at the concentration levels that usually prevail conductance varies nearly linearly with concentration, it is not difficult to predict behavior in conductometric titrations. To predict a titration curve we also need a listing of relative ionic conductances. Table 32.1 provides such values. Though the data are valid at infinite dilution, they provide a basis for comparison even at ordinary concentrations. The contribution of each ionic species is presumed independent of others. Figure 32.6b shows the result of applying this kind of analysis to the titration system of Fig. 32.6a. Figure 32.7 shows the procedure applied again.

With bridges like those in Figs. 32.4 and 32.5, we may expect a precision of $\pm 1\%$ or better under favorable circumstances. Where additional refinement in titration and thermostating is possible and we can arrange the use of a high-precision bridge like the Jones bridge manufactured by Leeds and Northrup, reliability can sometimes be of the order of $\pm 0.1\%$.

Limitations and Sources of Error. We may group the salient "chemical" limitations and sources of error of the conductometric method as follows:

1. an indistinguishable difference in slope of intersecting lines,
2. curvature in one or more conductance lines beyond end-point region,
3. volume increase during titration, and
4. temperature change.

Only the first two points require further discussion. In these comments we assume that a bridge or other instrumentation yielding results accurate to $\pm 1\%$ is used. With better thermostating and more precise instrumentation we may, of course, increase the acceptable range of conductance titrations.

The accuracy of locating the end point of a reaction depends significantly on how greatly intersecting conductance curves differ in slope. In this connection,

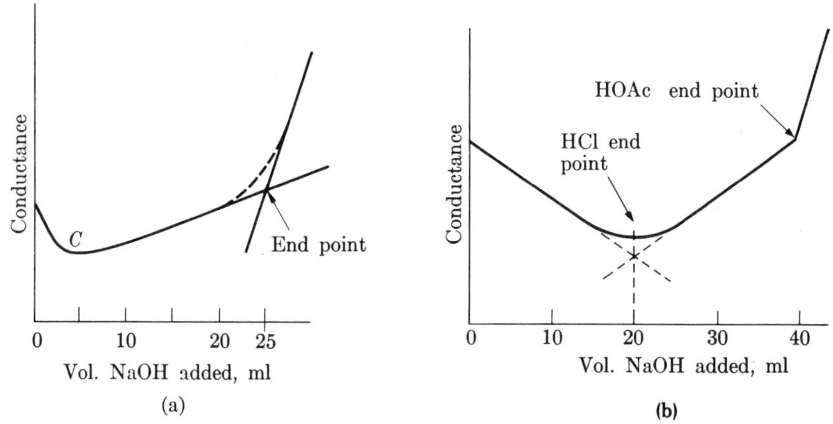

Fig. 32.7 Conductometric titrations involving a weak acid and strong base. (a) Titration of 0.01 *M* acetic acid by 0.1 *M* NaOH. In region *C* suppression of the ionization of weak acid by accumulating acetate ion becomes marked. In the end-point region the broken line indicates actual conductance; here curvature results from the hydrolysis of acetate ion. (b) Titration of an equimolar mixture of 0.01 *M* HCl and 0.01 *M* acetic acid by 0.1 *M* NaOH (titrant). The slight curvature at the acetic acid end point, resulting from hydrolysis, is not shown.

compare the curves in Fig. 32.6a and 32.7a. Accuracy is certain to be smaller in the latter case.

Indeed, if the change in slope is quite small an ionic reaction cannot be followed conductometrically. Usually this difficulty is inherent in a system and may arise for any of several reasons. First, it may be the result of a high concentration of a foreign electrolyte. For example, most redox reactions are ill adapted to conductometric monitoring. Often such solutions must be strongly acidic (as in dichromate and permanganate oxidations), basic, or contain added salt (as in many iodometric procedures). In a highly conducting solution a small change in conductance attributable to the desired reaction is difficult to detect precisely. Second, small differences in slope may arise because a sample is dilute. Third, there may be little change in slope because the sample is a very weak electrolyte. Figure 32.7a gives an illustration of this situation. In all such cases, unless instrumentation capable of better than $\pm 1\%$ precision is available, a conductometric titration will not be feasible.

The second source of chemical error, incompleteness of reaction or the occurrence of side reactions, needs only brief comment. If a product of the reaction undergoes substantial hydrolysis, ionization, dissolution, or indeed almost any side reaction, there will be pronounced curvature in the conductance curve and little possibility of determining the end point reliably. Often the disturbing effects may be suppressed.

In aqueous systems it is common to add ethanol to reduce ionization or to lower the solubility of a precipitate. When we can no longer locate the linear portion of a curve easily, we can sometimes apply a mathematical method of end-point calculation [13].

Mixtures of Acids or Bases. Often, we may analyze a mixture of a strong and a weak acid (or a strong and a weak base) by conductometric titration. The method is especially attractive when simple photometric and potentiometric methods do not give satisfactory results. In this situation, so long as one acid is strong and pK_a values differ by at least 5, the portion of the conductance curve attributable to each acid is definite. Figure 32.7b shows a curve for the titration of a mixture of acetic and hydrochloric acids. To interpret the graph, we should view the titration as a combination of the separate titrations of HCl and acetic acid already discussed.

Precipitation Titrations. Before deciding to follow a precipitation reaction conductometrically, we must study the growth of the crystals that would be involved and their tendency toward adsorption. Even though there is a sufficient change in conductance during the reaction, the slowness of precipitation, coprecipitation, appreciable solubility, and adsorption effects can greatly increase the uncertainty in the results. There are very favorable instances for the application of precipitation titration, for example, in determining some of the alkaline earths, by using sulfates. A particularly good illustration is the titration of solutions containing barium ions, using standard sulfuric acid solutions. The titration curve obtained resembles that in Fig. 32.7a.

32.7 OTHER APPLICATIONS

Direct conductance determinations of concentration are most often carried out when there is only a single electrolyte. The determination is possible because, as noted for conductometric titrations, an approximately linear relation is observed between specific conductance and concentration from 10^{-5} to 10^{-1} *M*. Provision for offsetting the effect of temperature becomes important in industrial determinations, where the temperature of a process stream may vary widely. Further, with a weak electrolyte allowance must be made for the effect of one or more equilibria.

Example 32.3. Conductance is widely used to monitor water purity, for example, in laboratory and industrial boiler water supplies. Pure water has a specific resistance of about 10 MΩ. Can conductance measurements also be applied successfully to estimation of salt levels in water?

For estimation of salt concentrations only an acceptable working curve is needed, a plot of specific conductance for a calibrated, thermostated cell for the analyte system. In oceanography salinity (really a total concentration of salts) can be registered on a scaled conductance meter.

Example 32.4. We can study the rate of the following reaction by monitoring the concentration of BH^+ and $CH_3CHNO_2^-$ conductometrically:

$$CH_3CH_2NO_2 + B = CH_3CHNO_2^- + BH^+.$$

Here B is a neutral organic base. Note that the only ionic species present are those formed during the reaction. In calculating the results, we assume the ions that are produced do not hydrolyze or associate at the concentration level used.

Example 32.5. When a conductance detector is used in ion chromatography what choices must be made about variables such as conductance cell constants, dc or ac measurement, mobile phase conductivity, and solution temperature?

As long as the chromatographic column resolves the ions in a sample, detection of species requires only that conductance is higher when they elute than it is for the mobile phase alone.

To preserve the resolution effected by the column the conductance cell should have a minute volume (about 2 μL). To ensure low detection limits the cell should also have a large electrode area.

Mobile phase conductivity can be electronically suppressed to achieve a stable chromatograph baseline. With that stability minimum changes in conductance of 10–40 parts in 10^5 can be detected. Conductance is commonly measured at 4–10 kHz.

When conductance is used to determine concentrations at trace levels, special care is required. Since the measurement is not selective, we must either first remove foreign conductors or make a correction.* In some instances, limiting ionic conductances can be used to calculate the concentration. More often we prepare a calibration curve in advance.

High-Frequency Methods: Oscillometry. Conductance measurements are also obtained at high audio and radio frequencies (rf) (about 10^5–10^7 Hz). Because substantially different instrumentation is required for rf measurements, these methods are usually collected under the name *oscillometry*. The advantage gained in oscillometry is that no metal electrodes need be in contact with the solution under investigation. For the measurements it is sufficient to place the glass cell or section of tubing holding the sample solution between the plates of a capacitor (or within the coil of an inductor) that is a part of the resonant circuit of an oscillator.

In instruments that handle separate samples, the oscillator itself serves as a detector since the frequency of its output depends on the conductance and dielectric constant of the solution. The capacitor type of arrangement is the more sensitive of the two means of coupling the solution to the oscillator circuit.

For *continuous monitoring* of the flow of a solution two loops or coils of wire are wrapped around the tube or pipe carrying the conducting solution far enough apart so that the conducting solution is the only current loop connecting them. Then the rf signal in the primary coil induces a signal in the second coil that is a measure of the conductance and capacitance of the solution in the tube and can be amplified and registered directly. A more precise arrangement is use of a circuit that provides for a null-balance readout. Oscillometry is useful in monitoring an electrolyte stream mainly when contact with metal electrodes is undesirable.

It is possible to carry out some nonaqueous titrations by oscillometer with surprising sensitivity if one component has a markedly different dielectric constant. Cells should be used that provide adequate shielding and are insensitive to fluid level when filled beyond a given point. For example, water is often determined in organic fluids by this method, and binary liquid pairs such as ethanol ($\epsilon = 24.3$)–nitrobenzene ($\epsilon = 34.8$) or benzene ($\epsilon = 2.27$)–chlorobenzene ($\epsilon = 5.62$) can be analyzed with precision.

REFERENCES

1. S. Glasstone, *Introduction to Electrochemistry*. New York: Van Nostrand, 1942.
2. J. F. Holler and C. G. Enke, Conductivity and conductometry, in *Laboratory Techniques in Electroanalytical Chemistry*, P. T. Kissinger and W. R. Heineman (Eds.). New York: Dekker, 1984. Chapter 8.
3. J. W. Loveland, Conductometry and oscillometry, in *Treatise on Analytical Chemistry*. Pt. I, Vol. 4. I. M. Kolthoff and P. J. Elving (Eds.). New York: Interscience, 1963.

*When the specific conductance of the solvent becomes as much as 0.5% of the value for the solution, a straight arithmetic correction is made. Since conductance (in dilute solutions, at least) is an additive property,

$$\kappa_{\text{electrolyte}} = \kappa_{\text{solution}} - \kappa_{\text{solvent}}.$$

4. E. Pungor, *Oscillometry and Conductometry*. New York: Pergamon, 1965.
5. R. A. Robinson and R. H. Stokes, *Electrolyte Solutions*. 2nd ed. New York: Academic, 1959.
6. T. Shedlovsky, Conductometry, in *Physical Methods of Organic Chemistry*. 3rd ed., Pt. IV. A. Weissberger (Ed.). New York: Interscience, 1960.

Special attention is given to bridges and detectors in:

7. B. Hague, *Alternating Current Bridge Methods*. 5th ed. London: Pitman, 1946.
8. Forest K. Harris, *Electrical Measurements*. New York: Wiley, 1952.
9. Melville B. Stout, *Basic Electrical Measurements*. New York: Prentice-Hall, 1950.

Conductometric and rf titrations are discussed in:

10. T. S. Burkhalter, High frequency conductometric titrations, in *Comprehensive Analytical Chemistry*. Vol. IIA. C. L. Wilson and D. W. Wilson, (Eds.). New York: Elsevier, 1964.
11. D. G. Davis, Potentiometric titrations and conductometric titrations, in *Comprehensive Analytical Chemistry*. Vol. IIA. C. L. Wilson and D. W. Wilson (Eds.). New York: Elsevier, 1964.
12. E. Grunwald, End point calculation in conductometric and photometric titrations, *Anal. Chem.*, **28,** 1112 (1956).
13. Walter Huber, *Titrations in Nonaqueous Solvents*. New York: Academic 1967.

Good theoretical treatments of conductance are provided on an intermediate to advanced level in:

14. R. M. Fuoss and F. Accascina, *Electrolytic Conductance*. New York: Interscience, 1959.
15. S. Petrucci (Ed.), *Ionic Interactions*. Vol. I. New York: Academic, 1971.

EXERCISES

When appropriate, refer to Table 32.1.

32.1 A cell is found to have a resistance of 4590 Ω when filled at 25°C with a 0.01-*M* solution of salt A of specific conductance $4.56 \times 10^{-4}\ \Omega^{-1}\ cm^{-1}$. When an 0.001-*M* solution of salt B is placed in the cell and a measurement made at 25°C, the resistance is found to be 25,230 Ω. Water of conductivity $2.5 \times 10^{-6}\ \Omega^{-1}\ cm^{-1}$ is used. What is the specific conductivity of the second salt?

32.2 A cell having a cell constant of $0.55\ cm^{-1}$ is filled with a dilute aqueous $NaNO_3$ solution and found to have a resistance of 3210 Ω at 25°C. If no other electrolyte is present, what is the approximate concentration of the salt? (Use limiting ionic conductances and refer to Section 32.2.).

32.3 Plot conductometric titration curves for the following systems, ignoring the ionization of any weak electrolyte and the solubility of any precipitate: (a) calcium chloride titrated with silver nitrate, (b) sodium acetate titrated with hydrochloric acid, (c) ammonia titrated with hydrochloric acid, and (d) acetic acid titrated with ammonia. Assume in each case a titrant concentration of 1 *M* and solution concentrations of 0.01 *M*. Calculate the specific conductivity when 0, 0.3, 0.7, 1.2, and 1.5 times the equivalent amount of titrant has been added.

32.4 In a conductometric titration, the more concentrated the titrant is, the more acute the angle of the conductance curve is, since the excess titrant line rises more steeply. Is the accuracy of the determination affected?

32.5 A precision conductance bridge with ratio arms in the ratio of 1 : 1 is to be used to measure solutions of specific conductances (a) 1×10^{-2}, (b) 1×10^{-4}, and (c) 1×10^{-6} with equal accuracy. What value of cell constant should be used with each solution?

32.6 An industrial system in which pure water ($R > 1$ MΩ) is furnished is monitored by identical conductivity cells at several points. The conductance of any cell may be checked at will, but ordinarily the cells are connected in parallel between an ac signal source and an indicating meter. Why is this arrangement effective in indicating ionic contamination at any of the check points?

32.7 In order to obtain a continuous record of a conductometric titration it is decided to record the unbalance current from a Wheatstone bridge, after amplification. Assume arm R_3 of the bridge is the titration cell, that arm R_4 is usually varied to obtain an initial balance and that resistances of ratio arms R_1 and R_2 are equal. (a) From the Thevenin analysis of the bridge in Appendix B, obtain an equation for the dependence of the unbalance current on the titration cell resistance R_3. (b) If R_{cell} is known from other measurements to vary from 200 to 100 Ω and back to 150 Ω should the bridge be balanced at the outset of a titration or come into balance midway if it is desired to obtain highest sensitivity of response to changes in cell conductance? (c) Assume that the bridge is balanced at the beginning of the titration just described. Graph the unbalance current against volume of reagent added.

ANSWERS

32.1 8.03×10^{-5}. *32.2* 1.4×10^{-3} *M*.

Appendix A

THE BINARY CODE AND OTHER NUMBER CODES

As is well known, the language of computers is the binary code. Digital computers are electronic devices that operate with two states. The essential idea was introduced in Chapter 5: in the binary system there are two states, HI and LO, or digits 1 and 0. Here some working aspects of the binary and the related codes, octal and hexadecimal, will be presented and the idea of interconversion with decimal code explored.

NUMBER CODES

Four *number codes*, decimal, binary, octal, and hexadecimal are in common use. The number of unique digits in each identifies the code and is said to be its *base*. Thus, there are 10 digits in the decimal system and 16 in the hexadecimal. In each code or system a *number* is a set of integers in which place is important. In a number each digit place, beginning at its left side or the most significant digit, and moving right, indicates that the integer in it must be multiplied by the base raised to a successively smaller power. An example in each code will illustrate the familiar principle involved. To distinguish numbers in different codes, each number will be followed by a subscript that indicates its base. For example, 15_8 is a number in octal notation. Special emphasis will be given the binary system. After interpretation of the different bases, interconversion to the binary system and some manipulations in that system will be very briefly explored.

Decimal. In decimal notation there are ten integers (0-9). For example, the number: 73.54_{10} stands for $7 \times 10^1 + 3 \times 10^0 + 5 \times 10^{-1} + 4 \times 10^{-2}$.

Binary. Similarly in binary notation, where 0 and 1 are available as digits, the number: 1011_2 stands for $1 \times 2^3 + 0 \times 2^2 + 1 \times 2^1 + 1 \times 2^0$, which equals 11_{10}.

Octal. In octal notation, 0, 1, 2, . . . , 7 are available as digits. For example, the number: 254_8 stands for $2 \times 8^2 + 5 \times 8^1 + 4 \times 8^0$. In decimal notation this number is $128 + 40 + 4 = 172$.

Hexadecimal. For hexadecimal notation the letters A–F serve as digits following the digit 9. Thus, its 16 digits are 0-9, A, B, . . . F. The decimal equivalents are: A = 10, B = 11, . . . F = 15. For example, the number $FA39_{16}$ stands for $F \times 16^3 + A \times 16^2 + 3 \times 16^1 + 9 \times 16^0$. In decimal notation this number is 61440 + 2560 + 48 + 9 = 64057.

INTERCONVERSION

To convert a decimal number to a binary number one need only divide the number and each successive dividend by 2. It is the remainders that are the binary digits. To form the binary number take the first remainder as the LSB. Successive remainders are progressively more significant bits. For example, converting 354_{10} to binary notation goes as follows (remainders are placed in parentheses):

$$354/2 = 177\ (0);\ 177/2 = 88\ (1);\ 88/2 = 44\ (0);\ 44/2 = 22\ (0);$$

$$22/2 = 11\ (0);\ 11/2 = 5\ (1);\ 5/2 = 2\ (1);\ 2/2 = 1\ (0);\ 1/2 = 0\ (1).$$

In binary code the number is 101100010. This approach to interconversion is generally useful in going from a higher base to a lower one.

What ways exist to convert numbers from binary code to the octal or hexadecimal equivalents, interconversions often needed in assembly code programming? A simple scheme works well. To obtain octal equivalents, group binary digits in sets of three beginning from the right. For the binary number 101100010 this approach yields the sets 101 100 010. Translate each set of digits into one octal digit. The octal number is 542_8 and the decimal equivalent, 354_{10}.

Similarly, to obtain the hexadecimal equivalent, group binary digits in packets of four beginning with the LSB. Then translate each set to a hexadecimal digit. For example, the binary number given yields the sets 1 0110 0010. The hexadecimal equivalent is 162_{16}.

How can one represent *negative numbers* simply in digital arithmetic? In particular, how can the use of minus and plus signs be avoided except in displaying results? Operations will be greatly speeded if *only* binary digits are employed. A protocol called the 2's *complement representation*, is probably the most widely used mode of representing negative numbers. This convention calls for a particular negative number to be stated as the 2's complement of its positive counterpart. Some advantages of the protocol are that:

1. The most significant bit gives the sign: if it is zero, the number is positive; if a one, the number is negative. Further,
2. Arithmetic is especially straightforward.

An example will provide the best explanation of how to obtain 2's complement of numbers.

EXAMPLE

Represent both 51 and −51 in binary notation. Obtain the negative number by using the 2's complement protocol. Demonstrate that use of the 2's complement makes the operation of subtraction as simple as addition.

Clearly $51_{10} = 110\ 011_2$. If a 16-bit number system is being employed, the number becomes 0 000 000 000 110 011. Since -51_{10} must be the 2's complement of this 16-bit binary number the process of complementing must be defined. In electronics to *complement* means to take the inverse. Thus, a 1's complement is obtained by writing for a given binary number a 0 for each 1 and a 1 for each 0. Then 1 is added to give the 2's complement.

For the binary number above the 1's complement is just 1 111 111 111 001 100. By adding 1 to that, one obtains 1 111 111 111 001 101 as the 2's complement. This number must represent -51_{10}. Note that the 2's complement representing -51_{10} might also have been obtained by counting down toward zero a total of 51 digits from 1 111 111 111 111 111.

ADDITION AND SUBTRACTION

A simple way to verify the result just obtained will be to *add* 51_{10} to -51_{10} in binary notation.

$$
\begin{array}{r}
0\ 000\ 000\ 000\ 110\ 011 \\
\underline{1\ 111\ 111\ 111\ 001\ 101} \\
0\ 000\ 000\ 000\ 000\ 000
\end{array}
$$

The expected result, zero, is generated along with a "carry," which is not registered. Actually, the operation just performed could also be interpreted as the subtraction of 51_{10} from 51_{10}. Any number to be subtracted is put in the form of its 2's complement and added.

Appendix B

APPLICATION OF THE THEVENIN MODEL TO A dc WHEATSTONE BRIDGE

By applying the Thevenin model (Section 2.4) to the Wheatstone bridge of Fig. 3.6 equations can be developed that will facilitate calculation of unbalance currents. Often such a bridge is used in this manner.

For the Thevenin model both the open circuit voltage at the output and the shorted value of the circuit resistance must be found. The open circuit voltage is just the difference voltage $V_B - V_C$. First, V_B is just

$VR_1/(R_1 + R_3)$ and V_C is $VR_2/(R_2 + R_4)$. Thus,

$$V_B - V_C = VR_1/(R_1 + R_3) - VR_2/(R_2 + R_4) = V_{th}$$

To obtain the value of Thevenin resistance, as discussed in Section 2.4, the regular voltage source must be short-circuited, that is, considered removed from the circuit. In Fig. B.1a the Wheatstone bridge is shown again, this time with the battery removed. Now points A and D are connected. As a result, arms R_1 and R_3 appear in parallel as do arms R_2 and R_4. Further, these two parallel combinations are in series. (We can ignore R_5 since it is between the output terminals.) The Thevenin resistance is given in Fig. B.1b.

To calculate the output current of the bridge, which might more appropriately be labeled the *unbalance current*, consider Fig. B.1b again. If a null detector of resistance R_5 is connected to the terminals of this Thevenin equivalent circuit, the current in it will be

$$I_{\text{null det}} = V_{eq}/(R_{eq} + R_5)$$

between points B and C. Since the voltage at B is $VR_1/(R_1 + R_3)$ and that at C is $VR_2/(R_2 + R_4)$, V_{eq} is

$$V_{eq} = V\left[\frac{R_1}{R_1 + R_3} - \frac{R_2}{R_2 + R_4}\right]. \tag{B.1}$$

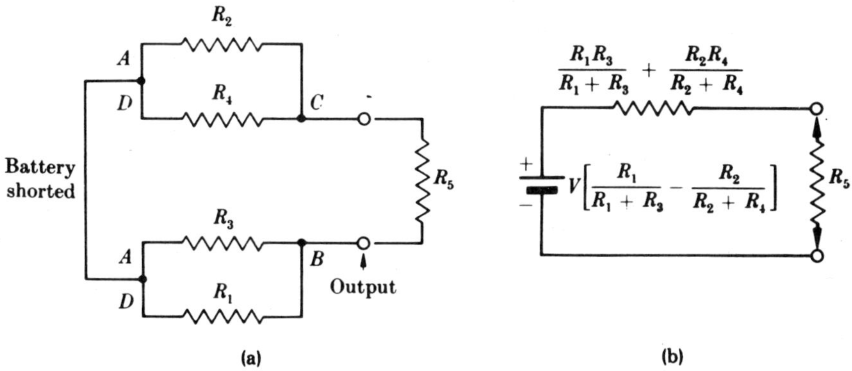

Fig. B.1 A Thévenin equivalent representation of a Wheatstone bridge. (a) Bridge circuit after shorting battery. (b) The final equivalent circuit with formulas for R_{th} and V_{th} and with the resistance of the null detector appearing as the load.

If a null detector of resistance R_5 is connected to the terminals of this Thevenin equivalent circuit, the current in it will be

$$I_{\text{null det}} = \frac{V_{eq}}{R_{eq} + R_5}. \tag{B.2}$$

INDEX